ELSEVIER GEO-ENGINEERING BOOK SERIES
VOLUME 2

Coupled Thermo-Hydro-Mechanical-Chemical Processes in Geo-Systems

Fundamentals, Modelling, Experiments and Applications

Ove Stephansson

GeoForschungsZentrum,
Telegrafenberg, Potsdam, Germany

Geo-Engineering Book Series Editor

John A. Hudson

Department of Earth Science and Engineering,
Imperial College of Science, Technology and Medicine,
London, UK

Lanru Jing

Department of Land and Water Resources Engineering,
Royal Institute of Technology (KTH),
Stockholm, Sweden

2004

ELSEVIER

Amsterdam – Boston – Heidelberg – London – New York – Oxford
Paris – San Diego – San Francisco – Singapore – Sydney – Tokyo

ELSEVIER B.V.
Sara Burgerhartstraat 25
P.O. Box 211, 1000 AE
Amsterdam, The Netherlands

ELSEVIER Inc.
525 B Street
Suite 1900, San Diego
CA 92101-4495, USA

ELSEVIER Ltd.
The Boulevard
Langford Lane, Kidlington,
Oxford OX5 1GB, UK

ELSEVIER Ltd.
84 Theobalds Road
London WC1X 8RR
UK

First edition 2004

Library of Congress Cataloging in Publication Data
A catalog record is available from the Library of Congress.

British Library Cataloguing in Publication Data
A catalogue record is available from the British Library.

ISBN: 0-08-044525-X
ISSN: 1571-9960

♾ The paper used in this publication meets the requirements of ANSI/NISO Z39.48-1992 (Permanence of Paper).
Printed in The Netherlands.

Series Preface

I am pleased to be able to introduce this book on "Coupled thermo-hydro-mechanical-chemical processes in geo-systems: fundamentals, modelling, experiments and applications" as the second publication in the Elsevier Geo-Engineering Book Series. The book contains the proceedings of the GeoProc2003 conference held at the Royal Institute of Technology in Stockholm, Sweden, in October 2003.

We do not anticipate that the Geo-Engineering Book Series will contain many such conference proceedings, but we have included these because of the extreme importance of the subject. In the past, different component disciplines within the overall geo-engineering subject area have separately used numerical modelling techniques with little interaction occurring between the disciplines. However, the principles of physics form a common basis for the component disciplines of radioactive waste disposal, general geo-systems, oil and gas reservoir engineering, geothermal energy engineering, geological systems, and geotechnical and environmental engineering. The physics is manifested in different ways according to the host rock conditions and the engineering perturbations. Thus, we should expect to see some convergence in approach in the numerical modelling and we should anticipate that cross-discipline interaction will provide fertile ground for moving the geo-engineering subject forwards.

In line with this thinking, the purpose of the GeoProc2003 conference was to present, for the first time, the state-of-the-art of numerical modelling in the component disciplines and to encourage interaction between the disciplines. Not only was this achieved successfully, but both the keynote and regular papers demonstrate the significant advances that have been made recently. Indeed, it is not too much to say that, through the information presented in this book, we are witnessing the birth of a new subject: i.e. the ability to incorporate thermal, hydrological, mechanical and chemical processes, together with their interactions, in numerical models and to apply these models across the spectrum of geo-engineering applications. Since the geo-engineer must have a predictive capability in order coherently to approach engineering design and that capability is now mainly provided by numerical modelling, the subject could not be more important — hence this publication.

We hope that you enjoy the book and we welcome proposals for new books. Please send these to me at the e-mail address below.

Professor John A. Hudson *FREng*
Series Editor
jah@rockeng.co.uk

PREFACE

This book presents the current state-of-the-art of coupled thermo-hydro-mechanical-chemical modelling in different geoscience disciplines and geo-engineering fields, and includes the associated scientific achievements reached in the recent international DECOVALEX III and BENCHPAR research projects. The material encompasses the different disciplines of geosciences and industrial applications, including geology, hydrogeology, rock and soil mechanics, geomaterial sciences, geotechnical and environmental engineering, geothermal and hydrocarbon reservoir engineering, mining, and radioactive waste disposal and management.

The contributions contained in the book were presented at the International GeoProc2003 Conference on Coupled T-H-M-C Processes in Geosystems: Fundamentals, Modelling, Experiments and Applications. This conference was the first international academic gathering in the growing field of modelling coupled thermal (T), hydraulic (H), mechanical (M) and chemical (C) processes in geological materials, geotechnical engineering and geological systems. The conference was held at the Royal Institute of Technology (KTH) in Stockholm, Sweden on October 13-15, 2003. The next T-H-M-C conference will be held in Nanking, China, in 2005.

The content of the book should attract significant interest from the international scientific and engineering community, both because of the leading edge nature of the subject and the high quality of the contributions. We believe that this book, published by Elsevier Science as the second in its new Geo-Engineering Book Series, will help to encourage further developments in the integration of multi-disciplinary geosciences and geo-engineering fields, thus assisting society through advanced scientific knowledge, efficient industrial applications and healthy and safe environments.

The editors are grateful to all the contributors and supporting organizations and individuals for their efforts which made this book possible. We are also grateful to Chin-Fu Tsang for writing such a useful introductory overview of the T-H-M-C processes.

Ove Stephansson, John A. Hudson and Lanru Jing

April 2004

About the Editors

Professor Ove Stephansson

For many years, Professor Stephansson headed the Engineering Geology Division at the Royal Institute of Technology in Stockholm, Sweden. He is a geologist with strong mathematical and engineering leanings and has contributed to many aspects of geo-engineering, especially rock stress, rock fractures and rock fracturing. He is the co-author with Professor Amadei of the seminal book "Rock Stress and Its Measurement". During the last 10 years, he has been one of the leaders of the international DECOVALEX program which advances the state-of-the-art of numerical modelling for radioactive waste disposal understanding and design. He recently retired to a position at the GeoForschungsZentrum in Potsdam, Germany.

Professor John A Hudson

Professor Hudson has spent his professional career in engineering rock mechanics (as it applies to civil, mining and environmental engineering) in consulting, research, teaching and publishing. From 1983 to the present, Professor Hudson has been affiliated with Imperial College, UK, as Reader and Professor. He edited the Elsevier 1993, five volume, 4407 page, "Comprehensive Rock Engineering" compendium, and currently edits the International Journal of Rock Mechanics and Mining Sciences. He is a co-author, with Dr John Harrison, of two Elsevier textbooks on "Engineering Rock Mechanics" published in 1997 and 2000. In 1998, he was elected as a Fellow of the Royal Academy of Engineering in the UK.

Dr Lanru Jing

Since 1991, Dr Jing has been based at the Royal Institute of Technology in Stockholm, Sweden and is currently an Associate Professor. He is an expert on rock fractures and in the techniques and applications of numerical modelling. His April 2003 review paper for the International Journal of Rock Mechanics and Mining Sciences, titled "A review of techniques, advances and outstanding issues in numerical modelling for rock mechanics and rock engineering", is the most popular paper — in terms of downloads — that the Journal has published in recent times. Dr Jing has been responsible for the Technical Secretariat of the DECOVALEX international programme for the last ten years.

INTERNATIONAL SCIENTIFIC ADVISORY COMMMITTEE

ORGANIZING COMMITTEE

Contents

Introductory Article

COUPLED THM PROCESSES IN GEOLOGICAL SYSTEMS AND THE DECOVALEX PROJECT

Chin-Fu Tsang,[1] Ove Stephansson,[2] Fritz Kautsky,[3] and Lanru Jing[4]

[1]*Lawrence Berkeley National Laboratory, Berkeley, California 94720, USA*
[2]*GeoForschungsZentrum, Telegrafenberg, D-14473 Potsdam, Germany*
[3]*Swedish Nuclear Power Inspectorate, SE-10658 Stockholm, Sweden*
[4]*Royal Institute of Technology, SE-10044 Stockholm, Sweden*

ABSTRACT: An overview is given of recent progress in the understanding, monitoring, and modeling of coupled thermo-hydro-mechanical (THM) processes in geologic systems, in the context of major practical applications. The progress has been made possible through individual research efforts, as well as international cooperative research projects. As an example of international cooperation, the DECOVALEX project is described. Initiated in 1992, the project has progressed successfully through three major stages. It has played a key role in the development of the field of mathematical modeling and testing of coupled THM processes in fractured rocks and buffer/backfill materials, a subject of importance for performance assessment of a radioactive waste geologic repository. The DECOVALEX project has been supported by a large number of radioactive waste management organizations and regulatory authorities, including those in Canada, Finland, France, Japan, Germany, Spain, Sweden, UK, USA, and EU. This paper presents a summary of the project, including the objectives, scope, problems investigated, scientific achievements and major outstanding issues, with emphasis on the science of the coupled THM processes.

1. INTRODUCTION

The last fifteen years have seen substantial progress in experimental and theoretical studies regarding the effects of coupling temperature gradient (T), hydrologic flow (H), and mechanical deformation (M) in fractured rocks. Much of the impetus behind these efforts is the concern over the role of such couplings in the performance and safety assessment of a heat-releasing nuclear waste repository in the subsurface (Tsang, 1987). However, the problem is of wider interest, ranging from coupled THM processes associated with geothermal energy extraction, gas production from coal beds, seismicity induced by fluid injection and the construction of underground openings, to guidelines on injection pressures needed for stimulating deep petroleum reservoirs with water colder than *in situ* fluids.

To understand and predict the effects of coupled processes in all these practical problems, models are being developed that are capable of simulating coupled thermo-hydro-mechanical (THM) processes. The term "coupled processes" implies that one process affects the initiation and progress of another. Thus, the rock mass behavior cannot be predicted with confidence by considering each process individually or in simple succession. Previously, binary TM and TH couplings have been studied in the context of, for example, rock mechanics and geothermal reservoir engineering. For many current applications, however, we must study the full triply-coupled THM processes. In some cases, the introduction of chemical processes (C) into the study of THMC couplings and its impact are also of important. Work has been initiated by different researchers, in various countries

The coupling of THM processes is a major challenge to the geoscience community, since the three processes have widely different characteristic time and spatial scales. The thermal effect in rock material has relatively long time and spatial scales. Mechanical effects, on the other hand, have a short time scale, since changes in the mechanical response can propagate through the rock mass with the speed of elastic waves, and deformability is often dominated by the presence of fractures of various size scales, such as joints, faults and fracture zones. Finally, groundwater flow and transport are sensitive to both small-scale heterogeneities and fracture system characteristics, but are characterized by longer flow and solute transport times.

Numerically, these processes can be modeled by different techniques, such as finite-difference methods (FDM), finite-element methods (FEM), and discrete-element methods (DEM), as well as discrete fracture network (DFN) methods. In addition, many of the coupled processes are nonlinear, and the constitutive equations typically contain certain parameter sets. Combining all these processes into an efficient model for the simulation of coupled THM processes in fractured rocks is a challenging task.

A key element in the study of coupled THM processes is the verification of numerical codes and validation of model results against well-conditioned field and laboratory experiments. Here, the challenge lies in providing a set of well-defined conditions for the boundaries, the rates of thermal and mechanical loading, the initial state (of stress, temperature and flow), and constitutive equations for coupling and material properties. Coupled THM experiments in the field require well-considered test designs, robust instrumentation, careful result interpretation, and often have durations of months and years.

In the next section, recent progress in the understanding of the various coupled THM processes in geo-systems for a wide range of problems is briefly reviewed, with references to other chapters in this volume. In the following section, we turn from a general discussion to one particular international cooperative project involving research teams from about ten countries to develop and study a set of coupled THM problems and data sets devoted to nuclear waste isolation. This is the so-called DECOVALEX project, within which significant advances on modeling THM processes related to nuclear waste repositories have been made over the past 12 years. The paper then concludes with some general remarks.

2. PROGRESS IN RESEARCH OF COUPLED THM PROCESSES IN GEOLOGICAL SYSTEMS

The progress made over the last decade in our understanding, monitoring, and modeling of coupled THM processes in geo-systems has been on many fronts, such as on problems related to nuclear waste geological disposal, development of geothermal and hot dry/wet rock systems, coal bed gas production and underground coal gasification, petroleum production and reservoir dynamics, and stability of large-scale civil constructions. Research in these fields is presented in the various chapters in this volume. The present section briefly describes a number of highlights and points out the potentials for cooperation and cross-fertilization among these fields.

In the study of coupled THM processes related to nuclear waste geological isolation, many laboratory experiments to investigate the constitutive relationships and individual processes have been performed, and a number of large-scale (~10–100m), long-term (1–8 years) field experiments have been conducted (see, for example, Alonso and Alcoverro, 2004; Datta et al., 2004; Chijimatsu et al., 2004). Modeling capability has advanced from studies of THM-induced permeability changes caused by fracture opening and closing, shear displacement and dilation in hard rocks, to simulation of flow and transport under coupled THM processes in fractured rocks accounting for complex geologic structures and heterogeneity. A number of attempts for homogenization of the discrete fracture network for simulation of coupled THM effects have been made (Liu H. et al., 2004; Guvanasen and Chan, 2004; Gomez-Hernandez and Cassiraga, 2004). Significant progress has also been made in coupled THM processes in bentonite, which is used as a buffer around a nuclear waste canister emplaced in a geologic repository, and in its interaction with neighboring rock (Thomas et al., 2004; Rutqvist et al., 2004). Increasingly, work has now turned to detailed studies of the role of gases (air from tunnel ventilation or gas produced from canister corrosion) and of coupled processes for softer rocks (Alonso and Olivella, 2004; Shao et al., 2004; Su et al., 2004). Soft rocks, such as clays and shales, are also being considered as potential host rocks for nuclear waste repositories. These rocks can experience large deformation, and their behavior is particularly sensitive to changes in moisture level or humidity in their environment, e.g., in the tunnels and boreholes. These cases under coupled THM processes represent challenging and exciting areas of continued research.

In the field of geothermal hot dry rock or hot wet rock (HDR/HWR) reservoirs, considerable progress in detecting and mapping of hydraulically created permeable structures has been made (Niitsuma, 2004). Much of the progress has been in the improved analysis of microseismic monitoring, which is currently the best available method for obtaining three-dimensional structural information away from boreholes. The maps of seismic clouds produced through conventional analysis of seismic events due to hydrofracturing involve significant location inaccuracy. Such errors can be much reduced through a suite of integrated methods, such as collapsing, doublet/multiplet analysis, clustering analysis and multiplet-clustering analysis. These methods are able to reveal, from the conventional seismic clouds, the macro- and microstructures, and these structures can be correlated with geologic and hydraulic structures identified from well logs. The method has been successfully applied to data from the cooperative European HDR research program at Soultz-sous-Forêts site in France. Besides this work, significant progress has also been made in modeling of coupled THM processes as well as their interaction with chemical processes in geothermal systems (see for example, Fomin et al., 2004b; Hosni et al., 2004).

In contrast to nuclear waste geologic isolation, in which disturbance to the geological formation is to expected to be minimized, petroleum exploration aims at maximizing production and large changes in pressure, temperature, stresses, deformation, flow rates and chemistry are expected. These have stimulated studies of mechanical-chemical coupling, quasi-static solid-liquid coupling, and dynamic solid-liquid coupling (Dusseault, 2004). Both the fundamental processes and practical implications of these couplings are investigated. Borehole integrity is an important issue in petroleum exploitation, often controlled by coupled THMC processes in shales present around the borehole (where the C in THMC refers to chemical effects). Shale expansion or contraction changes the stress states, causing potential for damage. Another issue is sand production, creating stress changes and the potential for further yield, channeling and dilation. Improved analysis and modeling of these types of problems have been conducted (e.g., Choi et al., 2004, Guimaraes et al., 2004; Han and Dusseault, 2004).

Underground coal gasification involves even more drastic coupled THM processes than petroleum exploitation. The high heat energy released in in-situ coal seam burning and the creation of new cavities in the coal seam dramatically modified the local permeability field, which in turn affects the gasification process. Such cavity evolution has been accounted for in a coupled THM model of the system (Liu J. et al., 2004). For the more moderate processes of coal mining and methane production from coal seams, there are the problems of gas pressure changes, opening or closing of coal cleats, and gas sorption or desorption from the coal matrix. These affect gas production and, in some cases, cause coal and gas outbursts. Modeling of these processes has made significant progress (Choi and Wold, 2004; Sun, 2004).

Advances in coupled processes research have also been made in the context of construction of major dams (Guo et al., 2004; Zhu et al., 2004), development of cold storage caverns (Lee and Lee, 2004), and grouting boreholes in permafrost (Fomin et al., 2004a).

The discussion above has centered on geological systems impacted by human activities. Much work has also been done on natural geological systems, in which coupled THMC processes play a key role in understanding the natural evolution of the systems. Such research includes modeling of sediment compaction during burial in sedimentary basins (Bjørlykke et al., 2004), lava dome collapse triggered by rainfall infiltration (Elsworth et al., 2004), and state of earth crust structure as a result of rock fracturing under high pressure-temperature conditions (Nikolaevskiy and Garagash, 2004).

It is interesting to note some of the common problems connecting the different fields. The reasons for such connections are the basic physics and the need to account for geological structures common to all these fields. For example, identification of faults and fractures away from boreholes is an important problem in nuclear waste geologic isolation, since faults or fractures may be the main conduits for leakage from the repository. In a recent multi-year heater test (Datta et al., 2004), microseismic events have been measured. The advances in methodology in analyzing these events made by HDR researchers can be applied to such data to great advantage. Another example is the interest of coupled processes in shales related to stability in petroleum boreholes. This interest is shared by researchers who are studying tunnel stability where shales are present, and by researchers considering the potential of storage of nuclear waste in shales. There are also overlaps between research on gas processes in coal seams and on gas transport in petroleum reservoirs. Often a computer code and model developed and tested in one field can be modified and applied to another. Then those computer codes and models, that have been successfully applied and tested in more than one field, will command a much higher confidence level.

A significant part of the progress would have been much delayed without major international cooperative programs. Cooperation brings together the best researchers from a number of countries and serves both to encourage and stimulate development and also to peer review each others' work. Two examples are the Soultz HDR project, supported by the European Commission, and the DECOVALEX project. To illustrate the importance and effectiveness of cooperative research, the rest of this paper will be devoted to an overview of the DECOVALEX project, and the major results pertaining to coupled THM processes that have emerged from this international cooperative project will be summarized.

3. THE DECOVALEX PROJECT

An international cooperative project DECOVALEX (acronym for DEvelopment of COupled THM models and their VALidation against EXperiments) was established in 1992 by national regulatory authorities and waste management organizations involved in nuclear waste disposal to cooperate in developing and testing models capable of simulating coupled THM processes. The participating organizations would share results from major field and laboratory experiments, results that would enhance understanding of coupled processes and provide data for model validations. Over the last twelve years, more than 15 research teams from 10 countries have participated in this joint effort. The project objectives include:

- To support development of computer simulators for THM modeling
- To investigate and implement suitable algorithms for THM modeling
- To compare model calculations with results from field and laboratory experiments
- To design new experiments to support code development
- To study the application of THM modeling to performance and safety assessment of nuclear waste repositories.

A large number of benchmark tests (BMT) and test cases (TC) have been studied within the project. The former are hypothetical problems used for investigating the behavior of individual or coupled THM processes with general complexity in processes, properties and parameters, and evaluating the extrapolations of results into time and spatial scales of interest to repository performance (which cannot be performed by experiments). As part of BMTs, sensitivity and scoping analysis will be conducted with alternative conceptual and numerical models by different teams. The TCs are laboratory and field experiments that were designed to advance our understanding of the THM processes, and numerical modeling was applied to both test the models and to help interpret the test results. A number of major, large-scale, multiyear experiments have been studied within the project

In addition to analysis of these BMTs and TCs, particular topics were selected for discussion and review among project participants. These range from the state-of-the-art of constitutive relations of rock fractures to the current international treatment of THM issues in repository performance assessment.

The activities of the DECOVALEX project are organized around the study and modeling of BMTs and TCs by multiple research teams using different mathematical models and computer codes. Both BMTs and TCs are carefully developed as initial-boundary value problems with proper thermal, hydraulic, and mechanical initial-boundary conditions and loading sequences. Based on the results from these studies, new experiments were proposed to provide more rational tests of concepts and models, and to advance the state of mathematical modeling for coupled THM processes in fractured rocks and buffer materials. Analytical and semi-analytical solutions to the coupled problems were also developed whenever possible. Representatives of the national regulatory authorities and radioactive waste management organizations participating in the project took an active part in the whole process, to ensure that the project was conducted equally from scientific, engineering, and managing points of view. The physical processes studied in the BMTs and TCs are listed in Table 1.

The next section describes the activities of the three stages of DECOVALEX I, II, and III, covering the years 1992–1995, 1995–2000 and 2000–2003, respectively. Following this, two specific examples are presented to illustrate, in different depths, the work of one BMT and one TC performed under the project.

Table 1. Physical phenomena studied in DECOVALEX	
Components	Phenomena
Physio-mechanical processes	Thermal expansion, diffusion, and convection in fractured rocks and buffer materials Fluid flow in fractured rocks and buffer materials Deformation of fractured rocks and buffer materials Constitutive laws for rock fractures, fractured rock masses, and buffer materials Swelling pressure and suction potential of the buffer materials
Geometrical factors and properties	Rock fracture networks and their characterization and representation Rock fracture properties (aperture, roughness, gouge production, filling, conductivity, storativity) Variability and representability of network connectivity

4. SUMMARY OF DECOVALEX STAGES I, II, AND III

(a) DECOVALEX I

In the DECOVALEX I project, modeling was conducted on three hypothetical BMTs and six TCs, involving three small laboratory tests of rock samples and fractures, and three large field tests. Details may be found in Jing et al. (1995). One of the BMTs simulated a Swedish KBS-3 disposal concept in a fractured granitic rock with a fracture network system and properties similar to those of the Stripa granite. This case will be presented in more detail in section 5.

(b) DECOVALEX II

In the second phase of the project, DECOVALEX II, studies were focused on two major large-scale *in situ* experiments, and also on evaluating how the studies conducted in the project could be applied to the performance and safety assessment of a potential repository. The following studies were undertaken:

- Task 1: Numerical study of NIREX's Rock Characterization Facility at Sellafield
- Task 2: Numerical study of the in-situ THM experiment in the Kamaishi Mine, Japan
- Task 3: Review of the state-of-the-art in constitutive relations of rock fractures
- Task 4: Current understanding on the coupled THM processes related to design and performance assessment of radioactive waste repositories.

Results of Task 1, 2, and 4 are presented in Stephansson et al. (2001). Of particular note is Task 1, under which an extensive data package on the geology, hydrology, and rock mechanics was distributed to the research teams, who were then free to select, develop, and parameterize their own conceptual models of the local site within the Sellafield system. One lesson learned from this procedure is the recognition of the significance of calibration procedures in predicting the response to pumping and shaft sinking. Another lesson learned is the importance, and also the difficulty, of performing the simulations assuming fully coupled processes. The simulations of the Kamaishi Mine heater experiment, Task 2, have also provided valuable experience in analyzing coupled THM processes in the near field of a waste canister-bentonite-rock system. Some details of this study are given in section 6.

(c) DECOVALEX III

In the third phase of the project, DECOVALEX III, the following tasks have been performed:

- Task 1: FEBEX experiment conducted by ENRESA in Grimsel Mines in Switzerland
- Task 2: The Drift Scale Test (DST) in the Exploratory Studies Facility (ESF) at Yucca Mountain, USA
- Task 3: Three benchmark tests about., (a) near-field repository performance (BMT1), (b) Material property homogenization and far-field repository performance (BMT2), and (c) Glaciation process effects on far-field repository performance (BMT3)

- Task 4: Survey and discussion of THM processes in performance assessment.

Below we shall give a brief description of these tasks, and detailed results are presented in some of the other chapters in this volume.

In the FEBEX experiment, a full-scale *in situ* experiment is being performed on a heater-buffer-rock system in the Grimsel Test Site in Switzerland, with a long period of heating followed by cooling. The aim of the project is to demonstrate the present capabilities for building bentonite barriers in conditions similar to actual repository design and providing monitoring data to understand coupled THM processes in the near field. A large quantity of monitoring data on stress, deformation, water content, water pressure, and temperature distributions and their histories were recorded at a large number of monitoring places *in situ*. Also, a large number of rock/buffer property parameters were measured in laboratory tests. Three subtasks are conducted within DECOVALEX III: (1) simulation of hydro-mechanical behavior in the fractured rock mass with respect to the tunnel excavation; (2) simulation of the THM processes of the heater-buffer system; and (3) simulation of coupled THM responses of the complete rock-buffer-heater system during the heating period. Ten research teams supported by nine national organizations participated in this task.

The second task, the Drift Scale Test (DST) in the Exploratory Studies Facility (ESF) at Yucca Mountain, is a large-scale thermal test conducted by the Yucca Mountain Site Characterization Office of the U. S. Department of Energy (DOE). It is part of DOE's program of characterizing the Yucca Mountain site to evaluate its suitability for a potential nuclear waste repository. The heating phase of the test started on December 3, 1997, and lasted for about four years, ending on January 14, 2002. The objective of the test is to help increase confidence in models of coupled THMC processes in the rock mass. These models will be employed to quantitatively assess the long-term performance of the potential repository. Heating is effected through nine cylindrical heaters placed on the floor of a 47.5 m drift and 50 wing heaters, each 10 m long, inserted into horizontal boreholes on either side of the drift. The purpose of this arrangement is to: (a) simulate the thermal field that an emplacement drift will experience from neighboring drifts, and (b) heat a large volume of rock mass to boiling temperatures within the experimental heating period of four years. Processes occurring in the DST include heat and fluid flow in unsaturated fractured rocks; heat-pipe effects and other heat transfer mechanisms; effects of temperature on permeability and conductivity; THM changes in concrete lining; THM processes in unsaturated fractured rocks with the presence of drifts; effects of thermo-mechanical processes on hydrologic characteristics; chemical changes under air-water-vapor flow in fractured rock; changes in Eh and pH chemical reactions under phase change; and effects of dissolution and precipitation on hydrologic characteristics. Six research teams supported by six national organizations participated in this task.

Task 3 consists of three benchmark tests: BMT1, BMT2, and BMT3. BMT1 focuses on near-field behaviour of a hypothetical repository in fractured hard rock from excavation to post closure. Databases developed at the Kamaishi experiment were used for the detailed technical definition of the BMT, with repository geometry similar to that of the Kamaishi mine experiment. The main performance assessment (PA) measures were the resaturation progress in buffer and rock, the mechanical effects on buffer and waste form, and the mechanical stability of the repository. The thermo-hydro-mechanical evolution of the whole configuration is simulated over a period of 1000 years. Comparison of the results predicted by fully coupled THM analysis with those from partially coupled TH, TM, and HM analysis (in terms of several predefined indicators) identifies the couplings that play a crucial role with respect to safety issues. Six research teams supported by six national organizations participated in this study.

The BMT2 is on homogenization (originally proposed as an upscaling problem) concerns the relationship between an equivalent continuum (which could be heterogeneous) and detailed discrete representations of fractured rocks. A key issue is the extrapolation of rock properties obtained from small-scale tests and observations to a large repository scale, with analysis for uncertainties. The main focus is on the method of deriving flow and deformation properties of the fractured rocks as equivalent continua, and its impact on prediction of large-scale changes in flow and deformability fields. The database developed at Sellafield for Task 1 of DECOVALEX II is used for the detailed technical definition of the BMT2. Eight research teams supported by eight national organizations participated in this study.

Finally, the glaciation test, BMT3, concerns mainly the hydro-mechanical processes during a cycle of glaciation and deglaciation for assessing the long-term performance (up to 100,000 years) of a hypothetical post-closure repository, without considering thermal effects. Many different alternative scenarios are included in the task, such as permafrost, different ice-rock interface conditions, 2D–3D transition, inland/coastal repository locations, or sea level changes. The main PA measures are the maximum deformation, changes in permeability fields, flow patterns and formation of critical flow paths, and groundwater pressure. Only long-lasting and large-scale changes in PA measures are significant. Four research teams supported by four national organizations participated in this study.

To better understand the relevance of THM coupling to PA and the associated uncertainties and the applicability ranges, Task 4 of DECOVALEX III was established as a platform for presentation, discussion, and documentation of the treatment of THM issues in the framework of PA analyses. The task contains two subtasks: (1) a state-of-the-art review on the current and past international treatment of THM issues in PA and (2) a forum on and documentation of THM treatment in the PA.

5. EXAMPLE OF A BMT STUDY

In this and next sections, we shall illustrate the type of research under DECOVALEX by two examples from DECOVALEX I and II, respectively. The first example is a BMT and the second example, a TC.

The nature of a BMT study is well demonstrated with BMT3 of the DECOVALEX I project. It was a problem associated with a near-field repository model, set up as a two-dimensional plane-strain problem in which a tunnel with a deposition hole was located in a fractured rock mass. The model is 50 × 50 m in size, and situated at 500 m below the ground level (Figure 1). The fracture network is a two-dimensional realization of 6,580 fractures from a realistic three-dimensional fracture network model of the Stripa Mine, Sweden (Figure 2). The problem is set up as a fully coupled THM near-field repository problem, with thermal effects caused by heat release from radioactive waste in the deposition hole (the heater). Heat output decreases exponentially with time. The rock matrix is assumed to be isotropic and linearly elastic, and its mechanical properties do not change with temperature variations. Its thermal conductivity and expansion are also assumed to be isotropic. The fractures are assumed to consist of parallel, planar, smooth surfaces at the macroscopic level, with an effective hydraulic aperture.

The initial and boundary conditions for the mechanical, thermal, and hydraulic effects are shown in Figure 1, with heating maintained for 100 years. (See Table 2 for the research teams and codes applied to this study). Figure 3 shows the different representations and simplifications in the models used by the various teams. This BMT was regarded as an excellent model for testing the capabilities of many alternative mathematical models and computer codes (Stephansson et al., 1996).

The major findings from this BMT study include the following:

- This BMT was a well-defined near-field problem, with both a realistic fracture network (which may likely be encountered in practice) and complete aspects of coupled THM processes.
- Agreement in temperature results from all research teams was remarkable.
- Heat convection had negligible effects on the temperature results.

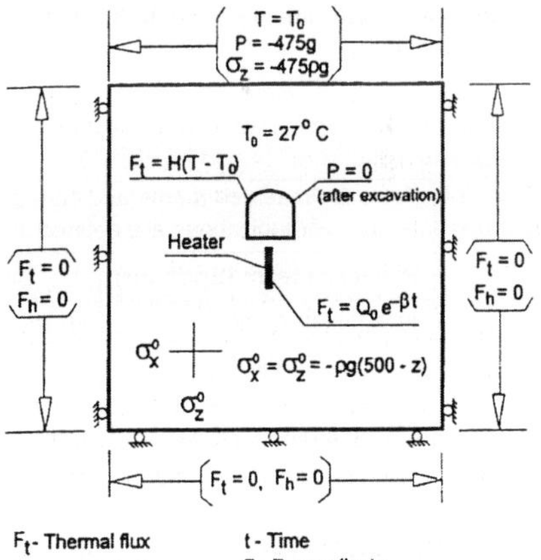

Figure 1. Definition of BMT3(DECOVALEX I): Model geometry and boundary conditions

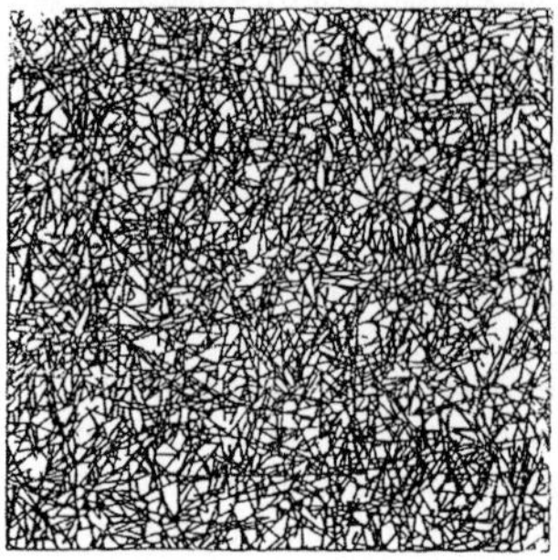

Figure 2. Reference fracture network from Stripa Mine for BMT3 (DECOVALEX I)

AEA (NAPSAC code, DFN model) Complete and explicit representation of 6580 fractures by DFN method, without heat and rock deformation, but with analytical model for stress approximation	
CEA/DMT (CASTEM 2000 code, FEM model) Equivalent continuum approach with FEM method (1344 elements, 4216 nodes), homogenization (scale 25 x 25 meters) by Crack Tensor theory (Oda, 1986)	
CNWRA (UDEC code, DEM model) Vertical symmetry and simplified fracture network with 295 fractures, 337 blocks, 1540 finite difference elements, and 10853 nodes. A small inner region with random fractures and lager outer region with regular artificial fractures	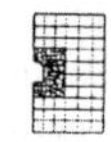
ITASCA (FLAC code, FDM model) Equivalent continuum approach with FDM (735 finite ifference elements and 792 nodes)without formal homogenization. Vertical symmetry and inhomogeneous permeability	
INERIS (UDEC code, DEM model) Simplified fracture network with inner region of random fractures and outer regions of regular artificial fractures (564 fractures, 677 blocks, 24391 finite difference elements, and 16791 nodes)	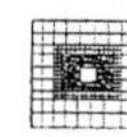
KPH (THAMES code, FEM model) Equivalent continuum approach with FEM method (674 elements, 2127 nodes). Homogenization (scale 10 x 10 meters) by Crack Tensor theory (Oda, 1986). Heat convection considered	
NGI (UDEC code, DEM model) Two models of vertical symmetry (one left half and one right half) with simplified fracture network without heating, about 512 fractures, 510 blocks, and 1580 finite difference elements	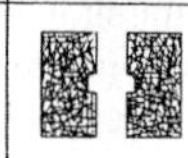
VTT (UDEC code, DEM model) Vertical symmetry (left half) with simplified fracture network, 814 fractures, 496 blocks, 1308 finite difference elements, and 2222 nodes.	

Figure 3. Different model representations and simplifications for BMT3 (DECOVALEX I)

- Displacement results agreed reasonably well among the research teams and between discrete and continuum approaches. Large discrepancies occurred only when monitoring points are located on loose blocks with large movements, close to the boundary of the tunnel and the deposition hole.
- Stress results agree less well among the teams than the displacement results, especially at points closer to the heat source and tunnel boundary, although a consistent general trend in the stress results can be observed.
- The flow rate results were highly divergent, both between the discrete and continuum approaches and among models in the same approach group.

Table 2. Research teams and codes applied to study BMT3 (the acronyms for the organizations are defined in Jing et al., 1995)

Research team (Sponsor)	**Code**	**Method**
AEA (NIREX, UK)	NAPSAC	DFN – fracture flow only
CEA/DMT (ANDRA, France; CEC)	CASTEM 2000	FEM – homogenization scale 25 m
CNWRA (NRC, USA)	UDEC	Dem with simplified fracture network
INERIS (ANDRA, France; CEC)	UDEC	DEM with simplified fracture network
ITASCA (SKB, Sweden)	FLAC	FDM without homogenization
KPH (PNC, Japan)	THAMES	FEM – homogenization scale 10 m
NGI (NIREX, UK)	UDEC	DEM with simplified fracture network
VTT (STUK, Finland)	UDEC	DEM with simplified fracture network

- Two homogenization scales were obtained by two teams using the same Crack Tensor theory, and the validity of representative elementary volume (REV) and its relationship to the size of finite element are unresolved, and remain an open problem.

6. EXAMPLE OF A TEST CASE STUDY

In this section, we present an example of a test case study to illustrate DECOVALEX activities. Task 2 of the DECOVALEX II project was the numerical modeling of the *in situ* THM experiment of a fractured rock-buffer-heater system at Kamaishi Mine, Japan (Figure 4). The Kamaishi Heater Test was an experiment conducted in a 5 × 7 m^2 alcove excavated from an existing drift, located at a depth of about 250 m. In 1995, a vertical test pit, 1.7 m in diameter and 5 m in depth, was drilled into the floor of the alcove (Figure 5). The hole was drilled with a gentle shot boring method, using a large-diameter boring machine to avoid mechanical disturbance of the rock. In 1996, an electric heater was installed into the test pit and surrounded by a buffer of bentonite clay. Bentonite was placed into the test pit in layers of 0.1 m, with compaction of each layer to a dry density of about 1.6 ton/m^3. After the entire test pit was filled with bentonite, a watertight concrete lid was placed on the drift floor, which in turn was supported by steel bars from the ceiling of the drift. Because the rock was not fully saturated immediately around the test pit, a flooding pool was set up on the drift floor above the test pit. At the end of 1996, the heater was turned on and the temperature was set to 100°C for 8.5 months, followed by a 6-month cooling period. System responses, including temperature, moisture content, fluid pressure, stress, strain and displacement, were measured in both the bentonite and surrounding rock mass. The experiment was completed in early 1998, and thereafter the monitoring sensors were recovered and recalibrated.

The task for the DECOVALEX research teams was to predict the THM effects in the buffer material inside the test pit and in the surrounding rock, both during excavation of the test pit and the heater testing. The test case was divided into three main tasks: Tasks 2A, 2B, and 2C. Task 2A was to predict the HM effects in the rock caused by the excavation of the test pit. Geometrical, mechanical and hydraulic rock properties, as well as hydraulic conditions before excavation, were given to the research teams, and they were asked to predict water inflow distribution in the test pit. Task 2B was a model calibration of rock and fracture properties and the hydromechanical boundary conditions, based on actual measured results predicted in Task 2A. Task 2C was to predict the THM effects in the rock and buffer during the heating experiment. The rock model was presumed to have properties based on the calibration in Task 2B, with the calibrated permeability distribution in the near-field rock. At

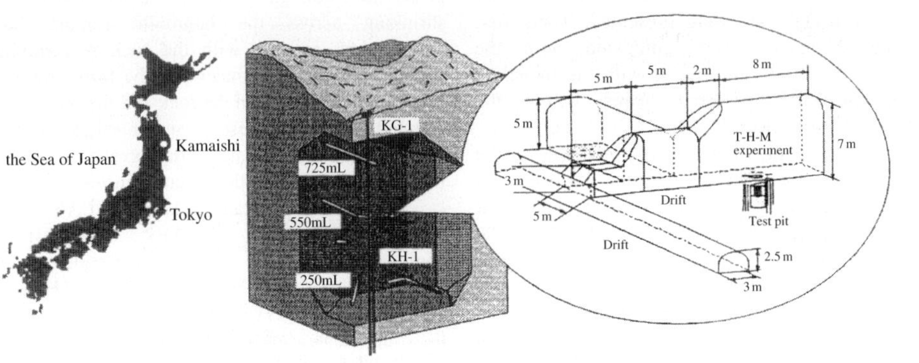

Figure 4. Location of the Kamaishi Main in Japan (a) and a 3D view of the test site in Kamaishi Main (b)

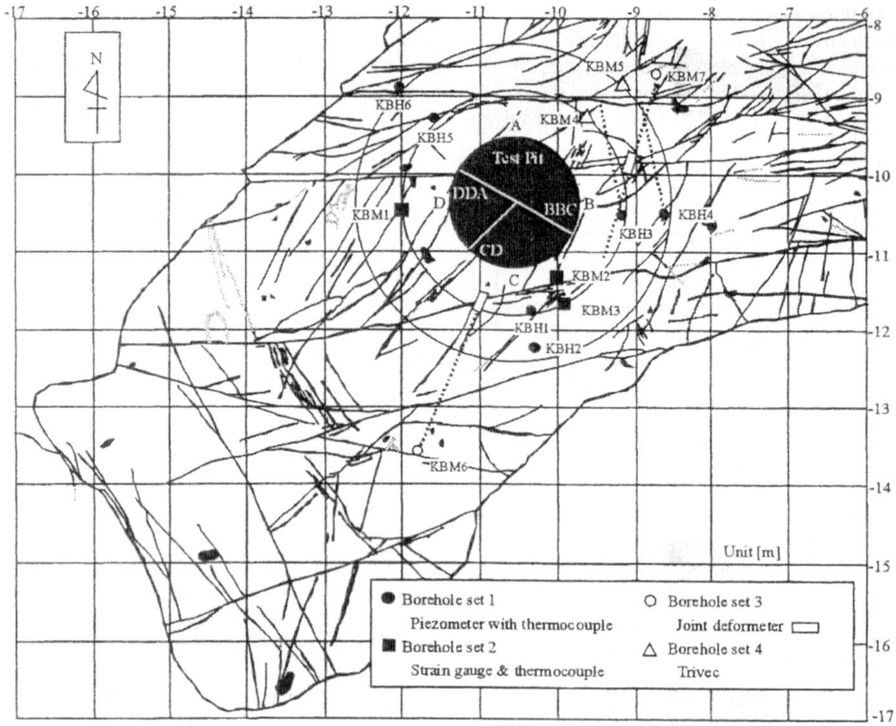

Figure 5. The fracture traces at the test drift floor and the locations of the instrumentation boreholes at the Kamaishi test site

every step, all the model predictions were made before completion of the respective test and before the experimental data were presented. Thereafter, the model results were compared with the experimental results, as well as with the modeling results of other research teams within the DECOVALEX project.

Processes being studied in the modeling of the Kamaishi Test included groundwater and heat flow in the rock matrix, fractures, buffer, and their interfaces under varying unsaturated conditions. Before emplacement of buffer and heater, the inflow of water into the test pit was affected, not only by the presence of fractures, but also by the unsaturated condition of the rock near the test pit. Strong variation in the areal distribution of inflow was observed on the walls of the test pit. After the heater and bentonite were emplaced, diffusion of water into the bentonite from the rock occurred simultaneously with drying of the bentonite near the heater. Multiphase flow in the bentonite region with phase transition gave rise to varying swelling or shrinking across the bentonite region. Such deformation interacts with the rock permeability, with open questions concerning the flow processes in the interface between the rock and the buffer. The coupled THM processes under such varying saturation conditions are complex and are at the leading edge of our modeling capabilities.

Four research teams—AECB, CLAY, KIPH and LBNL—studied the task with different computational models. The computer codes applied to the task were ROCMAS, FRACON, THAMES and ABAQUS-CLAY. All of them were based on the finite-element method (FEM). Figure 6 presents an overview of the geometry and the boundary conditions of respective models, including the near-field rock, bentonite buffer, concrete lid, and heater. The LBNL model is the largest and explicitly includes nearby drifts as well as three main fractures

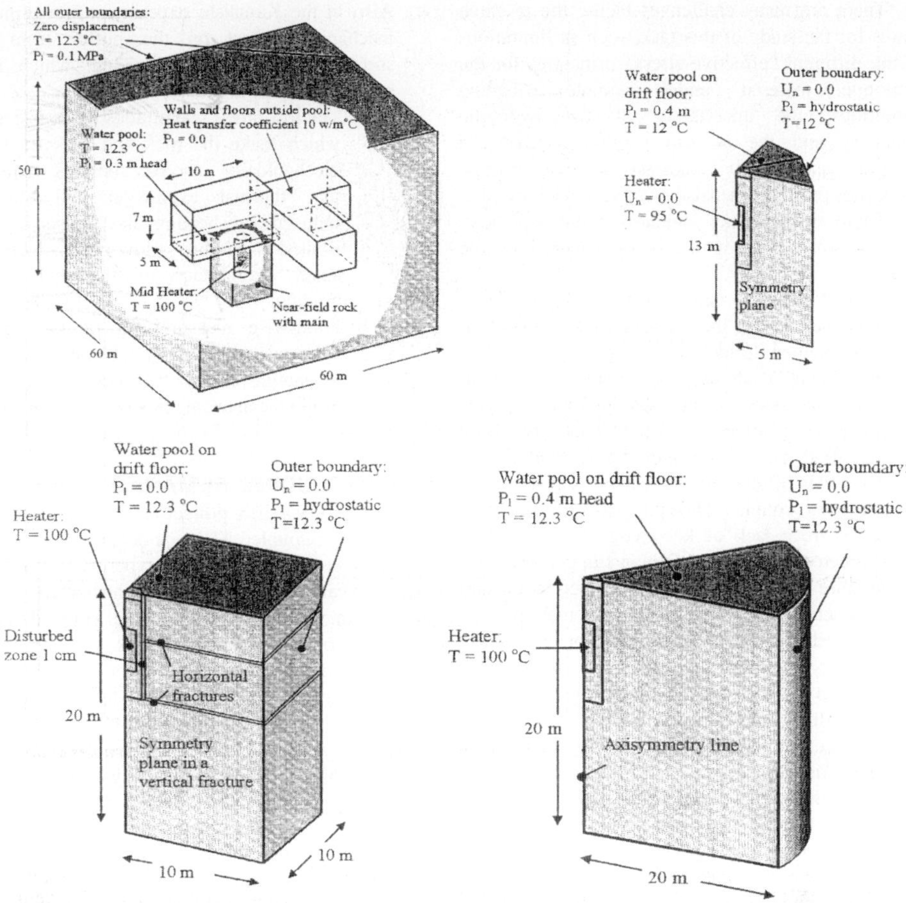

Figure 6.Geometry and boundary conditions of the computational models: (a) LBNL (11,000 elements), (b) AECB (288 elements), (c) CLAY (3,600 elements) and (d) KPIH (2800 elements)

in the near-field (on the drift floor in Figure 5). The fractures were included as discrete features because they are highly conductive and dominated the flow of water into the open test pit. The rock in between the main fractures is highly fractured (spacing 0.1 to 0.3 m) but has a much smaller hydraulic conductivity and was represented by an equivalent continuum. AECB, CLAY and KPIH teams reduced the size of their models and the drift system, because of the computer-intensive nature of the problem, which includes a year of simulation time and coupling of highly nonlinear processes in five degrees of freedom per nodal point.

A smaller model can also be justified if interest is limited to the behavior of the clay-buffer, and if the effect of the far-field rock mass can be represented by the boundary conditions on the near-field model. The axisymmetric geometry used in AECB and KPIH models was motivated through an exploratory two-dimensional modeling, with results showing that the resaturation of the bentonite buffer by wetting from the surrounding rock was uniform and axisymmetric, thus making it unnecessary to include discrete fractures in the model. The CLAY model is one-quarter symmetric.

There are many challenges facing the research teams for the study of this task, such as limitations in the different "effective stress" principles for the bentonite material under complex loading conditions; the uncertainty of the hydraulic boundary and the *in situ* conditions; and the complex and largely unknown *in situ* fracture properties (both geometrical and hydromechanical).

There were many scientific and operative lessons learned from the study, including the following:

- First of all, an in-depth understanding of THM processes in the heater-bentonite-rock system was obtained through the study of the Kamaishi experiment. Model results, when compared with test data, show that all the major THM processes involved in the system have been identified and their behavior described.
- The main limitations of current capability of modeling coupled HM processes in fractured rocks are a lack of knowledge about fracture geometry and uncertainty in *in situ* properties.
- Despite great progress in characterization and parameterization of bentonite, models and knowledge of the physical behavior of partially saturated swelling clays still need improvement in areas such as the effective stress behavior, vapor flow, and water retention.
- Our current knowledge of the rock-buffer interaction, especially the hydraulic interactions with the presence of rock fractures is limited and requires more study.
- In the Kamaishi experiment, very limited effect on the hydraulic behavior of the buffer from the surrounding rock and fractures could be observed. This may have resulted from the following:
    a) The *in situ* experiment was not maintained long enough such that possible larger hydro-mechanical interaction between the rock and buffer at the longer period could be observed.
    b) The hydraulic conductivity of the rock is much higher than that of the buffer, so that whether or not fractures in the rock were considered did not result in much difference in the hydraulic behavior of the buffer.
    c) The rock fractures near the buffer might have been sealed by buffer material during installation of the buffer material.
- Also in the Kamaishi experiment, very limited mechanical effect on the buffer from the surrounding rock was observed, which may have resulted from
    a) The low stress magnitudes in the test area, which make the rock and rock fractures mechanically inactive, besides the fact that thermal expansion and stresses induced by heating have also only a limited effect due to low power input and small magnitudes of the temperature gradient
    b) Point-wise measurement of mechanical behavior, which may not be nearly enough to capture essential aspects and patterns of the mechanical responses.

7. CONCLUDING REMARKS

The DECOVALEX project has played a key role in advancing coupled THM models of geomaterials and their testing against field experiments over the last ten years or so. New and sophisticated models have been developed to simulate coupled THM processes in heater-bentonite-rock systems. Their intercomparison through BMT studies, as well as their testing against major field tests in Europe, Japan and the U.S., has provided substantial insight into the effects and impacts on the coupled THM processes of a geological nuclear waste repository. It also gives us an understanding of our modeling capabilities, i.e., what we can and cannot predict well. The international cooperation under the DECOVALEX organizational framework has been very effective and profitable to all participants. Research teams have been able to share the results of very expensive major field experiments, as well as continually provide detailed ideas, technical suggestions, and peer review to each other's work. Such unique in-depth cooperation among multiple national teams has proved to be beneficial to all involved.

On a more general level, it can be seen from all the chapters of this volume, that substantial progress has been made in coupled THM processes in many related fields in the geosciences, ranging from nuclear waste isolation, to geothermal HDR energy extraction, civil engineering, mining projects, and petroleum reservoir dynamics. On the one hand, many challenging and important research issues remain, and interactions and discussions across these fields will be most useful. On the other hand, significant advances already made over the last decade have enabled us to model the relevant basic physics behind the coupled processes, and have allowed us to simulate these processes in realistic complex geologic systems. Now we are in a good position to study and evaluate the behavior of such complex systems under various THM scenarios relevant to practical applications.

ACKNOWLEDGEMENTS

The authors would like to express their gratitude, on behalf of all the research teams participated in the DECOVALEX Project, to all the Funding Organizations and EU for their financial support. They also thank John Hudson for discussion and encouragement, and Hui-Hai Liu and Jonny Rutqvist for their careful review of the manuscript. The first author would like to acknowledge that the paper is prepared with partial support of the Director, Office of Science, Office of Basic Energy Sciences, Geosciences Program of the U.S. Department of Energy under contract number DE-AC03-76SF00098.

REFERENCES

Alonso, E.E. and J. Alcoverro, *FEBEX benchmark test case definition and comparison of different modelling approaches*, this volume, 2004.

Alonso, EE. and S Olivella, *Gas Flow in Porous and Fractured Media*, this volume, 2004.

Bjørlykke, K., F Chuhan, A Kjeldstad, E Gundersen, O Lauvrak and K Høeg, *Modelling of sediment compaction and fluid flow during burial in sedimentary basins*, this volume, 2004.

Chijimatsu, M., L Jing, A Millard, TS Nguyen, A Rejeb, J Rutqvist, M Souley and Y Sugita, *Building confidence in the mathematical models by calibration with a T-H-M field experiment*, this volume, 2004.

Choi, SK., CP Tan and R Freij-Ayoub, *A coupled mechanical-thermal-physico-chemical model for the study of time-dependent wellbore stability in shales*, this volume, 2004.

Choi, SK. and MB Wold, *A coupled geomechanical-reservoir model for the modelling of coal and gas outbursts*, this volume, 2004.

Datta, R., D Barr and W Boyle, *Measuring thermal, hydrologic, mechanical, and chemical responses in the Yucca Mountain Drift Scale Test*, this volume, 2004.

Dusseault, M., *Coupled problems and petroleum geomechanics*, this volume, 2004.

Elsworth D., B Voight, J Simmons, S Young and B Winkler, *Vaporization-induced overpressures as a trigger for the hazardous collapse of lava domes*, this volume, 2004.

Fomin, S., Z Jing, and T Hashida, *The effect of thermal, chemical, hydrological, and mechanical factors on water/rock interaction in HDR geothermal systems*, this volume, 2004b.

Fomin, S. V Chugunov and T Hashida, *Mathematical modeling of the borehole grout off in permafrost*, this volume, 2004a.

Gómez-Hernández, J.J. and EF Cassiraga, *Impact of flow and transport coupling in the upscaling of transport parameters for performance assessment in the context of nuclear waste disposal*, this volume, 2004.

Guimaraes, LDN., A. Gens, and S. Olivella, *Coupled analysis of damage formation around boreholes*, this volume, 2004.

Guo, H., W Xu and Z Wu, *Study on the coupling influence of concrete dam base seepage, stress, and creep on structure behaviors of dam body*, this volume, 2004.

Guvanasen, V. and T. Chan, *Upscaling the THM properties of a fractured rock mass using a modified crack tensor theory*, this volume, 2004.

Han, G. and MB Dusseault, *Coupled analysis of sand stability in petroleum wellbores*, this volume, 2004.

Homepage for DECOVALEX III: www.DECOVALEX.com

Hosni A.,, S Gentier, A Genter, D Billaux and F Dedecker, *Coupled THM modeling of the access zones to the heat exchanger in the Soultz-sous-Forêts geothermal site (France)*, this volume, 2004.

Jing, L., C.-F. Tsang, and O. Stephansson. 1995. *DECOVALEX—An international cooperative research project on mathematical models of coupled THM processes for safety analysis of radioactive waste repositories*. Int. J. Rock Mech. & Mining Sci. and Geomech. Abstr. (Special Issue DECOVALEX I), 32, 389–398.

Lee, GS. and CI. Lee, *Thermo-hydrological analysis to predict the temperature distribution around a cold storage cavern*, this volume, 2004.

Liu, HH., J Rutqvist, Q Zhou and G.S. Bodvarsson, *Upscaling of normal stress-permeability relationships for fracture networks obeying the fractional levy motion, this volume. 2004.*

Liu, J., C Mallett, A Beath, D Elsworth and B Brady, *A coupled flow-transport-deformation model for underground coal gasification*, this volume, 2004.

Niitsuma, H., *Detection of hydraulically created permeable structures in HDR/HWR reservoir by high-resolution seismic mapping techniques*, this volume 2004

Nikolaevskiy, VN. and IA Garagash, *Earth crust structure as result of rock fracturing at high pt – conditions*, this volume, 2004.

Rutqvist, J., M. Chijimatsu, L. Jing, A. Millard, T.S. Nguyen, A. Rejeb, Y.Sugita and C.F. Tsang, *Evaluation of the impact of thermal-hydrological-mechanical couplings in bentonite and near-field rock barriers of a nuclear waste repository in sparsely fractured hard rock*, this volume, 2004.

Shao, JF, Y Jia, and D Kondo, *An elastoplastic damage model for unsaturated argillites*, this volume, 2004.

Stephansson, O., L. Jing and C.-F. Tsang. 1996. *Coupled Thermo-Hydro-Mechanical Processes of Fractured Media—Mathematical and Experimental Studies*. Development in Geotechnical Engineering, 79. Elsevier, Amsterdam. 575 p.

Stephansson, O., C.-F. Tsang, and F. Kautsky. 2001. *Foreword to special issue on DECOVALEX II*. Int. J. Rock Mech. & Min. Sci, 38, 1–4.

Su, K., O. Ozanam, and N. Hoteit, *Desiccation and rehumidification on the thermohydro-mechanical behaviour of the Callovo-Oxfordian argillaceous rock*, this volume, 2004.

Sun, PD., *Three-dimensional numerical simulation of coupled gas leak flow and coal-rock deformation in parallel coal seams*, this volume, 2004.

Thomas, HR., PJ Cleall, N Chandler, D Dixon and HP. Mitchell, *Analysis of the hydraulic interaction between clay buffers and host rock in large scale tests*, this volume, 2004.

Tsang, C.-F. 1987. *Coupled Processes Associated with Nuclear Waste Repositories*. Academic Press, San Diego, 801 p.

Zhu, Z., W Xu and A Zhang, *Coupling model of seepage field and damaging field of fractured rock mass and its application*, this volume, 2004.

Keynote Contributions

PREDICTING SOLUTE TRANSPORT IN FRACTURED ROCKS- PROCESSES, MODELS AND SOME CONCERNS

Ivars Neretnieks
Department of Chemical Engineering and Technology, Royal Institute of Technology, SE-100 44-Stockholm, Sweden

Abstract: Flow and solute transport in fractured crystalline rocks has gained increasing interest in the last decades because several countries plan to site final repositories for high level nuclear waste at large depths in rock formations. Water flow and solute transport by the seeping groundwater are slow and it is not possible to make experiments over the thousands of years and hundreds of meters of interest. Instead one has to rely on models that describe the processes and mechanisms that will be dominant over long times. It is found that most nuclides of interest are strongly retarded in relation to the water velocity because they enter the slightly porous rock matrix by molecular diffusion and sorb on the micropore surfaces in the rock matrix. One of the key questions is how large is the contact surface between the seeping water and the fractured rock.

1. INTRODUCTION AND SCOPE

In several countries there are plans to build a final repository for nuclear waste in deep lying geologic formations. In Sweden and Finland crystalline rock is the preferred geological medium. In these countries the fractures in rocks are water saturated from close to the ground surface and down to very large depths. The repositories will be located about 500 m below the ground. Canisters made of materials with extremely long lifetimes will contain the nuclear waste. They are expected to keep their integrity for hundreds of thousands to many millions of years. Nevertheless, it cannot be ruled out that one or a few canisters may have a small damage, which could allow groundwater to come into contact with the spent fuel. Nuclides would then be released and carried by the slowly seeping groundwater to the biosphere. The water flowrates are expected be very low and the residence times of the water to be long, tens to thousands of years, for a packet of water to arrive from repository depth to the ground. Even so long times are, however, short compared to the lifetimes of some nuclides. Fortunately there are mechanisms that retard most of the nuclides of concern and allow them to decay to insignificance.

The by far most important retardation mechanism is caused by the uptake of the nuclides from the flowing water in the fractures into the (slightly) porous rock matrix in the rock blocks between the conducting fractures. The stagnant water volume in the rock matrix is typically 10-50 times larger than that in the flowing fractures. If time were given for the nuclides in the flowing water to mix with the stagnant matrix water by molecular diffusion a nuclide pulse would be diluted and retarded by a factor of 10-50. Many of the nuclides are positively charged cat-ions and sorb on the negatively charged mineral surfaces within the rock matrix. This can cause an even much larger withdrawal of nuclides from the flowing water into the rock matrix. The retardation effect can be tremendous for the sorbing nuclides allowing them to decay to insignificance (Neretnieks 1980).

It is obviously not possible to make experiments with duration of hundreds of thousands of years and over distance of hundreds of meters in the tight rock formations of interest. It is therefore essential to understand the key processes so well that credible predictions can be made using models based on well-established laws of nature. The models must be supported by experiments that can credibly be extrapolated. The models used are based on the laws of mass and energy conservation and on laws of thermodynamics. The difficulties in applying these laws arise mainly from the fact that the rock mass cannot be described in detail. The location, orientation and detailed hydraulic properties of the fractures cannot be measured in detail. The diffusion and sorption properties of the interior of the rock mass under natural stress cannot be readily measured. Mixing processes of different water packages in fractures and at intersections are not fully understood. All this makes it difficult to build models that account

for all the processes. Simplifications are therefore made attempting to bring out the dominating processes.
Laboratory and field tests are performed to study the different processes. Some of the difficulties in interpreting observations and measurements will be discussed.
In this paper those processes that are deemed to be dominant over large distances and long times are emphasised. Processes that may dominate for flow and tracer transport over time scales of days to years are brought out only in the context of interpreting tracer experiments aimed at obtaining data for the long term processes. Some processes that may dominate experiments performed over short distances and times may become overshadowed by processes that dominate over longer times. The paper will try to bring out the latter.
A set of very thorough review papers have recently been published on processes and modelling discussed in this paper and these can be referred to for more detailed descriptions (Bodin et al. 2003a and 2003b).
One of the main aims of the site investigations for locating a nuclear waste repository is to obtain a coherent picture of all aspects of the rock in order to be able to predict the migration of any possibly leaking radionuclides from the repository.

2. FRACTURED ROCKS

Figure 1 shows a conceptual picture of fractured crystalline rock. The fractures are partly open and allow water to flow in them. At fracture intersections there may be additional conduits. The fracture zones can be much larger than the average fracture spacing and have smaller distances between the fractures within them. It is readily seen that there are considerable difficulties in characterizing rock volumes in detail. Lager fracture zones can be localised by drilling boreholes and making observations and measurements in the holes. The fracture frequency can be determined by observations on the bore cores as well as in the boreholes. Not all fractures and not all of a fracture conduct water.

The frequency of the flowing fractures (or channels) can be assessed by hydraulic testing (Crawford et al. 2003). The details in the channels such as apertures and aperture

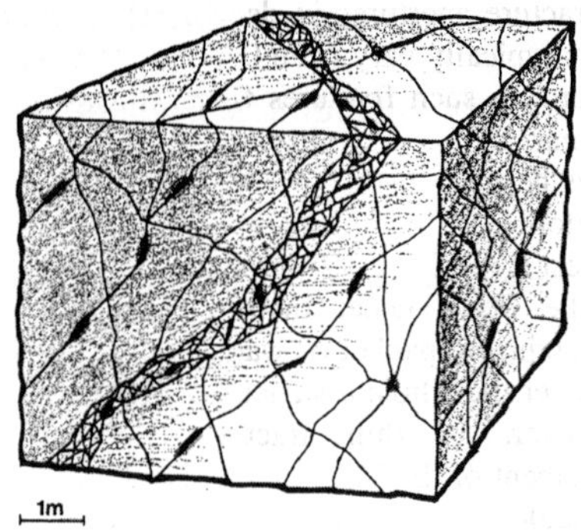

Figure 1. Cartoon of rock with irregular fractures and containing a fractured zone. Water flows in "channels" in the fractures

variations, connections between the channels at the fracture intersections cannot be determined with present day methods. Furthermore as only a limited number of boreholes can be drilled for the observations the measurements and observations sample only a very small fraction of all the fractures and channels. The models therefore must rely on a limited dataset and by necessity are forced to be simplified. Different models are based on different assumptions. Testing the different assumptions with various experiments is therefore a central part of many laboratory and field experiments.

3. PROCESSES AND MECHANISMS THAT AFFECT FLOW AND TRANSPORT IN FRACTURED ROCK

Bodin et al. (2003a) have recently made a thorough compilation of flow and transport processes in fractured rocks. Below only a short summary is made of processes deemed to be relevant for this paper.

We would wish to follow a small packet of contaminated water and study the fate of the mass of nuclides it carries. In the most idealised case the packet keeps its identity from start to finish. There are a number of mechanisms that disturbs this ideal picture. Mixing with other water packages and sorption on the solid material the packet passes are of main concern here.

3.1 Processes in a single fracture

Fracture apertures in deep lying crystalline rocks typically vary from 0.01 mm to fractions of mm. In such fractures flow will be clearly laminar. The water package (or stream) that flows between the two walls of a fracture will develop a velocity profile where the water nearest to wall is stagnant and the water in the middle flows faster than the average. The contaminated packet of water will thus spread out over the whole distance the packet travels. However, in thin fractures the random movement of the individual solute (and water) molecules will jump from one location to another location of the velocity profile. This effect, called Taylor dispersion, will considerably decrease the longitudinal spreading of the water package for the slowly flowing waters discussed here.

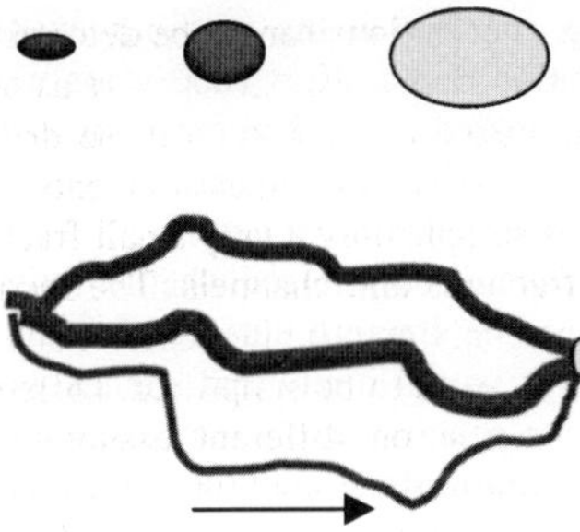

Figure 2. Two dispersion mechanisms in a fracture. Upper part of figure illustrated Fickian dispersion. Lower illustrates velocity dispersion. Flowrates and velocities are larger for thicker lines.

Collecting and mixing the streams would show a dispersal of the original package. This is illustrated in the lower part of Figure 2.

Much more important is the velocity distribution in a fracture caused by aperture variations. Abelin et al. (1991, 1994) have observed that in real fractures a large part of the fractures carry no water and that some or a few small parts carry the water and with different velocities. Should a water package split between a fast and a slow stream in the fracture this would cause part of the packet to arrive later than the other part. This would cause a spread in the Residence Time Distribution, RTD of the combined streams.

Repeated such divisions and later mixing will cause a packet of water to disperse in and across the flow direction. It has been observed that in granular porous media a pulse will spread similarly to spreading by molecular diffusion. This is illustrated in the upper part of Figure 2. Such dispersion can be characterised by a constant dispersion coefficient. In fracture media this is commonly not the case

The meandering channels in a fracture plane can be in contact with each other along part or all of the travel distance. When they are in contact there will exchange of solutes between them by molecular diffusion.

It is of some interest to consider how mixing takes place or does not take place in various situations. This will indicate whether the streams with different velocities and flowrates will mix and even out their concentrations. If they do then the fast paths would be retarded. It will also indicate when a higher spatial resolution in numerical models that discretize the fracture plane is not needed.

In laminar flow or in stagnant liquids, mixing takes place by molecular diffusion. Consider two streams of water, each W wide, that flow side by side during time t. For dimensionless time, Dt/W^2, less than 0.01 the diffusing species have penetrated a very short distance. For a dimensionless time $Dt/W^2 > 1$ the stream has practically equilibrated. This is illustrated in Figure 3. With the data in Table 1 the residence time of the streams is 2 years. The diffusivity in the water for small solutes is around $D=10^{-9}$ m^2/s. Very good mixing will take place when $Dt/W^2=1$ which occurs for W=0.25 m. Thus in this example streams narrower than a few tens of cm will be well mixed. Streams of meter width will be less well mixed but will still have had a noticeable exchange of solutes. Sorbing species will be retarded in their sideways diffusion as well as in their longitudinal movement. Therefore also the sorbing tracer pulse will mix sideways over comparable distances.

Observations of flow distribution in natural fractures show stream widths varying between cm and tens of cm (Abelin et al. 1991,1994). Measurements of aperture variations indicate correlation lengths in this range (Hakami and Larsson 1996). Simulations of flow in variable aperture fractures also suggest that stream widths from cm to tens of cm can be expected in fractured granitic rocks.

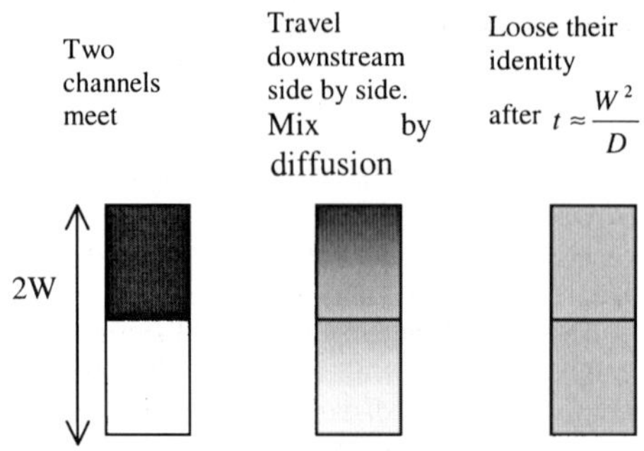

Figure 3. Diffusional mixing of two streams that travel side by side. After some time they loose their identity.

Matrix diffusion is another mechanism that causes mixing. The rock matrix is porous with a porosity ε_p in granites of typically 0.1-0. 5 %. The water in the rock matrix is stagnant. Mixing between the flowing water in the fractures and the matrix water takes place by molecular diffusion. A water package in the fracture that equilibrates with the matrix water by molecular diffusion will be diluted. A contaminant pulse travelling in a fracture will be retarded in relation to the advective velocity of the water.

For very long contact times the penetration by diffusion can reach far into the rock matrix whereas for short times the penetration depth will be small. From diffusion theory one can derive a simple expression for the penetration depth, here defined as the distance that leading edge of the solute has travelled. The concentration at the leading edge of the penetrating solute is here taken to be 1% of that at the fracture surface. The relation is

$$\eta_{0.01} = 4\sqrt{D_a t} \tag{1}$$

D_a is the apparent diffusion coefficient of the solute, t is the contact time. For small non-reactive solutes in water D_a is about 10^{-11} m^2/s. For a 3 year contact time a penetration depth of about 12 cm is reached. Somewhat larger penetration distances were actually found in a field experiment lasting 3.5 years (Birgersson and Neretnieks 1990). For a 1000 year contact time (A damaged canister could leak for a very long time) the penetration depth will be about 2 m.

The sorbing radionuclides will be retarded in their diffusion into the matrix by a factor *R* because they sorb to the micropore surfaces and are constantly withdrawn from the water. They accumulate on the surfaces and can attain high concentrations there. The volumetric sorption coefficient K differs between nuclides. For nonsorbing nuclides K=0 but for sorbing nuclides it can range from less than 1 to hundreds of thousands.

$$R = \frac{\varepsilon_p + K(1-\varepsilon_p)}{\varepsilon_p} = 1 + \frac{K(1-\varepsilon_p)}{\varepsilon_p} \tag{2}$$

For ε_p=0.005 a nuclide with *K*=50 will be retarded by a factor of 10 000 and have an apparent diffusivity 10 000 times lower than a nonsorbing species. In the previous example the penetration depth, after 1000 years would be 2 cm compared to the 2 m for a nonsorbing nuclide. A 100 000 year contact time will give 10 times longer penetration depths.

Questions have been raised whether only the rock nearest the fracture surface (1-5 cm) is porous. This would imply that solutes would not be able to penetrate more than a few cm (in rock under natural stress conditions.

Electrical conductivity measurements with long samples, up to tens of cm have shown that the rock is porous and the porosity is connected over these distances (Skagius and Neretnieks, 1986a,b, 1988, Ohlsson et al. 2001. Löfgren and Neretnieks (2003a) in laboratory experiments have shown that ions can be brought through long rock samples in a DC field clearly showing that the pore system is connected. They have compared electrical conductivity measurements in deep boreholes with diffusion and electrical conductivity measurement on cores from the same boreholes. The in-situ electrical measurements penetrate the rock up to several meters. The measurements show that the in-situ rock has 2 to 4 times lower electrical conductivity compared to the laboratory samples. This agrees well with the in-situ diffusion measurements of Birgersson and Neretnieks (1990) and those of Skagius and Neretnieks (1986b) where samples were tri-axially recompressed in the laboratory to several hundred bars. We no longer doubt that rock in-situ has a connected porosity extending several meters.

In some fractures alterations and somewhat increased porosity of the rock can be observed a few cm from the fracture surface. In this region molecular diffusion would be further facilitated. Some fractures contain fracture infill materials that also can add to the sorption and retardation effects.

Sorption coefficients are mostly measured on crushed rock samples because the sorbing nuclides would take extremely long time to fully penetrate even a one cm coupon of rock. This raises questions whether the sorption coefficients in intact rock can be derived from crushed material as the crushing induces new micro fractures in the grains and also creates new surfaces on the outside of the grains. Löfgren and Neretnieks (2003b) have proposed that a DC field could be used to induce electro-osmotic flow in rock pieces of rock and also at the same time the electro- migration will further speed up the ionic movement in the pores. Preliminary tests show that an increase of ionic migration velocity by several hundred to a thousand times can be achieved compared to that of molecular diffusion. This could potentially permit measurements of sorption coefficients in intact rock samples.

The exchange of solutes between the stagnant water in the matrix and the flowing water in the fractures takes place over the contact surface between the stream and the rock. This is often called the Flow Wetted Surface, FWS. Obviously, for a given packet or stream of water, the larger the FWS the faster will be the exchange. As fractures do not conduct water evenly over the surface of the fracture, and even are closed at the contact points, different paths in the fracture plane will have different flowrates and these flowrates will see different magnitudes of FWS. In a flowpath with average width W and length L the FWS is $2LW$. There will thus be a distribution of LW/q of the streams (or water packages) traversing a fracture. This is illustrated in the lower part of Figure 2.

For a channel with no dispersion where the incoming water has a concentration c_o the concentration at the outlet can be obtained from Equation (3) (Neretnieks 1980). The Residence Time Distribution, RTD. in each stream will be strongly influenced by the LW/q ratio.

This is illustrated in Figure 4 for a case for some typical data from Table 1

$$\frac{c}{c_o} = Erfc\left(\frac{LW}{q}\frac{MPG}{\sqrt{t-t_w}}\right) \quad (3)$$

The entity LW/q depends only on the geometry of the flow path and the flowrate whereas the second entity, called the Materials Property Group, MPG, depends on the rock matrix porosity, the pore diffusion coefficient and the retardation coefficient that is determined by the sorption coefficient.

$$MPG = \varepsilon_p \sqrt{D_p R} \quad (4)$$

The following data are used in the example. They are within the range for fractured rocks in Sweden and for long-term repository conditions at depth.

Table 1. Data used in the examples

$K_h=10^{-8}$ m/s
$L=6$ m
$i=0.003$ m/m
$D_p=2\ 10^{-11}$ m^2/s
Pe=10
K=1 and 50
$\varepsilon_f=3\ 10^{-4}$
$\varepsilon_p=5\ 10^{-3}$
$a_R=1$ m^2/m^3 (FWS/volume of rock)
$\sigma_{cLog10}=0.8$ (Standard dev. of flowrates)

Derived entities

$u_f=K*i/\varepsilon_f=10^{-7}$ m/s->3m/a
$q_{fracture}=3.6\ 10^{-10}$ m^3/s
$t_w=2$ years
$\delta=6*10^{-4}$ m
$LW/q=10^{11}$ s/m

Figure 4 illustrates the very strong influence on the breakthrough curve for a sorbing species by a moderate change in LW/q. After 100 years the stream that has a large LW/q will give a breakthrough concentration of a few % of that at the inlet, whereas the stream with a small LW/q will already have attained 80 % of the inlet concentration. Seen in another way: In the faster case the 0.01 % edge concentration arrives 0.65 years after the water residence time t_w, whereas for the slower case this happens after 53 years. In general a change

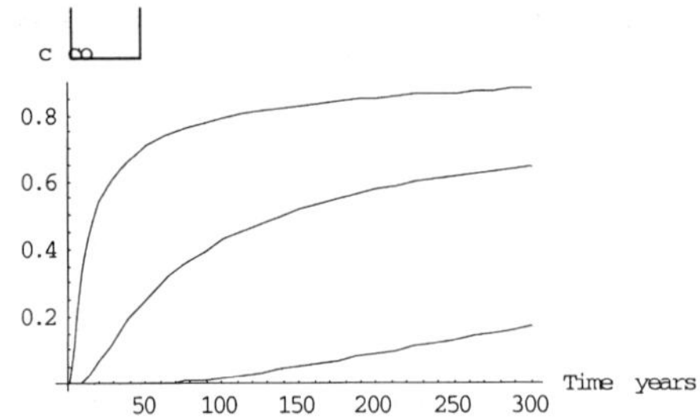

Figure 4. Breakthrough curves for three streams in a fracture. One stream has the average LW/q, one 3 times smaller, lower curve, and one three time larger, upper curve.

in *LW/q* by a factor x will change the time of arrival of a given fraction of the inlet concentration by x^2.

The Residence Time Distribution, RTD, in a pathway depends in a strongly non-linear way on *LW/q*. It may be noted that in the example all three streams were taken to have a water residence time of 2 years. The water residence time has a small influence on the RTD of a sorbing tracer in this case. Moderate changes in *LW/q* can cause dramatic changes in the RTD. The factor of nine difference in *LW/q* between the fast and the slow channel used in the example is by no means an upper bound of possible variations. This suggests that it is important to consider the presence of pathways (channels) with high flowrates and small FWS's. It may also be noted that a change in the retardation factor only influences the RTD linearly. On the other hand the retardation factor may vary from 1 for a nonsorbing nuclide to hundreds for weakly sorbing nuclides and millions for strongly sorbing nuclides.

3.2 Processes at fracture intersections and in networks

In one and the same fracture the apertures are correlated so that the aperture at one point in the fracture close to another point usually is much more alike than at points further away. This may not be the case for points in two different fractures that intersect each other. A potentially conductive channel in one fracture may meet a closed part in the other fracture it intersects. Flow would not be possible in this case. There are some observations that the intersection itself may form a conduit. When and where this occurs the water flowing in a channel in one fracture will flow along the conduit at the intersection until it reaches a channel in the other fracture. At present there are practically no investigations and observations in real fractures that can be used to quantify and illuminate these questions. There are several of numerical Channels Network models that can be used to simulate these processes but because of lack of data no firm conclusions as to the importance of these issues has been reached. In an attempt to bypass these problems (Moreno and Neretnieks 1993) formulated a simple channel network model that does not use detailed information on aperture variations and fracture orientations. It uses information obtained from hydraulic measurements in boreholes on the flowrate distribution and the frequency of conductive fractures found in the boreholes. The model assumes that every measured flowing fracture in a borehole represents a channel with constant properties

In order to follow a packet of water along the different fractures it traverses it is necessary to have information on the mixing and division processes at intersections. Two streams from different fractures that meet flat side by flat side at an intersection will most likely mix already in the intersection because although the residence time in the intersection is short, the distance to diffuse is also short. Should the mixing not be complete at the intersection it will occur as the streams flow along the fracture. Mixing in the direction perpendicular to the aperture will very quickly even out any concentration difference in this direction by the reasons discussed above for mixing of parallel streams. By the time the molecules in a stream have reached the next intersection they will have forgotten if they were near this wall or the other when they entered because they will have bounced back and forth between the walls so many times.

In the appendix the solution of the equations for flow and matrix diffusion in a channel with variable properties is derived. The solution shows that one can follow a constant stream *q* through a path with arbitrarily varying velocities, widths and material properties.

When one attempts to follow a "packet of water" through the network the packet, however small, will have to divide at fracture intersections. If it is a well-mixed stream it will divide into the outgoing streams in proportion to the total flowrates in the outgoing streams.

The original stream with flowrate q will be split into smaller and smaller fractions as it proceeds through the network. However, if we follow each increasingly smaller fraction through the system each fraction will experience the same residence time as the other water traversing a fracture. Therefore the residence time of each fraction will be the sum of the residence times in all the paths it has visited. In the same way the (little) stream q contacts its fraction of L_iW_i in the fracture with total flowrate Q_i. The LW/q for the whole path is the sum of L_iW_i/Q_i of all the channels that have been visited.

Figure 5 illustrates different paths that are taken by different parts of the original stream emanating from one point in a fracture network. All the paths will have a different RTD. Each path will have its own mean travel time. When matrix diffusion is active there will also be a spread in residence times along each trajectory such as seen in Figure 5. The mean arrival times for the different trajectories will be different giving rise to velocity dispersion which then is superimposed on the spreading along each trajectory caused by the matrix diffusion effects.

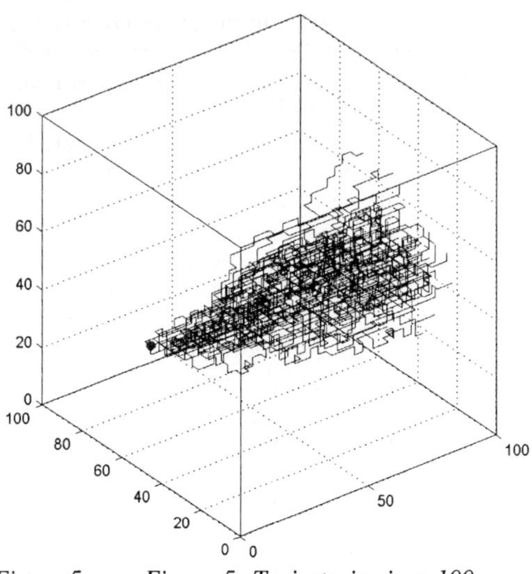

Figure 5. Figure 5 Trajectories in a 100 m cube

4. FRACTURE ZONES

Fracture zones can extend over hundreds of meters to many kilometres and vary in width from just a few fractures close to each other up to many tens of meters containing small rock blocks and degraded rock that has formed clays and other secondary minerals

The hydraulic transmissivities of the zones usually are larger, or much larger than the common smaller fractures in the rock although locally the zones may have lower conductivities in places containing much clay. Not atypical distances between the smaller zones are 100s of meters. The larger zones are much less frequent. The network of zones can often have a large influence on the flowrate distribution in the bedrock. The higher flowrates in the zones will tend to cause higher water velocities and although there will be a larger FWS because of the higher fracture frequency the entity LW/q may become smaller. Therefore also the sorbing nuclides will tend to move faster in the zones compared to the velocities in the smaller fractures. As in the smaller fractures the variations in local transmissivities can vary over many orders of magnitude. This implies that the water and nuclide movement must be viwed as a stochastic process.

5. MODELLING FRACTURED ROCK AS A CONTINUUM

Because of the stochastic nature of the system it has long been attempted to model the flow and transport in the fractured rocks as were it taking place in a porous medium with average properties. The Advection-Dispersion equation, AD-equation that has been found useful in beds of small particles such as sand has been used extensively. However, there is increasing concern that it may not be applicable. Observations in drifts and tunnels show that even over distances of hundreds of meters the flowrate distribution is extremely uneven. (Abelin et al.1991). Also the basic assumption that dispersion can be characterised by a constant dispersion coefficient does not agree with observations. The dispersion coefficient seems to increases with observation distance up to km distances at least (Gelhar et al. 1992).

Nevertheless, simulations using the AD-equation and Network models give not very dissimilar results provided the same LW/q is used. This is shown in Figure 6 where three models are compared. A single fracture, a network of fractures with stochastic transmissivities and the AD-equation. For the

AD-equation a Peclet number of Pe=10 is used. This gives a spread in the RTD (dispersion) that is within the range of observed values in the field. The standard deviation of the transmissivities of the channels in the network model was chosen such that the RTD of a nonsorbing tracer also has a Pe=10. In the time scale presented there seem not be dramatic differences between the model results for the Network and the AD-models. It should be noted, however, that the early arrival times are of special interest for radionuclides that decay. There the differences are considerable. Similar results were obtained in comparisons of different models using very similar data bases (Selroos et a. 2002). In that comparison a discrete fracture network model, a channel network model and a stochastic continuum model were used.

It is at first surprising that the AD equation gives an earlier arrival than the network model because the AD model uses and average *LW/q* or (a_R/u_o) which implies that even those channels with smallest flowrates contribute equally to the retardation of the solutes by the matrix interaction. However, the meandering paths as seen in Figure 5 often travel longer than what would be obtained superimposing the same hydraulic gradient over the loc modelled as is done for the AD model.

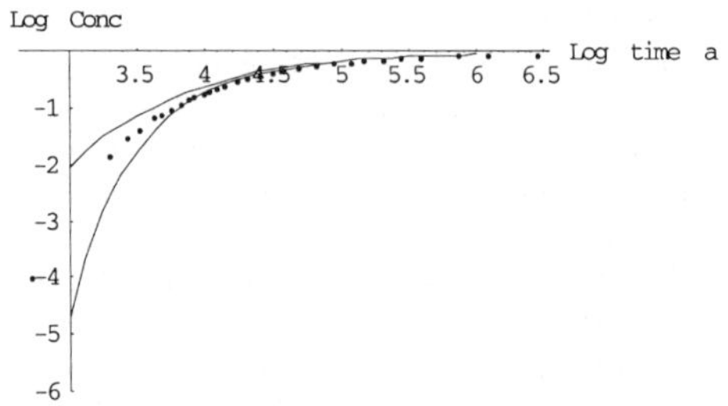

Figure 6. Break through curves for K=1 and Pe=10 for AD equation, CNM (dots) and single channel (from top down) for transport through a 100 m block. Data from Table 1.

6. DATA NEEDED AND HOW THEY ARE OBTAINED

It has been pointed out that both hydraulic data and materials property data are needed in order to simulate radionuclide transport in fractured rocks. Sorption coefficients are commonly measured in batch experiments on crushed rock samples. There are concerns that this may not represent values for intact rock.

Diffusion coefficients and rock matrix porosities are measured on rock samples in the laboratory. Commonly they are not re-stressed to conditions at repository depth. In-situ techniques are being developed to measure diffusivities in-situ.

The entities needed to assess *the LW/q* for the fracture networks are determined from down-hole measurements. The frequency of flowing fractures and the transmissivity (actually flowrate) distributions are obtained by injecting or withdrawing water at narrow and well defined locations along the whole lengths boreholes. An example of the transmissivity distribution from five boreholes at the ÄSPÖ rock laboratory is shown in Figure 7. The average spacing between flowing fracture (Conductive Fracture Frequency, CFF) gives information of how dense the fracture network is. The CFF gives information on the FWS per volume of rock a_R. Together with the flux u_o expected for a given hydraulic gradient it can give a first indicator of the retention properties of the rock mass. a_R/u_o can be seen as an average *LW/q* for the whole rock mass. It can be used in the AD equation or in Equation (3) for scoping calculations of the RTD.

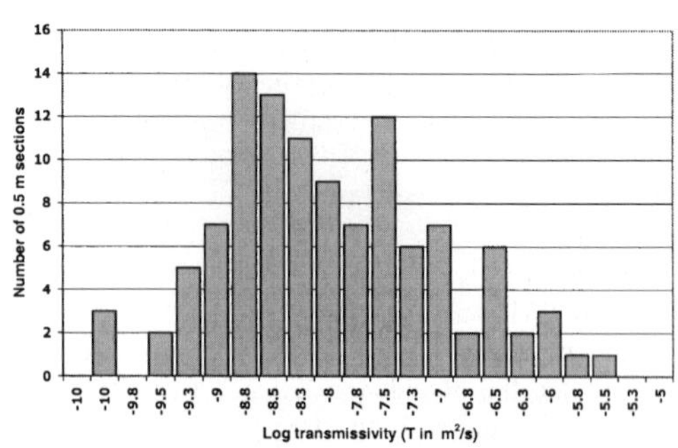

Figure 7. Histogram of transmissivities from boreholes at ÄSPÖ rock laboratory

7. INTERPRETING FIELD EXPERIMENTS

There have been a number of field experiments over distances ranging from few meters to 30-50 m in fractured crystalline rocks with the aim of testing the existing models for flow and transport, some of which are discussed in this paper (Abelin et al. 1991,Winberg et al. 2000, Andersson et al. 2002). Results from tracer experiments in the field are sometimes used to validate the concept underlying the models and to verify that independent data can be used to predict the outcome of the experiments. Some difficulties have been found in such attempts. One major difficulty is that the man made experiments are performed over months to years whereas the aim is to simulate processes that will have durations of thousands of years. Different retentions processes and mechanisms may dominate in the different time scales. One example related to the matrix diffusion is that in short term experiments the matrix diffusion will only penetrate the first few mm from the fracture wall. That part of the rock is sometimes altered and is more porous and the exchange can be faster than in unaltered rock. For sorbing tracers even a small amount of infill material can dominate the retardation in the short term whereas in the long term the infill will have equilibrated early and the main retardation would be due to uptake into the deeper parts of the matrix. In practice the pathways are unknown and it is not possible to follow them. In the Stripa 3D experiment that had a duration of several years and where 9 different tracers were used over distances of 11-40 m it was found that the non-recovery of part of the traces could be explained by uptake into matrix. The CFF that gave the FWS and the matrix property data obtained in laboratory could reasonably well explain this behaviour.

Converging flow and tracer tests at the ÄSPÖ Hard rock laboratory in Sweden using non-sorbing as well as sorbing tracers in fractured crystalline rocks were performed recently (Winberg et al. 2000). Laboratory data on diffusion and sorption properties were used to predict the Residence Time Distribution, RTD, of the sorbing tracers. The field explorations were mainly aimed at assessing the hydraulic properties of one specific fracture that was deemed to be the main flow conduit, the so-called Feature A. It was found that the residence time distributions could not be explained if it was assumed that all flow takes place in only Feature A. An approximately 30 times larger Flow Wetted Surface or 1000 times larger matrix diffusivity or sorption coefficients than what laboratory data indicated would be needed (Neretnieks and Moreno 2002).However, this could caused by inflow of water to the pumping hole from other fractures present in the rock mass which could invalidate the assumption of one major fracture carrying all the pumped water. The fracture frequency and the transmissivity distribution measured in the rock mass adjacent to the experimental site suggest that this could have occurred.

Another set of experiments was made over 14-33 m distances (Andersson et al. 2002). Similar problems in interpreting the test were experienced. The results suggest that what at first seemed to be prominent rather isolated fractures belong to a complex network of intersecting fractures that strongly influence the flow distribution in the rock. (Neretnieks 2003)

8. CONCLUSIONS

Solute transport in fractured crystalline rock will be strongly influenced by the exchange of solutes between the flowing water and stagnant water in the rock matrix. There is increasing evidence that the matrix porosity is connected over at least meter distances. For the slow natural flowrates the solutes will be retarded compared to the seeping velocity of the water. Sorbing species will be even more retarded because they are withdrawn from the moving water by sorption on the mineral surfaces on the fractures and in the rock matrix. There are some concerns that sorption coefficients measured on crushed rock may not be representative of whole rock pieces. The conductive fracture frequency will determine the contact surface between the seeping water and the rock. There is an increasing consensus on how flow and transport should be modelled in fractured rock systems. Fracture and channel network models are being increasingly more used compared to the traditional Advection-Dispersion models.

9. REFERENCES

Abelin H., Birgersson L., Moreno L., Widén H., Ågren T., Neretnieks I. *A Large Scale Flow and Tracer Experiment in Granite II. Results and interpretation.* Water Resources Research, 27(12), p 3119-3135, 1991

Abelin H., Birgersson L., Widén H., Ågren T., Moreno L., Neretnieks I. *Channeling experiments in crystalline fractured rocks.* J. Contaminant Hydrology. 15, p129-158, 1994

Andersson P., Byegård J., Winberg A., *Final report on the TRUE block scale project, 2Tracer tests in the block scale,* SKB TR-02-14, May 2002

Bodin J., Delay F., de Marsily G., *Solute transport in a single fracture with negligible matrix permeability: 1. fundamental mechanisms,* Hydrology J., 11,418-433, 2003a

Bodin J., Delay F., De Marsily G., *Solute transport in a single fracture with negligible matrix permeability: 2. mathematical formalism,* Hydrology J., 11,434-454, 2003b

Birgersson, L., and I. Neretnieks, *Diffusion in the matrix of granitic rock. Field test in the Stripa mine.* Water Resources Research 26 p 2833-2841, 1990

Crawford J., Moreno L., and Neretnieks I., *Determination of the flow-wetted surface in fractured media,* J. Contaminant Hydrology, 61, p 361-369, 2003

Gelhar L.W, Welty C., Rehfeldt K.R., *A critical review of data on field-scale dispersion coefficients,* Water Resources Research 29, p1955-1974, 1992

Hakami E., Larsson E. *Aperture measurements and flow experiments on a single natural fracture,* Int J Rock Mech Mineral Sci Geomech abstr 33(4) p 395-404, 1996

Löfgren M., Neretnieks I. *Formation factor logging by electrical methods: Comparison of formation factor logs obtained in situ and in the laboratory,* J. Contaminant Hydrology, 61, p 107-115, 2003a

Löfgren M., Neretnieks I. *On the pore connectivity in intrusive igneous rock. Formation factor measurements with direct and alternating current,* Paper presented at Migration 03, Gyeongju, Korea, September 2003b

Moreno L. and I. Neretnieks, *Fluid flow and solute transport in a network of channels.* J. Contaminant Hydrology 14, 163- 192, 1993

Neretnieks, I., *Diffusion in the rock matrix: An important factor in radionuclide retardation*? J. Geophys. Res. 85, p 4379-4397, 1980.

Neretnieks I, Moreno L. *Analysis of the TRUE experiments considering a 3D flow-pattern.* Paper presented at the international Groundwater symposium, Berkeley, California March 25-28, 2002, Proceedings. Ed. Findikakis A. N.

Neretnieks I. *Interpretation of some in-situ tracer test experiments in fractured crystalline rock at Äspö hard rock laboratory,* Paper presented at this congress, 2003

Ohlsson Y, Löfgren M, Neretnieks I., *Rock matrix diffusivity determinations by in-situ electrical conductivity measurements,* J Contam Hydrol 47, 117-125 , 2001

Skagius, K., Neretnieks I., *Porosities of and diffusivities of some non-sorbing species in crystalline rocks.* Water Resources Res. 22, p 389-398, 1986a

Skagius, K., Neretnieks I., *Diffusivity measurements and electrical resistivity measurements in rock samples under mechanical stress.* Water Resources Research 22 (4), p 570-580, 1986b

Skagius, K., Neretnieks I., *Measurements of cesium and strontium diffusion in biotite gneiss.* Water Resources Research 24 (1), p 75-84, 1988.

Selroos J-O., Walker D.D., Ström A., Gylling B., Follin S., *Comparison of alternative modelling approaches for groundwater flow in fractured rock,* J Hydrology 257,p 174-188, 2002

Winberg A., Andersson P., Hermanson J., Byegård J., Cvetkovic V., Birgersson L. *Final report of the first stage of the tracer retention understanding experiments,* Technical Report, TR 00-07 , Äspö Hard Rock Laboratory, March 2000.

APPENDIX

Solution of the equations for flow and matrix diffusion in a channel with variable properties

The equations describing transport of a solute with advective flow in a stream and matrix diffusion of the solute are

$$\frac{\partial c}{\partial t} + u\frac{\partial c}{\partial x} = a_w D_e \frac{\partial c_p}{\partial z}(z=0) \text{ for x>0 and t>0} \quad \text{(A1)}$$

$$\frac{\partial c_p}{\partial t} = D_a \frac{\partial^2 c_p}{\partial z^2} \text{ for all x>0 and z>= 0} \quad \text{(A2)}$$

a_w, D_a, D_e and u depend on x but not on t

Introduce a new variable into (A2)

$$Z = z\frac{1}{\sqrt{D_a}} \quad \text{(A3)}$$

(A2) becomes

$$\frac{\partial c_p}{\partial t} = \frac{\partial^2 c_p}{\partial Z^2} \quad \text{(A2a)}$$

and (A1) becomes

$$\frac{\partial c}{\partial t} + u\frac{\partial c}{\partial x} = \frac{a_w D_e}{\sqrt{D_a}}\frac{\partial c_p}{\partial Z}(Z=0) \quad \text{(A1a)}$$

Introduce new variables θ to follow the stream with its velocity $u(x)$ and a length variable X. Noting that we can chose to follow a stream with constant flowrate q in a streamtube X can be written as the RHS of (A4)

$$X = \int_0^x \frac{a_w D_e dx'}{\sqrt{D_a} u} = \frac{2}{q}\int_0^x W(x')\frac{D_e}{\sqrt{D_a}}dx'$$

$$= \frac{2x}{q}\left(\frac{W(x)D_e}{\sqrt{D_a}}\right)^{Average} \quad \text{(A4)}$$

$$dX = \frac{2WD_e dx}{\sqrt{D_a} q} \quad \text{(A5)}$$

Introduce the water residence time to travel from 0 to x

$$t_w = \int_0^x \frac{dx'}{u} \quad \text{(A6)}$$

and a new variable

$$\theta = t - X \quad \text{(A7)}$$

With these transformed variables (A1a) and (A2a) become

$$\frac{\partial c}{\partial X} = \frac{\partial c_p}{\partial Z}(z=0) \quad \text{(A1b)}$$

$$\frac{\partial c_p}{\partial \theta} = \frac{\partial^2 c_p}{\partial Z^2} \quad \text{(A2b)}$$

The Laplace transforms of (A1b) and (A2b) are

$$\frac{\partial \tilde{c}}{\partial X} = \frac{\partial \tilde{c}_p}{\partial Z}(z=0) \quad \text{(A1c)}$$

$$\frac{\partial^2 \tilde{c}_p}{\partial Z^2} = p\tilde{c}_p \quad \text{(A2c)}$$

(A2c) is solved with boundary conditions $\tilde{c}_p(z=0) = \tilde{c}$ and $\tilde{c}_p(z->\infty) = 0$ from which

$$\frac{\partial \tilde{c}_p}{\partial Z}(z=0) = -\tilde{c}\sqrt{p} \quad \text{(A8)}$$

and

$$\frac{\partial \tilde{c}}{\partial X} = -\tilde{c}\sqrt{p} \quad \text{(A9)}$$

The solution to (A9) with the BC that c=1 at x=0, which gives const=$1/p$, is

$$\tilde{c} = \frac{e^{-\sqrt{p}X}}{p} \quad \text{(A10)}$$

The inverse Laplace transform of (A10) gives

$$\frac{c}{c_o} = Erfc\left(\frac{X}{2(\theta)^{\frac{1}{2}}}\right) = Erfc\left(\left(\frac{D_e W}{\sqrt{D_a}}\right)^{Average}\frac{x}{2q(t-t_w)^{\frac{1}{2}}}\right) \quad \text{(A11)}$$

For a Dirac pulse the solution is

$$c\,q = \dot{m} = \frac{M \cdot G}{q\sqrt{\pi}}\frac{e^{\frac{G^2}{q^2(t-t_w)}}}{(t-t_w)^{3/2}} \quad \text{(A12)}$$

where M is the injected mass and MPG is called the Materials Property Group

$$G = 2\int_0^x MPG(x')W(x')dx' =$$

$$2\int_0^x W(x')\frac{D_e}{\sqrt{D_a}}dx' \quad \text{(A13)}$$

When W, L Q and MPG are piecewise onstant G/q is obtained from

$$\frac{G}{q} = 2\sum_i \frac{MPG_i W_i L_i}{Q_i} \quad \text{(A14)}$$

NOTATION

a_R FWS per volume of rock
a_w FWS surface per volume of water
c Concentration of solute
c_o Concentration at inlet
c_p Concentration in matrix pore water
D_a Apparent diffusion coefficient
D_e Effective diffusion coefficient
D_L Dispersion coefficient
D_p Diffusion coefficient in pore water
D_w Diffusion coefficient in free water
i Hydraulic gradient
K Volume based sorption coefficient
K_h Hydraulic conductivity
L Length of flowpath
M Mass of solute injected in a stream
R Retardation factor in matrix
q Constant flowrate stream
Q Flow rate in a channel
u Water velocity
u_f Water velocity in fracture
u_o Water flux (Darcy velocity)
$W(x)$ Local width of a fracture
t Time
t_w Water residence time
x Distance along flow direction
z Distance into matrix
δ Fracture aperture
ε_f Flow porosity
ε_p Matrix porosity

MODELLING GAS FLOW THROUGH DEFORMABLE FRACTURED ROCKS

S. Olivella and E.E. Alonso
[1]) Geotechnical Engineering Department, Technical University of Catalunya,
c/ Jordi Girona 1-3, Edificio D-2, 08034 Barcelona, Spain.

Abstract: This paper presents a model for the simulation of two phase flow phenomena in deformable fractured rocks. The main problem is that gas pressure may play an important role on the mechanical behavior specially if discontinuities exist or develop. Fracture aperture changes and fracture failure are accounted by the model. Aperture of the discontinuity is used as the main variable for permeability and capillary pressure variations. Injection pressures that show peak values before steady state regime is attained are obtained with the model as shown in some simulations performed.

1. INTRODUCTION

Gas generation due to corrosion of radioactive waste canisters and migration through the geological barriers is of major concern in the context of safety analysis. The generated gas may accumulate and, as the pressure increases, it will start flowing through the engineered barrier and eventually through the geological barrier (Gens et al 2001, Alonso et al 2002). The mechanisms of gas migration can be: diffusion, two phase flow and two phase flow coupled to mechanical effects which may lead to the fracturing of the porous media. While diffusion and two phase flow have been investigated in great detail, two phase flow coupled to deformation has received less attention.

The low permeability of the engineered and geological barriers implies that the capillary pressure to start desaturation (air entry pressure) will be high (small pore sizes). Then, gas pressure may increase reaching values that may lead to gas fracture processes or simply gas flow through existing discontinuities that may undergo changes in its aperture. Modelling gas flow under these conditions requires the introduction of mechanical formulations that allow taking into account fracture opening or fracturing formation plus subsequent opening. In such case fracture aperture is a key variable.

This paper describes a procedure to model in a unified manner two phase flow with concentrated gas flow through fractured porous rocks.

2. MODEL DESCRIPTION

The basic idea of the model presented here consists in the appropriate representation of a single fracture embedded in a continuous finite element. Figure 1 shows a fractured rock composed by a matrix and a single fracture.

Hydraulic and mechanical effects have to be included because gas flow depends critically on the mechanical interactions because they control fracture aperture or porosity changes.

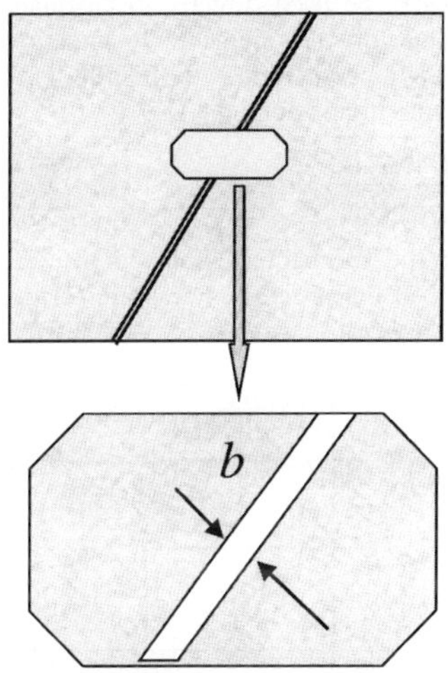

Figure 1. Rock with a single idealized fracture.

Consider first the flow phenomena through a single fracture. Liquid and gas flow will be calculated using Darcy's law. The most important parameter in this law is the intrinsic permeability which can be calculated, assuming laminar flow, as:

$$k_{fracture} = \frac{b^2}{12} \qquad (1)$$

where b is the aperture of the single fracture. When this fracture is included in a finite element, the intrinsic permeability of the element can be calculated as:

$$k = k_{fracture}\frac{b}{s} + k_{matrix}\frac{s-b}{s} \cong \frac{b^3}{12s} + k_{matrix} \qquad (2)$$

where s is the element size (width normal to flow direction) and k_{matrix} is the reference intrinsic permeability of the rock matrix, i.e. the rock without fractures. Permeability of the matrix will be important only for very low apertures; otherwise fracture permeability will dominate the total permeability and matrix permeability will be negligible.

The second hydraulic process that is included in the fracture is the variation of capillary pressure caused by changes in the aperture. According to Kelvin's law the capillary pressure necessary to desaturate a fracture is given by:

$$P_o = \sigma\left(\frac{1}{r_1}+\frac{1}{r_2}\right)=\frac{2\sigma}{b} \tag{3}$$

which is obtained when $(1/r_1) = 0$ and $r_2 = b/2$ (the wetting angle has been assumed to be 0). This equation can be used directly to calculate the air entry value of the element. If equation (3) is combined with equation (2) the capillary pressure to start desaturation is obtained as:

$$P = P_o \frac{\sqrt[3]{k_{fracture\_o}}}{\sqrt[3]{k_{fracture}}} \tag{4}$$

From the mechanical point of view, we are interested in the process of fracture formation and aperture.

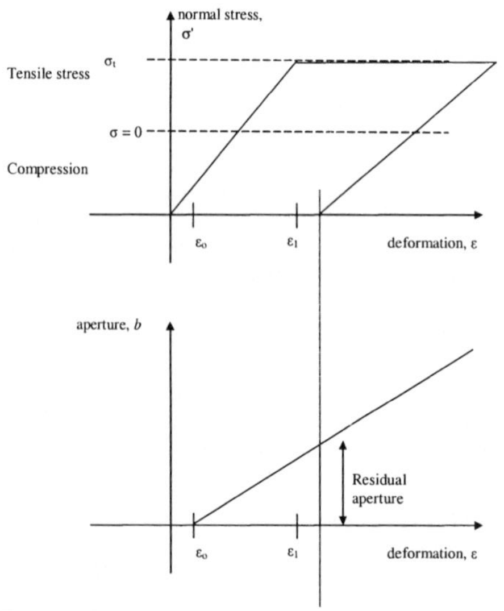

Figure 2. Stress-strain behavior coupled to fracture aperture changes.

The aperture of the fracture can be estimated as a function of deformation in the following way:

$$b = b_o + \Delta b$$
$$\Delta b = s\Delta\varepsilon = s(\varepsilon - \varepsilon_o) \tag{5}$$

The assumption is here that deformation is localized and results in changes in aperture. A threshold value (ε_o) is considered. Therefore the changes in aperture start when deformation reaches this value. Obviously, deformation perpendicular to the fracture plane should be used when aperture changes have to be obtained. The threshold value (ε_o) is associated with fracture initiation. It marks the initiation of the failure of the rock in tension. This parameter will be set to zero if the fracture already exists and has an initial aperture b_o. The initial aperture can be nearly zero when the fracture exists but is closed.

The stress-strain behaviour of the rock, including fracture formation, is a crucial component of the aperture changes. As mentioned before the deformation normal to the fracture is considered in equation (5) in order to obtain aperture changes. If an elastoplastic model is considered for the rock mass behaviour, fracture initiation can be associated with tension stresses. Fracture orientation is sensitive to the stress tensor orientation so the plane where the minimum principal stress (compression positive) occurs defines the plane of fracture formation.

Figure 2 shows the general stress-strain behaviour coupled to aperture changes. A threshold strain (ε_o) defines the initiation of fracture aperture. A strain corresponding to failure is indicated (ε_1). Failure is achieved when the normal stress reaches the tensile strength (σ_t). ε_o, ε_1 and σ_t are model parameters.

From this general model the following cases can be established:

a) Existing fracture initially closed:

$$\sigma_t = 0,\ \varepsilon_o \approx 0$$

b) Non existing fracture:

$$\sigma_t \neq 0 \text{ and } \varepsilon_o = \varepsilon_1 \neq 0$$

The first case (a) corresponds to a fracture that due to compression stresses is almost closed or has an initial aperture b_o. Normal extension will induce aperture opening right from the beginning of stressing. As the normal stress reaches a zero value, deformation will increase under constant stress (irreversible deformations). Unloading will imply fracture closure but a residual aperture is expected due to irreversible deformations. An appropriate elastic modulus could be used if complete closure had to be achieved.

The second case (b) implies that a tensile strength exists and, therefore, aperture cannot be initiated before failure by tension occurs. For this reason the threshold strain (ε_o) is set equal to the strain corresponding to failure (ε_1). This latter value depends on the tensile strength. Normally, failure would have to be associated with softening because the material is damaged. But this problem is beyond the scope of this paper.

This model has been implemented in CODE_BRIGHT (Olivella et al 1995) and some cases analysed to investigate the possible responses of the coupled behaviour.

Further improvements of the model may incorporate different aspects such as the influence of the formation of infilling material due to shear displacements between crack interfaces. Lee and Cho (2002) have investigated the variation of intrinsic permeability when fractured rocks (granite and marble) are subjected to normal and shear deformations. This work shows that mechanical and hydraulic apertures are not always in linear trend. For instance, deviation takes place for large shear strains due to formation of infilling materials. The overall effect is that permeability was bounded by a maximum value. This effect may justify the inclusion in the model of a maximum hydraulic aperture.

The fact that this model considers elastic and inelastic fracture aperture or closure has already been discussed in the literature. For instance, Renner et al (2000) investigated the behaviour of fractured argillaceous rocks including permeability variations induced by changes in confining pressures. In this work crack dimensions and permeability are correlated by means a model that takes into account elastic crack closure and crack closure controlled by inelastic processes. This later is explained by asperity indentation when rough crack walls contact each other.

3. SIMULATION OF HYDROFRACTURE TESTS

In order to study the possibilities of the model an hydrofract test is first simulated.

A rock sample is considered with an initial permeability assumed to be 10^{-19} m^2, both in the porous medium and in the element that will contain the fracture. Capillary pressure to start desaturation (air entry value) in the porous medium is assumed to be 0.3 MPa.

Confining stress is 1 MPa (horizontal) and the initial pressure of water is 0.1 MPa (atmospheric pressure). Tensile strength in the fracture is assumed to be 0.5 MPa which means that the fracture does not exist before fluid injection. Elastic Bulk Modulus was assumed to be K=1000 MPa and elastic shear modulus was G=1000 MPa.

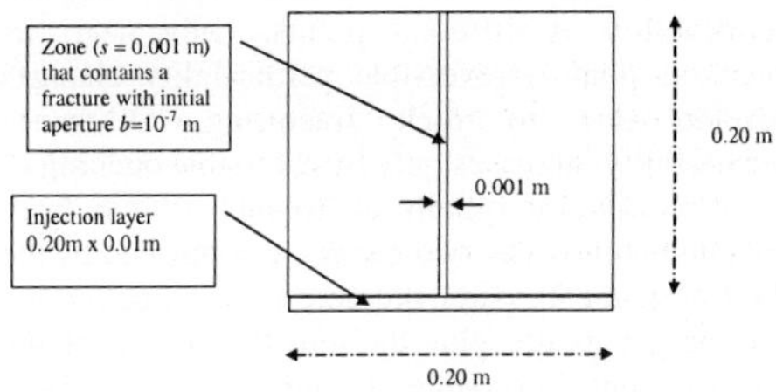

Figure3. Schematic representation of the 2-D representation of a sample containing a fracture initially closed (initial aperture 10^{-7} m)

The water flow rate injected is 3.6 g/h and injection was stopped at 1400 h. The initial strain to start discontinuity aperture is $\varepsilon_o = 0.0001$ (while $\varepsilon_1 = 0.001$ for this case). This means that before failure, increase of permeability will take place but in a reversible way.

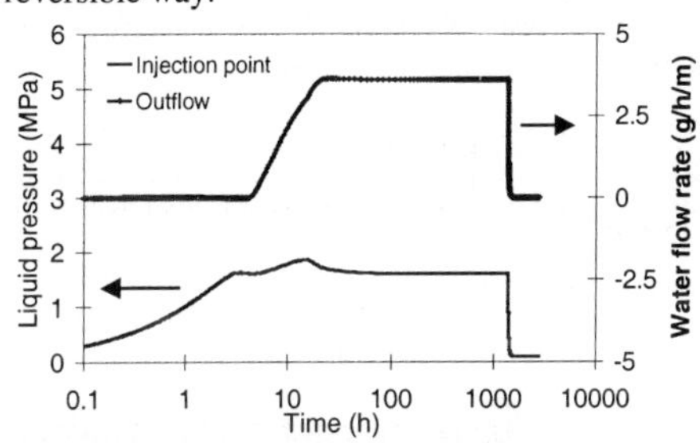

Figure 4. Liquid pressure at injection point and water flow rate at outflow point.

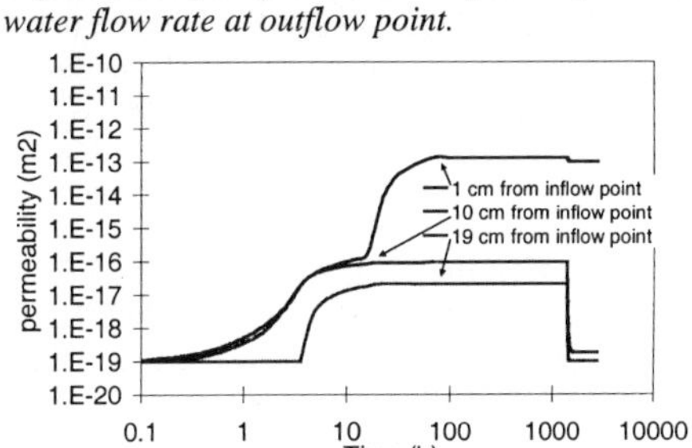

Figure 5. Equivalent intrinsic permeability of the elements that contain the fracture for different distances from injection point

Figure 4 shows the evolution of injection pressure and the flow rate at the outflow point. It can be observed that at 4.4 h a breakthrough takes place. A maximum injection pressure of 1.91 MPa is calculated and, due to mechanical effects, the pressure decreases. This is better explained in figure 5 which shows the evolution of intrinsic permeability at different points. Only near the injection point, irreversible permeability changes develop due to rock fracturing. Otherwise permeability increases only by reversible opening of the discontinuity (points at 10 and 19 cm from injection point). The two zones are controlled by the fluid pressure. If water pressure remains below the confining pressure plus the tensile strength of the rock then only reversible opening takes place. Near the injection point, permeability has increased by 7 orders of magnitude due to fracture opening

After breakthrough and fracture opening, steady state conditions are achieved. If the flow rate had been fixed at a rate higher than 3.6 g/h, more points would have reached the failure condition and therefore an increased permeability would have been developed (Table 1).

Table 1. Maximum water pressure as a function of the injected water flow rate

Flow rate (g/h/m)	Maximum pressure (MPa)	Comments
3.6	1.91	Hydro fracture initiation near injection point
5.4	2.44	Hydro fracture extends along the sample (except near outflow)

Finally, if a different value of ε_o is considered, the results would have been quite different. For instance if ε_o is set equal to 0.001 so $\varepsilon_o = \varepsilon_1 = 0.001$, then failure takes place along the sample for 3.6 g/h and the maximum gas pressure calculated is 2.88 MPa. This situation approximates better a fracture that forms in a completely undamaged medium.

4. SIMULATION OF GAS FLOW TESTS THROUGH AN EXISTING FRACTURE

The same rock sample is considered now in order to study the behavior of gas flow. In this first case, the fracture is assumed to be present before injection of the gas starts. Section 5 will deal with gasfract test, i.e. when the fracture is not present.

Confining stress is 1 MPa (horizontal) and the initial pressure of water and gas is 0.1 MPa (atmospheric pressure). Tensile strength in the fracture is assumed to be 0 MPa i.e. the fracture already exists at the beginning of the test. Fracture opening will start immediately when the normal stress decreases. Pressures cannot exceed the confining pressure along all the fracture because in that case the fracture can open freely (tensile strength is zero). However, since there is a pressure gradient, gas pressure can be locally higher than the confining pressure. Gas flow rate at injection point is fixed at 0.0432 g/h.

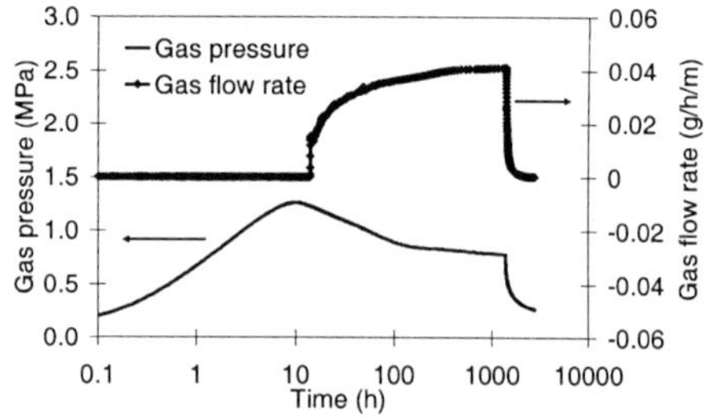

Figure 6. Gas pressure at injection point and gas flow rate at outflow point.

Figure 6 shows the evolution of gas pressure at the injection point and the gas flow rate at the outflow point. It can be observed that after a period of gas pressure increase corresponding to the desaturation of the injection layer, gas starts to penetrate in the medium. Figure 7 shows degree of saturation for both the fracture and the rock matrix. Gas flow takes place through the fracture because permeability is higher than in the matrix and because desaturation is easier than in the matrix. Due to fracture aperture, permeability increases and the air entry pressure decreases.

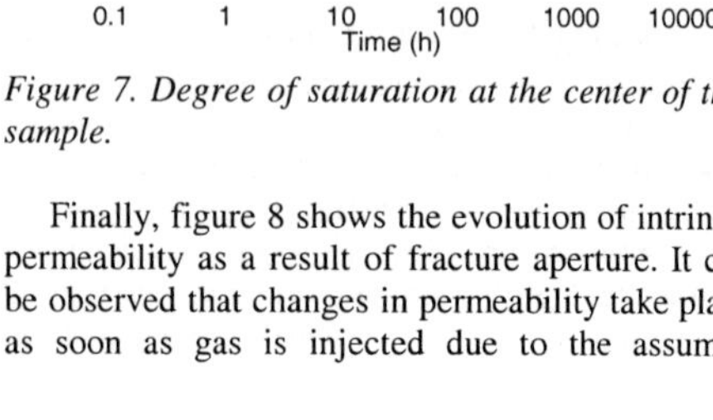

Figure 7. Degree of saturation at the center of the sample.

Finally, figure 8 shows the evolution of intrinsic permeability as a result of fracture aperture. It can be observed that changes in permeability take place as soon as gas is injected due to the assumed existence of the fracture. For this simulated test, steady state is practically reached after 1000 h although there is still some gas pressure dissipation.

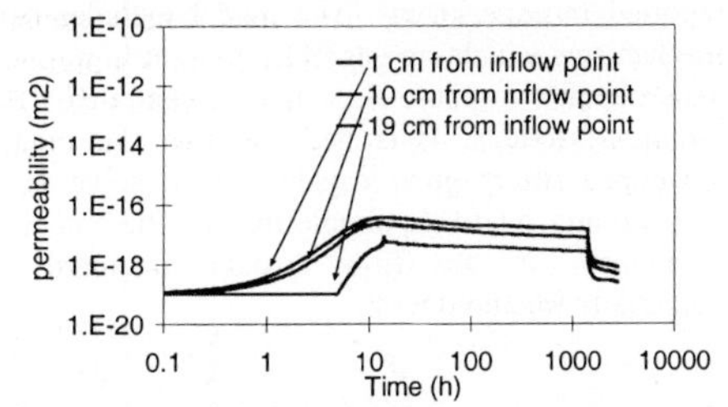

Figure 8. Equivalent intrinsic permeability of the elements that contain the fracture for different distances from injection point

This analysis has been performed for different values of the gas flow rate. The behaviour was similar and the maximum gas pressures achieved are included in Table 2. It can be observed that a flow rate of 0.0576 g/h produces failure, which implies large aperture of the fracture, large permeability increase, sharp drop of the gas injection pressure and, consequently, unstable numerical solution.

Table 2. Maximum gas pressure as a function of the injected gas flow rate

Flow rate (g/h/m)	Maximum gas pressure (MPa)	Comments
0.0144	0.78	Gas pressure below confining pressure
0.0288	1.14	Gas pressure below confining pressure (locally higher)
0.0432	1.26	Intermediate situation
0.0576	1.36	Gas opening of the fracture along the sample due to gas pressure higher than confining pressure

5. SIMULATION OF GASFRACT TESTS

The same rock sample is considered now in order to study the behavior of gas flow in a non-damaged rock. In this second case, the fracture does not exist at the initiation of the test.

Confining stress is 1 MPa (horizontal) and the initial pressure of water and gas is 0.1 MPa (atmospheric pressure). Tensile strength in the fracture is assumed to be 0.5 MPa and the initial strain to start fracture aperture is set to 0.0001. This value is probably too low and has been introduced instead of using the value of 0.001 used in section 3 for hydro-fracture simulation. This value permits gas flow before fracture failure. Gas was injected in this case at 0.216 g/h, i.e. higher flow rate than in the case of previous existing fracture.

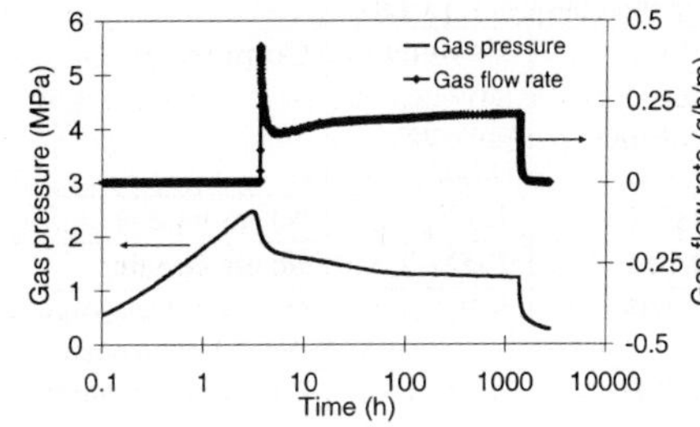

Figure 9. Gas pressure at injection point and gas flow rate at outflow point.

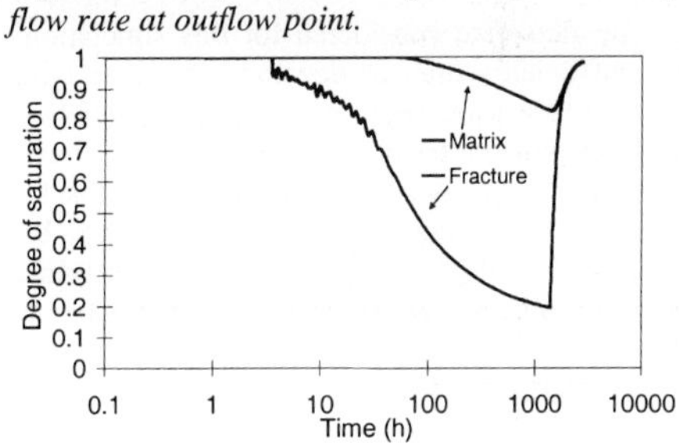

Figure 10. Degree of saturation at the center of the sample.

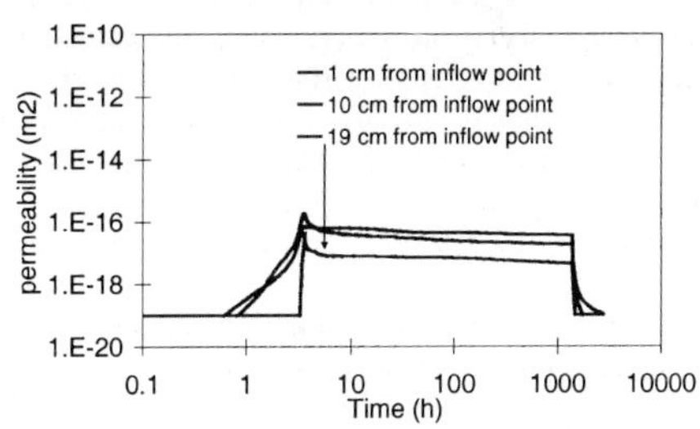

Figure 11. Equivalent intrinsic permeability of the elements that contain the fracture for different distances from injection point

Figure 9 shows the pressure at injection point and the gas flow rate at the extraction point. It can be observed that a sharp increase of flow rate develops which can be associated with the gas accumulation at the injection layer before the breakthrough takes place (3.7 h). This breakthrough is related to two phase flow and to fracture formation. While Figure 10 shows desaturation Figure 11 shows permeability changes due to fracture aperture. A preferential path has developed due to increase of gas permeability caused by desaturation and deformation.

Table 3. Maximum gas pressure as a function of the injected gas flow rate

Flow rate (g/h/m)	Maximum gas pressure (MPa)	Comments
0.072	1.57	No gas fracture
0.144	2.02	No gas fracture
0.216	2.50	Gas fracture initiation near injection point
0.288	2.84	Gas fracture along the sample

The flow rate considered for this simulation is not suffficienly high to develop a fracture failure because pressure does not reach the confining pressure plus the tensile strength along the fracture (except near the injection point). Increasing the flow rate will imply failure. The simulation has been performed for different flow rates and the maximum pressures obtained are summarized in table 3.

6. CONCLUSIONS

Using a few simple concepts of fracture development and the associated two phase flow phenomena along discontinuities, a continuum model capable of simulating flow in a fractured porous media has been developed. The calculation procedure is developed within the framework of a hydro-mechanical model for geological media which is based on a formulation and numerical program that can account for several types of simulations and problems. A realistic representation of irreversible fracture opening requires elastoplastic constitutive model for the soil or rock skeleton. Fracture opening is characterized by two strain parameters: the threshold strain which marks the opening of the fracture and the tensile strain which marks the fracture in tension of the rock. Both can be equal in a simplified situation.

A few computer simulations, inspired in tests conducted to study gas flow through compacted clays and rocks have been presented. The analyses show the sensitivity of results to a number of factors controlling the tests usually performed: the gas flow rate, the confining stress, and the mechanical parameters of the material. It is shown that the model reproduces in a natural way the peak discharge phenomena and peak gas pressure often reported in experiments. The model includes two mechanisms which are generally present in a given experiment: two phase flow and concentrated flow through fractures. It is believed that the model developed offers good capabilities to solve in a realistic and relatively simple manner the complex phenomena of gas flow through fractured or potentially fractured rock.

7. ACKNOWLEDGEMENTS

The authors would like to acknowledge NAGRA for funding this investigation

REFERENCES

Alonso, E.E., S. Olivella & C. Delahaye (2002) Gas Migration in Clays. *Environmental Geomechanics,* L. Vulliet, L. Laloui and B. Schrefler eds, Presses polytechniques et universitaires romandes, Lausanne, Switzerland : 83-94.

Gens, A., S. Olivella & B. Valleján (2001) Analysis of gas phase transport phenomena in compacted clay barriers. *Computer Methods and Advances in Geomechanics:* 735-742.

Lee, H.S. and T.F. Chou, (2002), Hydraulic Characteristics of Rough Fractures in Linear Flow under Normal and Shear Load, *Rock Mech.Rock Engng,* 35(4), 299-318.

Olivella, S., A. Gens, J. Carrera & E.E. Alonso (1995) Numerical formulation for a simulator (CODE_BRIGHT) for the coupled analysis of saline media.. *Engineering Computations*, 13: 87-112.

Renner, J. T. Hettkaamp, and F. Rummel (2000) Rock Mechanical Characterization of an Argilaceous Host rock of a Potential Radioactive Waste Repository. *Rock Mech. Rock Engng,* 33 (3), 153-178.

RESEARCH AND APPLICATION ON COUPLED T-H-M-C PROCESSES OF GEOLOGICAL MEDIA IN CHINA – A REVIEW

Xia-Ting Feng[1] , Jianjun Liu[1, 2] and Lanru Jing[1, 3]

[1)]Institute of Rock and Soil Mechanics, Chinese Academy of Sciences, Wuhan 430071, China
[2)] Institute of Mechanics of Porous Media, Wuhan University of Technology, Wuhan 430023, China
[3)]Royal Institute of Technology, Stockholm , Sweden

Abstract: Theoretical models of coupled T-H-M-C processes of geological media and the associated numerical solutions have become an attractive research focus in geomechanics and related fields in China. This paper provides a systematic overview of the past progress in the fundamental studies of the coupled THM models and numerical methods, and their applications in the fields of oil/gas reservoir, coal mining, and water resources engineering works. The key areas of weakness in research in this field are also outlined and possible directions for the future development are discussed.

1. INTRODUCTION

Although the coupling effects between different physical processes in rock and soil engineering problems have been long recognized and their numerical solutions have also been attempted in the early1970's, such as FEM solutions for problems of thermal stresses, soil consolidation, ground subsidence and dam-reservoir interactions, significant progress in the establishment of fundamental theories, development of reliable numerical methods and computer codes, extensive experimental investigations and large scale applications have only been made in the last two decades due mainly to the demands from nuclear waster disposal, oil/gas and geothermal reservoir engineering and environmental preservation in the world. Although started at one decade later compared with international communities in this field, the research and development in the field of coupled THMC processes of geological materials and systems in China have made impressive progresses due mainly to the demands from the rapid development of infrastructure since 1990's, and has become one of the most attractive focuses of research in the general field of geosciences.

However, the majority of scientific publications in these researches and applications are on domestic Chinese journals that cannot be readily reached by international readers. In order to compensate this shortcoming, this review presents systematically the scope and spectra of the research and applications in the field of coupled T-H-M-C processes published on Chinese journals, concentrating mainly on the fundamental studies and applications in oil/gas reservoir engineering, coal mining, engineering in cold regions and water resources engineering. The hydro-power and general civil engineering problems will not be included for space limit. Brief summary of key points of progress and the spots of weakness are identified in view of the international status, together with possible future directions of research and developments at the end.

2. FUNDAMENTAL STUDIES

The research works for fundamental studies of the coupled processes focused mainly on experimental investigations of coupled behaviours of single rock fractures and fractured rock masses, and development of numerical methods.

2.1 Experimental studies on coupled stress- flow of single rock fractures

The fundamental behaviour of rock fractures under combined mechanical loading and fluid flow occupies the central attention of the experimental study, with objectives of providing conceptual models of coupled hydro-mechanical behaviour of single rock fractures in order to support numerical modelling and design parameters. Some of the typical achievements are reported in Su et al. (1997) and Shen et al. (1998). The former reported an exponential relation between the fracture aperture and normal stress and the latter reported the results of the probabilistic density function of the roughness of fractures and its impact on the contact areas, stresses and flow rate. The permeability-strain relations of intact samples of sandstone during a complete stress-strain process, including post-peak behaviour, was conducted by Li et al. (1995). Experimental studies on effects of chemical corrosion on the micro-fracturing process of granite sandstone rocks under laboratory conditions were reported by Fang et al. (2002).

2.2 Theoretical studies on the coupled T-H-M behaviour of rock fractures

Theoretical models for coupled HM and THM behaviour of rock fractures were also established to support numerical modelling and design practices in engineering, with the focus on the constitutive models describing the interactions between aperture (or transmissivity) and stress (or deformability) of rock fractures. Typical works can be seen in Liu (1987), Zhou and Xiong (1996a) based on Hertzian contact mechanics principles of rough surfaces for both normal and shear deformation processes, Zhou and Xiong (1996b) on generalized cubic law for fracture flow and Liu et al. (2001) for a constitutive model describing coupled T-H-M behaviour of rock fractures using experimental test results from granite rock fracture samples with temperature up to 300 °C. Large block tests (420mm x 395mm x 200mm in size) with multiple fractures of different apertures were also reported in Zhang and Zhang (1997).

2.3 Mathematical models and numerical methods for the coupled processes of fractured rock masses, soils and general porous media

The establishment of mathematical models (including the governing equations of coupled processes and constitutive models) and numerical methods for the solutions of practical problems attracted extensive attention in this area since mid 1990's. Both equivalent continuum approach and discrete fracture system approach are used, with focus on coupled T-H, H-M, M-H-C and T-H-M processes of fractured rocks, soils and general porous media. Summarized below are some of the representative works:

1) Equivalent continuum models of coupled temperature-flow processes of fractured rock masses using FEM (Cai and Han, 1997);

2) Constitutive models and numerical methods for coupled stress/deformation-flow processes of fractured rocks, using seepage energy superposition (Zhou and Xiong, 1996c), lumped-parameter approach (Wu, 1997; Wu et al., 1994), dam foundations (Chen and Zhang, 1994; Gao and Xia, 1997), hydraulic fracturing processes (Yang et al., 2002), four-degree-of freedom approach for both elastic and elastoplastic deformations (Wang et al., 1998, 2000a), coupling effects of unsteady flow and rock damage (Yang and Zhou, 1991; Zhang et al., 2000; Zheng, 2000; Zhu and Sun, 1999), equivalent compliance and permeability tensors with coupled stress-flow effects (Wu, 1996), and dual porosity model with stress-flow coupling (Wu and Zhang, 1996);

3) Hybrid model of Discrete Fracture Network (DFN) and Discrete Element Method (DEM) for coupled stress-flow analysis of rock slope stability (Wang, 2000);

4) Models and methods for coupled thermo-hydro-mechanical processes of fractured rocks, in saturated conditions using crack-tensor approach (Liu et al., 2002), the equivalent continuum approach and FEM solution technique with consideration for two-phase flow of water/vapour due to phase change by evaporation for fully or partially saturated cases (Li et al., 2000).

5) Models and numerical solutions based on principles of mechanics of porous media for coupled THM processes, using closed-form solutions (Xu and Xu, 1999a,b,c; Xu et al., 1999; Zhang and Li, 1997; Cai & Wu, 2001), dynamic consolidation problems (Li et al., 2001a), with elastoplastic deformation and FEM (Liu & Gen, 2000), FEM formulation for coupled temperature, flow and deformation processes (Chen et al., 2001), Non-Fourier and Non-Fickian heat and mass transfer models (Wang et al., 2001a), constitutive models for swelling rocks and soils using a reduced suction technique (Zhan et al., 1998).

6) Models for coupled T-H-M-C problems in soils, considering heat-moisture-salt transfer in freezing and thawing processes (Yue, 1994) and electro-chemical hydrodynamic behaviour of saturated soils and clays (Guan and Fang, 2000).

7) Fractal structure of the time distribution of microfracturing in rocks under water chemical corrosion (Feng and Seto, 1999a) and neural network models for acoustic emission behaviors of rocks under different stress conditions with water chemical corrosion (Feng and Seto, 1998a,b,c, 1999b).

8) Coupled models of multiphase flow and solid deformation processes for oil/gas reservoirs, using mixture theory for saturated (Li et al., 2001b) or unsaturated porous media using FEM solutions (Xue and Song, 1999a,b; Dong & Xu, 1998; Dong et al., 2000).

3. APPLICATIONS IN OIL/GAS RESERVOIR ENGINEERING

3.1. Models and numerical technique on coupled flow-deformation effects

The effects of the coupled multiphase fluid flow and solid deformation/stress play an important role for characterization, analysis and operation of the oil/gas reservoirs and naturally becomes one of the most active research fields in China. Ran and Gu (1998a,1999) established three-dimensional three-phase (fluids, gas and solid) flow and transport models with elastoplastic rock deformation, using hybrid and staggered Finite Difference and Finite Element solution techniques. The methods were applied to dynamic simulation of single-well oil production processes, and revealed significant dynamic effects of the coupled flow and deformation fields on the oil production process (Fig. 1). Similar studies were also carried out by Fan et al. (2000) for coupled flow and fracturing effects of low-permeability gas reservoirs considering influences of time-dependent variation of stress, strain, porosity and permeability, using the Finite Difference technique.

An important aspect in Chinese oil/gas industry is that oil/gas formations of low-permeability play an increasing role in the overall production. The flows in low-permeability reservoirs are often Non-Darcian and fracturing plays a significant role for both production and modelling. An experimental study (Liu and Liu, 2001a) reported variations of porosity and permeability of rock samples of low permeability with the effective stress, with reductions of 13% in porosity and 87% permeability when an effective stress of 20 MPa was applied (Fig. 2). The constitutive models and a FEM solution method based on the above experimental results were reported in Liu et al. (2002), and applied to simulating processes of hydraulic fracturing and flooding prevention for the Mao-2 block, Daqing oil field in China.

Study on fracture systems characteristics and fracturing process becomes important tasks in such conditions. In this direction, Liu and Liu (2001b)

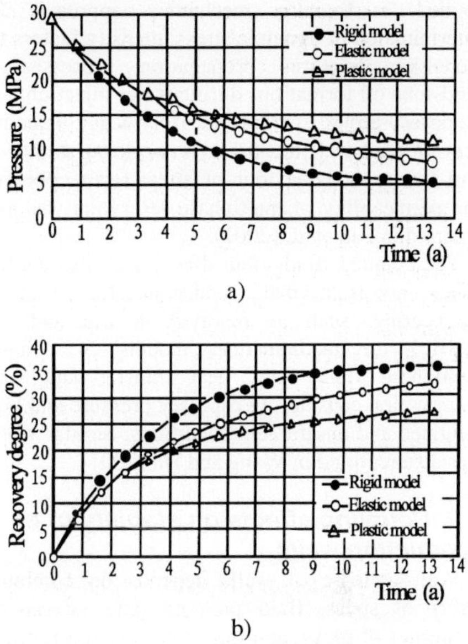

Figure 1. a) Effects of solid deformation on fluid pressure evolution; b) effects of solid deformation on oil recovery evolution (after Ran & Gu, 1999).

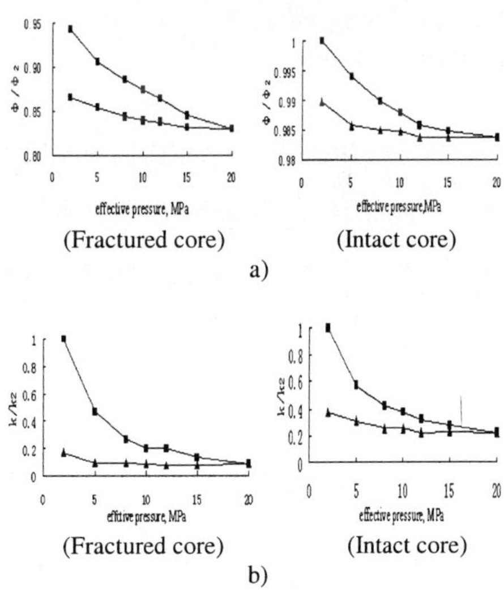

Figure 2. Effect of effective stress on porosity (a) and permeability (b) of the fractured and intact core samples (after Liu and Liu, 2001a).

applied a fracture mechanics approach for determination of dynamic stress intensity factors for modelling fractures propagation processes in sandstone oil formations during water injection, and proposed design parameters for water injection pressure and well locations. A work on studies of post-fracturing behaviour of stress-strain, porosity and permeability of multiphase reservoirs was also reported in Fan et al. (2001).

The coupled fluid-solid effects was also studied under non-isothermal conditions for reservoir engineering, such as reported in Liu and Liu (2001c) for mathematical models of coupled multiphase fluid flow, heat transfer and solid deformation considering capillary pressure and both saturated and unsaturated conditions. Similar works was also reported in Wang and Du (2001).

3.2. Coupling effects on stability of oil production wells

Wall collapse of wells depends on combined effect of stress, fluid pressure, and mechanical properties of rocks, and the latter is also a function of fluid-solid coupling through interactions of rock strength/deformability and porosity. Researches in this direction started in 90's in China, considering mainly the effects of creep deformation of rocks on well casings (Wang et al., 1994), especially for inclined wells, with coupled mechanical and chemical effects in shale formations (Huang, 1995; Lin, 1996; 1998; Liu et al., 2000; Sun et al., 1998). Figure 3 gives examples of the typical relationships between stress invariants, drilling fluid density, well inclination angle and degree of mineralization of the shale.

FEM methods have been extensively used for well collapse analysis, such as reported in Wang et al. (2000b), simulating the complete well collapse processes with effects of drilling fluid density, stabilized well shape and well enlargement rate. Closed-form solutions considering non-linear deformation of the rocks and coupling flow-stress effects were also developed due to the axi-symmetric geometry of the wells (Li & Li, 1997).

Damage mechanics principles have been applied extensively for well stability problems, especially for hard, brittle and fractured shale rocks. Constitutive models using damage concept, coupled with hydro-chemical swelling models, have been developed based on experimental results, and applied for design of drilling fluid density, combined with FEM methods (Liu, 1995).

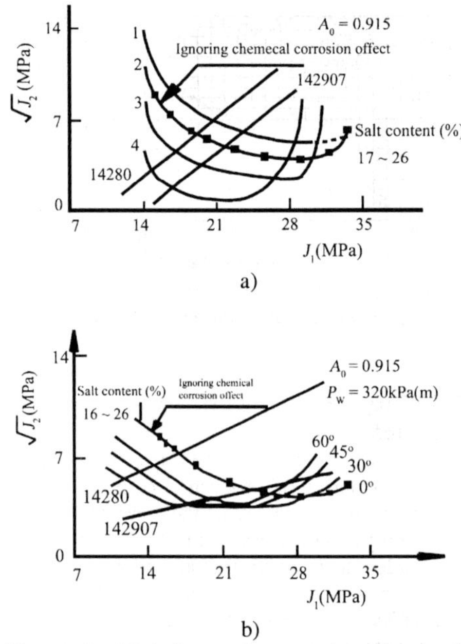

Figure 3. Variations of stress invariant and mineralization degree of shale with: (a) drilling fluid density and b) well inclination (after Sun et al., 1998.

3.3. Coupling effects on Casing failure

Casing failure in the oil/water production wells is one of the difficulties in petroleum reservoir engineering, and the rock deformation is one of the main reasons causing casing failures. The general understanding is that the stress change, fault reactivations or inter-layer slipping caused by swelling or compaction due to overpressure in the formation may be important factors. Based on a non-linear fluid flow and solid deformation coupling model and FEM solution technique, Liu et al. (2003) and Yin et al. (2003) investigated the tensile, compressive and shear failure mechanisms of casings during oil production, and derived the critical parameters for casing failure prevention. Figure 4 illustrates the obtained relations between the casing extrusion pressure and blowout flux under a pressure gradient of 12 MPa and in-situ stress of 30 MPa, for different formation permeability values.

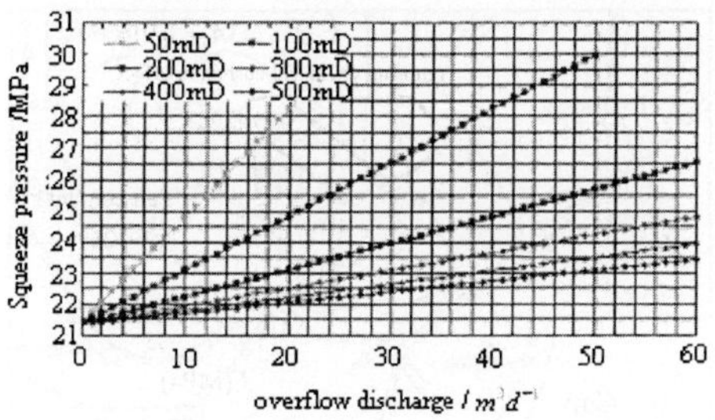

Figure 4. Relations between blowout flux and extrusion pressure (after Yin et al., 2003).

3.4. Coupling effects on sand production

Sand production in low-density sandstone formations during oil production depends on many factors, such as stresses in the formation beyond rock strength. The stresses in the formation include overburden, tectonic stresses and seepage pressure. The formation strength depends mainly on mechanical properties and cementation degree. The stress change and reduction of the cementation degree of the sandstone during the water injection and oil extraction controls the sand production from the formation. The studies in this direction in China focus mainly on numerical modelling, with typical works by Liu et al. (2003) for the models of sand production in low-density sandstones using coupled fluid flow -solid deformation, using the pipe-flow concept.

4. APPLICATIONS IN COAL MINING

Coupled coal-methane-water process has been an active research area in China for a long time due to its utmost importance to mine safety and productivity and the traditional importance of coal mining in Chinese energy industry. In this review only some of the recent works on mathematical modelling on methane migration and water burst problems are highlighted due to space limit.

4.1. Coupled models and solutions for methane migration

The methane migration during coal mining is caused by loss of the in-situ equilibrium between the coal skeleton stress and methane pressure. The interactions between coal deformation and methane migration have several aspects: effective stress change in coal seams due to excavation and mining, porosity change and methane content increase, methane pressure increase, change in the absorption capacity of the methane particles on the surfaces of coal solid skeleton, and changes in the gas permeability due to coal deformation.

Zhao (1994) presented a model of coupled coal deformation and methane migration based on a consolidation theory of elastic medium with Darcy fluid flow and the Terzaghi effective stress law, and its numerical solution technique and applications to practical problems. Works using similar approaches were also reported in Liang et al. (1995,1996), Sun and Xian (1999), Ding et al. (1999) and Wang et al. (2001). Extensions to consider coal-methane interactions under non-isothermal conditions were reported in Liang et al. (2000). These studies provided not only the models for methane migration analysis and prediction, but also software as design tools for considering dynamic interactions between the rock stresses and methane pressure around the working faces for coal mine safety management. Cheng (2000) investigated thermal effects on adsorption and release mechanisms of methane particles during coal seam fracturing with interior heat sources, and applied the models for interpreting the thermal behaviour of pre-burst of the coal seams.

4.2. Coupling effects on water burst in coalmines

Water burst is one of the main failure mechanisms of the water protection pillars in coalmines. A fracturing zone is usually formed to certain extent in the water protection pillars due to mining excavations. Under pore pressure, the strength and deformability of the pillars will change with time due to transient seepage process, even when mining is stopped. The time-dependent pillar strength reduction with seepage may reach a critical stage to cause sudden loss of pillar integrity and outburst of water. It is therefore also an issue of the long-term pillar stability.

The criteria for predicting critical conditions of water protection pillars for water bust are of fundamental importance. A research by Gao (1995) reported a criterion for water burst from the floors of working faces, considering coupled water flow and coal seam deformation, water chemistry, and fracture-water interaction in coal seams, based on a system energy concept. A dynamic analysis method was proposed by Yang and Zhang (1995) for water bust from working face floors, using numerical modelling techniques with coupled water flow and coal seam coupling.

5 APPLICATIONS IN LAND SUBSIDENCE

Land surface subsidence due to water extraction has becoming a special kind of geological hazards in large cities in China like Shanghai, Tianjin etc. (Duan 1998) due mainly to the fast economic and industrial development in the past two decades. Extensive research has been devoted to this subject and substantial progress has been made, especially in the areas of mechanism studies, modelling, experiments, predictions and control measures. Among the super-large cities in the world Shanghai established the first network for subsidence monitoring with 28 central stations and 1150 nodes covering the whole city area, using GPS and computerized online data collection, transmission, processing and decision-making systems. At present, the total extraction volume of groundwater is controlled within 50 million m^3/year and the rate of subsidence is kept within 5 mm/year (Gong, 1997).

It was found that the subsidence of the ground surface is a result of coupled effects between groundwater flow and rock/soil deformation, with the seepage and consolidation affected by changes in rock/soil porosity and permeability. Numerical modelling using coupled flow-deformation models become necessary.

Using a three-dimensional FEM model and Terzaghi's consolidation theory, Wei et al. (2000) predicted ground subsidence contours and groundwater flow in Jining city in North China, with reasonable agreement compared with measured data (Figure 5) in 1997 and predicted the future subsidence contours in 2000 and 2010. This works have contributed significantly for the urban planning of the city.

A dynamic coupling model of time-dependent subsidence of soft clay formations was developed by Ran and Gu (1998b) based on principles of poro-mechanics, considering three-dimensional seepage and consolidation (both primary and secondary), gravity, non-homogeneity and anisotropy, time- and stress-dependent changes of porosity and permeability, and different loading-unloading features. The model considered also multiple aquifers and water extraction well systems.

Ground surface subsidence is also an important environmental issue in mining industry. Liu et al. (1999) reported a numerical study on ground surface deformation and subsidence due to water drainage in a large-scale open-pit coalmine, using random media theory, Markov Chain principles and superposition techniques. The predicted and the measured subsidence curves agreed reasonably well along the chosen profiles (see Figure 6), which helped the mine planning and environmental management significantly.

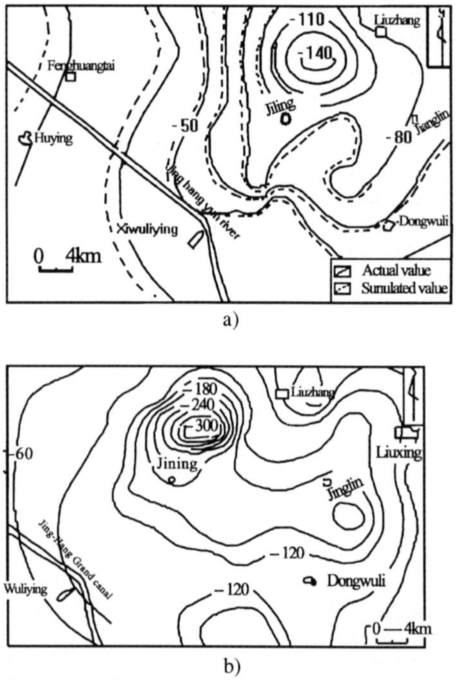

Figure 5. Simulated and measured subsidence contours for Jining city a) Measured and simulated subsidence contours in 1997. b) Predicted subsidence contours in 2000 (after Wei et al., 2000).

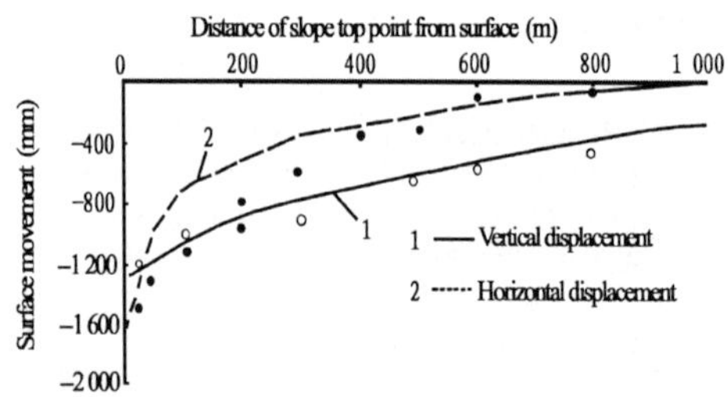

Figure 6. Comparison between the calculated and measured values of ground surface movements (after Liu et al., 1999).

6 APPLICATIONS IN RAIWAY AND HIGHWAY WORKS IN COLD REGIONS

There are large areas in Northeast, Northwest and Southwest China with cold climate. Rock/soil engineering works in such cold regions have special features of coupled T-H-M processes, induced mainly by permafrost phenomena, freezing-thawing cycles and changes of rock/soil/water properties in cold temperature regime and with time. Typical examples are the difficulties encountered in the highway and railway development in Qing-Zang (Tibetan) plateau, such as frozen pressure induced tunnel lining failure.

The research works in this direction has been mainly led by Lanzhou Institute of Glaciology and Geocryology, Chinese Academy of Sciences, with focus on numerical modelling of coupled T-H-M behaviours of railway tunnels and highway foundations. Based on the theories of Fourier heat transfer, Darcy fluid flow and elastic deformation of frozen and unfrozen porous media, mathematical models coupling the temperature, stress and fluid flow with phase changes and FEM solution techniques and computer codes were developed and applied to predict the frozen pressure induced stress, temperature and seepage fields in the surrounding near-field rocks of railway tunnels, and the lining failure processes (Lai et al., 1999) Creeping deformation and moving interfaces between frozen-unfrozen areas are included. Figure 7 compares the different frozen zones surrounding a tunnel when coupled seepage-temperature is considered or not, after 5 years of construction.

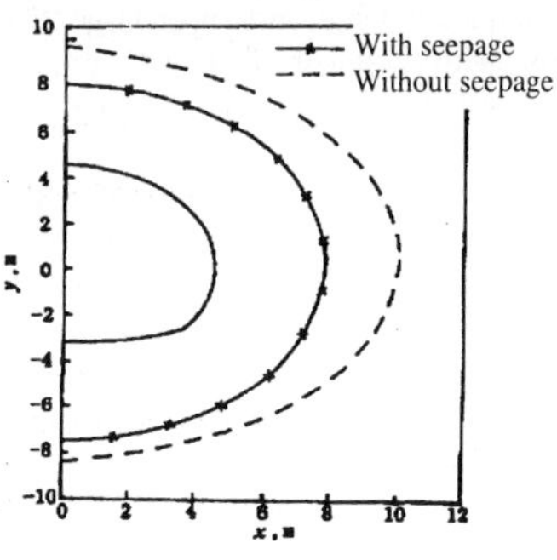

Figure 7. The maximum frozen zone surrounding a rock tunnel after 5 years of construction (after Lai et al., 1999).

Another study reported a coupled seepage-fluid flow effects on long-term behaviour of a highway foundation in a permafrost area on the Qing-Zang (Tibetan) plateau, Ling and Wu (1999), during a cyclic freezing-thawing process. The results show that the seepage from the foundation to the drainage ditches at the two sides of the highway, due to surface recharge, plays a significant role in decreasing mechanical stability of the road by increasing the thawing depth inside the foundation.

7 DISCUSSIONS AND CONCLUDING REMARKS

7.1 Regarding fundamental studies

Although the development of constitutive models, governing equations and FEM solution techniques are, in general, at the similar status compared with international levels, the non-proportional works between the experimental and theoretical and modelling studies pointed out the main weakness area: the needs for more comprehensive experimental works on fundamentals of coupled T-H-M-C processes in geo-materials and systems, especially large-scale in-situ experiments and more QA-directed and in-depth laboratory investigations.

In terms of numerical methods, the dominance by FEM with equivalent continuum approach might not be most suitable for sparsely or moderately fractured hard rocks and more advanced methods and codes using discrete approach are needed. The issue of applicability of the equivalent media approach, the associated scale effects, and uncertainty evaluations need to be fully explored. The processes are dominated by coupled stress-flow problems and effects of thermo-chemical effects need more attention. More works for soils, clays, sands and other similar media, which are equally, if not more, important in the fields of geo-engineering and environments, seem also needed.

Another important issue is the lack of documentation about the validation/verification of the developed computer codes with properly and internationally accepted QA-measures and well defined benchmark test problems so that the reliability of the software can be well established, especially for large-scale industrial applications.

7.2 Regarding applications

The extent and scope of the reported applications of the coupled approaches and methods on the domestic journals are impressive and appeared to be in an accelerating state. Many significant progresses were made in applying numerical models for predicting behaviour of geological materials and performance of the geological/engineering systems, especially in the fields of reservoir engineering, coal mining and land subsidence. However, lack of progress can still be detected in a number of important aspects, such as:

- Lack of adequate support of measured data to numerical predictions, due perhaps to the lack of in situ measurements and monitoring systems;
- Lack of more comprehensive descriptions of model conceptualisation, data acquisition and evaluation, system characterization, and uncertainty evaluation in quite a number of publications, especially regarding motivation and justification of the approaches (discrete vs. equivalent continuum) and models for fractured rocks;
- Lack of applications and publications on general environmental effects related to engineering activities, as demonstrated by much less works reported about contaminant transport and consequences, such as nuclear waste management, geothermal energy utilization, landfill performance evaluation and landslides;
- Lack of large or super-large scale computational case studies with full utilizations of numerical modelling potentials, due perhaps to lack of available large scale computing capacities in terms of software.

Although the above lack of progress is not unique in China, they placed significant constraints on further development of capabilities and reliabilities of the numerical modelling for coupled T-H-M-C processes for geological materials, systems and projects. Some of the future development directions may be predicted as given below.

7.3 Predicted future R&D directions

As in every other country, the research and development is driven by the needs of social, economic, scientific and technical needs of the society at large. As a developing country with rapid economy growth, environment protection and sustainable infrastructure development requires adequate support from science and technology in the field of coupled T-H-M-C processes of geological materials, systems and projects. Some of the most important areas for further extensive R & D in this field can be briefly summarized below.

1) More systematic and integrated research on basis of multi-fields, multi-process, multi-media, multi-scale and multi-discipline in the field of coupled T-H-M-C processes of geo-materials and systems. At present, the Newtonian and Non-Newtonian flow in fractured geological media at macro-, micro- and meso-scopes using deterministic, probabilistic, fuzzy, intelligent or chaotic approaches in both experimental and theoretical directions will occupy the primary attention.

2) Development of more advanced and reliable modern laboratory techniques and large scale, long-term in-situ monitoring systems, in combination with GPS, GIS, Remote Sensing and information technology to support data acquisition/transmission, site characterization, model/system evaluation and software development, interfaced with visual/virtual simulations, artificial intelligence and other newly emerged technologies. In-situ measurements and long-term monitoring technique in cold regions will be especially focused.

3) Development of capabilities and necessary software systems for large and super-large scale computations with coupled T-H-M-C process, with some core functions such as alternative material models, complex boundary/initial conditions, man-made engineering processes (e.g. sequential excavation, supporting, monitoring) and functions supporting evaluations of system/model uncertainty, variability, and sensibility. Special target areas of applications are nuclear waste disposal, landslides, land subsidence, methane migration and water burst in coal mining, dam break, landfill performance evaluation and geothermal energy exploration. Special attention on modelling coupled T-H-M-C processes using advanced numerical techniques for large scale analysis such as parallel processing and integrated multi-station simulations will also be made.

ACKNOWLEDGEMENTS

The financial supports from The Special Funds for Major State Basic Research Project under Grant no. 2002CB412708 and National Nature Science Foundation of China under Grant no.10072073 are gratefully acknowledged.

REFERENCES

Cai J and Han Q. 1997. *Continuum model of coupled seepage and temperature fields in rock masses.* Groundwater, 19(2), pp. 59-62.

Cai J and Wu Y. 2001. *Analysis of one-dimensional seepage model by coupling hydraulic head gradient, temperature gradient and stress.* Chinese J. Rock Mech. & Engng., 20(suppl), pp. 1026-1028.

Chen P and Zhang Y. 1994. *Coupling analysis of seepage/stress for jointed rock.* Chinese J. of Rock Mech. & Engng., 13(4), pp. 299-308.

Chen B, Li N and Zhuo R. 2001. *FEM analysis on fully coupled thermo-hydro-mechanical behaviour of porous media.* Chinese J. of Rock Mech. & Engng., 20(4), pp. 467-472.

Cheng W. 2000. *Study on heat effect mechanism of gas interstitial flow in coal seam.* J. of China Coal Society, 25(5), pp. 506-509.

Ding J, Ma Y, Zhao G, Lu W and Guo D. 1999. *Instability theory and numerical analysis of solid-fluid coupling on coal-methane outburst.* Engng. Mech. 16(4), pp. 47-53.

Dong P and Xu X. 1998. *Mathematical Models for Fluid-Solid Coupling in Reservoir and its Finite Element Method Equations.* ACTA Petrolesi Sinica, 1998,19(1):64-70

Dong P, Lang Z and Xu X. 2000. *The fully coup led fluid-solid analysis for the reservoir rock deformations due to oil withdrawal.* Chinese J. of Geomechanics, 6(2), pp. 6-10.

Duan Y. 1998. *Research status on land subsidence and sustainable development in the 21st century of China.* Chinese J. of Geological Hazard and Control, 9(2), pp. 1-5.

Fan X, Li X, Zhang S and Xu X. 2000. *Mathematical simulation of coupled fluid flow and geomechanical behaviour for full low permeability gas reservoir fracturing.* Petroleum Exploration and Development, 27(1), pp.76-83.

Fan X, Xu X and Zhang S. 2001. *Analysis on changes of stress, strain, porosity and permeability in multiphase reservoir with fluid solid coupled and geomechanical mathematic simulation.* Rock & Soil Mechanics, 22(1), pp. 47-50.

Feng X, Wang C and Chen S. 2002. *Testing study and real-time observation of rock meso-cracking process under chemical corrosion.* Chinese J. Rock Mech. & Engng., 21(7), pp. 935-939.

Gao Y. 1995. *Mechanical effects of water pressure on water burst at working face floors.* Coal Field Geology & Exploration, 24(6), pp. 506-509.

Gao H and Xia S. 1997. *Three-dimensional modelling of coupled seepage and stress fields of fractured rock masses.* Chinese J. of Geotech. Engng., 19(2), pp. 102-105.

Gong S. 1997. *Forecast of Shanghai land subsidence.* Forecast, Vol. 2, pp. 33-35.

Guan J and Fang W. 2000. *Coupling relationship for electrochemical hydrodynamics in water saturated sandy and clayey soils.* Chinese J. Geotech. Engng., 22(6), pp. 664-667.

Huang R. 1995. *study on shale stone well-bore stability coupled with mechanical and chemical. completing and drilling fluid.* Drilling and Completion Fluids,12(3), pp. 50-53.

Lai Y, Wu Z, Zhu Y and Zhu L. 1999. *Non-linear analysis for the coupled problem of temperature, seepage and stress fields in cold region tunnels.* Chinese J. of Geotech. Engng., 21(5), pp. 529- 533.

Li N, Chen B and Dang F. 2000. *A model of coupled temperature-seepage-deformation of fractured rocks and FEM solutions.* Progress in Natural Sciences, Vol. 8, pp. 722-728.

Li Y, Chen J and Zhou S. 2001a. *Numerical simulation for a foundation reinforcement procedure using dynamic consolidation* techniques. Rock & Soil Mech. 22(1), pp.6-11.

Li J and Li Z. 1997. *Rock elastic-plastic stresses around a wellbore and wellbore stability under permeation osmosis.* Engng. Mech., 14(2), pp. 131-137.

Li S, Li Y and Wu Z. 1995. *The Permeability-strain equation relating to complete stress-strain path of the rock.* Chinese J. of Geotech. Engng., 17(2), pp. 13-19.

Li S, Xu B and Duan Y. 2001b. *Flow through fissured reservoir with deformable media.* Chinese J. of Computational Mechanics, 18(2), pp.133-137.

Liang B, Zhang M, Pan Y and Wang L. 1995. *Theory of instability of flow fixation coupling for coal and gas outburst.* J. of China Coal Society, 20(5), pp.492-496.

Liang B, Zhang M and Wang Y. 1996. *Mathematical model and numerical method for coupled gas flow in coal seams and coal deformation.* Chinese J. of Rock Mech. & Engng., 15(2), pp.135-142.

Liang B, Liu J, Fan H and Zhang M. 2000. *The mathematical model and its numerical solution of gas flow under unequal temperature.* Chinese J. Rock Mech. & Engng., 19(1), pp.1-5.

Lin Y. 1996. *Experimental study on shale rock hydrate damaged under 3D stress.* Oil Drilling & Exploration,, 25(1), pp. 10-14.

Lin Y. 1998. *Coupling method of mechanics and chemistry for hole stability prediction.* Petroleum Drilling Technology, 26(3), pp. 19-22.

Ling F and Wu Z. 1999. *Numerical simulation of the influence of seepage on temperature field of roadbed in permafrost regions.* Chinese J. of Glasiology & Geocryology, 21(2), pp. 115-119.

Liu J. 1987. *Seepage relation in single fracture under normal stress effect.* J. of Hydrogeology & Engineering Geology, Vol. 2, pp. 22-27.

Liu Y. 1995. *Mud density for wellbore stability when formation rock is damaged.* Acta Petrolei Sinica,, Vol. 3, pp. 123-128.

Liu B, Yang J and Zhang J. 1999. *Surface subsidence and deformation caused by open pit mining and dewatering.* J. of China Coal Society, Vol. 1, pp. 46-52.

Liu Y, Cai Y, Liu Q and Wu Y. 2001. *Thermal-hydraulic-mechanical coupling constitutive relation of rock mass fracture interconnectivity.* Chinese J. of Geotech. Engng., 23(2), pp. 196-200.

Liu Y, Cai Y, Liu Q and Wu Y. 2002. *Study on mechanical behaviour of jointed rock mass affected by temperature and saturated water.* Chinese J. of Rock Mech. & Engng., 21(2), 233-237.

Liu J, Feng X and Yang C. 2003. *Numerical simulation of the effect water flooding well releasing pressure on casing extrusion stress.* Proc. of the Chinese National Conf. on Rock Mech. & Engng., (2003.5,Beijing), pp. 27-29.

Liu J and Gen W. 2000. *The solid-liquid coupling theory of the groundwater seepage and the numerical method.* Site Investigation Sci. & Tech. For Exploration, Vol. 4, pp.7-10.

Liu P, Zhang M, Sun M and Li R. 2000. *The coupling method research on the mechanics and chemistry of the shale hole stabilization.* Drilling and Production Technology, 23(6), pp.10-12.

Liu J and Liu X. 2001a. *The effects of effective pressure on porosity and permeability of low permeability porous media.* Chinese J. of Geomechanics, 7(1), pp.41-44.

Liu J and Liu X. 2001b. *Dynamic model of fracture propagation in Oil/gas reservoirs.* Engng Mech., (supp.), pp. 118-122.

Liu J and Liu X. 2001c. *Models for coupled fluid flow, temperature and stress fields for oil reservoir engineering.* Special Oil and Gas Reservoirs, 8(2), pp. 31-37.

Liu J, Zhang S and Liu X. 2002. *Theory of fluid-solid coupling theory and numerical simulation in fractured low permeability oil reservoir.* ACTA Mechanics Sinica,, 31(19), pp. 392-398.

Ran Q and Gu X. 1998a. *Coupling analysis of multiphase flow and stress for oil reservoir.* Chinese J. of Geotech. Engng., 20(2), pp. 69-73.

Ran Q and Gu X. 1998b. *A coupled model for land subsidence computation with consideration of rheological property.* Chinese J. of Geological Hazards and Prevention, 9(2), pp. 99-103.

Ran Q and Gu X. 1999. *Numerical simulation of multiphase flow in an elastoplastic deforming oil reservoir.* Chinese J. of Computational Mechanics, 16(1), pp. 24-31.

Shen H, Gao H and Xia S. 1998. *Experimental study on seepage characteristics of fissured rock mass with stress.* Journal of Yangtze River Scientific Research Institute, 15(3), pp.35-39.

Su B, Zhan M and Wang Y. 1997. *Experimental study on the coupled stress and fluid flow in fractured rock masses.* Chinese J. of Geotech. Engng., 19(4), pp. 73-77.

Sun Y, Liu P, Song W and Gai X. 1998. *Mechanics and Chemistry coupled model of shale stabilization in deviated hole.* Petroleum Drilling Technology, 26(4), pp. 51-53.

Sun P and Xian X. *Coupled models for coal seam deformation-gas leakage and its applications.* J. of China Coal Society, 24(1), pp. 60-64.

Wang M, Lu W and Hu J. 1994. *Casing damage mechanism of water and oil well and protection.* Oil industry Press, Beijing.

Wang Y, Xu Z. and Su B. 1998. *Four-freedom complete method for the seepage-stress coupled analysis in fractured rock masses.* J. of Hydraulic Engineering, Vol. 7, pp. 55-59.

Wang H. 2000. *Hybrid model of discrete fracture network and discrete element method for rock slope stability analysis.* J. of Hydrogeology & Engineering Geology, Vol. 2, pp. 30-34.

Wang Y, Xu Z and Su B. 2000a. *Complete coupled analysis of fluid flow and elastoplastic stress in complicated fractured rock masses.* Chinese J. of Rock Mech. & Engng, 19(2), pp. 177-181.

Wang J, Deng J and Li B. 2000b. *Numerical simulation of borehole collapse procedure and prediction on well-hole enlargement.* Petroleum Drilling Techniques 28(6), pp. 13-14.

Wang Z and Du Z. 2001. *Mathematical models and numerical simulations of multiphase fluid flow and solid deformation processes under non-isothermal conditions for elastoplastic oil reservoir formations.* Petroleum Exploration and Development, 28(6), pp. 68-72.

Wang S, Shi M and Yu W. 2001a. *Research on relaxation phenomena of heat and mass coupling in wet porous media.* J. of Applied Sciences, 19(4), pp. 349-352.

Wang Y, Liu J, Yang J and Zeng M. 2001b. *Simulation of fluid-solid coupling for flow of coal-bed methane.* J. of China Coal Society, 26(3), pp. 285-289.

Wei J, Cui Y and Shao J. 2000. 3-D *Finite Element method simulation of ground water and land subsidence in Jining City.* J. of Changchun Univ. of Sci. & Tech.,30(4), pp. 376-381

Wu Y. 1996. *The coupled stress-ground water flow model for jointed rock mass.* J. of Geological Hazards & Environmental Control, 7(1), pp. 31-34.

Wu Y. 1997. *Fundamentals of rock mass hydraulics (III)-lumped parameter model and continuum models of seepage and stress coupling of rock masses.* Journal of Hyrdogeology and Engineering Geology, Vol. 2, pp. 54-57.

Wu Y and Zhang Z. 1996. *An application of generalized double porosity media model for coupled stress and groundwater flow analysis in fractured rock mass.* J. of Engineering Geology, 1(3), pp. 40-46.

Wu Y, Zhang Z, Wang S and Huang R. 1994. *Lumped parameter model of the coupled flow and stress effects of fractured rock masses.* J. of Engineering Geology, 2(1), pp. 9-13.

Xu Z and Xu X. 1999a. *Liquid-solid coupled problem of the flow in confined strata under two dimensional in-situ stresses.* Chinese J. of Rock Mech. & Engng., 18(6), pp. 645-650.

Xu Z and Xu X. 19 99b. *Liquid-solid coupled problem of radial flow through porous media under general plain stress conditions.* Chinese J. of Geomechanics, 5(1), pp.12-16

Xu Z and Xu X. 1999c. *A fluid-solid coupling problem in saturated porous media with a fixed fluid extraction rate.* Chinese J. of Geotech. Engng., 21(6), pp. 737-741.

Xu Z, Xu X, and Shen L. 1999. *One-dimensional unsteady flow of fluid in deformable porous media.* Chinese J. of Applied Mechanics, 16(4), pp. 46-52.

Xue S and Song H. 1999a. *The theory of immiscible saturated two-phase flow in deformed porous media I-mathematical model.* Seismology & Geology, 21(3), pp. 243-252.

Xue S and Song H. 1999b. *The theory of immiscible saturated two-phase flow in deformed porous media ii-coupling solution and finite element equations.* Seismology & Geology, 21(3), pp. 253-260.

Yang X and Zhang Y. 1995. *Analysis on dynamics phenomena in mine water outburst.* Mine Pressure & Roof Management, Vol. 1, pp. 13-15.

Yang Y and Zhou W. 1991. *Coupling model of seepage field and damaging field in fractured rock mass and lts application.* J. of Hydraulic Engineering, Vol. 5, pp. 19-27.

Yang T, Tang C, Zhu W and Feng Q. 2002. *Coupling analysis of seepage and stresses in rock failure process.* Chinese J. of Geotech. Engng, 23(4), pp. 489-493.

Yin Z, Liu J and Wu Q. 2003. *Effect of water flooding well pressure releasing on casing extrusion stress.* Chinese J. of Rock Mech. & Engng. (in press)

Yue H. 1994. *Primary study on model for coupled heat-moisture-salt transfer in soil during freezing-thawing processes.* J. of Glaciology & Geocryology, 16(4), 308-313.

Zhan Z, Wen J, Wu G and Liao Y. 1998. *Research on constitutive model of wet swellling rock and soil masses.* J. of Geological Hazards & Environmental Preservation, 9(2), pp. 49-54.

Zhang W and Li X. 1997. *Analytic solution to Poro-mechanical problems in a hollow axi-symmetric domain.* Chinese J. of Rock Mech. & Engng., 16(5), pp. 83-489.

Zhang H, Zhou C and Yi L. 2000. *Damaging osmotic tensor of rock mass under hydraulic coupling conditions.* Geotech. Engng. World, 3(3), pp. 11-14.

Zhang Y and Zhang J. 1997. *Experimental study of the seepage flow-stress coupling in fractured rock masses.* Rock and Soil Mechanics, 18(4), pp. 59-62.

Zhao Y. 1994. *Mathematical models for coal strata and gas coupling and numerical method.* Chinese J. Rock Mech. Engng., 14(3), pp. 201-205.

Zheng S. 2000. *Theory of coupling analysis on seepage and damage field and its application.* Ph. D. Thesis. Institute of Rock and Soil Mechanics, Chinese Academy of Sciences, Wuhan, China.

Zhou, C and Xiong W. 1996a. *Study on the mechanism of coupling seepage and deformation in un-continuum fissure.* J. of Hydrogeology and Engineering Geology, Vol.3, pp. 14-17.

Zhou C and Xiong W. 1996b. *A generalized cubic law for percolation in rock joints.* Rock & Soil Mechanics, 17(4), pp. 1-7.

Zhou C and Xiong W. 1996c. *Permeability tensor for jointed rock.* Chinese J. of Rock Mech. & Engng., Vol. 15, pp. 338-344.

Zhu Z and Sun J. 1999. *Analysis models for unsteady flow in fractured rock masses coupled with damaged field.* J. of Hydrogeology & Engineering Geology, Vol. 2, pp. 35-41.

COUPLED PROCESSES AND PETROLEUM GEOMECHANICS

Maurice B. Dusseault
Earth Sciences, University of Waterloo, Waterloo Ontario

Abstract: Coupling in geomechanics arises in various natural and man-made systems, but seems ubiquitous in petroleum geomechanics because of large changes in pressures, temperatures, stresses, rates and chemistry. Practical petroleum geomechanics examples of mechano-chemical coupling, quasi-static solid-liquid coupling, and dynamic solid-liquid coupling are presented. Mechano-chemical coupling in shale stability shows how pressure, temperature, and concentration driven diffusion processes are coupled and lead to volume changes, which in turn lead to stress changes, yield, and changes in physical properties. Sand production from high porosity sandstones is an example of quasi-static solid-liquid coupling. Apparently, yield and dilation of the sand fabric must precede full liquefaction and sand efflux. Recent progress toward understanding dynamic-static coupling has opened a new applications area in geomechanics, and effects such as enhanced fluid flow rate are beginning to be exploited commercially.

1 INTRODUCTION

Coupling is fundamental to modern soil and rock mechanics; the first basis for coupling was Terzaghi's 1-D consolidation theory linking liquid diffusional flow (Darcy flow) to the stress partitioning between the solid and the matrix ("Terzaghi's Law", $\sigma_{ij} = \sigma'_{ij} + p_o$). A rich scientific and engineering literature emerged as Terzaghi's initial limiting assumptions were relaxed one by one by persons such as Biot (static consolidation), Booker (thermal consolidation), Mitchell (chemical consolidation), and others. Generalized consolidation theory now includes diffusion processes and stress effects such as the Mandl-Cryer effect, countercurrent flow of ions and solutes in mechano-chemical processes, in situ hydraulic fracturing generated by mineralogical changes, etc.

The need to formulate and solve coupled problems arises often in petroleum geomechanics:

- In porous solids, several fluid phases can interact in terms of flow, gas solubility, altering saturations, phase changes, etc.
- Problems may involve simultaneous transport and mechanical deformation; transport includes Darcian, Fickian, Fourier and Ohmic diffusion.
- Stress, pressure and temperature changes ($\Delta\sigma$, Δp, ΔT) can be huge; at 5 km deep, effective stress of 75 MPa to 0 MPa are induced during drilling or hydraulic fracturing. Mineral phases cannot be considered incompressible, and strong non-linear effects occur. Large Δp and ΔT generate large, coupled, non-linear and often irreversible effects.
- Pore fluid chemistry changes alter chemically sensitive rocks, particularly shales, so that their properties change substantially.
- Processes may lead to fabric changes, dilation, microfissuring, fracturing, loss of cementation, and so on, with massive effects on coupled processes, and in highly non-linear ways.

Several examples of coupling in petroleum geomechanics will be described. Such descriptions are necessary prerequisites to equations-based mathematical model development. In the examples chosen, strong predictive models generally do not yet exist because of the behavioural complexity.

2 ΔV AND COUPLING EFFECTS

Quantitative coupling of mechanical (stress-strain) and diffusion (transport) effects requires that the diffusion process be expressed in terms of the volume change that it generates. For heat flux, this is straightforward: thermoelastic volume changes are computed using the thermal compressibility coefficient $\beta = (\partial V/\partial T)/V$, which may also be non-linear and anisotropic (as for shales). Physically, a ΔV arising because of a temperature increase causes a "push" against the surrounding rocks, increasing the stress in the direction of expansion. In practical terms, a thermal expansion tensor is rarely highly deviatoric, and a single scalar value generally suffices for most problems.

To couple mechanical effects to Darcian and Fickian processes, effective stress compressibility is used for pore pressure changes and chemically-induced volume changes are used for ionic and salinity changes. Particularly for Fickian flux, the physical coupling parameters (likely non-linear) are generally unavailable, and the literature is deficient in addressing this area.

Shale may be the only "common" rock type where all four diffusion processes can co-exist as processes of first-order importance within typical engineering conditions (Dusseault, 2003). Coupling effects in shale are extremely complex because microfissuring and yield take place, giving rise to many high-order non-linearities. For example, during deep drilling, high borehole wall deviatoric stress (a mechanical factor) leads to microfissuring, and the permeability in the damaged zone increases by at least 4-6 orders of magnitude (from nanoDarcy to milliDarcy range). This leads to facilitated Darcian flux and Fickian flux, so concentration gradients are both diffusively and advectively controlled. Pore pressure gradients that act to support the wall disappear, shrinkage (or expansion) can take place, and the microfissuring propagates, changing rock behavior. Heat transfer rates are also affected if advective flow occurs.

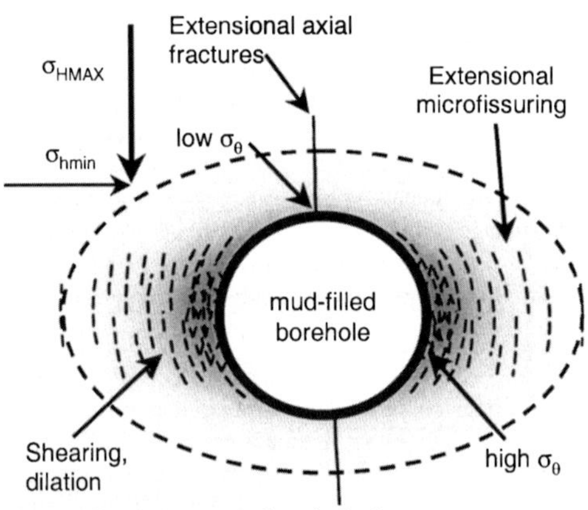

Figure 1. Anisotropic borehole damage

This may sound complicated, but reality is even more complex. The damaged skin usually sloughs off the wall during drilling; hence, in addition to non-linearities arising from weakening and microfissuring, an ill-defined moving geometric boundary condition exists. Furthermore, the process is inherently anisotropic because of natural shale fabric anisotropy and in situ stress anisotropy. This leads to different sloughing rates at different points around the borehole wall, generating borehole elongation in the direction of the least compressive stress acting normal to the borehole wall. These "borehole breakouts" can be small, but in highly stressed regions they can extend several diameters from the borehole wall.

Fickian diffusion also alters borehole yielding processes (Geilikman et al. 1996); the ionic change $Na^+ \rightarrow Ca^{++}$ leads to shrinkage of ductile smectitic shales. In stiff shale, a $\Delta V/V$ of much less than 0.01 will cause stresses to change by many MPa, leading to a self-reinforcing coupled process:

- Shrinkage reduces σ_θ until $p_w > \sigma_\theta$, usually at 90° to the direction of σ_{hmin} (**Fig. 1**).
- A short radial fracture develops, propagating to the point where $p_w \approx \sigma_\theta$.
- The fracture permeability is vastly larger than the intact shale, more rapidly exposing shale away from the borehole wall to the new cation.
- Newly exposed shale shrinks, fostering fracture growth, thus the process is self-reinforcing.
- Because of flux limits and scale effects (fixed lithological boundaries, flow path length), the process slows with distance.

For geochemical effects in shale, incorporating mechanochemical coupling requires $\Delta V/\Delta C$ data for Fickian processes (C is specie concentration); such data remain largely unavailable.

Clearly, mechanochemical coupling in the presence of coupled diffusion processes is extremely rich from a scientific and engineering point of view. The key geomechanics factor is to quantify ΔV as a function of T, C and p because ΔV drives $\Delta\sigma'_{ij}$, which leads to yield (shearing and fracturing), which leads to changes in diffusivity parameters. Methods must be developed for quick assessment of these properties. For all coupled geomechanics processes with yield and microfissuring, the associated strong non-linearities make analysis singularly challenging.

3 WEAKENING AND DIFFUSION

Diffusive processes alone can lead to internal weakening without a change in external macroscopic stress that causes yield. Consider the propagation of a front of ionic change, such as inward Ca^{++} diffusion with adsorption in smectitic shale. The front is relatively sharp because of adsorption, and there is a volume decrease associated with the $Na^+ \rightarrow Ca^{++}$ cation change, on the order of 1% for typical conditions. The sharpness of the front leads to high local strain gradients, rather than uniform general strain. A strain gradient is accompanied by a stress gradient, leading to high local deviatoric stresses, perhaps on the scale of a few millimetres. These stresses lead to rupture of brittle cohesive bonding, reducing material strength. Thus, even though shrinkage occurs, a process usually considered strengthening, diffusively propagating strain inhomogeneities can lead to weakening through generation of irreversible internal damage. Extensive strength tests on specimens after changing the ionic state while the specimens were maintained at a constant confining stress have clearly demonstrated this effect (Bostrøm et al. 1998, Horsrud et al. 1994).

Other diffusion processes can change strength and other physical properties. For example, in polycrystalline rocks subjected to cyclic ΔT, differential thermal expansion of different mineral grains and different thermal expansion coefficients in different crystallographic orientations generate local stresses high enough to nucleate and propagate grain boundary microfissures. This occurs even without a thermal gradient (Rosengren and Jaeger, 1968), although a sharp temperature front would contribute to the effects.

Thus, gradual cohesion loss in a polycrystalline rock can take place through Fourier diffusion. As damage accumulates, an interconnected microfissure network is generated, and the rock permeability, allowing water to enter the polycrystal. Water enhances microfissure propagation because of hydrolytic weakening (Hadizadeh and Law 1991), which is the provision for –OH groups to create silanol groups, rather than unsatisfied silane bonds (Griggs 1967, Swolfs, 1978). Furthermore, if the rock is heated, the speed of all chemical reactions is accelerated, another aspect of geochemical coupling.

4 SHALES

Many aspects of shale behavior are dealt with in another article in these Proceedings (Dusseault, 2003). Other coupling effects include:

- Temperature changes (Fourier) alter the viscosity of the permeant, changing the hydraulic conductivity (Darcian flux rate).
- Temperature changes (Fourier) affect mobility of electrostatically bound water, changing the hydraulic conductivity (and thus Darcian flux).
- Ionic and salinity changes (Fickian) alter the degree of electrochemical bonding of the interstitial water, changing the pore throat hydraulic conductivity (and thus Darcian flux).
- Darcian flux displaces the pore liquid, altering the ionic and salinity state, thereby changing the Fickian flux.

4.1 Origin of Q-I fractured shales

Borehole instability arises mainly in geochemically sensitive smectitic (S) shales, usually of higher porosity (>10%), and in low-porosity (<10%), highly fractured, quartz-illite (Q/I) shale. These rocks are encountered in most deep drilling operations in both tectonic and non-tectonic situations. The origin of the fractured shales is of geological and practical interest.

With increased burial depth and temperature, smectite undergoes mineralogical change: **smectite → quartz + illite + water**, with a 10-15% volume loss. This generates cemented, lower-porosity Q/I shales. Intense fracturing occurs because volume shrinkage reduces horizontal stress; when $\sigma_h < p_o$, fracturing starts, a process similar to columnar shrinkage joint formation in cooling basalt.

The **S→I** phase change begins at 2500-3000 m depth; below 5000 m, smectite is rare. Many intermediate S/I mixed-layer clay variants are generated during this transition, and these clays have a decreasing geochemical sensitivity that depends on the proportion of smectite layers.

Accompanying these mineralogical phase changes are horizontal stress state changes (σ_v is ~constant because of the overburden weight). Shrinkage reduces σ_h in the shale, and equilibrium dictates a concomitant gain in σ_h in bounding layers that have not experienced large volume changes (**Fig. 2**). Thus, attempts to control shale sloughing in Q-I fractured shale may simply lead to lost circulation through hydraulic fracturing.

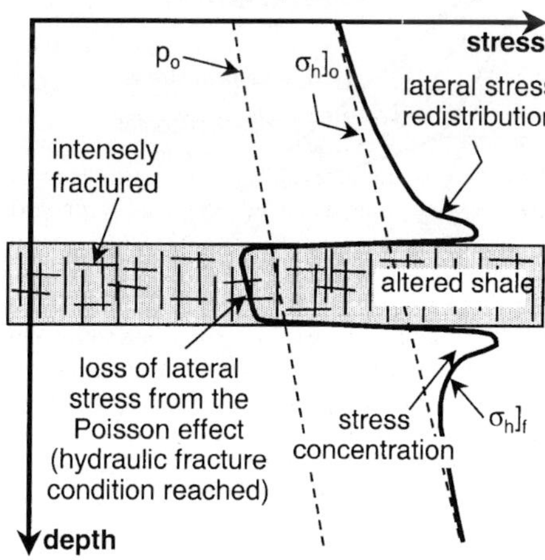

Figure 2. Diagenesis Causes Stress Changes

Not only are stresses altered, but fabric is also massively and irreversibly changed. The blocky nature of the fractured shale creates a scale issue: larger diameter boreholes are more likely to experience sloughing of the shale chunks (**Fig 3**).

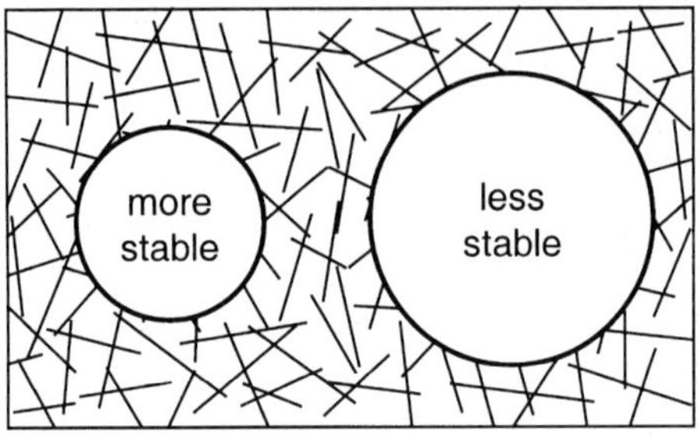

Figure 3. Scale Effect in Fractured Shales

Intense fracturing increases permeability by many orders of magnitude. The regional sealing Intense fracturing increases permeability by many orders of magnitude. The regional sealing effect of ductile shales is destroyed, and this is thought to be responsible for the pressure reversion that is found below overpressured zones (**Fig 4**).

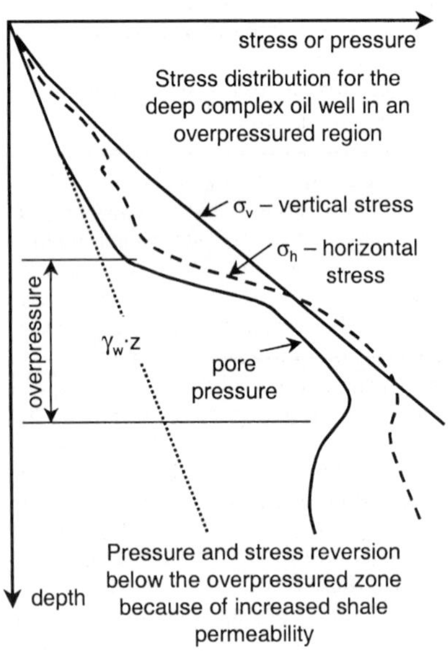

Figure 4. Deep Pressure and Stress Reversion

The permeable fracture network destroys the pressure drop across the borehole wall ($p_w - p_o$) that is a seepage support. Higher borehole pressures cause fractures to dilate, increasing flow capacity and reducing the ability to build a low permeability filter cake. This effect is exacerbated by bottom-hole cooling, as the drilling fluid at the bit can easily be 25-35°C cooler than the formation temperature. Cures to borehole instability in this case involve lowering (not raising) the borehole fluid pressure and adding carefully graded solids to the drilling fluid to bridge fractures in the shale.

4.2 Smectites and fluid chemistry

Smectites are geochemically sensitive because of a high surface area: if the electrostatically bound water layer thickness changes, ΔV takes place, with large stress changes. Several examples are given.

Using a Ca^{++}-based drilling fluid in a Na^{+}-based shale causes an adsorbed water thickness reduction. This has beneficial effects: the shrinkage causes a stress redistribution that is stabilizing (**Fig 5**).

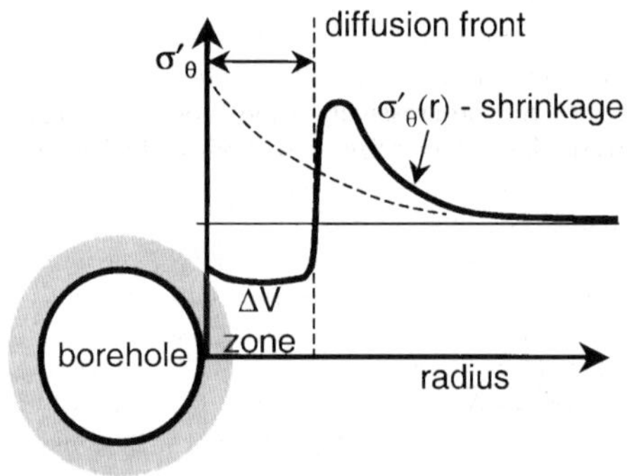

Figure 5. Reduced $\sigma'_\theta(r)$ from Shrinkage

Shrinkage capacity depends first on smectite content, exchangeable cations and porosity. Then, it depends on the nature of the drilling fluid filtrates to which the smectite is exposed. High salinity, potassium (K^+) rich, or cooler drilling fluids will each cause shrinkage. In contrast, heating and use of fresh water-based fluids can lead to expansion, accelerated sloughing and swelling. If shrinkage is sufficient to allow tensile fracturing, the potential for advective flow and diffusive ionic exchange into the shale may increase, introducing a type of non-linearity that is common in coupled processes. It appears that geochemical changes, advective flow, and heat flow can all couple strongly, generating large-scale stress changes that lead to fabric changes (shear yield, swelling, fracture…).

The need to account for shale properties as well as pore fluid chemistry has led to the suggestion of a descriptive dimensionless parameter, the Reactivity Coefficient (Fam et al. 2003), for shales.

4.3 Can water flow in smectitic shale?

Magnetic resonance probing of specimens of 15-20% porosity smectitic shales (Pierre II shale from Wyoming, ~40% smectite) indicated that ~5% of the water content was fully "free", not structured by electrostatic forces acting on the polar water molecules. Similar tests of a quartz-illite shale (Queenston Shale from Ontario) showed that ~50% of the interstitial water was "free", despite a much lower porosity (6-8%). Magnetic resonance relaxation curve analysis allows volumes of water at different binding energies to be quantified (Dolinšek et al., 1999).

Almost all the water in smectitic shales is to some degree electrostatically bound to clay particle surface or cations. However, all the water is not equally bound: molecules adjacent to the clay particle surface are tightly bound (H-H bonding), water molecules farther away from the mineral surface are progressively more loosely bound. Nevertheless, thermally activated diffusive exchanges can still take place (f(T)). Some conclusions at this stage are warranted:

- In smectitic shales of 10-20% porosity, true advective Darcian flow may not take place.
- Because water is "structured", there may be a threshold pressure gradient before flow starts.
- Large diameter hydrated ion flux is impeded, allowing high osmotic pressures (~1-2 MPa).
- Because of different binding levels, advective permeability is pressure-gradient-dependent.
- +ΔT will increase Brownian motion, reducing bound water layer thickness, giving an increase in hydraulic conductivity with temperature, and leading to thermal consolidation.
- Hydraulic conductivity in smectitic shales appears not to be a unique function of the pore and throat geometry; it depends on specific surface, the electro-active character of that surface, temperature, and of course the ionic concentration in the interstitial water.

4.4 Drilling long open-hole sections

Exceptionally, drilling may expose 2500-3500 m of rock along the borehole to a drilling fluid of constant density and chemistry (**Fig. 6**). It is usually necessary to control elevated pressures at depth, so that a large Δp between the borehole fluid and the pore pressure exists at shallow depth. This high Δp helps drive flux into the strata, and can help axial fracture propagation (**Fig. 1**).

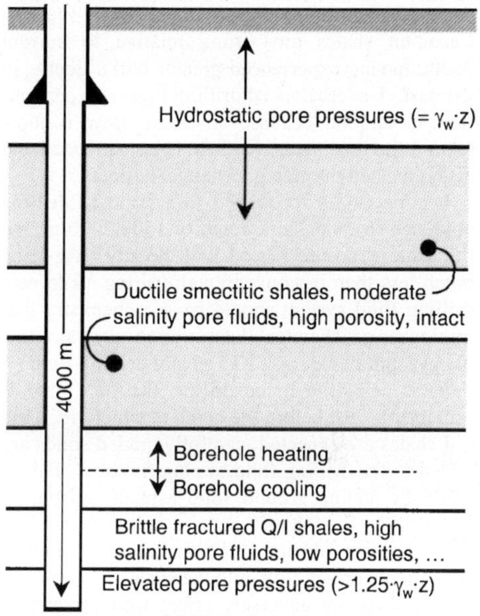

Figure 6. Mixed Conditions in Deep Drilling

Shales vary substantially in physical properties along the borehole so that corrective actions taken to control instability at the bit may trigger problems higher in the hole, or vice-versa. For example, in gas wells in western Alberta and north-eastern British Columbia, long open-hole sections simultaneously expose intact ductile smectitic shales in the upper portion, and intensely fractured Q-I shales in the lower part, several hundred metres above the target horizons at ~4000 m depth. Using an oil-based drilling fluid is effective in the ductile shales, but leads to sloughing and hole control problems in the deeper, fractured shales. Using a water-based drilling fluid leads to sloughing problems in the upper part of the hole because of prolonged exposure time to filtrate. Temperature effects over a long hole interval can also exacerbate sloughing high in the hole.

The conservative strategy is to place an intermediate casing string seated at the boundary between ductile and fractured shales, having drilled the ductile shale with oil-based drilling fluid, then converting to a water-based fluid to drill out of the intermediate casing. An intermediate casing string costs several hundred thousand dollars.

What is the "best possible approach" to cope practically with these coupled problems? First, drill as deeply as possible in an underbalanced condition ($p_o > p_w$), preferably with a carefully designed foam (Argillier et al., 1998). Western Canadian shales are strong relative to current depth, having experienced greater burial depths in the past. Underbalanced drilling improves penetration speed, and because advective flow is mow toward the hole, reactive shale is not corroded and disaggregated through geochemical effects.

It is necessary to convert to a standard drilling approach ($p_w > p_o$) at a depth of 1500-2000 m, but the shale exposure period will already have been much less than in conventional drilling. The new drilling fluid system should have a chemistry that minimizes geochemical deterioration, such as a low pH gypsum-based, or a KCl-glycol drilling fluid (if there is anhydrite to be drilled, the Ca^{++} mud is preferred). Just before the depth at which fractured Q-I shales are expected, carefully graded solids are added to the drilling fluid to plug fissures. This is aided by adding ground solid asphalt (gilsonite). The drilling fluid density is kept as low as possible to maximize drilling speeds and to reduce advectively driven flow into the shale.

If it does not adversely affect well production rates, may also drill a smaller diameter hole to improve penetration rates and reduce exposure time of shales to diffusion processes.

5 SAND PRODUCTION

5.1 Impact of sand production

In high flow rate oil and gas wells, sand influx from formation erosion (piping) is considered negative, though this view is slowly changing (Dusseault et al. 2000). Screen and granular filter installations are used if even a small sand influx risk exists, and materials and installation costs are high, particularly in offshore conditions. Also, reduced production rates attend the use of screens and gravel packs; wells are poorer producers because flow impediments and substrates for mineral precipitation have been placed in the fluid path. These devices also become blocked by fine-grained mineral particles, and frequent, expensive well interventions are used to restore productivity.

During 1990-2000, it became apparent that sand influx leads to better oil production rates (Dusseault and Santarelli 1989, Dusseault and El-Sayed, 2000). In heavy oil production from cohesionless sandstones (Geilikman et al. 1997), production rates with free sand influx can be 10-30 times higher (10-30 m^3/d rather than ~1 m^3/d) than in the same well with a screen installed. Even in high-rate offshore oil wells (1000-4000 m^3/d), permitting some sand influx has led to production rate increases of >35% (data from over 200 oil wells in different North Sea fields, Dusseault et al. 2000).

In Canada in 2002, ~650,000 bbl/d of oil (viscosity 800 - 12,000 cP) was produced using massive sand influx, generally with 0.5 to 6-8% sand by volume of the produced volume (liquids and solids). This represents ~5 billion dollars for Canadian producers at $20.00 US/bbl. Of course, 500,000 m^3/yr of produced sand is a challenging rock mechanics problem, as the most environmentally secure disposal methods are fracture injection and salt cavern placement (Dusseault and Bilak 1999, Duyvestyn et al. 1998).

Sand production has application in any reservoir where conditions are appropriate (low cohesion, high porosity, blockage or scaling problems, etc.).

5.2 Increased production in heavy oil

Figure 7 shows a "typical" sand and oil production history for a Canadian heavy oil well (400-700 m deep, 1000-3000 cP heavy oil). Sand influx declines and approaches a "stable" level that is a function of oil viscosity, but oil rates remain many times larger than sand-free oil production.

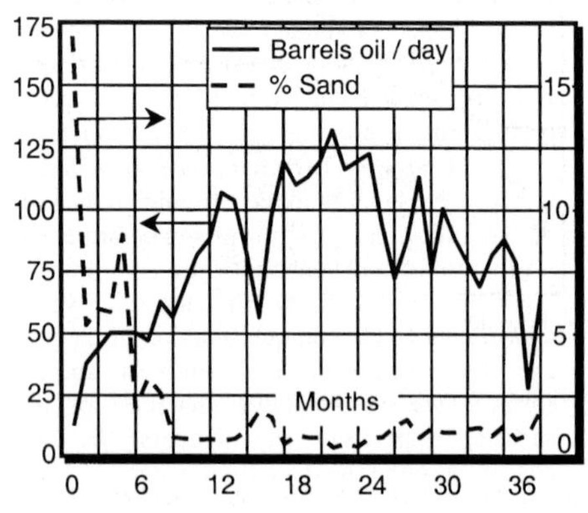

Figure 7. Oil and Sand Co-Production

What causes such large production rate increases? These five factors are of first-order importance:

- Darcy velocity is $v_f - v_s$, thus solid flux increases fluid flux, affecting flow rate by perhaps a factor of two. This may dominate rate enhancement during initial production.
- Removal of hundreds of cubic metres of sand leads to dilated, a high permeability zone that propagates outward, giving a "large wellbore" effect proportional to $\ln(r_o/r_w)$ (r_o/r_w = ratio of the high permeability zone to the original wellbore radii). This can enhance rates by up to ×5 and dominates late life flow.
- In heavy oil, exsolving gas stays as a dispersed bubble phase during flow, with many interesting effects (Geilikman and Dusseault, 1999). As bubbles expand down gradient, an "internal drive" is generated, causing fluids to flow more rapidly to the wellbore.
- If sand is being produced, pore blocking by fines or precipitated asphaltenes does not arise.
- Sand removal creates a softened zone, stresses are redistributed outward from the soft zone, and vertical stress concentrations, combined with lateral stress reduction because of sand removal, bring the force of gravity to bear, aiding destabilization and extrusion of sand toward the producing wellbore (**Fig. 8**).

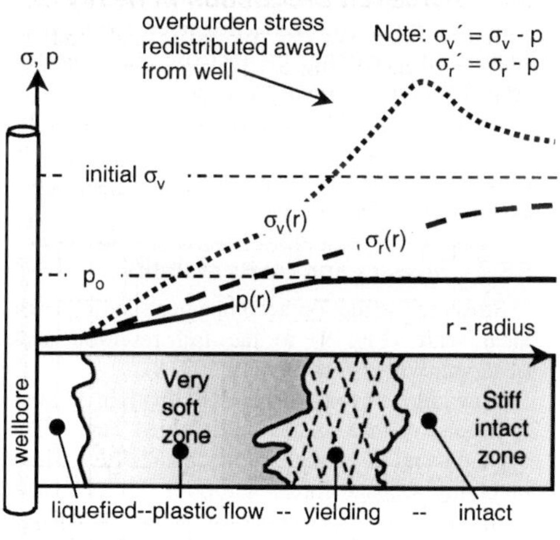

Figure 8. Sand Production Stress Redistribution

5.3 Increased rates in conventional oil

Sand management implementation in the Adriatic Sea (1995-2000) and in the North Sea (1996-2003) has clearly demonstrated that:

- Screens, gravel packs and other such devices reduce the well Productivity Index (PI).
- Screens and packs increase scaling and mineral plugging, with attendant PI reductions.
- Sanding in high-rate oil and gas wells occurs as "bursts" of 5-30 kg of sand that usually attenuate within an hour. Sand bursts are always accompanied by an increase in PI.
- Sand bursts are complex events that remain unpredictable, with many contributing factors (blockages, $\Delta\sigma'_{ij}$, Δp, saturation changes…).
- Erosion of surface facilities is about 10-15% as serious as expected, and risks can be easily reduced by facilities redesign and monitoring.
- Sand bursts are spontaneous "clean-up" events that improve local flow characteristics; they are triggered by high local pressure gradients.
- Wells not only produce oil faster, they require fewer expensive interventions.

Deliberate sand clean-up procedures have been developed to encourage sanding for a short time under controlled conditions, and this helps develop oil rate enhancements.

5.4 Liquefaction and sanding

Likely, the intact dense sands involved (porosities 25-32%), though cohesionless, cannot liquefy or erode from an intact face because of strong dilation and arching. Sanding is viewed as a liquid-solid coupling process that takes the geological material progressively from an intact dense state to a dilute particulate suspension.

Well perforaton damages a zone around each entry channel through crushing, cohesion destruction and porosity increase. When production starts, seepage forces dislodge the loosened, yielded sand in the damaged region and transport it into the wellbore, generating a "cavity". At depth the stresses are high, leading to shearing and yield of the cavity walls, generating a dilated yielded zone with little cohesion. Erosion enlarges the cavity, which must collapse at some size, with further porosity increases and softening.

Thus, sanding processes are viewed as yield and dilation before liquefaction (piping) and dilution. The final grain plucking occurs through the seepage force $\mathbf{F} = S \cdot A \cdot \partial p/\partial l$, where S is a shape factor. This small unbalanced pressure cannot overcome strong arching forces or cohesive resistance, but both of these have been destroyed by yield and dilation under low confining stress and high deviatoric stress (**Fig. 9**).

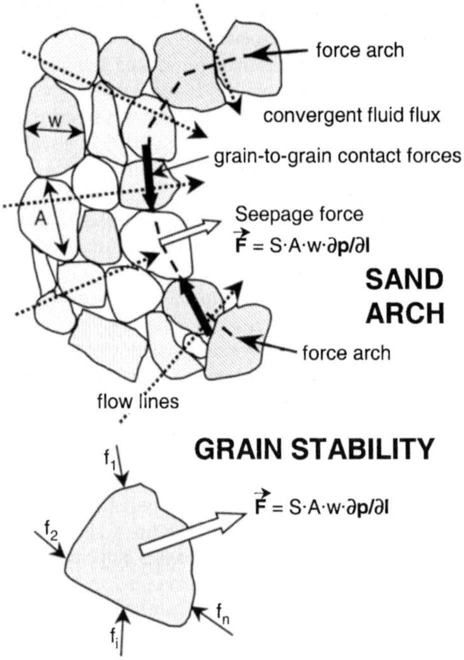

Figure 9. Granular Mechanics & Flow Coupling

Thus, intact sand (n ~30%) is yielded and dilated (n → 35-45%) by high shear stress, and confining stress is reduced until liquefaction occurs (n → ~50%). Also, because sand is easily retarded during flow to the well, dilution by faster flowing liquid further increases the porosity to the value seen at the wellhead (n → 94-99%).

Different processes occur for a conventional oil well evidencing sand bursts. During the sand-free period, near-perforation permeability is gradually restricted by mineral deposition or filtration, leading to local pressure gradient increases until the weakest "arch" behind the casing collapses, leading to a sudden influx of several kilograms of sand at high velocity through the port. This reduces the pressure gradient because of a large increase in local permeability, so the seepage force drops to a low value, and sand grains arch again. Effects related to multiphasic capillary cohesion and altered relative permeability during saturation changes can affect sand strength, effective stresses and seepage forces around a wellbore (Han and Dusseault 2003). Finally, large scale reservoir depletion changes effective stresses, and this can also lead to sand yield, cohesion loss, and erosion.

5.5 Coupling in sand production

Mathematical simulation of sanding requires that physical processes be well-understood and properly formulated. This section is an attempt to identify some of these mechanisms. Some models exist, but robust physical modelling remains a difficult goal because of multiphase flow, slurry flow, massive stress changes, porosity varying from 30% to >95%, and so on.

5.5.1 Gas behavior

Gas exsolves slowly in heavy oil, so that equilibrium thermodynamic assumptions, common to all reservoir simulators, are no longer valid: a kinetic model is required, coupling time with exsolution rates, pore pressures and saturations.

When gas evolves, bubbles grow and block pores, impeding liquid flow and increasing local pressure gradients. This process remains poorly understood, and involves issues such as bubble size and pore throat size distributions. Effects on liquid rate can be large, as much as a factor of 3 to 4 reduction because of bubble blockage.

Bubbles probably also aid sand destabilization by concentrating pressure gradients and therefore seepage forces. However, a dispersed bubble phase appears if anything to increase the apparent viscosity of an oil-water-gas slurry.

In a slurry (gas bubbles, sand, oil and water), the gas bubbles and sand are of substantially different densities, and somevertical segregation takes place during flow to the wellbore.

5.5.2 Stresses and cavity stability

Stress redistribution accompanies sanding at the local scale (**Fig. 9**) to the full reservoir scale (reservoir wide stress arching, **Fig. 8**).

How large a cavity can exist at depth in an intact sandstone or in a dilated and yielded sand? This question is a coupled mechanics-flow issue involving seepage forces acting at the local scale (cm), pore pressure variations acting at a larger scale (m), and stress redistribution at all scales.

Is sand bursting in conventional wells and shear events in conventional wells chaotic? Is it linked to stress redistributions and how do we predict long-term behavior? Microseismic monitoring could help quantify these mechanical coupling issues.

5.5.3 Sand liquefaction and slurry flow

Transition from a dense sand to a slurry, a phase change akin to melting of an alloy, with porosity playing the thermodynamic role of temperature, remains difficult to simulate. Earthquake liquefaction involves a contractile solid strain, whereas in sand production it is dilatant. Conventional liquefaction is treated empirically because of the complexity of cyclic loading, but in the case of sand production, inertial effects are irrelevant. A mathematical method of handling the disappearance of effective stress with concomitant liquefaction and porosity increase is needed.

Liquefaction occurs as porosity surpasses ~50%, but in a dense slurry there is also large energy consumption from collisions. As slurry becomes more and more diluted during flow to the well, sand capture and re-release can occur (Zhang and Dusseault 2000). Incorporating such a capture factor allows the mean fluid flow rate to exceed sand flux, thus progressive dilution from n = 50% to n > 95% can take place. Incorporating this coupled solid and liquid flux rigorously into stress-flow mechanics is beyond current capabilities.

5.5.4 Nonlinearities, moving boundaries

Sanding processes are accompanied by strong non-linearities. Cohesion loss takes place early in the dilation process because stresses are high and shear distortion occurs. Permeability changes from 0.5 – 5 Darcy to very high values, and properties such as matrix compressibility change from $\sim 0.50\times10^{-6}$ kPa^{-1} to "infinite" when the system is liquefied. Initial fabric eventually is fully destroyed by these plasticity processes.

Sand production involves a moving boundary value problem analogous to melting. Although the boundaries may be diffuse (as in melting of an alloy), asymmetric (intact permeability and stress anisotropy may have effects), and irregular, they must move outward from the producing well as a function of the amount of sand produced. Simulating this requires important assumptions about dilation and regularity that have not been tested in practice. It is not known if the boundaries are of constant dimension (thickness), or can be scaled to the disturbed zone size (self-similarity).

In sand production issues, many types of coupling exist under the most challenging of non-linear conditions and moving boundary situations. The economic incentive to quantitatively address these issues is large, given the value of the oil that can be produced by encouraging sand flux.

6 DIFFUSION-DYNAMICS COUPLING

Conventional recovery from sandstones is often as low as 20-40% of the oil in place, and improved recovery technologies can increase this by 10-20% (i.e. to 30-60% of the oil in place). Only exceptionally is more than 60-65% of the oil in place recovered. Various physical processes that are linked to Darcy flow, i.e. capillarity, viscosity permeability, heterogeneity, and pressure gradients, dictate the amount of oil recovered.

- Capillary pore throat blockage isolates bodies of oil as the capillary entry pressure cannot be overcome under static flow conditions.
- If the viscosity ratio of displacing to displaced fluids << 1, the less viscous phase channels through the other phase. This viscous fingering is endemic in water flooding of viscous oils.
- Water and gas coning to producing oil wells are forms of viscous instabilities and they lead to sudden declines in oil production rate.
- Zones of different permeability lead to flow channeling, where most or all the displacing fluids flow through the most permeable zone.

The last three processes are liked to pressure gradients and viscosity and are called advective instabilities.

The physical process in conventional oil exploitation is pressure-driven fluid flow. It has long been supposed that the physics of flow in porous media leads unavoidably to recovery limits in multiphase systems, but two major recent developments are altering this view. Gravity drainage can achieve recoveries as high as 99% in ideal cases, and pressure impulse flow enhancement has shown substantial flow rate improvements compared to conventional methods. Only the latter is discussed here; it is an interesting case of fluid-solid and static-dynamic coupling in geomechanics. The finding that dynamic (inertial) energy can affect flow processes in porous media is general, and will affect other industries such as environmental clean-up and deep well solid waste disposal processes for biosolids.

6.1 Darcy and Biot

Consider mechanical excitation frequencies in a liquid-saturated porous medium (**Fig. 10**). Two formalisms were conventionally used: for high frequencies, Biot-Gassmann wave mechanics (Biot 1956) is used; for low frequencies, Darcy theory is used (inertial effects assumed negligible). However, there clearly must be approximately three orders of magnitude where both diffusion and inertial effects are equally important. No coupled inertial-diffusion theory existed that permitted general analysis of the whole range of frequencies.

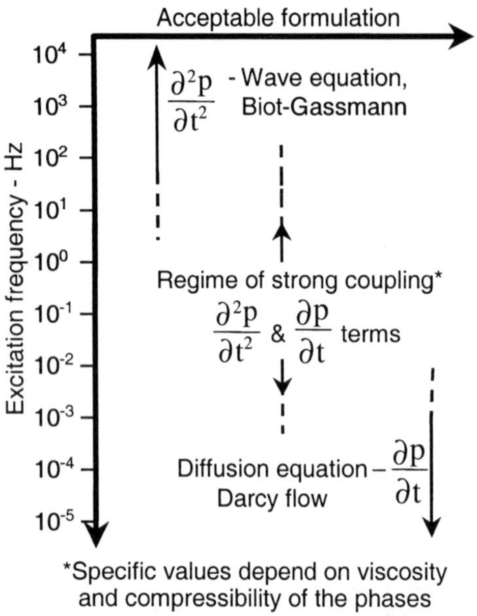

Figure 10. Excitation Frequencies and Theories

In Biot formalism, there exist several assumptions that restrict its generality and make true liquid-solid coupling impossible. Biot assumed that for a REV in a multiphasic porous medium, a single energy functional could be stipulated to define the energy state. It has been shown that for N continuous contiguous phases, N functionals are needed to fully describe behavior. (For example, simultaneous countercurrent flow of two immiscible liquids is evidence that at least two separate energy functionals are needed.)

Biot assumed that porosity (ϕ) was a fixed parameter; recent work shows that for porous media, ϕ is a thermodynamic variable similar to pressure and temperature, and it must be treated as a dynamic variable (Spanos et al. 2003). For the temporal evolution a porous medium, the porosity change must be stipulated as follows:

$$\frac{\partial \eta}{\partial t} = \delta_s \overline{\nabla} \bullet \overline{\mathbf{v}}_s - \delta_f \overline{\nabla} \bullet \overline{\mathbf{v}}_f \qquad 1$$

Porosity change with time (porosity diffusion) is specified by the divergence of the velocities of the phases, multiplied by their densities (δ_s, δ_f).

Because of the constant porosity assumption and a single energy functional, it is impossible to stipulate wave attenuation in Biot wave mechanics from first principles of thermodynamics, which should be part of any complete theory. Biot approached wave attenuation (not spatial spreading) empirically, rather than accounting for energy losses through compression and rarefaction cycles of the solid and liquid phases.

An implicit assumption in wave mechanics is that liquids deform by straining, and that no discrete flow takes place during dynamic excitation. Attempts to overcome this shortcoming through introduction of "squirt flow" concepts have not been entirely satisfactory.

At the other end of the excitation frequency spectrum, Darcy formalism deals with flow through porous media with these assumptions:

- Liquids are incompressible and strains are small (not the same as saying that the solid skeleton is incompressible). Modifications exist to analyze gas flow to wells.
- There are no dynamic (inertial) effects, therefore all motion can be described by a set of diffusion equations.

Thus, Darcy flow theory contains no dynamics: it is a "quasistatic" diffusion theory. It may be a reasonable construct for excitation frequencies less than ~10^{-4} Hz, as liquids do behave incompressibly in this range, leading to a pure displacement process through the pores (diffusion).

A complete theory of flow in porous media with dynamic-diffusive coupling requires that the governing equations contain both diffusion ($\partial \mathbf{u}/\partial t$) and inertial ($\partial^2 \mathbf{u}/\partial t^2$) terms (**Fig. 10**).

6.2 The porosity dilation (PD) wave

Biot and Darcy theory shortcomings have been largely overcome by development of a coupled diffusion-dynamic formalism (de la Cruz et al. 1993, Spanos 2001, Spanos et al. 2003). Porosity is treated as an explicit thermodynamic variable, so that $\partial\phi/\partial t$ and $\partial^2\phi/\partial t^2$ terms are found in the equations. The equations are mixed hyperbolic and parabolic, highly non-linear, and not amenable to easy numerical model development. Nevertheless, if they are solved subject to the assumption of the incompressibility of a liquid saturant, the existence of a slow wave is predicted. It is called the porosity dilation (PD) wave; it is not a strain wav, it is a coupled liquid-solid displacement wave, and it has some interesting properties.

The PD wave is a body wave of small elastic porosity dilation that propagates through a liquid-saturated porous medium. The wave cannot exist without liquid-solid coupling; in oil-water systems, it is generated through 0.1 – 1 Hz excitation, in which range the liquid evidences a transition to incompressible behavior. The PD displacement wave is analogous to a tsunami (a liquid displacement wave), and propagates at roughly $1/20^{th}$ the velocity of strain waves in the liquid.

6.3 PD wave micromechanisms

To clarify the effect of the PD wave on fluid flow, a micromechanics explanation is helpful.

If a porous solid is struck a direct blow at the right velocity, a compressional wave propagates outward; if an askance blow is used, the wave is dominated by shear strains. If a sudden dilation (or contraction) at the right frequency is applied to a liquid-saturated pore (achieved in practice by a sudden pressure pulse applied to the liquid phase), a coupled solid-liquid displacement wave is generated. Incompressible fluid expulsion (or uptake) takes place, causing adjacent pores to expand (or contract). During PD wave propagation, pore contraction forces liquid down the pressure gradient, and the "forward" pore dilates as liquid rushes into it. This causes a small elastic porosity change, thus transmitting the excitation and propagating the PD wave. If pores are substantially gas-filled, coupled liquid-solid dilation cannot occur, and rapid attenuation through gas compression (high energy loss) takes place. (This is is exploited in vehicle shock absorbers.)

If the porous medium is liquid saturated, the low-frequency, long-wave length PD wave propagates conservatively (low energy losses), just as a tsunami propagates great distances with little energy loss. Furthermore, it is kept within the porous and permeable zone and does not transit into impermeable or low permeability rocks because the liquids cannot move into and out of the pores at a sufficient velocity.

Repeated impulses generate more PD waves, and this leads to repeated accelerations of pore throat liquid as the waves transit. Sudden pressure pulses are the most effective way to generate the impulse, but a blow to the solid skeleton that has the right frequency content will generate the PD wave because of coupling; however, if high frequencies are used (e.g. seismic frequencies of 10-1000 Hz), very little energy will be converted to PD waves.

6.4 Some dynamic coupling effects

There are examples of the porosity dilation wave in practice. The human heart succeeds in passing red blood cells through capillaries barely larger in diameter than the cell diameter. This is possible because the heart is a pulsing device; if the heart were a rotary pump, all capillary flow would cease. The red blood cells continue to pass through the capillaries because porosity dilation waves transit, causing capillary walls to dilate, allowing cells to move forward in "steps". Of course, the frequency content of the heartbeat is just optimum for the "porous medium" of the capillary network. The velocity of the dilation wave as it travels toward the capillaries is slow (metres/sec) because of the high compressibility of blood and artery walls.

Post-earthquake flow rate enhancements or alterations in oil wells and water wells (Vorhis 1967) can be explained by the micromechanical effects associated with the PD wave. Groundwater table response after a large earthquake at a distance is well documented, but it is also known that the response generally occurs only long after the P, S and Rayleigh waves have passed. The delay is because the PD wave, which is responsible for the effects, is a slow wave (100-200 m/s velocity). Similarly, delayed triggering of sympathetic earthquakes (Geilikman et al. 1993), groundwater response to low-frequency vibrations (such as storm-wave induced inland flow in porous sediments), and other phenomena can be better understood by PD wave mechanics.

Capillary blockage occurs when the non-wetting phase is blocked at a pore throat by the interfacial tension (γ_{ow}) with the wetting phase (**Fig. 11**).

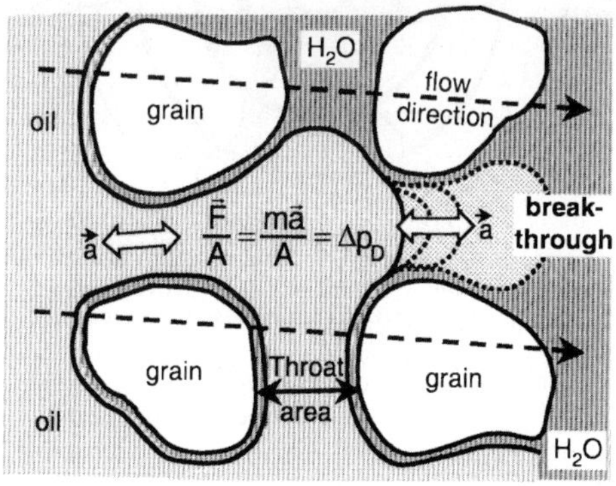

Figure 11. Dynamics and Capillary Blockage

The pressure barrier is $\Delta p = \gamma_{ow}/2r$, where r is the interface radius of curvature, set by the diameter of the throat. The PD wave causes acceleration ($\bar{a}$) of a small amount of the fluid mass (m) in the pore throat; this generates a force, $\mathbf{F} = m\bar{a}$. A dynamic pressure $\Delta p_D = \mathbf{F}/A$ is generated, and when $m\bar{a}/A > \gamma_{ow}/2r$, $\Delta p_D > \Delta p$, causing phase breakthrough. Once breakthrough is achieved, fluid flow can continue through the pore more easily, and oil production can continue with less impedance. If capillary blockage is reduced, there is less bypassed oil, increasing the ultimate oil recovery.

Figure 12 represents viscous fingering arising when a low viscosity liquid displaces a higher viscosity liquid. This instability bypasses oil, causes early breakthrough, results in poor placement of treatment liquids, etc. If the low viscosity liquid is injected with vigorous pulsing, the PD wave suppresses the viscous instability near the wellbore, but because waves spread radially and the amplitude decays by 1/r, the energy does not have as large an effect at a distance, and the characteristic length of the fingers is reduced. Figure 13 shows identical tests of oil displacement by water, showing different characteristics when pulsed externally (no pumping).

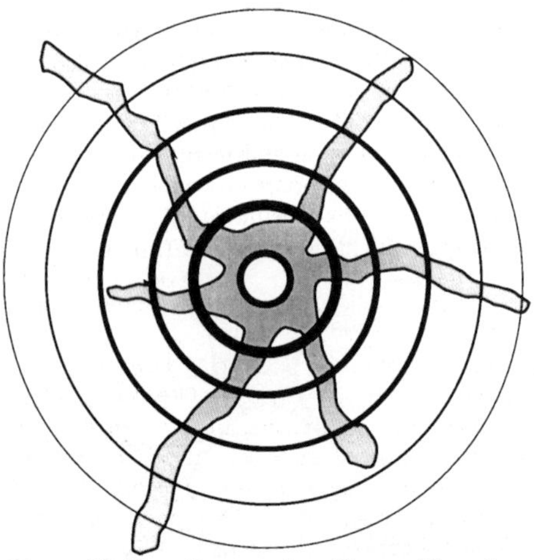

Figure 12. Wave Propagation, Viscous Fingering

Similar effects occur that reduce the magnitude of coning and the negative aspects of permeability channelling. However, the liquid-solid coupling does not totally overcome the instabilities, but reduces their impact. Even in single phase flow, high amplitude dynamic excitation at the right frequency will accelerate fluid flow rates and lead to internal pressure distributions that are different than predicted by quasi-static Darcy theory.

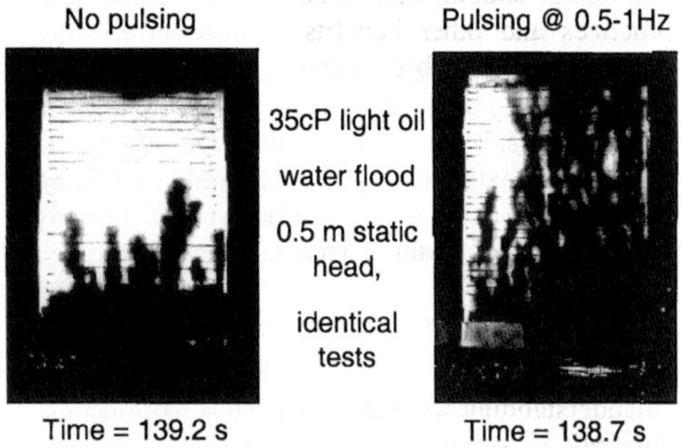

Figure 13. Dynamic Viscous Finger Suppression

Pressure pulsing can ameliorate restriction of pore throats by fine-grained fluid-transported minerals or precipitates such as asphaltenes that restrict flow to wellbores. The episodic accelerations at the pore throats that are characteristic of the passage of the PD wave serve to dislodge pore throat blockages, allowing the fine-grained materials to be flushed to the producing well (**Fig. 14**). This appears to be the same mechanism that causes groundwater wells to turn murky with colloidal substances some time after a distant earthquake has occurred.

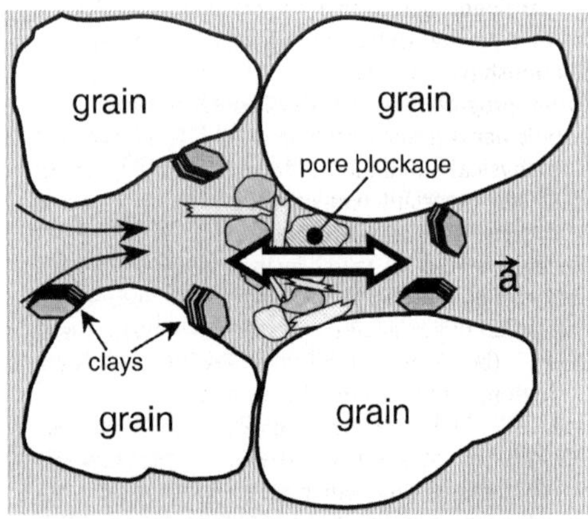

Figure 14. Fluid Accelerations Dislodge Blockages

PD wave excitation should be a highly efficient method of triggering compaction in appropriate reservoirs, and in cases where sand densification in situ is desired. This is because the porosity dilation effect causes cyclic changes in the grain-to-grain contact forces, perturbing the equilibrium and leading toward densification.

Pulsing should also lead to better grouting practices and other benefits such as accelerated clean-up of potable water aquifers that are contaminated by NAPLs and other materials. In these cases, the suppression of advective instabilities and the aid to overcoming capillary blockages will lead to more efficient displacement, less fingering of flushing agents, and so on.

7 CLOSURE

Understanding the consequences of coupling of physical processes is a leading-edge issue in petroleum geomechanics. A fundamentally scientific approach is required in petroleum geomechanics, as rarely do issues of civil society safety or personnel security enter into the design process, so the whole concept of safety factor is irrelevant to the design process. Thus, the issues do not permit lumping together the effects of various uncertainties into a single "safety" factor. The safety factor approach also tends to suppress the detailed scientific investigations that push forward our various disciplines in geomechanics. A common attitude is "If we can get away with simply increasing the design safety factor, we can then ignore the details of the physics." Ultimately, this is counterproductive.

Petroleum geomechanics continues to labour under a general deficiency of data and constitutive relationships. These would aid optimization and foster progress in analysis, but serious issues of sample damage and cost hamper testing and laboratory physical simulations. More monitoring seems to be an important part of understanding coupled processes. Pressures, temperatures and flow rates are the parameters taken by petroleum engineers; only through difficult inference can these be of first-order importance to petroleum geomechanics issues (borehole stability, sanding, dynamic excitation, compaction, fracture flow, cementation…). We need to measure displacements, transmission of acoustic (strain) or displacement waves, stresses, and other measures more closely linked to stress-strain of geomaterials.

In coupled processes, the issue of volume change arises repeatedly: shale expansion or contraction controls stress states around the borehole; withdrawal of sand creates stress changes and the potential for further yield, channeling and dilation; dynamic volumetric dilatation allows inertial effects to be brought to bear at the pore throats. Understanding the geomechanics implications means quantifying the volume changes and incorporating these into a numerical model.

Several economically important applications of coupling in Petroleum Geomechanics have been given: one each of mechano-chemical coupling, fluid-solid quasi-static coupling, and static-dynamic coupling. Better understanding, better modeling, and better measurements in all these cases will bring substantial economical benefits.

REFERENCES

Argillier, J.-F., Saintpère, S., Herzhaft, B. and Touré, A. 1998. *Stability and Flowing Properties of Aqueous Foams for Underbalanced Drilling.* Proc. SPE Annual Tech. Conf., New Orleans, SPE #48982.

Biot, M.A., 1956a&b. *Theory of propagation of elastic waves … porous solid.* Jour. of the Acoust. Society of America, 28, 168-178; 179ff.

Bostrøm, B., Svanø, G., Horsrud, P., Askevold, A., 1998. *The shrinkage rate of KCl-exposed smectitic North Sea shale simulated by a diffusion model.* SPE 47254, Proc. Eurock '98, Trondheim, 8-10 July.

de la Cruz, V., Sahay, P.N., and Spanos, T.J.T., 1993. *Thermodynamics of porous media.* Proc. Royal Society of London, A 443, 247-255.

Dolinšek, J., Bharatam, J., Dusseault, M.B. and Pintar, M.M. 1998 *Two-dimensional NMR study of surface water dynamics in hydrated silica spheres.* Physical Review B, 58(11), 7340-7346.

Dusseault, M.B. 2003. *Coupled Thermo-mechano-chemical Processes in Shales: The Petroleum Borehole.* Proceedings GeoProc 2003, Stockholm, Sweden, 8 p.

Dusseault, M.B. 2003. *Coupled Thermomechano-chemical processes in shale.* Proc. GeoProc 2003 Conference , Stockholm, Sweden, 8 pages.

Dusseault, M.B. and Bilak, R.A. 1999. *Regulatory controls and slurry fracture injection.* Jour of Canadian Petroleum Technology, December Special Volume. CD version

Dusseault, M.B. and El-Sayed, S., 2000. *Heavy oil well production enhancement by encouraging sand influx.* SPE/DOE Improved Oil Recovery Symposium, Tulsa, SPE #59276.

Dusseault, M.B. and Santarelli, F.-J. 1989. *A conceptual model for massive solids production in poorly-consolidated sandstones,* Proc. ISRM Int. Symp. on Rock at Great Depth, eds:

Maury, V. and Fourmaintreaux. D., Balkema, Rotterdam, V. 2, pp. 789-797.

Dusseault, M.B., Tronvoll, J., Sanfilippo, F., Santarelli, F.J. 2000. *Skin self-cleaning in high-rate oil wells using Sand Management.* Proc. SPE Int Conf on Formation Damage, Lafayette, SPE #58786

Duyvestyn GM, Davidson BC, Dusseault MB, 1998. *Salt solution caverns for petroleum industry toxic granular solid waste disposal.* Proc. SPE/ISRM Eurock'98, Balkema, Rotterdam, 239-246.

Fam, M. & Dusseault, M.B. 1999. *Determination of the reactivity of clay-fluid systems using liquid limit data.* Canadian Geotechnical Journal 36(1), 161-165.

Fam, M., Dusseault, M.B., & Fooks, J. 2003. *Drilling in mudrocks: rock behavior issues.* J. Petroleum Science and Eng., at the press.

Fam, M.A. & Dusseault, M.B. 1998. *Borehole stability in shales: a physico-chemical perspective.* Proc. SPE/ISRM Eurock '98, Balkema, Rotterdam, 461-470.

Geilikman, M.B. and Dusseault, M.B. 1996. *Mechano-chemical corrosion of shales.* Proc. 2nd NARMS '96, Montreal, Canada, 959-964.

Geilikman, M.B. and Dusseault, M.B. 1999. *Sand Production Caused by Foamy Oil Flow.* Transport in Porous Media, 35: 259-272.

Geilikman, M.B., Dullien, F.A.L. and Dusseault, M.B., 1997. *Erosional creep of fluid-saturated granular medium,* Journal of Eng. Mechanics of ASCE, v 123 (7), 653-659.

Geilikman, M.B., Spanos, T.J.T., and Nyland, E., 1993. *Porosity diffusion in fluid-saturated media.* Tectonophysics, 217, 11-113.

Griggs, D.T. 1997. *Hydrolytic Weakening of Quartz and Other Silicates*", Geophys. J. R. Astr., Soc., 14, 19-31, 1967.

Hadizadeh, J. and Law. R.D. 1991. *Water-Weakening of Sandstone and Quartzite Deformed at Various Stress and Strain Rates,* Int. J.Rock Mech. Min. Sci.&Geomech, Abstr, 28, 5, 431-439.

Han, G, and Dusseault, M.B. 2003. *Coupled Analysis of Sand Stability in Petroleum Wellbores.* Proc. GeoProc 2003 Conference , Stockholm, Sweden, 6 pages.

Han, G., Ioannidis, M., and Dusseault, M.B. 2002. *Semi-analytical solutions for the effect of well shut-down on rock stability.* Proc Can Int Petrol Conf, Calgary, June Paper 2002-50, 9 p.

Horsrud, P., Bostrøm, B., Sønstebø, E.F. & Holt, R.M. 1998. Interaction between shale and water-based drilling fluids: Laboratory exposure tests give new insight into mechanisms and field consequences of KCl contents. SPE 48986. SPE Annual Tech. Conf. & Exhibition, New Orleans, 27-30 September.

Horsrud, P., Holt, R.M., Sønstebø, E.F., Svanø, G. & Bostrøm, B. 1994. *Time dependent borehole stability: Laboratory studies and numerical simulation of different mechanisms in shale.* SPE/ISRM 28060, EUROCK'94, Delft, NL, Balkema, Rotterdam. 8pp.

MacGillivray, D., Davidson, B. & Dusseault, M.B. 1996. *One-dimensional thermal conductivity measurements in quartz-illite and smectitic shales.* Proc. Eurock '96, ISRM Int. Symp., Italy, Balkema, 107-113.

Rosengren K.J. and Jaeger, J.C. (1968). *The mechanical properties of an interlocked low-porosity aggregate. Géotechnique*, **18**(3), 317-326.

Schmitt, L., Forsans, T. & Santarelli, F.J. 1994. *Shale testing and capillary phenomena.* Int. J. Rock Mechanics Mining Sciences & Geomechanics Abstracts, 31, 411-427.

Sønstebø, E.F. & Horsrud, P. 1996. *Effects of brines on mechanical properties of shales under different test conditions.* Proc. Eurock '96, Barla (ed.), Balkema, Rotterdam.

Spanos, T.J.T. 2001, *The Thermophysics of Porous Media*, Boca Raton, Chapman & Hall/CRC Monographs and Surveys in Pure and Applied Mathematics.

Spanos, T.J.T., Dusseault, M.B. Udey, N. 2003. *Fundamental thermodynamics requirements for porous media description.* Proc. GeoProc 2003 Conference , Stockholm, Sweden, 6 pages.

Swolfs, H.S.: *Chemical Effects of Pore Fluids on Rock Properties*, Underground Waste Management and Environmental Implications, Proc. AAPG Symp., Memoir 18, 224-233, Dec.6-9, 1971.

Vorhis, R.C., 1967, Hydrologic effects of the earthquake of March 27, 1964, outside Alaska: U.S. Geological Surv. Prof. Paper 544-C, 54 p.

Wang, Y., Chen, C.C. and Dusseault, M.B. 2001. *An integrated reservoir model for sand production and foamy oil flow during cold production.* SPE Conf. on Thermal Operations and Heavy Oil Symp., Margarita, Venezuela, SPE 69714.

Zhang, L. and Dusseault, M.B. 2000. *Sand production simulation in heavy oil reservoirs.* Proc. SPE Int. Oil and Gas Exhibition and Far Eastern Technical Conf., Beijing, SPE 64747.

SOME THMC CONTROLS ON THE EVOLUTION OF FRACTURE PERMEABILITY

Derek Elsworth

Department of Energy and Geo-Environmental Engineering and The Energy Institute, Pennsylvania State University, University Park, PA 16802, USA

Abstract: The significant controls exerted on the evolution of fracture permeability by ambient conditions of stress, temperature, and chemical potential are illustrated for contrasting rock types. Natural and artificial fractures in Berea Sandstone, Arkansas Novaculite, and Bellefonte Limestone are confined within an x-ray transparent pressure cell and water flow-through tests conducted. The samples are monitored for change in permeability that evolves concurrently with the circulation of the water, and for net mineral efflux that accompanies this circulation. Periodic imaging by x-ray CT is used to define source areas of mineral efflux and its redistribution, and to provide constraint to phenomenological models that describe the observed behaviour. Vastly different behaviours are observed between the fractures present within the varied host materials – some gape with net mineral efflux, and some seal. These responses are viewed within the framework of the competition between mass mobilized by pressure solution, and net free-face dissolution or precipitation on the fracture walls, and in the matrix. Viewed in this context, the surprising range of behaviours, observed under similar THMC conditions in these three contrasting material types, may be unravelled.

1. INTRODUCTION

Paths of stress, temperature, and chemical potential are known to strongly influence the evolution of the transport properties of rocks. This is especially true where the materials are fractured, and permeability is strongly coupled to relatively small changes in fracture apertures, that may in turn be driven by changes in stress or chemical potential. Data constraining the role of stress-mediated chemical effects in fractures are sparse, but are available at elevated temperatures (>300°C) in granite [Moore et al., 1994], and at lower temperatures (50°C–150°C) in tuff [Lin et al., 1997]. These are augmented by results for available composite aggregates of quartz [Elias and Hajash, 1992], halite [Gratier, 1993], calcite [Zhang et al., 1994] and albite [Hajash et al., 1998], at moderate temperatures (23°C–150°C), and the same material suites at elevated temperatures and pressures [e.g., Zoback and Byerlee, 1975; Siddiqi et al., 1997]. The limited studies on fractures [Moore et al., 1994; Lin et al., 1997; Durham et al., 2001] suggest an increased sensitivity of their transport properties to thermal, hydraulic, mechanical, and chemical processes, over porous medium flows. This is apparent even at temperatures as low as 100°C, where the mobile dissolved species is silica, the test duration is of the order of a month [Elias and Hajash, 1992; Lin et al., 1997; Polak et al., 2003], and where permeability may be reduced by a factor of 10^4 [Lin et al., 1997; Polak et al., 2003]. Importantly, fracture permeabilities may either reduce or increase, in surprising ways, depending on the paths of stress or chemical potential. We illustrate this behaviour through observations during flow-through tests on samples of varied rock types. These include a fractured porous medium (Berea sandstone), and fractures in both silicic (Arkansas novaculite) [Polak et al., 2003; Yasuhara et al., 2004] and carbonate rocks (Bellefonte Limestone) [Polak et al., 2004].

2. EXPERIMENTAL METHOD

Flow-through experiments are conducted and reported on fractured samples confined within and x-ray transparent cell to evaluate the influence of stress-mediated changes in dissolution and precipitation on the evolution in transport properties. Cylindrical samples containing a single axial fracture are confined within a flexible membrane and end-to-end flow tests completed, as illustrated in Figure 1. Three independent measurements are made to constrain the processes promoting the redistribution of mineral mass within the fracture: hydraulic flux, mineral mass flux, and volumetric imaging by x-ray CT. First, permeability is monitored continuously by prescribing flow rate and measuring pressure drop across the sample. Effective stresses are retained within a narrow range by controlling the upstream back pressure. This measurement records permeability, and through this defines change in average aperture within the system [Piggott and Elsworth, 1992]. Second, the net efflux of dissolved mineral mass is measured periodically to provide a record of net mass removal, and to correlate this with observed changes in aperture, defined by the flow tests.

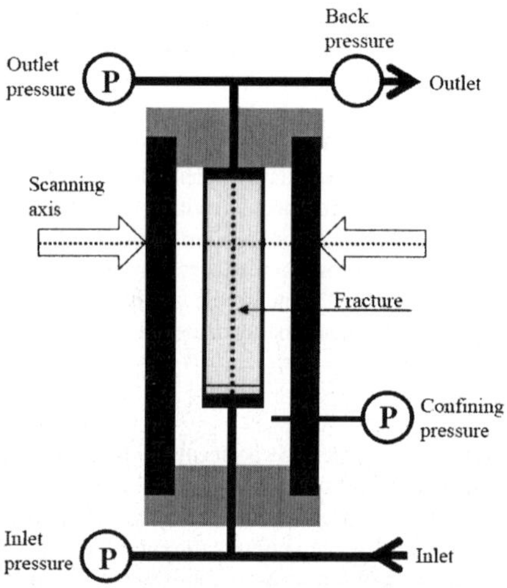

Figure 1. Experimental arrangement showing confining cell and sample with axial fracture. Flow at prescribed flow-rate and measured pressure drop records permeability. Constraint on process is provided by measurement of dissolved mass efflux, and periodic scanning by x-ray CT.[Polak et al., 2003]

Finally, periodic non-destructive imaging by x-ray CT is used to view redistribution of mineral mass within the heated sample, as illustrated in Figure 2. X-ray CT records changes in density within the sample as a proxy for mineral removal or redistribution, with the scanner used in this study capable of resolving down to ~1/1000th of the diameter of the sample. For sample diameters of 3-5 cm used in the following study, the minimum voxel size scales to the order of 30-50 μm. This resolution is at the limits of utility in defining mass redistribution – it is adequate to define mass redistribution in samples of Berea and limestone, but not for the relatively tighter fracture in novaculite.

Figure 2. Pressure vessel with heater tapes and flow couplings within OMNI-X scanner. X-ray source (left) penetrates aluminium pressure vessel, and is recorded on the detector (right).[Courtesy A.B. Polak].

These measurements provide unusual constraint on the evolving processes. Importantly, they allow the source of dissolved components to be determined: we need to discriminate whether the source is from free-face dissolution of the fracture wall, or from stress-mediated dissolution at contacting asperities. This distinction is crucial since these two mechanisms impart opposite effects in the sense of permeability change, under net dissolution: free-face dissolution increases permeability, and pressure solution reduces permeability.

3. OBSERVATIONS AND PROCESSES

Observations are reported on flow-through tests on porous-fractured and fractured samples to illustrate both the large magnitude and fast rate of change in permeability that evolves with the circulation of fluids that are not in chemical equilibrium with the host. The spectrum of materials (Table 1) include rocks with both high (Berea) and insignificant (novaculite and limestone) matrix porosity, and both high (limestone) and low (Berea and novaculite) dissolution and precipitation rate constants, allowing differing responses to be contrasted. The tests are conducted under invariant moderate stresses (~3.5 MPa), for different paths of chemical potential in the influent fluids, and at both isothermal and incremented temperatures. These contrasting experimental and material characteristics are itemized in Table 1.

	Berea	Novaculite	Limestone
Matrix porosity	~20%	<0.01%	<0.01%
Temperature, *°C*	20-120	20-150	20
Effective Stress, *MPa*	3.5	3.5	3.5
Permeants	DI	DI	G/W &DI
Dissolution rate, $Mol.m^{-2}.s^{-1}$	$\sim 10^{-9}$	$\sim 10^{-9}$	$\sim 10^{-6}$
Precipitation rate, $Mol.m^{-2}.s^{-1}$	$\sim 10^{-7}$	$\sim 10^{-7}$	

Table 1. Contrasting experimental and material characteristics of the component rock types.

Porous-Fractured Medium

The surprising changes in permeability that result from the circulation of water through Berea sandstone are illustrated in Figure 3. As temperature of the sample is raised from 20°C to 120°C, the bulk permeability (matrix and fracture combined) of the axially fractured core drops more than two orders-of-magnitude from 1200 mD to 5 mD. This occurs while mineral mass of principally *Si* is net removed from the effluent at ~25 ppm.

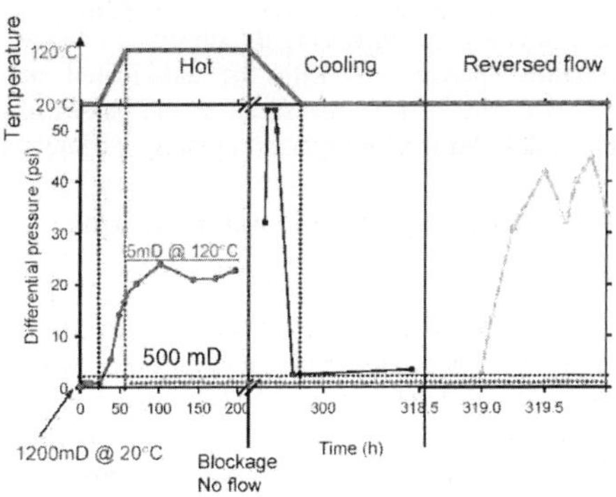

Figure 3. Change in differential pressure versus time for a flow-though test on Berea sandstone. Sample temperature incremented from 20°C to 120°C and then returned to 20°C. Matrix permeability 500 mD. [Courtesy A.B. Polak].

Permeability remains at this reduced magnitude while the core is at 120°C, and reduces further as the core is cooled, until it finally rises suddenly as it approaches 20°C. This behaviour, together with the pre-and post-test scans across the sample by x-ray CT (Figure 4) that show closure of the fracture, could be interpreted as the initial sealing and welding together of the fracture, by either pressure solution or precipitation, followed by abrupt thermal cracking as the sample is cooled. However, this interpretation is refuted by the observation that the bulk permeability of the sample dips below the porous medium permeability of 500 mD, and that the reduction in permeability is restored when flow within the cold sample is reversed (Figure 3). Rather, the interpretation is consistent with the migration of fines within the pore space, moving across the pore diameter to clog the pore throat. As the flow is reversed, the clay minerals move to the opposite side of the pore and re-clog after a small amount of flow, as suggested in SEM images (Figure 5) taken pre- and post-test.

Figure 4. Pre-test (left) and post-test (right) scans of the fracture in Berea sandstone. Dark tones are high density, and light are low density. Fracture runs one o'clock to seven o'clock. Note post-test closure of the fracture in the 5 cm diameter core. [Courtesy AB.. Polak].

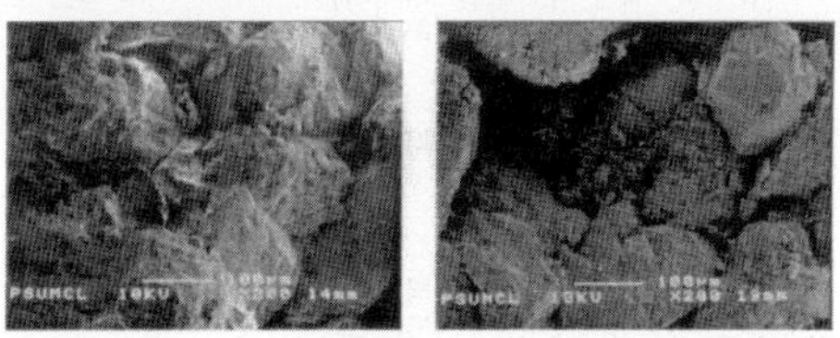

Figure 5. Pre-and post-test SEM of the Berea interior. Quartz present as light mineral and clays plugging pore throats as dark mineral, especially in central figure, post test. [Courtesy A.B. Polak].

Fractured Novaculite

Interpretations of the observed changes in pressure drop across the sample of fractured Berea sandstone, are ambiguous, largely because the permeability of the matrix is high, relative to the fracture. Hence the source of any permeability change cannot be isolated. This difficulty may be removed if the processes are observed where the ambiguities in interpretation that result from both the presence of clays, and high matrix permeability are both absent. Constant-stress flow-through experiments on a natural fracture in Arkansas Novaculite are conducted to observe the bounds of this curious phenomenon – where net dissolution reduces fracture permeability.

Pre-test scans of the sample, by x-ray CT with a resolution of 37 µm, recover an image of the void space present within the fracture, as illustrated in Figure 6. The "islands" of imaged fracture void are presumed connected via open conduits that are less than 37 µm in aperture, and are eliminated in the thresholding of the image. This is confirmed by the flow tests, initiated under a confining stress of 3.5 MPa at 20°C, where the initial mean hydraulic aperture is of the order of 12 µm, as illustrated in Figure 7. Fluid flow through the sample is at constant rate, Q, of 2 cc/min as the temperature is incremented and held, first to 80°, then to 120°, then to 150°C. Apparent during heating is the rapid reduction in aperture, and amplified reduction in fracture permeability that accompanies the incremental stages of heating [Polak et al., 2003]. Measured aperture initially drops precipitously, and then slows to an equilibrium magnitude, and occurs with net dissolution of *Si* of the order of 0.77 ppm throughout. These changes in fracture aperture are large, reducing aperture five-fold, and permeability by two orders-of-magnitude, and are surprisingly rapid – each thermally driven cycle completes in a few days, and is complete for the full range of temperatures in only 40 days.

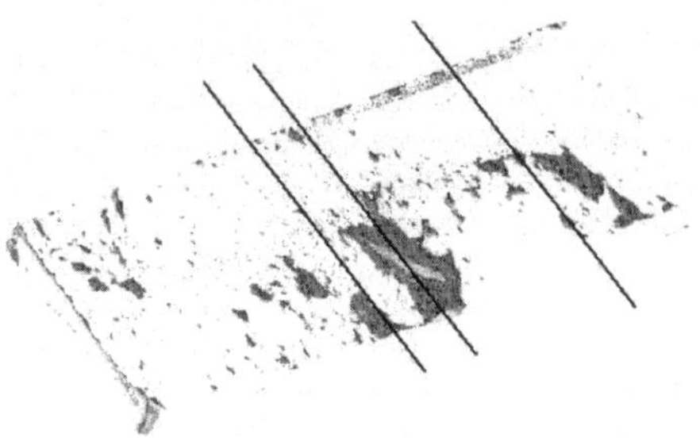

Figure 6. Reconstruction of the fracture within the sample of novaculite. Pixel resolution is 37 µm. [Courtesy A.S. Grader].

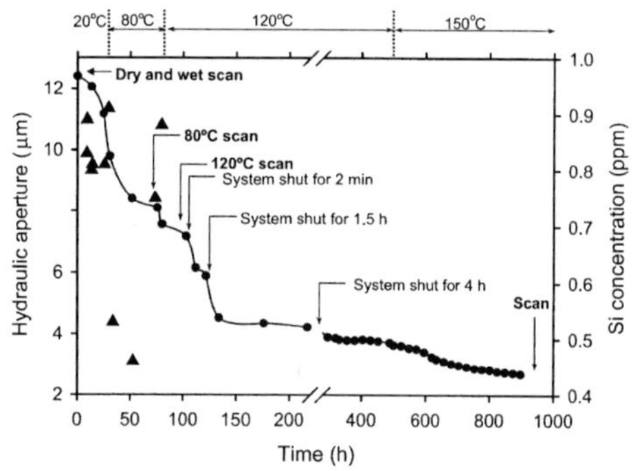

Figure 7. Change in aperture with time for a circulation test on novaculite. Temperatures incremented to 20°C, 80°C, 120°C, and 150°C. Triangles show Si concentrations recovered from the effluent fluid. [Polak et al., 2003].

Perhaps even more interesting than the large and rapid change in permeability, is the sense of the permeability change – permeability reduces despite net dissolution from the fracture. The presumed mechanism for this is through pressure solution at the contacting asperities. Locally amplified stresses at asperity contacts elevate the chemical potential of the silica that, in turn, is mobilized into the saturated fracture void, and removed from the system. The reasonableness of this mechanism may be evaluated by comparing the observed change in fracture aperture, b, with the anticipated mass removal rate, dM/dt, measured in the experiment. For a defined fracture contact area, A_c, and density of quartz, ρ_g, these are linked through the effluent concentration, C_p, as $db/dt = dM/dt \cdot 1/A_c\rho_g$ or $db/dt = QC_p/A_c\rho_g$ [Yasuhara et al., 2004a]. For the average effluent concentration of *Si*, of 0.77 ppm and fractional contact areas in the range 5-30% [Pyrak-Nolte et al., 1987], this results in calculated aperture closure rates of the order of 10^{-10} to 10^{-12} m/s, with measured closure rates, in the range 10^{-11} to 10^{-13}m/s (Figure 8). Correspondingly, removal of mineral mass from the contacting asperities is a plausible explanation for the observed behaviour, and is consistent with the activation of pressure solution. This behaviour is also consistent with observations of the sectioned fracture, post-test, where SEM micrographs show dissolution sites, and the blunting and welding of asperity contacts in a natural fracture that had been separated prior to the test (Figure 9)

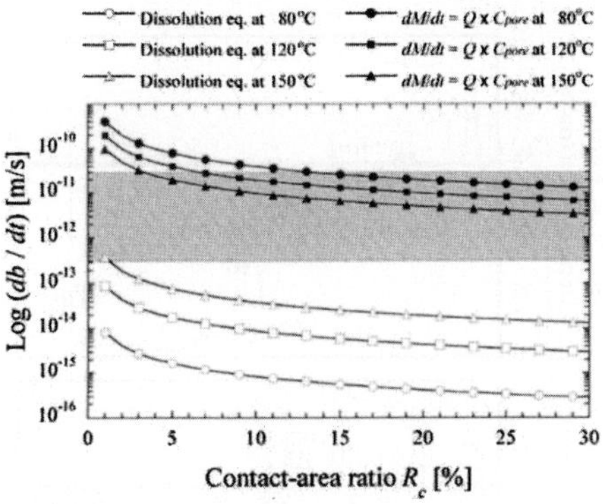

Figure 8. Correlation between aperture closure rate, and nominal contact area. Traces are calculated from measured efflux concentration of Si *(0.77 ppm) and the shaded area shows the range of rates evaluated from the hydraulic data. Open circles show rates predicted from unmodified rate constant data. [Yasuhara et al., 2004a].*

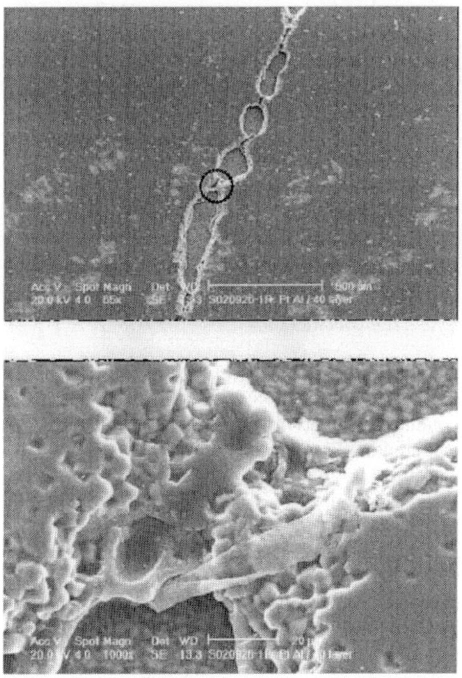

Figure 9. SEM of fracture in novaculite post-test, showing welded portions of the fracture that had originally been separated. Rounded grain surfaces indicative of dissolution. [Courtesy A.B. Polak].

The stress-mediated redistribution of mineral mass by pressure solution may be represented by the three serial processes of: mineral dissolution beneath the contacting asperity, transport along an interfacial water-film to the fracture void, and then redistribution by precipitation within the fracture void, or efflux from the sample. The asperities interpenetrate as mass is removed from contacting asperities, in a similar manner to the mechanisms that drive grain interpenetration and cementation in sandstones [Yasuhara et al., 2003], as illustrated in Figure 10.

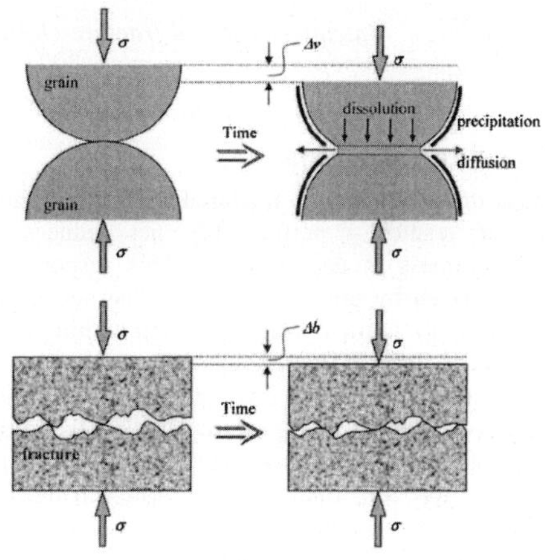

Figure 10. Model for asperity interpenetration follows that for grain interpenetration in granular materials. [Yasuahra et al., 2003].

For the open system considered here, reprecipitation on the fracture wall is ignored. Considering an assumed roughness for the fracture surface profile, enables the progress of fracture closure, driven by pressure solution to be followed, as illustrated for this particular experiment, in Figure 11. The model is able to follow the progress of aperture reduction in a systematic manner, but only where the surface area in contact is increased from the anticipated macroscopic contact area to represent the true reactive surface area at the microscale [Yasuhara, 2004a]. Regardless of this modification, the observed response is accurately tracked at multiple temperatures for a consistent suite of material parameters, adequately explaining this enigmatic response of net permeability reduction as a result of net mass dissolution.

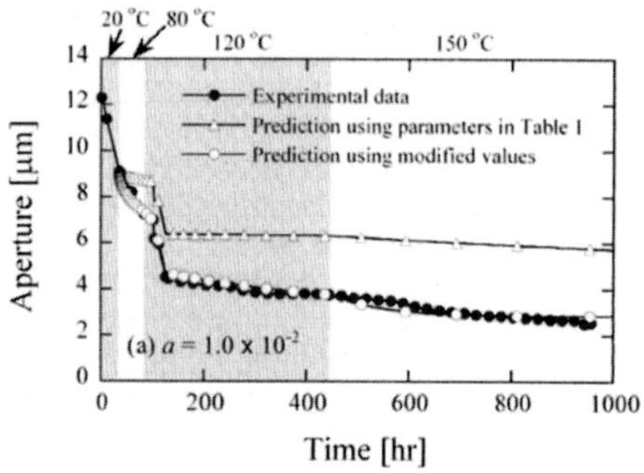

Figure 11. *Matching observed fracture closure with modelled response. [Yasuhara et al., 2004a].*

Fractured Carbonate

Flow through tests on fractured carbonates show similar results – permeability net reduces as mineral mass is net removed. This response is again driven by pressure solution. The net (outlet minus inlet) mineral concentrations within fluids circulated during a flow-though test on a fracture in limestone [Polak et al., 2004] are shown in Figure 12. Groundwater (pH~8) is circulated for the initial 935 h, followed by the circulation of distilled water (pH~6) for the final 515 h. The change in permeability recorded through this period is shown in Figure 13. This notes the steady monotonic decline in permeability that occurs with the circulation of groundwater – the fracture net closes despite a small net dissolution. Following the switching of the permeant to distilled water (pH~6), the reduction in aperture initially accelerates, driven by the amplification of pressure solution as the dissolution rate constant is elevated in the more acidic permeant. This change is manifest as a coherent signal in the elevated output of *Ca*, signalling the more active effect of pressure solution. However, following this increase in fracture closure rate, and with no change in the experimental conditions, permeability change spontaneously switches from net reduction to a rapid increase.

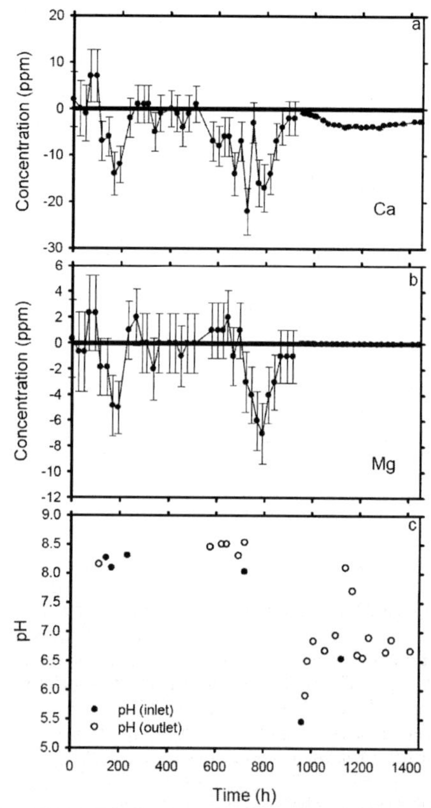

Figure 12. *Net efflux of Ca and Mg, and change in pH, from the flow-though test in limestone, with time. [Polak et al., 2004].*

The switch in response results from the observed localization of mass dissolution along a narrow conduit, evident in the fracture morphology, captured in CT images taken both pre- (Figure 14a) and post-test (Figure 14b). These images show the localization of mass removal along a "wormhole" that grows and ultimately dominates flow within the system. Since these experiments provide independent constraint on hydraulic behaviour and the redistribution of mass within the sample (through measurements of effluent concentrations and through imaging by x-ray CT), these measurements may be compared. If consistency is sought between observed the hydraulic and mineral mass balance data, then added constraint is applied to the problem, as evident in Figure 13.

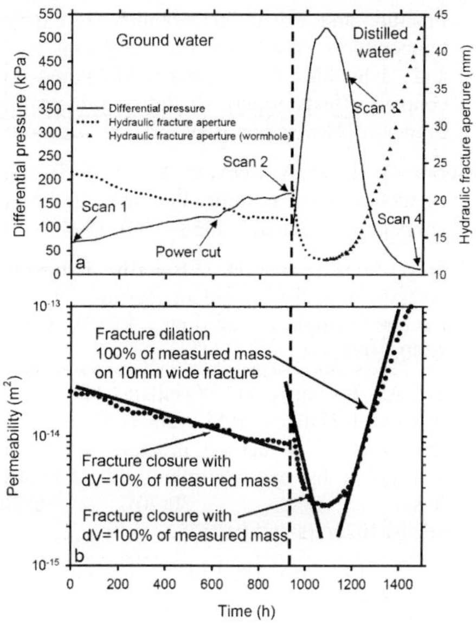

Figure 13. Change in (a) driving differential pressure (end-to-end) with time, and evaluation of fracture aperture, as a proxy for (b) permeability change. Overlain on (b) are evaluations of permeability change constrained only on initial permeability magnitude and mass removal rate dM/dt. [Modified from Polak et al., 2004].

Anticipated changes in permeability may be predicted from the mass removal rates provided an appropriate mechanistic model is prescribed. Assuming that mass is removed only from the contacting asperities, and that the contact area is typically in the range 5-30%, then permeability may be evaluated directly. Applying this procedure to the data in Figure 13, shows the difficulty in balancing measured mass removed with that observed in early time – the measured mass removed is a factor of 10 larger than that predicted from the change in permeability, suggesting that large measurement error in the mass concentration data is a major limitation [Polak et al., 2004]. Alternatively, additional dissolved mass from the sample interior, not contributing to the change in aperture could participate in the process, and result in the observed mismatch. Upon switching the influent to distilled water (pH~6), the observed change in permeability closely matches the change predicted from the mass efflux. This matches for an unmeasured but reasonable asperity contact area of 25%. Finally, as the aperture closes, and flow is progressively restricted to the continuous path of the largest initial apertures, one single conduit develops as the principal flow path – the observed mass removal rate is consistent with the evolution of a flow path 10 mm in width. This observation is consistent with the development of the wormhole, apparent in the scanning data of Figure 14. Although only diagnostic, these models are consistent with the findings from prognostic lumped-parameter models that enable this response to be followed *a posteriori* [Yasuhara et al., 2004b] and to match proposed mechanisms with observations.

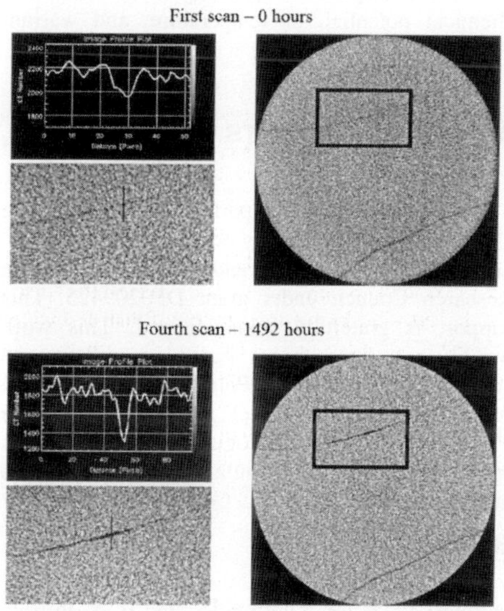

Figure 14. Scan of the fractured limestone sample pre- (above) and post-test (below). Growth in the fracture aperture occurs during the test. CT number measured across the sample confirms growth of the aperture as a "wormhole." [Polak et al., 2004].

4. CONCLUSIONS

The previous has documented the surprising influence that complex paths of temperature and chemical potential may exert on the transport properties of fractured rocks. Pressure solution is shown to be an important factor, which, under certain circumstances may strongly influence response. Pressure solution, representing a mechanism of stress-mediated (enhanced) dissolution, can be mobilized at relatively modest temperatures (~20°-80°C) and stresses (~3.5 MPa), and can act rapidly – on the order of a few to tens of days. These effects are shown to be surprisingly virulent– capable of changing permeabilities by many orders of magnitude within these short timeframes. Surprisingly, these effects may result in the net reduction of fracture aperture, and consequent reduction in permeability, even when mineral mass is net dissolved and removed from the sample, throughout. Correspondingly, these processes may exert a significant influence on the circulation of fluids in natural and engineered environments, especially those involving significant temporal or spatial gradients in stress, chemical potential, or temperature, and warrant close scrutiny.

ACKNOWLEGEMENTS

This work is a result of partial support from the U.S. Department of Energy under grant DOE-BES-DE-FG02-00ER15111, and the Australian Research Council under grant DP0209425. This support is gratefully acknowledged. This work would not have been possible without the contributions of Amir Polak, Hide Yasuhara, Avrami Grader, Phil Halleck, and Jishan Liu. Mike Howard (Arkansas Geological Commission) is thanked for providing a bulk sample of Arkansas novaculite.

REFERENCES

Durham, W. B., W. L. Bourcier, and E. A. Burton, Direct observation of reactive flow in a single fracture, Water Resour. Res., 37(1), 1 – 12, 2001.

Elias, B. P., and A. Hajash, Changes in quartz solubility and porosity due to effective stress: An experimental investigation of pressure solution, Geology, 20, 451–454, 1992.

Gratier, J. P., Experimental pressure solution of halite by an indenter technique, Geophys. Res. Lett., 20(15), 1647– 1650, 1993.

Hajash, A., T. D. Carpenter, and T. A. Dewars, Dissolution and timedependent compaction of albite sand: Experiments at 1000C and 1600C in pH-buffered organic acids and distilled water, Tectonophysics, 295(1– 2), 92– 115, 1998.

Lin, W., J. Roberts, W. Glassley, and D. Ruddle, Fracture and matrix permeability at elevated temperatures. Workshop on significant issues and available data. Near-field/Altered-zone coupled effects expert elicitation project, San Francisco, November, 1997.

Moore, D. E., D. A. Lockner, and J. A. Byerlee, Reduction of permeability in granite at elevated temperatures, Science, 1558–1560, 1994.

Piggott, A. R., and D. Elsworth, Laboratory assessment of the equivalent apertures of a rock fracture, Geophys. Res. Lett., 30(13), 1387 – 1390, 1993.

Polak, A., Elsworth, D., Yasuhara, H., Grader, A.S., and Halleck, P.M. (2003) Permeability reduction of a natural fracture under net dissolution by hydrothermal fluids, Geophys. Res. Lett., 30(20), 2020, doi:10.1029/2003GL017575.

Polak, A., Elsworth, D., Liu, J., and Grader, A. (2004) Spontaneous switching of permeability changes in a limestone fracture under net dissolution. Water. Resour. Res., Vo. 40, W03502, doi:10.1029/2003WR002717.

Pyrak-Nolte, L. J., L. Myer, N. G. W. Cook, and P. A. Witherspoon, Hydraulic and mechanical properties of natural fractures in low permeability rock, Proc. Int. Cong. Int. Soc. R. Mechs., Montreal, 225– 231, 1987.

Siddiqi, G., B. Evans, G. Dresen, and D. Freund, Effect of semibrittle deformation on transport properties of calcite rocks, J. Geophys. Res., 102(B7), 14,765– 14,778, 1997.

Yasuhara, H., Elsworth, D., Polak, A., Liu, J., Grader, A., and Halleck, P. (2004b) Spontaneous Permeability Switching in Fractures in Carbonate: Lumped Parameter Representation of Mechanically- and Chemically-Mediated Dissolution. Submitted for Publication. Transport in Porous Media. 30 pp.

Yasuhara, H., Elsworth, D., and Polak, A. (2004a) The evolution of permeability in a natural fracture: significant role of pressure solution. J. Geophys. Res., Vol. 109, B03204, doi:10.1029/2003JB002663.

Yasuhara, H., Elsworth, D., and Polak, A. (2003) A mechanistic model for compaction of granular aggregates moderated by pressure solution. J.

Geophys. Res. *108*(B11), 2530, doi:10.1029/2003JB002536.

Zhang, S., S. Cox, and M. Paterson, The influence of room temperature deformation on porosity and permeability in calcite aggregates, J. Geophys. Res., 99, 761–775, 1994.

Zoback, M., and J. Byerlee, The effect of microcrack dilatancy on the permeability of Westerly Granite, J. Geophys. Res., 80, 752–755, 1975.

DETECTION OF HYDRAULICALLY CREATED PERMEABLE STRUCTURES IN HDR/HWR RESERVOIR BY HIGH RESOLUTION SEISMIC MAPPING TECHNIQUES

Hiroaki Niitsuma
Graduate School of Environmental Studies, Tohoku University

Abstract: There have been considerable improvements in microseismic mapping technology in these ten years. Various high resolution mapping techniques to reduce location error have been developed which include collapsing, doublet/multiplet analysis, clustering analysis, and multiplet-clustering analysis. These high-resolution seismic mapping techniques have been applied to the microseismic data set obtained during the hydraulic stimulation at Soultz-sous-Forêts HDR field, France, in 1993. A seismic structure, which was growing linearly at the early stage of the stimulation, has been detected and analysed. It is estimated that this structure is enhanced permeable paths and consist of sub parallel microstructures, which are oriented to the directions most easy to slip by the stimulation.

1. INTRODUCTION

A Comprehensive understanding of the fracture distribution and hydro-geomechanical processes occurring during hydraulic stimulation is essential in both conventional and Hot Dry Rock (HDR)/ Hot Wet Rock (HWR) geothermal development. Some of this information can be obtained from well logs such as flow, temperature, pressure, BHTV, FMI, etc. However, they only provide restricted information near the well.

Microseismic monitoring is currently the best available method for obtaining three-dimensional information about reservoirs and fracture systems at locations remote from boreholes. The primary problem of the method is its location accuracy. When located with conventional location algorithms, the events are frequently seen to be distributed as a cloud with little macro- or microstructure. It is difficult to meaningfully correlate such images with logging data of existing wells. Therefore, information on detailed reservoir structure, fracture orientation, and hydraulic behavior cannot be obtained from the application of conventional microseismic location techniques.

There have been considerable improvements in microseismic mapping technology in these ten years (Niitsuma et al., 1999; Fehler et al., 2001). Various techniques to reduce location error have been developed which include collapsing, doublet/multiplet analysis, double-difference method, clustering analysis, and multiplet-clustering analysis. These high resolution mapping techniques demonstrate that the diffuse clouds obtained using conventional techniques are largely artefacts of location error, and reveal macro- and microstructures within the cloud that can be correlated with geological and hydraulic structures identified from well logs.

This paper overviews the high resolution microseismic mapping techniques and discusses a microseismic structure, as an example, which was observed during the Soultz '93 stimulation and revealed by the techniques (Niitsuma et al., 2002).

2. HIGH RESOLUTION SEISMIC MAPPING TECHNIQUES

Jones and Stewart (1997) proposed a method called "collapsing" to focus a seismic cloud. This method is based on the concept that there is probability that the location of an event can occur anywhere within some error ellipsoid surrounding the event. The error ellipsoid is a function of the geometry of the seismic network and the misfit between the measured arrival times and those predicted using the estimated event location and origin time. The collapsing method is applied by shifting locations within their error ellipsoids towards the centre of mass of the events that fall within the ellipsoid, assuming that actual physical seismic locations have a simpler structure than that of the calculated locations because of the random noise. Iterative shifting of event locations proceeds until the distribution of movement vectors for all events in the cloud approximates that for normally distributed location uncertainties. This method does not make the uncertainties in the data any smaller nor does it provide more accurate event locations; it simply highlights structures already inherent within the unfocused data set. An example of application of the collapsing method to the earthquakes in California, USA by Jones (1998) is shown in Figure 1. Structures in the seismic cloud have been revealed by this method.

Asanuma et al. (2001) proposed a modification to the collapsing method. In this modified method the movements of the locations are dependent on the shape of the distribution of the seismic events within the confidence ellipsoid, and not just the

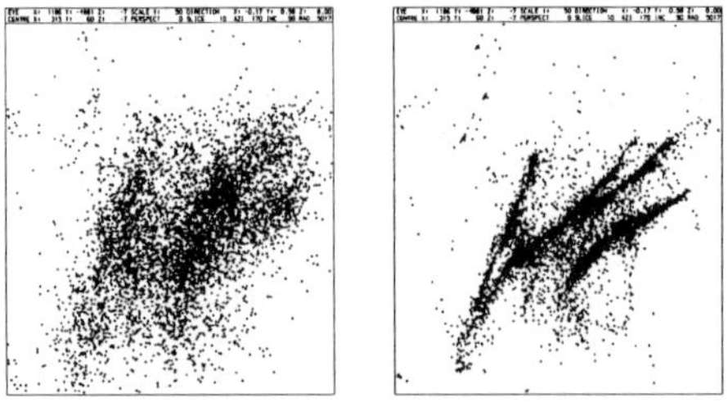

Figure 1: Results of collapsing (right) and conventional mapping (left) for microseismicity in Mammoth Lake, USA, by Jones (1998).

position of the centre of mass, as is the case in the original method. Additionally, whereas the original collapsing method implicitly assumes that all locations belong to point structures, in this modified version three types of structure are considered: point, line and plane. Principal component analysis of the locations is used to evaluate to which type of structure each location most probably belongs. The modified technique has been applied to microseismic events associated with hydraulic stimulation at the Fenton Hill HDR Field; NM, USA, and known small-scale structures were imaged (Asanuma et al., 2001).

The clustering analysis is another mean to find and evaluate structures in seismic clouds. Phillips *et al.* (1997) and Phillips (2000) proposed a repicking method, which produces the relative time difference of phases within a group of similar microseismic events, followed by master event relative source locations. For this analysis, waveforms are lined up after low pass filtering, with P- and S-wave arrival times are manually repicked from a similar portion of the waveform. The technique can be used to identify wave arrival times and improve the accuracy of relative source location determinations. Phillips (2000) applied this analysis to microseismic clouds created by hydraulic stimulation in the Soultz HDR reservoir and delineated two distinct intersecting planar structures.

A pair or group of events that have similar or almost identical waveforms is called a "doublet" or "multiplet". The similarity of waveforms makes it possible to precisely detect relative delays by means of the modern time-delay estimation methods. Poupinet *et al.* (1984) suggested that location accuracy is improved 10 times by using a cross-spectrum analysis, compared with conventional methods. Fréchet *et al.* (1989) applied a cross-spectrum analysis to similar induced microseismic events at *La Mayet de Montague* HDR geothermal site (France), to identify fluid-filled fracture pathways. Moriya *et al.* (2001, 2002) and Gaucher *et al.* (1998) also applied this method to the seismicity in Soultz HDR field and found micro structures in seismic cloud. Figure 2 shows a result of the doublet analysis applied to the AE/MS observed in Hijiori HDR field, Japan (Tezuka and Niitsuma, 2000). The left hand plot shows locations determined using conventional location technique and the right hand shows locations determined using the doublet analysis. Clear planar structures are found by this analysis.

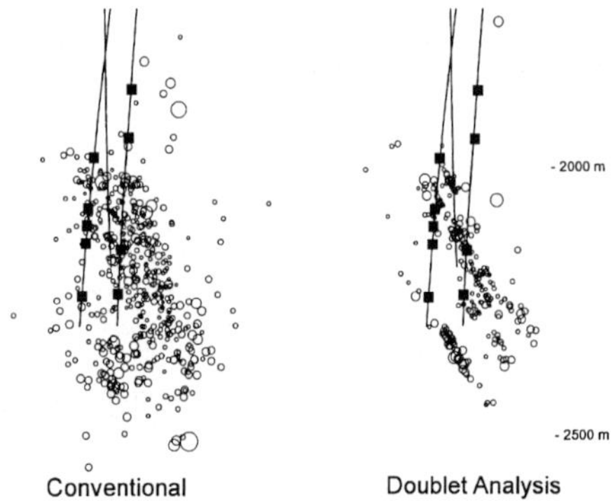

Figure 2: Result of doublet analysis (right) and conventional mapping for micorseismicity in the Hijiori HDR field, Japan, by Tezuka and Niitsuma (2000).

The multiplet analysis provides information on the micro structures in the seismic cloud. However, it does not give knowledge of absolute locations of individual multiplet clusters with the same order of precision to the relative location of events in a multiplet cluster. Moriya et al. (2002) proposed the "multiplet-clustering method" to reduce the location error of multiplet clusters using the technique of clustering analysis. In this method, the waveforms of events in a multiplet group are stacked, and similar phases of the stacked waveforms are compared in time domain. Then, relative arrival time differences (i.e. relative locations between the stacked events) are more precisely determined.

3. APPLICATION TO THE SOULTZ '93 SEISMICITY

3.1 Hydraulic fracturing in 1993 at the Soultz field

A major hydraulic fracturing experiment was performed in September and October 1993 at Soultz-sous-Forêts. In the September test, 25,000 m^3 of fresh water were injected between 2850-3350 m at progressively higher rates to 40 l/s and pressures of up to 10 MPa over a 17 day period. In October, a further 20,000 m^3 of water were injected at up to 50 l/s into the entire open hole section. Through the test, it was demonstrated that the fracture network in the basement rock was well developed, with enhanced permeability and a substantial increase in transmissivity (Baria et al., 1999; Evans, 2000).

Induced microseismic activity was detected using three downhole 4-component detectors and one hydrophone, installed in wells #4550, #4616, #4601 and EPS1 in the 1993 experiment. The 4-componet detector consists of accelerometers mounted in a housing and set in sand at the bottom of the borehole. The three 4-component seismic detectors and hydrophone were set at depths of 1,500 m, 1,420 m, 1,600 m and 2,850 m, respectively.

Using these detectors, source locations for more than 10,000 events have been determined (Baria et al., 1999). Figures 3 shows source locations for all of the analysed events. As shown in this figure, the seismic cloud has approximate dimensions of 0.5 by 1.2 by 1.5 km, striking N30° W and dipping nearly vertically (Phillips, 2000).

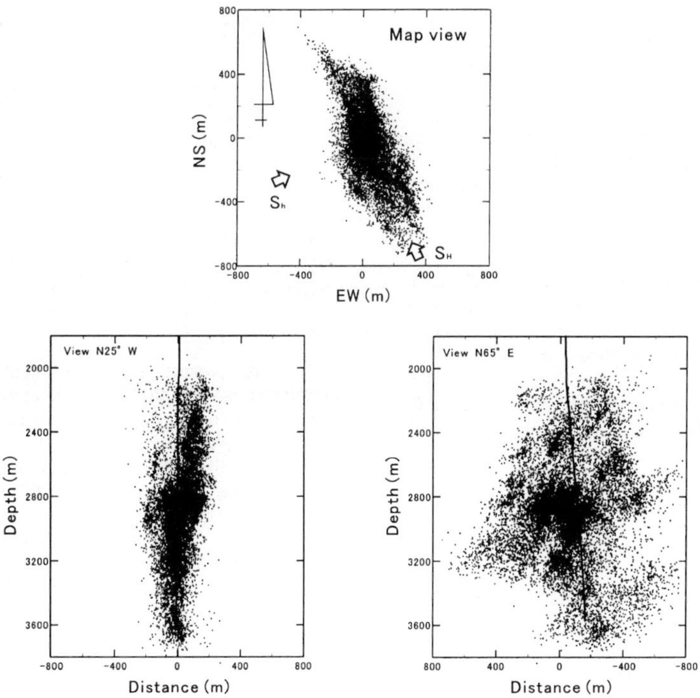

Figure 3: Seismic cloud associated with the '93 stimulation at Soultz HDR field.

3.2 Seismic line revealed by the collapsing method

By applying the collapsing method for the microseismic cloud developed during the September 1993 stimulation (Jones et al., 1995),

Evans et al. (1999) found a major linear structure, which developed early in the injection. This structure, shown in Figure 4, initiated at 2950 m near the well and extended to 3300 m prior to the evolution of the other principal structures.

Figure 5 shows the growth of this structure in time. It initiated at a depth where significant flow left the well, and it then grew downwards in a stepwise manner, the steps corresponding to step-increases in injection pressure and flow with a delay of about two hours. Analysis of flow and temperature logs during and following the stimulations indicated that the seismic structure is also a high-permeability, hydraulically-significant structure created/enhanced by the hydraulic stimulation (Evans, 2000, 2001).

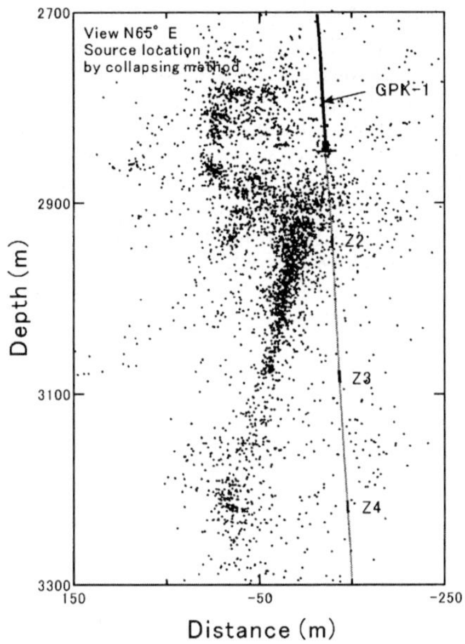

Figure 4: Seismic line revealed by the collapsing method.

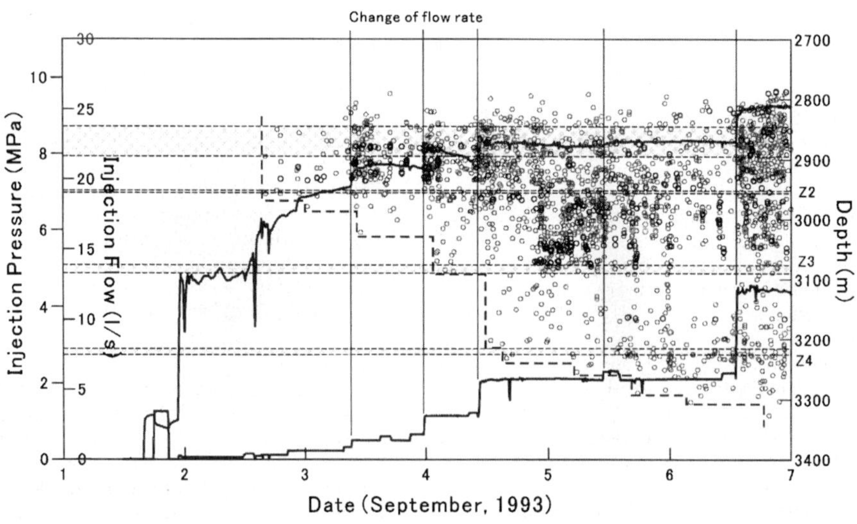

Figure 5: Growth of the seismic line during the stimulation.

3.3 Estimation of microstructure by the multiplet analysis

In the '93 experiment, about 60 % of located events belong to some of the multiplet groups. Figure 6 shows the result of precise relative mapping of a multiplet group (Moriya et al., 2002). The relative location error is estimated to be less than 1 m. The multiplet groups usually define planer structures with diameters of 2 - 12 m. Thus, the multiplet analysis sheds light on the micro structure of the total seismic cloud.

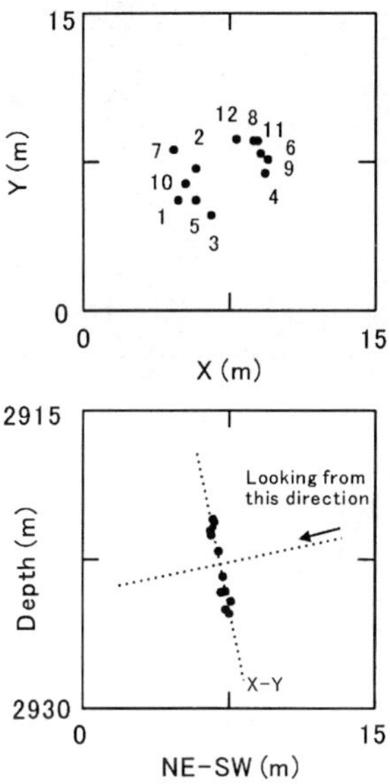

Figure 6: Planer structure of a seismic multiplet group.

97 multiplet groups, which consist of 713 events in the seismic line, have been analysed (Moriya et al., 2001). Figure 7 (a) shows the distribution of orientations of individual multiplet planes plotted in lower hemisphere of Schmidt net. Critical pore pressure for frictional slip based on Coulomb's law, using a representative stress field at the depth of 2900 m, is also shown in this figure, where friction coefficient was assumed to be 0.8. Figure 7 (b) shows a fracture distribution detected by FMI log at 2000 – 3800 m in GPK-1 (Genter and Dezayes, 1993).

As shown in Figure 7 (a), multiplet planes are almost vertical and their strikes are approximately NNW-SSE. Similar distribution is observed for the pre-existing fractures (Figure 7 (b)). The direction

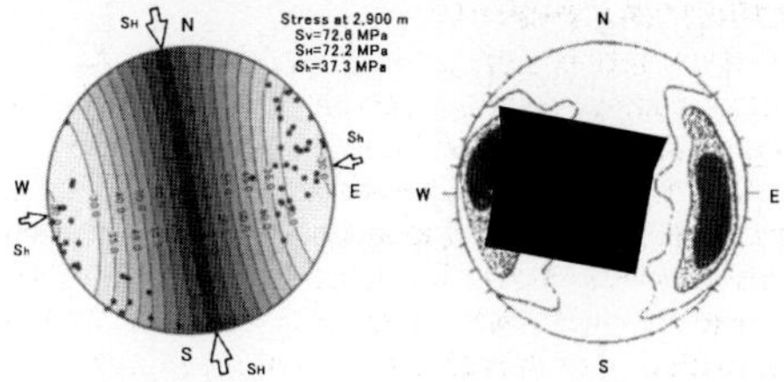

Figure 7: Orientation of multiplet structure and pre-existing joints. Orientation of multiplet structure and critical pore pressure calculated from stress data (left), and orientation of natural joints before the stimulation detected by FMI in GPK-1 (right).

of the multiplet planes mostly distributes in the area where the critical pore pressure is less than 30 MPa. Because the seismicity started when the wellhead pressure exceeded 5 MPa, the pore pressure would exceed the critical pressure during the growth of the seismic line for most of the plane (Moriya et al., 2001).

3.4 Structure of the seismic line estimated by the multiplet-clustering analysis

Figure 8 shows a result of the multiplet-clustering analysis applied to the multiplet groups in the microseismic structure shown in Figure 3. The 97 multiplet clusters are plotted in this figure. Relative locations between clusters are determined with error less than 6 m, while the location error for the JHD (Joint Hypocenter Determination) mapping shown in Figure 3 is about 15 m.

As shown in this figure, the structure of the seismic line has a thin volume by 20 x 80 x 400 m, which consists of the sub parallel multiplet clusters shown in Figure 7 (a). The cluster found by Phillips (2000) is included in this volume at the depth around 3200 m.

3.5 Discussion

The multiplet and multiplet-clustering analysis methods have been used to resolve the detailed microstructure of the seismic line which appeared at the early stage of the stimulation. The microstructure consists of the sub parallel multiplet planes. This result suggests that natural joints which are oriented to the directions most easy to slip, were stimulated by the injection. Because the structure developed stepwise corresponding to the

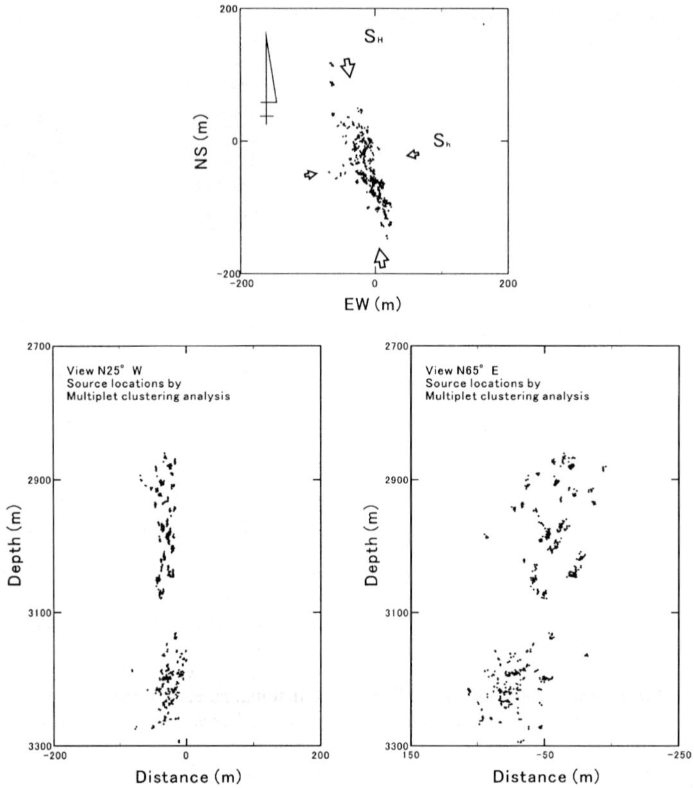

Figure 8: Detailed structure of the seismic line estimated by the multiplet-clustering analysis.

step of injection pressure and flow with delay, some permeable paths were created aseismically by the seismic events, which were observed as the multiplet clusters, and the structure developed iteratively. The fact that the structure is restricted within the thin layer suggests existence of a fractured layer in the region.

4. CONCLUSION

The recent high-resolution seismic mapping techniques which include collapsing, doublet/multiplet analysis, clustering analysis, and multiplet-clustering analysis are effective for a comprehensive understanding of the fracture distribution and hydro-geomechanical processes occurring during hydraulic stimulation in HDR/HWR geothermal development. These techniques have been applied to the microseismic data set obtained during the hydraulic stimulation at Soultz HDR field, France, in 1993. Detailed seismic structure of a seismic line, which was growing linearly at the early stage of the stimulation, has been analysed. It is estimated that this structure shows enhanced permeable paths and consist of sub parallel microstructures, which are oriented to the directions most easy to slip by the stimulation.

ACKNOWLEDGEMENT

This work was carried out as a part of the MTC/MURPHY International Collaborative Project (International Joint Research Grant) supported by NEDO and MESSC, and by a grant from the Industrial Technology Research Grant Program in 2000, supported by NEDO. We would also like to thank SOCOMINE for providing the data from the European HDR site at Soultz-sous-Forêts, which is supported mainly by the European Commission, BMBF (Germany), and ADEME (France).

REFERENCES

Asanuma, H., Ishimoto, M., Jones, R. H., Phillips, W. S. and Niitsuma, H. (2001), *A variation of the collapsing method to delineate structures inside a cloud,* Bull. Seism. Soc. Am., Vol.91, 154-160.

Baria, R., Baumgärdner, J., Gérard, A., Jung, R. and Garnish, J. (1999). *European HDR research programme at Soultz-sous- Forêts (France) 1987-1996.* Geothermics, Vol. 28, 655-669.

Evans, K. F., Jones, R., Soma, N. and Baria, R. (1999). *Correlation of near-borehole microseismic structures defined by the 'collapsed cloud' of events induced by the 1993 GPK1 stimulation with flowing fractures.*, Internal Report of MURPHY Project 1999, NEDO.

Evans, K. F. (2000). *The effect of the 1993 stimulations of well GPK1 at Soultz on the surrounding rock mass: Evidence for the existence of a connected network of permeable fractures.*, Proc. World Geothermal Congress 2000, Vol. 6, 3695-3670.

Evans, K. F. (2001). *Determining the effect of stimulation injections on the rock mass around the well GPK-1 from borehole data: Analysis of the 1993 injections.*, Report of BBW Project no.98.0008-1, 6-74.

Fehler, M., Jupe, A. and Asanuma, H. (2001). *More Than Cloud: New techniques for characterizing reservoir structure using induced seismicity*, The Leading Edge, 324-328.

Genter, A. and Dezayes, C. (1993). *Fracture evaluation in GPK1 borehole by using FMI data.* Field Report, BRGM Orléans.

Fréchet, J., Martel, L., Nikolla, L. and Poupinet, G. (1989). *Application of the cross-spectral moving-window technique (CSMWT) to the seismic monitoring of forced fluid migration in a rock mass*, Int. J. Rock Mech. Min. Sci. Geomech. Abstr., **26**, 221-233.

Gérard, A., Baumgärdner, J. and Baria, R. (1997). *An attempt towards a conceptual model derived from 1993-1996 hydraulic operations at Soultz.* Proc. NEDO International Symposium, Vol. 2, 329-341.

Gaucher, E., Cornet, F. H. and Bernard, P. (1998), *Induced seismicity analysis for structure identification and stress field determination*, Paper SPE 47324, Proc. SPE/ISRM, Trondheim, Norway.

Jones, R. H., Beauce, A., Jupe, A., Fabriol, H. and Dyer, B. C. (1995). *Imaging induced microseismicity during the 1993 injection tests at Soultz-sous-Forêts, France.* Proc. World Geothermal Congress 1995, Vol. 4, 2665-2669.

Jones, R. H. and Stewart, R. (1997). *A method for determining significant structures in a cloud of earthquakes.* J. Geophys. Res., Vol. 102, 8245-8254.

Jones, R.H. (1998). *Collapsing: A new method for improved earthquake location*, Proc. Geoscience 98.

Moriya, H., Nakazato, K., Evans, K. F., Niitsuma, H., Jones, R.H. and Baria, R. (2001). *Evaluation of the Soultz HDR system by precise mapping techniques of induced microseismic event.* Trans. Geothermal Resources Council 2000, Vol. 25, 187-190.

Moriya, H., Nakazato, K., Niitsuma, H. and Baria, R. (2002). *Detailed fracture system of the Soultz-sous-Forêts HDR field evaluated using microseismic multiplet analysis.*, Pure appl. Geophys., Vol. 159, 517-541.

Niitsuma, H. et al. (1999). *Current status of seismic and borehole measurements for HDR/HWR development*, Geothermics, Vol. 28, 475-490.

Niitsuma, H., Moriya, H., Asanuma, H., Evans, K., Jones, R., Jung, R. and Baria, R. (2002). *Detecting hydraulically created permeable structures in the Soultz HDR site by high resolution seismic mapping techniques*, Proc. of New Zealand Geothermal Workshop.

Phillips, W. S., House, L. and Fehler, M. (1997). *Detailed joint structure in a geothermal reservoir from studies of induced microearthquake clusters*, J. Geophys. Res., 102, 11745-11763.

Phillips, W. S. (2000). *Precise microearthquake locations and fluid flow in the geothermal reservoir at Soultz-sous-Forêts, France*, Bull. Seism. Soc. Am., Vol. 90(1), 212-228.

Poupinet, G., Ellsworth, W. L. and Fréchet, J. (1984). *Monitoring velocity variations in crust using earthquake doublets: An application to the Calaveras fault, California,* J. Geophys, Res., Vol.89, 5719-5731.

Tezuka, K. and Niitsuma, H. (1997). *Integrated interpretation of microseismic clusters and fracture system in a hot dry rock artificial reservoir*, Expanded Abstr., SEG 67th Ann. Intern. Meeting.

RECENT STUDY OF COUPLED PROCESSES IN GEOTECHNICAL AND GEOENVIRONMENTAL FIELDS IN CHINA

Wang Sijing[1] and Wang Enzhi[2]
[1]Institute of geology and Geophysics, CAS, Beijing, 100029
[2]Tsinghua University, Beijing, 100084

Abstract: This paper introduces the current state of research on the coupled Thermal-Hydraulic-Mechanical-Chemical (THMC) processes in geo-materials and applications for geotechnical and geo-environmental engineering in China. The paper puts equal emphasis on both the achievements of experimental results for fundamental studies, and the applications in rock mechanics and foundation engineering, tunnel engineering, geothermal mining, unsaturated soil and dynamic consolidation. At the end, it presents a model for chemical mining.

1. INTRODUCTION

To describe and predict the behaviors of coupled processes of the geo-materaials and geological systems, the coupling effect draws equal attention from both geologists and engineers. Recently the problems related to coupled processes in rock mechanics and engineering geology has been raised as one of main fields for further development (Wang Sijing, 2002, 2003).

In fact some preliminary results of relevant studies have been achieved in Chinese literature for the past decades. As early as in 1980's the study of coupled processes were started in soil and rock mechanics (Qian Jiahuan, 1986; Kou Shaoquan, 1987; Tao Zhenyu et al.,1988), but more focused attention was paid to the problem only in 1990's. More recent progress has appeared in different areas of civil engineering, water resource management and mining engineering (Sun Jun et al., 2000; Zhang Yuzhuo, 2001).

In viewing of the large number of references existing in the Chinese literature which is not readily accessible to the international communities, this paper focuses on the new progresses gained in the past few years in the research of coupled process and applications in the geotechnical and geo-environmental engineering in China. A new model for the coupled THMC processes for chemical mining is proposed and some suggestions for future developments are also given at the end.

2. COUPLED THMC PROCESSES IN GEO-SYSTEMS AND GEOTECHNICAL - GEOENVIRONMENTAL ENGINEERING

2.1 Interactions Between The Coupled Processes

The main processes involved in the geotechnical and geo-environmental engineering may be classified as thermal (T), hydraulic (H) , mechanical (M), and chemical (C). In a geo-engineering system these processes are related to each other through interaction. The properties and parameters in one process may be affected by the others. Mutual feedbacks of the influences then lead to a combined process of development, the so called coupled processes. However, in a geo-engineering system not all the processes involved may be equally important in controlling the total system behavior. In many cases the particularity of the studied problems has to be taken into account.

Figure 1 shows a sketch of interactions among the coupled THMC processes in a geo-engineering system. The main effects produced from one process to another are indicated in terms of increment of the controlling variables. The symbols are listed in the Appendix. The coupling effects in one direction may be different from that in the opposite direction. For example, when the thermal process plays a role of an agent process, its variation may cause the change of the object processes, such as the hydraulic one, in terms of the

change in the fluid viscosity and

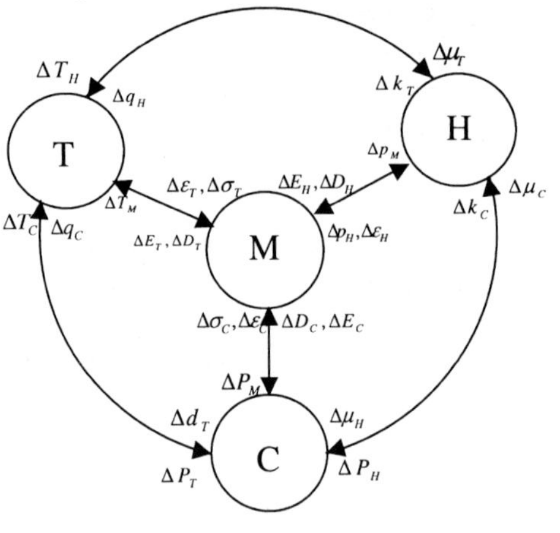

Figure 1 Main directions of Interaction Coupling between processes

Conductivity, while the change in the thermal process is the variations of temperature and heat energy (Table 1).

2.2 Typical Problems Involving Coupled Process

The types and numbers of processes involved in the coupled system are dependent upon the nature of the problem studied for solution.

In principle all the THMC processes may be involved in the geotechnical and geo-environmental problems. However, to simplify the problem for solution, only the major processes are considered for a particular problem. For example, the dam foundation problems are practically dominated by the coupled HM processes. The objectives of the solution are the interactions between the foundation stresses and deformation (the mechanical process), and the seepage pressure and flow rate (hydraulic process). Only in some special cases the thermal and/or chemical processes may also be involved, say, for dams built in cold region or on rock foundation of high solubility.

Some typical coupled processes in the geotechnical and geo-environmental engineering are listed in Table 2.

Table 1 Effect of interaction between coupling process

Object / Agent	T	H	M	C
T	○	Δk_T; $\Delta\mu_T$	$\Delta\sigma_T$; $\Delta\varepsilon_T$ ΔE_T; ΔD_T	ΔP_T; Δd_T; $\Delta\mu_T$
H	ΔT_H; Δq_H	○	Δp_H; $\Delta\varepsilon_H$ ΔE_H; ΔD_H	$\Delta\mu_H$; ΔP_H Δd_H
M	ΔT_M	Δk_M	○	ΔP_M
C	ΔT_C; Δq_C	Δk_C; $\Delta\mu_C$	$\Delta\sigma_C$; $\Delta\varepsilon_C$ ΔE_C; ΔD_C	○

(Note: Δ -increment; T -Temperature; q -Heat; k -Permeability; μ -Viscosity; p -Water pressure; σ -Stress; ε -Strain; E -Elasticity; D -Damage; P -Chemical potential; d -Density)

3. EXPERIMENTAL STUDIES

Laboratory experiment and in-situ monitoring are important measures in studying the coupled phenomenon for finding its mechanism controlling the development and evolution of the process. Due to complexity of the field observations, most experiments were conducted in laboratories. Some of the successful experiments conducted in China in the past are listed in Table 3. Most of studies are focused on coupling of two processes (TH, HM, CM), but some of them involve three coupling processes (THM). Some examples are given below.

Example 1: A study of effect of stress state on seepage conductivity of single natural rock fracture (MH) was conducted by one of authors of this paper. In Figure 2 the effect of normal stress on the permeability of rock joint in granite is shown. The original fracture transmissivity is reduced due to

Table 2 Typical geotechnical and geo-environmental issues involving coupled process

Type of coupling	Geotechnical and geo-environmental issues
H-M	Seepage and deformation of dam foundation; Drainage and stability of slopes; Soil consolidation, surface subsidence and cracking, due to extracting ground water; Hydraulic fracturing of rocks; Groundwater inflow into underground extraction
T-M	Ventilation of underground excavation; Frost or heat development in underground storage systems.
T-H-M	Deep seated tunnel and mining; Permafrost and freezing soil engineering; Injection of gas into ground for energy storage; Geothermal exploitation; Underground disposal of nuclear wastes.
H-M-C	Chemical grouting in foundation treatment; Ground pollution and treatment.
T-H-M-C	Chemical mining; Underground waste disposal.

cyclic normal loading, with the fracture roughness playing a certain role in the changing of transmissivity. For a monotonic loading the hydraulic aperture of fracture versus normal stress can fit with an exponential relationship, as obtained in the previous studies (Su Baoyu et al., 1997). However, in the case of cyclic loadings the final relationship between hydraulic aperture of a rough fracture and normal stress can be better expressed by a hyperbolic function, as following.

$$b = b_0 + \frac{C}{\sigma_0 + \Delta\sigma} \tag{1}$$

where c is an experimental constant.

Example 2: One of the most complicated coupling issues is to quantify the effect of chemical process in a coupled system. Only a few works are devoted to this study in Chinese literature (Feng et

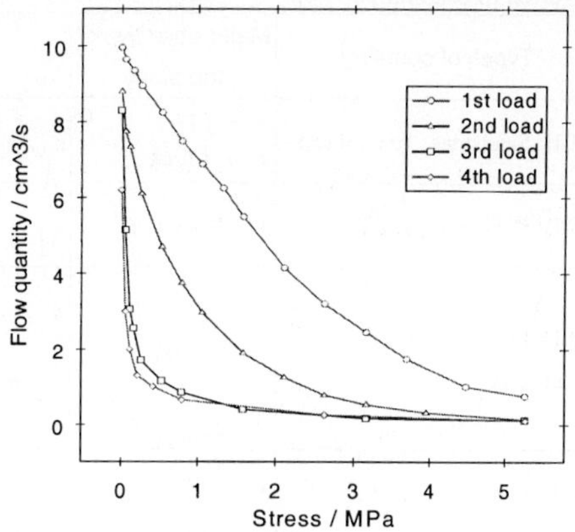

Figure 2 Curves of flow quantity through fracture of Miyun granite under normal stress

al., 2000, 2002; Yue, 1994; Guan Jiteng et al., 2000). The chemical effect in the coupled CM process has been recently studied by one of the authors of this paper in terms of chemical damage of rock (Tang Liansheng and Wang Sijing, 2002; Tang Liansheng et al., 2002a, 2002b). The degree of rock damage is given by following definitions.

a. Chemical damage index in terms of fracture toughness D^K

$$D^K = 1 - \frac{K_c}{K_{c0}} \tag{2}$$

b. Chemical damage index in terms of uniaxial compressive strength D^C.

$$D^C = 1 - \frac{\sigma_c}{\sigma_{c0}} \tag{3}$$

c. Chemical damage index in terms of increase of joint aperture D^b

$$D^b = 1 - \frac{b_j}{b_{j0}} \tag{4}$$

where K_C and K_{Co} –fracture toughness of rock affected and unaffected by chemical fluid, respectively; σ_C and σ_{Co}–uniaxial compressive strength of rock affected and unaffected by chemical fluid, respectively; b_j and b_{jo} –aperture of rock joint affected and unaffected by chemical fluid, respectively.

Table 3 Some experimental studies on the effect of coupling processes in Chinese literature (Note: All symbols are referred to original papers).

Types of coupling	Major objective of the study	Main results obtained by the study	References
M-H: Sandstone, Triaxial test	$k = f(\varepsilon)$ $\varepsilon = f(\sigma)$	k_{max}/k_0—1.5~2.0	Li Shiping et al., 1995
M-H: Rock joint, Triaxial test	$k_f = f(\sigma, p)$	$k_f = K_0 \sigma_{ne}^{-a}, \sigma_{ne} = \sigma_2 - \gamma(\sigma_1 + \sigma_3) - p$	Zheng Shaohe et al., 1999
M-H: Low cemented sandstone, Uniaxial	$k_f = f(\sigma)$	$k_f = ax^3 + bx^2 + cx + d$ $x = \frac{\sigma_a \cot^2(45° - \varphi/2) + 2C \cot(45° - \varphi/2)}{\sigma_a}$	Zhang Shouliang et al., 2000.
M-H: Gas flow in joints	$k_f = f(\eta, \sigma, p)$	$k_f = k_0 p^{-\eta} e^{[(a(\theta - bp)]}$	Zhao Yangsheng et al., 1999b
H-M: Sandstone and Granodiorite, Uniaxial	$k = f(w_s)$ $E = f(w_s)$	$\sigma_{CH} = \sigma_{Cdry}(A_1 \exp(-B_1 w_s) + C_1$ $E_H = E_{dry}(A_2 \exp(-B_2 w_s) + C_2$	Chen Ganglin et al., 1991
H-M: Flat fractures	$Q = f(\sigma_n)$	$Q / \Delta h = a \exp(-b\sigma_n)$	Shen Hongjun et al., 1998
H-M: rock joints	$\Delta b = f(\sigma_n)$ $Q = f(\sigma_n)$	$b = a \exp[(A + B/(1 - C\sigma_n))/3]$ $\ln(Q / \Delta H) = A + B/(1 + C\sigma_n)$	Su Baoyu, 1997
H-H: Gas-liquid two phase flow	$k_f = f(w_s)$	$Gas: k_{fg} = a_1 e^{(-b_1 w_s)}, Water: k_{fw} = a_2 w_s + b_2$	Zhao Yangsheng et al., 1999a
H-H: Sandstone	$k_f = f(p)$	$k_f = K_0(1 - p/(a + bp))^4$	Zhen Zuan et al., 1995
C-M: Rock fracturing	$k_I = f(C)$	Decrease of fracture toughness	Feng Xiating et al., 2000
C-M: Granite, limestone sandstone	$D_C = f(\Delta P_H, t)$	Increase of chemical damage	Tang Liansheng et al., 2002a,b,c
C-M: Microfracturing	Fracturing (σ, C)	Decrease of fracturing stress	Feng Xiating et al., 2002
T-M: Granite Fracture density	$\eta = f(t)$	$\eta \sim (t)$, linear increase	Kou Shaoquan, 1987
T-H-M: Granite experiment	$N_\mu = f(U)$	$N_U = a_1 \left(\frac{UL}{\mu}\right)^{a_1} \left(\frac{\mu}{\alpha}\right)^{\frac{1}{3}}$	Jian Zhao, 1999
T-H-M: Saturated joint Triaxial experiment	$\eta_r = f(T)$ $\sigma_{CT} = f(T)$	$\eta_r = a + bT + cT^2 + dT^3$ $\sigma_{CT} = \sigma_C \{1 + (a_1 + b_1 T_r + cT_r^2) T_r\}$ $T_r = (T - 70)/100$	Liu Yachen et al., 2002
T-H-M: Rock joint Triaxial test	$k_f = f(\sigma, T)$ $\Delta b = f(\sigma, T)$	$k_f = Ae^{(-a\sigma_n - bT)}$	Liu Yachen et al., 2001

In view of its general character the value of pH of the chemical fluids was also accepted as one of the variables of the chemical damage function. The preliminary results indicate that rocks are commonly sensitive to the pH value. Most types of rocks, such as granite, sandstone and limestone, etc., can be more prone to damage by chemical fluid of low or high pH values, but less damaged in fluid with medium pH(=7) value (Figure 3). The chemical damage of rocks leads to neutralizing of the fluid. Chemical damage of rock is a time-consume process, and commonly increases with time in a rate decreasing way (Figure 4).

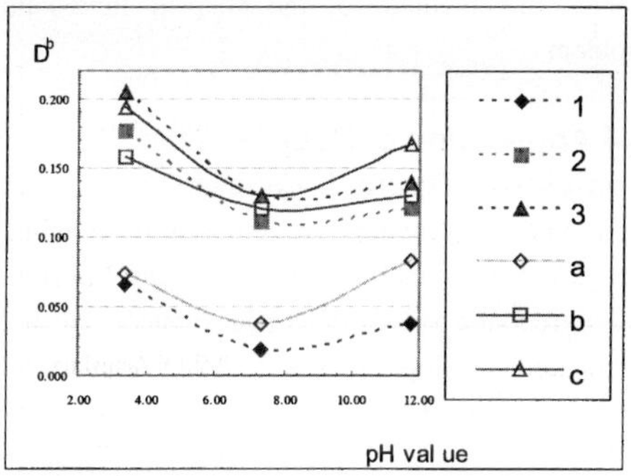

Figure 3 Chemical damage ratio of Db versus pH of chemical fluid.1, a-No flow, 2, b-Partially merged circulation, 3, c-Fully merged circulation.1, 2, 3, for 571 hours, a, b, c for 2418 hours.

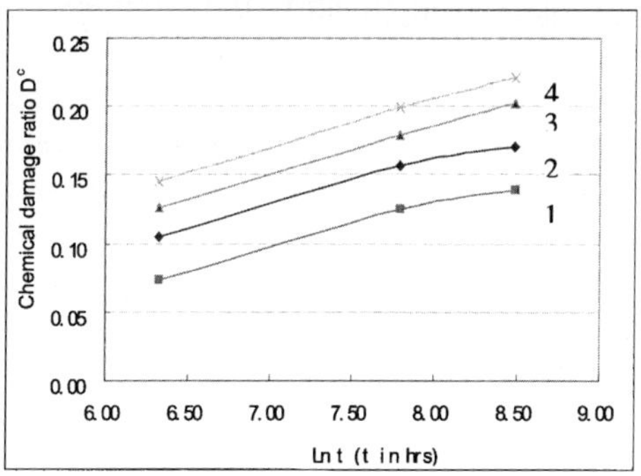

Figure 4 Chemical damage ratio(D^C) versus time(t). 1-Granite C; 2-Granite A; 3-Limestone C; 4-Limestone A

4. COUPLING ANALYSIS AND APPLICATIONS

4.1 General Remarks

The basic approach in the full coupling analysis is to solve a system of the equations (A) controlling the individual processes (T, H, M, C) and the equations (B) of relationship among coupling processes (TH, HM, MC, TM, etc.). In full coupling analysis the equations (A) controlling the coupling processes should contain the results of other processes, i.e.

$$T = f_t(T, H, M, C),$$

$$H = f_h(T, H, M, C), \text{ etc}$$

The relationship equations (B) should involve both direct action and inversion, for example,

$$T = f_t(H) \text{ and } H = f_H(T), \text{ and so on.}$$

To simplify the solution some authors proposed a partial coupling approach. In this case a part of inversion of interaction is not included, and the solution can be practically accepted for some special problems. Huang Tao (2002) used this approach for study of seepage field under influence of stresses and thermal action.

Another simplifying approach is the non-simultaneous coupling (Sun Yue, 2003), in which different numerical methods can be involved in the analysis, for example, the FEM is step by step incorporated with the Manifold method.

4.2 Application in Rock Mechanics and Foundation Engineering

In most cases of dam foundation problems the coupled HM processes are dominant for the foundation stability and/or seepage through foundations. A FEM model of seepage-rock damage coupling was first established by Yang Yangyi et al. (1991) for dam foundation analysis. Recently this model has been further developed by Zheng Shaohe et al. (2001) for rock engineering.

The jointed rock is treated as a hydraulically discontinuous mass. According to this model the rock matrix is not permeable and the seepage flow is limited in fractures. Taking into account the contribution of water saturation and seepage pressure, the total strain and constitutive equations of the rock mass was obtained.

The permeability tensor, being dependent upon the stress state, changes with evolution of rock damage. The expressions for permeability tensor can be obtained for fractures subjected to different stress states and failure modes.

Another FEM model of HM coupling for elasto-plastic rock mass was proposed by Wang Yuan et al. (2000). Both elasto-plasticity of rock matrix and fracture behavior are taken into consideration. A non-linear equation is obtained for controlling the coupled process

$$[M(\sigma)]\Delta\{\delta\} = \Delta\{F(\delta)\} \qquad (5)$$

where, $\{M(\sigma)\}$-the stiffness matrix depending on the stress and hydraulic pressure, $\{\delta\}$- the unknown displacement vector , $\{F(\delta)\}$ -the load vector depending on the displacement.

4.3 Application of Coupled HM Analysis in Dynamic Consolidation for Pavement Foundation

A study of numerical modeling of rutting in flexible pavement foundations under cyclic loading was reported by Huang Yu et al., 2001. On the basis of a coupled dynamic soil-pore water system theory a visco-elastic dynamic model has been developed to predict the non-linear response of pavement structure.

In this work the dependency of the maximum shear modulus upon the effective stress, and relationship between strain and damping ratio are studied. Based on the laboratory experiment, the function of pore water pressure buildup versus the number of stress cycles is obtained.

The main procedures of analysis are as follows: (1) Calculate initial static state, (2) Determine parameters of dynamic model, (3) Calculate dynamic state, (4) Calculate pore water pressure increment and undrained residual strain, (5) Repeat steps of 2-4 till the end of cyclic loading, (6) Continue post-cyclic static analysis until the full dissipation of pore water pressure.

Thus, the rutting depth and permanent deformation of pavement foundation can be obtained.

An application of dynamic consolidation in the compaction for foundation reinforcement was reported by Li Yaochen et al., 2001. The solution for an axi-symmetric problem was achieved by using FEM method for the coupled fluid-solid problem.

4.4 Application of Coupled THM Analysis in Tunnel Engineering

A number of works are devoted to coupling analysis in the tunnel engineering. Huang Tao et al. (1999) reported a study of seepage, temperature and stress field of slope, in which the Qinglin tunnel was excavated. The main equations used are as follows.

$$\left.\begin{array}{l}[K(\sigma,T)]\{H\} = \{B\} \\ \{\sigma\} + \{p(\Delta H)\} + \{R_b(\Delta T)\} = [D]\{\varepsilon\} \\ \{[\lambda(\sigma)] + [C_a(H)]\}\{T\} = \{T_b\}\end{array}\right\} \qquad (6)$$

where, $[K]$-permeability matrix, $\{H\}$-unknown water head vector, $\{B\}$ -boundary seepage condition vector, $\{\sigma\}$-stress vector, $\{p\}$-dynamic water pressure vector, $\{R_b\}$-thermal stress vector, $[D]$ -elasticity matrix, $\{\varepsilon\}$ -strain vector, $[\lambda]$ -thermal conductivity matrix, $[C_a]$ -thermal capacity matrix, $\{T\}$- unknown temperature vector, $\{T_b\}$-thermal boundary condition vector .

A cross iterative method is used in the FEM calculation. First, the initial seepage field is obtained, followed by calculation of coupled seepage and stress fields. The coupled temperature, stress and seepage fields are then obtained subsequently.

4.5 Application of THM Coupling Analysis of Unsaturated Soils

On the basis of existing studies on the coupled HM process Wu Wenhua et al. (2002) proposed a constitutive model for coupled THM behaviour of unsaturated soils. The thermal softening effects, defined to consider the decrease in value of the pre-consolidation pressure and decrease in critical value of the suction of the SI (suction increasing), are introduced, based on the formulation of a CAP model.

The total strain is defined as the sum of elastic-plastic strain from mechanical action, strain due to thermal effects and strain due to suction. Then, the constitutive relationship in a rate form for the coupled THM process in the unsaturated soils is obtained.

Comparison of the results obtained from the numerical modeling by using the constitutive model with existing experiment results (Benchmark Test 2.2) shows that the proposed model can be used satisfactorily.

Other works devoted to study of unsaturated soils can be referred to Li Xikui et al. (1998) on the coupled seepage, suction and stress fields and Yang Daiquan et al. (2000) on the transfer of pore moisture, air and heat.

4.6 APPLICATION OF THM COUPLING ANALYSIS IN GEOTERMAL ENGINEERING

A 3D FEM simulation for coupled THM process of fractured rocks for heat extraction in HDR was reported by Zhao Yangsheng et al. (2002).

The equations introduced to represent the coupling process are obtained by assuming that the fluid circulated in the fractures is water, and the heat exchange and transport are by modes of both convection and conduction.

In the equation for rock deformation the thermal stresses in rock blocks are considered, and the seepage pressure is included in calculating the deformation of fractures. The thermal exchange between rock block and seepage water is taken into account. Change of water density is accepted in the analysis.

5. FULL THMC COUPLED PROCESS AND APPLICATION—A DISCUSSION

A full THMC coupling method is useful for understanding and quantitatively describing many geo-processes. Although there a number of studies on coupled process appeared in China, the problem of fully coupled THMC process has not been well tackled yet.

To establish the governing equations the full interaction of coupling processes should be taken into consideration (Table 4). The system reactions should be included into the main coupled equations, in addition to some equations concerning the relationship between different parameters.

Chemical process is complicated because of the difficulty to establish a unified governing equation. In some cases large number of chemical reactions exists in one system, causing complicated interactions. Therefore, the study can be started from specified problems with a few chemical reactions involved.

In addition to the waste disposal, another application area of THMC coupling models is the chemical mining. In the Jingchuan Nickel Mine, Northwestern China, a program of study and experiment is proposed for extraction of metals by circulation of chemical fluid in ore blocks.

Since the ore of Jinchuan Mine mainly consists of oxides and sulfides of Ni,Cu and Co, the chemical procedure of oxidation of ore to separate the useful metals is adopted.

For oxides the chemical reaction is

$$MeO+2H_2SO_4(aq)\rightarrow Me^{2+}+2HSO_4^-$$

For low-grade sulfide ore the bioleaching technology should be used and the chemical reaction is following:

$MeS+2O_2+H_2SO_4$(bacteria)$\rightarrow Me^{2+}+2HSO_4^-+H_2O$
(with heat yielding)

The percolation of acid plus bacteria in the ore and rock can gradually extract the metal and destroy the structure of rock. As shown in Figure 6, the leaching rate of metals increases with time of percolation. In general, the increase of leaching rate becomes slow with decreasing of content of remaining metal in the ore.

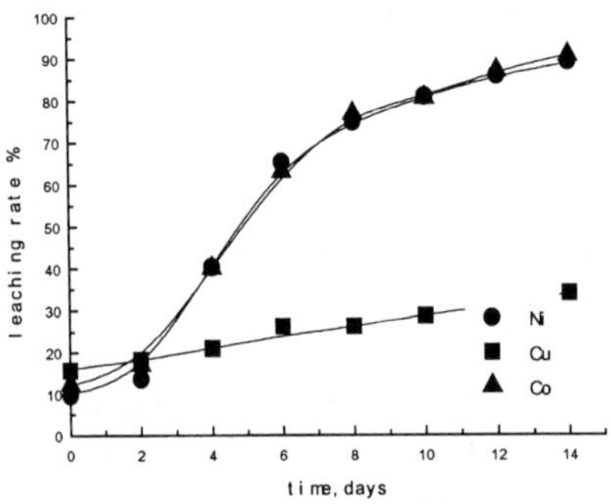

Figure5 Leaching rate of metals from ore rock versus time (Zhang Guangji,2002)

Accompanying with extraction of metals and other components and disintegration of ore rock, the mechanical strength and seepage conductivity should be affected severely. In addition the chemical process is associated with thermal release which may affect the mechanical and hydraulic processes, and accelerate the chemical reaction it self. All these complex coupling processes need a comprehensive modeling test prior to the technological design.

6. CONCLUSION

(1) Study of THMC coupled process becomes one of frontier on geotechnical and geo-environmental engineering in China.

(2) More experimental studies and in-situ coupled monitorings are essential for further development.

(3) More study works are needed to involve the chemical process.

(4) A scheme of chemical mining suggested in this paper is proposed for further development.

REFERENCES

Chen Ganglin and Zhou Rende, 1991, *An experimental study concerning the macro-scopic effect of water on the deformation and failure of loaded rocks*, Acta Geophysica Sinica, Vol.34, No.3, pp.335-342

Chen Zuan, Wu Xiangyang, Sun Demin and Yang Wei, 1995, *Study on relationship between permeability of sandstone and hydrostatic pressure*, Chinese J. Rock Mech. Eng., Vol.14, No.2, pp.155-159

Fang Hsai-Yang, 2000, *Environmental Geotechnology-perspective in the 21st century*, Chinese J. Geotech. Eng., Vol.22, No.1, pp.3-11

Feng Xiating and Masahiro Seto, 2000, *Rock fracturing behaviors under chemical corrosion—part I: Experimental study*, Chinese J. Rock Mech. Eng., Vol.19, No.4, pp.403-407

Feng X, Wang C and Chen S. 2002. *Testing study and real-time observation of rock meso-cracking process under chemical corrosion.* Chinese J. Rock Mech. & Eng., 21(7), pp. 935-939.

Guan Jiteng and Fang Wenjing, 2000, *Coupling relationship for electromechanical hydrodynamics in water saturated sandy soil and clayey soil*, Chinese J. Geotech. Eng., Vol.22, No.6, pp.664-667

Huang Tao, 2002, *Coupling study among seepage-stress-temperature in fractured rock mass*, Chinese J. Rock Mech. Eng., Vol.21,No.1, pp.83-87

Huang Yu, Stsphen F Browm and Glenn R Mcdowell, 2001, *Dynamic coupled analysis for rutting in flexible pavement foundation under cyclic loading*, Chinese J. Geotech. Eng., Vol.23, No.6, pp.757-762

Jiang Zhenquan and Ji Liangjun, 2001, *The laboratory study on behavior of permeability of rock along the complete stress-strain path*,

Chinese J. Geotech. Eng., Vol.23, No.2, pp.153-156

Kou Shaoqun, 1987, *Influence of thermal cracking on the deformation and failure of granite*, Acta Mechanica Sinica, Vol.19, No.6, pp.550-555

Li Caihua, Chen Congxin and Fu Shaolan,2002, *Testing Study on seepage characteristics of single fracture with sand under shearing displacement*, Chinese J. Rock Mech. Eng., Vol.21, No.10, pp.1457-1461

Li Shiping, Li yushou and Wu Zhenye, 1995, *The permeability-Strain equations relating to complete stress-strain path of the rock*, Chinese J. Geotech. Eng., Vol.17, No.2, pp.13-19

Li Xikui and Fan Yiqun,1998, *Finite element analysis of deformation and seepage process in unsaturated soils*, Chinese J. Geotech. Eng., Vol.20, No.4, pp.20-24

Li Yaochen, Chen Jie and Zhou shunhua, 2001, *Numerical simulation for a foundation reinforcement procedure using dynamic consolidation technique*, Rock and Soil Mechanics,Vol.22, No.1, pp.6-11

Liu Caihua , Chen Cunxin and Fu Shaolan, 2002, *Experimental study of law of seepage through single fracture in rock under normal stress*, Chinese J. Rock Mech. Eng., Vol.21,No.10, pp.1457-1461

Liu Yachen, Cai Yaoqing., Liu Quansheng and Wu Yushan, 2001, *Thermal-hydraulic-mechanical coupling constitutive relation of rock mass fracture interconnectivity*, Chinese J. Geotech. Eng., Vol.23, No.2, pp.196-200

Liu Yachen, Cai Yongqing, Liu Quansheng and Wu Yushan, 2002, *Study on mechanical behavior of jointed rock mass affected by temperature and saturated water*, Chinese J. Rock Mech. Eng., Vol.21,No.2, pp.233-237

Miao Linchang and Yin Zongze, 1999, *Shear strength of unsaturated soils*, Rock and Soil Mechanics. Vol.20, No.3, pp.1-6

Qian Jiahuan, Qian Xuede, Zhao Weibing and Shuai Fangsheng, 1986, *Theory and practice of dynamic consolidation*, Chinese J. Geotech. Eng., Vol.8, No.6, pp.1-17

Shen Hongjun, Gao Haiying, Xia Songyou, 1998, *Experimental study on seepage characteristics of fissured rock mass with stress*. Journal of Yangtze River Scientific Research Institute, Vol.15, No.3, pp.35-39

Sun Jun and Wang Sijing, 2000, *Rock mechanics and engineering in China: developments and current state-of-the art*, Int. J. Rock Mech. & Min. Sci., Vol. 37, pp.447-465

Sun Yue, 2003, *Application research on non-simultaneous coupling calculation of deformation and seepage with manifold method and FEM*, Chinese J. Rock Mech. Eng., Vol.22, No.6, pp.943-950

Su Baoyu, Zhan Meili and Wang Yuan, 1997, *Experimental study on the coupling characteristics of fracture seepage and stresses*, Chinese J. Geotech. Eng., Vol.19, No.4, pp.73-77

Tang Liansheng and Wang Sijing, 2002a, *Analysis on mechanism and quantitative methods of chemical damage in water-rock interaction*, Chinese J. Rock Mech. Eng., Vol.21, No.3, pp.314-319

Tang Liansheng, Zhang Pengcheng and Wang Sijing, 2002b, *Testing study on macroscopic mechanics effects of chemical action of water in rock*, Chinese J. Rock Mech. Eng., Vol.21, No.4, pp.526-531

Tang Liansheng, Zhang Pengcheng and Wang Sijing, 2002c, *Testing study on effects of chemical action of aqueous solution on crack propagation in rock*, Chinese J. Rock Mech. Eng., Vol.21, No.6, pp.822-827

Tao Zhenyu and Shen Xiaoying, 1988, *Coupled analysis of reservoir zone stress field*, Journal of Wuhan Hydro-electrical University, No.1, pp.8-14

Wang Sijing, 2002, *New era of rock mechanics and engineering*, Proc. of 7th Congress of Chinese Society of Rock Mechanics and Engineering, China Press of Science and Technology, Beijing, pp.1-3

Wang Sijing, 2003, *Century achievements and new historical mission of rock mechanics and engineering in China*, Chinese J. Rock Mech. & Eng., vol.22, No.6, pp.867-871

Wang Yuan, Xu Zhiying and Su Baoyu, 2000, *Full coupling analysis of seepage and elsto-plasticity stresses in complex rock masses*, Chinese J. Rock Mech. Eng., Vol.19, No.2, pp.177-181

Wang Yuping and Wang Yonghong, 1999, *The effects of soil-water interaction on the flow in soil fissure*, Chinese J. Rock Mech. Eng., Vol.18, No.5, pp.554-557

Wu Wenhua and Li Xikui, 2002, *Constitutive model and numerical simulation of thermo-hydro-mechanical behavior in unsaturated soils*, Chinese J. Geotech. Eng., Vol.24, No.4, pp.411-416

Yang Daiquan, Shen Zhujiang, Hariato Rahardjo and Leong Eng Choon, 2000, *Modelling fully coupled moisture, air and heat trasfer in unsaturated soils*, Chinese J. Geotech. Eng., Vol.22, No.3, pp.357-361

Yang Yanyi and Zhou Weiyuan, 1991, *A coupled seepage-damage analysis model for jointed rock masses and its application to rock engineering*, Chinese J. Water Resource, No.5, pp.19-27

Yue H. 1994. *Primary study on model for coupled heat-moisture-salt transfer in soil during freezing-thawing processes*. J. of Glaciology & Geocryology, 16(4), pp.308-313.

Zhang Yuzhuo and Yao Jianguo, 2001, *New progress and sustainable development of rock mechanics and engineering in China, in Frontiers of Rock Mechanics and Sustainable Development in the 21st Century*, ed: Wang Sijing et al., A.A.Balkema, pp.3-16

Zhao Jian, 1999, *Experimental study of flow-rock heat transfer in rock fractures*, Chinese J. Rock Mech. Eng., Vol.18,No.2, pp.119-123

Zhao Yangsheng, Hu yaoqing, Yang Dong and Zheng Shaohe, 1999a, *A simulation testing study on seepage law of gas-liquid two-phase fluid in fissure*, Chinese J. Rock Mech. Eng, Vol.18, No.3, pp.354-356

Zhao Yangsheng, Hu Yaoqing, Yang Dong and Wei Jingping, 1999b, *The experimental study on the gas seepage law of rock related to adsorption under 3-D stresses*, Chinese J. Rock Mech. Eng., Vol.18, No.6, pp.651-653

Zhao Yangsheng, Wang Ruifeng, Hu Yaoqing, Wang Zhijun and Xie Yaoshe, 2002, *3D numerical simulation for coupled THM of rock matrix-fractured media in heat extraction in HDR*, Chinese J. Rock Mech. Eng., Vol.21, No.12, pp.1751-1757

Zhang Guangji,2002, *Studyon mechanism and kinetics of the bioleaching of nicopyrite,*PhD Thesis of the Institute of process Engineering, Chinese Academy of Sciences,Beijing,1-108

Zhang Shouliang, Shen Chen and Deng Jingen, 2000, *testing study on the law of permeability variation in progress of rock deformation and damage*, Chinese J. Rock Mech. Eng., Vol.19, Supplement, pp.885-888

Zheng Shaohe, Zhao Yangsheng and Duan Kanglian, 1999, *An experimental study on the permeability law of natural fracture under 3-D stresses*, Chinese J. Rock Mech. Eng., Vol.18, No.2, pp.133-136

Zheng Shaohe and Zhu Weishen, 2001, *Theoretical analysis on a coupled seepage-damage model of fractured rock mass*, Chinese J. Rock Mech. Eng., Vol.20, No.2, pp.156-159

Zhou Chuangbing and Xiong Wenlin,1996, *Permeability tensor for jointed rock masses in coupled seepage and stress field*, Chinese J. Rock Mech. Eng., Vol.15, No.4, pp.345-352

Table 4 Interaction among THMC coupling processes

	Main parameters	Direct effect	System reaction	Main Equations
T	• ΔT • Δq	T-H: $\Delta k_T, \Delta\mu_T$ T-M: $\Delta\sigma_T, \Delta\varepsilon_T, \Delta E_T, \Delta D_T$ T-C : $\Delta P_T, \Delta d_T, \Delta\mu_T$	$\Delta T_H, \Delta q_H, \Delta T_M, \Delta T_C,$ Δq_C	$\Delta T = f(\Delta T_C, \Delta T_H, \Delta T_M)$ $\Delta q = f(\Delta q_H, \Delta q_C)$
H	• Δp • ΔQ	H-T: $\Delta T_H, \Delta q_H$ H-M: $\Delta p_H, \Delta\varepsilon_H, \Delta E_H, \Delta D_H$ H-C: $\Delta\mu_H, \Delta P_H, \Delta d_H$	$\Delta k_T, \Delta\mu_T, \Delta k_M, \Delta k_C,$ $\Delta\mu_C$	$\Delta k = f(\Delta k_C, \Delta k_H, \Delta k_M)$ $\Delta p = f(\Delta p_H)$ $\Delta k = f(\Delta\mu)$
M	• $\Delta\sigma$ • $\Delta\varepsilon$	M-T: ΔT_M M-H: Δk_M M-C: ΔP_M	$\Delta\sigma_T, \Delta\varepsilon_T, \Delta E_T, \Delta D_T,$ $\Delta p_H, \Delta\varepsilon_H, \Delta E_H, \Delta D_H,$ $\Delta\sigma_C, \Delta\varepsilon_C, \Delta E_C, \Delta D_C$	$\Delta\varepsilon = f(\Delta\varepsilon_H, \Delta\varepsilon_T, \Delta\varepsilon_C)$ $\Delta\sigma = f(\Delta\sigma_T, \Delta\sigma_C, \Delta p_H)$ $\Delta D = f(\Delta D_C, \Delta D_T, \Delta D_H)$ $\Delta D = f(\Delta p, \Delta\sigma)$
C	• Δd • ΔP	C-T: $\Delta T_C, \Delta q_C$ C-H: $\Delta k_C, \Delta\mu_C$ C-M : $\Delta\sigma_C, \Delta\varepsilon_C, \Delta E_C, \Delta D_C$	$\Delta P_T, \Delta d_T, \Delta\mu_T,$ $\Delta\mu_H, \Delta P_H, \Delta d_H, \Delta P_M$	$\Delta P = f(\Delta P_T, \Delta P_H, \Delta P_M)$ $\Delta d = f(\Delta d_H, \Delta d_T)$ $\Delta\mu = f(\Delta\mu_C, \Delta\mu_T)$ $\Delta d = f(\Delta T)$

Theme 1
Coupled T-H-M-C Processes in Radioactive Waste Disposal Systems

THE FEBEX BENCHMARK TEST. CASE DEFINITION AND COMPARISON OF DIFFERENT MODELLING APPROACHES

E.E. Alonso and J. Alcoverro
Geotechnical Engineering Department, Universitat Politècnica de Catalunya,
c/ Jordi Girona 1-3, Edificio D-2, 08034 Barcelona, Spain.

Abstract: The FEBEX (Full-scale Engineered Barriers Experiment in Crystalline Host Rock) "in situ" test was installed at the Grimsel Test Site underground laboratory (Switzerland) and is a near-to-real scale simulation of the Spanish reference concept of deep geological storage in crystalline host rock. A modelling exercise, aimed at predicting field behaviour, was divided in three parts. In Part A, predictions for both the total water inflow to the tunnel as well as the water pressure changes induced by the boring of the tunnel were required. In Part B, predictions for local field variables, such as temperature, relative humidity, pore water pressure, stresses and displacements at selected points in the bentonite, and global variables, such as the total input power to the heaters were required. In Part C, predictions for temperature, stresses, water pressures and displacements in selected points of the host rock were required. Eleven Modelling Teams from Europe, North America and Japan were involved in the analysis of the test. Differences among approaches may be found in the constitutive models used, in the simplifications made to the balance equations and in the geometric symmetries considered. Several aspects are addressed in the paper: the basic THM physical phenomena which dominate the test response are discussed, a comparison of different modelling results with actual measurements is presented and a discussion is given to explain the success of the various predictions.

1. THE FEBEX "IN SITU" TEST

The Febex "in situ" test is currently in operation at the Grimsel Test Site, located in the granitic rocks of the Aare Massif in central Switzerland.

In order to perform the FEBEX in-situ test, it was decided to excavate a new drift. Prior to it, two pilot boreholes (FEBEX 95.001 and FEBEX 95.002) were drilled in the area. Afterwards, the FEBEX drift was excavated between these pilot boreholes. It was parallel to FEBEX 95.002. Figure 1 shows a perspective of the FEBEX drift and associated boreholes.

Borehole FEBEX 95.002, with a diameter of 86 mm and a length of 132.36 m, was equipped with packers to isolate sections in which pore water pressures could be monitored.

During the period 25.09.95 to 30.10.95, the FEBEX drift was excavated using a TBM. The FEBEX drift has a diameter of 2.28 ± 0.01 m and a total length of 71.4 m.

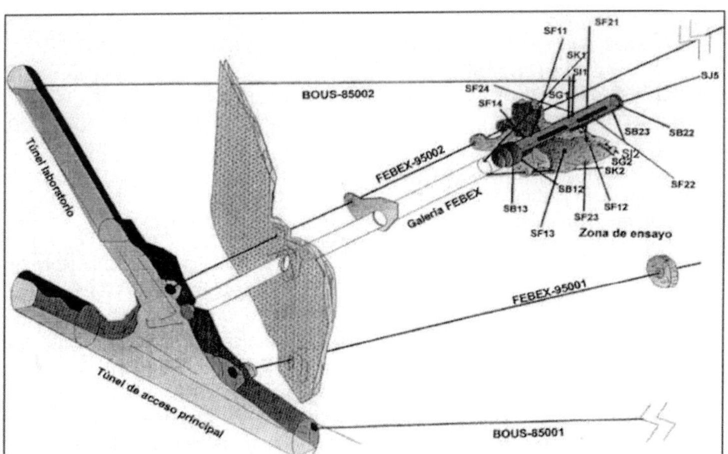

Figure 1. Layout of FEBEX test and associated boreholes (Pardillo et al, 1997).

Discrete flow measurements of water flow towards the excavated FEBEX tunnel were conducted in November 1995 using diapers (cellulose sheets) located at various points of the tunnel wall. Flow measurements were obtained by dividing the increase of weight of the diapers by the corresponding time. Bulk inflow measurements were also performed in January 1996 using a gauge. During the period 12.02.96 to 04.04.96, 19 in-drift boreholes (Ø 66 mm and Ø 146 mm) with a total length of 233 m were drilled from the test area of the FEBEX drift (tunnel meters 50-71 m). Initially, the borehole layout was planned to be strictly radial. However, the boreholes were re-oriented in order to intersect the most relevant geological features. (Figure 1) These boreholes were initially used to conduct hydraulic tests and later to monitor several variables during the test (water pressure, temperature, stresses, displacements).

Extensive geological and hydraulic characterizations have been performed. Relevant documents are Keusen et al. (1989), Pardillo et al. (1997) and Guimerà et al (1996, 1998).

Shear zones are of considerable thickness at the area (5 to 20 m). At the intersection with tunnels, they display major outflows indicating their relevance as preferential flow paths. Lamprophyre dykes have also considerable dimensions (thickness up to several meters), although their relevance as preferential flow paths is not as important as shear zones. A preferential flow path within these dykes is the contact surface between the lamprophyre and the host rock. Figure 2 provides a geological cross section of the Grimsel area and shows the position of the FEBEX drift, bounded by two main shear zones

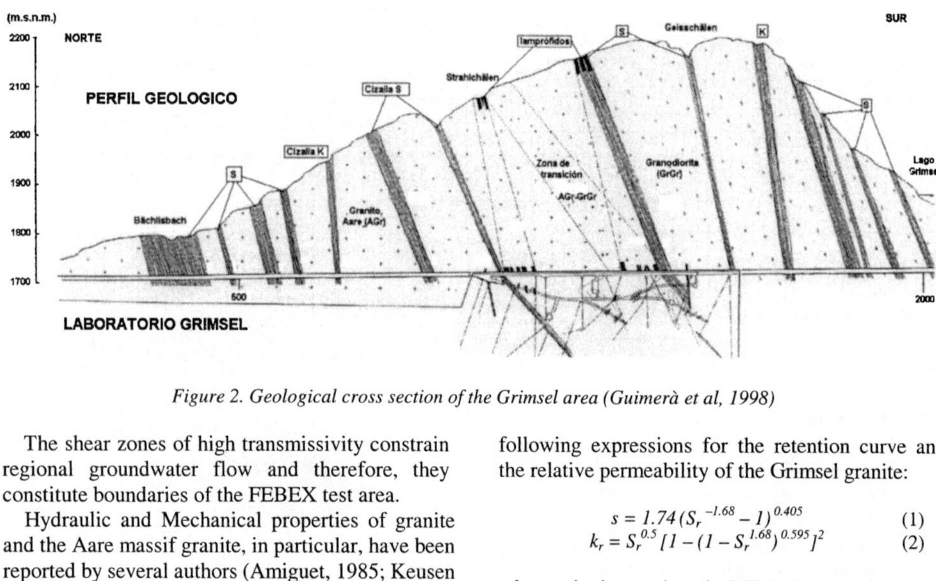

Figure 2. Geological cross section of the Grimsel area (Guimerà et al, 1998)

The shear zones of high transmissivity constrain regional groundwater flow and therefore, they constitute boundaries of the FEBEX test area.

Hydraulic and Mechanical properties of granite and the Aare massif granite, in particular, have been reported by several authors (Amiguet, 1985; Keusen et al, 1989; Bräuer et al, 1989; Pahl et al, 1989). Table 1 provides a set of significant parameters of intact Aare granite essentially based on core testing. In addition, the following properties are given for granite in Amiguet et al (1985):

coef. linear thermal expansion	$(5\text{-}12)\cdot 10^{-6}$	K^{-1}
coef. vol. thermal expansion	$(20\text{-}30)\cdot 10^{-6}$	K^{-1}
specific heat	*800-1250*	*J/kgK*

Finsterle and Pruess (1995) obtained by backanalysis of a ventilation experiment the following expressions for the retention curve and the relative permeability of the Grimsel granite:

$$s = 1.74\,(S_r^{\,-1.68} - 1)^{0.405} \qquad (1)$$

$$k_r = S_r^{\,0.5}\,[1 - (1 - S_r^{\,1.68})^{0.595}]^2 \qquad (2)$$

where s is the suction (in MPa), S_r is the degree of saturation and k_r is the relative permeability.

The vertical lithostatic stress at the level of the GTS test galleries is around 9-12MPa. However, the horizontal stresses are substantially higher than the depth-related overburden pressure. Significant horizontal forces have been measured in the main compression direction NW-SE. Maximum and minimum horizontal stresses are in the range 18/45 MPa and 15/32MPa. More details may be found in Bräuer et al (1989) and Pahl et al (1989)

Table 1: Rock mechanical parameters of granite at the Grimsel Test Site (Keusen et al., 1989). (refers to fractures).*

Parameter	*Central Aare granite*	*Units*
density	*2660±23.8*	*kg/m³*
porosity	*0.4-1.0*	*vol %*
uniaxial comp. strength	*169.1±37.1*	*MPa*
Young's modulus E_{50}	*53.3±11.0*	*GPa*
Poisson's ratio	*0.37±0.12*	-
	0.33±0.03	
tensile strength	*9.06±1.48*	*MPa*
triaxial comp. strength	*5.0 ; 263.0±29.9*	*MPa*
(σ_3 ; σ_1)	*10.0 ; 333.0±20.6*	
	20.0 ; 410.0±63.8	
friction angle	*(*) 33*	°
p-wave vel. (specimen)	*3111±278*	*m/s*
p-wave vel. (whole rock)	*5600±100*	*m/s*
therm. conductivity (wet)	*2.58±0.19*	*W/m·K*
therm. conductivity (dry)	*3.34±0.35*	*W/m·K*
permeability	$5\cdot10^{-12}$ *(10 MPa)*	*m/s*
	$3.5\text{-}4.5\cdot10^{-12}$ *(5-15 MPa)*	
	$5\cdot10^{-12}$ *(5-30 MPa)*	

The "in situ" test consists of a near to full-scale simulation of a HLW disposal facility, following the scheme shown in Figure 3. Two electrical heaters, of dimensions and weight equivalent to those of the nuclear canister were placed in the drift just described, the entire space surrounding the heaters being filled with blocks of compacted bentonite to complete the 17.4 m of barrier for the test section. This test zone was closed with a concrete plug.

In addition to the clay barrier and the heaters, 632 sensors of diverse types were installed. The sensors monitor the different thermo-hydro-mechanical processes that occur in both the clay barrier and the surrounding rock throughout the entire life of the test. A heating stage of more than three years was planned, followed by dismantling of the test. The present paper reports on the measurements recorded during the first 1000 days of operation.

Compacted bentonite blocks were manufactured by compacting statically in appropriate moulds a crushed bentonitic rock under vertical stresses of 40-45MPa. The average values of water content and dry density of all the manufactured blocks are 14.4% and 1.69 g/cm³, respectively.

Figure 4 shows a view of the test during the process of installation. Heaters were located inside a perforated clay liner as shown in the figure. A typical cross section of the test showing the relevant dimensions is shown in Figure 5.

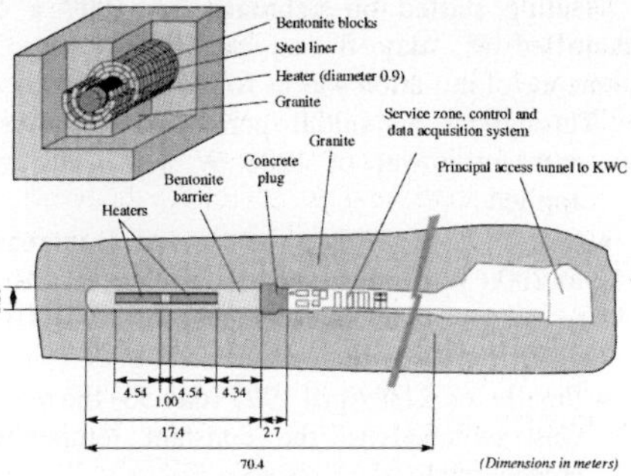

Figure 3. General scheme of the FEBEX in situ test

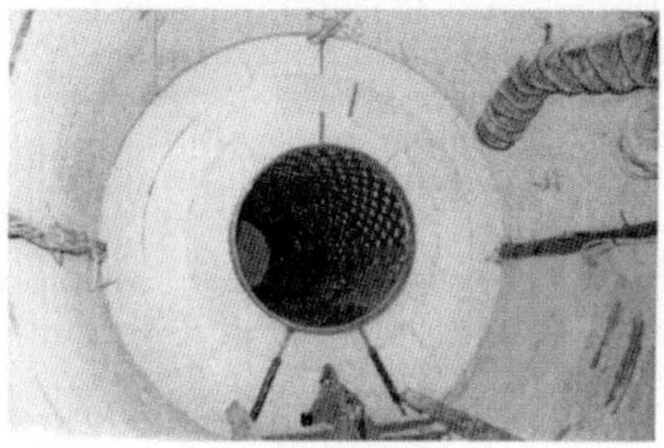

Figure 4. A view of FEBEX test construction

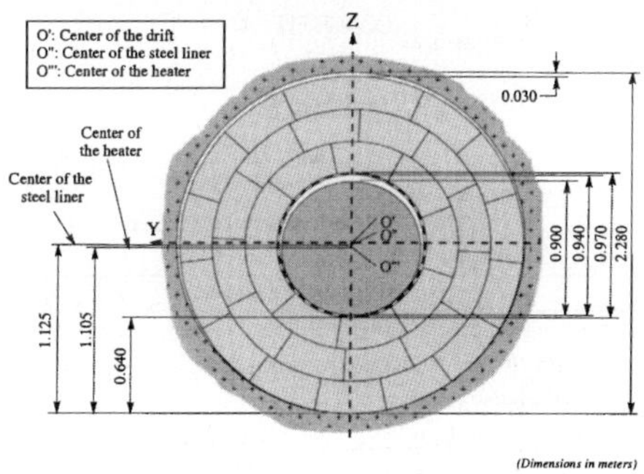

Figure 5. Typical cross section of the clay barrier

Sensors located inside the barrier were arranged in cross sections as shown in Figure 6. Cables were carried radially to outer grooves in the bentonite and taken eventually, through the concrete plug, to the service area of the tunnel.

A total volume of gaps of 5.53% of the emplacement volume was measured. Therefore, the average dry density of the barrier was established as 1.60t/m³. Typically, the dry density varied by +-0.25t/m³ among the different slices of assembled blocks.

Heating started on February 27, 1997 a date identified as "day 0" on the time scale. The sequence of initiation was as follows:

- Throughout an initial period of 20 days a constant power of 1200 W per heater was applied
- Over the next 33 days the power was increased to 2000 W per heater and maintained constant to approximate the temperature of 100 °C desired at the surface of the steel liner
- Finally, on 21st April 1997 (day 53) the system was switched to the constant temperature control mode.

1.1. The FEBEX bentonite

The properties of FEBEX bentonite have been extensively investigated over the past 8 yrs. Several recent publications describe the thermo-hydro-mechanical properties of the compacted material (Villar, 2002; Villar and Lloret, 2001, 2002, 2003; Villar et al, 2002; Huertas et al, 2000; Romero et al, 2001; Pintado et al, 2002; Lloret et al, 2003). Extensive information is also available for the bentonite "S-2", a material of essentially the same origin (a quarry in Serrata de Níjar, Almería, Spain) as the FEBEX one. A summary of relevant properties, taken from some of the mentioned references, is given here.

The conditioning of the bentonite in the quarry, and later in the manufacturing factory, was strictly mechanical (homogenization, rock fragment removal, drying, crumbling of clods, and sieving) to obtain a granulated material with the specified characteristics of grain-size distribution and water content: Fraction of particles of more than 5 mm, less than 5%, and fraction of particles smaller than 74 μm, greater than 85%; water content, after conditioning, between 12.5% and 15.5%. Table 2 provides a set of identification parameters as determined by two laboratories: Ciemat and UPC.

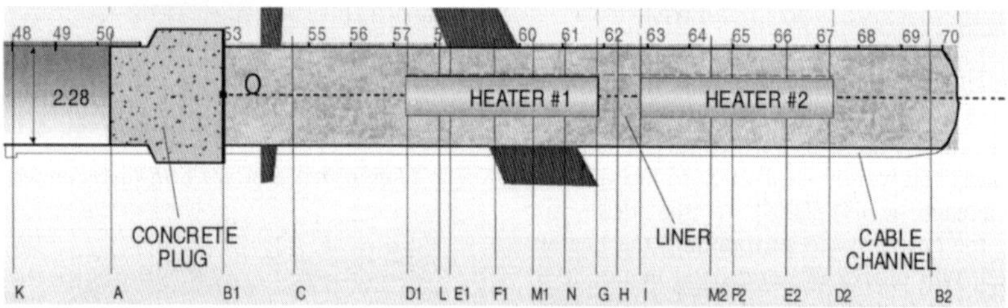

Figure 6. Arrangement of instrumented sections

Table 2: Identification properties.

Property	*CIEMAT*	*UPC-DIT*
Water content in equilibrium with the air in the laboratory, in %	*13.7 ± 1.3*	*13.3 ± 1.3*
Liquid limit, in %	*102 ± 4*	*93 ± 1*
Plastic limit, in %	*53 ± 3*	*47 ± 2*
Plasticity index	*49 ± 4*	*46 ± 2*
Specific weight	*2.70 ± 0.04*	—
Grain-size distribution, in %		
Fraction less than 74 μm	*92 ± 1*	*87*
Fraction less than 2 μm	*68 ± 2*	*45*
Specific surface, in m²/g		
Total	*725 ± 47*	—
External, BET	*32 ± 3*	—

Swelling pressure tests were performed using conventional oedometers on samples saturated with distilled water. A regression curve was developed as a function of dry density for the swelling pressure of the FEBEX bentonite, as shown on Figure 7 and expressed by the equation

$$P_s = exp\,(6.77\,\rho_d - 9.07) \qquad (3)$$

where P_s is the swelling pressure in MPa and ρ_d is the dry density in g/cm^3 (Figure 7).

A series of swelling under load tests, performed in oedometer tests, provided additional information on the expansivity and mechanical behaviour of the material. The following equation provides a relationship between 1D swelling strain upon saturation, applied confining stress (MPa) and dry density (g/cm^3):

$$\varepsilon = -46.9 - 19.4 \log \sigma + 36.6\,\rho_d \qquad (4)$$

Many determinations of saturated hydraulic conductivity are available. Some of them are collected in Figure 8.

Infiltration tests were also performed to determine the relative permeability. It was found that the relationship

$$k_r = S_r^n \qquad (5)$$

where "n" is a parameter could fit experimental data. A reference value for n is 3 although a significant variability is found.

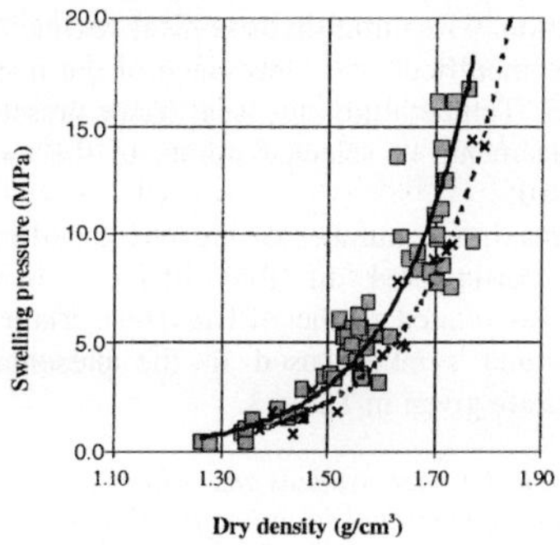

Figure 7. Swelling pressure of FEBEX bentonite (square dots). Crosses correspond to the "S-2" bentonite.

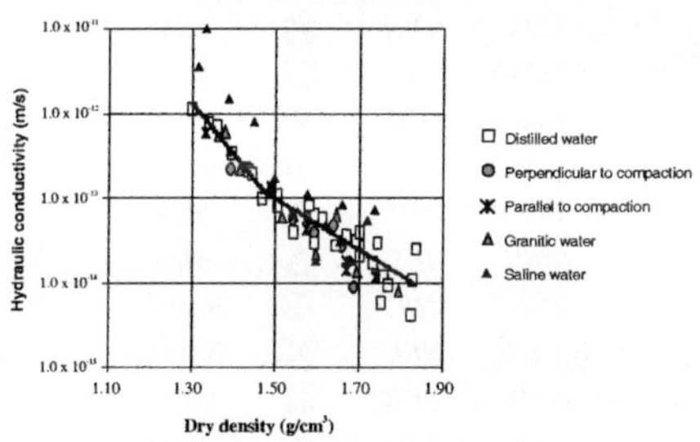

Figure 8. Saturated hydraulic conductivity as a function of dry density of FEBEX bentonite

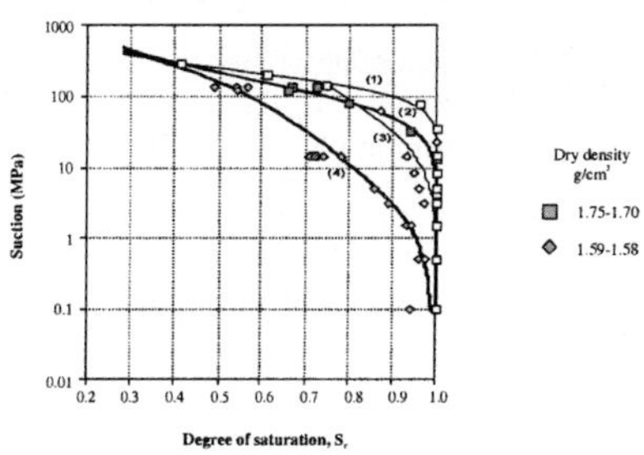

Figure 9. Water retention characteristics of FEBEX bentonite. Bold symbols correspond to wetting paths

Water retention curves were determined under free and constant volume conditions. The latter are more precise. Curves relating suction and degree of saturation for wetting-drying paths at different constant dry density are shown in Figure 9. It was found that the concept of a constant intrinsic permeability does not hold, even approximately, for an expansive material such as the FEBEX bentonite. Figure 10 shows values of "intrinsic" permeability determined in gas and water flow experiments. Extremely large differences are found, which are due to the change in microstructure of the compacted material when its water content changes. Thermal properties are also available. Specific heat has been determined only for bentonite S-2. The relationship between specific heat and temperature fits the following equation, in a range of temperatures between 45°C and 150°C:

$$c_s = 1.38 \cdot T + 732.5 \qquad (6)$$

where c_s is the specific heat, in J/kg°C, and T is the temperature, in °C. Measured thermal conductivities are given in Figure 11.

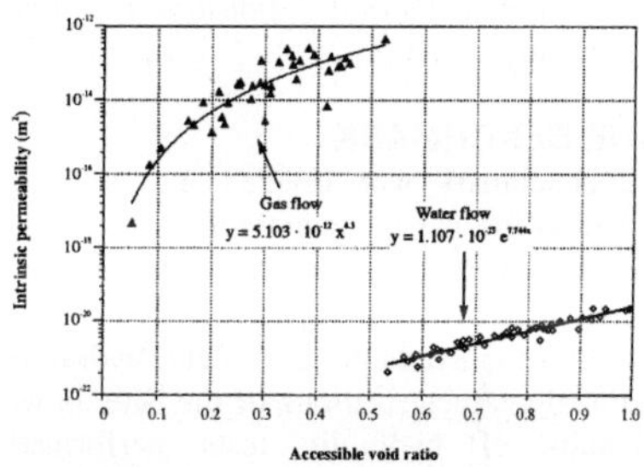

Figure 10. Intrinsic permeability of the compacted clay obtained from saturated water flow and from unsaturated gas flow tests. In gas flow tests, the accessible void ratio indicates the ratio between gas accessible pore volume and particle volume (e (1-S_r)).

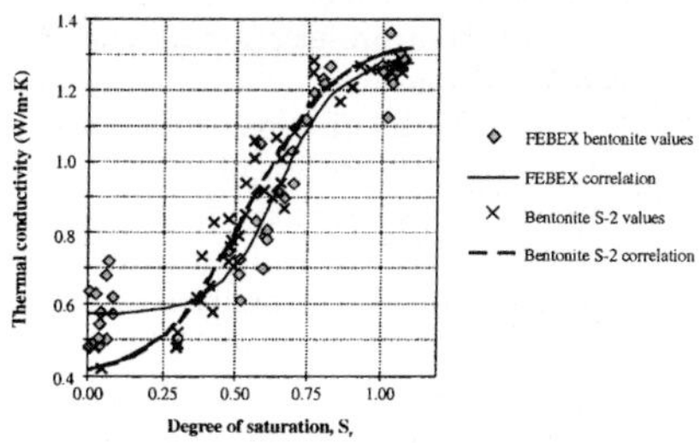

Figure 11. Thermal conductivity as a function of degree of saturation.

Measured thermal expansions for heating and cooling paths are given in Figure 12

More sophisticated suction controlled tests which explore the behaviour of the bentonite under specific stress and suction paths are described in the references given above. Tests on small scale cells involving simultaneous hydration and heating were also performed. They are boundary value problems and their analysis may provide a refined evaluation of constitutive parameters. Some of the research groups participating in the FEBEX benchmark test have used this information to their advantage. Their analysis is published elsewhere.

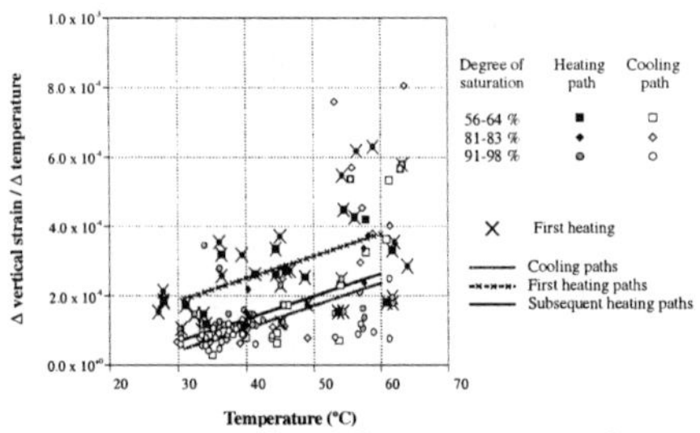

Figure 12. Linear thermal expansion as a function of temperature.

2. THE BENCHMARK

The benchmark was divided into three parts, described as follows in general terms:

- Part A: Hydro-mechanical modelling of the rock.

Based on the available geological, hydraulic and mechanical characterizations of the Site as well as on results of hydraulic tests performed on boreholes, a hydro-mechanical model for the zone around the FEBEX tunnel was to be prepared. Using this model, changes in water pressure induced by the boring of the FEBEX tunnel in the near vicinity, as well as the total water flow rate to the excavated tunnel was required.

- Part B: Thermo-hydro-mechanical analysis of the bentonite behaviour.

Based on the characterization of the bentonite and on the details of the process of test installation, a thermo-hydro-mechanical model for the bentonite barrier and the heaters was to be prepared. Using this model, the thermo-hydro-mechanical response of the bentonite barrier as a result of the heat released by the heaters and the hydration from the host rock was required. Local field variables such as temperature, relative humidity, pore water pressure, stresses and displacements, as well as global variables such as total input power to the heaters was required.

- Part C: Thermo-hydro-mechanical analysis of the rock.

Based on the characterization of the rock massif and on the details of the process of test installation and performance, the rock response in the immediate vicinity of the buffer was required. The rock is now subjected to the heat released by heaters and to swelling pressures resulting from bentonite hydration. The initial hydrological regime (Part A) is also modified by the presence of the impervious barrier. Temperature, stresses, water pressures and displacements in selected points of the rock were required.

A maximum number of eleven modelling teams have participated in the different benchmark activities related to the FEBEX test. Their names, codes and symbols used in the presentation of results are given in Table 3.

Table 3. Codes, symbols and colours assigned to participants and co-ordinator.(F.O: Funding Organization)

F. O.	*code*	*symbol*	*color*
ANDRA	*ANG*	■	*red*
ANDRA	*ANN*	□	*red*
BGR	*BGR*	◆	*green*
CNSC	*CNS*	◇	*green*
US DOE	*DOE*	▲	*blue*
IRSN	*IPS*	△	*Blue*
JNC	*JNC*	⊠	*Brown*
US NRC	*NRC*	✕	*Brown*
SKB	*SKB*	⊞	*Black*
SKI	*SKI*	+	*Black*
STUK	*STU*	●	*orange*
Coord	*UPC*	○	*orange*

Details on the modeling approaches used by the different participants in the Benchmark are described in companion papers to this Symposium. The performance of the test and a comparison of calculated and most significant measured variables are described in the next three sections. The Benchmark was conceived as a blind prediction exercise. The graphs presented include the prediction made by different teams. Some of the participants performed later additional analysis once the actual field data was officially released.

3. PART A: HYDRO-MECHANICAL MODELLING OF THE ROCK

Although the FEBEX experiment concentrates on the behavior of the bentonite buffer around waste canisters (Parts B and C address this fundamental aspect of the test) it turned out that some of the observations made during the excavation of the tunnel and immediately afterwards provided an interesting large scale experiment of a hydro-mechanical nature. Two types of measurements have been selected to develop the modelling exercise: the actual water inflow rates into the FEBEX tunnel and the water pressure response in the vicinity of the tunnel outer perimeter against the tunnel excavation by a Tunnel Boring Machine

Flow measurements into the tunnel provide an integrated variable, controlled by the problem geometry, rock mass fracture pattern, fracture anisotropy, rock matrix permeability and boundary conditions.

The second part asked for a prediction of the transient changes in water pressure recorded at two borehole intervals in the close proximity of the advancing tunnel. The pore water pressure record exhibited a marked transient behaviour when the tunnel face was close to the measuring section of the borehole. This transient behaviour was characterized (in one of the measuring segments) by a rather sharp increase in pressure followed by a slow decay as the tunnel moved away from the measuring section.

This behavior is related to the interaction between rock deformation and water pressure and provides an interesting record of hydro-mechanical interaction in the saturated granitic rock mass. Figure 13 is a Plan view of the Febex tunnel and the Borehole FEBEX 95.002 where observations were made in the intervals P3 and P4.

Total water inflow in the test zone (which extends along 17.40 m along tunnel axis, from coordinates 54.00 to 71.40) was measured by two techniques (absorbing pads on selected points of the tunnel wall and small gauge measuring of overall leaked water) at different dates in the period January-May 1996, once the tunnel was fully excavated.

The first technique involved discrete measurements at selected points on the FEBEX tunnel by means of absorbing pads. The absorbing pads were weighted before and after their placement in order to determine the volume of leaked water. With this information, it was possible to know the distribution of water input flow on the wall of the FEBEX tunnel.

The surface of the test zone of the FEBEX tunnel was divided into three types of zones: (1) the granite matrix, (2) fracture zones and (3) well-defined water inflow points. The total water inflow into the FEBEX tunnel was estimated to be 1.30×10^{-7} m^3/s (about 7.8 ml/min). The total water inflow is made of the contributions coming from: (1) the matrix with an inflow of 3.51×10^{-8} m^3/s, (2) the fractures with an inflow of 2.79×10^{-8} m^3/s and (3) the well-identified points with an inflow of 6.67×10^{-8} m^3/s. From these results, it may be concluded that: (1) about the 27% of the water inflow is through the matrix (this flow, usually not considered, is important in this case) and (2) about the 51% of the water inflow is through well-identified points (conventional methods measure typically this flow).

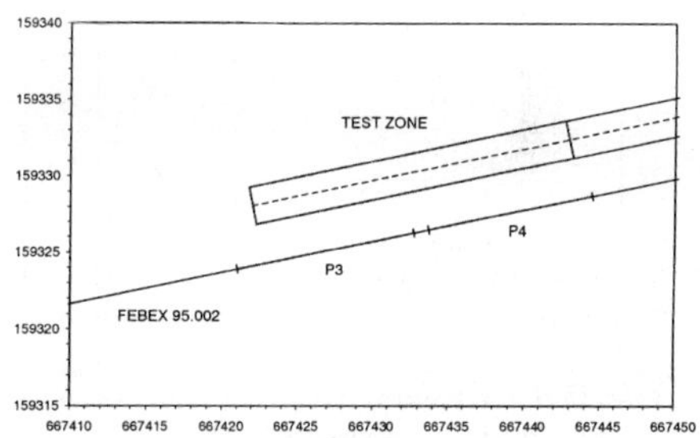

Figure 13. Plan view of the test zone of the FEBEX drift and the borehole FEBEX 95.002, showing the intervals P3 and P4. GTS coordinates (in m) are used (north is parallel to the y-axis).

Table 4. Predictions by participants at day 100. (*) Predictions by ANG refer to day 45.

Predictions	*Q (ml/min)*
ANG	*(*) 0.13*
ANN	*6.45*
CNS	*10.10*
DOE	*10.10*
IPS	*34.53*
JNC	*0.39*
SKB	*7.01*
SKI	*6.94*

The comparison between measured and predicted pore water pressure records will be presented in two plots (Figures 14a and 14b).

Pressure (in MPa) and tunnel drilling advance (in m) are plotted as a function of time. The position of the measuring interval (P4) along borehole FEBEX 95.002 is indicated by a vertical segment, which spans the appropriate tunnel metering. Predictions of modelling teams are indicated with the corresponding acronyms and selected format.

Pressure readings at P4 showed a fast increase in pore water pressure directly connected with the periods of tunnel drilling. The periods of drilling inactivity (night shifts, weekends can be identified) were immediately reflected in a transient decay in pore water pressure. Once the tunnel face had gone beyond the P4 interval, a progressive decay of pore water pressure was monitored.

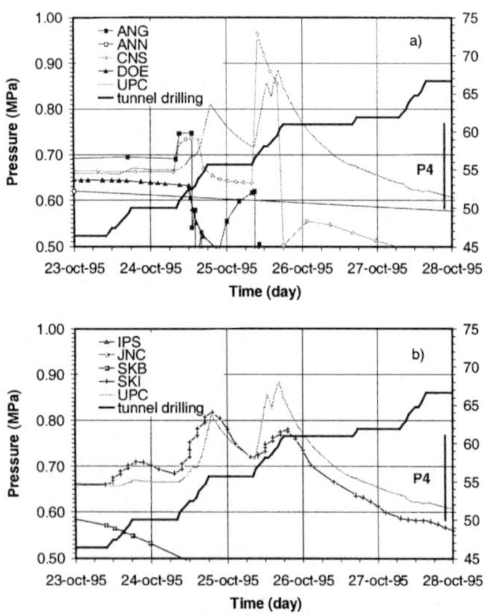

Figure 14a, b. Water Pressure Evolution (borehole FEBEX 95.002, interval P4).

The patterns of prediction are very different from team to team. These differences are explained by the nature of the models being used and other computational details, such as the orientation and intensity of assumed initial stress, which is a fundamental issue in this case.

Some computational models (such as ANN, DOE, IPS, JNC and SKB) do not show any increase in pore water pressure. ANG predicts sharp transient responses for a relatively extended period of time but their signs are the opposite of actual measurements: predicted pressure decrease during excavation times and increase during periods of inactivity. CNS predicts transient changes in pore water which follow the measured trends, although their model predicts a rapid dissipation and negative values of pore pressures once the tunnel goes beyond section P4. SKI presents more accurate predictions, although the selected "in situ" stress field does not correspond with available information.

The long term reaction of the pressure measured in the borehole section P4 is to show a steady decrease since the excavation of the tunnel implies a neighboring boundary at a given relative humidity (the RH prevailing in the FEBEX tunnel prior to the buffer and heater installation). Most of the models show this trend although the rates of water pressure decay may change.

3.1. Conclusions for Part A

Widely different models for water inflow were used. Some teams (ANN, CNS, and SKB) used uncoupled hydraulic transient models to solve the first part of the exercise, whereas others (ANG, DOE, SKI) used a coupled HM modelling. It does not seem that the mechanical coupling introduces any advantage in this case. In fact, the reason for some of the better predictions (such as SKB calculation) may be associated with previous calibration of the model using other hydraulic data in the same area. Some models describe water circulation in the rock by means of discrete features (tubes, channels such as ANN) equivalent porous medium for different zones (such as DOE, SKI) and others combine porous medium and discrete fractures (ANG, CNS, SKB). Again the overall results do not show a particular advantage of a given conceptualisation. Some of the calculations (such as SKI) provide the proportion of flow rates attributed to different origins (matrix, fracture zones). The distribution of measured water inflow rates along the tunnel axis provides a first approximation to the relative proportions of flow through discrete fractures/shear zones and the matrix or distributed flow (Figure 15).

Pore water pressure changes in the vicinity of the tunnel excavation are a direct consequence of changes in the volumetric strain of the rock. Later, pore water pressure dissipations are a consequence of the transition flow towards a new equilibrium, which now has a modified boundary condition (the tunnel surface) in the vicinity. Therefore, fully coupled hydro-mechanical analyses are required to try to capture actual measurement. In fact, one-way coupling (hydraulic parameters updated as the rock mass deforms) is not capable of reproducing the observed behaviour.

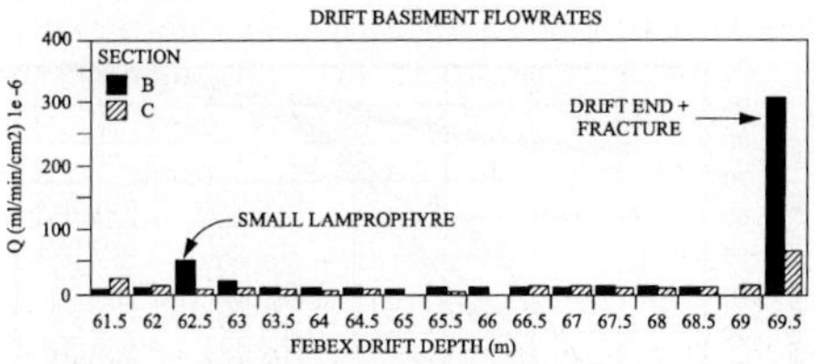

Figure 15. Results of inflow measurements in part of FEBEX test area arranged in rows (sections) and their relationships to geological structures (Guimerà et al, 1998).

However, the case has demonstrated that even if a fully HM coupled model is used, the difficulties to capture the actual pore pressure of the granitic mass are very high. It was well established that the volumetric behaviour of the rock in the vicinity of the tunnel depends critically on two aspects: the orientation and the intensity of the initial stress field. "In situ" stresses show often a large variability. Field determinations at Grimsel suggest that the major principal stress at the location of the FEBEX tunnel is horizontal (around 30 MPa), whereas the minor principal stress may be considered vertical and defined by geostatic conditions (around 10 MPa). The intermediate principal stress, also horizontal, may reach intermediate (around 15 MPa) values, but remains substantially higher than the vertical stress. It was shown that this particular distribution of initial stress leads to results which are opposite in trend to the actual measurements (dilation of the rock, instead of compression is computed at the P4 locations). In order to match the actual measurements, changes in the intensity of the vertical stress and on the direction of principal horizontal stresses had to be introduced. Moreover, the same initial stress field does not seem to be valid to reproduce results at P3 and P4. It should be added that local conditions at P3 and P4 do not seem to be the same since a more previous zone (which reduces the trend for a rapid initial increase in pore pressure) is present in P3. The finite length of the measuring intervals allows also an easy connectivity between pervious and impervious zones.

4. PART B: THERMO-HYDRO-MECHANICAL ANALYSIS OF THE BENTONITE BEHAVIOUR.

The performance of the test and the comparison with modeler's predictions will be based on the following variables:

- The evolution of the total heating power of one of the heaters. The starting time will be the "day 53" which corresponds to the beginning of the automatic control, once the hottest point at the contact between the heater and the bentonite has reached 100ºC. The heating power is a global performance variable which integrates the changing thermal conduction coefficient of the bentonite as the barrier experiences water content changes
- The distribution and evolution of relative humidity of the bentonite barrier. This is a key variable of the experiment since it shows the effects of inner heating and outer hydration from a host rock in a direct manner. Relative humidity variations are the result of water transport processes in liquid and vapour form. Two locations have been selected to show the radial distribution of RH: Section E1 (see Figure 6), which affects the heater directly, and section H that is a centered section located between the two heaters. The evolution in time of RH is examined in three points at increasing radial distance (in both sections E1 and H): a point close to the heater, "H", a mid point, "C" and a point close to the granite wall, "G".
- Field information on temperature is vast but it offers limited information on other relevant physical processes and it is essentially controlled by conduction phenomena. Couplings do not affect temperature in a significant way and experience indicates that predictions with most models lead to accurate results. Therefore, the comparison between measurements and temperature predictions will be made in one section (D1) in radial direction.
- Radial stress evolution at some points located at different radial distances in section E2. Measurement errors are often likely in the case of (normal) stresses. In addition, heterogeneous stress distributions and large variations over short distances are possible. Nevertheless the cells installed offered also an interesting field information to the development of swelling-induced stresses inside the barrier

Figure 16 shows the evolution of measured input electrical power on heater 1 and the set of predictions

The measured power decreases steadily with time for the first 400 days after the beginning of the automatic operation. This is consistent with the progressive drying of the inner annulus of the barrier and the associated decrease of thermal conductivity of the bentonite. The slight increase in power in the second part of the period represented is attributed to the progressive hydration of the barrier due to the incoming water. Some predictions reproduce accurately the observed behavior.

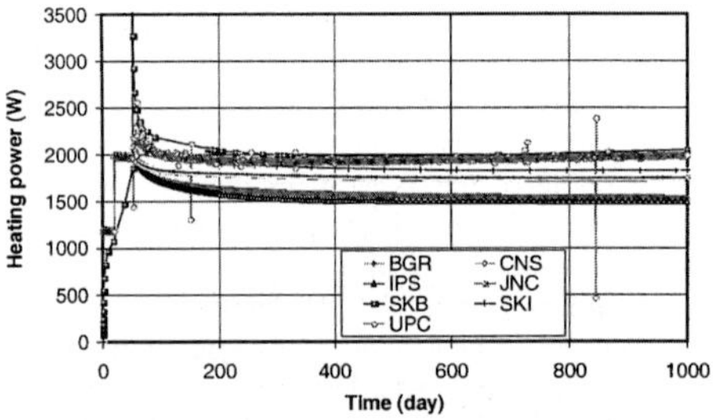

Figure 16. Measured and predicted evolution of heating power. Heater 1.

The variation of relative humidity with radial distance in Section E1 is shown, in Figure 17a, for t=1000days. Measured data of four radial directions (two vertical, two horizontal) are lumped together in the figure. The heater directly affects section E1. The measured data does not show a clear preferential direction of hydration. The test seems to have a cylindrical symmetry. This is an interesting observation since a lamprophyre dyke, that provided a non-uniform inflow of water into the tunnel, diagonally crossed Section E1. It seems that the low permeability of the saturated bentonite, ($1{\cdot}10^{-21}m^2$) compared with the granite matrix permeability ($5{\cdot}10^{-19}m^2$), leads to a fairly homogeneous hydration of the barrier. Some of the predictions capture well the distribution of RH. Others have difficulties to handle the change of phase of water at the hot areas and predictions are far from reality. In one case (SKI) the 3D analysis performed leads to small variations of relative humidity with the different radial orientations, a result that may also be supported by the experimental evidence.

A similar plot is given in Figure 17b for Section H, centered between the two heaters and, therefore, not directly affected by the heat released by them. This section is obviously wetter. Predictions require now a 2D axisymmetric (or 3D) geometry. Two of the models predict quite accurately the experimental data.

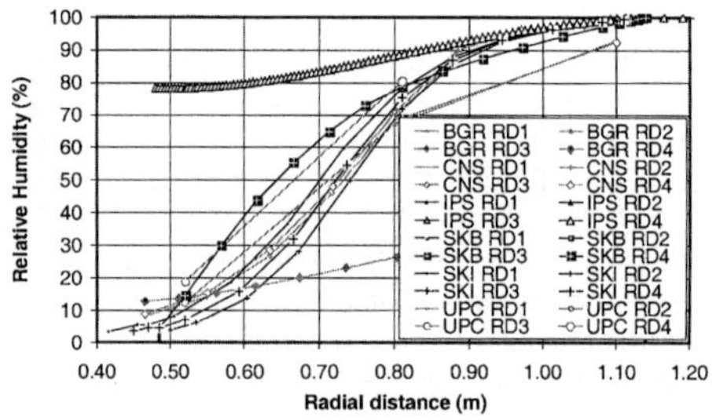

Figure 17a. Variation of relative humidity with radial distance in Section E1 for t=1000days

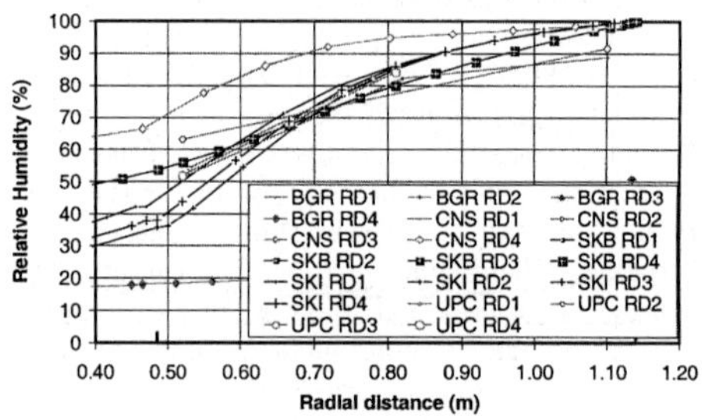

Figure 17b. Variation of relative humidity with radial distance in Section H for t=1000days

Evolution plots of relative humidity at some selected points (H, C and G) in Figures 18 and 19. Figure 18 corresponds to section E1, directly affected by the heating action of heater 1. Measurements exhibit a very distinctive pattern. The point close to the granite (G) becomes rapidly saturated. The point in the bentonite close to the heater (H) experiences an intense desiccation. The initial RH decreases to values as low as 10% (this is equivalent to a very high suction: more than 300MPa) 200-300days after the beginning of heating. Later, the humidity increases at a slow, progressive rate. The point in the center (C) displays a more complex behavior: it becomes initially wetter, reaches a peak value of RH, but partially dries immediately afterwards. These transient phenomena take place during the first 80 days of heating. Later, a progressive hydration is recorded. This behavior is explained by the role of water vapor and phase changes in water transfer inside the barrier. Water evaporates in areas of the bentonite close to the heater. Hot vapor in high concentration migrates radially outward and becomes eventually cooler. Then it condensates. This explains the peak in RH observed in Figure 18. But this is an expanding process. As the temperature increases inside the barrier, this phenomenon moves in an outward direction and the inner points of the barrier become drier again. At higher times the water inflow (in liquid phase) from the outer zones eventually dominates the barrier hydration. The model results reproduced in Figure 18b capture correctly this phenomenon. The models represented in Figure 18a find more difficulties. These difficulties are associated with incomplete formulations of the phase change of water and the transportation of vapor. One thousand days after the beginning of heating (100ºC) the barrier is still far from saturation.

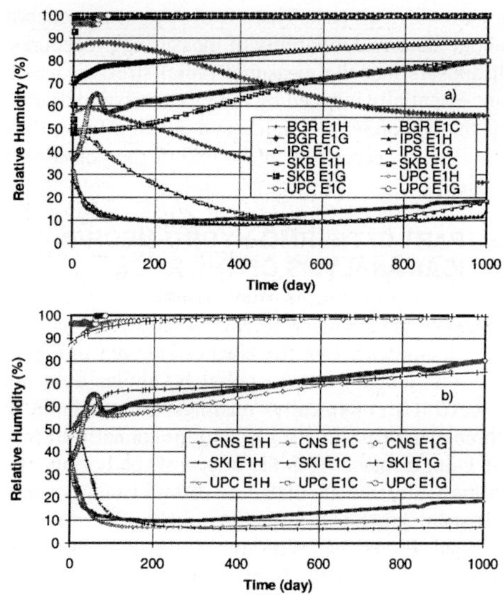

Figure 18a, b. Measured and predicted variation of RH at three radial distances in section E1

Figure 19 presents equivalent results for the central section H. A more complex initial transient is measured at the inner point (H) in this case. The central and outer point exhibits a progressive hydration from the early stages. One of the model predictions in Figure 19b (SKI) is remarkably accurate. The three additional models represented in Figure 19 meet more difficulties.

Only one figure regarding temperature distribution across the barrier is included here (Figure 20). At the particular location shown (section D1 at the edge of Heater 1) temperature decreases from a value above 80ºC at the heater to 30-35ºC at the granite contact. The lower figure (30ºC) corresponds to 90 days of heating and the upper one (35ºC) to 1000 days of heating. All the predictions follow the measured decay with reasonable accuracy. However, when comparing extreme predictions for a common time (1000 days) a maximum difference of 27ºC is found.

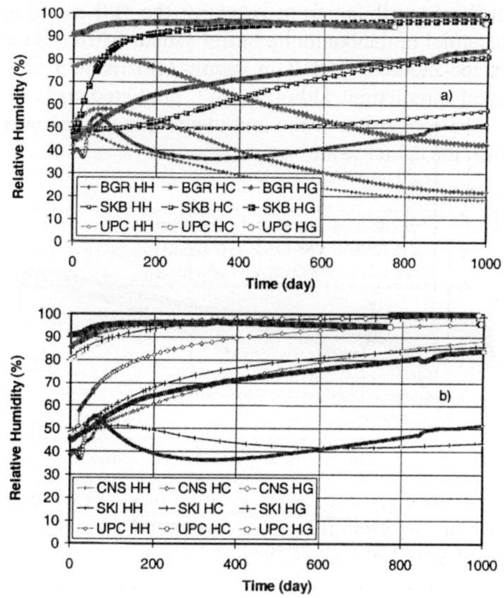

Figure 19a, b. Measured and predicted variation of RH at three radial distances in section H

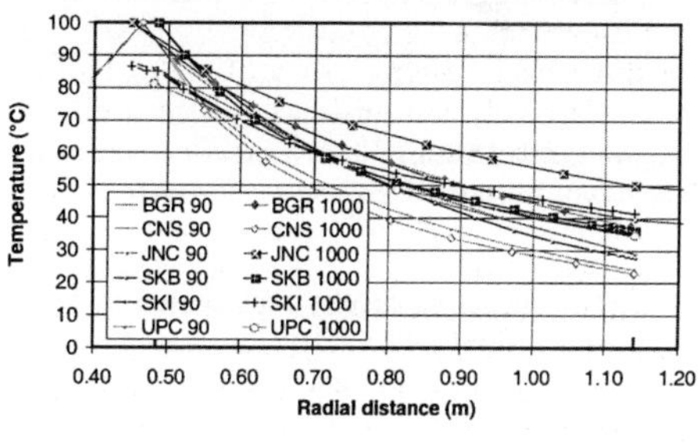

Figure 20. Measured and calculated temperature variation across the barrier, at section D1, for two times: 90 and 1000days.

Measured radial stresses are given in Figures 21a, b. Two of the selected measurement cells (G1 and G2) are at the contact between the bentonite blocks and the granite wall. The third one (H1) is closer to the heater. The outer cells develop consistently stronger radial stresses than the inner one during the considered time span. Stress build-up is faster in one of the G cells but the recorded radial stress in both cells is similar and close to 3.5 MPa at the end of the 1000 days period. The inner cell, however, shows a slower and steady development of the confining stress. The recorded value at the end of the period has not reached 1 MPa. It is clear that the buffer is still far from reaching the full swelling potential of the bentonite buffer (about 6 MPa). One of the modelers (SKB in figure 21a) made a very good prediction although the predicted rate of increase of stress at the end of the period is slower than the actual value.

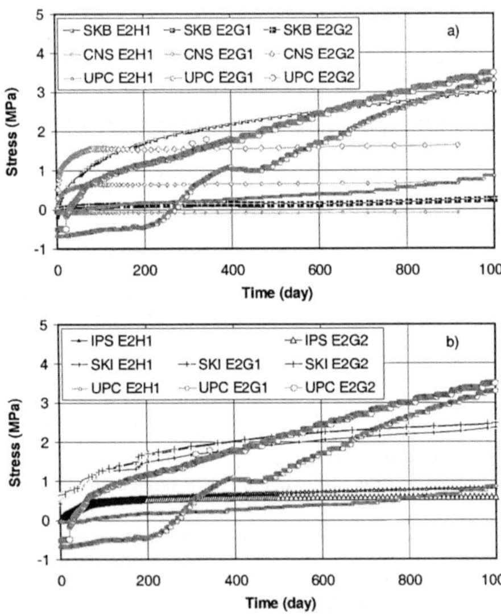

Figure 21a, b. Measured and calculated evolution of radial stresses at the three points indicated in section E2.

The two modeling results represented in Figure 21b predict a similar stress development in the three points considered. Both underpredict, to a different degree, the actual measurements, especially at later times

4.1. Conclusions for Part B

Only a reduced number of modeling teams participated in the blind prediction of Part B. As shown in the comparisons of RH and stress variables only three teams (CNS, SKB and SKI) were able to provide predictions for the full exercise. Modeling teams who used a 1D coupled model (IPS) provided approximate results for cases, which could be approximated by a radial symmetry. This is the case of sections normal to the test axis at the center of each one of the heaters. Models prepared to solve only the Thermo-hydraulic part of the problem (BGR) could not provide predictions for stress development. In some of the models, (BGR, IPS) phase change and vapor transfer was not considered and this limitation hampered the correct reproduction of measured variables. In fact, vapor transfer plays a dominant role in the early stages of the test.

The three fully coupled models (CNS, SKI, and SKB) behaved in general terms in a quite satisfactory manner. They predicted quite accurately the evolution of relative humidities inside the barrier. Stress prediction, however, has proved to be a more difficult task. There is always some concern about the actual reliability of measuring procedures. It appears that the measured radial stresses, which are essentially induced by the progressive hydration of the bentonite, are higher and develop faster than predictions, especially at the end of the considered period.

5. PART C: THERMO-HYDRO-MECHANICAL ANALYSIS OF THE ROCK

Several instruments were located, at increasing depth, in the auxiliary boreholes shown in Figure 1. An enlarged view of the borings is given in Figure 22. The length of these radial boreholes does not exceed 15m. Typically, readings are available in three or four positions: close to the tunnel wall (say at 1-3m distance from the origin of the boring, one or two intermediate distances and a distant position (13-14m deep)

Data is available on the following variables: temperature, water pressure in rock, water pressure in packers, stress state and radial displacements.

The Cartesian co-ordinate system shown in Figure 22 is used to orientate and locate the measuring devices. The origin is located at the intersection of the tunnel axis and the contact plane between the concrete plug and the bentonite buffer. Positive X axis is directed along the tunnel axis towards the other end of the test section. The Z-axis is vertical, pointing upwards and the Y-axis is perpendicular to the (XZ) plane in the position indicated in Figure 22.

Water pressure was measured in some boring intervals (a few meters long) separated by packers. The rubber packers were also filled with water and their pressure was maintained and measured externally. Water pressure gauges are located in the measuring area of the FEBEX tunnel. They are connected with the measuring intervals by means of steel tubing.

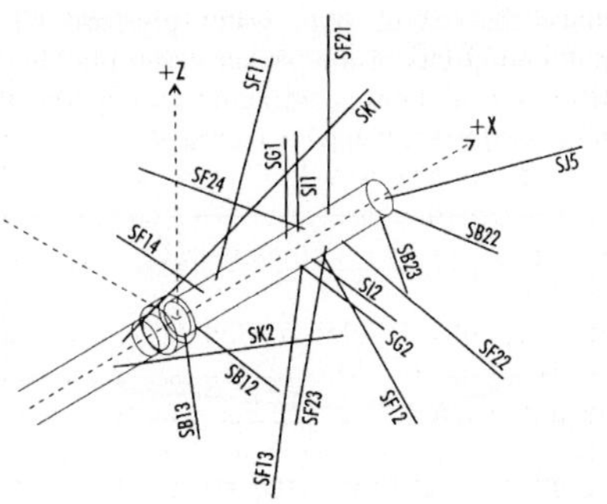

Figure 22. Auxiliary instrumentation boreholes around FEBEX test section

Three normal components of stress are measured by means of total pressure cells. Four sets of 3 cells were prepared and grouted "in situ" in boreholes SG1 and SG2. Each cell had five sensors oriented in different directions and fixed on a common support 2 m long. Each one of the sensors was a circular steel flat cell. An interpretation of the five readings provides the normal stress components in three directions: radial (with respect to the tunnel axis) (σ_r), axial (along the direction of tunnel axis) (σ_x) and circumferential (normal to the radial direction (σ_θ). Once the sets were in position the borehole was filled with a slightly expansive mortar. Once the mortar was cured, cells were pressurized against the mass of surrounding mortar to guarantee a good initial contact. Note that this procedure provides increments of stress over the moment of cell installation. It will record therefore stress increments due to the performance of the test (temperature effects, modification of pore water pressures, swelling of the bentonite)

Radial displacements were measured by means of borehole extensometers installed in borings SI1 and SI2. They were located close to the position of the stress cells. Each extensometer consists of graphite rods with independent anchoring points located at a depth of 1.0, 3.0 and 7.0 m into the borehole.

Measured and computed evolutions of temperature in Boring SF 21, at a radial distance of 1.20m are shown in Figure 23. Most predictions are quite accurate. Two of them, however, depart in a significant way. Prediction of temperature depends essentially on a correct estimation of thermal conduction coefficients and boundary conditions.

Water pressures are represented in Figure 24a, b. Both figures show evolutions of water pressures recorded in Borehole SF 21 at radial distances of 3m (Figure 24a) and 13.6m (Figure 24b). These radial distances locate the center point of the measuring intervals, which have lengths of 2.34 m and 2.82 m respectively.

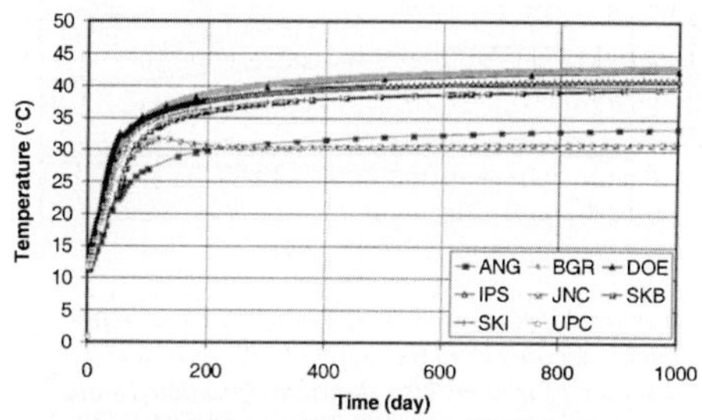

Figure 23. Measured and predicted evolution of temperature in borehole SF 21 at a radial distance of 1.20m

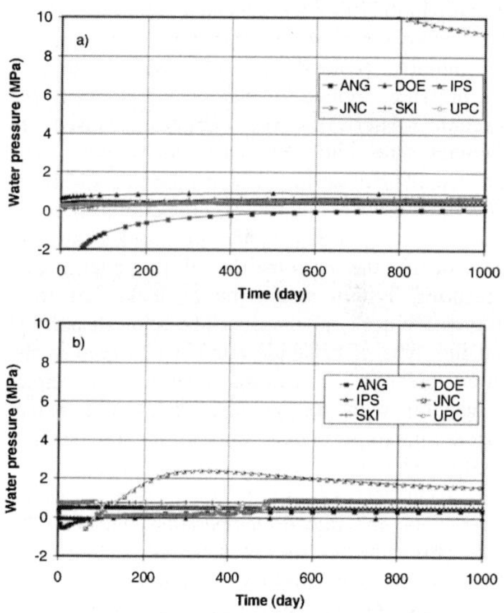

Figure 24a, b. Measured and predicted evolution of water pressures in borehole SF 21 at radial distances of 3.03 m and 13.6 m respectively.

Measured water pressures close to the tunnel wall (data in Figure 24a) increase slowly with time from an initial value close to 0.5MPa to a value in the vicinity of 0.7MPa, 1000day after the beginning of heating. Some of the predictions made are very good and reproduce the absolute measured values and the increasing trend (IPS, DOE, and SKI). Measured water pressures are stable and close to 1MPa at the more distant point (Figure 24b). A very good prediction is made by SKI.

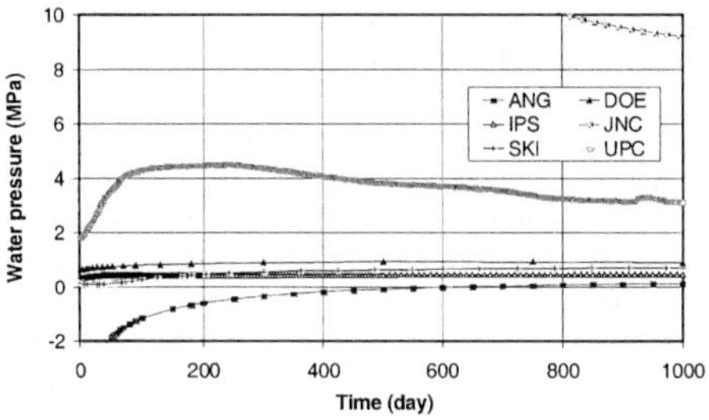

Figure 25. Measured evolution of pressure in first packer located in radial borehole SF 21 at a radial distance of 1.86 m. Also shown are predicted water pressures for a radial distance of 3.03 m.

It is interesting at this point to observe the data of the packer pressures. The evolution of water pressure in the first packer located in borehole SF21 at a radial distance of 1.86 m is shown in Figure 25. The initially recorded water pressure is related to the pressure applied to the packer to seal the borehole. Afterwards the circuit is closed, to maintain constant volume conditions. The substantial increase in pressure observed in Figure 25 is most probably due to the temperature-induced dilation of the water inside the packer system. Presumably, the deformation of the packer and measuring system and/or minor leaks led to a progressive decay afterwards. The relevant point is that this type of behavior was not observed in the water pressure measuring intervals in direct connection with the granitic rock. A transient increase in temperature of a porous medium such as a saturated rock induces a transient increase and subsequent decrease in pore water pressure, which depends on the rate of temperature increase, the rock permeability, the rock porosity, the rock stiffness and the geometry (drainage conditions). Water pressure increases because water dilates more than the rock skeleton. In parallel, a drainage process reduces excess pore pressures towards zero. The higher the porosity, the higher peak pore water pressure observed. The lower the permeability and the stiffness of the rock the higher the pore water pressure peak and the slower the subsequent dissipation. The fact that no transient excess pore water pressures were detected in the rock is probably due to the relatively high permeability of the rock: Excess pore water pressures dissipate immediately after they are induced. In addition, recorded water pressures are average values over the measuring interval. This effect contributes also to reduce the excess pore water pressures. The different conditions of the packer system (a "high" porosity and a rigid confinement) explain the completely different behavior recorded.

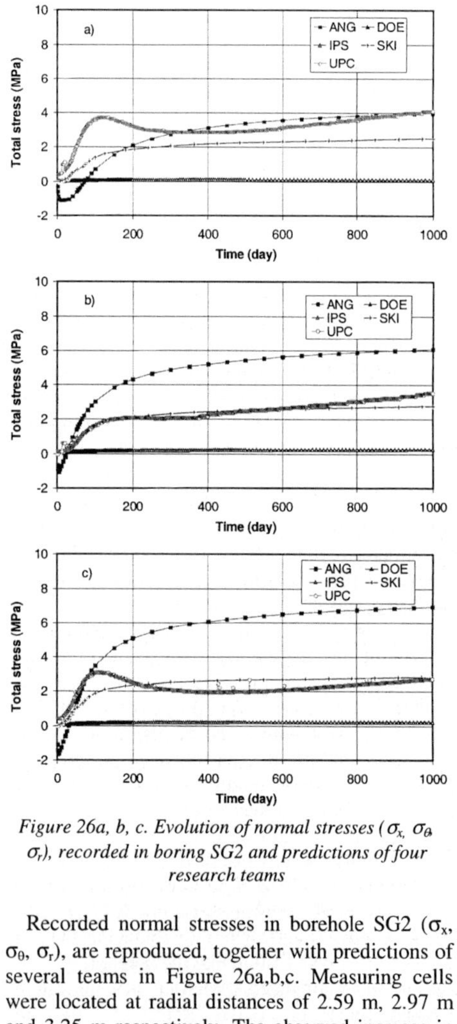

Figure 26a, b, c. Evolution of normal stresses (σ_x, σ_θ, σ_r), recorded in boring SG2 and predictions of four research teams

Recorded normal stresses in borehole SG2 (σ_x, σ_θ, σ_r), are reproduced, together with predictions of several teams in Figure 26a,b,c. Measuring cells were located at radial distances of 2.59 m, 2.97 m and 3.25 m respectively. The observed increase in total stress is due to differential rock dilation effects induced by the varying temperature field and by the increasing swelling pressure at the tunnel wall.

Recorded values at a larger radial distance (r=7.10-7.80 m), not shown here, were significantly lower. Interestingly, a peak is recorded at early stages in all the three stress components, followed by a transient decay and a progressive increase at later dates. The stress cells measure total stress and the observed behavior is consistent with the expected change in pore water pressures, explained before. The stress predictions shown in the figures do not reproduce the measured transient, however. Some of the calculated values (especially for SKI) are quite close to actual absolute values.

Radial displacements measured in one of the extensometer rods installed are shown in Figure 27. The extensometers measure the relative motion of the anchoring point with respect to the tunnel wall. The data in Figure 27 refers to a 4m long bar. Measured displacements (away from the tunnel wall) are very small, in the order of 0.1 mm. They increase rapidly at early stages due to the stressing and dilation of the rock and remain essentially constant for the rest of the measuring period (despite the increasing swelling pressure; probably the increase in radial stress at the tunnel boundary has a very limited effect). Some of the predictions are quite accurate (IPS, SKI). Some uncertainties remain, however, because the measured values are close to the precision of the measuring system.

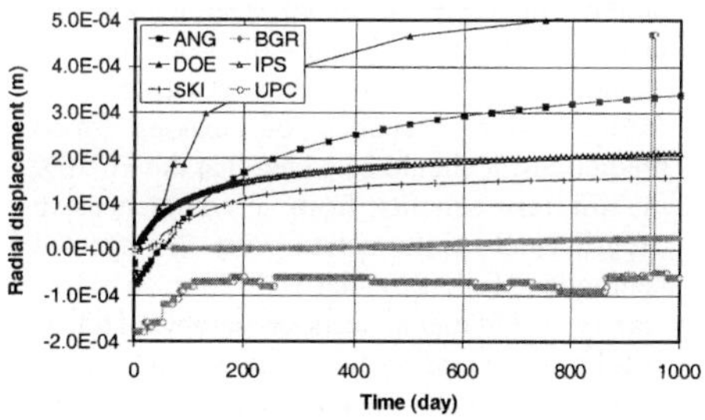

Figure 27. Evolution of radial displacements recorded in boring SI1, extensometer fixed at a radial distance of 8m and predictions of research teams

5.1. Conclusions for Part C

As in Part B, only a reduced number of modelling teams provided blind predictions for the rock behavior, once the expansive bentonite barrier was in place. Coupled THM models are also required for this part of the Benchmark although the temperature increase plays a dominant effect on the rock behavior. As it is frequently the case, temperature changes are well reproduced in general terms. Rock water pressures development integrates two separate phenomena: the modification of the hydrogeological regime in the vicinity of the tunnel due to the presence of the barrier and the temperature effects. Temperature effects, in turn, depend on a number of rock properties: rock dilation coefficient, porosity, stiffnes and permeability. The actual development of excess pore pressures are additionally controlled by the rate of temperature change and by the general drainage conditions in the area. It has been suggested that the limited transient reaction of the pore water pressure in the Grimsel host rock is a natural consequence of the high permeability of the rock and, to a lesser extent, of the averaging effect of the measuring interval (a few meters of borehole). The records of packer water pressures have provided interesting complementary evidence of the transient pore water pressure development. Also, the evolution of total stresses, 3 m away from the tunnel wall, shows a transient initial peak which has also been attributed to excess pore water pressure behavior.

Long-term water pressures increase slowly with time in the tunnel immediate vicinity (a few meters). Measured water pressures after 1000 days of test operation are, however, relatively small (1MPa).

A calculation of the evolution of effective mean stress and the octahedral shear stress in borehole SG2 at a radial distance r= 3m has been made. Pore pressures were approximated by data collected in neighbouring boreholes. It was also assumed that normal stresses on the three normal planes are principal stresses. The results are given in Table 5

Table 5. Mean effective and octahedral shear stresses recorded at a radial distance of 3m in borehole SG2

Time (days)	*Incremental mean effective stress (MPa)*	*Incremental octahedral shear stress (MPa)*	*Incremental max. shear stress (MPa)*
0	*0*	*0*	*0*
100	*2.56*	*0.82*	*1*
300	*1.96*	*0.37*	*0.4*
700	*2.26*	*0.53*	*0.65*
1000	*2.8*	*0.53*	*0.65*

Note that the values given are increments over the initial state of stress once the FEBEX tunnel was open. They represent therefore the combined effect of temperature effects (the dominant phenomenon), the generation and dissipation of pore pressures and the swelling pressure effects. At the radial distance of 3 m the increments of mean effective stress dominate. Incremental shear stresses tend to be a fraction of the incremental mean stresses. The maximum difference of effective normal stresses is slightly higher, as shown also in the same table. Given the expected strength properties of the granite, the effective stress path corresponding to these values separates from the failure envelope. It is also interesting to note that the worst condition (in terms of proximity to a failure envelope) is met when total stresses reach a peak (at t=100days approximately). At later times mean effective stresses dominate.

Very small incremental displacements were recorded in the 1000-day period (a tenth of a millimetre in 4 and 8m long intervals).

Rock water pressures were reasonably well predicted by three of the research teams (IPS, DOE, and SKI). More limited success was achieved in the prediction of stresses and displacements, with the exception of SKI.

6. CONCLUDING REMARKS

The FEBEX test is one of the few large-scale tests available to gain an integrated perspective of the behaviour of current concepts for nuclear waste disposal in crystalline rock. The comprehensive instrumentation installed in the rock and in the compacted bentonite buffer has yielded vast amounts of data over the past six years. Part of this data, the data corresponding to the first three years of heating, has been used to conduct a Benchmark exercise to evaluate the capabilities of a number of finite element codes developed to handle coupled problems in geological and porous media. This paper provides an account of the main results achieved during the performance of the exercise. Some selection of the large amount of results has been unavoidable. A description of the hypotheses and specific features of the different codes participating is also outside the scope of this paper. However, a few companion papers provide a detailed insight into some of the models and computer tools participating in the Benchmark.

For the purposes of the organization of the exercise into specific tasks the Benchmark was divided into three main parts: A Rock behaviour during the excavation of the FEBEX tunnel, B Buffer behaviour and C Rock behaviour during the heating and (partial) hydration of the buffer. This distribution has been maintained in the paper.

Specific conclusions for each of the mentioned parts have been given before. Only a few concluding remarks will be added here:

- The best predictions of the water inflow into the excavated tunnel are found when the hydrogeological model is properly calibrated on the basis of other known flow measurements in the same area. The particular idealization of the rock mass (equivalent porous media, discrete fractures) plays a secondary role
- The development and dissipation of excess pore water pressures in the vicinity of the advancing tunnel (at the time of the FEBEX tunnel excavation) was a clear example of hydro-mechanical interaction. It was concluded that the development of pore pressures was controlled by the initial stress field state, by the rate of excavation and by the permeability and drainage properties of the granite. However, the available information on the intensity and direction of principal stresses in the area was found inconsistent with the actual measurements. The problem posed by this discrepancy was essentially unsettled since a precise determination of the initial stress state in the vicinity of the FEBEX tunnel was not available.
- Predicting the behaviour of the buffer under the combined heating and wetting actions requires a fully coupled THM formulation, which incorporates all the necessary physical processes controlling the bentonite behaviour. Only a partial set of codes could offer the required features. Particularly relevant to predict the early stages of heating was the inclusion of phase changes of water and the vapour transport. Codes incorporating these features were capable of making good predictions. It should be added that the FEBEX in situ test benefits from a comprehensive experimental information on compacted bentonite properties derived from a large variety of laboratory tests on samples and on small-scale hydration and heating cells.
- It has been shown that the hydration of the bentonite buffer was essentially independent of the heterogeneous nature of the rock hydraulic conductivity features. This is explained by the fact that the rock matrix permeability is higher than the saturated bentonite permeability. Some 3D analyses performed, where the heterogeneous permeability features of the rock have been included, tend to support also this conclusion.
- The heating of the rock resulted in a significant increase in rock stresses in the vicinity of the FEBEX tunnel. Water pressures remained however essentially unchanged. The relatively

high rock permeability explains the absence of significant pore water pressure transients. Only one of the participating modelling teams was capable of achieving a consistent prediction of all the measured variables in the rock: temperature, water pressures, rock stresses and radial displacements

7. ACKNOWLEDGEMENTS

The authors wish to acknowledge and thank the support provided by the Research Project Decovalex III during the development of the work partially reported in this paper. Discussions with all the modelling teams participating in the benchmark exercise have been very useful to increase our understanding of the observed behaviour of the FEBEX "in situ" test. A particular appreciation is extended to ENRESA, the Spanish National Agency for Nuclear Waste Disposal, owner and manager of the FEBEX test, for his permission to release the data for the purposes of the benchmark test and for his support during the test operation and analysis. The European Community has economically supported the research activities in FEBEX by means of Projects funded within the 4th and 5th Research Framework Programs.

8. REFERENCES

Amiguet, J.L., *Grimsel Test Site. Felskennwerte von intaktem Granit. Zusammenstellung felsmechanischer Laborresultate diverser granitischer Gesteine.* NAGRA, NIB 85-08, Sep. 1985.

Bräuer, V., Kilger, B., and Pahl, A., *Grimsel Test Site. Engineering geological investigations for the interpretation of rock stress measurements and fracture flow tests.* NAGRA, NTB 88-37E, Apr. 1989.

Finsterle, S., and Pruess, K., *Solving the estimation-identification problem in two-phase flow modeling.* Water Resour. Res., 31(4), 913-924, 1995.

Guimerà, J., Ortuño, F., Vázquez, E., Ruiz, B., Martínez, L., Carrera, J., and Meier, P., *Pulse tests at "in drift" boreholes. Performance and evaluation.* UPC, 70-UPC-L-0-1001, 1996.

Guimerà, J., Carrera, J., Martínez, L., Vázquez, E., Ortuño, F., Fierz, T., Bülher, C., Vives, L., Meier, P., Medina, A., Saaltink, M., Ruiz, B., and Pardillo, J., *FEBEX Hydrological characterization and modeling.* UPC, 70-UPC-M-0-1001, 1998.

Huertas, F., Fuentes-Cantillana, J.L., Jullien, F., Rivas, P., Linares, J., Fariña, P., Ghoreychi, M., Jockwer, N., Kickmaier, W., Martinet, M.A., Samper, J., Alonso, E., and Elorza, F.S., *Full scale engineered barriers experiment for a high-level radioactive waste in crystalline host rock (FEBEX Project). Final Report.* European Commission. Report nº EUR 19147 EN, 2000.

Keusen, H.R., Ganguin, J., Schuler, P., and Buletti, M., *Grimsel Test Site. Geology.* NAGRA, NTB 87-14E, 1989.

Lloret, A., Villar, M.V., Sánchez, M., Gens, A., Pintado, X., and Alonso, E.E., *Mechanical behavior of heavily compacted bentonite under high suction changes.* Géotechnique, 53(1): 27-40, 2003.

Pahl, A., Heusermann, St., Bräuer, V. and Glöggler, W., *Grimsel Test Site. Rock stress investigations.* NAGRA, NTB 88-39E, 1989.

Pardillo, J., Campos, R., and Guimerà, J., *Caracterización geológica de la zona de ensayo FEBEX (Grimsel - Suiza).* CIEMAT, 70-IMA-M-2-01, 1997.

Pintado, X., Ledesma, A., and Lloret, A., *Back-analysis of thermohydraulic bentonite properties from laboratory tests.* Engineering Geology, 64: 91-115, 2002.

Romero, E., Villar, M.V., and Lloret, A., *Thermo-hydro-mechanical behavior of two heavily overconsolidated clays.* 6th Int. Workshop on Key Issues in Waste Isolation Research. Barcelona, 2001.

Villar, M.V., *Thermo-hydro-mechanical characterization of a bentonite from Cabo de Gata. A study applied to the use of bentonite sealing material in high level radioactive waste repositories.* PhD Dissertation and Enresa Technical Publication 01/2002. Madrid.

Villar, M.V., and Lloret, A., *Variation of the intrinsic permeability of expansive clays upon saturation: measurement with gas and water.* Clay Science for Engng. Balkema: 259-266, 2001.

Villar, M.V., Martín, P.L., Lloret, A., and Romero, E., *Second report on thermo-hydro-mechanical laboratory tests.* CIMNE-CIEMAT Report nº 70-IMA-L-0-97, 2002.

Villar, M.V., and Lloret, A., *Temperature influence on the hydromechanical behavior of a compacted bentonite.* Proc. Int. Meeting on Clays in Natural and Engineered Barriers for Radioactive Waste Confinement. Reims 2002.

Villar, M.V., and Lloret, A., *Temperature influence on mechanical behavior of a compacted bentonite.* Proc. GEPROC. Int. Conf on Coupled THMC Processes in Geosystems, 2003.

MODELLING THE RESPONSE OF THE BENTONITE IN THE FEBEX HEATER EXPERIMENT

T.S. Nguyen[1], A.P.S. Selvadurai[2] and G. Armand[3]
[1)]Canadian Nuclear Safety Commission, Canada
[2)] McGill University, Canada
[3)] formerly McGill University, Canada; currently, ANDRA, France

Abstract: The Canadian Nuclear Safety Commission (CNSC) used the finite element code FRACON to perform blind predictions of the FEBEX heater experiment. The FRACON code numerically solves the extended equations of Biot's poro-elasticity. The rock was assumed to be linearly elastic, however, the poro-elastic coefficients of variably saturated bentonite were expressed as functions of net stress and void ratio using the state surface equation obtained from suction-controlled oedometer tests. In this paper, we will summarize our approach and predictive results for the Thermo-Hydro-Mechanical response of the bentonite. It is shown that the model correctly predicts drying of the bentonite near the heaters and re-saturation near the rock interface. The evolution of temperature and the heater thermal output were reasonably well predicted by the model. The trends in the total stresses developed in the bentonite were also correctly predicted, however the absolute values were underestimated probably due to the neglect of pore pressure build-up in the rock mass.

1. INTRODUCTION

The FEBEX experiment is an in-situ experiment performed at the Grimsel site in Switzerland. The FEBEX experimental gallery is circular in cross section, with diameter 2.28 m and length 70.4 m. It was excavated at a depth of approximately 450 m in granitic rock. Two electrically powered heaters, with nominal diameter 0.9 m and length 4.54 m, were emplaced at the end of the gallery. Blocks of compacted bentonite were emplaced around the heaters to form a 17.4 m test section, isolated from the rest of the gallery by a concrete plug.

The bentonite and heaters were installed between July 1, 1996 and October 15, 1996. Heating commenced on February 27, 1997 (set at time=0), in the following three phases:

i) Constant power output of 1200 W per heater for 20 days.

ii) Constant power output of 2000 W per heater for the next 33 days

iii) Constant temperature control (between 95 to 100 °C) on the surface of the heaters from day 53.

The research teams participating in the DECOVALEX III project were asked to predict the Thermo-Hydro-Mechanical (THM) response of both the bentonite and the rock mass after the commencement of the heating (from day 0). This paper summarizes the approach and results from the CNSC team for the bentonite response.

2. GENERAL MODELLING APPROACH

Based on a review of the information (UPC, 2000 and ENRESA, 1998) provided by the UPC(Polytechnical University of Catalonia, Barcelona) and the authors' experience with similar type of in-situ experiments (Nguyen, 2000-a and b) the following approach was adopted:

2.1 Conceptualisation

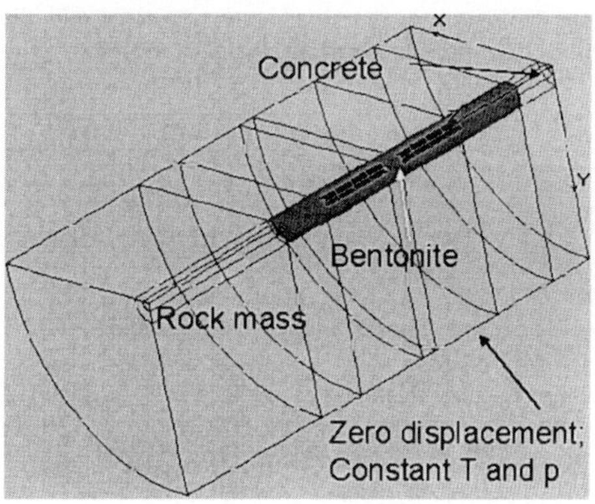

Figure 1. Conceptual representation of the FEBEX heater experiment

The conceptual model is schematically represented in Figure 1. We assume radial symmetry around the axis of the FEBEX tunnel, with a radius of influence of approximately 11 m from the centreline of the tunnel. The FEBEX gallery is intersected by several structural features, such as fracture zones and lamprophyres zones, which are more hydraulically conductive than the relatively intact rock mass. However, in this paper, the above features were ignored and an equivalent homogeneous porous medium approach was adopted for the rock mass. The above assumption originates from past experience gained from observations made at the Kamaishi Mine in Japan (Nguyen, 2000-a and b) where it was found that the THM behaviour of the bentonite was not influenced by the presence of discrete fractures in the rock mass but rather by the contrast in the permeability of the bentonite, and the bulk permeability of the rock mass. In a our effort to model the rock mass response (Selvadurai, Nguyen and Armand, 2003), these structural features needed to be explicitly represented.

2.2 Mathematical model and computer code

In order to predict the T-H-M response of the bentonite, a coupled T-H-M transient analysis was performed with the Finite Element Code FRACON. The governing equations incorporated in the FRACON code were derived from an extension of Biot's (1941) theory of poro-elasticity to include the T-H-M behaviour of the unsaturated FEBEX bentonite. The model formulation(Nguyen, Selvadurai and Armand, 2003) resulted in three governing equations where the primary unknowns are temperature, the displacement vector and the pore fluid pressure, as follows:

$$\frac{\partial}{\partial x_i}\left(\kappa_{ij}\frac{\partial T}{\partial x_j}\right)+q=\rho C\frac{\partial T}{\partial t} \tag{1}$$

$$\begin{aligned}&\frac{\partial}{\partial x_i}\left(\frac{\rho_f k_{ij} K_r}{\mu}\left(\frac{\partial p}{\partial x_j}+\rho_f g\, j\right)\right)-\frac{\partial}{\partial x_i}\left(D_T\frac{\partial T}{\partial x_i}\right)\\&+\rho_f\left[\frac{-nS}{K_f}+G_{sub}(1-n)\frac{dw}{dp}+S\left(S-\frac{G_{sub}w}{n}\right)\frac{n-\alpha}{K_s}\right]\frac{\partial p}{\partial t}\\&+\rho_f(S\alpha)\frac{\partial}{\partial t}\left(\frac{\partial u_k}{\partial x_k}\right)+\rho_f S\big((1-\alpha)\beta-n\beta_f-(1-n)\beta_s\big)\frac{\partial T}{\partial t}=0\end{aligned} \tag{2}$$

$$G\frac{\partial^2 du_i}{\partial x_j\partial x_j}+(G+\lambda)\frac{\partial^2 du_j}{\partial x_i\partial x_j}+\alpha S\frac{\partial dp}{\partial x_i}-\beta K_D\frac{\partial dT}{\partial x_i}+dF_i=0 \tag{3}$$

Equation (1) is the equation of conservation of heat, where heat conduction is assumed to be the only mechanism of heat transport. In this equation, κ_{ij} is the thermal conductivity tensor (W/m /^{0}C), ρ is the density of the bulk medium (kg/m^3), C is the bulk specific heat of the medium (J/kg/ ^{0}C) and q accounts for distributed heat generation in the poro-elastic medium (W/m^3).

Equation (2) is the equation of pore water flow in the saturated-unsaturated porous medium, derived from considerations of mass conservation:

- The first term of the equation results from a generalization of Darcy's law of water flow in variably saturated porous media. In this term, k_{ij} is the saturated permeability tensor (m^2); K_r (non-dimensional) is the relative permeability of unsaturated media and is a function of the degree of saturation S (For S=1, K_r=1); μ (kg/m/s) is the viscosity of water; ρ_w is the density of water, which are both functions of temperature.
- The second term represents vapour flow due to thermal gradients. In this term D_T (m^2/s) is the coefficient of thermal vapour diffusivity.
- The third term represents water retention due to the unsaturated state of the medium. In this term, w is the gravimetric water content, n is the porosity and G_s is the specific gravity of the solid particles. The water content for the unsaturated FEBEX bentonite can be expressed as an empirical function of suction:
  w (%) = 36.1 -12log(s) (4)
  s (MPa) = $p_g - p$ is suction;
  p_g is the gas pressure, assumed constant in our model;
  log is the decimal logarithm.
- The fourth term represents water retention due to compressibility of the water and the solid phase, where $1/K_w$ is the coefficient of compressibility of water (Pa^{-1}) and $1/K_s$ is the coefficient of compressibility of the solid phase (Pa^{-1})
- The fifth term represents water retention due to the consolidation of the porous medium
- The sixth term represents water flow due to the difference in thermal expansion between the water and the solid material, where β_w and β_s (1/^{o}C) are the coefficient of volumetric thermal expansion of the water of the solid material, respectively.

Equation (3) is the equation of equilibrium of the porous medium. In this equation, it is assumed that the medium is non-linearly elastic, and G (Pa) and λ (Pa) are Lame's constants of elasticity and β is the coefficient of volumetric thermal expansion of the solid matrix. G and λ , and also K_D the bulk modulus can also be expressed as functions of the more commonly used Young's modulus E (Pa) and Poisson's ratio ν (non-dimensional). α is the pseudo-Biot's coefficient. In this work, the coefficients of elasticity are assumed constant when the material is saturated. When the material is unsaturated, as described in Nguyen, Selvadurai and Armand (2003), the coefficients of elasticity and the coefficient α are expressed as functions of suction and the void ratio, using a state surface equation (Figure 2) derived from suction controlled oedometric tests. The bulk modulus and the pseudo-Biot's coefficient resulting from the state surface equation approach are shown in Figures 3 and 4, respectively.

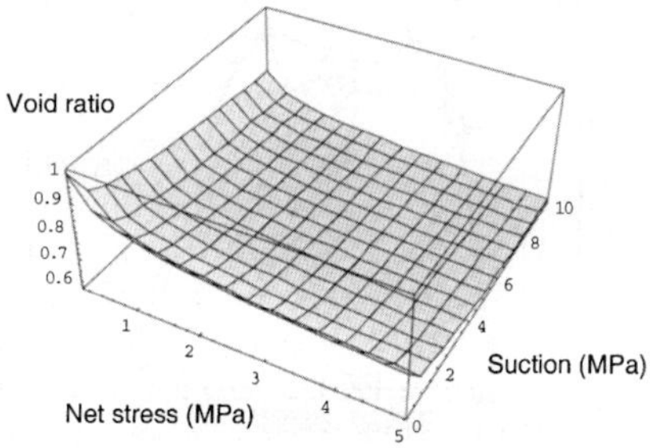

Figure 2. State Surface for FEBEX bentonite

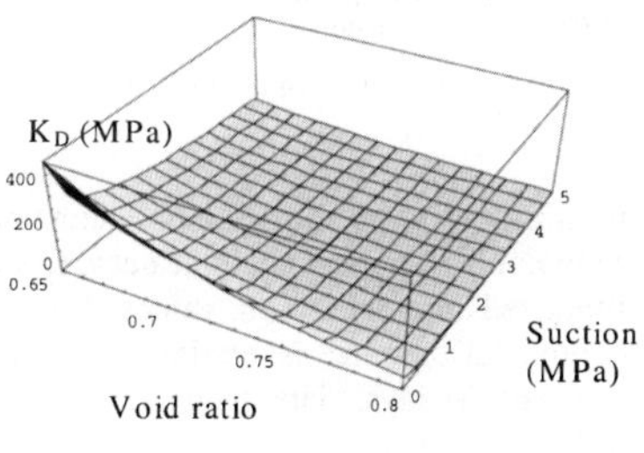

Figure 3. Bulk Modulus for FEBEX bentonite

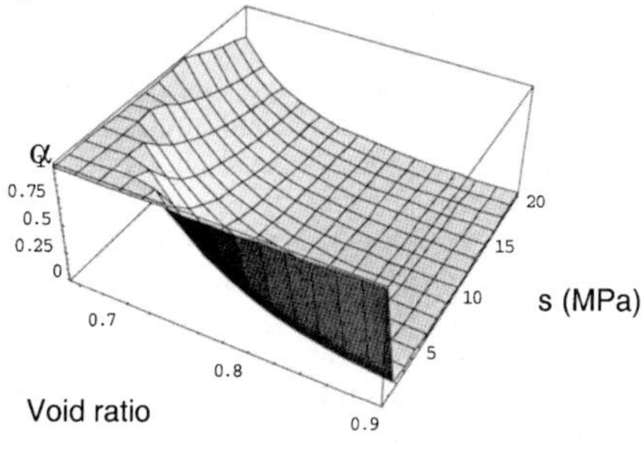

Figure 4. Pseudo-Biot's coefficient for FEBEX bentonite

2.3 Sequence of transient analysis using the FRACON code

In modelling the response of the bentonite, we mathematically simulate as closely as possible the actual test sequence. Each of the following phases were modelled with the FRACON code, with the results of the last time step of each phase being used as starting conditions for the subsequent phase:

t= -180 day to t=0: the emplaced bentonite was allowed to re-saturate isothermally.

t=0 to t= 20 days: the heaters were set at an output of 1200 W each

t= 20 to 53 days: the heaters were set at an output of 2000 W each

t=53 to 450 days: the surface of the heaters was set at 100 °C.

3. FINITE ELEMENT MODEL

The geometric configuration shown in Figure 1 was discretised into 1876 isoparametric solid elements joined by 5093 nodes. Three material types are represented in the finite element model:

i)the bentonite with constitutive equations and properties as described in Nguyen, Selvadurai and Armand (2003) .

ii)the rock mass which is assumed to be homogeneous and linear elastic, as justified in section 1. The following values of the elastic constants and the permeability were based on our previous modelling of the excavation of the FEBEX gallery (Nguyen, 2001) : E=10 GPa, ν= 0.37 ,$k_{radial}=5\times10^{-19}$ m^2; $k_{axial}=5\times10^{-18}$ m^2.

iii)The concrete which is also assumed to be homogeneous, isotropic and linear elastic, with the following assumed properties: E=30GPa, ν=0.2, k=10^{-21} m^2.

The outside boundaries of the rock mass are assumed to be fixed against normal displacement, and are specified to remain at a constant temperature (12°C). The measured steady state pressure in the rock in the vicinity of the gallery varies between 0 (atmospheric) and less than 1 MPa. These values are very small to influence the hydraulic response of the bentonite. For the sake of simplicity, we assumed that the outside boundary of the rock mass is maintained at a constant pressure of 0 MPa. The heaters are not explicitly represented; instead, the boundaries between the heaters and the bentonite are assumed to be fixed at zero normal displacement and zero fluid flux (very rigid and impermeable heaters), and an imposed power output is specified at either a constant rate or constant temperature depending on the sequence of heating. The initial conditions are T=12°C throughout the model, p=0 in the rock and concrete, and initial water content of 13.4 % .

4. COMPARISON OF PREDICTED AND MEASURED RESULTS

We now present the results for different T-H-M output parameters at points and along sections where these parameters were experimentally measured.

The relative humidity Hr is a function of suction and temperature as given in the Kelvin's equation:

$$Hr = Exp\,(-\,s m_v v_w / R / (T+273.16)) \qquad (5)$$

where m_v is the specific volume of water (0.001 m^3/kg); v_w is the molecular mass of water vapour (18.016×10^{-3} kg/mol); R is the universal gas constant (8.31432 J/(mol. °K)

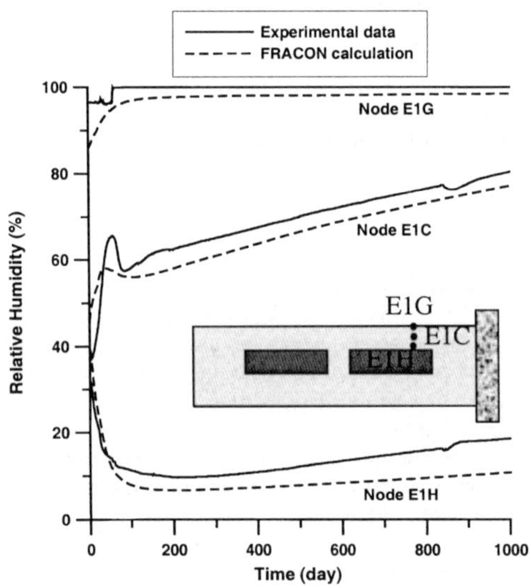

Figure 5. Evolution of relative humidity

The evolution of relative humidity at three points E1H, E1C, E1G, is shown in Figure 5. This figure expectedly shows that resaturation is faster near the rock (E1G). Near the heater (point E1H) the results indicate drying effects which persist up to and probably well beyond 1000 days. At point E1C, midway between the heater and the rock, the relative humidity increases in the early time up to approximately 15-20 days, due to vapour condensation. After that time, the relative humidity decreases when the thermal pulse reached point E1C, until approximately 100 days, when it increases again, due to water input from the rock mass. Figure 5 also shows excellent agreement between the predicted and experimental results for the evolution of the relative humidity.

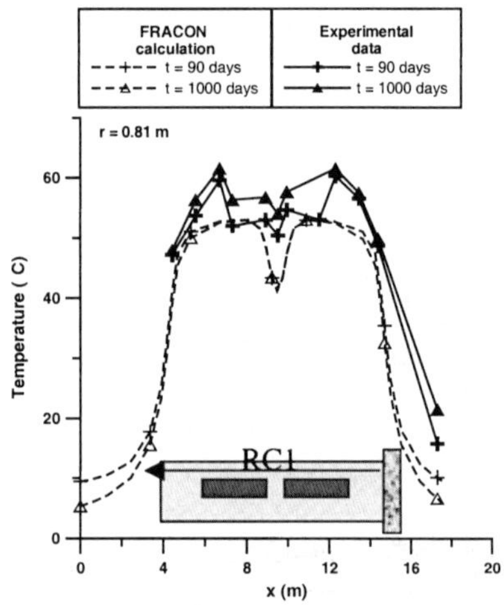

Figure 6. Temperature distribution along axial direction RC1

The distribution of temperature along one axial section is shown in Figure 6. It could be seen, from both the predicted and measured values, that a steady state thermal regime is attained not too much later than 90 days. The model generally underpredicts the values of temperature, due to the constant temperature conditions of 12°C imposed on the outer boundaries of the finite element model. The predictions could be improved if these boundaries were defined at larger distances.

The evolutions of total radial stress at points E2G1, E2G2 (near the rock interface) and E2H1 (near the heater 2) are shown in Figure 7. Compressive stress values of up to 1.4 MPa were calculated at the rock interface (E2G2). Near the heater the stresses are found to be small and attaining a maximal value of 0.1 MPa and originating from thermal expansion of the bentonite. The model predicts that a steady state stress regime is attained very rapidly in the bentonite, due to swelling of the bentonite from resaturation. The measured values however show that the stress is still increasing at 1000 days. We believe that heating of the surrounding rock mass results in thermal expansion of the porewater and the solid matrix. Since the coefficient of thermal expansion of the water is higher than the one of the solid matrix, this results in an increase in the porewater pressure. That water pressure increase was in fact predicted in our simulation of the rock mass response (Selvadurai, Nguyen and Armand, 2003) and would be transmitted to the total stress measured in the bentonite. That phenomenon is neglected in our model since we impose a zero pore pressure in the rock mass.

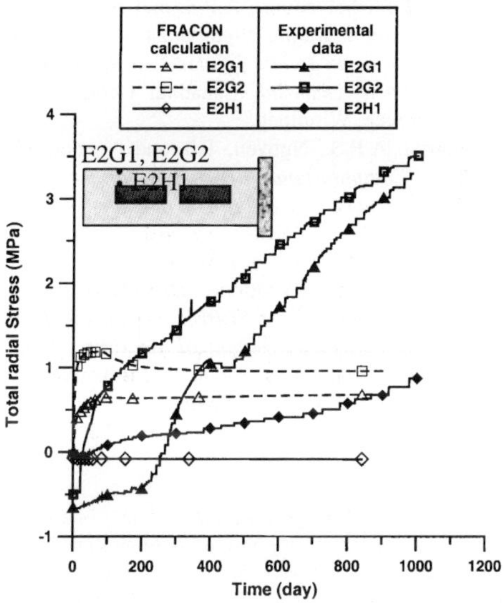

Figure 7. Evolution of radial stresses in bentonite

The power output from the heaters is shown in Figure 8. The output is predicted to be almost the same for the two heaters. During the temperature control phase, the power would have to be jacked up initially to increase the temperature at the surface of the heaters to 100 °C; this power will then decrease to an asymptotic value. The power output predicted by the model agrees very well with the measured value, although the measured output from heater 2 (near the concrete block) is higher that from heater 1.

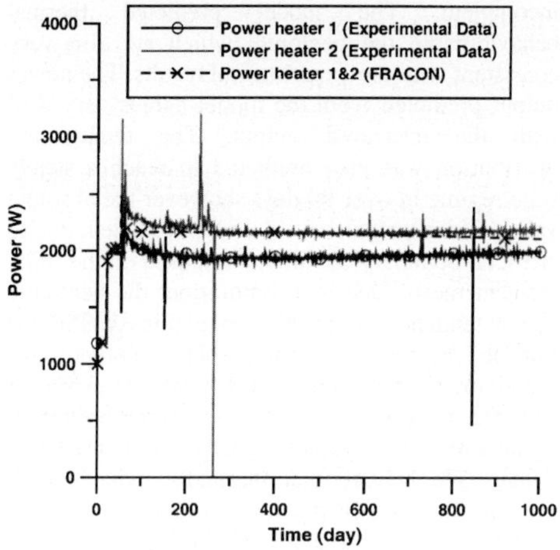

Figure 8. Power output from heater

5. CONCLUSIONS

The FEBEX T-H-M experiment is a valuable and important project which should lead to an improvement in the understanding of the behaviour of the bentonite barrier around heat-emitting Nuclear Fuel Waste(NFW) containers. Such large field experiments should always be undertaken with the simultaneous development of constitutive and computational models to interpret the experiments. The FEBEX bentonite possesses strong nonlinear behaviour in the unsaturated state. In order to simulate that behaviour, we have adopted a nonlinear poro-elastic approach. In this approach, the coefficients of the poroelastic equations are assumed to be functions of suction and the void ratio. These functions are derived from the state-surface equation which has been experimentally obtained from suction-controlled oedometric tests performed by the Spanish research organizations UPC and CIEMAT.

The equations of poro-elasticity were solved with the finite element code FRACON. The FRACON code was used to predict the in-situ T-H-M experiment at the FEBEX gallery in Grimsel, Switzerland. The FRACON correctly predicted that the bentonite would resaturate from the rock interface. Near the heater, it also correctly predicted that initial drying will take place, followed by a slow resaturation. The model correctly predicted that at the end of 1000 days of heating, resaturation of the bentonite was still incomplete. The model predicted thermal behaviours in the bentonite which are also very consistent with the experimental results. The power output predicted from the model agrees very well with the measured output. The temperature distribution was also predicted to reach a steady state regime in over 90 days; however the absolute values of temperature were under-predicted, due to artificial boundary conditions imposed on the finite element mesh. Due to re-saturation, the bentonite has a tendency to swell. Since this swelling is impeded by the surrounding rock, a compressive radial stress of the order of 1.5 MPa is predicted near the rock interface. The measured values of radial stress show values higher than the predicted values. We believe that thermally induced pore pressure increase in the surrounding rock mass is an additional component contributing to increases in stress. That component was neglected in our computations.

ACKNOWLEGGEMENT

The authors sincerely thank the Canadian Nuclear Safety Commission (CNSC) for its financial support and ENRESA/UPC for providing experimental data from the FEBEX project. The opinion expressed in this paper is the authors' and does not necessarily reflect the opinion of the CNSC or McGill University.

REFERENCES

Biot, M.A. 1941. *General theory of three dimensional consolidation*. J. Appl. Phys., 12:pp 15-164

ENRESA. 1998. *FEBEX full-scale engineered barriers experiment in crystalline host rock-Bentonite: origin, properties and fabrication of blocks*. Publicacion Tecnica 05/98. ENRESA. Emilio Vargas, 7-28043, Madrid

Lewis, R.W. and Schrefler, B.A., 1987. *The finite element method in the deformation and consolidation of porous media*, John Wiley and Sons.

Nguyen, T.S., 2000-a. *Mathematical and laboratory simulation of the thermo-hydro-mechanical behaviour of unsaturated bentonite*, Proc. Canadian Geotechnical Conference, Montreal, Bitech Publishers Ltd., Canada.

Nguyen T.S., 2000-b. *Mathematical Modelling of an in-situ Experiment for the T-H-M Behaviour of Unsaturated Bentonite and Granitic Rock Mass*, Proc. Canadian Geotechnical Conference, Montreal, Bitech Publishers Ltd., Canada

Nguyen, T.S., 2001. *Modelling of the excavation of the FEBEX gallery*, Canadian Nuclear Safety Commission, Internal Report.

Nguyen, T.S. and Selvadurai, A.P.S. 1995. *Coupled thermal-hydrological-mechanical processes in sparsely fractured rock.* Int. J. Rock Mech. Min. Sci., 32: pp 465-480.

Nguyen, T.S., Borgesson, L., Chijimatsu, M., Rutqvist, J., Fujita, T., Hernelind, J., Kobayashi, A., Onishi, Y., Tanaka, M. and Jing, L. 2001.*Hydro-mechanical response of a fractured granitic rock mass to excavation of a test pit-the Kamaishi Mine experiment in Japan.* Int. J. Rock Mech. Min. Sci., 38(1): pp 79-94.

Nguyen, T.S., Selvadurai, A.P.S. and Armand, G. 2003. *Computational Predictions of the Buffer Response in an In-situ Heater Test for Nuclear Fuel Waste Disposal*, Canadian Geotechnical Conference, Winnipeg.

Selvadurai. A.P.S., Nguyen, T.S. and Armand, G. 2003. *Computational Predictions of the Rock Mass Response in an In-situ Heater Test for Nuclear Fuel Waste Disposal*, Canadian Geotechnical Conference, Winnipeg.

UPC, 2000, Task definition: *DECOVALEX III, Task 1: Modelling of FEBEX in-situ test, Part B: Thermo-hydro-mechanical analysis of the bentonite behaviour.* Polytechnical University of Catalonia, Barcelona, Spain.

THM SIMULATION OF THE FULL-SCALE IN-SITU ENGINEERED BARRIER SYSTEM EXPERIMENT IN GRIMSEL TEST SITE IN SWITZERLAND

Yutaka SUGITA[1], Masakazu CHIJIMATSU[2], Akira ITO[1], Hiroshi KURIKAMI[1], Akira KOBAYASHI[3] and Yuzo OHNISHI[3]
[1]) Japan Nuclear Cycle Development Institute
[2]) Hazama Corporation
[3]) Kyoto University

Abstract: Coupled THM simulation of the FEBEX, which is the full-scale in-situ Engineered Barrier System Experiment performed in Grimsel Test Site in Switzerland, is one Task in the international cooperation project DECOVALEX III. In the Task, the simulation of the thermal, hydraulic and mechanical behaviour in the buffer during heating phase is required, e.g. the evolutions and the distributions of stress, relative humidity and temperature at the specified points in bentonite buffer material.

This report presents our approach for calculation of the Task. Our numerical code THAMES is the three-dimensional finite element simulator of fully coupled processes. First, we defined the input data for THAMES from the supplied properties of FEBEX bentonite. After calibrations of some parameters such as thermal vapour diffusivity, the analysis that treats fully coupled thermal, hydraulic and mechanical processes was carried out.

1. INTRODUCTION

The FEBEX is the full-scale in-situ Engineered Barrier System (EBS) Experiment performed in Grimsel Test Site (GTS) in Switzerland (enresa (2000)). The simulation of the coupled thermal, hydraulic and mechanical (THM) behaviour of FEBEX is the task of the DECOVALEX (DEvelopment of COupled models and their VALidation against EXperiments) III.

We have developed and improved the numerical code THAMES to predict and assess the coupled THM behaviour in and around the EBS of the geological disposal of high-level radioactive waste. THAMES is a finite element code for analysis of coupled thermal, hydraulic and mechanical behaviour of a saturated-unsaturated medium. Then, we applied THAMES to calculate this task of the DECOVALEX III. In this task, the evolutions and the distributions of stress, relative humidity and temperature at the specified points in bentonite buffer material are required.

First, we identified the input parameter for THAMES on properties of FEBEX bentonite, because the fundamental properties of FEBEX bentonite had been obtained by various laboratory tests to identify the input data for the numerical code CODE BRIGHT (enresa (1998)). After calibrations of the all required parameters for THAMES, such as thermal vapour flow diffusivity and intrinsic permeability, the coupled THM simulations were carried out.

2. NUMERICAL CODE

THAMES was developed originally by Ohnishi et al. (1987) to simulate the coupled THM behaviour in rock media, and later extended to simulate the coupled THM behaviour of a saturated-unsaturated buffer by adding function of taking into account unsaturated water movement by heat gradient in the buffer by Chijimatsu et al. (2000) and swelling characteristics by Komine et al. (2003).

2.1 Conservation law

Conservation law of momentum represents the equilibrium equations as

$$\Delta\sigma_{ij,j} = 0 \qquad (1)$$

where $\Delta\sigma_{ij}$ is tensor of total stress increment.

Law of fluid mass conservation can be written as

$$\frac{\partial \theta\rho_w}{\partial t} + q_{i,i} + Q = 0 \qquad (2)$$

where θ is volumetric water content, ρ_w is water density, t is time, q_i is groundwater flux vector and Q is sink/source.

Law of heat energy conservation can be described as

$$\frac{\partial\{(\rho c)_m T\}}{\partial t} + q^h_{i,i} + Q^h = 0 \qquad (3)$$

where $(\rho c)_m$ is specific heat of mixture of gas, liquid, and solid, T is temperature, q^h_i is heat flux vector and Q^h is heat sink/source.

2.2 Constitutive equations for bentonite

In this paper, only constitutive equations in unsaturated bentonite are explained because the other constrains are almost general. Assuming elasticity of bentonite, increment of total stress in unsaturated media can be written as follows.

$$\Delta\sigma_{ij} = \frac{1}{2}C_{ijkl}\left(\Delta u_{k,l} + \Delta u_{l,k}\right) - \Delta\sigma_{sw}\delta_{ij} - \beta\Delta T\delta_{ij} \quad (4)$$

where C_{ijkl} is elastic modulus tensor, u_i is displacement, β is the parameter regarding to thermal stress and δ_{ij} is Kronecker's delta. The second term represents swelling pressure.

Groundwater flux is expressed by

$$q_i = -\rho_w(D_\theta)_{ij}\theta_{,j} - \rho_w(D_T)_{ij}T_{,j} \quad (5)$$

where D_θ is water diffusivity and D_T is thermal water diffusivity. D_θ is dependent of both saturation degree and temperature and shows concave up when it is plotted against volumetric water content. It is because moisture movement in unsaturated media happens both in liquid phase and in vapour phase, where the former is dominant in high saturation, while the latter is dominant in low saturation. D_θ of bentonite in Japan is given by following experimental equation, which is combination of moisture movement in vapour and liquid phase.

$$(D_\theta)_{ij} = \delta_{ij}(D_{\theta l} + D_{\theta v}) \quad (6)$$

where subscript l represents liquid and v represents vapour phase. And isotropy is assumed.

Heat flux, q^h, consists of the advective term, q^a, and the dispersive term, q^d, as

$$q_i^h = q_i^a + q_i^d \quad (7)$$

where

$$q_i^a = S_r q_i T \quad (8)$$

$$q_i^d = -\lambda_m T_{,i} \quad (9)$$

where λ_m is thermal conductivity of mixture.

THAMES has been validated and improved through applying to the large-scale laboratory test such as the DECOVALEX project, the Kamaishi in-situ experiment, the Prototype Repository Project in SKB and so on (Fujita et al. (1996); Rutqvist et al. (2001); Sugita et al. (2002)).

3. IDENTIFICATION OF INPUT DATA

3.1 FEBEX bentonite

Fundamental properties of FEBEX bentonite are supplied as the data set by enresa (e.g. enresa (2000)). However, the supplied data set of FEBEX bentonite is short of the information for the input data of THAMES. Therefore, the input data should be identified from the supplied data set.

a) Thermal properties

(a) Specific heat

Specific heat is given from the laboratory experiments and the specific heat capacity is given as

$$(\rho C)_m = (1\text{-}n)\,\rho_s C_s + nS_r\rho_w C_w \quad (10)$$

where C_s is the specific heat of solid particle [J/kgK], C_w is the specific heat of water [J/kgK], n is the porosity [-], S_r is the degree of saturation, ρ_s is the density of solid particle [kg/m^3], ρ_w is the density of water [kg/m^3]. Specific heat of solid particle C_s is a function of temperature as follows,

$$C_s = 1.38\,T + 732.5 \quad (11)$$

(b) Thermal conductivity

Thermal conductivity is a function of the degree of saturation as follows,

$$\lambda = 1.28 - \frac{0.71}{1 + e^{(S_r - 0.65)/0.10}} \quad (12)$$

where λ is the thermal conductivity [W/mK]. It is a function of the degree of saturation.

(c) Linear expansion

Linear expansion is $1.0\cdot10^{-4}$[1/K] which is given from the laboratory experiments.

b) Hydraulic conductivity

Regarding FEBEX bentonite, hydraulic conductivities [m/s] are given as a function of dry density [kg/m^3] (enresa (2000)).

On the other hand, THAMES requests intrinsic permeability K [m^2] as an input parameter.

Assuming that these values were obtained under 25°C, they can be converted into the equation (13).

$log(K) = -0.00600\rho_d - 11.10$ (ρ_d=1300-1470) (13)
$log(K) = -0.00296\rho_d - 15.61$ (ρ_d=1470-1840)

c) Water retention curve

Modified van Genuchten expression is applied to water retention curve.

$$\frac{S_r - S_{r\min}}{S_{r\max} - S_{r\min}} = \left[1 + (\psi / P_0)^{\frac{1}{1-\lambda}}\right]^{-\lambda} \left[1 - \psi / P_s\right]^{\lambda_s} \quad (14)$$

where S_{rmin} and S_{rmax} are the residual and maximum degree of saturation and P_0 [MPa], P_s [MPa], λ and λ_s are material parameters. ψ is suction [MPa]. Table 1 shows the parameters fitted for the wetting paths, that are plotted together with the experimental points.

Table1.Parameters fitting of the drying paths after saturation

P_0	λ	S_{rmin}	S_{rmax}	P_s	λ_s
35	0.30	0.01	1.00	4000	1.5

d) Water diffusivity

Philip and de Vries (1957) identified water flow velocity q as equation (15) assuming that water potential is function of volumetric water content and temperature.

$$q_i = -\rho_w \left[(D_\theta)_{ij}\theta_{,j} + (D_T)_{ij}T_{,j}\right] \quad (15)$$

where D_θ is water diffusivity, D_T is thermal vapour diffusivity. D_θ is defined as equation (16) using relative permeability r_k and the gradient of water retention curve $\partial\psi/\partial\theta$.

$$(D_\theta)_{ij} = r_k \frac{\rho_{w0} g k_{ij}}{\mu_w} \frac{\partial \psi}{\partial \theta} \quad (16)$$

Figure 1 shows the identified water diffusivity as functions of volumetric water content. Water diffusivity is U-shaped. This is considered that water moves in the vapour phase at the dry condition and does in the liquid phase at the wet condition.

e) Thermal vapour diffusivity

Transport of water under thermal gradient is expressed with the thermal vapour flow diffusivity in THAMES. The parameter for thermal vapour flow diffusivity was determined by back analysis of temperature gradient test by enresa (2000) (see Figure 2). Three specimens of bentonite compacted at a dry density of 1680 kg/m^3 and with initial water contents of 15.3, 16.9 and 17.3% were tested. Figure 3 shows the comparison between monitoring data and simulation results.

$$D_T = 1.0E\text{-}12 \ [m^2/s°C] \quad (17)$$

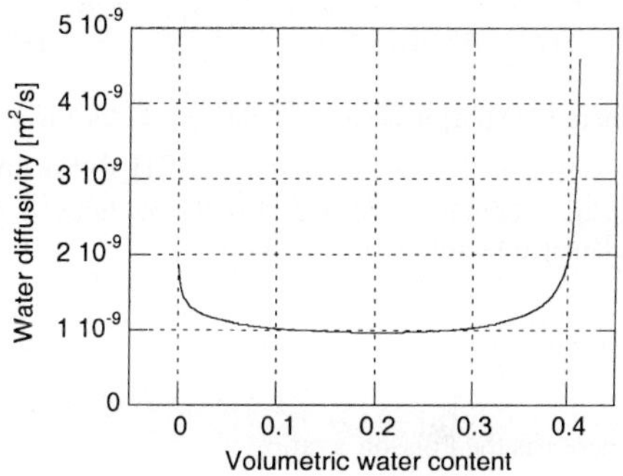

Figure 1. Identified water diffusivity for THAMES

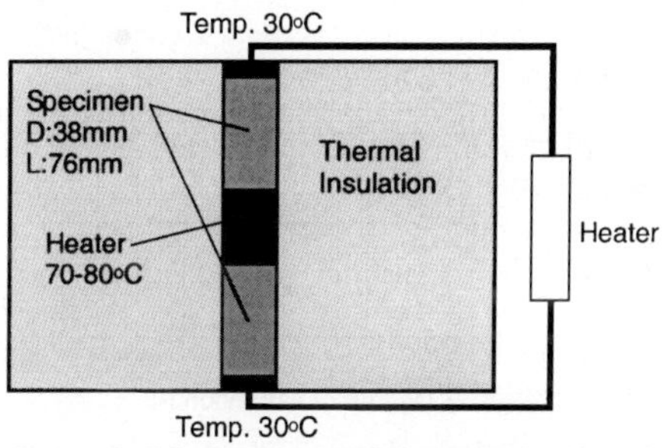

Figure 2. Schematc view of thermohydraulic cell

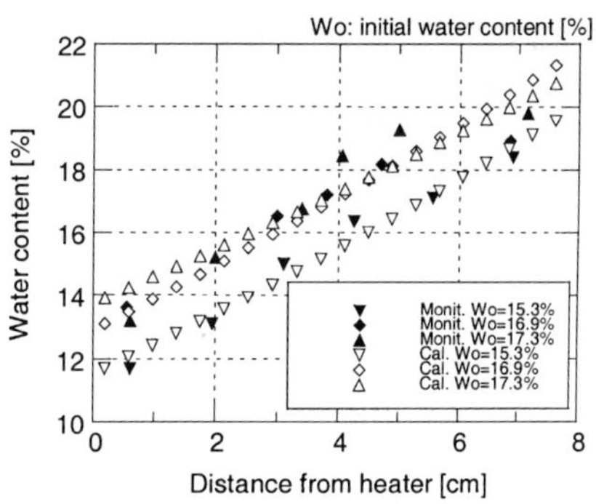

Figure 3. Comparison of water content distribution in the specimen between TH analysis and measured results of temperature gradient test

f) Elastic modulus

In data set of FEBEX bentonite, shear modulus G is measured with different saturation conditions, different dry density and different confining pressure. The shear modulus was identified as a function of the degree of saturation and the dry density. Equation (18) is defined by means of recurrence analysis of the data set of FEBEX bentonite (see Figure 4).

$$G = -234.7 + 0.176\rho_d + 2.1S_r \quad (18)$$

where G [MPa] is shear modulus, ρ_d [kg/m^3] is dry density , S_r [%] is saturation. For THAMES, the elastic modulus is needed. Elastic modulus E is calculated by following equation.

$$G = \frac{E}{2(1+\nu)} \quad (19)$$

where ν is the Poisson's ratio.

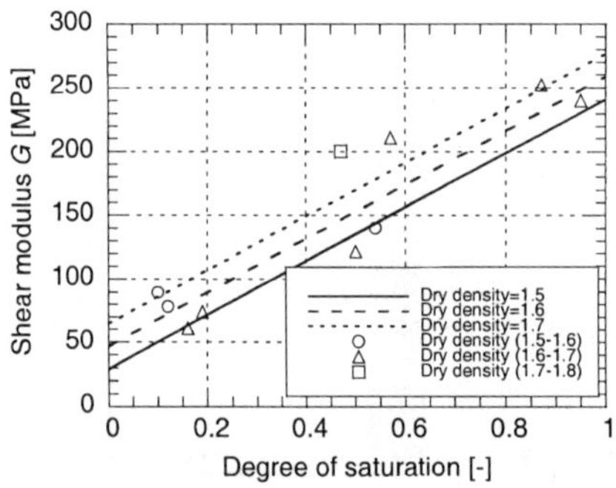

Figure 4. Relation between Shear modulus and degree of saturation

g) Swelling pressure

Maximum swelling pressure of FEBEX bentonite is defined as function of dry density as shown following equation

$$\sigma_{swmax} = exp(0.00677\rho_d - 9.07) \quad (20)$$

where σ_{swmax} [MPa] is maximum swelling pressure, ρ_d [kg/m^3] is dry density.

During the re-saturation process of buffer, swelling pressure under unsaturated conditions is important, though the above equation is limited to saturated conditions.

Free energy change is said to be equal to potential swelling pressure (Nakano et al.(1984)) and is approximately equal to the suction in the case of the buffer material. Because gravity potential of buffer is negligible compared with the osmotic and the matrix potential that are the main contributions to the suction. Therefore, it can be said that swelling pressure is expressed by suction. The suction, however, is consumed during re-saturation process not only by swelling but also by soil structure changes. We define the function F that illustrates the ratio of suction contributing to swelling pressure.

$$\Delta\sigma_{sw} = -F\Delta\psi \quad (21)$$

And F is defined as function of saturation.

$$F = 2\sigma_{sw\max} S_r \frac{\partial S_r}{\partial \psi} \quad (22)$$

3.2 Boundary conditions

Figure 5 shows the schematic view of the FEBEX. FEBEX has two heaters. Vertical sections, as D, E1, E2 and G, in the test tunnel are instrumented sections that are selected for comparison between the prediction and the monitored data for part B. Sections of E1 and E2 are for relative humidity simulation. Sections of D and G are for temperature simulation. Section of E2 is for total pressure simulation.

Figure 6 shows the numerical model geometry for the simulation. The model includes the two heaters, bentonite, plug, tunnel and host rock. Numerical area is decided considering influence of boundary conditions. Since the site can be assumed line symmetric with the xz-plane, the model is made only in the positive part of y direction. The number of elements is 5,760. The number of nodes is 26,401. The host rock is assumed to be a homogenous media.

The simulation started after the installation of heaters, bentonite and plug. The influences against water pressure and temperature in the host rock due to excavation of the tunnel did not considered. In this analysis, initial hydraulic head is specified 100m in the host rock. The initial temperature is 10°C in the all region, and the initial water content of buffer is 14%.

The boundary condition of the surface of heaters is specified as the heat flux from start to 53 days. Each heater power is set 1200w from 0 until 20 days and 2000W until 53 days. After that, temperature of the heater is fixed 100°C. All outer boundaries of the model were fixed at the initial value.

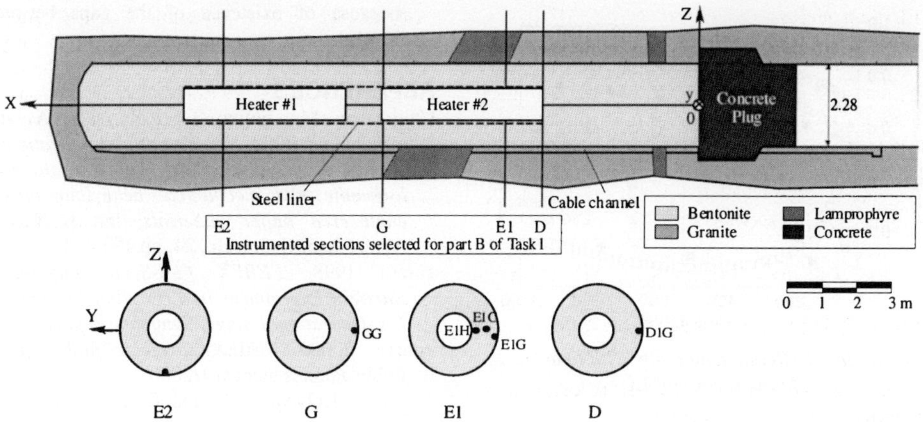

Figure 5. Schematic view of FEBEX and instrumented sections

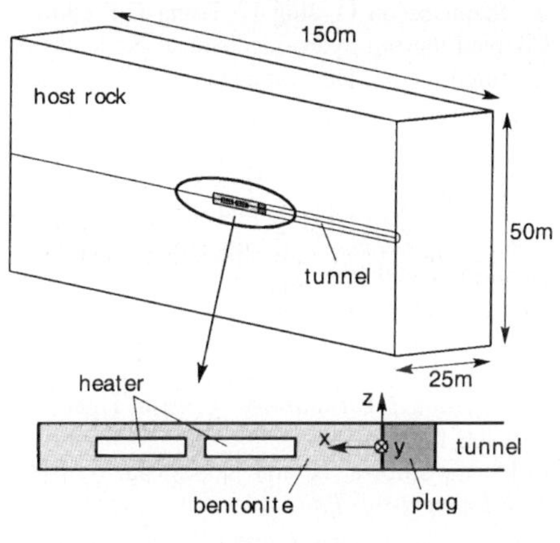

Figure 6. Numerical model geometry

4. RESULTS

Figure 7 shows the variation in temperature as function of time in the section of D and G in bentonite. The measured temperature increased rapidly with time and increased gradually with time after 200 days. The measured temperature at the point of GG is higher than the point of D1G, because the section G is affected by two heaters. The simulation values agree well with the measurement values.

Figure 8 shows the variation in relative humidity as function of time in the section of E1 in bentonite. In the section of E1, the measured relative humidity near the point of the surface of heater decreased once, and then increased. The measured relative humidity at the points of the center of bentonite and near host rock increased. There is great similarity in the tendency between simulation and monitored data. The simulation values agree almost well with the measurement values.

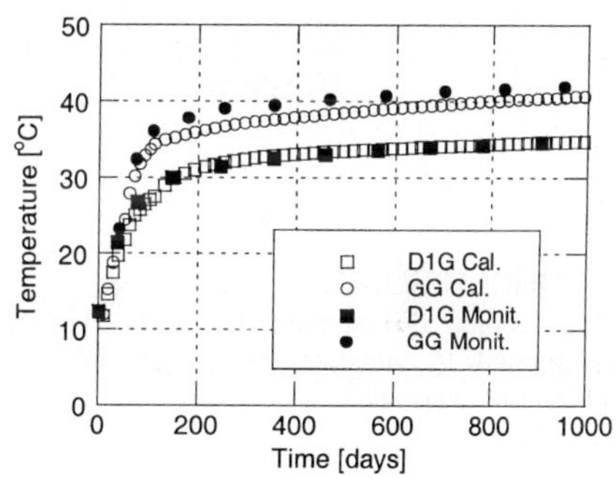

Figure 7. Temperature with time
in the sections of D and G

Figure 9 shows the variation in total pressure as function of time in the section of E2. The measured total pressure increased with time. Although the value around 1000 days is permissible, the process differs from the monitored, especially before 400 days. One of the reasons of the difference might be the existence of initial gaps within the bentonite blocks or those between the blocks and host rock. Modelling of initial gaps may be necessary.

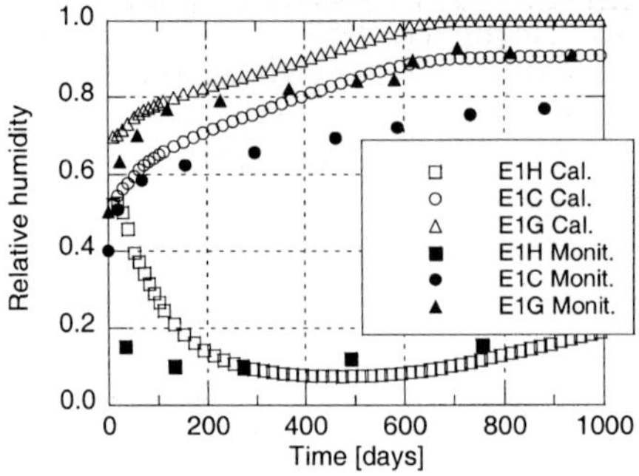

Figure 8. Relative humidity with time In the section of E1

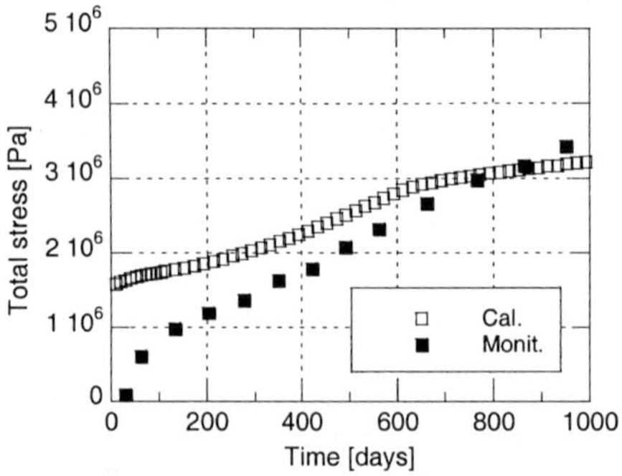

Figure 9. Total pressure with time in the section of E2

5. CONCLUSIONS

The coupled T-H-M numerical code THAMES was applied to simulate the in-situ full scale engineered experiment performed in GTS in Switzerland.

1) The identified water diffusivity is U-shaped as functions of volumetric water content. This is considered that water moves in the vapour phase at the dry condition and does in the liquid phase at the wet condition.
2) Simulated temperature showed good agreement with measured one.
3) Philip & de Vries model can be applied to simulate the unsaturated water movement in not only Japanese bentonite but also FEBEX bentonite.
4) Total pressure showed good agreement with measured except for early phase. The discrepancy at the early phase is probably because of existence of the gaps between blocks.

REFERENCES

Chijimatsu, M., Fujita, T., Kobayashi, A. and Nakano, M. 2000. *Experiment and validation of numerical simulation of coupled thermal, hydraulic and mechanical behaviour in the engineered buffer materials*, Int. J. Numer. Anal. Meth. Geomech., 24, pp.403-424.

enresa 1998. *FEBEX Full-Scale Engineered Barriers Experiment in Crystalline Host Rock, Pre-Operational Stage, Summary Report*

enresa 2000. *FEBEX project finil report*, Publicaciones tecnica 1/2000.

Fujita T., Kobayashi A. and Börgesson L. 1996. *Experimental investigation and mathematical simulation on coupled T-H-M processes of the engineered buffer materials*, the TC3 problem. In: Stephansson O, Jing L, Tsang C-F editors. Coupled thermo-hydro-mechanical processes of fractured media, Developments in Geotechnical Engineering, No.79, Amsterdam: Elsevier, pp.369-392.

Komine H., Kurikami, H., Chijimatsu, M., Kobayashi, A., Sugita Y. and Ohnishi, Y. 2003. *Coupled thermal, hydraulic and mechanical simulation with theoretical model for swelling characteristics*, GeoProc 2003.

Nakano, M., Amemiya, Y., Fujii, K. & Ishida, A. 1984. *Water Penetration and Swelling Pressure of Restrained Unsaturated Clay*(in Japanese). JSIDRE 112: pp.55-66.

Ohnishi, Y., Shibata, H. and Kobayashi, A. 1987. *Development of finite element code for the analysis of coupled thermo-hydro-mechanical behavior of a saturated-unsaturated medium.* In: Tsang, C.-F. editor, Coupled processes associated with nuclear waste repositories, Academic Press, Orlando, pp.551-557.

Philip, J.R. and de Vries, D.A. 1957. *Moisture movement in porous materials under temperature gradient*, Transactions, Aemerican Geophysical Union, vol.38(2), pp.229-237.

Rutqvist J., Börgesson L., Chijimatsu M., Nguyen S., Jing L., Noorished J. and Tsang C-F. 2001. *Coupled thermo-hydro-mechanical analysis of a heater test in fractured rock and bentontie at Kamaishi Mine – comparison of field results to predictions of four finite element codes*, IJRMMS, Vol38, No.1, pp.129-142.

Sugita Y., Ito A., Chijimatsu M. and Kurikami H. 2002. *Prediction analysis A for the PRP with numerical code THAMES*, SKB IPR-02-24.

HYDROMECHANICAL RESPONSE OF JOINTED HOST GRANITIC ROCK DURING EXCAVATION OF THE FEBEX TUNNEL

Steven Sobolik[1], Stephen Webb[1], Akira Kobayashi[2], and Masakazu Chijimatsu[3]
[1]) Sandia National Laboratories, Albuquerque, New Mexico, USA *
[2]) Kyoto University, Grad. School of Agricultural Science, Kyoto, Japan
[3]) Hazama Corporation, Tokyo, Japan

Abstract: The FEBEX experiment is a coupled T-H-M test conducted at the Grimsel Test Site in Switzerland designed to study various processes occurring near a radioactive waste storage repository. The experiment uses cylindrical heaters placed along the axis of a drift to simulate the presence of radioactive waste canisters. An engineered barrier made of bentonite blocks is placed in the surrounding annular space. Several research teams performed blind predictions of changes in water pressure and flow rate induced by the excavation of the FEBEX tunnel. This report outlines two numerical approaches to these predictions. One, the compliant joint model, captures the average response of a jointed rock mass by distributing the response of the individual joints throughout it. The other, a modified Crack tensor theory, treats the non-linear mechanical and hydraulic behavior of fractures. The difference of results between the models are discussed and compared to the actual test data.

1. INTRODUCTION

The DECOVALEX III project (DEvelopment of COupled models and their VALidation against Experiments) is an international, multi-disciplinary interactive and co-operative research effort. The purposes of the project are the study of the various coupled thermo-hydro-mechanical (THM) and geochemical processes occurring in the near field of a high activity radioactive waste storage facility, a demonstration of the feasibility of the construction of engineered barriers, and how they affect the role of performance assessment for radioactive waste storage. Thirteen international funding organizations from industry and regulatory authorities participate in the project, from Canada, Finland, France, Germany, Japan, Spain, Sweden, Switzerland, the United Kingdom, and the United States.

Task 1 of the DECOVALEX III project is the analysis of the FEBEX (Full-scale Engineered Barriers Experiment in Crystalline Host Rock) test, an underground heated drift test at the Grimsel Test Site, Switzerland. The FEBEX project is co-financed by the Spanish nuclear agency ENRESA and the European Commission. The purpose of the project is the study of the various processes occurring in the near field of a high activity radioactive waste storage. The experiment uses two cylindrical heaters to simulate radioactive waste canisters. The heaters are placed along the axis of a drift drilled in a rock formation and an engineered barrier is placed in the annular space left between them. The engineered barrier is made of high density compacted bentonite blocks that swells due to water input from the host rock, providing an impervious sealing.

The FEBEX in situ test is well suited to be the object of a modelling exercise due to the detailed geological and hydrogeological characterization of the Grimsel Test Site (GTS), the bentonite used as the engineered barrier, and the monitoring performed during the drilling of the FEBEX tunnel and during the test. For the DECOVALEX III project, the modelling exercise was divided into three parts. This report considers Part A of the FEBEX analysis task, for which hydro-mechanical models for the zone around the FEBEX tunnel were prepared. Using these models, predicted changes in water pressure induced by the boring of the FEBEX tunnel as well as the total water flow rate to the excavated tunnel were required. This report summarizes the results produced by two of the participating organizations: the US Dept. of Energy (DOE), represented by the team from Sandia National Laboratories (SNL); and the Japan Nuclear Cycle Development Institute, Japan (JNC), represented by the team from Kyoto University and Hazama Corporation. The difference of results between the models are discussed and compared to the test data.

2. DOE MODEL

The DOE approach for solution of the FEBEX excavation problem included a mechanical calculation of the mining of the drift and the resulting changes in permeability, followed by a hydrological calculation setting permeability as a function of time during the mining process. The finite element code JAS3D (Blanford, 1999) was chosen to perform the mechanical calculations, and the two-phase integrated finite difference flow code TOUGH2 (Pruess, 1991) was selected to perform

* Sandia is a multiprogram laboratory operated by Sandia Corporation, a Lockheed Martin Company, for the United States Department of Energy's National Nuclear Security Administration under contract DE-AC04-94AL85000.

the hydrological calculations. These codes are used by the Yucca Mountain Site Characterization Project of DOE for performance assessment and design-related analyses.

2.1 Model description

The continuum mechanics modeled by JAS3D are based on two fundamental governing equations. The kinematics is based on the conservation of momentum equation, which can be solved either for quasi-static or dynamic conditions (a quasi-static procedure was used for these analyses). The stress-strain relationships are posed in terms of the conventional Cauchy stress. JAS3D includes at least 30 different material models.

The Compliant Joint Model (CJM) was chosen for the mechanical calculations for this analysis. The CJM uses an equivalent continuum approach to model the behavior of jointed media. An equivalent continuum approach captures the average response of a jointed rock mass by distributing the response of the individual joints throughout the rock mass. The CJM in JAS3D can model up to four joint sets of arbitrary orientation, with the fractures in each set assumed to be parallel and evenly spaced. The intact rock between joints is treated as an isotropic linear-elastic material. More detailed descriptions of the CJM model can be found in Chen (1991).

The CJM models both the normal and shear displacement along joints. The joint normal response in the CJM is non-linear elastic and is based on a rational polynomial fit to experimental data. For the CJM, joint normal behavior is characterized by

$$T_{nn} = \frac{A\nu_n}{V_{max} + \nu_n}, \tag{1}$$

where T_{nn} is the joint normal stress, ν_n is the joint normal displacement, and V_{max} is the maximum possible closure of the joint. The maximum closure, V_{max}, is usually assumed to be the average joint normal opening for the joint set in an unstressed condition. The parameter A is called the half-closure stress, because $-A$ is the compressive stress required to close the joint to a joint normal displacement of $\nu_n = -V_{max}/2$. As the joint opening increases, the joint normal stress approaches A asymptotically. The quantity $V_{max} + \nu_n$ measures the amount of separation between the two faces of the joint.

The CJM assumes bilinear behavior of the joint shear stress across the joint as a function of a scalar measure, δ, of joint shear displacement. Joint shear is treated as linear elastic in shear stress (τ) versus shear displacement (δ) below a critical shear stress magnitude governed by a Mohr-Coulomb failure criterion (joint shear stiffness, G_s). Beyond this critical shear stress magnitude, a linear relation analogous to strain-hardening plasticity governs the joint shear stress versus joint shear displacement relation. The onset of inelastic response (slip) is assumed to be governed by a linear Mohr-Coulomb failure criterion. The slip function is given by

$$F = |\tau| + \mu T_{nn} - C, \tag{2}$$

where μ is the coefficient of friction and C is the joint cohesion. The joint behavior is elastic if $F < 0$ and inelastic if $F \geq 0$. The scalar measure of joint shear displacement, δ, can be interpreted as being made up of two parts: (1) any unrecovered elastic displacement, and (2) the total arc length of the path traced on one joint surface by a point on the opposite joint surface while the joint is sliding.

The current model for converting fracture deformation behavior to changes in fracture permeability utilizes only the normal deformation component. The fracture permeability is calculated using the predicted fracture apertures with the cubic law:

$$K_f = \frac{b^3}{12s} \tag{3}$$

where K_f is the fracture permeability in m², b is the fracture aperture, and s is the fracture (joint) spacing. The fracture aperture b is obtained from the predicted fracture closure from normal deformation in Eq. 1, where $b = V_{max} + \nu_n$.

The boundary conditions shown in Figure 1 were determined from FEBEX in situ stress data (Pahl et al., 1989). An additional boundary condition was used to model the time-dependent stress induced by the alpine miner during excavation of the drift. A pressure value of 25 MPa was applied to the wall in successive sections of the drift when that material was "excavated" numerically in the calculation.

The two-phase integrated finite difference flow code TOUGH2 was selected to perform the hydrological calculations for this FEBEX analysis. TOUGH2 is a general-purpose program for the simulation of multi-dimensional, multiphase fluid flow in porous and fractured media. Fluid advection is calculated by a multiphase extension of Darcy's law, and diffusion is included. Local thermodynamic equilibrium is assumed in all phases. The Equivalent Continuum Model (ECM) for matrix/fracture flow was chosen for the rock fluid flow conceptual model. The ECM is based on pressure equilibrium between the fractures and matrix. Pressure equilibrium does not imply equal saturation. Two-phase characteristic curve functions relating liquid saturations and liquid relative permeability to capillary pressure are necessary to perform the calculations. The two-phase characteristic curves defined in UPC (1999) for capillary pressure and liquid phase relative permeability are used for both the fractures and the matrix.

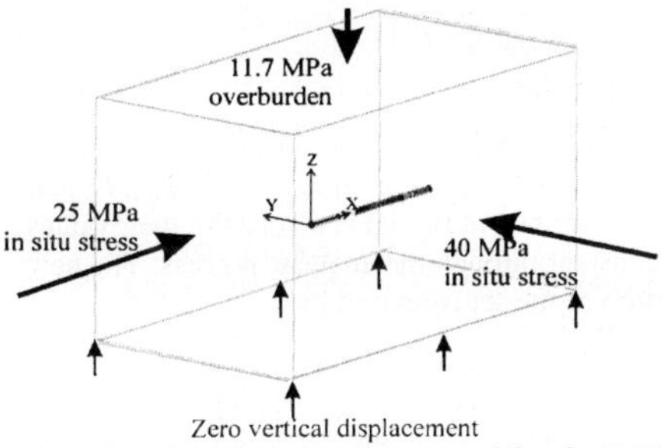

Figure 1. Boundary conditions used for the DOE mechanical calculations (Pahl et a.

2.2 Parameters

Two types of rock were modeled for the mechanical calculations: granite and lamprophyre. Most of the required input parameters for these analyses came directly from GTS data. The rock properties used for this analysis are listed in Table 1 (Keusen et al., 1989). These properties include measured intact rock mechanical properties such as elastic modulus, and a measured friction angle for their fractures. Two vertical joint sets were modeled in the mechanical calculations. Their orientations with respect to the axis of the FEBEX drift were calculated and used in the input parameters. Table 2 lists the measured and derived in situ properties related to fracture development in three regions of the 71.4m-long FEBEX drift: granite with low fracture density (primary rock); granite in the shear zone; and lamprophyre. These in situ properties include bulk permeability K_b (Guimerà et al., 1998) and fracture frequency data (Pardillo and Campos, 1996) taken from several nearby boreholes.

The average value of porosity for the granite was 0.007 (Keusen et al., 1989); this value was also used for the lamprophyre due to lack of specific information. Similarly, the granite matrix permeability was estimated as $5.\times10^{-19}$ m^2, which was also used for lamprophyre. The compressibility of the rock was estimated from mechanical properties for intact rock as 1.46×10^{-11} Pa^{-1} for granite and 2.41×10^{-11} Pa^{-1} for lamprophyre. The permeability of the fractures was calculated using the cubic law.

Table 1. Rock properties used for mechanical calculations.

Property	Granite	Lamprophyre
Elastic modulus E, intact rock (GPa)	53.3	42.4
Poisson's ratio ν	0.37	0.33
Coeff. of friction $\mu=\tan(\phi)$ (fracture)	0.65	0.64
Cohesion *C* (MPa)	20	28.5
Half-closure stress *A* (MPa)	1.0	1.0
Elastic joint shear stiffness G_s (MPa/m)	1.0E+6	1.0E+6
Inelastic joint shear stiffness G_s' (MPa/m)	1.0E+3	1.0E+3

Table 2. In situ fracture properties used for mechanical calculations.

Property	Granite	Granite in shear zone	Lamprophyre
Bulk permeability K_b (m²)	4.7E-18	1.8E-15	5.0E-17
Fracture freq. n (frac./m)	1.5	5.6	5.5
Initial fracture aperture *b* (μm)	3.4	15.7	4.8
Max. closure of the joint V_{max} (μm)	S1/S2: 138 S3/K2: 87	S1/S2: 643 S3/K2: 408	S1/S2: 196 S3/K2: 124

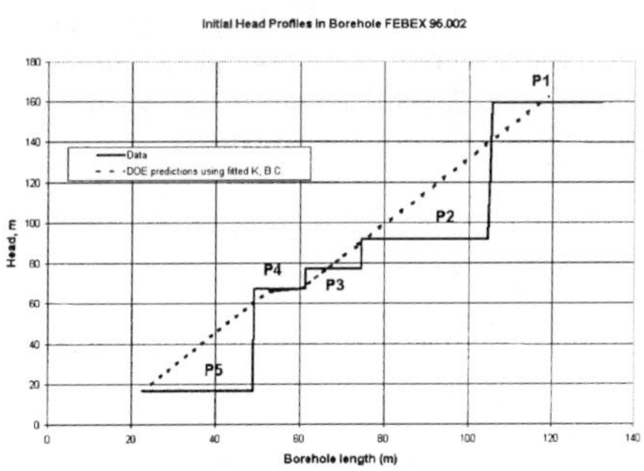

Figure 2. Initial head conditions used for the DOE hydrological calculations

The initial conditions and permeabilities of the in situ rock in the TOUGH2 model were adjusted to reflect the head profiles measured in a nearby, roughly parallel borehole, FEBEX 95.002 (Meier et al., 1995). The P4 and P3 intervals are the packed sections of the borehole for which pressure predictions during drift construction were required; the drift construction proceeds in a direction going from P5 to P1. The initial head profile used in the calculations, and a comparison with the measured values, are shown in Figure 2. At the ends of the mesh, hydrostatic boundary conditions were assumed, which resulted in a pressure distribution within the model. The flow is generally parallel to the borehole during this initial phase, which is in general agreement with the pressure contours shown in Meier et al. (1995).

3. JNC MODEL

The equivalent continuum model used for the JNC model is developed by combining the crack tensor theory proposed by Oda (1986) and the constitutive model by Barton et al. (1985). This model simulates the anisotropy of the media by crack tensor theory and the non-linear mechanical and hydraulic behavior by the Barton-Bandis (BB) model. Thus, the properties are changed heterogeneously and anisotropically through the excavation process. The constitutive law is incorporated into the coupled THM code, THAMES (Ohnishi, et al., 1985).

3.1 Model Description

The Barton-Bandis (BB) model represents the change in the normal and shear stiffness due to the stress change by using *JRC* (Joint Roughness Coefficient) and *JCS* (Joint Compression Strength). By using the JRC_n^I and JCS_n^I (MPa) which is the mean values for each set *I*, the initial normal stiffness of joint, K_{ni} (MPa/mm), is calculated for each set *I*,

$$K_{ni}^I = -7.15 + 1.75 JRC_n^I + 0.02\left(JCS_n^I / a_j^I\right) \tag{4}$$

$$a_j^I = \frac{JRC_n^I}{5}\left(0.2\frac{UCS}{JCS_n^I} - 0.1\right) \tag{5}$$

where a_j(mm) is the initial joint aperture, and *UCS* is the uniaxial compression strength of rock. The normal stress, σ'_n, is calculated for each element. The normal and shear stiffness of the joint are revised by using the normal stress at each time step.

$$K_n^I = K_{ni}^I\left[1 - \frac{\sigma'^I_n}{V_m K'^I_{ni} + \sigma'^I_n}\right]^{-2} \tag{6}$$

$$K_s^I = \frac{100}{L^I}\sigma_n^I \tan\left[JRC_n^I \log_{10}\left(\frac{JCS_n^I}{\sigma'^I_n}\right) + \phi'^I_r\right] \tag{7}$$

V_m in Eq. 6 is assumed to be the same as a_j^I of Eq. 5. L^I is the mean fracture length of the set (mm) and ϕ_r is the residual friction angle. By using $h^I = K^I_n L^I$, $g^I = K^I_s L^I$, the elastic compliance C_{ijkl} and hydraulic conductivity tensor k_{ij} are defined by

$$C_{ijkl} = \sum_I \left(\frac{1}{h^I} - \frac{1}{g^I}\right)F_{ijkl}^I + \frac{1}{4g^I}\left(\delta_{ik}F_{jl}^I + \delta_{jk}F_{il}^I + \delta_{il}F_{jk}^I + \delta_{jl}F_{ik}^I\right) \tag{8}$$

$$k_{ij} = \sum_I \frac{1}{12} e^{I^3}\left(P_{kk}^I \delta_{ij} - P_{ij}^I\right) \tag{9}$$

where the summation is carried out for sets. δ_{ij} is the Kronecker delta. The equations for F_{ijkl}, F_{ij} and P_{ij} are described in Barton et al. (1985); their values are constant through the analysis process. The new aperture of the set is revised by

$$e^I = e^I_{\text{int}} - \frac{\Delta\sigma_n^I}{K_n^I}, \quad e^I_{\text{int}} = \sqrt[3]{\frac{18 \times \mu_l \times K_{mean}}{9.8 \times P_{KK}^I}} \tag{10}$$

where μ_l is the viscosity of water., and K_{mean} is the mean hydraulic conductivity.

By using the above non-linear equivalent model, the coupled equations are solved in three dimensions in the coupled THM code THAMES. The continuity equation includes gravity, porosity, the degree of saturation, and the compressibility of water. The equilibrium equations are written as

$$\left[\frac{1}{2}T_{ijkl}^{-1}\left(u_{k,l} + u_{l,k}\right) + \chi T_{ijkl}^{-1} C_{kl} \rho_l h\right] + \rho b_i = 0 \tag{11}$$

where $T_{ijkl} = (M_{ijkl} + C_{ijkl})$, in which M_{ijkl} is the elastic compliance of the rock matrix. $C_{ij} = C_{ijkl}\,\delta_{kl}$..

3.2 Parameters

The geometric information of the fracture sets is mainly obtained from Keusen et al. (1989). Table 3 shows the data used in this analysis. The uniaxial compression strength is set at the average of the values indicated in UPC (1999). JCS is assumed to be 80% of the uniaxial compression strength.

The upper boundary has a prescribed total water head of 0.8MPa and the other boundaries are set at no flow condition. The mechanical boundary condition was a roller condition for all boundaries. The intact rock mechanical properties were similar to those used for the DOE analysis. The transmissivities in UPC (1999) were used in a kriging process to develop a heterogeneous hydraulic conductivity profile throughout the computational mesh. Eq. 10 was applied to get the initial aperture distribution. Hydraulic conductivities in the range of 4×10^{-12} to 6×10^{-9} m/s were calculated using this method.

Table 3. Fracture geometry used in the JNC analysis

	Set	Joint System	Density (N/m^3)	Length (m)	Dip dir./ slope	JRC	JCS (MPa)	UCS (MPa)
Mat. 1	1	S1+S2	1.00	2.44	155/70	2.99	82.4	103
	2	S3	0.83	1.63	188/81			
	3	K2	1.05	1.83	210/75			
	4	K3	0.72	2.41	258/80			
Mat. 2	3	Lamprophyre	10.5	1.83	210/75	2.81	63.5	79.4

Two cases were calculated. Case 1 used the non-linear anisotropic mechanical and hydrologic properties. Case 2 used a linear isotropic elastic mechanical model and a heterogeneously anisotropic hydrologic model. In Case 2, the crack tensor is not considered, so the deformation is carried out based on the properties of the intact rock. Cases 1 and 2 have homogeneous mechanical properties in the lamprophyre zone and the granite zone at the initial conditions. In Case 2, the properties are constant through the excavation process, while the mechanical and hydraulic properties are changed though the excavation process in Case 1.

4. RESULTS

The analytical results were obtained from "blind" analyses; i.e., the measured flow rate and pressure histories were not known to the analysts prior to performing these calculations. The analyses were then modified slightly for this report to solve problems identified early by the analysts. The predicted flow rates into the FEBEX drift 100 days after excavation, compared with the measured value, are shown in Table 4. The predicted pressure histories in sections P3 and P4, compared to the measured values, are shown in Figure 3.

The results from the DOE calculations were very encouraging. The predicted flow rate was close to the measured value. The initial pressures in P3 and P4 match the data very closely, and the pressure drops match the data well temporally. However, the calculations overpredict the magnitude of the pressure drop. In the investigation of this problem, it was noted that there is an apparent mass-balance problem with TOUGH2; the mass of the fluid is based on the fracture and matrix porosities and saturations, and there is no mechanism to change the porosities and conserve water mass. Additionally, there is a slight increase in pressure in P4 and P3 as the alpine miner approaches them. The DOE analysis did not capture this pressure increase.

Table 4. Measured and Predicted Flow Rate in the FEBEX Drift After 100 Days

	Flow rate (ml/min)
Measured	7.8
DOE	10.1
JNC, Case 1	0.39
JNC, Case 2	1.04

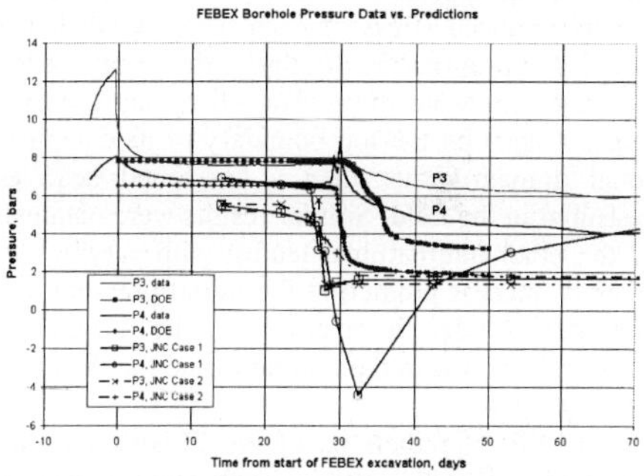

Figure 3. Predicted and measured pressure histories in sections P3 and P4 of borehole FEBEX 95.002.

The predicted flow rates from the JNC analyses are less than the measured flow. The flow rate for Case 1 is smaller than that of Case 2, which infers that the non-linear anisotropic model predicts a greater rock deformation and thus a greater reduction of the hydraulic conductivity around the drift. The predicted pressure at P4 is drastically changed by excavation in Case 1. The pressure becomes negative and then recovers gradually to the initial state. The effects of the drift construction are predicted to happen two to three days before they are actually witnessed in the data. On the other hand, there is little difference in behavior at P3 for the two cases. This means that the influence of excavation at P3 is very small and the non-linear effect there is negligible. The overall pressure values in P3 are less than the data, which is likely due to the choice of hydrologic boundary conditions.

5. CONCLUSIONS

The results of the two analyses described in this report illustrate the difficulties inherent in implementing a coupled hydrological-mechanical model for repository-related processes. The DOE analysis using a compliant joint model effectively modeled the changes in permeability, but identified continuity problems in the hydrologic portion of the code. The DOE team has been developing a fully-coupled THM code based on the TOUGH2 and JAS3D codes, and the issue of mass continuity is being dealt with so that the mechanical changes to porosity will be appropriately included in the hydrological calculation. The inability of the DOE analysis to capture the pressure increases coincident with the approach of the alpine miner indicates that the induced stress of excavation is not being modeled appropriately. Similarly, the results of the JNC analyses were more affected by the use of a constant head on the top boundary of its computational domain rather than a horizontal head as measured in the field. Similar results were obtained by the other international teams, with varying degrees of success predicting the initial pressure, the pressure rise due to excavation, the asymptotic pressures reached in the borehole, and the flow rate after 100 days.

These blind predictions of the FEBEX data do not make a strong case that, for this particular geomechanical situation, a coupled analysis is entirely necessary. The granite in this case is sparsely fractured, and most of the inflow occurs at the lamprophyre and other more fractured areas. Also, the rock mass is sufficiently nonporous and saturated that inelastic deformation of the rock matrix is not a significant issue for repository performance. However, the exercise was very valuable for developing rationale for modeling the more complex coupled problems associated with the introduction of the bentonite barrier and the heat of the simulated waste.

6. REFERENCES

Barton, N., Bandis, S., and Bakhtar, K. 1985. “Strength, Deformation and Conductivity Coupling of Rock Joints.” *International Journal of Rock Mechanics and Mining Sciences & Geomechanics Abstracts*, 22(3), pp. 121-140. Oxford, United Kingdom; New York, New York: Pergamon Press.

Blanford, M.L. 1999. *JAS3D - A Multi-Strategy Iterative Code for Solid Mechanics Analysis, Users' Instructions, Release 1.6.* Internal SNL document. Albuquerque, New Mexico: Sandia National Laboratories.

Chen, E.P. 1991. *A Constitutive Model for Jointed Rock Mass with Two Intersecting Sets of Joints.* SAND89-0592. Albuquerque, New Mexico: Sandia National Labs.

Guimerà J., Carrera J., Martínez L., Vázquez E., Ortuño F., Fierz T., Bülher C., Vives L., Meier P., Medina A., Saaltink M., Ruiz B. and Pardillo J. (1998). *FEBEX Hydrological characterization and modelling.* UPC, 70-UPC-M-0-1001, Jan. 1998.

Keusen H.R., Ganguin J., Schuler P. and Buletti M. (1989) *Grimsel Test Site. Geology.* NAGRA, NTB 87-14E, Feb. 1989.

Meier P., Fernández P., Carrera J. and Guimerà J. (1995) FEBEX - PHASE 1. *Results of hydraulic testing in boreholes FBX 95.001, FBX 95.002, BOUS 85.001 and BOUS 85.002.* UPC, Dec. 1995.

Oda, M., 1986. An Equivalent Continuum Model for Coupled Stresses and Fluid Flow Analysis in Jointed Rock Masses, *Water Resour. Res.* 22, 13, pp.1845-1856.

Ohnisi, Y. Shibata, H. Kobayasi, A., 1985. Development of finite element code for the analysis of coupled thermo-hydro-mechanical behaviors of saturated-unsaturated medium, *Proc. Int. Symp. Coupled Processes Affectting the Performance of a Nuclear Waste Repository*, Berkeley, 263-268.

Pahl A., Heusermann St., Bräuer V. and Glöggler W. 1989. *Grimsel Test Site. Rock stress investigations.* NAGRA, NTB 88-39E, Apr. 1989.

Pardillo J. and Campos R. (1996). *FEBEX - Grimsel Test Site (Switzerland) Considerations respect to the fracture distribution.* CIEMAT, 70-IMA-L-2-05, Mar. 1996.

Pruess, K. 1991. *TOUGH2 – A General-Purpose Numerical Simulator for Multiphase Fluid and Heat Flow.* LBL-29400. Berkeley, California. Lawrence Berkeley Laboratory.

UPC (Technical University of Catalonia), 1999. *DECOVALEX III, Task 1: Modeling of FEBEX in-situ test: General Specifications.* Barcelona , Spain.

ANALYSES OF COUPLED HYDROLOGICAL-MECHANICAL EFFECTS DURING DRILLING OF THE FEBEX TUNNEL AT GRIMSEL

J. Rutqvist[1], A. Rejeb[2], M. Tijani[3], C.-F. Tsang[1]
[1]) Ernest Orlando Lawrence Berkeley National Laboratory (Berkeley Lab), USA
[2]) Institut de Radioprotection et de Sûreté Nucléaire, France
[3]) Ecole des Mines de Paris, France

Abstract: This paper presents analyses of coupled hydrological-mechanical (HM) processes during drilling of the FEBEX tunnel, located in fractured granite at Grimsel, Switzerland. Two and three-dimensional transient finite-element simulations were performed to investigate HM-induced fluid-pressure pulses, observed in the vicinity of the FEBEX tunnel during its excavation in 1995. The results show that fluid-pressure responses observed in the rock mass during TBM drilling of the FEBEX tunnel could not be captured using current estimates of regional stress. It was also shown that the measured pressure responses can be captured in both two and three-dimensional simulations if the stress field is rotated such that contraction (compressive strain rate) and corresponding increases in mean stress occur on the side of the drift, where increased fluid pressure spikes were observed.

1 INTRODUCTION

When tunnels are excavated in saturated rocks, peculiar responses in the fluid pressure have been observed in surrounding rocks. One example is the excavation of the FEBEX tunnel, located in fractured hard rock at the Grimsel Test Site (GTS), in Switzerland (McKinley et al., 1996). During tunnel-boring-machine (TBM) excavation of the FEBEX tunnel, distinct increases in fluid pressure were observed in boreholes located a few meters away from the drift wall (Figure 1). The increased fluid pressure typically occurred during ten-hour-long working shifts of active tunnel boring. Fluid pressure decreased during time periods of no tunnelling activity. These pressure responses are believed created by a rapid squeezing of the porous rock under increasing mechanical load in a coupled hydrological-mechanical (HM) process.

As a part of the international DECOVALEX III project, several independent research teams have analysed coupled HM effects during TBM drilling of the FEBEX tunnel. This paper presents the analyses conducted by two different research teams: Institut de Radioprotection et de Sûreté Nucléaire (IRSN) in collaboration with Ecole des Mines de Paris (EMP), who conducted a two-dimensional numerical modelling, and the Lawrence Berkeley National Laboratory (Berkeley Lab), who conducted a three-dimensional transient numerical modelling. The paper focuses on the pressure responses in borehole section P4, which are most distinct.

2 TWO DIMENSIONAL MODELLING

A two-dimensional finite-element modelling was carried out by the IRSN/EMP to analyse HM effects during excavation of a circular hole in a homogeneous, isotropic, porous elastic medium. The finite-element coupled codes VIPEF (mechanic) and HYDREF (hydraulic) was used for this HM analysis (Tijani, 1996).

2.1 *Model of the FEBEX tunneling*

In this analysis, the transient tunnelling process was simulated in a two-dimensional section across the FEBEX tunnel. A coupled HM analysis was conducted using a Biot (1941) model with Young's modulus E = 24.68 GPa, Poisson's ratio $\nu = 0.37$, Biot's coefficient b = 1 (Terzaghi assumption), and a Biot's modulus M equal to infinity (the storage phenomena is caused only by skeleton strain). The hydraulic permeability was set to 7×10^{-18} m^2 after model calibration against observed water inflow into the FEBEX tunnel.

The TBM excavation process was modelled by a decrease of the effective force and pore pressure on the wall of the tunnel. This decrease depends on the time and the distance to the front of the excavation. The excavation was simulated according the excavation sequence shown in Figure 1 for an initial effective stress of: $\sigma_H = \sigma_h = 29.4$ MPa (horizontal maximum and minimum compressive), and $\sigma_v = 7.14$ MPa (vertical).

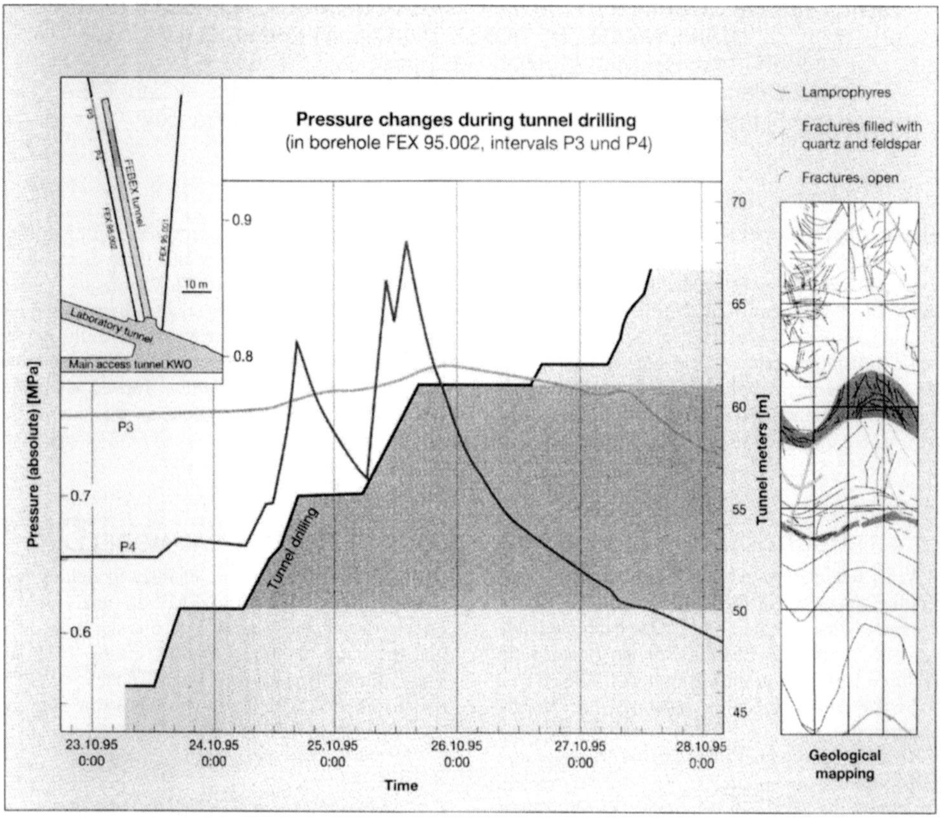

Figure 1. Pressure changes in borehole FEX 95.002 induced by the TBM drilling of the FEBEX tunnel at the Grimsel Test Site (McKinley et al., 1996).

2.2 Strain rate

Figure 2 shows the calculated distribution of average volumetric strain rate for the assumed stress field. The location of six points (θ= 30° to θ = 69°) where fluid pressure changes were evaluated and interval P4 of borehole FEX 95.002 are also shown. Simulation results, as well as analytical solutions, reveal that a pore-pressure increase will only occur at the four points located in the contracted zone of compressive strain rate (θ > 45° in Figure 2). However, the P4 interval (θ = 14° in Figure 2) is located in the zone of extensional strain rate, and therefore no pore pressure increase can occur in at that location for the assumed stress field.

2.3 Pore pressure variations

Figure 3 presents the simulated pore pressure versus time at the 6 points described above and the measured pore pressure at the P4 interval of the FEBEX.95002. The pore pressure measured in the P4 interval (θ = 14°) agree well with the numerical results obtained from the point located at θ = 53°. This means that a good agreement between measured and simulated pressure response at P4 could be achieved merely by changing the orientation of the initial stress.

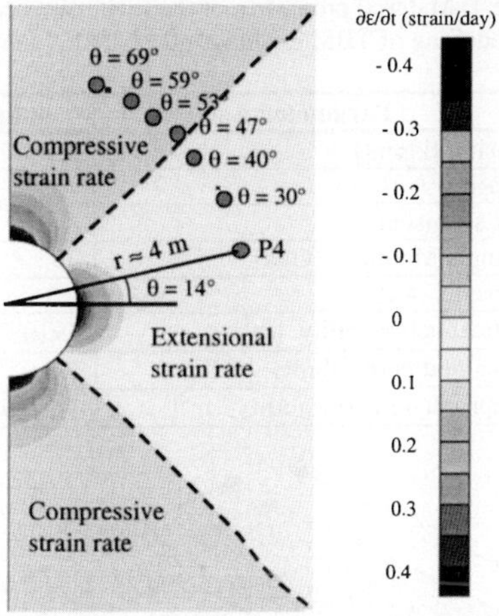

Figure 2. Simulated strain rates distribution indicating zones of contraction (compressive strain rate) and extension (extensional strain rate) and locations for evaluation of pore-pressure variations.

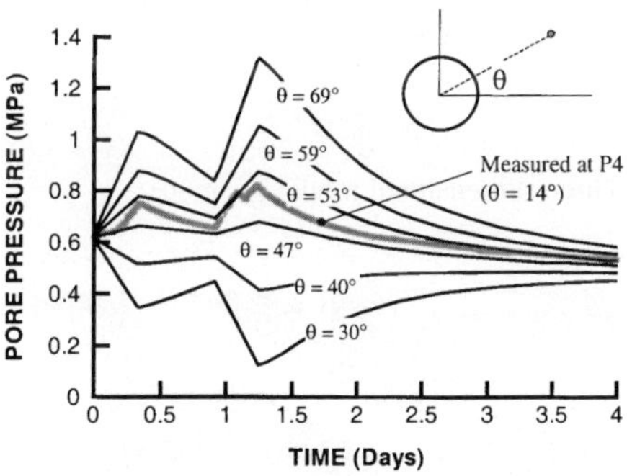

Figure 3. Simulated pore-pressure responses at points with locations shown in Figure 2, compared with measured pressure response in interval P4 of borehole FEX 95.002. Time 0 in this figure corresponds to 7:00 on October 14 in Figure 1.

2.4 *Influence of permeability and excavation duration*

The influence of rock permeability and excavation duration on the pore-pressure evolution was investigated. In this context, relative time is defined as $\tau = t / T$, where t is time and T is the excavation duration. Figure 4 presents the pore-pressure evolution during excavation at the point located at $\theta = 53°$ (Figure 2) with variation of permeability and excavation duration. The permeability is varied according to $k = \lambda k_0$, where $k_0 = 7\times10^{-18}$ m^2 (reference case) and λ takes the values 0.1, 0.2, 1, 2, 5, 10, 100, and infinity. In addition, it can be shown that the curves in Figure 4 also represent variations in excavation duration according to $T = \lambda T_0$, where $T_0 = 0.35$ day. Significantly, for $k \leq 7\times10^{-18}$ m^2 ($\lambda \leq 1$), a maximum pressure increase of 0.8 MPa was obtained. On the other hand, for a permeability two orders of magnitude higher $\lambda = 100$) no significant pressure increase occurred. Thus low rock permeability is necessary to simulate fluid pressure responses observed during TBM drilling of the FEBEX tunnel.

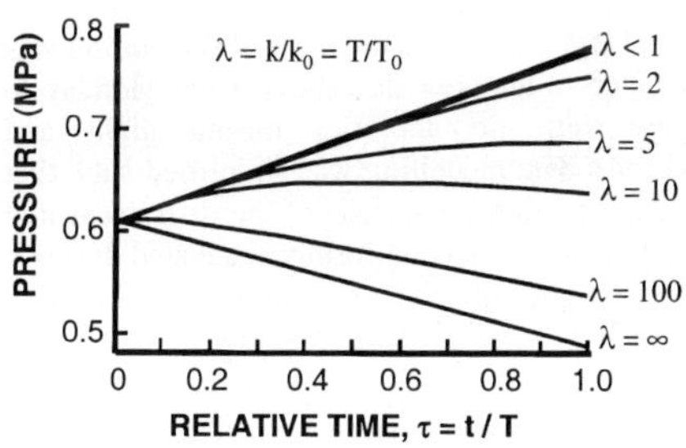

Figure 4. Simulated pore pressure evolution with variation of permeability and excavation duration.

3 THREE-DIMENSIONAL MODELLING

A fully coupled three-dimensional HM modelling of the TBM drilling for the FEBEX tunnel was conducted by Berkeley Lab, using the finite-element code ROCMAS (Rutqvist et al., 2001).

3.1 *Model conceptualization of FEBEX tunnelling*

A finite-element mesh was constructed as shown in Figure 5. The hydrological properties of the different units within the finite-element model was first estimated from local geohydrological field data and then calibrated to match observed head distribution and inflow into the open drift. An initial stress was assigned according to σ_v = 10 MPa, σ_h = 15 MPa, and σ_H = 30 MPa, where σ_H is oriented 45° from the axis of the FEBEX tunnel (Figure 5b). These values are within the range of stress measurements in the GTS area (Pahl *et al.*, 1989).

The material properties are given in Table 1. Except for permeability, the properties of the high-permeability zones (Lamprophyres) and the surrounding rock are the same. The mechanical rock-mass properties are obtained from the geological description of the GTS (Keussen et al., 1989). Significantly, the Young's modulus of the rock mass was reduced to 70% of its value for intact rock.

In the Berkeley Lab's numerical modelling, the FEBEX tunnel was excavated according to the actual TBM boring schedule. The TBM boring was conducted in ten-hour day shifts, from Monday to Friday, with no activities during night and weekends. The modelling was simplified such that one daily section—a section of the drift excavated during one day—was uniformly excavated during a ten-hour work shift. The excavation process in each drift section was simulated as follows:

1. Young's modulus of elements within the excavated tunnel section was set to zero.
2. Internal stress and fluid pressure within the tunnel were linearly reduced with time over about ten hours.
3. Final condition at the end of the work shift was zero internal stress and zero fluid pressure within the excavated tunnel.

The entire one-month TBM drilling of the FEBEX tunnel was simulated.

Table 1. Material properties of the rock mass used in modelling of TBM drilling of the FEBEX tunnel

Parameter	Value
Density, [kg/m^3]	2,700
Porosity	0.01
Biot's constant, b	1.0
Young's Modulus, [GPa]	35
Poisson's ratio,	0.3
Vertical permeability, [m^2]	5×10^{-18}
Horizontal permeability, [m^2]	5×10^{-19}
Lamprophyre permeability, [m^2]	1.1×10^{-17}

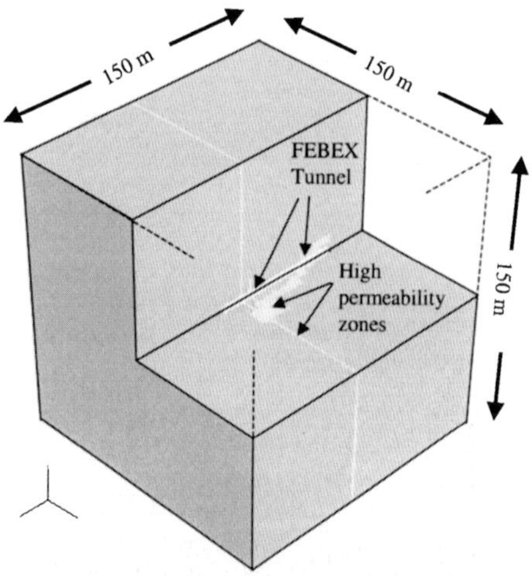

(a) Three-dimensional model geometry

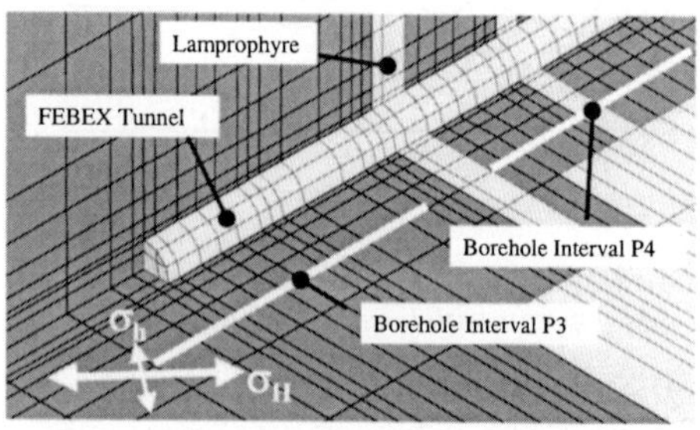

(b) Detailed view of the near field of FEBEX

Figure 5. Finite-element model for a fully coupled simulation of the one-month TBM drilling for the FEBEX tunnel.

3.2 Result of fluid pressure variation

Figure 6 presents simulated changes in mean stress and fluid pressure at 18:00 on October 25, which is at the end of one-day shift, when the excavation had reached 61 m (Figure 1). Figure 6b shows that the fluid pressure has increased in two zones: a zone around the front-left side of the excavation and a zone above the excavation, near its front. Figure 6a shows that near these two zones of pressure increase; the mean stress has increased as a result of the excavation. In contrast, the fluid pressure on the side of the drift has decreased where the mean stress has decreased. This shows that HM coupled processes induce these changes in fluid pressure.

Figure 6 also illustrates that the HM-induced changes in fluid pressure are temporal. For example, the HM-induced fluid-pressure increase above the drift is diminished 5 to 10 m behind the excavation front. Thus, away from the active TBM drilling, the open boundary at the tunnel wall controls the fluid pressure.

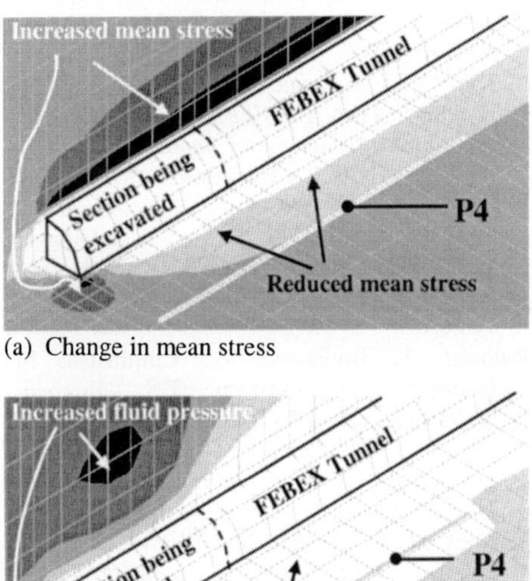

(a) Change in mean stress

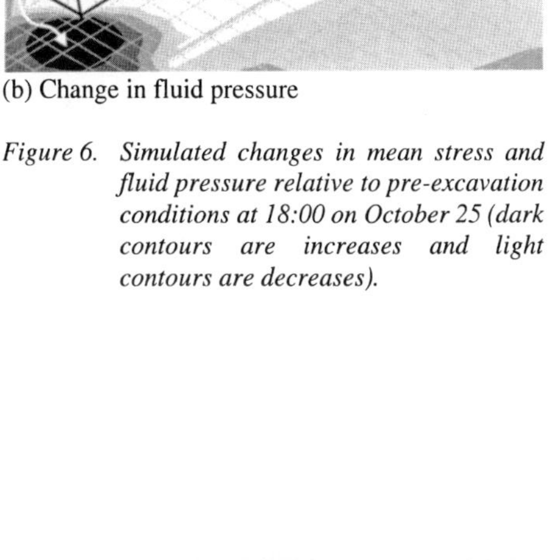

(b) Change in fluid pressure

Figure 6. Simulated changes in mean stress and fluid pressure relative to pre-excavation conditions at 18:00 on October 25 (dark contours are increases and light contours are decreases).

3.3 Predicted pressure evolution

Figure 7 presents predicted and measured pressure responses in interval P4. During three consecutive day shifts of active tunnel boring, both predicted and measured pressure respond when the excavation front passes parallel to P4 (between 7:00, October 23 to 18:00, October 25, in Figure 1). However, the predicted pressure responses are the opposite of what was measured. The measured pressure shows two peaks at 714 and 738 hours. In the simulation, on the other hand, the pressure decreases at 714 and 738 hours.

3.4 Calibrated pressure evolution

A parameter variation showed that the HM-induced pressure responses depend on many parameters, including rock-mass deformation modulus, Biot's coupling constants, hydraulic permeability, and the magnitude and orientation of the *in situ* stress field. The three material parameters affect only the magnitude of the HM-induced pressure response. On the other hand, the magnitude and direction of the *in situ* stress field are important factors that determine where and when the fluid pressure will increase or decrease.

An attempt was made to match the field test responses by adjusting the initial stresses in the model. It was concluded that it is possible to produce increasing pressure peaks at 714 and 738 hours in interval P4 by changing orientation and magnitude of the stress field. Basically, the vertical stress had to be substantially increased such that the TBM drilling produces a mean stress increases on the side of the tunnel. Figure 8 presents one example in which the *in situ* stress field has been rotated in order to match simulated and measured fluid pressure responses.

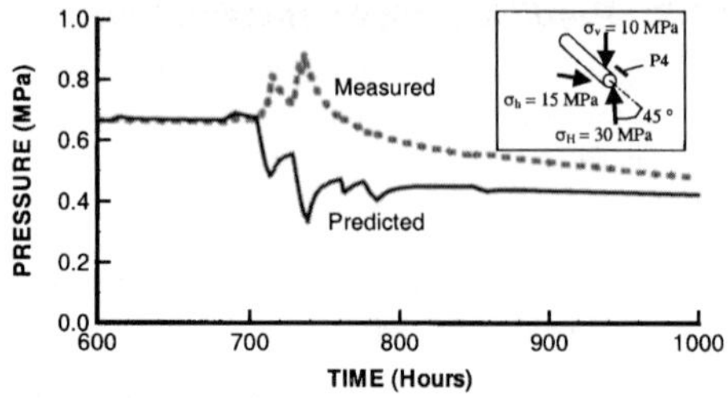

Figure 7. *Predicted and measured fluid pressure evolution in interval P4 of borehole FEX 95.002, for estimated regional stress field at GTS area.*

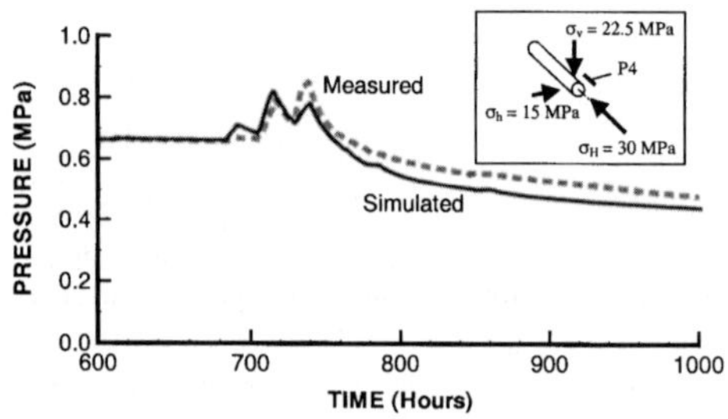

Figure 8. *Simulated and measured fluid pressure evolution in interval P4 of borehole FEX 95.002, for adjusted local stress field at the FEBEX tunnel.*

4 CONCLUDING REMARKS

The results from two and three-dimensional numerical analyses show that fluid-pressure responses observed in the rock mass during TBM drilling of the FEBEX tunnel could not be captured using current estimates of regional stress. It was shown that the measured pressure responses can be captured in both two and three-dimensional simulations if the stress field is rotated such that contraction (compressive strain rate) and corresponding increases in mean stress occur near borehole FEX 95.002 on the side of the drift. From the results of the two-dimensional analysis, it appears that good agreement between measured and simulated evolution of fluid pressure could be obtained if the maximum principal stress were rotated about 40° (53° minus 14° in Figure 2) from the horizontal. Such a rotation of the local *in situ* stress field is not unrealistic, especially considering the presence of the Lamprophyre zones and other geological features.

ACKNOWLEDGEMENTS

Review and comments by Dan Hawkes, Lawrence Berkeley National Laboratory are much appreciated. Financial support for the Berkeley Lab's research team was provided by a grant from the Swedish Nuclear Power Inspectorate.

REFERENCES

Biot, M. A. 1941. *General theory of three-dimensional consolidation.* J. Applied Physics 12: 155-164.

Keusen H.R., Ganguin J., Schuler P. & Buletti M, Grimsel. 1989. *Test Site-Geology.* NAGRA, NTB 87-14E.

McKinley, I., Kickmaier, W., del Olmo, C., & Huertas F. 1996. *The FEBEX project: full-scale simulation of engineered barriers for a HLW repository.* NAGRA Bulletin No. 27, NAGRA, Switzerland.

Pahl, A., Heusermann, St., Bräuer, V. & Glöggler, W. 1989. *Grimsel Test Site. Rock stress investigations.* NAGRA, NTB 88-39E, Apr. 1989.

Rutqvist, J., Borgesson, L., Chijimatsu, M., Kobayashi, A., Nguyen, T.S., Jing, L., Noorishad, J. & Tsang, C.-F. 2001. *Thermohydromechanics of partially saturated geological media: governing equations and formulation of four finite element models.* Int. J. Rock. Mech. Min. Sci. 38(1): 105-127.

Tijani, S.M. 1996. Short description of VIPLEF code. *Coupled thermo-hydro-mechanical processes of fractured media.* Developments in Geotechnical Engineering, 79: 507 – 511.

THERMOMECHANICAL MODEL FOR COMPACTED BENTONITE

Petri Jussila
Helsinki University of Technology
Institute of Mathematics

Abstract: A general thermomechanical model is derived for an arbitrary mixture. The system is described by proper choices of free energy and dissipation function. The general theory is applied to a TH model for a mixture of compacted bentonite, liquid water, vapour, and air. The free energy of the system is chosen to take into account the individual behaviours of the components and their mutual interactions through adsorption, vaporisation, and mixing of the gaseous components. An important tool is the equilibrium condition for the water species in different combinations of the components. The resulting thermodynamically consistent macroscopic model is fitted to a suction experiment and applied to a simple 1D TH simulation of the bentonite buffer of the Febex in situ test. The results calculated with FEM are compared to measurements.

1. INTRODUCTION

A buffer of compacted bentonite is planned to be used to prevent the movement of groundwater and the consequential escape of material from a geological repository for spent nuclear fuel. Fluid flow, phase changes, mechanical behaviour of the buffer, rock, and the waste canisters, and the heat produced by the waste constitute a coupled thermo-hydromechanical system. The aim of the study is to derive a general thermodynamically consistent THM model for an arbitrary mixture. The general theory is applied to the thermohydraulic modelling of a mixture of compacted bentonite, liquid water, vapour, and air.

The system is described by proper choices of free energy and dissipation function. The free energy of the system is chosen to take into account the individual behaviours of the components and their mutual interactions through adsorption, vaporisation, and mixing of the gaseous components. The choice is based on the equilibrium conditions for the water species in different combinations of the components.

The model is fitted to a suction experiment for Febex bentonite and applied to the TH simulation of the bentonite buffer of the Febex in situ test, which is considered in the international Decovalex 3 project. The present approach is to describe the essential features of the TH behaviour of the buffer in a simple 1D geometry. The results calculated with FEM are compared to the measurements.

2. THE GENERAL MODEL

The system consists of solid skeleton (s), liquid water (l), water vapour (v), and air (a). Vapour and air occupy the same gaseous (g) volume fraction. The molar volume fraction $\xi_k = \beta_g x_k$ for $k \in \{a, v\}$ and $\xi_k = \beta_k x_k$ for $k \in \{s, l\}$ of component k is defined by means of the volume fraction $\beta_j = V_j/V$ of the phase $j \in \{s, l, g\}$ and the molar fraction $x_k = n_k/(n_a+n_v)$ for $k \in \{a, v\}$ and $x_k = 1$ for $k \in \{s, l\}$ of component k, where n_k [mol] is the mole number of component k. Apparently, we have $\sum_k \xi_k = 1$. The molar volume fractions relate the apparent densities $\rho_k = m_k/V$ [kg/m^3] to the intrinsic densities $\tilde{\rho}_k = m_k/(x_k V_j)$ [kg/m^3], for which j = g for $k \in \{a, v\}$ and j = k for $k \in \{s, l\}$ by the equation $\rho_k = \xi_k \tilde{\rho}_k$.

We also define porosity $\eta = 1 - \xi_s$, liquid saturation $\chi = \xi_l/(1 - \xi_s)$, and vapour fraction $\zeta = \xi_v/(\xi_a+\xi_v)$ as alternative variables.

In the general model the state variables are chosen to be the molar volume fractions ξ_k, intrinsic densities $\tilde{\rho}_k$, and strains ε_k [-] of the components, and the common temperature T [K]. The indicator function taking care of the restriction for the molar volume fractions is

$$I(\xi_s,\xi_l,\xi_v,\xi_a) = \begin{cases} 0, & (\xi_s,\xi_l,\xi_v,\xi_a) \in C, \\ +\infty, & \text{otherwise,} \end{cases} \tag{1}$$

where the feasible set is

$$C = \left\{ (\xi_s,\xi_l,\xi_v,\xi_a) \in \mathbb{R}^4 \,\middle|\, \sum_k \xi_k = 1, \xi_k \geq 0, k \in \{s, l, v, a\} \right\} \tag{2}$$

The free energy [J/m^3] of the system is

$$\psi = \sum_{k \in \{s, l, v, a\}} \rho_k \psi_k = \sum_{k \in \{s, l, v, a\}} \rho_k \tilde{\psi}_k + TI(\xi_s,\xi_l,\xi_v,\xi_a), \tag{3}$$

where $\tilde{\psi}_k$ [J/kg] is the specific free energy of component k without the restriction (2). Dissipative behaviour of the system is characterised by the dissipative variables, which in the general work are chosen to be the rates of deformation $\mathbf{D}_k$ [1/s], the heat fluxes $\mathbf{q}_k$ [W/m^2], and the relative velocities $\mathbf{V}_k$ [m/s]. The dissipation function ϕ [W/m^3] is a function of the state variables and the dissipative variables.

We follow the procedure introduced by Hartikainen and Mikkola (1997) of using the principle of maximal rate of entropy production (Ziegler and Wehrli, 1988) with the orthogonality relations

$$\phi = \nu\left[\sum_k (\partial\phi/\partial\mathbf{D}'_k):\mathbf{D}'_k + (\partial\phi/\partial(\mathrm{tr}\mathbf{D}_k))\,\mathrm{tr}\mathbf{D}_k + (\partial\phi/\partial\mathbf{q}_k)\cdot\mathbf{q}_k + (\partial\phi/\partial\mathbf{V}_k)\cdot\mathbf{V}_k\right], \tag{4}$$

$$\nu = \phi\left[\sum_k (\partial\phi/\partial\mathbf{D}'_k):\mathbf{D}'_k + (\partial\phi/\partial(\mathrm{tr}\mathbf{D}_k))\,\mathrm{tr}\mathbf{D}_k + (\partial\phi/\partial\mathbf{q}_k)\cdot\mathbf{q}_k + (\partial\phi/\partial\mathbf{V}_k)\cdot\mathbf{V}_k\right]^{-1}, \tag{5}$$

As a result we get the following general constitutive relations

$$\boldsymbol{\sigma}'_k = \rho_k \frac{\partial\tilde{\psi}_k}{\partial\boldsymbol{\varepsilon}'_k} + \nu\frac{\partial\phi}{\partial\mathbf{D}'_k}, \tag{6}$$

$$p_k = -\rho_k\frac{\partial\tilde{\psi}_k}{\partial(\mathrm{tr}\boldsymbol{\varepsilon}_k)} - \nu\frac{\partial\phi}{\partial(\mathrm{tr}\mathbf{D}_k)} + p_k^{\mathrm{th}}, \tag{7}$$

$$s_k = -\frac{\partial\tilde{\psi}_k}{\partial T} - \frac{I}{\tilde{\rho}_k}, \tag{8}$$

$$-\frac{\nabla T}{T} = \nu\frac{\partial\phi}{\partial\mathbf{q}_k}, \tag{9}$$

$$-\mathbf{m}_k - \sum_j\left(\rho_k\frac{\partial\tilde{\psi}_k}{\partial\xi_j} + \xi_k B\right)\nabla\xi_j + \sum_j\left(\rho_j\frac{\partial\tilde{\psi}_j}{\partial\xi_k} + \xi_j B\right)\nabla\xi_k = \nu\frac{\partial\phi}{\partial\mathbf{V}_k}, \tag{10}$$

$$p_k^{\mathrm{th}} = \xi_k\tilde{\rho}_k^2\frac{\partial\tilde{\psi}_k}{\partial\tilde{\rho}_k} - \xi_k T I, \tag{11}$$

$$G_q - G_p = 0, \tag{12}$$

where $\boldsymbol{\sigma}'_k$ [Pa] is the deviatoric Cauchy stress, p_k [Pa] is the pressure,

$$p_k^{\mathrm{th}} = \xi_k\left(B + \sum_j \rho_j\frac{\partial\tilde{\psi}_j}{\partial\xi_k}\right) \tag{13}$$

is the relative thermodynamical pressure [Pa], B [Pa] is the relative hydrostatic pressure for which

$$B = \left\{B_k \forall k \in \{\mathrm{s, l, v, a}\} \,\middle|\, (B_s, B_l, B_v, B_a) \in T\partial I(\xi_s, \xi_l, \xi_v, \xi_a)\right\}, \tag{14}$$

s_k [J/(kgK)] is the specific entropy, $\mathbf{m}_k$ [N/m^3] is the rate of production of linear momentum, and

$$G_k = \tilde{\psi}_k - \frac{1}{2}\mathbf{U}_k\cdot\mathbf{U}_k + \frac{p_k^{\mathrm{th}}}{\rho_k} \tag{15}$$

is the generalised specific Gibbs function [J/kg], where $\mathbf{U}_k$ [m/s] is the absolute velocity of component k.

3. THERMOHYDRAULIC MODELLING OF BENTONITE

3.1 Constitution

The general theory is applied to thermohydraulic modelling of bentonite buffer with an assumption of a rigid skeleton. We get the constitution from the general constitutive relations with appropriate choices of the free energies and the dissipation function. The chosen specific free energies of the components are the following

$$\psi_s(T,\tilde{\rho}_s,\xi_j) = -c_s T\ln\frac{T}{T_0} + \frac{T}{\tilde{\rho}_s}I(\xi_j), \tag{16}$$

$$\psi_l(T,\tilde{\rho}_l,\xi_j) = -c_l T\ln\frac{T}{T_0} + \frac{RT}{M_v}h(\xi_s,\xi_l) + \frac{T}{\tilde{\rho}_l}I(\xi_j), \tag{17}$$

$$\psi_v(T,\tilde{\rho}_v,\xi_j) = -c_v T\ln\frac{T}{T_0} + \frac{RT}{M_v}\ln\frac{\tilde{\rho}_v}{\tilde{\rho}_{v,0}} + \frac{RT_0}{M_v}\frac{\tilde{\rho}_{v,0}}{\tilde{\rho}_v} + f(T) + \frac{RT}{M_v}g_v(\xi_j) + \frac{T}{\tilde{\rho}_v}I(\xi_j), \tag{18}$$

$$\psi_a(T,\tilde{\rho}_a,\xi_j) = -c_a T\ln\frac{T}{T_0} + \frac{RT}{M_a}\ln\frac{\tilde{\rho}_a}{\tilde{\rho}_{a,0}} + \frac{RT_0}{M_a}\frac{\tilde{\rho}_{a,0}}{\tilde{\rho}_a} + \frac{RT}{M_a}g_a(\xi_j) + \frac{T}{\tilde{\rho}_a}I(\xi_j), \tag{19}$$

where c_k [J/(kg K)] is the specific heat at constant volume, M_k [kg/mol] is the molar weight, R [J/(molK)] is the universal gas constant, f [J/kg] is a vaporisation function, h [-] is an adsorption function, g_k [-] is a mixing interaction term for the gaseous components, and $\tilde{\rho}_{k,0}$ [kg/m^3] is the intrinsic density at the reference state $(P,T) = (P_0,T_0)$, where P [Pa] is the (mixture) pressure.

The chosen forms for the adsorption function, mixing terms, and vaporisation function, respectively, are

$$h(\xi_s,\xi_l) = b\left(\frac{\xi_s}{\xi_l} - \frac{1}{\chi_{\max}}\frac{1-\eta_0}{\eta_0}\right)^2 = a\left(\frac{1}{\chi} - \frac{1}{\chi_{\max}}\right)^2, \tag{20}$$

$$g_k(\xi_v,\xi_a) = \ln\left(\frac{\xi_k}{\xi_a+\xi_v} \Big/ \frac{\xi_{k,0}}{\xi_{a,0}+\xi_{v,0}}\right), \quad k \in \{a, v\}, \tag{21}$$

$$f(T) = -\frac{RT}{M_v} - L_0\frac{T-T_0}{T_0}, \tag{22}$$

where $\chi_{\max}$ [-] is the maximum saturation, $a = b((1-\eta_0)/\eta_0)^2$ is a constant material parameter, and L_0 [J/kg] is a constant related to the latent heat.

The dissipation function is chosen to be

$$\begin{aligned}\phi(\mathbf{q}_j,\mathbf{V}_j,\mathbf{V}_{vg}) = &\sum_{k\in\{s,l,v,a\}}\left(\frac{1}{\xi_k\lambda_k T}\mathbf{q}_k\cdot\mathbf{q}_k\right) + \\ &+\sum_{k\in\{l,g\}}\mu_k\mathbf{k}_k^{-1}(\beta_k\mathbf{V}_k)\cdot(\beta_k\mathbf{V}_k) + \\ &+\tilde{\rho}_v\frac{RT}{M_v}\frac{1}{D}\frac{\beta_g}{\zeta(1-\zeta)}(\zeta\mathbf{V}_{vg})\cdot(\zeta\mathbf{V}_{vg}).\end{aligned} \tag{23}$$

where λ_k [W/(Km)] is the heat conductivity, μ_k [kg/(sm)] is the dynamic viscosity, $\mathbf{k}_k$ [m^2] is the permeability tensor, D [m^2/s] is the diffusivity, $\mathbf{V}_g = \zeta\mathbf{V}_g + (1-\zeta)\mathbf{V}_a$ is the molar weighted velocity of gas, and $\mathbf{V}_{vg} = \mathbf{V}_v - \mathbf{V}_g$.

The result for the relative intrinsic pressure denoted as $\tilde{p}_k = p_k/\xi_k$ gets the form

$$\tilde{p}_k = B + \tilde{\rho}_l\frac{RT}{M_v}\xi_l\frac{\partial h}{\partial \xi_k}, \quad k \in \{s, l, v, a\}, \tag{24}$$

$$\tilde{p}_k = \tilde{\rho}_k\frac{RT}{M_k} - \tilde{\rho}_{k,0}\frac{RT_0}{M_k} - TI = P - P_0 - TI, \quad k \in \{v, a\}. \tag{25}$$

Here, the relative hydrostatic pressure $B = P - P_0$ is found to equal to the mixture pressure.

The Darcy law for liquid, the Darcy law for gas, the Fick law for the relative velocity of vapour, and the Fourier heat conduction law, respectively, get the form

$$\xi_l\mathbf{V}_l = -\frac{\mathbf{k}_l}{\mu_l}\left[\nabla\left(\frac{p_l}{\xi_l}\right) - \tilde{\rho}_l\mathbf{g} + \tilde{\rho}_l\frac{RT}{M_v}\nabla h\right], \tag{26}$$

$$\beta_g\mathbf{V}_g = -\frac{\mathbf{k}_g}{\mu_g}\left[\nabla P - \frac{\rho_a+\rho_v}{\beta_g}\mathbf{g}\right], \tag{27}$$

$$\zeta\mathbf{V}_{vg} = -D\left[\nabla\zeta - \zeta(1-\zeta)\frac{M_v-M_a}{RT}\mathbf{g}\right], \tag{28}$$

$$\mathbf{q}_k = -\xi_k\lambda_k\nabla T\ . \tag{29}$$

Component interactions are studied by considering the Gibbs equilibrium (12) of the water species in different mixtures. The resulting Clausius-Clapeyron equations "outside" and "inside" the porous medium, respectively, are

$$\ln\frac{\zeta^0P^0}{\zeta_0P_0} = \frac{M_v}{RT}\left[L_0\frac{T-T_0}{T_0} + \left(c_v^p - c_l^p\right)T\ln\frac{T}{T_0} + \frac{P^0-P_0}{\tilde{\rho}_l}\right], \tag{30}$$

$$\begin{aligned}\ln\frac{\zeta P}{\zeta_0P_0} = &\frac{M_v}{RT}\left[L_0\frac{T-T_0}{T_0} + \left(c_v^p - c_l^p\right)T\ln\frac{T}{T_0} + \frac{P-P_0}{\tilde{\rho}_l}\right] + \\ &+h+\xi_l\frac{\partial h}{\partial \xi_l}\ ,\end{aligned} \tag{31}$$

where ζP, ζ^0P^0, and ζ_0P_0 denote the partial pressure of saturated vapour from the gaseous phase inside the porous medium, outside the porous medium, and at the reference state, respectively. This is compatible with the conventional equilibrium conditions for ideal vapour with the assumptions $c_l^p = c_l$, $c_v^p = c_v + R/M_v$, and $L = L_0 + (c_v^p - c_l^p)T$, where c_k^p [J/(kgK)] is the specific heat at constant pressure and L [J/kg] is the latent heat.

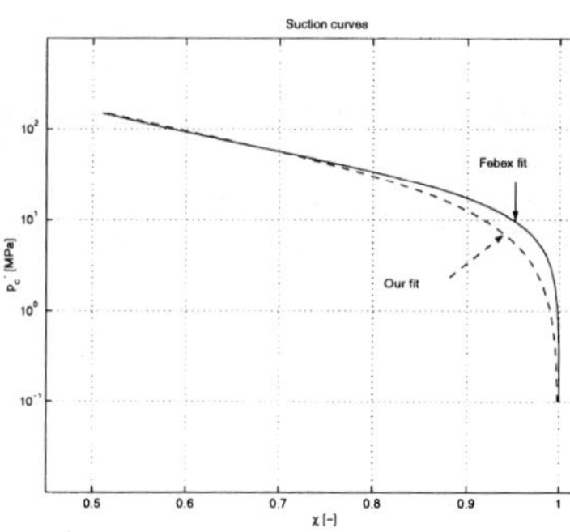

Figure 1. The suction curve (dashed) and the Febex fit (solid). The wetting experiment of a confined sample with ρ_{dry} = 1600...1650 kg/m^3.

By combining the Gibbs equilibria (30, 31) at the same pressure and temperature we get the relative humidity inside the porous medium as

$$\ln RH = \ln\left(\frac{\zeta P}{\zeta^0 P^0}\right) = \ln\left(\frac{\zeta}{\zeta^0}\right) = h + \xi_1 \frac{\partial h}{\partial \xi_1} \tag{32}$$

and the Kelvin law for the suction pressure p_c by means of the adsorption function as

$$p_c = -\tilde{\rho}_1 \frac{RT_0}{M_v}\left(h + \xi_1 \frac{\partial h}{\partial \xi_1}\right), \tag{33}$$

which form is also presented by Frémond and Nicolas (1990). The fitted suction curve by means of (20, 33) along with a van Genuchten fit used in the Febex project is plotted in Figure 1.

3.2 The thermohydraulic model

The final state variables are saturation χ, temperature T, and vapour fraction ζ. Inertial terms and gravitation are neglected and the total pressure is assumed to be constant. The model consists of the conservation of the mass of the water species, and the thermal energy conservation, i.e.,

$$\frac{\partial}{\partial t}(\rho_1 + \rho_v) + \nabla \cdot [\tilde{\rho}_1 \xi_1 \mathbf{V}_1 + \tilde{\rho}_v \xi_v \mathbf{V}_v] = 0 \ , \tag{34}$$

$$c\frac{\partial T}{\partial t} - L\left(\frac{\partial \rho_1}{\partial t} + \nabla \cdot (\rho_1 \mathbf{V}_1)\right) + \nabla \cdot \mathbf{q} = 0 \ , \tag{35}$$

where $c = \sum_k \rho_k c_k$. The constitutive relations are reduced to the Clausius-Clapeyron equation, vapour state equation, Darcy's law for liquid, the result of Fick's law for vapour, Fourier's law for heat flow, and to the latent heat of vaporisation, i.e., to

$$\ln\frac{\zeta}{\zeta_0} = \frac{M_v}{RT}\left[L_0 \frac{T - T_0}{T_0} + \left(c_v^p - c_1^p\right) T \ln\frac{T}{T_0}\right] + h + \xi_1 \frac{\partial h}{\partial \xi_1}, \tag{36}$$

$$P = \tilde{\rho}_v \frac{RT}{M_v} \ , \tag{37}$$

$$\xi_1 \mathbf{V}_1 = \frac{\mathbf{k}_1}{\mu_1} \tilde{\rho}_1 \frac{R}{M_v} \frac{2a}{\chi^3} \left[\chi(1 - \chi/\chi_{\max}) \nabla T - T \nabla \chi\right], \tag{38}$$

$$\xi_v \mathbf{V}_v = -\beta_g D \nabla \zeta \ , \tag{39}$$

$$\mathbf{q} = \sum_k \mathbf{q}_k = -\sum_k \xi_k \lambda_k \nabla T \equiv -\lambda \nabla T \ , \tag{40}$$

$$L(T) = L(T_0) + \left(c_v^p - c_1^p\right)(T - T_0) = L_0 + \left(c_v^p - c_1^p\right) T \ , \tag{41}$$

respectively. The relation between saturation and relative humidity is given by equations (20, 32).

4. THERMOHYDRAULIC SIMULATION OF THE FEBEX IN SITU TEST

The thermohydraulic aspects of the Febex in situ test (Figures 2 and 3) are simulated in 1D polar coordinate $r \in [R_1, R_2] = [0.485, 1.14]$ m.

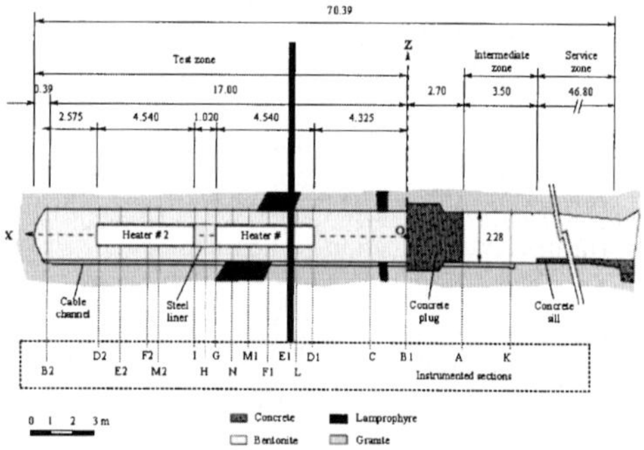

Figure 2. The Febex test profile and section E1.

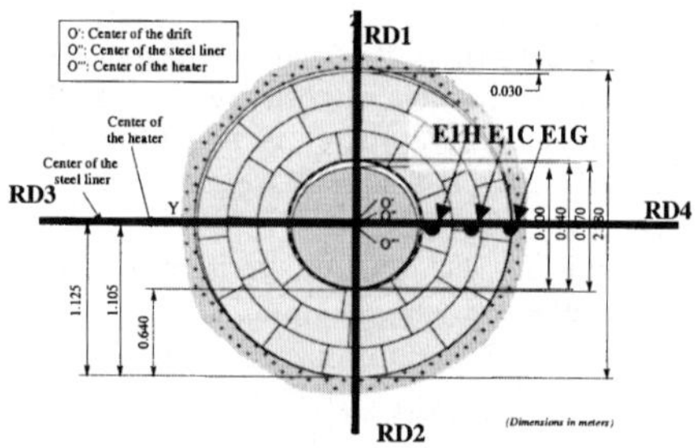

Figure 3. The simulation points and directions.

The presented simulation covers the radial distributions of relative humidity and temperature for the section E1 in the direction RD4, and the evolution of relative humidity in 3 points (E1H, E1C, and E1G) of the section E1. The initial conditions are $\chi_{\text{init}} = 0.54$, corresponding to RH = 38 %, and $T = T_r = 285.15$ K. The simulated sequence of –180…1000 d is divided in three periods. The moisture boundary conditions are no-flow at the heater and full saturation at the rock.

1) Days between −180…0 d correspond to the isothermal saturation period between the construction and the starting of the heaters.
2) Days between 0…53 d correspond to the heater adjustment period, which is approximated by a linear increase of the heater temperature to the final value assessed by the measurements. The temperature boundary condition at the rock is of the Robin type $\mathbf{q}=H(T-T_r)$ with a calibrated value of the heat transfer coefficient H [W/(m^{2}K)].
3) Days between 53…1000 d correspond to the period of controlled heating during which the temperature of the hottest point on each heater is kept at 373.15 K. The period is simulated by keeping the heater temperature constant. The other boundary conditions are the same as for the period 2.

Table 1. Initial, reference, and literature values.

Initial and Reference		Literature		
name	value	name	value	source
T_0	293.15 K	μ_l	$1.0 \cdot 10^{-3}$ kg/(sm)	Schmidt, 1989
P_0	0.1013 MPa	ζ_0	0.023	Schmidt, 1989
χ_{init}	0.54	D_1	$2.2 \cdot 10^{-5}$ m^2/s	Vargaftik, 1975
η_0	0.45	T_1	273 K	Vargaftik, 1975
χ_{max}	0.999	n	1.8	Vargaftik, 1975
ρ_{dry}	1690 kg/m^3	c_v^p	$1.87 \cdot 10^3$ J/(kgK)	Schmidt, 1989
T_r	285.15 K	c_l^p	$4.18 \cdot 10^3$ J/(kgK)	Schmidt, 1989
		$L(T_0)$	$2.45 \cdot 10^6$ J/kg	Schmidt, 1989
		$\tilde{\rho}_l$	998 kg/m^3	Schmidt, 1989

The results are calculated with the FEM program ELMER (CSC, 2003) in co-operation with CSC. The simple 1D mesh consists of 40 linear elements. A time step of 1 d is used for every case.

For heat capacity, heat conductivity, and liquid permeability, respectively, the following empirical relations are employed

$$c = c_{eff}(\chi,T) = \eta_0 \chi \tilde{\rho}_l c_l + \rho_{dry}(1.38(T-273.15)+732.5), \quad (42)$$

$$\lambda = \lambda_{eff}(\chi) = A_2 + \frac{A_1 - A_2}{1+e^{(\chi-x_0)/dx}}, \quad (43)$$

$$k_l = k_l(\chi) = k_u \chi^3, \quad (44)$$

where A_1 and A_2 [W/(Km)] are the thermal conductivities for dry and fully saturated medium, respectively, x_0 [-] is the saturation for which the thermal conductivity is the average value between the extreme values, dx [-] is a parameter, and k_u [m^2] is the permeability of the fully saturated medium. For diffusivity of vapour-air mixture we use the relation $D = D_1(T/T_1)^n$ (Vargaftik, 1975). The parameter values are given in Tables 1 and 2.

Table 2. Laboratory and calibrated values.

Laboratory		Calibration		
name	value	name	value	case
A_1	0.57 W/(Km)	a	0.4	Suction test
A_2	1.28 W/(Km)	H	1.9 W/(m^{2}K)	
x_0	0.65	$T(R_1)$	266.15 K	E1, RD4
dx	0.10			
k_u	$2.0 \cdot 10^{-21}$ m^2			

The simulated and measured results for the evolution of relative humidity at the points E1H, E1C, and E1G are illustrated in Figure 4. Figures 5 and 6 show the radial distribution of relative humidity and temperature, respectively, for section E1 in the direction RD4.

5. CONCLUSIONS

The presented TH simulations are essentially simplified 1D predictions of the relative humidity by means of calibration of the temperature boundary conditions. The actual geometry, behaviours of the heater, liner and rock, and the deformation of the buffer are not considered. Convection heat transfer and variations of pressure, viscosity of liquid, or liquid density are neglected, and the gas diffusion is of a simple form. The main simulational simplifications are the approximation of the heater adjustment period by a linear time dependence of temperature, and the use of a full saturation boundary condition at the rock surface.

Because the heater performance is simulated by means of temperature boundary conditions, the lack of temperature data constitutes a source of uncertainty. Other main uncertainties are related to the initial moisture distribution in the buffer and to the effect of the construction gaps and the instrumentation.

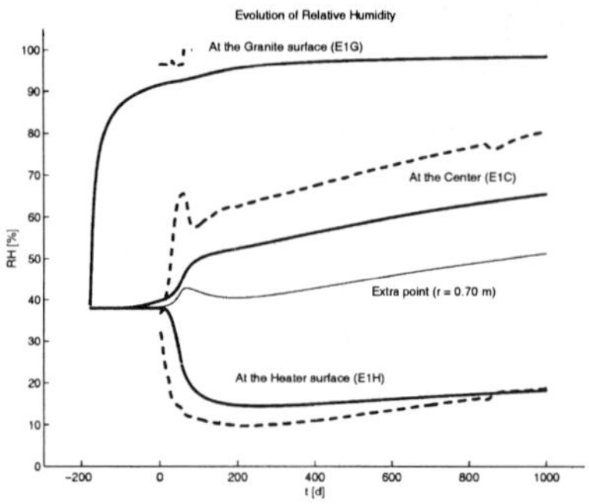

Figure 4. *Simulated (solid) and measured (dashed) evolution of RH at the points E1H, E1C, and E1G. The thin solid curve shows the simulated extra point.*

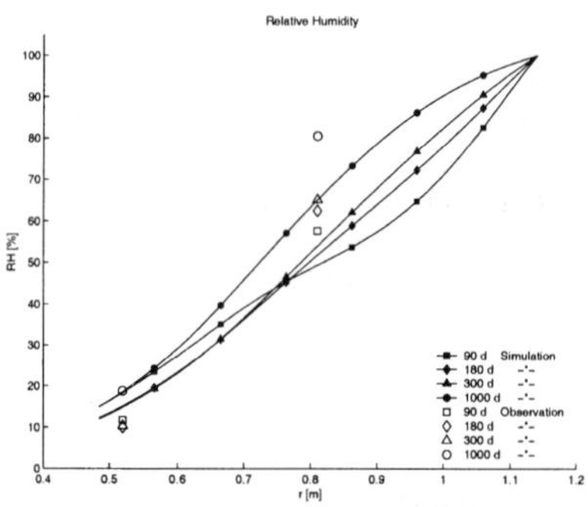

Figure 5. *Simulated and measured radial distribution of RH at 90, 180, 300, and 1000 d for the section E1 in the direction RD4.*

Fitting of the model against the suction experiment is done only by means of a single wetting curve with shortage of data at the dry region.

The largest discrepancy between the measured and simulated results can be seen in the case of relative humidity evolution at the centre of the buffer after the starting of the heaters. First a rapid condensation is measured followed by a rapid vaporisation after a sufficient temperature increase. In the simulation the same effect of consecutive wetting and drying is seen in considerably less extent at the region nearer the heater, e.g., at the extra point depicted in Figure 3. We suggest that this discrepancy be due to the gaps and instrumentation present in the actual case.

Despite the simplifications and uncertainties the simulations and the observations are fairly consistent. The model succeeds in its aim to describe the essential features of the TH behaviour, i.e., heat transfer, vaporisation, fluid flow, gas diffusion, and suction induced by adsorption.

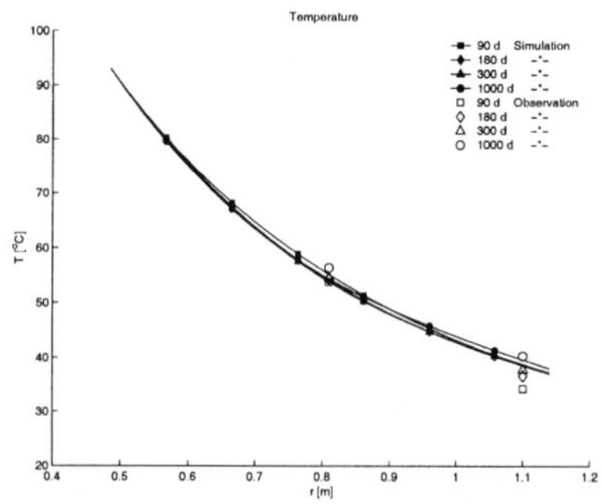

Figure 6. *Simulated and measured radial distribution of T at 90, 180, 300, and 1000 d for the section E1 in the direction RD4.*

REFERENCES

CSC - The Finnish IT Center for Science, 21st January 2003, *ELMER,* [WWW document] http://www.csc.fi/Elmer/.

Frémond, M., and Nicolas, P., *Macroscopic thermodynamics of porous media*, Continuum Mechanics and Thermodynamics 1990, 2:119-139.

Hartikainen, J., and Mikkola, M., *Thermomechanical model of freezing soil by use of the theory of mixtures*, In Proceedings of the 6th Finnish Mechanics Days, Aalto, J., and Salmi, T., (eds), 1997, 1–26.

Schmidt, E., *Properties of Water and Steam in SI-Units*, Springer, Berlin, 1989.

Vargaftik, N. B., *Tables on the thermophysical properties of liquids and gases. - 2. ed*, Wiley, New York, 1975.

Ziegler, H., Wehrli, C., *The derivation of constitutive relations from the free energy and dissipation function*, In *Advances in Applied Mechanics*, volume 25, pages 183-238, New York, 1988.

A FULLY COUPLED THREE-DIMENSIONAL THM ANALYSIS OF THE FEBEX IN SITU TEST WITH THE ROCMAS CODE: PREDICTION OF THM BEHAVIOR IN A BENTONITE BARRIER

J. Rutqvist and C.-F. Tsang
Earnest Orlando Lawrence Berkeley National Laboratory (Berkeley Lab)
Earth Sciences Division, Berkeley, CA 947 2, USA

Abstract: This paper presents a fully coupled thermal-hydrological-mechanical analysis of FEBEX—a large underground heater test conducted in a bentonite and fractured rock system. System responses predicted by the numerical analysis—including temperature, moisture content, and bentonite-swelling stress—were compared to field measurements at sensors located in the bentonite. An overall good agreement between predicted and measured system responses shows that coupled thermal-hydrological-mechanical processes in a bentonite barrier are well represented by the numerical model. The most challenging aspect of this particular analysis was modeling of the bentonite's mechanical behavior, which at FEBEX turned out to be affected by gaps between prefabricated bentonite blocks. At FEBEX, the swelling pressure did not develop until a few months into the experiment when moisture swelling of bentonite blocks had closed the gaps completely. Moreover, the wetting of the bentonite took place uniformly from the rock and was not impacted by the permeability difference between the Lamprophyres dykes and surrounding rock.

1 INTRODUCTION

As part of the DECOVALEX III project, several independent research teams have used various numerical models to analyze a full-scale engineered barrier experiment (FEBEX), currently conducted at the Grimsel Test Site in Switzerland. This paper presents the analysis conduced by the Berkeley Lab research team, using the fully coupled thermal-hydrological-mechanical (THM) numerical model ROCMAS. Specifically, the paper presents model predictions of coupled THM responses at FEBEX that are compared to field measurements.

2 FEBEX *IN SITU* TEST

The FEBEX *in situ* test has been conducted as a multi-national project coordinated by the Spanish organization ENRESA. In essence, the experiment involves the installation of two cylindrical heaters (4.54 m long and 0.97 m in diameter) centered in a tunnel (diameter 2.27 m) and surrounded by a barrier (thickness 0.64 m) made of highly compacted unsaturated blocks of bentonite (Figure 1). The tunnel is located in a mountainous area of massive saturated granite about 420 m below ground surface.

The FEBEX *in situ* test started in 1995 with rock-mass characterization and tunnel excavation. After the installation of heaters, the bentonite barrier, and monitoring equipment, the heating began on February 27, 1997. The heater power was first increased stepwise to reach a targeted maximum heater temperature of 100°C. After 53 days, the system was switched to a constant-temperature mode, keeping the maximum temperature at 100°C for the next several years. System responses—including temperature, moisture content, fluid pressure, stress and displacements—were monitored by several hundred sensors located in both the buffer and surrounding rock mass.

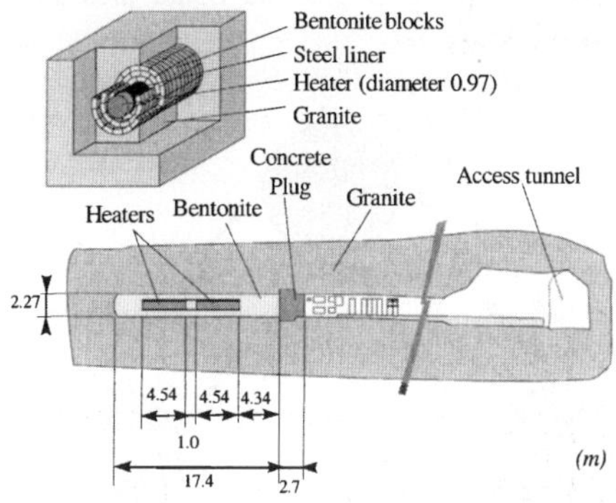

Figure 1. The set up of the FEBEX in situ test located at the Grimsel Test Site about 420 m below the ground surface.

3 ROCMAS

ROCMAS is a finite-element code for analysis of coupled THM processes in partially saturated geological media (Noorishad and Tsang, 1996, Rutqvist et al., 2001). In ROCMAS, the Biot (1941) formulation is extended to partially saturated media through Philip and de Vries' (1957) theory for heat- and moisture-flow in soil. This results in a comprehensive coupled THM formulation for partially saturated geological media that includes the coupled processes shown in Figure 2.

In ROCMAS, solid, liquid, and gas phases are considered. However, it is assumed that the gas pressure P_g is constant and equal to atmospheric pressure throughout the porous medium. As a consequence, vapor transport occurs only through molecular diffusion driven by a gradient in vapor concentration. The heat transfer may take place through conduction over all phases and by advection with water in both liquid and gas phases.

For the analysis of the FEBEX *in situ* test, a soil mechanics state-surface model was implemented. This state-surface approach provides a better representation of bentonite behavior under partially saturated conditions than a single effective stress approach. The logarithmic state surface model proposed by Lloret and Alonso (1985) was adopted in this analysis. In their model, void ratio (e) is a function of both net mean stress, ($\sigma_m'' = \sigma_m - p_g$, where σ_m = total mean stress and p_g is gas pressure) and suction ($s = p_g - p_l$, where p_g and p_l is gas and liquid pressures, repectively).

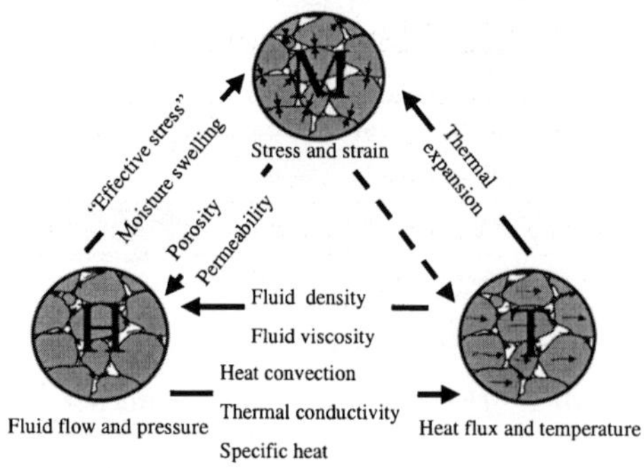

Figure 2. Coupled THM processes in unsaturated geological media simulated by the ROCMAS code

4 FEBEX MODEL

A three-dimensional fully coupled THM analysis of the FEBEX *in situ* test was carried out for the first 1,000 days of heating. Model geometry and material properties are presented in the next two subsections.

4.1 Geometry

A half-symmetrical three-dimensional model was discretized as shown in Figure 3a. The model includes two highly permeable lamprophyre dykes that intersect the FEBEX tunnel (Figure 3). These zones are included in the model because wetting rate of the bentonite might depend on the local permeability of the surrounding rock. The engineered materials included in the model are a concrete plug, the bentonite barrier, a steel liner (around the heaters), and the heater canisters (Figure 3b).

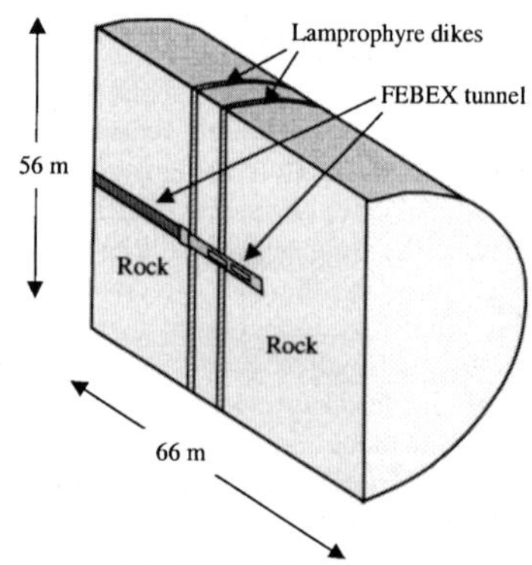

(a) Entire model of FEBEX

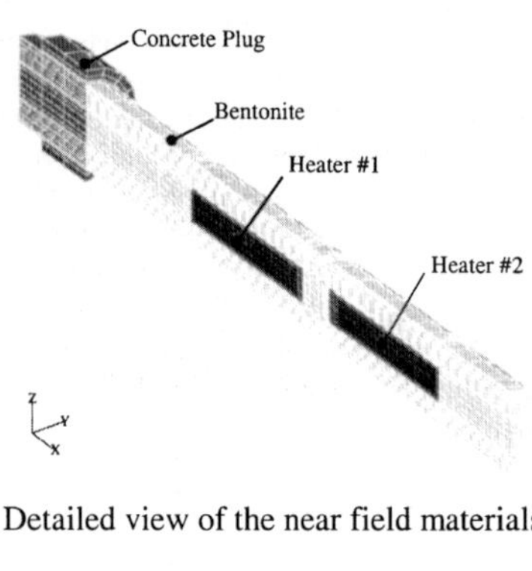

(b) Detailed view of the near field materials

Figure 3. Finite element model of FEBEX

4.2 *Material properties*

The material properties were obtained from various field and laboratory tests that were performed before the emplacement of the buffer and heaters. For example, the inflow into the open drift was utilized to calibrate the *in situ* hydraulic properties of the rock mass and Lamprophyre dykes. Moreover, several laboratory tests were utilized for numerical back-analyses of coupled THM properties of the bentonite. The properties of the bentonite and rock are listed in Tables 1 and 2.

Some of the material properties listed in Table 1 are given in the form of functions. The water-retention curve of the bentonite is described by a modified van Genuchten (1980) function as:

$$S = 0.01 + (0.99)\left[1 + (s/35)^{1.43}\right]^{-0.30}\left[1 - s/4000\right]^{1.5} \quad (1)$$

where S is liquid saturation and s is suction.

The thermal conductivity is a function of saturation according to:

$$\lambda_m = 1.28 - \frac{0.71}{1 + e^{(S-0.65)/0.1}} \quad (2)$$

and specific heat is a function of temperature according to:

$$c_s = 1.38T + 732.5 \quad (3)$$

The state surface model for this particular bentonite's mechanical behavior is defined as:

$$e = 1.03 - 0.49\log\sigma''_m - 0.17\log s + 0.23\log\sigma''_m \log s \quad (4)$$

As shown in Table 2, the properties of the Lamprophyre properties and the surrounding rock are the same except for permeability and thermal conductivity. The mechanical rock-mass properties are obtained from the geological description of the Grimsel Test Site (Kneussen et al., 1989). Significantly, the Young's modulus of the rock mass was reduced to 70% of its value for intact rock.

Table 1. Material properties of the bentonite barrier used in modeling the FEBEX heater test

Parameter	Value
Dry density, [kg/m^3]	$1.6\cdot10^3$
Porosity, [-]	0.41
Saturated permeability, [m^2]	$2.0\cdot10^{-21}$
Relative permeability, k_{rl}	$k_{rl} = S^3$
Water retention	Equation (1)
State surface	Equation (4)
Poisson ratio, [-]	0.35
Thermal expan. coeff., [1/°C]	$1.0\cdot10^{-5}$
Dry specific heat, [J/kg·°C]	Equation (3)
Thermal cond., [W/m·°C]	Equation (2)
Flow times tortousity factor	0.8
Thermal diffusion factor	2.0

Table 2. Material properties of the rock mass used in modeling the FEBEX heater test

Parameter	Value
Density, [kg/m^3]	2700
Porosity, [-]	0.01
Biot's constant, α [-]	1.0
Young's Modulus, [GPa]	35
Poissons ratio, [-]	0.3
Specific heat, [J/kg·°C]	800
Thermal conductivity, [W/m·°C]	3.6
Thermal expan, coefficient [1/°C]	$8.21\cdot10^{-6}$
Vertical permeability, [m^2]	5×10^{-18}
Horizontal permeability, [m^2]	5×10^{-19}
Van-Genuchten's parameter, P_0 [MPa]	1.47
Van Genuchten's parameter, β	2.47
Lamprophyre permeability, [m^2]	1.1×10^{-17}
Lamprophyre thermal cond. [W/m°C]	2.45

5 SIMULATION RESULTS

The simulation of the FEBEX *in situ* test included both pre-heating and heating periods, as shown in Figure 4. The simulation began eight months before heating to take into account the wetting of the benonite during the experimental setup. After the initiation of heating, the heater power was increased step-wise during the first 53 days. Then the power of each heater was individually controlled by a constant heater temperature of 100°C. The simulated heater power shown in Figure 4 is slightly lower than the measured one.

In the following subsections, simulated and measured system responses are compared at selected monitoring points within the bentonite barrier.

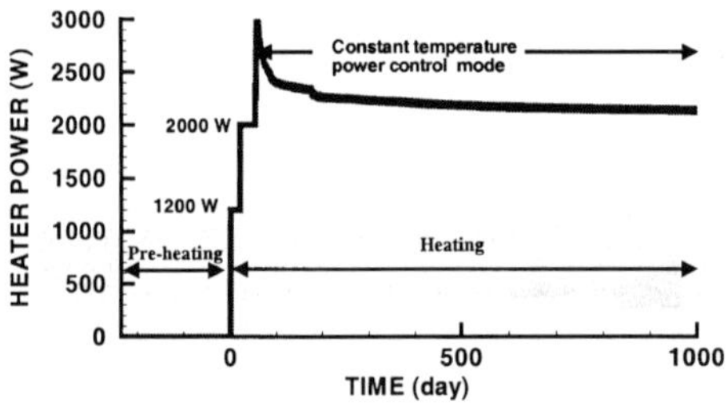

Figure 4. Simulated heater power as a function of time for each heater. A constant temperature mode is implemented after 53 days

5.1 Results of temperature

Figure 5 presents simulated temperature in the bentonite barrier after 1,000 days. The figure shows the location of the hottest point on the heater surface, which is kept at a constant temperature of 100°C. At the drift wall the temperature reaches a maximum of about 40–50°C, giving rise to a thermal gradient of about 80 to 90°C/m.

In Figure 6, simulated and measured temperature evolutions are compared at monitoring point D1G, located near the drift wall. The good agreement between simulated and measured temperature shown in Figure 6 is typical. This confirms that temperature can be very well predicted by numerical analysis in this type of environment.

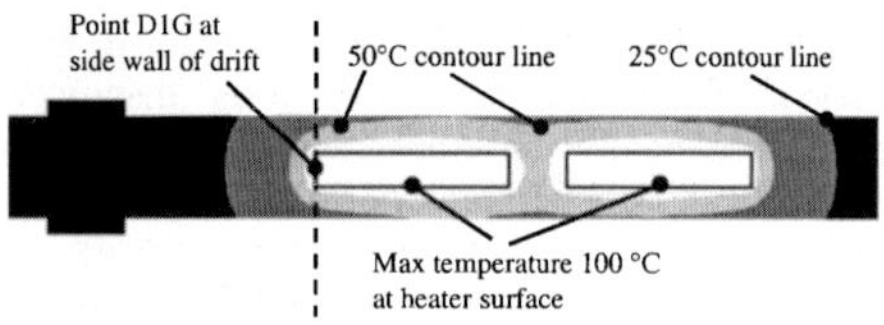

Figure 5. Simulated temperature contour after 1,000 days of heating

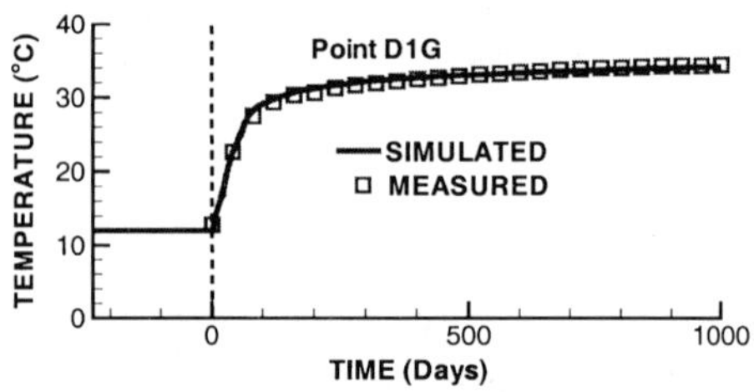

Figure 6. Simulated and measured temperature evolution in point D1G located at the drift wall

5.2 Results of moisture content

Figure 7 shows simulated liquid saturation after 1,000 days of heating. The dark contours in Figure 7 indicate zones where wetting of the bentonite have taken place. The simulated wetting has begun already before heating through water infiltration from the fully saturated rock. The light contours in Figure 7 indicate that the drying zone has taken place near each of the heaters. This drying is caused by evaporation and transport of vapor along the thermal gradient away from the heaters. Another observation in Figure 7 is that the wetting of the bentonite progresses uniformly all around the drift without any visible effects from the high permeability Lamprophyres.

Figure 8 presents simulated and measured evolutions of the moisture content (relative humidity) at three points located along a vertical monitoring section between the two heaters (HG, HC and HH in Figure 7). Figure 8 shows that the wetting of the bentonite starts at the drift wall (Point HG located a few centimeters from the rock surface) as soon as the bentonite is installed in that section. At this monitoring point, the relative humidity has increased to about 80–90% at heater turn-on (time = 0 in Figure 8).

At point HH, condensation of vapor caused a slight wetting during the first 100 days. For the period between 100 to 300 days, the water content decreased slightly because of drying induced by the heat. After 300 days, the relative humidity increased again as a result of capillary driven liquid flow from the outer regions of the bentonite.

At point HC—located halfway between the heater and the drift wall—the moisture content increased gradually as a result of infiltration from the fully saturated rock.

Figure 8 indicates that the evolution of relative humidity was reasonable well predicted by the numerical analysis.

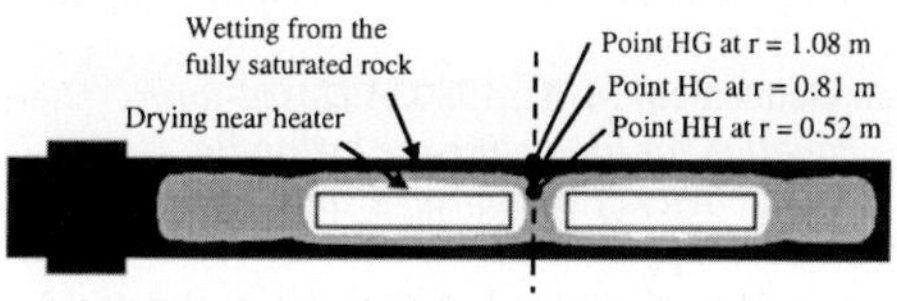

Figure 7. Simulated contour of liquid saturation in the bentonite buffer after 1,000 days of heating: Dark and light contours indicate wetting and drying, respectively

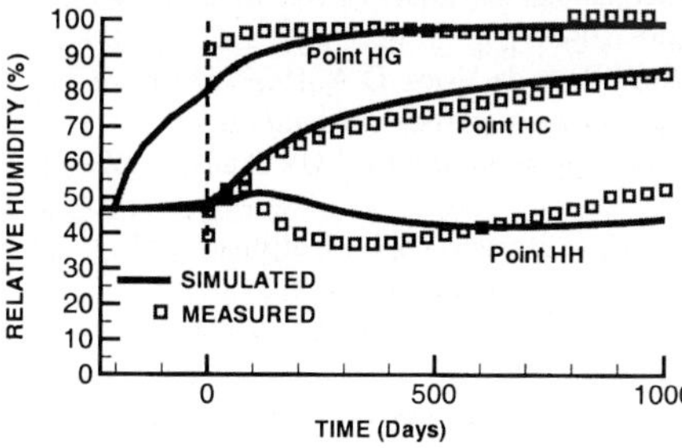

Figure 8. Simulated and measured evolution of relative humidity at points HG, HC and HH with locations shown in Figure7

5.3 Results of Stress

Figure 9 presents simulated contours of intermediate compressive-principal-stress at 1,000 days. The figure shows that compressive stress caused by swelling pressure has developed at the drift wall, especially at the front end of the tunnel, where the bentonite blocks were first installed.

Figure 10 presents simulated and measured evolution of stress normal to the drift wall at two locations (E2G2 and B2G in Figure 9). The simulated stress began to develop as soon as the wetting commenced and increased to about 2 to 2.5 MPa at 1,000 days. The measured stress indicates that the swelling stress might not have begun to develop until several months after heater turn-on. This delay in the development of swelling stress was a common observation at many monitoring points in the bentonite barrier at FEBEX.

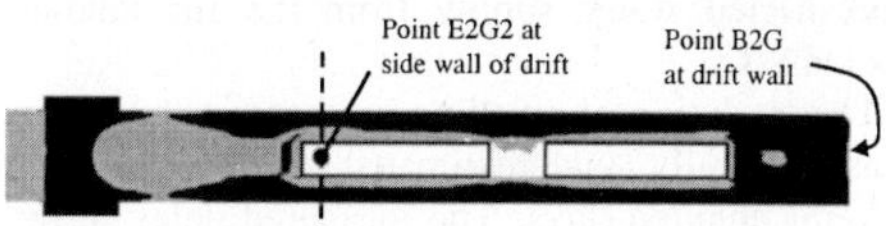

Figure 9. Simulated contour of intermediate principal stress in the bentonite buffer after 1,000 days of heating: dark contour indicates the highest compressive stress (2 to 3 MPa), whereas light contour indicates the lowest compressive stress (0 to 1 MPa)

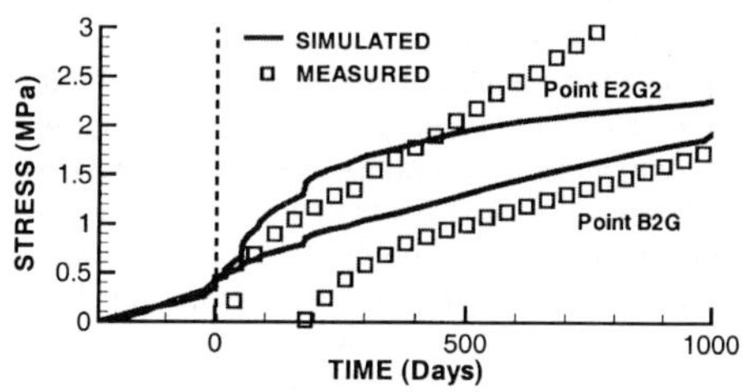

Figure 10. Simulated and measure evolution of radial stress at points E2G2 and B2G, with locations shown in Figure 9

6 DISCUSSION

The general agreement between simulated and measured THM responses at FEBEX indicates that coupled THM processes are well represented in the ROCMAS code.

The good agreement between simulated and measured temperature indicates that thermal responses can be predicted with high confidence. This is a consequence of heat conduction being the dominant mode of heat transport. If the thermal conductivity and specific heat are accurately represented in the model, the temperature responses should be well predicted.

The evolution of the moisture content was reasonably well predicted both in trends and magnitudes. Moreover, results from the modeling and field measurements at FEBEX indicate that the presence of highly permeable Lamprohyres (Figure 3a) did not have a significant impact on the wetting of the buffer. This shows that the wetting of buffer was controlled by the hydraulic properties of the bentonite barrier with unrestricted water supply from the surrounding rock mass.

The evolution of swelling pressure in the buffer was generally underestimated during the first several hundred days. The measured delay in the swelling stress at FEBEX was probably caused by the existence of gaps between the bentonite blocks. It appears that the swelling stress in the buffer begun to develop after a few months when the gaps had closed. These gaps were not considered in our numerical model, and consequently the simulated swelling stress developed much earlier.

7 CONCLUSIONS

The numerical model ROCMAS was applied to predict coupled THM processes in a bentonite barrier at the FEBEX *in situ* test. The results indicate that numerical modeling can provide highly reliable predictions for temperature distribution, and reasonably reliable predictions for moisture flow and stress in a bentonite barrier. Moreover, field observations and modeling shows that resaturation of the buffer was controlled by the properties of the bentonite barrier whereas the permeability of the rock was sufficiently high to act as an unrestriced water source. Therefore, the wetting of the bentonite took place uniformly from the rock and was not impacted by the permeability difference between the Lamprophyres dykes and surrounding rock.

The evolution of stress in the bentonite barrier at FEBEX was affected by the existence of gaps between the pre-fabricated bentonite blocks. The swelling pressure did not develop until moisture swelling of the bentoninte blocks had closed the gaps completely.

ACKNOWLEDGEMENTS

Reviews and comments by Hui-Hai Lui and Dan Hawkes, Lawrence Berkeley National Laboratory are much appreciated. Financial support was provided by a grant from the Swedish Nuclear Power Inspectorate (SKI).

REFERENCES

Biot, M. A. 1941. *General theory of three dimensional consolidation.* J. Applied Physics 12: 155-164.

van Genuchten, M. T., 1980. *A closed-form equation for predicting the hydraulic conductivity of unsaturated soils.* Soil Sci Soc Am J 44: pp. 892-898.

Keusen H.R., Ganguin J., Schuler P. & Buletti M, Grimsel. 1989.*Test Site-Geology.* NAGRA, NTB 87-14E.

Lloret A. & Alonso E.E. 1985. *State surfaces for partially saturated soils.* Proc 11th Int Conf Soil Mech Fdn Engng, San Francisco, Vol. 2, pp. 557-562.

Noorishad, J. & Tsang, C.-F. 1996. *ROCMAS simulator; A thermohydromechanical computer code.* In O. Stephansson, L, Jing & C.-F. Tsang, (Eds), Coupled thermo-hydro-mechanical processes of fractured media: 551-558. Elsevier.

Philip, J.R. & de Vries, D.A. 1957. *Moisture movement in porous material under temperature gradients.* EOS Trans., AGU. 38: 222-232.

Rutqvist, J., Borgesson, L., Chijimatsu, M., Kobayashi, A., Nguyen, T.S., Jing, L., Noorishad, J. & Tsang, C.-F. 2001. *Thermohydromechanics of partially saturated geological media: governing equations and formulation of four finite element models.* Int. J. Rock. Mech. Min. Sci. 38(1): 105-127.

A DISCRETE APPROACH TO MODELLING HYDROMECHANICAL ROCK RESPONSE OF FEBEX TUNNEL EXCAVATION (GRIMSEL UNDERGROUND RESEARCH LABORATORY, SWITZERLAND)

Véronique MERRIEN-SOUKATCHOFF[1], Imad KADIRI[1], K. SU[2], Yves GUGLIELMI S.[3]
[1]) LAboratoire Environnement Géomécanique Ouvrages, Ecole des Mines, Nancy, France
[2]) ANDRA, Châtenay-Malabry, France
[3]) Geosciences Azur, Université de Nice-Sophia Antipolis, France

Abstract: LAEGO and ANDRA have participated in the modelling of the task 1A of the "DECOVALEX" III international research project, which was devoted to modelling the hydromechanical effects of FEBEX tunnel excavation at the Grimsel Test Site in Switzerland. After presenting the site, this paper will focus on the 3D discontinuous modelling performed by associating 3DEC and 3FLO codes, which leads to a discontinuous semi-coupled approach. The modelling tools, choice of geomodel, the mechanical boundary and related initial conditions, the hydraulic boundary and related initial conditions, as well as the calibration process of hydraulic properties will all be discussed herein. A purely-hydraulic modelling first yields a rather good estimation of both total flow leading towards the tunnel and the general shape of pressure evolution in a nearby borehole. A hydromechanical computation representing the final stage of excavation reveals a difference of just 5% on the evaluation of total water flow rate and does improve the prediction of water head in the boreholes surrounding the tunnel.

1 INTRODUCTION

Within the framework of the international research project entitled DECOVALEX III ("International co-operative project for the DEvelopement of Coupled models and their Validation against Experiments in nuclear waste isolation"), we have participated in the modelling of Task 1A, which focuses on modelling the hydromechanical effects of tunnel excavation in crystalline rock. The case examined herein was the FEBEX tunnel, situated at the Grimsel Test Site (GTS), an underground laboratory in Switzerland operated by NAGRA. The purpose of the FEBEX (Full-scale Engineered Barriers Experiment in Crystalline Host Rock) project is to study the various processes occurring in the vicinity of high-activity radioactive waste storage.

The specific task undertaken has consisted of predicting changes in water head induced by boring the FEBEX tunnel (pressure variations vs. time) as well as in the total water flow rate into the excavated tunnel over the last 17.40 meters. The data accessed for this purpose were the available geological, hydraulic and mechanical characterizations of GTS, along with the results of hydraulic tests performed on boreholes. In order to accomplish this task, we employed a discontinuous approach. The next section of this paper will describe the test site and subsequent sections will discuss in greater detail the modelling tools, methodology and results.

2 SITE DESCRIPTION

The GTS is located at an elevation of 1,730 m within the granite rocks of the "Aare Massif" in central Switzerland, about 450 m beneath the eastern flank of the Juchlistock Mountain.

This experimental site is set up in a tunnel system (see Figure 1) that branches off from a main access tunnel towards the KWO (Kraftwerke Oberhasli AG) underground power station. The GTS tunnel system comprises a laboratory tunnel with a total length of almost 1,000 m plus a central building that houses all the operating facilities (Ortuño, 2000).

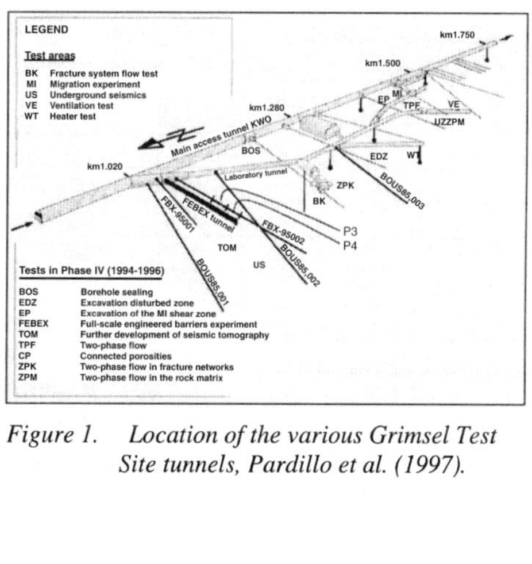

Figure 1. Location of the various Grimsel Test Site tunnels, Pardillo et al. (1997).

2.1 Geographical and geological context

The FEBEX drift excavation is located in the northern part of the Grimsel Test Site (GTS) and was completed with a TBM from September 25 to October 30, 1995. Four predrilled pilot boreholes have been positioned in the surrounding area (BOUS 85.001, BOUS 85.002, FEBEX 95.001 and FEBEX 95.002.). The FEBEX tunnel displays a circular section with an average diameter of 2.28 m and a total length of 70.4 m. It was excavated with an ascending slope of 1% to allow for natural drainage. The rock mass is predominantly granite and granodiorite and has been affected by several fracturing phases. Keusen *et al.* (1989) identified 12 possible discontinuity systems, with the most significant discontinuity families being listed in Figure 2.

2.2 Hydrogeological data

The GTS has been excavated within a fractured rock mass of low permeability. Observation in the laboratory tunnel has revealed that water flow is concentrated almost exclusively in discontinuities (Keusen *et al.*, 1989). Figure 2 demonstrates the relative importance of water-bearing discontinuities. The S families are schistosity regrouped into shear zones of considerable thickness (5 to 20 m) displaying major outflows at the intersection with tunnels, hence indicating their relevance as preferential flow paths. L is a regional structure of large-sized lamprophyre dikes (with thickness reaching several meters). Although their relevance as preferential flow paths is not as pronounced as the shear zones, their contact with the host rock yields planes of major transmissivity.

A regional numerical model produced by Voborny *et al.* (1991) has provided an estimate of the main flows within a vicinity of several kilometres around the GTS, while a second model has allowed deriving more details of the site area. These models have been used as references to infer boundary conditions for the more heavily-localized model performed. Two shear zones (K and S) located within the vicinity of the FEBEX tunnel constrain groundwater flow due to their high transmissivity and therefore constitute FEBEX environment boundaries; in subsequent model, they have been considered as imposed head limits (see Figure 3).

Numerous hydraulic tests were carried out, during the various project phases, either in an isolated borehole or between boreholes. Hydraulic experiments consisted of pulse tests, constant head injection and/or extraction and constant rate injection/extraction using borehole intervals delimited in FBX 95001, FBX 95002, BOUS 85.001 and BOUS 85.002 (Figure 1). Both water flow and piezometric level measurements for borehole intervals are available. Continuous pressure monitoring has allowed detecting cross-hole responses during pulse testing; these have then been used to calibrate the hydraulic conductivities of fractures.

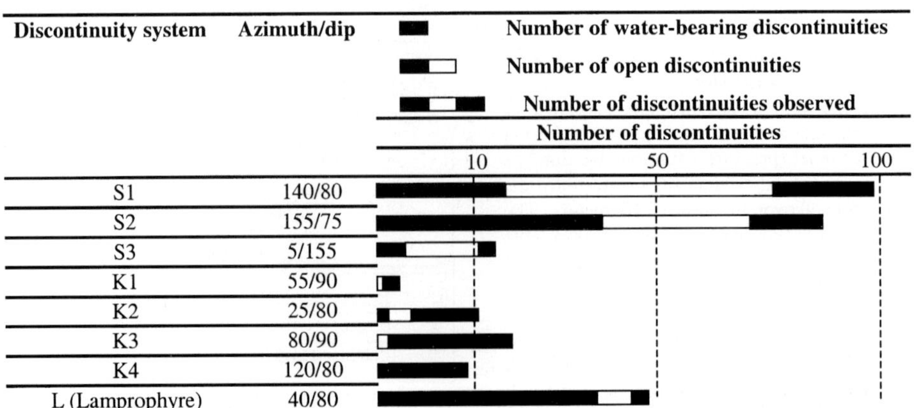

Figure 2. Discontinuity systems in the laboratory tunnel and water flow for each set (according to Keusen et al., 1989).

2.3 Geomechanical data

The principal mechanical properties of granite rock matrix were provided for the purposes of this task; no mechanical characterization of discontinuities was available. The stress measurements presented by Pahl *et al.* (1989) show that the stress field is triaxial with horizontal stresses 4 to 5 times higher than the lithostatic pressure and a difference of greater than 10 MPa between the minimum and maximum horizontal stresses (see Figure 3).

3 DISCONTINUOUS MODELLING TOOLS EMPLOYED

As seen in the previous section, water flow is driven by discontinuity patterns. We have elected to use discontinuous modelling tools to run the simulation of this case and more precisely to combine a 3-code application, known as HM3D. HM3D allows modelling the hydromechanical behaviour of a rock mass by means of successively employing the three following codes: RESOBLOK, a code developed by LAEGO to enable constructing the site's geometrical representation (the "geomodel"); mechanical calculations performed with the 3DEC distinct element code (ITASCA), flow version 1.34; and the 3FLO flow-modelling code (ITASCA), version 0.90. Special procedures allow for transferring information from one code to another. The hydraulic computation is influenced by mechanical results via the variation in joint apertures due to the mechanical displacements. Due to the deterministic selection of discontinuities in the FEBEX case, only 3DEC and 3FLO were successively employed. Application of RESOBLOK would have been of interest in the case of statistical knowledge of the fractures. The geometrical representation step, which leads to setting the joints in place, is essential for the mechanical and hydraulic steps and exerts however a sizable impact on subsequent results (Kadiri, 2002; Kadiri *et al.*, 2002).

The mechanical phase of the computation has made use of the 3DEC distinct element code, which relies upon an explicit solution of motion equations based on small time steps; the ensuing computation cycles lead to achieving equilibrium. The blocks separated by discontinuities were considered as elastic in the FEBEX case. Joints respect the Coulomb slip criterion characterised by normal and shear stiffnesses, as well as friction, cohesion and tensile strength properties assigned to each joint.

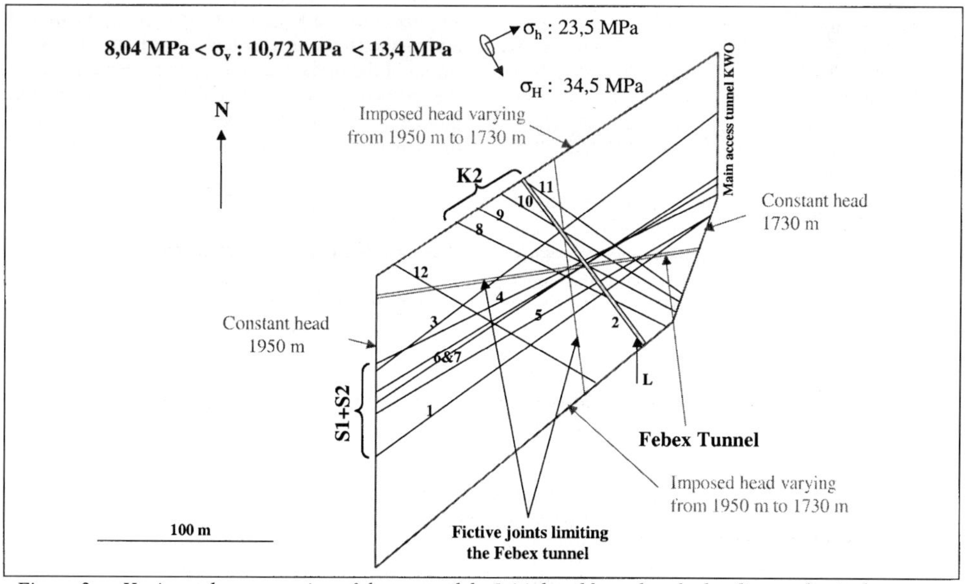

Figure 3. Horizontal cross-section of the geomodel - Initial and boundary hydraulic, mechanical conditions.

For the hydraulic phase, each discontinuity is considered to be crossed by a set of channels (1D hydraulic element) generated either deterministically or statistically. At the intersection of two joints, a hydraulic conduit, called a tube, is produced. Flow is limited to fractures and the blocks are impervious. The diffusion equation is resolved throughout the finite differences. Initial conductivities are assigned to both channels and tubes (combined within the denoted pipe). Displacements computed with the 3DEC code influence the conductivity of hydraulic elements by the following relationship:

$$\frac{C}{C_0} = \left(\frac{a}{a_0}\right)^{exp} \qquad (1)$$

where C_0 is the initial conductivity (before stress variation) of the hydraulic element, C the hydraulic element conductivity once displacements have been taken into account, a_0 the initial aperture (for a channel) or initial radius (for a tube), a the aperture or radius once displacements have been taken into account, and default exp is 3 for a channel or 4 for a tube. This behaviour gives rise to a semi-coupled approach (i.e. the influence of mechanical computation on hydraulic properties).

4 MODELLING OF THE FEBEX CASE

4.1 The "geomodel"

In order to perform hydromechanical computations, the choice of which discontinuities to represent must be a judicious one. We only retained the 3 sets of fractures with the most significant hydraulic impact (S1+S2; K2 and L), as shown in Figure 2. Selection of specific fractures to be introduced into the model consisted of:

- indexing the fractures identified in boreholes within the vicinity of FEBEX and the FEBEX tunnel itself;
- selecting open fractures that could play a hydraulic role according to descriptive criteria (we have retained those fractures considered as being open by the geophysical log and indicated by the TLV probe);
- keeping discontinuities, whose hydraulic role was demonstrated by hydraulic tests and water flow measurements conducted in the boreholes.

This choice was also influenced by the fact that HM3D does not allow introducing a large number of fractures. Fractures in the same area with the same orientation were considered as a single fracture (with dip and azimuth being averaged from all individual fractures taken into account): a joint in the model represents a set of *in situ* fractures.

A horizontal view at the altitude of the FEBEX tunnel (1,730 m) of the geomodel obtained is presented in Figure 3. The horizontal dimensions of the model were dictated by the hydraulic boundaries, as will be explained below. The model is 150 m wide in the north-south direction and 200 m wide east-west. Moreover, it is 200 m high (100 m above and below the tunnel altitude).

4.2 Mechanical boundaries and initial conditions

The initial stresses and boundary stress conditions were introduced in accordance with the *in situ* data described in Section 2.3. Vertical stress varies between 8.04 MPa and 13.4 MPa, depending on the lithostatic pressure (with the overburden varying from 300 to 500 m). The maximum horizontal stress is constant, equal to 34.5 MPa (mean of the measured values) and oriented N 130° (i.e. perpendicular to S2 schistosity), whereas the minimum horizontal stress is 23.5 MPa (mean of the measured values) and oriented N 40° (i.e. parallel to S2 schistosity).

4.3 Hydraulical boundary conditions, initial conditions and property calibration

On the whole, flow is moving from within the massif towards the laboratory tunnel (from west to east). Before excavation of the FEBEX tunnel, the equipotentials were roughly parallel to the wall of this main tunnel. A zero pressure has been assumed in the laboratory tunnel. To the north and south, the model is constrained by two major shear zones S1+S2, as specified in Section 2.2. Figure 3 displays the hydraulic boundary conditions adopted, in a plane view. The head and gradient adopted were deduced both from measurements conducted within the boreholes intersecting our model boundary and from the regional model developed by Voborny *et al.* (1991). In the third dimension, the upper and lower surfaces have been considered as impervious.

The conductivity and specific storativity of the fracture were obtained by calibration in a purely-hydraulic computation. In each fracture, two perpendicular sets of channels 2 meters apart were generated. At first, the same initial aperture of 10^{-5} was assigned to all fractures, but since a joint in 3FLO represents several *in situ* discontinuities (see Section 4.1), the conductivity Cond was calculated as follows:

$$Cond = n\frac{\rho \cdot g}{12\mu}a^3 \qquad (2)$$

where a is the aperture, n the number of discontinuities represented by a joint in the model, ρ the water mass density (1,000 kg/m3), g the gravity constant (9.81 m.s-2) and μ the water dynamic viscosity ($\mu = 10^{-3}$ Pa.s).

The calibration process was performed iteratively by simulating the test (in general, an imposed zero pressure) for each borehole interval until simulated flow equals measured flow. The first calibration step consists of considering each interval separately and leads to allotting an initial conductivity to each fracture and each test. The iteration then makes it possible to consider the whole intervals and the whole fractures, even those that are not directly in connection with the borehole interval targeted for calibration. Furthermore, in order to simulate the head variation observed in the four boreholes (BOUS 85.001, BOUS 85.002, FEBEX 95.002 and FEBEX 95.002), the conductivity assigned to the fractures was reduced within the western part of the model. The calibration step leads to assigning hydraulic apertures varying from 10 to 20 μm.

Let's note herein that the proposed model is not unique: the calibration of another "geomodel" also compatible with the observations could have led to a different set of parameters in good accordance with local information.

Pressure Static Recovery tests were modelled (within a purely-hydraulic computation) in order to calibrate fractures storativity. The difference between computed storativity, which takes only joint aperture into account, and calibrated storativity reveals the major role played by rock mass on the storativity and/or the importance of weak joint stiffness. Since further computations were merely conducted under steady-state conditions, storativities are not subsequently used.

4.4 Purely-hydraulic computation

As part of the "DECOVALEX" project, we were asked to predict the total water flow rate to the excavated tunnel over the last 17.40 meters of FEBEX tunnel section (between 54.00 and 71.40 m). A purely-hydraulic computation, using the previous calibration, yields an estimated steady-state flow of between 6 and 7.6 ml/min. This size scale is to be compared with total water inflow estimated from a semi-quantitative hydrogeological map of the FEBEX tunnel: 7.8 ml/min (with an estimation of about 27% of inflow water coming through the matrix). In comparison with the continuous prediction (Alonso *et al.*, 2001), this estimation is rather good and the discontinuous approach allows localizing water output with greater precision. (This kind of comparison was not included within the "DECOVALEX" task.)

For the prediction of water head changes induced by FEBEX tunnel boring, it was impossible to simulate a transient evolution due to changes in geometry with HM3D. We chose to model tunnel boring in 4 excavation phases and then simulate the steady state corresponding to the end of each phase.

Figure 4 shows the comparison carried out between the measured and simulated pressures. The general shape of the curves obtained with hydraulic simulation is quite similar to measurement results, except there is a constant difference of about 1 bar between the P4 measurements and the simulation. This point will be discussed in Section 4.5 below. Our simulation approach (a hydraulic steady flow analysis with four steady-state steps) was not able to reproduce pore pressure increase (led to a higher horizontal stress than vertical stress) as excavation neared the monitored borehole intervals and as post-tunnel face dissipation was completed.

4.5 Hydromechanical computation

A hydromechanical computation, which takes both the mechanical boundary and initial conditions described in Section 4.2 plus the hydraulic initial conditions, boundary conditions and calibration described in Section 4.3 into account, has been performed. Only the final stage of the excavation was simulated and this led to a steady-state flow estimate of between 5.7 and 7.2 ml/min, i.e. 5% less than the flow predicted by a purely-hydraulic computation.

The difference in head between the hydraulic and hydromechanical computations is more significant (see Table 1): 6 m, or 0.7 bar in pressure. This discrepancy may explain the observed difference between measurements and simulation in the purely-hydraulic computation (see Figure 4).

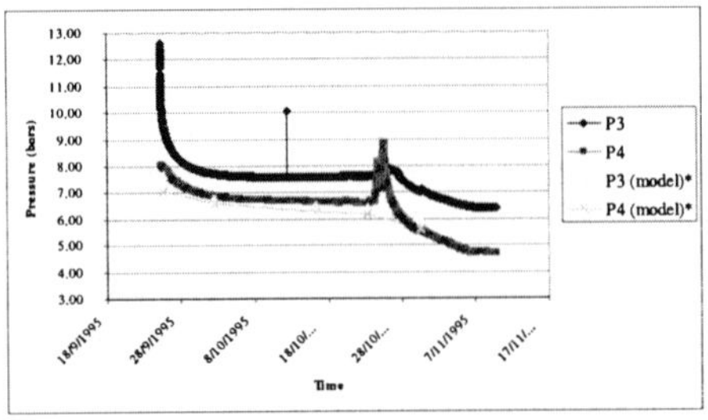

Figure 4. *Comparison of measured and simulated (by hydraulic computation) pressures over intervals P3 and P4 of the FBX 95002 borehole.*

Table 1. Comparison of measured and modelled heads over the P3 and P4 intervals in both a purely-hydraulic model (**H**) and a hydromechanical (**HM**) computation

FBX 95002 Intervals	Measured head (m)	Simulated head (m) H	Simulated head (m) HM
P3	1,787	1,781	1,787
P4	1,769	1,778	1,784

5 CONCLUSION

FEBEX excavation modelling using discontinuous codes has proven the ability of this conceptual approach to reproduce both the global flow rate towards the excavated tunnel and the general shape of head variation in the surrounding area. The refinement introduced by a hydromechanical computation, in comparison with a purely-hydraulic computation, only mildly influences (5%) flow estimation, yet remains more sensitive with respect to pore pressure. Due to the partial coupling along with the inability of tools to simulate transient hydromechanical processes, the precise variations associated with tunnel progression have not been reproduced. The result however incites use of the discontinuous approach even if there is a need of improving the computer codes because they enables to reproduce accurately the localisation of water outflow witch is not the case for equivalent porous medium codes.

REFERENCES

Alonso E.E. & Alcoverro J. 2001. *DECOVALEX III, Task 1: Modelling of FEBEX in-situ test, Introduction report and comparison of results by UPC.* Internal Report.

Kadiri I., Merrien-Soukatchoff V. & Su K. 2002, *Discontinuous Hydromechanical Modeling of the impact of the FEBEX tunnel (Grimsel, Switzerland).* 5th European Conference on Numerical Methods in Geotechnical Engineering (NUMGE) 2002, 4-6 September 2002, Paris, France, pp. 585-590.

Kadiri I. 2002. *Modélisation hydromécanique des milieux fracturés*, Ph.D thesis, LAEGO, Ecole des Mines, INPL, France, 28 October 2002, 143 p.

Keusen H.R., Ganguin J., Schuler P. & Buletti M. 1989. *Grimsel Test Site. Geology.* NAGRA, NTB 87-14E, Feb. 1989.

Ortuño F. 2000. *Discrete water inflow measurements at the FEBEX tunnel. DECOVALEX III, Task 1: Modelling of FEBEX in-situ test.* Internal Report. July 2000.

Pahl A., Heusermann St., Bräuer V. & Glöggler W. 1989. *Grimsel Test Site. Rock stress investigations.* NAGRA, NTB 88-39E.

Pardillo J., Campos R. & Gimerà J. 1997. *Caracterización geológica de la zona de ensayo FEBEX (Grimsel - Suiza).* CIEMAT, 70-IMA-M-2-01, May 1997.

Voborny O., Adank P., Hürlimann W., Vomvoris S. & Mishra S. 1991. *Grimsel Test Site. Modelling of groundwater flow in the rock body surrounding the underground laboratory.* NAGRA, NTB 91-03.

MEASURING THERMAL, HYDROLOGICAL, MECHANICAL, AND CHEMICAL RESPONSES IN THE YUCCA MOUNTAIN DRIFT SCALE TEST

Robin Datta[1], Deborah Barr[2], William Boyle[2]
[1]Bechtel SAIC Company, 1180 Town Center Dr., Las Vegas, NV 89144, USA
[2]Office of Repository Development, U.S.Dept. of Energy, POB 30307, N. Las Vegas, NV 89036 USA

Abstract: The Drift Scale Test (DST) in the Exploratory Studies Facility at Yucca Mountain, Nevada, USA is an in-situ heater test to study coupled thermal, hydrological, mechanical, and chemical processes caused by the decay heat from high level nuclear waste emplaced in the proposed geologic repository. The objective of the test is to develop, refine, and verify/validate the coupled-process models to be employed in evaluating the performance of the proposed repository. The four-year heating phase of the test, started in December 1997, was terminated in January 2002, leading into the on-going cooling phase measurements. The DST is also the benchmark experiment central to Task 2 of the DECOVALEX III international research project for the "DEvelopment of COupled models and their VALidation against EXperiments." Several other papers in this GeoProc2003 conference discuss the thermal-hydrologic (TH), thermal- hydrologic-mechanical (THM), and thermal-hydrologic-chemical (THC) processes in the DST. To provide the context to those presentations, this paper describes the DST outlining its configuration, the geologic setting, the instrumentation and measuring systems, and other physical and temporal features of the test. Typical measured results relative to the four processes of interest are presented and discussed.

1. BACKGROUND

The United States Department of Energy is conducting the Drift Scale Test in Yucca Mountain, Nevada as part of its program of site investigation for a proposed geologic repository for the permanent disposal of spent nuclear fuel and high-level radioactive waste. The objective of the test is to better understand the coupled processes that would occur in the rock around the proposed repository because of the decay heat from the emplaced waste. The four principal processes of concern, occurring in the rock, are thermal(T), hydrological(H), mechanical(M), and chemical(C). The most challenging task in this regard is to understand the interactions or couplings between the individual processes; to translate the understandings mathematically; and to incorporate the mathematics into the numerical tools which are employed to exercise the coupled process models. The results contribute to the knowledge base upon which the total system performance assessment (TSPA) of the proposed repository is built.

2. TEST DESCRIPTION

The DST is located in a drift off the main tunnel of the Exploratory Studies Facility (ESF) at Yucca Mountain, some 135 kilometers north west of Las Vegas, Nevada. It involves heating a 5m diameter, 47.5m long drift by electric heaters for four years, and measuring the thermal, hydrological, mechanical, and chemical responses before, during, and after the heating.

2.1 Geologic Setting

Yucca Mountain is part of a series of north-trending ridges in the Great Basin physiographic province of North America. The mountain is underlain by 1000 to 1500 meters of Tertiary volcanic tuffs, formed from the ash of eruptions occurring between 8 and 16 million years ago. The tuffs are generally layered, separated into beds which range from non-welded to densely welded; in addition, some are devitrified and others are vitric. The proposed repository horizon is within a group of beds, up to 350 meters thick, of moderately to densely welded devitrified tuff known as the Topopah Spring Tuff of the Paintbrush group. The sub-division of the Topopah Spring tuffs is largely based on the relative abundance of lithophysae which are cavities with dimensions on the order of millimeters to hundreds of millimeters formed because of the presence of gases in the cooling ash flows. The presence or absence of lithophysae can have a measurable effect on the mechanical and hydrological properties of the rocks. There are four sub-units of Topopah Spring Tuff: the upper and lower lithophysal units represented by Tptpul and Tptpll respectively, and the middle and lower nonlithophysal units represented by Tptpmn and Tptpln respectively. Over two-thirds of the proposed repository will be in Tptpll, the rest in Tptpmn and Tptpln. The Drift Scale Test block is entirely in Tptpmn. The sub-units of Topopah Spring Tuff Tptpul, Tptpmn, Tptpll, and Tptpln are also represented by Tsw33, Tsw34, Tsw35, and Tsw36 respectively in discussing hydrologic processes.

2.2 Configuration

The DST consists of a 5m diameter, 47.5m long drift heated by 9 cylindrical electric heaters placed on the floor. Each heater, 1.7m diameter and 4.6m long, is capable of generating a maximum of 15kW. Additional heating is applied by 50 rod heaters, referred to as "wing heaters", emplaced in horizontal bore holes drilled into either side wall of the drift (Figure 1).

The drift cross-section and the cylindrical heaters are approximately the sizes of emplacement drifts and waste packages respectively, being considered for the proposed repository. The wing heaters are employed to simulate the heat that would come from adjoining drifts in the proposed repository, and thus to provide better test boundary conditions. Each wing heater, 10m long, has inner and outer segments capable of generating 1145W and 1719W respectively. The heater outputs can be varied over a range of 0% to 100% of the designed capacity of the heaters.

An Access/Observation (AOD) drift parallel to the Heated Drift and an orthogonal Connecting Drift (CD) delineates the periphery of the test block (Figure 1). The AOD and the HD are 30 meters apart, center to center, on plan. Rockbolts and wire-mesh are installed as ground support in the Heated Drift (HD) and the other excavations. Also, the final 12.5m length of the Heated Drift is supported by a cast-in-place concrete liner, approximately 12 inches thick, in order to observe the performance of such a concrete-lined drift at elevated temperatures. An insulated thermal bulkhead isolates the heated length of the HD from the CD. The bulkhead is not a pressure bulkhead, and allows some heat exchange by convection. A small alcove or niche is excavated into one side wall of the drift just outside the bulkhead for Plate Loading Test(s) to be conducted to investigate the load-deformation responses of heated and unheated rock. (Figure 1).

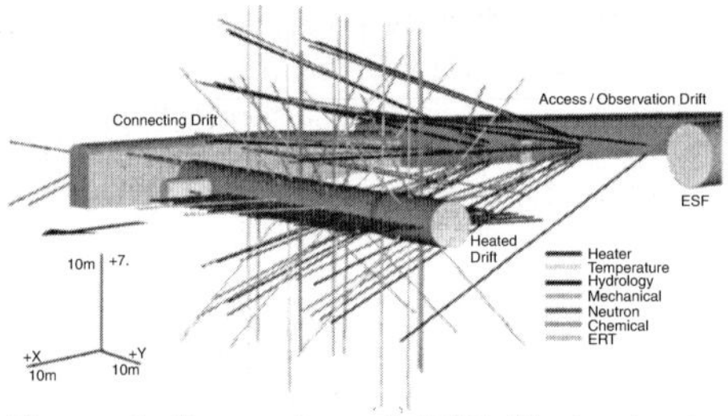

Figure 1 Perspective of DST block showing multiple boreholes to house heaters and sensors

The HD was circular in cross-section when excavated. The concrete invert on the floor is 1.25 meters thick. The origin (0,0,0) of the co-ordinate system employed to specify locations in the DST is the trace of the center line of the HD as excavated on the hot side surface of the bulk head. The center line is also the y-axis, with positive x going away from the AOD, positive y going into the HD and positive z going up.

As shown in Figures 1, a large number of bore holes are drilled from the drifts into the test block to house the heaters, instruments, and sensors. Approximately 3500 sensors of various types are installed in some 3300 meters of bore holes and elsewhere. At the end of four years of heating in January 2002 the volume of rock heated to above 100^0C exceeded 20,000 cubic meters. The DST is described in detail in "Drift Scale Test Design and Forecast Results" (CRWMS M&O, 1997).

2.3 Measurements Made

The measurements made in the DST can be grouped under thermal, hydrological, mechanical, and chemical processes.

Thermal measurements include heater power, and temperatures of rock, air, water, and at various other surfaces. Rock temperatures are measured by both resistance temperature devices (RTDs) and thermocouples. The RTDs are grouted in holes drilled into the rock to house them. Figure 2 shows the RTD holes in the DST. Thermocouples are attached to surfaces such as the drift wall, wing heaters, cable-trays etc.

A total of approximately 2650 temperature sensors are installed in the DST. Thermal conductivity and diffusivity of the rock measured by REKA (rapid evaluation of K and alpha) probe (Danko and Mousset-Jones, 1989) can also be categorised as thermal measurements.

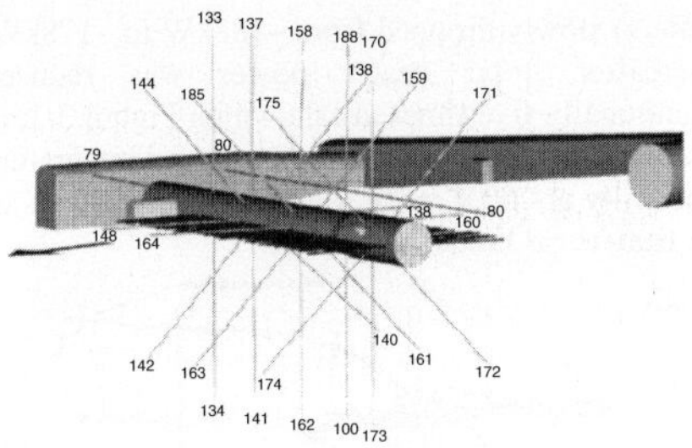

Figure 2 RTD holes in the DST

Moisture content of the rock, temperature and relative humidity of air, fluid pressure, fracture permeability, and gas tracer tests are some of the hydrologic measurements made in the DST. Rock moisture content is measured by three different geophysical methods; these are neutron logging, electric resistivity tomography (ERT), Ramirez and Daily, (2001) and ground penetrating radar (GPR) Hubbard et al, (1997). Fracture permeability is measured by air-injection testing in the hydrology holes drilled from the AOD to above and below the HD. Segments of these holes are isolated by inflatable packers. Air is injected and pressure responses are monitored to generate data from which permeabilities are computed, Cook (2000), Tsang et al (2000). There are twelve hydrology holes in two transverse arrays of five holes each and one array of two holes as shown in Figure 3. Each hydrology hole has typically four segments in which air-injection and pressure monitoring are performed.

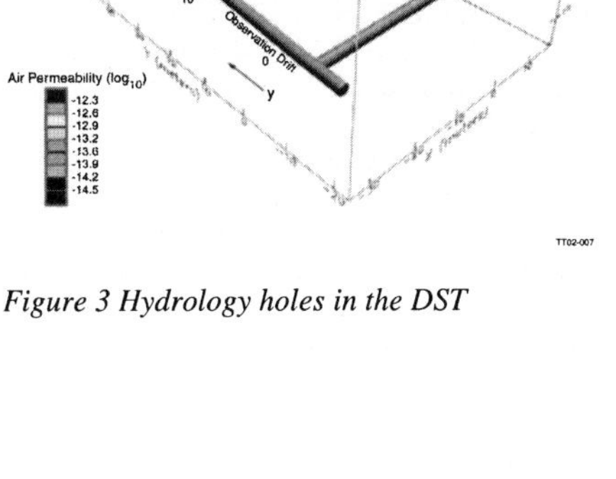

Figure 3 Hydrology holes in the DST

The mechanical measurements include displacement in rock by multiple point borehole extensometers (MPBX), deformation of concrete lining by convergence monitors, and strain in concrete by electrical strain gages. Rock mass modulus measurement(s) in the Plate Loading Test niche and monitoring of acoustic emissions by 16 accelerometers installed at different locations in the test block are the other mechanical measurements. There are 17 MPBX holes each with four or more anchors. Twelve of these are radial holes drilled from the HD in three transverse arrays. The displacement sensors in these holes are LVDTs (linear variable differential transducer) capable of functioning at elevated temperatures. The LVDTs are mounted at the collar of the hole and the measured displacements at the anchors 1 through 4 are relative to the LVDT. Three MPBX holes drilled down from the AOD toward the HD before it was excavated contain vibrating wire type sensors. These record the rock movement at ambient temperature caused by the mining of the HD as well as the movement caused subsequently by the heating.

Analyses of gas and water samples collected from the DST are the principal chemical measurements in the DST. Periodic collection of samples of rock from the test block and analysing the mineralogic-petrologic changes with time are also in the chemical measurements category. Ten chemistry holes, drilled from the AOD and equipped with inflatable flexible liners, water-absorbing pads, and tubing failed their intended purpose of yielding water and/or gas samples.

Miscellaneous measurements include periodic video photographing inside the HD along its entire length; planned measurements include examination of coupons of candidate materials for the waste containers and batches of native microbes left inside the HD to study the effect of the heating and cooling on them.

2.4 Automated Data Collection

An automated data collection system (DCS) of approximately 6000 channels continuously records the readings of the various sensors on an hourly basis. The DCS scans and records the readings of temperature, humidity, pressure, MPBXs, strain gages, convergence monitors, and current and voltage for heater power. The measurements made by the DCS are referred to as passive monitoring. Other measurements such as ERT, neutron logging, GPR, acoustic emissions, REKA probe, and others are made periodically (every few weeks to few months) using individual data collection system. These non-DCS measurements are referred to as 'active testing'.

2.5 Test Schedule

Although conceptual planning for a large scale field thermal test to characterise the Yucca Mountain site continued following the submission of the Site Characterisation Plan (SCP) U.S. DOE (1988), the concrete outlines of the DST as implemented were conceived in the early 1995 time frame. Initiation of the four-year heating phase in December 1997 was preceded by detailed planning, test designs, construction, installation, and pre-test characterisation of the test block. The cooling phase, which began in January 2002, is expected to span approximately four years and will be followed by post-test characterisation, analyses of results, documentation, and a final report in the 2007/2008 time frame.

2.6 Changes

Two of the ten original chemistry holes (Section 2.3) were re-equipped like the hydrology holes approximately one year after the start of heating to collect more water and gas samples from below the HD. Two other chemistry holes were used to periodically collect rock cores by side-wall sampling to study mineralogic changes, if any, as the heating progressed. Approximately three years after the start of heating one new chemistry hole was drilled through the boiling zone to study the changes in the rock in that region.

3. OBSERVATIONS, SAMPLING, ANALYSES, AND MEASUREMENTS

These are described herein under the four processes of concern. Most of the measurements and testing in the DST listed in section 2.3 above can be grouped under, and are performed during, one or more of five largely chronological phases. They are the pre-test characterization (of the block), the pre-heating baseline, heating, cooling, and post-test characterization phases. A fairly comprehensive cataloging of the various measured data collected during the heating phase of the DST can be found in BSC (2002).

3.1 Thermal Response

The actual applied heater power is shown in Figure 4 for the entire heating phase. Over the first 820 days the total power (canister heaters + wing heaters) slowly dropped from ~188kW to ~178kW. Thereafter, total heater power was reduced intentionally five times, as shown on Figure 3-1, in an attempt to maintain the drift wall temperature nominally at 200^{0}C prior to the switching off of all the heaters on January 14, 2002.

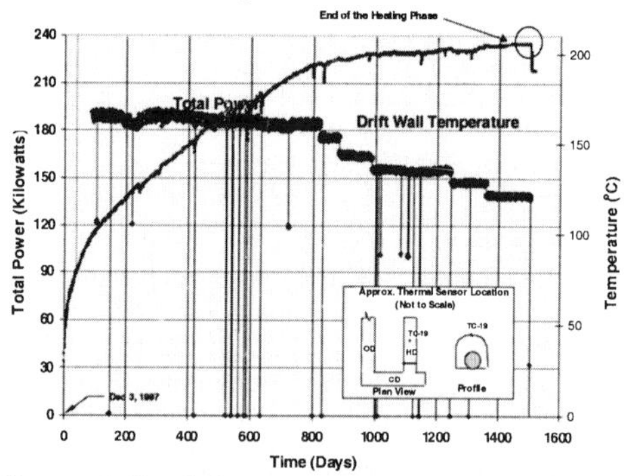

Figure 4 Total heater power and drift wall temperature during DST heating phase

Figure 4 also shows the temperature history of the sensor at the crown at approximately the mid-length of the drift. Figure 5 shows the temperature histories at selected sensors in the radial RTD holes 158 and 164 located at the middle of the HD. Five of the fifteen temperature history plots in Figure 5 clearly illustrate the heat-pipe effect at approximately 96^{0}C (the temperature at which water boils at Yucca Mountain). Heat transfer in the welded unsaturated tuffs in Yucca Mountain is more or less entirely by conduction in the sub-boiling regime. As the rock temperature reaches the boiling point, the thermal pulse goes toward the latent heat of vaporization causing a hiatus in the rise in temperature; once all the available water is boiled off, the conduction-dominated heat transfer process resumes and temperature rises again.

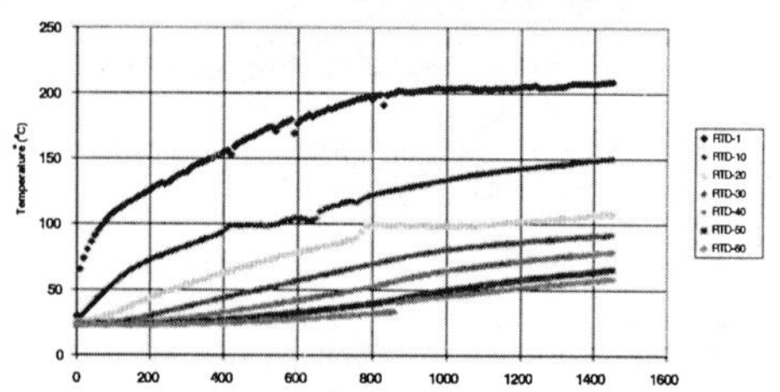

Figure 5a Temperature histories at selected sensors in DST borehole 158

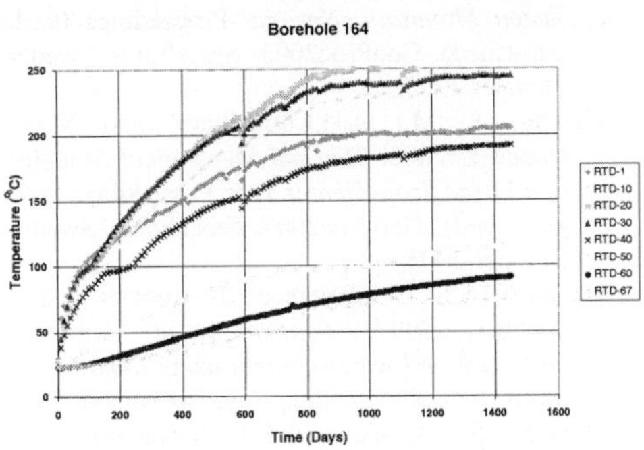

Figure 5b Temperature histories at selected sensors in DST borehole 164

3.2 Hydrological Response

The pneumatic permeabilities of the fractures measured after the start of heating generally decreased relative to the pre-heating baseline measurements as shown in Figure 6. This decrease is interpreted to be due to the condensation of the moisture in the fractures in the cooler regions at a distance from the heat source, as the pore water in the vicinity of the source was vaporised and driven away. The increased fracture saturation restricted the air flow decreasing the air permeability.

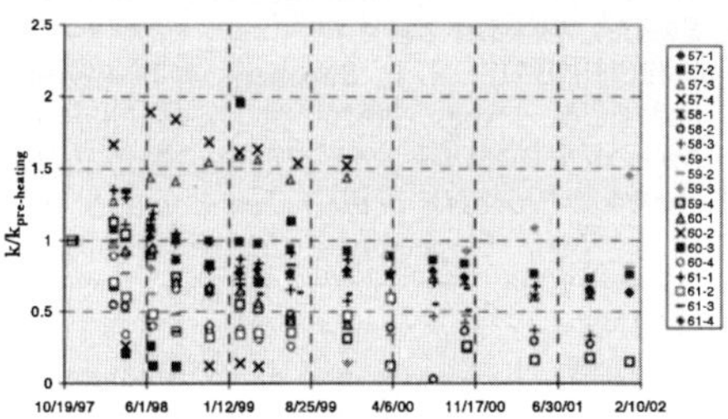

Figure 6 Changes in fracture permeabilities in DST boreholes 57-61 during heating

The data in Figure 6 represent the measured permeability at a specific segment of a Hydrology hole at a particular time during the heating phase expressed as a ratio of the baseline permeability at that segment measured before the start of heating. Fracture permeabilities evolve due to complex interactions between thermal, hydrological, and mechanical processes. These observed changes are analysed in-depth by Rutqvist et al, 2003. Green and Painter, 2003 and Olivella et al, 2003 also discuss the modeling of TH/THM processes during the heating phase of the DST.

3.3 Mechanical Response

Displacements measured by MPBXs constitute the bulk of the mechanical measurements in the DST. The DCS continuously scanned the displacement sensors and recorded the readings at a frequency of four per 24 hours. Figure 7 shows the measured displacement histories in one of the 12 radial holes drilled from inside the HD. Hole 155 is inclined up at 30 degrees to the vertical in a transverse plane at y = 21m. Millard and Rutqvist, 2003 and Hsiung et al, 2003 discuss the thermal-mechanical modeling of the DST.

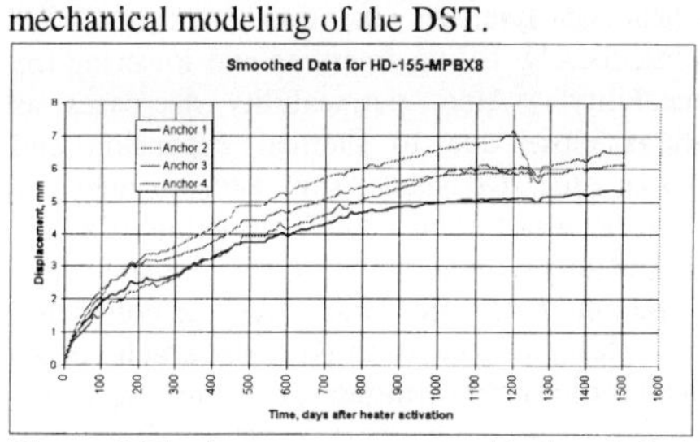

Figure 7 Displacements measured in borehole 155

3.4 Chemical Response

Evolution of the composition of the gas and water samples from the DST with the progress of heating provides a variety of interesting data. The exsolution of CO_2 from the pore water with temperature rise soon after the start of heating causes a rapid increase in the CO_2 concentration. Dissolution of the CO_2 in the cooler region depresses the pH of the water compared to that of the baseline pore water. Sonnenthal et al, (2003) discusses the modeling of the thermal-hydrologic-chemical processes in the DST and compares the predicted and observed results. Water chemistry (cation and anion analyses), gas chemistry including stable isotopes, strontium and uranium isotopic compositions of water samples and mineralogical and petrologic analyses of rock samples collected before and during the test constitute a large volume of chemical data from the DST.

4. CONCLUDING REMARKS

The Drift Scale Test at Yucca Mountain, Nevada provides a rare opportunity of examining the heat-driven coupled processes in a high-level nuclear waste repository. The observed data from the test are the results of the interactions or combined effects of multiple processes. Although, fielding this large and complex DST capable of functioning effectively at relatively high temperatures over many years, was not a small accomplishment, the real challenge of the DST is the sound and reliable interpretation and analyses of the large volumes of data of various kinds the test has generated and will continue to generate over next several years.

The DST confirmed that the heat transfer process is conduction dominated, although pore water plays an important role especially in the sub-boiling regime. Vaporized pore water travels away from the heat via fractures and condenses in cooler regions, thereby filling fractures and lowering the permeability. Also, permeability decreases as fractures close due to thermal expansion and increases due to shear slip accompanied by dilation. Rutqvist et al (2003) show how the TH and THM effects may be separated. Sonnenthal et al (2003) show how the THC process is dominated by the release of CO_2 from the rock which begins soon after the start of heating.

REFERENCES

BSC (Bechtel SAIC Company) 2002. *Thermal Testing Measurements Report.* Report # ANL-NBS-HS-000041 REV 00. Bechtel SAIC Company, Las Vegas, Nevada, October 2002.

Cook, P., 2000. *In situ pneumatic testing at Yucca Mountain.* Int. J. of Rock Mech & Mining Sci. 37: 357-367.

CRWMS M&O, 1997. *Drift Scale Test Design and Forecast Results.* Report # BAB000000-01717-4600-00007, Rev.01, prepared by the Civilian Radioactive Waste Management System Management and Operating Contractor for the U.S. Department of Energy Yucca Mountain Site Characterisation Office, Las Vegas, Nevada, December 1997.

Danko, G., and P.F. Mousset-Jones. *An In Situ Technique for Rock Thermophysical Property Measurement and Site Characterisation.* Proceedings, 31st Heat Transfer and Fluid Mechanics Institute, California State University, Sacramento, 1989, pp.397-417.

Green, R.T., and S. Painter, 2003. *Numerical Simulation of Thermohydrological Processes Observed at the Drift Scale Heater Test at Yucca Mountain, Nevada.* Proceedings (to be published), GeoProc2003, Stockholm, Sweden, October 2003.

Hsiung, S. M., A.H.Chowdhury, and M. S. Nataraja, 2003. *Thermal-Mechanical Modeling of a Large Scale Heater Test.* Proceedings (to be published), GeoProc2003, Stockholm, Sweden, October 2003.

Hubbard, S.S., J.E.Peterson, J. Roberts and F. Wobber, 1997. *Estimation of permeable pathways and water content using tomographic radar data.* The Leading Edge, November 1997.

Olivella, S., A. Gens and C. Gonzalez, 2003. *Coupled THM Modeling in a Fractured Permeable Rock.* Proceedings (to be published), GeoProc2003, Stockholm, Sweden, October 2003.

Millard, A. and J. Rutqvist, 2003. *Comparative Analyses of Predicted and Measured Displacements During the Heating Phase of the Yucca Mountain Drift Scale Test.* Proceedings (to be published), GeoProc2003, Stockholm, Sweden, October 2003.

Ramirez, A. and Daily, W., 2001. *Electrical Imaging of the Large Block Test —Yucca Mountain, Nevada.* Journal of Applied Geophysics, Vol 46, pp85-100.

Rutqvist, J., C.-F.Tsang, and Y.Tsang, 2003. *Analysis of Stress and Moisture-Induced Changes in Fractured Rock Permeability in the Yucca Mountain Drift Scale Test.* Proceedings (to be published), GeoProc2003, Stockholm, Sweden, October 2003.

Sonnenthal, E. L., N. F. Spycher, M. E. Conrad, and J. A. Apps, 2003. *A Conceptual and Numerical Model for Thermal-hydrologic-chemical Processes in the Yucca Mountain Drift Scale Test.* Proceedings (to be published), GeoProc2003, Stockholm, Sweden, October 2003.

Tsang, Y.W., K. Huang, and G. S. Bodvarsson. *Estimation of the Heterogeneity of Fracture Permeability by Simultaneous Modeling of Multiple Air-Injection Tests in Partially Saturated Fractured Tuff.* Dynamics of Fluids in Fractured Rock. Geophysical Monograph 122, American Geophysical Union. 2000

U.S. DOE, 1988. *Site Characterization Plan: Yucca Mountain Site, Nevada Research and Development Area, Nevada.* Report DOE/RW-0199, Office of Civilian Radioactive Waste Management, U.S. Department of Energy, Washington, D.C., 1988.

ANALYSIS OF STRESS AND MOISTURE INDUCED CHANGES IN FRACTURED ROCK PERMEABILITY AT THE YUCCA MOUNTAIN DRIFT SCALE TEST

J. Rutqvist, C.-F. Tsang and Y. Tsang
Ernest Orlando Lawrence Berkeley National Laboratory (Berkeley Lab), USA

Abstract: This paper presents a coupled thermal-hydrological-mechanical analysis of a large-scale underground heater test conducted in unsaturated fractured welded tuff at Yucca Mountain, Nevada. Key processes in this study are thermal-hydrological (TH) changes in fracture moisture content and thermal-mechanical (TM) changes in fracture aperture, which both contribute to changes in local fracture-permeability. The soundness of the modeling approach and in particular the model for thermal-mechanical induced changes in fracture permeability is demonstrated through good agreement between simulated and measured changes in air-permeability in several boreholes at DST. This shows that a continuum modeling approach is appropriate and indicates that fracture opening/closure caused by changes in fracture normal stress is the dominating mechanism for TM-induced changes in permeability at the site.

1. INTRODUCTION

The Yucca Mountain Drift Scale Test (DST) is a large underground heater test designed to study coupled thermal-hydrological-mechanical-chemical behavior in an unsaturated, fractured, and welded tuff rock. A coupled thermal-hydrological-mechanical (THM) analysis of the DST was conduced with TOUGH-FLAC (Rutqvist et al., 2002), which is a simulator based on a coupling of two established computer codes: TOUGH2 (Pruess et al., 1999) and FLAC3D (Itasca Consulting Group, 1997). This paper focuses on fracture-permeability changes caused by thermal-mechanical (TM) and thermal-hydrological (TH) processes at the DST. The results of TM-induced rock-mass deformations are presented elsewhere (Millard and Rutqvist, 2003).

2. THE YUCCA MOUNTAIN DRIFT SCALE TEST

The Drift Scale Test (DST) is carried out at Yucca Mountain, Nevada, in a side alcove of an underground tunnel, the Exploratory Studies Facility (ESF) at a depth of about 250 m in the middle nonlithophysal zone (Tptpmn) of the Topopah Spring Tuff formation. The DST centers around a heated drift, that is 5 m in diameter and has a 47.5 m long heated section (Figure 1). Heating is provided by nine floor heaters within the heated drift, as well as 50 rod heaters, referred to as "wing heaters," placed into horizontal boreholes emanating from the heated drift. The dimensions of the heated drift are similar to the current design of waste emplacement drifts for a repository at Yucca Mountain.

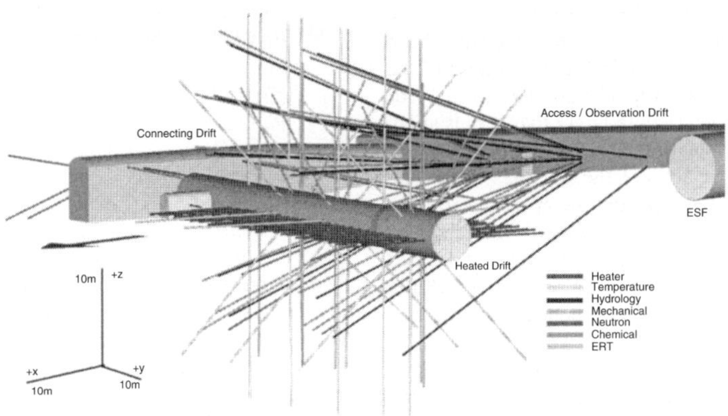

Figure 1. Three-dimensional view of the Yucca Mountain Drift Scale Test

The heaters of the DST were activated on December 3, 1997. The heating phase continued for approximately four years, till January 14, 2002, when heater power was turned-off. Currently, the DST is in the midst of a four-year cooling period.

The DST test block has been instrumented with thousands of sensors to monitor thermal, mechanical, hydrological, and chemical changes. For the purposes of this paper, the focus is on periodic testing of fracture permeability changes through air injection (Tsang et al., 2000) conducted in several-meters-long packed-off sections of so called hydrological boreholes.

The periodic air-permeability measurements provide an excellent opportunity to study the evolution of TM-induced changes in fracture permeability, and TH-induced changes in fracture moisture content, throughout the four-year heating period and the subsequent cooling period.

3. MODEL CONCEPTUALIZATION OF THE DST

The DST is simulated with TOUGH-FLAC in a two dimensional cross section oriented normal to the drift axis (Figure 2a). The highly fractured rock mass at the test site is modeled as a dual-permeability medium, which consists of interacting matrix and fracture continua (Birkholzer and Tsang, 2000). Fracture mapping at the site shows three dominating fracture sets (Nieder-Westermann, 1998, Section 7.4.5):

1) One prominent vertical, southeast trending
2) One less prominent vertical, southwest trending
3) One less prominent subhorizontal

The above-mentioned fracture sets have been derived through line mapping along sections of the ESF at Yucca Mountain. The average spacing for mapped fractures of length larger than 1 m is about 0.3 to 0.4 m. However, detailed cell mapping has shown that about 80% of the fractures at the site are less than one meter, and therefore the fracture spacing counting all fractures would be less than 0.3 m. Moreover, air-permeability tests conducted in short-interval (0.3 m) packed-off boreholes sections show that hydraulic conducting fractures exist at least every 0.3 m (Wang et al., 2001, Section 6.1). This evidence of a highly fractured rock justifies the using a continuum modeling approach.

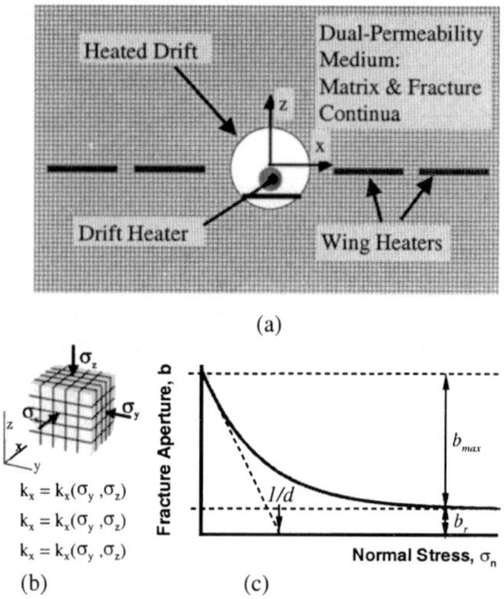

Figure 2. TOUGH-FLAC simulation of the DST: (a) schematics of two-dimensional model geometry; (b) conceptual model for stress-permeability coupling; (c) normal stress versus aperture relationship for fractures

For the Yucca Mountain site, incorporation of stress effects into hydraulic properties is based on a conceptual model of a highly fractured rock mass that contains three orthogonal fracture sets, as shown in Figure 2b. Porosity correction factor (F_ϕ) and permeability correction factors (F_{kx}, F_{kx} F_{kx}) calculated from the initial and current apertures (b_{1i}, b_{2i}, b_{3i} and b_1, b_2, b_3, respectively) in fracture sets 1, 2, and 3, according to:

$$F_\phi = \frac{b_1 + b_2 + b_3}{b_{1i} + b_{2i} + b_{3i}} \tag{1}$$

$$F_{kx} = \frac{b_2^3 + b_3^3}{b_{2i}^3 + b_{3i}^3},\ F_{ky} = \frac{b_1^3 + b_3^3}{b_{1i}^3 + b_{3i}^3},\ F_{kz} = \frac{b_1^3 + b_2^3}{b_{1i}^3 + b_{2i}^3} \tag{2}$$

where fractures in sets 1, 2, and 3 are assumed to be equally spaced and oriented normal to x, y, and z directions, respectively, and a parallel-plate fracture flow model is adopted. The capillary pressure is corrected for porosity and permeability changes according to a Leverett (1941) type of relationship:

$$F_{Pc} = \sqrt{\frac{F_k}{F_\phi}} \quad (3)$$

where

$$F_k = \sqrt[3]{F_{kx} \times F_{ky} \times F_{kz}} \quad (4)$$

In this study, the current fracture aperture b depends on the current effective normal stress σ'_n, according an exponential function:

$$b = b_r + b_m = b_r + b_{max}\left[\exp(d\sigma'_n)\right] \quad (5)$$

where b_r is a residual aperture, b_m is mechanical aperture, b_{max} is the maximum mechanical aperture, and d is a parameter related to the curvature of the function (Figure 2c). Equation (5) can be inserted into Equation (2) to derive expressions for rock-mass permeability correction factors in x, y, and z directions.

4. MATERIAL PROPERTIES AND INITIAL CONDITIONS

The material properties of the DST test block are presented in Table 1. These include both matrix and fracture hydrological properties for the dual-continuum approach as well as equivalent continuum rock-mass mechanical properties. Hydrological properties include water-retention curves and relative permeability functions based on van-Genuchten's (van-Genuchten, 1980) model. Mechanical properties include a fractured rock-mass deformation modulus of 14.77 GPa, which is about 50% lower than the Young's modulus of intact rock. A temperature-dependent thermal expansion coefficient is adopted (See also Millard and Rutqvist, 2003). The parameters b_{max} and d defining the stress-aperture relationship in Equation (5) have been estimated by model calibrations against various field experiments at Yucca Mountain (described by Rutqvist and Tsang, 2003). In the analysis presented here, a b_{max} of 150 μm and d of 0.6 Pa^{-1} are adopted.

The initial conditions for the simulation are presented in Table 2. At Yucca Mountain, vertical stress calculated from the weight of the overburden is the maximum principal stress, whereas the horizontal stresses are estimated to be about half the magnitude of the vertical stress. At an ambient temperature of 25°C, the matrix is almost fully saturated with water, whereas fractures are almost dry (as a result of matrix water imbibition).

Table 1. Rock properties of the fractured welded tuff (Tptpmn unit) containing the DST test block

Property	**Value**
Matrix Hydraulic and Thermal Properties	
Permeability	1.24E-17 m^2
Porosity	0.11
Van Genuchten, α_m	2.25E-6 Pa^{-1}
Van Genuchten m_m	0.247
Residual saturation	0.18
Rock grain density, ρ	2,530 kg/m3
Rock grain specific heat	953 J/kg K)
Dry thermal conductivity	1.67 W/m K
Wet thermal conductivity	2.0 W/m K
Fracture Hydrological and Hydro-Mechanical Properties	
Permeability, k	1.00E-13 m^2
Porosity	0.263E-3
Van Genuchten, α_f	9.73E-5 Pa^{-1}
Van Genuchten, m_f (-)	0.492
Residual saturation	0.01
Fracture frequency	4.32 m^{-1}
b_{max} for Equation (5)	150 μm
Exponent d for Equation (5)	0.6 MPa^{-1}
Rock Mass Mechanical Properties	
Young's Modulus	14.77 GPa
Poisson's ratio	0.21
Thermal Expan. Coeff.	5+0.0583×T 10^{-6}/°C

Table 2. Initial conditions

Parameter	**Approximate value at the level of DST**[1]
Vertical stress, σ_{zi}	≈ 5.7 MPa
Min horizontal stress, σ_{yi}	≈ 2.9 MPa
Max horizontal stress, σ_{xi}	≈ 3.4 MPa
Initial temperature, T_i	≈ 25°C
Initial Fracture saturation	≈ 9%
Initial matrix saturation	≈ 90%
Initial gas pressure	≈ 0.9 bar

1) Approximate values of the depth-dependent parameters at the drift level.

5. RESULTS OF SIMULATION

The simulated total thermal input power and the temperature at the drift wall for the entire 49.5-month heating period are presented in Figure 3. The simulated thermal input is based on average values of measured thermal-power input over discrete time periods. During the test, the thermal input was manually reduced with time to limit the rising temperature at the drift wall to about 200ºC. Calculated results after 12 months of heating are presented in Figures 4 to 7.

Figure 4 shows that after 12 months of heating, the temperature has risen above the boiling point around the heated drift and near the wing heaters. High temperature induces strongly coupled TH processes with evaporation of liquid water and drying near the heat source (Figure 5, dryout zone). The evaporated water is transported as vapor away from the heat source toward cooler regions, where it is condensed to liquid water (Figure 5, dark zone). As a result, a dry-out zone is created near the heat source and a condensation zone moves progressively away from the heat source. In the condensation zone, increases in fracture liquid saturation should result in a decreased air permeability.

On the other hand, high temperature gives rise to thermal expansion of the rock mass with associated thermal stresses (Figure 6). Near the heat source, the horizontal compressive stress increases strongly, with a maximum increase at the drift wall and near the wing heaters. Such an increase in compressive stress tends to tighten fractures to smaller aperture, leading to a reduction in air permeability. Away from the heat source, however, the horizontal stress decreases slightly (Figure 6, top). This reduction in horizontal stress will allow pre-existing vertical fractures to open to a larger aperture leading to an increase in air permeability in this area.

Figure 7 presents the calculated THM-induced changes in air permeability. These changes are caused by the combined effect of TH-induced changes in fracture moisture content (Figure 5) and TM-induced changes in fracture aperture (Figure 6). Near the heat source, permeability decreases mainly because of fracture closure, but is also affected by TH-induced wetting and drying. Away from the heat source, a zone of increased permeability has developed as a result of the opening of vertical fractures (Figure 7, near borehole section 74:4).

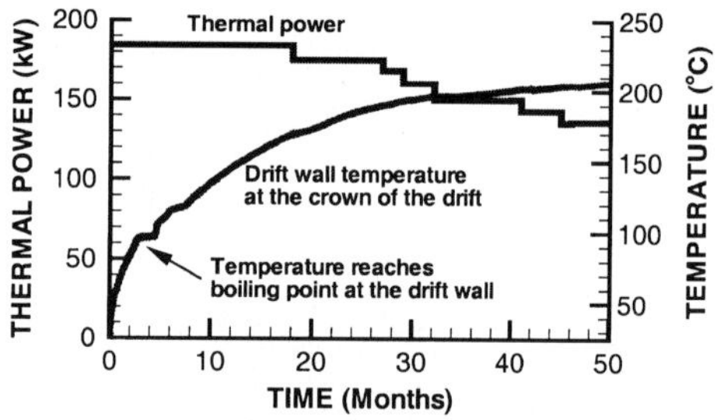

Figure 3. Simulated thermal power and drift-wall temperature at the crown of the drift during the 49.5-month heating period

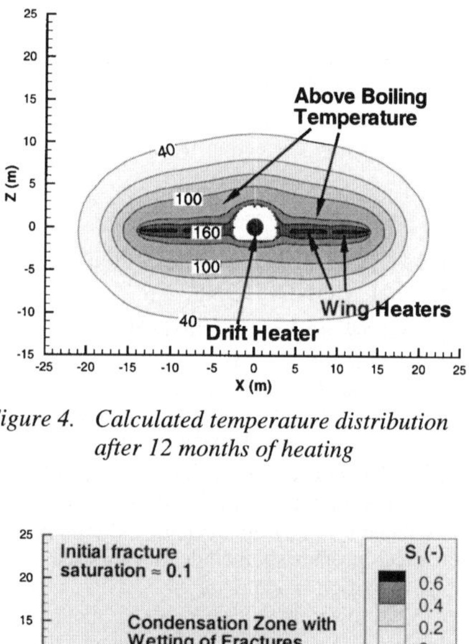

Figure 4. Calculated temperature distribution after 12 months of heating

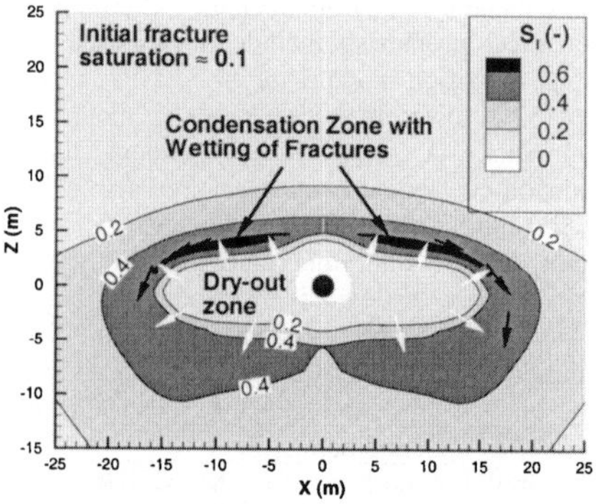

Figure 5. Calculated liquid saturation (S_l) in fractures after 12 months of heating

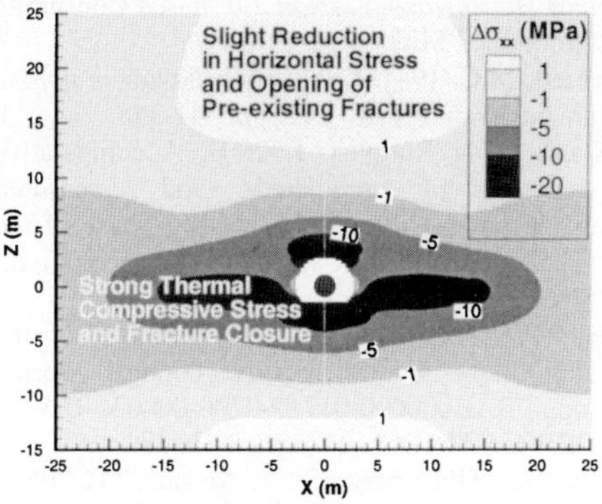

Figure 6. Calculated TM-induced horizontal stresses ($\Delta\sigma_{xx}$) after 12 months of heating

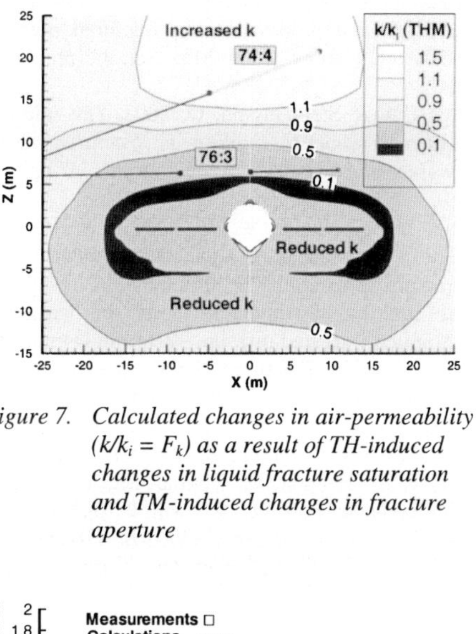

Figure 7. Calculated changes in air-permeability ($k/k_i = F_k$) as a result of TH-induced changes in liquid fracture saturation and TM-induced changes in fracture aperture

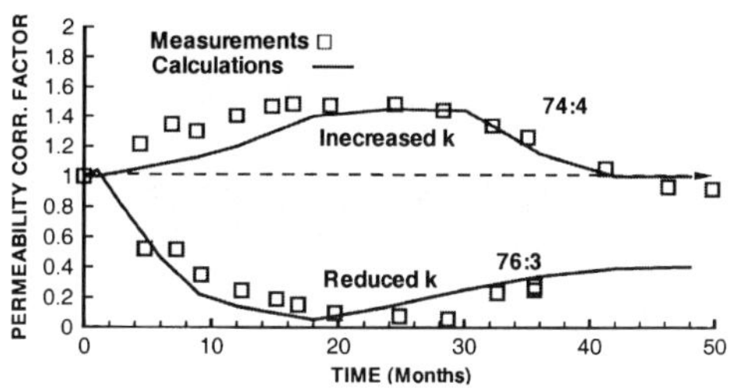

Figure 8. Calculated and measured permeability correction factor F_k *(see Equation 4 for definition of* F_k*)*

6. COMPARISON OF SIMULATED AND MEASURED AIR-PERMEABILITY CHANGES

Figure 8 presents a comparison of calculated and measured air-permeability at two borehole sections. The calculated changes in air-permeability are in good agreement with the measured ones both in trends and magnitudes. The good agreement between the calculated and measured air-permeability shows that the adopted conceptual model for TM-induced changes in permeability (shown in Figure 2) is sound and that the numerical values of b_{max} and d defining the normal stress-aperture function in Figure 2c are appropriate.

7. DISCUSSION

The conceptual model in Figure 2, combined with a continuum model approach, is shown to be appropriate for the analysis of THM processes at the DST because the rock mass is highly fractured, forming a dense, well-connected fracture network for fluid flow. This differs from many other fractured rock sites in Canada, Europe, and Asia, where underground tests have been conducted in sparsely fractured crystalline rocks (Rutqvist and Stephansson, 2003). In those formations, fluid flow is dominated by a few widely spaced fractures, which means that a continuum approach may not apply on the drift scale. In relation to other fractured rock sites, the rock mass at Yucca Mountain is relatively homogenous (ubiquitously fractured), with much less variability in rock-mass mechanical and hydrological properties.

Air-permeability measurements during the cooling phase of the DST will be extremely helpful for confirming the stress-aperture function and for observing potential irreversible (permanent) permeability changes. For an elastic rock mass, mechanical deformations and permeability changes would be fully reversible, meaning that after the temperature declines to ambient, air permeability would return to their original values. However, irreversible changes in permeability will almost certainly occur after the temperature has declined to ambient. The question is, how large will these irreversible changes be? So far, trends in displacement and permeability changes at the DST indicate that irreversible changes in permeability will be small, most certainly not larger than one order of magnitude.

8. CONCLUSIONS

The good agreement between simulated and measured air permeability indicates that the adopted conceptual model is sound and that the model for stress-versus-permeability coupling is appropriate for predicting TM-induced permeability changes at Yucca Mountain. This leads to the following specific conclusions regarding the coupled THM :

- A continuum model approach is appropriate for simulating relevant coupled THM processes at the DST.
- Fracture closure/opening caused by changes in normal stress across fractures is the dominating mechanisms for TM-induced changes in fracture permeability, whereas fracture shear dilation does not appear to be significant at the DST.

The conclusion that fracture shear dilation is not significant at the DST is pending on measurements during the cooling phase of the DST.

ACKNOWLEDGEMENTS

Technical review and comments by Sumit Mukhopadhyay and Dan Hawkes, Lawrence Berkeley National Laboratory are much appreciated. The work was partly supported by the Director, Office of Civilian Radioactive Waste Management, U.S. Department of Energy, through Memorandum Purchase Order EA9013MC5X between Bechtel SAIC Company, LLC and the Ernest Orlando Lawrence Berkeley National Laboratory (Berkeley Lab) through the U.S. Department of Energy Contract No. DE-AC03-76SF00098, and partly by a grant from the Swedish Nuclear Power Inspectorate (SKI).

REFERENCES

Birkholzer, J.T. & Tsang, Y.W. 2000. Modeling the thermal-hydrologic processes in a large-scale underground heater test in partially saturated fractured tuff. Water Resources Research, 36(6): pp. 1431-1447.

van Genuchten, M. T., 1980. A closed-form equation for predicting the hydraulic conductivity of unsaturated soils. Soil Sci Soc Am J 44: pp. 892-898.

Itasca Consulting Group Inc. 1997. *FLAC-3D Manual: Fast Lagrangian Analysis of Continua in 3 Dimensions–Version 2.0.* Itasca Consulting Group Inc., Minnesota, USA,

Leverett, M.C. 1941. *Capillary behavior in porous media.* Trans AIME. 142: pp. 341-358.

Millard, A. & Rutqvist, J. 2003. A comparative analysis of predicted and measured displacements during the heating phase of the yucca mountain drift scale test. Geoproc Conf., Stockholm, Sweden.

Nieder-Westermann G.H. 1998. Geology of the exploratory studies facility Topopah Spring loop. BAB000000-01717-0200-00002 REV01. CRWMS M&O, Las Vegas, Nevada.

Pruess, K., Oldenburg, C., & Moridis., G. 1999. *TOUGH2 User's Guide, Version 2.0*, Report LBNL-43134, Lawrence Berkeley National Laboratory, Berkeley, California.

Rutqvist, J., Wu, Y.-S., Tsang, C.-F. & Bodvarsson, G. 2002. A modeling approach for analysis of coupled multiphase fluid flow, heat transfer, and deformation in fractured porous rock. Int. J. Rock Mech. Min. Sci. 39: pp. 429-442.

Rutqvist, J. & Stephansson, O. 2003. *The role of hydromechanical coupling in fractured rock engineering.* Hydrogeology Journal, 11: pp. 7-40.

Tsang, Y.W., Huang, K. & Bodvarsson, G. S. 2000. Estimation of the heterogeneity of fracture permeability by simultaneous modeling of multiple air-injection tests in partially saturated fractured tuff. pp. 99-114 in Dynamics of Fluids in Fractured Rock, Edited by B. Faybishenko, P.A. Witherspoon, S. Benson, American Geophysical Union, Washington, DC.

Wang, J., Cook, P., Trautz, R,, Flexser, S., Hu, Q., Salve, R., Hudson, D., Conrad, M., Tsang, Y., Williams, K., Soll, W., & Turin, J. 2001. *In-situ field testing of processes ANL-NBS-HS-000005 REV01.* Las Vegas, Nevada: Bechtel SAIC Company.

THERMAL-MECHANICAL MODELING OF A LARGE-SCALE HEATER TEST

S.M. Hsiung[1], A.H. Chowdhury[1], and M.S. Nataraja[2]
1) Center for Nuclear Waste Regulatory Analyses, U.S.A.
2) U.S. Nuclear Regulatory Commission, U.S.A.

Abstract: Modeling of the drift-scale heater test at the Exploratory Studies Facility at Yucca Mountain, Nevada, U.S.A. was performed. The objectives of the analysis were to investigate the (i) temperature effects on mechanical deformation surrounding the heated drift and (ii) thermal-mechanical effects on rock-mass permeability. The continuum representation of a deformation-permeability relationship based on fracture normal stress was developed to assess rock-mass permeability variations because of temperature changes. The estimated rock-mass displacements and permeability variations as a function of heating time were compared with field measurements. The estimated trend of permeability responses using a normal stress-based deformation-permeability relationship compared reasonably to that measured.

1. INTRODUCTION

DECOVALEX (acronym for the **DE**velopment of **CO**upled models and their **VAL**idation against **EX**periments in nuclear waste isolation) is an international cooperative project that has been designed to increase understanding of coupled thermal-hydrological-mechanical processes as they affect rock-mass responses and radionuclide release and transport from a repository to the biosphere. One of the tasks in Phase III of this cooperative project was modeling of the drift-scale heater test at the Exploratory Studies Facility at Yucca Mountain, Nevada, U.S.A. The object of the drift heater test was to study the coupled behavior of the thermal-mechanical-hydrological-chemical processes in an unsaturated environment. This paper presents the modeling results on (i) temperature-induced rock-mass deformation surrounding the heated drift and (ii) thermal-mechanical effects on rock mass permeability and their comparison with the measured data.

2. MODELING APPROACH

2.1 Description of drift-Scale heater test

The host rocks for the proposed repository at Yucca Mountain include Topopah Spring Upper Lithophysal (Tptpul), Topopah Spring Middle Nonlithophysal (Tptpmn), Topopah Spring Lower Lithophysal (Tptpll), and Topopah Spring Lower Nonlithophysal (Tptpln) units. The drift-scale heater test facility is located in the Tptpmn unit (CRWMS M&O, 1997). The Tptpmn unit is 30–40 m thick at the location of the drift-scale test area. This unit is overlain by the Tptpul and underlain by the Tptpll units. The heater drift is approximately 5 m in diameter and 47.5 m long, and its entrance is closed by a thermal bulkhead.

Thermal sources for the heated drift consisted of 9 canister heaters, placed end to end on the concrete inverts of the heated drift, and 50 wing heaters (25 on either side) placed in horizontal boreholes drilled into the sidewalls of the heated drift approximately 0.25 m below the springline. The wing heaters were spaced 1.83 m apart and separated from the heater drift by 1.5 m.

2.2 Numerical model

The analyses were conducted using a continuum approach using the FLAC computer code (Itasca Consulting Group, Inc., 2000). A two-dimensional vertical cross section was configured for numerical analysis. This cross section intersected the axis of the heated drift at middistance between the thermal bulkhead and its terminus. The FLAC models had a dimension of 1,000 m in width and 740 m in height, with the origin of the coordinates located at the center of the heated drift. A 5-m diameter circular drift was located with its center 500 m from the left boundary and 500 m from the bottom of the model domain.

Table 1. Rock-mass mechanical properties.

Unit	RMQ*	Young's modulus, GPa	Poisson's Ratio	Bulk Density, kg/m^3
Tptpul	2	14.28	0.23	2,160
Tptpmn	2	12.02	0.21	2,250
Tptpll	2	12.02	0.21	2,250

* Rock-Mass Quality

Fixed horizontal displacement boundaries were applied to the sides and a fixed vertical displacement boundary to the bottom of the model. The top of the model coincided with the ground surface. Initial stresses consistent with overburden depth were applied as initial *in-situ* conditions.

The FLAC models included three thermal-mechanical lithologic units: Tptpul unit on the top, Tptpmn unit in the middle, and Tptpll unit on the bottom. The Tptpmn unit was 36 m thick, and its bottom was at 482 m from the bottom of the model domain. Note that the heater drift was located in the midheight of the Tptpmn unit.

The rock-mass mechanical and strength properties used for the three units were obtained from a CRWMS M&O report (CRWMS M&O, 1999) provided by the technical monitoring team and are listed in Tables 1 and 2. The acronym RMQ in Tables 1 and 2 means Rock-Mass Quality. The dilation angle for each unit was taken as half the corresponding friction angle. The rock-mass friction angle for the Tptpmn unit provided in CRWMS M&O (1999) was greater than the intact rock friction angle. The intact-rock friction angle was used in the analysis instead of the rock-mass value for the Tptpmn unit. To study the sensitivity of rock-mass properties on the thermal-mechanical effects, two variations of rock-mass properties also were modeled for each rock unit (Tables 3 and 4).

Table 2. Rock-mass strength properties.

Unit	RMQ	Cohesion, MPa	Tensile Strength, MPa	Friction Angle, Degree
Tptpul	2	1.4	1.13	46.00
Tptpmn	2	2.3	1.36	48.15
Tptpll	2	1.4	1.13	46.00

Table 3. Rock-mass mechanical properties for sensitivity analyses.

Unit	RMQ	Young's modulus, GPa	Poisson's Ratio	Bulk Density, kg/m^3
Tptpul	1	9.03	0.23	2,160
	5	20.36	0.23	2,160
Tptpmn	1	8.98	0.21	2,250
	5	24.71	0.21	2,250
Tptpll	1	8.98	0.21	2,250
	5	24.71	0.21	2,250

The measured temperature data were obtained from the technical monitoring team. These data were presented in three-dimensional grids in a two-day interval. Data selected as input for analyses were at 3, 6, and 9 months and 1, 2, 3, and 4 years of heating, and at the cross section 23 m from the thermal bulkhead. A temperature distribution in the rock mass after 4 years of heating is shown in Figure 1. When conducting the analyses, temperature changes between heating times were applied to simulate the heating process as a function of time. The thermal expansion coefficients, based on laboratory measurements (CRWMS M&O, 1999), for three rock units modeled are shown in Figure 2. It can be observed that the thermal expansion coefficients vary more than a factor of 5–7 from 25 to 300 EC. Two rock-mass failure criteria were used in the analyses to examine their effect on the rock-mass responses to the heating process: Mohr-Coulomb and ubiquitous failure criteria. When the ubiquitous failure criterion was used, fracture sets with dip angles (counterclockwise from the positive horizontal axis) of 82, 83.5, and 80.5 degrees were assumed for the Tptpul, Tptpmn, and Tptpll units, respectively (CRWMS M&O, 2000). The fracture cohesion, tensile strength, friction angle, and dilation angle were assumed to be 0.1 MPa, 0.0 MPa, 41 degrees, and 20.5 degrees for all three units (CRWMS M&O, 2000).

Table 4. Rock-mass strength properties for sensitivity analyses.

Unit	RMQ	Cohesion, MPa	Tensile Strength, MPa	Friction Angle, Degree
Tptpul	1	1.1	0.90	44.00
	5	2.9	2.26	47.00
Tptpmn	1	1.9	1.16	48.15
	5	3.9	2.22	48.15
Tptpll	1	1.1	0.90	44.00
	5	2.9	2.26	47.00

3. RESULTS AND DISCUSSIONS

3.1 Thermally induced rock-mass mechanical responses

Thermally induced failure of rock mass

After heating started, the rock mass began to deform in response to the thermally induced stresses. At some point of heating, a yield zone developed for the models with the Mohr-Coulomb failure criterion in the Tptpul unit near the interface between the Tptpul and Tptpmn units. This yield zone continued to grow upward throughout the heating process. The modeling results indicated that the timing of occurrence and the ultimate size of this yield zone were rock-mass property dependent. This yield zone tended to develop earlier and bigger for models using larger RMQ designations (e.g., RMQ 5 versus RMQ 2). The reason for this behavior is obvious because the differences between the Young's modulus among the RMQs are significant but the differences for the strength properties are relatively small. Thermally induced stresses are directly proportional to the Young's modulus. Consequently, it is more likely for the models with higher RMQ designations to develop yield zones larger than those with lower RMQ designations. Similar trend could be observed for development of a yield zone beneath the heated drift in the Tptpll unit at the interface between the Tptpmn and Tptpll units. This yield zone, however, developed later and relatively smaller in size than the yield zone above the heated drift [Figure 3(a)].

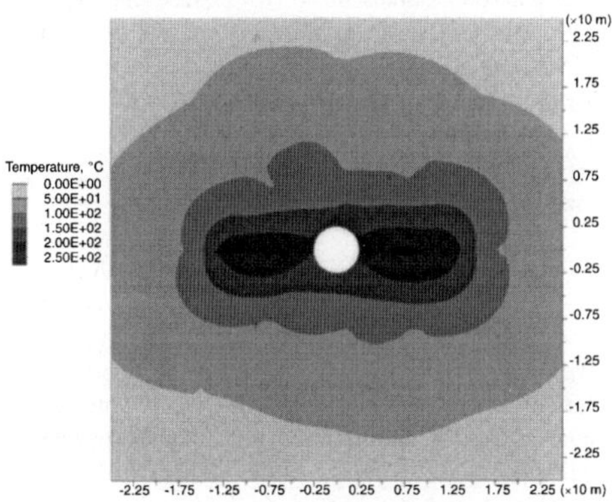

Figure 1. Temperature distribution in the rock mass after 4 years of heating.

The majority of rock yielding for models using the ubiquitous fracture failure criterion took place along fractures (fracture slip) because of fracture shear failure. As a result, the yielding behavior for the models using the ubiquitous failure criterion was substantially different from that for the models using the Mohr-Coulomb failure criterion. The fracture slip zones started first in the Tptpmn unit at the early stage of heating, mainly at a diagonal direction, forming an approximately 45-degree angle counterclockwise from the vertical axis and more than 50 degrees from the fracture dip angles [Figure 3(b)]. These fracture slip zones formed long shear bands stretching along the diagonal direction discussed earlier. As heating proceeded, new shear bands continued to develop vertically either upward or downward with respect to the heated drift. Also, these shear bands joined together. Another group of shear bands, at an approximately 45-degree angle clockwise from the vertical axis [Figure 3(b)], started to develop later and expand outward from the heated drift at a much slower rate than the shear bands at the perpendicular direction. Similar to the yielding trend observed for the models with the Mohr-Coulomb failure criterion, the shear bands tended to develop earlier and bigger for models using larger RMQ designations than those for models using smaller RMQ designations.

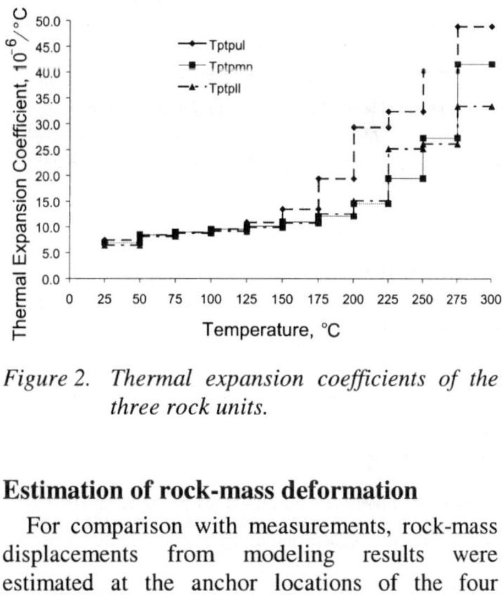

Figure 2. Thermal expansion coefficients of the three rock units.

Estimation of rock-mass deformation

For comparison with measurements, rock-mass displacements from modeling results were estimated at the anchor locations of the four multiple-position extensometers (MPBXs) installed at the cross section approximately 21 m from the thermal bulkhead. For these four MPBXs, two were vertically oriented (one in the crown, MPBX9, and one in the invert, MPBX10). The remaining two (MPBX7 and MPBX8) were inclined at approximately 30 degrees to MPBX9 on either side of MPBX9. Figures 4, 5, 6, and 7 show the estimated displacements as a function of time for the model using the RMQ 2 rock-mass properties and the ubiquitous failure criterion. Note that each MPBX contained four anchors. Anchor 1 was located closest to the heated drift, while

Anchor 4 was the farthest. Also included in the figures were the measured displacements. Letters E and M in the legends denote the estimated results and the measured data.

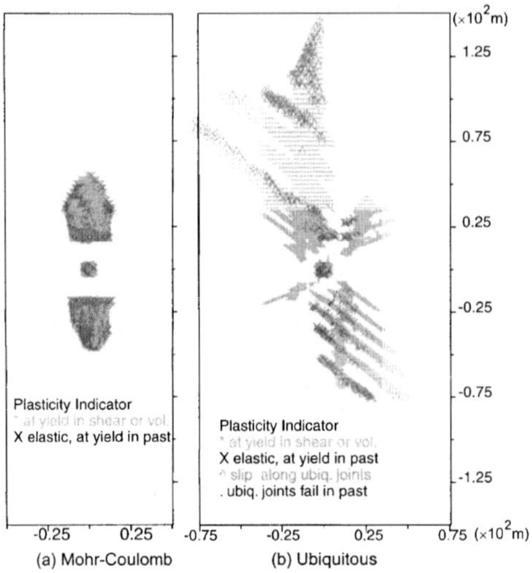

Figure 3. Thermally induced rock-mass yielding.

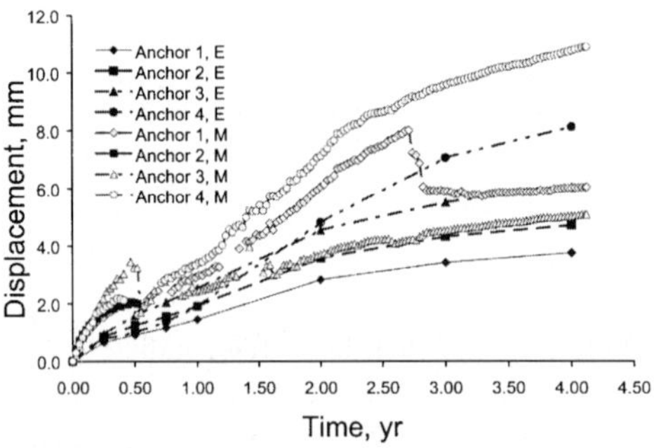

Figure 4. Anchor displacements for MPBX7.

Other than in the early stage of heating and regardless of the rock-mass properties or failure criteria used, the anchor displacements relative to the MPBX assembly head, located at the collar of the heated drift, were larger for the anchors located farther away from the heated drift. This displacement pattern indicated that the neighboring anchors of an MPBX continued to move away from each other as heating proceeded, suggesting an expansion of the rock mass between the two anchors. This behavior was not strictly observed in the measured data where some compression zones could be identified in the rock mass between anchors; for example, between Anchors 1 and 3 of MPBX7 (Figure 4) and between Anchors 2 and 3 of MPBX9 (Figure 6) and MPBX10 (Figure 7).

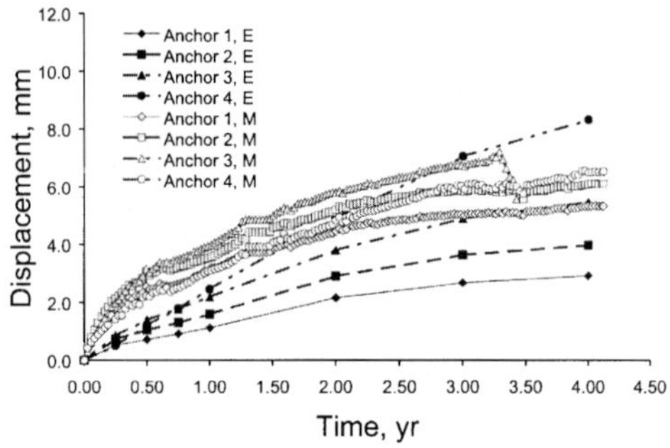

Figure 5. Anchor displacements for MPBX8.

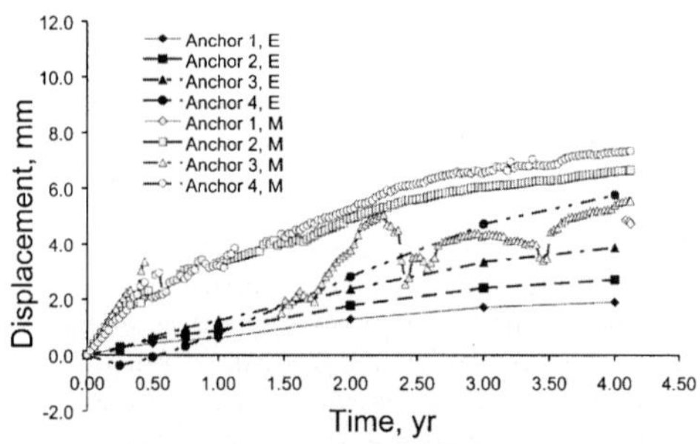

Figure 6. Anchor displacements for MPBX9.

In general, the estimated anchor displacements were larger when the rock-mass properties for larger RMQ designations were used. During the early stage of heating, Anchor 4 for both vertical MPBXs moved closer to the heated drift before Anchor 4 and the heated drift moved away from each other as heating proceeded further. This phenomenon was not observed in the measured data. Instead, the measured data indicated that all anchors in an MPBX had almost identical displacements for the first few months of heating, suggesting that only the rock mass between Anchor 1 and the assembly head deformed while there was not much deformation in the rock-mass between anchors.

The modeling results underestimated displacements for most anchors being evaluated. The measured anchor displacements for MPBX8 appeared similar in magnitude, suggesting that rock-mass expansion took place mostly in the region between Anchor 1 and the assembly head, while the deformation between anchors was relatively small. This behavior might be a result of heterogeneity of the rock mass in the region not reflected in the modeling results.

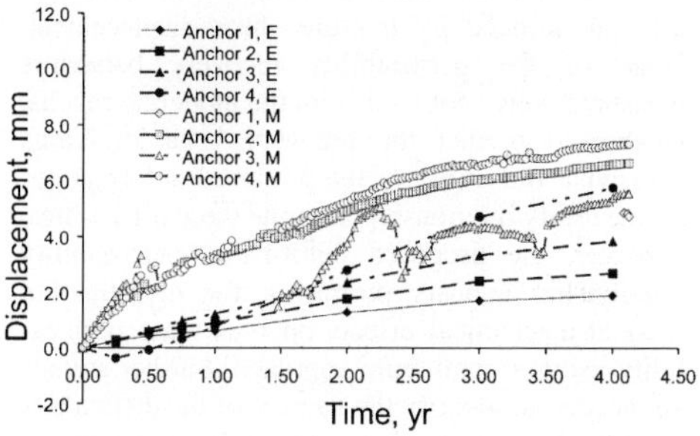

Figure 7. Anchor displacements for MPBX10.

The measured displacements also suggested potential localized fracture slip. For example, fracture slip in the rock mass between Anchors 2 and 3 of MPBX7 after approximately 6 months of heating caused Anchor 3 to move relatively closer to the assembly head. Another example was the slip of a fracture located possibly between Anchor 1 and the assembly head of MPBX7 at approximately 2.7 to 2.8 years of heating that caused Anchor 1 to move closer to the assembly head. These slips caused a displacement reduction. This distinct behavior represented one form of heterogeneity of the rock mass studied. A continuum analysis, such as that used in this study, could not capture this behavior. Consequently, the estimated displacement from modeling results would not be able to predict a displacement reduction representing fracture slip.

3.2 Thermal-Mechanical effects on rock-mass permeability

Permeabilities at specified locations

To assess the thermal-mechanical effects on rock-mass permeability, a deformation-permeability relationship based on fracture normal stress was developed.

$$k_f = k_{fr}\left(\frac{a}{c(\sigma_n+1)}\sqrt{\frac{\phi_{fr}}{12k_{fr}}} - \frac{e_{ftp}+e_{fsp}\tan\psi_f}{\phi_{fr}}\right)^3 \quad (1)$$

where k_f and k_{fr} are permeability at time of interest and reference permeability; a and c are constants, σ_n is the normal stress; ϕ_{fr} is the reference fracture porosity; e_{ftp} and e_{fsp} are rock-mass or fracture tensile and shear plastic strains; and ψ_f is the rock-mass or fracture dilation angle. Constant c can be determined from

$$c = \frac{-1 \pm \sqrt{1+4a\sigma_{nr}\sqrt{\frac{\phi_{fr}}{12k_{fr}}}}}{2\sigma_{nr}} \quad (2)$$

where σ_{nr} is the reference normal stress, and constant a is assumed to be the reciprocal of the initial normal stiffness (Bandis, et al., 1983). In this study, the normal stress and the fracture permeability and porosity before heating were used as reference. The reference fracture permeability and porosity for Tptpul were $5.50 \times 10^{\&13}$ m^2 and 0.0066, for Tptpmn $2.76 \times 10^{\&13}$ m^2 and 0.01, and for Tptpll $1.29 \times 10^{\&12}$ m^2 and 0.011 (CRWMS M&O, 2001). The initial fracture stiffness was 2.01 $\times 10^5$ MPa/m (CRWMS M&O, 1999).

As a part of the heater tests, several hydrologic holes were drilled and pressure sensors installed in the holes to measure permeability variations during the heating process. In this paper, the estimated permeability variations at the pressure sensor locations in Holes 57 and 59, at the cross section 10 m from the thermal bulkhead, and Holes 74 and 76, at the cross section 30 m from the thermal bulkhead (Figure 8), were compared with the measured data. Figures 9 and 10 show the variations of permeability as a function of time for Holes 57 and 59 along with field measurements. Letters E and M in the legends denote estimated results and measured data. Vertical axes in the figures are the ratio of the current permeability to the reference permeability. In general, the trends of permeability response to heating for the models with different RMQs and failure criteria were similar. The difference among the models studied was mainly the magnitude of variation.

Figures 9 and 10 suggested that the continuum analyses modeled reasonably the trending of permeability responses to the heating process although there were obvious discrepancies between the estimated and the measured trends. One discrepancy was the variations in permeability estimated at the location of Pressure Sensor 1 (PS1) in Holes 57 and 59. The measured data at this location indicated a permeability reduction (smaller than the reference permeability) trend for the heating duration, whereas the estimated trend was a permeability enhancement (larger than the reference permeability) at the beginning of the heating. The extent of the enhancement decreased as heating proceeded and, for some models analyzed, this enhancement disappeared and, eventually, led to a permeability reduction. The estimated reduction, however, was small in magnitude when it is compared with the measurements. Also, the measured data and modeling results of all models showed different trending in permeability variations at PS1 in Hole 59.

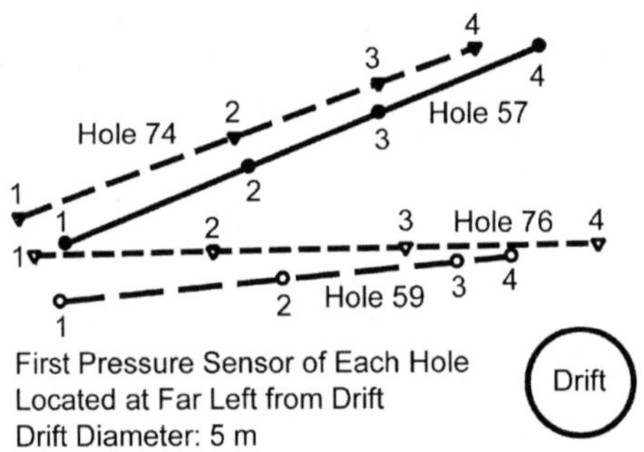

Figure 8. Location of hydrologic holes.

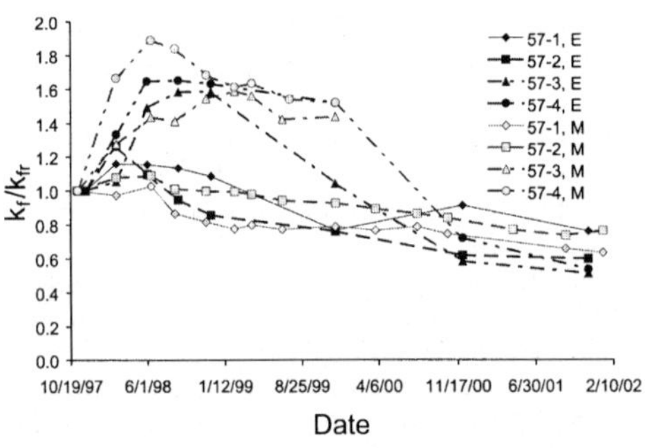

Figure 9. Permeability variations for Hole 57.

The measured data showed permeability recovery at some pressure sensor locations in the middle of the heating process after some large reduction of permeability earlier in the heating process. These pressure sensor locations included PS2 and PS3 in Hole 59 (Figure 10) and PS3 and PS4 in Hole 76 (no figure shown). The permeability at PS3 in Hole 59 not only recovered completely but also started to experience a permeability enhancement between 3 and 4 years of heating. There were two possibilities causing permeability to recover. One possibility was a decrease in fracture normal stress that resulted in a reduction in fracture closure, which, in turn, increased the fracture aperture. When the fracture normal stress became smaller than the reference fracture normal stress, permeability enhancement would result. The other possibility was related to dilation induced by fracture shear displacement. None of the permeability recovery behaviors measured were observed from the modeling results. Unable to predict the recovery behavior limits somewhat the utility of the proposed deformation-permeability relationship. Despite these difficulties, however, the proposed deformation-permeability relationship appears promising for investigating thermal-mechanical effects on rock-mass permeability using a continuum approach. Further studies are needed to identify the sources of the difficulties so that the deformation-permeability relationship can be calibrated, if necessary.

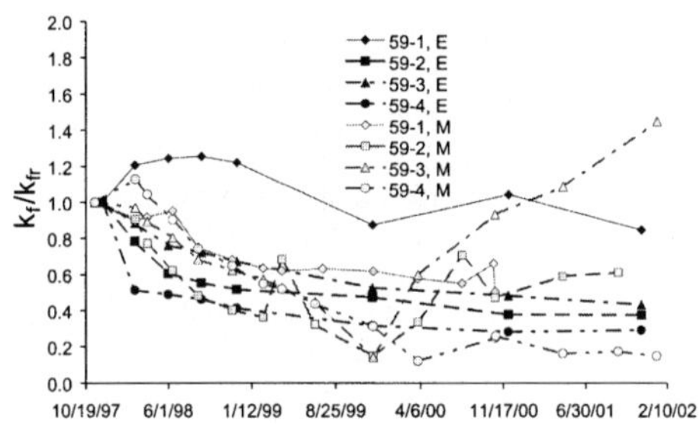

Figure 10. Permeability variations for Hole 59.

Spatial permeability variations

Spatial permeability variations in the region 60 m from all sides of the heated drift after 4 years of heating are shown in Figure 11 for the model with the RMQ 2 rock-mass properties and ubiquitous failure criterion. The contours in the figure are ratios of the current permeability to the reference permeability. The figure indicated, in general, a permeability reduction zone surrounding the heated drift regardless of the failure criteria used. This zone extended diagonally outward from the top-left and low-right corners of the heated drift. In this permeability reduction zone, the region with the extensive permeability reduction was in the rock mass immediately surrounding the heated drift.

Notice the presence of zones with high permeability enhancement. Locations of these high-permeability enhancement zones were consistent with those of the shear bands shown in Figure 3(b) [Note the scale difference between Figure 3(b) and Figure 11.]. Notice also the presence of isolated permeability enhancement zones in the essentially permeability reduction zone.

If the same failure criterion was used, the spatial permeability pattern would be the same regardless which rock-mass properties were used. The magnitude of change, however, would be different. In general, the larger the RMQ designations, the greater the variation magnitude.

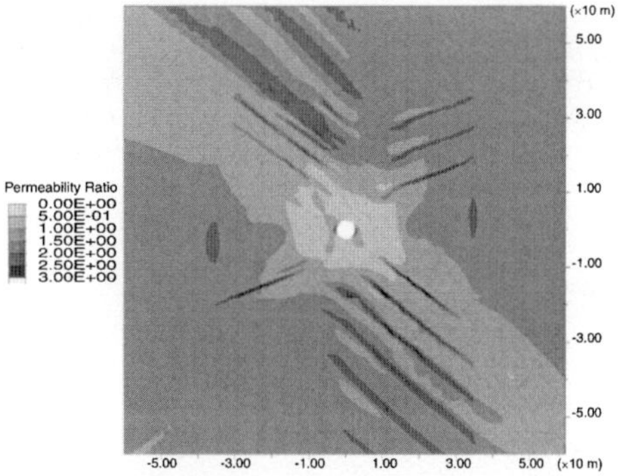

Figure 11. Permeability contour plot for a model using the ubiquitous failure criterion.

4. CONCLUSION

A continuum approach was adopted to assess thermally induced rock-mass responses of a drift-scale heater test. The modeling results underestimated the displacements of the rock mass surrounding the heated drift. One possible cause for the underestimation might be related to the temperature data used. This temperature data set was constructed using a kriging procedure that had the tendency to introduce uncertainties by smoothing out the temperature distribution. These uncertainties could affect the displacement estimation around the heated drift if the temperature in the area was underestimated. Another possible cause might be that the seven heating times used in the analyses were too coarse. For all models studied, large yield zones were developed in the rock mass. Because plastic deformation is path dependent, it may be necessary to use a smaller heating timestep to simulate more realistically the plastic deformation behavior in the analyses.

A normal stress-based deformation-permeability relationship was proposed to investigate the thermal-mechanical effects on rock-mass permeability. The estimated trend of permeability responses to heating compared reasonably to that measured. The modeling results, however, were not able to predict the permeability recovery observed at certain locations.

REFERENCES

Bandis, S.C., A.C. Lumsden, and N.R. Barton. 1983. Fundamentals of Rock Joint Deformation. *International Journal of Rock Mechanics and Mining Sciences & Geomechanics Abstracts.* Vol. 20, No. 6. pp. 249–368.

CRWMS M&O. 2001. *Multiscale Thermo-hydrologic Model.* ANL–EBS–MD–000049. Rev. 00 ICN 02. Las Vegas, Nevada: CRWMS M&O.

CRWMS M&O. 2000. *Ground Control for Emplacement Drifts for SR.* ANL–EBS–GE–000002. Rev. 00. Las Vegas, Nevada: CRWMS M&O.

CRWMS M&O. 1999. *TBV-332/TBD-325 Resolution Analysis: Geotechnical Rock Proper ties.* B00000000–01717–5705–00134. Rev. 00. Las Vegas, CRWMS M&O.

CRWMS M&O. 1997. *Ambient Characterization of the Drift-Scale Test Block.* BADD00000–01717–5705–00001. Rev. 01. Las Vegas, Nevada: CRWMS M&O.

Itasca Consulting Group, Inc. 2000. *FLAC—Fast Lagrangian Analysis of Continua.* Version 4.0. Minneapolis, Minnesota: Itasca Consulting Group, Inc.

ACKNOWLEDGEMENTS

The work described in the paper was performed at the Center for Nuclear Regulatory Analyses, Southwest Research Institute®, San Antonio, Texas, U.S.A., on behalf of the U.S. Nuclear Regulatory Commission (NRC) Office of Material Safety and Safeguards, Division of Waste Management. This paper is an independent product of the CNWRA and does not necessarily reflect the views or regulatory position of the NRC. The NRC staff views expressed herein are preliminary and do not constitute a final judgment of the matter addressed or of the acceptability of a license application for a geologic repository at Yucca Mountain.

NUMERICAL SIMULATION OF THERMAL-HYDROLOGICAL PROCESSES OBSERVED AT THE DRIFT-SCALE HEATER TEST AT YUCCA MOUNTAIN, NEVADA

Ronald T. Green[1] and Scott Painter[1]
[1])Center for Nuclear Waste Regulatory Analyses, U.S.A.

Abstract: Results from the four-year long heating phase of the Drift-Scale Heater Test at the Exploratory Studies Facility at Yucca Mountain, Nevada, provide a basis to evaluate conceptual and numerical models used to simulate thermal-hydrological coupled processes expected to occur at the proposed repository. A three-dimensional numerical model was built to perform the analyses. All model simulations were predicated on a dual (fracture and matrix) continuum conceptualization. A 20-percent reduction in the canister heat load to account for conduction and radiation heat loss through the bulkhead, a constant pressure boundary condition at the drift wall, and inclusion of the active fracture model to account for a reduction in the number of fractures that were hydraulically active provided the best agreement between model results and observed temperatures. The views expressed herein are preliminary and do not constitute a final judgment of the matter addressed or of the acceptability of its use in a license application

1. INTRODUCTION

Results from the Drift-Scale Heater Test (DST) at the Exploratory Studies Facility at Yucca Mountain (YM) were used to evaluate the capability of a multiphase code to simulate thermal-hydrological processes in partially saturated fractured rock. The DST is a planned eight-year long *in situ* heater test, with four years of heating followed by four years of cooling. The four-year heating phase was completed in January 2002. The test has been monitored using a vast array of instrumentation to measure, or infer, changes in temperature, saturation, stress, and chemistry.

For the modeling study presented in this paper, temperature and saturation distributions in the rock were simulated for the four-year heating phase of the DST. The model results were compared with measured temperatures and saturations at the onset of the test and with measured temperatures after three months, one year, and four years of heating to guide refinement of the model. These modeling studies are an important part of the process of assessing and developing confidence to independently evaluate the safety case for the proposed geologic repository at YM. The predictive modeling analyses reported in this paper were conducted as part of Task 2 of the DECOVALEX III program.

2. DRIFT-SCALE HEATER TEST DESCRIPTION

The DST is located in the Topopah Spring middle nonlithophysal (Tptpmn) unit, within the horizon of the proposed YM repository. The Tptpmn unit is approximately 30–40 m thick at the location of the DST, overlain by the Topopah Spring upper lithophysal (Tptpul) and underlain by the Topopah Spring lower lithophysal (Tptpll) units. The DST test block was characterized prior to the onset of heating. On-site characterization of the local geology and *in situ* hydrology was supplemented with laboratory tests of thermal-hydrological properties.

A plan-view schematic of the relative placement of the heated drift to the observation drift is shown in Figure 1. The 5-m diameter, 47.5-m long heated drift is closed at the east end by a thermal bulkhead. Approximately 12 m of the west end of the heated drift is lined with cast-in-place concrete ground support. Concrete inverts were placed along the entire floor of the heated drift to provide a flat surface. Thermal sources for the heated drift consist of nine canister heaters, placed end to end on the floor of the heated drift, and 50 wing heaters (25 on either side), emplaced in horizontal boreholes drilled into the sidewalls of the heated drift about 0.25 m below the springline. The wing heaters are spaced 1.83 m apart. Each wing heater has two segments, both 5 m long, separated by 1.67 m with a larger power output from the outer segment. The inner wing heater segment is separated from the heated drift wall by 1.5 m.

3. CONCEPTUAL MODEL

A three-dimensional numerical model was assembled to perform the analyses. A variably sized grid in the xz-plane, commonly referred to as an

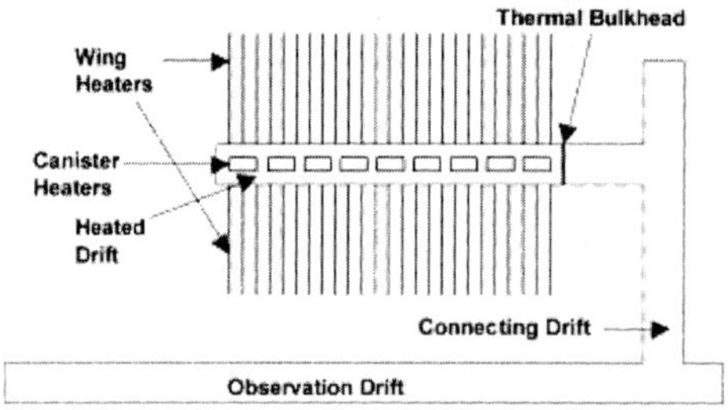

Figure 1. Plan-view schematic of the drift-scale heater test (DST).

unstructured grid, was used to construct the numerical model. This allowed for higher resolution near the heated drift and the wing heaters, areas with the greatest gradients in temperature and saturation. The xz-plane was extruded as 14 planes in the y-direction to give the model its third dimension. Themedium was represented as a dual continuum, one for the matrix and one for the fractures. The fracture continuum implemented the active fracture model (Liu et al., 1998). Following are descriptions of these conceptualizations.

3.1 Dual continuum model

The dual continuum model (DCM) formulation is comparable to the dual permeability model (DKM) formulation (TRW Environmental Safety Systems, Inc., 2000). The DCM and DKM conceptualizations provide separate continua for the matrix and the fractures. The dual continua are coupled throughout the model domain by transfer functions for heat and mass transfer between the fractures and matrix. Use of a DCM increases the complexity of the numerical model used in the simulations, but offers the potential to realistically partition flow between matrix and fractures. Mass flow across the matrix/fracture interface is directionally dependent. When liquid pressure in the matrix exceeds the pressures in the fractures (i.e., $P_{lm} > P_{lf}$), liquid flow, Q_l, from the matrix to the fracture continuum is defined by

$$Q_l = \frac{A_{fm}A_{\text{mod}}}{\mu_f} k_{harmonic} k_{rl,m\to f} \frac{P_{lm} - P_{lf}}{d} \quad (1)$$

where the m, f, l, and r subscripts denote matrix, fracture, liquid, and relative. d is the distance from the matrix block center to the fracture center and $k_{harmonic}$ is the harmonic mean of the absolute permeability. A_{mod} is a modifier term included to allow a reduction, but not an increase, in the interfacial area between the matrix and fracture continua. For $P_{lf} > P_{lm}$, flow from the fractures to the matrix is defined by an equation analogous to Equation 1, with $k_{rl,m\to f}$ replaced by $k_{rl,f\to m}$. The matrix liquid-phase relative permeability is used for the matrix-to-fracture relative permeability $k_{rl,m\to f}$. For flow from fracture to matrix, a separate representation $k_{rl,f\to m}$ is used, thereby providing additional flexibility to the model. Similarly, an analogous form of Equation 1 defines mass flow of gas, Q_g, between matrix and fracture continua.

$$Q_g = \frac{A_{fm}A_{\text{mod}}}{\mu_g} k_{harmonic} k_{rg} \frac{P_{gf} - P_{gm}}{d} \quad (2)$$

The upstream value for the relative permeability is used to define gas flow between the matrix and fracture continua.

3.2 Active fracture model

The active fracture model for unsaturated flow through fractured rocks proposed by Liu et al. (1998) was used to modify flow through fractures. This model is based on the hypothesis that only a portion of connected fractures is active in conducting water. The hypothesis stipulates that: (i) all connected fractures are active if the system is fully saturated, (ii) all fractures are inactive if the system is at residual saturation, and (iii) the fraction of fractures that is active is related to water flux through the fractures. Liu et al. (1998) proposed that the fraction of active fractures be a power function of effective water saturation in connected fractures.

The liquid-phase relative permeability function for the fracture continuum defined in the Mualem relationship is modified to

$$k_{rl} = \left(S_l^{eff}\right)^{\frac{1+\gamma}{2}} \left\{1 - \left[1 - \left(S_l^{eff}\right)^{\frac{1+\gamma}{2}}\right]^m\right\}^2 \quad (3)$$

to describe fracture to matrix flow. S^{eff} is effective saturation. Similarly, the van Genuchten relationship between effective saturation and capillary pressure is modified to

$$S_l^{eff} = \left[1 + \left(\alpha|P_c|\right)^{\frac{n}{1-\gamma}}\right]^{-m} \quad (4)$$

where α is the inverse of the air-entry pressure, P_c is capillary pressure, n and m are fitting parameters, and γ is a positive constant depending on the properties of the fracture network.

4. NUMERICAL SOLUTION

The multiphase simulator MULTIFLO was used to perform the thermal-hydrological simulations (Painter et al., 2001). MULTIFLO is based on a fully implicit formulation using a variable substitution approach. Space discretization is based on a block-centered grid using an integral finite-volume difference scheme, an approach suitable for an unstructured grid with arbitrary interblock grid connectivity and any polygon block boundary.

The numerical model grid is aligned with the axis of the heater drift. Vertical symmetry is assumed along and orthogonal to the drift axis. One plane of symmetry is placed mid-distance between the bulkhead and the terminus of the heated drift. The other plane is colinear with the drift axis. This allows modeling only ¼ of the DST block volume. The modeled area extends 200 m in the vertical direction and 100 m in the horizontal direction. The center of the heated drift is vertically placed at the midpoint of the model. Planes that intersect the heated drift have 1,123 nodes for each continuum. Planes that do not intersect the heated drift have additional nodes to account for the drift (i.e., 1175 nodes). The entire model, therefore, has 16,068 nodes in each continuum. Finer mesh resolution is included in areas expected to experience large temperature, saturation, and pressure gradients (Figure 2).

Three hydrostratigraphic units are included in the model: Tptpul (TSw33), Tptpmn (TSw34), and Tptpll (TSw35) of the Topopah Spring welded unit. Values for key properties assigned to the three units are presented in Table 1 (CRWMS M&O, 2001). Property values are assigned uniformly to each unit.

4.1 Boundary and initial conditions

The vertical boundaries of the model were specified as adiabatic with no fluid flow. The top boundary was prescribed as a mixed boundary with specified flux and constant temperature and pressure. The bottom boundary was prescribed as a Dirichlet type with specified pressure, temperature, and gravity drainage. The mixed boundary condition at the top allows gas and heat transport in or out of the model while maintaining pressure and temperature as specified. The heater drift was not explicitly included in the model; instead, the heater drift wall was specified as constant pressure (at atmospheric), allowing for heat and mass loss out of the drift.

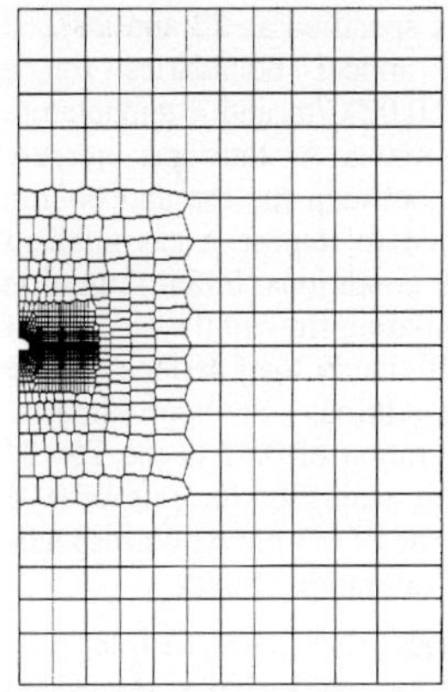

Figure 2. Unstructured grid assigned to the numerical model

Table 1. Key property values assigned to modeled units.

Unit	Tptpul	Tptpmn	Tptpll
Fracture permeability (m^2)	5.50e-13	2.76e-13	1.29e-12
Fracture porosity (-)	0.0066	0.010	0.011
Matrix Permeability (m^2)	3.08e-17	4.04e-18	3.04e-17
Matrix Porosity (-)	0.154	0.110	0.131
Thermal conductivity (dry)(W/m-K)	0.79	1.56	1.20
Thermal conductivity (wet)(W/m-K)	1.68	2.33	2.02

The temperature was specified as 22 and 26°C at the top and bottom model boundaries for a geothermal gradient of 0.02°C/m and a temperature of 24°C at the DST horizon. A static gas pressure difference of 2,156 Pa between the top and bottom boundaries was specified to impose a gas gradient consistent with ambient conditions. Initial saturation was determined by simulating flow in the absence of heat at the DST for sufficiently long periods of time that steady-state flow conditions were approximated. An ambient matrix saturation of 0.92 in the TSw34 (Tptpmn) is predicted at an infiltration rate of 0.06 mm/yr. An infiltration rate of 0.3 mm/yr corresponds with an ambient matrix saturation of 0.99.

4.2 Model heat source

Heat was introduced into the model at the heated drift and at the inner and outer wing heaters. The heated drift cavity was not explicitly included in the model to avoid difficulties associated with representing the air space within the drift, radiative and convective heat transfer between the heater canisters and the drift wall, and the physics of heat and mass transfer at the drift-cavity/drift-wall boundary. The disadvantage to this simplification is that coupled thermal-hydrologic processes at the drift wall cannot be directly or easily investigated using this model.

The heat-source levels were applied uniformly to the drift boundary elements at the drift wall. The cylindrical wing heaters were represented as rectangular slabs, thereby smearing the heat deposition in the y-direction of the model. The DST experienced measured heat loads that deviated significantly (less) from the levels of the design heat loads since energized in December 1997. Piece-wise linear heat loads assigned to the model are compared with measured wing and canister heat loads in Figure 3. It was assumed that the decrease in cumulative heat load for the wing heaters occurred uniformly over both the inner and outer wing heaters. The canister heat load was reduced by 20 percent to account for heat loss through the thermal bulkhead.

5. MODEL RESULTS

Heat and mass transfer were simulated and compared to the DST data for the four-year heating phase of the DST. Temperature was directly measured throughout the DST affected volume. Saturation for the matrix is inferred using

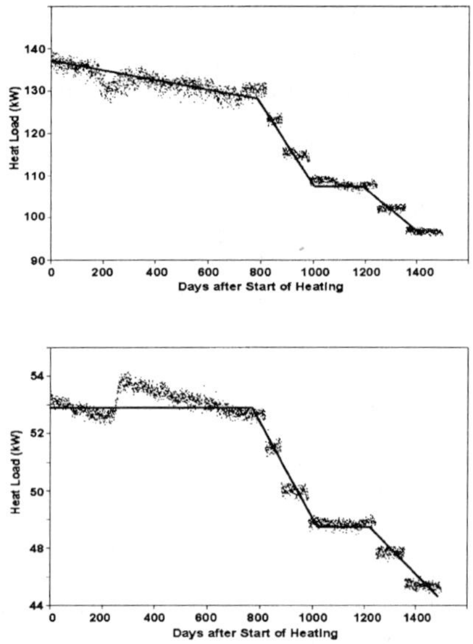

Figure 3. Measured (dots) and simulated (lines) heat load for the wing (top) and canister (bottom) heat sources.

geophysical methods (i.e., electrical resistivity tomography, ground-penetrating radar, and neutron logging). Air permeability tests were conducted to provide a measure of fracture saturation. These measurements only provide a qualitative measure of saturation at the DST. Therefore, only temperature simulation results at this time, although inferred saturation measurement techniques may promote a clearer interpretation of the DST.

Temperatures calculated using a 2D model were measurements are used for comparison with compared with similar results using a 3D model to evaluate the effect of dimensionality. The comparison was made for the vertical plane located at the midpoint in the heated drift. The difference in temperatures calculated at the end of the four-year heating phase is illustrated in Figure 4. The 2D model calculated temperatures a maximum of 10.5 C higher at boiling isotherm located above the outer wing heater. Also calculated was a slight decrease in temperature in the halo formed just within the boiling isotherm. The slightly lower temperatures calculated

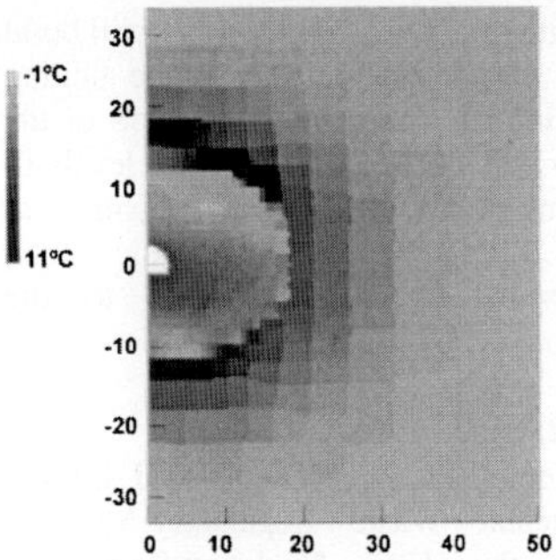

Figure 4. Temperature difference between 2D and 3D models after four years of heating.

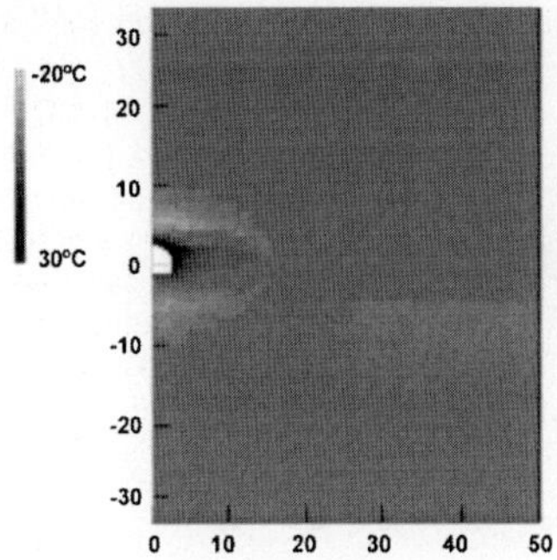

Figure 5. Temperature difference between open and closed boundary at the drift wall.

in the 2D model are interpreted to be a consequence of increased condensed water caused by the slightly higher temperatures outside the boiling isotherm.

The effect of open versus closed drift-wall boundary conditions on calculated temperature was evaluated. Temperatures calculated for a model with the drift wall boundary treated as a Neumann boundary with no fluid and heat flow were compared with results from a model with a Dirichlet boundary condition at the drift wall. Temperature and saturation were allowed to vary while pressure was kept constant at one atmosphere at this boundary. The open boundary condition simulations resulted in reduced saturation at the drift wall. This lower saturation lowered the saturation-dependent thermal conductivity near the drift wall, resulting in higher temperatures in this region. As a consequence, temperatures calculated by the model with an open boundary at the drift wall were as much as 30°C higher near the drift and 20°C lower at the outer edge of the boiling isotherm (Figure 5). The open boundary at the drift wall is considered to more closely approximate the effects of heat and mass loss through the thermal bulkhead.

Matrix temperature, matrix saturation, and fracture saturation are calculated after four years of heating using the 3D model with dual continua, an active fracture model, and an open boundary at the drift wall (Figures 6-8). This model conceptualization provided the best agreement with temperatures measured during the DST at boreholes 158, 160, and162. Temperatures calculated for the fracture continuum are the same as those calculated for the matrix continuum. Maximum temperatures of 122, 185, and 275°C were calculated near the wing heaters after three months, one year, and four years of heating.

6. CONCLUSION

The multiphase code MULTIFLO was used to simulate the thermal-hydrological coupled processes observed during the four-year heating phase of the DST at Yucca Mountain, NV. The effects of different conceptual models on calculated temperature and saturation were compared. Conceptual models for the DST were evaluated by comparing simulated temperatures with temperatures measured at three boreholes. The models with the best agreement between calculated and measured temperatures included dual continua (matrix and fracture) and active fracture. to reflect actual heat loads. The canister heat load was decreased by 20 percent to account for radiative and conductive heat loss through the thermal bulkhead.

The effect of dimensionality (i.e., 2D versus 3D) on temperature was evaluated. A maximum reduction in temperature of about 10°C near the wing heaters was calculated after four years of heating when a 3D model was used compared with a 2D model. Changing the drift wall from a closed boundary to an open boundary increased temperatures near the drift by a maximum of 30°C and decreased temperatures at the outer extent of the boiling isotherm by about20°C. Higher temperatures at the drift wall are attributed to lower thermal conductivities caused by drier rock.

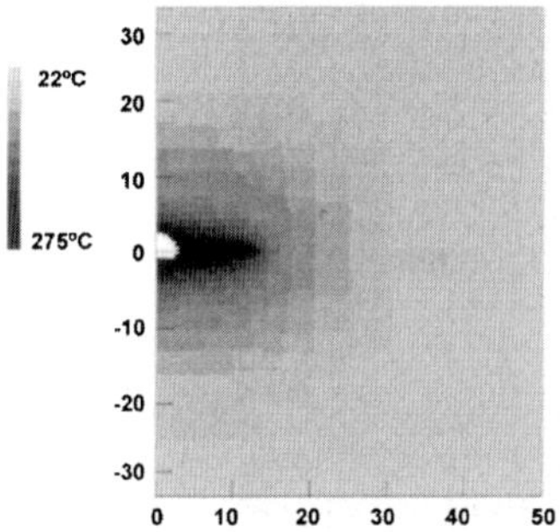

Figure 6. Calculated temperature after four years of heating.

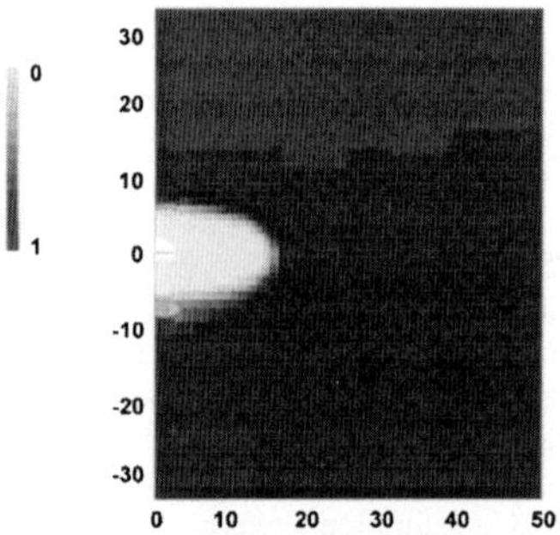

Figure 7. Calculated matrix saturation after four years of heating.

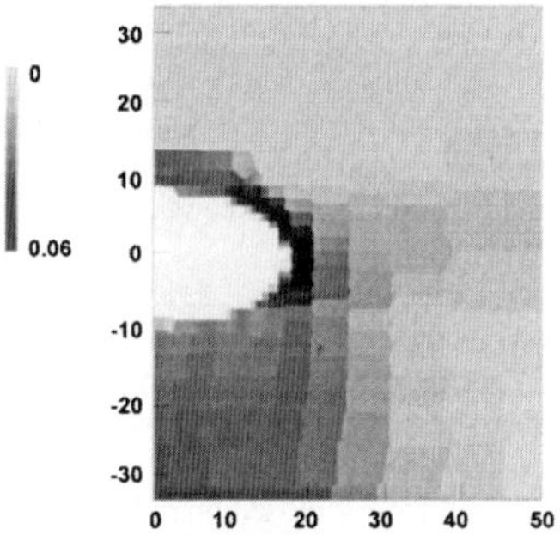

Figure 8. Calculated fracture saturation after four years of heating.

The homogeneity of property assignment is reflected in the uniform nature of predicted matrix temperature and saturation. Saturation predicted for the fracture continuum indicated that water will build up in the areas above and outside of the boiling isotherm. Shedding of water off of the side of the boiling isotherm was observed. Saturation levels in the fracture continuum were highly sensitive to physical (i.e., porosity) and hydraulic (i.e., permeability) property values assigned to the fractures.

7. ACKNOWLEDGMENTS

This paper documents work performed by the Center for Nuclear Waste Regulatory Analyses (CNWRA) for the U.S. Nuclear Regulatory Commission (NRC). Activities reported here were performed on behalf of the NRC Office of Nuclear Material Safety and Safeguards, Division of Waste Management. The report is an independent product of the CNWRA and does not necessarily reflect the views or regulatory position of the NRC

REFERENCES

CRWMS M&O, *Multiscale Thermohydrologic Model, ANL-EBS-MD-000049, Revision 00, ICN02,* North Las Vegas, NV, Civilian Radioactive Waste Management System, Management and Operating Contractor, 2001.

Liu, H.H., C. Doughty, and G.S. Bodvarsson. *An active fracture model for unsaturated flow and transport in fractured rock.* Water Resources Research 34(10): 2,633–2,646. 1998.

Painter, S.L, P.C. Lichtner, and M.S. Seth. *MULTIFLO User's Manual MULTIFLO Version 1.5–Two-Phase Nonisothermal Coupled Thermal-Hydrologic-Chemical Flow Simulator.* Revision 3. Change 0. San Antonio, TX: Center for Nuclear Waste Regulatory Analyses. 2001.

TRW Environmental Safety Systems, Inc. *Thermal Tests Thermal-Hydrological Analyses/Model Report.* Las Vegas, NV: TRW Environmental Safety Systems, Inc. 2000.

THM ANALYSIS OF A HEATING TEST IN A FRACTURED TUFF

S. Olivella, A. Gens and C. Gonzalez
[1]) Geotechnical Engineering Department, Technical University of Catalunya, c/o Jordi Girona 1-3, Edificio D-2, 08034 Barcelona, Spain.

Abstract: This paper contains THM analyses simulating the drift scale test (DST) at Yucca Mountain. By introducing appropriate hydrological parameters the analysis is performed in single porosity-permeability. The predicted temperatures and degrees of saturation are quite close to the measurements obtained in the test. The THM calculation is used to investigate gas permeability variations that were measured at different locations during the test. These gas permeability variations may be associated to hydraulic effects but also to mechanical effects. Finally, a sensitivity analysis is performed to study the influence of permeability changes induced by deformations on the shape and size of the dried zone.

1. INTRODUCTION

The in situ heating test DST performed in Yucca Mountain is a challenging modeling problem because it involves several coupled phenomena, concerning multiphase fluid flow in deformable rock.

The in situ DST (Drift Scale Test) consist in a 47.5 m long, 5 m diameter drift heated by 9 heaters simulating waste canisters placed on the floor. Additional heat is supplied by 50 wing heaters inserted into horizontal boreholes drilled into each side wall (Datta, 2002).

Thermal-hydrological simulations have been performed by Birkholzer and Tsang (2000) in three dimensions using the finite volume simulator TOUGH. In other reports containing modeling efforts of the DST test, thermo-mechanical calculations were performed in two dimensions using the calculated temperatures from the thermo-hydrological models. The coupling considered in those calculations was only in one direction, i.e. temperatures were influenced by the hydrological processes, but the thermo-hydrological problem was not influenced by deformations. However, one can expect that water and air flow, which is controlled by intrinsic permeability, will be influenced by deformations.

This paper contains a first attempt to perform a coupled thermo-hydrological-mechanical (THM) analysis based on the DST in situ test configuration. The coupled THM analysis has been performed in two dimensions.

2. CODE DESCRIPTION AND THM FORMULATION

The governing equations for the THM problem are balance equations and constitutive equations. Mass conservation equations apply to water and air. When the porous medium is deformable, the momentum balance equation (mechanical equilibrium) is also taken into account. In non-isothermal problems, the internal energy balance for the total porous medium must be considered. The basic equations solved by the finite element code CODE_BRIGHT are:

Mechanical equilibrium equations (1, 2 or 3 dimensions):

$$\nabla\cdot\boldsymbol{\sigma}+\mathbf{b}=\mathbf{0} \qquad (1)$$

Water mass balance:

$$\frac{\partial}{\partial t}\left(\omega_l^w\rho_l S_l\phi+\omega_g^w\rho_g S_g\phi\right)+\nabla\cdot(\mathbf{j}_l^w+\mathbf{j}_g^w)=f^w \qquad (2)$$

and a similar equation for air balamce. Energy balance:

$$\frac{\partial}{\partial t}\left(E_s\rho_s(1-\phi)+E_l\rho_l S_l\phi+E_g\rho_g S_g\phi\right)+ \qquad (3)$$
$$+\nabla\cdot(\mathbf{i}_c+\mathbf{j}_{Es}+\mathbf{j}_{El}+\mathbf{j}_{Eg})=f^Q$$

where ϕ is porosity, ρ_α is density, ω^i_α is mass fraction, S_α is degree of saturation, $\mathbf{j}$ are fluxes, and f^α is a source/sink term.

More details of the formulation and numerical method can be found in Olivella et al (1995).

3. SIMPLE VS DOUBLE STRUCTURE

The specific nature of the host rock (fractured tuff) makes it necessary to recognize that a double porosity and double permeability play a role in the hydrological problem. Properties that correspond to the tuff in the vicinity of the drift (tptpmn) are summarized in Table 1. Properties for tptpul and tptpll are in the same range and can be found elsewhere.

The intrinsic permeabilities together with relative permeability (calculated using the Van Genuchten parameters for retention curve) permit to obtain a function of permeability versus capillary pressure which is represented in figure 1. The curve is obtained by adding the contribution of the matrix and the fracture. This permeability function is approximated in a single structure analysis by the following combination:

- Retention curve: parameters from matrix
- Intrinsic permeability: value from fracture
- Relative permeability: $m = 0.04$

Table 1. Tuff properties near the drift (tptpmn).

Property	Matrix	Fractures
Porosity (-)	0.11	0.0026
Int. permeability (m^2)	$0.124x10^{-16}$	$0.1x10^{-12}$
P_o in VG retention (MPa)	0.444	0.01027
m in VG retention	0.247	0.492

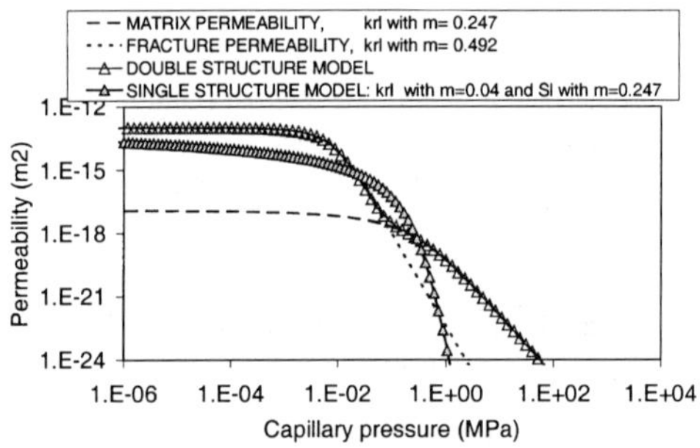

Figure 1. Liquid permeability as a function of capillary pressure.

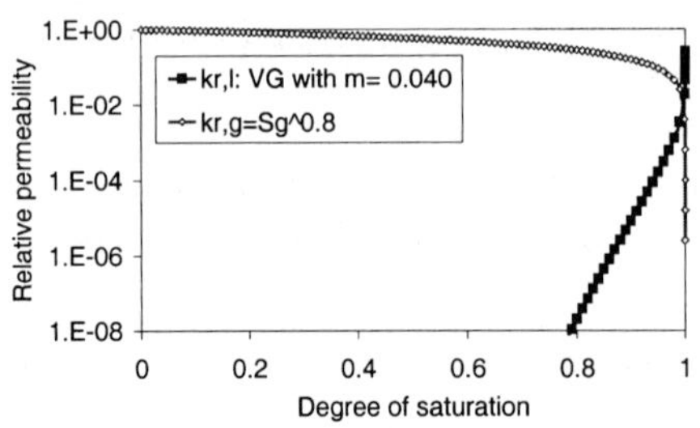

Figure 2. Liquid and gas relative permeability functions considered for the single structure

Therefore for a single structure model the retention curve is taken from the matrix (because most of water is stored in the rock matrix), intrinsic permeability is taken from the fracture (because most of water flows through the fracture) and relative permeability should be a different function from the ones corresponding to the matrix and to the fracture.

Figure 2 shows the relative permeability functions considered for liquid and for gas. These curves recognize two aspects:

- Gas flow is only reduced when the fractures saturate (near full saturation of the rock)
- Liquid flow reduces strongly as the fractures desaturate (fractures desaturate before the matrix).

A series of double structure themo-hydrological calculations have been performed and are reported elsewhere (Olivella et al, 2003). The results obtained with this single structure model are qualitatively similar to the ones obtained considering explicitly the double structure.

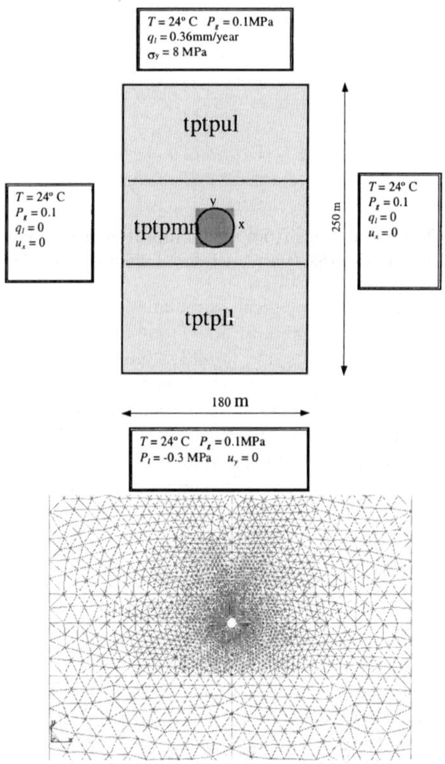

Figure 3. Geometry and boundary conditions for THM calculations (above). Detail of the finite element mesh (below).

The heat power introduced in the calculations in two dimensions corresponds to both the heaters in the drift and the wing heaters in the sidewall boreholes. Since the model is 2D (per meter in the axial direction), the following values have been considered:

– Heat power in canister-heaters:
36400 W/can/(9 canister x 4.7 m) = 860 W/m

– Heat power in inner wing-heaters:
37520 W/wings/(50 wings x 1.87 m) = 401 W/m

– Heat power in outer wing-heaters:
55300 W/wings/(50 wings x 1.87 m) = 591 W/m

which correspond to 70% of the nominal power. This reduction is based on a 3D-2D comparison for the heat flow problem (Olivella et al 2003).

Figure 3 shows the geometry, boundary conditions and a detail of the finite element mesh used in the calculations.

4. THERMAL AND HYDROLOGICAL RESULTS

The thermo-hydrological part of the THM analysis provides a temperature distribution that can be compared with measurements available for the 4 years calculation. Figure 4 shows the calculated temperature distribution after 4 years of heating and figure 5 shows the desaturated zone around the drift. Measurements of temperature are available at different locations from which representative points in boreholes 158 (vertical from drift wall) and 160 (horizontal from drift wall) have been chosen (figure 6).

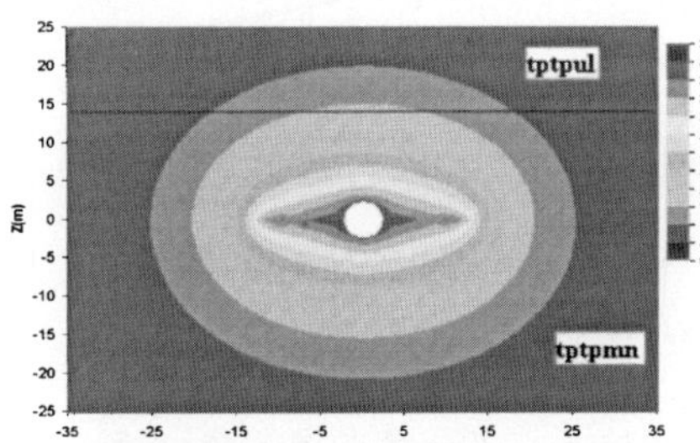

Figure 4. Calculated temperature at 1460 days.

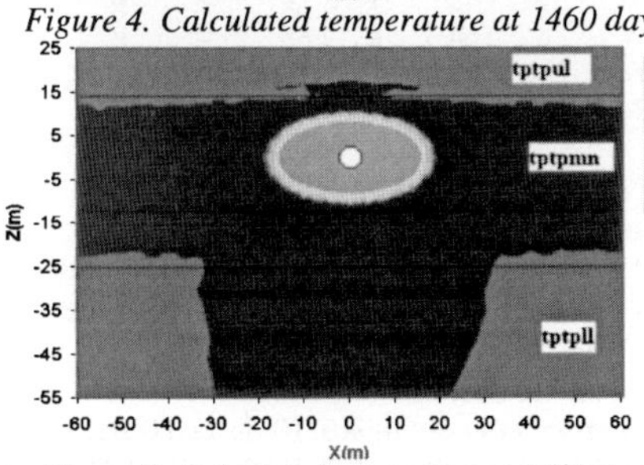

Figure 5. Calculated saturation at 1460 days.

The over prediction of temperatures after 2 years is caused by the constant power considered in the analysis. The actual power was reduced to limit the maximum temperature in the heaters.

Finally, the fluxes of liquid and gas are represented in Figure 7. From these fluxes it can be observed that the dried zone is not static. Liquid water continuously flows towards the dry zone due to capillary pressure gradients. This flux is compensated by water evaporation and gas flow in the opposite direction. The downward regional flux is diverted by the dried zone. This indicates that dissolved substances would tend to be confined in the dried zone because, although water is flowing, the flow in the outward direction is gas flow and does not transport solutes while the flow in the inward direction is liquid flow. So, solutes would tend to precipitate in the dry zone.

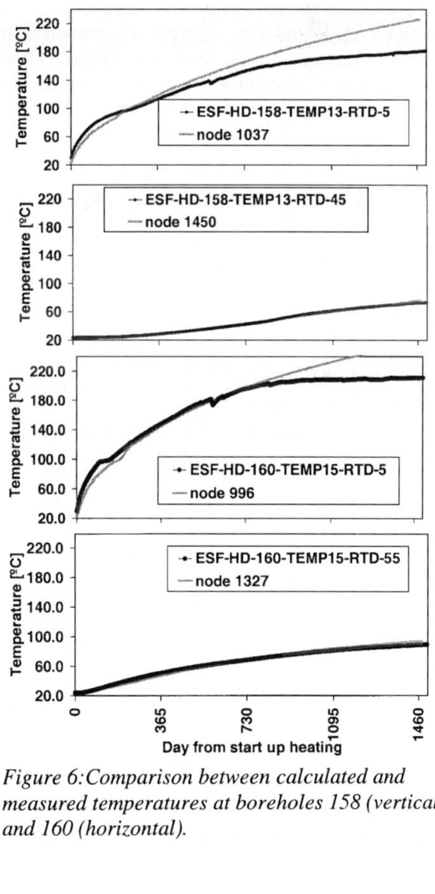

Figure 6: Comparison between calculated and measured temperatures at boreholes 158 (vertical) and 160 (horizontal).

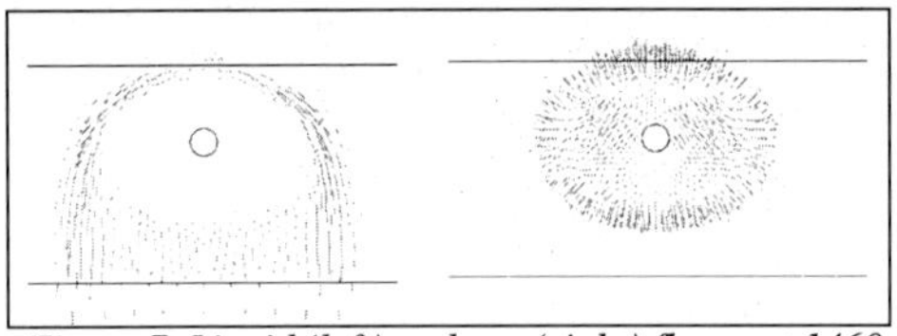

Figure 7. Liquid (left) and gas (right) fluxes at 1460 days (not with the same scale).

5. MECHANICAL EFFECTS ON THM

In order to solve the coupled THM problem, mechanical properties should be considered for the rock and a constitutive law is also necessary to account for permeability changes induced by deformations. A simple way that permits coupled calculations is using the following properties for the rock: Elastic modulus: 36800 MPa; Poisson ratio: 0.2; and thermal expansion: 0.00002. As heating of the rock takes place, volumetric expansion in the confined rock induces changes in porosity. In reality, fractures will tend to close. These porosity/aperture changes can be transformed into permeability variations. A exponential law has been used to model permeability variations:

$$k_{\text{intrinsic}} = k_o k_{\text{mech}}(\phi) = k_o \exp\left(-b\left(\phi_o - \phi\right)\right) \qquad (4)$$

For which the value b=1000 has been adopted (see figure 8).

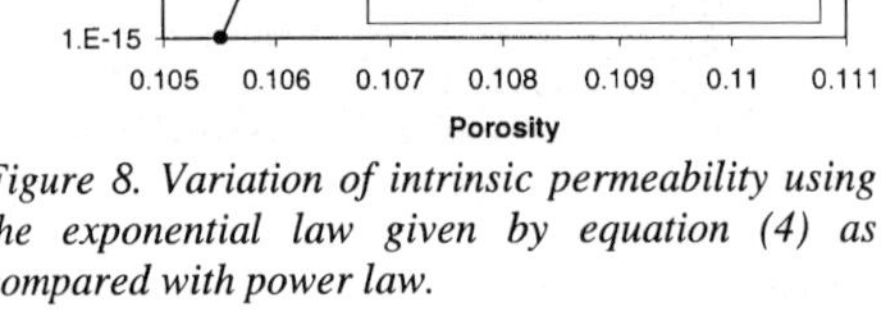

Figure 8. Variation of intrinsic permeability using the exponential law given by equation (4) as compared with power law.

Figure 8 shows the variation of permeability in the range of porosity variations that take place in the THM calculations as compared with what would be obtained using cubic law. It can be observed that for b=1000 the maximum reduction of permeability is about 2 orders of magnitude. Using this permeability function, the effects on the thermal and hydrological solutions are small. In other words, the mass and heat transport are affected but the change in temperature and degree of saturation distributions are small. Other values of b have been considered in Section 6 (see below).

An interesting result from THM calculation is gas permeability which was measured in the test. The combined effect of hydraulic and mechanical changes is that gas permeability may reduce or increase depending on time and location. Hydraulic conductivity to gas is calculated as:

$$K(\phi, S_g, T) = k_{\text{intrinsic}} k_{\text{rg}} \frac{\rho_g g}{\mu} = k_o k_{\text{mech}} k_{\text{rg}} \frac{\rho_g g}{\mu} \qquad (5)$$

where k_{mech}(deformation) and k_{rg} (gas saturation) are functions that modify intrinsic permeability, respectively, when mechanical and hydrological effects take place. The product of these two functions ($k_{\text{mech}}k_{\text{rg}}$ i.e. hydro-mechanical effects) has been represented in figures 9 and 10 for selected points where measurements were available. In order to see the relative importance of each contribution, the hydrological effect is also included in a separated curve (k_{rg}). The functions are represented in the figures normalized by the first measurement. Figure 11 shows the location of points.

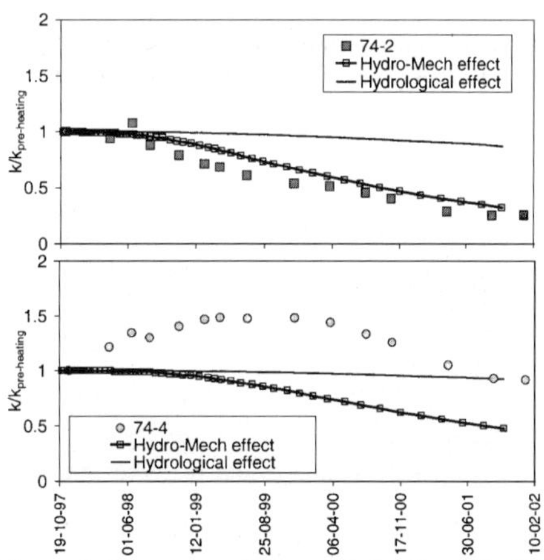

Figure 9. Gas permeability evolution for point 74.2 and 74.4. Symbols alone are measurements.

The effect of volumetric deformation is that permeability decreases slowly as for instance in point 74.2. Clearly, the hydrological effect was not enough to explain the permeability decrease for point 74.2.

The strong decrease of permeability (calculated) observed initially in figure 10 is associated with water condensation and, therefore, gas saturation decrease. However, the measurements in point 76.1

indicate that gas permeability increases instead of decreasing which would be associated to either fracture aperture or gas saturation increase due to drying. Points 76.1 and 76.4 correspond to two points symmetrically situated with respect to the drift and the measurements would imply heterogeneous behavior.

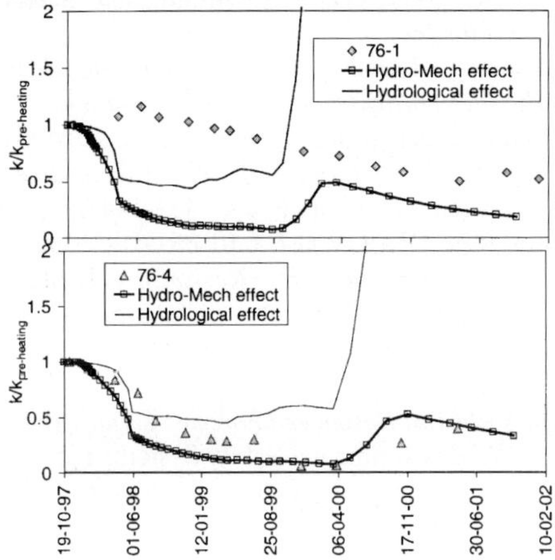

Figure 10. Gas permeability evolution for point 76.1 and 76.4. Symbols alone are measurement.

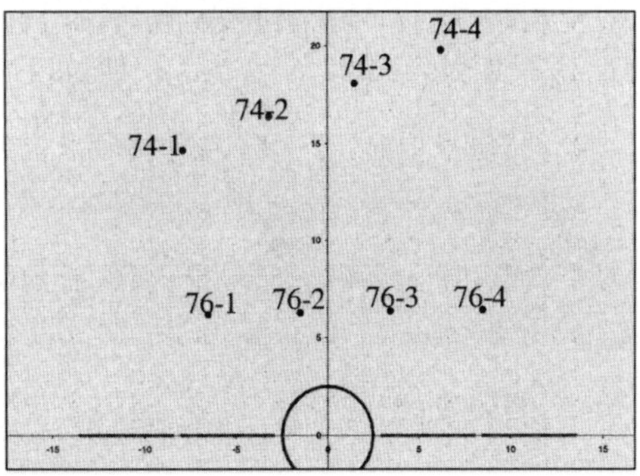

Figure 11. Location of points for gas permeability measurement.

6. SHAPE OF THE DRIED ZONE DEPENDING ON PERMEABILITY

In order to investigate the possible influence of permeability changes on water movement, the shape of the dried zone has been compared for different permeability variations. The following cases have been defined: Case CIP: Constant intrinsic permeability; Case DIP: Decreased permeability by four orders of magnitude near the drift induced by volumetric deformations (b=2000 in equation (4); Case IIP: Increased permeability by two orders of magnitude near the drift induced by shear deformations (b=-1000 in equation (4)). Using a negative b in equation (4) is based on the concept that mean stress (inducing volumetric deformation) and shear stress (inducing shear deformation) undergo similar variations during heating. Of course, this would require a constitutive model that was able to induce dilatancy when shear stresses change.

Figure 12 shows that the shape of the dried zone is different after 4 years depending on the case. If permeability is constant or decreases moderately (with b=500 and b=1000, the results were similar), the shape is elliptical.

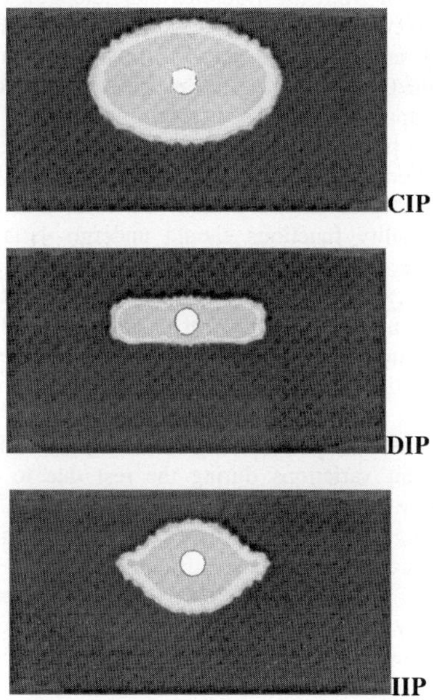

Figure 12. Different shape of the dried zone depending on the intrinsic permeability variations induced by deformations.

If intrinsic permeability decreases strongly (maximum of 4 orders of magnitude near the drift) due to volumetric deformation induced by compression, the dried zone becomes elongated horizontally but it is flattened in the vertical direction. The volume of the dried zone is smaller than in the case of constant permeability so the drying due to vapor flow is less efficient. This can be due to the lower gas phase permeability.

If permeability increases, the reduction of the dried zone takes place both horizontally and vertically. Again, the volume of the dried zone is smaller but now the reason can be that liquid water flow is more efficient so the vapor migration is compensated in a more efficient way by the liquid flow towards the drift.

7. CONCLUSIONS

A single structure calculation has been considered in order to investigate the THM behavior of the DST test at Yucca Mountain.

The thermo-hydrological calculations have indicated that it is possible to choose appropriate hydrological parameters in order to obtain a distribution of saturation similar to the one prevailing in the in situ test. Intrinsic permeability was taken from the fractures and retention curve was taken from the matrix. Relative permeability for gas and for liquid had to be modified. None of the functions valid for the matrix or the fracture were appropriate. The problem in fact, is that relative permeabilities are controlled by degree of saturation in the fracture and this model used a global degree of saturation. Therefore, relative permeability functions should undergo variations near full saturation because the fractures desaturate for low capillary pressures compared to the matrix.

The mechanical problem is considered elastic but the intrinsic permeability variations induced by deformations have been introduced. This permits a coupling from mechanical to hydraulic and in turn to thermal. Gas permeability measurements undergo important variations during the test due to both moisture redistribution and to deformations. The hydrological effect is more important than the mechanical. It is in fact, difficult to separate both effects. The model is only able to reproduce correctly gas permeability variations at some points. Heterogeneity may play an important role but it has not been introduced in this model.

Finally, a sensitivity analysis has been presented concerning the shape of the dried zone for different possibilities of intrinsic permeability evolutions induced by deformations. It is interesting that the dried zone shows low sensitivity to the strong intrinsic permeability variations introduced which confirms that hydrological effects are more important than the mechanical couplings, and secondly, that the dried zone tends to reduce in size regardless of the intrinsic permeability variation introduced. This later can be explained by the possible influence of intrinsic permeability on the liquid phase or on the gas phase.

ACKNOWLEDGEMENTS

The authors would like to acknowledge ENRESA (Empresa Nacional de Residuos, S.A.) for funding the participation in Decovalex III.

REFERENCES

Ambient Characterization of the Drift Scale Test Block, 1977, TRW Environmental Safety Systems Inc.

Birkholzer, J.T. and Y.W. Tsang, 2000, *Modelling the thermal-hydrologic processes in a large-scale underground heater test in partially saturated fractured tuff.* Water Resources Research, vol 36, no 6, pp: 1431-1447.

Datta, R.N. DECOVALEX III PROJECT, *Task 2A, Interim Report (Revised). Thermal-Hydrological Predictive Simulation of the Yucca Mountain Project Drift Scale Test*, February 2002.

Drift Scale Test Design and Forecast Results, 1977, TRW Environmental Safety Systems Inc.

Olivella, S., Gens, A., Carrera, J., Alonso, E. E. 1995. *Numerical formulation for a simulator (CODE_BRIGHT) for the coupled analysis of saline media.* In Engineering Computations, n. 13, pp. 87-112.

Olivella, S. Gonzalez, C. Gens, A., Luna, M *Progress report DECOVALEX III*, Task 2A, February 2003.

COMPARATIVE ANALYSES OF PREDICTED AND MEASURED DISPLACEMENTS DURING THE HEATING PHASE OF THE YUCCA MOUNTAIN DRIFT SCALE TEST

Alain MILLARD[1], Jonny RUTQVIST[2]
[1])CEA/DM2S/LM2S, Saclay, FRANCE
[2]) Lawrence Berkeley National Laboratory, Earth Sciences Division, Berkeley, USA

Abstract: As a part of the DECOVALEX III project—model predictions were carried out of thermo-mechanical (TM) rock-mass responses at the Yucca Mountain drift scale test (DST), Nevada. This paper presents model predictions of TM-induced rock displacements at the DST carried out by two independent research teams using two different approaches and two different numerical models. Displacements predicted by the two independent analyses compare reasonably well to the measured ones, both in trends and average magnitude. The analyses indicate that the rock mass behaviour is essentially elastic and that the in situ rock mass thermal expansion coefficient is well represented a temperature-dependent thermal-expansion derived from laboratory tests on intact rock.

1. INTRODUCTION

This paper presents the effort of two research teams, Lawrence Berkeley National Laboratory (LBNL) and Commissariat a l'Energie Atamique de Cadarache (CEA), to independently predict Thermo-mechanical (TM) responses at a large-scale heater test, performed at Yucca Mountain, Nevada (Datta et al., 2003). The work was carried out as part of the DECOVALEX III project, an international co-operative research project for the development of coupled models and their validation against experiments. The heater test, denoted the Yucca Mountain Drift Scale Test (DST) is conducted at a depth of about 250 m in partially saturated, fractured tuff. Along the 50 m length of the drift, heat is supplied by nine floor heaters and 50 horizontal wing heaters. At the DST site, mechanical displacements are measured along radial boreholes in three cross sections along the drift axis (y=13.7, 21.0, and 41.1 m), with the intermediate section close to the middle of the drift.

From the geotechnical investigations performed at Yucca Mountain, it appears that the rock mass where the DST is located is highly fractured, with three orthogonal fractures sets orientated close to the vertical axis (z) as well as along the (drift) y- and x-axes. Because of the scale of the DST and the surrounding highly fractured rock mass, an equivalent continuum approach is appropriate. Moreover, since the drift is long and the wing heaters are spaced about 1.80 m apart, one may reasonably assume that the temperature profile will be rather uniform along the drift, close to the mid-point. These considerations led the two research teams, LBNL and CEA, to restrict their analysis to a vertical cross section of the drift, normal to its axis, by using two-dimensional equivalent continuum approaches. The predicted displacements were compared to measured ones either at the three sections (LBNL) or in the mid-section (CEA).

2. LBNL MODEL

The LBNL team performed a fully coupled THM analysis, using an explicit sequential coupling of two established computer codes: TOUGH2 and FLAC3D (Rutqvist et al., 2002). In this approach, a fully coupled multiphase fluid and heat transport analysis is conduced with TOUGH2, while a simultaneous rock-mechanics analysis is performed with FLAC3D. The two codes were coupled through coupling modules: (1) a module that transfers temperature changes to the rock-mechanics analysis, and (2) a module that transfers stress-induced changes in hydraulic properties to the fluid and heat transport analysis (Rutqvist et al., 2002).

2.1 Calculation model

For symmetry considerations, only half of the cross section is considered, with a vertical symmetry plane along the drift axis. To be free from boundary effects, the model domain extends vertically from z = -150 m to z =100 m, with the origin at the centre of the drift, and horizontally, from x = 0 to x = 100 m, as indicated in Figure 1a.

Mechanical boundary conditions are depicted in Figure 1b, as well as the initial state of stresses in the rock mass, estimated using an average density of 2,200 kg/m^3 (calculated from the weight of overlying rock using density values given in CRWMS M&O (1999)).

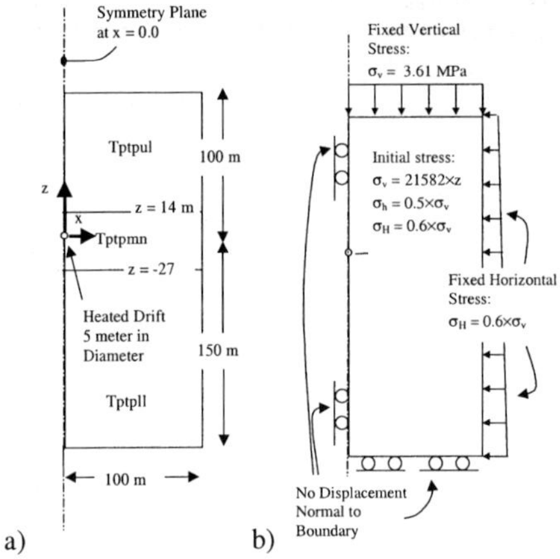

Figure 1. LBNL model domain (a) and boundary conditions (b)

2.2 Material properties

For the equivalent continuum, the LBNL research team used a linear elastic material model. A rock-mass Young's modulus 14.77 GPa and a rock-mass Poisson's ratio of 0.21 were adopted from CRWMS M&O (1999). These elastic parameters, which represent the bulk rock mass (including the effect of fractures) have been estimated using an empirical method based on the Geological Strength Index (GSI). The adopted rock-mass Young's modulus is about 50% lower than the Young's modulus of intact rock determined on core samples from the site.

To determine the thermal expansion coefficient, which plays a crucial role in the analysis, the LBNL research team fitted a temperature-dependent function to measured values for intact rock sample extracted from the DST area. As a result, a linear function is derived according to the following expression:

$$\alpha = 5.0 + 0.0583 \times T \ (10^{-6}/°C) \quad (1)$$

The comparison between adopted and measured values is shown in Figure 2.

3. LBNL RESULTS

In three cross sections of the heated drift, displacements were measured along four boreholes located as shown in Figure 3. The calculated vertical displacement field after 1 year of heating is also shown.

The measured values correspond to the radial displacement component, measured at 4 anchors, and relative to the collar placed at the drift wall. Distances to the collar are respectively 1, 2, 4, and 15 m for Anchors 1, 2, 3, and 4.

To display typical trends of the field data and their variability, the LBNL research team chose to compare the calculated values to measurements that have been grouped into equivalent categories, based on their location relative to the heated drift.

The comparison is plotted in Figure 4 for 60° inclined boreholes (BH 147, 148, 154, 155, 178 and 179). The noticeable spread of the results may be attributed to the local rock mass heterogeneities, such as fractures. With the exception of the very early time, simulated displacements are within the range of the measured ones. The best agreement is obtained for Anchors 2 and 3, whereas the

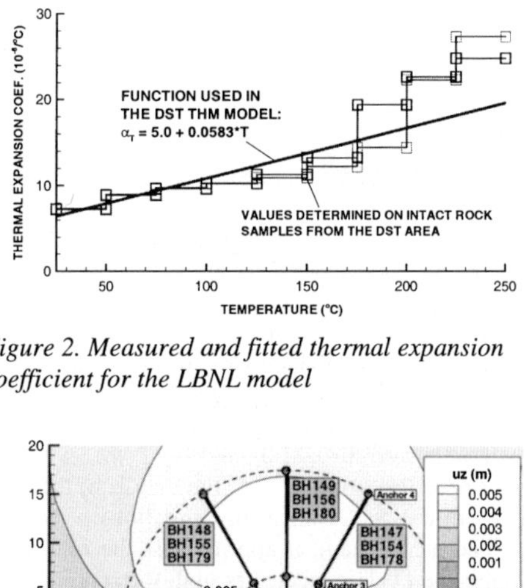

Figure 2. Measured and fitted thermal expansion coefficient for the LBNL model

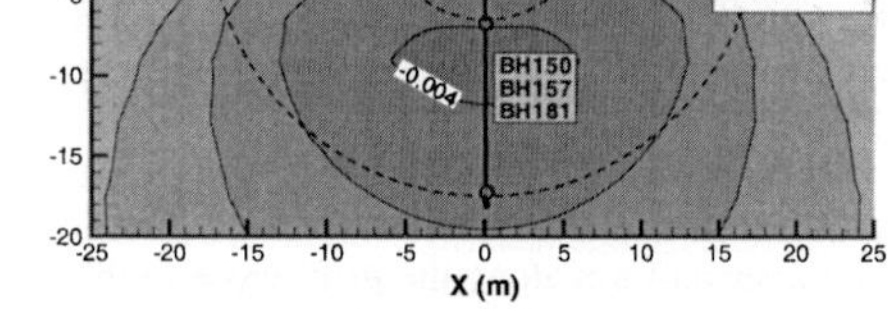

Figure 3. Orientation of mechanical boreholes and magnitude of calculated vertical displacement after 1 year of heating

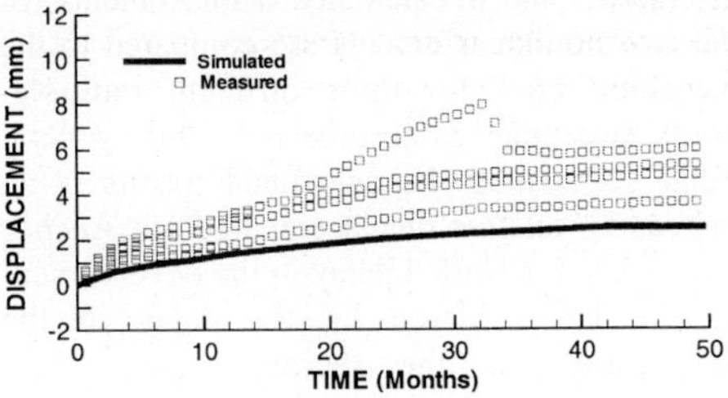

(a) Anchor 1

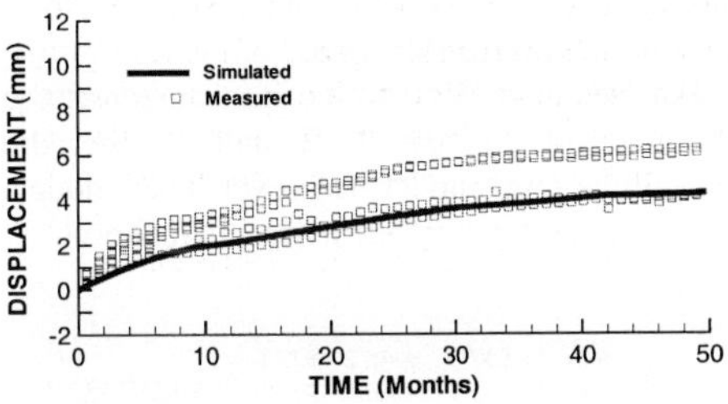

(b) Anchor 2

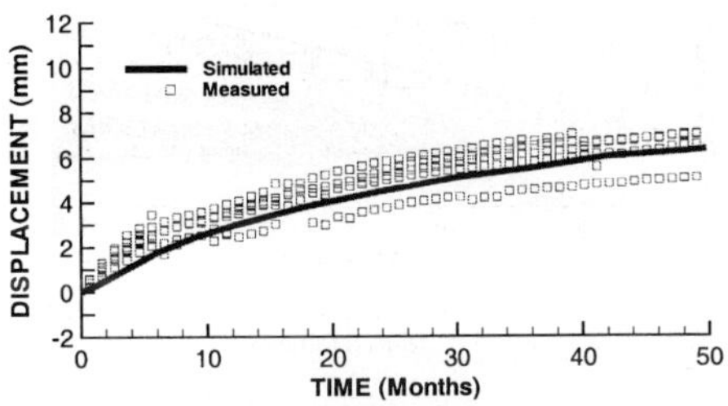

(c) Anchor 3

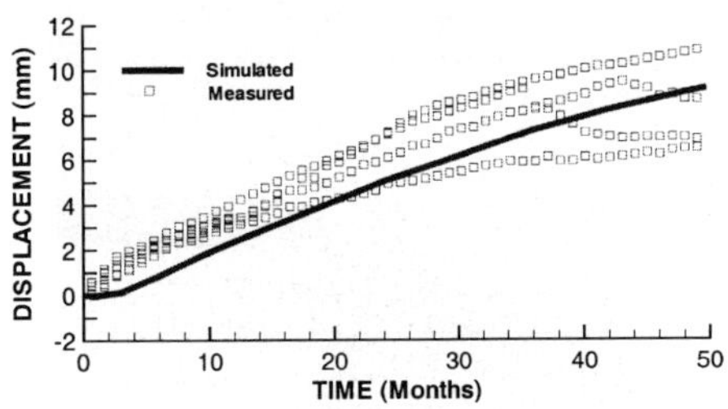

(d) Anchor 4

Figure 4. Measured and LBNL's simulated displacements for 60° inclined boreholes.

calculated displacement at Anchor 1 is a lower bound prediction. In Anchor 4, the early time displacements are slightly underpredicted, while the final average displacement at 50 months is well predicted. These results tend to indicate that the intact-rock thermal expansion coefficient determined from core samples was appropriate for the *in situ* conditions at the DST. This point is further investigated in the analysis conducted by the CEA research team as presented below in sections 4 and 5.

4. CEA MODEL

While the LBNL team calculated the temperature field using their fully coupled THM model, CEA used measured *in situ* temperatures for their analysis. The temperature field was imported at regular time intervals into their model grid for a TM analysis. The mid-section y = 21 m was selected for the two-dimensional TM analysis. Calculations were performed using CEA's computer code Castem2000. (Verpeaux et al (1989)).

4.1 Calculation model

As in LBNL's analysis, only half of the cross section was modelled. The domain extended from z = -150 m to z = 150 m and from x = 0 to x = 285 m. Initial stresses and boundary conditions were similar to LBNL (Figure 1b), except on the lateral vertical boundary, where zero normal displacements were prescribed.

4.2 Material models and properties

At Yucca Mountain, rock properties such as Young's modulus and thermal expansion coefficient have been estimated for both intact rock (based on laboratory tests on samples) and fractured rock mass (based on empirical methods or *in situ* tests). All calculations conduced by the CEA research team were performed using both sets of parameters (Millard 2003) and show that in fact a better agreement between calculated and predicted displacements is obtained when using intact-rock properties. This is in agreement with the finding by the LBNL team (Section 3), which concluded that the displacement was well predicted using the thermal expansion coefficient derived from intact-rock specimens.

Different properties were introduced for each stratigraphic unit, according to Table 1. The thermal expansion coefficients are variable with temperature, with the functions defined by the experimental points (see also Figure 2 for examples of intact rock thermal expansion data measured at DST).

For these calculations, three different material models were investigated: a linear elastic, an elasto-brittle, and elasto-plastic ubiquitous joint model. The ubiquitous joint, elasto-plastic model, is defined by a two-dimensional yield criterion, composed of two Mohr-Coulomb criteria, along two predefined directions characterised by their normal vectors n_1 and n_2.

$$F_i(\sigma) = F_i(\sigma_{ni},\tau_i) = \sigma_{ni} \cdot tg\varphi_i + | \tau_i | - C_i \leq 0 \quad (2)$$

where $i = 1$ or 2.

In this expression, σ_{ni} and τ_i represent, respectively, the normal stress and the shear stress acting on the plane orthogonal to n_i; φ_i is the friction angle; and C_i is the cohesion. The plastic flow is nonassociated and derives from a potential G_i, which has a similar expression, where ψ_i is the dilation angle:

$$G_i(\sigma_{ni},\tau_i) = \sigma_{ni} \cdot tg\psi_i + | \tau_i | \quad (3)$$

The elasto-brittle law is also defined according to predefined orthogonal failure directions. The criterion is defined by a limit on the normal stress:

$$F_i(\sigma_{ni}) = \sigma_{ni} - R_t \leq 0 \quad (4)$$

where R_t is the uniaxial traction resistance of the material. Once this limit is reached, the stress decreases linearly down to 0., corresponding to the rupture strain ε_r.

Table 1. Material properties in the CEA model

Geological Unit	Tptpmn	Tptpll	Tptpul
Young's Modulus (GPa)	32.93	27.54	20.36
Thermal Expansion Coeff. $*10^6$	7.6 to 53.17	6.41	6.29 to 34.24
Traction Strength (MPa)	3.95	3.30	2.44
Friction Angle	30°	30°	30°

5. CEA RESULTS

In Figure 5 and 6, calculated displacements for the two nonlinear models are compared to the measured ones for three different multiple-point borehole extensometers. One general trend concerning the calculated results is an increasing relative displacement, from Anchor 1 to Anchor 4, with a delay in the development of the latter caused by the delay in the propagation of the thermal front. Non-linearities are more pronounced when using the ubiquitous joint model, whereas the brittle model leads to weakly nonlinear behaviour, close to the pure elastic one.

For borehole BH155, the best agreement is obtained for Anchors 1, 2, and 3 using the ubiquitous joint model. However, both models overestimate the displacements in Anchor 4.

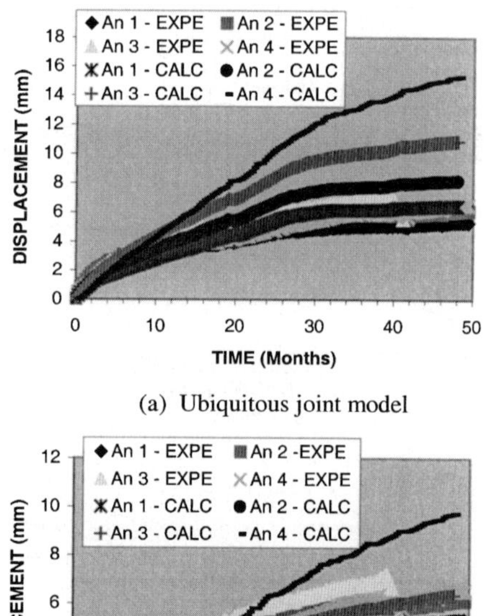

(a) Ubiquitous joint model

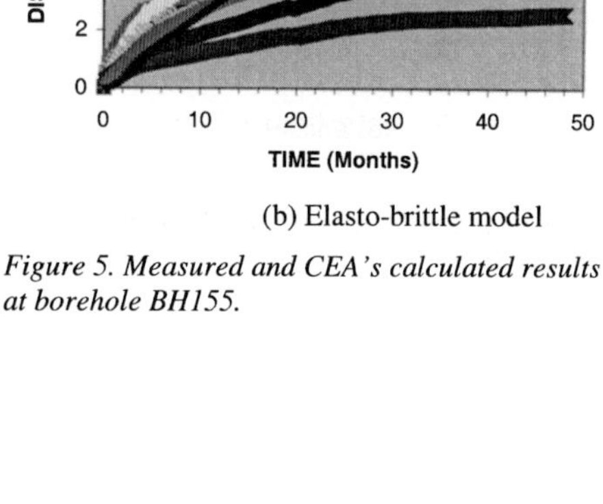

(b) Elasto-brittle model

Figure 5. Measured and CEA's calculated results at borehole BH155.

For extensometers BH156, a difference can be observed in the response of Anchor 4 at very early times: the ubiquitous joint model predicts, as is observed in the experiment, a very limited negative displacement, smaller than the elastic model, which indicates that some plastic deformation occurs. This is confirmed by looking at calculated contours of plastic zones: plastic shear develops at the drift boundary as early as 3 months. For this extensometer, the right order of magnitude is predicted for Anchor 3 and 4, with an excellent agreement for Anchor 4 using the ubiquitous joint model.

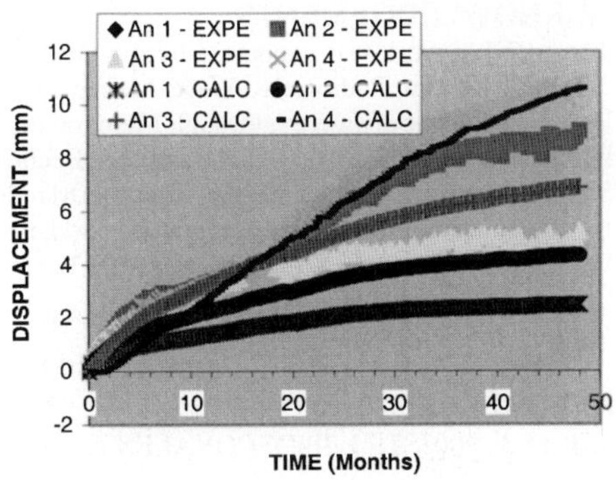

(a) Ubiquitous joint model

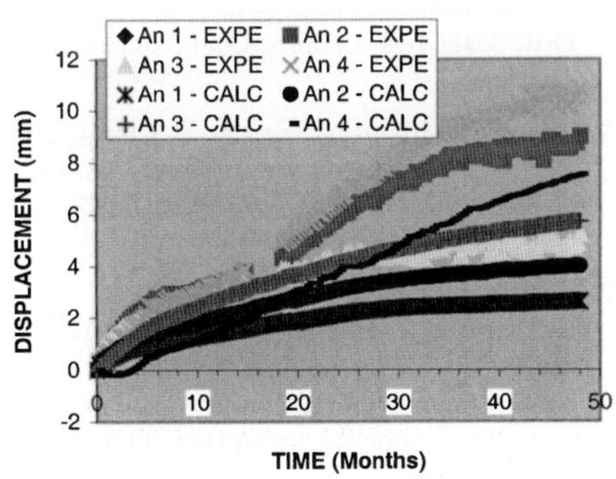

(b) Elasto-brittle model

Figure 6. Measured and CEA's calculated results at borehole BH156

6. COMPARISON TO OVERALL MEASURED DISPLACEMENTS

Figure 7 provides a comparison of model predictions by both LBNL and CEA to measured displacements along two boreholes after 48 months of heating (close to the end of the heating period). The figure shows that the predictions by LBNL's elastic model and CEA's elasto-brittle model are almost identical. Note that the CEA elasto-brittle model did not induce significant inelastic behaviour resulting in an essentially elastic response. Nevertheless, the close agreement between these two independent model predictions gives confidence that the TM process model, input data, and boundary conditions are correctly implemented both numerical models.

Figure 7 shows that LBNL's elastic and CEA's elasto-brittle models are within the range of maximum and minimum measured values, except in areas close to the drift wall. Near the drift wall, there appears to be an additional shift in the displacement field, possibly caused by inelastic behaviour near the open wall surface. The CEA's ubiquitous joint model overpredicts the displacement in the 60° inclined boreholes (Figure 7a) possibly by an overprediction of near-wall inelastic deformations.

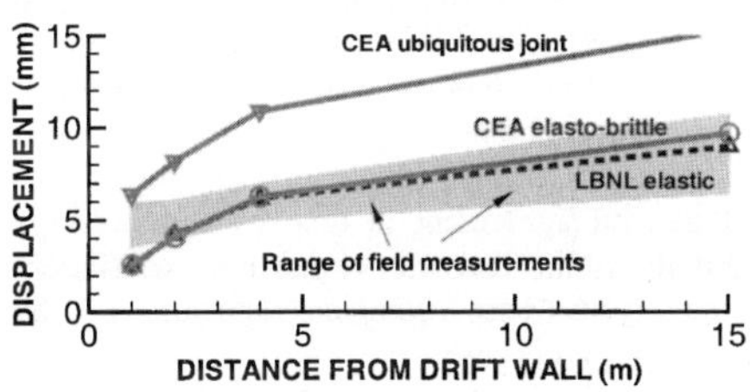

(a) 60° inclined boreholes

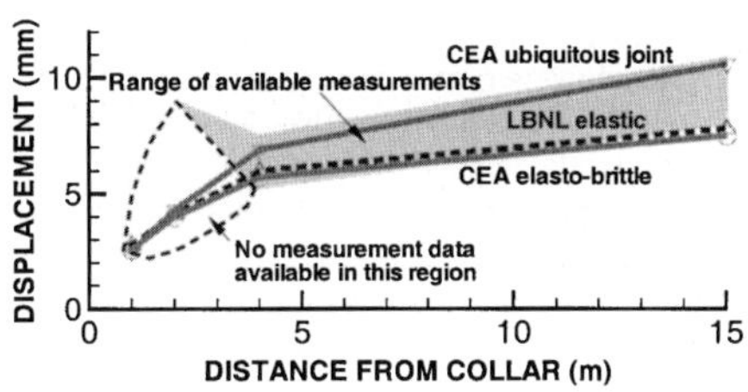

(b) Upper vertical boreholes

Figure 7. Comparison of model predictions and measured displacement after 48 month of heating.

7. DISCUSSION

The model predictions presented in this paper were obtained by two independent analyses, using two different approaches and two different numerical models. The approached used by LBNL and CEA differ in several aspects:

(1) LBNL conducted a fully coupled THM analysis to calculate temperature and TM responses, whereas CEA used measured temperature data, which was imported into their TM analysis at regular intervals.
(2) LBNL used an elastic material model, whereas CEA used two different elasto-plastic models—an ubiquitous joint and elasto-brittle material model
(3) LBNL applied a constant stress lateral boundary at x = 100 m, whereas CEA used a no-displacement boundary at x = 250 m.

Despite these differences, the predicted displacements along the boreholes are almost identical for LBNL and CEA for the case of elastic rock mass behaviour (Figure 7). The agreement between the two independent analyses by LBNL and CEA provide confidence that both approaches were appropriately implemented.

The most important parameter for the prediction of TM-induced displacement is the thermal expansion coefficient. Both LBNL and CEA used a temperature-dependent rock-mass thermal expansion coefficient derived from measurements on intact rock samples (Figure 2). Judging from a good general agreement in displacement average magnitude, the thermal expansion coefficient derived from core samples appears to be appropriate in this case.

In general, the predicted displacement using both LBNL's elastic and CEA's elasto-brittle (weakly inelastic) models are within the ranges of field measurements, except for very close to the drift wall. However, in a few individual anchors, displacement values are more than 50% larger than predicted by the elastic material behaviour. The increased displacement in these anchors may be explained by inelastic responses leading to a better agreement with the ubiquitous joint model (e.g. Anchor 4 in Figure 5a).

8. CONCLUSIONS

The following specific conclusions can be drawn from the analysis of thermally induced displacement at the Yucca Mountain Drift Scale Test:

- Displacements predicted by the two independent analyses compare reasonably well to the measured ones, both in trends and average magnitude.
- In situ rock mass thermal expansion coefficient is appropriately represented by a temperature-dependent thermal-expansion coefficient derived from intact rock samples.
- The rock mass behaviour is essentially elastic although in a few instances inelastic behaviour can be observed near the drift wall.

ACKNOWLEDGEMENTS

The LBNL's work was supported by the Director, Office of Civilian Radioactive Waste Management, U.S. Department of Energy, through Memorandum Purchase Order EA9013MC5X between Bechtel SAIC Company, LLC and the Ernest Orlando Lawrence Berkeley National Laboratory (Berkeley Lab) through the U.S. Department of Energy Contract No. DE-AC03-76SF00098.

REFERENCES

Millard A. 2003. Task 2B DECOVALEX report

Datta R., Barr, D, Boyle, W., Jing, L. 2003. Measuring the thermal, hydrologic, mechanical and chemical responses of the Yucca Mountain Drift Scale Test. Geoproc Conf., Stockholm, Sweden

Verpeaux P., Millard A., Charras T., Combescure A. 1989. *A modern approach of large computer codes for structural analysis*. Proc. SmiRT conf., Los Angeles, USA

Rutqvist, J, Y-S. Wu, C-F Tsang and G. Bodvarsson, 2002. A modeling approach for analysis of coupled multiphase fluid flow, heat transfer, and deformation in fractured porous rock, *Int. J. Rock Mech. Min. Sci.* 39, 429-442.

CRWMS M&O (Civilian Radioactive Waste Management System Management & Operating Contractor). 1999. *TBV-332/TBD-325 Resolution Analysis: Geotechnical Rock Properties*. B00000000-01717-5705-00134 REV 00. Las Vegas, Nevada.

BUILDING CONFIDENCE IN THE MATHEMATICAL MODELS BY CALIBRATION WITH A T-H-M FIELD EXPERIMENT

M. Chijimatsu[1], L. Jing[2], A. Millard[3], T.S. Nguyen[4], A. Rejeb[5], J. Rutqvist[2,6], M. Souley[7] and Y.Sugita[8]
[1]) Hazama Corporation, Japan
[2]) Royal Institute of Technology (KTH), Sweden
[3]) Commissariat a l'Energie Atomique (CEA), France
[4]) Canadian Nuclear Safety Commission (CNSC), Canada
[5]) Institut de Radioprotection et de Sûreté Nucléaire (IRSN), France
[6]) Lawrence Berkeley National Laboratory (LBNL), USA
[7]) INERIS-LAEGO, France
[8]) Japan Nuclear Cycle Development Institute (JNC), Japan

Abstract: Geological disposal of nuclear fuel wastes relies on the concept of multiple barrier systems. In order to predict the performance of these barriers, mathematical models have been developed, verified and validated against analytical solutions, laboratory tests and field experiments within the international DECOVALEX project. These models in general consider the full coupling of thermal (T), hydrological (H) and mechanical (M) processes that would prevail in the geological media around the repository. This paper shows the process of building confidence in the mathematical models by calibration with a reference T-H-M experiment with realistic rock mass conditions and bentonite properties and measured outputs of thermal, hydraulic and mechanical variables.

1. INTRODUCTION

Geological disposal of nuclear fuel wastes relies on the concept of a multiple barrier system. It is generally recognized that groundwater is the main agent of transport of contaminants from the wastes to the biosphere. In order to minimize the movement of contaminants in groundwater, a series of barriers is provided: the rock mass, which should have low permeability and high strength; the buffer and backfill that should have low permeability and high sorption capacities; the container which should have high mechanical strength and corrosion resistance. However, the excavation of the repository and the heat generated by the emplaced wastes would affect the performance of the above barriers. In order to predict the performance of these barriers, mathematical models have been developed, verified and validated against analytical solutions, laboratory tests and field experiments within the international DECOVALEX project during the last ten years (Jing et al, 1996, 1999). These models in general considered the full coupling of thermal (T), hydrological (H) and mechanical (M) processes that would prevail in the geological media around the repository. These models are complex, require considerable effort to develop, and need a large number of input data. This paper shows the process of confidence building in the mathematical models by calibration with a reference T-H-M experiment with realistic rock mass conditions and bentonite properties and measured outputs of thermal, hydraulic and mechanical variables. The reference experiment chosen for this work is the coupled in-situ THM experiment at Kamaishi Mine, Japan. Five research teams (Table 1) simulated the Kamaishi in-situ experiment. A number of improvements to previous work performed during DECOVALEX II (Rutqvist et al., 2001 a and b) were suggested and tested in this study. The suggested improvements were tested using a simplified axisymmetric model of the in-situ experiment. The present work constitutes Phase A of a benchmark test (BMT1A) of the international DECOVALEX III project (Jing et al., 2003).

Table 1 Research team and simulation code

Research team	Simulation code
CNSC (Canada)	FRACON
JNC (Japan)	THAMES
KTH/SKI (Sweden)	ROCMAS
CEA/IRSN (France)	Castem 2000
INERIS/ANDRA (France)	FLAC

2. OUTLINE OF THE KAMAISHI IN SITU T-H-M EXPERIMENT

A Schematic view of the Kamaishi in-situ THM experiment is shown in Figure 1 (Chijimatsu et al, 2001). A granulated bentonite was compacted directly into the test pit, by layers of 10 cm in thickness. The initial water content (by weight) of the bentonite was 15%. A heater was installed in the test pit, within the bentonite. When bentonite was compacted within the last 50 cm of the pit, a concrete lid was installed in the remaining part; this lid was restrained by vertical steel bars connected to the ceiling of the drift, in order to restrict vertical movements. A 40 cm deep water pool was then formed on the floor of the test drift. The temperature at the center of the heater was maintained at 100°C during a heating phase of 258 days. The heater was then turned off, and the natural cooling phase took place for approximately 180 days. Measurements from the installed sensors were recorded during the complete heating and cooling phases.

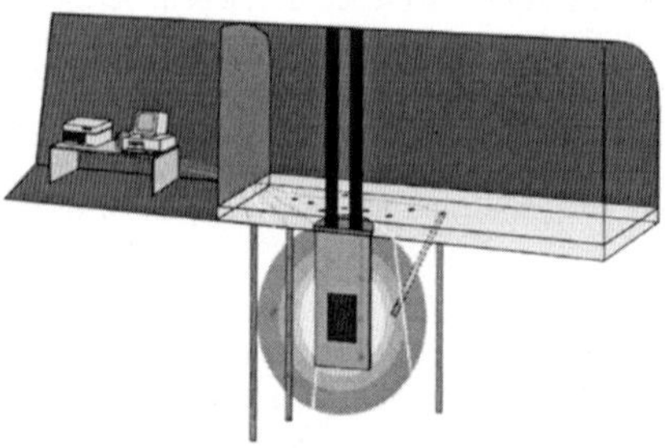

Figure 1. Schematic view of Kamaishi T-H-M Experiment.

3. THE CALIBRATION OF THE T-H-M IN-SITU EXPERIMENT USING A SIMPLIFIED AXISYMMETRIC MODEL

A simplified calibration test case of the in-situ experiment was proposed and defined below. The case focuses on the THM behaviour of a radial line (with a radial distance r as the coordinate) from the centre of the heater, with the axisymmetric geometry shown in Figure 2.

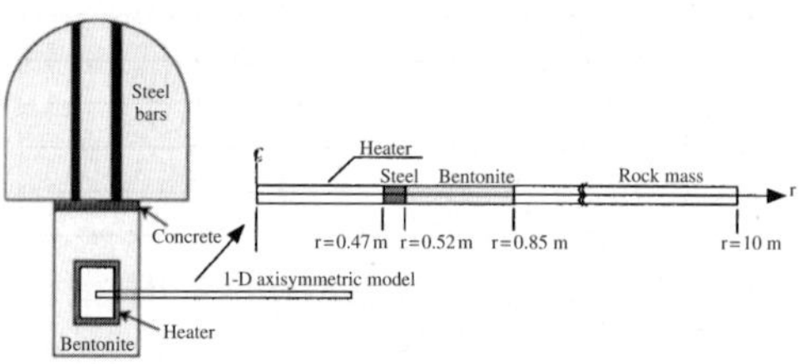

Figure 2. Geometry of the simplified axisymmetric model.

The desired output parameters are: T-temperature (°C); u-radial displacement (m): p-pore pressure (Pa); ω-water content by weight (%); σ-radial total stress (Pa); and ε_ρ; ε_τ-radial and tangential strain, respectively. Time histories of these output parameters are calculated for both the heating and cooling phases.

The boundary conditions are defined as follows:

At r = 0.47 m:

·T=100°C (heating) and free temperature (cooling), Free displacement, Impermeable

At r = 3 m

·T = 12 °C, u=0, p=3.9 kPa (equivalent to 0.4 m of water)

The initial conditions are as follows:

·T = 12 °C and zero displacements and stresses everywhere; p =3.9 kPa in the rock; ω =15% in the bentonite

4. PHYSICAL PROCESSES AND CONSTITUTIVE MODELS

4.1 The CNSC model

In the FRACON code developed and used by the CNSC, the bentonite was assumed to be a poroelastic continuum of a generalized Biot's type. The physical processes considered are the heat conduction, pore water flow in saturated /unsaturated porous media, vapor flow driven by temperature gradient, and mechanical deformation of the skeleton. These processes are described by the following governing equations.

The energy conservation equation is given as

$$\frac{\partial}{\partial x_i}\left(\kappa_{ij}\frac{\partial T}{\partial x_j}\right) + q = \rho C\frac{\partial T}{\partial t} \tag{1}$$

where T is temperature, κ_{ij} is the heat conductivity tensor, q the heat source (or sink), ρ the bulk density and C is the specific heat.

The continuity equation of groundwater is given as

$$\frac{\partial}{\partial x_i}\left[\frac{\rho_w k_{ij} K_r}{\mu}\left(\frac{\partial P}{\partial x_j} + \rho_w g_i\right)\right] - \frac{\partial}{\partial x_i}\left(D_{Tv}\frac{\partial T}{\partial x_i}\right) + \rho_w n\frac{dS_r}{dP}\frac{\partial P}{\partial t} - S_r B_w\frac{\partial P}{\partial t} - \rho_w S_r\frac{\partial}{\partial t}\left(\frac{\partial u_r}{\partial x_r}\right) + \rho_w S_r[n\beta_w + (1-n)\beta_s]\frac{\partial T}{\partial t} = 0 \tag{2}$$

where k_{ij} is the saturated permeability tensor; K_r is the non-dimensional relative permeability of the unsaturated medium; S the function of the degree of saturation; μ the water viscosity; ρ_w the water density; D_{Tv} the thermal diffusivity of the vapor; n the porosity; and B_w the coefficient of water compressibility. The symbol β_w and β_s are the thermal expansion of water and solid, respectively. P is the water pressure.

The equilibrium equation is given as

$$G\frac{\partial^2 u_i}{\partial x_j \partial x_j}+(G+\lambda)\frac{\partial^2 u_j}{\partial x_i \partial x_j}+\chi(S)\frac{\partial P}{\partial x_i}-\beta K_d\frac{\partial T}{\partial x_i}+F_i=0 \qquad (3)$$

where, u_i is the displacement vector, G and λ are Lame's elasticity constants, β is the volumetric thermal expansion coefficient of the solid matrix and F_i is the body force.

4.2 The JNC model

The behavior of the buffer material is influenced by the interdependence of thermal, hydraulic and mechanical phenomena. To simulate the water/vapor movement and heat induced water movement, the continuity equation used in the extended THAMES code is as follows;

$$\left\{\xi\rho_w D_\theta\frac{\partial\theta}{\partial\psi}(h,_i-z,_i)+(1-\xi)\frac{\rho_w{}^2 gK}{\mu_l}h,_i\right\}+\{\rho_w D_{Tv}T,_i\},_i \qquad (4)$$
$$-\rho_{wo}nS_r\rho_w g\beta_P\frac{\partial h}{\partial t}-\rho_w\frac{\partial\theta}{\partial\psi}\frac{\partial h}{\partial t}-\rho_w S_r\frac{\partial u_{i,i}}{\partial t}+\rho_{wo}nS_r\beta_T\frac{\partial T}{\partial t}=0$$

where D_θ is the isothermal water diffusivity, θ the volumetric water content, ψ the water potential and K the intrinsic permeability. The symbol ξ is the relative saturation parameter so that ξ=0 when the medium is fully saturated, ξ=1 otherwise. μ_l is the viscosity of water, n the porosity, S_r the degree of saturation, β_P the compressibility of water, β_T the thermal expansion coefficient of water and z the elevation head. The variable u_i is the displacement vector, T is temperature, h is the total head and t is time. The subscript 0 means the reference state.

The energy conservation equation takes into account, heat transfer by conduction and convection, as well as the energy change by evaporation. The equation is given as

$$(\rho C_v)_m\frac{\partial T}{\partial t}+nS_r\rho_w C_{vl}V_{li}T,_i-K_{Tm}T,_{ii}+L\left\{D_{\theta v}\frac{\partial\theta}{\partial\psi}(h,_i-z,_i)\right\},_i$$
$$+nS_r T\frac{\beta_T}{\beta_P}\left\{\xi D_\theta\frac{\partial\theta}{\partial\psi}(h,_i-z,_i)+(1-\xi)\frac{\rho_w gK}{\mu_l}h,_i+D_T T,_i\right\},_i \qquad (5)$$
$$+\frac{1}{2}(1-n)\beta T\frac{\partial}{\partial t}(u_{i,j}+u_{j,i})\delta_{ij}=0$$

where $(\rho C_v)_m$ is the specific heat of the material consisting of water and the solid particles, C_{vl} the specific heat of water, V_{li} the velocity vector of water, K_{Tm} the thermal conductivity of the medium consisting of water and the solid particles, L the latent heat of vaporization per unit volume, $D_{\theta v}$ the vapor diffusivity.

The equilibrium equation takes the swelling behavior of the partially saturated bentonite into account.

$$\left[\frac{1}{2}C_{ijkl}(u_{k,l}+u_{l,k})-F\pi\delta_{ij}-\beta\delta_{ij}(T-T_o)+\chi\delta_{ij}\rho_w g(h-z)\right]+\rho b_i=0 \qquad (6)$$

where C_{ijkl} is the elastic matrix, ρ the density of the medium and b_i the body force. χ is the parameter for the effective stress, χ =0 for the unsaturated zone and χ=1 for the saturated zone. The symbol F is the coefficient relating to the swelling pressure process and β=(3G+2λ)α_s, where α_s is the thermal expansion coefficient.

4.3 The KTH/SKI model

The governing equations of ROCKMAS are as follows. The mass conservation of water and vapor, energy conservation and momentum conservation equations are given as

$$\alpha[S_r\rho_w+(1-S_r)\rho_v]$$
$$+\left[nS_r\rho_w^0\beta_p+n(\rho_w-\rho_v)C_s+n(1-S_r)\frac{\rho_v}{\rho_w RT}\right]\frac{\partial P}{\partial T} \qquad (7)$$
$$+\left[nS_r\rho_w^0\beta_T+n(1-S_r)\left(h\frac{\partial\rho_{vs}}{\partial T}+\frac{\rho_v P}{RT^2}\right)\right]\frac{\partial T}{\partial t}$$
$$=\nabla\cdot\left[\left(\frac{\rho_w k_w}{\mu_w}+D_{Pv}\right)\nabla P+\frac{\rho_w k_w}{\mu_w}\gamma_w\nabla z\right]-\nabla\cdot(D_{Tv}\nabla T)$$

$$\left[(1-n)\rho_s C_{vs}+\rho_w C_{ws}nS_r+\rho_v C_{pv}n(1-S_r)+H_l\left(h\frac{\partial\rho_{vs}}{\partial T}-\frac{\rho_v P}{RT^2}\right)\right]\frac{\partial T}{\partial t}$$
$$+\beta T_0\frac{\partial e}{\partial t}+\nabla\cdot(C_{vw}q_w^m+C_{pv}q_v^m)\nabla T \qquad (8)$$
$$=\nabla\cdot[(K_M+LD_{Tv})\cdot\nabla T+LD_{pv}\nabla P]$$
$$-H_2\frac{\partial(nS_r)}{\partial t}-\left(H_1\frac{\rho_v}{\rho_w RT}+H_2\frac{\partial(nS_r)}{\partial P}\right)\frac{\partial P}{\partial t}$$

$$\frac{\partial}{\partial x_j}\left[C_{ijkl}e_{kl}-\delta_{ij}\beta(T-T_0)-\delta_{ij}\zeta(S_r-S_{r0})-\delta_{ij}\alpha S_r\overline{P}\right]+\overline{\rho}_s f=0 \qquad (9)$$

where, α is the Biot's coupling constant, γ_w the fluid density by weight, ξ the coefficient of swelling, the average $\overline{\rho}$ solid mass density, ρ_v the vapor mass density, ρ_{vs} the saturated vapor mass density, C_s the moisture capacity, e the volumetric elastic strain, R the specific gas constant, f the body force per unit mass, k_w the water permeability, K_M the thermal conductivity of solid-fluid mixture, D_v the effective molecular diffusivity of water vapor in air, L the latent heat of vaporization, and the C_{vs}, C_{vw}, C_{Pv} the volumetric specific heat capacity of solid, water and vapor, respectively.

4.4 The IRSN/CEA model

The model that has been implemented in the

computer code Castem 2000 (Verpeaux et al., 1989) for the simulation of coupled THM responses of unsaturated soils is based on the following governing equations. The equations governing the evolution of the chosen primary variables, which are the displacements ξ, the liquid water pressure p_l and the temperature T, are classically obtained from conservation equations together with state laws.

The water mass conservation equation is

$$\begin{aligned} \frac{\partial}{\partial t}(\rho_l\, S_l\, \varphi) &= - div\, \underline{q_l} - div\, \underline{q_v} \\ &= + div\left[\frac{K\, k_{rl}\, \rho_l}{\eta_l}\left(\underline{grad}\; p_l - \rho_l\, \underline{F}\right)\right] \\ &\quad + div\,(D_{T_v}\, \underline{grad}\, T) \end{aligned} \tag{10}$$

where K is the intrinsic permeability, k_{rl} the permeability relative to liquid water and η_l the dynamic viscosity of water. $\underline{F}$ stands for body forces such as gravity.

The conservation of momentum is written classically as

$$div\, \underline{\underline{\sigma}} + \rho\, \underline{F} = 0 \tag{11}$$

where $\underline{\underline{\sigma}}$ stands for the total stress in the porous medium, and ρ its equivalent density.

Finally, the conservation of energy is simply taken in a simplified form:

$$\rho C \frac{\partial T}{\partial t} - div\left(K_T \cdot \underline{grad}\, T\right) = q \tag{12}$$

where ρC stands for the equivalent porous medium heat capacity, K_T is the heat conductivity and q represents a heat source.

The temperature equation is solved separately, whereas the water mass conservation and momentum conservation equations are solved together. Because of the strong non-linearity that are present in these equations, a fully implicit time integration scheme is used.

4.5 The INERIS/ANDRA model

The two-phase flow option in FLAC allows numerical modelling of the flow of two immiscible fluids through porous media. A description of the concepts involved in the mathematical description of multi-phase flow may be found in reference books such as "Fundamentals of Numerical Reservoir Simulation" (Peaceman, D. W., 1977). Some of these concepts are addressed below.

In a two-phase flow problem, the void space is completely filled by two fluids. One of the fluids (the liquid fluid, identified by the subscript lq) wets the porous medium more than the other (the gas fluid, identified by subscript gz). As a result, the pressure in the non-wetting fluid will be higher than the pressure in the wetting fluid. The pressure difference P_{gz}-P_{lq} is the capillary pressure P_c, which is a function of saturation S_{lq}. Darcy's law is used to describe the flow of each fluid. The effective intrinsic permeability is given as a fraction of the single-fluid intrinsic permeability. The fractions (or relative permeability) are functions of saturation, S_{lq}.

In addition to the mechanical balance of momentum, the mechanical constitutive equations and compatibility equation, some additional equations are used for coupled calculations, as follows.

The transport equations of liquid and gas components are described by the Darcy's law:

$$\begin{cases} q_i^{\,lq} = -k_{ij}^{\,lq} \kappa_r^{\,lq} \dfrac{\partial}{\partial xj}\left(P_{lq} - \rho_{lq} g_k x_k\right) \\ q_i^{\,gz} = -k_{ij}^{\,lq} \kappa_r^{\,gz} \dfrac{\partial}{\partial xj}\left(P_{gz} - \rho_{gz} g_k x_k\right) \end{cases} \tag{13}$$

where k_{ij} is saturated mobility tensor, κ_r the relative permeability for the fluid, and μ the dynamic viscosity.

The constitutive equations are

$$\begin{cases} S_{lq} \dfrac{\partial P_{lq}}{\partial t} = -\dfrac{K_{lq}}{\phi}\left[\dfrac{\partial \zeta_{lq}}{\partial t} - \phi \dfrac{\partial S_{lq}}{\partial t} - S_{lq} \dfrac{\partial \varepsilon}{\partial t}\right] \\ S_{gz} \dfrac{\partial P_{gz}}{\partial t} = -\dfrac{K_{gz}}{\phi}\left[\dfrac{\partial \zeta_{gz}}{\partial t} - \phi \dfrac{\partial S_{gz}}{\partial t} - S_{lq} \dfrac{\partial \varepsilon}{\partial t}\right] \end{cases} \tag{14}$$

where K_{lq} , K_{gz} are liquid and gas bulk modules.

5. COMPARISON OF RESULTS

Figure 3 and Figure 4 compare the temperature results by five models, at r=0.685m and r=1.45m, respectively. The calculated and measured values agree well and demonstrate that conductive heat transfer is the main heat transport mechanism. All models reproduced correctly the temperature behaviour during the complete heating-cooling phases, and the reasons for the discrepancies are perhaps the differences in model geometry, mesh size, moisture-heat interaction models with different thermal diffusion laws and values of thermal diffusivities.

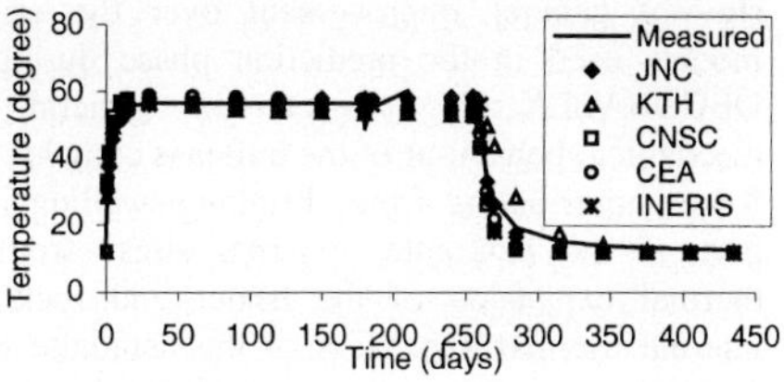

Figure 3. Time histories of temperature at r=0.685m.

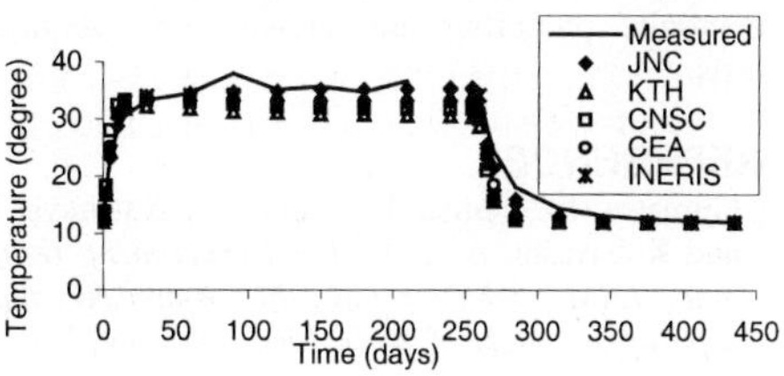

Figure 4. Time histories of temperature at r=1.45m.

Figures 5 to 7 compare the water contents at r=0.54m, r=0.685m and r=0.85m, respectively, as functions of time. The calculated results agree well with the measured ones at r=0.54m and r=0.85m, both qualitatively and quantitatively, for all teams except the INERIS/ANDRA model during both heating and cooling periods, which may be caused mainly by the treatment of steel of the canister as a porous medium and the fact that the thermally driven water transport was ignored in the INERIS/ANDRA model. The major discrepancies exist at r=0.685m in all models except for the KTH/SKI model. The possible causes for such major discrepancies could be many, especially coupled porosity-permeability models, which was ignored in all models, and condensation-evaporation coupling effects, which was ignored in CEA/IRSN, CNSC, JNC and INERIS/ANDRA models.

Figure 8 compares the radial stress at r=0.54m as a function of time, where experimental data are available. For the CEA/IRSN, CNSC, KTH/SKI and JNC models, although large discrepancies still exist, the general trend among the models agrees fairly well with the measured data at this point. This indicates that the thermal expansion phenomenon is modelled fairly well by these teams, but not the swelling process, which may be the reason for the discrepancies.

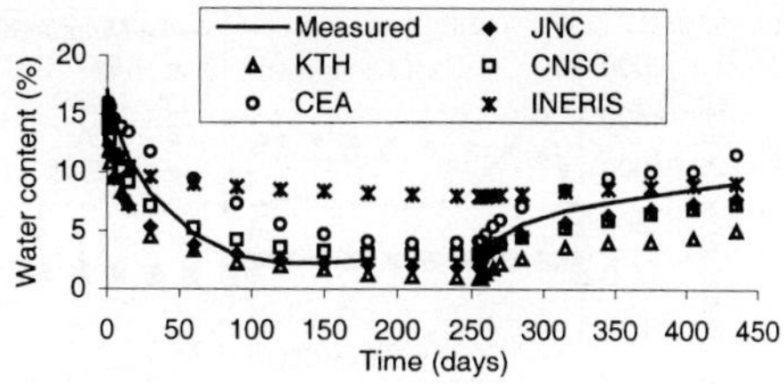

Figure 5. Water contents at r=0.54m.

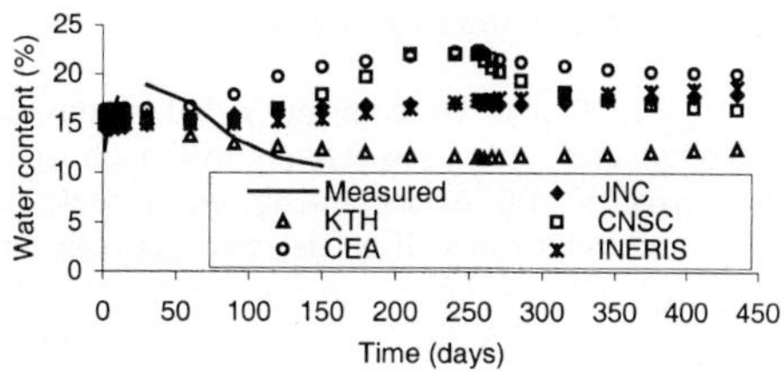

Figure 6. Water contents at r=0.685m.

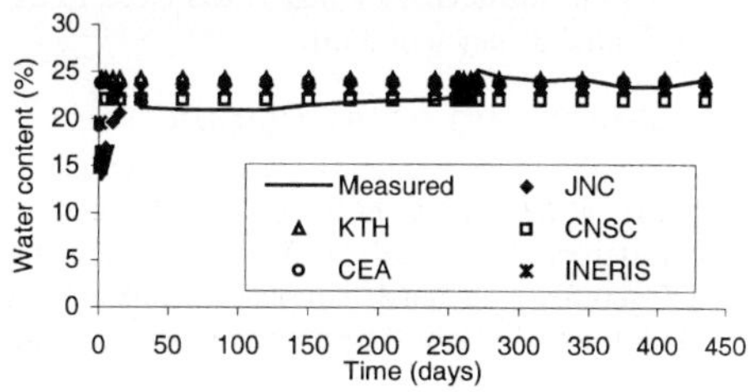

Figure 7. Water contents at r=0.85m.

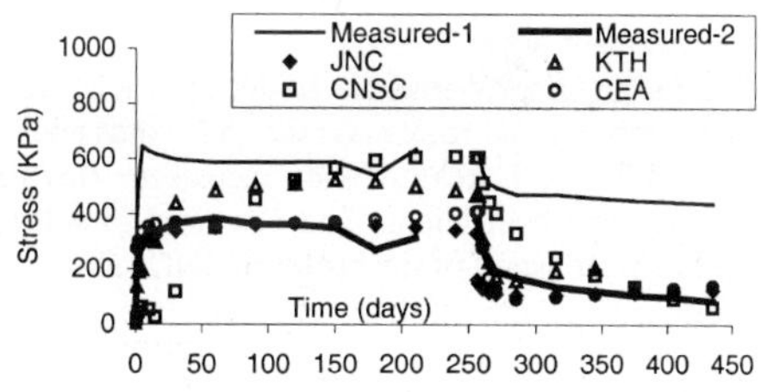

Figure 8. Radial stress at r=0.54m.

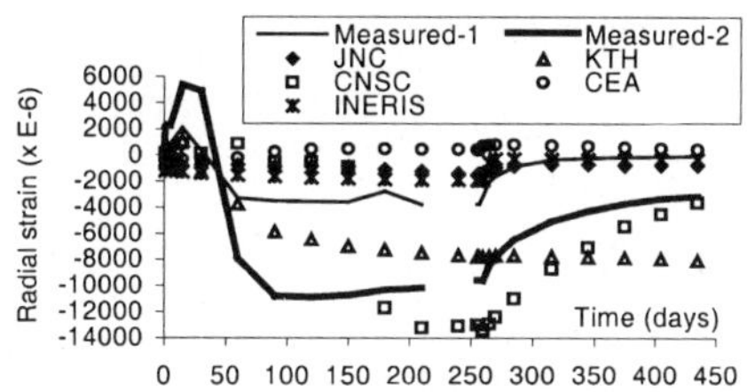

Figure 9. Radial strains at r=0.685m.

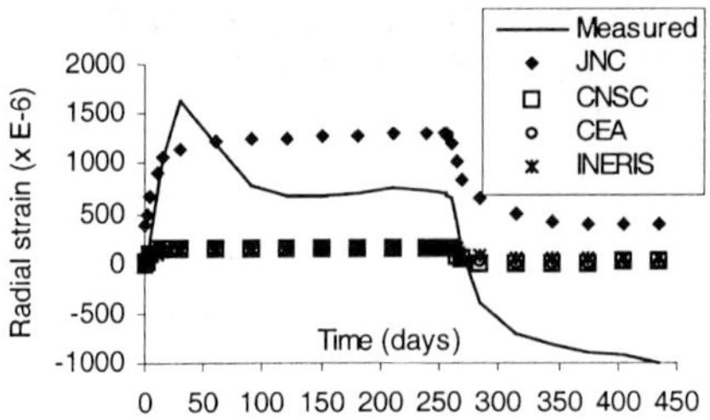

Figure 10. Radial strains at r=1.45m.

Figures 9 and 10 compare radial strains at *r*=0.685m and 1.45m, respectively. Although large discrepancies still exists among the models, a consistent trend can still be detected between the measured and calculated strains at *r*=0.685m. At *r*=1.45m in the rock, the JNC results of the strain is consistent with the experimental data in trend, and CNSC model results were under-predicted, due to the fixed outside boundary that is too close to the center (radial distance of 3 m).

6 SUMMARY AND CONCLUSIONS

A simplified axisymmetric model of the in situ THM test at the Kamaishi mine was simulated by five different numerical models. Although the model geometry is much simplified compared to the field test conditions, improved simulation of the general THM responses were obtained as compared to previous blind predictions performed within the DECOVALEX II project. The measures taken for improvement were:

·Parameter improvements (reduced rock mass permeability and rock mass thermal expansion by the KTH/SKI team, and increased thermal expansion coefficient and reduced swelling pressure constant of the buffer by JNC team)

·Inclusion of the sealing of rock fractures by penetrating bentonite by the KTH/SKI team, which can explain the uniform (axisymmetric) wetting of the bentonite.

·An improved swelling/shrinking strain function combined with an increased thermal expansion of the bentonite giving a good match of the mechanical (stress, strain) behavior of the buffer by the KTH/SKI team.

·Use of higher *E* (Young's modulus) and ν (Poisson's ratio) of the bentonite near the heater, and use of a "sealed" layer of rock around the bentonite by the CNSC team.

As a results of the above measures, the results from the simplified axisymmetric model used in the re-evaluation of the Kamaishi mine experiment showed general improvement over the original models used in the prediction phase during the DECOVALEX II project. In general, the mechanical behaviour of the buffer is complex with forces contributing from shrinking/swelling in all part of the bentonite, external stress from the thermal expansion of the heater and rock, and internal thermal expansion of the bentonite itself. However, a reasonable prediction of the mechanical behaviour can be done if all relevant bentonite properties are known from laboratory tests.

REFERENCES

Chijimatsu, M., Fujita, T., Sugita, Y., Amemiya, K. and Kobayashi, A. 2001. *Field experiment, results and THM behavior in the Kamaishi mine experiment.* Int.J. of Rock Mechanics and Mining Sciences, Vol. 38, No.1, pp.67-78

Jing, L., Stephansson, O., Tsang, C-F. and Kautsky, F. 1996. *DECOVALEX -Mathematical Models of Couples T-H-M Processes for Nuclear Waste Repositories.* Executive Summary for Phases I, II and III, SKI Report 96:58.

Jing, L., Stephansson, O., Börgesson, L., Chijimatsu, M., Kautsky, F. and Tsang, C-F. 1999 *DECOVALEX II, Technical Report –Task 2C.* SKI Report 99:23.

Jing, L and Nguyen, T.S. (eds.). 2003. *Report of BMT1A/WP2 - Implications of T-H-M coupling on the near-field safety of a nuclear waste repository.* SKI (in Press).

Peaceman, D. W., 1977. Fundamentals of Numerical Reservoir Simulation, *Developments in Petroleum Science*, 6. Elsevier.

Rutqvist, J., Börgesson, J., Chijimatsu, M., Kobayashi, A., Jing. L., Nguyen. T.S., Noorishad. J.and Tsang, C.-F. 2001a. *Thermohydromechanics of partially saturated geological media: governing equations and formulation of four finite element models.* Int.J. of Rock Mechanics and Mining Sciences, Vol. 38, No.1, pp.105-128

Rutqvist, J., Börgesson, J., Chijimatsu, M, Nguyen. T.S, Jing. L., Noorishad. J. and Tsang, C.-F. 2001b. *Coupled thermo-hydro-mechanical analysis of a heater test in fractured rock and bentonite at Kamaishi mine – comparison of field results to predictions of four finite element codes.* Int.J. of Rock Mechanics and Mining Sciences, Vol. 38, No.1, pp.129-142

Verpeaux,P.,A. Millard, T. Charras, A. Combescure 1989. *A modern approach of large computer codes for structural analysis.* Proc. SMIRT 10 Conf. Los Angels, USA.

NUMERICAL SIMULATION OF VARIABLY COUPLED THERMO-HYDRO-MECHANICAL PROCESSES IN FRACTURED POROUS MEDIA

Martin Kohlmeier, René Kaiser, Werner Zielke

Institute of Fluid Mechanics and Computer Applications in Civil Engineering
University of Hannover, Appelstraße 9a, 30167 Hannover, Germany

Abstract: This contribution deals with the modeling of coupled thermal (T), hydraulic (H) and mechanical (M) processes in subsurface structures or barrier systems. We assume a system of three phases: a deformable fractured porous medium fully or partially saturated with liquid and a gas which remains at atmospheric pressure. Consideration of the thermal flow problem leads to an extensively coupled problem consisting of an elliptic and parabolic-hyperbolic set of partial differential equations. The resulting initial boundary value problems are outlined. Their finite element representation and the required solving algorithms and control options for the coupled processes are implemented using object-oriented programming in the finite element code RockFlow/RockMech.

Model performance is demonstrated for a simplified three-dimensional test case that takes into consideration both the thermo-hydro-mechanical interaction and a stress dependent permeability. The functionality of the described control concept for the coupled processes is illustrated by some selected simulations of Benchmark Test 1B (BMT1B) of DECOVALEX III.

1. INTRODUCTION

The application of numerical modeling in water resource and waste disposal management is constantly increasing. An important factor is the requirement to predict environmental impact. As the processes are usually very complex, the substantial interest is directed towards diverse process factors, possibly interacting with each other. Unfortunately, a separate and reliable analysis of a coupled multi-field problem requires well-founded assumptions and restrictions of the problem itself. Consequently, the analysis of coupled processes is a challenging application of the finite element method.

The possibility to classify processes and couplings according to their importance is the main feature of the developed coupled numerical model. Thus, the execution of assessment procedures and the supply of decision criteria for public waste disposal authority can be simplified with these tools.

Coupled process modeling is requested in the DECOVALEX international project in order to study the influence of process interactions. Therefore a process oriented strategy and the implementation of a control concept was demanded, which allows the selective activation of processes. Furthermore, it was necessary to have selectable couplings in order to perform sensitivity analyses. Thus, the importance of different kinds of couplings or dependencies of material properties can be estimated.

Concerning the DECOVALEX benchmarks our part of the work is restricted to isothermal conditions. Multiphase flow modeling considering non-isothermal effects is performed by the Center for Applied Geosciences (ZAG) in Tübingen, Germany.

2. THE ROCKFLOW/ROCKMECH CODE

The RockFlow finite element code was originally developed for groundwater flow and transport problems. Additionally, multiphase flow problems (Thorenz (2001)) and grid adaptation (Kaiser (2000)) were taken into account. The ongoing development of RockFlow/RochMech focuses on the thermo-mechanical (TM) behavior of subsurface systems coupled to fully or partially saturated flow (H). We are benefiting from the applicability of the code to perform reactive transport processes (C), Habbar (2001). The strategy of object-oriented implementation of all these processes is the basis to perform coupled or separate analyses of multi-field problems.

3. APPROACH AND ASSUMPTIONS FOR THE THM COUPLED PROBLEM

In this section the governing equations of the THM coupled problem are presented in the framework of partially saturated porous media, see e.g. Lewis and Schrefler (1998). Here, we are assuming an incompressible liquid in a moving porous solid and negligible gas pressure gradients (Richards' approximation). Effects due to vapour transport are not incorporated in the presented formulation.

3.1 Balance equations

Solid sub-problem. The balance of linear momentum for the solid is given by

$$\nabla \cdot (\boldsymbol{\sigma}' - S_l p \mathbf{I}) + \rho_m \boldsymbol{g} = \mathbf{0} \quad (1)$$

where $\boldsymbol{\sigma}'$ is the effective stress, S_l is the liquid saturation, p is the liquid pressure and $\mathbf{I}$ is the second-order unit tensor. The density of the mixture and the acceleration due to gravity are denoted by ρ_m and $\boldsymbol{g}$, respectively.

Fluid sub-problem. Here the balance of mass for the liquid phase is given in its volumetric representation:

$$\nabla \cdot \boldsymbol{q} + n \frac{\partial S_l}{\partial t} + \alpha \nabla \cdot \dot{\boldsymbol{u}} = 0 \quad (2)$$

where $\boldsymbol{q}$ is the relative fluid flux, n is the porosity, α is Biot's coefficient and $\dot{\boldsymbol{u}}$ is the solid displacement velocity.

Heat sub-problem. The balance of heat energy is given by

$$\nabla \cdot \boldsymbol{q}_\mathrm{h} + ((1-n) c_s \rho_s + n c_l \rho_l) \frac{\partial T}{\partial t} + c_l \rho_l \; \boldsymbol{q} \nabla T = 0 \quad (3)$$

where $\boldsymbol{q}_\mathrm{h}$ is the heat flux, T is the temperature, c_s, ρ_s, c_l and ρ_l are the specific heat capacity and the density of the solid and the liquid phase, respectively.

3.2 Constitutive equations

The stress strain relation (generalized linear Hooke's law) incorporating the thermal expansion of the solid is

$$\boldsymbol{\sigma}' = \lambda \, \mathrm{tr}(\boldsymbol{\varepsilon} - \boldsymbol{\varepsilon}_\mathrm{T}) \mathbf{I} + 2G(\boldsymbol{\varepsilon} - \boldsymbol{\varepsilon}_\mathrm{T}) \quad (4)$$

with $$\boldsymbol{\varepsilon} = \frac{1}{2}(\nabla \boldsymbol{u} + (\nabla \boldsymbol{u})^\mathrm{T}) \quad (5)$$

$$\boldsymbol{\varepsilon}_\mathrm{T} = \alpha_\mathrm{T} (T - T_\mathrm{ref}) \mathbf{I} \quad (6)$$

where λ is the Lamé constant, G is the shear modulus and α_T is the thermal expansion coefficient of the solid.

The liquid flux (modified Darcy's law) is

$$\boldsymbol{q} = n S_l \boldsymbol{v}_l = \frac{k_\mathrm{rel}^l \boldsymbol{k}}{\mu_l} (-\nabla p + \rho_l \boldsymbol{g}) \quad (7)$$

where k_rel^l and μ_l are the relative permeability and the dynamic viscosity of the liquid phase. The intrinsic permeability tensor of the porous medium is denoted by $\boldsymbol{k}$.

The heat conduction is

$$\boldsymbol{q}_\mathrm{h} = \boldsymbol{D}_m (-\nabla T) \quad (8)$$

where $\boldsymbol{D}_m$ is the averaged heat conductivity tensor of the mixture.

3.3 Material properties and nonlinear material laws

The intrinsic permeability $\boldsymbol{k} = \boldsymbol{k}(\varepsilon_\mathrm{v}, \boldsymbol{\sigma}')$ depends on the volumetric strain or in the case of fractured rock it can be formulated as a function of the effective normal stress using e.g. the following definition of the void aperture b_v

$$b_\mathrm{v} = b_\mathrm{vr} + \Delta b_\mathrm{v} \quad (9)$$

where b_vr is the residual void aperture which defines the minimum void aperture of a completely compressed fracture and $\Delta b_\mathrm{v} = \Delta b_\mathrm{v}(\sigma'_\mathrm{n})$ is the stress induced void aperture, e.g. defined by the Barton-Bandis' hyperbolic normal closure model (Bandis et al. (1983))

$$\Delta b_\mathrm{v} = \frac{k_\mathrm{n0} V_\mathrm{m0}^2}{\sigma'_\mathrm{n} \, k_\mathrm{n0} \, V_\mathrm{m0}} \quad (10)$$

where σ'_n, k_n0 and V_m0 are the effective normal stress, the initial normal stiffness at zero normal stress and the maximum normal closure. Finally, the transmissivity of a parallel fracture can be approximated by

$$T = \frac{b_h^3 \rho g}{12\mu} \tag{11}$$

where b_h is the hydraulic aperture, which can be derived by $b_h = b_v^2 / JRC_0^{2.5}$, where JRC is the joint roughness coefficient, see Bandis et al. (1983).

Multiphase flow modeling in non-saturated soils requires the incorporation of the relative permeability saturation relation $k_{rel}^l = k_{rel}^l(S_l)$ and the capillary pressure saturation relation, e.g. defined by van Genuchten (1980).

3.4 Set of coupled field equations

The finite element formulation is obtained using the Standard Galerkin method.

The final algebraic equations are set up by assembling the element contributions: the stiffness matrix and the coupling matrix for each desired process or coupling, respectively. Additionally, the boundary conditions have to be incorporated. Consequently, the resulting equations are strongly or weakly coupled with each other and it has to be decided whether they can be solved iteratively in a partitioned scheme or in one step in a directly coupled system of equations, see Eq. (12).

$$\begin{bmatrix} \mathbf{K}_u & \mathbf{C}_{up} & \mathbf{C}_{ut} \\ \mathbf{C}_{pu} & \mathbf{K}_p & \mathbf{C}_{pT} \\ \mathbf{0} & \mathbf{C}_{Tp} & \mathbf{K}_T \end{bmatrix} \begin{bmatrix} \Delta\mathbf{u} \\ \mathbf{p} \\ \mathbf{T} \end{bmatrix} = \begin{bmatrix} \mathbf{R}_u \\ \mathbf{R}_p \\ \mathbf{R}_T \end{bmatrix} \tag{12}$$

3.5 Solving coupled equations

The coupled linear system in Eq. (12) represents the algebraic formulation of the interacting processes. As mentioned in the previous section, the composition of this set is achieved by a flexible assembling strategy, which needs several pieces of information. This means that the pre- and post-processing of the solving step needs access to the data of each unknown involved in the linear system. They are summarized in a data object consisting of sub-objects for each solver needed for a multi-field problem. The resulting structure of the data object is arranged in Table 1.

Table 1. Object-oriented data set for solving sets algebraic equations.

OBJECT: **DATA** for all SOLVERS*		
Object: **SOLVER**	• number of primary unknowns • solver properties • Object UNKNOWN	
	Object: **UNKNOWN**	• ICs • BCs • element contribution • assemble function • number of degrees of freedom • nodal data

*of linear or nonlinear type.

4. CONTROL CONCEPT

The RockFlow/RockMech code provides several finite element algorithms representing hydrologic thermal, or reactive processes with diffusive and advective transport. New developments for mechanical analyses and the required couplings concerning THM modeling demand a facility to pick a variable amount of processes and to decide whether they are coupled or not. Using the presented control concept, the desired process can be specified by keywords, see Table 2. The processing can be started or inactivated separately for each process. In addition, these specifications can be used to decide which processes shall interact with another.

Table 2. Selected processes and their access keywords within the control concept.

Process*	Keyword for specification of desired process
T	HEAT_TRANSPORT (HT)
H	FLUID_FLOW (FF)
M	SOLID_DEFORMATION (SD)
C (Transp. and reaction of chemical substances)	MASS_TRANSPORT (MT)

*Further processes can easily be added.

The coupling terms of interacting processes are affected by the parameter values of the respective material groups. The quantity of these functional dependencies is predetermined by the formulation of the processes themselves. Some of these terms are listed in the matrix of Tab. 3.

Table 3. The matrix of coupling terms.

	Influencing Process		
Process	T	H	M
T		$q_h = q_h(q)$	–
H	$\rho_l = \rho_l(T)$		$q = q(\alpha \dot{\varepsilon}_v)$ $q = q(k(\sigma_n))$
M	α_T	p	

5. APPLICATIONS

Important applications for coupled finite element strategies can be found in the theory of consolidation of partially saturated soils. Some test cases were performed, (see Kohlmeier et al. (2002)) using the Richards' formulation.

5.1 Density dependent THM modeling

This numerical example is concerned with density dependent problems like brine filled solution-mined cavities or the thermo-hydraulic interaction of thermally expanding fluids. In order to demonstrate this coupling phenomenon we assume a fully saturated flow of a thermally expanding liquid within a porous torus (see Fig. 1).

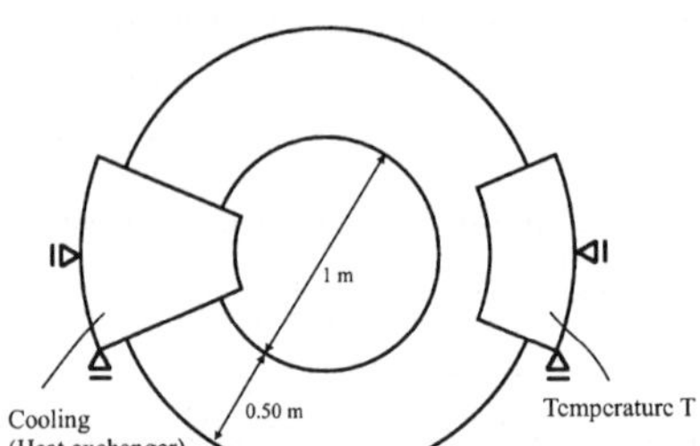

Figure 1. Geometry and boundary conditions.

Due to density gradients in the liquid a flow field is established. The resulting counter-clockwise movement of the liquid forces an advective transport of heat energy, which can be seen in the temperature field depicted in Fig. 2. The resulting displacement field is shown in Fig. 3.

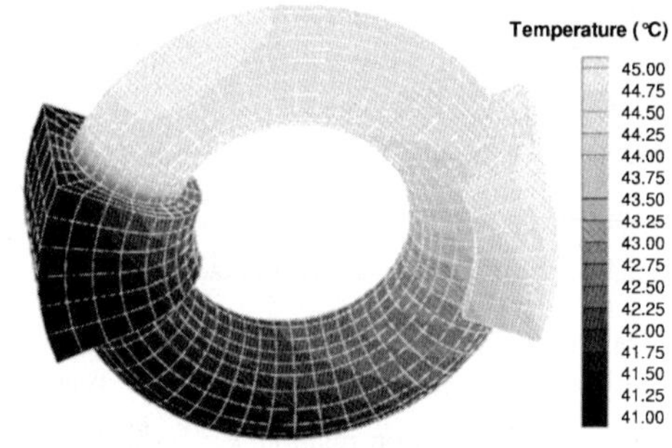

Figure 2. Temperature field.

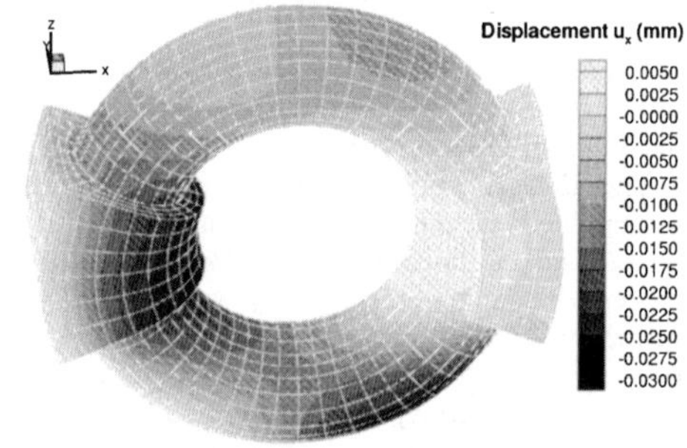

Figure 3. Displacements u_x.

5.2 Thermally influenced fracture transmissivity

Here, we assume a dipole fluid flow initialized by the infiltration of cold water in a hot rock and its extraction after the heating-up, see Fig. 4. A fracture with a stress dependent transmissivity is located in the front part of the domain. Due to the thermal contraction of the rock the transmissivity of the fracture increases. Fig. 5 shows the different velocity curves in the fracture and in the non-affected matrix. Fig. 6 shows the final temperature field in the deformed configuration.

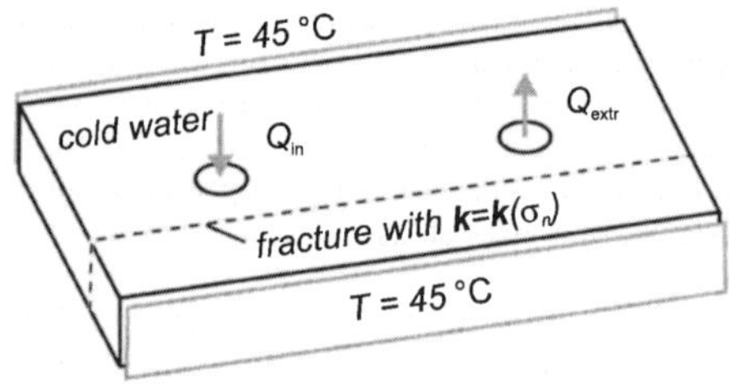

Figure 4. Geometry and boundary conditions.

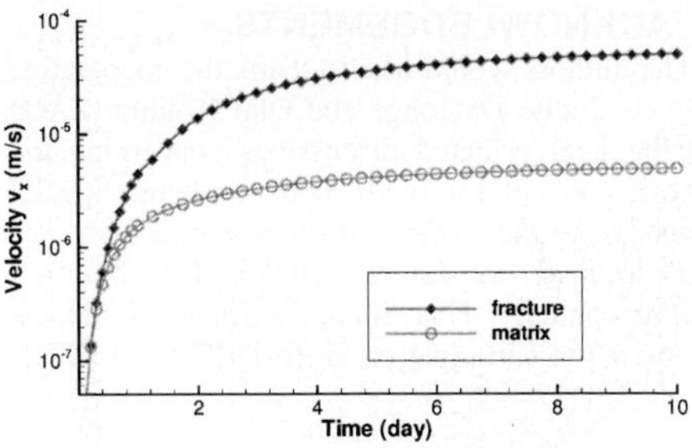

Figure 5. Velocity curves in fracture and matrix across the centreline of the domain.

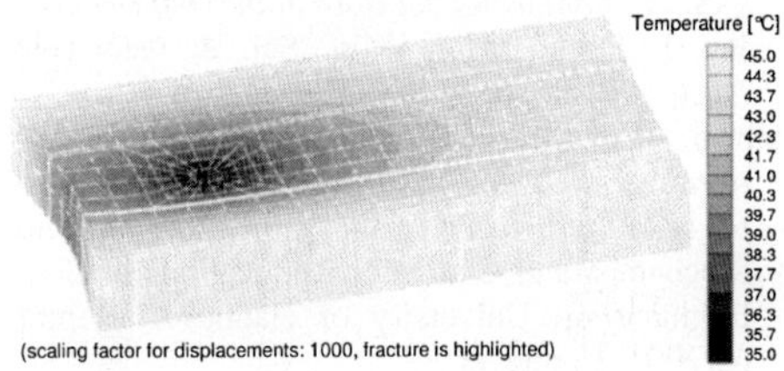

Figure 6. Temperature field in the deformed configuration (t=10 days).

5.3 Thermo-mechanical modeling of the DECOVALEX BMT1B F0/F1

In this section we present some results of BMT1B. Its definition is based on a hypothetical set up of a nuclear waste repository. A discretized quarter of the symmetric domain gives a flavor of the geometrical model and is depicted in Fig. 7. It consists of the following materials: fractured rock, backfill, buffer, waste and overpack. The material properties are listed in Table 3.

Additionally, a heat output curve, boundary and initial conditions are prescribed in order to complete the description of the initial boundary value problem to be solved.

We are presenting a choice of results of the thermo-mechanical (TM) analysis. The resulting temperature and stress curves of some selected points in the buffer are given in Fig. 8 and Fig. 9. The heat producing waste affects the temperature field. Starting with an assumed initial temperature of 45 °C a maximum temperature is predicted and the influence on the mechanical behavior of the solid structure can be seen. Effects of vaporization of water or vapor transport are not incorporated.

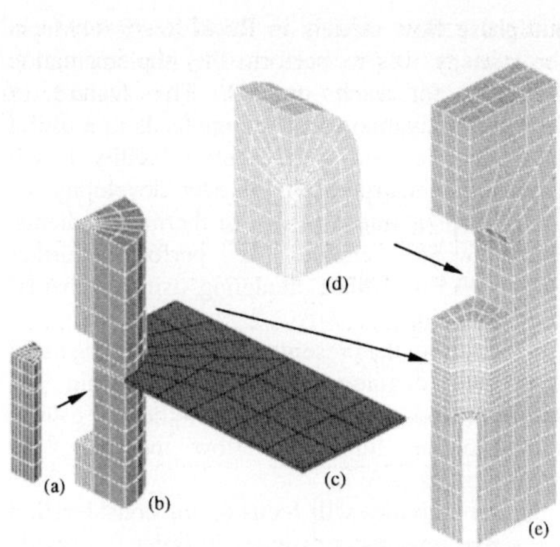

Figure 7. Partitions of the mesh for the fractured rock case (F1): waste and overpack (a), buffer (b), fracture (c), backfill (d), rock mass (e).

Table 3. Material properties.

Material parameter	Rock mass	Backfill	Buffer	Waste and overpack
Young´s modulus, *E* (MPa)	61.0× 10^3	5.0	100.0	200.0×10^3
Poisson´s ratio, ν	0.303	0.4	0.4	0.3
Thermal expansion coefficient α_T (K^{-1})	8.21× 10^{-6}	8.21×10^{-6}	8.21× 10^{-6}	1.64×10^{-6}
Effective porosity, *n* (%)	0.379	30	20	0.0379
Density, ρ (kg/m³)	2746	1750	1800	7800
Specific heat, *c* (J/(kgK))	900	826.087	826.087	460
Thermal conduc-tivity, λ (W/(mK))	2.6	1.46213	1.36760	53

6. CONCLUSION

The objective of this work has been the development of finite element algorithms for hydraulically coupled three-dimensional thermo-mechanical processes to be applied to geological problems. A further aim was the realization of couplings to existing finite element algorithms for multiphase flow models in RockFlow/RockMech. The strategy was to perform the implementation separately for each process. The feature of piecewise activation of couplings leads to a useful and extendible control and steering facility. It will provide the incorporation of new developments, e.g. the vapour transport due to thermal gradients, which would enable to perform further DECOVALEX, BMT1 modeling using Richards' approximation.

In the future, the presented coupling strategy will incorporate ongoing developments within the RockFlow code concerning non-isothermal effects in multiphase (liquid/gas) flow modeling, De Jonge, Kolditz (ZAG).

Further activities will focus on the consideration of reactive transport processes in order to provide radionuclide release and transport modeling.

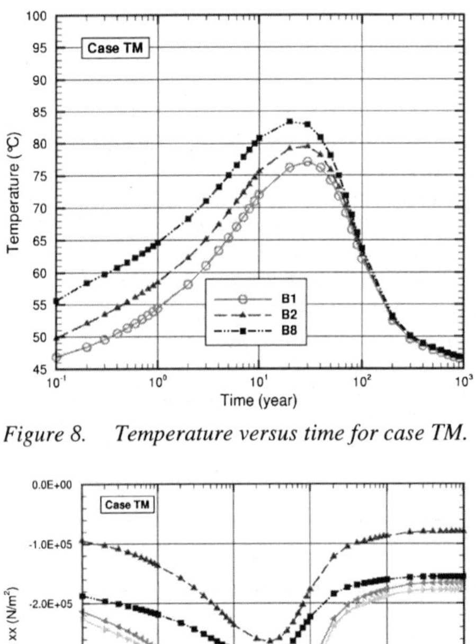

Figure 8. Temperature versus time for case TM.

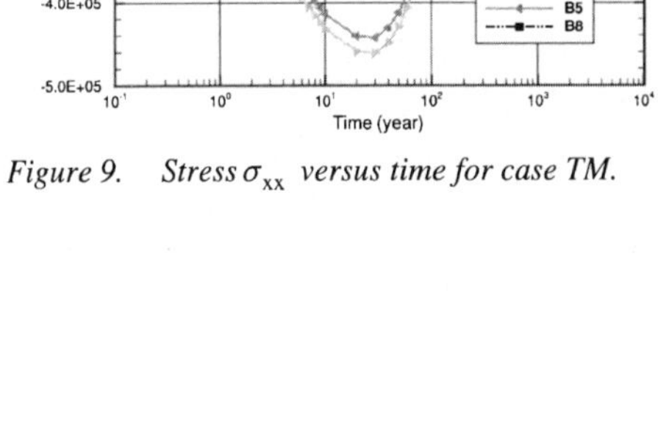

Figure 9. Stress σ_{xx} *versus time for case TM.*

7. ACKNOWLEDGEMENTS

The authors would like to thank the cooperative partners Joelle De Jonge and Olaf Kolditz (ZAG) for the goal oriented discussions concerning the control concept for multi-field problems. Special thanks go to the *Federal Institute for Geosciences and Natural Resources* (BGR), Dr. Manfred Wallner and Dr. Hua Shao, for financial support implying the participation in the DECOVALEX III project.

REFERENCES

Bandis, S. C., Lumbsden, C. A. & Barton, N. R. 1983. *Fundamentals of Rock Joint Deformation.* Int. J. Rock. Mech. Min. Sci. & Geomech. Abstr., **20**, No. 6, pp 249-268.

Habbar, A. 2001. *Direkte und Inverse Modellierung reaktiver Transportprozesse in klüftig-porösen Medien.* Ph.D. thesis, Institute of Fluid Mechanics and Computer Applications in Civil Engineering, University of Hannover, Report 65/2001, Hannover.

Kaiser, R. 2000. *Automatic Grid Adaptation for Multidimensional Coupled Processes in Subsurface Hydrosystems.* In: Groundwater Updates, International Symposium 2000 on Groundwater (IAHR), Sato, K. & Iwasa, Y. (eds.), pp. 339 - 344, Sonic City, Omiya, Japan, May 8-10, 2000. Springer-Verlag, Tokyo, Berlin, New York.

Kohlmeier, M., Kaiser, R., Kolditz, O., Zielke, W. 2002. *Finite element simulation of consolidation and bentonite swelling in the framework of unsaturated porous media.* In Hasanzadeh, S. M. et al. (eds.), *Developments in Water Science, 47, Vol. 1, pp. 57-64, Proc. XIVth Int. Conf. on Computational Methods in Water Resources, June 23-28, 2002, Delft.*

Lewis, R. W., Schrefler, B. A. 1998. *The Finite Element Method in the Static and Dynamic Deformation and Consolidation of Porous Media.* Wiley, Chichester.

Thorenz, C. 2001. *Model Adaptive Simulation of Multiphase and Density Driven Flow in Fractured and Porous Media.* Ph.D. thesis, Institute of Fluid Mechanics and Computer Applications in Civil Engineering, University of Hannover, Report 62/2001, Hannover.

van Genuchten, M. 1980. *A closed-form equation for predicting the hydraulic conductivity of unsaturated soils. J. Soil Science Society of America*, **44**: 892-98.

NUMERICAL IMPLEMENTATION OF THERMALLY AND HYDRAULICALLY COUPLED PROCESSES IN NON-ISOTHERMAL POROUS MEDIA

Joëlle De Jonge, Mingliang Xie, Olaf Kolditz
Centre for Applied Geoscience (ZAG) / Section Geohydrology/Geoinformatics
University of Tübingen, Germany

Abstract: To better understand the coupling of thermal (T), hydraulic (H) and mechanical (M) processes (T-H-M processes) and their influence on the system behaviour, models allowing T-H-M coupling are developed. These models allow simulations in the near-field of the system. Such a model has been developed within the simulator RockFlow/RockMech. This paper concentrates on the thermal and hydraulic processes. For those processes, the material parameters and state variables are highly non-linear and mostly functions of temperature, saturation and pressure. This paper shows how these dependencies are formulated mathematically and are implemented into RockFlow/RockMech. The implementation allows phase changes between the fluid phases (gas and liquid) to occur explicitly. The model allows the simulation of very low permeability clays with high capillary pressures. An example for code validation is shown, where low permeability clay is desaturated, lastly, current work on the calculations performed in the near field study (BMT1) of the DECOVALEX III project is outlined.

1. INTRODUCTION

For deep geological disposal of nuclear waste, engineered barrier systems have to be developed. To develop appropriate barriers, the disposal scenario has to be simulated numerically, as a prediction and engineering tool in the near-field of the disposal site. The numerical model has to be able to reproduce the various processes. Figure 1 shows a plan view of a possible disposal scenario, with nuclear waste packed into the core at the centre, a bentonite buffer surrounding it. Waste canister and buffer are placed into the host rock. The processes arising in this situation are coupled processes, composed of thermal processes (heating, cooling, evaporation, condensation), hydraulic processes (water intrusion, saturation, desaturation), mechanical processes (elastic and plastic deformation), and chemical processes (radiation, reactions). This paper concentrates on the coupling of thermal and hydraulic processes.

The engineered buffer used in deep geological disposal sites is often pure bentonite or a bentonite – sand mixture. Bentonite is a material with very low permeability and swelling properties. The swelling properties are used as a seal against water intrusion.

RockFlow is a simulator that is used to model near-field situation of the type described above (Kolditz et al. (2003)). It is a finite element code, programmed using object-oriented techniques. Object-orientation allows a wide spectrum of applications for the code, and also simultaneous code development (Kolditz (2002)). This paper will concentrate on the model within RockFlow, that is used for simulating the coupling of thermal and hydraulic processes (TH). The governing equations of the model are shown, as well as how these equations are treated numerically. A code application example for the desaturation of bentonite will be presented. The first example is based on an example in Olivella and Gens (2000) and partially serves as model verification. Furthermore, the code development finds application in bench mark test 1 (BMT1) of the DECOVALEX III (International co-operative project for the DEvelopment of COupled models and their VALidation against EXperiments in nuclear waste isolation) project (Jing et al. (1999)). Current work on this project is presented.

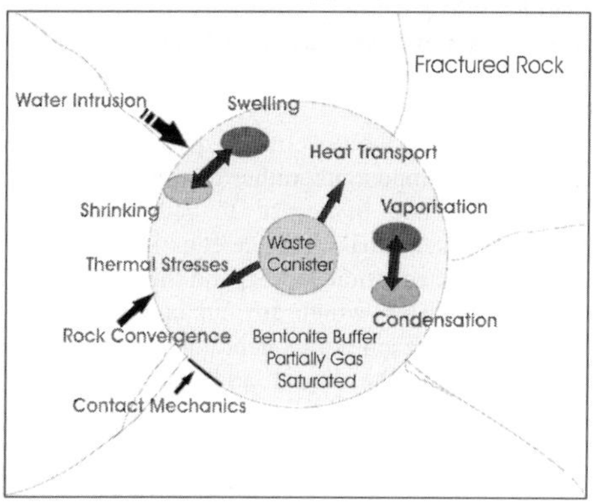

Figure 1. THM processes in deep geological disposal sites.

2. GOVERNING EQUATIONS

We use a componental formulation as applied by Emmert (1997) for modelling of steam injection for soil remediation strategies. The TH code in RockFlow has three primary variables, gas phase pressure, liquid phase saturation, and temperature. The set of three governing equations result from two balance equations, the general mass balance equation and the energy balance equation. The general mass balance is shown in equation (1), where n [-] is porosity, S [-] is saturation, ρ [kg/m^3] is density, $\boldsymbol{J}$ [kg/(ms)] is flux, and Q [m^3/s] is a source-sink term. The superscripts l, g, and s denote the liquid phase, gas phase, and solid phase respectively. The subscript k denotes the component, which is air (a) or water (w). The first term of equation (1) is a mass storage term, the second term shows mass movement and the right hand side accounts for mass leaving or entering the system.

$$\frac{\partial}{\partial t}(nS^l\rho_k^l + nS^g\rho_k^g) + \nabla\cdot(\mathbf{J}_k^{ls} + \mathbf{J}_k^{gs}) + \nabla\cdot(\mathbf{J}_k^g + \mathbf{J}_k^l) = Q_k \tag{1}$$

Equation (2) shows the general energy balance equation, where u [J/kg] is internal energy and $\boldsymbol{J}$ [J/(ms)] is heat flux. The structure of the equation is similar to that of equation (1).

$$\begin{aligned} & \frac{\partial}{\partial t}\left((1-n)\rho^s u^s + nS^g\rho^g u^g + nS^l\rho^l u^l\right) \\ + & \left(\nabla\cdot\mathbf{J}_h^s + \nabla\cdot\mathbf{J}_h^g + \nabla\cdot\mathbf{J}_h^l\right) = Q_h \end{aligned} \tag{2}$$

Equations of state and material laws give the necessary specifications to transform the balance equations into governing equations for the model. The equations of state and material laws used in the TH model are thoroughly described in De Jonge and Kolditz (2002).

In the TH code the governing equations are split according to components, rather than phases. This choice was made in regard to phase changes. Hence, the general mass balance equation results in the component equation for air, and the component equation for water, which are our components in the TH model. Each component equation contains terms for both, the liquid and the gas phase. Equation (3) shows the governing equation for air or water, (k = a, w), where X [-] is the mass fraction, P [Pa] is pressure, k_{rel} [-] is the relative permeability, $\boldsymbol{k}$ [m^2] is the reference permeability, μ [Pas] is viscosity, P_c [Pa] is capillary pressure, $\boldsymbol{g}$ [m/s^2] is the gravity vector, and D [m^2/s] is the diffusion coefficient.

$$\begin{aligned} & nX_k^g\left(S^g\frac{\partial\rho^g}{\partial p^g} + S^l\frac{\partial\rho^l}{\partial p^l}\right)\frac{\partial p^g}{\partial t} \\ - & \nabla\cdot\left(\rho^g X_k^g\frac{k_{\text{rel}}^g\mathbf{k}}{\mu^g}\nabla p^g\right) - \nabla\cdot\left(\rho^l X_k^l\frac{k_{\text{rel}}^l\mathbf{k}}{\mu^l}\nabla p^g\right) \\ + & n(-\rho^g X_k^g + \rho^l X_k^l)\frac{\partial S^l}{\partial t} \\ = & Q_k \\ + & (nS^l X_k^l\frac{\partial\rho^l}{\partial p^l})\frac{\partial p_c}{\partial t} - \nabla\cdot\left(\rho^l X_k^l\frac{k_{\text{rel}}^l\mathbf{k}}{\mu^l}\nabla p_c\right) \\ - & \nabla\cdot\left(\rho^g X_k^g\frac{k_{\text{rel}}^g\mathbf{k}}{\mu^g}\rho^g\mathbf{g}\right) + \nabla\cdot\left(\rho^l X_k^l\frac{k_{\text{rel}}^l\mathbf{k}}{\mu^l}\rho^l\mathbf{g}\right) \\ + & \nabla\cdot(nS^g\rho^g D_k^g\nabla X_k^g) \end{aligned} \tag{3}$$

Equation (4) shows the governing equation resulting from the energy balance equation, where c [J/(kgK)] is heat capacity, λ [J/(Kms)] is thermal conductivity, and h [J/kg] is enthalpy.

$$\begin{aligned} & \left((1-n)\rho^s c^s + nS^g\rho^g c^g + nS^l\rho^l c^l\right)\frac{\partial T}{\partial t} \\ + & \left(\rho^g c^g\frac{k_{\text{rel}}^g\mathbf{k}}{\mu^g}(\nabla p^g - \rho^g\mathbf{g})\right. \\ & \left. + \rho^l c^l\frac{k_{\text{rel}}^l\mathbf{k}}{\mu^l}(\nabla p^l - \rho^l\mathbf{g})\right)\nabla T \\ - & \nabla\cdot\left(((1-n)\lambda^s + nS^g\lambda^g + nS^l\lambda^l)\nabla T\right) \\ - & \nabla\cdot(nS^g\rho^g D_a^g h^g\nabla X_a^g) \\ - & \nabla\cdot(nS^g\rho^g D_w^g h^g\nabla X_w^g) \\ = & \rho Q_T \end{aligned} \tag{4}$$

3. NUMERICAL IMPLEMENTATION

Figure 2 illustrates the numerical implementation used in RockFlow. The method of weighted residuals is applied to the governing equations, resulting the weak formulation of the algebraic equations. The order of spatial derivatives is reduced by using the Gauss-Ostrogradski theorem, thus resulting in the algebraic equation system.

Equation (5) shows the resulting algebraic equation for the water and air components, where (k = a, w) and θ is a value between [0,1]. C_P, C_S, K_P, and K_S are element matrices, that are dependent on element dimension and shape. Development for the case of 1-D elements is shown in Kolditz and De Jonge (2003), and 2-D triangular elements in De Jonge and Kolditz (2002).

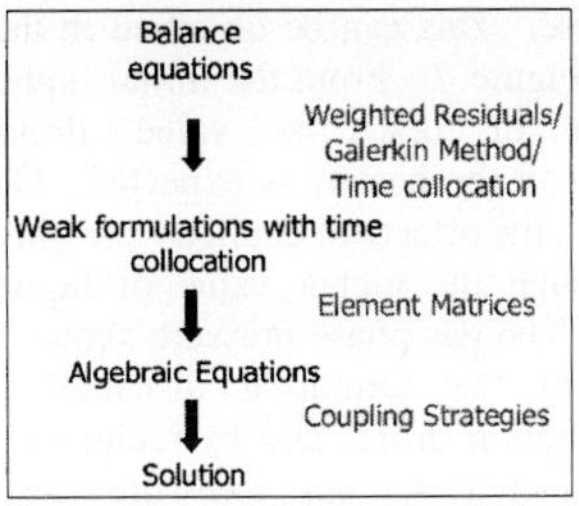

Figure 2. Numerical implementation

$$
\begin{aligned}
& \left[\frac{1}{\Delta t}(\mathbf{C}_{P_k}^g + \mathbf{C}_{P_k}^l) + \theta(\mathbf{K}_{P_k}^g + \mathbf{K}_{P_k}^l)\right] [\hat{\mathbf{p}}^g]^{n+1} \\
+ & \left[\frac{1}{\Delta t}(\mathbf{C}_{S_k}^l - \mathbf{C}_{S_k}^g)\right] [\hat{\mathbf{S}}^l]^{n+1} \\
= & \left[\frac{1}{\Delta t}(\mathbf{C}_{P_k}^g + \mathbf{C}_{P_k}^l) - (1-\theta)(\mathbf{K}_{P_k}^g + \mathbf{K}_{P_k}^l)\right] [\hat{\mathbf{p}}^g]^{n} \\
+ & \left[\frac{1}{\Delta t}(\mathbf{C}_{S_k}^l - \mathbf{C}_{S_k}^g)\right] [\hat{\mathbf{S}}^l]^{n} \\
+ & \left[\frac{1}{\Delta t}\mathbf{C}_{P_k}^l\right] ([\hat{\mathbf{p}}_c]^{n+1} - [\hat{\mathbf{p}}_c]^{n}) \\
+ & [\mathbf{K}_k^{ls}]\,(\theta[\hat{\mathbf{p}}_c]^{n+1} + (1-\theta)[\hat{\mathbf{p}}_c]^{n}) \\
- & [\mathbf{K}_k^{g}]\,(\theta[\hat{\mathbf{X}}_k^g]^{n+1} + (1-\theta)[\hat{\mathbf{X}}_k^g]^{n}) \\
+ & (\mathbf{K}_k^{gs} + \mathbf{K}_k^{ls})\mathbf{g}
\end{aligned} \tag{5}
$$

Equation (6) shows the algebraic equation for the energy component in matrix form. As above, the element matrices for the 1-D and 2-D triangular elements can be found in De Jonge and Kolditz (2002).

$$
\begin{aligned}
& \left[\frac{1}{\Delta t}\mathbf{C}_T + \theta\mathbf{A}_T + \theta\mathbf{K}_T\right] [\hat{\mathbf{T}}]^{n+1} \\
= & \theta[Q_T]_\tau^{n+1} + (1-\theta)[Q_T]^n \\
+ & \left[\frac{1}{\Delta t}\mathbf{C}_T - (1-\theta)\mathbf{A}_T - (1-\theta)\mathbf{K}_T\right] [\hat{\mathbf{T}}]^{n}
\end{aligned} \tag{6}
$$

In the implementation, each algebraic equation is linked to a primary variable. Thus, the algebraic equation for air is linked to gas pressure, the equation for water is linked to liquid phase saturation and the energy equation is linked to temperature. The system of equations can be solved in two ways in the TH model; either they are solved iteratively, gas pressure and liquid saturation are solved simultaneously, followed by temperature, as shown in the system of equations (7).

$$
\begin{aligned}
\left[\frac{1}{\Delta t}\mathbf{C}_{Pa} + \mathbf{K}_{Pa}\right] [\hat{\mathbf{p}}^g]^{n+1} + \left[\frac{1}{\Delta t}\mathbf{C}_{Sa}\right] [\hat{\mathbf{S}}^l]^{n+1} &= \mathrm{RHS}_a \\
\left[\frac{1}{\Delta t}\mathbf{C}_{Pw} + \mathbf{K}_{Pw}\right] [\hat{\mathbf{p}}^g]^{n+1} + \left[\frac{1}{\Delta t}\mathbf{C}_{Sw}\right] [\hat{\mathbf{S}}^l]^{n+1} &= \mathrm{RHS}_w \\
\left[\frac{1}{\Delta t}\mathbf{C}_T + \mathbf{K}_T + \mathbf{A}_T\right] [\hat{\mathbf{T}}]^{n+1} &= \mathrm{RHS}_T
\end{aligned} \tag{7}
$$

4. EXAMPLE – DESATURATION

The example presented here is based on the one shown in Olivella and Gens (2000) with slightly modified material properties and temperature boundary condition values. The example portrays the desaturation of bentonite due to heating in a closed system. The set-up is as illustrated in Figure 3. A bentonite block is heated to 100°C on the left side. Dimensions and initial conditions are as shown on Figure 3. The only boundary condition is the temperature boundary condition at the heater.

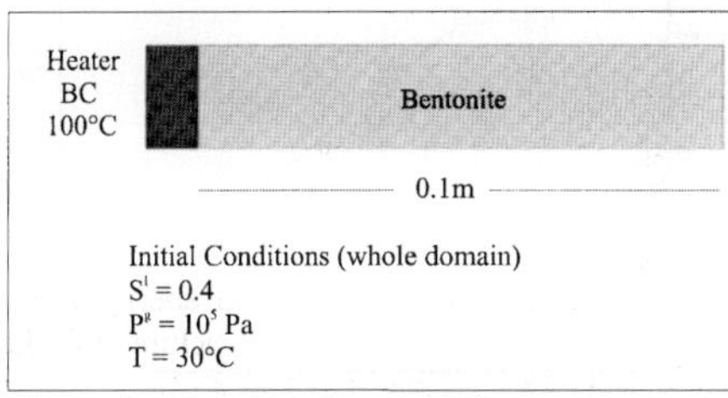

Figure 3. Desaturation example set-up

The material parameters are as summarised in Table 1. Because the capillary pressures in bentonite are very high (up to 10^{10} Pa, as shown in Figure 5), a large suction pressure is created, that strongly favours saturation of the sample. Vapour movement, and hence desaturation due to sample heating can only be modelled if vapour flow is encouraged. This can be achieved by including a vapour diffusion term into the balance equation or

Table 1. Material parameters for the desaturation example.

Parameter	Value
Porosity [-]	0.44
Tortuosity [-]	0.1
Permeability [m^2]	5 x 10^{-19}
Relative permeability [-]	Figure 4
Capillary pressure	Figure 5
Density [kg/m^3]	1650
Heat capacity [J/(kg*K)]	1605
Heat conductivity [J/(K*m*s)]	0.5

alternatively, by choosing a dual relative permeability – saturation relationship so that vapour movement is favoured at low liquid phase saturations (when capillary pressure is highest). Olivella and Gens (2000) developed such a model. It is based on experimental observations of bentonite properties. The reasoning behind it is, that in the dry bentonite pore spaces are higher, resulting in a higher permeability. When bentonite is wet however, it swells and pore spaces are accordingly small. Figure 4 shows an example of such a dual relative permeability – saturation model.

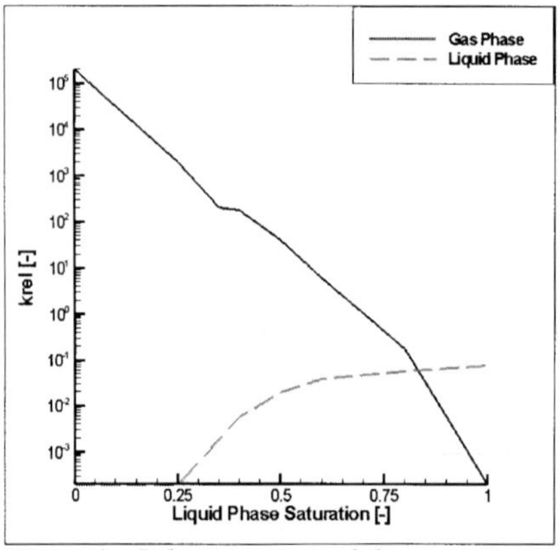

Figure 4. Relative permeability – saturation relationship – dual model

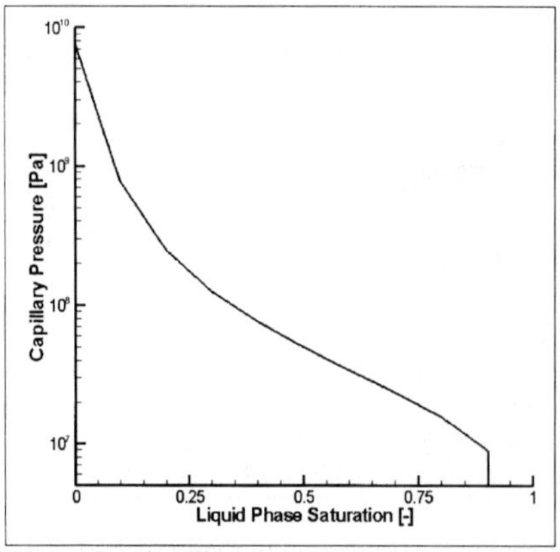

Figure 5. Capillary pressure – saturation relationship.

Because there are no saturation and pressure boundary conditions in the example, saturation is free to fall and rise. This can be observed in the results shown in Figure 7. From the initial liquid phase saturation of 0.4, the value drops progressively next to the heater, as expected. On the right boundary, the effects of capillary pressure can be seen through the higher value of liquid phase saturation. The gas phase pressure shows a small response to the saturation distribution. Where liquid saturation drops, and by reciprocity gas saturation increases, gas pressure also increases. At the right boundary, the decrease in gas pressure corresponds to the increase in liquid saturation. One can see that the stationary temperature is reached in less than ½ day. The vapour mass fraction responds to the temperature distribution immediately.

The results thus show successful modelling of a desaturation process in bentonite, by using the RockFlow TH model.

5. EXAMPLE – BMT1 DECOVALEX III

This section describes the current work on the near-field bench mark test (BMT1) of the DECOVALEX project. The results shown in this section are preliminary and serve to outline the direction of current code development and to further illustrate the working of the RockFlow TH model.

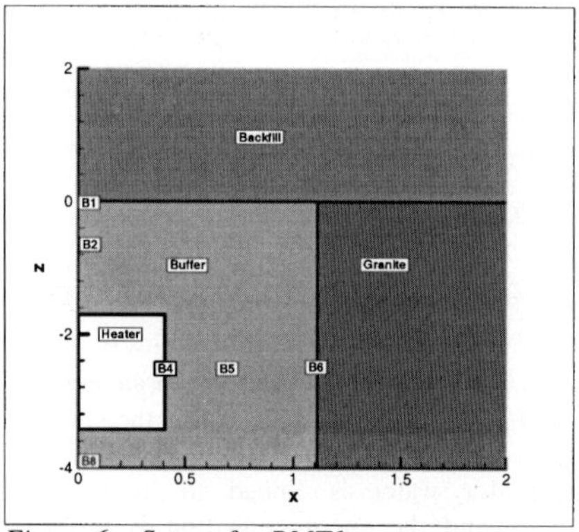

Figure 6. Set-up for BMT1.

In BMT1 a typical deep geological waste repository set-up is studied. Figure 6 shows a detail of the set-up. The waste canister is surrounded by a bentonite buffer material, and the backfill for the tunnel is made of a bentonite-sand mixture. The host rock is granite. Material

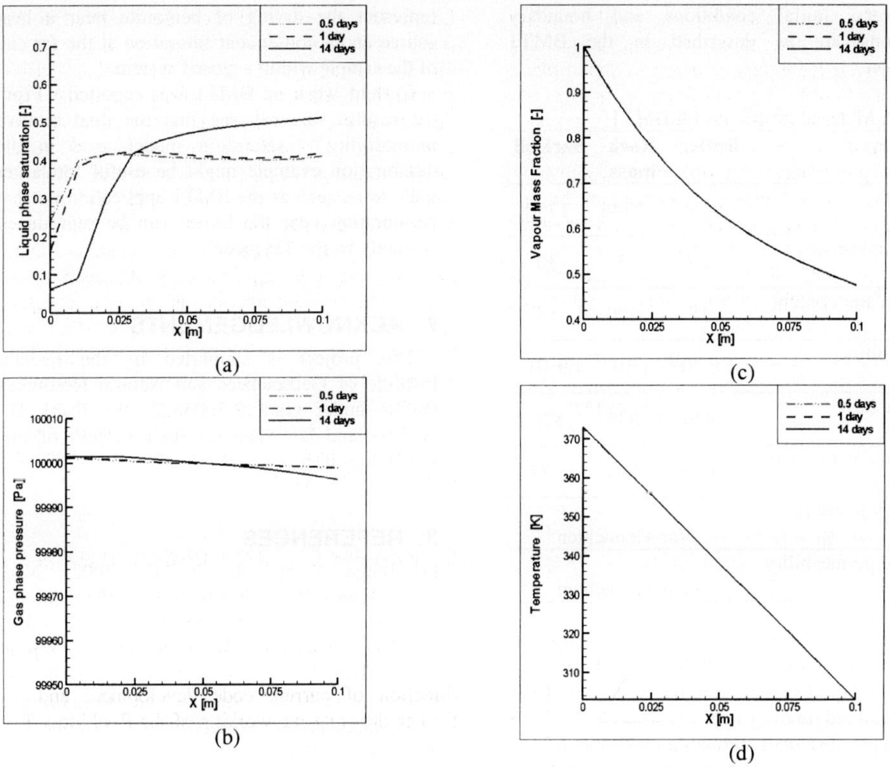

Figure 7. Results of desaturation example. (a) liquid phase saturation, (b) gas phase pressure, (c) vapour mass fraction, (d) Temperature.

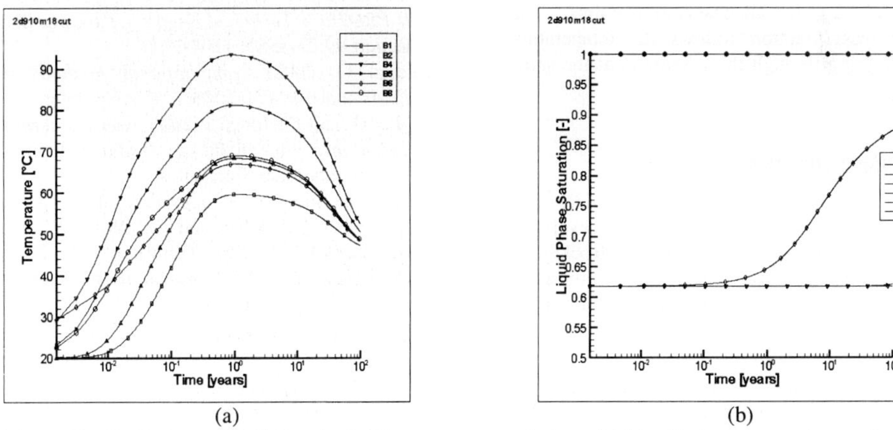

Figure 8. Preliminary BMT1 Results (a) Temperature evolution , (b) Liquid phase saturation evolution in the observation points.

parameters, initial conditions and boundary conditions are as described in the BMT1 definition.

Table 2. Material properties for BMT1

Property	Buffer	Rock mass	Backfill
Density [kg/m3]	1650	2746	1750
Effective Porosity [%]	20	3	30
Initial Water content [%]	15	100	15
Permeability [m/s]	$1.6\ 10^{-20}$	$1\ 10^{-17}$	$1.9\ 10^{-19}$
Specific heat [J/kg/K]	426	833	826
Thermal conductivity [W/m/K]	1.26	2.71	1.37
Capillary pressure [Pa]		Van Genuchten	
Relative permeability [-]		Van Genuchten	

The liquid saturation rises steadily in observation point B6. At all the other observation points the liquid saturation remains the same over the considered time.
The temperature goes through a maximum after a period of 1.5 years. At observation point B4, the maximum temperature is 78°C. After 100 years, the temperature decreases to a range between 48°C and 52°C for all observation points. The vapour mass fraction follows the temperature pattern, going through the maximum at the same time.

6. CONCLUSIONS

This paper presented the thermo-hydraulic (TH) coupled finite element model that we developed within the object-oriented simulator RockFlow. The governing equations, their numerical implementation, and the algebraic equations, as well as available coupling schemes for the RockFlow TH model were presented.

The model can now be used for the analysis of flow and heat transport processes in the near field of nuclear waste disposal. The model benefits of the componental approach, and is not restrained by the Richards approximation.

The bentonite evaporation example showed that the RockFlow TH model is able to faithfully represent the drying of bentonite near a heat source and a consequent saturation at the far end of the sample within a closed system.

Current work on BMT1 was reported. From the results, one can see that the dual relative permeability – saturation model used in the desaturation example might be useful for larger scale tests, such as the BMT1 application, so that desaturation near the heater can be reproduced correctly by the TH model.

7. ACKNOWLEDGEMENTS

This project is supported by the Federal Institute of Geosciences and Natural Resources (BGR) under grant 2-1003822. We thank Dr. Wallner and Dr. Shao for their support of this research project.

8. REFERENCES

De Jonge, J. and Kolditz, O. (2002): *Non-Isothermal Flow Processes in Porous Media, Part I: Governing Equations and Finite Element Implementation,* Rockflow Preprint [2002-1].

Emmert. M. 1997. *Numerische Modellierung nichtisothermer Gas-Wasser Systeme in porösen Medien.* PhD thesis, Insitut für Wasserbau, Universität Stuttgart.

Jing L. Stephansson O., Börgesson L. Chijimatzu M. Kautsky F. Tsang C-F. 1999. *DECOVALEX II PROJECT Technical Report – Task 2C.* SKI Report 99:23. ISSN 1104-1374

Kolditz. O. 2002. *Computational Methods in Environmental Fluid Mechanics.* Springer.

Kolditz O. and De Jonge J. 2003. *Non-Isothermal Two-Phase Flow in Porous Media.* Submitted to Computational Mechanics.

Kolditz O., Bauer S., Beinhorn M., de Jonge J., Kalbacher T., McDermott C., Wang W., Xie M., Kaiser R., and Kohlmeier M. 2003. *ROCKFLOW - Theory and users manual, release 3.9.* Groundwater modeling group, Center for Applied Geosciences, University of Tübingen, Institute of Fluid Mechanics, University of Hannover.

Olivella S. and Gens. A. 2000. *Vapour transport in low permeability unsaturated soil with capillary effects.* Transport in Porous Media, 40:219–241.

EVALUATION OF THM COUPLING ON THE SAFETY ASSESSMENT OF A NUCLEAR FUEL WASTE REPOSITORY IN A HOMOGENEOUS HARD ROCK

A. Millard[1], A. Rejeb[2], M. Chijimatsu[3], L. Jing[4], J. De Jonge[5], M. Kohlmeier[6], T.S. Nguyen[7], J. Rutqvist[8], M. Souley[9], Y. Sugita[10]
[1])CEA, France
[2]) IRSN, France
[3]) Hazama Corporation, Japan
[4]) Royal Institute of Technology (KTH), Sweden
[5]) ZAG,University of Tubingen, Germany
[6]) ISEB,University of Hannover, Germany
[7]) Canadian Nuclear Safety Commission (CNSC), Canada
[9]) Lawrence Berkeley National Laboratory (LBNL), USA
[9]) INERIS-LAEGO, France
[10]) Japan Nuclear Cycle Development Institute (JNC), Japan

Abstract: An evaluation of the importance of the thermo-hydro-mechanical couplings (THM) on the performance assessment of a deep underground storage design has been made as part of the international DECOVALEX III project. It is a numerical study that simulates a generic repository configuration in the near field in a homogeneous hard rock. A periodic pattern comprises a single vertical borehole, containing a canister surrounded by an over-pack and a bentonite layer, and the backfilled upper portion of the gallery. The thermo-hydro-mechanical evolution of the whole configuration is simulated over a period of 100 years. The importance of the rock mass intrinsic permeability has been investigated through three values : 10^{-17}, 10^{-18} and 10^{-19} m^2. Comparison of the results predicted by fully coupled THM analysis as well as partially coupled TH, TM and HM analysis, in terms of several predefined indicators, enables us to identify the couplings, which play a crucial role with respect to safety issues. The results demonstrate that temperature is hardly affected by the couplings. In contrast the influence of the couplings on the mechanical stresses is considerable.

1. INTRODUCTION

The general benchmark exercise BMT1, performed in the framework of DECOVALEX III project, has for objective the evaluation of the Thermo-Hydro-Mechanical couplings on the safety assessment of a high level nuclear wastes repository (Nguyen et al., 2003). In the present paper, attention is focused on part B of this exercise, which consists of studying the relative importance of these couplings in the near field of a nuclear waste repository, first during excavation and then over a period of 100 years of heating. A hypothetical repository has been considered, based on the one hand on a Japanese design, and on the other hand, on rock mass characteristics derived from the Canadian Shield. The repository, located at a depth of 1000 m, consists mainly of parallel galleries, in which vertical boreholes are dug in order to emplace the waste canisters (see Nguyen et al., 2003).

For this exercise, the rock mass is considered as homogeneous, the presence of fractures being taken into account in part C of the BMT1 exercise (Rutqvist et al., 2003).

With respect to the safety assessment, the importance of the couplings will be appraised from the following indicators: the evolution of the THM state of the buffer (in particular the maximum temperature, the time to re-saturation and the maximum stresses), and the rock mass permeability changes and possible failure.

2. SPECIFICATION OF THE TASK

2.1 Geometry

The homogeneity of the rock mass as well as the periodicity of the galleries and of the boreholes enables to consider only one quarter of the generic pattern, which consists in the vertical borehole and the above portion of the gallery. The dimensions are reported on Figure 1.

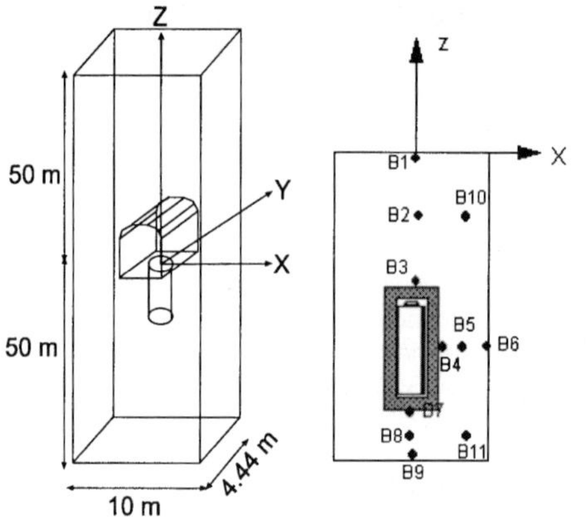

Figure 1. Dimensions of a generic pattern and position of points of interest

2.2 Initial and boundary conditions, modelling sequences

As detailed in Nguyen et al. (2003) the initial and boundary conditions are dictated by typical THM conditions for Japanese conditions, as well as symmetry arguments. The analysis was performed in two phases: an excavation phase performed to steady state, and then the phase after waste emplacement. For this latter period, the wastes, the buffer and backfill were assumed to be emplaced instantaneously at time zero. The THM response of the engineered and geological barriers would be profoundly influenced by the heat output from the wastes (see Nguyen et al. 2003).

2.3 Model and Material properties

This paper presents modelling results from five research teams for task BMT1-B: CNSC (Canada), CEA-IRSN (France), JNC (Japan), SKI (Sweden) and ISEB/ISAG (ISEB/)(Germany).

Table 1 Research team, code and model type

Research team	Code	Model type
CEA/IRSN	CASTEM 2000	3D
CNSC	FRACON	3D
JNC	THAMES	2D Axi
SKI	ROCMAS	3D
ISEB/ZAG	RF/RM	3D

All teams used a fully coupled model. The detailed descriptions of the models are given in Chijimatsu, 2003. Their main features are :

- the gaz pressure is supposed constant, equal to the atmospheric pressure;
- liquid water flow is governed by a generalized Darcy's law (or equivalent) where the conductivity depends on the degree of saturation;
- vapor flow is induced by the thermal gradient and, for SKI, also by the pressure gradient;
- the bentonite swelling is accounted for by either a 'Bishop like' effective stress or by a swelling stress function (JNC).

All teams assumed linear elastic behaviour for the different media; in addition, SKI used a no-tension model for the buffer and the backfill.

Because of the crucial role of the rock mass permeability, three values of its initial intrinsic permeability have been considered: 10^{-17}, 10^{-18} (base case) and 10^{-19} m^2. Moreover, a variation of the permeability with respect either to the porosity φ or to the effective mean stress σ_M (SKI) has been assumed:

$$K(\varphi) = 2.186\ 10^{-10}\ \varphi^3 - 5.185\ 10^{-18} \quad (1)$$

$$K(\sigma_M) = 1.5\ 10^{-15}\ \sigma_M^{-1.92} \quad \text{or}$$

$$K(\sigma_M) = 2.0\ 10^{-18}\ \sigma_M^{-0.9} \quad (2)$$

The bentonite properties have been calibrated from various tests results during DECOVALEX II (Rutqvist et al., 2001) and have been used, together with the rock mass properties, for verification in Task A of the BMT1 exercise (Chijimatsu et al., 2003).The backfill material consists of a mixture of bentonite (15%), sand and gravel.

All parameters can be found in Jing et al., 2003. Here, attention is restricted to properties which differ from one team to another, and which may thus partly explain differences in the results. These properties are listed in tables 2 and 3.

Table 2 Main different properties for buffer

Team	Young's Modulus (MPa)	Thermal expansion(°C)
CEA	150	3. 10^{-5}
CNSC	150	3. 10^{-6}
JNC	250	3. 10^{-6}
SKI	26	3. 10^{-5}
ISEB	100	8.21 10^{-6}

Table 3 Main different properties for rock mass

Team	Permeability range (m^2)	Biot's coefficient
CEA	10^{-17} - 10^{-19}	1.
CNSC	10^{-18} - 10^{-19}	0.2
JNC	10^{-18} - 10^{-20}	1
SKI	10^{-17} - 10^{-19}	0.5

Some crucial data, such as water retention characteristics and relative water permeability functions were represented by the different teams according to their engineering judgment.

3. THM RESULTS

In this section, some of the results predicted by a fully coupled THM analysis are presented, in terms of the predefined indicators.

3.1 Temperature in the buffer

The maximum temperature is reached at the contact with the over-pack, after about 30 years of heating. Three-dimensional models predict a value within the range of 75 to 80°C, whereas the axisymmetric model (JNC) leads to about 90°C. The evolution of temperature at point B4 is summarized on Figure 2. On this figure, the ISEG curve relates to a purely TM calculation.

As might be expected, the value of the initial intrinsic rock mass permeability has a low influence on these results (Figure 3).

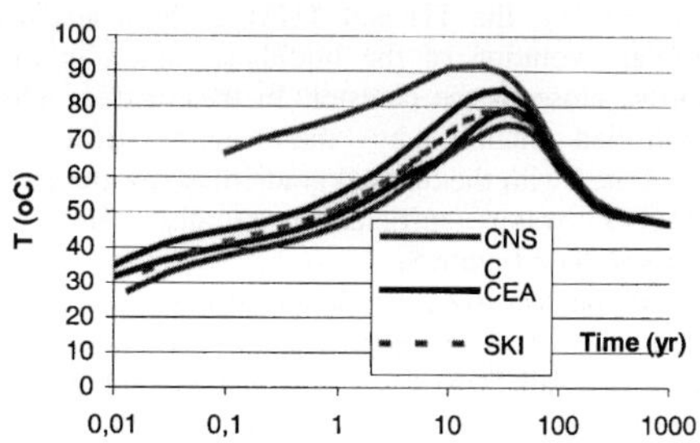

Figure 2 . Evolution of temperature at B4 (base case, k = 10^{-18} m^2)

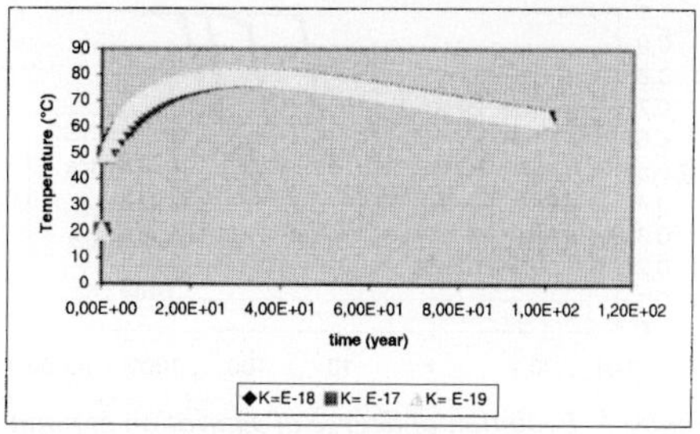

Figure 3. Evolution of temperature at B4 for different rock mass permeability (CEA)

3.2 Degree of saturation in the buffer

The thermal gradient induced by the heat generated from the wastes causes first a de-saturation close to the heater, followed by a re-saturation controlled by the rock mass permeability. For the base case, the predicted time necessary for a full re-saturation of the buffer, varies between 7 years (CEA) and 55 years (SKI) for the three dimensional models. The evolution of the degree of saturation at point B4 is plotted on Figure 4.

As expected, it strongly depends upon the rock-mass permeability, as can be seen on Figure 5 for the low permeability case.

3.3 Stresses in the buffer

A good indication of the stress level at B4 is given by the horizontal total stress σ_{XX} (radial direction), depicted on figure 6. This stress incorporates the effects of the temperature increase, the shrinking and swelling due to de-saturation and re-saturation, and finally the pore pressure build-up. In fact, this latter effect is clearly the largest one, as indicated by the results from ISEB, which do not take into account the pore pressure effect, and are less than 1 MPa.

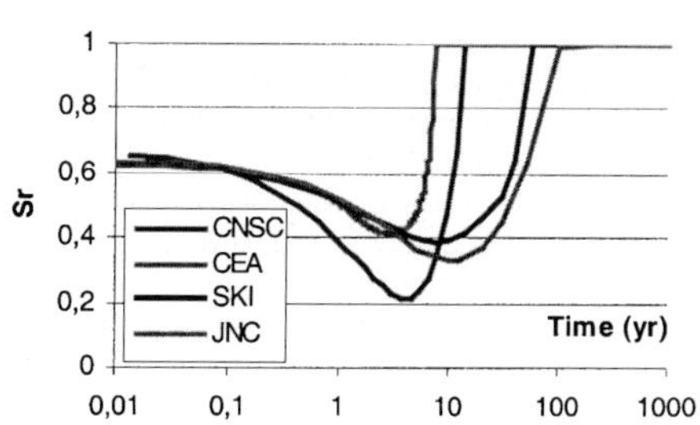

Figure 4. Evolution of degree of saturation at point B4 (base case, k = 10^{-18} m^2)

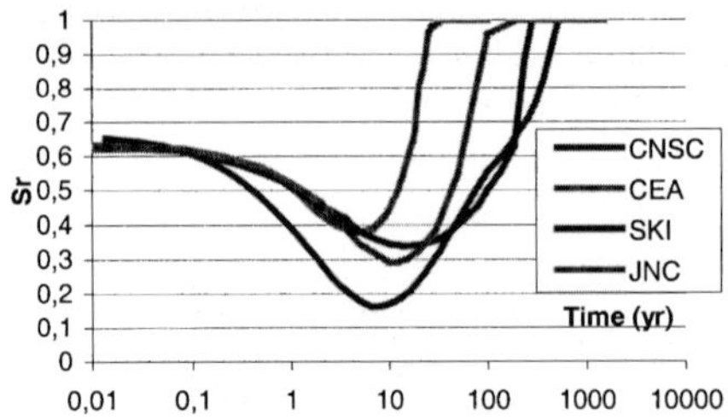

Figure 5. Evolution of degree of saturation at point B4 for low rock mass permeability (10^{-19} m^2)

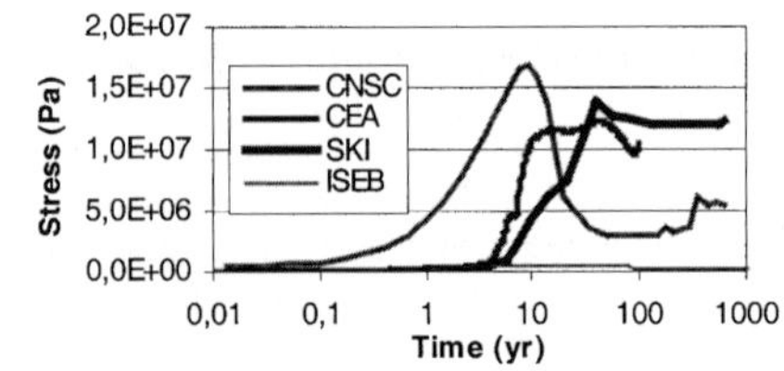

Figure 6. Evolution of total radial stress σ_{XX} at point B4 for base case

3.4 Rock failure

Possible rock failure has been investigated by means of the Hoek and Brown's failure criterion:

$$F = \sigma'_1 - \sigma'_3 - (m\,\sigma_c\,\sigma'_3 + s\,\sigma_c^2)^{0.5} \quad (3)$$

Where σ'_1 and σ'_3 are the maximum and minimum principal effective stress (in the above formula, compression is taken positive, and the principal effective stresses are supposed to be positive), σ_c is the uniaxial compressive stress equal to 123 MPa, and m and s are two constants, m = 17.5, s = 0.19.

CEA, CNCS and SKI teams predicted failure at excavation, mainly due to traction stresses. Moreover, compressive failure was also found later on, by CEA from 40 years on, associated with the thermal peak, and by SKI at about 100 years.

4. EVALUATION OF COUPLINGS

In this paragraph, the importance of the couplings on the safety indicators is evaluated by comparing the THM calculations to the partially decoupled ones TH, TM and HM. However, all calculations start from the same initial conditions. The influence of the initial rock mass permeability on the results has been investigated as well.

4.1 Effect on temperature

A very limited effect is found on temperature: the maximum difference, arising close to the canister (point B4) and for the low permeability, is due to the variation of thermal conductivity induced by the changes in saturation. An illustration is given on Figure 7 for points B4 and B6, for the base case. From these results, it appears that, in the present analysis, the temperature distribution can be reasonably predicted by a single uncoupled thermal analysis.

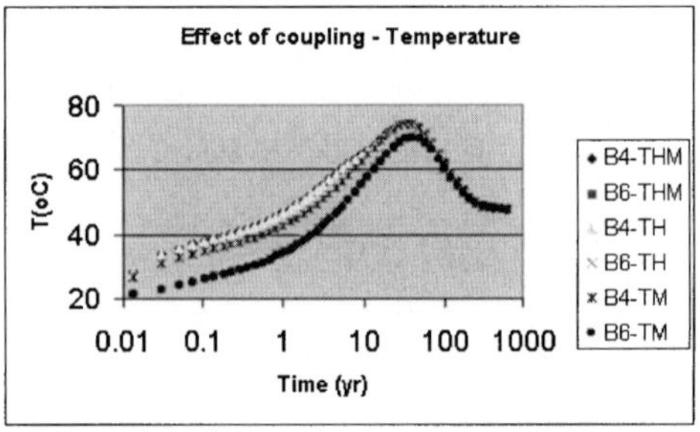

Figure 7. Effect of couplings on temperature at points B4 and B6 (base case, CNSC)

4.2 Effect on saturation on pore pressure

The situation is completely different for the saturation. In fact, for medium or high rock mass permeability, the TH and THM analysis lead to similar evolution of the liquid saturation in the buffer, close to the canister. In these cases, it is controlled mainly by the vapour diffusion associated with the thermal gradient, while the rock supplies water sufficiently rapidly. This is illustrated on Figure 8.

For the low rock permeability case, a correct prediction of the pore pressure in the rock mass requires a full THM analysis, as shown on Figure 9 for point B6 of the rock adjacent to the buffer. At early times, due to its low permeability, the rock could not supply water to the buffer at a sufficient rate, leading to a drop in pore pressure. Rebuilding of the hydrostatic pressure starts at about 30 years for the THM case, and about 60 years for the HM case, while it is still unsaturated after 100 years in the TH case.

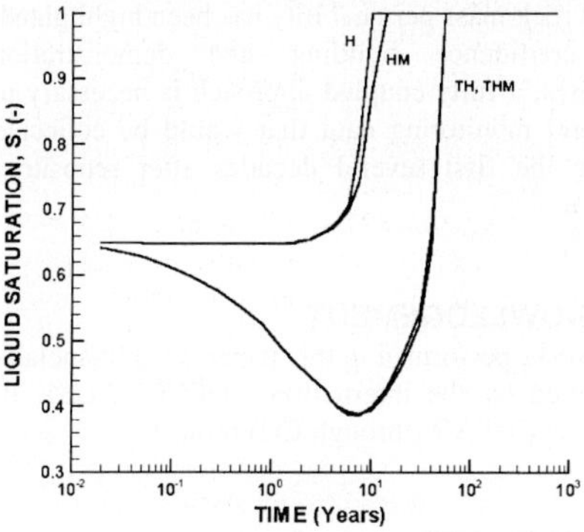

Figure 8. Evolution of liquid saturation at B4, for the base case (SKI).

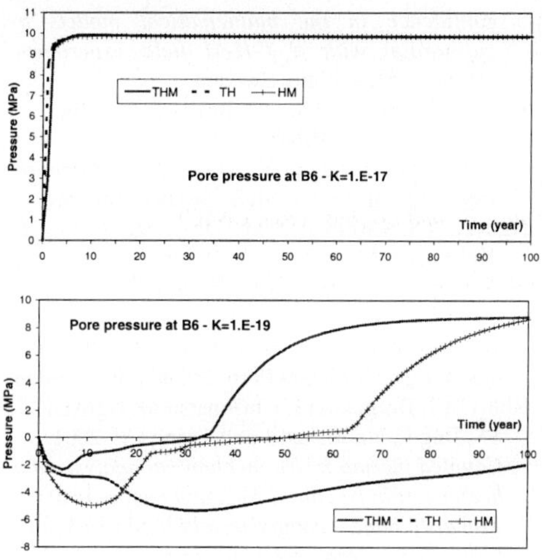

Figure 9. Pore pressure at point B6 (CEA). Influence of rock permeability.

4.3 Effect on total stresses

As might be expected, couplings have a strong effect on the total stresses, on one hand because of the thermal stresses (of the order of magnitude of 0.1 MPa), and on the other hand because of the swelling pressure and the hydrostatic pressure contributions; the predominance of this latter effect (of the order of magnitude of 10 MPa), is confirmed in Figure 10.

The same conclusion can be drawn for all three values of permeability, when looking for example at point B6 (Figure 11).

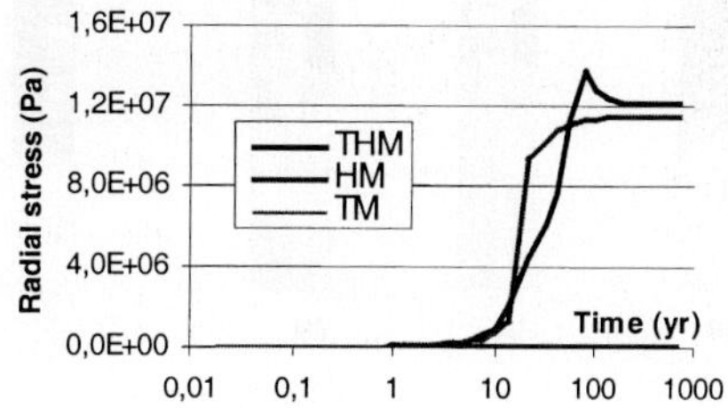

Figure 10. Total radial stress at point B4 (base case SKI)

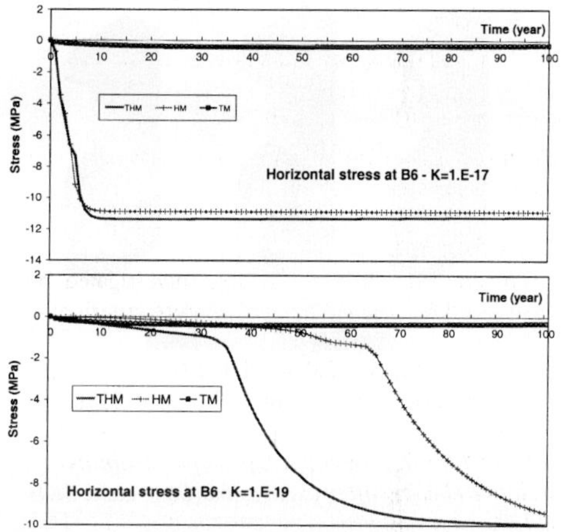

Figure 11. Horizontal total stress at point B6 (CEA). Influence of rock permeability. (Here, tractions are considered positive)

4.4 Effect on rock failure and rock permeability changes

Failure is conditioned to the effective stress state, according to equation (3). Figure 12, which compares the safety factor (defined by the ratio F/σ'_1) after excavation, shows the necessity to consider full THM couplings when studying the rock mass stability (Failure corresponds to a safety factor lower than 1, in dark blue on Figure 12).

Concerning the rock permeability changes, the influence of couplings is not so pronounced for medium and high initial permeability, the variations being within one order of magnitude (Figure 13). Conversely, in the case $K=10^{-19}$ m^2, a large influence is obtained.

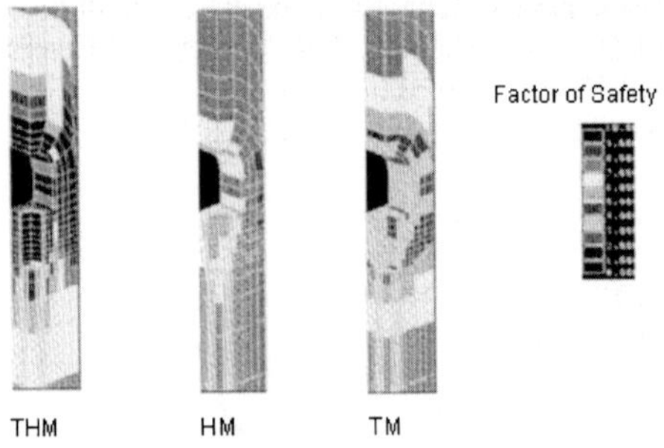

Figure 12. Effect of couplings on rock mass stability at excavation (base case, CNSC)

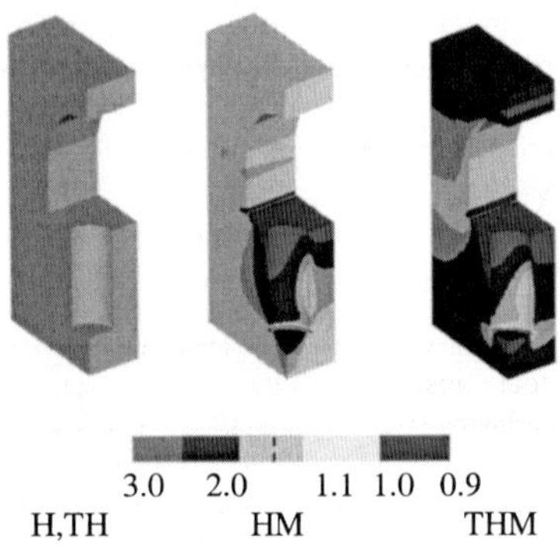

Figure 13. Effect of couplings on permeability relative change, after excavation (base case, SKI)

5 CONCLUSION

From these comparative studies, and in the particular case considered here, it has been possible to evaluate the importance of the various T-H-M couplings on the safety assessment of typical repository, in the near field. The major findings are summarized in table 4.

Table 4 Effect of T-H-M couplings on near field repository response

Indicator	THM	TH	TM	HM
Temperature	low	low	low	-
Resaturation	medium/ high	medium	-	low
Swelling stress	high	-	low	high
Rock failure	high	-	low	high
Permeability change	medium/ high	medium	medium	low

In addition to this table, the importance of the initial rock mass permeability has been highlighted. For confidence building and demonstration purposes, a fully coupled approach is necessary to interpret monitoring data that would be collected during the first several decades after repository closure.

ACKNOWLEDGEMENT

The works performed in this paper were financially supported by the international DECOVALEX III and BENCHPAR (through EU) projects.

REFERENCES

Chijimatsu M, Giraud A, Jing L, Millard A, Nguyen TS, Rejeb A, Rutqvist J, Souley M, Sugita Y and Thoraval A. 2003. *Building confidence in the mathematical models by calibration with a T-H-M field experiment.* Geoproc, Stockholm, Sweden

Jing L, Nguyen TS, Chijimatsu M, Jing L, Kohlmeier M, Millard A, Rejeb A, Rutqvist J, Souley M and Sugita Y. 2003. *BMT1-B Reportl.* Decovalex III report, SKB, Sweden. (in press).

Nguyen TS, Chijimatsu M, De Jonge J, Jing L, Kohlmeier M, Millard A, Rejeb A, Rutqvist J, Souley M and Sugita Y. 2003. *Implications of coupled Thermo-Hydro-Mechanical processes on the safety of a hypothetical nuclear fuel waste repository.* GeoProc, Stockholm, Sweden

Rutqvist J, Börgesson J, Chijimatsu M, Nguyen TS, Jing L, Noorishad J and Tsang CF. 2001. *Coupled thermo-hydro-mechanical analysis of a heater test in fractured rock and bentonite at Kamaishi mine – comparison of field results to predictions of four finite element codes.* Int. J. of Rock Mechanics and Mining Sciences, Vol. 38, No.1, pp.129-142

Rutqvist J, Chijimatsu M, De Jonge J, Jing L, Kohlmeier M, Millard A, Nguyen TS, Rejeb A, Souley M, Sugita Y. 2003. *Scoping calculation for a fractured rock medium.* GeoProc, Stockholm, Sweden (this proceedings).

EVALUATION OF THE IMPACT OF THERMAL-HYDROLOGICAL-MECHANICAL COUPLINGS IN BENTONITE AND NEAR-FIELD ROCK BARRIERS OF A NUCLEAR WASTE REPOSITORY IN SPARSELY FRACTURED HARD ROCK

J. Rutqvist[1,2], M. Chijimatsu[3], L. Jing[2], A. Millard[4], T.S. Nguyen[5], A. Rejeb[6], Y.Sugita[7] and C.F. Tsang[1]

1) Lawrence Berkeley National Laboratory, USA
2) Royal Institute of Technology, Sweden
3) Hazama Corporation, Japan
4) Commissariat a l'Energie Atomique, France
5) Canadian Nuclear Safety Commission, Canada
6) Institut de Radioprotection et de Sûreté Nucléaire, France
7) Japan Nuclear Cycle Development Insititute, Japan

Abstract: As part of the international DECOVALEX III project and the European BENCHPAR project, this paper evaluates the impact of thermal-hydrological-mechanical (THM) couplings on the performance of a bentonite back-filled nuclear waste repository in sparsely fractured hard rock. The significance of THM coupling on the performance of a hypothetical repository is evaluated by several independent coupled numerical analyses. Moreover, the influence of a discrete fracture intersecting a deposition hole is discussed. The analysis shows that THM couplings have the most impact on the mechanical behaviour of bentonite-rock system, which is important for repository design considerations.

1 INTRODUCTION

This paper evaluates the impact of thermal-hydrological-mechanical (THM) couplings on the performance of a bentonite-back-filled nuclear waste repository in sparsely fractured hard rock. The analysis was carried out as a part of the DECOVALEX III and BENCHPAR projects. DECVALEX III is an international co-operative research project for the development of coupled models and their validation against experiments, whereas BECHPAR is a project supported by the European Union aimed at improving the THM coupled processes content of radioactive waste repository performance assessment. In BENCHPAR/DECOVALEX III, seven research teams independently performed coupled THM analysis of a hypothetical repository, defined in Nguyen et al., (2003). The hypothetical repository consists of horizontal drifts, which are back-filled with a rock-bentonite mixture and vertical deposition holes, in which the waste canisters are embedded in bentonite (Figure 1).

This paper focuses on a repository located in sparsely fractured rock with a hydraulic conducting horizontal fracture intersecting the vertical deposition hole (Figure 1b). The analysis for the case of a homogenous intact rock (Figure 1a) is presented in Millard et al., (2003). This paper present results of coupled THM analyses conducted by four research teams: Royal Institute of Technology (KTH), Canadian Nuclear Safety Commission (CNSC), Commissariat a l'Energie Atomique de Cadarache (CEA) and Japan Nuclear Fuel Cycle Development Institute (JNC). Computer codes used and their sources are listed in Table 1.

Table 1. Research Teams, codes and their sources

Team	Code	Reference
KTH	ROCMAS	Rutqvist et al. (2001a)
CEA	Castem2000	Verpeaux et al (1989)
CNSC	FRACON	Nguyen (1996)
JNC	THAMES	Ohnishi and Kobayashi (1996)

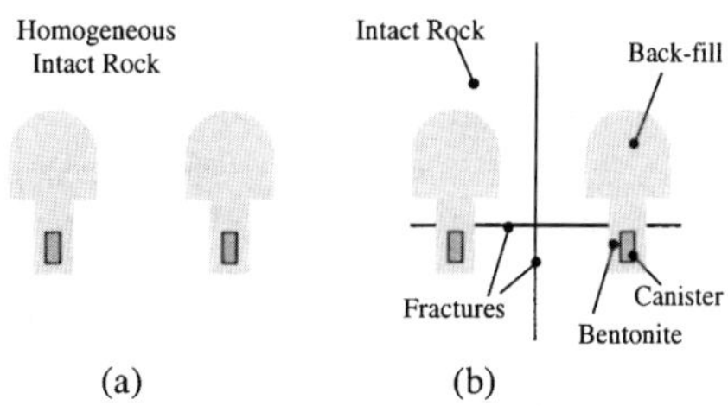

Figure 1. (a) Intact homogenous rock versus (b) sparsely fractured rock cases.

2 MODEL CONCEPTUALIZATION

The repository geometry is based on the Japanese H12 project (JNC, 2000). Because of repetitive symmetry, the simulations were conducted on a one-quarter symmetric model containing one deposition hole (Figure 2). The upper and lower boundaries are placed at vertical distances of 50 m from the drift floor according to the BMT1 definition (Nguyen et al., 2003).

Most of the material properties for bentonite and the rock were extracted or developed during DECOVALEX II for modelling of the Kamaishi Mine heater test (Rutqvist et al., 2001b). Bentonite properties were further calibrated for improved model representation of *in situ* THM responses at Kamaishi Mine (Chijimatsu et al., 2003). The rock matrix represents intact granite with a permeability varying between 1×10^{-19} and 1×10^{-17} m^2, a Young's modulus of about 60 GPa, and a thermal expansion coefficient of about 8×10^{-6} 1/°C.

The mechanical and hydromechanical properties of the horizontal rock fractures were estimated using the Barton-Bandis' Joint model (Barton and Bakhtar, 1983). In this context, the void aperture, b_v, is defined as the accessible volume per unit area of a fracture. The void aperture at a given effective normal stress is

$$b_v = b_{vr} + \Delta b_v \quad (1)$$

where b_{vr} is a residual void aperture when the fracture is completely compressed from a mechanical point of view, and Δb_v is mechanically induced void aperture for an incompletely compressed fracture.

The mechanically induced void aperture can be related to the current effective normal stress, σ'_n, using Barton-Bandis' hyperbolic normal closure model according to

$$\Delta b_v = \frac{k_{n0} V_{m0}^2}{\sigma'_n + k_{n0} V_{m0}} \quad (2)$$

where k_{n0} and V_{m0} is the normal stiffness and maximum normal closure at the zero stress intercept (Barton and Bakhtar, 1983). The parameters k_{n0} and V_{m0} were estimated using the basic parameter values $JRC_0 = 9$, $JCS_0 = 105$ MPa and $\sigma_c = 123$ MPa (extracted from the Kamaishi Mine data set). Using formulas presented in Barton and Bakhtar (1983), $k_{n0} = 56$ GPa/m and $V_{m0} = 65$ μm were derived for the fourth loading cycle.

For the hypothetical case presented in this analysis we assume $b_h = 0.85\times b_v$, where b_h is the hydraulic aperture defined from the parallel plate flow relationship

$$T_f = \frac{b_h^3 \rho g}{12\mu} \quad (3)$$

Figure 3 presents the resulting relationship between fracture transmissivity and effective normal stress. This function represents a fracture whose initial aperture is 10 μm at an initial effective stress of 17 MPa, corresponding to the initial vertical effective stress across the horizontal fracture.

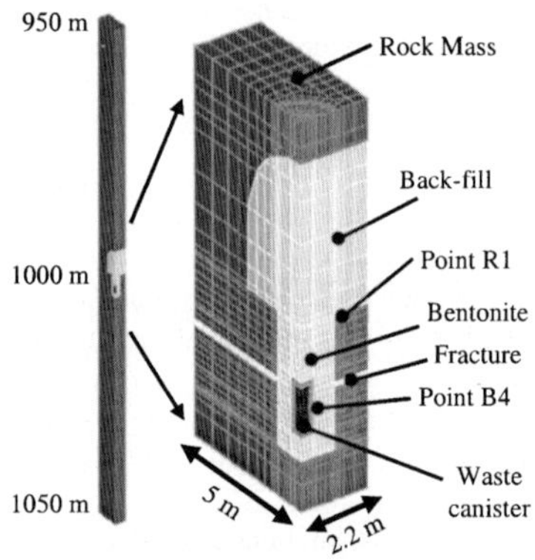

Figure 2. Quarter symmetric model of the hypothetical repository located at 1000 meters depth (KTHs model).

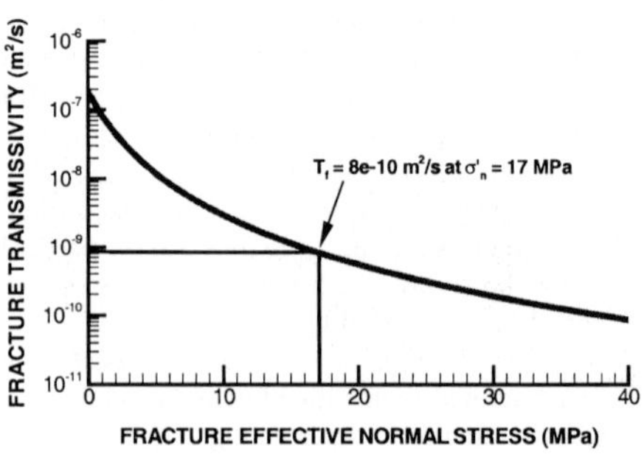

Figure 3. Fracture transmissivity versus effective fracture stress derived for $k_{n0} = 56$ *GPa/m and* $V_{m0} = 65$ *μm.*

The mechanical integrity of the rock mass is estimated using Hoek and Brown's failure criterion, which can be written as (Hoek and Brown, 1997):

$$\sigma_1' = \sigma_3' + \sigma_c \left(m_b \frac{\sigma_3'}{\sigma_{ci}} + s \right)^a \tag{4}$$

where σ'_1 and σ'_3 are the maximum and minimum effective stresses at failure respectively, m_b is the values of Hoek-Brown constant m for the rock mass, s and a are constants which depends upon the characteristics of rock mass, and σ_c is the uniaxial compressive strength of the intact rock pieces. In this simulation, the empirical constants have been estimated to $a = 0.5$, $m_b = 17.5$ and $s = 0.19$, which represents a very good quality rock mass of sparsely fractured granite.

3 THM MODEL RESULTS

Figure 4 summarises the general results of THM analyses by CEA, CNSC, KTH and JNC. Results of time evolution at selected points (B4 in the buffer and R1 in the rock) are shown in Figures 5 and 6.

Figure 5a shows that the maximum temperature of about 75 to 80°C is reached at the surface of the waste canister after about 30 to 40 years. The slightly lower peak temperature obtained by CNSC appears to be related to a lower initial temperature.

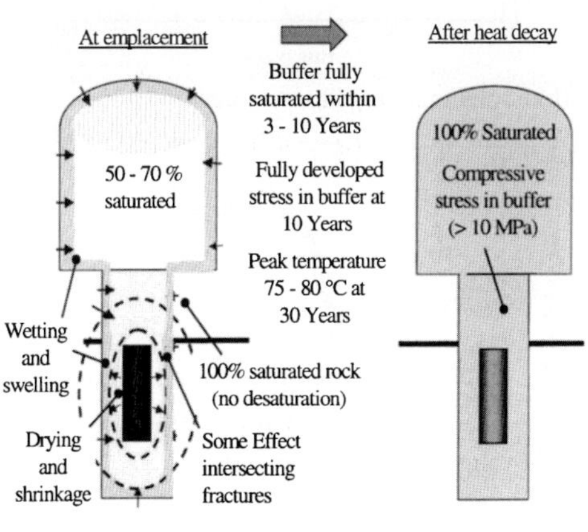

Figure 4. General results of coupled THM simulations of a hypothetical repository in sparsely fractured rock

Figure 5b shows that the time to full resaturation of the buffer varies between 3 to 10 years. The different resaturation times among these models are in line with simulation results obtained for the well-defined axisymmetric problem (Chijimatsu et al., 2003) and for the case of a homogenous rock (Millard et al., 2003). This indicates that the variation of the resaturation time among these models are caused by slightly different input hydraulic properties of the buffer, such as relative permeability and water retention curves.

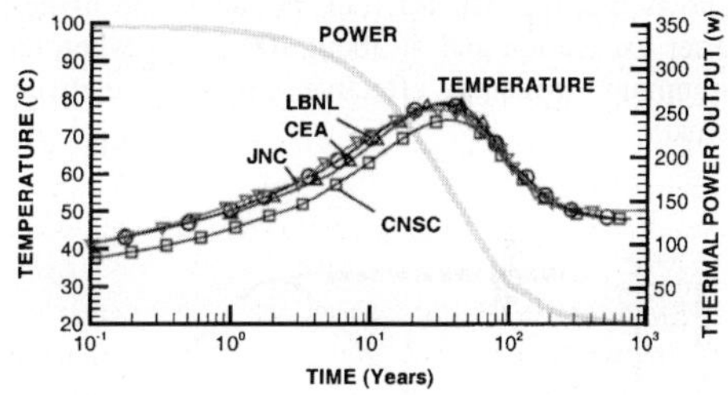

a) Heat power and temperature evolution

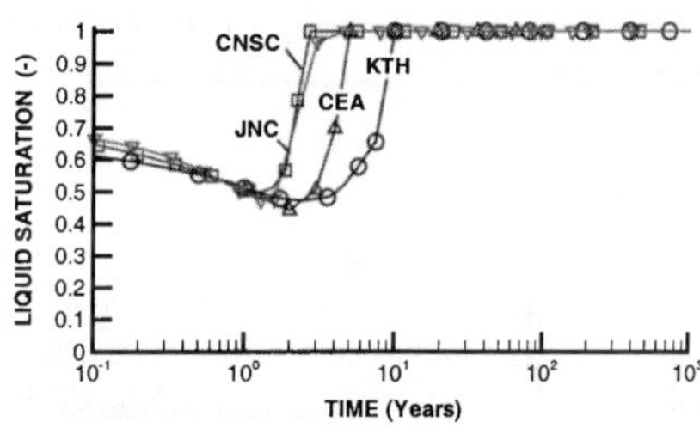

b) Evolution of liquid saturation

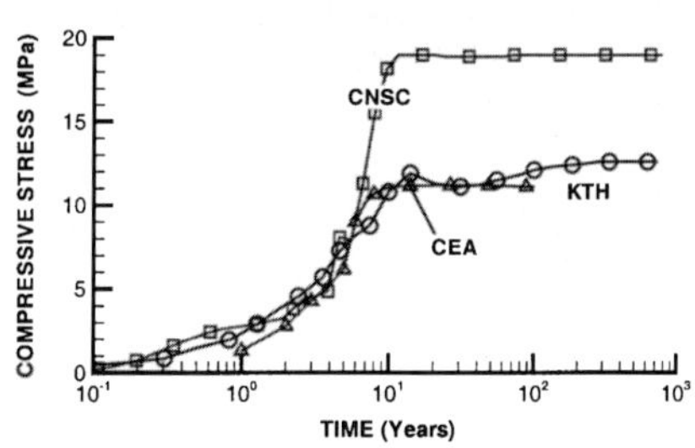

(c) Evolution of compressive stress (total stress)

Figure 5. Simulation results by CEA, CNSC, JNC and KTH for point B4 located in the buffer at the surface of the waste canister as shown in Figure 2.

The simulated evolution of compressive stress in the buffer is very similar for CEA and KTH, whereas CNSC obtained a substantially higher compressive stress (Figure 5c). The results by CEA and KTH indicate that the impact of the fluid pressure restoration is a dominant component contributing about 90% of the total compressive stress.

Figure 6 presents the evolution of stress in the rock at point R1, which is a critical location for the evaluation of rock failure. Figure 6a shows that the maximum compressive principal effective stress is 93 MPa at about 40 years. However, Figure 6b shows that the risk for rock failure is the highest after excavation and at about 100 years, when the minimum principal effective stress is relatively small.

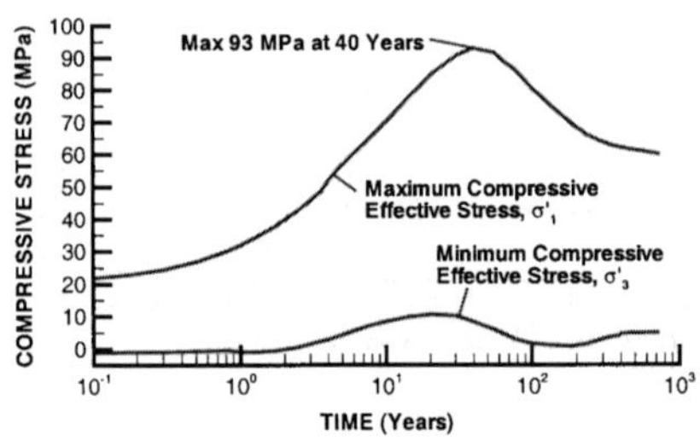

(a) Time evolution of σ'_1 and σ'_3

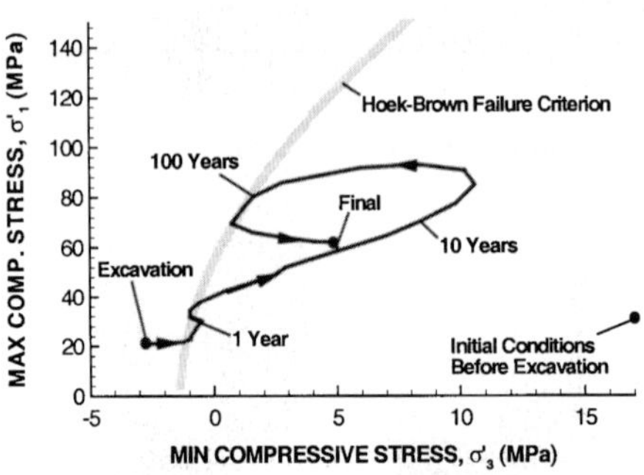

(b) Principal stress path and failure envelop

Figure 6. Simulation results by KTH showing evolution of maximum and minimum principal compressive effective stresses for point R1 located in the rock just below drift floor as shown in Figure 1.

4 IMPACT OF THM COUPLINGS

Comparisons partially coupled TH, HM and TM solutions with a fully coupled THM solution are utilised to study the impact of various couplings. Figure 7a shows that there is a small impact of TH couplings on the temperature field until the buffer is fully saturated. This slight impact of TH coupling is caused by the dependency of thermal conductive on liquid saturation. However, there is no visible impact of THM couplings on the peak temperature.

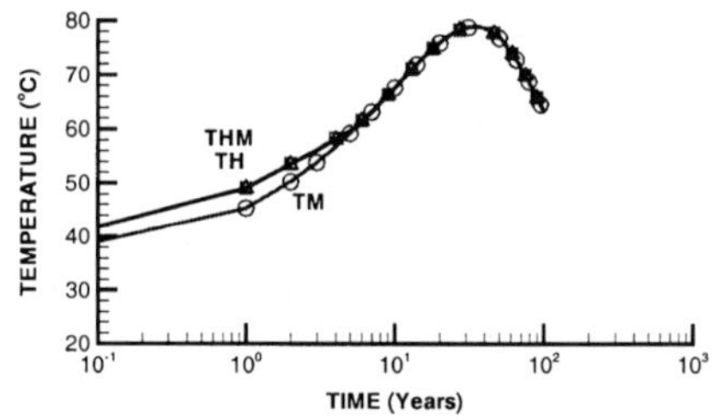

(a) Temperature evolution

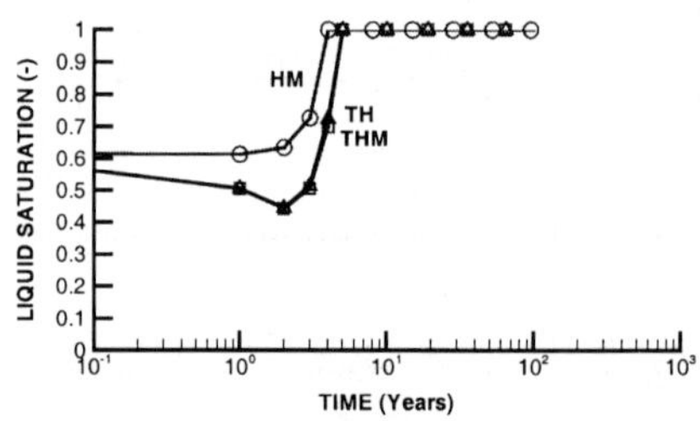

(b) Evolution of liquid saturation

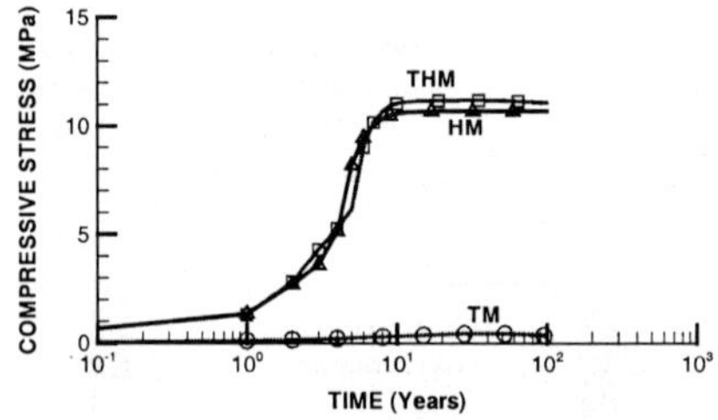

(c) Evolution of compressive stress

Figure 7. Simulation results by CEA showing the impact of HM, TH, TM and THM coupling for point B4 located in the buffer at the surface of the waste canister as shown in Figure 2.

There is a slight impact of TH coupling on the resaturation of the buffer (Figure 7b). In TH and THM simulations, the resaturation of the bentonite buffer is slightly delayed because of thermally induced drying near the canister surface.

Figure 7c shows that there is a strong impact of HM coupling on the total stress in the buffer. The total stress developed in this buffer material is dominated by fluid pressure, which at 1,000 meters depths develops to 10 MPa for final hydrostatic conditions. The magnitude of TM-induced stresses is less than 1 MPa.

Figure 8a shows that the TM coupling has a dominant impact on the evolution of the maximum principal compressive stress at point R1. The evolution of the minimum principal stress is more complex, with some degree of impact from both TM and HM couplings. Nevertheless, the results in Figure 8 shows that full THM analysis must be conducted for an appropriate evaluation of the evolution of stress in the near field rock.

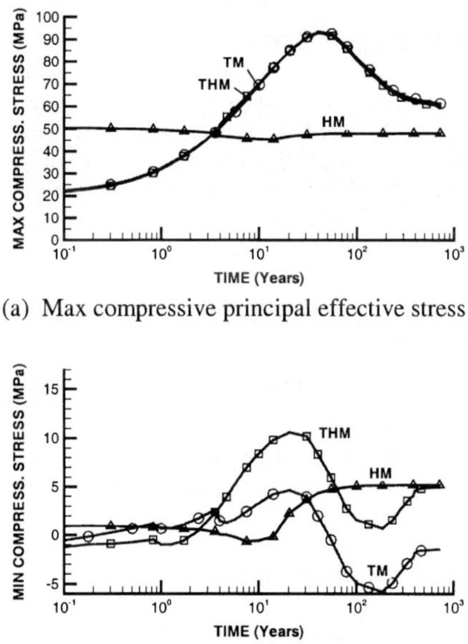

(a) Max compressive principal effective stress

(b) Min compressive principal effective stress

Figure 8. Simulation results by KTH showing the impact of HM, TM and THM couplings on the principal effective stress evolution for point R1 located in the rock just below drift floor as shown in Figure 1.

5 DISCUSSION ON THE IMPACT OF THE INTERSECTING FRACTURE

A comparison of simulation results for the homogenous intact rock (Figure 1a) and sparsely fractured rocks (Figure 1b) shows that the main effect of the fractures is an accelerated resaturation of the buffer leading to a much shorter resaturation time. For the case of an unfractured low permeability rock (k=1e-19 m^2), the time to full resaturation ranged from 40 to 200 years (Millard et al., 2003). By introducing hydraulic conducting fractures, the resaturation time is reduced to 3 to 10 years (Figure 7b).

The impact of fractures is illustrated by the simulation results of KTH shown in Figure 9. For the case of a low rock matrix permeability, desaturation of the rock surrounding the repository delays the resaturation of the buffer (Figure 9a). As seen in Figure 9b, there is no such desaturation of the rock when the nearby fractures are included.

Figure 10 presents the evolution of the maximum principal compressive stress in the buffer and back-fill. Figure 10 shows that stress begins to develop near the intersecting fracture and then in the lower parts of the buffer. This result shows that an intersecting fracture impacts the stress evolution and its spatial distribution in the buffer.

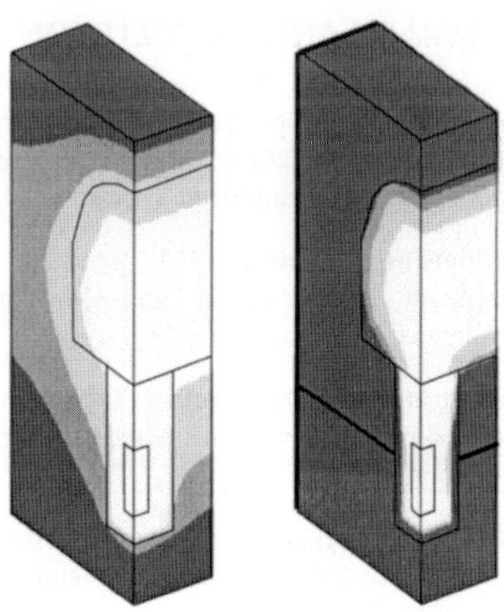

(a) Intact rock (b) Fractured rock

Figure 9. Liquid saturation at 1 year from KTH simulation with ROCMAS. Darkest contour indicates fully saturated conditions whereas lightest contour indicates less than 70% saturation.

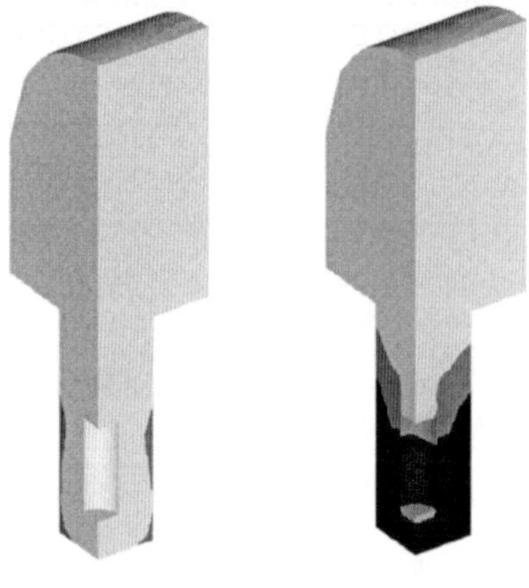

(a) 1 year (b) 10 years

Figure 10. Evolution of maximum compressive stress in the buffer and back-fill from KTH simulation with ROCMAS. Light contour indicates almost stress free conditions whereas the darkest contour indicate a fully developed compressive stress close to 10 MPa.

6 SUMMARY AND CONCLUSION

This analysis aimed to evaluate the impact of THM couplings on the performance of a repository located in sparsely fractured rock. The results of this analysis can be summarised as follows:

- **Temperature evolution (T process)** : No significant effect of H and M coupling (conduction dominates)
- **Resaturation of the buffer (H process)** : Affected by T coupling but not significantly by M coupling
- **Stress evolution in the buffer (M process)**: Strongly affected by H coupling and slightly affected by T coupling.
- **Stress evolution in rock for stability and design considerations (M process)**: Strongly affected by both T and H coupling.

It is clear that the temperature can be predicted accurately without consideration of coupling to hydraulic and mechanical processes. It is also clear that mechanical behaviour, that is, evolution of stress in the buffer-rock system, cannot be appropriately predicted without consideration of temperature effects and effects of fluid pressure. It is not clear at this point whether the hydraulic behaviour (for example resaturation of the buffer and radioactive nuclide transport) can be significantly impacted by T and M processes. For the parameter set adopted in this analysis, the resaturation time is slightly impacted by the effect of temperature whereas the mechanically induced changes in permeability does not significantly impact the resaturation process.

The general results of the impact of various THM couplings for sparsely fractured rocks conducted in this paper are in line with those of a homogenous low permeability rock (Millard et al., 2003). The main difference is that the hydraulic conducting fractures provide an additional water supply that prevents desaturation of the rock and accelerates the buffer resasuration process.

ACKNOWLEDGEMENTS

The following organisations are gratefully acknowledged for their financial support (alphabetical order): the Canadian Nuclear Safety Commission; the Commissariat a l'Energie Atomique; the European Commission through the BENCHPAR project under Contract FIKW-CT-2000-00066; the Institut de Radioprotection et de Sûreté Nucléaire; the Japan Nuclear Fuel Cycle Development Institute, the Swedish Nuclear Power Inspectorate.

REFERENCES

Barton, N.R. & Bakhtar, K. 1983. *Rock joint description and modeling for the hydrothermomechanical design of nuclear waste repositories.* Technical Report 83-10, TerraTek Engineering, Salt Lake City, Utah.

Chijimatsu, M., Giraud, A., Jing, L., Millard, A., Nguyen, T.S., Rejeb, A., Rutqvist, J., Souley, M., Sugita, Y. & Thoraval, A. 2003. Building *Confidence in the mathematical models by calibration with a T-H-M field experiment.* GEOPROC 2003, Stockholm 13-15 October, 2003.

Hoek, E. & Brown, E.T. 1997. *Practical estimates of rock mass strength.* Int. J. Rock mech. Min. Sci. 34: pp. 1165-1186.

JNC. 2000. *H12: Project to Establish the Scientific and Technical Basis for HLW Disposal in Japan. Second progress report on research and development for the geological disposal of HLW in Japan.* Five volumes. Japan Nuclear Cycle Development Institute (JNC).

Millard, A., Rejeb, A., Chijimatsu, M., Jing, L., De Jonge, J., Kohlmeier, M., Nguyen, T.S., Rutqvist, J., Souley, M. & Sugita, Y. 2003. *Evaluation of thm coupling on the safety assessment of a nuclear fuel waste repository in a homogeneous hard rock.* GEOPROC 2003, Stockholm 13-15 October, 2003.

Nguyen, T.S. 1996. Description of the computer code FRACON. In Stephansson, O., Jing, L., and Tsang, C.-F. editors. *Coupled Thermo-hydro-mechanical Processes of Fractured Media.* Developments in Geotechnical Engineering, Elsevier, 79: pp. 539–544.

Nguyen, T.S., Chijimatsu, M., De Jonge, J., Jing, L., Kohlmeier, M., Millard, A., Rejeb, A., Rutqvist, J., Souley, M. &. Sugita,Y. 2003. *Implications of coupled thermo-hydro-mechanical processes on the safety of a hypothetical nuclear fuel waste repository.* GEOPROC 2003, Stockholm 13-15 October, 2003.

Ohnishi, Y. and Kobayashi, A. 1996. THAMES. In Stephansson, O., Jing, L., and Tsang, C.-F. editors. *Coupled Thermo-hydro-mechanical Processes of Fractured Media.* Developments in Geotechnical Engineering, Elsevier, 79: pp. 545–549.

Rutqvist, J., Börgesson, L., Chijimatsu, M., Kobayashi, A., Nguyen, T.S., Jing, L., Noorishad, J., & Tsang, C.-F. 2001a. *Thermohydromechanics of Partially Saturated Geological Media – Governing Equations and Formulation of Four Finite Element Models.* Int. J. Rock mech. Min. Sci. 38: pp. 105-127.

Rutqvist, J., Börgesson, L., Chijimatsu, M., Nguyen, T.S., Jing, L., Noorishad, J., and Tsang, C.-F. 2001b.*Coupled Thermo-hydro-mechanical Analysis of a Heater Test in Fractured Rock and Bentonite at Kamaishi Mine – Comparison of Field Results to Predictions of Four Finite Element Codes.* Int. J. Rock mech. Min. Sci. 38: pp. 129-142.

Verpeaux, P., Millard, A., Charras, T. & Combescure, A. 1989. *A modern approach of large computer codes for structural analysis.* Proc. SmiRT conf., Los Angeles, USA

IMPLICATIONS OF COUPLED THERMO-HYDRO-MECHANICAL PROCESSES ON THE SAFETY OF A HYPOTHETICAL NUCLEAR FUEL WASTE REPOSITORY

T.S. Nguyen[1], M. Chijimatsu[2] , J. De Jonge[3] , L. Jing[4], M. Kohlmeier[5], A. Millard[6], A. Rejeb[7], J. Rutqvist[8], M. Souley[9] and Y. Sugita[10]
1)Canadian Nuclear Safety Commission, Canada
2) Hazama Corporation, Japan
3) University of Tuebingen, Germany
4) Royal Institute of Technology, Sweden
5) University of Hannover, Germany
6) Commissariat a l'Energie Atomique, France
7) Institut de Radioprotection et de Sûreté Nucléaire, France
8) Lawrence Berkeley Laboratory, USA
9) Ecole des Mines de Nancy, France
10) JNC, Japan

Abstract: In Bench Mark Test no. 1 (BMT1) of the DECOVALEX III international project, we looked at the implications of coupled thermo-hydro-mechanical (THM) processes on the safety of a hypothetical nuclear waste repository. The research teams first calibrated their models with the results of an in-situ heater experiments to obtain confidence in the capability of the models to simulate the main physical processes. Then the models were used to perform scoping calculations for the near-field of the hypothetical repository, with varying degrees of THM coupling complexity. The general conclusion from the BMT1 exercise is that it would be prudent to perform full THM coupling analyses for two main reasons. First, several safety features might be overlooked with lesser degrees of coupling. Second, the ability to predict and interpret several physical processes, during the post-closure monitoring period, is important for confidence building and public acceptance. Such ability is attainable only with fully coupled THM models.

1. INTRODUCTION

The DECOVALEX project is an international co-operative project, initiated by SKI, the Swedish Nuclear Power Inspectorate, to support the development of mathematical models of coupled THM processes in the host rock for potential nuclear fuel waste repositories. The DECOVALEX project has been running for more than a decade; the third phase of DECOVALEX called DECOVALEX III is now being finalized with two main objectives. The first objective is the validation of codes by simulating large-scale experiments. The second objective is to determine the relevance of THM processes on the safety of a repository. To achieve the second objective, benchmark tests are proposed, where typical repository designs, engineered barriers and host rocks are examined. The benchmark problem called BMT1 defined in this paper looks at the implications of coupled THM processes on the near-field performance of a repository. The performance of a nuclear waste repository is dependent on two main components:

i) the groundwater flow field since groundwater is recognized as the main agent of contaminant transport from the repository to the biosphere.

ii) the structural integrity of the engineered and geological barriers.

The performance assessment (PA) of a repository is often performed with probabilistic assessment computer codes, using Monte-Carlo simulations. Typically thousands of runs are performed in order to estimate the probabilistic distribution of a dose to potential receptors. It is thus necessary to simplify the processes involved in the different barriers to contaminant migration. The rock mechanics expert, the engineered barrier expert, and the hydrogeologist usually provide information to the PA analyst on the structural integrity of the rock mass and engineered barriers and the groundwater flow field; this information is then synthesized and simplified by the PA analyst before he (she) defines the input to the PA model. Up to the present time, the above information are provided separately, without considering the coupling between thermal, mechanical processes (implication on structural integrity), and hydraulic processes (implications on flow field). In this benchmark problem, we propose that scoping calculations be performed in order to determine how THM processes can influence the flow field, as well as the structural integrity of the geological and engineered barriers in the near-field of a typical repository .

The present paper is the first in a series of companion papers on bench mark test 1(BMT1) of DECOVALEX III. In this paper we introduce the subject, thoroughly define the problem and summarize the main conclusions. BMT1 is divided into three phases that will be discussed in more details in the following companion papers:

i) In Phase A (Chijimatsu et al. 2003), the research teams calibrate their T-H-M mathematical models with the measured output from an in-situ heater experiment, at the Kamaishi Mine in Japan. The objective of the calibration is to build confidence in the models used by different research teams in the subsequent scoping calculations.

ii) In Phase B (Millard et al. 2003), the research teams perform scoping calculations for a hypothetical repository in homogeneous rock. The bentonite considered as an engineered barrier in the hypothetical repository is the same as the one used in the Kamaishi Mine experiment. However, the permeability and strength characteristics of the rock mass are based on data obtained from sparsely fractured rock of the Canadian Shield.

iii) In Phase C (Rutqvist et al. 2003), the research teams perform scoping calculations for a hypothetical repository similar to the one in Phase B, with the additional consideration of discrete fractures intersecting the repository.

2. GENERAL DEFINITION OF THE PROBLEM

A hypothetical case is defined in which the feasibility of constructing a nuclear waste repository in a granitic rock formation at a depth of 1000 m is studied. No particular site has been selected although investigations have been performed at some experimental areas. For example at one experimental site, galleries were excavated down to a depth of 600 m, and a variety of hydraulic, mechanical, geochemical tests have been performed (Figure 1-a). Of particular relevance to the assessment proposed in BMT1, is a THM experiment which replicates the near field behaviour of the rock mass and buffer around a single waste container (Figure 1.b). The type of bentonite used in the above experiment and the dimensions of the experimental borehole are comparable with the parameters of the conceptual design of the repository shown in Figure 2.

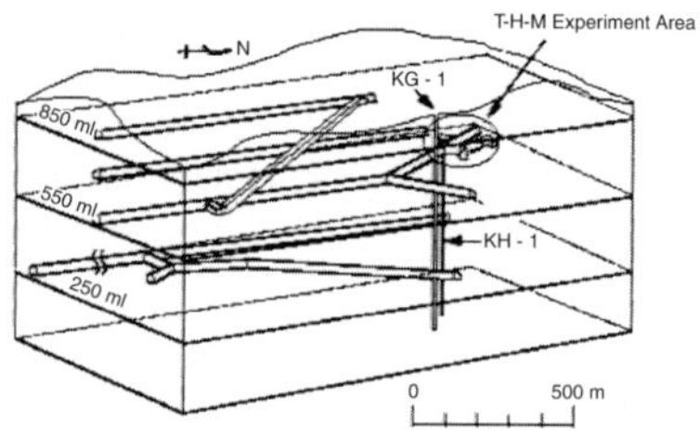

a) Underground Experimental Site

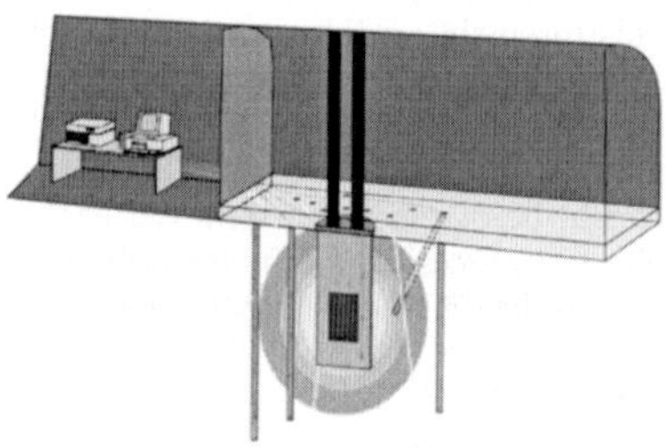

b) In-situ heater experiment (details)

Figure 1. Site and details for in-situ heater experiment

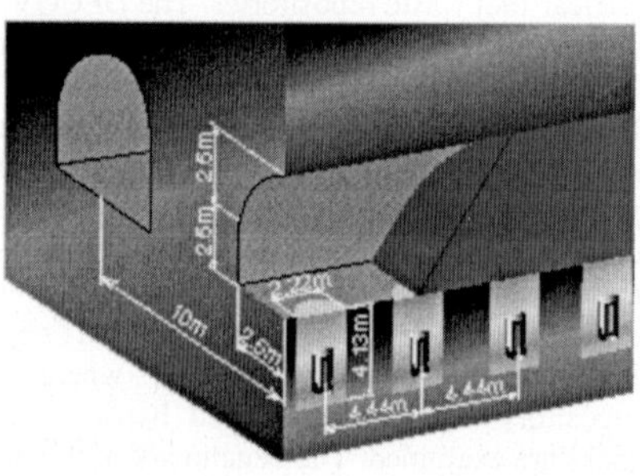

Figure 2. Conceptual design of hypothetical repository (JNC, 2000)

The conceptual design of the repository (Figure 2) consists of a series of parallel tunnels, where the wastes would be emplaced in boreholes excavated in the floors of the gallery. The centreline distance between adjacent tunnels is 10 m and the centreline distance between adjacent inground boreholes for the wastes is 4.44 m. The depth of each borehole is 4.13 m and the diameter is 2.22 m. The overpack for vitrified wastes would be emplaced into the borehole, and a bentonite buffer material would be compacted around the overpack. The tunnels would also be backfilled with a mixture of gravel and clay.

The PA modeller in charge of studying the near field behaviour would use a Monte Carlo assessment code in order to assess the transport of radionuclides through the engineered barriers to the surrounding rock. The PA modeller would like to have feedback from the engineering experts on the following key points:

1. What is the temperature evolution in the near-field?

2. How long would it take for the buffer to resaturate?

3. What are the stresses on the overpack and the buffer ? Will they be structurally stable?

4. How will the permeability and the flow field of the rock mass in the near-field evolve?

5. Is there a potential for rock mass failure in the near-field?

6. Last but not least, is THM coupling important and what combinations of coupling can be ignored?

The research teams involved in BMT1 would act as engineering experts for this hypothetical case. They would use THM mathematical models, implemented in computer codes, as tools to address the questions raised by the PA modeller. The research teams would divide their work in two phases:

1. Calibration of the code with the T-H-M experiment performed at the 550 m-Level gallery at the experimental site, illustrated in Figure 1. This step is considered as crucial in building confidence in the codes' capabilities to take into account the main physical processes.

2. Use of the codes to perform scoping calculations of the T-H-M near field behaviour, of the generic design shown in Figure 2, in order to specifically give feedback to the PA modeller.

The hypothetical repository defined in the above has rather composite features: it is in fact based on one of the Japanese conceptual designs (Figure 2). The buffer material is the same pure bentonite used in a real heater experiment at the Kamaishi Mine in Japan (Figure 1). The in-situ state of stress and the natural thermal gradient are also based on Japanese geological data; however as we will illustrate in the next section, the rock mass properties in term of strength and permeability are based on typical Canadian Shield's data.

3. CALIBRATION OF THE HEATER EXPERIMENT

Detailed descriptions of the experiment is provided by Chijimatsu et. al (2001). A general description is given here.

A circular test pit was excavated in the floor of the experimental drift. A granulated bentonite was compacted directly into the borehole, by layers of 10 cm in thickness. The initial water content (by weight) of the bentonite was 15%. A heater was installed in the test pit, within the bentonite. When bentonite was compacted within the last 50 cm of the pit, a concrete lid was installed in the remaining part; this lid was restrained by vertical steel bars connected to the ceiling of the drift, in order to restrict vertical movements. A dam was then built on the floor, and a water pool of 40 cm was created. The temperature of the water pool was maintained at 12.3 °C during the duration of the experiment. The heating phase of the experiment started when the temperature at the centre of the heater was set to 100°C; this heating phase lasted 258 days. The heater was then turned off, and the cooling phase started; measurements from the sensors were recorded during a period of approximately 180 days.

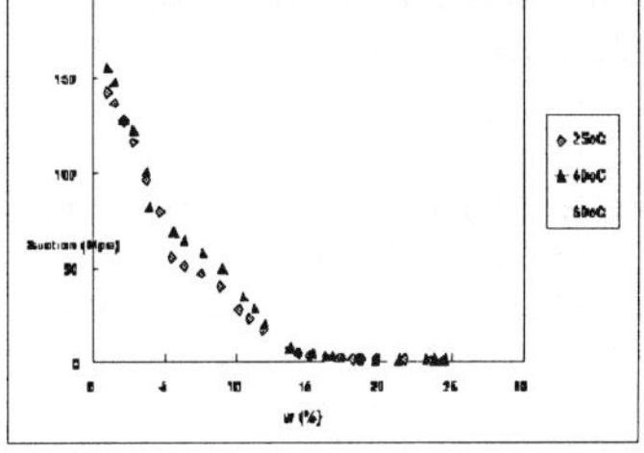

Figure 3. Water retention characteristics of bentonite

The basic THM properties of the buffer material were determined by laboratory tests. These properties include: saturated permeability, thermal conductivity, Young's modulus, water retention curves, isothermal infiltration tests, swelling pressure, and moisture flow under thermal gradient tests. The details of the tests and the results are given by JNC (2000). Typical water retention characteristics are shown in Figure 3 for illustration purposes.

The basic properties of the rock matrix were determined from laboratory tests on intact samples and were provided by JNC (2000). Typical values are: effective porosity: 0.379 %; Young's modulus: 61 GPa; Poisson's ratio: 0.303; coefficient of linear thermal expansion: 8.21x 10^{-6} $^{o}C^{-1}$; thermal conductivity: 2.54 to 2.71 W/m^{o} K; specific heat: 900 J/kg^{o} K.

The THM response of the buffer and rock mass were monitored by a network of instruments. For example, sensors to measure water content, temperature, pore pressure, stress and strain in the bentonite were installed at three radial sections. In the calibration exercise, the evolution of the temperature, water content, pore pressure, stress and strain along one radial axis intersecting the centre of the heater will be calculated and the calculated results will be compared and adjusted to the measured results.

4. SCOPING CALCULATIONS FOR THE NEAR FIELD

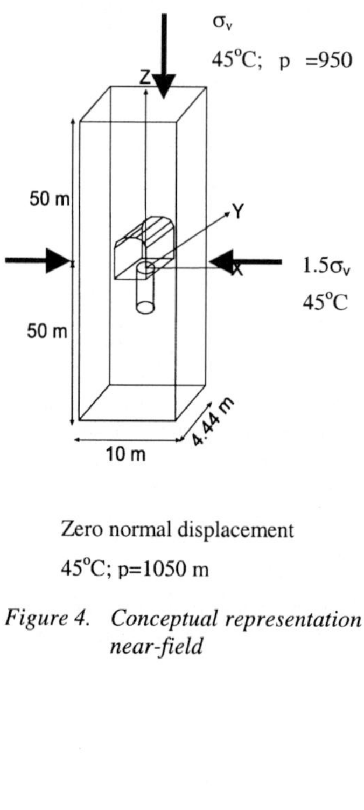

Figure 4. Conceptual representation of repository near-field

From the conceptual design of the repository as shown in Figure 2, and assuming repetitive symmetry, one web of the system, comprising one borehole and a slice of rock and backfill as shown in Figure 4 is considered.

For the rock mass, two main cases were considered:

i) The rock mass is considered to be homogeneous. The scoping calculations for that case constituted Phase B of BMT1.

ii) One discrete horizontal fracture intersects the centre of the emplacement borehole. A combination of this horizontal fracture and other intersecting fractures could also be considered. The scoping calculations for that case constituted Phase C of BMT1.

4.1 Rock mass properties

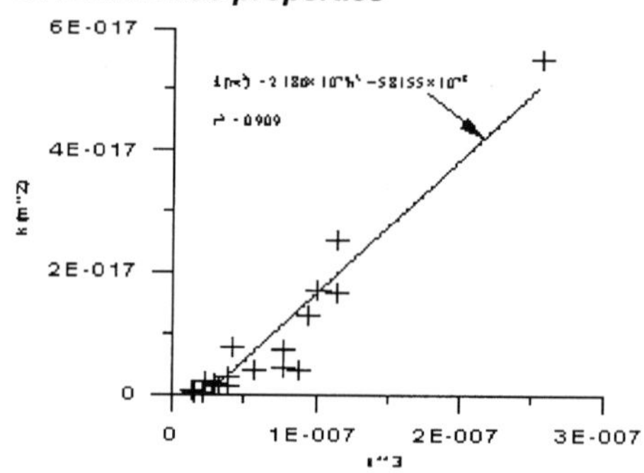

Figure 5. Permeability-porosity function for rock mass

Initial rock mass permeability values of 10^{-18}, 10^{-17} and 10^{-19} m^2 will be considered by the research teams, with the median value of 10^{-18} defined as the base case. Additionally, the rock mass permeability is assumed to vary with the effective porosity according to the following function:

$$k = 2.186x10^{-10} n^3 - 5.8155x10^{-18} \quad (1)$$

where k is the permeability (m^2) and n is the effective porosity.

Equation (1), illustrated in Figure 5, is derived from experimental data obtained from sparsely fractured rock of the Canadian Shield (Katsube and Kamineni, 1983), with a permeability range between 10^{-19} to 10^{-17} m^2.

The rock mass failure is assumed to be governed by the Hoek and Brown's criterion:

$$\sigma'_{1f} = \sigma'_1 - \sigma'_3 - (m\sigma_c\sigma'_3 + s\sigma_c^2)^{0.5} \quad (2)$$

where $\sigma'_{1f}, \sigma'_1, \sigma'_3$ are respectively the major effective principal stress at failure, the major effective principal stress and the. minor effective principal stress; m=17.5 and s=0.19 are empirical constants typical of granitic rock of the Canadian Shield (AECL, 1992); σ_c = 190 MPa is the uniaxial compressive strength of intact rock samples.

4.2 Analysis sequence

For both Phase B and Phase C, the research teams will perform the following sequence of simulation:

i)firstly, determine the effects of excavation by performing a steady-state analysis, with the boundary conditions as illustrated in Figure 4. The output of this analysis consists of the distribution of temperature, pore pressure, permeability, and factor of safety for rock mass failure. The ambient temperature of 45°C is consistent with Japanese thermal conditions at a depth of 1000m. A temperature of 20°C is assumed inside the tunnel in order to simulate cooling.

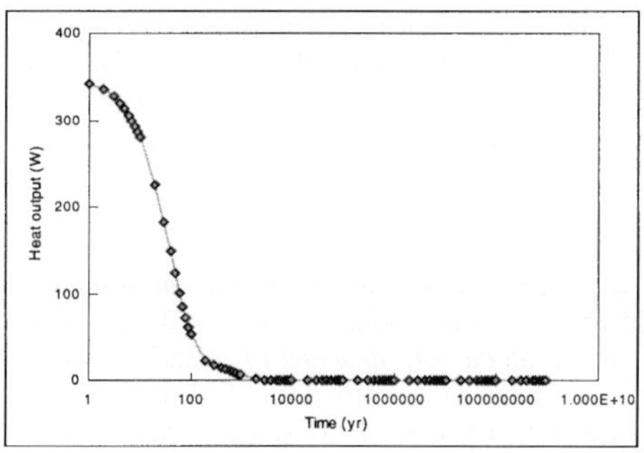

Figure 6. Heat output from waste canister

ii) secondly, perform transient analyses assuming that the buffer, backfill and heater are emplaced instantaneously at time t=0. The analysis shall be conducted for a period of at least 100 years. The results from the steady state analysis for excavation effects for temperature, stresses, and pore pressure in the rock mass will be used as initial conditions. For the buffer and backfill, the initial stresses are assumed to be zero, the initial temperature is 20°C, and the initial water content is 15 %. The heat output from the waste is shown in Figure 6 and is representative of Japanese spent fuel. The boundary conditions for the transient analysis are similar to the ones during excavation, except for the lateral boundaries where zero normal displacement will now be imposed to simulate periodic symmetry. The time evolution of temperature, water content and total stresses in the bentonite will be calculated at specific points. In the rock mass, the distribution of temperature, pore pressure, factor of safety against failure and permeability are required at specific times.

4.3 Evaluation of coupling effects

For both the excavation phase, and the long term phase with emplaced buffer, backfill and heater, the analyses shall be performed with increasing degree of complexity of coupling. A comparison matrix will be established in order to compare the implications of various orders of complexity of the coupling, as shown in table 1.

Table 1. Comparison of output for different degrees of T-H-M coupling

Coupling / Output	THM	TH	TM	HM
Temperature				NA*
Water content			NA	
Pore pressure			NA	
Stress/safety factor		NA		
Permeability		NA		

* NA: not applicable

5. CONCLUSIONS

In this paper, we provided the rationale and definition of a benchmark test called BMT1 to look at the implications of THM couplings on safety parameters in the near field of a hypothetical repository. This hypothetical repository possesses composite features since it is based on a Japanese design, with a Japanese bentonite used as buffer material and the heat output characteristics of Japanese spent fuel. However, the permeability and strength characteristics of the rock mass are based on typical properties of granites of the Canadian Shield.

In Phase A of BMT1, the T-H-M models used by the research teams are calibrated against the measured THM output of an in-situ heater experiment. All models could reasonably simulate the fundamental physical phenomena, with thermal output being best simulated and swelling stress in the bentonite being the most difficult parameter to simulate (Chijimatsu et al. 2003).

In Phase B, the research teams used the T-H-M models to perform scoping calculations for a repository in a homogeneous rock mass. For that case the fully coupled THM analysis predicted localized rock mass failure and also important features related to the buffer/backfill resaturation, the rock mass permeability evolution and the swelling stress development in the buffer which might be overlooked by more simple analyses. The effects of coupling on the above features seem to be amplified when the rock mass permeability is lower. Temperature is the only output parameter which is not significantly influenced by coupling (Millard et al. 2003).

In Phase C (Rutqvist et al. 2003), the research teams performed calculations for the case where one or several water-bearing discrete fractures intersect the repository. As in Phase B, the temperature field shows nearly no difference between THM, TH, TM calculations and is very similar to the one in Phase B. The fracture(s) accelerates the resaturation of the buffer/backfill and prevents the desaturation of the rock mass. From a mechanical point of view, the fracture(s) constitutes a zone of weakness and results in a more extended zone of damage as compared to the homogeneous scenario. TH or THM calculations produce very similar pore pressure fields. With respect to the stresses in the buffer, the conclusions are similar to the homogeneous case, with a predominant effect of pore pressure on total stresses, compared to the thermal stresses.

From the results of the present work, it appears that from a technical point of view the effect of coupling will be either short lived (several decades to 100 years) and would not impact on long term (thousands to hundred of thousand years) safety issues, or could be rectified by adequate design and operation methodology (e.g. avoid over-cooling the galleries). The influence of the host rock properties (e.g. permeability) on the long term safety seems to be much more important than coupling, since one has much less control over these properties. However, the short term period where coupled processes are important corresponds to the post closure monitoring period for most disposal systems. This post closure period is crucial for confidence building, demonstration purposes, and public acceptance. In order to interpret and assess the monitoring data collected during that post closure period, we believe a fully coupled approach is necessary.

ACKNOWLEDGEMENTS

The following organisations are gratefully acknowledged for their financial support (alphabetical order): the Canadian Nuclear Safety Commission; the Commissariat a l'Energie Atomique; the European Commission through the BENCHPAR project; the Federal Institute for Geosciences and Natural Resources, Germany; the Institut de Radioprotection et de Sûreté Nucléaire; the Japan Nuclear Fuel Cycle Development Institute, the Swedish Nuclear Power Inspectorate.

REFERENCES

AECL (Atomic Energy of Canada Ltd) 1992. *Used fuel disposal centre- a reference concept*, AECL Whiteshell, Pinawa, Manitoba, Canada.

Chijimatsu M, Giraud A, Jing L, Millard A, Nguyen TS, Rejeb A, Rutqvist J, Souley M, Sugita Y and Thoraval A. 2003. *Building confidence in the mathematical models by calibration with a T-H-M field experiment.* Geoproc, Stockholm, Sweden

JNC (Japan Nuclear Cycle Development Institute) , 2000. *H12: Project to establish the scientific and technical basis for HLW disposal in Japan.*

Katsube TJ and Kamineni DC. 1983. *Effect of alteration on pore structure of crystalline rocks: core samples from Atikokan, Ontario.* Canadian Mineralogist, Vol. 21, 637-646.

Millard A, Rejeb A, Chijimatsu M, Jing L, De Jonge J, Kohlmeier M, Nguyen T.S, Rutqvist J, Souley M, Sugita Y. 2003. *Evaluation of coupled THM coupling on the safety assessment of a nuclear fuel waste repository in a homogeneous hard rock.* GeoProc, Stockholm, Sweden.

Rutqvist J, Chijimatsu M, De Jonge J, Jing L, Kohlmeier M, Millard A, Nguyen TS, Rejeb A, Souley M, Sugita Y. 2003. *Scoping calculations for a fractured rock medium.* GeoProc, Stockholm, Sweden.

DEVELOPMENT OF A METHODOLOGY TO QUANTIFY THE IMPORTANCE OF HYDRO-MECHANICAL PROCESSES IN RADIONUCLIDE MIGRATION ASSESSMENTS

Philipp Blum, Rae Mackay & Michael S. Riley
School of Geography, Earth and Environmental Sciences,
University of Birmingham, UK

Abstract: A methodology for quantifying the contributions of hydro-mechanical processes to fractured rock hydraulic property distributions has been developed and tested. Simulations have been carried out on discrete fracture networks to study the sensitivity of hydraulic properties to mechanical properties, stress changes with depth, mechanical boundary conditions, initial mechanical apertures and fracture network geometry. The results indicate that the most important (and uncertain) parameters for modelling HM processes in fractured rock are fracture density and rock/fracture mechanical properties. Aperture variation with depth below ground surface is found to be of second order importance.

1. INTRODUCTION

A methodology for quantifying the contributions of hydro-mechanical (HM) processes to fractured rock hydraulic property distributions is presented. The purpose is to assess the importance of modelling HM processes for the calculation of large space, long time scale radionuclide migration through a repository host rock. The methodology is based on the identification of statistical models of fracture set properties and their use to generate multiple realisations of discrete fracture networks (DFN) numerically. The behaviour of DFN has been studied under different stress and fluid pressure conditions using numerical models of the hydraulic and mechanical processes to explore large scale flow behaviour. The work has been carried out using data from the Sellafield site in the United Kingdom provided by United Kingdom Nirex Limited. These data have been used by several research teams to investigate the upscaling of thermo-hydro-mechanical (THM) processes in two dimensions as part of the DECOVALEX III, an international co-operative programme. It is recognised that several options for the long-term management of radioactive waste, one of which is to place these wastes in a repository excavated in stable rock formations. This methodology appears to be useful for intentifying the issues affecting HM processes in performance assessment of a nuclear waste repository for this option.

2. METHODOLOGY

The main objective of this paper is to present the methodology developed to determine the large scale contributions to flow and transport arising from local scale HM processes and local scale fracture network heterogeneity (*Figure 1*). The methodology does not consider THM processes in the near-field of deep waste repositories, instead it examines the variations in the hydraulic behaviour of the host rock as a function of stress and mechanical properties. A similar methodology has been used by Martin *et al.* (1990), who applied it to investigate a major shear zone in granite and Eidsvig (2000) who used it to describe the stress-dependent permeability of fractured chalk. An incentive for the development of the methodology is that, potentially, it makes it possible to examine fault zones and to explain the depth distributions of flowing features observed during the Sellafield site investigation programme (Nirex 1997a). The methodology comprises 4 main stages. The first two stages and the results of their applications to the Nirex data sets are considered in this paper. The development and application of stages 3 and 4 are considered in Blum *et al.* (2003).

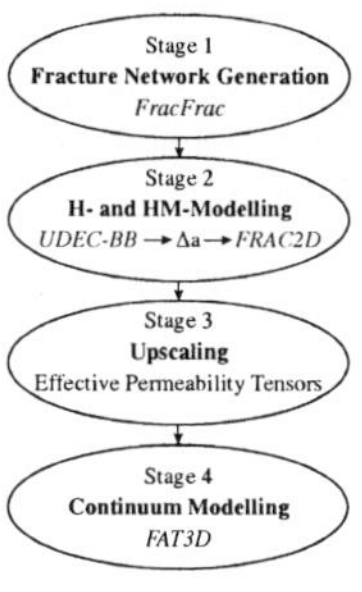

Figure 1. Overall model strategy (Δa = hydraulic aperture change).

Stage 1 concerns the DFN generation, which was performed with the code *FracFrac*. The uncertainties in the statistical parameterisation of the fracture length and density data are examined and show the importance of careful fitting of the power-law length distribution to the data.

Stage 2 comprises both hydraulic analysis and hydro-mechanical analysis. The purpose of stage 2 is to establish the hydro-mechanical contributions to an equivalent continuum permeability tensor, the minimum size of averaging volume (denoted here as the approximate REV) required to model large scale flow and the statistical variability in the upscaled parameter values. Several methods exist to evaluate the 2D permeability tensor for fractured rock (e.g. Long *et al.* 1982, Zhang *et al.* 1996). For this application the evaluation is based on the work by Jackson *et al.* (2000), which models flow through the DFN in several directions using boundary conditions set beyond the REV boundary and corresponding to a uniform head gradient. The impact of HM-modelling on the permeability tensor is explored by considering the change in hydraulic aperture distribution and establishing an HM-modified permeability tensor. Sensitivity studies using the distinct element code, UDEC, incorporating the Barton Bandis model for mechanical fracture closure (UDEC-BB) were carried out to examine the impact of mechanical properties, stress changes with depth, mechanical boundary conditions, initial mechanical apertures and fracture geometries.

3. APPLICATION

The data for the first trial of the methodology are from the Borrowdale Volcanic Group (BVG) rock formation, Cumbria, UK.

3.1 Fracture Network Generation

The most abundant data for the DFN generation (stage 1) are related to fracture density and length. A power-law fracture length distribution has been fitted to the available data as shown in *Figure 2*. The number of fractures per km^2 (N_F) longer than the length (L) is given by:

$$N_F = CL^{-D} \tag{1}$$

where C is the density constant and D is the fractal dimension. The data include only fractures with a length greater than 0.5 m, hence the lower cut-off length is assumed to be 0.5 m, which governs the fracture density for a given C and D. The data set in *Figure 2* can be modelled using alternative values of C and D. Three cases have been considered (low-density: C = 1.25, D = 2.0; medium-density: C = 3.23, D = 2.08; high-density: C = 4.0, D = 2.2), which all sensibly fit the data (*Figure 2*). The mean fracture densities (P_{21}) of 5.0 m/m^2, 13.2 m/m^2 and 16.9 m/m^2 for the low, medium and high density cases are illustrated in *Figure 3*. This shows the sensitivity of the fracture data analysis to the choice of power-law length distribution. Realisations of fracture length were constructed by sampling the power-law cumulative distribution function:

$$F = 1 - \left(\frac{L_{\min}}{L}\right)^D \tag{2}$$

where $L_{\min}$ is the lower-cut-off length and L is the fracture length.

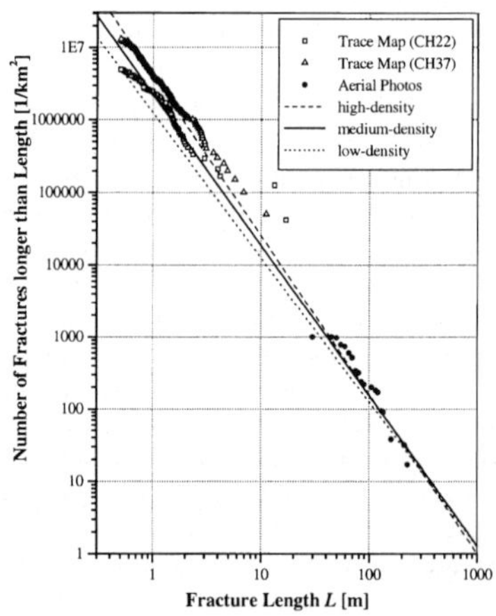

Figure 2. Power-law distribution of fracture length (based on Nirex 1997b).

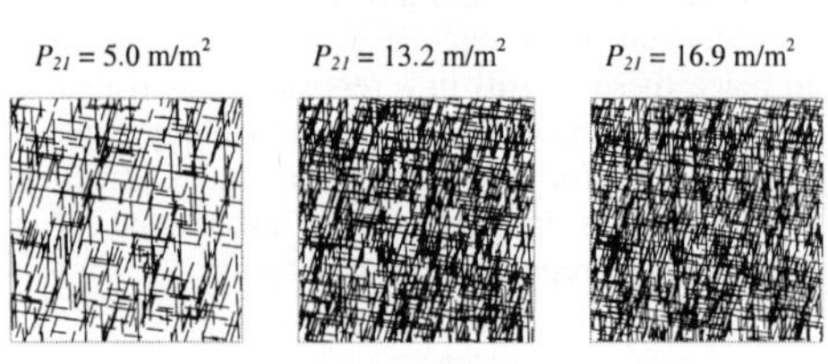

Figure 3. DFN with different fracture densities and a power-law length distribution (10 m × 10 m).

3.2 H-Analysis

The initial hydraulic analysis incorporating a baseline REV determination was performed with the fracture flow code FRAC2D using DFN with constant hydraulic aperture (a_h = 130.7 μm). To evaluate the 2D permeability tensor for square domains the boundary conditions were adjusted to give head gradient directions between 0° and 150° in 30° steps. The flows across the domain boundaries were calculated for each gradient direction with a head gradient of one. By assuming Darcian flow and minimising the expression:

$$\sum_{i=1}^{6}\left[q_x + \left(K_{xx}\cos\theta_i + K_{xy}\sin\theta_i\right)\right]^2 + \left[q_y + \left(K_{xy}\cos\theta_i + K_{yy}\sin\theta_i\right)\right]^2 \tag{3}$$

the three permeability tensor coefficients K_{xx}, K_{yy} and K_{xy} were calculated, where θ_i is the orientation of the hydraulic gradient. Fracture network generation regions of 100 m × 100 m were adopted from which different size DFN could be sampled. Numerous simulations were performed for each domain size and the principal permeability components (K_{max}, K_{min}) and principal direction (θ_{max}) were calculated using equations from Zhang *et al.* (1996). *Figure 4* illustrates graphically the results for one domain size. The H-analysis was performed for increasing domain sizes (side length 5-100 m). For the low-density case 100 simulations were required to achieve acceptable convergence of the sample mean and standard deviation of the permeability coefficients and direction, while only 50 simulations were necessary for the medium- and high-density case. *Figure 4* illustrates the permeability anisotropy, which is caused by the two vertical fracture sets with the highest fracture density.

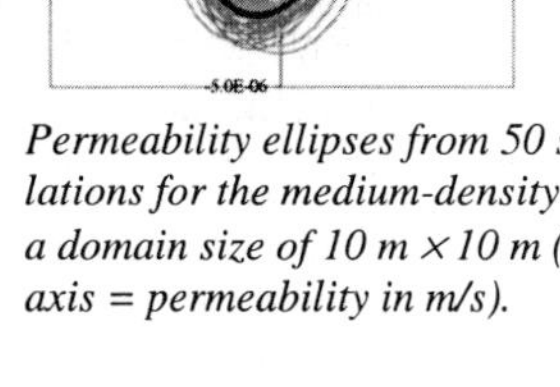

Figure 4. Permeability ellipses from 50 simulations for the medium-density case with a domain size of 10 m × 10 m (x- and y-axis = permeability in m/s).

REV-Analysis

Subjective criteria for an REV were set such that if the cumulative variances of the principal permeability coefficients and the principal directions are less than 5 % of their cumulative averages, then the domain size is equal to or greater than the minimum REV. The application of these criteria showed that for the medium- and high-density cases all domain sizes greater than 10 m × 10 m were above the minimum REV size. However for the low-density case no REV was achieved up to a domain size of 100 m × 100 m (the largest size that could be accurately modelled with FRAC2D). This is explained by the decreased connectivity in the minor flow direction for this case. The change of connectivity also causes a rotation of the principal directions of conductivity with increasing domain size. The baseline analysis of the minimum REV size was used to provide a starting point for the analysis of the HM interactions described in the next section.

3.3 HM-Analysis

HM-analysis has been only considered for the medium-density case. The HM-modelling was performed with the UDEC-BB model. The empirical relations in the UDEC-BB model are described in Bandis *et al.* (1983) and Barton *et al.* (1985). The in-situ stress field (MPa) is described by the following equations (Nirex 1997c):

$$\begin{aligned}\sigma_V &= 0.0294D + 0.26622\\ \sigma_H &= 0.03113D + 1.88747\end{aligned} \tag{4}$$

where σ_V is vertical stress, σ_H is horizontal stress and D is depth below ground level [m]. For HM-modelling, data sets are required that describe the rock block material and the joint material. Isotropic elastic behaviour is assumed for the rock blocks and values are listed in Table 1.

Table 1. Block data for the UDEC-BB model.

Block Data	
Bulk density ρ [kg/m^3]	2750
Young's modulus E_{block} [GPa]	84.0
Poisson ratio ν [−]	0.24
Uniaxial compressive strength [MPa]	157.0

Table 2. Joint Properties.

Joint Data	
Joint normal stiffness K_n [GPa/m]	434.0
Joint shear stiffness K_s [GPa/m]	434.0
Joint roughness coefficient JRC_0 [-]	3.85
Joint compressive strength JCS_0 [MPa]	112.2
Laboratory-scale joint length l_0 [m]	0.03
Residual friction angle [°]	27.2
Initial mechanical aperture a_{ini} [μm]*	138

* Calculated with equation 5.

The non-linear empirical Barton-Bandis (BB) joint model describes the joint behaviour. The mean input values are given in Table 2. The initial mechanical aperture (a_{ini}) at zero normal stress is calculated with the following empirical relationship (Bandis *et al.* 1983):

$$a_{ini} = \frac{JRC_0}{5}\left(0.2\frac{UCS}{JCS_0} - 0.1\right) \tag{5}$$

where *UCS* is the uniaxial compressive strength.

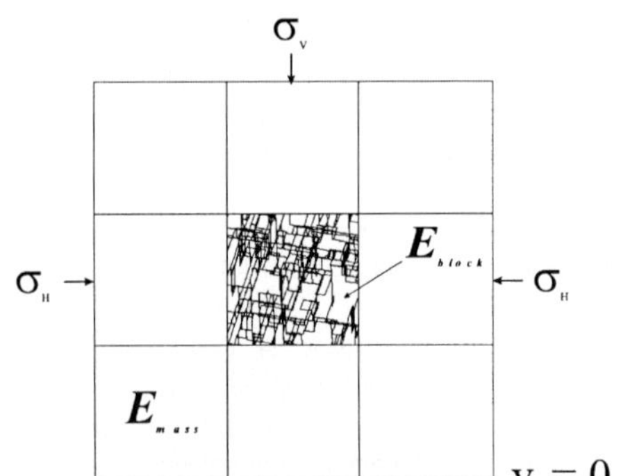

Figure 5. Mechanical boundary conditions for the HM analysis with the block-in-block method.

Figure 5 illustrates the applied stress field for the HM-modelling including both mechanical stresses and the hydrostatic fluid pressure. In this figure the bottom boundary is rigid in the *y*-direction and all other mechanical boundaries are free to move. An embedded domain method has also been tested to reduce the bottom boundary constraint. The 5 m × 5 m DFN was located in the centre of eight blocks of with uniform material properties characteristic of a disturbed (jointed) rock mass. The Young's modulus (E_{mass}) assigned to the surrounding rock mass was 65.0 MPa about 77 % of the undisturbed rock mass (E_{block}). Stresses were then applied across the outer boundary. The name block-in-block method has been coined for this approach. The resulting median hydraulic apertures are similar. The single block method yielded a mean aperture of 22.4 μm, while the block-in-block method produced a value of 24.3 μm. The block-in-block method has been used for all calculations given the better representation of the boundary conditions on the modelled domain.

Sensitivity studies have been carried out on a 10 m × 10 m model domain in a 30 m × 30 m block-in-block simulation region. The data set defining the joint parameters *JRC* and *JCS* show major variations (*Figure 6*) and no obvious relationships is observed between *JRC*, *JCS* and depth. The four pairs of *JRC*/*JCS* values identified in *Figure 6* have been used to assess the sensitivity of the BB-model to mechanical properties.

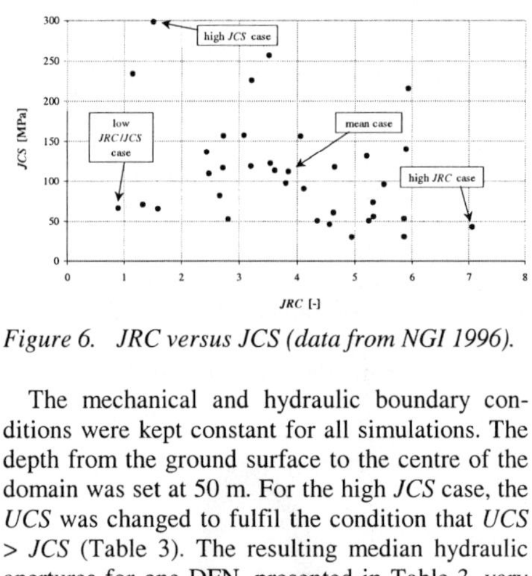

Figure 6. JRC versus JCS (data from NGI 1996).

The mechanical and hydraulic boundary conditions were kept constant for all simulations. The depth from the ground surface to the centre of the domain was set at 50 m. For the high *JCS* case, the *UCS* was changed to fulfil the condition that *UCS* > *JCS* (Table 3). The resulting median hydraulic apertures for one DFN, presented in Table 3, vary between 6.1 μm and 226.9 μm. The proportion of the total fracture length with apertures less than the hydraulic aperture a_h is shown in *Figure 7*. Fracture aperture is significantly controlled by fracture orientation, as illustrated in *Figure 8* and leads to enhanced anisotropy of the upscaled hydraulic conductivity. The high *JRC* case produces the largest hydraulic apertures as anticipated while the smallest hydraulic apertures are observed for a high *JCS* value and low *JRC* value.

Table 3. Input parameters (in bold) for the 4 *JRC/JCS* cases and the resulting median aperture values.

	Mean	High *JRC*	High *JCS*	Low *JRC/JCS*
JRC_0 [-]	3.9	7.1	1.51	0.9
JCS_0 [MPa]	112.2	43.1	298.4	66.4
UCS [MPa]	157.0	157.0	308.8	157.0
a_{ini} [μm]	138.5	887.5	32.3	67.1
a_h[1] [μm]	23.7	226.9	30.4	6.1
a_h[2] [μm]	18.1	1.4	57.8	6.3

[1] median hydraulic aperture using the calculated a_{ini}, [2] median hydraulic apertures calculated with a_{ini} = 77 μm.

To test the significance of the calculated initial mechanical apertures, a constant initial aperture of 77 μm was used and the simulations repeated. The results produced different results to those for the calculated hydraulic apertures but the trend was similar (Table 3). Thus, the initial mechanical apertures emerge to have an impact on the resulting hydraulic apertures but, for the results presented, the significance appears to be less than the impact of the variation of the mechanical properties. From these results the importance of the mechanical properties and their spatial distribution in the rock mass to the estimation of hydraulic aperture appears to be strong.

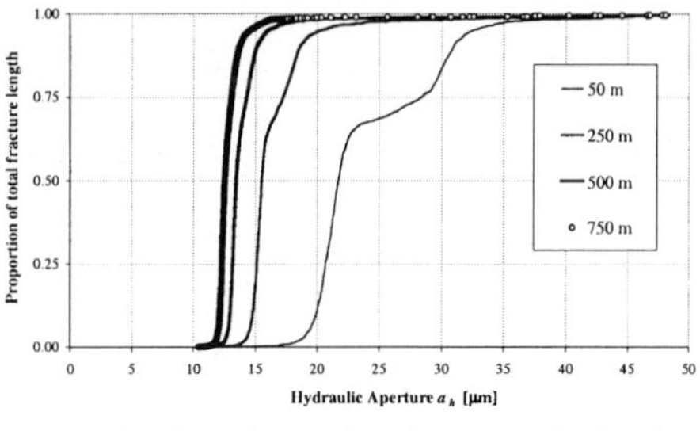

Figure 7. Cumulative distributions of hydraulic aperture for one DFN for different depth below ground level.

The effect of increasing stresses on the hydraulic aperture has been tested by simulating increasing depth of burial of the DFN for the median fracture frequency case. The calculated median hydraulic aperture was 23.7 μm for a burial depth of 50 m and 12.7 μm at 1000 m (*Figure 7*). This is less than a factor of 2 reduction and assuming the validity of the cubic law, potentially represents a reduction of hydraulic conductivity of less than one order of magnitude. The aperture variations due to mechanical property differences are significantly larger.

Further the results indicate that changes to the distribution of the hydraulic apertures progressively decrease with increasing depth (*Figure 7*). There is almost no difference between the calculated aperture distributions at 750 m and at 1000 m depth: a residual hydraulic aperture distribution is achieved at around 750 m depth. The anisotropy in the aperture distribution is removed between 250 m and 500 m depth. Above 250 m depth the anisotropy can have a substantial impact on the permeability and the principal direction, which is further discussed in detail in Blum *et al.* (2003).

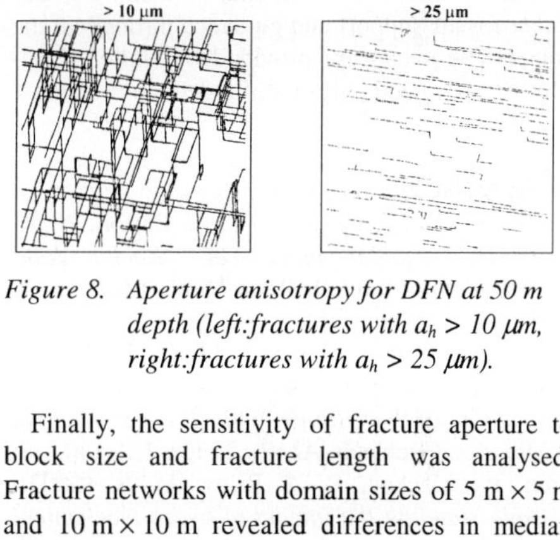

Figure 8. Aperture anisotropy for DFN at 50 m depth (left:fractures with a_h > 10 μm, right:fractures with a_h > 25 μm).

Finally, the sensitivity of fracture aperture to block size and fracture length was analysed. Fracture networks with domain sizes of 5 m × 5 m and 10 m × 10 m revealed differences in median hydraulic apertures of less than 0.5 μm. Even with a block size of 15 m × 15 m and effectively infinitely long fractures, the change in median hydraulic aperture remained less than 1 μm. It can be concluded that the size of the REV determined for flow only is also suitable for mechanical calculations, and that, for the assumed spatial distribution and orientation of fracturing, fracture length has only a minor impact on the hydraulic aperture distribution.

4. CONCLUSIONS

A key uncertainty for modelling fracture patterns arises from the inference of the parameters of the power-law fracture length distribution. Several models have been fitted to the BVG data, each yielding different hydraulic results. The main geometric effect on hydraulic behaviour arises from the fracture density but the orientation of individual fracture sets does impact on the anisotropy of REV scale hydraulic properties.

The sensitivity studies using the UDEC-BB model have shown that the HM-coupling (stress changes with depth) depends most strongly on the magnitude and variation of the rock and fracture mechanical properties and less on the fracture geometry. Thus understanding the spatial variation of the mechanical properties appears to represent the key aspect to understanding the impact of HM processes for fractured rock hydraulic characterisation.

ACKNOWLEDGEMENTS

The authors are grateful to all our colleagues from the DECOVALEX project for the many fruitful discussions. We would like to thank Dr. Les Knight for his support and guidance throughout the project. The data were provided by United Kingdom Nirex Limited, who also funded the research.

REFERENCES

Bandis, S. C., Lumsden, A. C. & Barton, N. R. (1983). *Fundamentals of Rock Joint Deformation*. Inter. J. Rock Mech. Min. Sci. & Geomech. Abstr., 20 (5), 249-268.

Barton, N., Bandis, S. & Bakhtar, K. (1985). *Strength, Deformation and Conductivity Coupling of Rock Joints*. Inter. J. Rock Mech. & Min. Sci. Geomech. Abstr., 22 (3), 121-140.

Blum, P., Mackay, R. & Riley, M. S. (2003). *Understanding the impact of hydro-mechanical coupling on performance assessment*. In GeoProc 2003, this issue.

Eidsvig, U. M. K. (2000). *Stress-dependent Permeability in Fractured Rock*. In 3rd Euro-conference on Rock Physics and Rock Mechanics, Bad Honnef, Germany.

Long, J. C. S., Remer, J. S., Wilson, C. R. & Witherspoon, P. A. (1982). *Porous Media Equivalents for Networks of Discontinous Fractures*. Water Resources Research, 18 (3), 645-658.

Martin, C. D., Davison, C. C. & Kozak, E. T. (1990). *Characterizing normal stiffness and hydraulic conductivity of a major shear zone in granite*. In Barton & Stephansson (eds.), Rock Joints, Balkema, Rotterdam, 549-556.

NGI (1993). *Geotechnical (CSFT) laboratory testing of BVG joint samples from boreholes RCF1 and RCF3*. NGI Report No. 931005-102/2.

Nirex (1997a). *An Assessment of the Post-closure Performance of a Deep Waste Repository at Sellafield*. Nirex Science Report, S/97/012, Harwell, UK.

Nirex (1997b). *Evaluation of heterogeneity and scaling of fractures in the Borrowdale Volcanic Group in the Sellafield area*. Nirex Report SA/97/028, Harwell, UK.

Nirex (1997c). *Assessment of the in-situ stress field at Sellafield*. Nirex Report S/97/003, Harwell, UK.

Jackson, C. P., Hoch, A. R. & Todman, S. (2000). *Self-consistency of a heterogeneous continuum porous medium representation of fractured media*.- Water Resources Research, 36 (1), 189-202.

Zhang, X., Sanderson, D. J., Harkness, R. M. & Last, N. C. (1996). *Evaluation of the 2-D Permeability Tensor for Fractured Rock Masses*. Inter. J. Rock Mech. Min. Sci. & Geomech. Abstr., 33 (1), 17-37.

UNDERSTANDING THE IMPACT OF HYDRO-MECHANICAL COUPLING ON PERFORMANCE ASSESSMENT OF DEEP WASTE DISPOSAL

Philipp Blum, Rae Mackay & Michael S. Riley
School of Geography, Earth and Environmental Sciences,
University of Birmingham, UK

Abstract: A methodology for understanding the importance of modelling hydro-mechanical (HM) processes in performance assessment of a radioactive waste site in fractured rock has been established. Results of HM-modelling performed with UDEC-BB, the universal distinct element code incorporating the empirical Barton-Bandis model, revealed large variations in hydraulic aperture distribution depending on the applied mechnical properties and the stress conditions. Continuum modelling was undertaken using mean upscaled hydraulic conductivity and porosity values for two cases - a hydraulic only analysis and a hydro-mechanical analysis for a hypothetical repository setting. The modelling of these cases showed that the most significant HM factors for performance assessment at the regional scale are the aperture distributions that depend on the variations and the spatial distribution of the mechanical properties. Uncertainties in the fracture density and the spatial distribution of different fracture densities appear to be less important than knowledge of the mechanical properties.

1. INTRODUCTION

Studies of the stress-dependent permeability of discrete fracture networks (DFNs) have been undertaken to examine the importance of hydro-mechanical (HM) processes for the performance assessment of nuclear waste repositories. Rock stress/fluid stress interactions have been modelled to build upscaled permeability relationships to support far-field continuum modelling. The HM-calculations at the small scale (< 30 m) have been performed using the two-dimensional (2D) universal distinct element code incorporating the empirical Barton-Bandis model (UDEC-BB) and the fracture flow and transport code, FRAC2D. Upscaled HM-modified, effective hydraulic conductivity tensors (k_{eff}) been evaluated at a scale at which the variability of the upscaled results is sufficiently small for the values to be used in a regional continuum model. The magnitude of the effective porosity has been determined from a calculation of the mean advective velocity through the discrete fracture network (DFN). The same averaging scale has been adopted for both the effective porosity and the effective hydraulic conductivity. The continuum flow and transport code FAT3D has been used for the far-field studies and particle tracking has been undertaken to examine the flow paths and travel times through the repository host rock under different physical conditions.

Two models are presented and compared in this paper: (1) a homogeneous hydraulic base case assuming constant hydraulic apertures throughout the model domain and (2) a non-homogeneous hydro-mechanical base case with hydraulic apertures determined from analysis of HM coupling using the mean mechanical properties for each formation. These are hereafter referred to respectively as Case H and Case HM. The regional groundwater system assumes flow in a 2D cross section through a hypothetical host repository. The rock mass consists of three contrasting geological formations (two laterally extensive formations and a vertically extensive fault zone). The rock property data for each formation are based on the Sellafield (Cumbria, UK) site investigation programme performed by United Kingdom Nirex Limited but the spatial distribution of the formations has been modified from that found at Sellafield.

2. SMALL-SCALE ANALYSIS

An extensive examination of the fracture network and mechanical data has been undertaken to determine models of the fracture characteristics of the three formations, the uncertainties in the parameterisation of the models, and the sensitivity of the upscaled flow properties to the underlying parameter variations. The methodology used to calculate effective hydraulic conductivity values and their sensitivity to the small-scale model is described in Blum *et al.* (2003). The study undertaken to obtain the effective hydraulic conductivity under different stress conditions and presented in Blum *et al.* (2003) revealed that the important parameters in modelling HM processes in the fractured rock mass are the fracture density, the mechanical (M) properties and the M property variations through the rock mass.

While alternative fracture densities were studied for the effective hydraulic conductivity analysis, only the intermediate density case for each formation has been considered here. Figure 1 presents an example of a DFN with medium density for Formation 2. The results presented here are based upon just one DFN for each formation to illustrate the methodology to be employed in a more extensive and longer term numerical investigation.

Figure 1. An example of a DFN for Formation 2 with a medium fracture density of 13.2 m/m^2. The block size is 10 m × 10 m.

Figure 2 shows the effective hydraulic conductivity ellipses for the three rock formations for Case H. The ellipses are closely aligned since the fracture set orientations of the three formations are similar. The maximum principal directions vary from 67.8° to 79.0° and the anisotropy ratios (K_{max}/K_{min}) vary from 1.9 to 3.5.

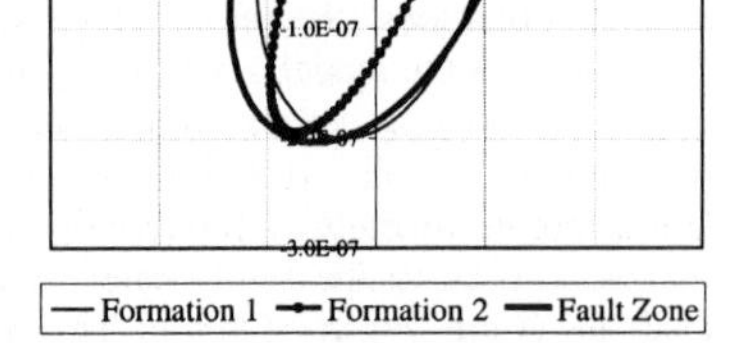

Figure 2. Hydraulic conductivity ellipses for the three rock formations (network size = 10 m × 10 m, hydraulic aperture (a_h) = 50 μm, x- and y-axis = k_{eff} in m/s).

For Case HM, pairs of joint roughness coefficient (*JRC*) and joint compressive strength (*JCS*) values were used to represent the range of mechanical properties observed in all three formations (Formation 1 see Blum *et al.* 2003; Formation 2: *JRC* (*1-4*) = 4.28, 5.98, 4.18, 2.29; *JCS* (*1-4*) = 39.3, 31.9, 90.9, 43.1 in MPa and the unchanged uniaxial compressive strength for all 4 cases *UCS* = 120.0 MPa; Fault Zone: *JRC* = 4.22, *JCS* = 105.9 MPa and the *UCS* = 128.4 MPa). In case of Formation 2 only four pairs were available, thus the entire data set is provided here. Stress conditions corresponding to five depths were also applied to the DFN. Table 1 summarizes the hydro-mechanical modelling results in terms of the median hydraulic apertures. Values range between 0.3 μm and 180.7 μm.

Considerable variability of fracture apertures is simulated even at depths of 500 m - 1000 m as observed in actual fractured rock at depth (e.g. Brace 1980, Armitage *et al.* 1996).

Table 1. HM modelling results (median hydraulic apertures in μm).

		25 m	75 m	175 m	375 m	750 m
F1	*1*	27.3	22.1	18.0	15.0	13.2
	2	-	-	-	-	180.7
	3	-	-	-	-	1.7
	4	-	-	-	-	18.2
F2	*1*	3.9	1.8	0.8	0.3	-
	2	93.5	68.4	46.0	31.2	-
	3	10.3	8.1	6.6	5.6	-
	4	5.1	2.4	1.1	0.4	-
FZ	*1*	18.2	14.9	12.6	10.9	9.8

F1 = Formation 1, F2 = Formation 2, FZ = Fault Zone, *1* (mean *JRC* and *JCS*), *2* (high *JRC* and mean *JCS*), *3* (mean *JRC* and high *JCS*), *4* (low *JRC* and low *JCS*).

Table 2. Anisotropy ratios (top row) and maximum principal directions (bottom row in degrees).

	25 m	75 m	175 m	375 m	750 m
F1	1.1	1.1	1.2	1.2	2.1
	25.8	74.3	58.6	54.7	65.7
F2	2.5	3.3	3.1	3.6	-
	37.1	43.7	54.7	58.7	-
FZ	2.0	1.3	1.6	1.4	1.5
	11.7	38.6	75.3	65.2	68.7

Mean *JRC* and mean *JRC* mechanical properties.

In addition, the maximum principal directions and anisotropy ratios for the hydraulic conductivity are presented for the mean mechanical property case in Table 2. Figure 3 shows the changes to the hydraulic conductivity ellipse with depth for the Fault Zone. At shallow depths (below 100 m) the maximum principal direction is aligned with the sub horizontal fractures. At the greatest depths, the shape and orientation of the ellipse approximates those seen in the constant hydraulic case (Figure 2) due to the convergence of the fracture apertures towards a uniform value approximately equal to the residual aperture. The principal directions and anisotropy ratios for the other two formations display similar trends (Table 2).

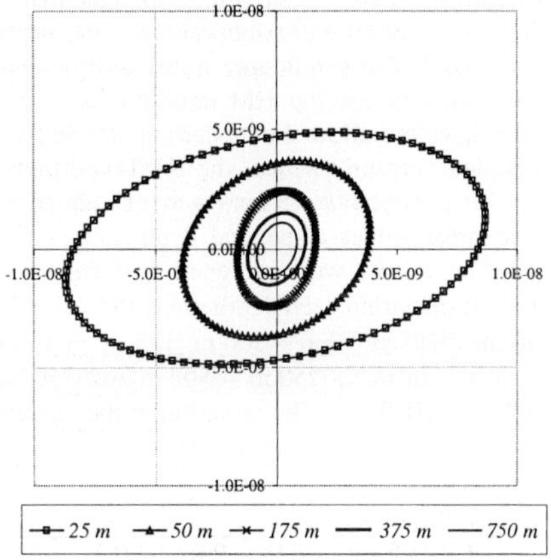

Figure 3. HM-modified hydraulic conductivity ellipses of the Fault Zone (x- and y-axis = k_{eff} in m/s).

To calculate travel times at large scales using the continuum model, the effective porosity (n_e) of the fractured rock is required. For this study the effective porosity has been assumed to be independent of the flow direction (Endo *et al.* 1984) although anisotropy has been observed in the calculations. The effective porosity in fractured media has been calculated using the following equation (e.g. Carrera *et al.* 1997):

$$n_e = \frac{q}{v} = \frac{t_f q}{L} \qquad (1)$$

where q is specific discharge, v is seepage velocity, L is travel distance and t_f is mean travel time. The mean travel time is calculated across the DFN considering all pathways through the fracture network. The number of pathways increases rapidly with network size. Particle tracking approaches are commonly used but these tend to consider only the most likely paths as the number of particles is usually orders of magnitude smaller than the number of alternative pathways. To avoid the approximation required by particle tracking, an algorithm has been developed that integrates travel time calculations calculated over connected sub-domains to minimise the size of the calculation whilst sampling all pathways. This algorithm is able to return the mean travel time accurately and is sufficient for the calculation of an effective porosity but does not provide data on particle dispersion through the fracture network. Table 3 summarises the calculated effective porosities for case *1* of the mechanical properties.

Table 3. Summary of the effective porosities for the mean mechanical property case.

	25 m	75 m	175 m	375 m	750 m
F1	1.71×10^{-4}	$1.46\ 10^{-4}$	1.21×10^{-4}	1.04×10^{-4}	9.05×10^{-5}
F2	5.93×10^{-6}	4.83×10^{-6}	2.63×10^{-6}	1.51×10^{-6}	-
FZ	1.34×10^{-4}	1.30×10^{-4}	1.15×10^{-4}	1.19×10^{-4}	9.24×10^{-5}

3. LARGE-SCALE ANALYSIS

The upscaled continuum modelling was performed with the flow and transport code FAT3D. Figure 5 shows the hypothetical repository and the three host rock formations, which are based on the Borrowdale Volcanic Group (BVG) in Cumbria, UK. The model domain is 5 km long and 1 km deep and extends under the sea from a low lying area on land. The orientation of the 2D cross section through the host rock is NNW/SSE (159/339°). The repository is located at 520 m depth and 10 m NNW from the 10 m wide Fault Zone, which penetrates from ground surface to the base of the section. The hydraulic boundary conditions are identical for both the H and HM cases. The base and the sides of the model domain are prescribed by no flow boundaries. The top boundary is prescribed by a constant pressure set equal to atmospheric pressure. This boundary condition corresponds to the water table being at ground surface or sea level as appropriate. For the hydraulic case, 3 particle trajectories are used to illustrate the flow system. Two starting points are located at the ground surface at the right hand side of the model domain to present the general flow regime in the system. The third trajectory is initiated at the NNW corner of the repository. For the HM case an additional particle trajectory starting at the top of the Fault Zone is presented. All particle trajectories are calculated assuming advection only.

3.1 Hydraulic base case

For the hydraulic base case, constant hydraulic apertures and fracture densities were considered. Homogeneous hydraulic conductivity tensors and porosities were applied to each formation. The case with a constant hydraulic aperture of 10 μm and medium fracture density for all formations is illustrated in Figure 5, where the streamlines and the particle travel times are shown. The time of travel between each marker is 1000 days. Owing to the low effective porosity values and the relatively high fracture density, particle travel times through the host rock are fast. The mean particle travel time from the repository to the seabed is only 123 years, when a constant hydraulic aperture of 10 μm is used.

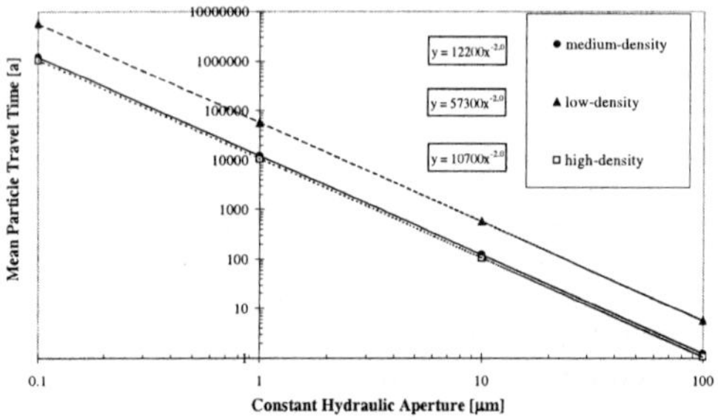

Figure 4. Summary of the mean particle travel times from the repository to the seabed for various constant hydraulic apertures and fracture densities.

The results for the particle travel times for various constant hydraulic apertures and fracture densities from the repository to the seabed are presented graphically in Figure 4. For the low fracture density case, the hydraulic conductivity estimated for a block size of 25 m × 25 m has been used even though this size does not correspond to the REV, which is estimated to be greater than 100 m × 100 m. This block size does allow, however, a calculation of the mechanical closure of the fracture apertures for the HM case and therefore a comparison of the results between the two cases for low-density conditions. The results of the continuum model based on constant hydraulic apertures display very rapid mean particle travel times from the repository to the seabed. For example, for the low and high fracture density networks adopting a constant hydraulic aperture of 10 μm, particle travel times from the repository to the seabed are 580 years and 106 years respectively. A doubling of the aperture increases the conductivity by a factor of 8 and increases the porosity by a factor of 2 if the cubic law is adopted. This leads to an overall increase in velocity of a factor of 4. Errors in velocity are therefore related to the square of the errors in the estimation of the hydraulic aperture. This sensitivity is clearly important in determining the accuracy of any travel time modelling and appropriate ranges for hydraulic aperture.

3.2 Mean HM base case

For the HM base case the mean mechanical properties have been used to calculate the hydraulic aperture distributions over the depth of the model. The continuum model and the applied methodology for the HM coupling in fractured rock does not allow the modelling of a fully HM-coupled system, hence the HM-modified hydraulic conductivity tensors were calculated at the mid point values of several depth ranges (Table 1). The results were assigned uniformly to the formation within each depth range (25 m ⇒ 0 m – 50 m, 75 m ⇒ 50 m – 100 m, 175 m ⇒ 100 m – 250 m, 375 m ⇒ 250 m – 500 m and 750 m ⇒ 500 m – 1000 m). The variation in the calculated aperture values decreases as depth increases, which allows for the larger depth bands at the base of the model.

The result for the HM base case is shown in Figure 6. The streamlines reveal a quite different geometry to the H base case. The impact of the anisotropy of Formation 2 is apparent. The increase to 4,860 years for the particle travel time from the repository to the seabed from the H base case is predominantly caused by the tight upper Formation 2 with an average hydraulic aperture for the entire formation of around 1.0 μm. The low-conductivity upper formation significantly reduces the total flow through the whole system, which consequently lowers the fluid velocities through the host rock.

The mean particle travel time down the higher conductivity Fault Zone from the ground surface to the top of Formation 1 only takes around 50 years, showing that the Fault Zone can act as a conductive pipe connecting the surface to the lower formation and influencing the flow geometry when the upper formation is tight. The very low flow contributions from the right hand side of the model to the flow regime are illustrated by the closely spaced, near vertical trajectories adjacent to the left hand boundary of the particles starting from the right hand side. This further indicates the strong influence of the tight Formation 2 coupled to the conductive fault.

4. DISCUSSIONS

The results of the HM modelling using the UDEC-BB calculated hydraulic aperture distributions under different stress conditions demonstrate the importance of the mechanical properties and their spatial variations in the rock mass on its hydraulic behaviour. The large-scale studies show the strong dependency of the travel times within the different formations on the effective porosities and hydraulic conductivities. At this stage, only the mean mechanical property values have been used in the continuum model, thus the impact of the spatial distribution of the entire range of observed mechanical properties has still to be assessed. The lack of precise knowledge about the distribution of the mechanical properties in the host rock and the large variability observed in the field data indicate that a stochastic approach is essential. However, unless verifiable models of the spatial patterns of the mechanical properties can be developed and sufficient data collected to support the use of these models, it is likely that very large uncertainties in the modelling of travel times in fractured formations will remain in the future.

One outcome of this study demonstrates the impact of different fracture densities on the large-scale analysis. The comparison between the medium- and high-density case with a constant aperture of 10 μm revealed a relatively minor change of only 17 years (123 years – 106 years) to the particle travel time from the repository to the seabed, which in comparison to the impact of changing mechanical properties and therefore apertures on the particle travel times appear to be very modest. Variations in the mechanical properties can change the aperture distribution of a formation by up to 2 orders of magnitude and thus can change the particle travel time by up to 4 orders of magnitude. This change in particle travel times is not possible by only changing the fracture density within the range of uncertainty identified by the data. Hence the results suggest that in the context of a performance assessment of deep waste disposal the impact caused by the uncertainty in the spatial distribution of mechanical properties is probably of much greater significance than the uncertainty caused by the spatial distribution of the fracture densities. Although potentially reasonable for the mean density cases, for the low-density case, where no REV could be obtained (Blum *et al.* 2003), this conclusion may be invalid due to the potential for the total loss of connectivity of the fracture network at the large-scale.

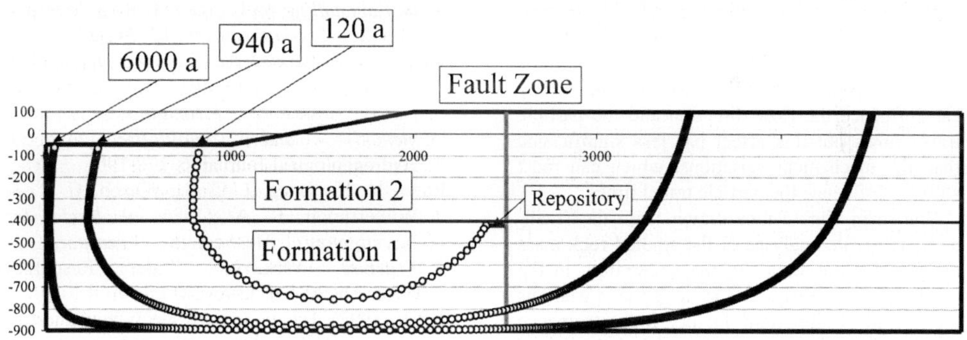

Figure 5. Hypothetical host rock with streamlines and particle travel times for the H base case (medium-density DFN with constant hydraulic aperture of 10 μm).

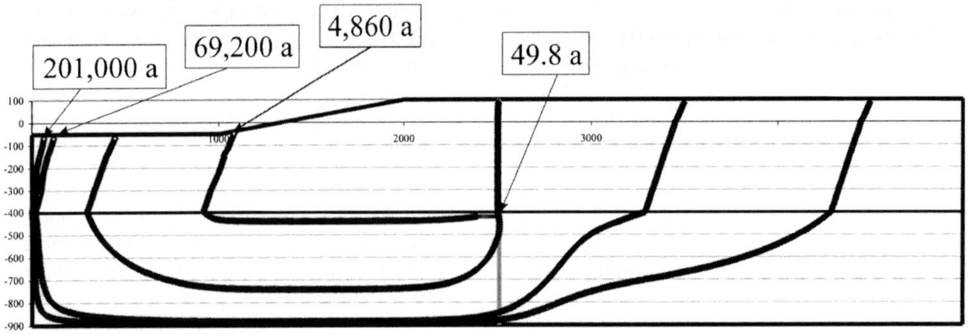

Figure 6. Hypothetical host rock with streamlines and particle travel times for the mean HM base case.

5. CONCLUSIONS

The methodology for upscaling HM processes from the small-scale (metre-scale) to the large-scale (km-scale) has been used to assess the importance of HM processes in performance assessment of deep waste disposal. The results reveal a wide range of potential hydraulic aperture distributions in the rock mass and a strong sensitivity of the hydraulic apertures to the mechanical properties of the rock mass and the fractures. The median hydraulic apertures ranged between 0.3 μm and 180.7 μm depending on the mechanical properties and the applied stresses. The HM modelling of all 3 formations showed a non-linear decrease in the hydraulic apertures with increasing depth. The aperture anisotropy can influence the flow direction and the particle travel times but this effect has less significance than the mechanical variations, which can most strongly influence the particle travel times.

The main conclusions drawn from the small- and large-scale analyses of the studied rock mass are that for this data set, the uncertainty in the mechanical properties and their spatial variations are the most important factors in performance assessment of deep waste disposal, followed by the uncertainty of the fracture density and the spatial distribution of the fracture density.

ACKNOWLEDGEMENTS

The authors are grateful to all our colleagues from the DECOVALEX/BENCHPAR project for the many fruitful discussions. We would like to thank Les Knight for his support and guidance throughout the project. The data were provided by United Kingdom Nirex Limited, who also funded the research.

REFERENCES

Armitage, P., Holton, D., Jefferies, N. L., Myatt, B. J. & Wilcock, P. M. (1996). *Groundwater flow through fractured rock at Sellafield.* European Science, Nuclear Science and Technology, EUR 16870 EN.

Blum, P., Mackay, R. & Riley, M. S. (2003). *Development of a methodology to quantify the importance of hydro-mechanical processes in radionuclide migration assessments.* In GeoProc 2003, this issue.

Brace, W. F. (1980). *Permeability of argillaceous and crystalline rocks.* Inter. J. Rock Mech. & Min. Sci. Geomech. Abstr., 17, 241-251.

Carrera, J., Sanchez-Vila, X., Benet, I., Medina, A., Galarza, G. & Guimera, J. (1998). *On matrix diffusion: formulations, solution methods and qualitative effects.* Hydrogeological Journal, 6, 178-190.

Endo, H. K., Long, J. C. S., Wilson, C. R. & Witherspoon, P. A. (1984). *A Model for Investigating Mechanical Transport in Fracture Networks.* Water Resources Research, 20(10), 1390-1400.

Nirex (1997). Data summary sheets in support of gross geotechnical predictions. Nirex Report, SA/97/052, Harwell, UK.

IMPACT OF FLOW AND TRANSPORT COUPLING IN THE UPSCALING OF TRANSPORT PARAMETERS FOR PERFORMANCE ASSESSMENT IN THE CONTEXT OF NUCLEAR WASTE DISPOSAL

J. Jaime Gómez-Hernández, Eduardo F. Cassiraga
Universidad Politécnica de Valencia, 46071 Valencia - Spain

Abstract: The disparity of scales between measurements and models requires a decision on how to extrapolate information collected at a given scale (generally small) onto the scale at which performance assessment is evaluated (generally orders of magnitude larger than the measurement scale). This problem only applies for those parameters with a considerable degree of spatial variability. If the spatial variability is small, the problem of upscaling is alleviated since a simple average value could be sufficient to represent adequately the parameter at the larger scale. But it is when the spatial variability is large, that there is a need to evaluate the impact of replacing the complex of the spatial variability of a given parameter by a homogeneous value over a large scale. It is the case of hydraulic conductivities. On a synthetic experiment based on data gathered from the Sellafield site it is shown that proper upscaling of hydraulic conductivities is not enough to reproduce the transport behaviour of the formation due to the non-linear coupling between flow and transport. Only after introducing a fictitious retardation factor in the upscaled model it is possible to reproduce some of the transport response of the formation observed in the small scale modelling.

1 INTRODUCTION

One of the workpackages in the EC-funded project BENCHPAR addresses the problem of upscaling in the context of performance assessment. This paper presents the results of the work performed by the team from the Universidad Politécnica de Valencia.

The need for upscaling rules originates from the impossibility of carrying out groundwater flow and mass transport modelling at the scale of measurements in the framework of performance assessment. There is a need to extrapolate the information collected at the measurement scale about the parameters controlling the mass transport processes up to a modelling scale, generally orders of magnitude larger than the measurement scale.

Upscaling rules are needed only for those parameters that display an important degree of spatial variability. During the upscaling process, not all of the information contained about the spatial distribution of the parameters can be transferred to the modelling scale; thus it is important to decide which is the information that should be transferred between scales. As a consequence of this decision, the upscaling rules will change depending on the objective pursued. For example, if the upscaled model should be able to reproduce the early arrival time of the contamination released from a given source, the upscaling rule will be different from that chosen to match the total amount of mass that crosses a given boundary during a specific time lapse.

In the example discussed in the text, the focus will be in the upscaling of conductivities with the aim of reproducing the early arrival times of a contamination release to a control plane; it will be seen that, for this purpose, not only a rigorous conductivity upscaling is needed but also the introduction of a new "calibration" parameter that could be interpreted as a porosity is needed.

2 EXAMPLE DESCRIPTION

We consider a synthetic example that mimics the release of radionuclides from a deep underground repository built in low permeability, fractured rock. The example is two-dimensional and its geometry is shown in Figure 1. It represents a vertical cross-section, and displays two horizontal formations with a major vertical fracture. Groundwater flow is driven by topography and it occurs from the recharge area to the left of the model to the sea.

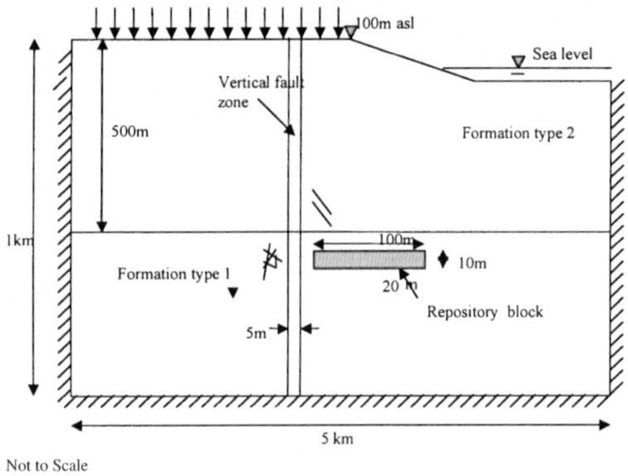

Figure 1. Geometry definition. Not to scale

Table 1 shows average hydraulic conductivities and their standard deviations based on similar data collected by NIREX at the Sellafield site.

Table 1. Hydraulic conductivy based on NIREX data from the Sellafield site (means and standard deviations).

Rock Unit	mean $K(ms^{-1})$	mean $\log_{10}K$	std $\log_{10}K$
Formation Type 2	$8.5\ 10^{-7}$	-6.0	0.61
Vertical Fault in Formation Type 2	$2.7\ 10^{-6}$	-5.6	0.61
Formation Type 1	$5.7\ 10^{-11}$	-10.2	0.48
Vertical Fault in Formation Type 1	$4.5\ 10^{-11}$	-10.3	0.87

Given that the mean hydraulic conductivities for the faults were too similar to those of the formations, it was decided to increase in two orders of magnitude the conductivities for the fault in the Formation Type 1, and four orders of magnitude in the fault in Formation Type 2. It was also decided to make a sensitivity analysis of the upscaling procedure as a function of the standard deviation of $\log_{10}K$, so the parameters in Table 1 were modified as indicated in Table 2.

Table 2. Hydraulic conductivity used in the example.

Rock Unit	mean $\log_{10}K$	set of std $\log_{10}K$ used in the sensitivity analysis
Formation Type 2	-6.0	0.0, 0.2, 0.4, 0.6, 0.8, 1.0
Vertical Fault in Formation Type 2	-3.6	
Formation Type 1	-10.2	
Vertical Fault in Formation Type 1	-6.3	

The degree of spatial correlation for $\log K$ was decided on the basis of scattered information from short pumping tests, which pointed to a correlation of a few tens of meters, and on the discussion with some experts knowledgeable of the Sellafield site. It appears that most of the fracturing is subvertical, thus a strongly anisotropic vertical correlation was assumed for the conductivities in all formations. After some preliminary tries, in which a horizontal range of 10 m and a vertical one of 50 m appeared as a too strong vertical correlation, it was decided to reduce the anisotropy ratio to 2, and to use ranges of 10 m and 20 m in the horizontal and vertical directions.

The repository is assumed filled with bentonite clay of mean $\log_{10}K$ –13.2 ($\log_{10}$ ms^{-1}) and standard deviation 1.7 ($\log_{10}$ ms^{-1}).

3 ANALYSIS PROCEDURE

Establishing upscaling rules requires knowledge of the behaviour of the system at the original scale, and at the target scale, in order to verify that these upscaling rules are indeed valid.

Furthermore, one can never expect to have perfect knowledge of the spatial distribution of conductivities in the are being analyzed, thus a stochastic approach will be taken, in which many realizations of conductivity are generated and the upscaling rules are defined to work, on average, for all realizations.

3.1 Computing flow resistance

Flow resistance can be interpreted as the time that a particle would take in travelling a given distance assuming that it travels with a velocity equal to the specific discharge (i.e., as if the local porosity is equal to 1).

Computing flow resistance is done similarly as how advective travel times are computed by particle tracking. One hundred particles are distributed evenly along the floor of the repository and their trajectories are determined. Flow resistance f [T] is then given by:

$$f = \int_L \frac{dl}{q} \tag{1}$$

in which q [LT^{-1}] is the specific discharge and L [L] represents the trajectory. The specific discharge is obtained by applying Darcy's law ($q = \mathbf{K} \cdot \nabla h$) after the solution of the groundwater flow equation at steady-state, with the boundary conditions as shown in Figure 1.

$$\nabla \cdot (\mathbf{K} \cdot \nabla h) + W = 0 \tag{2}$$

where $\mathbf{K}$ is the hydraulic conductivity tensor [LT^{-1}], h is piezometric head [L], W represents discharges/recharges per unit volume of aquifer [T^{-1}]

3.2 Numerical model

For the solution at the small scale, the system is discretized in square cells of 5 m by 5 m, for a total of 200 000 cells. Each cell is assigned a scalar conductivity value generated stochastically according to the conductivity variance and its range.

For the solution at the large scale, an upscaling procedure is applied over the small-scale model to determine the best conductivity value to be assigned to each square block of 10 by 10 cells. This procedure results in a conductivity tensor that is assigned to each block. The large-scale model has only 2 000 cells of 50 m by 50 m.

In both cases the flow equation (2) is solved by block-centred finite-differences using the code *flowxy* (Gómez-Hernández, 1991), capable to handle full conductivity tensors with arbitrary local principal directions.

3.3 Monte Carlo

Multiple realizations of conductivities at the small scale of 5 m by 5 m cells are generated. For each realization, the numerical modelling described in the previous subsection is applied. The realizations are generated independently for each formation type and for the vertical fracture in each formation type by sequential simulation using the code *gcosm3d* (Gómez-Hernández, 1991). For each formation the same spatial correlation (10 m range horizontal and 20 m range vertical) is assumed, but different means and standard deviations, according to the values in Table 2.

A typical realization at the small scale can be seen in Figure 2. In this figure the contrast between the two formations is apparent as well as the vertical correlation.

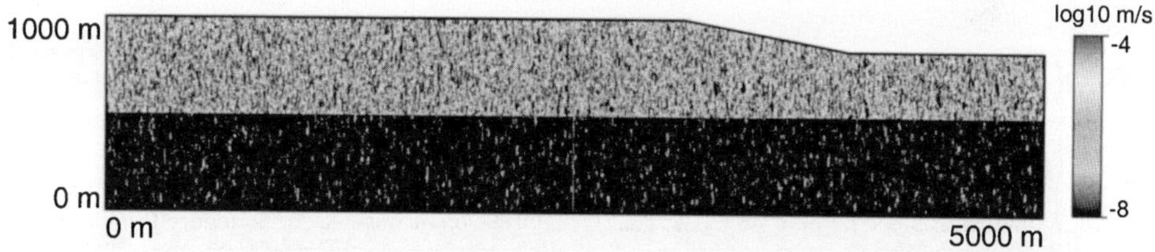

Figure 2. Typical realization of scalar conductivities (log_{10} m/s) at the small scale. Cells are 5 m by 5 m.

3.4 Block upscaling

For each block of 10 by 10 cells in the small scale model, a block conductivity tensor is determined by solving the flow equation (2) for a set of boundary conditions and then determining which is the best conductivity tensor that is able to match the total flow crossing the block at the small scale (Gómez-Hernández and Journel, 1994). The procedure is sketched in Figure 3.

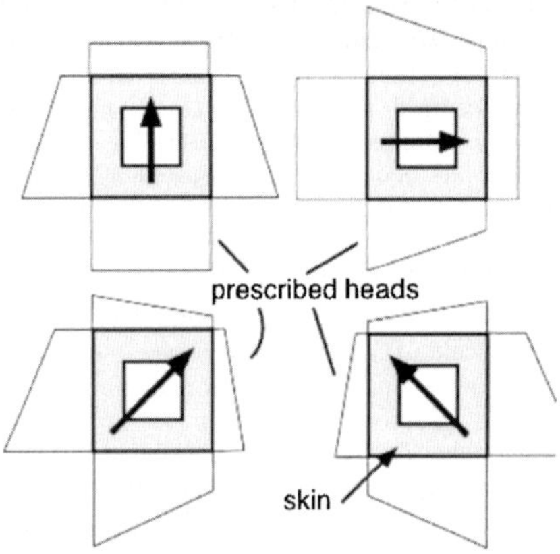

Figure 3. Set of four prescribed head conditions applied around a skin centered on the block being upscaled

conductivity tensors after upscaling a typical realization are shown.

Particle trajectories are then computed for both the small and the large-scale models in order to computing the flow resistances. For this purpose, the programa *3dtransp* was used (Hendricks, 2001). Figure 5 shows the particle trajectories in a small-scale realization and in the associated upscaled model. The shape of the trajectories is very similar in both models; the strong conductivity contrast between the two formations produces the observed refraction: the lower formation acts almost as an impermeable boundary to flow from/to the upper formation.

4 RESULTS

The following procedure is taken to define an upscaling rule that will not only reproduce total flows at the 50 m by 50 m block scale, but also at

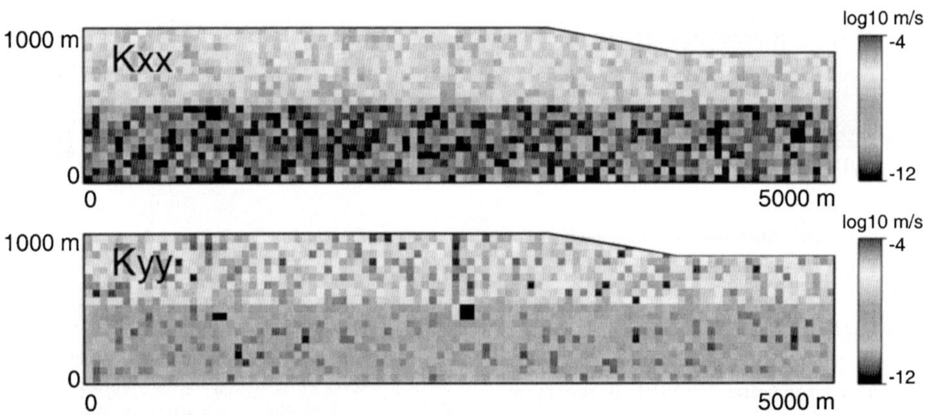

Figure 4. Diagonal components of the log_{10}hydraulic conductivity tensor (log_{10} m/s) after upscaling a small scale realization. The blocks are 50 m by 50 m in size.

The block of 10 by 10 cells to be upscaled is isolated along with a skin. The flow equation is solved using the boundary conditions shown in the Figure. These boundary conditions force flow in four different directions. Then, a simple least-square procedure is used to determine the single conductivity tensor that is able to reproduce best the flows crossing the block for all four boundary conditions. The result is a full hydraulic conductivity tensor with arbitrary principal directions, not necessarily parallel to the Cartesian axes. In Figure 4, the diagonal components of the the system scale and more importantly with regard to its transport behaviour as measured by the flow resistance between the repository and the seabed.

For each standard deviation listed in Table 2 (going from no variability, to substantial variability) 100 realizations of the system were generated at the 5 m by 5 m cell scale.

For each set of 100 realizations, first flow and then advective transport were solved for the boundary conditions shown in Figure 1. With the advective transport solution, the total flow resistance is computed and compared between at the two scales. The discrepancies found could be corrected introducing a fictitious retardation factor into the calculations as shown below.

Only some figures for the set of 100 realizations with a standard deviation of 0.6 are shown. Figure 6 shows the cumulative distributions of flow resistances obtained with the one hundred realizations, flow in the upscaled fields (darker color) travels faster than in the small-scale fields, basically due to the smearing effect of upscaling.

It appears that introducing a fictitious retardation factor (generally associated to sorption process) it could be possible to delay the flow resistance cumulative distribution in the upscaled field to match that in the small-scale field. However, if this factor is computed for different values of the cumulative distribution, this retardation factor is not constant. Indeed, by computing the ratio between the cumulative distribution in the small-scale field and that in the

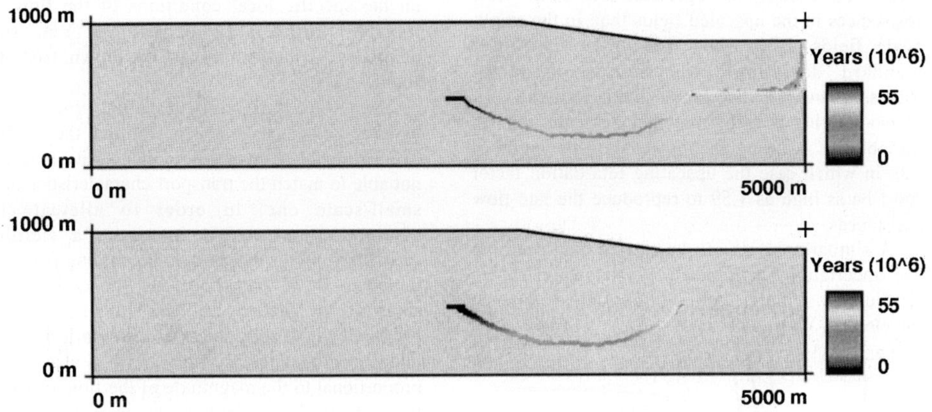

Figure 5. Trajectories from the repository to the seabed in the small scale model (top) and in the upscaled model (bottom) for a realization and its corresponding upscaled realization.

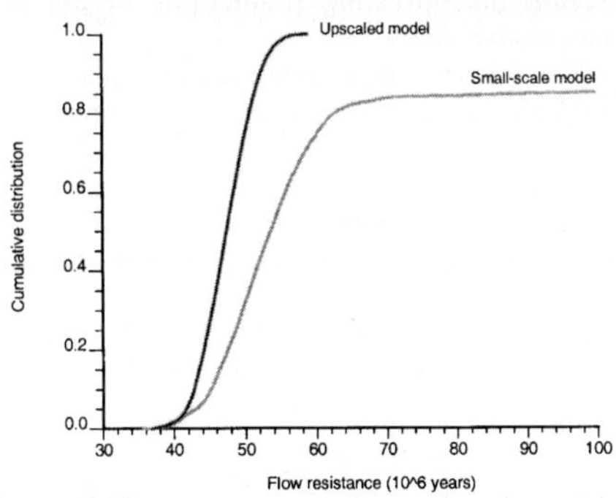

Figure 6. Flow resistance in the small-scale model and in the upscaled model, for the set of realizations with a standard deviation of 0.6

upscaled field, and doing so for all standard deviations we find a clear trend as displayed in Table 3.

Table 3. Fictitious retardation factor to be used in the upscaled field to match the flow resistance on the small-scale fields

Std	Cumulative distribution	Upscaling retardation factor
0.2	10	1.05
	50	1.05
	80	1.05

0.6	10	1.05
	50	1.14
	80	1.23
1.0	10	1.14
	50	1.41
	80	1.59

For the standard deviation of 0.2, it appears that a constant upscaling retardation factor of 1.05 would be enough to reproduce the same flow resistances in the upscaled fields than in the small-scale fields. But, when moving up to higher standard deviations, the retardation factor increases, and it is also larger, the later is the flow resistance that is to be matched, with the highest variabilities occuring for the standard deviation of 1.0, in which case the upscaling retardation factor must be as high as 1.59 to reproduce the late flow resistances.

A similar analysis can be carried out regarding the uncertainties associated with flow resistance predictions. Figure 7 shows the flow resistance envelops

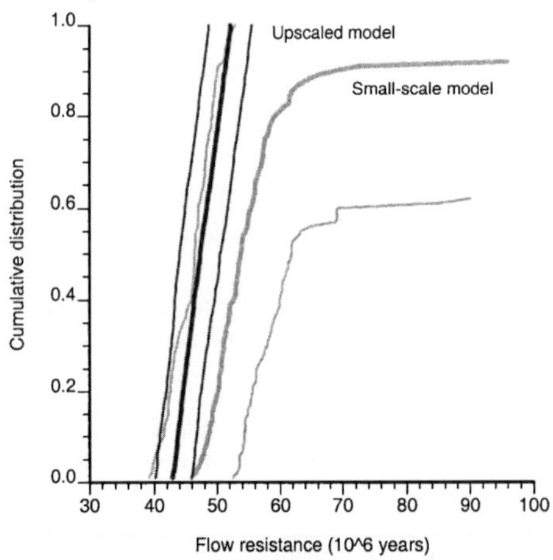

Figure 7. Uncertainty envelops for the 100 realizations of conductivity with a standard deviation of 0.6. Ligther curves for the small-scale realizations, darker curves for the upscaled ones. The thick lines represent the median, and the two other lines the 80% confidence interval

From the flow resistance envelops a similar conclusion is reached, it is possible to introduce a factor (that could not be called retardation) in the upscaled model in order to match the uncertainty about the predictions measured by the width of the confidence interval for different values of the cumulative distribution. However, as it happens in Table 3, there is no single factor that would be able to match all uncertainties for all levels of the cumulative distribution.

5 CONCLUSIONS

It has been recognized for a long time that the problem of upscaling is non-local and thus depends on the specific local conditions of the flow and transport problems under study. Yet, some qualitative conclusions can be drawn from this study.

Even when a rigorous upscaling procedure for hydraulic conductivities is performed, the resulting overall model of hydraulic conductivity tensors is not able to match the transport characteristics of the small-scale one. In order to alleviate this discrepancy we could introduce a fictitious retardation factor that acts as a coupling parameter between flow, transport and upscaling. This retardation factor appears to be directly proportional to the standard deviation of the underlying conductivity field and it is also directly proportional to the magnitude of the flow resistance that one wishes to match. In any case, for the specific case considered in this study, the upscaled fields are conservative since they predict earlier arrival than the actual value. At the same time, the uncertainty associated with the predictions in the upscaled field is smaller than in the actual field, leaving the upscaling predictions in the non-conservative side.

The full procedure described here may not be applicable in a real case due to computational limitations, but at least such a study should be performed on a reduced portion of the study area to try to establish the trends between the predictions in the upscaled field and those in the original one.

Acknowledgments

Financial aids from the EC through contract FIKW-CT2000-00066 and from the Spanish Nuclear Waste Management Agency, ENRESA, through contract 0770140 are gratefully acknowledged

REFERENCES

Gómez-Hernández, J.J., 1991, *A Stochastic Approach to the Simulation of Block Conductivities Conditioned upon Data Measured at a Smaller Scale*, Ph.D. Thesis, Stanford University, UMI, Ann Harbor, 428 pp.

Gómez-Hernández, J.J., and Journel, A.G., 1994, Stochastic Characterization of Gridblock permeabilities, *SPE Formation Evaluation, June 94*, 93-99.

Hendricks Franssen, H.-J., *Inverse Stochastic Modeling of Groundwater Flow and Mass Transport,* Ph.D. Thesis, Universidad Politécnica de Valencia, Proquest Information and Learnging, Ann Harbor, 363 pp.

UPSCALING THE THERMOHYDROMECHANICAL PROPERTIES OF A FRACTURED ROCK MASS USING A MODIFIED CRACK TENSOR THEORY

Varut Guvanasen[1] and Tin Chan[2]

[1]HydroGeoLogic Consulting, Inc.

[2]Atomic Energy of Canada Limited (AECL)

Abstract: A methodology based on the modified crack tensor theory of Oda (1986) was developed for the purposes of upscaling the thermohydromechanical properties for fractured rock mass. The methodology was applied to the Bench Mark Test Case 2 (BMT2) of DECOVALEX III, which is based on a hypothetical site with three fractured rock formations. It was found that the upscaled equivalent rock mass permeability, hydraulic and mechanical porosities, and Young's modulus approached respective asymptotic values at an upscale length of less than 10 meters. The equivalent properties are highly dependent on effective stress. The theoretically determined values of Young's modulus and permeability are generally very similar to, or on the same order of magnitude as, those obtained by Nirex and cited in the BMT2 Problem Definition and Description document.

1. INTRODUCTION

A methodology has been developed to generate homogenized or upscaled thermohydromechanical properties using field and laboratory data obtained from field investigations and samples of discrete rock joints and rock matrix from the three formations in DECOVALEX III (2001) Bench Mark Test 2 (BMT2). Details of upscaling methodology and application results of the upscaling approach to the three formations at the hypothetical site are described in this paper.

2. THEORETICAL FUNDAMENTAL

General assumptions and approach

The theoretical framework is based on a crack tensor theory modified from the original work of Oda (1986) to handle thermohydromechanical properties. The general assumptions based on the modified crack tensor theory are as follows. 1) Each fracture may be represented by a circular disk of finite and uniform aperture. However, the distribution of fracture diameter, orientation, and aperture is governed by the site-based statistics of fractures. 2) Scale dependency is incorporated through some parameters such as JRC, JCS (joint roughness coefficient, joint compressive strength), fracture area within the homogenization volume (L_s^3, where L_s is the length scale of homogenization), and indirectly through PDF (Probability Distribution Function) of fracture aperture, orientation, and diameter (different scales may have different PDFs). 3) There are enough fractures for statistical averaging to be applicable. 4) The dilatancy and contractancy effects of fractures under shear deformation are not included. 5) The hysteresis of repeated fracture deformation is ignored. 6) Continuum theory is applicable to the effective (averaged) hydroelastic and thermoelastic rock mass. 7) The permeability of the rock matrix is negligible in comparison with that of the fractures. 8) Contributions from various fractures and fracture sets can be superimposed. 9) Conversion between mechanical energy and thermal energy is negligible. 10) Fractures do not contribute to thermoelastic strain and the matrix does not contribute to hydroelastic strain. 11) Biot's (1955) theory of coupled hydroelastic processes is valid. 12) Porosities are based on mechanical and hydraulic apertures. Other assumptions are provided in Oda (1986), and Guvanasen and Chan (2000).

The material properties to be homogenized and upscaled are drawn from Biot's theory extended for non-isothermal consolidation (Guvanasen and Chan 2000). The modified crack tensor theory of Oda (1986) has been modified to include other transport and thermoelastic properties specified by Guvanasen and Chan (2000).

2.2 Upscaling of thermohydromechanical properties

Generalized equations describing isothermal hydroelastic equilibrium in poroelastic materials have been given by Biot (1955). These equations can be extended to describe thermohydroelastic phenomena by assuming that: 1) the fluid and solid are in thermodynamic equilibrium, and 2) a change in fluid density has no effects on the solid strain. Based on these assumptions, the following equation can be written tensorially (Guvanasen and Chan 2000):

$$e_{ij} = \left(C_{ijkl}\right)^{-1}\left(\tau_{kl} - \tau_{kl}^{o}\right) + \Delta p U_{ij} + \Delta T W_{ij} \qquad (1)$$

with i, j, k, l = 1, 2, 3; where: e_{ij} is the strain tensor, u_i the deformation in the i-th direction, C_{ijkl} the elastic constant tensor, τ_{ij} the stress tensor, τ_{ij}^0 the initial stress tensor, U_{ij} the isothermal hydroelastic constant tensor, W_{ij} the thermoelastic constant tensor, p the fluid pressure, p_{ref} the reference fluid pressure, $\Delta p = p - p_{ref}$, T the temperature, T_{ref} the reference temperature, and $\Delta T = T - T_{ref}$.

Using the assumptions stated in Section 2.1, Equation (1) may be re-written as:

$$e_{ij} = \left(C_{ijkl}^{(m)-1} + C_{ijkl}^{(f)-1}\right)\left(\overline{\tau}_{kl} - \overline{\tau}_{kl}^{o}\right) + \Delta\overline{p}\left(C_{ijkl}^{(f)-1}\delta_{kl}\right) + \Delta\overline{T}W_{ij}^{(m)} \tag{2}$$

so that:

$$\tau_{ij} = C_{ijkl}^{(B)}\overline{e}_{kl} + \Delta\overline{p}\left(C_{ijmn}^{(B)}C_{mnkl}^{(f)-1}\delta_{kl}\right) + \Delta\overline{T}C_{ijkl}^{(B)}W_{kl}^{(m)} + \overline{\tau}_{ij}^{o} \tag{3}$$

where $C^{(B)}_{ijkl}$ is the bulk elastic constant tensor $[C^{(m)}{}_{ijkl}{}^{-1} + C^{(f)}{}_{ijkl}{}^{-1}]^{-1}$, $C^{(m)}{}_{ijkl}{}^{-1}$ the compliance tensor for background rock mass, $C^{(f)}{}_{ijkl}{}^{-1}$ the compliance tensor for fractures, and $W^{(m)}{}_{ij}$ the thermoelastic tensor for background rock mass.

The compliance tensor for background rock matrix is a general expression; however, in the current work, it is defined by elastic constants. For an assumed transversely anisotropic material, the tensor is defined by five elastic constants (E_1, E_2, ν_1, ν_2, and G_2 - Young's modulus in the horizontal plane, Young's modulus in the vertical plane, Poisson's ratio in the horizontal plane, Poisson's ratio in the vertical plane, and shear modulus in the vertical plane of the background rock mass, respectively). The compliance tensor for fractures is defined by:

$$C_{ijkl}^{(f)-1} = \left(\frac{1}{\overline{h}} - \frac{1}{\overline{g}}\right)F_{ijkl} + \frac{1}{4\overline{g}}\left(\delta_{ik}F_{jl} + \delta_{jk}F_{il} + \delta_{il}F_{jk} + \delta_{jl}F_{ik}\right) \tag{4}$$

where:

$$F_{ij..k} = \rho_{Fr}^{L} R_{NFr} L_s^2 \int_{\Omega} \int_0^{d_{max}} \int_0^{E_{max}} d\, n_i n_j ... n_k \Xi(\underline{n}, d, E_M) \cdot \frac{A_f}{L_s^2} d\Omega\, dd\, dE_M \tag{5}$$

$$\overline{h} = d \int_{\Omega} \Xi(\underline{n})\, H\, d\Omega \tag{6}$$

$$H = \left[K_{no} + \frac{\overline{\tau}'_{ij} n_i n_j}{V_m}\right] \tag{7}$$

$$\overline{g} = G\frac{d}{\overline{\sigma}'_n}\overline{\tau}'_{ij} \int_{\Omega} \Xi(\underline{n})\, n_i n_j\, d\Omega \tag{8}$$

$$\overline{\tau}'_{ij} = \overline{\tau}_{ij} + p\,\delta_{ij} \tag{9}$$

$$R_{NFr} = \frac{\rho_{Fr}^{V}}{\rho_{Fr}^{L_s}} \tag{10}$$

$$A_f = Min\left(\tfrac{1}{4}\pi d^2, L_s^2\right) \tag{11}$$

and G is the joint shear stiffness, H the joint normal stiffness, V_m the joint maximum closure, K_{no} the initial joint normal stiffness, $\rho_{Fr}{}^{Ls}$ the lineal fracture density, $\rho_{Fr}{}^{V}$ the volumetric fracture density, R_{NFr} the fracture density ratio, $\underline{n}$ the unit vector normal to a fracture plane, Ω the range of solid angle in which $\underline{n}$ is defined, E_M the joint mechanical aperture, E_{max} the maximum value of mechanical aperture, d the joint diameter, d_{max} the maximum value of joint diameter, A_f the joint area and Ξ is the PDF.

The hydroelastic and thermoelastic tensors are derived based on assumptions that fractures do not contribute to thermoelastic strain, and that the rock matrix does not contribute to hydroelastic strain. The bulk hydroelastic and thermoelastic tensors (Uij(B), Wij(B), respectively) are given below.

$$U_{ij}^{(B)} = C_{ijmn}^{(B)} C_{ijmn}^{(f)-1} \delta_{kl} \tag{12}$$

$$W_{ij}^{(B)} = C_{ijkl}^{(B)} W_{kl}^{(m)} \tag{13}$$

2.3 Upscaling of hydraulic and transport parameters

Permeability Tensor A study by Long et al. (1982) suggested that rock mass behaves more like a porous medium if many fractures make a sufficient number of flow paths. Using a modified form of Oda's crack tensor theory (Oda 1986), the permeability of the homogenized material is given by:

$$k_{ij} = \lambda \tfrac{1}{12}\left(P_{kk}\delta_{ij} - P_{ij}\right) \tag{14}$$

where:

$$P_{ij} = \rho_{Fr}^{L_s} R_{NFr} L_s^2 \int_{\Omega} \int_0^{d_{max}} \int_0^{E_{max}} \frac{A_f}{L_s^2} E_M'^3 n_i n_j \Xi(\underline{n}, d, E_M) \cdot$$

$$\cdot F(E'_M, e'_h)^3 \, dd \, dE_M \, d\Omega \tag{15}$$

$$E'_M = E_M + \frac{\overline{\sigma}'_n}{H} \leq E_M - V_M \tag{16}$$

$$F(E'_M, e'_h) = \frac{1}{E'_M} Min\left(\frac{E'^2_M}{JRC^{2.5}}, E'_M \right) \tag{17}$$

and e_h is the joint hydraulic aperture, E_M the joint mechanical aperture, e'_h the joint hydraulic aperture with deformation, E'_M joint mechanical aperture with deformation, F the transformation function from deformed mechanical aperture to deformed hydraulic aperture with both apertures in microns, V_m the maximum joint closure, JRC the joint roughness coefficient, and λ the connectivity factor ranging from 0 to 1. The connectivity factor is calculated using the expression proposed by Oda (1986).

Storage Term Using the hydrolastic and thermoelastic terms in Equations (12) and (13), the storage term for the flow equation (Guvanasen and Chan 2000) is given by:

$$\begin{aligned} \varsigma = & C^{(B)-1}_{ijmn} C^{(f)-1}_{mnkl} e_{kl} + \theta_o C_f \Delta p \\ & - \left(C^{(B)-1}_{ijmn} C^{(f)-1}_{mnkl} W^{(m)}_{kl} - \theta_o \beta_f \right) \Delta T \end{aligned} \tag{18}$$

where C_f is the compressibility of fluid, θ_o the initial porosity, and β_f the thermal expansion coefficient of fluid.

Porosities The unstressed original porosities (due to mechanical and hydraulic apertures) attributed to unstressed fractures are given by:

$$\theta_{oM} = \rho^{L_s}_{Fr} R_{NFr} L^2_s \int_\Omega \int_0^{d_{max}} \int_0^{E_{max}} \frac{A_f}{L^2_s} E_M \, \Xi(\underline{n}, d, E_M) \cdot$$

$$\cdot dd \, dE_M \, d\Omega \tag{19a}$$

$$\theta_{oh} = \lambda \rho^{L_s}_{Fr} R_{NFr} L^2_s \int_\Omega \int_0^{d_{max}} \int_0^{E_{max}} \frac{A_f}{L^2_s} E_M \, \Xi(\underline{n}, d, E_M) \cdot$$

$$\cdot F(E_M, e_h) \, dd \, dE_M \, d\Omega \tag{19b}$$

where θ_{oM} is the initial porosity based on unstressed mechanical apertures, and θ_{oh} the initial porosity based on unstressed hydraulic apertures. Porosities based on deformed apertures are given by.

$$\theta'_M = \rho^{L_s}_{Fr} R_{NFr} L^2_s \int_\Omega \int_0^{d_{max}} \int_0^{E_{max}} \frac{A_f}{L^2_s} E'_M \, \Xi(\underline{n}, d, E_M) \cdot$$

$$\cdot dd \, dE_M \, d\Omega \tag{20a}$$

$$\theta'_h = \lambda \rho^{L_s}_{Fr} R_{NFr} L^2_s \int_\Omega \int_0^{d_{max}} \int_0^{E_{max}} \frac{A_f}{L^2_s} E'_M \, \Xi(\underline{n}, d, E_M) \cdot$$

$$\cdot F(E'_M, e'_h) \, dd \, dE_M \, d\Omega \tag{20b}$$

where θ'_M is the porosity based on deformed mechanical apertures, and θ'_h the porosity based on deformed hydraulic apertures. The initial porosity in Equation (19a) is used for terms that require initial porosity. For terms relating to convective components of the transport equations, the porosities in Equations (20a) and (20b) are used.

Effective Thermal Capacity Term The effective thermal capacity term, $\theta'_M \rho_f c_f + (1 - \theta'_M) \rho_s c_s$, can be determined directly from the porosity term in Equation (20a). In the expression above, ρ_f is the fluid density, c_f the thermal capacity of fluid, ρ_s is the solid density, and c_s the thermal capacity of solid. Because the porosity term is normally very small for fractured media, the thermal capacity term may be approximated using the unstressed mechanical porosity in Equation (19a), thereby removing the dependency of the effective thermal capacity term on stress, pressure, and temperature.

Effective Thermal Conductivity Term Assuming that the heat conduction term is much larger than the mechanical dispersion term, the hydrodynamic dispersion term can be approximated by:

$$H_{ij} = \lambda^T_{ij} \tag{21}$$

where H_{ij} is the thermal hydrodynamic dispersion tensor, and $\lambda_{ij}{}^T$ the thermal conductivity tensor. The above assumption has been used to simulate heater experiments at the Stripa mine in Sweden (Chan et al 1980). Simulation results suggested that the above assumption was technically sound. The validity of the assumption is partly due to the fact that the rock mass porosity in which the experiments took place was relatively small.

2.4 Homogenization scale and scale dependency

The homogenization of the thermohydromechanical, hydraulic, and transport properties is performed within a prescribed volume of L^3_S A number of joint parameters (e.g., shear and normal stiffnesses) are dependent on joint roughness coefficient and joint compressive strength (JRC and JCS, respectively). These two parameters are, in turn, dependent on the length scale of fractures.

3. UPSCALING PROCEDURE

3.1 Thermohydromechanical properties

Upscaling of thermohydromechanical properties is based on the assumption that all the properties are linear. The upscaling process is performed on a single fracture set at a time. Results from all fracture sets are aggregated toward the end of the process to provide unified bulk thermohydromechanical properties that are reflective of the presence of all fracture sets and the interaction between the fractures and the background rock matrix. The upscaling procedure for thermohydromechanical properties is briefly described below.

For a given fracture set i, determine the equivalent compliance tensor for fractures, $C^{(f)}{}_{ijkl}^{-1}$, using Equation (4). Repeat this process for all fracture sets. Combine results to obtain aggregated $C^{(f)}{}_{ijkl}^{-1}$. Construct the rock matrix compliance tensor $C^{(m)}{}_{ijkl}^{-1}$. Combine the fracture and rock-mass compliance tensors, and invert the resulting tensor to obtain $C^{(B)}{}_{ijkl}$ (see Equation (3)). Construct the hydroelastic and thermoelastic tensors using the previously obtained tensors and Equations (12) and (13).

3.2 Upscaling of hydraulic and transport properties

The upscaling procedure for upscaling hydraulic properties is similar to that described in the above subsection. For a given fracture set i, determine the permeability tensor using Equation (14), adjusting for deformation using Equations (15) to (17). Repeat the process for all fracture sets. Combine results to obtain aggregated permeability tensor. Similar to the determination of the permeability tensor, porosities are estimated setwise and aggregated using Equations (19a) to (20b).

4. AN UPSCALING EXAMPLE

The theory and procedure described in Sections 2 and 3 above were applied to the data given for BMT2. There are three formations: Formation 1 (the Bulk Longhands Farm Member), Formation 2 (the Altered Longhands Farm Member), and Formation 3 or the Fault Zone (the Faulted Longhands Farm Member). All the parameter data cited below were either extracted from or analyzed using the raw data given in the DECOVALEX III (2001) document.

Elastic constants, obtained from core or wireline data, are presented in Table 1. For the sake of brevity, only parameters specific to Formation 1 are presented in this paper (Table 2). Parameters for Formations 2 and 3 are not significantly different from those of Formation 1. In the table, the lineal fracture densities (P10 or ρ_{Fr}^{Ls}, number of fractures/ m), dip angles, and dip azimuth angles were obtained from core logs. The shear stiffness constant (g_{Sh}, dimensionless) and the initial joint normal stiffness (K_{no}, Pa/m) were obtained from the reported joint shear and normal deformation test results. Reported JRC values (adjusted from the 300-mm test scale to the 100-m homogenization scale) were used in the analysis. The normals of fracture planes in each fracture set were assumed to conform to Fisher's distribution with the given BMT2 Fisher coefficient K_w. The injection packer test data over 100 intervals, each of 1.56 m length, were used to obtain a distribution of mechanical apertures, E_M, based on the assumption that all fractures in each interval have the same aperture. It was also assumed that all fracture sets intersect the test interval. The total number of fractures in each interval was obtained from lineal fracture densities corrected by dip angles of respective fracture sets.

Table 1.Elastic properties.

Parameter	Formation		
	1	2	3
Young's modulus (GPa)	84.6	57.4	60
Poisson's ratio	0.24	0.26	0.23
Thermal expansion coefficient *(1/°C)	1×10^{-5}	1×10^{-5}	1×10^{-5}

* Assumed value

Table 2.Properties of fracture sets for Formation 1 (Bulk Longhands Farm Member).

Parameter	Fracture set			
	1	2	3	4
ρ_{Fr}^{V} (P30)	4.99	5.56	5.16	4.66
ρ_{Fr}^{Ls}(P10)	3.45	3.85	3.57	3.23
R_{NFr}(P30/P10)	1.45	1.45	1.45	1.45
P32/ P30	0.60	0.60	0.60	0.60
g_{Sh} (dimensionless)	1.35 $\times 10^4$	1.35 $\times 10^4$	1.35 $\times 10^4$	1.35 $\times 10^4$
K_{no} (Pa/m)	9.43 $\times 10^{10}$	9.43 $\times 10^{10}$	9.43 $\times 10^{10}$	9.43 $\times 10^{10}$
V_{m50} (m)	2 $\times 10^{-5}$	2×10^{-5}	2×10^{-5}	2×10^{-5}
JRC	3.2	3.8	3.1	4.6
Dip angle (°)	8	88	76	69
Dip azimuth angle (°)	145	148	21	87
Fisher K_w	5.9	9	10	10
d_{50} (m)	0.87	0.87	0.87	0.87
E_{M50} (m)	2.1 $\times 10^{-5}$	2.1 $\times 10^{-5}$	2.1 $\times 10^{-5}$	2.1 $\times 10^{-5}$

The calculation of apertures took into account the in situ effective stresses within every packer test interval using an empirical relationship (Barton et al. 1985). Owing to the lack of additional data, it was further assumed that the same set of packer test data was applicable to all the three formations. However, the interpretation for each formation differed due to formation-specific fracture characteristics. The 50th percentiles of the fracture apertures derived from the above-described packer tests for all fracture sets in Formation 1 are given in Table 2. Note that the full PDFs for all the distributed parameters were used in the upscaling calculation although only the 50th percentile values are given in the table. The surficial fracture trace length data, along with an empirical relationship between the total trace length per unit area and fracture cut-off length (DECOVALEX 2001), numbers of fractures observed in respective length intervals, dip and dip azimuth angles were used to estimate volumetric fracture densities (ρ_{Fr}^{V} or P30, number of fractures/ m^3) and fracture intensities (P32) for all fracture sets in Table 2. Details of the calculations will be presented in a publication in the near future.

Preliminary results for Formation 1 are presented in Table 3.

Table 3.Upscaled parameters for formation 1 (bulk longhands farm member).

Depth (m)	E (Pa)	k (m^2)	θ_M	θ_h
0	1.03-1.28×10^4	0.79-1.12 ×10^{-14}	1.30×10^{-4}	1.29×10^{-4}
250	0.76-10.2×10^9	.33-2.15 ×10^{-16}	2.76×10^{-5}	1.98×10^{-5}
500	13.1-17.3×10^9	4.67-7.23 ×10^{-17}	1.66×10^{-5}	1.07×10^{-5}
750	18.3-23.6×10^9	2.84-4.17 ×10^{-17}	1.23×10^{-5}	7.55×10^{-6}
750*	58.5-65.5×10^9	4.49-6.61 ×10^{-18}	1.95×10^{-6}	1.28×10^{-6}

*Assumed fracture density reduced by a factor of 3.

In the table, Young's moduli, fractured rock permeabilities, and mechanical and hydraulic porosities at different depths are presented. Effective stresses in fractures were obtained from the lithostatic stresses specified in the DECOVALEX III (2001) document corrected by corresponding hydrostatic water pressures. In the table, at each depth, a range of Young's moduli and a range of permeabilities are presented. The ranges were obtained from the three diagonal terms of the tensors of Young's modulus and permeability.

Owing to increasing effective stresses with depth, Young's moduli increase rapidly while permeabilities and porosities decrease rapidly with depth. Note that matrix porosities were not included in the table. At the 750-m depth, the properties were also calculated assuming that only 1/3 of the fractures observed in the rock mass contribute to the upscaled properties (Öhman et al 2003). This reduction in fracture densities causes Young's moduli to increase and permeabilities and porosities to decrease. Results for Formations 2 and 3 (not presented in this paper) share the same characteristics with those for Formation 1. Convergence tests, exemplified in Table 4, showed that the effective upscaled properties approach their asymptotic values at a scale length less than 10 m.

Table 4.Convergence analysis formation 1 (at the 500 m depth).

Scale length (m)	E (GPa)	k_{11} (m^2) ×10^{17}	θ_M ×10^5	θ_h ×10^5
0.5	9.41	64.5	20.9	13.7
1	10.3	34.5	12.3	7.69
2	12.2	9.45	3.53	2.44
3	13.1	4.67	1.66	1.07
10	13.1	4.67	1.66	1.07
100	13.1	4.67	1.66	1.07

A comparison between the theoretical upscaled Young's moduli and permeabilities and those upscaled by Nirex (DECOVALEX 2001) are presented in Tables 5 and 6, respectively. Nirex upscaled porosities were not available. The theoretical values are similar to the Nirex values. However, the theoretical Young's moduli tend to be consistently smaller than the Nirex values. In the two tables, ranges of Young's moduli and permeabilities are given for the case of all active fractures and the case of one-third active fractures for comparison purposes. Although inconclusive, the results suggest that not all fractures contribute to the upscaled properties.

During the upscaling process, it was observed that effects of very long fractures could not be effectively included using the proposed homogenization scheme described herein. Because the probability of occurrence (or frequency) of these fractures tends to be extremely low compared with smaller fractures, the impacts of extremely long fractures were lost in the homogenization process, which is based on the probability of occurrence of fractures of different spatial scales. It is recommended that extremely long fractures with very low probability of occurrence be treated deterministically. In other words, the domain of interest should consist of two components: the background rock mass, and the extremely long and large fractures. The homogenization process should be utilized to provide upscaled properties for the background rock mass. The extremely large fractures should be modelled as discrete fractures.

Table 5.Comparison between upscaled permeabilities (k in m^2) and Nirex values.

Formation	Upscaled k	Nirex k	Remarks
1	$4.67\text{-}7.23\times10^{-17}$	5.67×10^{-18}	(1)
	$0.74\text{-}1.15\times10^{-17}$		(2)
2	$2.16\text{-}3.74\times10^{-14}$	8.51×10^{-14}	(3)
	$5.30\text{-}8.82\times10^{-15}$		(2)
3	$2.54\text{-}4.07\times10^{-12}$	2.69×10^{-13}	(3)
	$5.55\text{-}8.82\times10^{-13}$		(2)

Notes: (1) Assumed calculated at 500 below ordnance datum; (2) Fractured density reduced by a factor of 3; (3) Calculated at ground surface.

Table 6.Comparison between upscaled Young's moduli (E in GPa) and Nirex values.

Formation	Upscaled E	Nirex E	Remarks
1	13-17	65	(1)
	32-39		(2)
2	15-16	44	(1)
	32-44		(2)
3	11-16	44	(1)
	25-33		(2)

Notes: (1) Depth at which E values were calculated is not reported; (2) Fractured density reduced by a factor of 3.

5. SUMMARY AND CONCLUSIONS

The crack tensor theory of Oda (1986) was modified and enhanced in order to upscale or homogenize physical properties of fractured rock mass within a predetermined upscaling volume. Parameters upscalable by the revised methodology include: thermohydroelastic and storage properties based on Biot's theory, and stress-dependent flow and transport properties. The revised methodology was applied to DECOVALEX III Bench Mark Test Case 2 (BMT2) which is based on a hypothetical site with three fractured rock formations. It was found that the upscaled equivalent rock mass permeability, hydraulic and mechanical porosities, and Young's modulus approached respective asymptotic values at an upscale length of less than 10 meters. The upscaled properties are highly dependent on effective stress and, therefore, depth below surface. The theoretically determined values of Young's modulus and permeability are generally very similar to, or on the same order of magnitude as, those obtained by Nirex and cited in the BMT2 Problem Definition and Description document. Field-observed values of other thermohydroelastic properties, and porosities were not available for comparison. The data also suggested that extremely long and large fractures with very low probability of occurrence (or frequency) should be modelled as discrete fractures.

ACKNOWLEDGEMENT

The work described in this paper was funded by Ontario Power Generation as part of the Deep Geologic Repository Technology Program.

REFERENCES

Barton, N., Bandis, M. & Bakhtar, K. 1985. *Strength, deformability and conductivity coupling of rock joints.* Int. J. Rock Mech. Min. Sci. & Geomech. Abstr., **22**, 121-140.

Biot, M.A. 1955. *Theory of elasticity and consolidation for a porous anisotropic solid.* Journal of Applied Physics, 26: 182-5.

Chan, T., Javandel, I., & Witherspoon, P.A. 1980. *Heat transfer in underground heating experiments in granite, Stripa, Sweden.* In Heat Transfer in Nuclear Waste Disposal, ASME monograph HTD Vol. 11, 1980, Ed. F.A. Kulacki and R.W. Lyckowski, pp. 1-8.

DECOVALEX III. 2001. *Benchmark Test 3 and BENCHPAR Work Package 3 – Problem definition and description. Version 6 posted on www.decovalex.com*

Guvanasen, V., & Chan, T. 2000. *A three-dimensional numerical model for thermohydromechnical deformation with hysteresis in a fractured rock mass,* Int. J. Rock Mech and Mining Sc., 37: 89-106.

Long, J.C.S., Remer, J.S., Wilson, C.R. & Witherspoon, P.A. 1982. *Porous media equivalents for networks of discontinuous fractures.* Water Resources Research, Vol. 18, No. 3, pp: 645-658.

Oda, M., 1986. *An equivalent continuum model for coupled stress and fluid flow analysis and jointed rock masses,* Water Resources Research, V. 22, No. 13, pp. 1845-1856.

Öhman, J., Antikainen, J, & Niemi, A. 2003. *Thermo-mechanical effects of hydraulic conductivity in a nuclear waste repository setting.* In the Proceedings of this conference.

EFFECT OF THE FRACTURE GEOMETRY ON THE COUPLED PHENOMENA IN LARGE SCALE

Akira KOBAYASHI [1], Yutaka SUGITA[2] and Masakazu CHIJIMATSU[3]
1) Kyoto University
2)Japan Nuclear Cycle Development Institute
3) Hazama Corporation

Abstract: For the performance assessment of the high level radioactive waste repository, the transport analysis in large region will be carried out. For this aim, the up-scaling of the permeability has to be examined. In this paper, the permeability is up-scaled by the pixel method and the effect of the fracture geometry on up-scaled parameter is evaluated. Then, thermal and hydraulic coupled analysis is carried out and then the performance of the far field is investigated by a particle tracking method. By discussing the change in the breakthrough curve due to the difference of the fracture geometry, it is found that the fracture geometry has a relatively large effect.

1. INTRODUCTION

For the performance assessment of radioactive waste disposal repository, the solute transport in the large region has to be examined. Probably, the effect of the heat generation from the repository has a significant role on the flow behaviour, if the leakage is caused at the early stage of the disposal. When a finite element method is used for the assessment, the region is divided by the elements, which have a parameter value as a material property. In this process, the elemental value is decided from some homogenization method because the scale of element is, in general, larger than that of in-situ test and many fractures are included into the volume of an element. As a homogenisation method, the crack tensor theory (Oda, 1986) is useful while the method cannot treat the effect of the heterogeneous distribution of the fractures. To treat the heterogeneous fracture geometry, the discrete fracture network method (e.g., Dershowitz et al., 1991) is effective. The method is, however, constrained by the computer capacity and so the selection of the hydraulically effective fracture has to be needed. At present, the objective criterion of the selection has not been examined well. To avoid the difficulty of both methods, a pixel method is used in this study, which method can treat the fractures from small to large in the same manner. However, the smallest area called pixel has to be decided arbitrarily. To obtain the homogenized property of a given area consisting of many pixels, the averaged property is calculated under some boundary condition giving one-dimensional flow field.

There may exist a scale effect of homogenized value, which scale effect is inherent for fracture system. On the other hand, the size of element is dependent on the size of structure and objective of the analysis, e.g., near or far field examination. Thus, the parameter value of an element has to be properly set for the size of element. In this study, a concept to treat the scale effect of the element is introduced.

The variation of the permeability distribution can be used for the analysis of the dispersion behaviour by ergodic hypothesis. The degree of the variation is dependent on the fracture geometry. In this study, the effect of the fracture geometry on the flow regime in the coupled behaviour is examined by considering some different fracture systems.

2. MODELING STRATEGY

Firstly, the fractures are generated in 60m cube according to the fracture geometry information, which was obtained from the data of Sellafield, Cumbria, UK. The central horizontal section is considered as an original plane. The density of fracture is determined from the total fracture length per unit area, P_{21}, on the plane. The original plane is divided by 0.2m squares, i.e. pixel. The number of fracture existing in each pixel is counted and the permeability proportional to the number of fracture is provided to each pixel. The size of homogenized area is 5m, 10m, 15m, 20m, 30m and 60 m square, which consists of pixels. The samples of each size are derived from the original plane. The directional permeability of the multiple samples is examined and the average and standard deviation are obtained. The effect of the scale is examined by the variation of the average homogenized permeability and its standard deviation. To examine the effect of fracture geometry on the scale effect, some parametric studies are carried out.

From the results of up-scaling examination, the concept to treat the scale effect of permeability is considered. Then, by using the concept, the large scale analysis is carried out by considering the coupled thermal and hydraulic phenomena. At that time, some parametric studies are also carried out.

2.1 Fracture length

Cumulative density of fracture trace length *d(L)* is given by (Nirex, 1997a)

$$d(L) = 4 \times L^{-2.2} \tag{1}$$

in which L is the trace length (m). By using this equation, the total number of fractures of all sets per square meters is

$$P_{20} = 4 \times (0.5)^{-2.2} - 4 \times (250)^{-2.2} = 18.4 (m^{-2}) \tag{2}$$

In this process, the maximum length of 250 m and the minimum length of 0.5 m are assumed. Thus, the probability density function, PDF, of fracture length is written by

$$PDF = \frac{4 \times (2.2) \times L^{-3.2}}{P_{20}} \tag{3}$$

From this equation, the length is generated by

$$L = \left\{ \left(4 \times (0.5)^{-2.2} - R_0^1 P_{20}\right) \frac{1}{4} \right\}^{-\frac{1}{2.2}} \tag{4}$$

in which $R_0^{\,l}$ is the random number between 0 and 1. The mean fracture length is 0.92m.

To examine the uncertainty related to the fracture length, the case that the mean length is 2.13m and the exponent of L is 1.2 is examined. Moreover, the case having the lognormal distribution of the fracture length with the mean length and variance from 4,000 realizations of Equation (4) is also examined. The latter case is to examine the uncertainty for the interpretation of the observation result.

2.2 Fracture density

Fracture density by which the number of generated fracture is determined is defined by P_{21}. P_{21} is assumed to be given by the 2-D fracture density (Nirex, 1997b)

$$P_{21} = 2.4 \times (0.5)^{-1} = 4.8 \ (m^{-1}) \tag{5}$$

However, the other consideration can be made by obtaining the total fracture length through the mean fracture length and total number of fracture.

$$P_{21} = 0.92 \times P_{20} = 0.92 \times 18.4 = 16.9 (m^{-1}) \tag{6}$$

Since this is considered as an uncertainty for the fracture information, the difference of the up-scaling result for both definitions are examined.

2.3 Permeability

The permeability of the pixel is calculated by

$$\log(K_{Pixel}) = \frac{N_{frac} - \mu_{frac}}{\xi_{frac}} \xi_{perm} + \mu_{perm} \tag{7}$$

where μ_{fac} is the average number of fractures in pixel, ξ_{frac} is its standard deviation, μ_{perm} is the average of measured log(K) and ξ_{perm} is its standard deviation. The statistical information of permeability is given in the UK data (Nirex, 1997c). In this examination, μ_{perm} of –7.07 and ξ_{perm} of 0.604 for Formation 2 of Sellafield site are used. Thus, the distribution of the permeability of the pixel is log-normal.

3. SCALE EFFECT OF PERMEABILITY

3.1 Methods

Figure 1 shows the distribution of the pixels having more than one fracture on the original plane. The size of the plane is 60 x 60 m. By dividing this plane by a given size, the samples are produced. The sample has a distribution of the permeability of the pixel according to Eq. (7). The sample is tested by imposing the boundary condition shown in Figure 2. The horizontal and vertical average permeability is tested. Since multiple samples are obtained from the reference plane, the average and the standard deviation are calculated for different size.

3.2 Examination cases

To investigate the effect of fracture geometry on the scale effect of the permeability, some parametric study is carried out. Table 1 shows the parameters examined for this aim. Case 1 is the reference case in which the original measured data at Sellafield are used. Case2 is the case to examine the effect of uncertainty on the fracture density, in which the density is defined by Eq. (6). Thus the number of fracture in this case is about 4 times that in Case 1. Case 3 is related to the uncertainty on the

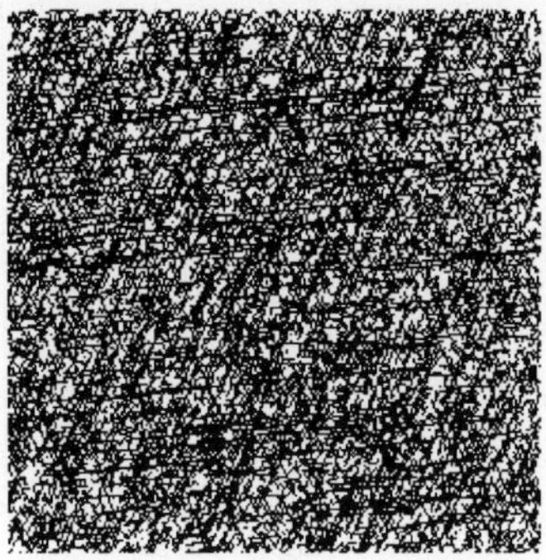

Figure 1. Distribution of pixel having more than one fracture

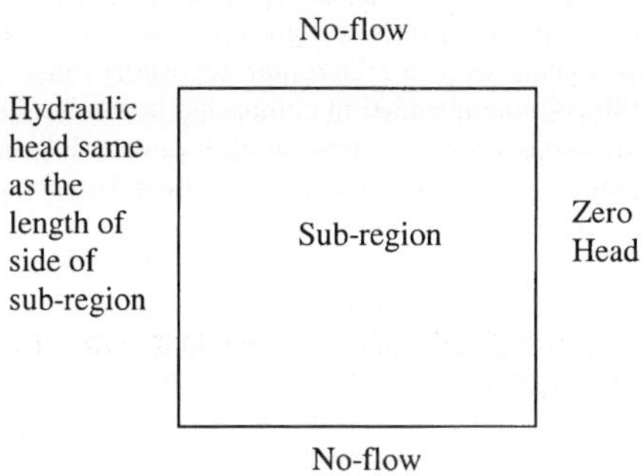

Figure 2. Boundary condition for examination of average directional permeability

fracture length information, in which the average fracture is set at double as mentioned in Section 2.1. Case 4 is related to the uncertainty on the fracture length modelling. In this case, the probability density function of fracture length is assumed to be a lognormal having the same average and standard deviation as those from 4,000 realizations of Eq. (4). This is related to the uncertainty about the interpretation of field data.

Table 1 Examination cases for up-scaling

Case	Remarks
1	Reference case, Eqs. (4) and (5)
2	Fracture density by Eq. (6)
3	Double mean fracture length of Case 1
4	Different fracture length modelling

3.2 Results and conclusions

Figures 3-6 show the examination results of each case. It is seen that the average vertical permeability is larger than average horizontal one for all cases, while the standard deviation does not have anisotropic tendency. Moreover, as a tendency of all cases, the scale effect of the average permeability is smaller than that of standard deviation. The average permeability shows a very small change with scale and can be considered a constant value as a practical matter. It can be, thus, concluded as a scale effect model that the average permeability is a constant, while the standard deviation becomes small with increase in scale. This indicates that the large and small elements have the same average permeability, while the standard deviation of small element is larger than that of large element. Therefore, the small elements have a large variation in the distribution, although the large element has mostly constant permeability.

For the effect of the fracture density, it is found that the difference is not seen between Figures 3 and 4. This is probably because the density larger than 4.8 m^{-1} gives the enough connectivity of the flow paths.

About the fracture length information, if the average fracture length becomes doubled, the scale giving the stable permeability is also doubled. The length of side giving the stable average vertical permeability in Case 1 is about 15 m while that in Case 3 is about 30 m as shown in Figure 5. Moreover, the average permeability and standard deviation of Case 3 become a little larger than those of Case 1. On the other hand, if the probability density function of the fracture length is lognormal, as shown in Figure 6, the scale effect becomes smaller than that in Case 1 where the function is a power law as shown in Eq. (3). The average and standard deviation of permeability of Case 4 becomes a little smaller than those of Case 1

It can be concluded from the above discussion that the uncertainty of the fracture length is more important than that of fracture density.

4. FLOW IN LARGE SCALE

4.1 Model description

Thermal and hydraulic coupling analysis is carried out to examine the effect of uncertainty related to the up-scaling method on the performance assessment. The hypothetical region is shown in Figure 7. At the depth of 530m, the high level

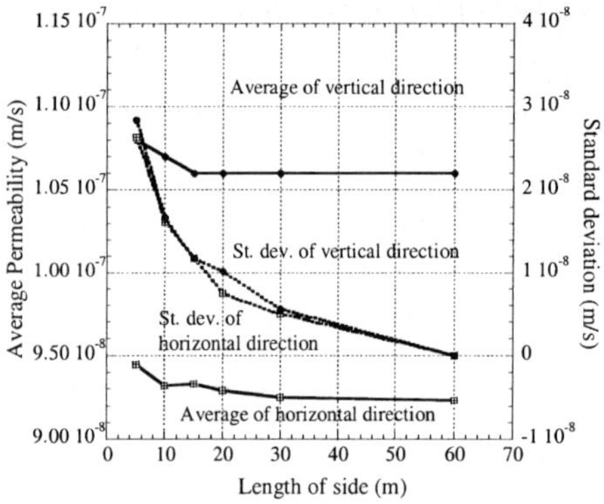

Figure 3. Scale effect of Case 1

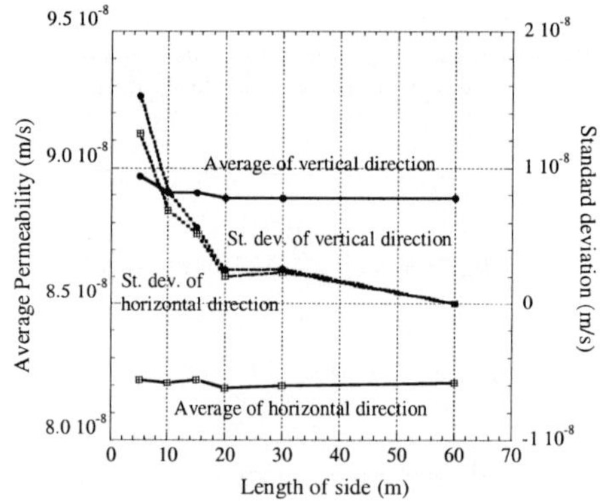

Figure 6. Scale effect of Case 4

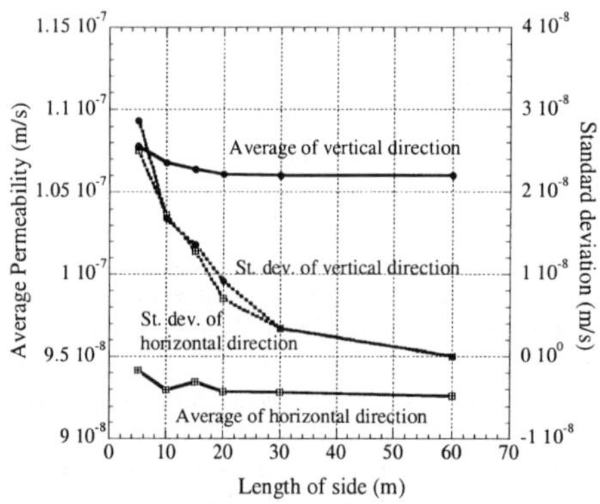

Figure 4. Scale effect of Case 2

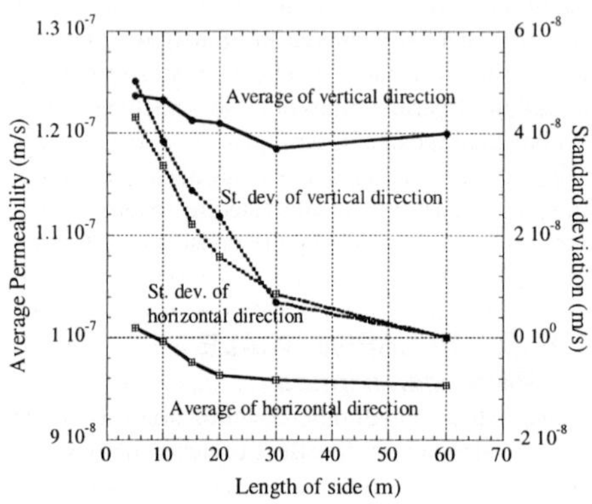

Figure 5. Scale effect of Case 3

radioactive waste disposal repository is located and comprise 60 canisters of high level waste uniformly distributed over a horizontal repository area of 100x100 m embedded in compacted bentonite. The heat evolution versus time for each canister is given from a realistic consideration. The boundary condition for thermal and hydraulic analysis is also shown in Figure 7. Since the water table is fixed at the ground surface level, the groundwater moves from the high hill part at left side to the sea at right side. The left hill is flat and 100 m above sea level.

Figure 8 shows the finite element mesh. The above-mentioned treatment of the scale effect of permeability is applied to this mesh. Although the results of particle tracking method have a mesh dependency, only the mesh shown in Figure 8 is used. As mentioned above, the small element has a large variance of the lognormal distribution, while the large one has a mostly constant permeability. For 30 realizations of the field, the coupled thermal and hydraulic analysis is carried out and the velocity distribution is obtained. Then, the particles from the repository to the sea are traced by using the velocity distribution.

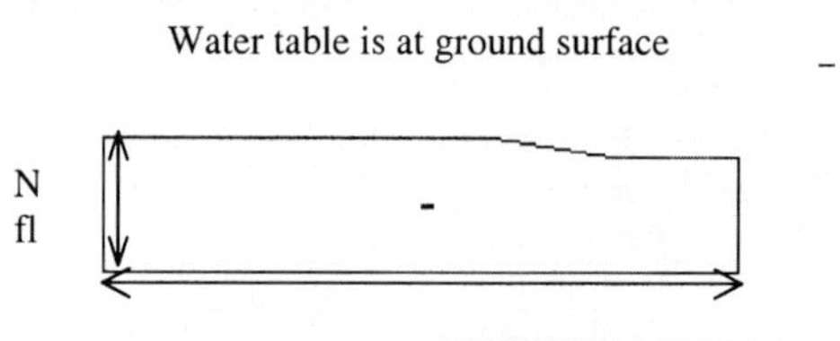

Figure 7. Model for large region analysis

The break through curve is obtained by summing up the results of 30 realizations. The results are examined with the break through curve at sea floor.

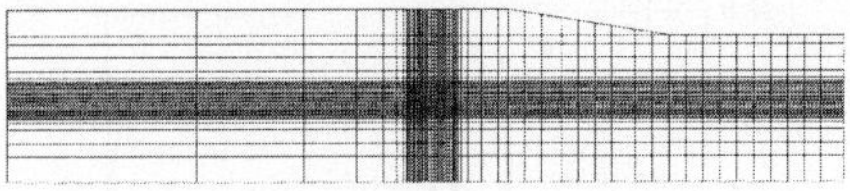

Figure 8. Finite element mesh

4.2 Conditions and cases

Table 2 shows the properties used for the coupled analysis. Thermal properties are the same for rock mass and bentonite. It is assumed that the repository consists of bentonite having very low permeability. To examine the uncertainty of upscaling process on the performance assessment, the cases shown in Table 3 are examined.

Case A is the reference case where the data measured at Sellafield site is used. The scale effect of standard deviation and the average permeability are obtained from the results of Case 1.

Case B is the case in which mean fracture length is doubled, which case is corresponding to Case 3 in the above section. It is expected from this case that the uncertainty related to the fracture length data is examined.

In Case C, the horizontal and vertical permeability is set vice versa. This is concerning to the uncertainty about the fracture direction. If the fracture direction is measured incorrectly, the anisotropy of the permeability may be differently evaluated. In this study, while the degree of anisotropy is not so large, the effect of the small difference of the interpretation on the anisotropy is examined.

In Case D, the relation of the thermal disturbance and value of permeability is discussed. While the heat generation at the repository is assumed by the realistic estimation, the thermal disturbance around the repository is very dependent on the permeability of rock mass. When the permeability is low, the buoyancy effect by heat generation can be emphasized because the boundary condition does not have a large effect on the entire groundwater movement.

When the particle is released from the repository, two cases can be assumed. One is the case when the particle does not go through the bentonite. The other is the case when the particle goes through the bentonite. The former case indicates

Table 2 Properties for large scale analysis

Terms	Values
Thermal conductivity	2.29 W/mK
Specific heat	798 J/kg/K
Effective porosity of rock	9.67%
Effective porosity of bentonite	40.9%
Permeability of bentonite(-logK)	13.59(m/s)

Table 3 Cases for large scale analysis

Case	Remarks
A	Reference case
B	Double mean fracture length
C	Converse anisotropy
D	10^{-4}-fold Mean permeability

the direct effect of the modelling of the rock mass. On the other hand, although the latter case is realistic, the effect of the low permeability of the bentonite is emphasized on the result. In this study, the results of both cases are presented.

4.3 Results and discussions

Figure 9 shows the break through curve at the sea floor for the case when the particle does not go through the bentonite. Since the permeability is set at 10^{-4} times in Case D, the peak time is delayed at the mostly same order. From Case B, if the mean fracture length is about 2 m, the resultant peak time becomes faster by a few hundreds years than that in the case where mean length is about 1 m. Moreover, it is found from Case C that the error related to the anisotropy has a relatively large effect on the peak time in spite of the case where the geometry does not show the clear anisotropy as shown in Figure 1.

Figure 10 shows the results of the case when the particle goes through the bentonite. It is found that the difference among Cases A, B and C is hardly seen. This is because time taken for going through the bentonite is quite large. It is concluded from this result that the error about the interpretation of the fracture geometry does not have an effect on the performance assessment if the buffer material works well. Moreover, the difference of peak time between Cases A and D is also smaller than that of permeability. It is found that the effect of the error related to the evaluation of the permeability value of rock mass becomes small when the buffer material is well constructed.

Figures 11 and 12 show the velocity distributions around the repository at the time of 70 years from heating for Cases A and D. The effect of buoyancy is small in Case A because the large permeability of rock mass induces intensively the groundwater flow due to boundary condition. On the other hand, the buoyancy effect can be seen in Case D, although the upward velocity is very small. However, this buoyancy effect on the break through curve is very small.

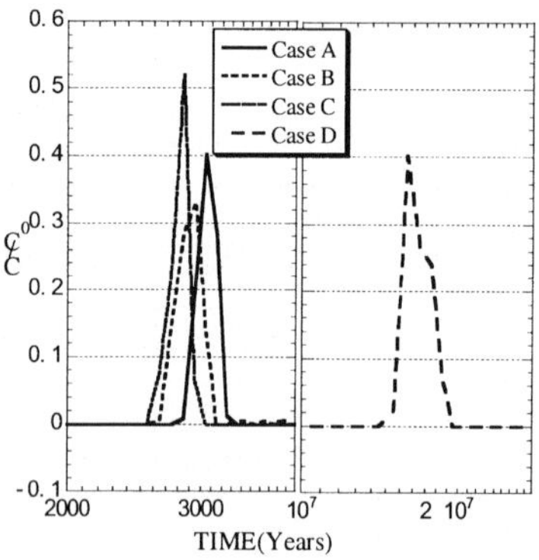

Figure 9. Break through curves for the case when particle does not go through bentonite

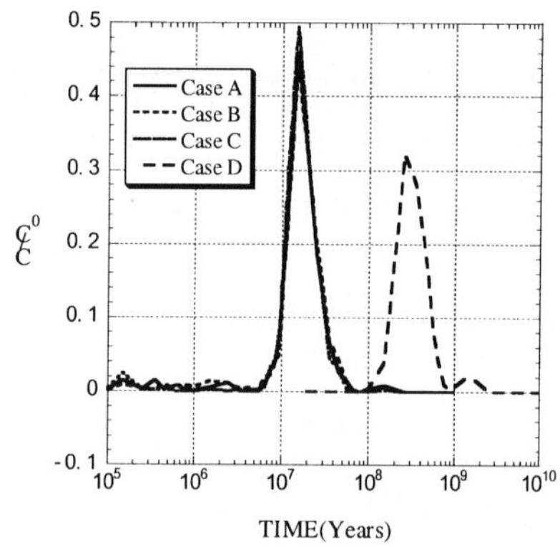

Figure 10. Break through curves for the case when particle goes through bentonite

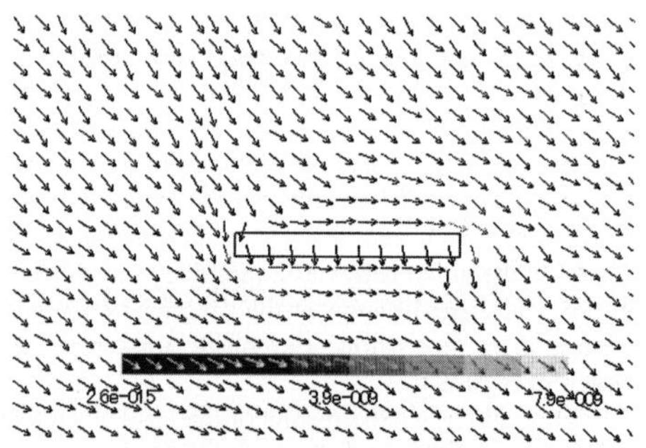

Figure 11. Velocity distribution around repository of Case A at the time of 70 years from start

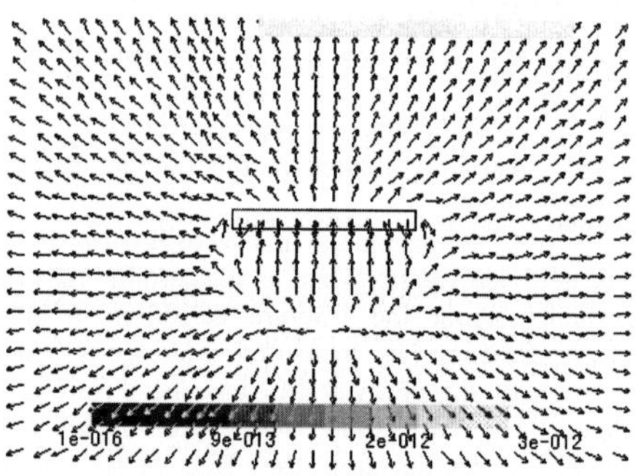

Figure 12. Velocity distribution around repository of Case D at the time of 70 years from start

REFERENCES

Dershowitz W., Wallmann P., Kindred S. 1991. *Discrete fracture modelling for the Stripa site characterization and validation drift inflow prediction*, Stripa Project Technical Report 91-16, SKB.

Nirex 1997a. *The lithological and discontinuity characteristics of the Borrowdale Volcanic Group at outcrop in the Craghouse Park and Latterbarrow areas*, Nirex Report SA/97/029.

Nirex 1997b. *Evaluation of heterogeneity and scaling of fractures in the Borrowdale Volcanic Group in the Sellafield area*, Nirex Reports SA/97/028.

Nirex 1997c. *An assessment of the post-closure performance of a deep waste repository at Sellafield: Volume 2 –Hydrogeological model development- Effective parameters*, Nirex Reports S/97/012.

Oda M. 1986. *An Equivalent Continuum Model for Coupled Stresses and Fluid Flow Analysis in Jointed Rock Masses*, Water Resources Research, Vol. 22, No. 13:1845-1856.

UPSCALING OF NORMAL STRESS-PERMEABILITY RELATIONSHIPS FOR FRACTURE NETWORKS OBEYING FRACTIONAL LEVY MOTION

Hui-Hai Liu, Jonny Rutqvist, Quanlin Zhou and Gudmundur S. Bodvarsson

Earth Sciences Division, LBNL, Berkeley, California

Abstract: Understanding and modeling coupled thermal, hydrological, mechanical, and chemical (T-H-M-C) processes in fractured rocks are of interest to many areas of active research, including geological disposal of nuclear waste. A key parameter for modeling these processes is the relationship between *in situ* stress and permeability. Since this relationship is generally measured at small scales, upscaling is needed for large-scale models. Few studies on the upscaling of this relationship have appeared in the literature. The major objective of our present effort is to develop closed-form upscaled normal stress-permeability relationships for fracture networks. Focusing on data from Sellafield site, UK, we demonstrate that measured permeability data form a distribution very well described by fractional Levy motion (fLm). This is consistent with findings reported in the literature indicating that fractures are spatially clustered and that the clustering patterns could be described by fractals. Assuming several different correlations between local permeability and fracture apertures, we develop two upscaling relationships between normal stress and permeability for fracture networks. These relationships capture the relevant large-scale effects of normal stress change on rock permeabilities, at least for the case of permeability distributions characterized by fLm.

1. INTRODUCTION

Geological disposal of nuclear waste generally involves coupling between thermal, hydrological, mechanical, and chemical (T-H-M-C) processes. This coupling can have a significant effect on rock mass stability, groundwater flow, and subsurface contaminant transport. Understanding and modeling the coupled T-H-M-C processes in fractured rocks is, therefore, important for assessing the performance of a geological disposal site. This study focuses on several important issues related to modeling large-scale coupled processes.

T-H-M-C processes are significantly affected by subsurface heterogeneity, which results in scale-dependence of the related parameters. To handle this scale-dependent behavior, we need to characterize this heterogeneity and consider its effects at different scales. In this study, we demonstrate that the measured permeability data from Sellafield site, UK, are very well described by fractional Levy motion (fLm), a stochastic fractal. This finding has important implications for modeling large-scale coupled processes in heterogeneous fractured rocks.

A key parameter for modeling T-H-M-C processes is the relationship between *in situ* stress and permeability. Since this relationship is generally measured at small scales, upscaling is needed for large-scale models. Few studies on upscaling of this relationship have appeared in the literature. In this study, we develop closed-form upscaled normal stress-permeability relationships for fracture networks that can be used for large-scale modeling studies.

2. THE FLM AND A PERMEABILITY DATA SET

While a number of researchers have used fLm to characterize subsurface heterogeneities in porous media (e.g., Painter, 1996; Liu and Molz, 1997; Molz and Liu, 1997; Liu et al., 2003), the usefulness of fLm for dealing with heterogeneity in fractured rocks has not been investigated in the literature. In this section, we present a brief introduction to fLm and then demonstrate the consistency between measured permeability data from the Sellafield site and fLm.

2.1 Fractional Levy Motion

Levy-stable distribution associated with the fLm is a generalization of the well-known Gaussian distribution, given by

$$f(x) = \pi^{-1} \int_0^\infty \exp(-|Ck|^\alpha) \cos(kx) dk \qquad (1)$$

where α is the Levy index and C is the width parameter. The Levy stable distribution will be reduced to the Gaussian distribution for α =2.

Width parameter C corresponds to the standard deviation of Gaussian distribution. Compared with Gaussian distribution, Levy distribution is characterized by long tails and infinite theoretical second (and higher) moments. We further define increments of log(K), where K refers to permeability, as

$$\Delta(h) = \log K(z+h) - \log K(z) \qquad (2)$$

where z is the spatial coordinate and h (m) is the lag. If a log(K) distribution is consistent with fractional Levy motion, the probability distributions of log(K) increments for different lags follow Levy-stable distributions (Equation 1), with the same Levy index and different width parameters given by

$$C(h) = C_0 h^H \qquad (3)$$

where C_0 and H are constants, with H representing the Hurst coefficient.

The parameters characterizing fractional Levy motion include α, C, C_0, and H. To determine these parameters from observed data, we used the quantile-based estimators for α and C for a given lag h (Fama and Roll, 1972). The scaling parameter H is calculated by fitting Equation 3 to the estimated C as a function of lag h.

2.2 Analysis of a Permeability Data Set

A permeability data set for fractured rocks was collected from the short interval tests carried out at the Sellafield site (United Kingdom Nirex Limited, 1997). These tests were conducted over a 156 m length interval and typically tested 1 to 2 m lengths of the borehole. They are comprised of 100 pulse injection tests and six production tests.

Figure 1 shows that the Levy-stable distribution fits the data much better than the Gaussian distribution, suggesting a strong non-Gaussian behavior. Figure 2 shows the match between Equation 3 with C values estimated from the short-interval testing data, indicating a fractal scaling relationship. Analysis results are supported by a number of studies reported in the literature. For example, based on observed data, Hirata (1987) and Yamamoto et al. (1993) found that fractures are spatially clustered and that the spatial clustering patterns could be described by fractals. Obviously, spatial clustering patterns of fractures should be closely related to spatial distributions of fracture permeability.

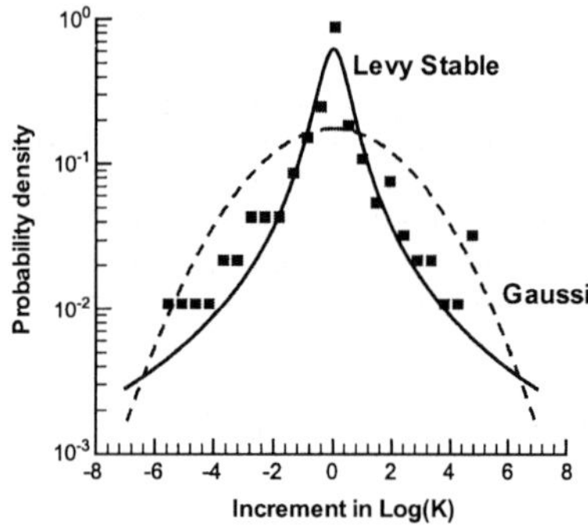

Figure 1. A comparison between a Gaussian distribution with variance determined from the data and a Levy-stable distribution with C and α values estimated from the data for h=1.56 m.

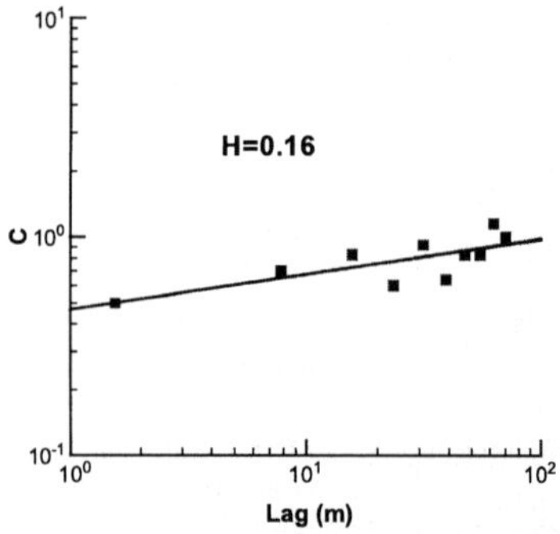

Figure 2. Levy width factor (C) as a function of lags (Levy structure function).

3. UPSCALING NORMAL STRESS-PERMEABILITY RELATIONSHIPS

Stress-induced changes in fracture aperture can give rise to significant changes in the rock mass permeability, through a sensitive "cubic" relationship between fracture aperture and fracture flow. This section discusses how to determine this relationship for large-scale heterogeneous fractured media (characterized by fLm) from small-scale laboratory measurements. The determination is based on the assumptions that an equivalent heterogeneous-porous medium could approximately describe flow processes within a subgrid fracture network (Jackson et al., 2000), and stress distributions at subgrid scale are uniform.

3.1 Laboratory Data and Upscaling Procedure

A small-scale stress-permeability (or aperture) relationship is generally obtained from laboratory tests on single fractures. As an example, Figure 3 shows measured normal stress-aperture relations for a rock sample collected from the Shellafield site. The observed function can be fitted by:

$$b = b_r + b_m = b_r + b_{\max} \exp(-\alpha' \sigma) \qquad (4)$$

where b is the total hydraulic aperture, b_r is the hydraulic residual aperture, b_m is the "mechanical aperture", b_{max} is the maximum "mechanical aperture", α' is a parameter related to the curvature of the function, and σ is the normal stress across the fracture. The stress-aperture relationship can be easily translated into a stress-permeability relationship (see Section 3.2).

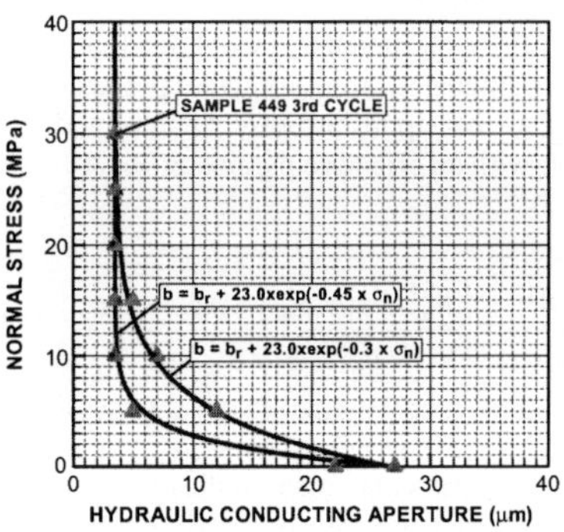

Figure 3. Stress vs. aperture function (Equation 4) fitted to laboratory experimental results

Upscaling of the stress-permeability relationship is implemented by a stochastic simulation of stress and fluid flow through a heterogeneous domain corresponding to a gridblock for a large-scale model (Figure 4). As an example, we consider the size of local scale to be 1.56 m, corresponding to the average measurement interval for the short-interval well tests. The initial permeability field is generated from fLm with parameter values determined from the data (Section 2), using the method of Liu et al. (2003). Then, we simulate the effective permeability (K_{eff}) for the gridblock consisting of a number of 1.56 m x 1.56 m blocks, using the TOUGH2 code (Pruess, 1991). A stress change is then induced, and a new flow simulation is conducted to derive a new effective permeability. For simplicity, we use a single stress-permeability relationship for both loading and unloading conditions. An average α' value (0.38), determined from Figure 3, is used for the stress-permeability relationship at the local scale.

An important step for upscaling the stress-permeability relationship is to relate local permeability (at a 1.56 m scale) to parameters b_r, b_{max}, and fracture frequency f. We assume that at the 1.56 m scale, local permeabilities can be considered to result from horizontal and vertical fractures. Different assumptions regarding relations among local permeability and the relevant parameters (f, b_r, and b_{max}) will give rise to different upscaling relationships. In this paper, we present two useful relationships.

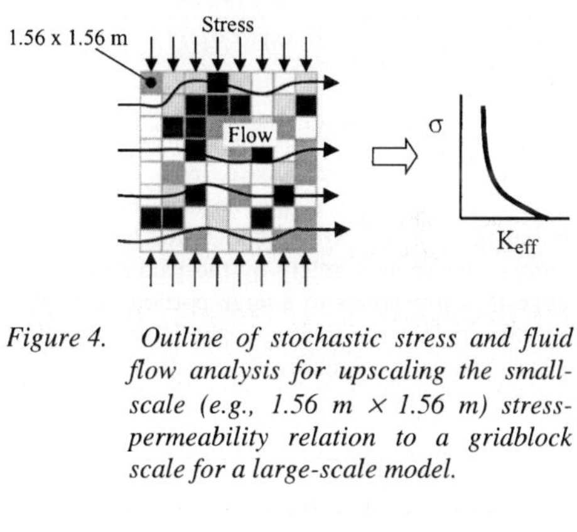

Figure 4. Outline of stochastic stress and fluid flow analysis for upscaling the small-scale (e.g., 1.56 m × 1.56 m) stress-permeability relation to a gridblock scale for a large-scale model.

3.2 Upscaling Relationship I

Local permeability variation is expected to result from both fracture frequency f and aperture variations. In this model, we assume that all fractures are identical and that permeability variation results purely from changes in f. Based on the "cubic" law, the relation among fracture aperture, frequency and local permeability K is (CRWMS M&O, 1999):

$$b = (\frac{12K}{f})^{1/3} \tag{5}$$

Combining Equations 4 and 5 gives

$$\frac{K(\sigma_{in},\Delta\sigma)}{K(\sigma_{in})} = \left(\frac{b}{b_{in}}\right)^3 = \left[\frac{R_b \exp(\alpha'\sigma_{in}) + \exp(-\alpha'\Delta\sigma)}{R_b \exp(\alpha'\sigma_{in}) + 1}\right]^3 \tag{6}$$

where σ is the (effective) normal stress (MPa), subscript "in" refers to the initial (*in situ*) conditions, $R_b = b_r/b_{max}$, and $\Delta\sigma$ is the stress change ($\sigma - \sigma_{in}$). Since all fractures are assumed to be identical, R_b will be a constant through the model domain.

Note that local permeability ratio (Equation 6) does not depend on local permeability and frequency. Therefore, the same relation (Equation 6) is also valid for effective permeability (K_{eff}) ratio at the gridblock scale:

$$\frac{K_{eff}(\sigma_{in},\Delta\sigma)}{K_{eff}(\sigma_{in})} = \left[\frac{R_b \exp(\alpha'\sigma_{in}) + \exp(-\alpha'\Delta\sigma)}{R_b \exp(\alpha'\sigma_{in}) + 1}\right]^3 \tag{7}$$

In other words, the permeability ratio is not scale-dependent in this case, and a numerical upscaling procedure is not needed. For a given initial stress and a stress change, the permeability ratio is determined solely by the parameter R_b. A large R_b corresponds to a large portion of residual aperture (b_r) in the total aperture b, and therefore gives rise to a relatively weak response to mechanical process.

3.3 Upscaling Relationship II

Although Relationship I gives a simple, closed-form stress-permeability relation at gridblock scale, it is not consistent with field observations from the unsaturated zone of Yucca Mountain, Nevada. This inconsistency shows that the permeability ratio depends on local permeability. A smaller initial local permeability generally corresponds to a larger permeability ratio. Motivated by this observation, we developed the second relationship by assuming that single vertical and horizontal fractures exist within a 1.56 m × 1.56 m block (Figure 4). The permeability variation is assumed to result from b_r variation while b_{max} is fixed (b_{max} = 2.3E-5 m as given in Figure 3).

The numerical upscaling procedure described in Section 3.1 above is used for a 50 m × 50 m gridblock containing a heterogeneous permeability distribution characterized by fLm (Figure 4). Within a 1.56 m × 1.56 m block, the initial aperture is calculated from Equation 5, using the corresponding local permeability. Then b_r at that small block is calculated from Equation 4 for a given initial stress. Note that Equation 6 is still valid for the local scale permeability ratio, but Equation 7 does not hold for the 50 m × 50 m gridblock, because R_b is not constant anymore.

For a given mean initial log(K) for the 1.56 m x 1.56 m blocks (Figure 4), we can simulate effective permeability ratios for the gridblock for different initial stresses and stress changes. Based on Equations 4 and 5, the local scale R_b is given by

$$R_b = \left(\frac{K(\sigma_{in})}{K^*}\right)^{1/3} - \exp(-\alpha'\sigma_{in}) \tag{8-1}$$

$$K^* = \frac{b_{max}{}^3 f}{12} \tag{8-2}$$

Numerical experiments (see Section 3.1) indicate that simulated stress-effective permeability relations can be approximately represented by Equation 7, with an effective parameter R_b^* :

$$\frac{K_{eff}(\sigma_{in},\Delta\sigma)}{K_{eff}(\sigma_{in})} = \left[\frac{R_b^* \exp(\alpha'\sigma_{in}) + \exp(-\alpha'\Delta\sigma)}{R_b^* \exp(\alpha'\sigma_{in}) + 1}\right]^3 \tag{9}$$

Numerical experiments also indicate that the effective parameter R_b^* can be approximated by a modified form of Equation 8:

$$R_b^* = \left(\frac{K_{eff}(\sigma_{in})}{K^*}\right)^{1/3} - \exp(-\alpha'\sigma_{in}) \tag{10}$$

As an example, Figure 5 shows a comparison between results calculated from Equations 9 and 10 with numerical simulation results (data points) for initial mean log(K) =-15. A good agreement is obtained. Note that both parameter R_b* and permeability ratio (Equation 9) are scale-dependent here because K_{eff} depends on the measurement scale (Equation 10).

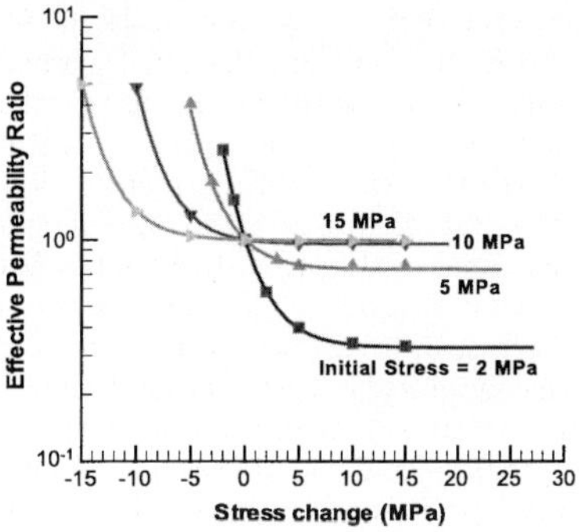

Figure 5. A comparison between results calculated from Equations 9 and 10 and numerical simulation results (data points).

3.4. Additional Remarks

Two methods are developed for calculating gridblock-scale stress-permeability relationships. Although the first method is developed by assuming identical properties for individual fractures and spatially variable fracture densities, it can be mathematically shown that the model is also valid for individual fractures with different aperture parameters (e.g., b_r and b_{max}) as long as R_b is the same for individual fractures.

Another interesting finding from this upscaling exercise is that local permeability ratio depends on R_b for a given initial stress and a stress change. R_b is an important parameter for characterizing the effects of the mechanical process on hydraulic properties. An effective R_b for large-scale stress-permeability relationships, R_b*, seems to exist. How to relate R_b (or R_b*) to local (or large-scale) permeability is of interest for coupled-process modeling, and deserves a further study.

4. CONCLUSIONS

While it has been shown that fLm is useful for characterizing subsurface heterogeneity in natural porous media, we demonstrate, for the first time, the consistency between fLm and permeability distributions in fractured rocks by analyzing the data set collected from the Sellafield site. This finding is also supported by results reported in the literature indicating that fractures are spatially clustered and that the clustering patterns could be described by fractals.

A key parameter for modeling the T-H-M-C processes is the relationship between *in situ* stress and permeability. Assuming several different relations between local permeability and fracture apertures, we developed two upscaling relationships between normal stress and permeability for fractured rock characterized by fLm. While more rigorous upscaling relationships are desirable, the proposed relationships capture relevant large-scale effects of normal stress changes on rock permeabilities.

ACKNOWLEDGMENTS

We are indebted to Tianfu Xu and Dan Hawkes at Lawrence Berkeley National Laboratory for their careful reviews of a preliminary version of this paper. This work was supported by the Director, Office of Civilian Radioactive Waste Management, U.S. Department of Energy, through Memorandum Purchase Order EA9013MC5X between Bechtel SAIC Company, LLC, and the Ernest Orlando Lawrence Berkeley National Laboratory (Berkeley Lab). The support is provided to Berkeley Lab through the U.S. Department of Energy Contract No. DE-AC03-76SF00098.

REFERENCES

CRWMS M&O 2000. *Analysis of Hydrologic properties data.* Las Vegas, Nevada.

Fama, E. & Roll, R. 1972. *Parameter estimates for symmetric stable distributions.* J. Am. Stat. Assoc. 66: pp. 331-338.

Hirata,T., Saton, T. & Ito, K. 1987. *Fractal structure of spatial distribution of micro fracturing in rock.* Geophys. J. R. Abstr. Soc. 90: pp. 369-374.

Jackson, P. C., Hoch, A. R. & Todman, S. 2000. *Self-consistency of a heterogeneous continuum porous medium representation of a fracture continuum.* Water Resources Research 36: pp.189-202.

Liu, H. H. & Molz, F. J. 1997. *Comment on "Evidence for non-Gaussian scaling behavior in heterogeneous sedimentary formations" by Scott Painter*, Water Resources Research 33: pp. 907-908.

Liu, H.H., Bodvarsson, G. S., Molz F. J. & Lu, S. 2003. *A generalized successive random additions algorithm for simulating fractional Levy motion.* Mathematical Geology (in review).

Molz, F.J. & Liu, H. H. 1997. *Fractional Brownian motion (fBm) and fractional Gaussian noise (fGn) in subsurface hydrology.* Water Resources Research 33(10): pp. 2273-2286.

Painter, S. 1996. *Evidence for non-Gaussian scaling behavior in heterogeneous sedimentary formations.* Water Resources Research 32(5): pp.1183-1195.

Pruess, K. 1991. *TOUGH2-A general purpose numerical simulator for multiphase fluid and heat flow.* Rep. LBNL-29400, Lawrence Berkeley National Laboratory, Berkeley, CA.

United Kingdom Nirex Limited 1997. *An Assessment of the post-closure performance of a deep waste repository at Sellafield.* Science Report S/97/012.

Yamamoto, H., Kojima K. & Tosaka, H. 1993. *Fractal clustering of rock fractures and its modeling using cascade process.* In Pinto da Cunha (ed.), Scale Effects in Rock Masses 93, Balkema, Rotterdam.

A BLOCK-SCALE STRESS-PERMEABILITY RELATIONSHIP OF A FRACTURED ROCK DETERMINED BY NUMERICAL EXPERIMENTS

Ki-Bok Min[1], Jonny Rutqvist[2], Chin-Fu Tsang[2], Lanru Jing[1]
[1] Engineering Geology and Geophysics Research Group, Royal Institute of Technology (KTH), Stockholm, Sweden
[2] Earth Science Division, Lawrence Berkeley National Laboratory (LBNL), Berkeley, CA, USA

Abstract: We present the stress induced permeability change in a fractured rock with special focuses on 1) fracture closure/opening process, 2) anisotropic stress conditions, and 3) fracture dilation. In order to overcome the limitations from the experimental study on a single fracture or analytical study on orthogonal and/or persistent fracture models, realistic Discrete Fracture Network (DFN) models were used as a geometrical basis. A series of numerical experiments on fluid flow are conducted under different stress boundary conditions for calculation of equivalent permeability. From the analysis on the realistic DFN model, this paper demonstrates the decrease of permeability with the fracture normal closure, increased anisotropy in permeability with anisotropic stress and increase of permeability and significant channelling effect due to fracture dilation. It is shown that block-scale permeability change in fractured rock depends not only on stress magnitude but also on stress orientation and differential stresses.

1. INTRODUCTION

Stress induced permeability change is of crucial importance in various kinds of applications such as nuclear waste disposal in deep geological formations, geothermal energy utilization and underground excavations. In particular, coupling between the stress and permeability is a key element in understanding the nature of flow in the fractured rock (Rutqvist and Stephansson, 2003). This is because fractures, which are the main pathways of fluid flow in fractured hard rocks, are heavily dependent on the stress conditions for their deformations.

Field observations indicate that the stress condition affects the flow characteristics of fractured rock mass (Barton et al., 1995; Ito and Hayashi, 2003; Pusch, 1989). Many laboratory investigations on single fractures also prove that the normal closure and shear dilation can significantly change the transmissivity of fractures (Makurat et al., 1990; Olsson et al., 2001). When it comes to the 'block scale' stress-permeability relationship, the analytical models based on the orthogonal and/or persistent fracture model are available (Bai and Elsworth, 1994). However, analytical solutions do not generally exist for more 'realistic' fracture systems. Furthermore, to the authors' knowledge, 'block-scale' study about the effect of shear dilations of fractures on the change of permeability has not been conducted so far.

We investigate the stress induced permeability change in fractured rocks by a discrete element approach considering interactions between the fluid flow in fractures and stress-deformation in rock blocks. Special focuses are given to 1) permeability change caused by fracture closure/opening 2) induced anisotropy of permeability under anisotropic stress conditions, and 3) permeability change and channelling flow effect induced by fracture dilation. These processes are investigated by properly considering the mechanical deformation of a realistic block/fracture system under various stress conditions and calculating the equivalent permeability through numerical experiments using the Distinct Element Method.

2. MECHANISMS OF STRESS INDUCED PERMEABILITY CHANGE

Figure 1 presents the three basic mechanisms of stress induced permeability change in fractured rock: (a) normal closure/opening, (b) shear dilation/contraction and (c) induced anisotropy due to different orientations of fractures and anisotropic stress condition.

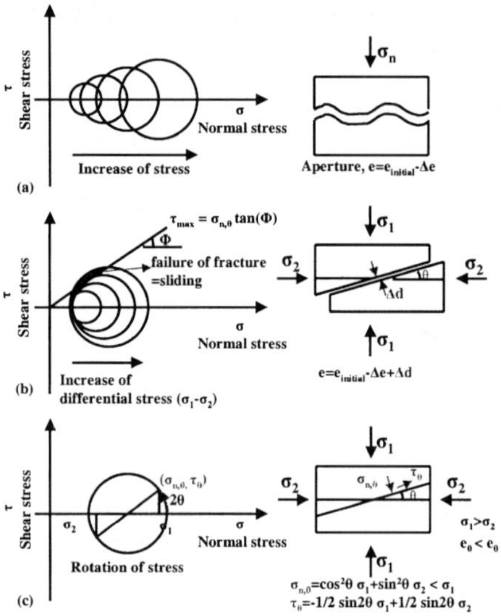

Figure 1. Basic mechanisms of stress induced permeability change in fractured rock. (a) normal closure/opening, (b) shear dilation/contraction, (c) effects of fracture orientation in multiple fractures.

The normal closure/opening mechanism as shown in Figure 1(a) is relatively well understood. This mechanism causes a decrease/increase of fracture transmissivity with an increase/decrease of normal stress magnitude. This mechanism plays a dominant role in fluid permeability of fractured rocks when failure does not occur in the fracture and deformation is not close to residual state.

The second mechanism is dilation/contraction induced permeability change during shear (Figure 1 (b)). Fractures dilate when fracture failure occurs under high differential stresses. This mechanism can play a major role in the flow field since the dilated fracture tend to have much higher transmissivity than the fracture under elastic deformation.

The third mechanism can be seen as a combination of first and second basic mechanisms (Figure 1(c)). In a fractured rock with multiple fractures, anisotropy in fluid permeability may be significant due to the different orientations of fractures and anisotropic stresses. Each fracture is under different contact normal or shear stresses mobilized through deformations and this make the directional permeability anisotropic.

Stress conditions for this study are selected in two ways, based on the above conceptual mechanisms of hydro-mechanical coupling in fractured rock.

Firstly, a horizontal and a vertical normal stress component are applied incrementally to the model boundaries with a fixed ratio, k, defined as the ratio of horizontal to vertical normal boundary stresses. The ratio, k, in this study is set to be 1.3 (Decovalex, 2000) and it is expected that no significant shearing will occur in fractures under this stress condition.

Secondly, horizontal normal boundary stress is increased incrementally while keeping the vertical normal stress constant to produce increased shear stress in the fractures in various orientations. Through this process, anisotropic permeability and the effect of dilation on the permeability change is observed.

After the application of the various stress boundary conditions, differential hydraulic heads are applied to opposite model boundaries to introduce fluid flow through the models, and the equivalent permeability of the model is calculated as the function of the applied stress boundary conditions.

3. PROCEDURE OF NUMERICAL EXPERIMENTS

The two-dimensional distinct element code, UDEC (Itasca, 2000) is applied for the modeling of mechanical and hydraulic behaviours. UDEC simulates the response of a fractured rock mass represented as an assemblage of discrete deformable blocks, subjected to the mechanical stress and hydraulic pressure boundary condition. Numerical experiment consists of 1) generation of a Discrete Fracture Network (DFN) as a geometric model, 2) application of various stress conditions, and 3) application of fluid boundary condition and calculation of equivalent permeability.

3.1 Generation of DFN

A realistic Discrete Fracture Network is generated based on statistical information on fracture by Monte Carlo simulations (Min and Jing, 2003). Data for this analysis is taken from the result of a site investigation at Sellafield undertaken by Nirex, UK (Table 1, Decovalex 2000).

Table 1 DFN parameters.

Set	Dip/Dip direction	Fisher constant (K)	Fracture density (m^{-2})	Mean trace length (m)
1	8/145	5.9	4.6	0.92
2	88/148	9.0	4.6	0.92
3	76/21	10.0	4.6	0.92
4	69/87	10.0	4.6	0.92

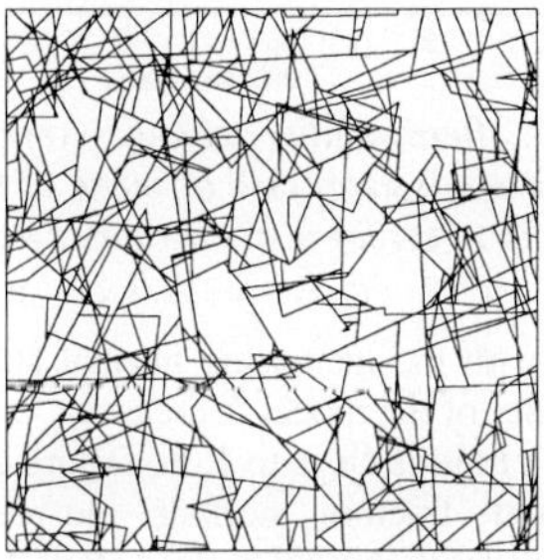

Figure 2. Geometry of DFN model.

Four fracture sets are used and their orientation show near-random patterns due to their low Fisher constants. The size of the model is as 5 m × 5 m (Figure 2). The fracture trace lengths are characterized by a fractal scaling law as follows

$$N = 4 \times L^{-2.2} \quad (1)$$

where N is the number of fractures greater than a given fracture length L (m) per unit area (m^2). The minimum and maximum cut-off trace lengths are chosen as 0.5 m and 250 m, respectively.

3.2 Application of stress boundary conditions

Each block is modeled as linear, isotropic, homogeneous and elastic medium and subdivided with a mesh of constant-strain triangle finite-difference elements. Key factors affecting the hydraulic behaviour of fractures such as opening, closure, sliding and dilation of fractures are modeled by an elasto-perfectly plastic constitutive model of a fracture. A step-wise non-linear normal stress-normal closure relationship is adopted with a linear Mohr-Coulomb failure for shear (Figure 3).

Dilation of a fracture occurs as it starts to slide (Figure 4, Itasca, 2000). This dilation continues until the critical shear displacement (Ucs) and dilation behaviour after Ucs is not modeled for simplicity.

Table 2 shows the model parameters (Decovalex, 2000). These are chosen to demonstrate the significance in stress-flow behaviour of fractured rocks and do not necessarily reflect the in-situ conditions at Sellafield. Figure 5 shows the boundary conditions for the numerical experiments.

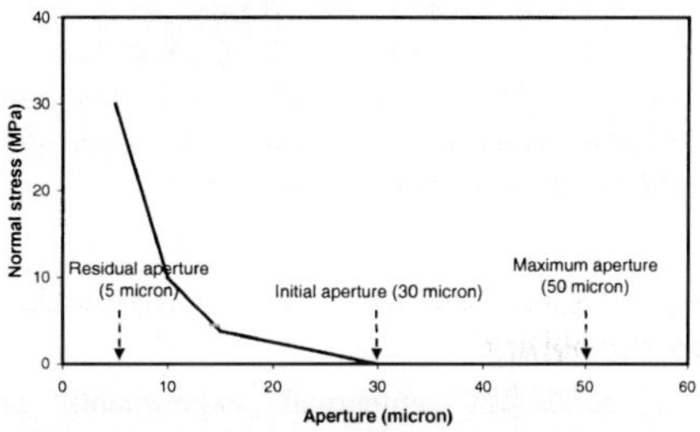

Figure 3. step-wise non linear behaviour of fractures under normal loading

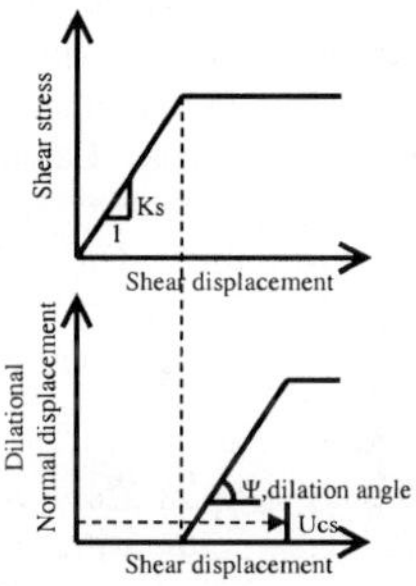

Figure 4. Shear and dilational behaviour of fracture.(after Itasca,2000)

Table 2 Model parameters.

	Properties	
Intact rock	Elastic modulus (GPa)	84.6
	Poisson's ratio	0.24
Fractures	Normal stiffness (GPa/m)	434
	Shear stiffness, Ks (GPa/m)	434
	Friction angle (°)	24.9
	Dilation angle (°)	5
	Cohesion (MPa)	0
	Critical shear displacement for dilation, Ucs (mm)	3
	Initial aperture (μm)	30
	Residual aperture (μm)	5
	Maximum aperture (μm)	50

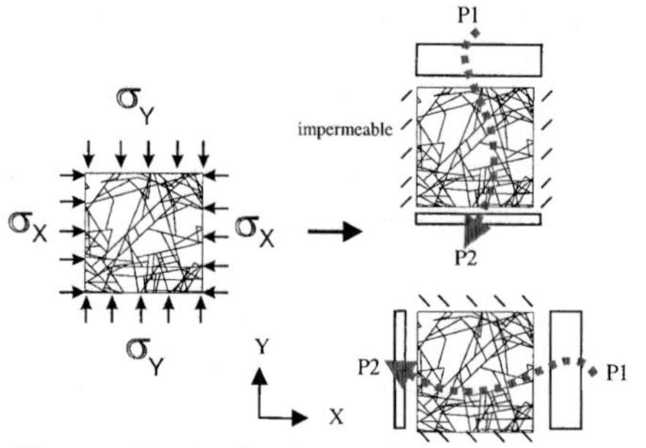

Figure 5. Application of stress boundary condition and calculation of equivalent permeability in x- and y- directions.

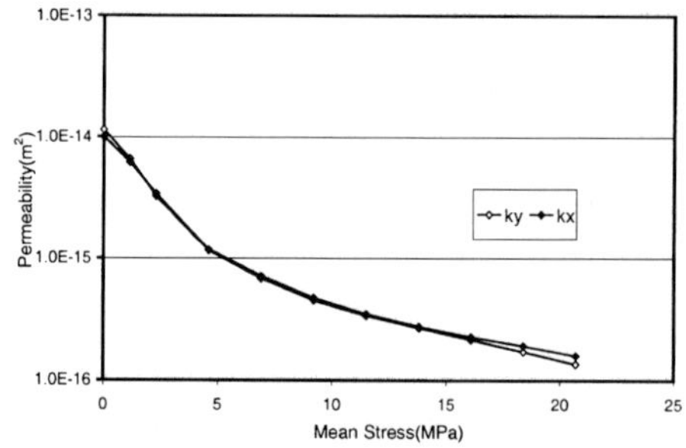

Figure 6. Permeability change versus stress change. Stress is expressed as mean of horizontal and vertical stress for a stress ratio, k= 1.3.

3.3 Calculation of equivalent permeability

A series of numerical experiments are conducted to calculate the equivalent permeability of the model under various stress conditions. The flow through fractures is calculated by the Cubic law and blocks are treated as impermeable medium. Mass conservation is assured at each fracture intersection and in domains between them, through an iterative process. Lateral sides of the model are set impermeable and x- and y-directional permeability with the specified hydraulic pressure boundary conditions are calculated by the following equations.

$$Q_x = A\frac{k_x}{\mu}\frac{\partial P}{\partial x}, Q_y = A\frac{k_y}{\mu}\frac{\partial P}{\partial y} \qquad (2)$$

where Q_x *and* Q_y = flow rate in x and y directions; A = cross-section area; k_x and k_y = permeability in x and y directions; μ = dynamic viscosity; and P = hydraulic pressure.

4. RESULTS

4.1 Permeability change versus stress change with fixed ratio of horizontal to vertical stress

Figure 6 shows the permeability change with the stress changes when the ratio, k is fixed as 1.3. Both horizontal and vertical boundary stresses are incrementally increased at each step while keeping the k ratio constant. Due to the closure of fractures with the increasing normal stresses, the permeability of the model decreases.

As the k ratio is close to 1, no failure of fracture occurs and fracture normal closure is the dominating mechanism for fluid flow. Non-linear behaviour of fracture normal stress-deformation makes the permeability change at the lower stress levels more dramatic than that at the higher stress levels. The reduction of permeability is about two orders of magnitude and anisotropy in permeability is not significant, mainly due to the fact that the constant k ratio that is close to an isotropic boundary stress condition.

4.2 Permeability change versus increasing differential stresses

Figure 7 presents the permeability change with the increasing k ratio simulated by increasing the horizontal boundary stress from 2.5 MPa to 25 MPa with the fixed vertical boundary stress of 5 MPa (the k ratio changes from 0.5 to 5). To evaluate the effect of shear dilation, the results are compared with the pure elastic fracture model that does not consider the failure and dilation.

Both pure elastic and elastoplastic models show similar response until the k value reaches approximately 2.5. With the increase of horizontal stress, the anisotropy in the permeability becomes significant. This is because the sub-vertical fractures are more vulnerable to the horizontal stress and its effect on y-directional permeability is larger. The difference between x- and y-directional permeability is about a factor of 2 for k ratios of 0.5 and 2.

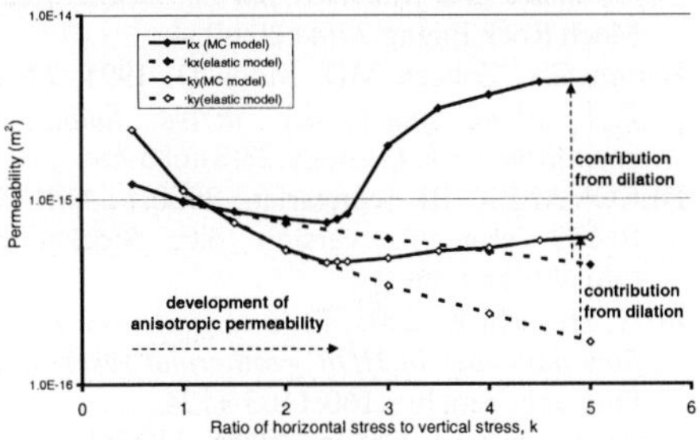

Figure 7. Permeability change due to the change of differential stress.

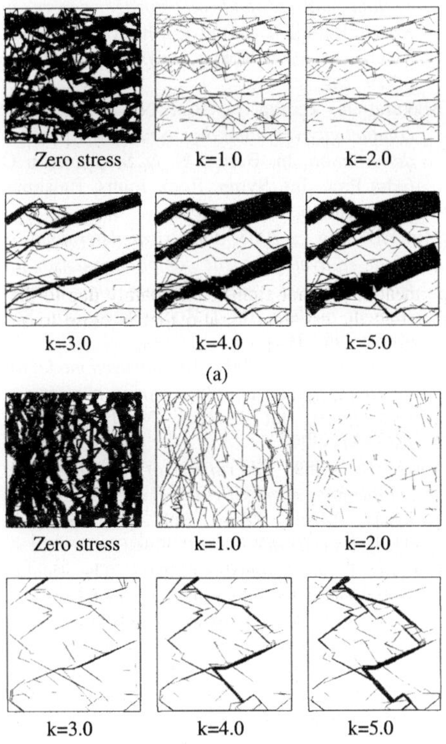

Figure 8. Fluid pathways during stress applications with the direction of hydrualic pressure gradient (a) from right to left. (b) from top to bottom. Thickness of the lines represent the flowrates and k is defined as the ratio of horizontal to vertical normal boundary stress.

When the k ratio is beyond 2.5 some fractures start to fail with continuous dilation and notable differences from elastic models are observed. Interesting aspects from these results are 1) the increase of permeability with the increase of differential stresses, and 2) dilation induced channelling effect.

The increase of permeability is due to the increased apertures from fracture dilation. For this study, k=2.45 is the starting point of shear failure of fractures inclined at about 33 degrees from the horizontal plane. As horizontal boundary stress increases, the range of possible failure orientation angle also becomes larger and this gives increased permeability. The increase of permeability seems to stabilize after a certain k ratio because the dilation does not continue after the critical shear displacement.

Figure 8 shows some examples of the channelling flow effect due to the fracture dilation. Note that all models have the same reference legend. One line indicates the flowrate of 10^{-9} m^3/sec and the flowrates smaller than this value are not drawn in Figure 8.

Dilation induced channelling may be explained in the following ways. As the dilation is limited to the fractures that have preferential orientations and trace lengths, many fractures, especially the sub-vertical fractures, still undergo the normal closures without any dilation. This situation makes a high contrast of apertures between neighbouring fractures. Therefore only the well-connected fractures of large apertures become the major pathways of fluid flow. Furthermore, dilated fractures tend to be those inclined more horizontally with increasing horizontal normal boundary stress and this made the channelling effect with horizontal hydraulic pressure gradient (Figure 8(a)) more manifest than the vertical pressure gradient (Figure 8(b)).

5. CONCLUSIONS AND DISCUSSION

1) When the ratio of two principal boundary normal stresses is not favorable to the shear failure of the fractures, the fracture normal closure/opening is the dominating factor for the change of permeability. Permeability decreases with the increase of stress and the reduction of permeability was about two orders of magnitude under a stress increase from 5 to 20 MPa for the model parameters assumed in this study.

2) Anisotropy in the permeability begins to be more significant due to the different orientation of fractures and increasing differential stresses. The maximum anisotropy in permeability has a factor of 2 for the stress ratio 2. The model for this study has near-randomly distributed orientations of fractures. If the model has more unidirectional orientations of fractures, this effect of anisotropic stress will be more significant.

3) Permeability increase due to the fracture dilation. When the ratio k is beyond the point to induce shear failure of fractures, dilation of fracture is the dominating factor for the change of permeability. Under this condition, permeability can increase with the increase of boundary normal stresses. The contribution of dilation can be one order of magnitude higher than pure elastic analysis. This increased permeability is analogous to the experiment on the intact rock with similar decrease/increase of permeability during increasing differential stress (Lee and Chang, 1995). The occurrence of failure and dilation depends on the orientation and connectivity of the fractures. The range of orientation for the dilation is widened with the increased differential stress. In this study the range of 20 –40 degrees were the orientations for the fracture dilation occurrence at the k ratio of 5. The dilated fractures are the major pathways of the fluid flow and this result is in line with the field observations (Barton et al, 1995; Ito and Hayashi, 2003).

4) Prominent channelling effect is observed during the increase of differential stresses. This is due to both anisotropic stress and the dilation of the fractures. Results show that well-connected fractures of larger apertures are the major pathways for fluid flow in fractured rock and pathways can be significantly changed after application of stress.

ACKNOWLEDGEMENTS

The financial supports by the European Commission through the BENCHPAR project (FIKW-CT-2000-00066) and the Swedish Nuclear Inspectorate (SKI) through the DECOVALEX III project are greatly acknowledged.

REFERENCES

Bai M, Elsworth D, 1994, *Modeling of subsidence and stress-dependent hydraulic conductivity for intact and fractured porous media*, Rock Mech Rock Engng;27(4):209-234

Barton CA, Zoback MD, Moos D, 1995, *Fluid flow along potentially active faults in crystalline rock*, Geology;23(8):683-686

DECOVALEX III secretariat, 2000, TASK 3, BMT2 protocol, version 6.0, Stockholm. (unpublished report).

Ito T, Hayashi K, 2003, *Role of stress-controlled flow pathways in HDR geothermal reservoirs*, Pure appl geophys;160:1103-1124

Itasca Consulting Group, 2000, UDEC user's guide, Ver. 3.1, Minnesota.

Lee CI, Chang KM, 1995, *Analysis of permeability change and groundwater flow around underground oil storage cavern in Korea*, In:Fujii T (ed) Proc 8th Int Congr Rock Mechanics, 1995, Tokyo, Balkema, Rotterdam, p.779–782

Makurat A, Barton N, Rad NS, 1990, *Joint conductivity variation due to normal and shear deformation*, In: Barton N. & Stephansson O (eds) Proc. Int. Symp. Rock Joints, Balkema, Rotterdam, p.535-540

Min KB, Jing L, 2003, *Numerical determination of the equivalent elastic compliance tensor for fractured rock masses using the distinct element method*, Int J Rock Mech Min Sci. 40(6): 795-816.

Olsson R, Barton N, 2001, *An improved model for hydromechanical coupling during shearing of rock joints*, Int J Rock Mech Min Sci; 38(3):317-329

Pusch R, 1989, *Alteration of the hydraulic conductivity of rock by tunnel excavation*, Int J Rock Mech Min Sci & Geomech Abstr;26(1):79-83

Rutqvist J, Stephansson O, 2003, The role of hydromechanical coupling in fractured rock engineering, Hydrogeology Journal;11:7-40.

HYDRO-MECHANICAL UPSCALING OF A FRACTURED ROCKMASS USING A 3D NUMERICAL APPROACH

Thoraval Alain, Renaud Vincent
INERIS, ENSMN, Parc de Saurupt, 54042 Nancy Cedex, France.

Abstract: A new upscaling method has been developed using 3D numerical tools (RESOBLOK & 3DEC). This method has been successfully compared with standard analytical approaches in the case of a simple fracture network. This method has been applied to determine the equivalent permeability, stiffness and Biot tensor of a real fracture rock-mass at different scales. The effects of the fracture network properties and of the state of stress on the result have been investigated.

1. CONTEXT OF THE WORK AND OBJECTIVES

This work is part of the INERIS contribution to work package 3 (WP3) of the European project BENCHPAR. The purpose of WP3 is the better understanding the impact of upscaling on the THM processes on Performance Assessment of nuclear waste disposal problems.

The INERIS contribution to WP3 consist in:

- The development and evaluation of a new upscaling method, and its application to WP3 fracture data;
- Comparison between H, HM, THM simulations considering an equivalent homogeneous medium with FLAC3D (whose properties will have been determined previously by the upscaling process).

This paper presents only the upscaling work.

2. DESCRIPTION OF THE UPSCALING APPROACH & VERIFICATIONS

A new upscaling method has been proposed by INERIS to determine the equivalent hydro mechanical properties of a fractured rock-mass. This method is based on the 3D numerical simulations of the behaviour of a "sample" of fractured rock-mass submitted to different hydromechanical boundary conditions. The simulations are defined in order to determine the equivalent permeability, stiffness and Biot tensor of a fractured rock-mass.

Two numerical tools are used: RESOBLOK [Heliot, 1988] to generate the 3D fracture network; 3DEC [Itasca Consulting Group, 1994; Damjanac, 1994] to make the hydro-mechanical computations.

2.1 Generation of the fracture network

The code used to generate the fracture network (RESOBLOK) is based on the assumption that the fracture can be considered as polygon. This code is able to make determinist or stochastic simulations. To make the stochastic simulations done for the WP3 exercise (figure 1), the following assumptions are done:

- Fracture density (number of fracture per unit volume of rock - P_{31}) is assumed to follow a Poisson law;
- Fracture orientation is assumed to follow a Fisher law. Average dip & dip direction as well as the Fisher coefficient (k_f) has to be given for the different fracture sets;
- Fracture length is assumed to follow a power law distribution of the form $N=CL^{-D}$, where N is the number of fractures which are equal to or greater than a given trace length L, D is the fractal dimension and C is a constant.

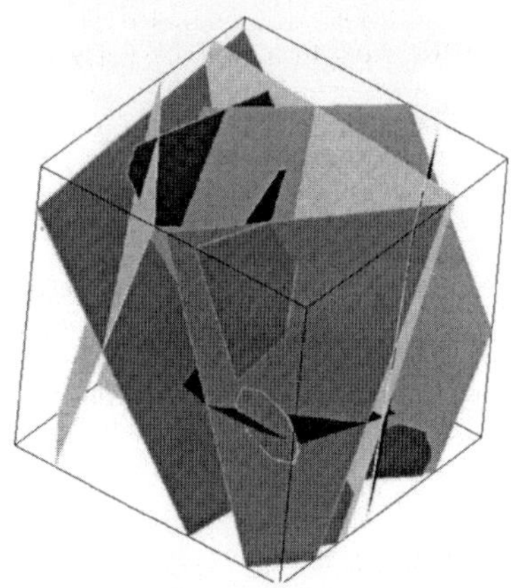

Figure 1. Stochastic fracture network generated by RESOBLOK.

2.2 Determination of the equivalent permeability tensor Kij

The hydraulic gradient J and the fluid flow Q are linked by the relationship:

$$Q_i = -K_{ij} J_j \quad (1)$$

To determine all the tensor terms with 3DEC, 3 sets of (hydraulic) boundary conditions have to be defined (figure 2).

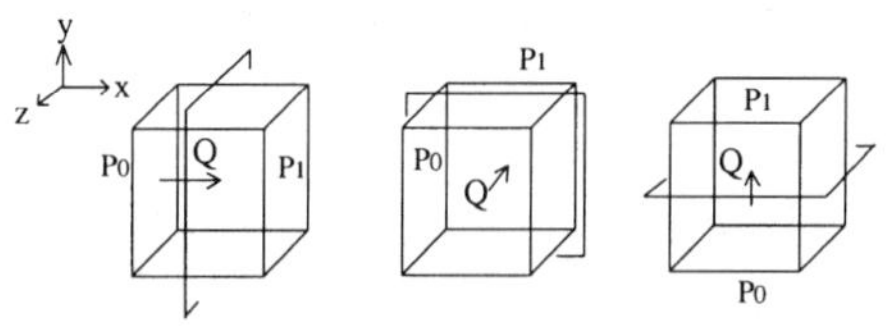

Figure 2. Boundary conditions used to determine the permeability tensor with 3DEC.

After each computation, the "equivalent" vectorial flowrate is averaged from the flowrate value at different locations of the fracture network. The different terms of the equivalent permeability tensor are then computed from equation (1).

The results of this new upscaling approach have been compared with an analytical solution given for the simple geometry (figure 3).

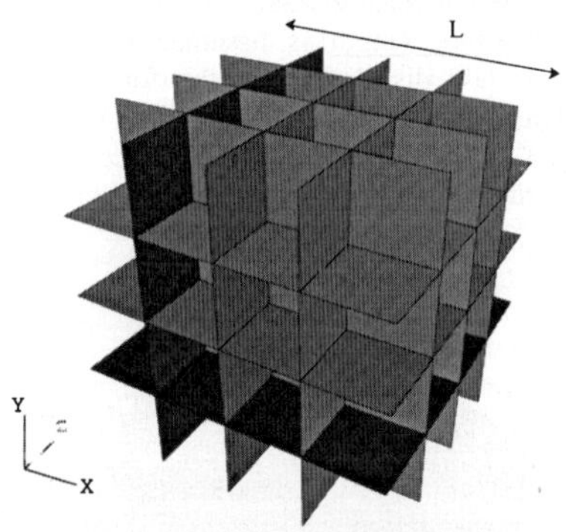

Figure 3. Fracture network used for comparison with analytical upscaling approach.

The analytical equivalent hydraulic conductivity tensor can be expressed in this case as following:

$$K_{ij\ analytical} = (\rho\, g\, a^3 / 2\, \mu\, L)\, \delta_{ij} \quad (2)$$
$$= 2.45\ 10^{-7}\ m/s$$

where: a (= 10^{-4} m) is the hydraulic aperture of the fracture; μ (= 10^{-3} Pa s) is the dynamic viscosity of the fluid; (L = 20 m) is the edge length of the model; ρ = 1000 kg/m^3; g =9.81 m/s^2.

The equivalent permeability given by 3DEC is:

$$K_{ij\ 3DEC} = (6\, \rho\, g\, a\, q_{3DEC} / \Delta P)\, \delta_{ij} \quad (3)$$
$$= 2.35\ 10^{-7}\ m/s$$

where q_{3DEC} = 4 10^{-2} m/s is the discharge (m/s) computed with 3DEC for $\Delta P = 10^6$ Pa.

We can see that the 3DEC result is very close to the analytical solution.

2.3 Determination of the equivalent Stiffness tensor Tijkl

The stress tensor σ_{ij}, the elastic strain tensor ε_{kl}, the Biot's tensor B_{ij} and the Biot coefficient G are linked by the Biot's equation:

$$\sigma_{ij} = T_{ijkl}\, \varepsilon_{kl} - B_{ij}\, P \quad (4)$$

$$P = -G(B_{kl}\, \varepsilon_{kl} - \xi) \quad (5)$$

where P is the fluid pressure and ξ the fluid production.

To determine all the tensor terms with 3DEC, 6 sets of (mechanical) boundary conditions have to be defined (Figure 4 and 5).

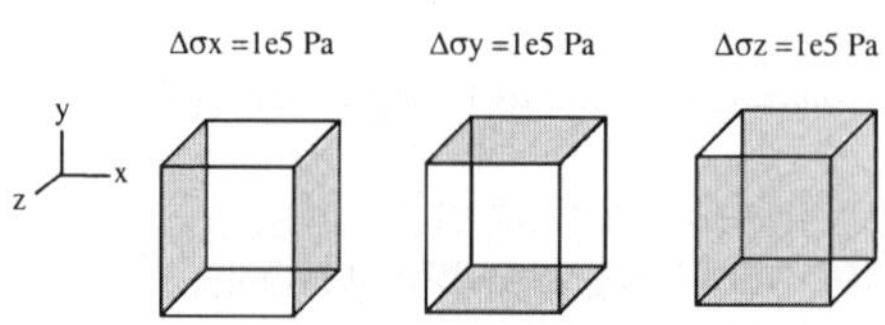

An initial state of stress (σ = 1e6 Pa) is applied into the model; a constant stress increment (Δσ = 1e5 Pa) is apply on the grey faces

Figure 4. First set of boundary conditions used to determine Tijkl with 3DEC.

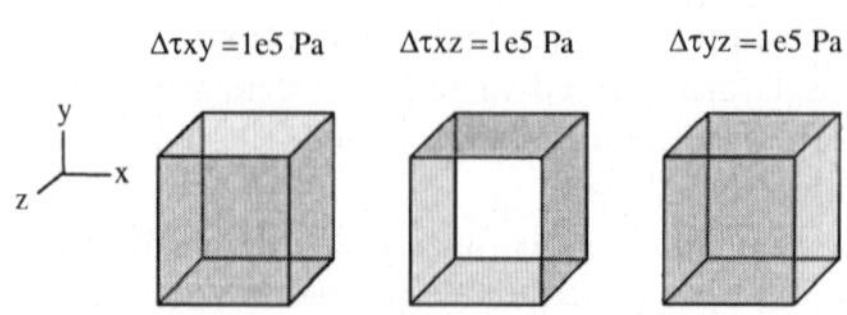

An initial state of stress (σ = 1e6 Pa) is applied into the model; a constant shear stress increment (Δτ = 1e5 Pa) is apply on the grey faces

Figure 5. Second set of boundary conditions used to determine Tijkl with 3DEC.

During each 3DEC runs, the average strain tensor is computed from displacement differences between opposite face of the model (100 pairs of point is used for each strain term).

A verification has been done again for the same simple geometry (figure 3). From Oda theory [1986], it can be shown that, in this case, $(T_{ijkl})^{-1}$ is equal to:

$$\begin{bmatrix} \frac{1}{E}+\frac{n}{LK_n} & -\frac{\nu}{E} & -\frac{\nu}{E} & 0 & 0 & 0 \\ -\frac{\nu}{E} & \frac{1}{E}+\frac{n}{LK_n} & -\frac{\nu}{E} & 0 & 0 & 0 \\ -\frac{\nu}{E} & -\frac{\nu}{E} & \frac{1}{E}+\frac{n}{LK_n} & 0 & 0 & 0 \\ 0 & 0 & 0 & \frac{1+\nu}{E}+\frac{n}{LK_s} & 0 & 0 \\ 0 & 0 & 0 & 0 & \frac{1+\nu}{E}+\frac{n}{LK_s} & 0 \\ 0 & 0 & 0 & 0 & 0 & \frac{1+\nu}{E}+\frac{n}{LK_s} \end{bmatrix}$$

Assuming E (Young's modulus)=5.294 GPa; ν (Poisson ratio) =0.324; Kn & Ks (normal &shear joint stiffness) =1 GPa/m; n (number of fracture in each fracture set) = 3; L = 20 m, we get (in GPa):

$$\text{Tijkl}_{\text{analytical}} = 3.205 \begin{bmatrix} 1 & 0.220 & 0.220 & 0 & 0 & 0 \\ 0.220 & 1 & 0.220 & 0 & 0 & 0 \\ 0.220 & 0.220 & 1 & 0 & 0 & 0 \\ 0 & 0 & 0 & 0.780 & 0 & 0 \\ 0 & 0 & 0 & 0 & 0.780 & 0 \\ 0 & 0 & 0 & 0 & 0 & 0.780 \end{bmatrix}$$

The determination of this matrix with the numerical approach gives the following result:

$$\text{Tijkl}_{\text{3DEC}} = 3.205 \begin{bmatrix} 1.000 & 0.220 & 0.220 & -0.014 & -0.014 & -0.005 \\ 0.220 & 1.000 & 0.220 & -0.014 & -0.005 & -0.014 \\ 0.220 & 0.220 & 1.000 & -0.005 & -0.014 & -0.014 \\ -0.001 & -0.002 & 0.000 & 0.780 & 0.000 & 0.000 \\ 0.001 & 0.000 & 0.000 & 0.000 & 0.780 & 0.000 \\ 0.001 & 0.003 & 0.003 & 0.000 & 0.000 & 0.780 \end{bmatrix}$$

It can be noticed that, in all cases, the differences between the analytical and the 3DEC solution are very small (less than 1.5%).

2.4 Determination of the equivalent Biot tensor Bij

The numerical procedure is to impose a fluid pressure built up from P to P + ΔP within the model, assuming no mechanical deformation at the boundaries. In this case the Biot's equations (4 & 5) can be rewritten as follows:

$$\Delta\sigma_{ij} = -B_{ij}\,\Delta P \;\;\&\;\; \Delta P = G\,\Delta\xi. \qquad (6)$$

with $\Delta\xi = V_1 - V_0/V_t$, where V_0 the initial fluid volume (under P), V_1 is the final fluid volume (under P+ΔP) and V_t the total model volume.

For the geometry of figure 3, it can be shown:

$$B_{ij} = \frac{nE}{LK_n(1-2\nu)+nE}\,\delta_{ij} \;\; and \;\; G = \frac{LK_n}{3n\delta_{ij}(\delta_{ij}-B_{ij})} \qquad (7)$$

With 3DEC, we determine the average stress variation $\Delta\sigma_{ij}$ and volumetric water content variation Δξ induced by the hydraulic pressure ΔP.

We have found again in this case that the difference between analytical and numerical solution are below 2%.

3. APPLICATION OF THE UPSCALING TECHNIQUE TO WP3

The objective of this section is to compute, with the numerical approach already presented and validated, the equivalent permeability tensor of the fracture network of the WP3 case. The set of computation presented here concerned the fracture network of formation 1 (computation has also been done for formation 2 and the fault zone).

The size of the zone of interest where the fracture network is simulated has been progressively increased from 2 m to the greater possible size according to the numerical problem encountered. Five simulations have been done for each model size to get an idea of the standard deviation.

3.1 Input of the fracture network geometry into 3DEC

3.1.1 RESOBLOK simulations

The fracture network is generated by RESOBLOK from the data gathered in Table 1.

Table 1: Joint set properties for formation 1

Fracture set	dip (°)	dd (°)	k_f	P_{11} density (FLT = 0.5 m)	D
1	8	145	5.9	0.16	1.2
2	88	148	9	0.31	0.7
3	76	21	10	1.0	1.1
4	69	87	10	0.5	1.1

The simulations have been done considering initial value of P_{31} equal to the value of P_{11} (for a Fracture Length Threshold of 0.5 m) given in Table 1. Average values of P_{22}, computed from several RESOBLOK simulations, were compared to P_{22} measurements (5m/m^2 for a fracture length threshold of 0.5 m) to evaluate the realism of the simulations. The results are gathered in Table 2.

We can notice that the computed P_{22} value at 3 m or 4 m are close to the measured value (5m/m2).

Table 2: Determination of fracture density from RESOBLOK fracture network simulations.

«Sample » size	Set	Computed P22 density (m/m^2)
3×3×3 m3	1	0.11
	2	1.37
	3	1.87
	4	0.97
	Total	**4.39**
4×4×4 m^3	1	0.26
	2	1.27
	3	2.57
	4	0.97
	Total	**5.01**

3.1.2 Interfacing RESOBLOK to 3DEC

A specific command of RESOBLOK is used to generate a 3DEC file including all information regarding fracture network geometry.

3DEC does not allow generating fractures that stop inside a block. During the 3DEC cutting process, the plane where the RESOBLOK polygonal fracture is lying is continued up to the next fracture (or block face). However, 3DEC automatically distinguish 2 zones (from information output from RESOBLOK): an "active" zone inside the polygon area (real joint), an "inactive" zone outside the polygon (fictitious joint). So 3DEC is able to "give" different properties to "real" joint and fictitious joint.

3.1.3 Checking the effect of fictitious joints on the equivalent permeabity tensor

We have shown, from various 3DEC runs, that the prolongation of joint up to the next block face (that artificially increased the fracture network connectivity) could lead to overestimate the permeability by a factor of 2 or 3.

The hydraulic aperture used for the real joint is supposed to be 6.5 10^{-5} m. We have considered a hydraulic aperture of 10^{-6} m for the fictitious joint. We have checked that the equivalent hydraulic permeability tensor computed considering a fictitious joint network is 5 orders of magnitude smaller than if we consider a real joint network.

3.1.4 Checking the effect of fictitious joints on the equivalent stiffness tensor

The joint stiffness used for the real joint is supposed to be 4.34 10^{-11} Pa/m. We have considered a joint stiffness of 10^{-12} Pa/m for the fictitious joint. We have checked that, for this value, the equivalent stiffness tensor is very close to the intact rock stiffness tensor.

We have evaluated the effect of considering the fictitious joints as real joints on the equivalent stiffness tensor T_{ijkl}. We can notice that the terms of the equivalent stiffness tensor terms are smaller if we consider 3DEC fictitious joints are real. For the WP3 case (where high joint stiffness values are considered) the differences remain small. It increased up to 30 % for Kn = 4.34 10^{-10} Pa/m.

3.1.5 Limitation of 3DEC code

We have noticed that some problems occur during the cutting process into 3DEC (version 3.00.073), when the fracture number becomes important (and maybe when the angle between fracture become small). I happens that in some cases, the cutting is rejected or an error appears later in the process of block geometry recognition. Some efforts have been done (in collaboration with ITASCA) to overcome those limitations but, up to now, no satisfying answer has been found

The consequence of this limitation is that the determination of K_{ij} and T_{ijkl} was not possible (for WP3 data) for a model size higher than 2×2×2 m^3 if we want to avoid artificial prolongation of joint into 3DEC. The limit is push back to 5×5×5 m^3 for T_{ijkl} (and unchanged for K_{ij}) a artificial joint prolongation is considered.

3.2 Determination of the WP3 "equivalent" permeability tensors (Kij)

3.2.1 Assuming constant hydraulic aperture

We have assumed that the hydraulic apertures were constant and equal to 6,5 10^{-5} m. The equivalent permeability tensor is given below at 2 m scale for the fracture network of formation 1 and considering a fracture length threshold of 0.5 m:

$$K_{ij} = \begin{bmatrix} 5.06 & -0.44 & 0.25 \\ -0.54 & 6.95 & -0.40 \\ 0.36 & -0.19 & 4.53 \end{bmatrix} \pm \begin{bmatrix} 1.65 & 0.62 & 0.52 \\ 0.81 & 1.88 & 0.62 \\ 0.33 & 0.56 & 1.60 \end{bmatrix} (10^{-8}\ m/s)$$

Because of 3DEC limitations, the upcaling was done only at 2 m scale; so the existence of a REV (and its value) could not be established at the present time.

Fracture network simulations were done considering the relation (established from measurements by Nirex (Andersson, 2000) between the number of fractures per meter and the fracture length threshold. With a view to the fracture network simplification, we have tried to determine the effect of the fracture length threshold (used to make the RESOBLOK simulations) on the upscaling process. Unfortunately, we have shown that, at 2 m scale, it was not possible to conclude about the effect of the fracture length threshold on K_{ij}. The reason is that the connection of the fracture network to the model boundaries, where the pressure gradient is applied, is artificially increased when the model size is small in comparison with the fracture length threshold.

3.2.2 Assuming stress dependent hydraulic aperture

For HM computation, 3DEC consider a relation between the hydraulic aperture "a" and the mechanical aperture "u" that can be written: $a = a_0 + \Delta u$, where a_0 is the zero stress aperture. A maximum and residual aperture is considered for numerical stability reason. We have assumed that $a_{max} = a_0 = 6.5\ 10^{-5}$ m and that $a_{res} = 1.8\ 10^{-5}$ m. Figure 6 represent the evolution of the diagonal terms of the equivalent permeability tensor with stress applied on the model boundaries (at 2 m scale).

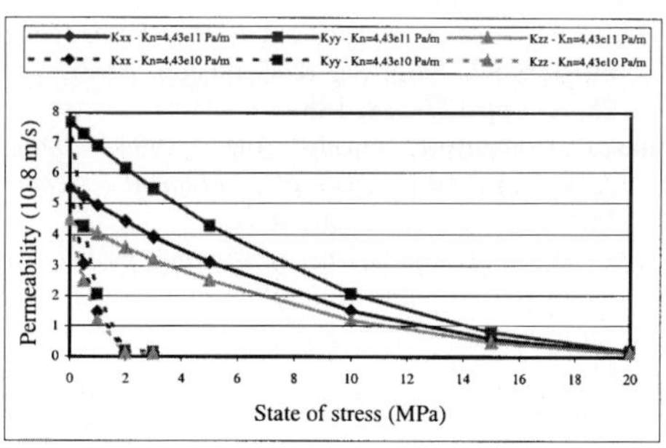

Figure 6 Equivalent permeability tensor variation with stress.

We can see a decreasing of the permeability when the applied stress increases. This decreasing is much faster if the normal joint stiffness is assumed to be smaller ($4.43\ 10^{10}$ Pa/m instead of $4.43\ 10^{11}$ Pa/m). It can be notice that there is a residual equivalent permeability (around 10^{-9} m/s) that has to be related to the residual hydraulic aperture. That residual permeability is reached for $\sigma > 20$ MPa (≈ 800 m) if the joint normal stiffness is $Kn = 4.43\ 10^{11}$ Pa/m and for $\sigma > 2$ MPa (≈ 80 m) $Kn = 4.43\ 10^{10}$ Pa/m.

In this work we have assumed no relation between the joint stiffness and the stress or with the fracture length. This assumption could be revisited later on. We have also considered an isotropic state of stress. Others stress conditions could be applied and would probably involve other kinds of evolution of the equivalent permeability tensor.

3.2 Determination of the WP3 "equivalent" stiffness tensor (T_{ijkl})

T_{ijkl} has been first determined considering a prolongation of the joint up to the next joint (all the cutting plane generated by 3DEC are assumed to be active). The results for formation 1 are gathered in Table 3 (we have distinguish the average values from the standard deviation of the different terms).

Table 3: Equivalent stiffness tensor for formation 1 at 2 m scale

a – average values

67,12	15,71	13,44	-4,61	-2,54	-0,91
15,73	72,86	14,38	-5,13	-1,08	-2,51
13,38	14,41	60,97	-1,49	-2,32	-2,75
-1,97	-2,07	-0,60	54,50	-0,79	-0,89
-1,51	-0,44	-1,25	-1,17	49,99	-0,66
-0,46	-1,55	-1,21	-0,77	-1,38	52,00

5,77	2,70	1,71	4,69	3,47	1,35
2,66	7,58	2,10	5,21	1,68	3,76
1,75	2,17	4,42	1,62	3,04	3,74
2,25	2,52	0,75	3,56	1,06	1,06
1,55	0,58	1,38	0,99	2,06	1,52
0,61	1,95	1,65	1,12	1,34	2,73

b- standart deviation

Figure 7 and figure 8 gives the evolution of the average value and standard deviation of the diagonal terms of T_{ijkl} terms from 2 m to 5 m scale.

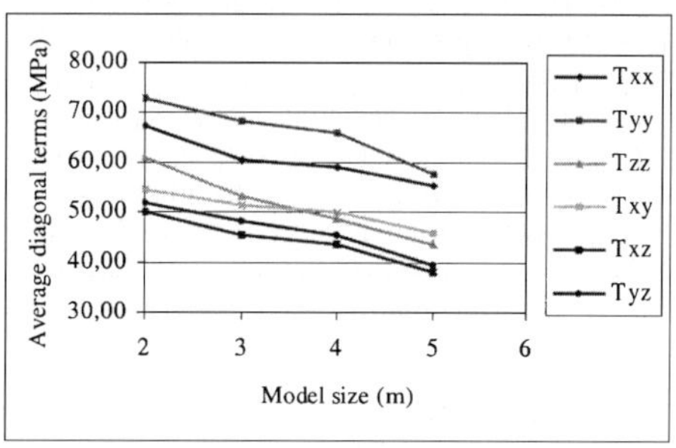

Figure 7. Evolution of average T_{ijkl} diagonal terms for formation 1.

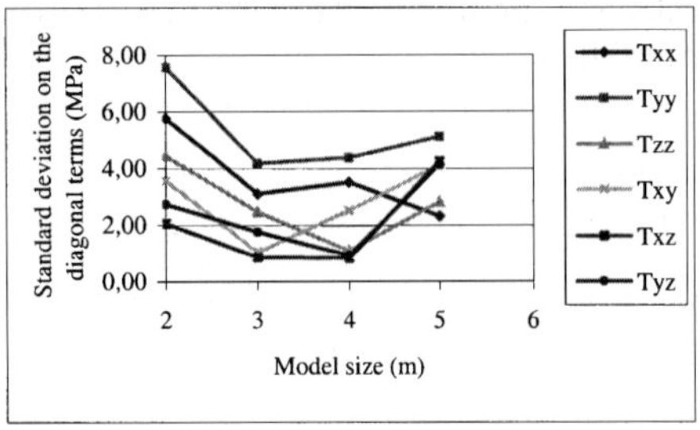

Figure 8. Evolution of standard deviation on T_{ijkl} diagonal terms for formation 1.

From this result, we can see that the REV is higher than 5 m. Model sizes become too important (≈ 250 Mo) to allow us to check if stabilization is about to be reached.

T_{ijkl} has also been determined at 2 m scale considering the real length of the joint. Differences with the previous simulation remain below 10 %. This could lead us to think that this factor is not very important for the equivalent mechanical behaviour. This result is consistent with the fact that no significant fracture length threshold effect has been shown on T_{ijkl}.

The results is highly related to the choice of the fracture stiffness value. This invites us to investigate later on the effect of stress on the result considering stress dependent stiffness.

4. CONCLUSIONS

An upscaling approach based on the 3D numerical simulation has been set up. Very good correlations with analytical solutions have been found for a network made of 3 sets of parallel fractures.

This approach has then been applied to fracture data given in the WP3 specifications in order to get the equivalent hydromechanical properties of the fractured rockmass.

The equivalent permeability tensor K_{ij} has been computed at 2 m scale considering the real length of the joint (we have shown the importance to avoid artificial joint prolongations). The REV could not be estimated due to code limitation reasons. The relation between permeability and stress (considering an isotropic state of stress) has been determined and is highly dependent on the joint stiffness value.

The equivalent stiffness tensor T_{ijkl} has been computed up to 5 m scale. No clear conclusion can be given for the mechanical REV. We have notice that the fracture network connectivity does not seem to have a strong influence on T_{ijkl}.

The new method presented in this paper is challenging because it is based on 3D explicit simulations of the hydromechanical behaviour of a fractured rock-mass. Some improvements are still to be done to make this method really operational. However those 3D results could contribute (through comparisons) to the evaluation of the more operational 2D methods used in BENCHPAR WP3. Accordingly this contribution could be profitable to the upscaling world.

5. ACKNOWLEDGEMENT

We wish to thank Ki-Bok Min from Royal Institute of Technology (KTH), Sweden for a helpful collaboration about mechanical upscaling using UDEC and 3DEC, and Johan Andersson from JA Streamflow AB, Sweden for many fruitful advices. We acknowledge with gratitude the financial support of the European Community and of the MEDD and MEFI french ministries.

6. REFERENCES

Damjanac, B. 1994. *A three-dimensional numerical model of water flow in a fractured rock mass.* Doctoral Thesis proposal, March 1994, University of Minnesota.

Heliot, D. 1988. *Generating a Blocky Rock Mass.* Int. J. Rock. Mini Sci & Geomech. Abstr. Vol 25, N°3 pp 127-138, 188.

Itasca Consulting Group, Inc., 1994. *3DEC 3-Dimensional Distinct Element Code.* Minneapolis, Minnesota: ICG.

Oda, 1986. *A equivalent continuum model for coupled stress and fluid flow analysis in joint rock masses.* Water Ressources Research, 22, pp 1845-1856.

Andersson, J. 2000. *Understanding the impact of upscaling THM processes on performance assessment.* version 6 (08/12/2000).

Vuillot, E. 1995. *Modélisation thermo-hydro-mécanique de massifs rocheux fracturés. Application au stockage de déchets radioactifs.* Thèse de Doctorat de l'INPL, Nancy, France.

THERMO-MECHANICAL EFFECTS ON HYDRAULIC CONDUCTIVITY IN A NUCLEAR WASTE REPOSITORY SETTING

Johan Öhman[1], Juha Antikainen[2] and Auli Niemi[1]
[1])Uppsala University
[2]) Helsinki University of Technology

Abstract: Understanding hydrogeological conditions around a future repository is crucial for safe nuclear waste disposal. Strong heterogeneity, even for *in-situ* fractured media, makes hydraulic conductivity predictions challenging. Furthermore, thermo-mechanical changes from repository excavation and waste heat emission will alter near-field hydrogeological conditions. This study estimates the magnitude of thermo-mechanical effects on performance of a hypothetical nuclear repository. First, complex 3D fracture networks are probabilistically analysed by traditional hydrological upscaling with fracture transmissivity based on *in-situ* hydraulic data. The resulting effective conductivity distribution is used as input for large scale stochastic continuum models. Next, separate thermo-mechanical modelling evaluates stress field changes due to repository excavation, bentonite swelling and heat emission. Obtained stress field changes, which are related to fracture aperture and transmissivity changes, alter near-field upscaled effective conductivity. Finally, applying the latter to the large scale model estimates the significance of thermo-mechanical effects on particle transport in relation to hydraulic *in-situ* uncertainty.

1. INTRODUCTION

Understanding the hydrogeological conditions for a future repository is crucial for safe disposal of high level nuclear waste. It is well known that only a part of geologically observable fractures are hydraulically conductive. Due to heterogeneity, the prediction of hydraulic conductivity of fractured media is a challenging task.

In a repository setting the hydrogeological conditions also change, due to both mechanical (M) changes, such as tunnel excavation and swelling of back filling material, and thermo-mechanical (TM) effects, due to heat emission from the waste. The question addressed here is how significant these changes are in comparison to the initial heterogeneity induced uncertainty, in predicting radionuclide travel times for a hypothetical repository. The model scenario and input data come from one of the test cases in the international collaboration project DECOVALEX (Development of Coupled Models and their Validation against Experiments), Decovalex (2000).

2. APPROACH

In our approach a detailed hydraulic analysis is carried out, where the flow and transport properties at the small scale are analysed by means of the fracture network software FracMan, Derschowitz et al. (1998), and MAFIC, Miller et al. (1999), that handle complex fracture geometries and fracture transmissivity distributions. The approach is probabilistic. A large number of network realizations is generated and the conductivity characteristics of small scale fracture network blocks are studied by determining directional hydraulic conductivity versus angle plots, which allows examination of the validity of continuum approximation, Öhman and Niemi (2003a). The upscaled conductivity statistics is correlated by an upscaled variogram, originating from borehole data using GSLIB, by Deutsch and Journel (1998), and used as input for the large scale stochastic continuum model TOUGH2, by Pruess et al. (1999). The resulting large scale flux and head fields are then used for particle tracking, using transport statistics obtained from small scale fracture network simulations.

Next, TM effects, caused by repository excavation, swelling of backfilling material and heating from the nuclear waste, are modelled separately using UDEC, by Itasca (2000). Our evaluation of the TM effect on hydraulic properties focuses on the repository near-field. The modelling results give changes in fracture apertures at various locations and for different main fracture orientations. These results are transferred into the previous hydrological model to examine the impact on near-field upscaled hydraulic properties.

Finally, the TM altered effective conductivity distribution is applied to the large scale model, as a sensitivity study, to compare the significance of the TM effects in relation to the underlying hydrology-induced uncertainty.

3. MODELLING

In order to perform the large scale hydraulic flow and transport simulations using a continuum media based model, the first step is to examine the possibility of approximating hydraulic fracture network flow as continuum media at some small scale [Long et al. (1982) and Cacas et al. (1992)].

3.1 Discrete fracture network data

Three-dimensional (3D) fracture networks were generated with the BART Beacher model, Beacher et al., (1977), using the FracMan software. Input data for these networks are distributions of geologically observed fracture orientations and lengths, density, transmissivity and intersection termination percentage. The fracture trace length distribution, reported by Nirex (1997a), is corrected to a 3D fracture radii distribution according to the method of Pigott (1997). Two-dimensional (2D) fracture density and fracture mean perpendicular spacing in boreholes, Nirex (1997b and c), are used to calculate 3D fracture area per volume intensity, P_{32}, Dershowitz and Herda (1992). A fracture transmissivity distribution was obtained from a probabilistic analysis of hydraulic borehole data, Osnes, et. al. (1988), implemented in the FracMan software. Non-conductive and low-transmissive fractures are excluded from the networks.

3.2 Small scale hydraulic upscaling

Flow simulations were performed for 150 fracture network realizations, each examined in various directions in a vertical plane parallel to the large scale modelling plane, by a gradually rotated hydraulic gradient. The approach is described in detail, Öhman and Niemi (2003a). A cubic flow region with 7.5 m sides was found large enough for the upscaling, yet small enough for computational demands of fine meshing for a high accuracy. The hydraulic conductivity in each direction was evaluated by the well known Darcy's law, yielding a distribution of effective conductivities at the 7.5 m scale, $G(K_{eff})$.

The validity of a continuum approximation for the fractured media was evaluated by a second order symmetric conductivity tensor, Harrison and Hudson (2000), fitted to the simulated directional conductivities of each fracture network realization. These fits were evaluated by root-mean-squares and classified as continuum or discontinuum-type. It was found that 5% of the fracture networks could not be described by a conductivity tensor, 15% have acceptable fits, and 80% fitted very well.

3.3 Small scale transport properties

In order to obtain distributions of transit times, τ, and transport resistance, β, Cvetkovic (1998), for the small scale continuum blocks, we performed particle tracking through 30 generated fracture networks in a similar way as for upscaling the hydraulic flow. In each realisation a gradually rotated $15 \times 7.5 \times 7.5$ m^3 flow region was examined by a hydraulic gradient of 0.05, corresponding to the average regional gradient. No-flow conditions were applied to the boundaries parallel to the large scale modelling plane. A cubic transport region, with side length 7.5 m, was defined at the centre of the flow region, as shown in Figure 1. The purpose for a somewhat larger flow field than the transport region in the transverse direction is to have 'guard zones', Jackson, et. al. (2000), that reduce the risk of allowing shortcuts to the adjacent boundaries.

10 000 particles were released from a 2×2 m^2 surface at the centre of the 15×7.5 m^2 inflow boundary and collected at the remaining sides of the transport region. In total, over 3 million particle travel times were obtained. Assuming the effective porosity to be proportional to $K_{eff}^{1/3}$, following 'the cubic law', each travel time was scaled by the directional conductivity of the transport region and joined into one single probability distribution, $G(\tau \times K_{eff}^{2/3})$. The purpose is to obtain a conductivity independent transit time distribution to be used as input for heterogeneous conductivity fields in the large scale particle tracking. Similarly, a linearly conductivity scaled distribution for travel resistance, $G(\beta \times K_{eff})$, was also obtained. The details of the transport modelling are given in Öhman and Niemi (2003b).

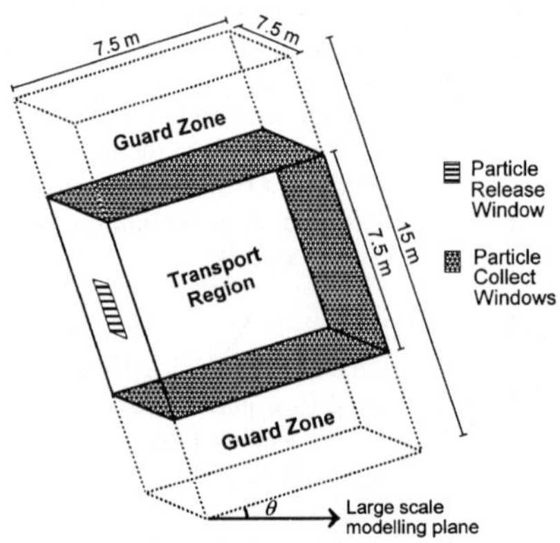

Figure 1. Flow and transport regions vertically rotated in the large scale model plane.

3.4 Large scale hydraulic modelling

The large scale modelling region, shown in Figure 2, was discretized into cubic blocks with side length 7.5 m, for consistency with the upscaled hydraulic input data. Element conductivity was stochastically generated from the probability distribution $G(K_{eff})$, and correlated with an upscaled exponential variogram. The region to the left of the fault zone, $0 < x < 2475$ m, was discretized with coarser elements, side length 45 m, and assigned a separate conductivity distribution, upscaled to 45 m. The reason for this is to reduce simulation time, since no particle tracking will take place within this region. The coarse grid has little impact on the large scale flow due to the geometry of the modelling region and the conductive fault zone.

No-flow was assigned for boundaries x = 5000 m, x = 0 m and z = -1000 m, and constant hydraulic heads of 100 m and 0 m for z = 0 m and the sea, respectively. A conductivity depth trend was super imposed on the correlated conductivity field. The depth trend was taken from *in-situ* M modelling and is a function of overburden weight.

3.5 Large scale particle tracking

The large scale transport simulations use results from the small scale transport simulations and head fields and vertical and horizontal flux fields extracted from steady state solutions of large scale hydraulic simulations. The particle tracking scheme is explained more in detail in Öhman and Niemi, (2003b). A large number of particles are released uniformly within the repository region and move proportionally to the local flux field until reaching the sea, as shown in Figure 2. The total travel time for a pathway is a sum of transit times, which are randomly taken from the distribution $G(\tau \times K_{eff}^{2/3})$ and scaled with the local hydraulic gradient and local conductivity. Transport resistance was obtained in a similar way.

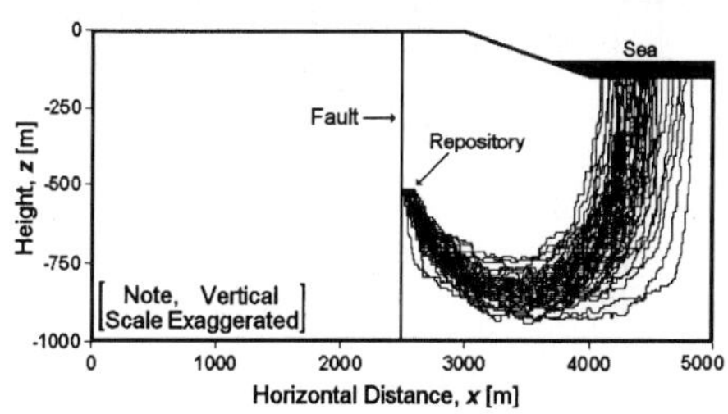

Figure 2. Simulated particle pathways for one correlated stochastic conductivity realization on the large scale.

3.6 Thermo-Mechanical modelling

The M effect of repository excavation, swelling pressure of filling material and TM effects from heat production of deposited waste was evaluated by separate 2D TM modelling using UDEC. The repository excavation was modelled by decreasing the rock stiffness and stress by 25% within the 100 x 10 m^2 repository area. The swelling pressure was modelled by applying 2 MPa pressure inside the repository. The heating of the repository was modelled as uniformly distributed, decaying heat source with initial power of 0.6 W/m^3. To reduce simulation time, one-directional coupling was used for heating times exceeding 3 years. The temperature field was first calculated with a separate continuous model (Figure 3) and then applied for the fracture network model.

The obtained fracture normal stresses, σ_n (MPa), were related to hydraulic apertures, b (μm), using an approximate empirical relationship based on the laboratory loading-unloading tests of core data in Equation 1. These hydraulic apertures, at 100 years of heating, were in turn related to fracture transmissivity, using 'the cubic law', and compared to the original, *in-situ,* fracture transmissivity.

$$b = \frac{27}{(\sigma_n + 1)^{0.66}} \qquad (1)$$

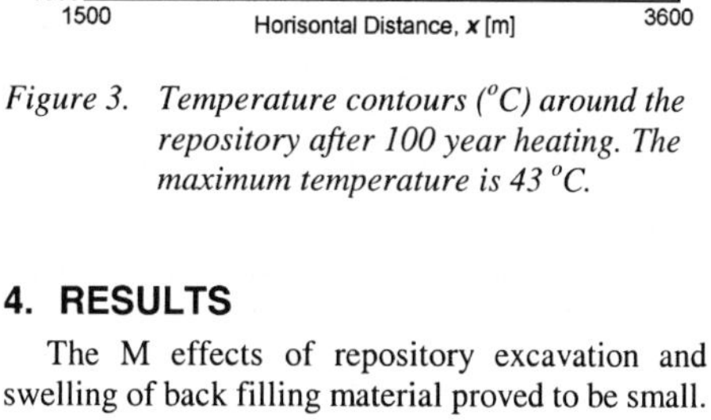

Figure 3. Temperature contours (°C) around the repository after 100 year heating. The maximum temperature is 43 °C.

4. RESULTS

The M effects of repository excavation and swelling of back filling material proved to be small. Our results focus on TM effects at 100 years of repository heating in the lower formation, as described in section 3.6. The changes in fracture transmissivity as a function of radial distance from the repository centre, r, are shown in Figure 4.

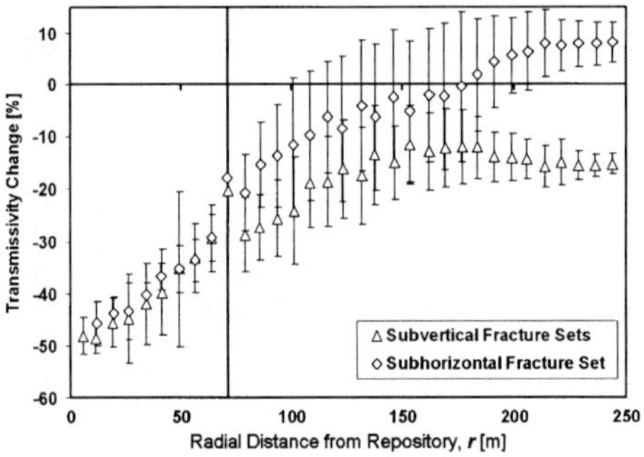

Figure 4. Fracture transmissivity change due to 100 years of heating, for repository distance divided into 7.5 m intervals.

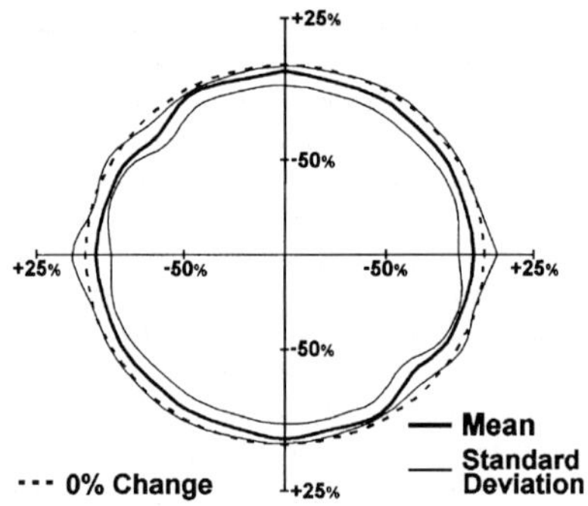

Figure 6. Change in directional conductivity for 10 small scale realizations at r =180 m.

A distortion can be seen at r = 70 m, which is caused by two fractures, partly inside the fault zone. The most interesting locations for further study was decided to be at r = 15 m, where transmissivity is reduced by a factor 2, and at r = 180 m, where sub-vertical fracture transmissivity is slightly decreased, while sub-horizontal fractures show a slight increase.

The statistics for transmissivity change at these two locations were transferred to 10 small scale fracture network realisations, each selected so as to represent different anisotropies and continuum classes. The small scale upscaling procedure was then repeated for these modified fracture networks, as described in section 3.2, in order to evaluate the impact on effective conductivities. Since the 2D TM modelling excluded the fracture set parallel to the large modelling plane, no changes were applied for this set in the 3D hydraulic modelling. The results are shown in Figures 5 and 6.

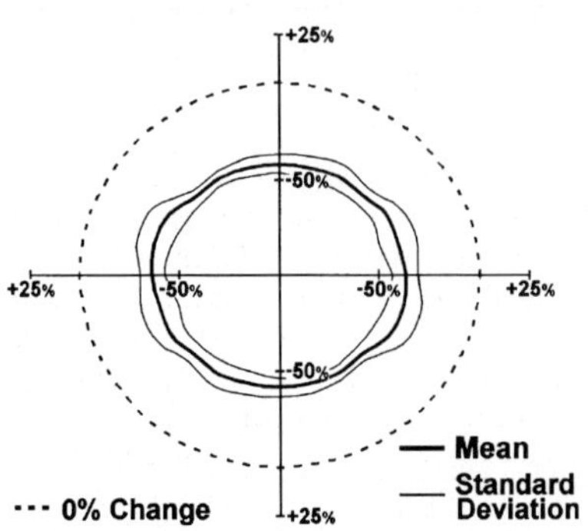

Figure 5. Change in directional conductivity for 10 small scale realizations at r = 15 m.

The effect of 100 years heating on effective conductivity at the 7.5 m scale was reduced by 40% near the repository and by 5% at 180 m distance. Close to the repository, the results only show small changes in continuum behaviour. The most probable reason for the slight discontinuum increase is that transmissivity was halved for three sets, but left unchanged for the fourth. At 180 m distance no changes in continuum behaviour are observed. Even though sub-horizontal and sub-vertical fractures are given different transmissivity changes at 180 m distance, as shown in Figure 4, no anisotropy increase can be observed in Figure 6. This is probably due to the fact that fracture sets have a large deviation around the mean angle.

The effective conductivity seems to follow the overall fracture transmissivity reduction. Figure 7 shows changes in effective conductivity, average fracture transmissivity change (disregarding orientations for fracture sets) and a fitted trend.

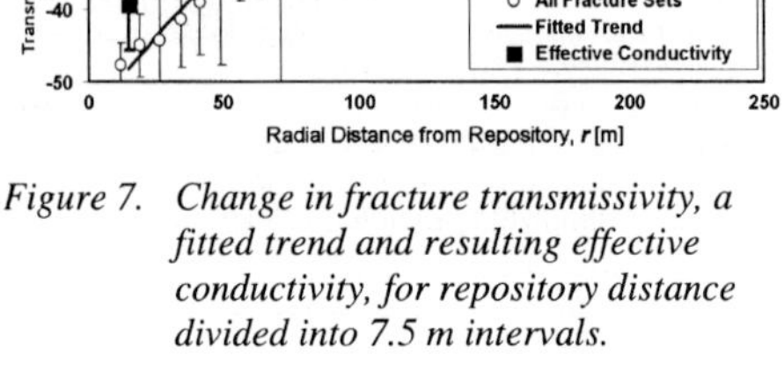

Figure 7. Change in fracture transmissivity, a fitted trend and resulting effective conductivity, for repository distance divided into 7.5 m intervals.

Assuming the excluded fracture set to undergo similar transmissivity change, as do the modelled sets, it seems likely that the changes in effective conductivity should follow the trend in Figure 7. Transferring effective conductivity changes as functions of r into large scale transport simulations as a "permanent conductivity change", yields an upper estimate of TM impact on travel times and transport resistance. Average travel times for the quantiles: 0-10%, 10-20%, 20-40%, 40-60% and 60-100% were calculated for 100 000 released particles. Our results for 10 realizations indicate that the transit time and the transport resistance increase at most by, approximately 5% and 8 %, respectively, as shown in Figures 8 and 9. Transport resistance calculated at a 50 m radial distance from the repository, on the other hand may increase as much as 25 %, as shown in Figure 10.

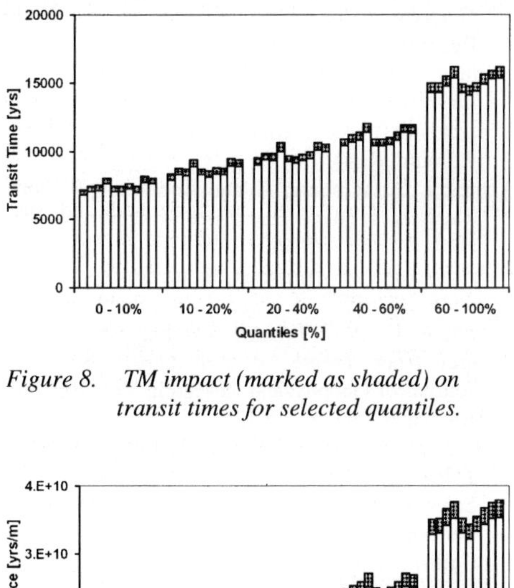

Figure 8. TM impact (marked as shaded) on transit times for selected quantiles.

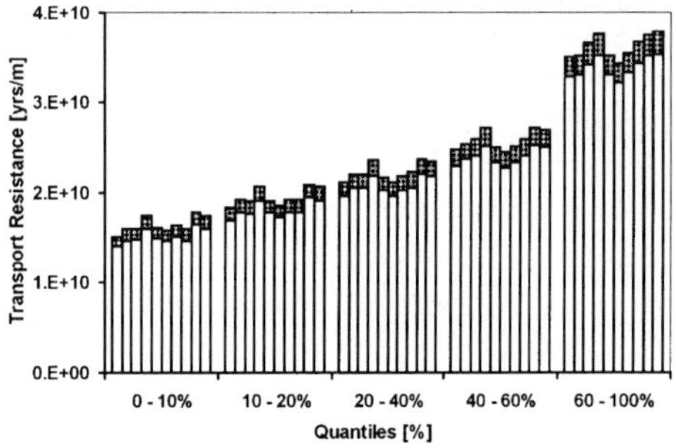

Figure 9. TM impact (marked as shaded) on transport resistance for selected quantiles.

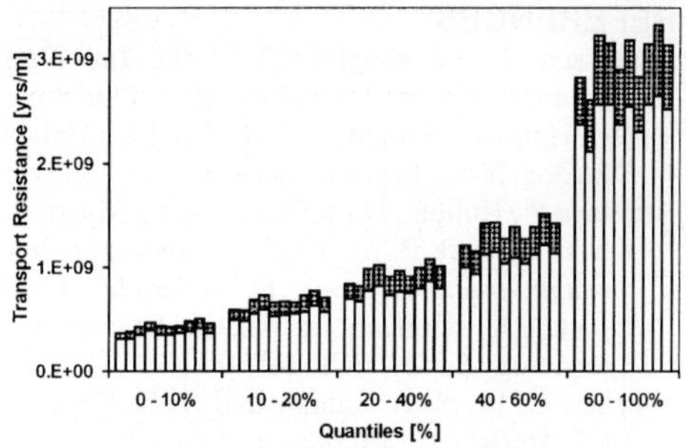

Figure 10. TM impact (marked as shaded) on transport resistance for selected quantiles 50 m from the repository.

5. CONCLUSIONS

The M effects of excavation and bentonite swelling proved to be negligible. The TM effects caused by repository heating seem to have reached a maximum at 100 years. The impact is a reduction in effective conductivity by a factor 2 for the repository near-field, while at a radial distance of 180 m from the repository the reduction is negligible. Transferring these changes into our large scale transport simulations as a permanent change in conductivity, yields an upper estimate of TM impact on transit time and transport resistance. Our results indicate that the transit time and the transport resistance increase, at the most by 5% and 8 %, respectively. Transport resistance calculated at a 50 m radial distance from the repository, on the other hand, may increase as much as 25 %.

Considering the long travel times (on the order of 10 000 years) in combination with peak temperatures at 100 years, seems to indicate that the need for full THM coupling in large scale simulations is not critical, in comparison to the intrinsic hydraulic uncertainties.

ACKNOWLEDGMENTS

We wish to thank the Finnish Radiation and Nuclear Safety Authority (STUK) and especially Dr. Esko Eloranta for the financial support and interest in this work. Further thanks go to the collaborators within the Decovalex group, especially Dr. Chin-Fu Tsang, Mr. Philipp Blum and Mr. Ki-Bok Min for their valuable advice and comments during the course of our work.

Finally we wish to thank Golder Associates for making the FracMan and MAFIC codes available for this study.

REFERENCES

Andersson, J. and Knight J. L. 2000. *The THM Upscaling Bench Mark Test 2 - Test Case Description.* Knight, J. L. (ed.), United Kingdom Nirex Limited, Harwell.

Armitage P., Holton., D., Jefferies, N. L., Myatt, B. J. and Wilcock P. M. 1996. *Groundwater flow through fractured rock at Sellafield,* Final report, E.C. publications, Nuclear Science and Technology Series, EUR 16870 EN.

Baecher, G. B., N. A. Lanney and H. H. Einstein. 1977. *Statistical Description of Rock Properties and Sampling*, Proceedings of the 18th U.S. Symposium on Rock Mechanics, American Institute of Mining Engineers, 5C1-8.

Cacas, M. C., Ledoux, E., de Marsily, G., Tillie, B., Barbreau, B., Durand, A., Feuga B. and Peaudecerf, P. 1990. *Modeling fracture flow with a stochastic discrete fracture network: Calibration and validation, 1, The flow model*, Water Resour. Res., 26(3), pp. 479-489.

Cvetkovic, V., Selroos, J. O. and Cheng, H. 1999. *Transport of reactive tracers in rock fractures*, J. Fluid Mech., vol 378, pp. 335-356.

Decovalex (2000): *Decovalex III, Bench Mark Test 2, Understanding the impact of Upscaling THM processes on Performance Assesment*, Test Case Description. 6.12.2000. (www.decovalex.com/)

Dershowitz, W, Lee, G., Geier, J., Foxford, T., LaPointe, P. and Thomas, A. 1998. *FracMan Interactive Discrete Feature Data Analysis, Geometric Modelling and Exploration Simulation*, User Documentation, Golder Associates Inc., Seattle, Washington.

Dershowitz, W. S. and Herda H. H. 1992. *Interpretation of fracture spacing and intensity*, Rock Mech., pp. 757-766.

Deutsch, C. V. and Journel, A. G. 1998. *Geostatistical Software Library and User's Guide*, Second edition, Oxford University Press.

Harrison J. P. and Hudson, J. A. 2000. *Engineering Rock Mechanics: Part 2. Illustrative examples*, Pergamon, Elsevier Science Ltd., The Boulevard, Longford Lane, Kidlington, Oxford OX5 1 GB, UK.

Itasca Consulting Group, Inc. 2000. *UDEC user's Guide*, Minneapolis, Minnesota, USA, (www.itascacg.com).

Jackson, C. P., Hoch, A. R. & Todman, S. 2000. *Self-consistency of a heterogeneous continuuum porous medium representation of fractured media.* Water Resour. Res., 36 (1), 189-202.

Long, J. C. S., Remer, J. S., Wilson, C. R. and Witherspoon, P. A. 1982. *Porous Media Equivalents for Networks of Discontinuous Fractures*, Water Resour. Res., 18, pp. 645-658.

Miller, I., Lee, G. and Dershowitz, W. 1999. *MAFIC Matrix/Fracture Interaction Code With Heat and Solute Transport User Documentation*, Version 1.6, Golder Associates Inc., Redmond, Washington, November 30.

Nirex. 1997a. *The lithological and discontinuity characteristics of the Borrowdale Volcanic Group at outcrop in the Craghouse Park and Latterbarrow areas*, Nirex Report SA/97/029.

Nirex. 1997b. *Evaluation of heterogeneity and scaling of fractures in the Borrowdale Volcanic group n the Sellafield area.* Nirex Report SA/97/028.

Nirex. 1997c. *Data summary sheets in support of gross geotechnical predictions.* Nirex Report SA/97/052.

Öhman, J. and Niemi, A. 2003a. *Upscaling of Fracture Hydraulics by Means of an Oriented Correlated Stochastic Continuum Model*, Water Resourc. Res. (Accepted).

Öhman, J. and Niemi, A. 2003b. *A new method for large scale stochastic continuum based particle tracking* (in preparation for submission to Water Resourc. Res.).

Osnes, J. D., Winberg, A. and Andersson, J. 1988. *Analysis of Well Test Data – Application of Probabilistic Models to Infer Hydraulic Properties of Fractures*, Topical Report RSI-0338, RE/SPEC Inc., Rapid City, South Dakota.

Pigott, A. R. 1997. *Fractal relations for diameter and trace length of disc-shaped fractures.* Journal of Geophysical Research, Vol. 102, No. B8, pages 18 121 - 18 125.

Pruess, K., Oldenburg, C. and Moridis, G. 1999. *TOUGH2 User's Guide*, Version 2.0, Earth Science Division, Lawrence Berkeley National Laboratory University of California, Berkeley, California 94720.

A FINITE-ELEMENT STUDY OF POTENTIAL COUPLED HYDROMECHANICAL EFFECTS OF GLACIATION ON A CRYSTALLINE ROCK MASS

Tin Chan[1], Frank W. Stanchell[1], Thomas Wallroth[2], Jan Hernelind[3] and Geoffrey Boulton[4]
[1]Applied Geoscience Branch, Atomic Energy of Canada Limited (AECL)
[2]Department of Geoengineering, Chalmers University of Technology (CTH)
[3] FEM-Tech AB
[4]School of GeoSciences, University of Edinburgh

Abstract: A number of studies related to past and on-going deep repository performance assessments have identified glaciation/deglaciation as major future events in the next few hundred thousand years capable of causing significant impact on the long term performance of the repository system. Bench Mark Test 3 (BMT3) of the international DECOVALEX III project has been designed to study the coupled hydro-mechanical (H-M) impacts of glaciation and deglaciation on the long-term (up to 100 000 years), post-closure performance of the geosphere in which a hypothetical repository is located. The BMT3 is a generic exercise based on simplified geological, hydrogeological and rock mechanical characteristics of a crystalline rock research area in the Canadian Shield. This paper presents the site-scale coupled hydro-mechanical finite-element modelling studies conducted by the AECL and CTH teams. Interim results suggest that coupled hydro-mechanical effects, transient effects and fracture zone structural geometry are important.

1. INTRODUCTION

Geological evidence has indicated that mid- to high-latitude locations in the Northern Hemisphere have experienced glaciation/deglaciation cycles in recent geological history. These cycles are likely to recur in the future within a time frame of several hundred thousand years and have to be considered in performance assessments of deep geological repositories of long lived nuclear wastes.

Bench Mark Test 3 (BMT3) of the international DECOVALEX III project, which is also Work Package 4 (WP4) of the European BENCHPAR project, is a numerical study with the following objectives (Chan et al. 2002): 1) to study the long-term evolution of a fractured rock mass in which a generic repository is located, as it undergoes a glacial cycle; and 2) to assess the associated coupled hydro-mechanical (H-M) impacts on the long-term waste isolation performance of the geosphere.

There are three types of numerical model studies, i.e., continental scale ice-sheet/drainage modelling, site-scale permafrost modelling and coupled H-M rock mass modelling. This paper pertains to the last type of modelling, as conducted by the AECL and CTH teams. Further details on other aspects of BMT3 can be found in Boulton and Hartikainen (2003) and Boulton et al. (2003).

Previous numerical studies of the effects of glaciation on the rock mass in which a repository is located have been either 2D coupled hydromechanical modelling based on simple, highly idealized geological structure (Nguyen et al. 1993, Chan et al. 1998) or rock mechanical modelling based on complex 3D geological structural models (Hansson et al. 1995). In this study time-dependent ice mechanical load and water pressure at the ice/bedrock interface, calculated by the ice-sheet/drainage model, were applied as transient boundary conditions for the coupled H-M models.

2. CONCEPTUAL MODEL

The conceptual model for HM modelling has been constructed from a simplification of the geological, hydrogeological and rock mechanical characteristics of AECL's Whiteshell Research Area (WRA) near the edge of the Canadian Shield in Manitoba, Canada. The model domain encompasses a variably fractured body of crystalline rock approximately 25 km x 37 km x 4 km deep. An interconnected 3D network of fracture zones traverses the rock mass. A generic spent-fuel repository measuring 2km x 2km x 10m was assumed to be located at approximately 500-m depth. The network of major fracture zones has been idealized so that there are 17 fracture zones, 12 at the vertical boundaries and 5 in the interior of the conceptual model. Of the five interior fracture zones two are vertical, one is horizontal and two are dipping at 45°. Coupled HM modelling has been performed in two vertical sections, Section 2 parallel to the maximum horizontal in situ stress (also the expected ice flow direction) and Section 1 perpendicular to this direction. The two configurations designated "Configuration 3" and "Configuration 6" in Figure 1 have been utilised for coupled HM modelling. The dipping and horizontal fracture zones are connected in Configuration 3 but not in Configuration 6. The structures are similar in the two sections. In Section 2 the horizontal fracture zone extends all the way to the vertical boundary. Further, the dipping fracture zone extends below the horizontal fracture zone.

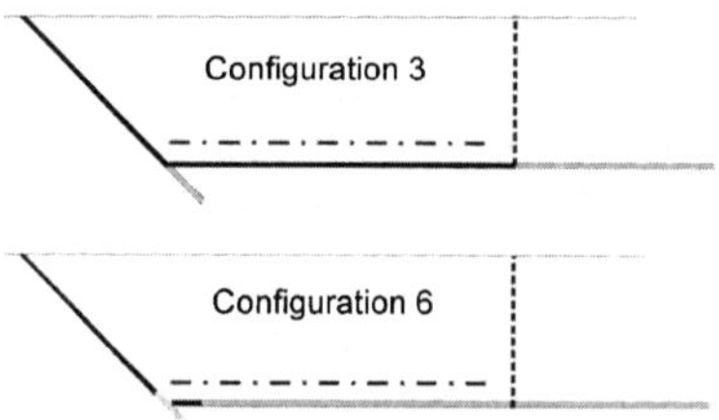

Figure 1. Central portion of fracture zone configurations for site-scale H-M simulations. Black is Section 1; grey is Section 2; dotted is vertical fracture zone; dashed is repository.

Hydraulically the background rock mass was represented by 13 layers with horizontal permeability, k_H, decreasing with depth from 10^{-15} m^2 near surface to 3.35 $x10^{-20}$ m^2 at the assumed repository level just below 500-m depth; below 750-m depth $k_H \approx 10^{-21}$ m^2. Vertical permeability was assumed to be 10 k_H for the top 300 m; below this depth the rock was assumed to be isotropic hydraulically. Porosity was assumed to be 0.003 for the background rock mass. For the vertical fracture zones k was assumed to be 10^{-13} m^2 from surface to 400-m depth, below which it decreases rapidly downward. The dipping fracture zone was assumed to have k value of 10^{-13} m^2 at all depths.

The following mechanical properties were assumed for the rock mass: density = 2650 $kg.m^{-3}$, Young's modulus = 35 GPa, Poisson's ratio = 0.22, cohesion = 5 MPa and internal friction angle = 30°. For the fracture zones Young's modulus = 3.5 GPa, cohesion = 3 MPa and friction angle = 25°. Biot's hydroelastic coefficient, α, was assumed to be 1 everywhere. In situ stresses were assumed to correspond to mean values for the Canadian and Fennoscandian Shields. These increase with depth, with the maximum principal stress in the horizontal NE-SW direction.

Transient mechanical and hydraulic boundary conditions calculated at the ice/bedrock interface by the ice-sheet/drainage models (Boulton and Hartikainen 2003) were prescribed to the top of the site-scale HM model, as illustrated in Figure 2.

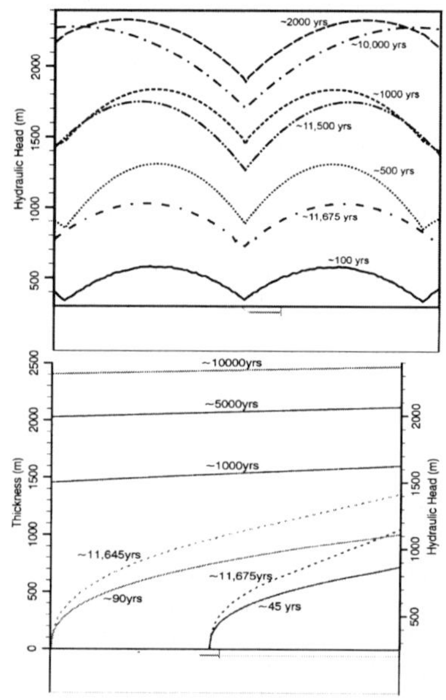

Figure 2. Bedrock surface boundary conditions. Top: Section 1 head; bottom: Section 2 head and ice thickness.

3. NUMERICAL MODELLING

The AECL team used an in-house MOTIF finite-element code (Guvanasen and Chan 2000), which is based on an extension of the classical poroelastic theory of Biot (1941). This code has undergone extensive verification and validation (Chan et al. 2003). The CTH team employed the commercially available, general-purpose finite-element code ABAQUS/Standard 6.3 (ABAQUS manuals). This code adopts a macroscopic thermodynamic approach. The porous medium is considered as a multiphase material, and an effective stress principle is used to describe its behaviour. ABAQUS allows the value of bulk modulus of the mineral grains as an input parameter. In order to select an appropriate value for this low-permeability, low-porosity rock, the CTH team compared the ABAQUS solution with Biot's (1941) analytical solution for 1D consolidation in the form presented by Chan et al. 2003).

4. RESULTS AND DISCUSSIONS

4.1 Hydraulic response

During the glaciation the evolution of subsurface hydraulic head follows the advance/growth and retreat/decay of the ice sheet, e.g., Figure 3 for two locations at approximately 500-m depth below ground surface – Point 1 in the dipping fracture zone and Point 4 in the repository. The hydraulic head rises rapidly from the pre-glacial value of ~265 masl (equivalent fresh water head) to ~2300 masl in about 1000 years after the onset of glaciation over the WRA and returns to pre-glacial level in as much time. Simulation results from the two teams generally agree well. There are, however, important subtle differences. According to the AECL model, in the first few hundred years the head in the dipping fracture zone, e.g., Point 1, can be several hundred metres higher than that in the repository, e.g., Point 4, or the surrounding rock mass. This is due to the rapid transmission of high head values from the top boundary down the very permeable ($k \sim 10^{-13}$ m^2) fracture zone, Figure 4. In contrast, there is only marginal difference between the two CTH curves for the two selected locations.

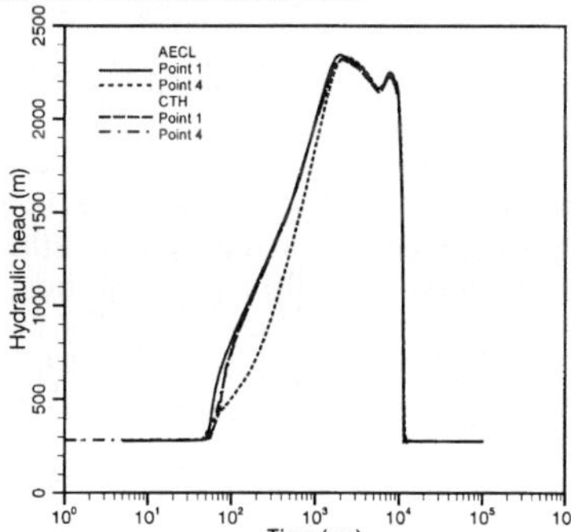

Figure 3. Evolution of hydraulic head calculated by AECL and CTH for Section 2 Configuration 3 in the dipping fracture zone (Point 1) and in the repository (Point 4).

During ice sheet advance there are horizontal hydraulic gradients all the way down to the bottom of the model (4km). This is definitely a mechanical-hydraulic (M-H) effect due to the compression of pore space by the ice loading. Otherwise, it would not be possible for the hydraulic gradient to penetrate to such depth in the very low-permeability rock mass. Head contours for Section 2 Configuration 6 during ice sheet advance (see Chan and Stanchell 2003) reflect the discontinuity between the dipping and horizontal fracture zones.

When the ice sheet covers the entire area, the hydraulic head is high (corresponding to the head boundary condition) down to a depth of about 650m, Figure 5. A very large downward hydraulic gradient (3-5 m/m) is seen below this depth. This is associated with the large permeability contrast in the bedrock. At depth there is an excess pore pressure equivalent to ~1/3 of the weight of the ice sheet. When the ice sheet retreats from the site a reversal of the vertical hydraulic gradient can be observed. The upward hydraulic gradient lasts for tens of thousands of years after the ice has disappeared from the WRA.

The head in the fracture zones is lower than in the surrounding rock mass during the retreat (Chan and Stanchell 2003; Wallroth et al. 2003). When the trailing edge of ice sheet leaves the site (at 11675 years), the largest residual heads exist at about 700 m depth. The position of the maximum residual head moves downwards with time as the excess pressure gradually dissipates. Around 9500 years after the ice sheet has retreated from the model domain, the highest residual excess pressure (~250 m) can be found at about 800-m depth (see 21200-year curve in Figure 5). At the end of the modelling period (100000 years) the residual pressure head in the repository is about 25 m.

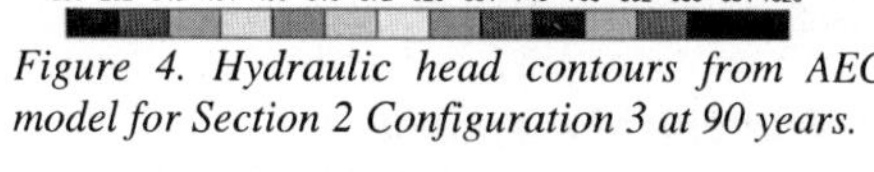

Figure 4. Hydraulic head contours from AECL model for Section 2 Configuration 3 at 90 years.

Maximum Darcy velocities are in the order of 1-10 cm/year in the fracture zones and 0.1-1 mm/year in the repository and the surrounding rock mass. These velocities are two orders of magnitude higher than their values in the non-glacial initial state. For Configuration 3, in which the fracture zones are connected, there is a distinct upward flow just before the ice front passes the site at Point 1 located in the dipping fracture zone. For Configuration 6, in which the dipping and horizontal fracture zones are disconnected, there is still upward flow with much reduced (by a factor of ~50) velocities in the AECL simulation, possibly due to M-H effects, but not in the CTH simulation. In general, the velocities calculated by the two teams do not agree as well as the hydraulic heads.

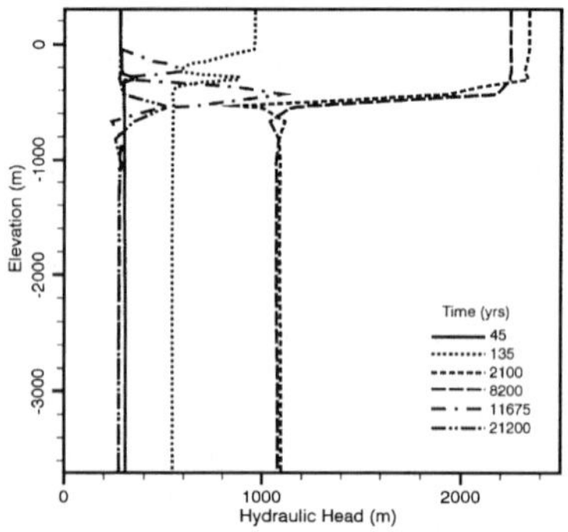

Figure 5. Vertical profile of hydraulic head (masl) through centre of repository from AECL Section 2 Configuration 3 simulation.

In Section 1 the evolution of subsurface hydraulic head during the glaciation follows the growth and decay of the ice sheet. A salient feature in the head distribution (not illustrated) is that shortly after the onset of glaciation, areas of lower head values appear beneath the three sub-glacial channels (troughs in Figure 3). Groundwater flows towards these sinks in the upper permeable rock mass. The side channels in the subsurface disappear as soon as they disappear from the top boundary.

During glaciation the much higher glacially induced regional flow in both sections overshadows the local flow that is due to topographic variations.

4.2 Mechanical response

Figure 6 depicts the evolution of mechanical displacement from the two teams for Section 2 Configuration 3 in the dipping fracture zone (Point 1) and in the repository (Point 4).

The downward displacement follows primarily the time varying ice loading at the top boundary, with minor time dependence due to the slow variation of pore pressure with time. From contour plots (Chan et al. 2003; Wallroth et al. 2003) it can be seen that there are larger downward displacements beneath the channels than in the surrounding rock. This is due to water drainage. The hanging wall and footwall of the dipping fracture zone are offset by a few cm of displacement because the 20-m thick fracture zone is much softer than the surrounding rock. Maximum downward displacement occurring at glacial maximum is about 1.35 m for the AECL and 1.7 m for the CTH model. Horizontal displacements are much smaller.

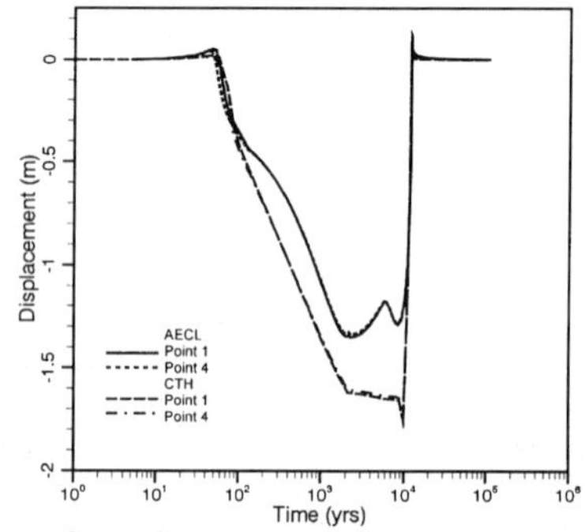

Figure 6. Evolution of mechanical displacement calculated by AECL and CTH for Section 2 Configuration 3 in the dipping fracture zone (Point 1) and in the repository (Point 4).

Effective stresses are affected in two different ways, an increase in total stress due to the mechanical loading of the ice sheet and a decrease in effective stress due to the increase in pore pressure. Consequently, changes in effective stresses are much less than expected from the mechanical stresses due to the weight of the ice sheet. Even in the dead-ended horizontal fracture zone in Section 2 Configuration 6 there is no hydraulic jacking, i.e., no effective tensile stress. No shear failure has been predicted. There is practically no rotation of principal effective stresses.

In general, the head and displacement results from the two teams are in better agreement than the Darcy velocity and effective stress results. This is not surprising. Head and displacement are primary unknown variables calculated by both MOTIF and ABAQUS, whereas Darcy velocity and stress are derivatives so that any inaccuracies arising from the extremely high material property contrast between the rock matrix and the fracture zones would be accentuated.

5. IMPORTANCE OF COUPLED AND TRANSIENT ANALYSES

A sequence of static equilibrium analyses was performed by applying the mechanical ice loading from the ice-sheet model at various times. Resulting maximum displacement is about 1.5 times as high as the coupled H-M analysis. Spatial and temporal variations of stress are qualitatively different from the H-M analysis. For example, the maximum change in the minimum principal stress in the horizontal fracture zone is about an order of magnitude higher than the corresponding effective stress change in the coupled H-M analysis. Principal stress axes rotate with space and time. Without the counter-balancing effect of pore pressure, a mechanical stress analysis of glacial loading predicts the maximum/minimum principal stress ratio to be much closer to 1, which leads to a much higher factor of safety, than the coupled H-M analysis. It would be unwise to base a mechanical stability assessment under glaciation condition on purely mechanical analysis.

Each team performed an uncoupled transient flow simulation. The spatial distribution and evolution of the simulated head are qualitatively similar to the coupled H-M simulations above a certain depth, which varies with time from near surface to 900-m depth. Below this depth, due to the very low rock permeability there is hardly any hydraulic response to glaciation. Residual excess pressures similar to the coupled H-M model results are predicted 9500 years after glacier is gone. In this case, however, the values (350-400 m of pressure head) are almost twice as high.

Several steady state hydraulic and coupled H-M simulations were performed. The results were found to be highly dependent on the particular steady state approximation chosen. None of the results resembles the coupled H-M simulation qualitatively or quantitatively. For example, the non-glacial steady state analysis under-predicts the flow velocities by two orders of magnitude. One coupled H-M simulation with steady state glacial boundary conditions predicted surface water recharging through sparsely fractured rock to 1200-m depth. A second steady state approximation predicted uniformly high hydraulic head throughout nearly the entire model domain. A third steady state approximation predicted a large hydraulic gradient at the glacier terminus situated over the model domain for 100 000 years. None of the steady state simulations was able to capture the upward hydraulic gradient during and after deglaciation. In short, no steady state analysis has been found to be a reasonable approximation to the coupled H-M analysis. Hence, it would be prudent to re-examine the validity of the steady state flow field assumption made in most safety assessments.

6. MODEL UNCERTAINTIES

There are a number of uncertainties in the H-M models due to the various assumptions and approximations including: 1) the one-way interfacing of the continental scale ice-sheet/drainage model to the site-scale H-M models through transient pressure head and ice (mechanical) loading boundary conditions; 2) limited testing of mesh refinement; 3) no numerical testing to assess whether the model domain is large enough to capture the complexity of hydromechanical response beneath and outside the ice terminus; 4) a very approximate representation of permafrost conditions as a no-flow boundary at ground surface not covered by ice rather than a more realistic representation as modified hydraulic and mechanical properties in the (time varying) frozen parts of the rock mass; 5) possibly insufficient discretization along the top boundary to capture the steep hydraulic gradient in the vicinity of sub-glacial channels since head values computed at 200-m spacing were transferred from the ice-sheet/drainage model; 6) generic, rather than site specific, input data have been used for the geometry and connectivity of the fracture zones, in situ stresses and mechanical strength parameters together with a simple failure criterion; 7) omission of influence of salinity on flow; 8) omission of effect of stress on permeability and 9) omission of large-scale crustal downwarping and rebound so that only displacements relative to the bottom of the model are simulated. Furthermore, uncertainties in the ice-sheet/drainage model (Boulton and Hartikainen 2003) would also be passed to the site-scale H-M models.

7. CONCLUSIONS

Two independent H-M models have yielded generally similar results indicating the possibility of high hydraulic gradients, high flow velocities and flow reversal during deglaciation, and high residual pressure long afterwards. Large hydraulic gradients can appear at the glacier terminus or near the sub-glacial channels (i.e. eskers). The extensive and thick ice-sheet may diminish the influence of topographic gradients, so that the groundwater flow regime beneath a continental ice-sheet may become more regional. Mechanical responses, though complicated, are relatively mild. No hydraulic jacking or hydraulic shearing and no rotation of principal effective stresses are indicated at nominal repository horizons. Implications of the Glaciation BMT study for performance assessment are discussed at some length in an overview paper (Boulton et al. 2003).

The transient coupled hydro-mechanical modelling in this study represents a step forward in advancing the state of the science for modelling geosphere response to glaciation. No uncoupled or steady state approximation has yet been found that captures all the essential hydro-mechanical responses of the geosphere that may be potentially important to long-term performance assessment.

While the ultimate model would be one that couples the simulations of the development and movement of the ice sheet, the associated hydraulics, the development of permafrost and the hydraulic and mechanical responses of the geosphere; worthwhile improvements in the interim include: 1) three-dimensional simulations; 2) conceptual model based on synthesis of site-specific data; 3) including stress-dependent permeability and 4) representing the effects of permafrost as modified, time dependent hydraulic and mechanical properties in the H-M model.

ACKNOWLEDGEMENTS

The authors are grateful to their respective funding organizations (Ontario Power Generation to AECL and SKB to CTH) for their financial support and technical advice. Financial support from the European Commission to CTH in the BENCHPAR project is also acknowledged.

REFERENCES

ABAQUS/Standard *User's Manual Version 6.3.* ABAQUS, Inc .

Biot, M.A. 1941. *General theory of three-dimensional consolidation.* Journal of Applied Physics 12: pp. 155-64.

Boulton, G., Chan, T., Christiansson, R., Ericsson, L., Hartikainen, J., Jensen, M.R., Stanchell, F.W. and Wallroth, T. 2003. *Thermo-Hydro-Mechanical (T-H-M) impacts of glaciation on the geosphere and its implications for deep geologic disposal of nuclear fuel waste.* This conference.

Boulton, G. and Hartikainen, J. 2003. *Ice sheet and permafrost modelling.* This conference.

Chan, T., P.A. O'Connor and F.W. Stanchell. 1998. *Finite-element modelling of effects of past and future glaciation on the host rock of a used nuclear fuel waste vault.* Ontario Hydro Nuclear Waste Management Division Report No. 06819-REP-01200-0020 R00. Written by Atomic Energy of Canada Limited for Ontario Hydro (now Ontario Power Generation), Toronto.

Chan, T., Christiansson, R., Andersson, J., Boulton, G.S., Stanchell, F.W. and Vidstrand, P. 2002. *DECOVLAEX III, task 3, BMT3 BENCHPAR Work Package 4: Glaciation Bench Mark Test, Rev. R1a,* Nov. 2002, available on www.decovalex.com

Chan, T. Guvanasen, V. and Stanchell, F.W. 2003. *Verification and validation of a three-dimensional finite-element code for coupled Thermo-Hydro-Mechanical and Salinity (T-H-M-C) modelling in fractured rock mass.* This conference.

Chan, T. and Stanchell, F.W. 2003. *DECOVLAEX III BMT3 (glaciation effects) Phase 2 progress report.* Ontario Power Generation Nuclear Waste Management Division Report. Atomic Energy of Canada Limited for Ontario Power Generation, Toronto. In preparation.

Guvanasen, V., and Chan, T., 2000. *A three-dimensional numerical model for thermohydromechnical deformation with hysteresis in a fractured rock mass,* Int. J. Rock Mech. and Mining Sc., 37: 89-106.

Hansson, H., O. Stephansson and B. Shen. 1995. *Far-field rock mechanics modelling for nuclear waste disposal (SITE-94),* SKI Report 95:40, Swedish Nuclear Power Inspectorate, Stockholm.

Nguyen, T.S., V. Polischuk and A.P.S. Selvadurai. 1993. *Effects of glaciation on a nuclear fuel waste repository.* In: Proceedings of the 46th Annual Canadian Geotechnical Conference, 1993 Sept., Saskatoon, Saskatchewan, 79-88.

Wallroth, T. Ericsson, L.O., Hernelind, J. 2003. *BENCHPAR WP4/DECOVALEX III BMT3 Phase II progress report.* Department of Geology, Chalmers University of Technology, Goteborg. In preparation.

THERMO-HYDRO-MECHANICAL IMPACTS OF COUPLING BETWEEN GLACIERS AND PERMAFROST

Geoffrey Boulton[1], Juha Hartikainen[2]
[1]) School of GeoSciences, University of Edinburgh
[2]) Laboratory of Strucural Mechanics, Helsinki University of Technology

Abstract: Cycles of glaciation in the recent past have been associated with the extension of deep permafrost and ice sheets into the middle latitudes of the northern hemisphere. Coupling between them creates a distinctive regime in which thermal, hydraulic and mechanical effects interact in ways that have the potential to change significantly geosphere conditions to depths of the order of 1000m. Ice sheet and shallow lithosphere models are coupled to investigate these impacts for a site with rock properties similar to those of the Äspö site in Sweden. An ice sheet model is forced by a realistic climate function for Europe during the last glacial cycle. It produces a 10,000 period of ice sheet occupancy over the site at the last glacial maximum. Before ice sheet arrival, permafrost extends to a depth of about 800m, but after glacier overriding of the site, it decays because of a subglacial thermal disequilibrium. Extension of permafrost beneath and beyond the ice sheet changes the hydraulic geometry, inhibits discharge to the surface of meltwater flowing as subglacial groundwater, increases heads and head gradients and leads to energetic groundwater circulation. Large stresses can be generated, with high stress and hydraulic gradients just below the base of the permafrost, with the potential to create failure under appropriate conditions of rock and fracture geometry.

1. INTRODUCTION

The last million years of Earth history have been characterised by cold, "Glacial Periods" of about 100,000 years in duration, separated by relatively warm "Interglacial Periods" of about 10,000 years in duration. Glacial periods were characterised by strong climatic variance, which included relatively short (5-15 ka duration) but intensely cold periods during which permafrost rapidly expanded to cover large areas of the middle latitudes of the Northern Hemisphere, and large ice sheets grew and progressively overrode the permafrost, as far south as latitude 40° in North America and 52° in north-west Europe. Modelling has suggested (Boulton et al, 1995) that glaciation comprehensively reorganises the geometry of groundwater flow and creates high subglacial water pressures. The growth of proglacial and subglacial permafrost inhibits the discharge of subglacial meltwaters, and thereby drives up subglacial groundwater head gradients and heads, an effect that may penetrate to depths greater than 700m, and create energetic sub-permafrost circulation compared with the weak circulation normally characteristic of sub-permafrost groundwater.

The last glacial period ended about 10,000 years ago, and given the typical length of interglacial periods of about 10,000 years, onset of glacial conditions would normally be imminent in geological terms. Indeed, until 20 years ago, the next major climate shift was expected to be glaciation, an expectation that has been replaced by that of global warming, as the scientific community became aware of the probable implications of the rapid increase of atmospheric greenhouse gases to levels unprecedented in the recent geological past. However, climate modelling (Burgess et al, 2000) has suggested that although an imminent phase of global warming is likely to delay the onset of further glacial periods, the natural tendency of climate would re-assert itself within 50,000 years. This is well within the time-frame for which assessments of the safety of potential geosphere repositories for long lived radio-nuclides are required in countries such as those of northern Europe and North America which have been subject to repeated glaciation in the past, and are likely to be so in future. In view of the energy of sub-glacial processes, it is important to assess the potential impacts of glacier and permafrost growth in the recent past, as a guide to the inclusion of these processes in safety assessments that take future climate and environmental change into account. However, although the extent and thickness of former ice sheets can be reconstructed on the basis of geological evidence and robust geophysical theories (e.g., Boulton and Payne, 1994), permafrost leaves little geological evidence that permits the reconstruction of permafrost depths. There is thus no alternative but to model its development, evolution and impacts. This paper presents an approach to the modelling of the past, coupled evolution of glaciers and permafrost and the implications for groundwater flow and transport, sub-surface stress and stability.

2. THE PROBLEM

Permafrost is perennially frozen ground with sufficient water content to cement the materials in which it is developed. The greatest reported depth of permafrost is in excess of 1400m in Siberia (Washburn, 1979). It is sustained by a thermal regime in which the Earth's surface, lying beneath a cold atmosphere, is a heat sink, and the geothermal flux a heat source. If the mean annual freezing point isotherm lies below the depth of seasonal temperature fluctuation, permafrost can exist. It is much more difficult to sustain permafrost beneath a glacier. In addition to the geothermal flux, the glacier sole is also a heat source, as a consequence of shear heating in the basal ice. If the upward temperature gradient in basal ice is unable to conduct both the frictional and geothermal fluxes away from the bed and through the ice, permafrost will be unstable and will decay. These conditions will almost invariably occur beneath a fast-flowing glacier.

As a result, pre-existing permafrost will tend to decay if it is overridden by a glacier. This has three consequences:

a) The glacier sole and decaying permafrost are groundwater sources, and water is driven outwards beneath the glacier by the glacier pressure gradient.
b) The marginal permafrost wedge (figure 1) is a zone of reduced hydraulic conductivity so that the outward groundwater flux is forced to flow through a more restricted section.
c) These circumstances will tend to generate higher groundwater flow velocities and larger heads and head gradients than in a non-glacial permafrost zone.
d) The rapid growth and decay of permafrost has the potential to create large transient overpressures or suction pressures as a consequence of the ice/water phase change.

Glacier-permafrost coupling creates a distinctive hydraulic regime, dominated by transient processes, which must be understood for long term safety assessments. In principle, the problem should be solved by a thermo-mechanically coupled model of the whole ice-water-rock system, driven by an external climate function. As an interim step, we use two separate models, one for the glacier and the other for the sub-glacial zone, which are weakly coupled together.

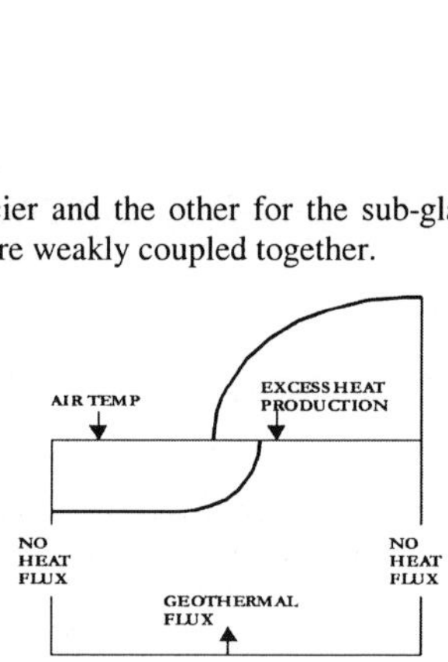

Figure 1. Geometry of the ice sheet/permafrost system and assumed thermal boundary conditions.

3. THE MODELS

3.1 The Ice Sheet

The model is based on that described in Boulton and Payne (1994).

a) Ice sheet form. The evolution of ice sheet form is determined using the continuity equation:

$$\frac{\partial H}{\partial t} = -\nabla_H .(\bar{v}H) + M - S \qquad (1)$$

where ∇_H is the two-dimensional horizontal gradient, $\bar{v}$ the vertically averaged velocity, M the mass balance (accumulation minus surface melting rate) and S the melting rate at the base of the ice column.

b) Ice flow. Strain rates of ice are related to the stress tensor by:

$$\dot{\varepsilon}_{ij} = A(T^*)\tau_*^{(n-1)}\tau_{ij} \qquad i, j = x, y, z \qquad (2)$$

where τ_* is the effective shear stress defined by the second invariant of the stress tensor, n the flow law exponent (typically 3-4) and A the temperature dependent flow law coefficient. T^* is the absolute temperature corrected for the dependence of the melting point on pressure.

The horizontal velocity vector is found by vertical integration of the flow law (Eq. 2):

$$v(z) - v(h) = -2(\rho_i g)^n (\nabla(H + h).\nabla(H + h))^{\frac{n-1}{2}}$$

$$\nabla(H + h)\int_h^z A(T^*)(H + h - z)^n dz \tag{3}$$

Integrating equation (3) from the ice base to the surface gives the vertically averaged velocity $\bar{v}$.

c) Ice sheet temperature. The temperature evolution of the ice sheet is described by:

$$\frac{\partial T}{\partial t} = \frac{k}{\rho_i c_p}\nabla^2 T - v.\nabla T + \frac{\Phi}{\rho_i c_p} \tag{4}$$

T is the absolute temperature, k is the thermal conductivity, c_p is the specific heat capacity and Φ is the heat generated due to friction. Thus heat is transferred by both diffusion and advection (first and second term of the RHS of Eq. (4), respectively).

The basal temperature gradient (figure 2) determines the extent to which heat can be removed from the base of the glacier, and the "excess heat" that may be generated that will increase the temperature and cause melting of subglacial permafrost.

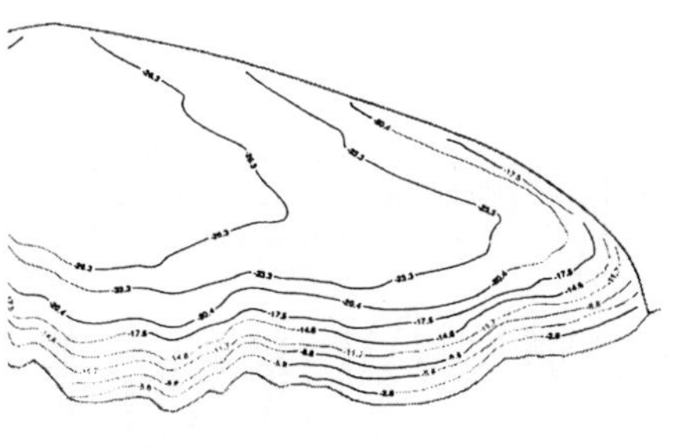

Figure 2. *Simulated temperature field in a X-section of the European ice sheet at its maximum last glaciation extent. The basal thermal gradient conducts away heat from the bed. The horizontal extent is 1200 km and the vertical 2,100m.*

d) Earth flexure. The evolution of the ice sheet is strongly coupled to the isostatic response of the Earth. The Earth model is based on that of Lambeck et al (1998) incorporating an elastic lithosphere and fluid asthenosphere.

3.2 The Subglacial Zone

The model is based on the approach described in Mikkola and Hartikainen (2001). The evolution of the subglacial zone is determined by the balance equations

$$\begin{aligned}
&\frac{\partial}{\partial t}\left(\beta^s\rho^s\right) + \nabla\cdot\left(\beta^s\rho^s\mathbf{v}\right) = 0 \\
&\frac{\partial}{\partial t}\left(\beta^w\rho^w + \beta^i\rho^i\right) + \nabla\cdot\left[\left(\beta^w\rho^w + \beta^i\rho^i\right)\mathbf{v}\right] + \nabla\cdot\left(\rho^w\mathbf{J}^w\right) = 0 \\
&-\nabla\cdot\left(\boldsymbol{\sigma}^* - B\mathbf{I}\right) = \left(\beta^s\rho^s + \beta^w\rho^w + \beta^i\rho^i\right)\mathbf{g} \\
&\frac{\partial \mathrm{E}}{\partial t} + \nabla\cdot\left[\mathrm{E}\mathbf{v} + \rho^w\left(C^wT + L\right)\mathbf{J}^w\right] + \nabla\cdot\mathbf{q} = 0
\end{aligned} \tag{5}$$

and the constitutive relations

$$\begin{aligned}
&\mathbf{J}^w = -\frac{k}{\rho^w|\mathbf{g}|}\left[\nabla\left(B - 2\rho^w\frac{lT}{T_o}f\right) - \rho^w\mathbf{g} + \rho^w\frac{lT}{T_o}\nabla f\right] \\
&\boldsymbol{\sigma}^* = G\left(\nabla\mathbf{u} + (\nabla\mathbf{u})^{\mathrm{T}}\right) + \left(K - \frac{2}{3}G\right)\nabla\cdot\mathbf{u}\mathbf{I} \\
&\rho^{\mathrm{k}} = \rho_o^{\mathrm{k}}\exp\left[\kappa^{\mathrm{k}}\left(B + \rho^w\frac{lT}{T_o}\frac{\partial f}{\partial\beta^{\mathrm{k}}}\right)\right] \\
&\mathrm{E} = \left(\beta^s\rho^sC^s + \beta^w\rho^wC^w + \beta^i\rho^iC^i\right)T + \beta^w\rho^wL \\
&\mathbf{q} = -\left(\beta^s\lambda^s + \beta^w\lambda^w + \beta^i\lambda^i\right)\nabla T \\
&\left(C^w - C^i\right)T\ln\left(\frac{T}{T_o}\right) + L\frac{T - T_o}{T_o} - \left(\frac{1}{\rho^w} - \frac{1}{\rho^i}\right)B + \frac{lT}{T_o}f = 0
\end{aligned} \tag{6}$$

β^{k}, ρ^{k}, C^{k}, λ^{k} and κ^{k}, k in $\{s,w,i\}$ are the volume fraction, the bulk density, the specific heat capacity, thermal conductivity and the compressibility of skeleton, water and ice, respectively. $\mathbf{J}^w$ and B denote the Darcy flux and the water pressure and $\mathbf{v}$, $\mathbf{u}$, $\boldsymbol{\sigma}^*$, G, K, E and $\mathbf{q}$ are the velocity, the displacement vector, the effective stress tensor, the shear modulus, the bulk modulus, the internal energy and the heat flux vector of the rock, respectively. $\mathbf{g}$ is the gravitational acceleration, T is the thermodynamic temperature and the constant L is related to the latent heat of fusion l such that $l = (C^w - C^i)T_o + L$, where T_o = 273.15 K.

The function f characterizes the adsorption of water in the porous medium. It is defined as $f = a[(1-\eta)/\eta/\chi - (1-\eta_o)/\eta_o]^2$, where $\chi = \beta^w/(\beta^w+\beta^i)$ is the relative water content, $\eta = 1 - \beta^s$ is the porosity and the parameter a depends on the nature of the porous medium. The hydraulic conductivity k reduces as a function of the water content by $k = k_o\chi^5$, where k_o is the hydraulic conductivity of the unfrozen rock.

4. THE SIMULATION EXPERIMENT

4.1 Approach and Boundary Conditions

We couple together the ice sheet/permafrost model by considering the thermodynamics at the ice/bed interface. Shear heating will be a maximum at the base of the glacier, because shear stresses are maximal at the bed, because of the large exponent in the flow law (equation 2) and also because of basal sliding when the ice/bed interface is unfrozen. The balance between heat production and heat conducted up through the glacier to its surface is analysed. If the heat conduction capacity exceeds the heat production capacity, freezing will occur beneath the ice sheet. If heat production exceeds conduction, the subglacial surface will become warmer. A variety of ice sheet simulations driven by a range of climate functions suggest that the range of variation of basal thermal gradient is relatively small. We have therefore used this to estimate the excess heat at the ice/bed interface.

We have modelled the evolution of the ice sheet /permafrost system and the associated changes in hydraulic and mechanical states along a 2D section, 60 km in length, parallel to ice flow and 4 km in depth. Rock properties vary one-dimensionally with depth and are those characteristic of the Äspö site in Sweden: λ^s = 3 W/m/K, C^s = 750 J/kg/K, ρ^s_o = 2650 kg/m^3, E = 35 GPa, ν = 0.22, η_o = 0.005, a = 0.00004 and k_o (m/s) varies with depth z as shown in Table 1.

Table 1.Hydraulic conductivity at the Äspö site.

Z (km)	0 – 0.1	0.1 – 1.1	1.1 -
$\log_{10}(k_o)$	-6	-7 - 3(z - 0.1)	-10 - (z - 1.1)/4.9

We then prescribe boundary conditions as follows (figure 1).

Up-glacier boundary. Groundwater flux, no heat flux:

$q_n = 0$, $J_n^w = \phi^H$, $u_n = 0$;

Down-glacier boundary. Free boundary for groundwater flow, no heat flux:

$q_n = 0$, $B = \rho^w gz$, $u_n = 0$;

Upper boundary. The heat flux from the ice, no flow across the boundary:

(a) in the glacier free part $T = T_s$, $B = 0$, $\boldsymbol{\sigma_n} = \mathbf{0}$,

(b) in the glacier covered part when no basal melting occucrs $q_n = \phi^T$, $B = 0$, $\boldsymbol{\sigma_n} = \boldsymbol{\rho^i g h}$,

(c) in the glacier covered part when basal melting occurs $T = T_m$, $J_n^w = (q_n - \phi^T)/(\rho^w l)$, $\boldsymbol{\sigma_n} = \boldsymbol{\rho^i g h}$;

Lower boundary. The geothermal heat flux, no flow across the boundary:

q_n = 0.04 W/m^2, $J_n^w = 0$, $u_n = 0$.

In addition, the water pressure B on the upper boundary is constrained not to exceed the maximum value of $\rho^w gh$, where h is the ice sheet thickness. ϕ^T is the excess heat from the ice sheet.

4.2 Glacier and Climate Forcing

Mean annual air temperatures that cause growth and decay of permafrost under non-glacial conditions are inferred from Johnsen et al (1992) in a way that is consistent with regional palaeoclimate (figure 3).

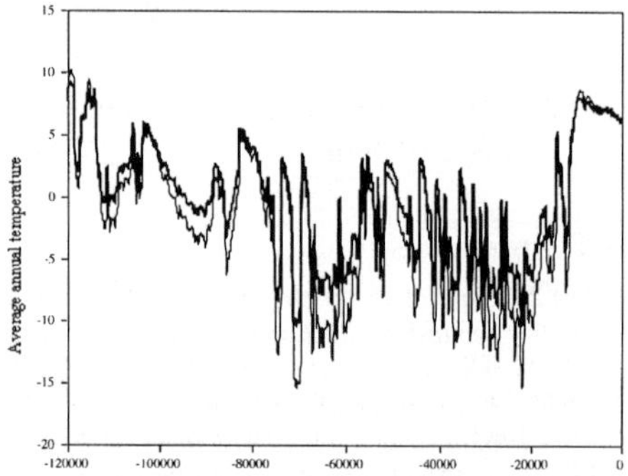

Figure 3. Warmer (upper) and colder (lower) temperature scenarios through the last glacial cycle based on Johnsen et al (1992).

A simplified form of climate function for the last glacial maximum has been used to drive a simple cycle of glacier advance and decay over a site that has a location similar to that of Äspö in relation to the maximum extent of glaciation. It produces an ice sheet residence time over the site of about 10,000 years. The variation in ice thickness through time along the transect is shown in figure 4, and the availability of excess heat through time along the transect is shown in Fig. 5.

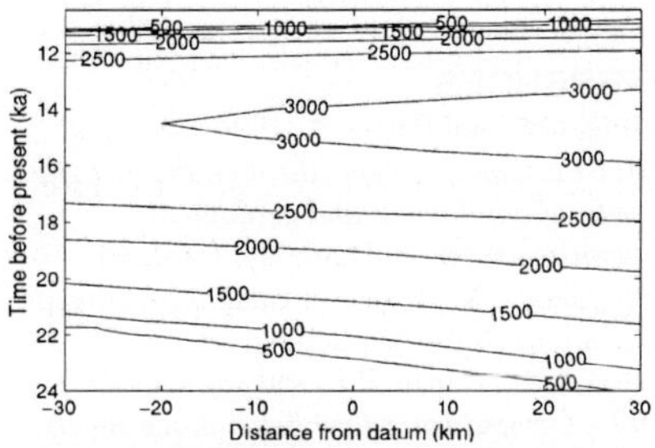

Figure 4. Evolution of ice sheet thickness h (m) in time along the transect.

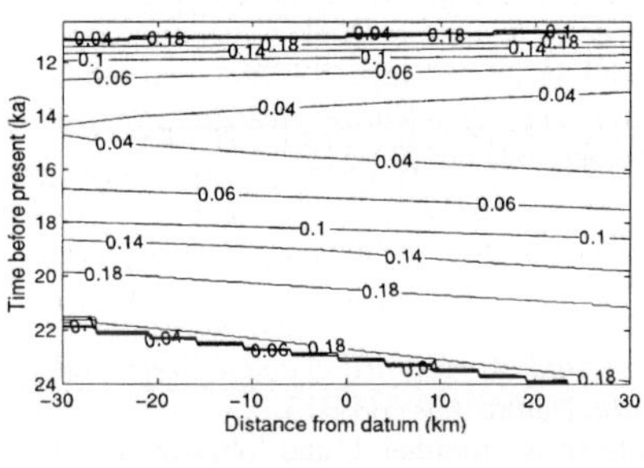

Figure 5. Evolution of excess heat production ϕ^T (W/m^2) through time along the transect.

4.3 Thermal Response

The thermal evolution of the sub-surface along the transect has been simulated by the model using the extra-glacial temperatures shown in figure 3 to determine the depth of permafrost prior to arrival of the glacier, and the subglacial boundary conditions shown in figure 5 to determine the progressive subglacial melting of this permafrost. Prior to the arrival on the transect of the ice sheet, which crossed the right hand margin at about 24 ka and the left hand margin at about 22 ka, freezing temperatures and associated permafrost developed during a long period of pre-glacier cooling, extended to a depth of about 800m. As the glacier overrode the transect, the warming effect of the ice sheet progressively increased subglacial temperatures and caused permafrost to decay. Figure 6 shows a snapshot of the temperature field along the transect at 19 ka before present, and figure 7 shows the progressive reduction in permafrost extent at several times during the maximum of glaciation. It shows that at datum 0, permafrost thickness decreased from 800m to almost zero in 6,000 years.

Figures 6-7 utilise the colder of the two scenarios shown in figure 3 as the extra-glacial boundary condition. This develops permafrost that survives for 40 km beneath the advancing glacier. In the warmer scenario, permafrost survives beneath the ice sheet for a distance of only 10 km.

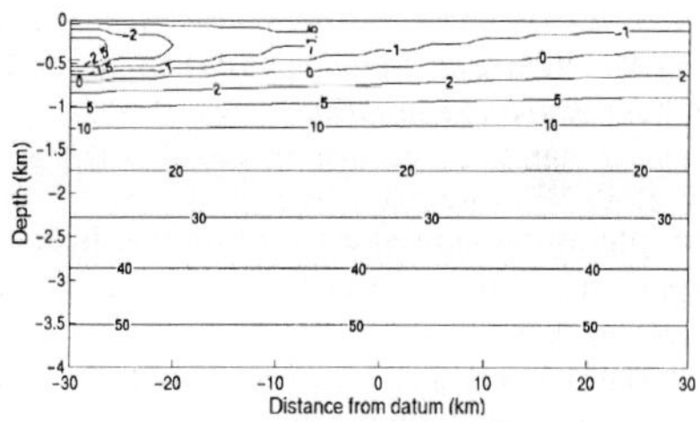

Figure 6. Temperature field (°C) along the transect in the subglacial zone at 19 ka before present.

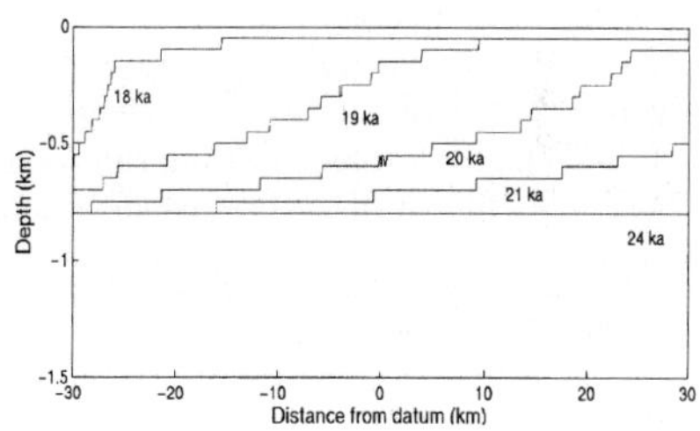

Figure 7. Progressive degradation of subglacial permafrost along the transect at several times during the glacial maximum.

4.4 Hydraulic Response

Figure 8 shows a snapshot of hydraulic heads at 19 ka, for the permafrost distribution at the same time shown in figure 7. The main flow is horizontal, being driven to the left by the leftward pressure gradient generated by the ice load shown in figure 4. Head gradients increase leftwards partly as a consequence of the boundary conditions, and partly as a consequence of thickening of the permafrost. Large heads and high gradients within the permafrost reflect increase in the amount of unfrozen water as the permafrost warms from both above and below, and because the permeability of partially frozen rock remains low.

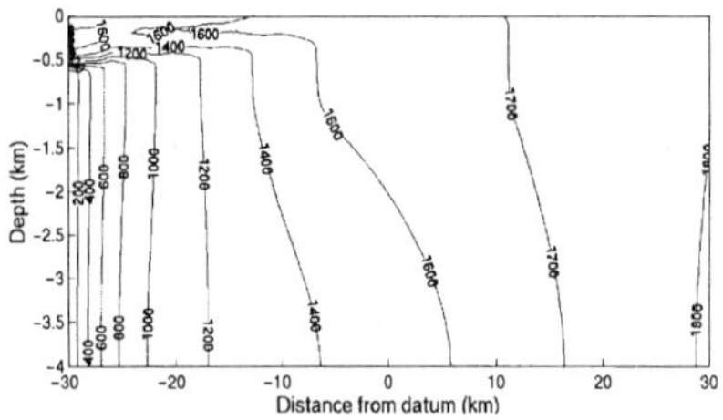

Figure 8. Hydraulic head contours in (m) at 19 ka before present.

An interesting consequence of thermo-hydraulic coupling is shown in figure 7. A thin, residual permafrost mass is shown after the main mass of permafrost has melted. This occurs because of the pore-pressure reduction in the unfrozen water in the thawing permafrost, which causes an increase in the melting point. Even as this is happening, melting continues at the glacier/bed interface.

4.5 Mechanical Response

Figure 9 shows the octahedral shear stresses generated by ice sheet loading and permafrost/hydraulic evolution in the sub-surface. We have not shown the corresponding strain response in view of the fact that earth properties were assumed to vary only vertically. It is clear however that the stresses could be large enough to generate significant strains under certain conditions of rock and fracture geometry.

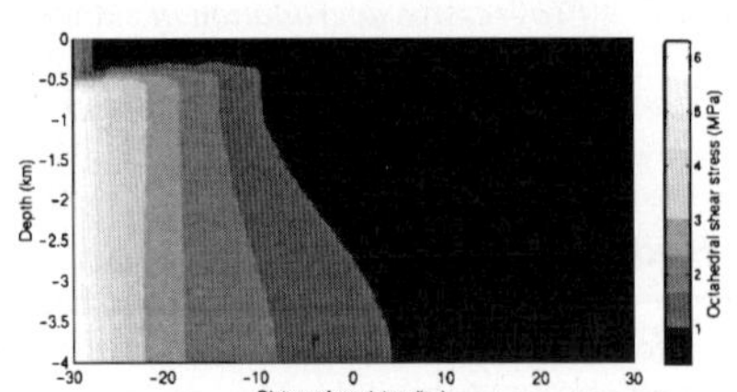

Figure 9. Octahedral shear stress field at 19 ka before present.

ACKNOWLEDGEMENTS

This work was undertaken as part of the collaborative programme involving the international project DECOVALEX III and the BENCHPAR project funded by the European Unions 5th Framework Programme. We acknowledge helpful discussions with our colleagues Tin Chan, Rolf Christiansson, Thomas Wallroth and Patrik Vidstrand.

REFERENCES

Boulton, G.S. and Payne, A. 1994. *Mid-latitude ice sheets through the last glacial cycle: glaciological and geological reconstructions*. In Duplessy, J.-C. and Spyridakis, M.-T. (eds.): Long-term climatic variations. Nato ASI Series I 22, pp. 177-212.

Boulton, G.S., Caban, P.E. and van Gijssel, K. 1995. *Groundwater flow beneath ice sheets: Part I – Large large scale patterns*. Quaternary Science Reviews, Vol. 14, pp. 545-562.

Burgess, P.E., Palutikof, J.P. and Goodess, C.M. 2000. *Investigations into long-term future climate changes*. In: Linking Climate Change to Land Surface Change. S.J. McLaren and D.R. Kniveton (eds.), Kluwer Academic Publishers, Netherlands pp. 231-246.

Johnsen, S.J., Clausen, H.B., Dansgaard, W., Fuhrer, K., Gundestrup, N., Hammer, C.U., Iversen, P., Jouzel, J., Stauffer, B. and Steffensen, J.P. 1992.*Irregular glacial interstadials recorded in a new Greenland ice core*. Nature *359*. 311-313.

Lambeck, K., Smither, C and Johnston, P. 1998. *Sea-level change, glacial rebound and mantle viscosity for northern Europe*. Geophysical Journal International, 125, pp. 340-354.

Mikkola, M. & Hartikainen, J. 2001. *Mathematical model of soil freezing and its numerical implementation*. Int. J. Num. Meth. Eng. 52, pp. 543-557.

Washburn, A.L. 1979. *Geocryology*. Arnold, UK.

THERMO-HYDRO-MECHANICAL (T-H-M) IMPACTS OF GLACIATION AND IMPLICATIONS FOR DEEP GEOLOGIC DISPOSAL OF NUCLEAR WASTE

Geoffrey Boulton[1], Tin Chan[2], Rolf Christiansson[3], Lars O. Ericsson[4], Juha Hartikainen[5], Mark R. Jensen[6], Frank W. Stanchell[2] and Thomas Wallroth[4]
[1]School of GeoSciences, University of Edinburgh; [2]Atomic Energy of Canada Limited; [3]Svensk Kärnbränslehantering; [4]Chalmers University of Technology; [5]Institute of Mathematics, Helsinki University of Technology; [6]Ontario Power Generation

Abstract: The thermo-hydro-mechanical impacts of extreme climate change on the lithosphere down to depths at which deep repositories might be sited have been simulated. The effects of glaciation, including ice sheet and permafrost development, have been studied using site-specific data by combining four models. A climate model provides the forcing function, and ice sheet, permafrost, and coupled hydromechanical models are used to assess impacts. It is concluded that glaciation occurs on a timescale and has impacts on a depth scale that require it to be analysed in a safety analysis for deep lithosphere disposal of long–lived radionuclides in areas that have been prone to glaciation in the past. The simulations have provided valuable insight about processes and mechanisms likely to influence the long-term performance of a repository, the geosphere, or both. The key impacts are discussed, and appropriate methods identified.

1. WHY SAFETY ASSESSMENTS MUST ADDRESS GLACIATION

Safety assessments of the disposal of long lived radioactive wastes in the middle to high latitudes of the northern hemisphere must recognise that these areas have been repeatedly glaciated in the recent geological past, that such areas have suffered glacial conditions for 90% of the last million years and that were it not for the prospect of human induced global warming, would expect an imminent descent into glaciation. Given the uncertainties about the Earth's capacity to buffer itself against enhanced greenhouse warming, e.g. Berger et al., 1996, glaciation scenarios should be considered for all timeframes between 1,000 and 100,000 years, well within the range for which safety assessments are required for wastes containing long lived radionuclides.

Glaciation has the potential to influence strongly the geosphere to the preferred depths for deep disposal sites of between 500 and 1000m. The strongest impacts in periods of glaciation are associated with the extension of ice sheets and perennial ground freezing to create "permafrost" to depths of several hundred metres. Their effects are:

***Ice sheets*:**

- can significantly alter the mechanical and hydraulic boundary conditions to a proposed repository site;
- can directly erode the bed over which they flow by the order of 10m during a glacial period (Clayton, 1994);
- cause large scale isostatic flexure of the lithosphere and influence regional seismicity (Stewart et al., 2000) by suppressing activity during loading (Johnston, 1987) and enhancing it during unloading, with possible reactivation of fault zones (Stanfors and Erikson, 1993), and an earthquake may possibly trigger secondary slip along larger fractures or induce fracturing at large depth (SKB 1997);
- melt basally and can stimulate groundwater flows that are more energetic than in non-glacial periods (Boulton et al.,1995);
- can discharge oxygenated, corrosive meltwaters into the substratum with the potential to influence the chemical conditioning of the near field around repositories (King-Clayton et al., 1997);
- produce channelised sub-glacial meltwaters that can erode deep channels (up to 400m in an extreme case) in the course of a single glacial period (Ehlers et al., 1984); subglacial channels are low water pressure zones and can draw up deep groundwaters towards them (Boulton et al., 2001).

Permafrost:

- can extend to depths in excess of 1000m (Washburn, 1979);
- dramatically reduces the hydraulic conductivity of the rocks in which it forms,

and creates an impermeable extension beyond a glacier that inhibits outflow of subglacial waters and generates sub-permafrost overpressures (Boulton et al., 1995) with the potential for hydraulic jacking (Pusch et al., 1990).

All these processes are the product of a system driven by the Earth's climate and characterised by strong thermo-hydro-mechanical coupling, in which both chemical processes and transient phenomena are important.

A "Bench Mark Test" (BMT) has been undertaken as part of the DECOVALEX and EU BENCHPAR projects to explore the impact of coupled glaciation processes on the lithosphere, and to determine the likely magnitudes of some effects at repository depths.

2. IDENTIFYING ISSUES FOR REPOSITORY PERFORMANCE ASSESSMENT

The geosphere effects of glaciation summarised in section 1 suggest that priorities for performance assessments should be to understand the long-term impacts of system changes on mechanical stability, hydraulic conductivity, groundwater flow and hydrochemistry. The objectives of the bench mark test are therefore:

a) to study, by analytical and/or numerical modelling the impact of a 100 ka glaciation-deglaciation cycle on the long-term evolution of a fractured rock mass in which a generic repository is located;
b) to assess the impact of the glaciation/deglaciation cycle on the coupled thermo-hydro-mechanical responses of the far field system around a repository and on its long-term performance in waste isolation;

A performance assessment (PA) of a deep repository consists of an analysis of the changes through time in the disposal facility as a consequence of both internal and external forces. Groups of coupled processes are linked together in a description of integrated evolution through time. The primary purpose of the benchmark test presented here is to develop modelling tools at a site scale for simulation of climate driven boundary conditions (ice sheet loading, groundwater hydraulics and permafrost) and to illustrate the magnitude of some T-H-M impacts in the far field in the context of a PA.

3. MODELLING APPROACH

The only way in which models of slow or rapid change between states that do not currently occur in Earth (i.e. there is currently no ice sheet glaciation in the middle latitudes) can be tested is through their capacity to simulate geologically known attributes of the past. Such geological testing of predictions is a vital complement to modelling. It gives confidence that we understand the system in question and provides a basis for use of models in scenarios of future geosphere evolution. We have therefore focused on simulating glaciation in a way that can be tested by geological observations, and applied the model to suggest subsurface impacts at a generic site to explore the implications for safety assessments.

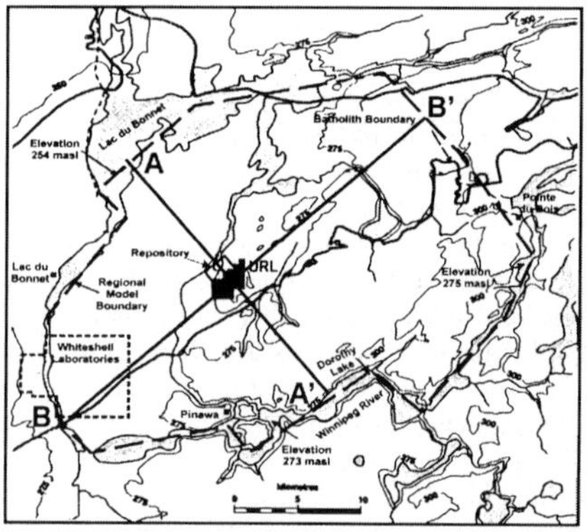

Figure 1. The Whiteshell site in Manitoba, Canada which provides site data for the modelling reported in this paper.

Site attributes have largely been based on those of the Whiteshell site in Manitoba (Figure 1), on the eastern part of the Canadian Shield at approximately 285 m.a.s.l., but with some simulations using properties of the Äspö site in Sweden, and assuming that the sites always lie above the marine limit. Simulations are designed to explore impacts of the growth and decay of ice sheets and permafrost during the last glacial cycle. The successive steps in our simulations are:

Step 1 – Simulation of the climate drive.

Our simulations are concerned with the whole of the last glacial period from 120,000 years ago (120 ka) to the present. The pattern of climate change is derived from the record from the Greenland ice sheet (Johnsen et al., 1992), adapted to the region using palaeo-climatic data from southern Canada and the northern USA and synoptic extrapolations.

Step 2 – Ice sheet loading and basal thermal and hydrological regime

A thermo-mechanically coupled, transient ice sheet model (Boulton and Payne, 1994), coupled with the Earth model of Lambeck et al., (1998) has been driven by the climate function over a prescribed topography of North America with a 10 km resolution. The model computes the temperature at the base of the ice sheet and the rate of basal melting in time and space. This is used to compute the spacing between channels that are required to exist at the ice/bed interface to discharge meltwater that cannot be discharged by groundwater flow (Boulton et al., 2001), and the head distribution at the ice/bed interface.

Step 3 – Permafrost distribution

A transient model of permafrost development based on the approach of Mikkola and Hartikainen, (2001) is driven directly by the climate/temperature function when there is no ice sheet present. When the ice overrides the site, the temperature at the base of the ice sheet is used as a boundary condition for permafrost development.

Step 4 – Groundwater flow, pressures and states of geosphere stress

Coupling between the permafrost and ice sheet are used to determine the transient response of the groundwater system and the state of rock stress along a 2D section parallel to ice flow using rock properties from the Åspö site. (Boulton and Hartikainen, this volume). Investigation of groundwater flow and geosphere stresses and strains have also been undertaken for steady state and transient conditions along sections both parallel and transverse to ice flow using the ABAQUS and MOTIF codes (Chan et al., this volume). Parallel simulations for the same boundary conditions were undertaken using the two codes.

4. MODELLED RESULTS

4.1 Climate through the glacial cycle

Figure 2 shows output from a climate/temperature simulation for the last glacial cycle for the ground surface at the Whiteshell site. It consists of two components. The first for periods when the site is ice-free, based on the Greenland ice core record and compatible with palaeo-climate and modern synoptic data. The second is the temperature condition at the ice bed interface, based on thermodynamic analysis of the ice sheet. Note that temperatures at the ice/bed interface, at 60ka and between 22.5ka and 11ka are higher and show less variance than extra-glacial temperatures because of the insulating and heating effect of the ice sheet.

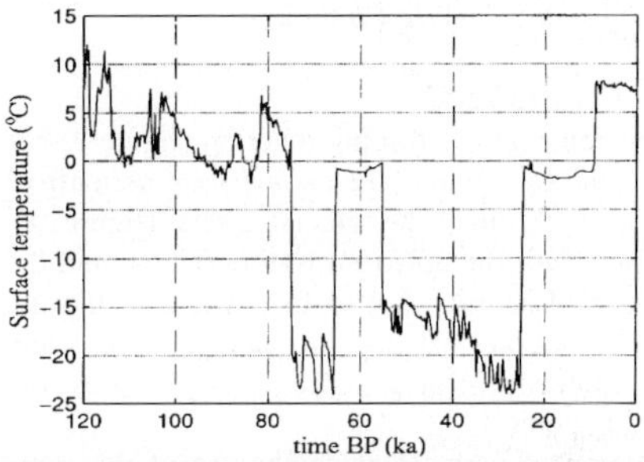

Figure 2. Modelled ground surface temperature at Whiteshell through the last glacial cycle.

4.2 Conditions at the base of the ice sheet

The climate/temperature function is used to drive a glaciological model of the Laurentide ice sheet through the last glacial cycle. It suggests that the Whiteshell site was glaciated at about 60ka and during the glacial maximum, between about 22.5ka and 11ka, which is compatible with geological evidence from the region. The maximum ice sheet thickness at the site is modelled as 3000m, which is likely to be an over-estimate. The model computes basal melt rates, and from a simplified, 1D description of hydraulic conductivity, computes the spacing of subglacial channels that would be required to drain the ice sheet bed. The channels are low pressure zones that draw down heads at the ice/bed interface. Figure 3a shows a time slice of the computed surface form of the advancing glacier at Whiteshell and Figure 3b the contemporary disposition of heads at the ice/bed interface. The longitudinal head gradients associated with the ice front and the transverse gradients associated with channels are much greater than the modern gradient, and create flow velocities one to two orders of magnitude faster than modern values with strong vertical flow components.

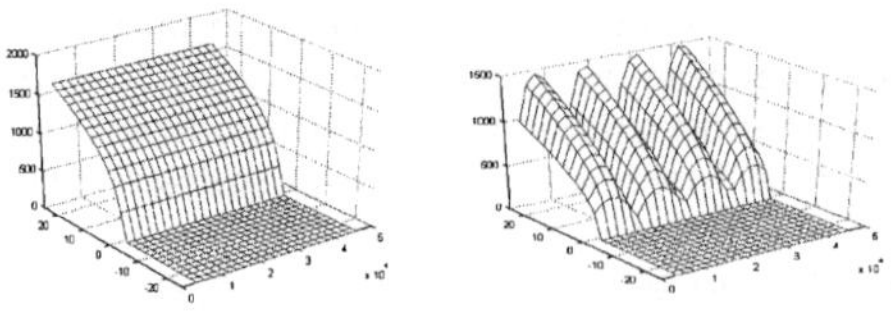

Figure 3a-b. a) The modelled form of the glacier surface with the glacier front lying along the line of A-A' (figure 1). b) Form of the heads due to channels at the ice/bed interface.

4.3 Permafrost

The temperature forcing function (figure 2) has been used to compute the evolution of permafrost thickness through the glacial cycle (figure 4), together with the unfrozen water content, and the increase of salinity due to freezing, but ignoring the insulation effects of snow cover and vegetation. Increasing salinity acts to depress the freezing point.

Computed permafrost depths are of the same order as anticipated repository depths, but ice sheet advance over the site leads to rapid permafrost decay (see Boulton and Hartikainen, this volume).

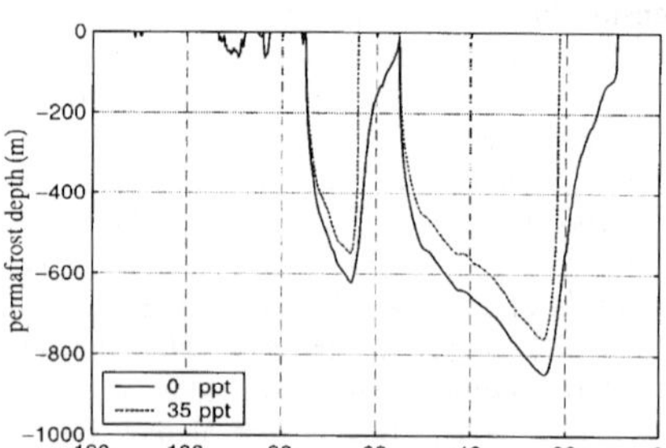

Figure 4. Modelled permafrost thickness through the last glacial cycle. One simulation ignores the salinity effect, the other includes it.

4.4 Groundwater flow

Strong groundwater flows, up to 2 orders of magnitude greater than in the non-glacial state, are generated beneath the glacier and beneath permafrost that extends beneath the glacier. Figure 5 shows modelled groundwater heads beneath the ice sheet along section B-B' (figure 1) at Whiteshell. Where permafrost is thin, significant water overpressures can develop and are enough to generate hydraulic jacking of bedrock. The consequences of glaciation at greater depth are (see Chan et al., this volume):

- A rapid increase of head during the first 1000 years of glaciation.
- A rapid transmission of these heads through the fracture systems, producing much higher early, transient heads than in the repository zone.
- During the glacier advance over the site, there is a large horizontal hydraulic gradient partly due to compression of pores by ice loading.
- As the area is completely covered by the ice sheet, a strong downwards hydraulic gradient develops , of as much as 3-5 m/m.
- At depth the excess water pressure is about 1/3 of the ice pressure.
- During ice sheet retreat, the gradient reverses, and is sustained, together with residual excess pressures that could be considerable if the hydraulic conductivity is low.
- Pressures in fracture zones decay rapidly after deglaciation.

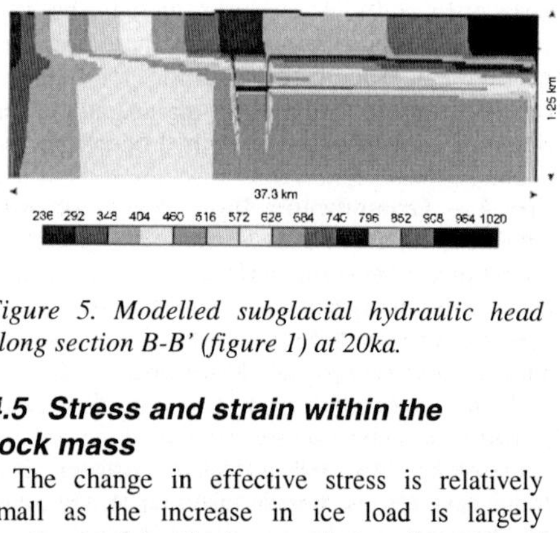

Figure 5. Modelled subglacial hydraulic head along section B-B' (figure 1) at 20ka.

4.5 Stress and strain within the rock mass

The change in effective stress is relatively small as the increase in ice load is largely compensated by the increased groundwater head (however, there is a transient effect, as the former is instantaneous whilst the latter diffuses through the system). There is very little rotation of principal effective stresses as the shear stress exerted by a glacier is small compared with the stresses due to the ice load. Even in dead-end horizontal fractures, there is no generation of tensile stresses and therefore no hydraulic jacking at depth. No shear failure is predicted. Figure 6 shows modelled downward displacements at repository depths both in the repository mass and in adjacent fractures (see also Tin Chan et al., this volume).

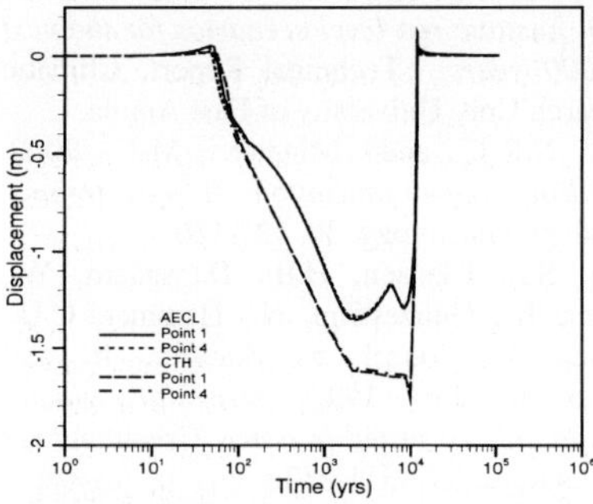

Figure 6. Evolution of mechanical displacements. Calculated using MOTIF (AECL) and ABAQUS (CTH) in a Fracture Zone (Point 1) and in the Repository zone (4).

5. IMPLICATIONS FOR PERFORMANCE ASSESSMENT

Although models of glacier-groundwater, glacier-permafrost-groundwater, glacier-groundwater-shallow failure systems have been presented (e.g. Boulton et al, 1995), this work is one of the first to assess impacts at repository depths using site specific data. The results provide valuable insights into the magnitude and rate of change of site-specific hydrogeologic and geomechanical properties in response to external, transient climate forcing. They clearly demonstrate the importance of glaciation scenarios in safety and performance assessments, the reality of effects that result from H-T-M coupling and underline the need for transient analyses of these coupled phenomena.

Although our method and approach has provided a systematic and structured framework for assess the relative importance of coupled H-T-M impacts on geosphere performance, there are four important further issues that need to be addressed further if such models are to be robust components of safety assessments:

a) Coupling. Modelling has used four separate components, a climate model, an ice sheet-earth model, a permafrost model and an earth hydro-mechanical earth model. The ice sheet-permafrost models are weakly coupled but the climate and hydro-mechanical earth models are uncoupled from other components. Information is merely passed in one direction between them. The development of a model in which the system is fully coupled and driven only by global climate, with feedbacks between the ice sheet and local climate is necessary if the full consequences of coupling are to be understood. This should comprise a highly simplified model, designed to explore system attributes, as a context for site specific models similar to that used in this project.

b) Transience. Transient effects are classically important in systems in which the response time is similar to, or longer than the rate of change in the forcing function, and where there are thresholds across which the system moves rapidly from one state to another. Both these types of transience occur in the glacier-permafrost system.

c) Testing. The non-linearity of some of the strong couplings in our models (e.g. the thermo-mechanical ice sheet coupling) and the inevitable uncertainty about known processes that have not been included (e.g. natural gas generation) and as yet unknown processes, require that models such as this are tested rigorously against their capacity to simulate non-trivial aspects of past geological evolution. Such testing is of fundamental importance and remains to be done.

d) Sensitivity testing. These transient boundary conditions have served as a basis for illustrative simulations in which groundwater movement, residence times and flow path evolution are shown to be sensitive to fracture network dimensionality, interconnectivity and permeability, and non steady-state H-M boundary conditions. The depth of permafrost formation was demonstrated as sensitive to ground surface periglacial temperatures, the insulating effects of ground conditions (vegetation, snow) sub-surface salinities and hydrological processes at the ice/bed interface. Unfortunately, the range of sensitivity testing that is required makes a large demand upon current computational resources.

6. CONCLUSIONS

The projects most important general conclusions are that geosphere impacts of glaciation occur:

- on a depth scale that is relevant to the safety of repositories buried several 100m beneath the surface;
- on timescales that are relevant to safety assessments for long lived waste,

and that:

- transient, coupled processes must be considered in safety assessments.

Site specific analysis for the Whiteshell site concludes that glaciation leads to:

- large heads and head gradients;

- increased groundwater velocities (about factor of 100), reduced travel times, faster recharge from surface and potentially higher water fluxes through a repository;
- hydraulic gradient reversal upon deglaciation;
- high residual pore pressure for 1000s of years after glacier has retreated from the site;
- the local flow field beneath the ice sheet becoming part of a much larger regional ice sheet flow system;

and that:

- permafrost may develop at repository depths;
- hydraulic jacking at depth is unlikely to be important;
- the impact on stress and mechanical stability at depth is minor;
- specific fracture zone geometry should be incorporated in site specific analyses.

ACKNOWLEDGEMENTS

This work was undertaken as part of the international DECOVALEX III programme and the EU BENCHPAR project under contract FIKW-CT-2000-00066. It has been supported by the EU 5th Framework programme, by Ontario Power Generation (OPG) and by Svensk Kärnbränslehantering AB and by STUK – Radiation and Nuclear Safety Authority of Finland.

REFERENCES

Boulton, G.S. and Payne, A. 1994. *Mid-latitude ice sheets through the last glacial cycle : glaciological and geological reconstructions.* In Duplessy, J.-C. and Spyridakis, M.-T. (eds.): Long-term climatic variations. Nato ASI Series I 22, 177-212.

Boulton, G.S., Caban, P.E. and van Gijssel, K. 1995. *Groundwater flow beneath ice sheets : Part I - Large scale patterns.* Quaternary Science Reviews, Vol. 14, 545-562.

Boulton, G.S., Zatsepin, S., and Maillot, B. 2001. *Analysis of groundwater flow beneath ice sheets.* SKB Technical Report, TR-01-06.

Clayton, K. 1994. *Glaciation of the British isles: an approach seeking to determine the role of glaciation in landform development over the last million years.* Safety Studies. Nirex Radioactive Waste Disposal NSS/R337.

Ehlers, J., Mayer, K.D. and Stephan, H.J. 1984. *Pre-Weichselian glaciation of North-West Europe.* Quaternary Science Reviews, 3, 1-40.

Goodess, C.M., Watkins, S.J. and Palutikof, J.P. 2000. *Eustatic sea-level scenarios for the next 150,000 years.* Technical Report, Climatic Research Unit, University of East Anglia.

Hulton, N.R.J. and Mineter, M.J. 2000. *Modelling self-organisation in ice streams.* Annals of Glaciology, 30, 127-136.

Johnsen, S.J., Clausen, H.B., Dansgaard, W., Fuhrer, K., Gundestrup, N., Hammer, C.U., Iversen, P., Jouzel, J., Stauffer, B. and Steffensen, J.P. 1992: *Irregular glacial interstadials recorded in a new Greenland ice core.* Nature, 359. 311-313.

Johnston, A.C. 1987. *Suppression in earthquakes by large continental ice sheets.* Nature 330, 467-469.

King-Clayton L., Chapman, N., Ericsson, L.O. and Kautsky, F. (eds) 1997. *Glaciation and Hydrology Workshop on the Impact of Climate Change and Glaciations on Rock stresses, Groundwater flow and Hydrochemistry.* Past, Present and Future Workshop Proceedings SKI Report 97:13

Lambeck, K., Smither, C. and Johnston, P. 1998. *Sea-level change, glacial rebound and mantle viscosity for northern Europe.* Geophysical Journal International 134, 102-144.

Mikkola, M. and Hartikainen, J. 2001. *Mathematical model of soil freezing and its numerical implementation.* Int. J. Num. Mech. Eng. 52, 543 - 557.

Pusch, R., Börgesson, L. and Knutsson, S. 1990. *Origin of silty fracture fillings in crystalline bedrock.* Geologiska Föreningens I Stockholm Förhandlingar, 112, (3), 209-213.

SKB, 1997. *SR 97 – Deep repository for spent nuclear fuel. SR-97 – Post-closure safety. Main report – Vol. I, Vol. II and Summary.* SKB TR 99-06.

Stanfors R., Eriksson, L.O. (editors),. 1993. *Post-glacial faulting in the Lansjärv area, Northern Sweden. Comments from the expert group on a field visit at the Molberget post-glacial fault area, 1991* SKB TR-93-11.

Stewart, S., Sauber, J. and Rose, J. 2000. *Glacio-seismotectonics: ice sheets, crustal deformation and seismicity.* Quaternary Science Reviews, 19 (14-15), 1367-1391.

Washburn, A.L. 1979. *Geocryology A survey of periglacial processes and environments.* Arnold IBSN: 0-7131-6119-1.

TEMPERATURE INFLUENCE ON THE MECHANICAL BEHAVIOUR OF A COMPACTED BENTONITE

María Victoria Villar [1], Antonio Lloret [2]

[1]) Centro de Investigaciones Energéticas, Medioambientales y Tecnológicas (CIEMAT), Madrid, Spain

[2]) Universitat Politècnica de Catalunya (UPC), Barcelona, Spain

Abstract: The design of high level radioactive waste (HLW) repositories in deep geological media –in which bentonite is proposed as sealing material– leads to the necessity of deepening the study of the behaviour of clays when subjected to hydraulic and thermal changes. The paper presents the results of an experimental study on the effects of temperature on the volumetric behaviour of a compacted bentonite used in FEBEX, which is a project for the study of the near field for a HLW repository according to the Spanish concept. The experimental programme includes swelling under load tests, swelling pressure tests and oedometric compression tests with and without control of suction. The results indicate that temperature reduces the swelling capacity and the swelling pressure and increases the volumetric strains induced by vertical loads due to a reduction of the yielding stress.

1. INTRODUCTION

The coupling between thermal, hydraulic and mechanical processes in the bentonite barrier of HLW repositories is recognised as a crucial aspect to evaluate their behaviour. The heating induced by the radioactive waste and the hydration with water supplied by the surrounding rock affect important properties of the compacted bentonite. In particular, the temperature changes affect the mechanical response of the clay in several aspects: swelling pressure, swelling and collapse, thermal dilatation and contraction, compressibility, yielding and effects on time dependent behaviour.

In relation to radioactive waste disposal, during last years a number of laboratory results referring to thermal effects on saturated soils have been presented (Baldi *et al.* 1988, Towhata *et al.* 1993, Tanaka *et al.* 1997, Sultan *et al.* 2002, Burghignoli *et al.* 2000). However, results focused on the thermal influence on volume change behaviour of unsaturated soils are still limited (Wiebe *et al.* 1998, Romero *et al.* 2003). In particular, information concerning the temperature effects on the mechanical behaviour of highly expansive clays in unsaturated conditions is still scarce (Lingnau *et al.* 1996, Romero *et al.* 2001).

The work presented here is being performed in the framework of FEBEX (Full-scale Engineered Barriers Experiment in Crystalline Host Rock), which is a project for the study of the near field for a HLW repository in crystalline rock according to the Spanish concept: the waste canisters are surrounded by a clay barrier constructed from highly compacted bentonite blocks (ENRESA 2000). In this paper, experimental results obtained on compacted specimens subjected to temperature, suction and stress are presented, as well as an attempt to interpret the cause of the temperature effects observed.

2. BENTONITE PROPERTIES

The tests have been performed with the FEBEX bentonite, which is the clay used for the FEBEX Project in the *in situ* (Grimsel, Switzerland) and the mock-up (Madrid, Spain) tests (ENRESA 2000).

The FEBEX bentonite has a content of montmorillonite higher than 90 percent. The cation exchange capacity (CEC) is of 102±4 meq/100g, and the major exchangeable cations are: Ca (42 %), Mg (33 %), Na (23 %) and K (2 %). The liquid limit of the bentonite is 102±4 percent. The hygroscopic water content in equilibrium with the laboratory atmosphere (relative humidity 50±10 %, temperature 21±3 °C, suction 120 MPa) is 13.7±1.3 percent.

The pore size distribution is bi-modal, as it is characteristic of this type of materials. The volume of intra-aggregate pores (smaller than 0.006 μm) is very similar for samples compacted at different dry densities, and represents 73-78 percent of total pore volume when the bentonite is compacted at a dry density of 1.7 g/cm^3.

Retention curves for the FEBEX bentonite at constant and free volume conditions for various dry densities and temperatures have been reported by Villar (2002) and Villar & Lloret (2002). Figure 1 shows the water retention curves for the bentonite compacted at dry density 1.65 g/cm^3 in wetting paths performed at 20, 40 and 60 °C (Villar & Lloret 2002). The retention capacity for the same dry density is slightly lower the higher the temperature.

The saturated permeability of the bentonite compacted at a dry density of 1.7 g/cm^3 is of the order of 10^{-14} m/s.

Other termo-hydro-mechanical and geochemical characteristics of the FEBEX bentonite are detailed in ENRESA (2000) and Villar (2002).

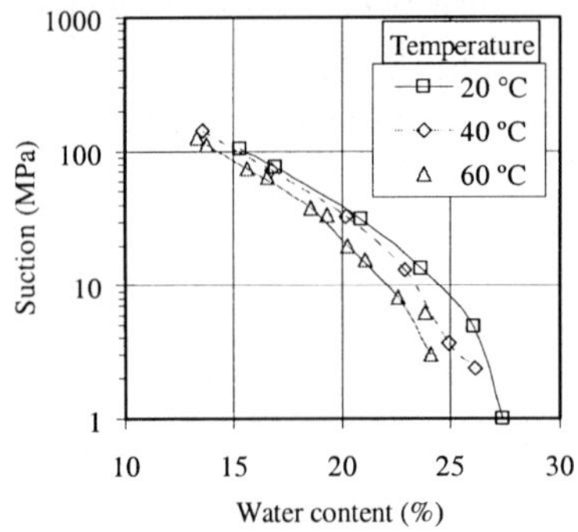

Figure 1. Water retention curves at constant volume of FEBEX bentonite compacted at dry density of 1.65 g/cm^3.

3. EXPERIMENTAL PROCEDURES

Five types of tests have been performed in oedometric conditions at various temperatures following different stress paths and suction changes.

3.1 Compression tests with hygroscopic water content

The influence of temperature on the compressibility of unsaturated clay has been studied by compression tests in oedometric conditions. The initial dry density of the samples has been 1.70 g/cm^3, and the clay was compacted uniaxially with its hygroscopic water content inside the oedometer ring (diameter of 5.0 cm, length of the specimen of 1.2 cm). A vertical pressure between 22 and 26 MPa was applied to manufacture the specimens. Three different samples were heated up to 25, 60 and 80 °C under a low vertical load (0.4 MPa). Afterwards, the samples were progressively loaded by steps up to 20 MPa. The tests were performed in an oedometric cell located inside a thermostatic bath (Figure 2). The two inlets of the cell were closed during the tests to avoid any water exchange with the environment, consequently the samples remained unsaturated.

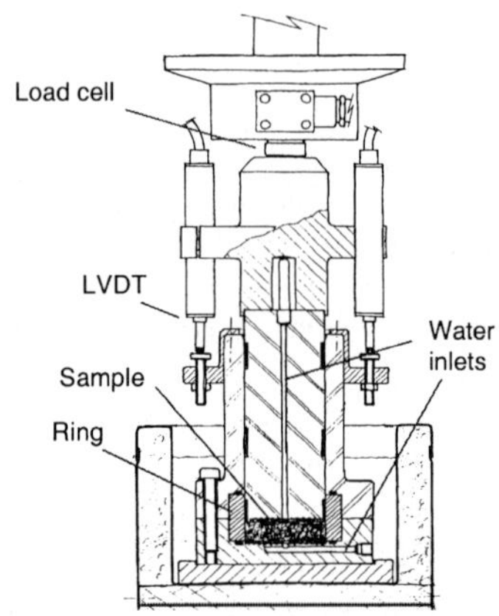

Figure 2. Schematic layout of the oedometric cell inside the thermostatic bath.

3.2 Soaking under vertical load tests

The influence of temperature on the swelling capacity of the clay was checked by soaking under load tests. They have been performed in an oedometer whose cell is placed in a thermostatic bath. Specimens of dry density 1.60 g/cm^3 were obtained by uniaxial compaction of the FEBEX clay with its hygroscopic water content. Vertical stresses of 15±1 MPa were applied for the manufacturing of the specimens, whose final height was 1.2 cm and diameter 5.0 cm. The swelling strain experienced by the clay upon saturation with distilled water has been determined for temperatures between 30 and 80 °C. The tests have been performed under vertical loads of 0.5, 1.5 and 3.0 MPa, after having reached the stabilisation of the target temperature.

3.3 Swelling pressure tests

The determination of the swelling pressure as a function of temperature in the interval between 25 and 80 °C has been performed in the oedometers shown in Figure 2. The clay has been uniaxially compacted with its hygroscopic water content at initial dry densities of 1.50 and 1.60 g/cm^3. Vertical stresses of 11 and 16 MPa, respectively, were applied to obtain specimens of 5.0 diameter and 1.2 cm height.

Once the temperature has stabilised, the sample is hydrated at constant volume through the bottom face with deionised water injected at a pressure of 0.6 MPa, while the upper outlet remains open to atmosphere. At the same time, the swelling pressure exerted by the clay is measured by a load cell installed in the loading frame and the vertical deformation of the specimen is measured by two LVDT's. The values of load, strain and water exchange are automatically recorded. The final density may differ slightly from the nominal one due to the small displacement allowed by the equipment.

3.4 Compression tests after saturation

In this type of tests the combined effect of the temperature and the hydration on the behaviour of the clay was investigated. The tests were performed at 25 and 80 °C in the oedometer shown in Figure 2, and the procedure to manufacture the samples was the same described in section 3.1. After the initial heating under a low vertical stress, the samples were saturated with distilled water under a vertical load of 1 MPa. Afterwards, the vertical load was progressively increased up to 20 MPa.

3.5 Wetting and compression tests under controlled suction

In these tests, bentonite was hydrated and loaded controlling the suction applied to the samples by using the axis translation technique. Figure 3 shows the layout of the oedometer cell, which was immersed inside a thermostatic bath (Romero *et al.* 2001). Suction is applied by changing the pressure of the gas phase in the pores of the sample by injecting nitrogen in the cell to the desired pressure. The bottom of the sample is in contact with water at atmospheric pressure through a cellulose membrane permeable to water but not to air.

Three tests on bentonite compacted at an initial dry density of 1.7 g/cm^3 with its hygroscopic water content have been performed at different temperatures (20, 40 and 60 °C), following the suction and stress paths indicated in Figure 4. In the first step, under a vertical load of 0.1 MPa, the suction of the sample was equalised to 14 MPa (due to mechanical limitations of the cell, this is the maximum suction that can be controlled). Afterwards, keeping this suction constant, the vertical load was increased up to 5 MPa. Under this vertical load, suction was decreased by steps until reaching a null suction state. Finally the sample was unloaded.

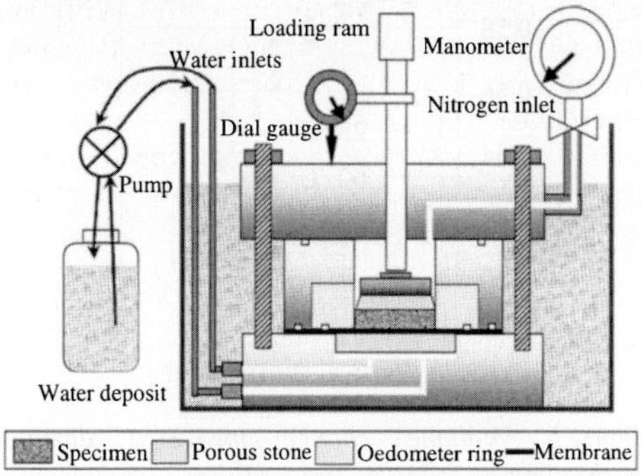

Figure 3. Schematic layout of the suction controlled oedometer cell inside the thermostatic bath.

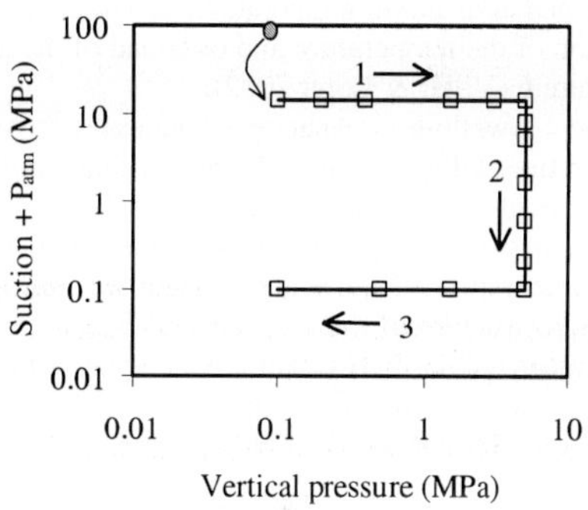

Figure 4. Stress path followed in wetting and loading tests under controlled suction at different temperatures.

4. RESULTS

4.1 Compression tests with hygroscopic water content

Figure 5 shows the volume change measured in the samples loaded at different temperatures keeping their water content constant. As it has been observed in saturated materials, the temperature increases the compressibility of the bentonite. A certain reduction in the size of the elastic domain with the temperature can be observed (Hueckel & Borsetto 1990), despite the fact that the vertical stresses applied are smaller than the compaction load.

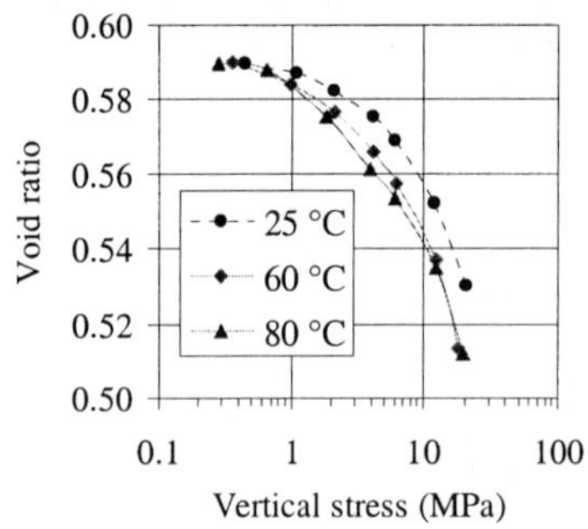

Figure 5. Volume change measured during loading at different temperatures and constant water content (initial dry density 1.70 g/cm³).

4.2 Soaking under vertical load tests

The final strains reached in soaking tests performed at different temperatures are plotted as a function of the temperature and overload of the test in Figure 6 (Villar & Loret 2002).

The swelling capacity decreases with temperature, although the influence of temperature is less evident when the overload is high. This behaviour can be explained if we consider that at high temperatures the amount of adsorbed water in the microstructure of the clay is smaller that at low temperatures (Ma & Hueckel 1992), what reduces the interlamellar swelling capacity, which is the prevailing mechanism in the swelling of a Ca-bentonite.

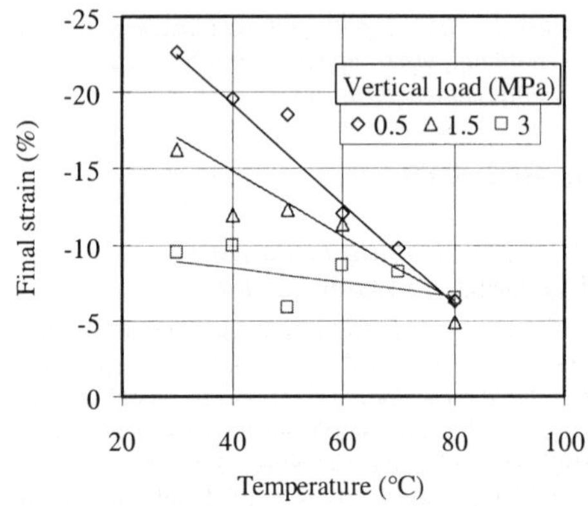

Figure 6. Final strain of samples of initial dry density 1.60 g/cm³ saturated under different vertical loads and temperatures.

4.3 Swelling pressure tests

The results of swelling tests are shown in Figure 7, in which the dispersion of data can be mostly attributed to the variations in dry density caused by the small displacement allowed by the equipment. A clear decrease of swelling pressure as a function of temperature is observed. This behaviour is well-matched with the observed increase in the compressibility of the bentonite in the compression tests and the reduction of swelling strains in the soaking tests. Lingnau *et al.* (1996) and Romero *et al.* (2003) found also a reduction in swelling pressure with temperature for a sand/bentonite mixture and for a moderately expansive clay, respectively.

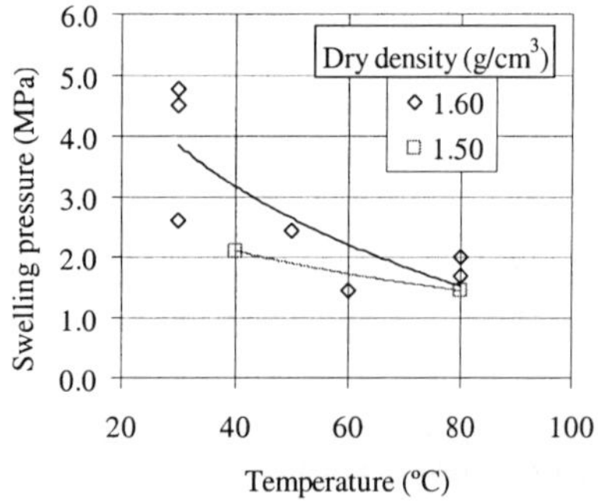

Figure 7. Effect of temperature on swelling pressure for two different dry densities.

4.4 Compression tests after saturation

Figure 8 shows the evolution of the swelling during saturation under a vertical load of 1 MPa at two different temperatures for samples of initial dry density 1.70 g/cm³. The increase of water permeability and the decrease of swelling capacity of the bentonite with temperature can be observed.

After saturation the samples were loaded under oedometric conditions. The volume changes measured during loading are plotted in Figure 9.

Hydration induces a significant increase of pore volume in both samples, but this swelling is greater in the sample tested at 25 °C (Figure 8), following the same trend of the soaking under load tests. This increase of volume is associated to irreversible changes in the macrostructure of the material and reduces the apparent preconsolidation stress of the bentonite (Gens & Alonso 1992, Alonso *et al.* 1999). These structural changes prevail over the increase of the compressibility of the bentonite with temperature, and the cold sample presents the smallest preconsolidation pressure and the biggest volume change under load application.

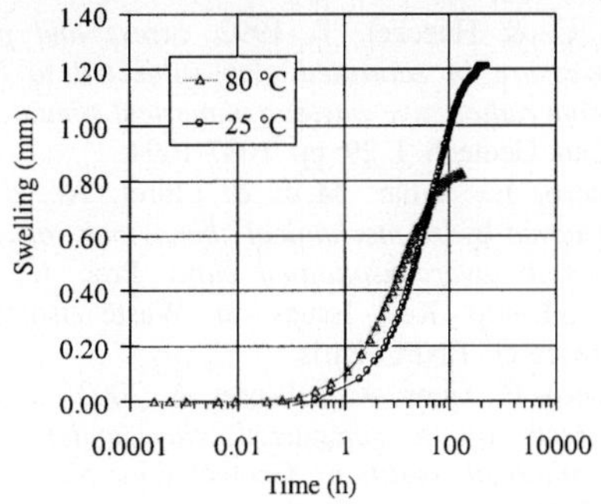

Figure 8. Swelling during saturation of bentonite under a vertical stress of 1 MPa.

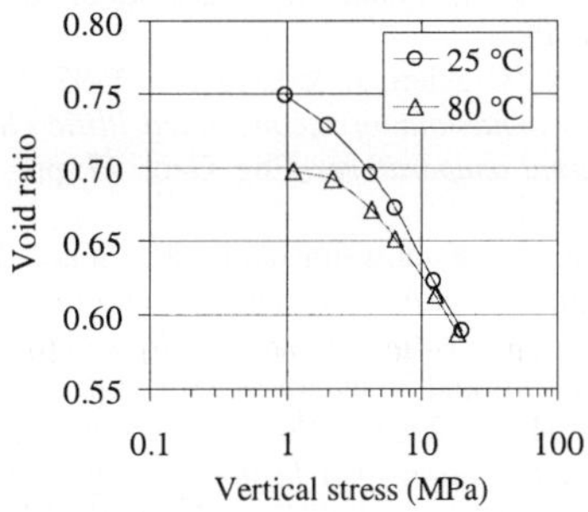

Figure 9. Volume change due to loading after saturation under a vertical stress of 1 MPa.

4.5 Wetting and compression tests under controlled suction

The results of the oedometric tests performed with control of suction are summarised in Figure 10 (Romero *et al.* 2001, Villar 2002). The swelling that occurs initially when suction decreases from the 120 MPa corresponding to the hygroscopic conditions of the clay to the 14 MPa experienced in the oedometer is greater the lower the temperature. Afterwards, the stiffness of the bentonite on loading from 0.1 to 5 MPa apparently increases with temperature (the results of this part of the path are detailed in Figure 11). It must be taken into account that after the volume changes due to the first suction reduction, the structure of the three samples is not the same and, therefore, their mechanical behaviour is strongly affected: the sample tested at 20 °C presents the higher void ratio and the more compressible structure. In fact, the yield point at 20 °C is obtained around a vertical stress of 1.5 MPa, whereas the samples tested at 40 and 60 °C present this point at higher stresses. In the subsequent wetting from 14 to 0 MPa, again the higher swelling strains are observed at laboratory temperature (Figure 10). In the final unloading steps the strain changes are more relevant in the samples tested at high temperatures.

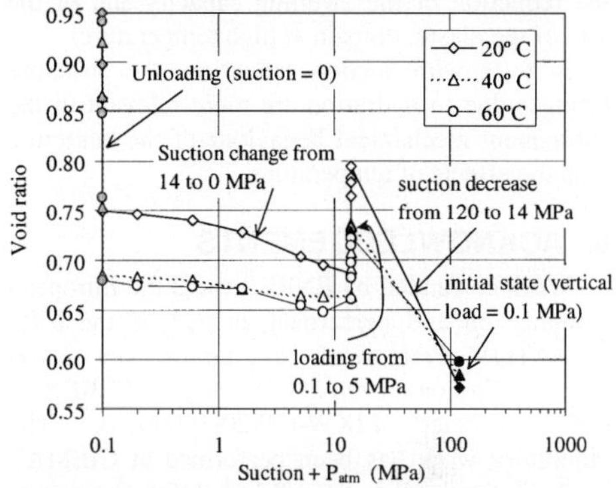

Figure 10. Results of isothermal oedometric tests with control of suction and temperature.

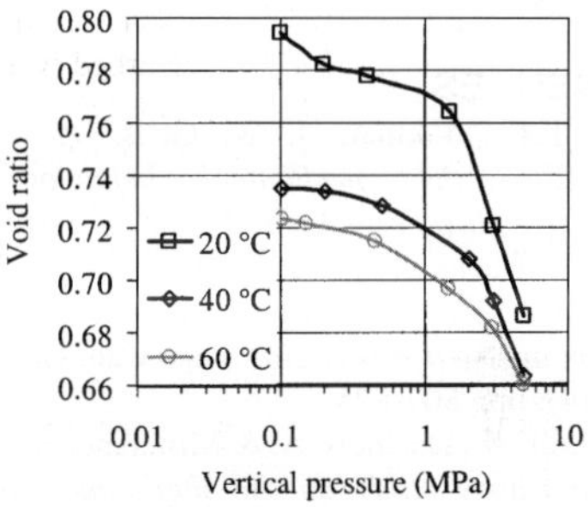

Figure 11. Evolution of void ratio during loading under suction 14 MPa (detail of Figure 10).

5. CONCLUSIONS

Different experimental techniques and equipments to study the influence of the temperature on the mechanical behaviour of the FEBEX bentonite under saturated and partially saturated states have been presented.

Under unsaturated conditions the compressibility of the bentonite increases with the temperature of the test. This result is coherent with the thermally induced reduction of the size of the yield surface proposed by several authors.

A decrease of the final swelling strain on isothermal suction reduction has been observed at elevated temperatures in both suction controlled and soaking tests. This effect can be explained by transfer induced by temperature between intra-aggregate adsorbed water and inter-aggregate free water. Swelling pressure also decreases with temperature. This decrease may be explained by the reduction of the swelling capacity and of the size of the elastic domain at high temperatures.

In tests with suction reduction, the structure changes due to hydration are more relevant in the subsequent mechanical behaviour of the bentonite than the effects of temperature.

6. ACKNOWLEDGEMENTS

Work co-funded by ENRESA and the European Commission and performed as part of the Fifth EURATOM Framework Programme, key action Nuclear Fission (1998-2002), Project FEBEX II (EC Contract FIKW-CT-2000-00016). The laboratory work has been performed at CIEMAT by R. Campos and J. Aroz. The helpful discussions with P. Rivas and P.L. Martín (CIEMAT) and E. Romero and A. Gens (UPC) are greatly acknowledged.

REFERENCES

Alonso, E.E., Vaunat, J. & Gens, A. 1999. *Modelling the mechanical behaviour of expansive clays*. Eng. Geol. 54: pp. 173-183.

Baldi, G., Hueckel, T., & Pellegrini, R. 1988. *Thermal volume changes of mineral-water system in low-porosity clay soils*. Can. Geotech. J. 25(4): pp. 807-825.

Burghignoli, A., Desideri, A. & Miliziano, S. 2000. *A laboratory study on the thermomechanical behaviour of clayey soils*. Can. Geotech. J. 37: pp. 764-780.

ENRESA. 2000. *FEBEX Project. Full-scale engineered barriers experiment for a deep geological repository for high level radioactive waste in crystalline host rock. Final Report*. Pub. Téc. 1/2000. 354 pp.

Gens, A. & Alonso, E.E. 1992. *A framework for the behaviour of unsaturated expansive clays*. Can. Geotech. J. 29: pp. 1013-1032.

Hueckel, T. & Borsetto, M. 1990. *Thermoplasticity of saturated soils and shales: constitutive equations*. J. Geotech. Eng. ASCE 116(12): pp. 1765-1777.

Lingnau, B.E., Graham, J., Yarechewski, D., Tanaka, N. & Gray, M.N. 1996. *Effects of temperature on strength and compressibility of sand-bentonite buffer*. Eng. Geol. 41(1-4): pp. 103-115.

Ma, C. & Hueckel, T. 1992. *Stress and pore pressure in saturated clay subjected to heat from radioactive waste: a numerical simulation*. Can. Geotech. J. 29: pp. 1087-1094.

Romero, E., Villar, M.V. & Lloret, A., 2001. *Thermo-hydro-mechanical behaviour of two heavily overconsolidated clays*. Proc. 6th Int. Workshop Key Issues in Waste Isolation Research. ENPC, Paris.

Romero, E., Gens, A. & Lloret, A. 2003. *Suction effects on a compacted clay under non-isothermal conditions*. Géotechnique 53 (1): pp. 65-81.

Sultan, N., Delage, P. & Cui, Y.J. 2002. *Temperature effects on the volume change behaviour of Boom clay*. Eng. Geol. 64: pp. 135-145.

Tanaka N., Graham, J. & Crilly, T. 1997. *Stress-strain behaviour of reconstituted illitic clay at different temperatures*. Eng. Geol. 47: pp. 339-350.

Towhata, I., Kuntiwattanakul, P., Seko, I. & Ohishi, K. 1993. *Volume change of clays induced by heating as observed in consolidation tests*. Soils and Foundations 33(4): pp. 170-183.

Villar, M.V. 2002. *Thermo-hydro-mechanical characterisation of a bentonite from Cabo de Gata*. Pub. Téc. ENRESA 01/2002. 258 pp. Madrid.

Villar, M.V. & Lloret, A. 2002. *Temperature influence on the hydro-mechanical behaviour of a compacted bentonite*. Proc. Int. Meeting Clays in Natural and Engineered Barriers for Radioactive Waste Confinement. ANDRA, Reims.

Wiebe, B., Graham, J., Tang, G. X. & Dixon, D. 1998. *Influence of pressure, saturation and temperature on the behaviour of unsaturated sand-bentonite*. Can. Geotech. J. 35: pp. 194-205.

IMPACT OF IN-SITU PARAMETERS AND BOUNDARY CONDITIONS ON THE THERMAL-HYDRO-MECHANICAL BEHAVIOUR OF A CLAY ENGINEERED BARRIER SYSTEM.

N. Barnel [1], T. Lassabatère [2], C. Le Potier [1] & P. Sémété[2]
[1]) CEA (Commissariat à l'Energie Atomique) / Centre de Saclay
[2]) EDF R&D (Electricité de France) / Site des Renardières

Abstract: This paper deals with the saturation kinetics of a clay-engineered barrier system. The thermal-hydraulic couplings of such a system are first identified on a reference thermal-hydraulic calculation. The saturation phenomenon kinetics is governed by darcean liquid flow, which is accelerated while heating. To complete this study, the impact of mechanics is taken into account. It tends to delay the saturation time, but darcean liquid flow remains the dominating flux. A second part is devoted to the impact of uncertainties concerning in-situ parameters or in-situ boundary conditions. On the one hand, the uncertainties concerning intrinsic conductivity can change the order of magnitude of saturation time. On the other hand, neither gas relative permeability nor gas or mechanical boundary condition influence significantly the saturation time. Finally, no heat-pipe effect can be created in clayed materials, whose kinetics is governed by darcean water flow in any case.

1. INTRODUCTION

In a common concept of radioactive waste repository, wastes are placed horizontally or vertically in cylindrical chambers. They are surrounded by an expansive clay, called engineered barrier (EB). When this clay is being saturated by the site water, it develops a swelling potential closing the technological gasps and insuring the mechanical stability of the system. For this reason, a good approximation of this resaturation phenomenon is fundamental in order to evaluate performance assessment of such a repository concept.

Previous works have been performed to evaluate the impact of a thermal radioactive waste on the thermal-hydraulic behaviour of a clay engineered barrier system. The results suggest that the saturation kinetics is governed by the darcean water flow. This flow can be accelerated by water dynamic viscosity decrease while heating, Barnel and al. (2002[a]). Furthermore, increasing the initial heat power has only a local effect and does not modify the saturation time. Finally, when increasing the time of heating, the gas pressure becomes strong enough to activate another coupling between suction and gas pressure. In this case, the kinetics is not governed by darcean water flow anymore, Barnel and al. (2002[b]).

This paper deals with a reference case, representing the wastes whose saturation kinetics seems to be governed by darcean water flow. In a first part, the thermal-hydraulic case is completed by studying the impact of mechanics. The thermal-hydro-mechanical (THM) couplings are identified. In a second part, we wonder if the uncertainties regarding in-situ initial and boundary conditions can affect significantly the saturation time.

More precisely, we first pay attention to intrinsic conductivity. Indeed, its measurement is very difficult, so the values commonly used vary in a large range, covering several orders of magnitude. Concerning the gas, we study the impact of the conductivity value which is much more difficult to evaluate than the liquid one. Particularly, we wonder if this parameter can induce a heat-pipe effect. We also determine the influence of gas boundary condition. An hypothesis consists in considering the gas confined in the repository chamber, another one permits it to escape. Finally, we tried to evaluate the influence of mechanical boundary condition, which depends on the width of void between the EB and the canister.

2. THE REFERENCE CASE

2.1 Geometry and thermal-hydraulic modelling

The canister, the engineered barrier and the host rock are modelled with a 1D-axisymetric geometry. The 0.24 m thick canister is not modelled. Next to it stands the 0.8 m thick engineered barrier (EB). The canister and the EB are placed in the host rock. The extension of the whole system is 50 m. The initial conditions are; 72.5 MPa of suction in the EB, whereas the host clay is initially saturated at 5 MPa of water pressure upon the hydrostatic level. Numerically, 1% of residual gas content is initially considered in the site. Physically, it is equivalent to assume that complete saturation can not be reached in both clays. The gas is assumed to be initially at atmospheric pressure. The initial temperature is 28°C everywhere.

Due to axisymetric geometry, there are no lateral fluxes. In the reference case, the contact between the canister and the clay is supposed to be perfect, which implies non-existence of gas and water flow between them. To model the canister heating, a time dependent heat flux, expressed in W, is imposed at that boundary. It follows equation 1:

$$Q(t) = 1070 \exp\left[-4 \ln 10 / 200 \times (t + 30)\right] \quad (1)$$

The materials chosen are Foca clay for the EB and Bure argillite for the host rock. Materials main properties are detailed in table 1.

Table 1.Materials main properties

	Foca clay	Bure argillite
Φ^*	30%	5%
$k_{int}{}^*$ in m^2	10^{-20}	10^{-19}
$\psi_0{}^*$ in MPa	1.5	10
n^*	0.06	1-(1./1.7)

* where Φ stands for porosity, k_{int} for intrinsic conductivity, ψ_0 and n are the two parameters of Van-Genuchten retention law (see equation 2):

$$S_l = \left[1 + \left(\frac{P_{cap}}{P_{cap}{}^0} \right)^{\frac{1}{1-n}} \right]^{-n} \quad (2)$$

In equation 2, S_l and P_{cap} denote respectively liquid saturation and capillary pressure, $P_{cap}{}^0$ =1.5 MPa and n = 0.06.

The model describing the partially saturated porous materials is composed of three balance equations. They express water mass conservation, air mass conservation and energy conservation. They can be found in Chavant (2001). The capillary pressure, the gas pressure and the temperature have been chosen as the main variables of the system. The three conservation equations are non-linear and coupled. As an example, these notions are explained on liquid conductivity expression, written as equation 3:

$$K_l = \rho_w \times g \times \mu_l{}^{-1}(T) \times K_{rg}(P_{cap}) \quad (3)$$

In equation 3, ρ_w stands for water density, μ_l for liquid dynamic viscosity and K_{rg} for relative conductivity. The liquid conductivity is associated to darcean liquid flow, in water mass conservation equation. This term is non-linear because water relative conductivity depends on capillary pressure, which is the main variable associated with water mass conservation equation. Furthermore it is coupled to thermal effects because liquid dynamic viscosity depends on temperature, which is the main variable associated with energy conservation equation.

The model is solved numerically by means of the finite element code ASTER. The equations are discretized within a finite element formulation. The time discretization is implicit and the coupling is solved by means of a global inversion of the system, as explained in Chavant (2001).

This referential thermal-hydraulic calculation have been studied in details by Barnel and al (2002^b). The main results were the following. Gas pressure does not seem to influence the saturation process. In fact, the saturation kinetics is governed by the darcean liquid flow. This flow is coupled with temperature, as can be seen in equation 3. It is accelerated when the heat source is taken into account. More precisely, this acceleration is entirely due to dynamic viscosity decrease while heating.

2.2 Impact of mechanics on the thermal-hydraulic behaviour

To get better the understanding of the system, the mechanical behaviour of the clay has been taken into account. The initial conditions are null total stresses everywhere. So, in-situ mechanical stresses are not taken into account. The results of our THM calculation show only stresses induced by thermal-hydro-mechanical couplings. The contact between the EB and the canister is once again supposed to be perfect, so that no radial displacement of the clay is allowed at that boundary. Biot's poroelastic model is chosen to represent clay behaviour. It takes partial saturation into account via an equivalent pressure which includes capillary effects, involving both gas and liquid, Dangla (1998). Biot's model is added as fourth equation to the system. The associated main variable is total stress state. The couplings with thermal-hydraulics behaviour are introduced by possible porosity variations.

Three calculations have been performed, denoted by $C_{OM}^{\mu w}$, $C_{IM}^{\mu w}$ and $C_{IM}^{\mu T}$. In the first one ($C_{OM}^{\mu w}$), the thermal source is not activated; only the hydro-mechanical couplings are present. In the second calculation ($C_{IM}^{\mu w}$), the source is activated but the liquid dynamic viscosity is taken as a constant. Finally, in the third calculation ($C_{IM}^{\mu T}$), the heat source is activated and the liquid dynamic viscosity depends on temperature.

Table 2. Mechanical results of C_{xM}^{x} caculations

	$C_{OM}^{\mu w}$	$C_{IM}^{\mu w}$	C_{IM}^{μ}
Δx^{max} in cm*	2.7	2.7	2.7
σ' in MPa*	33	32	33
σ^{fl} in MPa*	-64	-65	-65
σ^{tot} in MPa*	-31	-33	-32

* Δx^{max} represents the swelling of the EB at the interface with the site, σ' σ^{fl} and σ^{tot} denote respectively the net stress, the equivalent fluid stress and the total stress at the contact EB-canister. A negative stress is compressive.

Table 2 gives the main mechanical results of those three calculations. As no displacement is allowed at the interface EB-canister, saturation phenomenon develops stresses. A compressive stress of about 32 MPa is supported at the contact between the EB and the canister. The clay skeleton is there submitted to a net stress of about 33 MPa.

Furthermore, the EB swells in the direction of the site; the interface between the EB and the site is moved 2.7 cm away. These results show a quite significant coupling between hydraulics and mechanics.

Table 3. Saturation time of C_{xM}^{x} caculations

	saturation time years/months/days
$C_{OM}^{\mu w}$ and $C_{IM}^{\mu w}$	11 y / 11 m/ 20 d
$C_0^{\mu w}$ and $C_1^{\mu w}$	09 y / 05 m/ 21 d
$C_{IM}^{\mu T}$	05 y / 11 m / 25 d
$C_1^{\mu T}$	05 y / 00 m / 26 d

The saturation times obtained with these calculations are given in table 3. The results are compared with similar calculations that does not take mechanics into account, respectively denoted by $C_0^{\mu w}$, $C_1^{\mu w}$ and $C_1^{\mu T}$.

Obviously, mechanics delays saturation phenomenon. Theoretically, mechanics can affect hydraulics by a change in porosity. The chosen model does not consider conductivity changes associated to porosity variations. As the EB swelled, clay porosity has been increased. Thus, the volume to be filled by liquid is bigger when mechanics is taken into account. That's why the saturation time is delayed, even if the conductivity is not affected. The retroactive coupling of mechanics on hydraulics is present on these calculations.

Furthermore, according to water mass conservation equation, porosity does not affect liquid fluxes ratio. Therefore, darcean liquid flow still governs saturation kinetics. As can be seen on table 2, $C_{OM}^{\mu w}$ and $C_{IM}^{\mu w}$ calculations give the same saturation time, whereas in $C_{IM}^{\mu T}$ the saturation phenomenon is accelerated. As for thermal-hydraulics calculations, full saturation of the EB is reached earlier when both the heating source is being activated and liquid dynamics viscosity depends on temperature. Once again, this acceleration is only due to water dynamic viscosity decrease while heating.

$C_{OM}^{\mu w}$ and $C_{IM}^{\mu w}$ calculations give the same results. As an example final stresses, displacement and saturation time at the key points are given in table 2 and 3. It implies that, according to the chosen model, the thermal-mechanical coupling is negligible. This result can be criticized; the model is not really predictive. Nevertheless, we decided not to discuss the model here.

The THM couplings of the reference calculation have now been identified. We wonder now if this identification is still pertinent, when considering in-situ parameters and boundary conditions uncertainties.

3. INFLUENCE OF INTRINSIC CONDUCTIVITY

One of the most important uncertainty about in-situ parameter concerns materials intrinsic conductivity. They can vary by one or even two order of magnitude. This is quite important because, according to equation 3, darcean liquid flow is directly proportional to this parameter. Theoretically, if the conductivities ratio between the EB and the site remains constant, and, if darcean water flux is dominating, the saturation time should be proportional to EB intrinsic conductivity. To check this, three hydraulic calculations were performed, with EB intrinsic conductivity of respectively 10^{-19}, 10^{-20} and $10^{-21}m^2$. We chose a constant conductivities ratio between the EB and the site. Its value is 0.1. The calculations saturation times are included in figure 1.

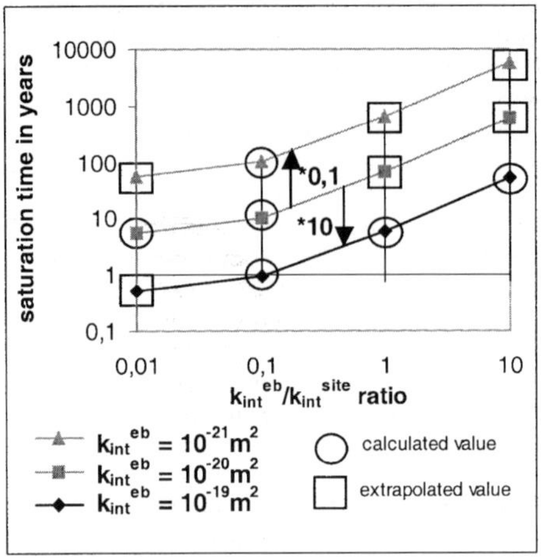

Figure 1. Impact of material intrinsic conductivity on EB saturation time.

As expected, the darcean water flux remains the dominating flux. Multiplying, respectively dividing, by a ten factor the intrinsic conductivity of both the EB and the site exactly multiplies, respectively divides, by a factor ten the saturation time.

In a second time, we try to change only one material intrinsic conductivity. As a consequence, the conductivities ratio between the EB and the site is modified. Three other hydraulic calculations have then been performed for different values of this ratio, respectively 0.01, 1 and 10. Taking into account the link between conductivities ratio on the saturation time, we extrapolate six other saturation times. They are only hypothetical saturation times, still they suppose that darcean liquid flow remains dominating. This assumption seems to be correct till saturation time is less than a thousand years (characteristic time of saturation when fickian flows dominate). But rigorously this should be checked by really performing the calculations. Nevertheless, these last six hypothetical times are reported on figure 1 to give a first approximation of saturation times. One can notice that those hypothetical times concern only slowly saturating cases, which are not the critical ones.

In each curve, in which k^{eb} is fixed, a slope break is visible for a k^{eb}/k^{site} ratio of 0.1. When this ratio is lower than 0.1, the site is further more permeable than the EB. Physically, the site loses its liquid retention ability. It is equivalent to consider the EB directly in contact with liquid water. On the contrary, when k^{eb}/k^{site} ratio is upper than 0.1, the saturation of the EB is limited by liquid water flow through the site.

As can be seen on figure 1, uncertainties on in-situ intrinsic conductivity values strongly affect the saturation time. This time would vary by the same order of magnitude than the intrinsic conductivity of material.

4. INFLUENCE OF GAS RELATIVE CONDUCTIVITY AND BOUNDARY CONDITION

In the reference case, the gas has no influence on liquid kinetics and saturation time. But we have to check this result, if the gas can escape from the chamber. In Pollock (1986), the heating of a partially saturated alluvium induces a heat-pipe effect. Liquid water vaporises near the heating source. The vapour is carried away in colder areas by means of gas darcean flux. There, vapour is liquefied. The reference case does not show any significant vaporization near the heating source, still the model we use and Pollock's one are quite similar. The non-existence of this phenomenon can not be imputed to the model. Two other explanations are then possible. It can be due to clay and alluvium behaviour differences, or to a bad evaluation of hydraulics parameters. Hence, tests have been performed to see if gas conductivity or retention curve parameters uncertainties can activate a heat-pipe phenomenon.

Six thermal-hydraulic calculations have been done. They are denoted by C_l^{Pg}, C_l^{Krg} , $C_l^{Pg\text{-}Krg}$, C_{le}^{Krg} , C_{le}^{Pg} and C_{le}^{Pg}. The difference between C_l^{Pg} and the reference case is the gas boundary condition; the gas pressure is maintained at the atmospheric level at the contact EB-canister for each calculation time. C_l^{Krg} differs from the reference case, because the relative conductivity law is replaced by the constant value of 10^8. This value is not realistic, but two reasons have motivated this choice. First, several orders of magnitude uncertainties concern both intrinsic and gas relative conductivity. We have chosen not to discuss intrinsic conductivity concept pertinence, so we have preferred to report the global uncertainties on a single parameter; the relative conductivity. Second, if the heat-pipe effect, observed by Pollock (1986), can be activated, the darcean water flux would not govern the saturation kinetics anymore. Taking an exaggerated value is a security to validate our results. In $C_l^{Pg\text{-}Krg}$, both modifications done in C_l^{Pg} and C_l^{Krg} are taken into account. Finally, C_{le}^{Krg} , C_{le}^{Pg} and C_{le}^{Pg} are respectively the same that C_l^{Pg}, C_l^{Krg} and $C_l^{Pg\text{-}Krg}$, but the retention curve parameters are P_{cap}^0 = 17 MPa and n = 0.27.

Table 4.Results of C_l^x calculations

	P_{cap}^{max}* MPa	S_l^{min}*	saturation time years/months/days
C_l	72.50	0.77	05y/00m/28d
C_l^{Pg}	75.23	0.77	05y/01m/08d
C_l^{Krg}	81.67	0.76	08y/00m/12d
$C_l^{Krg\text{-}Pg}$	81.42	0.76	07y/10m/08d

Table 5. Results of C_{le}^x calculations

	P_{cap}^{max}* MPa	S_l^{min}*	saturation time years/months/days
C_{le}	72.50	0.56	07y/08m/04d
C_{le}^{Pg}	76.96	0.55	07y/10m/08d
C_{le}^{Krg}	97.98	0.51	09y/01m/08d
$C_{le}^{Krg\text{-}Pg}$	96.55	0.51	09y/00m/20d

*P_{cap}^{max} and S_l^{min} denote respectively the maximum of capillary pressure and the minimum of liquid saturation degree observed at the interface EB-canister

The main results of the calculations, compared to referential cases, are reported in table 4 and 5. No visible heat-pipe effect is present; the saturation degree does not decrease much more than 0.05 compared to the reference value. Thus, the desaturation of the EB at the contact with the canister can be neglected in every case. The absence of a heat-pipe effect is related to clay specific behaviour.

Concerning the saturation time, gas mobility exaggeration can delay the saturation time. In our cases this schedule varies from 15% of delay to 60% in accordance with the parameters chosen for the retention law. The gas boundary condition, allowing the gas to escape the chamber, tends to delay the saturation phenomenon, but it is not significant. The delay is about 2% whichever saturation curve is chosen. This case is much more realistic than the ones where gas conductivity have been virtually increased. Thus we assume that it is very weakly probable that gas could delay darcean liquid flow.

5. INFLUENCE OF MECHANICAL BOUNDARY CONDITION

Two main issues are fundamental to predict the thermal-hydro-mechanics of the clay. First, the mechanical behaviour of swelling clays, like Foca, is very difficult to understand and to model. Many models exist, as mention in Barnel (2002). As a first approximation, we chose Biot's poroelastic model. Once again, the purpose here is not to discuss the model, which is a huge issue. We prefer to focus on boundary conditions. The contact between the EB and the canister can not be perfect. The width of the void between the EB and the canister is not known. So we decided to test five different boundary conditions to see the influence of mechanics on the thermal-hydraulic behaviour of the clay.

Five thermal-hydro-mechanical calculations have been performed, denoted by C_l^0, $C_l^{0.4}$, $C_l^{1.0}$, $C_l^{1.6}$ and C_l^{dx}. In C_l^0, the contact between the EB and the canister is supposed to be perfect, no displacement is allowed at that point. On the contrary, C_l^{dx} calculation considers that the void between the EB and the canister is large enough to allow free displacement of the EB in the direction of the canister. $C_l^{0.4}$, $C_l^{1.0}$ and $C_l^{1.6}$ are intermediary cases; The canister and the EB are supposed to be separated by a void of 4 mm, 1 cm and 1.6 cm respectively.

Such a boundary condition is an hypothetical state for the materials, and does not exactly reflect reality for at least two reasons. The EB is allowed to move in direction of the canister, but no fluids fluxes are allowed. If there were a liquid flow in the void, it would probably vaporize, because of gas circulation in the neighbourhood of the chamber. The flux boundary condition would neither be a null flux, nor be a constant capillary pressure. Moreover, the dependence between capillary pressure and time at this point would be very hard to determine. We then assume that, if there is a fluid flux in the void, it would not be important and can be neglected.

Furthermore, a space between the EB and the canister in axisymetric conditions is equivalent to a ring of void in a plane. This is not realistic at all so long as the canister is placed horizontally in the chamber, but it is a satisfying approximation if the canister is placed vertically in the chamber.

Although their boundary conditions are not totally satisfying, $C_l^{0.4}$, $C_l^{1.0}$ and $C_l^{1.6}$ calculations have been performed. A void of 1.7 cm is sufficient to allow free displacement of the EB during the whole saturation process. Comparing the five calculations and regarding the model, final values of total and effective stress are a linear function of radial displacement of the EB near the canister. This point is shown on figure 2. It is a consequence of the linearity of poroelastic Biot's model.

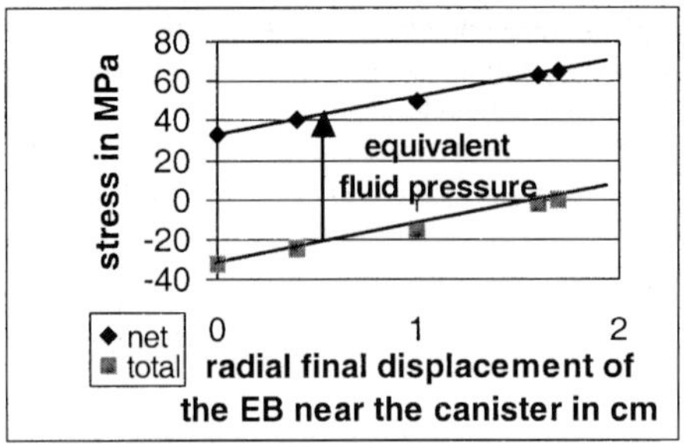

Figure 2. Relation between final stress and displacement at the nearest point from the canister in the EB.

The role of mechanics on liquid darcean flux does not depend on the mechanical boundary condition; the saturation time is 5 years 11 months and 26 days, like in the thermal-hydro-mechanical reference calculation. The maximal radial displacement is observed, when the saturation is achieved, at the contact EB-site; there the EB expands. The site is contracted in response. The maximal values for EB extension are available on table 6. It reveals that the boundary condition at the interface EB-canister only has a very weak effect on the EB-site interface behaviour, 80 cm away. Like gas boundary condition, mechanical boundary condition does not disturb the saturation phenomenon, whose kinetics remains dominated by darcean water flux.

Table 6. Results of C_l^x THM calculations; final radial displacement at the inerface EB-site (Δx^{max})

	C_l^0	C_l^{2m}	C_l^{1y}	C_l^{5y}	C_l^{dx}
Δx^{max}	2.7 cm	2.7 cm	2.6 cm	2.6 cm	2.5 cm

6. CONCLUSIONS

According to the chosen model, the saturation of a clay engineered barrier is governed by darcean water flow. The EB swells when possible or develops swelling pressure on materials in its neigbourhood. The expansion of the EB increases porosity, which increases the space to be filled by liquid during the saturation process. The saturation time is then delayed. Darcean water flow is accelerated when the heat source is activated, because dynamic liquid viscosity decreases while heating. Temperature increase seems to have a negligible effect on mechanical behaviour of the clays.

The uncertainties regarding in-situ intrinsic conductivity measurement and evaluation, are a huge issue. The saturation time is very sensitive to this parameter. It can change the order of magnitude of saturation time. On the contrary, the saturation time is not significantly affected by a change of gas or mechanical boundary condition. Moreover, gas conductivity uncertainties are not sufficient to induce a heat-pipe effect in clayed materials.

In this paper, we have not discussed the model. Still, the absence of thermal-mechanical coupling is directly linked to the model, so as the linear relation between final stress and displacement at the interface EB. Next step would be to discuss and evaluate the physical pertinence of the model.

REFERENCES

Barnel, N., Lassabatère, T., Le Potier, C., Maugis, P. & Mouche, E., 2002[a]. *Impact of a thermal radioactive waste on the thermal-hydraulic behaviour of a clay engineered barrier system.* In Auriault, J.-L., et al. (ed.), Poromechanics II, Balkema.

Barnel, N., Lassabatère, T., Le Potier & Sémété, P., 2002[b]. *Sensitivity analysis on the thermal-hydraulic parameters governing the saturation of a clay engineered barrier system.* Proc. Clays in natural and engineered barrier systems, Reims, 5-12 December 2002.

Barnel, N., 2002. *Le gonflement dans les argiles; des concepts aux modelisations.* Technical report, SFME/MTMS/RT/02-017/A.

Chavant, C., 2001. *Modèles de comportement THHM.* Technical report, EDF Clamart, MTI/MMN/R7.01.11-A.

Dangla, P. & Coussy, O., 1998. *Non linear poroelasticity for unsaturated porous materials: an energy approach.* In Thimus et al. (eds), Poromechanics, Balkema.

Pollock, D., 1986. *Simulation of fluid flow and energy transport processes associated with high-level radioactive waste deposal in unsaturated alluvium.* Water Resources Research, Vol 22, N°. 5, p. 765-775.

ANALYSIS OF THE THMC BEHAVIOUR OF COMPACTED SWELLING CLAY FOR RADIOACTIVE WASTE ISOLATION

A. Gens[1], L. do N. Guimarães[2], S. Olivella[1], M. Sánchez[1]

[1]) Geotechnical Engineering Department, Technical University of Catalunya, c/ Jordi Girona 1-3, Edificio D-2, 08034 Barcelona, Spain.

[2]) Federal University of Pernambuco. Av. Acadêmico Hélio Ramos, s/n, Centro de Tecnologia e Geociências, Departamento de Engenharia Civil, Cidade Universitária, CEP: 50740-530, Recife – PE, Brazil

Abstract: A fully coupled formulation combining reactive transport and an existing thermo-hydro-mechanical (THM) code is briefly described. Special attention has been given to phenomena likely to be encountered in clay barriers used as part of containment systems of nuclear waste. The types of processes considered in the chemical formulation include hydrolysis, complex formation, oxidation/reduction reactions, acid/base reactions, precipitation/dissolution of minerals and cation exchange. Both kinetically controlled and equilibrium-controlled reactions have been incorporated. The formulation has been implemented in the numerical code CODE_BRIGHT. An application is presented concerning the performance of a large scale in situ heating test simulating high-level nuclear waste repository conditions.

1. INTRODUCTION

Compacted swelling clays are often envisaged as the main component of engineered barriers for radioactive waste disposal. These barriers are subjected to thermal loading due to the heat emitted by the waste and to hydration from water coming from the adjacent rock. As a consequence of these thermo-hydraulic phenomena, mechanical and chemical changes arise that, in turn, may affect all other aspects of behaviour. A correct understanding and prediction of these barriers would require, therefore, the performance of fully coupled thermo-hydro-mechanical and chemical (THMC) numerical analyses.

Although there have been a number of coupled THM formulations proposed and the corresponding numerical codes have been available for some time (e.g. Gawin et al., 1995; Olivella et al., 1994, 1995; Thomas and He, 1995), coupled THMC formulations have lagged behind. Only recently, a number of proposals have been put forward in this field (Guimarães et al., 1999; Thomas et al., 2001; Gens et al., 2002). The research is now reaching a state that, through the incorporation into efficient computer codes, it is possible to perform coupled THMC numerical analyses to tackle engineering problems of a high level of complexity.

In this paper a fully coupled THMC formulation is briefly described that allows the performance of numerical analyses considering all four aspects of behaviour: thermal, hydraulic, mechanical and chemical and their interactions. The formulation takes into account the possibility that the material is partly saturated. Therefore, the term hydraulic should be interpreted in a wide sense, i.e. including both liquid and gas flow phenomena. Although the formulation is general, especial attention has been paid to the phenomena (especially chemical phenomena) that are likely to occur in swelling clays in an engineered barrier for nuclear waste isolation. To demonstrate that the tool developed is now capable of performing fully coupled THMC analysis involving real engineering problems, the formulation is applied to the analysis of a large scale in situ heating test simulating nuclear repository conditions.

2. DESCRIPTION OF THE FORMULATION

2.1 THM formulation

The THM formulation presented here is based on the one proposed in Olivella et al. (1994) applied to clays. The formulation uses a multi-phase, multi-species conceptual model. The porous medium is assumed to be made up of three phases: solid, liquid and gas. The main part of the solid phase is an inert mineral that does not dissolve but it may contain precipitated species. The liquid phase contains water, dissolved air and dissolved chemical species whereas the gas phase is made up of dry air and water vapour. Note that dry air is considered as a single species in spite of the fact that it is a mixture of gasses.

Using the compositional approach, the formulation is based on the establishment of balance equations for each species: mineral, water and air, instead of using the phases. Thus, the contributions of phase changes automatically cancel. In the following, variables may include a subscript and a superscript. The subscript refers to the phase (s for solid, l for liquid and g for gas), whereas the superscript indicates the species (w for water and a for dry air).

As an example, the balance equation for the mass of water is expressed as:

$$\frac{\partial}{\partial t}(\phi S_l \rho_l w_l^w + \phi S_g \rho_g w_g^w) + \nabla \cdot (\mathbf{j}_l^w + \mathbf{j}_g^w) = f^w \qquad (1)$$

where ϕ is the porosity, S_l and S_g the liquid and gas degree of saturation, ρ_l and ρ_g, the liquid and gas densities in Kg/m^3 of phase, w_l^w the mass fraction of water in the liquid (close to one in dilute solutions) and w_g^w the mass fraction of water vapour in the gas phase. The term $\phi S_l \rho_l w_l^w$ is the water content in the liquid phase per unit volume and $\phi S_g \rho_g w_g^w$ the water content in the gas phase. $\mathbf{j}_l^w$ and $\mathbf{j}_g^w$ are the mass flux of water in the liquid and gas phases respectively. The term f^w may represent, in the context of this paper, the water production or loss associated with precipitation or dissolution of minerals.

Adding the mass balance equations for water and air to the equation for energy balance and momentum balance (equilibrium), the governing equations for the THM problem are obtained. Theses equations are solved simultaneously at every time step of the analysis. To close the THM formulation it is necessary to specify the relevant constitutive equations that describe the set of phenomena considered. Phase changes (liquid water/vapour and gaseous air/ dissolved air) are assumed to be in equilibrium. The corresponding equilibrium restrictions are also incorporated in the formulation. More details are given in Olivella et al. (1994)

2.2 THMC formulation

Now a number of additional chemical species are considered. The mass continuity of each chemical species is expressed by the reactive transport equation. In addition to the usual homogeneous reactions occurring in the liquid phase (aqueous complex formation, acid/base and oxidation/reduction), heterogeneous reactions that exchange matter between the liquid and the solid phase are also taken into account. The main heterogeneous reactions considered are dissolution/precipitation of minerals and cation exchange. In general the chemical processes are faster than the characteristic times of the THM problem, so it is assumed that chemical equilibrium prevails. Only some dissolution/precipitation processes are assumed to be kinetics-controlled. The transport equations describe the mass continuity of each chemical species in the porous medium. In the chemical equilibrium problem, a relationship between the species concentrations is established that allows a reduction in the degrees of freedom of the reactive transport problem (Guimarães, 2002).

Let us consider the reactive transport of N chemical species in a deformable unsaturated porous medium. The transport of every one of those species can be expressed as:

$$\frac{\partial}{\partial t}(\phi S_l \rho_l c_i) + \nabla \cdot \mathbf{j}_i = R_i \quad (i = 1, \ldots, N) \qquad (2)$$

where c_i is the concentration of species i in mols/Kg of solution and R_i is the total production rate of species i due to chemical reactions, in mols/m^3/s. $\mathbf{j}_i$ is the total flux of species i expresses mols/m^2/s. As in Olivella et al. (1994), this flux is considered as the sum of advective and non-advective fluxes. Advective flux is, in turn, the sum of the movement of the liquid phase with respect to the solid phase (governed by generalized Darcy's law) and of the solid phase with respect to the reference configuration. The non-advective flux is the consequence of molecular diffusion and mechanical dispersion. For simplicity, it is assumed that the mechanical dispersion tensor is the same for all solutes. This has the advantageous consequence that electrical charge is conserved in the reactive transport equations when chemical reactions are charge balanced (Lichtner, 1985).

After several transformations, the final form of the reactive transport equations:

$$\frac{\partial}{\partial t}(\phi S_l \rho_l U_j) + \nabla \cdot \begin{pmatrix} \rho_l U a_j \mathbf{q}_l - \mathbf{D}_l \nabla U a_j \\ + \phi S_l \rho_l U_j \dot{\mathbf{u}} \end{pmatrix} + \sum_{m=1}^{N_m} \nu_{jm} r_m = 0 \quad (j = 1, \ldots, N_c) \qquad (3)$$

$$U_j = C_j + \sum_{i=1}^{N_x} \nu_{ij} X_i \quad (i = 1, \ldots, N_c) \qquad (4)$$

$$Ua_j = \lambda_j C_j + \sum_{i=1}^{N_x} \nu_{ij} \lambda_i X_i \quad (i = 1,\dots,N_c) \qquad (5)$$

where U_j is the total analytical concentration and Ua_j is the total aqueous concentration of the primary species j. C_j and X_i are the concentrations of the primary and secondary species respectively. λ_j and λ_i are the mobilities of the primary and secondary species. r_m is the rate of precipitation or dissolution of mineral m under kinetics conditions and N_m is the number of minerals in kinetics conditions. ν_{jm} is the number of moles of primary species j in a mol of mineral m.

In the present formulation, the total analytical concentrations U_j are the unknowns of the transport equations (3), and Ua_j and r_m are considered non-linear functions of U_j. The link between the unknowns U_j and the dependent variables Ua_j and r_m is the chemical equilibrium model described in Guimarães (2002). This type of unknown exhibits the interesting property of being independent of chemical equilibrium. Total analytical concentration changes in time are exclusively due to transport and kinetics reactions. Figure 1 illustrates the meaning of these variables in a particular example. The total analytical concentration, U_{Ca}, for the calcium takes into account its presence in the liquid phase (as dissolved Ca^{2+} ion and $CaSO_4{}^0$) and in the solid phase as precipitated gypsum and as exchangeable cation in the clay mineral. The aqueous concentration, Ua_{Ca}, only considers the calcium in the liquid phase.

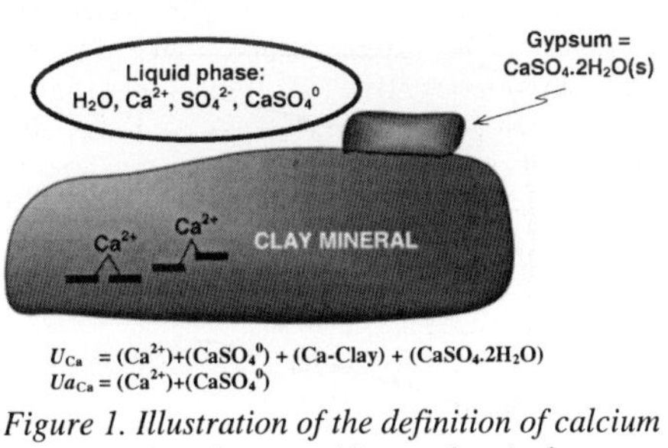

$U_{Ca} = (Ca^{2+}) + (CaSO_4{}^0) + (Ca\text{-}Clay) + (CaSO_4.2H_2O)$
$Ua_{Ca} = (Ca^{2+}) + (CaSO_4{}^0)$

Figure 1. Illustration of the definition of calcium concentrations in a specific geochemical system.

3. EXAMPLE OF APPLICATION

3.1 The FEBEX in situ heating test

A large scale in situ heating test is being performed at the Grimsel Test Site, an underground research laboratory located in the Swiss Alps. The rock in the site is mainly granite. The test tries to reproduce the main features of the current Spanish concept for deep underground disposal of high level radioactive waste (Febex, 2000).

The test layout (Fig. 2) consists in placing two heaters at the end of a drift of 2.28 m diameter and 70.4m length especially bored for this purpose. The diameter and length of the heaters (0.9m and 4.54m respectively) corresponds to the actual dimensions of the canister envisaged for radioactive waste storage. The heaters are placed in the axis of the drift at a 1m distance from each other.

The space between the drift and the heaters is filled by blocks of compacted bentonite with a smectite content in the range of 88%-96%. The test is heavily instrumented with measurements of temperatures, relative humidity (equivalent to total suction), pore pressures, displacements, and stresses.

The test schedule involves switching on the heaters an applying increasing power until the temperature reaches a value of 100°C at some point in the bentonite. From that moment on, the power of the heaters is constantly adjusted in order to keep the maximum temperature in the bentonite barrier at the 100°C mark. The test was run in this way for five years until one of the heaters was switched off and dismantled. The experiment continues with the other heater without a planned finishing date.

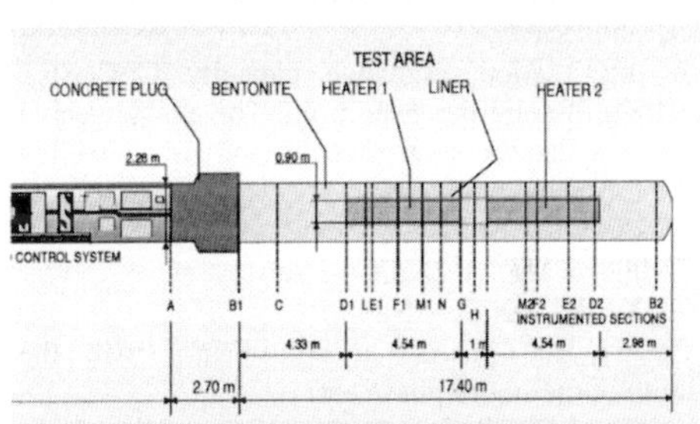

Figure 2. Lay out of the Febex in situ heating test

3.2 Features of analysis and chemical model

Because of space limitations only a brief overview of the analysis characteristics can be given. More information is reported in Febex (2000) and Gens et al. (2002). The numerical computations have been carried out using the fully coupled THMC version of CODE_BRIGHT. Radial symmetry has been assumed resulting in a 1-D axisymmetric analysis that is quite appropriate for examining the behaviour of the bentonite barrier and immediate adjacent rock.

After a short initial period of applying increasing power to the heaters, a constant 100° C temperature has been applied to the contact between the heater and the buffer, in correspondence with the test protocol. Although the full heating test has lasted for five years, the analysis has been run up to 100 years to check and predict long term THMC conditions. The rock thermal, hydraulic and mechanical boundary conditions have been based on the results of the comprehensive site investigation carried out at the site.

The initial conditions of the bentonite were as follows: dry density 1,7 g/cm^3 and water content 14.4%. This results in an initial degree of saturation of 0.65 and an initial suction of 115 MPa. Both the bentonite (solid phase and interstitial water) and the hydration water were subjected to a full chemical characterization. The thermo-hydro-mechanical parameters were determined in an independent laboratory testing programme.

The selected primary species are H_2O, Ca^{2+}, K^+, Na^+, Mg^{2+}, Cl^-, SO_4^{2-}, HCO_3^-, H^+, $SiO_2(aq)$, NaX. They are all in the liquid phase except adsorbed sodium, NaX, where X^- stands for the clay mineral. The secondary species, in equilibrium with the primary ones, are: OH^-, CO_3^{2-}, $CO_2(aq)$, $CaCO_3(aq)$, $CaHCO_3^+$, $CaSO_4(aq)$, $CaCl^+$, $MgSO_4(aq)$, $MgHCO_3^+$, $MgCl^+$, $NaHCO_3(aq)$, $NaSO_4^-$, $NaCl(aq)$, KSO_4^- and $HSiO_3^-$.

A total cation exchange capacity CEC=86.2 meq/100g of solid has been used. The exchangeable cations are also secondary species and are controlled by the following reactions:

$$CaX_2 = Ca^{2+} + 2NaX - 2Na^+$$
$$MgX_2 = Mg^{2+} + 2NaX - 2Na^+$$
$$KX = K^+ + NaX - Na^+$$

Finally, the phases in equilibrium with the solution of the liquid phase are:

Calcite: $CaCO_3(s) = Ca^{2+} + HCO_3^- - H^+$
Gypsum: $CaSO_4.2H_2O(s) = Ca^{2+} + SO_4^{2-} + 2H_2O$
Anhydrite: $CaSO_4(s) = Ca^{2+} + SO_4^{2-}$
Chalcedony: $SiO_2(s) = SiO_2(aq)$
Carbonic gas at $10^{-3.5}$atm:
$CO_2(g) = HCO_3^- - H_2O + 2H^+$

As a result of these assumptions the number of unknowns per node was 13: 1 temperature, 1 liquid pressure, 1 displacement and 10 primary chemical species (excluding water).

3.3 Results of the THMC analysis

Only a small selection of results will be shown here. They have been chosen to demonstrate the capabilities of the formulation and to point out some of the significant THMC features of the problem. THM results are given for three points of the buffer: one near the heater, one in the central part of the barrier and one closer to the rock (external buffer).

Figure 3 shows the variation of temperature at those points. Close to the heater the temperature is maintained at 100° C. At the other two points there is, after the initial transient period, a gentle temperature rise over time. Figure 4 shows the evolution of degree of saturation for the first five years. It can be observed that there is a strong drying close to the heater followed by a milder hydration afterwards. The part of the barrier close to the rock becomes quickly saturated, but the central part is still unsaturated after five years. In fact, the long term calculations (Figure 5) indicate that more than 35 years would be required to achieve complete barrier saturation. More information on the THM results is given in Gens et al. (2002).

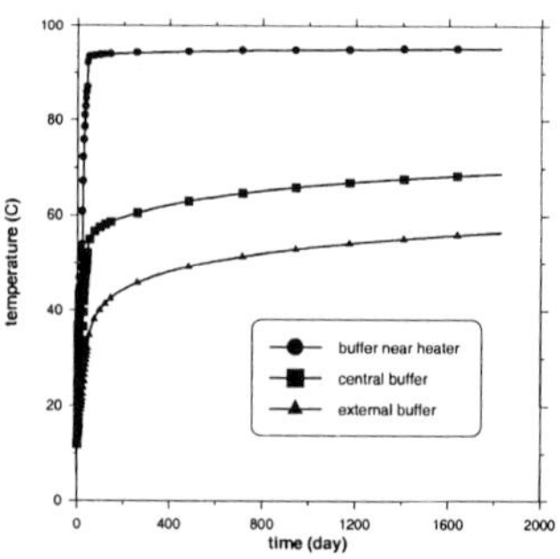

Figure 3. Variation of temperature with time (5 years) at three points of the bentonite barrier

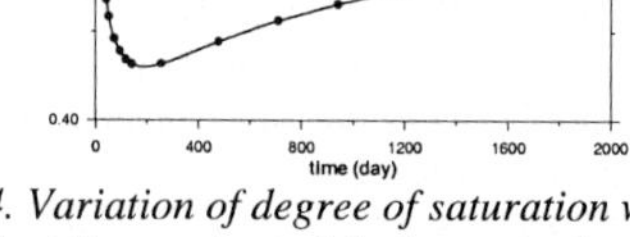

Figure 4. Variation of degree of saturation with time (5 years) at three points of the bentonite barrier

Figure 6 presents the ionic strength distribution in the buffer and in the rock at various times up to 100 years. Naturally the ionic strength in the granitic water is much lower than that of the bentonite water. It can be noted that the diffusion process is slow, after 100 years the solutes have only penetrated about 8 m in the rock. The ionic strength near the heater increases because of evaporation but the value reached is not very high.

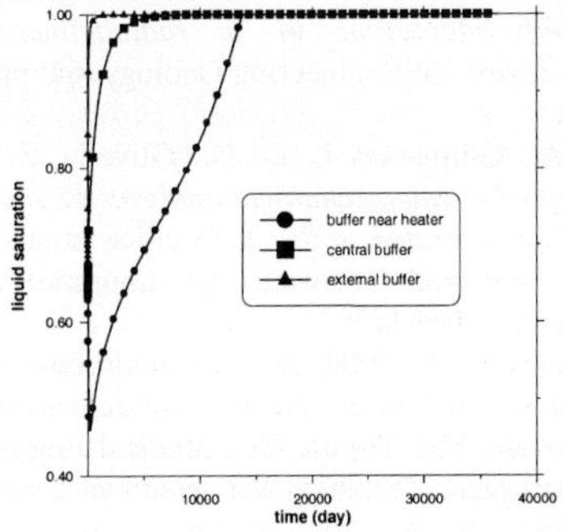

Figure 5. Variation of degree of saturation with time (100 years) at three points of the bentonite barrier

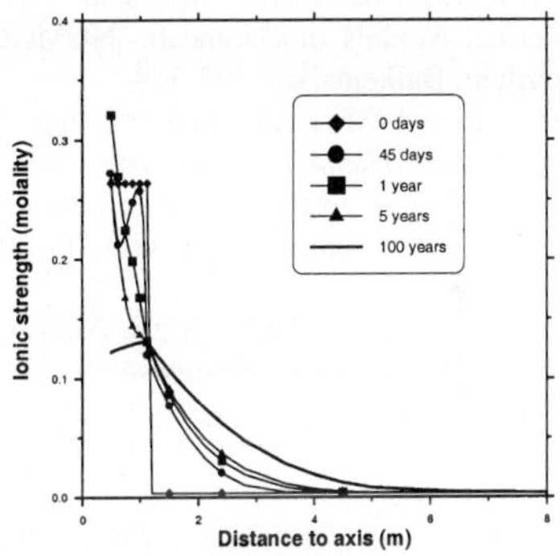

Figure 6. Distribution of ionic strength in the bentonite barrier and rock at various times

The computed concentrations for some of the individual cations can be examined in Figures 7 and 8. The tendency is towards a dilution of the aqueous species in the barrier, especially, at shorter times, near the rock (Figure 7). Concentrations in the rock increase progressively due to diffusion. The exchangeable cations in the buffer clay do not vary much (Figure 8). There are some initial changes due to the impact of heating but the final cation content in the solid phase after 100 years is rather similar to the initial one. It should be stressed, however, that there are some uncertainty over the details of cation exchange processes are elevated temperature.

To show the type of results concerning the behaviour of precipitated minerals, Figure 9 is presented showing the distribution of chalcedony concentration at various times. It can be noted that the mineral is dissolved at the two ends of the barrier. Near the rock the mineral dissolves because of the entry of very dilute granitic water. In contrast, near the heater there is initially some precipitation due to water evaporation induced by heating. However, in the long term, the mineral disappears from the inner zone of the barrier due to the higher solubility of the chalcedony leading to an increase in the concentration of $SiO_2(aq)$. This originates a molecular diffusion transport that leads to the development of a dissolution front (clearly visible in the 100 years results) that travels towards the cooler outer part of the barrier.

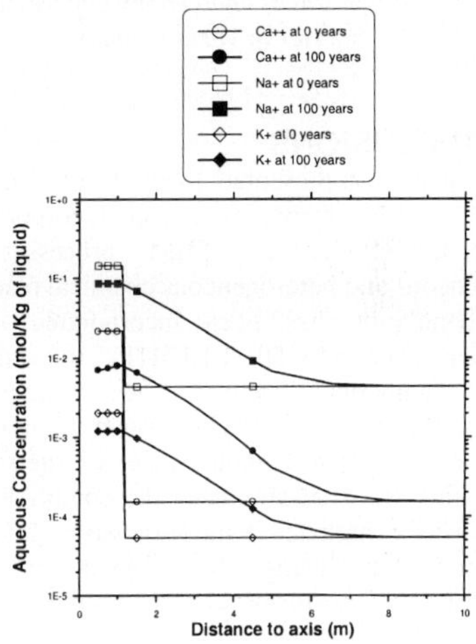

Figure 7. Distribution of the aqueous concentration of cations Ca2+, Na+ and K+ in the bentonite barrier and rock at various times

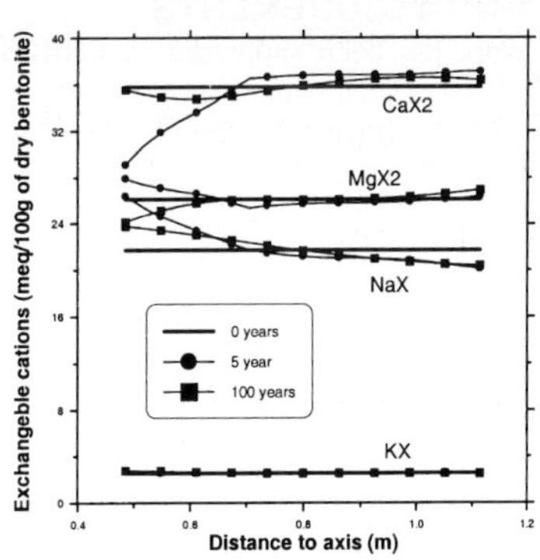

Figure 8. Distribution of exchangeable cations concentration in the bentonite barrier at various times

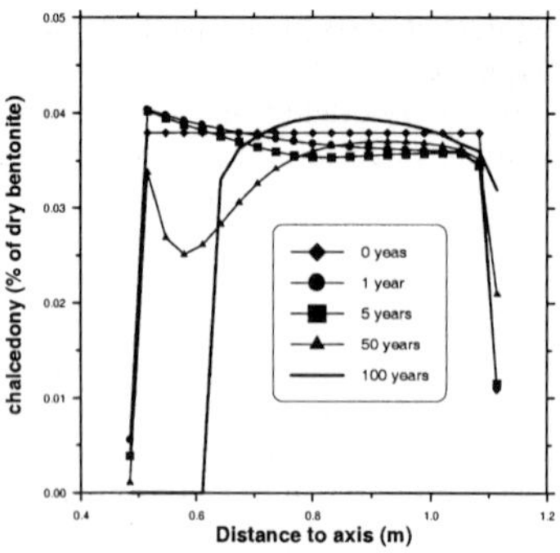

Figure 9. Distribution of chalcedony concentration in the bentonite barrier at various times

4. CONCLUSIONS

The paper has presented a very general and fully coupled THMC formulation that incorporates a significant number of THM processes and homogeneous and heterogeneous chemical reactions. The formulation has been incorporated into a computer code (CODE_BRIGHT) in order to perform numerical analysis. An example of application has been briefly described involving the simulation of a quite complex case: a large scale in situ heating test that simulates the conditions of a repository for high level nuclear waste. It can be concluded that, using the computational tool presented, the performance of coupled THMC analysis or real engineering problems is already a feasible proposition.

5. ACKNOWLEDGEMENTS

This work has been supported by ENRESA and the European Commision. The authors are also grateful for the financial support given by CNPq (Conselho Nacional de Desenvolvimento Cientíco e Tecnológico) and the assistance of the Ministerio de Ciencia y Tecnología of Spain through research grant BTE2001-2227.

6. REFERENCES

FEBEX, 2000. *Full-scale engineered barriers experiment for a deep geological repository for high level radioactive waste in crystalline host rock.* Final Report. In Technical Publication 1/2000. Madrid: ENRESA.

Gawin, D., Baggio, P. & Schhrefler, B.A. 1995. *Coupled heat, water and gas flow in deformable porous media.* In Int. J. Num. Meth. Fluids, 20: pp. 967-987.

Gens, A., Guimarães L. do N., García-Molina, A., Alonso, E.E. 2002. *Factors controlling rock-clay buffer interaction in a radioactive waste repository.* In Engineering Geology. 64: pp. 297-308.

Gens, A., Guimarães L. do N., Olivella, S. 2002. *Coupled chemomechanical analysis for saturated and unsaturated soils.* In Vulliet et al. (ed.), Environmental Geomechanics, Lausanne: EPFL Press: pp. 109-123.

Guimarães L. N. 2002. *Análisis multi-componente no isotermo en medio poroso deformable no saturado.* Phd Thesis, Geotechnical Engineering Department, Technical University of Catalunya, Spain.

Guimarães L. do N., Gens, A., Olivella, S. 1999. *THM and reactive transport coupling in unsaturated porous media.* In 7th Int. Symp. on Numerical Models in Geomech., NUMOG VII. Rotterdam: Balkema, pp. 303-308.

Lichtner, P. 1985. *Continuum model for simultaneous chemical reactions and mass transport in hydrothermal systems.* In Geochimica et Cosmochimica Acta, 49: pp 779-800.

Olivella, S., Carrera, J., Gens, A., & Alonso, E. E. 1994. *Nonisothermal multiphase flow of brine and gas through saline media.* In Transport in Porous Media, 15: pp. 271-293.

Olivella, S., Gens, A., Carrera, J., Alonso, E. E. 1995. *Numerical formulation for a simulator (CODE_BRIGHT) for the coupled analysis of saline media.* In Engineering Computations, n. 13, pp. 87-112.

Thomas, H.R. & He, Y. 1995. *An analysis of coupled heat, moisture and air transfer in a deformable unsaturated soil.* In Géotechnique, 45: pp 667-689.

Thomas, H.R., Cleall, P.J. & Hashm, A.A. 2001. *Thermal/hydraulic/chemical/mechanical (THCM) behaviour of partly saturated soil.* In Desai et al. (ed.) Computer Methods and Advances on Geomechanics, Rotterdam: Balkema, 1: pp. 743-748.

A NEW MECHANISTIC APPROACH TO SIMULATING SWELLING PROCESSES IN BENTONITE MATERIALS

Mingliang Xie, Wenqing Wang, Joëlle De Jonge and Olaf Kolditz
Center of Applied Geosciences(ZAG)/University of Tübingen, Germany

Abstract: Swelling and shrinking processes are crucial characteristics of bentonite. Laboratory and in-situ tests have shown that swelling pressure depends not only on the variation of water content and dry density, but also on the chemical composition of fluid, and on the microstructure of the expansive minerals, especially for the montmorillonite. Swelling pressure decreases also with the increase of solute (ionic) concentration. For the purpose of numerical modelling of such phenomena, a new model, the Chemical-Swelling-Model (CSM) is presented, which is based on the unique microstructure of bentonite and on the Gouy-Chapman DDL theory. According to this model the swelling effect is a potential of interlayer porosity change, which reveals itself to be a volume change in case of free expansion, or turns out to be a swelling pressure in case of constrained swelling. After the implementation of this model in the object-oriented simulator RockFlow, non-isothermal moisture transport processes and the corresponding swelling effects can be simulated.

1. INTRODUCTION

The basic elements of many waste repository concepts in geological formation are engineered barrier systems (EBS). Since bentonite possesses a remarkable swelling due to water intrusion and therefore restricts the migration of water and contaminants through it, bentonite is widely used as barrier material for waste disposal, especially for radioactive waste repositories. Hence, a fundamental understanding of the occurring processes within the bentonite buffer material and also in the near-field zone is essential for qualitative and quantitative evaluation of the efficiency of the barrier. These processes are highly complex and include fluid flow, heat transport, interaction between water/solution and bentonite, as well as mechanical reaction of the bentonite itself. Additionally, they are fully coupled processes, for instance, thermo-hydro-mechanical coupling phenomena (Gens et al. 1998). Numerical modelling can be an effective way to improve the understanding of those processes and to evaluate the long-term performance of engineered barrier systems.

The origin of the swelling effect of bentonite has been widely studied by many researchers, in the effort to reach a fundamental approach to relate swelling potential to basic particle-water-cation interactions (Norrish 1954, Sridhara & Jayadeva 1982, Low 1987, Quirk 1997). These approaches are based on the net negative total charge on the surface of montmorillonite particles, the Poisson-Boltzmann (PB) Theory and the Gouy-Chapman Diffusion Double-Layer (DDL) Theory. However, they are pure microscopic theories or semi-empirical correlation equations derived from theory and/or laboratory tests. Upscaling methods from microscopic theory to macroscopic effects are very important especially for real-world simulation.

Numerical simulation of the complex processes in bentonite as buffer materials has been received increasing attention for waste repository (Börgesson et al. 1995, Jing et al. 1999, Kolditz 2002). These models are mainly developed for the special case of a rigid soil skeleton.

This paper presents a numerical simulation of the swelling/shrinking processes using the numerical model RF-TH/M - a fully coupled thermo-hydraulic model with mechanical coupling - which has been developed for the purpose of modelling non-isothermal multiphase flow in swelling porous media. In this paper details of the mathematical and numerical multiphase-multicomponental formulation for bentonite, as well as code implementation are described. As an example a test case is investigated.

2. SWELLING MECHANISM

The swelling mechanism of bentonite results from the unique microstructure and its net negative charge (Low 1987).

Bentonite is a mixture of clayey minerals, in which smectite is the main mineral and proved to be the most important one for swelling. The structure of smectite like montmorillonite is composed of two structural units, the silica tetrahedron and the alumina octahedron. These units are interconnected and extend to form layers only about 10 Å thick. Several such layers are stacked one above the other, to form a crystal particle (Hicher et al. 2000).

The swelling effect results from additional embedding of water molecules into these thin layers. With the increase of the water content, the thickness of the water film increases. If the water is not pure, ion concentration results in a decrease of the film, which may be described by the electric double layer theory. As a consequence of this process, porosity and permeability will change. Swelling/shrinking is therefore a strong hydraulic-mechanic-chemical (HMC) coupling phenomenon.

3. CHEMICAL SWELLING MODEL

According to the concept mentioned above, the chemical swelling model (CSM) was set up based on a microstructural model and the electric double layer theory. The main equations of the swelling model were derived from the microstructural model with intrinsic parameters such as specific surface S_0, density ρ_0 and number of layers in one particle m. The Gouy-Chapman diffusion-double-layer theory (Stumm & Morgan 1996) was adopted to describe the thickness of the surrounding water film. The equations include those for interlayer porosity change (volume change) in case of free swelling and those for swelling pressure in case of constrained swelling.

3.1 Interlayer porosity

In the CSM model the total porosity n is divided into porosity between particles n_{IP} and that within the particles or interlayer porosity n_{IL}.

$$n = n_{IP} + n_{IL} \tag{1}$$

With this definition n_{IP} represents the coarse pore and can be assumed to correspond to the original void ratio of a dry sample. During the saturation process the solution will be absorbed and transported through the pores. In the meanwhile it can also enter the interlayer space of the particles. Thus, the particles are expanding with fluid intrusion. Consequently, the sample as a whole tends to swell. If the sample is not confined, the particles can expand freely. Diffusion double layers are easy to build up around the particles and within the interlayer space. In this case the maximal interlayer void ratio can be derived according to the CSM model as (Xie et al. 2003):

$$n_{IL} = mS_0\rho_0\delta \tag{2}$$

where m is the sheet numbers within one expansive mineral particle, S_0 is the specific surface of the bentonite in the state of full expansion, ρ_0 is the dry density of the sample, and δ is the thickness of the double layer, which depends on the fluid properties. It is inversely proportional to the ionic strength I of the fluid. Based on the Gouy-Chapman diffusion double layer model (Stumm & Morgan 1996), δ can be calculated as:

$$\delta = \left(\frac{\varepsilon\varepsilon_0 RT}{2F^2 I \times 10^3}\right)^{\frac{1}{2}} \tag{3}$$

where ε is the relative dielectric constant of water, ε_0 the permittivity of free space, R the molar gas constant, T the absolute temperature, F the Faraday constant, and I the ionic strength.

Substituting equation (3) into equation (2) yields the maximal interlayer porosity n_{ILmax}:

$$n_{ILmax} = mS_0\rho_0\left(\frac{\varepsilon\varepsilon_0 RT}{2F^2 I \times 10^3}\right)^{\frac{1}{2}} \tag{4}$$

As the swelling effects occur chiefly in the expansive minerals like montmorillonite, therefore the apparent interlayer porosity should be modified with the expansive mineral fraction β. For an absolutely dry sample, the interlayer porosity should be zero. The maximal interlayer porosity can only be reached in the free swelling case. Therefore, it is reasonable to assume that n_{IL} is a function of water saturation S^l. The simplest correlation is linear function $f(S^l) = S^l$.

$$n_{IL} = \beta \cdot f(S^l) \cdot n_{ILmax} \tag{5}$$

This model was verified by desiccation tests performed by the Japan Nuclear Cycle Development Institute (JNC 1999) with compacted samples of bentonite/sand mixture. A reversed correlation was derived to be $f(x) = (S^l)^{1.50}$ (s. Example 1 in Xie et al. 2003).

3.2 Swelling pressure

If free swelling is restricted, swelling pressure will occur (Figure 1). In practice the swelling pressure can be measured under a rigid boundary condition. That means that the sample volume remains constant during the test procedure (strains constant during the test procedure (strains are zero),

$$\Delta\varepsilon = 0 \tag{6}$$

In order to calculate swelling pressure, we propose the following algorithm:

(i) free swelling increment resulting in volume change, according to equation (5), $\Delta\varepsilon = \varepsilon^{fs}$.

(ii) pressing the free swelling increment back, $\Delta\varepsilon = -\varepsilon^{fs}$.

Using this algorithm, mechanical parameters for swelling pressure can be determined as:

$$d\sigma = E \cdot d\varepsilon \tag{7}$$

where σ is swelling stress, E is modulus of swelling processes.

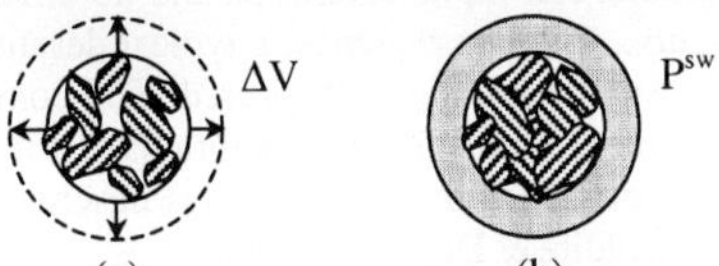

Figure 1. Schematic concept of free swelling (a) and constrained swelling (b)

In practice the samples used are usually highly compacted. Samples for swelling pressure tests are prepared through compaction or consolidation. During swelling tests no volume change can occur. However, the effective void ratio (interparticle porosity) in bentonite decreases with the intrusion of water, which results in the swelling pressure. This "swelling compaction" effect can be treated as a further consolidation process. In this case the history of the preconsolidation should be taken into consideration (Nishimura 2001). Thus the standard compaction curve will be employed as the basic relation for the effective void ratio change to the swelling pressure.

The overconsolidation (OC) line represents the primary phase of the swelling pressure test. The equation for the OC line can be derived as:

$$\Delta e = C_s log(\frac{\sigma_{v0} + \Delta\sigma_{v0}}{\sigma_{v0}}) \tag{8}$$

$$\Delta\sigma_{v0} = \sigma_{v0}(10^{\Delta e/C_s} - 1) \tag{9}$$

where Cs is the slope of the straight line or the compression index.

The change of the effective void ratio due to moisture change can be calculated as the interlayer porosity variation:

$$\Delta e = e - e_0 \tag{10}$$

$$e = \frac{n_{IL}}{1 - n_{IL}} \tag{11}$$

Swelling processes affect porosity and permeability. The effective porosity n will change between the maximal porosity n_{max} (at residual saturation) and minimal porosity n_{min} (at saturation S^l=*1*):

$$n = n_{max} - (n_{max} - n_{min}) \cdot \beta \cdot (S^l)^{3/2} \tag{12}$$

Permeability changes with the void ratio variation and is calculated with the Börgesson model (Börgesson et al. 1995).

4. MODEL EQUATIONS

The system of model equations to determine the chosen field variables: fluid gas pressure p^g, fluid liquid saturation S^l and equilibrium temperature T is based on the balance equations for fluid mass and heat in combination with the flux terms and the equations of state (Kolditz & De Jonge 2003).

4.1 Fluid component equation

The fluid component equation in terms of primary variables was derived from the fluid mass balance equation.

$$\begin{aligned}
& nX_k^g\left(S^g\frac{\partial\rho^g}{\partial p^g} + S^l\frac{\partial\rho^l}{\partial p^l}\right)\frac{\partial p^g}{\partial t} \\
- \;& \nabla\cdot\left(\rho^g X_k^g\frac{k_{\text{rel}}^g\mathbf{k}}{\mu^g}\nabla p^g\right) - \nabla\cdot\left(\rho^l X_k^l\frac{k_{\text{rel}}^l\mathbf{k}}{\mu^l}\nabla p^g\right) \\
+ \;& n(-\rho^g X_k^g + \rho^l X_k^l)\frac{\partial S^l}{\partial t} \\
= \;& Q_k \\
+ \;& (nS^l X_k^l\frac{\partial\rho^l}{\partial p^l})\frac{\partial p_c}{\partial t} - \nabla\cdot\left(\rho^l X_k^l\frac{k_{\text{rel}}^l\mathbf{k}}{\mu^l}\nabla p_c\right) \\
- \;& \nabla\cdot\left(\rho^g X_k^g\frac{k_{\text{rel}}^g\mathbf{k}}{\mu^g}\rho^g\mathbf{g}\right) + \nabla\cdot\left(\rho^l X_k^l\frac{k_{\text{rel}}^l\mathbf{k}}{\mu^l}\rho^l\mathbf{g}\right) \\
+ \;& \nabla\cdot(nS^g\rho^g D_k^g\nabla X_k^g)
\end{aligned} \tag{13}$$

4.2 Thermal energy equation

Based on the heat balance equation for the porous medium consisting of three phases (gas, liquid and solid phases), the following thermal energy equation was given.

$$
\begin{aligned}
& \left((1-n)\rho^s c^s + nS^g \rho^g c^g + nS^l \rho^l c^l\right) \frac{\partial T}{\partial t} \\
+ & \left(\rho^g c^g \frac{k^g_{\mathrm{rel}} \mathbf{k}}{\mu^g} (\nabla p^g - \rho^g \mathbf{g}) \right. \\
& \left. + \rho^l c^l \frac{k^l_{\mathrm{rel}} \mathbf{k}}{\mu^l} (\nabla p^l - \rho^l \mathbf{g}) \right) \nabla T \\
- & \nabla \cdot \left(\left((1-n)\lambda^s + nS^g \lambda^g + nS^l \lambda^l\right) \nabla T\right) \\
- & \nabla \cdot (nS^g \rho^g D^g_a h^g \nabla X^g_a) \\
- & \nabla \cdot (nS^g \rho^g D^g_w h^g \nabla X^g_w) \\
= & \rho Q_T
\end{aligned}
\tag{14}
$$

4.3 Numerical scheme

The method of weighted residuals is adopted to derive the weak forms of the differential field equations for gas fluid pressure $\hat{\mathbf{p}}^g$, liquid saturation $\hat{\mathbf{S}}^l$ and temperature $\hat{\mathbf{T}}$.

Fluid component equation (primary variables $\hat{\mathbf{p}}^g$ and $\hat{\mathbf{S}}^l$), where e.g

$$
(\hat{\mathbf{p}}^g)^T = [\hat{p}^g_1 \, \hat{p}^g_2 \, \dots \, \hat{p}^g_i]
\tag{15}
$$

$$
\begin{aligned}
& \left[\int_\Omega \mathbf{N}^T \left(nS^g X^g_k \frac{\partial \rho^g}{\partial p^g} + nS^l X^l_k \frac{\partial \rho^l}{\partial p^l}\right) \mathbf{N}\, d\Omega\right] \left[\frac{d\hat{\mathbf{p}}^g}{dt}\right] \\
+ & \left[\int_\Omega \nabla \mathbf{N}^T \left(\rho^g X^g_k \frac{k^g_{\mathrm{rel}} \mathbf{k}}{\mu^g} + \rho^l X^l_k \frac{k^l_{\mathrm{rel}} \mathbf{k}}{\mu^l}\right) \nabla \mathbf{N}\, d\Omega\right] [\hat{\mathbf{p}}^g] \\
- & \left[\int_\Omega \mathbf{N}^T \left(n\rho^g X^g_k - n\rho^l X^l_k\right) \mathbf{N}\, d\Omega\right] \left[\frac{d\hat{\mathbf{S}}^l}{dt}\right] \\
= & \int_\Omega \mathbf{N}^T Q_k \, d\Omega \\
+ & \left[\int_\Omega \mathbf{N}^T \left(nS^l X^l_k \frac{\partial \rho^l}{\partial p^l}\right) \mathbf{N}\, d\Omega\right] \left[\frac{d\hat{\mathbf{p}}\mathrm{c}}{dt}\right] \\
+ & \left[\int_\Omega \nabla \mathbf{N}^T \left(\rho^l X^l_k \frac{k^l_{\mathrm{rel}} \mathbf{k}}{\mu^l}\right) \nabla \mathbf{N}\, d\Omega\right] [\hat{\mathbf{p}}\mathrm{c}] \\
- & \left[\int_\Omega \nabla \mathbf{N}^T (nS^g \rho^g X^g_k D^g_k) \nabla \mathbf{N}\, d\Omega\right] \left[\hat{\mathbf{X}}^g_k\right] \\
- & \int_\Omega \nabla \mathbf{N}^T \left(\rho^g X^g_k \frac{k^g_{\mathrm{rel}} \mathbf{k}}{\mu^g} \rho^g \mathbf{g} + \rho^l X^l_k \frac{k^l_{\mathrm{rel}} \mathbf{k}}{\mu^l} \rho^l \mathbf{g}\right) d\Omega
\end{aligned}
\tag{16}
$$

Heat energy equation (primary variable $\hat{\mathbf{T}}$)

$$
\begin{aligned}
& \left[\int_\Omega \mathbf{N}^T \left((1-n)\rho^s c^s + n\rho^g c^g + n\rho^l c^l\right) \mathbf{N}\, d\Omega\right] \left[\frac{d\hat{\mathbf{T}}}{dt}\right] \\
+ & \left[\int_\Omega \mathbf{N}^T \left(n\rho^g \mathbf{J}^{gs} + n\rho^l \mathbf{J}^{ls}\right) \nabla \mathbf{N}^T \, d\Omega\right] \left[\hat{\mathbf{T}}\right] \\
+ & \left[\int_\Omega \nabla \mathbf{N}^T \left((1-n)\lambda^s + n\lambda^g + n\lambda^l\right) \nabla \mathbf{N}\, d\Omega\right] \left[\hat{\mathbf{T}}\right] \\
= & \int_\Omega \mathbf{N}^T Q_t \, d\Omega
\end{aligned}
\tag{17}
$$

The resulting algebraic equations, as well as details of the numerical solution algorithms are given in Kolditz & De Jonge (2003).

5. EXAMPLE

5.1 Example definition

This example is to test the swelling effects under capillary pressures up to 10^{10}Pa occurring in extremely low-permeable bentonite materials. For this purpose, a simple 1-D case is set up. A one meter long bentonite column is heated on the left hand side. Element discretization length is 0.01m. The initial conditions of the system are: atmospheric gas pressure, full liquid saturation and a temperature of 12°C. The heater has a constant temperature of 100°C. Flow boundary conditions on the left side are gas pressure of 10^6Pa and 15% liquid saturation. On the right side we have atmospheric pressure, full liquid saturation and no diffusive heat flux. As a consequence, a typical desaturation process of bentonite is triggered. The complete set of initial and boundary conditions and the material properties for this example was described in detail by Kolditz & De Jonge (2003).

5.2 Results

In order to illustrate the swelling effect on the hydraulic transport processes, two cases are studied: with and without swelling properties. The samples are compacted to a dry density of 1.83kg/m^3 with an initial total porosity 0.41. With the intrusion of fluid, the samples are supposed to be saturated after certain time. During this process the interlayer porosity will increase and fluid permeability will decrease during swelling. For the non-swelling case the interlayer porosity is supposed to be constant (zero). After that, the samples undergo a non-isothermal desaturation process mentioned above. The series of Figures 2, 3 show calculated profiles along the bentonite column at several time stages, t = 10^3 to 10^5 seconds for the swelling effect on the drying processes with the monolithic scheme of the multiphase-multicomponental (MPC) model.

The results show clearly the swelling effect on the drying processes. The gas pressure front at the same simulation time (t = 1×10^{5}s) reaches only about 0.08m into the probe in case of swelling (Figure 2(a)) but 0.11m in case of non-swelling (Figure 3(a)). The water is thus more difficultly to

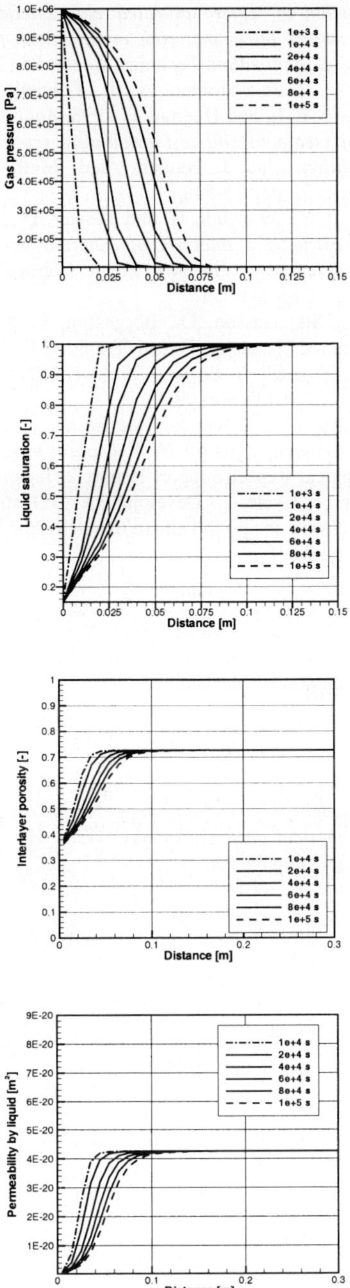

(a)

(b)

(c)

(d)

Figure 2. Computed profiles with swelling
(a) Gas pressure, (b) Liquid saturation
(c) Interlayer porosity, (d) Permeability

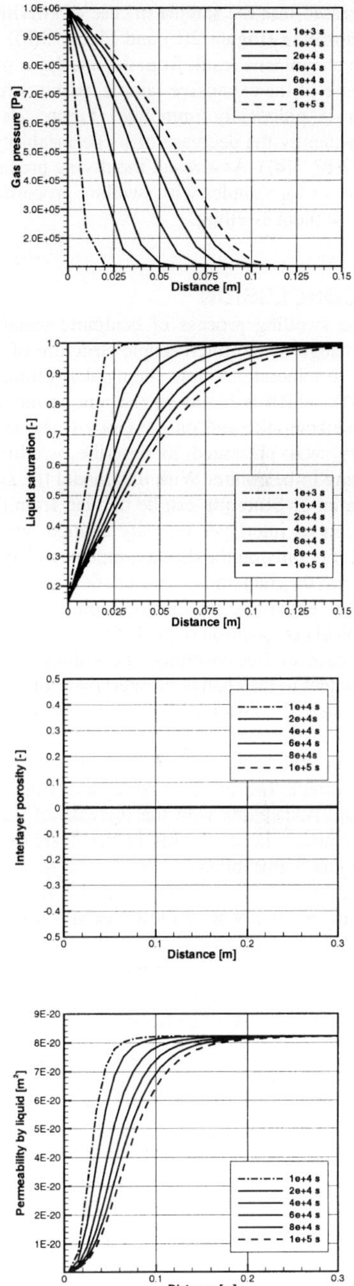

Figure 3. Computed profiles without swelling
(a) Gas pressure,(b) Liquid saturation
(c) Interlayer porosity, (d) Permeability

be pressed into the sample in case of swelling than non-swelling (Figure 2(b) and Figure 3(b)). All of these phenomena result from the interlayer porosity increase of the expansive sample, within which the water is practically immobile (Figure 2(c)) and consequently the decrease of its permeability (Figures 2(d), 3(d)). As a result the drying process will be slower for samples with swelling properties than those without swelling.

6. CONCLUSION

The swelling process of bentonite results from the change of the microscopic structure of the expansive minerals. A new chemical swelling model (CSM), which is based on the unique microstructure of bentonite and on the Gouy-Chapman DDL theory, was presented to describe swelling phenomena in bentonite. With this model the swelling potential of bentonite can be derived from the calculation of interlayer porosity change, which depends not only on the density, moisture change, but also on the mineral characteristics (e.g. specific surface, expansive mineral fraction etc.) and fluid chemical composition (eqn. 4, 5).

In case of free swelling, the volume change is equivalent to the change of interlayer porosity.

In case of constrained swelling, the potential of interlayer porosity change turns out to be swelling pressure owing to the further "swelling compaction" effect. The effective void ratio (interparticle porosity) decreases with the increase of the moisture content, hence results in the decrease of the bentonite permeability.

This model is implemented in the novel object-oriented simulator RockFlow-TH/M – a fully coupled thermo-hydraulic model with mechanical coupling, which can be used to simulate non-isothermal moisture transport processes and the corresponding swelling effects.

7. ACKNOWLEDGEMENTS

This study was supported by the BMBF (Federal Ministry of Education and Research), Germany. The authors would like also to acknowledge Tom Schanz at the University of Weimar for providing suggestions for the research.

8. REFERENCES

Börgesson, L., Johannesson, L-E., Sanden, T. and Herneling, H.. 1995. *Modelling of the physical behavior of water saturated clay barriers - laboratory tests, material models and finite element application.* Technical Report 95 (20).

Gens, A., Garcia-Molina, A. J., Oliviella, S., Alonso, E. E. and Huertas, F. 1998. *Analysis of a full scale in situ test simulating repository conditions.* Int. J. Numer. Anal. Meth. Geomech., 22:pp.515-548.

Hicher, P. Y., Wahyudi, H. and Tessier, D. 2000. *Microstructural analysis of inherent and induced anisotropy in clay.* Mech. Cohes.-Frict. Mater., 5:pp.341-371.

Jing, L., Stephansson, O., Börgesson, L., Chijimatzu, M. Kautsky, F. and Tsang, C-F. 1999. *DECOVALEX II Project, Technical Report – Task 2C,* SKI Report 99:23.

JNC. 1999. *H12:project to establish the scientific and technical basis for HLW disposal in Japan.* Technical Report Support Report 2, Japan Nuclear Cycle Development Institute. http://www.jnc.go.jp/kaihatu/tisou/zh12/h12/s02/pdf/b-09.pdf.

Kolditz, O. 2002. *Computational Methods in Environmental Fluid Mechanics.* Springer.

Kolditz, O. & J. De Jonge. 2003. *Non-isothermal two-phase flow in porous media.* (Submitted to Computational Mechanics).

Low, P.F. 1987. *Structural component of the swelling pressure of clays.* Langmuir, 3:pp.18-25.

Murad, M.A. & Cushman, J.H. 1997. *A multiscale theory of swelling porous media: II. dual porosity models for consolidation of clays incorporating physical chemical effects.* Trans. Por. Media, 28(1):pp.69-108.

Nishimura, T. 2001. *Swelling pressure of a compacted bentonite subjected to high suction,* pages 109-114. Clay Science for Engineering, Proceedings of the international symposium on suction, swelling, permeability and structure of clays - Is-Shizuoka. Balkema, Rotterdam.

Norrish, K. 1954. *Manner of swelling of montmorillonite.* Nature, 4397:pp.256-257.

Quirk, J.P. 1997. *Application of double-layer theories to the extensive crystalline swelling of Li-montmorillonite.* Langmuir, 13:pp.6241-6248.

Stumm, W. & Morgan, J.J. 1996. *Aquatic Chemistry, 3rd edition.* John Wiley and Sons, New York.

Xie, M., Kolditz, O., Tripathy, S. and Schanz, T. 2003. *Numerical modelling of swelling processes in compacted bentonite.* ROCKFLOW-Preprint 2003-4, Center of Applied Geosciences, University of Tübingen, Germany.

APPLICATION OF A THM-COUPLED CODE TO TRANSPORT PROCESSES IN A SWELLING BENTONITE BUFFER

Thomas Nowak, Hua Shao, and Manfred Wallner,
Federal Institute for Geosciences and Natural Resources (BGR), Hanover (Germany)

Abstract: The ability to model coupled thermal, hydraulic and mechanical processes is of importance for investigations on different concepts for permanent repositories. The radioactive waste heats the surrounding system of barriers. This process shows strong interactions with hydraulic and mechanical processes. Therefore BGR develops together with University of Hanover and University of Tübingen a code for modelling these processes in a coupled way.

Up to now two-phase two-component flow under non-isothermal conditions and coupled THM (one-phase flow) have been implemented in this code and validated against different experimental results. For the modelling of water penetration into unsaturated bentonite or clay, a swelling model is available in the code. If the test sample is confined within a constant volume, then a swelling pressure will build up which causes changes to the pore structure and reduces the porosity. A small change in porosity can, however, create a considerable reduction in permeability.

1. INTRODUCTION

Engineered barriers constitute a basic element in the conceptual design of repositories for radioactive waste in deep geological media. For the safety performance of a repository, it is very important to understand, on the one hand, individual processes in the barrier and the host medium in the near-field zone and the coupled mechanism, on the other hand, to parameterise all physical variables for the long-term modelling.

Clay-like materials swell and shrink with wetting and drying processes. Swelling and shrinking are hydro-mechanical coupling phenomena. They result in a dramatic alteration of material properties. The swelling effect results from additional embedding of water molecules into the solid matrix. The water becomes immobile. As a consequence, effective porosity for fluid flow and intrinsic permeability decrease. With drying of the material, the porosity will be recovered to a certain degree. This moisture transport as well as phase transitions (evaporation and condensation) are controlled by heat exchange processes. Swelling behaviour additionally depends on the mineral composition of the material. If the volume expansion of the material is restrained, swelling pressure is build up. This pressure can be measured in the laboratory by use of oedometers or triaxial cells. The relationship between swelling pressure and void ratio can then be used as a measure for the expansion or compression properties of the material.

This paper will focus on the hydraulic properties of the swelling materials, especially on the bentonite that is used in the Full-scale Engineered Barrieres Experiment in Crystalline Rock (FEBEX). The influence of swelling will be shown in a simulated laboratory experiment and in the application on the FEBEX *in situ* experiment.

2. PROCESS ANALYSIS

In principle there are a lot of variables that affect the total barrier system (Figure 1). All parameter variables are normally dependent on temperature and pressure. A fully coupled analysis requires equation coupling, constitutive coupling, and parameter coupling, Kolditz & Kohlmeier (2001).

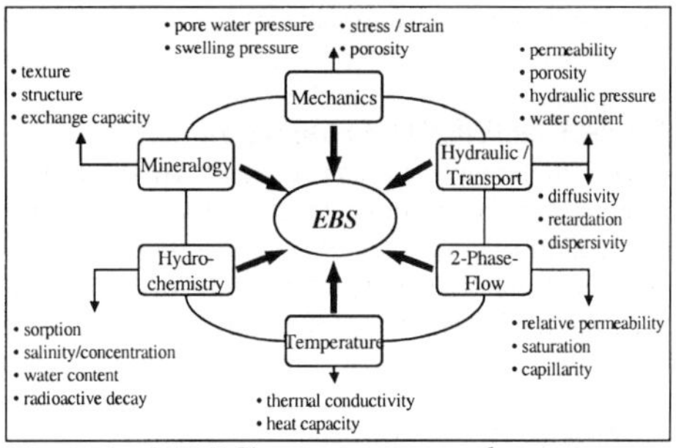

Figure 1. THM(C) processes and parameters associated with engineered barrier system.

The barrier response and the rock mass response to disposal of radioactive waste is a coupled phenomenon involving thermal, hydraulic, mechanical and chemical processes. Coupled processes imply that one process affects another (Figure 2). For simplifying the problem, effects between two processes can be analysed as follows.

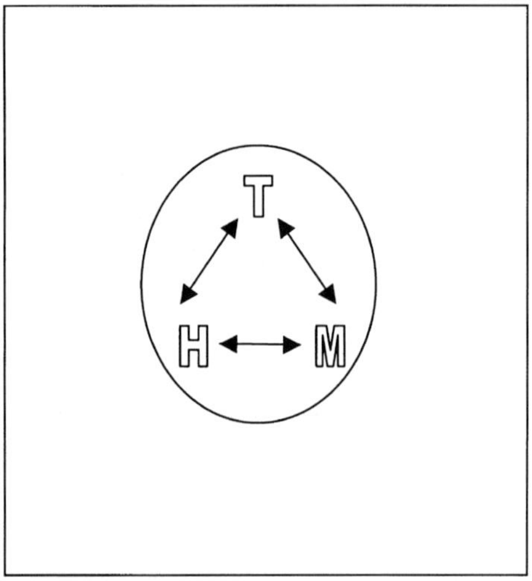

Figure 2. THM interactions.

The temperature rising in the vicinity of the heaters causes an evaporation of liquid. Due to thermal and pressure gradients it comes to a vapour diffusion, which can lead to a strong modification of the humidity distribution in the barrier system. Modifications in the humidity distribution on the other hand directly affect the hydraulic characteristics, since the permeability depends strongly and nonlinearly on the saturation. By cooling it comes to a condensation of water vapour and thus to the increase of water saturation. For the permeation of water from the adjacent rock into the barrier system the characteristics of the multiphase flow are crucial in the contact area of both materials - the contrast in permeability, relative permeability and capillary pressure. For the water flow into the barrier, the subsequent delivery of water is further important due to regional hydraulic gradients.

Further hydro-thermal effects concern the flow characteristics of the fluids (dependency of the viscosity on the temperature) and heat transport characteristics (dependency of heat conductivity and heat capacity on the water saturations).

The admission of aqueous solutions leads to swelling of clay-like materials and thus to a strong modification of the hydraulic and mechanical characteristics, like effective porosity, permeability, rigidity and shearing strength of the material. A reduction in humidity, e.g. by drying process or evaporation, and associated shrinkage of the clay can involve a material damage with appropriate porosity modification.

The consideration of thermal effects leads to the development of thermoplastic models. Furthermore also transient effects will play a role both in the barrier material and in the host rock under elevated temperature conditions. This requires an extension of the material models regarding visco-elasticity or visco-plasticity.

3. FINITE ELEMENT CODE

BGR used the finite element code RockFlow/ RockMech for the analysis of the experiment. Originally this code was used for analysis of uncoupled thermal (T), hydraulic (H), mechanical (M) and chemical (C) processes. A development strategy for the extension to a fully coupled tool has been elaborated, but at present the realisation of coupling of processes and the implementation of mechanical behaviour more complex than linear elasticity is limited, Kolditz & Kohlmeier (2001).

For the calculations presented in this paper we consider the flow of two phases (water and gas) and heat conduction and advection.

4. HYDRAULIC PROPERTIES OF FEBEX-BENTONITE

In the FEBEX project blocks made of Ca-bentonite are tested for their suitability in radioactive waste isolation. This material has been tested in various laboratory experiment to identify the material parameters. The modelling of gas and water flow in engineered barriers demands beneath the intrinsic permeability and porosity constitutive equations for capillary pressure and relative permeabilities.

For the calculations that are presented in this paper the identified material parameters from these laboratory experiments have been used. Capillary pressures can be very high at low water saturations (Figure 3).

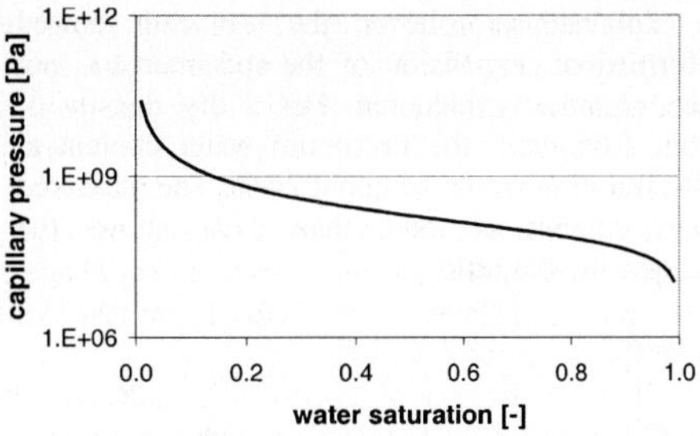

Figure 3. Capillary pressure vs. water saturation for FEBEX bentonite, DECOVALEX III (2001)

An additional phenomenon of this bentonite are the differences in the measured permeability whether gas or water is used as test fluid (Figure 4). As gas does not have an influence on the pore structure of bentonite the gas permeabilities are assumed to represent the intrinsic permeability of bentonite. Accessible void ratio means in the case of gas permeability the void ratio multiplied with the gas saturation. The curve for water permeability has been developed from tests at full water saturation.

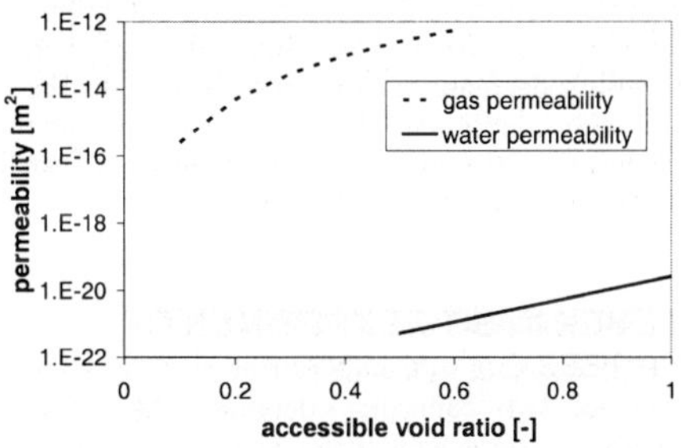

Figure 4. Permeability for water and gas vs. accessible void ration for FEBEX bentonite, DECOVALEX III (2001)

The infiltration of water into clay-like materials causes changes in the pore structure of the material. Water molecules place in the solid matrix and become immobile. In consequence the effective porosity and the intrinsic permeability decrease. If the volume expansion of the material is restrained, swelling pressure is observed which increases linearly with the degree of water saturation, Studer et al. (1984), and Börgesson (1984). The correlation between swelling pressure p_s and void ratio e can be expressed best with the following empirical relationship, Börgesson et al. (1995):

$$e = e_0 \left(\frac{p_s}{p_{s0}} \right)^{\beta} \tag{1}$$

The index 0 denotes a reference value for the respective parameter. The exponent β can be calculated as follows:

$$\beta = \frac{\Delta(\ln e)}{\Delta \ln(p_s)} \tag{2}$$

A small change in void ration can create a considerable reduction of the permeability. The correlation between permeability k and void ratio can be expressed with the following empirical relationship, Johannesson et al. (1995):

$$k = k_0 \left(\frac{e}{e_0} \right)^{\eta} \tag{3}$$

The index 0 denotes a reference value for the respective parameter. The exponent η can be calculated as follows:

$$\eta = \frac{\Delta(\ln k)}{\Delta \ln(e)} \tag{4}$$

5. LABORATORY EXPERIMENT

In laboratory experiments specimens from the FEBEX bentonite with a height of 0.13 m and a diameter of 0.15 m have been tested. In a series of tests the specimens have been subjected to different combinations of heating and water injection. In one of these tests the specimen had an initial dry density of about 1.65 g/cm^3 and an initial water content of 13.1 %. At the bottom end of the specimen water has been injected with a pressure of 1 MPa and after 148 d the distribution of water content along the height of the specimen has been measured, DECOVALEX III (2001).

This test has been modeled with RockFlow/ RockMech in 1D to check the chosen hydraulic parameters for the modeling of the FEBEX *in situ* experiment. The capillary pressure curve has been set as shown in Figure 3. Relative permeability for gas has been developed from the experimental results shown in Figure 4. Relative permeability for water has been assembled from the dependency on saturation and swelling pressure and an additional factor c.

$$k_{relw} = k^{S}_{relw} * k^{\varepsilon}_{rel} * c \tag{4}$$

The additional factor c meets the observation that measurements of permeability with gas result in values that are orders of magnitude higher than those measured with water (compare Figure 4). This formulation yields in two different permeabilities for water and gas. Table 1 contains the parameters and Figure 5 shows the relative permeabilities.

Table 1. Hydraulic parameter of FEBEX bentonite in the model.

Parameter	Value
intrinsic permability	$8\cdot 10^{-13}\ \mathrm{m}^2$
porosity	0.39
rel. permeability for gas	(see Figure 5)
rel. permeability for water	(see Figure 5) $k^{S}_{relw} = S_w{}^3$ $k^{\varepsilon}_{relw} = f(\beta,\eta)$, $\beta = -0.1$ $\eta = 6.2$ $kk_{relw}(S_w = 1) = 2.5\cdot 10^{-21}\mathrm{m}^2$ $c = 3\cdot 10^{-9}$
capillary pressure	see Figure 3

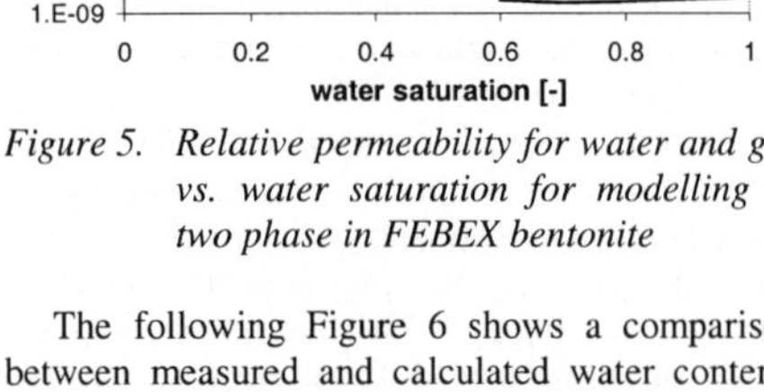

Figure 5. Relative permeability for water and gas vs. water saturation for modelling of two phase in FEBEX bentonite

The following Figure 6 shows a comparison between measured and calculated water contents (diamond shaped symbols respective the solid black line). The deviation between the measured and the calculated water content at the injection end of the specimen (distance from top 0.13 m) can be explained as follows: the test cell showed deformations, expansion of the specimen has not been completely hindered. For a dry density of about 1.65 g/cm^3 the maximum water content at full saturation would be about 24 %. The measured water content of more than 27 % shows the increase in void ratio.

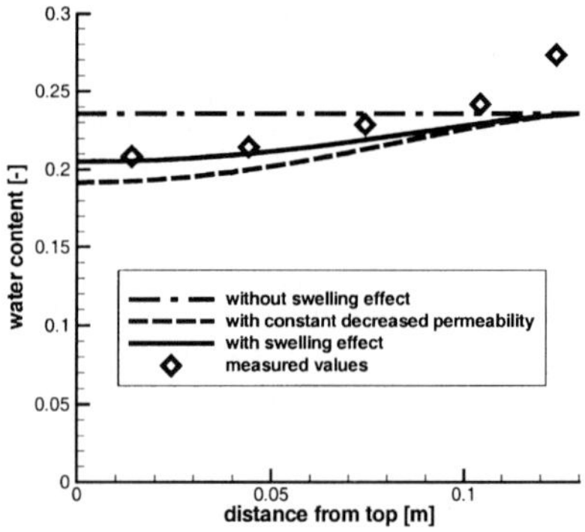

Figure 6. Comparison between calculated and measured water contents in a laboratory experiment

Two additional calculation have been performed to show the influence of k^{ε}_{rel}. The dash-dotted line shows the result without swelling effect, i.e. no permeability decrease. The dashed line shows the result with decreased permeability, i.e. low permeability as if complete swelling ($S_w = 1$) has already taken place in the complete specimen.

6. FEBEX IN SITU EXPERIMENT

The FEBEX *in situ* experiment simulates the disposal of two canisters containing high-level radioactive waste at real scale. The heaters which maintain a temperature of about 373 K are surrounded by bentonite blocks in a horizontal gallery, according to the Spanish reference concept of Deep Geological Disposal in Granite. The averaged dry density of the buffer is 1.6 g/cm^3 and the averaged initial water content is 14.4%, DECOVALEX III (2001). Figure 7 shows the layout of the FEBEX *in situ* experiment.

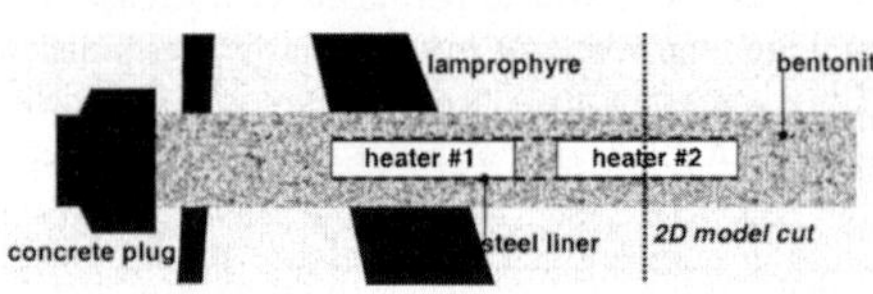

Figure 7. Layout of the FEBEX in situ experiment

Based on the characterization of the bentonite and the granite a 2D model which represents a cut through heater #2 has been applied to model the saturation process of the bentonite. The evaporation of water in the vicinity of the heaters has been modelled with a boundary condition for saturation evaluated from the measured relative humidity evolution. As this approach does not represent the phase transitions this model can just give an impression of the saturation process. Figure 8 shows the calculated distributions of saturation and temperature in the bentonite 1000 d after the heaters have been turned on.

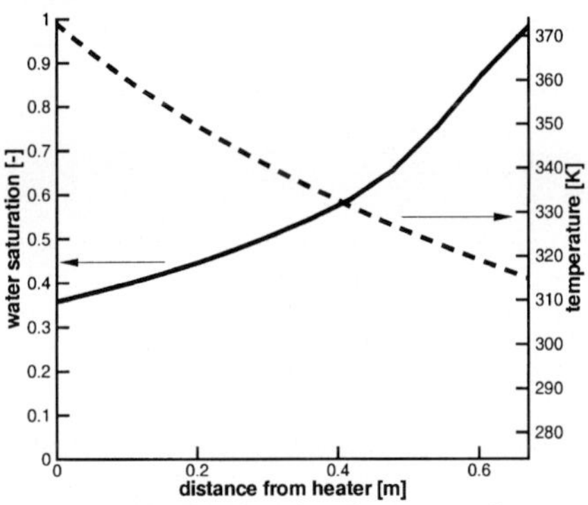

Figure 8. Calculated distribution of water saturation in the buffer of the FEBEX in situ experiment

The buffer in this experiment is made of bentonite blocks. Because there are gaps of up to 30 mm at the upper contact to the host rock the swelling of the blocks is not hindered at these blocks. The differences in back pressure at each block leads to different changes of the pore structure and permeability at each block.

7. CONCLUSIONS

The code RockFlow/RockMech that has been used in this study allows to take into account several couplings, for example the dependency of water permeability on swelling pressure (see chapter 4). The empirical relationship bases on experimental data of swelling material under confined conditions. With an additional factor c this relationship is suitable to reproduce the qualitative distribution of water content in the laboratory experiment (see chapter 5). More close to reality would be the formulation of this factor within the relative permeability depending on the swelling pressure (k_{rel}^{ε}).

The small permeability for water depends on changes in the pore structure caused by swelling. An important factor on the alteration of the pore structure is the back pressure, i.e. whether the bentonite is under confined conditions or not. In the FEBEX *in situ* experiment the back pressure is hardly to know as the barrier is made of bentonite blocks with gaps between the blocks and the host rock. The description of water permeability with the described empirical relationship meets its limitations at this point as it is not suitable to describe volumetric change of the swelling material (change of density) under unconfined conditions. For the latter, a HM-coupled code with a constitutive law for the mechanical behaviour of swelling material is necessary. For this reason the development of the code RockFlow/RockMech to a fully THM-coupled code is ongoing.

8. REFERENCES

Börgesson, L. 1984. *Water flow and swelling pressure in non-saturated bentonite-based clay barriers.* In Clay Barriers for Isolation of Toxic Waste; Proc. Intern. Symp., Stockholm, Sweden, 28-30 May 1984.

Börgesson, L., Johannesson, L.E., Sanden, T. & Hernelind, J. 1995. *Modelling of the physical behaviour of water saturated clay barriers.* SKB Technical Report 95-20, Stockholm, Sweden.

DECOVALEX III 2001. *DECOVALEX III, Task 1, Modelling of FEBEX in-situ test, Part B, Thermo-Hydro-Mechanical Analysis of the Bentonite Behaviour, Rev.* 2, Stockholm, Sweden.

Johannesson, L.E., Börgesson, L. & Sanden, T. 1995. *Compaction of Bentonite Blocks – Development of technique for industrial production of Blocks which are manageable by man.* SKB Technical Report 95-20, Stockholm, Sweden.

Kolditz, O. & Kohlmeier, M. 2001. *A Fully Coupled T-H-M Model for Non-isothermal Flow and Deformation Processes in Porous Media (Preprint).* Institute for Fluid Mechanics, University of Hanover, Germany.

Studer, J., Ammann, W., Meier, P., Müller, C. & Glauser, E. 1984. *Verfüllen und Versiegeln von Stollen, Schächten und Bohrlöchern, Band 2 Anhänge.* Nagra Technical Report NTB 84-33, Baden, Switzerland.

DRYING AND RESATURATION OF THE BENTONITE BARRIER IN A NUCLEAR WASTE REPOSITORY. ANALYSES BASED ON AN ANALYTICAL SOLUTION

Johan Claesson, Chalmers University, Sweden

Abstract: The bentonite layer, which surrounds the canisters in a nuclear waste repository deep down in rock, experiences an initial drying and a resaturation from the outer rock side. These complex processes interact and a key question is the degree of initial drying and the time it takes to resaturate the bentonite. It is shown that the highly non-linear coupled equations may, when formulated in a special way, be linearized with a loss of accuracy of some 10% only. The paper presents an analytical solution for the linearized case. The solution involves two key parameters only, a time scale and thermodiffusive parameter *a*. The largest drying at the canister wall and the resaturation are obtained from a single set of curves with a dimensionless time and *a* as parameter.

1. INTRODUCTION

The bentonite layer that surrounds and protects the canisters in a nuclear waste repository deep down in rock experiences a complex coupled heat and moisture flow process. The released heat from the canister will cause an initial drying. The water in the rock will on the other hand cause successive saturation of the bentonite from the outer rock side. These processes will interact. Key questions are the degree of initial drying and the time it takes to saturate the bentonite under various scenarios.

The considered radial process in the bentonite annulus is a complicated one with coupled, highly nonlinear flows that involve many things. There are liquid flow and vapor flow as well as conductive and convective heat flow depending on gradients in pressure, water vapor density and temperature. The flow coefficients depend on water properties such as saturation water vapor pressure and dynamic viscosity of water. They also depend on the properties of bentonite: water retention curve, hydraulic conductivity and water vapor diffusion coefficient, and thermal conductivity, all of which are functions of degree of water saturation.

The highly non-linear coupled equations may, in the form the equations are formulated below, be linearized with a loss of accuracy of around 10%. The equation for the moisture becomes of a convective-diffusive form. An analytical solution involving Bessel functions may be obtained for the transient drying and saturation of the bentonite.

The paper presents this analytical solution for the time-dependent radial water and heat flux through the bentonite annular region around the canister. The linearized solution involves two parameters.

The first key parameter t_0 gives a basic time scale for the whole process of drying and resaturation. The whole coupled process is expressed by a single thermodiffusive parameter a.

2. EQUATIONS AND SOLUTION

We will use the degree of water saturation S $(0 < S \leq 1)$ in the pores and the temperature T as basic state variables.

2.1 Flow equations

The moisture flux g (kg/(m^2,s)) in the bentonite has a liquid and a vapor component. The liquid flux g_ℓ is proportional to the gradient of the pore water pressure P with a hydraulic conductivity $k(S)$ that is a function of the degree of water saturation S. The flux is inversely proportional to the viscosity $\eta(T)$. The water vapor flux g_v is proportional to the gradient of the water vapor density ρ_v in the gas phases in the pores with a vapor conductivity factor $D_v(S)$ that is a decreasing function of S. The heat flux q (W/m^2) has a conductive part with a thermal conductivity $\lambda(S)$. There is also a negligible convective part. We have

$$g_\ell = -\frac{\rho_w k(S)}{\eta(T)} \cdot \frac{\partial P}{\partial r}, \quad g_v = -D_v(S) \cdot \frac{\partial \rho_v}{\partial r},$$

$$g = g_\ell + g_v \qquad q = -\lambda(S) \cdot \frac{\partial T}{\partial r}. \tag{1}$$

The water retention curve $P(S)$ is an important material property. We use the following functions for the bentonite: $k(S)$, $D_v(S)$, $\lambda(S)$ and $P(S)$.

We also need data for water. Kelvin's equation relates the water pressure P to the relative humidity. The water vapor density as function of P and T is

$$\rho_v(P,T) = e^{P/[R_w(273+T)\rho_w]} \cdot \rho_{v,sat}(T). \quad (2)$$

The moisture flux g and the total flux $G(r,t)$ (kg^3/s) over the height H may now be written in the following general form

$$g(r,t) = \frac{G(r,t)}{2\pi rH} = -K_S(S,T)\frac{\partial S}{\partial r} - K_T(S,T)\frac{\partial T}{\partial r}. \quad (3)$$

The moisture flow coefficients K_S and K_T are functions of the state variables S and T:

$$K_S(S,T) = \frac{\rho_w k(S)}{\eta(T)} \cdot \frac{dP}{dS} + D_v(S) \cdot \frac{\partial \rho_v}{\partial P} \cdot \frac{dP}{dS},$$

$$K_T(S,T) = D_v(S) \cdot \frac{\partial \rho_v}{\partial T}. \quad (4)$$

All above functions for bentonite and water are represented by explicit formulas with an error below 1% in the interval $10 < T < 100\,°C$ and $0.3 < S \le 1$. The mathematical program Mathcad is used. The coefficient functions $K_S(S,T)$ and $K_T(S,T)$ for moisture flux are readily obtained.

2.2 Basic equation

The moisture balance equation for water saturation $S(r,t)$ in the radial case for the annular region between canister wall ($r = r_c$) and rock (r_r) reads

$$2\pi rH \cdot \frac{\partial}{\partial t}\left[V_p \rho_w S\right] = -\frac{\partial G}{\partial r}, \qquad r_c < r < r_r. \quad (5)$$

Here, V_p is the porosity or pore volume per unit volume of bentonite. The total radial moisture flux $G(r,t)$ is given by (3).

The second flow function $K_T(S,T)$ determines the moisture flow due to the thermal gradient. The vapor flow is zero when the pores are totally blocked at full saturation $S = 1$. This means that the function contains the factor $1 - S$. We have $K_T(S,T) = K'_T(S,T) \cdot (1-S)$. Here, $K'_T(S,T)$ is different from zero for $S = 1$.

The total heat release from the canister is Q_c (W). It varies in time with a time scale of years. The time scale to establish a steady-state temperature profile over the bentonite annulus is below 24 hours. We may therefore with very good accuracy consider the temperature as independent of time. But the value of Q_c and the temperature level and profile in the bentonite will vary slowly with time. In any particular moderate time span, we have a constant value of Q_c and a constant temperature T_r at the rock boundary. The radial temperature profile $T(r)$ satisfies the equation

$$Q_c = -2\pi rH \cdot \lambda(S) \cdot \frac{\partial T}{\partial r}, \qquad T(r_r,t) = T_r. \quad (6)$$

Combining (3) and (6) we may eliminate the temperature derivative. We may now write the moisture balance equation (5) in the following way

$$rV_p\rho_w \cdot \frac{\partial S}{\partial t} = \frac{\partial}{\partial r}\Big[K_S(S,T) \cdot \left(r \cdot \frac{\partial S}{\partial r} - \frac{Q_c}{2\pi H} \cdot \frac{K'_T(S,T)}{K_S(S,T)\lambda(S)} \cdot (1-S)\right)\Big] \quad (7)$$

We obtain a single equation for $S(r,t)$. Any variation of the temperature profile with time is determined from (6) with a very slowly varying $Q_c(t)$. The equation is of a convective-diffusive character. The second term involving $1-S$ is caused by the water vapor diffusion due to the temperature gradient.

2.3 Linearized equation for $S(r,t)$

The variation of $K'_T(S,T)/[K_S(S,T)\lambda(S)]$ and $K_S(S,T)$ in, for example, the region 0.8<*S*<1 and 70<*T*<80 have been calculated for typical bentonite data. There is a variation of some $\pm 20\%$. This is quite typical for the type of cases that we consider. It may be noted that the factor $1-S$ varies much more (by a factor 20 between *S*=0.8 and *S*=0.99).

A drying and resaturation process will then have a variation of the flow coefficients $K_S(S,T)$ and $K'_T(S,T)/[K_S(S,T)\lambda(S)]$ in the governing equation (12) of up to $\pm 20\%$. We may with some caution approximate the two functions by suitable constant mean values. We use the approximations:

$$K_S(S,T) \approx K_S(S_0,T_0), \qquad \lambda(S) \approx \lambda(S_0),$$
$$K'_T(S,T) \approx K'_T(S_0,T_0). \qquad (8)$$

Here, S_0 is a mean water saturation level and T_0 the mean temperature in the bentonite. The error involved in this approximation is judged to be below 10 %, since the flow functions have values both above and below the approximate values.

We now introduce two new parameters:

$$D_0 = \frac{K_S(S_0,T_0)}{V_p \rho_w}, \quad a = \frac{Q_c}{4\pi H} \cdot \frac{K'_T(S_0,T_0)}{K_S(S_0,T_0)\lambda(S_0)}. \qquad (9)$$

Equation (7) with the approximations (8) may now be written

$$\frac{1}{D_0} \cdot \frac{\partial S}{\partial t} = \frac{\partial^2 S}{\partial r^2} + \frac{1+2a}{r} \cdot \frac{\partial S}{\partial r}, \quad r_c < r < r_r. \qquad (10)$$

The initial degree of saturation in the bentonite is S_{in}. Full saturation *S*=1 is maintained at the rock boundary $r = r_r$. The moisture flux is zero at the canister boundary. The boundary conditions and the initial condition are:

$$\left.\frac{\partial S}{\partial r}\right|_r = \frac{2a}{r_c}[1 - S(r_c,t)], \quad S(r_r,t) = 1, \quad S(r,0) = S_{in}. \qquad (11)$$

We have to solve the above problem for $S(r,t)$. It has a diffusive-convective character. It involves one essential dimensionless parameter a only. The time is scaled with the diffusivity factor D_0.

2.4 Solution of linearized equation

The solution of the problem (10)-(11) is reported in detail in Claesson (2003). We use a new function $\sigma(r,t)$ for the degree of saturation:

$$S(r,t) = S_{in} + (1 - S_{in}) \cdot \sigma(r,t). \qquad (12)$$

The function $\sigma = (S - S_{in})/(1 - S_{in})$ is zero when the saturation equals the initial level: $S = S_{in}$. The value is negative when the degree of saturation is below the initial value: $S < S_{in}$, and it lies in the interval $0 < \sigma \le 1$ for $S_{in} < S \le 1$.

The problem for $\sigma(r,t)$ satisfies the equations (10)-(11) except that the initial condition is $\sigma(r,0) = 0$. We get rid of the parameter S_{in}. The problem for $\sigma(r,t)$ has the parameters a, D_0, and r_c / r_r.

The solution for $\sigma(r,t)$ is

$$\sigma(r,t) = 1 - \left(\frac{r_r}{r}\right)^a \cdot \sum_{n=1}^{\infty} A_n u_n(r) e^{-\alpha_n^2 \cdot D_0 t / r_r^2} \qquad (13)$$

The derivation of the solution and mathematical details are reported in the separate study Claesson (2003). The difference $1-\sigma$ has homogeneous boundary conditions. Multiplication with r^a removes the factor $2a$ in (10). The method of separation of variables gives Bessel functions as solution. The eigenfunctions $u_n(r)$ are orthogonal, and the initial condition is satisfied by a suitable choice of coefficients A_n. We have:

$$J_a(\alpha)\cdot Y_{a-1}(\alpha r_c / r_r) - Y_a(\alpha)\cdot J_{a-1}(\alpha r_c / r_r) = 0$$
$$\Rightarrow \quad 0 < \alpha_1 < \alpha_2 < \ldots .$$

$$A_n = \left(\alpha_n J_{a-1}(\alpha_n') - 2aJ_a(\alpha_n)(r_c / r_r)^{a-1}\right)\cdot$$
$$\cdot\frac{\pi}{\alpha_n}\cdot\frac{J_a^2(\alpha_n)Y_{a-1}(\alpha_n')}{J_{a-1}^2(\alpha_n') - J_a^2(\alpha_n)}, \qquad \alpha_n' = r_c\alpha_n / r_r,$$

$$u_n(r) = \frac{J_a(\alpha_n r / r_r)}{J_a(\alpha_n)} - \frac{Y_a(\alpha_n r / r_r)}{Y_a(\alpha_n)}. \tag{14}$$

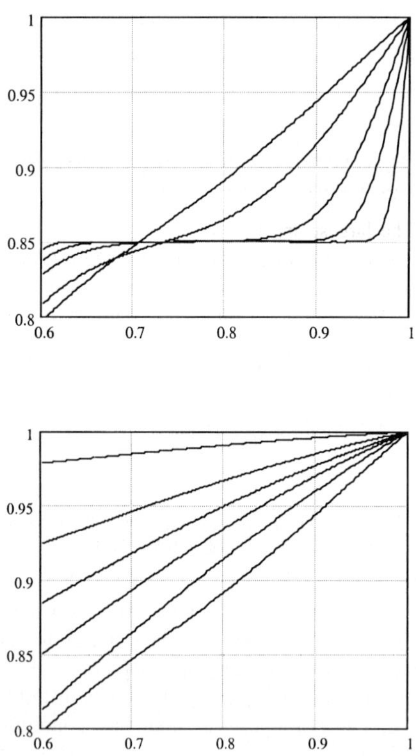

Figure 1. The degree of saturation S for a=0.9 as function of r / r_r for different τ. Top diagram: τ =0.002, 0.01, 0.03, 0.1 and 0.216 (maximum drying). Bottom diagram: τ =0.216 (bottom curve), 0.4, 0.694 (S=0.85), 1, 1.5 and 3.

The eigenvalues α_n, which are the positive roots of the top equation of (14), may be written as $\alpha_n = \nu_n \cdot \pi r_r / [2(r_r - r_c)]$. The values ν_n are quite close to $2n-1$ for $n \geq 2$. The exponent involving time *t* in (13) may be written:

$$\frac{\alpha_n^2 D_0 t}{r_r^2} = \nu_n^2 \tau \quad \tau = \frac{t}{t_0}, \quad t_0 = \frac{4(r_r - r_c)^2}{\pi^2 D_0}. \tag{15}$$

Here, $\tau = t / t_0$ is the dimensionless time, and t_0 a basic time scale for the considered process.

The above solution has been implemented in the mathematical computer program Mathcad. The solution is readily calculated. The result for *a*=0.9 and $S_{in} = 0.85$ is shown in Figure 1. The degree of saturation $S(r,\tau)$ is shown as a function of the radius r / r_r for a number of dimensionless times τ. The top graph shows short times from τ =0.002 to τ = 0.216, and the bottom graph values from τ = 0.216 to τ =3.

The curve for the shortest time τ =0.002 shows a small drying near the canister and increased saturation from the initial value *S*=0.85 to *S*=1 in a small region near the rock boundary. The intermediate region is unchanged. The dried region and the region of increased saturation both increase in time. Water from the dried region is displaced by vapor diffusion through the undisturbed intermediate region to the outer parts. At $\tau \approx 0.05$ (between the curves τ =0.03 and τ =0.1) the two regions meet at r / r_r =0.8. The drying in the inner region continues until the time of largest drying τ =0.216. Then the wetting process becomes dominant also at the canister wall. There is from this time a steady wetting of all the bentonite as shown in the bottom graph. The initial saturation at the canister wall (*S*=0.85) is restored at the time τ = 0.694.

2.5 Dependence on saturation and temperature level

The above solution involves the two key parameters t_0 and *a* given by (15) and (9). They depend on the choice of average saturation level S_0 and average temperature level T_0 in (8). We have from (15) and (9)

$$t_{0y}(S_0,T_0)=\frac{4(r_r-r_c)^2\cdot V_p\rho_w}{\pi^2\cdot K_S(S_0,T_0)\cdot t_y},$$
$$a(S_0,T_0)=\frac{Q_c}{4\pi H}\cdot\frac{K_T'(S_0,T_0)}{K_S(S_0,T_0)\lambda(S_0)}. \tag{16}$$

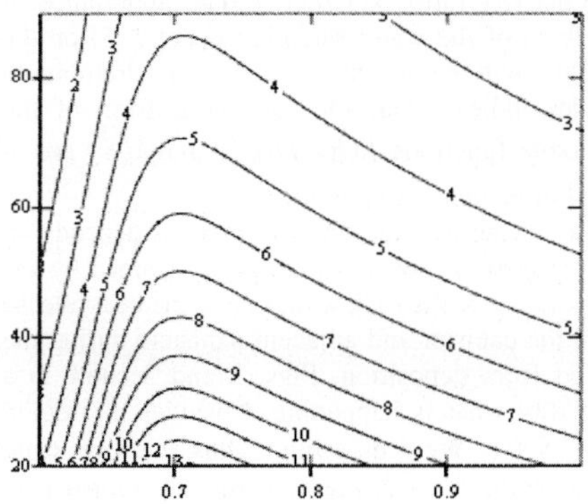

Figure 2. The time-scale function $t_{0y}(S_0,T_0)$ using reference bentonite data. Intervals in degree of saturation and temperature: $0.6\le S_0\le 1$, $20\le T_0\le 90$ °C.

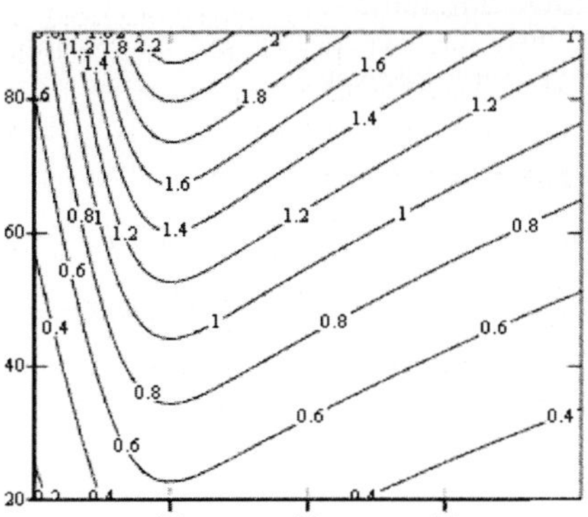

Figure 3. The thermodiffusive function $a(S_0,T_0)$ using reference bentonite data. Intervals in degree of saturation and temperature: $0.6\le S_0\le 1$, $20\le T_0\le 90$ °C.

The function $t_0(S_0,T_0)$ is divided by the time of a year $t_y = 3600\cdot24\cdot365$ s, which means that t_{0y} is the time scale in years. These two functions have been calculated for a wide range of saturation and temperature levels for a reference case taken from Rutqvist et al (1999). See Fig. 2 and Fig 3, respectively. The Mathcad program generates such a plot in no time for any bentonite (and water) data.

The time scale t_0, (15)-(16), is determined by the average diffusivity D_0, or by $K_S(S_0,T_0)$. The dimensionless parameter *a*, (9), is proportional to the heat flux Q_c/H. It is also proportional to the ratio between the flow coefficients for the temperature and saturation gradients. It is a thermo-diffusive parameter. It should be noted that the time scale t_0 is inversely proportional to the flow coefficient K_S, while *a* only depends on the ratio between the two basic flow coefficients.

3. DRYING AND RESATURATION AT THE CANISTER WALL

The largest drying and the slowest resaturation occur at the canister wall $r=r_c$. The degree of saturation at the wall is from (12), (13) and (15)

$$S_c(t)=S_{in}+(1-S_{in})\cdot\sigma_c(\tau),$$
$$\sigma_c(\tau)=\sum_{n=1}^{\infty}(r_r/r_c)^a A_n u_n(r_c)\cdot e^{-\nu_n^2\cdot\tau}. \tag{17}$$

The eigenvalues and the expansion coefficients, (14), are functions of the dimensionless quantities *a* and r_c/r_r only: $\sigma_c(\tau)=\sigma_c(\tau;a,r_c/r_r)$. In our case with $r_c/r_r=0.6$, the dimensionless quantities depend on *a* only. This means that the function $\sigma_c(\tau)$ involves the single parameter *a*.

The function $\sigma_c(\tau;a)$ has been calculated for different values of *a*. The curves are shown in Figure 4. The curves decrease to a minimum below zero, which gives the largest or maximum drying. Then the curves increase steadily up to 1 at full saturation.

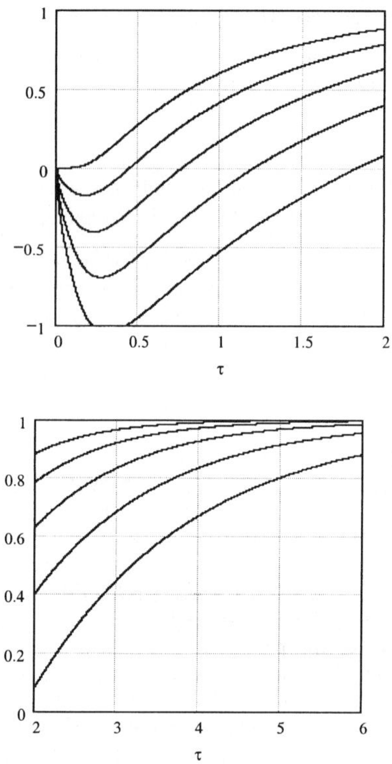

Figure 4. The function $\sigma_c(\tau;a)$ for the degree of saturation at the canister wall for diffent a-values: a=0 (top curve), 1.0, 1.5 and 2.0

4. CONCLUSIONS

The above solution provides a good picture of the complex coupled moisture and heat flow process. The whole solution, with the time-scale function $t_{0y}(S_0,T_0)$ and the thermodiffusive function $a(S_0,T_0)$, have been implemented as Mathcad programs. The whole program with formulas covering a few pages is presented in Claesson (2003). Any calculation requires a few seconds up to a minute only. The formulas may readily be implemented in any mathematical program.

The solution may be used for sensitivity studies. The level of the conductivity function $K_S(S,T)$ determines the time scale of the drying and resaturation process. The heat flux Q_c/H times the level of the moisture conductivity ratio $K'_T(S,T)/[K_S(S,T)\lambda(S)]$ gives the parameter *a*, (16). These are the key factors that influence the process.

The conductivity functions $K_S(S,T)$ and $K_T(S,T)$ depends on the bentonite data. Important data are the conductivity $k(1)$ at saturation and the conductivity ratio $k(1)/D_v(0)$. The importance of the form of the water retention curve *P*(*S*) on the process will be investigated further. Preliminary studies indicate that the functional form of the bentonite functions $k(S)$, $D_v(S)$ and $\lambda(S)$ are of small or moderate importance.

The drying and resaturation process depends on the temperature level, which is determined by the temperature at the rock wall, i.e. by the heat release from the canister and adjacent canisters during the period from deposition. Figs. 2 and 3 show in a nice way what is happening. Consider the region $0.8 < S \leq 1$. We see in Fig. 2 that the time scale varies from 8 to 3 years when the temperature increases from 20 to 90 °C. The thermodiffusive parameter increases from 0.4 to 1.4. The process is more rapid and intensive when the temperature increases. The main reason for this is that the vapor flow and vapor saturation pressure increase strongly with the temperature.

Acknowledgement.
This study was supported by SKI, the Swedish Nuclear Power Inspectorate.

REFERENCES

Abramowitz M. & Stegun I., 1972. H*andbook of Mathematical Functions.* Dover Publ. N.Y.

Carslaw H.S. & Jaeger J.C. 1965. *Conduction of Heat in Solids.* Oxford Sc. Publ., Oxford.

Claesson J. 2003. *Drying and Resaturation of a Bentonite Annulus. An Analytical Solution.* Report, Building Physics, Chalmers, Sweden.

Rutqvist J., Noorishad J. & Tsang C-F. 1999. *Coupled Termohydromechanical Analysis of a Heater Test in Unsaturated Clay and Fractured Rock at the Kamaishi Mine.* SKI report 99:50, the Swedish Nuclear Power Inspectorate.

FABRIC CHANGES OF A PELLET-BASED BENTONITE BUFFER MATERIAL AND THEIR EFFECTS ON MECHANICAL BEHAVIOUR

C. Hoffmann, E.E. Alonso, E. Romero
Geotechnical Engineering Department, Universitat Politècnica de Catalunya,
c/ Jordi Girona 1-3, Edificio D-2, 08034 Barcelona, Spain.

Abstract: The mechanical behaviour of granular aggregates of highly compacted bentonite has been investigated by means of suction controlled wetting tests. It was found that specimen behaviour is controlled by the rate of application of moisture at the specimen boundaries. Two extreme cases have been investigated: the slow or the fast application of suction changes. Measured results have been interpreted within the framework of a double structure model. Away from equilibrium, the capillary suction prevailing at the large macropores, is different from the microstructural suction, which controls pellet expansion (or shrinkage). A local transfer of water between both levels of structure and the associated changes in macro and microporosity explains the nature of observed phenomena.

1. INTRODUCTION

Artificially prepared aggregates of highly compacted bentonite are being considered as a suitable buffer material, which offers some advantages over the use of manufactured compacted blocks. Pneumatic machines can easily handle the granulated mixture and the process of buffer installation and compaction may be automated. A large-scale demonstration project, (the EU financed 'EB Project' Engineered Barrier Emplacement Experiment ; AITEMIN, 2001), is being performed in order to explore the benefits of using this material in barrier installation and performance.

Suction controlled oedometer tests were performed to gain insight into the characterization of the material, which exhibits a marked double structure: a microstructure, which describes the dense aggregates of clay platelets and a macrostructure which includes the inter-aggregate and inter-pellet voids. An important feature of this material is the evolution of the microstructure as the mixture hydrates.

The first part of the paper is devoted to the characterization of the pore structure and to the presentation of test results which show the importance of the double structure features. In the second part a double structure constitutive formulation is presented. Measured specimen performance is finally interpreted with the model in qualitative and quantitative terms.

2. TESTED MATERIAL

Pellets were prepared from FEBEX bentonite powder (w_L= 93%, w_P = 47%, % of particles< 75 μm = 85% γ_s = 2.70 Mg/m^3). Further details of this bentonite are described in ENRESA (2000). In order to achieve the emplacement dry density, an adequate pellet size distribution and a high dry density of the pellets (around 1.9 Mg/m^3) were required. Bentonite powder was compacted using a roller press. In order to facilitate this operation, the bentonite powder was pre-heated to 120 °C. As a result of this process, granules with a dry density ranging between 1.97 and 2.03 Mg/m^3 and a reduced water content of 3-4% were obtained. Finally, the fraction of granules with sizes larger than 10 mm was used as the basic material for the production of pellets with a jaw crusher. Figure 1 shows the grain size distribution of the tested specimens.

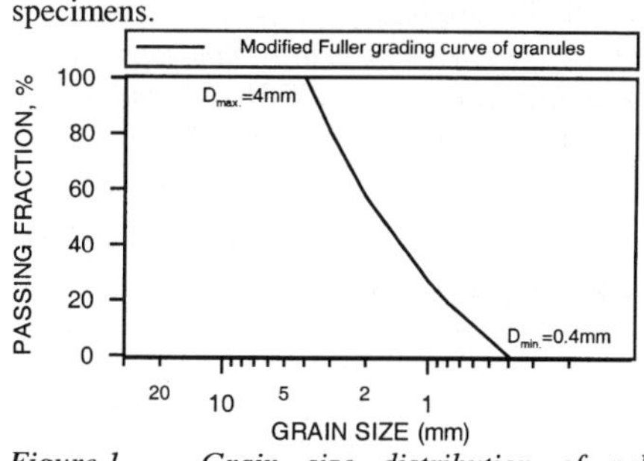

Figure 1. Grain size distribution of pellet mixtures.

3. PORE SIZE DISTRIBUTION

Mercury intrusion porosimetry (MIP) tests were performed to characterize the multiple-porosity network of the artificially prepared packing at different dry density values ranging from 1.2 to 1.95 Mg/m^3.

To avoid excessive shrinkage a freeze drying process was used to prepare the partially saturated samples (Delage *et al.* 1982). Figure 2 shows the measured pore size density functions for the as-compacted state, which correspond to a packing having ρ_d = 1.35 Mg/m^3, for pellets having ρ_d = 1.95 Mg/m^3.

The pore size distribution curve of the as-compacted material at a dry density of 1.35 Mg/m^3 displays two modes, at pore sizes of 13 nm and 40 μm. The porosimetry of the high-density pellets (Figure 2) also shows two modes at 13 nm and 11 μm. These modes correspond to the intra-aggregate and inter-aggregate pores (pores inside and between aggregations contained in a pellet). MIP technique was not adequate for measuring the pore size between pellets, which is larger than 400 μm.

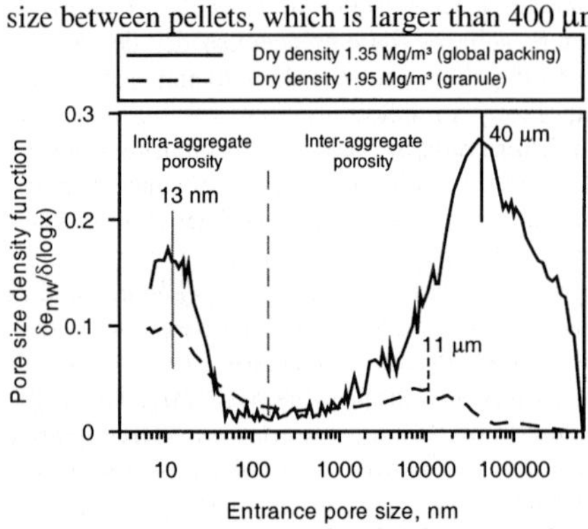

Figure 2. Pore size density functions for the as-compacted state (ρ_d = 1.35 Mg/m^3) and for the high-density pellets (ρ_d = 1.95 Mg/m^3).

4. HYDROMECHANICAL TESTS ON PELLET MIXTURES

Two oedometer tests following the same wetting under load path were performed on identical samples, using different techniques. A special oedometer cell was designed and built in order to perform the tests. The cell may handle two suction control techniques: vapour equilibration and liquid water transfer. In the second case an axis translation procedure may be introduced by means of gas and water pressure controls on the upper and lower porous disks bounding the specimen (Romero, 2001).

In the test involving wetting with liquid water, the stored water volume was controlled and suction values were derived from the measured water retention curve of the mixture. In the second test performed, the wetting path was followed in a step-by-step process by varying the chemical potential of different types of aqueous solutions used to control the relative humidity of the air circulated through the specimen. Each suction step was maintained until no volume change was registered in the sample. Once a minimum suction value of 3 MPa was reached (using copper sulphate saturated solution), liquid water was finally injected to achieve the saturated condition.

When the sample was initially wetted by liquid water (Figure 3a), an initial collapse (volumetric compression) was observed immediately after inundation. Then, the expanding granules started to swell and a net swelling deformation (volumetric expansion) was measured. At a final stage, however, when the specimen was close to saturation, additional collapse deformations were registered.

In the vapour transport case (Figure 3b), a net swelling deformation (volumetric expansion) was measured from the very beginning of the test and for all the subsequent vapour equilibrium stages until a suction value of 3 MPa was reached. Then, at the final stage, when liquid water was injected to bring the suction to zero, a sudden collapse was initially observed followed by some final swelling deformations.

In order to compare the kinematics of both tests, a normalized time was adopted in Figure 3c. Normalized time was calculated by dividing the actual real time by the total time length of each test. The vapour transfer test was much longer than the liquid transfer one due to the lower equilibration times associated with vapour transfer.

Differences in observed behaviour can be interpreted as follows. When wetting is forced by means of liquid water, water flows through the sample and fills initially the macropores. The macrostructural level becomes saturated, the capillary intergranular forces are destroyed and a structural collapse of the specimen, is produced. Immediately after, a water exchange between macro and microstructural levels starts and the potentially expanding pellets begin to swell. This explanation includes implicitly the assumption that two different water potentials exist, at a given time, inside the specimen. The water filling the macropores has a macro potential, s_M, not necessarily in equilibrium with the water potential inside the aggregates, s_m. The high initial permeability of the specimen macrostructure explains its rapid saturation when liquid water under some pressure is applied at the specimen boundaries.

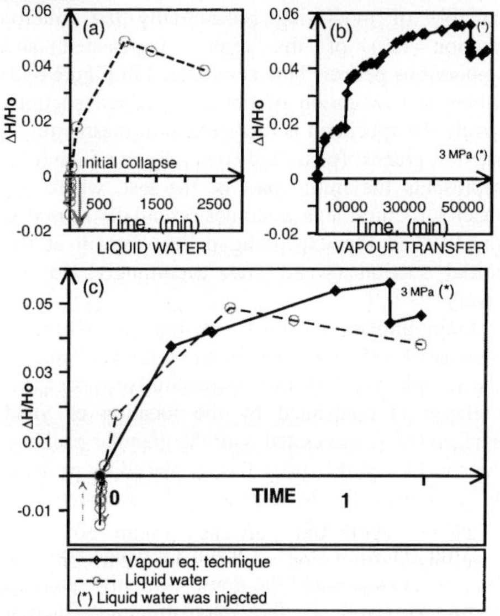

Figure 3. Evolution of wetting under load tests performed with liquid water (a) and vapour equilibrium technique (b). Test results plotted against the normalized time (c).

However, when a vapour equilibrium technique is used to wet the sample, the large pores remain essentially unsaturated and water is slowly transferred to the bentonite pellets. Under these conditions the macro and micro water potentials have the opportunity to remain essentially equal through the wetting process. Capillary forces holding together the macrostructure change slowly and no collapse is produced. In fact, an opposite phenomenon is more likely: the dense aggregates (inside the pellets) begin to swell and they will progressively occupy the macropores. The overall (net) specimen behaviour is an expansion, which starts at the early stages of the test.

What figure 3 shows is that different rates of moisture transfer lead to different results in terms of constitutive behaviour. Constitutive behaviour, even at the scale of a specimen (or even at the scale of a representative elementary volume) cannot be explained without a reference to the rate and phenomena of water transfer. From the perspective of conventional unsaturated soil mechanics these experimental observations pose a complicated problem for model formulation. The idea presented here to derive a working model is to accept that: 1) Two water potentials (macro and micro) control the mechanical behaviour of the material; 2) Both potentials may become locally different. As a consequence, different laboratory techniques lead to different histories of macro and micro potential changes and therefore to different specimen responses.

5. MODELLING THE EXPERIMENTAL RESULTS

5.1 Features of the model

Using concepts of hardening plasticity a double-porosity model was proposed by Gens and Alonso (1992) and Alonso, Gens & Gehling, (1994) to model expansive soil behaviour. This model extends the classical Barcelona Basic Model (BBM) for unsaturated soils (Alonso, Gens and Josa, 1990) to definitely expansive unsaturated soils. These developments are the starting point of the analysis presented here. "Conventional" unsaturated soil behaviour has been successfully modelled in the past considering two independent stress fields: the net stress (σ_{ij}-$u_a\delta_{ij}$), and the suction (u_a-u_w)δ_{ij} where: σ_{ij} =total stress; u_a=pore air pressure; u_w= pore water pressure.

The discussion above suggests that a single suction field is not enough to interpret the behaviour of expansive materials with a marked bi-modal porosity distribution. Therefore, in addition to the net stress, two additional water potential fields are considered: s_M and s_m. s_M may be interpreted as a capillary potential, but the concept of water potential, linked to the physico-chemical mechanisms of water bounding to the clay minerals, is more appropriate for s_m. For the purposes of discussing the experimental behaviour observed it is convenient to work with two stress spaces which integrate the net stress field and the two independent water potential variables: The space (σ_{ij},s_M) associated with the macrostructure and the space (σ_{ij},s_m) associated with the microstructural level. (A reference zero air pressure will be considered, for simplicity).

Two void ratios (macro and micro) are defined in the sample. In a given soil state, these void ratios can be quantified if a pore size distribution is available. Total volumetric deformation is computed as the sum of the micro and macro components. Based on the adopted phases the total volumetric deformation can be expressed as:

$$\dot{\varepsilon}_v = \frac{de_T}{1+e} = \frac{de_M + de_m}{1+e} = \dot{\varepsilon}_{vM} + \dot{\varepsilon}_{vm} \quad (2)$$

where

$$d\varepsilon_{vM} = \frac{de_M}{1+e} \text{ and } d\varepsilon_{vm} = \frac{de_m}{1+e} \quad (3)$$

The terms $\dot{\varepsilon}_{vM}$ and $\dot{\varepsilon}_{vm}$ are the volumetric deformations induced by net mean stress and suction changes at both macro and microstructural levels respectively.

For isotropic or one-dimensional conditions, Figure 4 shows the two stress spaces mentioned before. Also indicated are the assumed positions of the yield locus which mark the initiation of irreversible strains: The Loading –Collapse (LC) curve in the case of the macrostructure and the Suction-Decrease (SD) curve in the case of microstructural changes. More details of the model may be found in Alonso (1998) and Alonso et al (1999)

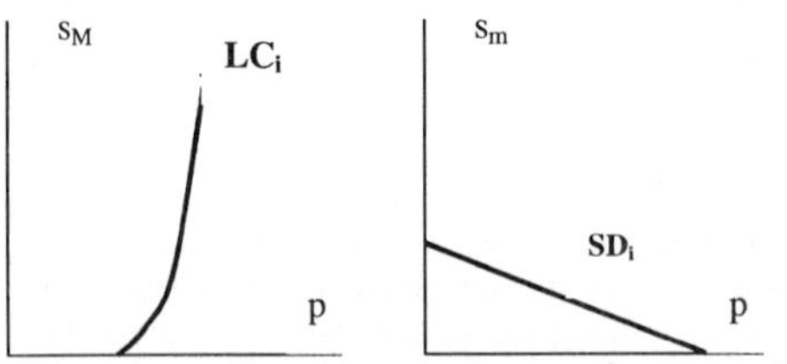

Figure 4. Stress- suction spaces for isotropic (or one-dimensional) conditions and associated yield surfaces.

5.2 Modelling results

A striking characteristic of the two tests was the widely different results (in terms of volumetric behaviour) observed in identical samples subjected to the same overall stress-suction path (Figure 5). Application of the model requires the knowledge of the evolution of the controlling suctions during wetting. The liquid wetting is characterized by an almost instantaneous reduction to zero of the macrostructural (capillary) suction and a delayed reduction of the microstructural suction. In the vapour transfer wetting case local equilibrium of both suctions is maintained. This information is enough to perform the analysis, as shown immediately.

Liquid water test

The liquid wetting reduces the Macro suction (s_M) at the early stages of the wetting because of the high open porosity of the specimen but it is not capable of modifying substantially the (micro) suction (s_m) of the highly compacted and impervious pellets. This is indicated in Figure 5a,b where the evolution of the two defined suctions inside the specimen is represented in mean stress - suction planes: (p,s_M) and (p,s_m). Figure 5a and 5b represents the initial part of the test, where the macrostructural level becomes essentially saturated ($s_M \cong 0$), and the expanding pellets remain at the initial suction ($s_m \cong s_o$) and, accordingly, do not swell.

Intergranular (Macro) suction is therefore dissipated and conditions for a potential collapse of the sample develop. In the constitutive model this collapse is controlled by the location of yield surface (LC_1) associated with the granular packing. As the LC_1 yield surface is displaced to position LC_2, dragged by the decreasing Macro suction, a collapse is predicted.

After this first stage, a water exchange between both levels starts and the dense bentonite granules begin to swell as the microstructural suction decreases. The microstructural suction decreases in search of the "boundary condition" imposed by the low macrostructural suction, s_M. Two different processes could be recognized. First, as it is shown in Figure 5d, the wetting path takes place in an elastic regime. Then, when the SD_1 yield surface is "activated" (Figure 5f), additional irrecoverable deformations occur at the macrostructural level due to the coupling between both levels. These deformations were computed by means of the interaction function f_D. = $\dot{\varepsilon}_{vM} / \dot{\varepsilon}_{vm}$ (Table 1)

Table 1.Model parameters

Virgin consolidation: $\lambda(0)=0.19$ $r=0.7$ $\beta=0.1$
Yield Curve LC_i: $\kappa=0.075$ $p_0^*=220\ kPa$ $p_c=75\ kPa$
Interaction function: $f_D = 1.577 * p/p_o - 2.04$
Yield curve SD_i: p+s = 18.3 MPa
Microstructure: $\hat{\kappa}_s=0.09$; Initial void ratio, $e_m=0.3$
Macrostructure: $K_{SM} \rightarrow \infty$, no elastic macro swelling was considered to occur within the elastic region. Initial void ratio, $e_M=0.75$

Model parameters were derived from this test and are collected in Table 1. They were derived from the measured test results and therefore no "predictive" capabilities are claimed here.

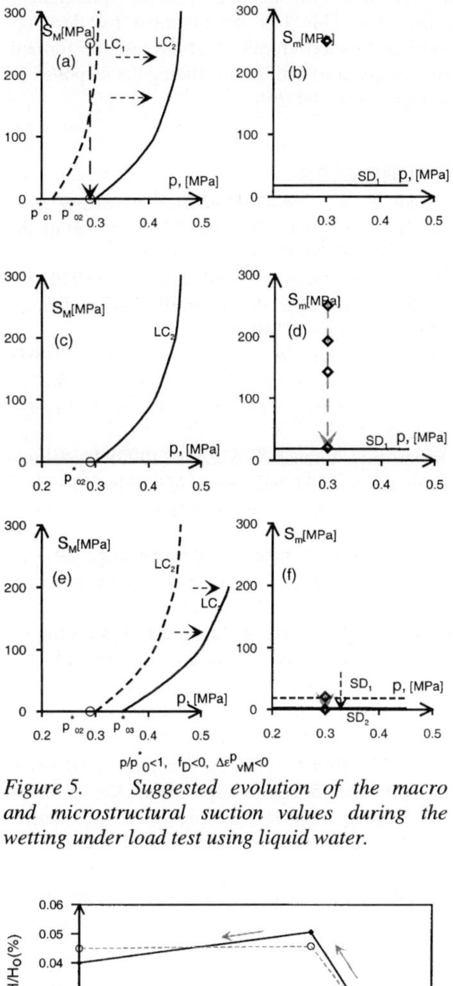

Figure 5. Suggested evolution of the macro and microstructural suction values during the wetting under load test using liquid water.

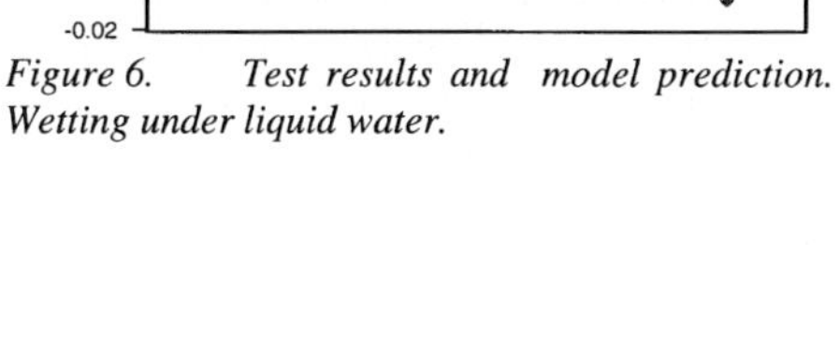

Figure 6. Test results and model prediction. Wetting under liquid water.

Test results and model predictions are presented in Figure 6. Measured (and modelled) volumetric deformation is represented as a function of the microstructural suction since the macrostructure can be considered in an almost saturated condition from the beginning of the test. The proposed model seems to provide an adequate framework to explain the observed behaviour.

Vapour transfer test

In this case net swelling deformations (volumetric expansion) were recorded from the very beginning of the wetting process. The vapour equilibrium technique favours the local equilibrium of macro and micro suctions inside the specimen. Using a similar representation in the (p,s_M) and (p,s_m) planes, the interpretation of this test has been represented in Figure 7. Two different stages could be identified. During the first stage (Figure 7b) collapse deformations, (associated with s_M changes) and swelling deformations (associated with s_m changes) take place simultaneously. The net overall volume change depends on the relative intensity of both mechanisms. Using the model parameters derived for the liquid wetting test, a net volumetric expansion was computed.

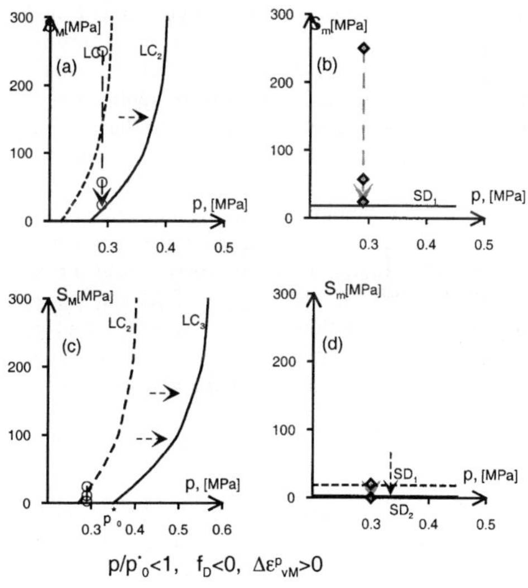

Figure 7. Evolution of the macro and microstructural suction values during the wetting under load test. Vapour equilibrium technique.

As the microstructural suction decreases to low values the stress path crosses the yield surface, SD, (Figure 7.b) and some plastic re-arrangements take place. In the model this phenomenon is described by the interaction function f_D. A net volumetric compression was observed (and computed) at this stage.

Experimental results and model predictions obtained for this test are presented in Figure 8.

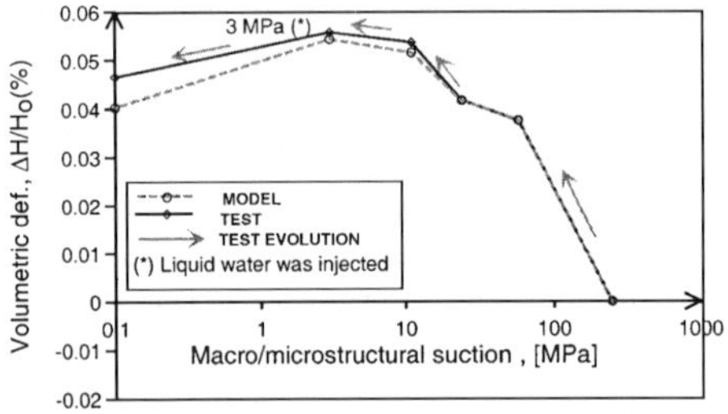

Figure 8. *Test results and model prediction.*

6. CONCLUSIONS

The behaviour of bentonite pellet mixtures has been investigated by means of wetting under load tests conducted in an special suction controlled oedometer cell prepared to handle liquid and vapour transfer processes. The experimental results reported in this paper show that specimen behaviour is controlled by the rate of application of suction changes. Two extreme cases have been investigated: the slow and the fast application of suction (or humidity) changes. Liquid water injection and a vapour transfer technique were used for a fast and slow wetting respectively. It was found that a liquid water injection results in an initial collapse of the sample followed by a subsequent expansion. If a vapour transfer technique is used, swelling strains are measured from the start of the test. Yet the apparent stress-suction paths of both types of test are identical.

Measured results are properly interpreted if due account is given to the evolution of the soil structure. Porosimetry determinations show that two distinct pore sizes are present. Their significance to explain the mechanical behaviour of the mixture has been explored in the paper. Two key aspects help to understand the measured behaviour: Two different water potentials (suctions) may coexist at a given time at a "point" of the soil and the rate of suction changes is controlled by the hydraulic boundary conditions of the test.

Experimental results have been interpreted within the framework of a double structure model, which seems to capture the essential features of observed behaviour. The proposed model is capable also of making precise quantitative predictions. This has been shown by deriving model parameters from one of the tests performed (liquid injection) and by calculating the response of the vapour transfer test.

REFERENCES

Alonso, E. E. (1998) *Modelling expansive soil behaviour.* Proc. 2nd Int. Conf. Unsaturated Soils. 2: 37-70. Beijing, China.

Alonso, E.E, Gens, A. and Josa, A. (1990). *A constitutive model for partially saturated soils.* Géotechnique, N° 40, 405-430.

Alonso, E.E, Gens, A. and Gehling, W.Y.Y. (1994) *Elastoplastic model for unsaturated expansive soils.* 3rd Conf. Num. Meth. Geotech. Engng: 11-18

Alonso, E, J Vaunat & A Gens (1999) *Modelling the mechanical behaviour of expansive clays.* Engineering Geology. 54: 173-183.

AITEMIN (2001). *Engineered Barrier Emplacement. Experiment in Opalinus Clay "EB" Experiment.* Internal Report. AITEMIN, Madrid

Delage, P., Tessier, D. & Audiguier, M.M. (1982). *Use of the cryoscan apparatus for observation of freeze-fractured planes of a sensitive Quebec clay in scanning electron microscopy.* Can. Geotech. J., 19: 111-114.

ENRESA (2000). FEBEX Project. *Full-scale engineered barriers experiment for a deep geological repository for high-level radioactive waste in crystalline host rock.* Final report. Madrid.

NAGRA (2001). *Short report on technical trials for producing bentonite granulations.* Internal Report. NAGRA, Switzerland.

Romero, E (2001) *Controlled-suction techniques.* 4° Simpósio Brasileiro de Solos Nâo Saturados. W.Y.Y. Gehling & F. Schnaid (eds). Porto Alegre, Brasil: 535-542.

Romero, E., Castellanos, E. and Alonso, E.E. (2002) *Laboratory tests on compacted sand/bentonite/lead nitrate buffer material for the GMT emplacement project.* Project Report GMT/IR 02-01. NAGRA, Switzerland.

A CONCEPTUAL AND NUMERICAL MODEL FOR THERMAL-HYDROLOGICAL-CHEMICAL PROCESSES IN THE YUCCA MOUNTAIN DRIFT SCALE TEST

Eric L. Sonnenthal, Nicolas F. Spycher, Mark Conrad, and John Apps
Lawrence Berkeley National Laboratory,
Berkeley, CA 94720 USA

Abstract: A numerical model was developed to predict the coupled thermal, hydrological, and chemical (THC) processes accompanying the Drift Scale Test (DST) at Yucca Mountain, NV. The DST has been closely monitored through the collection of gas, water, and mineral samples as well as thermal, hydrological, and mechanical measurements. A two-dimensional dual permeability model was developed to evaluate multiphase, multicomponent, reaction-transport processes in the fractured tuff. Comparisons between results using the TOUGHREACT code and measured water (e.g., pH, SiO_2(aq), Na^+, K^+) and gas (CO_2) compositions show that the model captures the chemical evolution in the DST. Non-reactive aqueous species (e.g., Cl) show strong dilution in fracture waters, indicating little fracture-matrix interaction. Silica concentrations are higher than in the initial pore water and show a trend of increasing reaction with fracture-lining silicates at higher temperatures. The narrow precipitation zone of predominantly amorphous silica observed above the heaters was also captured.

1. INTRODUCTION

The long-term performance of the repository at Yucca Mountain will be affected by the coupling of thermal, hydrological and chemical (THC) processes in the rock around the emplacement drifts. The transport of heat, fluid, and vapor will result in changes in water and gas chemistry, as well as mineral dissolution and precipitation which may lead to permanent changes in porosity, permeability and unsaturated hydrological properties. The purpose of this contribution is to describe the approach used to model reaction-transport processes in the Drift Scale Test (DST) with comparisons of simulation results to measured geochemical data on water, gas, and minerals.

The DST is the second underground thermal test to be carried out in the Exploratory Studies Facility (ESF) at Yucca Mountain, Nevada. The purpose of the test is to evaluate the effect of coupled thermal, hydrological, chemical, and mechanical processes that take place in unsaturated fractured tuff over a range of temperatures (approximately 25°C to 200°C). The DST heaters were activated on December 3, 1997, with a planned period of 4 years of heating, followed by 4 years of cooling. After just over 4 years, the heaters were switched off and since that time the system has been slowly cooling.

The DST consists of an approximately 50 m long drift that is 5 m in diameter. Nine electrical floor canister heaters were placed in this drift (the Heated Drift) to simulate nuclear-waste-bearing containers. Electrical heaters were also placed in a series of horizontal boreholes (wing heaters) drilled perpendicular outward from the central axis of the Heated Drift. These heaters were emplaced to simulate the effect of adjacent emplacement drifts. Details regarding the DST layout, borehole orientations, operation of the test, and measurements performed are discussed in BSC (2002), CRWMS M&O (1998), and in Datta et al. (2003, this volume).

The major driving forces for THC processes in the thermal tests and in the potential repository are the thermal heat load and (over long time periods) boundary conditions such as infiltration rate. For the short time scale of the DST the infiltration rate is negligible, being on the order of 1 to 10 mm/year (Sonnenthal and Bodvarsson, 1999).

A recurring theme in this work is the importance of the interaction of flow and transport between fractures and the adjoining porous matrix. We note here that large-scale chemical equilibrium cannot be maintained in a system where mineral precipitation and dissolution are taking place at different rates in fractures and matrix (which have different mineral assemblages and abundances), the advective transport in fractures is faster than diffusive transport in the matrix.

The development of a conceptual model and the strategy of incorporating the important THC processes for the thermal tests are based originally on studies done on TH processes for the Single Heater Test and the DST (e.g., Tsang and Birkholzer, 1999; Birkholzer and Tsang, 1998) and a predictive model for THC processes in the DST (Sonnenthal et al., 1998; Xu et al., 2001). The following sections discuss some of the important aspects of the THC model, followed by comparisons to measured geochemical data.

2. DRIFT SCALE TEST THC MODEL

2.1 Conceptual model for THC processes

The evolution of the chemical regime in the unsaturated zone surrounding a heat source is closely related to the spatial distribution of temperature and the transport of water and vapor. An important aspect of the unsaturated fractured tuff at Yucca Mountain is that the highly permeable fractures are nearly dry, whereas the low permeability and porosity rock matrix has a water saturation of about 90 percent. Thus, heating of the rock induces boiling of the matrix pore water, transport of water vapor into fractures, and condensation along fracture walls.

The numerical model for reaction-transport processes in the fractured welded tuffs must account for the different rates of transport in fractures, compared to a much less permeable rock matrix. Transport rates greater than the rate of equilibration via diffusion leads to disequilibrium between waters in fractures and matrix. Because the system is unsaturated, and undergoes boiling, the transport of gaseous species, especially CO_2, is an important consideration. The model must also capture the differences in initial mineralogy in fractures and matrix and their evolution.

Although mechanical processes were certainly active during the DST, measured changes in air permeability were typically less than one order of magnitude (BSC, 2002). Permeability reductions could also be attributed to increases in fracture liquid saturation in the condensation zones. In addition, variations in fracture permeability are at least a few orders of magnitude around the DST and therefore smaller mechanical effects would not have a major effect on changes in water and gas chemistry resulting from coupled THC processes.

2.2 Numerical grid and properties

The two-dimensional dual-permeability grid for the DST represents a vertical cross section through the Heated Drift at a distance approximately 30 m from the bulkhead, separating the Heated Drift from the Access Drift (Figure 1; adapted from Birkholzer and Tsang, 1998).

The top boundary is approximately 99 m above the drift center, with the bottom boundary located approximately 157 m below the center. Small grid blocks are employed adjacent to the wing heaters and drift wall to capture the strong gradients in temperature and liquid saturation in these regions.

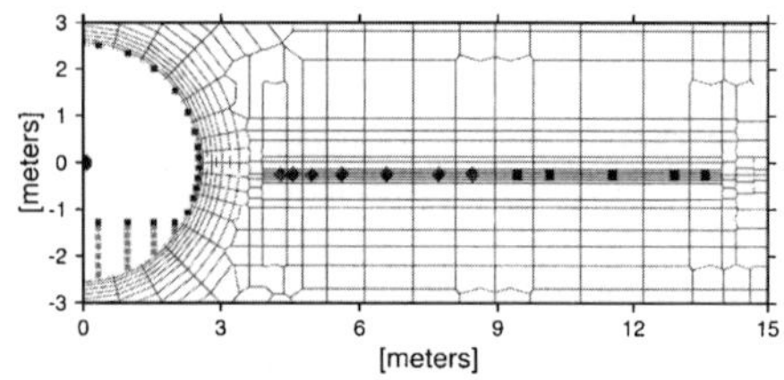

Figure 1. Partial view of 2-D numerical mesh for the DST (adapted from Birkholzer and Tsang, 1998). Symbols represent the central canister heater (red circle), wing heater locations (purple diamonds and red squares), concrete invert (green squares), and drift wall and invert top blocks (small red squares).

The Heated Drift grid block is connected directly to the Heater Test Alcove grid block and the connection area and distance were adjusted so that heat loss from the drift resulted in roughly similar crown temperatures to the maximum observed values. This was done to simulate heat and mass losses through the bulkhead, instead of reducing power by a factor. In the approximate location of the Observation Drift, the grid block volumes were increased to represent connection to the atmosphere. Details regarding the hydrological and thermal properties; thermodynamic and kinetic data, and; time-dependent heat input are discussed in Spycher et al. (2003).

2.3 Geochemical model

In order to handle separate yet interacting processes in fractures and matrix, the dual permeability method has been adopted, such that each grid block is divided into matrix and fracture continua, characterized by their own pressure, temperature, liquid saturation, water and gas chemistry, and mineralogy. Simulations of THC processes include coupling between heat, water, and vapor flow; aqueous and gaseous species transport; kinetic and equilibrium mineral-water reactions; and feedback of mineral precipitation/dissolution on porosity, permeability, and capillary pressure.

Aqueous species included in the model are H^+, Ca^{2+}, Na^+, K^+, $SiO_2(aq)$, Mg^{2+}, Al^{3+}, Fe^{3+}, SO_4^{2-},

HCO_3^-, Cl^-, F^-, and NO_3^-. The initial pore water chemistry is based on a sample that was ultracentrifuged from the proposed repository host rock (Tptpmn). Minerals considered include silica phases (α-cristobalite, quartz, tridymite, amorphous silica, and opal-CT), calcite, feldspars, smectites, illite, kaolinite, sepiolite, zeolites, fluorite, hematite, and gypsum. Treatment of CO_2 includes gas-water equilibration, diffusion, and advection.

2.4 TOUGHREACT code

Simulations of THC processes were performed using the TOUGHREACT code (Xu and Pruess, 2001; Xu et al., 2003). TOUGHREACT is based on TOUGH2 (Pruess, 1991) and in addition to the coupling between heat, water, and vapor flow for multiphase systems, it solves aqueous and gaseous species transport, kinetic and equilibrium mineral-water reactions, and feedback of mineral precipitation/dissolution on porosity, permeability, and capillary pressure.

TOUGH2 uses the integral finite difference method, with the equations for the multiphase system solved simultaneously, primarily with an iterative solver. Reaction-transport equations are solved using the sequential non-iterative operator splitting method, whereby each aqueous and gaseous component is transported separately by diffusion and advection, followed by solving the local kinetic and equilibrium mineral-water-gas reactions at grid block using Newton-Raphson iteration.

The rate of flow between fluids in fractures and in the rock matrix is a crucial factor for transport and reaction in fractures. For consistency with the formulation for flow between fractures and matrix used in the Yucca Mountain Project, the reactive surface area for minerals in unsaturated fractures has been related to the fracture-matrix interaction area based on a modified form of the "Active Fracture Model" (Liu et al., 1998; Sonnenthal et al., 2003). In this way, the wetted surface area for mineral-water reactions is consistent with that for flow and diffusion.

3. MODEL RESULTS AND COMPARISONS TO DATA

3.1 Gas-phase CO_2 evolution

The concentration of CO_2 in the gas phase is a function of temperature, pressure, aqueous-phase chemistry, mineral-water reactions, and advective and diffusive transport. From a model validation standpoint, the strong effect of CO_2 partial pressure on water pH and the final brine composition formed upon evaporation highlight the importance to the analysis of CO_2 concentrations. Simulation results are compared to concentrations measured in gas samples taken from boreholes during the heating phase of the DST. CO_2 concentrations in gases collected from the DST also provide a qualitative measure of the influence of atmospheric gas on the system, because of the relatively low and constant value in the atmosphere (≈400 ppmv).

The modeled evolution of CO_2 has been validated by comparison to over 4 years of measurements from the DST. Simulated CO_2 concentrations in the fracture gas after 3 years of heating (Figure 2) show the general outward migration of elevated CO_2 concentrations with the boiling front. The peak in CO_2 concentrations takes place at approximately 60°C, and at higher temperatures the concentrations generally decline as a result of degassing and transport with water vapor to cooler regions.

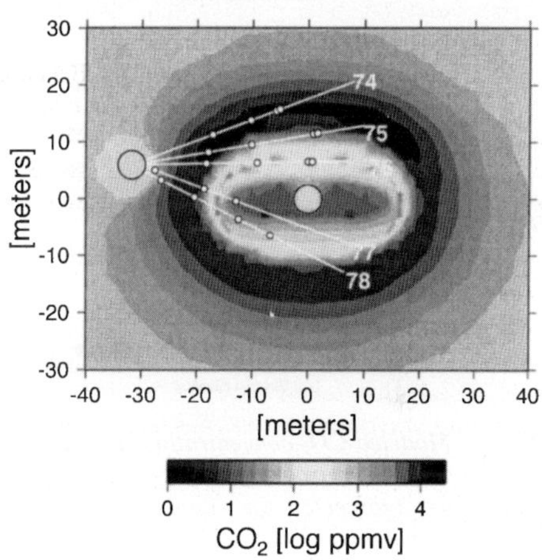

Figure 2. Modeled CO_2 concentrations in fractures after 3 years of heating. Note locations of numbered boreholes collared in the Observation Drift (circular region at left).

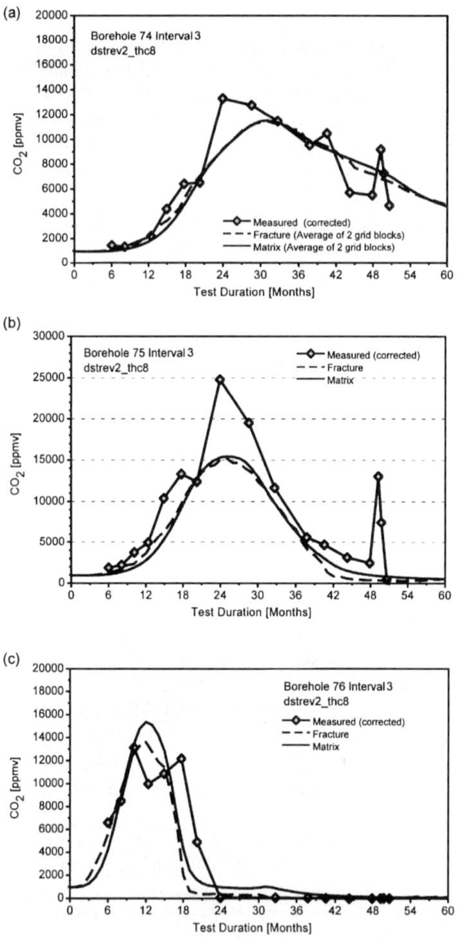

Figure 3. Modeled CO_2 concentrations in fractures and matrix compared to measured values from boreholes (corrected for vapor condensation) (a) Borehole interval 74-3 (average of bounding grid blocks); (b) Borehole interval 75-3; (c) Borehole interval 76-3.

Comparisons of modeled CO_2 to measurements performed on gas samples from boreholes (shown in Figure 2) are presented in Figure 3 (above). Samples were collected from zones a few meters (borehole interval 76-3) to about 15 meters away from the Heated Drift (borehole interval 74-3). Measured concentrations were corrected for water vapor condensation that took place as part of the procedure for gas sampling. Zones closest to the heaters (interval 76-3) exhibit narrower and earlier peaks in concentration compared to zones further out in the rock (interval 74-3).

Simulated and measured concentrations are close in magnitude and in their trends. There is little difference between fracture and matrix concentrations, because of rapid equilibration by advection and diffusion of gas species and their local equilibration with pore water.

3.2 Aqueous species evolution

The evolution of aqueous species in waters collected in the DST exhibit small reductions in pH (from about pH 8 in the pore water to about 6-8 in condensate waters. The drop in pH is related to the local increases in CO_2 concentrations. Figure 4 shows an example of the initial drop in pH during vapor condensation, followed by increasing pH as the zone is further heated and CO_2 is diluted by water vapor.

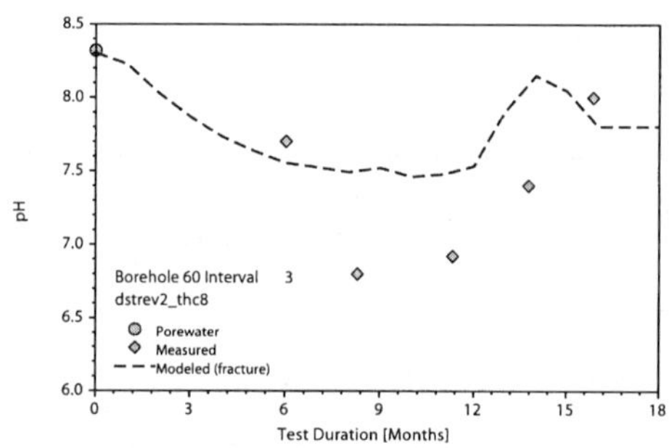

Figure 4. Measured and modeled pH (in fractures) for samples collected from borehole interval 60-3, located below the heaters.

Simulated and measured concentrations of conservative species in the fractures, such as chloride, are much lower than in the initial matrix pore water, indicating that fracture-matrix interaction has been negligible. However, reactive species, such as silica and potassium show significant effects of reaction with fracture-lining silicate minerals.

3.3 Mineral precipitation/dissolution

Model predictions, followed by analyses of *in-situ* sidewall core samples, showed that amorphous silica, calcite and lesser amounts of gypsum are the dominant phases expected to precipitate in the boiling regions. The greatest amount of mineral precipitation is predicted to be above the heaters where reflux of water condensed in fractures dissolves fracture-lining minerals and is boiled.

Simulations and measurements of amorphous silica and calcite, are shown in Figures 5 and 6. Amorphous silica forms only where strong evaporation by boiling takes place. In comparison, calcite that is originally present in fractures, dissolves in the lower pH waters in condensation zones around the boiling zone and in the drainage zones below the heaters. It also forms with amorphous silica and gypsum in the boiling zones.

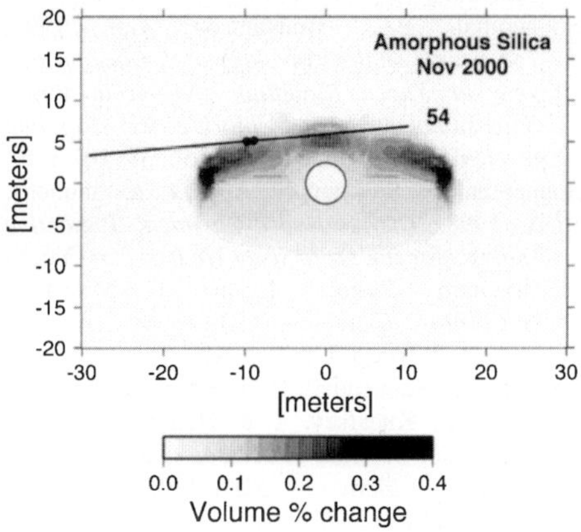

Figure 5. Volume percent change in amorphous silica abundance in fractures. Filled circle indicates sidewall core sample locations where it was observed.

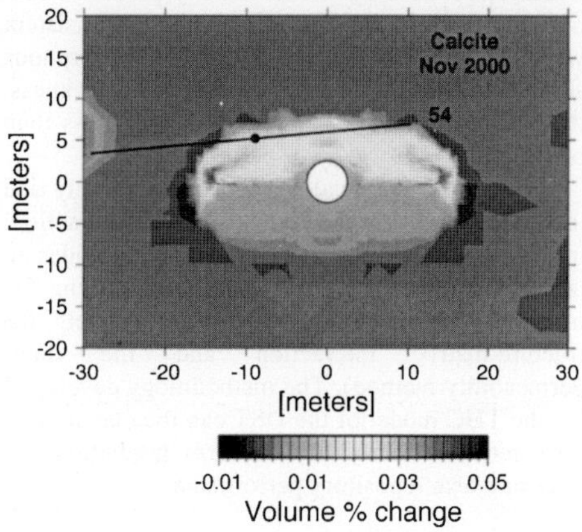

Figure 6. Volume percent change in calcite abundance in fractures. Filled circle indicates sidewall core sample locations where it was observed.

4. CONCLUSIONS

Prior to this study, little work had been done to model THC processes in unsaturated fractured rock under boiling conditions. The DST presented an unprecedented opportunity to test the conceptual models and codes on a well-constrained system, over time and spatial scales greater than simple lab experiments.

Comparisons of modeled concentrations and the chemistry of water (e.g., pH, $SiO_2(aq)$, Na^+, K^+) and gas samples (CO_2) collected over 5 years show that the THC model captures the trends and magnitude of chemical changes. CO_2 concentrations in gas samples increase from an initial value of about 1000 ppmv to greater than 25,000 ppmv as a result of exsolution from matrix pore water and transport into fractures. The CO_2 dissolves into condensate waters at lower temperatures, resulting in a drop in pH and also dissolution of calcite.

Based on the strong dilution trends of non-reactive species, such as chloride and sulfate, the extent of interaction of fracture condensate water with matrix pore water must have been very limited. However, reactive species, such as silica, show increasing concentrations owing to reaction with predominantly fracture-lining silica polymorphs and feldspars at higher temperatures. A precipitation zone of secondary mineral phases such as amorphous silica, calcite, and gypsum in fractures has also been successfully predicted in a narrow reflux zone above the heaters. Consistent with the observation of coatings of amorphous silica that are tens of microns or less in thickness, the predicted fracture porosity change is less than one percent.

Because the chemical changes are so closely tied to the TH processes and fracture-matrix interaction, modeling several chemical components provides an independent method for validating and refining TH models (e.g., the "Active Fracture Model" for fracture-matrix interaction and the dual-permeability method). The methodology developed for the THC model of the DST can then be applied with more certainty to long-term predictions of nuclear waste repository performance.

5. ACKNOWLEDGEMENTS

Aqueous geochemical data from the Drift Scale Test were obtained by Laura DeLoach (Lawrence Livermore National Lab). Schon Levy (Los Alamos National Lab) determined the mineral abundances in fractures. Sumit Mukhopadhyay and Jens Birkholzer developed the initial TH grid from which the THC model was developed. Sonia Salah worked on the thermodynamic and kinetic databases. This work was supported by the Director, Office of Civilian Radioactive Waste Management, U.S. Department of Energy, through Memorandum Purchase Order EA9013MC5X between Bechtel SAIC Company, LLC, and the Ernest Orlando Lawrence Berkeley National Laboratory (Berkeley Lab). The support is provided to Berkeley Lab through the U.S. Department of Energy Contract No. DE-AC03-76SF00098.

6. REFERENCES

BSC (Bechtel SAIC Company) 2002. *Thermal Testing Measurements Report.* Las Vegas, Nevada: Bechtel SAIC Company.

CRWMS M&O (Civilian Radioactive Waste Management System Management and Operating Contractor) 1998. *Drift Scale Test As-Built Report.* Las Vegas, Nevada: CRWMS M&O.

Birkholzer, J. T. & Tsang, Y. W. 1998. *Interpretive Analysis of the Thermo-Hydrological Processes of the Drift Scale Test.* In: Drift Scale Test Progress Report, Chapter 6. Yucca Mountain Project Level 4 Milestone SP2930M4. Berkeley, California: Lawrence Berkeley National Laboratory.

Datta, R., Barr, D., & Boyle, W. 2003. *Measuring thermal, hydrological, mechanical, and chemical responses in the Yucca Mountain Drift Scale Test.* Proceedings (to be published), GeoProc2003, Stockholm, Sweden, October 2003.

Liu, H.H., Doughty, C., & Bodvarsson, G.S. 1998. *An active fracture model for unsaturated flow and transport in fractured rocks.* Water Resources Research, 34(10), 2633–2646.

Pruess, K. 1991. TOUGH2: *A general purpose numerical simulator for multiphase fluid and heat flow.* Lawrence Berkeley National Laboratory Report LBL-29400, Berkeley, California.

Sonnenthal, E.L., Bodvarsson, G.S. 1999. *Constraints on the hydrology of the unsaturated zone at Yucca Mountain, NV from three-dimensional models of chloride and strontium geochemistry.* J. Contam. Hydrol., 38, 107-156.

Sonnenthal, E., Spycher, N., Apps, J., & Simmons, A. 1998. *Thermo-Hydro-Chemical Predictive Analysis for the Drift-Scale Heater Test.* Yucca Mountain Project Level 4 Milestone SPY289M4. Berkeley, California: Lawrence Berkeley National Laboratory.

Spycher, N., Sonnenthal, E., Dobson, P., Apps, J., Salah, S., Kneafsey, T & Haukwa C. 2003. *Drift-Scale Coupled Processes (DST and THC Seepage) Models.* MDL-NBS-HS-000001 REV02. BSC: Las Vegas, NV.

Tsang, Y., Birkholzer, J. 1999. *Predictions and observations of the thermal-hydrological conditions of the Single Heater Test.* Journal of Contaminant Hydrology, 38: pp. 385-425.

Xu, T., & Pruess, K. 2001. *Modeling multiphase non-isothermal fluid flow and reactive geochemical transport in variably saturated fractured rocks: 1. Methodology.* American Journal of Science, 301: pp. 16-33.

Xu T., Sonnenthal, E., Spycher, N., Pruess, K., Brimhall, G., & Apps, J., 2001. *Modeling multiphase non-isothermal fluid flow and reactive geochemical transport in variably saturated fractured rocks: 2. Applications to supergene copper enrichment and hydrothermal flows.* American Journal of Science, 301: pp. 34-59.

Xu, T., Sonnenthal, E., Spycher, N., & Pruess, K. 2003. *TOUGHREACT: A new code of the TOUGH family of non-isothermal multiphase reactive geochemical transport in variably saturated media.* Proceedings, TOUGH Symposium, 2003.

A RESEARCH PROGRAM FOR NUMERICAL EXPERIMENTS ON THE COUPLED THERMO-HYDRO-MECHANICAL AND CHEMICAL PROCESSES IN THE NEAR-FIELD OF A HIGH-LEVEL RADIOACTIVE WASTE REPOSITORY

Akira ITO[1], Mikazu YUI[1], Yutaka SUGITA[1], Susumu KAWAKAMI[1]
[1]) Japan Nuclear Cycle Development Institute

Abstract: In the near-field of a high-level radioactive waste (HLW) repository, the coupled thermo -hydro -mechanical and chemical (T-H-M-C) processes will occur. Observation periods by laboratory and in-situ experiments are very short and information by natural analogue studies is very limited, so numerical experiments are the only available approach to predict the near-field long-term evolution. We have initiated a research on the coupled T-H-M-C processes and are trying to carry out numerical experiments on the coupled T-H-M-C processes in order to predict the near-field long-term evolution. This paper presents our current status and future activities on the research for numerical experiments on the coupled T-H-M-C processes.

1. INTRODUCTION

The engineered barriers are composed of vitrified waste encapsulated in a canister, metal overpack and buffer material in a geological disposal system of high-level radioactive waste (HLW) adopted by JNC (2000). Highly compacted bentonite is considered as the candidate buffer material for the system. Figure 1 shows the schematic view of the geological disposal system and expected processes after emplacement of the engineered barriers.

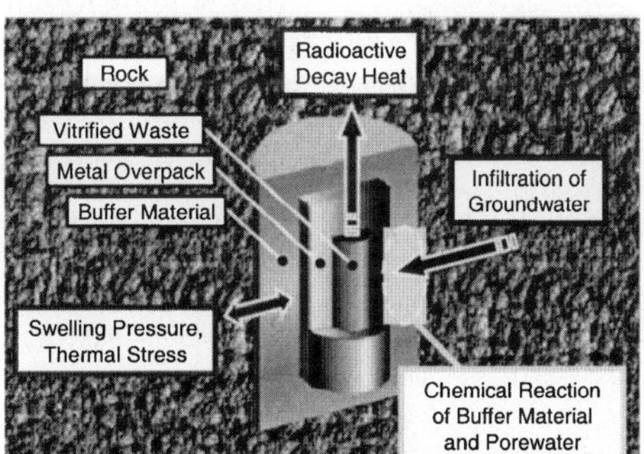

Figure 1. Schematic view of the geological disposal system and expected processes after emplacement of the engineered barriers

In the near-field of a HLW repository, the coupled thermo -hydro -mechanical and chemical (T-H-M-C) processes will occur, involving the interactive processes among radioactive decay heat from the vitrified waste, infiltration of groundwater into bentonite, swelling pressure of bentonite due to saturation and chemical reaction between bentonite and porewater. Observation periods by laboratory and in-situ experiments are very short and information by natural analogue studies is very limited, so numerical experiments are the only available approach to predict the near-field long-term evolution.

The coupled T-H-M model has been already developed through laboratory, in-situ experiments and DECOVALEX project (Stephansson et al. (1995) and Stephansson et al. (2001)), and is to predict maximum temperature, resaturation time and mechanical stability during the resaturation of engineered barriers. To understand and assess the long-term performance of a HLW repository, relevant predictions of the coupled T-H-M-C processes are required. The near-field chemistry, which is one of the most important factors for radionuclides migration and overpack corrosion, should be predicted under consideration of initial transient state, because the chemical process is strongly dependent on temperature, water movement and stress after the emplacement of engineered barriers. From a long-term viewpoint, the chemical process can lead to the changes in porosity, permeability and swelling pressure, and can affect the integrity of the near-field.

We have initiated a research on the coupled T-H-M-C processes and are trying to carry out numerical experiments on the coupled T-H-M-C processes in order to predict the near-field long-term evolution. The main objective of numerical experiments is to provide short-term confidence in performance assessment by pre-closure monitoring, and to predict precisely the near-field behaviour for long-term performance analysis by inclusion of the near-field chemical evolution. These approaches decrease the gap among the laboratory, in-situ experiment and the performance assessment, and will increase confidence in performance assessment.

2. DEVELOPMENT OF THE COUPLED T-H-M-C MODEL

2.1 Conceptual model

Interactions among thermal, hydrological and mechanical processes are well studied through DECOVALEX project. From a long-term viewpoint, chemical process can lead to the changes in thermal, hydrological and mechanical properties, and can affect the integrity of the near-field.

Some experimental studies are on-going by JNC to build the database for the interaction from chemical process to other processes, such as permeability and swelling properties of buffer material saturated by saline water and of buffer material reacted by cement. In the near future, we will introduce these databases to this research.

To develop the coupled T-H-M-C model, we sorted out the interactions among each process and built the conceptual model for the coupled T-H-M-C processes (Figure 2). Although many kinds of chemical process are assumed in the near-filed of a HLW repository, our current focal point is to take into account the interaction between minerals and porewater in the compacted bentonite.

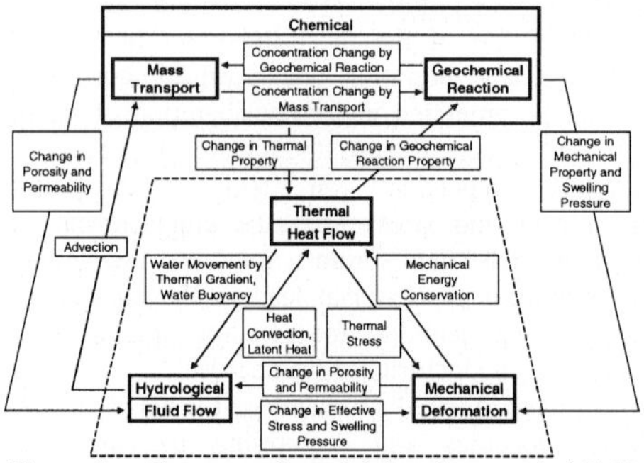

Figure 2. Conceptual model for the coupled T-H-M-C processes

2.2 Governing equations

From the conceptual model, we have developed the coupled T-H-M-C model which is based on the coupled T-H-M model (inside the broken line in Figure2; Ohnishi et al. (1985) and Chijimatsu et al. (2000)) and reactive-mass transport model (inside the box named "Chemical" in Figure 2). This is a system of governing equations composed of Equations (1)-(9), which couple heat flow, fluid flow, deformation, mass transport and geochemical reaction in terms of following primary variables: temperature T, pressure head ψ, displacement u_i, total dissolved concentration of the nth master species $C_{<n>}$ and total dissolved and precipitated concentration of the nth master species $T_{<n>}$. Here we set master species as the linear independent basis for geochemical reactions, and speciation in solution and dissolution/precipitation of minerals are calculated by a series of governing equations for geochemical reaction. Now we adopt equilibrium model for geochemical reaction (Parkhurst et al. (1980)), mainly because of reliability and abundance of thermodynamic data for geochemical reaction.

$$\frac{\partial}{\partial x_i}(\rho_l C_l q_l T)-\frac{\partial}{\partial x_i}\left(\lambda_m \frac{\partial T}{\partial x_i}\right)+\frac{\partial}{\partial t}(\rho_m C_m T)=0 \tag{1}$$

$$-\frac{\partial}{\partial x_i}\left\{\xi\rho_l D_\theta \frac{\partial \theta}{\partial x_i}+(1-\xi)\frac{\rho_l{}^2 gK}{\mu_l}\frac{\partial(\psi+z)}{\partial x_i}+\rho_l D_T \frac{\partial T}{\partial x_i}\right\} +\frac{\partial}{\partial t}(\rho_l nS)=0 \tag{2}$$

$$\frac{\partial}{\partial x_j}\left\{\frac{1}{2}C_{ijkl}\left(\frac{\partial u_k}{\partial x_l}+\frac{\partial u_l}{\partial x_k}\right)-\pi'\delta_{ij}-\frac{E}{1-2\nu}\delta_{ij}\alpha_s(T-T_0) +\chi\delta_{ij}\rho_l g\psi\right\}+\rho_m b_i=0 \tag{3}$$

$$\frac{\partial}{\partial x_i}\left(\rho_l q_l C_{\langle n\rangle}\right)-\frac{\partial}{\partial x_i}\left(\rho_l nSD_{ij}\frac{\partial C_{\langle n\rangle}}{\partial x_j}\right)+\frac{\partial}{\partial t}\left(\rho_l nST_{\langle n\rangle}\right)=0 \tag{4}$$

$$\sum_{a=1}^{A}(z_a m_a)=0 \tag{5}$$

$$\sum_{a=1}^{A}(v_a m_a)=OPV+\sum_{p=1}^{P}(u_p MIN_p) \tag{6}$$

$$C_{\langle n\rangle}=\sum_{a=1}^{A}(c_{a,n}m_a) \qquad C_{\langle n\rangle}=T_{\langle n\rangle}+\sum_{p=1}^{P}(b_{p,n}MIN_p) \tag{7}$$

$$\sum_{n=1}^{N}\{b_{p,n}\log(a_n)\}=\log(K_p) \tag{8}$$

$$\log(a_a)=\log(K_a)+\sum_{n=1}^{N}\{c_{a,n}\log(a_n)\} \tag{9}$$

Table 1 shows the nomenclature in Equations (1)-(9). Equation (1), (2) and (3) are the governing equations; of energy conservation for thermal process; of mass conservation of fluid for hydrological process; and of momentum conservation for mechanical process, respectively.

Equations (4)-(9) are governing equations for chemical process as follows. Equation (4) is for mass transport, and the governing equation of mass conservation of each master species through mass transport. A series of Equations (5)-(9) are for geochemical reaction, and the governing equations; of electrical neutrality; of conservation of electrons; of mass conservation of each master species through geochemical reaction; of mass action for each mineral; and of mass action for each aqueous species, respectively.

Table 1. Nomenclature in Equations (1)-(9)

A	Total number of aqueous species
a_a	Thermodynamic activity of the ath aqueous species
a_n	Thermodynamic activity of the nth master species
b_i	Body force
$b_{p,n}$	Stoichiometric coefficient of the nth master species in the pth mineral
$c_{a,n}$	Stoichiometric coefficient of the nth master species in the ath aqueous species
C_{ijkl}	Elastic matrix
C_l	Specific heat of liquid
C_m	Specific heat of medium
$C_{(n)}$	Total dissolved concentration of the nth master species
D_{ij}	Dispersion tensor
D_T	Thermal water diffusivity
D_θ	Isothermal water diffusivity
E	Young's modulus
g	Acceleration of gravity
K	Intrinsic permeability
K_a	Thermodynamic equilibrium constant for mass action equation for the ath aqueous species
K_p	Thermodynamic equilibrium constant for mass action equation for the pth mineral
m_a	Molality of the ath aqueous species
MIN_p	Moles of the pth mineral transferred into aqueous phase
N	Total number of master species
n	Porosity
OPV	Sum of operational valence of the speciation in the initial solution
P	Total number of mineral
q_l	Flux of liquid phase
S	Degree of saturation
T	Temperature
T_0	Initial temperature
$T_{(n)}$	Total dissolved and precipitated concentration of the nth master species
u_i	Displacement
u_p	Sum of operational valence of the constituents in the pth mineral
v_a	Sum of operational valence of the constituents in the ath aqueous species
z	Elevation head
z_a	Charge of the ath aqueous species
α_s	Linear thermal expansion coefficient of solid
δ_{ij}	Kronecker's delta
θ	Volumetric water content
λ_m	Thermal conductivity of medium
μ_l	Viscosity of liquid
ν	Poisson's ratio
ξ	Unsaturated parameter (equal to 0 at saturated zone and 1 at unsaturated zone)
π'	Swelling pressure contributing to swelling stress
ρ_l	Density of liquid
ρ_m	Density of medium
χ	Effective stress parameter
ψ	Pressure head

So as to predict precisely the near-field behaviour for long-term performance analysis by inclusion of the near-field chemical evolution, we think it important to take into account the interactive model from thermal, hydrological and mechanical processes to chemical process as the first step of this research. After this step, we can really introduce the interactive model from chemical process to other processes, and then we can realize the fully coupled T-H-M-C model.

3. PRELIMINARY ANALYSIS ON WATER MOVEMENT AND MASS TRANSPORT IN UNSATURATED BUFFER MATERIAL UNDER THERMAL GRADIENT

3.1 Analysis conditions

To demonstrate coupling between T-H-M and mass transport, we carried out preliminary analysis on water movement and mass transport in unsaturated buffer material under thermal gradient (Ito et al. (2002)). We assume a sample of bentonite-sand mixture 0.1 m in height (Kunigel V1 : Silica Sand = 70 wt% : 30 wt%, dry density ρ_d = 1.60 g/cm^3), and simulated saline water (containing NaCl at 3.3wt% per liquid mass) infiltrates from the top of the sample, with temperature at the top and bottom of the sample kept at 80 and 100 ^{O}C, respectively (Figure 3). Initial and boundary conditions for analysis are as shown in Table 2, and the initial condition for total concentration of NaCl $T_{<NaCl>}$ is set at 0.07 wt% as impurities in buffer material.

In this analysis, we used properties of buffer material obtained by laboratory experiments using distilled water, because unsaturated properties of buffer material with NaCl solution have not been well studied yet.

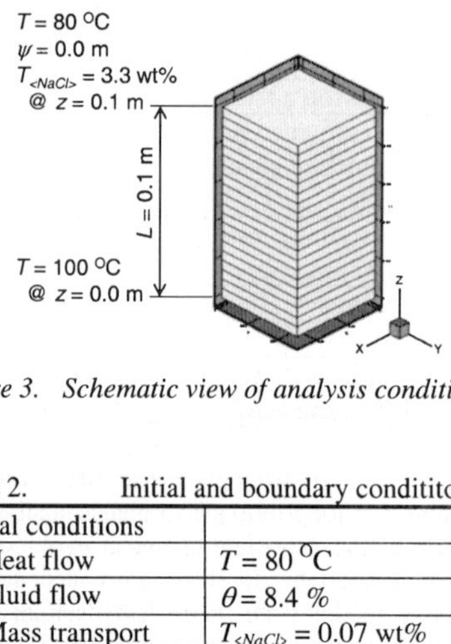

Figure 3. Schematic view of analysis conditions

Table 2. Initial and boundary condititons

Initial conditions	
1) Heat flow	$T = 80\ ^{O}C$
2) Fluid flow	$\theta = 8.4$ %
3) Mass transport	$T_{<NaCl>} = 0.07$ wt%
Boundary conditions	
1) Heat flow	$T = 80\ ^{O}C$ at $z = 0.1$ m $T = 100\ ^{O}C$ at $z = 0.0$ m Closed at vertical side
2) Fluid flow	$\psi = 0.0$ m at $z = 0.1$ m Closed at $z = 0.0$ m and vertical side
3) Mass transport	$T_{<NaCl>} = 3.3$ wt% at $z = 0.1$ m Closed at $z = 0.0$ m and vertical side

3.2 Analytical results

It is well known that water moves as both liquid and vapor in unsaturated porous medium under thermal gradient. The aqueous species is transferred only through liquid phase, so we have to distinguish liquid flux from water flux. Here we divide water flux q into liquid flux q_l and vapor flux q_v using the isothermal water diffusivity D_θ and thermal water diffusivity D_T as follows.

$$q = -D_\theta \frac{\partial \theta}{\partial x_i} - D_T \frac{\partial T}{\partial x_i} \tag{10}$$

$$q_l = -D_\theta \frac{\partial \theta}{\partial x_i} \tag{11}$$

$$q_v = -D_T \frac{\partial T}{\partial x_i} \tag{12}$$

Figure 4 shows the analytical results. As shown in Figure 4(a), volumetric water content in the middle and upper parts of the sample (z = 0.03 - 0.10 m) increases with time due to infiltration of saline water from the top of the sample. On the other hand, thermal gradient shown in Figure 4(c) generates the vapor movement from the bottom to top. So volumetric water content in the lower part of the sample (z = 0.00 - 0.03 m) decreases during about 5 days after loading heat at the bottom, and shifts to increase afterwards due to infiltration from the top (Figure 4(b)).

Figure 4(d) shows the liquid velocity V_l (equal to liquid flux q_l divided by volumetric water content θ), and we carried out the advection-dispersion transport analysis with this liquid velocity field. In this analysis, we considered only dissolution/precipitation of NaCl as geochemical reaction.

As shown in Figure 4(e), dissolved concentration of NaCl increases in all parts of the sample with time, and is concave downwards. This indicates that the diffusive transport is dominant in unsaturated buffer material. Dissolved concentration of NaCl in the lower part of the sample increases due to decrease in volumetric water content caused by vapor movement from the bottom to top (Figure 4(f)), but is not so large compared to that of infiltrating saline water at the top. Both precipitation of NaCl and local strong increase in dissolved concentration of NaCl are not found in all parts of the sample.

Based on this analysis, vapor movement due to thermal gradient does not show significant influence on dissolved concentration of NaCl in

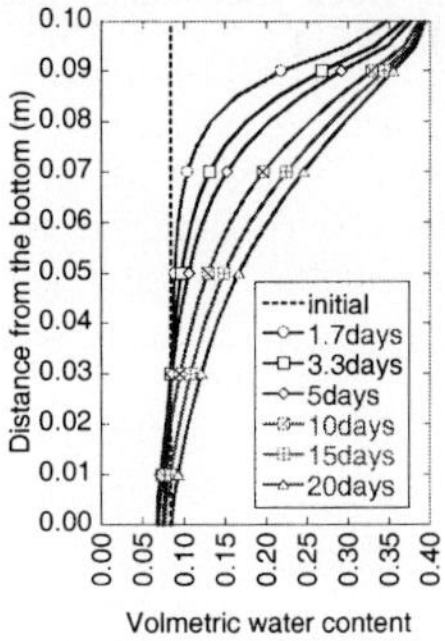

(a) Volumetric water content

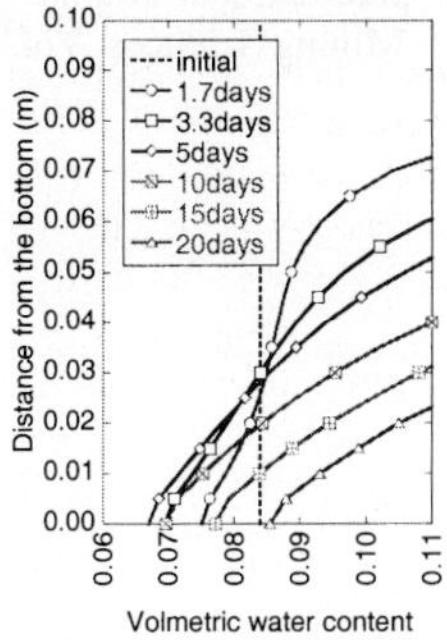

(b) Volumetric water content (Zoom up around the bottom)

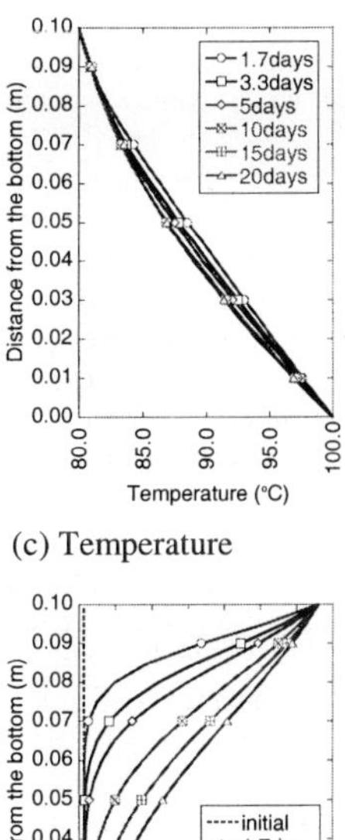

(c) Temperature

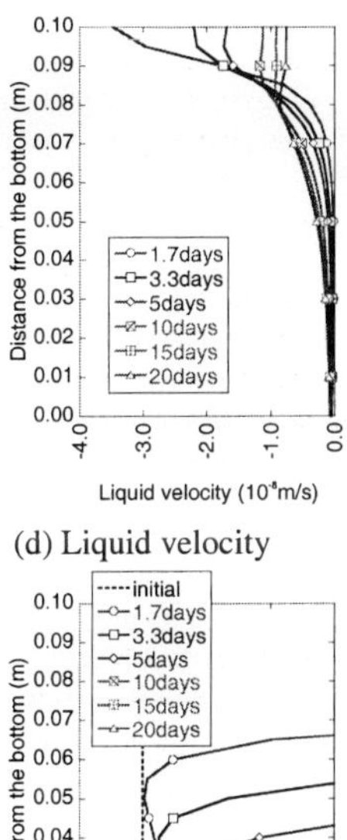

(d) Liquid velocity

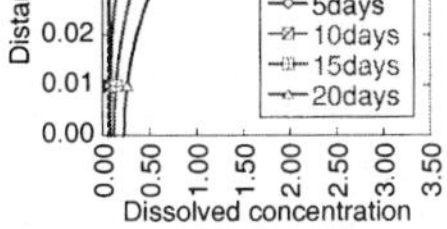

(e) Dissolved concentration of NaCl

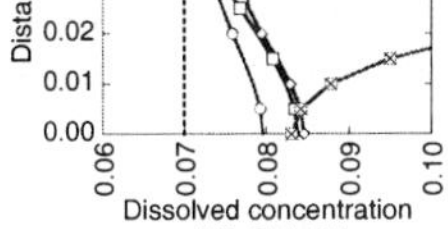

(f) Dissolved concentration of NaCl (Zoom up around the bottom)

Figure 4. Analytical results of volumetric water content, temperature, liquid velocity and dissolved concentration of NaCl

unsaturated buffer material. This indicates that (1) salt accumulation in buffer material is not so significant, if properties of buffer material with saline water are nearly the same as that with distilled water; and that (2) local increase in dissolved concentration of salt due to decrease in volumetric water content caused by vapor movement is expected to dissipate, if diffusive transport is dominant in unsaturated buffer material. So we infer that even if properties of buffer material with saline water are different from that with distilled water, salt accumulation in buffer material will not be significant as long as the transport is dominated by diffusion in unsaturated buffer material.

4. CONCLUSION

The coupled T-H-M processes are well studied through DECOVALEX project. To tighten the linkage between the research on the coupled processes and performance assessment, we think it necessary to put more attention to chemical process and have initiated the research on the coupled T-H-M-C processes. Observation periods by laboratory and in-situ experiments are very short and information by natural analogue studies is very limited, so numerical experiments are the only available approach to predict the near-field long-term evolution.

In this paper, we present the coupled T-H-M-C model and carry out the preliminary analysis on water movement and mass transport in unsaturated buffer material under thermal gradient. Based on the preliminary analysis, it is suggested that salt accumulation in buffer material will not be significant as long as the transport is dominated by diffusion in unsaturated buffer material. As shown by this analysis, numerical experiments on the coupled T-H-M-C processes can support our understanding of the near-field evolution in a HLW repository.

Our activities on development of the coupled T-H-M-C code and further demonstration analysis are presented by Neyama et al. in GeoProc 2003 conference, entitled "Prototype Code Development for Numerical Experiments on the Coupled Thermo-Hydro-Mechanical and Chemical Processes in the Near-field of a High-level Radioactive Waste Repository". The prototype code is based on the coupled T-H-M code, mass transport code and geochemical code which have been well verified and validated through benchmark tests and various experiments, and some of them are adopted in the second progress report on research and development for the geological disposal of HLW in Japan (H12 report; JNC (2000)).

So as to predict precisely the near-field behaviour for long-term performance analysis by inclusion of the near-field chemical evolution, we need to extend the interactive model from chemical process to other processes. In the near-future, we will introduce databases focused on chemical effects, e.g. high-pH plume by cementitious materials, on other processes through experimental studies on-going by JNC. We strongly believe numerical experiments on the coupled T-H-M-C processes will decrease the gap among the laboratory, in-situ experiment and performance assessment, and will increase confidence in performance assessment.

REFERENCES

Chijimatsu, M., Fujita, T., Kobayashi A. & Nakano, M. 2000. *Experiment and Validation of Numerical Simulation of Coupled Thermal, Hydraulic and Mechanical Behaviour in the Engineered Buffer Materials.* International Journal for Numerical and Analytical Methods in Geomechanics. Vol. 24: pp. 403 - 424.

Ito, A., Kawakami, S., Yui, M. & Chijimatsu, M. 2002. *Numerical Analysis about the Mass Transport in Unsaturated Buffer Material under Thermal Gradient (in Japanese).* Proc. the 57th Japan Society of Civil Engineers Annual Meeting. CS10-052: pp. 485 - 486.

JNC 2000. *Second progress report on research and development for the geological disposal of HLW in Japan, Project overview report.* JNC TN1410 2000-001.

Ohnishi, Y., Shibata, H. & Kobayashi, A. 1985. *Development of Finite Element Code for the Analysis of Coupled Thermo -Hydro - Mechanical Behaviors of a Saturated-Unsaturated Medium.* Proc. Int. Symp. on Coupled Process Affecting the Preformance of a Nuclear Waste Repository, Berkeley: pp. 263 - 268.

Parkhurst, D.L., Thorstnsen, D.C. & Plummer, L.H. 1980. *PHREEQE - A Computer Program for Geochemical Calculations.* U.S. Geological Survey. Water-Resources Investigations 80-96.

Stephansson et al., O. 1995. *Thermo -Hydro - Mechanical Coupling in Rock Mechanics.* International Journal of Rock Mechanics and Mining Sciences. Vol. 32. No. 5.

Stephansson et al., O. 2001. *DECOVALEX II.* International Journal of Rock Mechanics and Mining Sciences. Vol. 38. No. 1.

GEOMOD - AN INTEGRATED GEOSCIENTIFIC MODEL OF THE ÄSPÖ HARD ROCK LABORATORY, SWEDEN

R. Christiansson[1], J.A. Hudson[2], J. Berglund[3], M. Laaksoharju[4], H. Hakami[5], P. Vidstrand[6], J Sundberg[7]
[1]) SKB, Stockholm, Sweden [2]) Rock Engineering Consultants, UK
[3]) SwedPower, Gothenburg, Sweden [4]) Geopoint, Stockholm, Sweden
[5]) Itasca, Solna, Sweden [6]) Bergab, Gothenburg, Sweden,
[7]) Geo Innova, Linköping, Sweden

Abstract: This paper describes the GeoMod project, the aim of which is to update the Rhén *et al.*, 1997 geoscientific model of the Äspö Hard Rock Laboratory in Sweden. The update is based on geological, rock mechanics, thermal properties, geohydrological and hydrogeochemical models using results of all the investigations and experiments carried out at the Laboratory during its experimental phase since 1995. The project and this paper emphasise the integration of geoscientific disciplines and the extent to which individual model results support each other for the site description.

1. INTRODUCTION

The Äspö Hard Rock Laboratory (HRL) for studying various aspects of radioactive waste disposal is located in granitic rock on the south-east coast of Sweden. The facility is sited on an island in the Misterhult archipelago, just north of the nuclear power units. The underground access tunnel starts on the mainland in a direction towards the island of Äspö. At 200 m depth, the access ramp forms a spiral with 2.5 turns down to 450 m depth. The diameter of the spiral is 300 metres. A number of niches and shorter tunnels have been developed from the main tunnel. Most of the openings were developed by the drill and blast method, but the lower 50 m of the main tunnel (approx. 400 m of tunnel) has been excavated by TBM. Most of the openings have an areal size range of 25–35 m^2. Investigations from the surface began in 1986. By 1990, the site characterization was sufficient to provide a prediction of the conditions to be expected when tunnelling started later the same year. The geoscientific model was updated based on data obtained during construction, and finally completed in 1997 by Rhén *et al.* The model was based on data from a semi-regional study, deep boreholes and data collected during tunnelling. The site model integrated geology, hydrogeology, hydrogeochemistry and rock mechanics aspects. The engineering measures (grouting, rock support, etc) were also summarised.

From 1995, the Äspö HRL continued in an operational stage with various experiments for research and demonstration of disposal technologies. Additional characterisations were carried out via boreholes, often shorter than 100m. In addition, an extensive borehole monitoring programme has been in place and monitoring of ground water pressure and ground water chemistry has been carried out since 1990. These data, collected for various purposes, have contributed to a series of new, local geoscientific models (cf. Äspö HRL annual reports). This work has produced incremental geoscientific data and further site understanding, for particular locations at the site. The disciplines included are geology, rock mechanics, thermal properties, hydrogeology and hydrogeochemistry.

2. OBJECTIVES OF THE GEOMOD PROJECT

The objectives of the GeoMod project are to:

- provide an update of the geoscientific understanding of the Äspö HRL;
- set the boundary conditions for future experiments;
- develop a methodology for integration of the geoscientific disciplines; and
- provide a benchmark site descriptive model and associated methodology that could be useful for the on-going SKB site characterization work.

Thus, the work will provide on-going and future experiments at the Äspö HRL with an updated geoscientific model as a basis for identifying suitable experimental rock volumes and for establishing boundary conditions for the experiments. Most of the new data has been collected during the operational phase of different experiments conducted in the tunnel in the lower part of the Äspö HRL.

The updated model focuses on a volume which includes the tunnel spiral and will be contained within a cube having a 1 km side length, extending from the surface to 1000 m depth. This volume is to be linked to the earlier model (Rhén *et al.* 1997). Additionally, the development and refinement of the methodology and tools for geoscientific model construction will support the geoscientific characterization of a future repository site.

3. MODELLING STRUCTURE

In order to clarify the assumptions behind the models, the project has used a proposal by Olsson *et al.*, 1994 relating to the method for structuring these assumptions. The structure is presented in Table 1.

The basis for the model description is to specify the intended use of the model and the next step is to decide which physical processes should be included in the model. In some cases, these processes may be represented by constitutive equations.

Finally, the concepts required to solve the problem are defined, as specified below:

- the type of geometrical framework and framework-related parameters have to be defined;
- the types of material properties to be assigned to the domains defined by the geometrical framework must be decided upon;
- the spatial assignment method of the material properties within a domain has to be described;
- the model volume has a limited extent and the boundary conditions have to be defined in order to compute or judge the effects within the model; and
- the method for integrating different scientific disciplines into one logical and coherent model has to be established, so that synergy is achieved.

If required, a computational tool, numerical or analytical, that handles the processes and concepts should be chosen. Output parameters of interest for model purposes must then be defined.

Table 1. Structure for a condensed description of models in the Äspö HRL GeoMod project (Olsson *et al.*, 1994).

MODEL NAME	
Model scope or purpose Specification of the intended use of the model **Process description** Specification of the processes included in the model, definition of constitutive equations	
CONCEPTS	DATA
Geometrical framework and parameters	
Dimensionality and/or symmetry of model. Specification of what the geometrical (structural) units of the model are and the associated geometrical parameters (the ones fixed implicitly in the model and the variable parameters).	Specification of the size of the modelled volume. Specification of the source of data for geometrical parameters (or geometrical structure). Specification of the size of the geometrical units and resolution.
Material properties	
Specification of the material parameters contained in the model (it should be possible to derive them from the process and geometrical units).	Specification of the source of data for material parameters (could often be the output from some other model). Specification of the value of material parameters.
Spatial assignment method	
Specification of the principles for the way in which material (and if applicable geometrical) parameters are assigned throughout the modelled volume.	Specification of the source of data for model, material and geometrical parameters, as well as stochastic parameters. Specification of the results of the spatial assignment.
Boundary conditions	
Specifications of (type of) boundary conditions for the modelled volume.	Specification of the source of data on boundary and initial conditions. Specification of the boundary and initial conditions.
Numerical or mathematical tool Computer code used. **Output parameters** Specification of the computed parameters and possibly derived parameters of interest.	

4. MODELLING APPROACH

The modelling approach follows the strategies outlined for the on-going site investigations that SKB is currently conducting at two locations in Sweden. The strategies include modelling of geology (Munier *et. al.*, 2003), hydrogeology (Rehn *et. al.*, 2003), rock mechanics (Andersson *et al.*, 2002; Hudson *et al.*, 2002), hydrogeochemistry

(Smellie *et al.*, 2002) and thermal properties (Sundberg, 2003).

Modelling steps

The Site Descriptive Model should be a multidisciplinary interpretation of the relevant disciplines using available data. The modelling comprises the following iterative steps (Andersson, 2003):

- evaluation of primary data;
- descriptive and quantitative modelling in 3D; and
- overall confidence evaluation.

The measured (primary) data consist of a wide range of different measurement results. These data need to be checked for consistency and interpreted into a format amenable to three-dimensional modelling. Special attention is required here because some data at the Äspö HRL have been collected for specific purposes, sometimes using unique equipment for an individual experiment. The data are first evaluated within each discipline and then the evaluations are checked between the disciplines.

The actual three-dimensional modelling (i.e. estimating the distribution of parameter values in space) is made in a sequence where the geometrical framework is taken from the geological model and used by the rock mechanics, thermal and hydrogeological modelling. The hydrogeochemical description is to some extent developed independently, but its consistency with the hydrogeological description is checked. Even if most of these first steps in the development chain are made within each discipline, the other disciplines are kept sufficiently informed about the progress of these evaluations, e.g. through project group meetings.

After the individual discipline modelling and uncertainty assessment, a phase of overall confidence evaluation follows. Relevant members of the different modelling teams assess the suggested uncertainties and evaluate the feedback, in particular to the suggested geology, hydrogeology and hydrogeochemistry descriptions. After revisions, the teams reassess the model description with its uncertainty and confidence statements. These discussions should assess overall confidence by checking that:

- all relevant data are used;
- the different kinds of uncertainty are addressed; and
- suggested alternatives make sense and whether there is potential for additional alternatives.

Assessment of uncertainty and confidence

There are always uncertainties in interpreting measurements and rock parameters which vary in space. The three-dimensional description should present the parameters with their spatial variability over a relevant and specified scale, with the uncertainty included in this description. Different alternative descriptions have been discussed. In addition, it is necessary to describe the confidence in the model predictions.

For the GeoMod project, the first attempt was to increase the detail in the previous model by increasing the discrete descriptions of the geoscientific conditions. It has been found that the results using the unevenly distributed data are better explained using stochastic descriptions rather than discrete ones.

The confidence in the descriptive model is essentially a qualitative entity, although judgement of confidence can be based on quantitative analyses.

5. MODELLING RESULTS

The modelling of regional and local deformation zones in three dimensions at Äspö was conducted on the basis of their ductile history. A ductile precursor has been documented for all modelled zones but one (and may be valid for that one as well). The assumption is that ductile high strain zones of the scale considered (half a metre or more) are normally persistent over distances of more than a few tens of metres and are commonly the location of later reactivations. Ductile deformation was therefore normally a pre-requisite for interpolating structures with a distinct strike and dip over distances larger than a few tens of metres and also to extrapolate structures beyond the point of observation. Four deformation zones penetrate the central tunnel spiral at Äspö in the new model.

Although a good conceptual understanding of water-bearing structures at all scales has been developed through the studies performed at the Äspö site, the new model incorporates small-scale water bearing structures only to a minor extent. They have been considered, but the conceptual model only rarely lends support for interpolation between observation points, given the density of observations present in the Laboratory today. Compared to the older Äspö model (Rhen *et al.*

1997a,b,c) the new updated version is more comparable with the contemporary geometrical frames used by e.g. geology and rock mechanics. This resulted in the updated version containing fewer hydraulic conductors than previously. For the statistical description of both the rock and conductor volumes, the amount of data available in 1996 was as useful as the data now available, even though the amount has approximately doubled since 1996. However, some new knowledge is available. For example, one intriguing result is that the correlation between hydraulic conductivity and specific storage is not as simple as previously described, but exhibits a large variability and a lower lowest limit for the specific storage than previously prescribed based on the old rock mechanics interpretations.

In terms of hydrogeochemistry, the modelled present-day groundwater conditions at the Äspö HRL consist of a mixture in varying degrees of the following water types: brine, glacial, marine and meteoric water types. The HRL tunnel construction caused hydraulic changes, with the location, depth and hydraulic properties of the rock determining the degree of groundwater mixing. In the upper 150 m of the bedrock, the dominating component is meteoric water. At greater depths, 150-600 m, a brackish-saline water is present, consisting of proportions of present and ancient marine water and glacial melt water. Below this level, the saline water contains proportions of brine water of which the major portion has been stagnant for millions of years. The mixing portions, sinks and sources of the elements due to chemical mass-balance reactions such as calcite dissolution/precipitation, ion exchange, sulphate reduction and redox reactions have been quantified. Indications of upconing of saline water and calcite sealing of fractures have been observed and modelled. The important information concerning the origin, evolution and hydrodynamics of the groundwater at the Äspö site has been quantified, and the conceptual post-glacial hydrogeological model has been verified.

Rock mechanics investigations were focused on characterizing the spatial distribution of *in situ* stresses within the GeoMod area by conducting a number of numerical (3DEC) analyses. Results from pointwise stress measurements in boreholes were used as a basis for setting up the 3DEC model. Modelling the shearing behaviour of the major fracture zones, which were incorporated in the 3DEC model block, was an essential part of the analyses performed. The trajectories of the *in situ* principal stresses change direction when traversing fracture zones; the analyses clearly showed rotations of the stress tensor. Within a certain limited volume of rock neighbouring a major fracture zone, these stress changes should be included in the stability evaluations of a tunnel planned to pass through that volume of rock.

Thermal conductivities were measured and modelled from reference values of the thermal conductivity of different minerals and from the mineral composition of all Äspö samples in the SKB SICADA database. The thermal conductivity database has been analysed in terms of frequency, type of distribution, spatial distribution, variogram analyses, etc. The data have also been analysed according to the different rock types. However, there are uncertainties relating to the rock classification, mainly due to problems in distinguishing between the Äspö diorite and Ävrö granite, but also because different classification systems have been used.

6. MODEL INTEGRATION

The interactions between the component disciplines of the GeoMod project were studied by compiling an interaction matrix (Hudson, 1992). In this matrix, the five disciplines (geology, rock mechanics, hydrogeology, hydrogeochemistry and thermal properties/processes) are placed along the leading diagonal of the matrix from the top left to the bottom right, see Figure 1. Note that there is a clockwise interaction convention in the matrix so that, for example, the influences of rock mechanics processes on hydrogeology processes are located in Box 2,3; whereas the influences of hydrogeology processes on rock mechanics processes are located in Box 3,2. Working Groups were established for each discipline within the GeoMod project and asked to consider the main interaction processes.

The next step was to obtain an idea of how significant the interactions are. Each Discipline Working Group was then asked to assign a significance rating to the off-diagonal boxes in the row and column through their leading diagonal discipline box. The ratings used were 'high', 'medium', 'low' and 'no significance'. The significance of the interactive influences in each off-diagonal box was thus assessed by the two disciplines involved.

The conclusions relating to integration of the five disciplines that can be drawn from the interaction matrix compilation and analysis are as follows.

1,1 Geology	1,2 (High) Rock inhomogeneity and anisotropy, & fractures (geometry, etc.) affects rock properties/stresses	1,3 (High) Rock porosity, rock mass permeability, and water-rock interaction affects water flow	1,4 (High) Rock mineralogical composition and geometry of fractures affects hydrogeochemistry	1,5 (High) Mineralogical composition, porosity, textural and structural anisotropy effects
2,1 (Low) Stress data affecting the geo-interptretation of facture systems and rock mass	2,2 Rock Mechanics	2,3 (Medium) Spatial distribution of in situ stress and EDZ influences the hydrogeological regime	2,4 (High) Stress changes near fractures zones affecting flow may change precipitation	2,5 (Low) Rocks & fracture zones subjected to higher stresses may be more thermally conductive
3,1 (Medium) Hydrogeological tests and measures can affect the geo-interpretation of permeable features	3,2 (Medium) Water pressure changes the effective stress	3,3 Hydro-Geology	3,4 (Medium) The flow pattern affects dilution and mixing	3,5 (Low) The flow pattern will affect the temperature due to convection effects
4,1 (Low) Hydrogeochemistry interpretation affecting the interpretation of fracture min.rel.	4,2 (Medium) Precipitation in fractures affects the fracture stiffnesses, strengths and creep properties	4,3 (Medium) Effects of ground-water age, density, viscosity, and dissolution and precipitation	4,4 Hydro-geo-chemistry	4,5 (No effect)
5,1 (Medium) Thermal anisotropy and measurement affecting the geo-interpretation	5,2 (High) Change in temperature can change the local stress, possibly leading to failure	5,3 (Medium) Temperature gradients and thermal expansion in rock/fractures affect the water flow	5,4 (Medium) Dissolution and precipitation enhanced by thermal gradients	5,5 Thermal Properties

Figure 1. The discipline interaction matrix content, plus the significance allocated to each off-diagonal box, as 'high', 'medium', 'low', or 'no effect', indicated in parentheses in each box.

- The identification of the interactions was straightforward, with the two respective disciplines contributing to each of the separate 20 interactions being in agreement. This indicates a consistent inter-disciplinary understanding of the complete interactive model system.
- Of the 20 interaction boxes in the 5x5 interaction matrix in Figure 1, there was only one box in which no significant interaction was identified. Thus, there is a large degree of interaction between the five disciplines.
- There was also inter-disciplinary agreement about the significance ratings, which indicates that the different disciplines have a consistent view of the significances of the different interactions.
- 15 of the 20 interactions were rated as having a 'High' or 'Medium' significance, indicating that the majority of interactions are significant.
- The model system as a whole has a relatively high interaction intensity.

7. DISCUSSION AND CONCLUSIONS

One of the major contributions from the technical development in site investigation tools during recent years is the Boremap system for core logging that has been used at Äspö HRL. The use of regular core mapping,together with photographic images of the core in Boremap, delivers high quality oriented data from the boreholes, making 3D modelling much simpler. This kind of data has also shown that ductile shear zones in the dm to m scale are more common than previously realised. The narrow shear zones at the Äspö laboratory and their relatively minor extent do not seem to have influenced the stability of the tunnel systems, but they most certainly have played a significant role in the geological history during pressure cycles. Local strain has produced local failures and thus local fracture systems, within the zones themselves and in the volumes of rock between them.

In the context of hydrogeology, as clearly indicated by the interaction matrix in Fig. 1, the geological geometrical description in particular has a large impact on a 3D site descriptive hydrogeological model. Thus, this description should preferably be finalised before the start of analyses in the other disciplines. However, the integrated work (implemented by e.g. subject presentations, personal discussions, and direct observations) involving different teams has helped to establish good mutual understanding, both on specific characteristics of individual rock volumes and the behaviour of the hydraulic boundaries.

Multi-component mixing has occurred, together with water-rock interaction processes, to modify the groundwater composition. The many limitations of existing geochemical models used at the Äspö site, and the need to decode the complex groundwater information in terms of origin, mixing and reactions at site scale, necessitated the development of a new modelling tool. This new modelling concept was named M3. This code should be used together with standard groundwater modelling and evaluation techniques.

It has been informative to attempt the correlation of the findings of different disciplines, which were all obtained from the specific volume of rock at Äspö HRL. Attempts in outlining integrated systems, based on how geological, hydrogeological, geochemical, and rock mechanics features interact should, however, continue beyond the scope of this project in order to increase our understanding of the methodology. This will support rock mass characterisation for siting a repository in the future. As an example, knowledge of the spatial distribution of *in situ* stresses is not only useful in determining the conditions for siting a repository and later in stability analyses during the construction phase, but it also helps in understanding interactive aspects such as the characteristics of the water conducting structures, the pattern they occupy within a rock mass, and ways of limiting or re-routing the flow through those structures.

An interesting relation has been observed between thermal conductivity and density for the rock types at Äspö, and this may be used to evaluate the spatial distribution of the thermal properties from density logging. There is currently insufficient knowledge concerning the variation of thermal properties at different scales. If the whole observed variation within a rock type is based on the cm-m scale, the thermal influence at the canister scale is small. This is due to the fact that the small-scale variation in thermal properties is mainly averaged out on the 5-10 m scale. If the main variation within rock types is on the 5-10 m scale, there is probably a significant effect on the canister temperature. However, it is likely that the observed variation occurs on both these scales.

REFERENCES

Andersson, J. 2003. *Site Descriptive Modelling. Strategy for integrated evaluation.* SKB TR-03- SKB, Stockholm, Sweden.

Andersson J., Christiansson R. & Hudson J. A 2002. *Site investigations. Strategy for rock mechanics site descriptive model.* SKB TR-02-01. SKB, Stockholm, Sweden.

Hudson J. A. 1992. *Rock engineering systems: theory and practice.* Ellis Horwood Ltd., Chichester, UK.

Hudson J. A. (ed.). 2002. *Strategy for a rock mechanics site descriptive model. A test case based on data from the Äspö HRL.*. SKB R-02-04. SKB, Stockholm, Sweden.

Munier R., Stanfors R., Stenberg L., Milnes A. G., Triumf C.-A. 2003. *Geological site descriptive model – a strategy for the model development during site investigations.* SKB R-03-07. SKB, Stockholm, Sweden.

Olsson O., Bäckblom G., Gustafson G., Rhén I., Stanfors R. & Wikberg P., 1994. *The structure of conceptual models with application to the Äspö HRL Project.*. SKB TR 94-08. SKB, Stockholm, Sweden.

Rhén I. (ed), Bäckblom G., Gustafson G., Stanfors R. and Wikberg P., 1997a. *Äspö HRL - Geoscientific evaluation 1997/2. Results from pre-investigations and detailed site characterization. Summary report.* SKB TR 97-03. SKB, Stockholm, Sweden.

Rhén I., Gustafson G., Wikberg P., 1997b. *Äspö HRL - Geoscientific evaluation 1997/4. Results from pre-investigations and detailed site characterization. Comparison of predictions and observations. Hydrogeology, groundwater chemistry and transport of solutes.* SKB TR 97-05. SKB, Stockholm, Sweden.

Rhén I. (ed), Gustafson G., Stanfors R. and Wikberg P., 1997c. *Äspö HRL - Geoscientific evaluation 1997/5. Models based on site characterization 1986-1995.* SKB TR 97-06. SKB, Stockholm, Sweden.

Rhén I., Follin S., Hermansson J., 2003. *Hydrogeological site descriptive model – a strategy for its development during site investigations.* SKB R-03-08. SKB, Stockholm, Sweden.

Smellie J., Laaksoharju M., Tullborg, E.-L., 2002. *Hydrogeochemical site descriptive model - a strategy for the model development during site investigations.* SKB R-02-49. SKB, Stockholm, Sweden.

Sundberg J. 2003. *Site Investigations - Thermal site descriptive model. A strategy for model development during site investigations. Version 1.0.* SKB R-03-10. SKB, Stockholm, Sweden.

PROTOTYPE CODE DEVELOPMENT FOR NUMERICAL EXPERIMENTS ON THE COUPLED THERMO-HYDRO-MECHANICAL AND CHEMICAL PROCESSES IN THE NEAR-FIELD OF A HIGH-LEVEL RADIOACTIVE WASTE REPOSITORY

Atsushi NEYAMA[1], Akira ITO[2], Masakazu CHIJIMATSU[3], Yoshinao ISHIHARA[4], Tomoyuki HISHIYA[5], Mikazu YUI[2], Yutaka SUGITA[2], Susumu KAWAKAMI[2]
[1]) Computer Software Development Co., Ltd.
[2]) Japan Nuclear Cycle Development Institute.
[3]) Hazama Corporation.
[4]) Mitsubishi Heavy Industries Ltd.
[5]) Dia Consultants Company Limited.

Abstract: Numerical experiment is the only available approach to demonstrate long-term complex evolution of a high-level radioactive waste repository. It is very difficult to construct the code for coupled processes by using individual process codes in a short time, because detailed modification for coupling is needed in each code. We have developed the coupling system "COUPLYS" that is very effective to improve the efficiency and quality for linkage of existing individual process codes. Using this coupling system "COUPLYS", the prototype code for the coupled T-H-M-C processes has been developed.

1. INTRODUCTION

Geological disposal of high-level radioactive waste (HLW) in Japan is based on a multibarrier system composed of engineered and natural barriers. The engineered barriers are composed of vitrified waste encapsulated in a canister, metal overpack and buffer material. Highly compacted bentonite is considered as the candidate buffer material, mainly because of its low hydraulic conductivity and high sorption capacity of radionuclides.

In a HLW repository, the coupled thermo -hydro -mechanical and chemical (T-H-M-C) processes will occur, involving the interactive processes among radioactive decay heat of the vitrified waste (thermal processes), infiltration of groundwater (hydrological processes), swelling pressure of buffer material due to saturation (mechanical processes) and chemical evolution of buffer material and porewater (chemical processes).

Japan Nuclear Cycle Development Institute (JNC) has already developed the coupled thermo - hydro and mechanical (T-H-M) model and has initiated a research on the coupled T-H-M-C processes to predict the chemical evolution of buffer material and porewater chemistry, and the chemical effects on other (thermal, hydraulic and mechanical) processes. In this research, numerical experiment system for the coupled T-H-M-C processes is developed in order to predict the long-term evolution of the near-field (engineered barriers and surrounding host rock) for various repository designs and geological environments.

It is very difficult to construct the code for coupled processes by using individual process codes in a short time, because detailed modification for coupling is needed in each code. We have developed the coupling system "COUPLYS" (Neyama et al. (2002)) that is very effective to improve the efficiency and quality for linkage of individual process codes. Using this coupling system "COUPLYS", the prototype code for the coupled T-H-M-C processes has been developed. The coupled T-H-M-C processes in the prototype code are analysed by the coupled T-H-M code "THAMES" (Chijimatsu et al. (1999)), mass transport code "Dtransu-EL" (Nishigaki et al. (2001)) and geochemical code "PHREEQE" (Parkhurst et al. (1980)). This paper presents the prototype code development and typical results of exercise by the prototype code.

2. DEVELOPMENT OF PROTOTYPE COUPLING SYSTEM

The coupling technique was selected under consideration of use of the existing analysis codes, from the viewpoint of the efficiency and quality to develop the coupling code. And we develop the coupling technique which is not dependent on the programming language of the existing analysis codes.

2.1 Memory management technique

Coupling code can be developed efficiently by this concept. An end-user defines the block of data definition based on the information of coupling data. This data definition includes variable name, type, dimension and size for each variable by text format. This technique uses "Data Definition Language (DDL)" to describe the coupling data among each analysis code. Module of shared memory management prepares the reference table to access coupling data, and this module ensures memory area based on data definition by "DDL". Reference table controls the correlation between variable name and address in the shared memory. Interface program has a function to access coupling data between shared memory and each analysis code by using command of "get" and "set" described in source program of each analysis code. And then interface program transfers coupling data among each analysis code through shared memory.

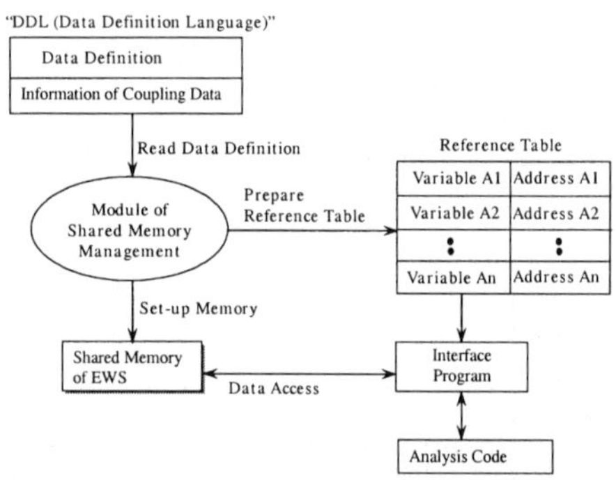

Figure 1. Memory management technique by "DDL"

2.2 Process management technique

This process management program controls the start-up / termination and pause / restart of process, based on the input data of process management in text format, in which the executive order of the processes and termination conditions are described. This program adopted "semaphore" to control the process of each analysis code. The "semaphore" is implemented on the UNIX machine by the standard, and it is the general-purpose function, which aims synchronous among each process. Figure 2 shows the binary "semaphore" for simple example. The "semaphore" is the flag for execution and termination of process. Process A and B refer to this flag, so the execution and termination of process are done alternately. And as the characteristics of "semaphore", a process during standing-by doesn't influence CPU. "COUPLYS" adopted "semaphore" set, which could use value from 0 to 32767 to take plenty of processes into consideration.

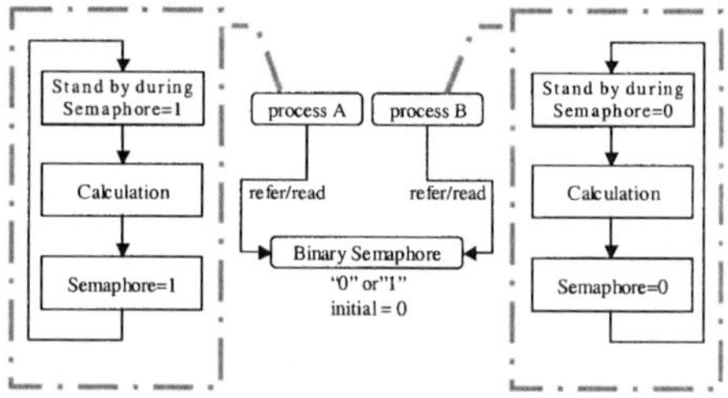

Figure 2. Process management technique by "semaphore"

3. PROTOTYPE CODE DEVELOPMENT

3.1 Coupling of existing analysis code

Existing codes of THAMES, Dtransu and PHREEQE are used in the prototype code based on the concept of the Figure 1 and Figure 2. Prototype coupling system has a function of start-up / termination and pause / re-start of each code, and command of "get" and "set" for coupling data is described in source program of each analysis code (Figure 3). This prototype code is executed on a SUN ULTRA, Linux and SGI workstation.

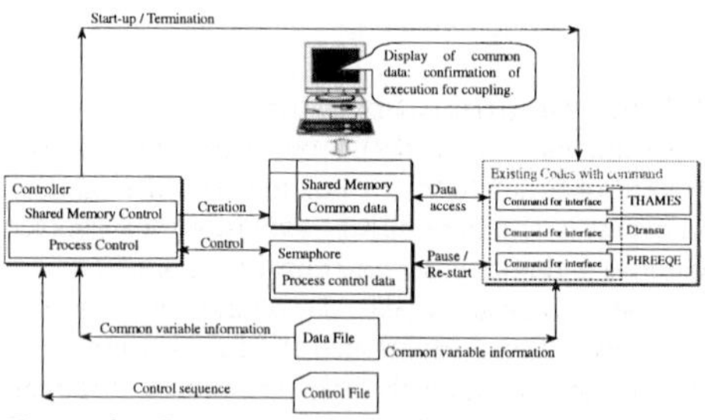

Figure 3. Prototype coupling system by "COUPLYS"

Data file in Figure 3 defines variable name, type, dimension and size for coupling data by text file. And control file defines procedure of coupled T-H-M-C analysis by each analysis code and coupling module. Coupling module contains the treatment necessary for coupling analysis.

For example, as shown in Figure 4, the result of geochemical calculation is delivered to mass transport code, since the unknown variable of mass transport equation is related to the dissolved and precipitated concentration $T_{\langle n \rangle}$. The mass conservation in mass transport is confirmed by introducing the factor γ in the time discretization of mass transport equation. The factor γ is calculated from the result of geochemical analysis and the dissolved concentration $C_{\langle n \rangle}$ is calculated by using the factor γ in mass transport analysis during the numerical iteration between mass transport analysis and geochemical analysis. This procedure is dealt with not in the existing code but in the additional coupling module.

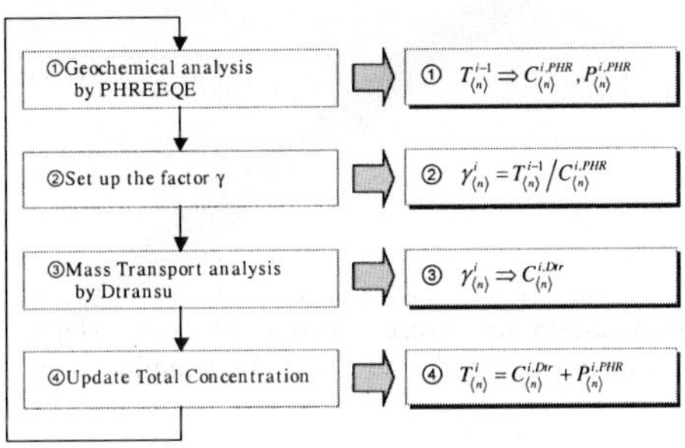

Figure 4. Coupling procedure between mass transport analysis and geochemical analysis

3.2 Verification of prototype code

Comparison with the existing code HYDROGEOCHEM, which couples mass transport and geochemistry, and the prototype code is carried out in order to confirm the applicability of the prototype code. A verification analysis assumes that calcite is filled within one-dimensional column and dissolved from one side. Mass transport is dealt with as diffusion in one-dimensional column. Geometry of one-dimensional column is shown in the Figure 5. And properties of the medium, initial condition and boundary condition are shown in the Table 1. In this verification analysis, calcite dissolves gradually in the end part of column and it is exhausted, because boundary is a pure solution in the end part (nodal point 21 and 22) of the column. Since calcite is exhausted at the end part, aqueous concentration of Ca and C become smaller than that in dissolution equilibrium of calcite and pH also becomes smaller than that in dissolution equilibrium of calcite. This interpretation is shown in Figure.6, 7, 8 and 9.

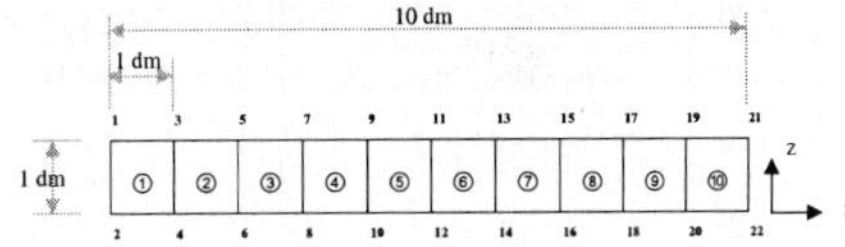

Figure 5. Geometry of one-dimensional column for verification analysis

Table 1. Conditions for verification analysis

Parameter	Unit	Value
Diffusion Coefficient	m^2/s	3×10^{-10}
Porosity in the column	-	0.33
Boundary total concentration Nodal point 1, 2 [Ca, C]	mol/l	1×10^{-2}
Boundary total concentration Nodal point 21,22 [Ca, C]	mol/l	1×10^{-10}
Initial total concentration Nodal point 3 ~ 20 [Ca, C]	mol/l	1×10^{-3}

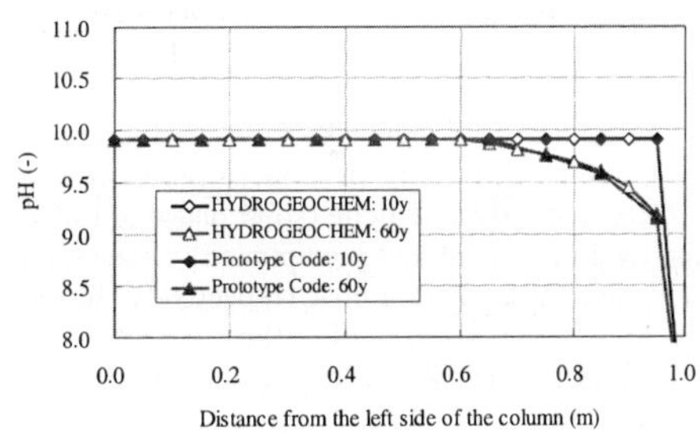

Figure 6. Distribution of pH

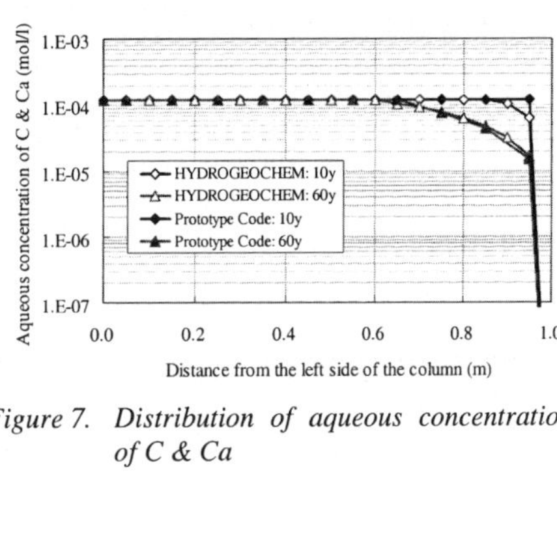

Figure 7. Distribution of aqueous concentration of C & Ca

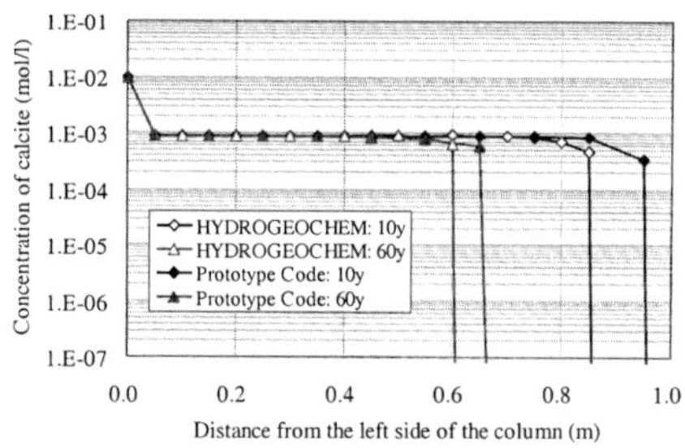

Figure 8. Distribution of concentration of calcite

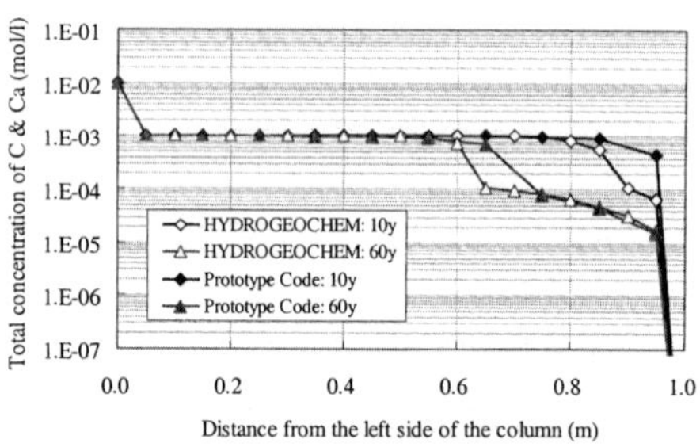

Figure 9. Distribution of total concentration of C & Ca

Figure 6, 7, 8 and 9 show the results of distribution of pH, aqueous concentration of C and Ca, concentration of calcite (solid phase of $CaCO_3$) and total (dissolved and precipitated) concentration of C and Ca in the column, respectively. In these figures comparison of the results with HYDROGEOCHEM and the prototype code at 10 years and 60 years are shown. While HYDROGEOCHEM calculates the geochemistry at each node, the prototype code calculates geochemistry at each finite element in order to treat the multi-medium easily. This is because we have to treat multibarrier system in the near-field of the HLW repository. We have confirmed agreement between the results of HYDROGEOCHEM and the prototype code. Little difference between them is caused by the difference between geochemical calculation at each node and that at each finite element.

4. DEMONSTRATION ANALYSIS

In this study, we have tried the one-dimensional and three-dimensional demonstration analysis in order to confirm the applicability of prototype code on "COUPLYS".

4.1 One-dimensional demonstration analysis

The near-field geometry of one-dimensional demonstration analysis is shown in the Figure 10. This test case is Ca-C-Mg-Na-K-Fe-Al-Si system because smectite, calcite and chalcedony are assumed in the buffer material region, and calcite and chalcedony are assumed in the hard rock region.

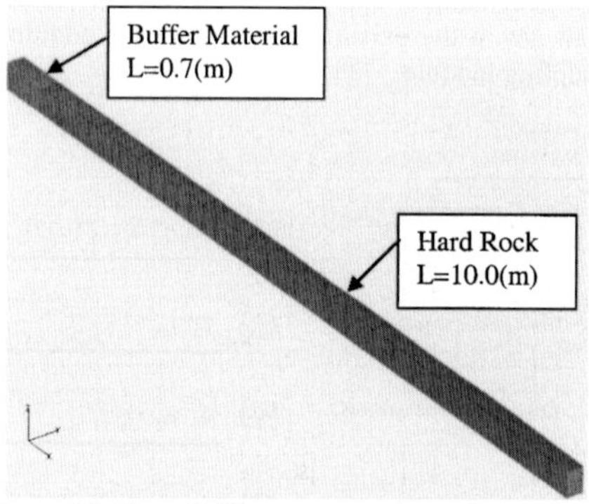

Figure 10. Near-field geometry of one-dimensional demonstration analysis

Properties of the medium, initial condition and a boundary condition for thermal process, hydrological process, mass transport and geochemistry are shown in the Table 2. Temperature is fixed at 80°C in the inner boundary of buffer material and the outer boundary of hard rock is assumed adiabatic condition. And the buffer material is unsaturated in initial condition; on the other hand hard rock is saturated in initial condition.

Figure 11, 12, 13 and 14 show the result of the temperature, saturation, pH and concentration of Ca(aq) in buffer material and hard rock, respectively. In these figures, results at 1 year and 10 years are shown and vertical line at x= 0.7 m shows the buffer material / hard rock boundary. As shown in Figure 11, 12, 13 and 14, the results at 1 year is the stage of heat conduction from the inner boundary of buffer material to the outer boundary of hard rock and groundwater infiltration into buffer material from hard rock. And pH and concentration of Ca(aq) at 1 year show the transient state influenced by temperature. At 10 years, heat conduction is over and buffer material is resaturated. And pH and concentration of Ca(aq) at 10 years show the steady state, because temperature becomes constant in all region. Since solubility of minerals depends on temperature, the results of porewater chemistry are highly dependent on temperature.

Table 2. Conditions for one-dimensional demonstration analysis

Parameter	Unit	Buffer Material	Hard Rock
Permeability	m^2	4×10^{-20}	1×10^{-15}
Porosity	-	0.403	0.005
Thermal Conductivity	$Wm^{-1}K^{-1}$	1.0(Init. Cond.) ~ 2.0(Sat. Cond.)	2.8
Specific Heat	$kJkg^{-1}K^{-1}$	0.7(Init. Cond.) ~ 1.1(Sat. Cond.)	1
Diffusion Coefficient	m^2/s	5×10^{-10}	5×10^{-10}
Initial Temperature	^oC	45	45
Initial Saturation	%	40	100
Thermal Boundary	-	T=80°C	adiabatic
Hydraulic Boundary	-	No flow	h=50m
Minerals	-	Calcite Chalcedony Smectite	Calcite Chalcedony

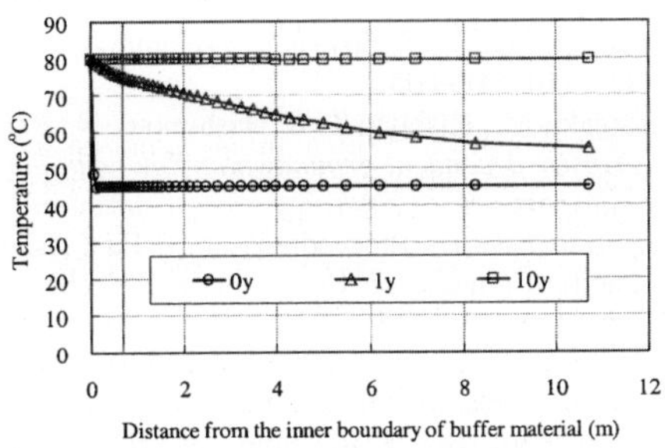

Figure 11. Distribution of temperature

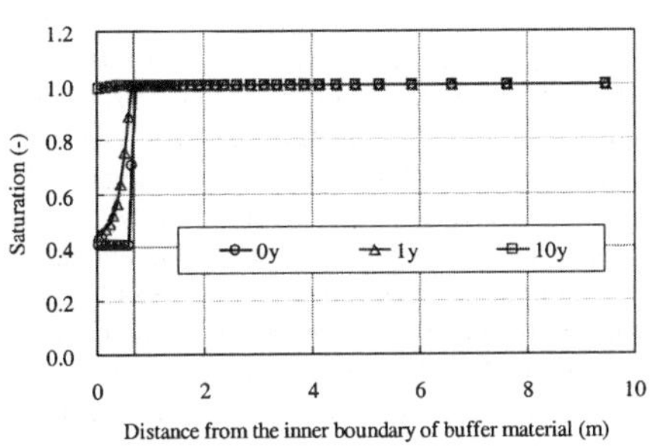

Figure 12. Distribution of saturation

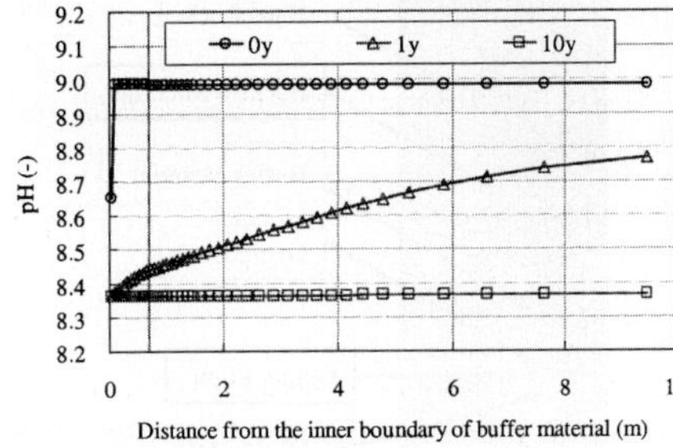

Figure 13. Distribution of pH

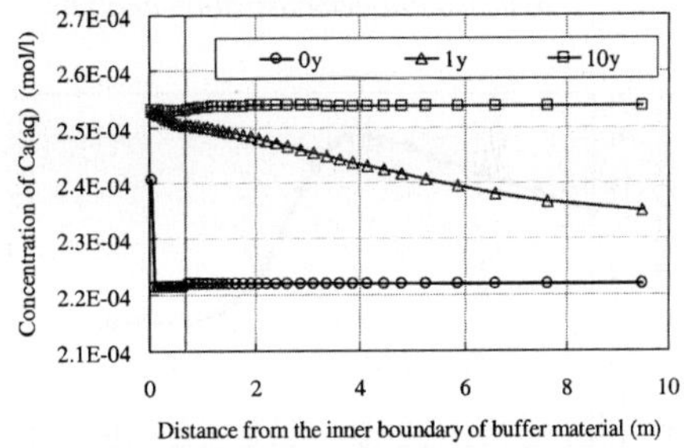

Figure 14. Distribution of concentration of Ca(aq)

4.2 Three-dimensional demonstration analysis

The near-field geometry of three-dimensional demonstration analysis is shown in the Figure 15. This test case is Ca-C-Mg-Na-K-Fe-Al-Si system because smectite, calcite and chalcedony are assumed in the buffer material region and the backfill region. And calcite and chalcedony are assumed in the hard rock region. Heat is generated from the region of vitrified waste based on the second progress report (JNC (2000)). All side surfaces are assumed adiabatic and impermeable boundary. Boundary temperature and hydraulic head on the upper and lower surface are fixed.

Figure 16 and 17 show time history of temperature and pH inside buffer material (at the boundary between overpack and buffer material as shown in Figure 15), respectively. The pH inside buffer material highly depends on temperature and becomes constant at 1,000 years. Through this type of numerical experiments, we can predict near-field chemistry for radionuclides migration and overpack corrosion.

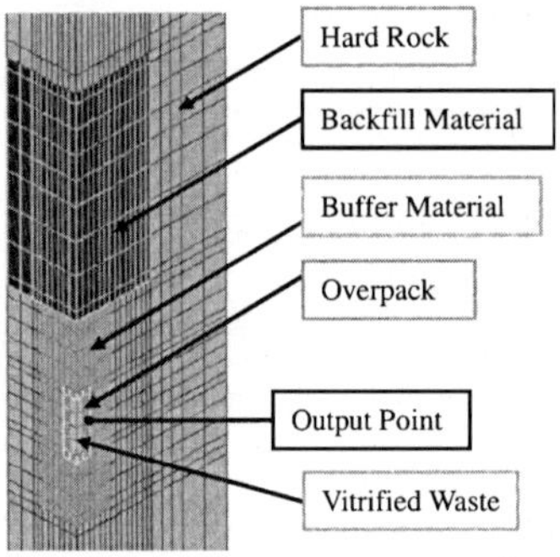

Figure 15. Near-field geometry of three-dimensional demonstration analysis

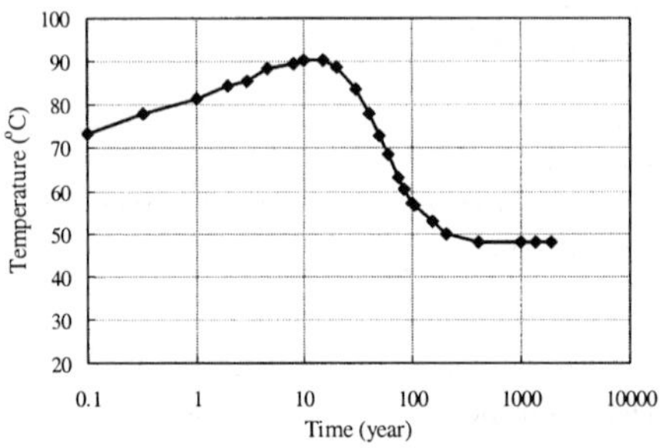

Figure 16. Time history of temperature inside buffer material

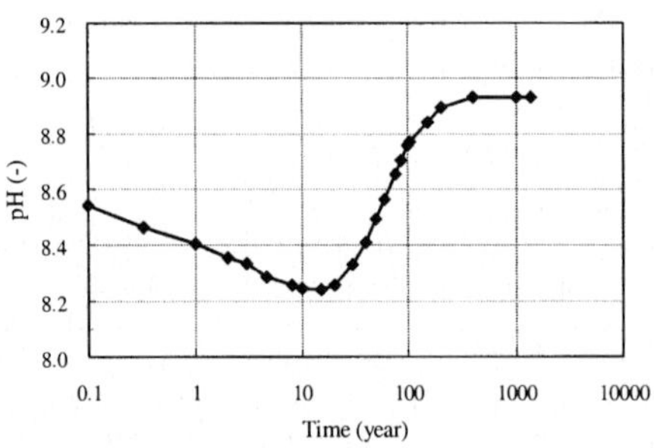

Figure 17. Time history of pH inside buffer material

5. CONCLUSION

The applicability of the prototype coupling system "COUPLYS" by "DDL" and "semaphore" to develop the coupled T-H-M-C code was shown by verification and demonstration analysis. The use of the prototype coupling system like "COUPLYS" is very effective to improve the efficiency and quality to develop the coupling code. This system provides us the basis of numerical experiment of the long-term evolution of a HLW repository.

The preliminary conclusions through demonstration analysis on the coupled T-H-M-C processes are as follows.

(1) The pH in buffer material highly depends on temperature. This is because solubility of calcite, which influences porewater pH, depends on temperature.

(2) When equilibrium model is adopted for geochemical reaction, porewater chemistry is not influenced by water saturation of medium.

On mass transport and geochemical reaction, we will advance our model to take into account degassing, reaction of ionic exchange, surface complexation and kinetic reaction of dissolution/precipitation.

REFERENCES

Chijimatsu, M., Fujita, T., Kobayashi, A. & Ohnishi, Y. 1999. *Coupled Thermo-Hydro-Mechanical Experiment at Kamaishi Mine, Technical Note 16-99-03, Analyses of Task 2C, DECOVALEX II*. JNC TN8400 99-031.

JNC 2000. *Second progress report on research and development for the geological disposal of HLW in Japan, Supporting Report 2, Repository Design and Engineering Technology*. JNC TN1410 2000-003.

Nishigaki, M., Hishiya, T. & Hashimoto, N. 2001. D*ensity dependent groundwater flow with mass transport in saturated - unsaturated porous media*. Proceedings of the First Asian-Pacific Congress on Computational Mechanics: pp. 1375 - 1380.

Neyama, A., Ishihara, Y., Yui, M. & Ito, A. 2002. *Development of Process Coupling System for the Numerical Experiment of High Level Radioactive Waste*. WSEAS Transactions on Mathematics. Volume 1. ISSN 1109-2769: pp. 186 - 191.

Parkhurst, D.L., Thorstnsen, D.C. & Plummer, L.H. 1980. *PHREEQE - A Computer Program for Geochemical Calculations*. U.S. Geological Survey. Water-Resources Investigations 80-96.

Yeh, G.T. & Tripathi, V.S. 1990. *HYDROGEOCHEM: A Coupled Model of HYDROlogic Transport and GEOCHEMical Equilibria in Reactive Multicomponent Systems*. ORNL-6371. Environmental Sciences Division Publication No.3170. Oak Ridge National Laboratory.

MODELLING THREE PHASE HYDRO-MECHANICAL COUPLING IN POROUS MEDIA: APPLICATION TO A REAL SCALE EXPERIMENT

Georg Klubertanz[1], Jean Croisé[1], Michel de Combarieu[2], Ken Ando[3]
[1]) Colenco Power Engineering, Baden, Switzerland
[2]) Agence nationale pour la gestion des déchets radioactifs (ANDRA), France
[3]) RWMC, Radioactive Waste Management Funding and Research Center,Tokyo,Japan

Abstract: The numerical code MHERLIN is being applied to the Gas Migration Test (GMT) at the Grimsel rock laboratory/ Switzerland. The ongoing GMT experiment, by the Japanese Radioactive Waste Management Funding and Research Center RWMC, is located at the Grimsel test site in Switzerland. It studies the gas migration through a sand-bentonite engineered barrier system under conditions that are close to those expected in a repository for radioactive waste in a crystalline host rock when gas is generated by the waste through steel corrosion. The model used is based on the continuum theory of mixtures and treats the soil as a three phase porous medium (solid, liquid and gas). The principal field variables are solid deformation, liquid pressure and gas pressure.In this paper we report the comparison between calculations and measured results for the resaturation of the engineered barrier.

1. INTRODUCTION

In recent years, the hydromechanical coupling in unsaturated soils as well as their deformation behaviour has been subject to increased research. At this time, the theoretical base for the hydromechanical coupling in deformable porous media with two pore fluids is known, Lewis & Schrefler (1999). Nevertheless, application of these theories and studies concerning their implementation are rather young, e.g. Klubertanz (1999), Lewis & Schrefler (1999).

The application of the hydro-mechanically coupled three phase models to real scale experiments is however still a challenge. This paper presents a successful application of such a model to a complex real scale experiment involving numerous materials with very different properties.

2. THREE PHASE MODEL

We briefly recall a model for hydromechanical coupling in unsaturated soils that is implemented in the used computer code. The relations and their application in the present context is extensively discussed in Klubertanz (1999) or Klubertanz et al. (1999), a broad discussion of the employed material laws for saturation can be found in Seker (1983).

The numerical model for unsaturated porous media is based on the continuum theory of mixtures and treats the soil as a three phase porous medium (solid, liquid and gas). The principal field variables are solid deformation, liquid pressure and gas pressure. The specific density of the solid grains is considered constant, the gas phase is governed by the ideal gas law. The two fluid phases flow freely and a non-linear pore pressure-saturation relation is used. The resulting system of equations is discretized in space using the finite element technique and in time by the Θ-method.

2.1 Momentum balance equation

The momentum balance of the three phase mixture reads, neglecting inertia terms reads (see e.g. Hutter et al. (1999))

$$\nabla \bullet \boldsymbol{\sigma} + \rho \mathbf{g} = 0 \qquad (1)$$

where $\boldsymbol{\sigma}$ is the (total) stress tensor, $\mathbf{g}$ the gravity vector and ρ the density of the mixture with $\rho = (1-n)\,\rho^s + n\,S^w\,\rho^w + n\,S^a\,\rho^a$ and n the porosity. Note that here ρ^s, ρ^w and ρ^a are the densities of solid, water and air; S^w and S^a are the degrees of water and air saturation defined as the volume ratio of water and air, respectively, divided by the total pore volume of the control volume. Please note that throughout this paper, quantities related to the solid, water or air phase are denoted by the superscripts s, w and a, respectively. Furthermore $\nabla\bullet$ denotes the divergence operator.

The total stress tensor $\boldsymbol{\sigma}$ may be divided in stresses in the fluid and in the solid matrix. It is essential to choose an appropriate formulation for the effective stresses in the solid skeleton. Here a classical formulation for those effective stresses is chosen Klubertanz (1999).

$$\boldsymbol{\sigma} = \boldsymbol{\sigma}' - S^w\,p^w\,\mathbf{I} - (1-S^w)\,p^a\,\mathbf{I} \qquad (2)$$

where $\boldsymbol{\sigma}$ and $\boldsymbol{\sigma}'$ are the total and effective stresses respectively, p^w the water and p^a the air pressure, $\mathbf{g}$ the gravitation vector and $\mathbf{I}$ is the identity matrix. Combination of (1) and (2) yields the form

$$\nabla \bullet (\boldsymbol{\sigma}' - S^w p^w \mathbf{I} - (1-S^w) p^a \mathbf{I}) + \rho \mathbf{g} = 0 \tag{3}$$

2.2 Mass balance equations

The mass balance equations for the solid phase is:

$$\frac{\partial[(1-n)\rho^s]}{\partial t} + \nabla \bullet [(1-n)\rho^s v^s] = 0 \tag{4}$$

where $\mathbf{v}^s$ is the solid velocity.

For the water phase the mass balance equation reads:

$$\frac{\partial[nS^w \rho^w]}{\partial t} + \nabla \bullet [nS^w \rho^w v^w] = 0 \tag{5}$$

or, more explicitly

$$\begin{aligned} & S^w \rho^w \frac{\partial n}{\partial t} + n\rho^w \frac{\partial S^w}{\partial t} + nS^w \frac{\partial \rho^w}{\partial t} \\ & + nS^w \rho^w \nabla \bullet [\mathbf{v}^w] + \mathbf{v}^w \nabla [nS^w \rho^w] = 0 \end{aligned} \tag{6}$$

where $\mathbf{v}^w$ is the (absolute) water velocity.

Finally, the air mass balance is:

$$\frac{\partial nS^a \rho^a}{\partial t} + \nabla \bullet [nS^a \rho^a \mathbf{v}^a] = 0 \tag{7}$$

or in detail

$$\begin{aligned} & S^a \rho^a \frac{\partial n}{\partial t} + n\rho^a \frac{\partial S^a}{\partial t} + nS^a \frac{\partial \rho^a}{\partial t} + nS^a \rho^a \nabla \bullet [\mathbf{v}^a] \\ & + \mathbf{v}^a \nabla [nS^a \rho^a] = 0 \end{aligned} \tag{8}$$

where $\mathbf{v}^a$ is the (absolute) air velocity.

Simplifying (6) and (8) by dividing with $S^w \rho^w$ or $S^a \rho^a$ respectively and (4) one obtains mixed mass balances for water and solid

$$\begin{aligned} & \frac{n}{S^w}\frac{\partial S^w}{\partial t} + \frac{n}{\rho^w}\frac{\partial \rho^w}{\partial t} + n\nabla \bullet \mathbf{v}^w + \frac{\mathbf{v}^w}{S^w \rho^w} \\ & \nabla [nS^w \rho^w] + (1-n)\nabla \bullet \mathbf{v}^s + \mathbf{v}^s \bullet \nabla (1-n) = 0 \end{aligned} \tag{9}$$

as well as for air and solid

$$\begin{aligned} & -\frac{n}{S^a}\frac{\partial S^w}{\partial t} + \frac{n}{\rho^a}\frac{\partial \rho^a}{\partial t} + n\nabla \bullet \mathbf{v}^a + \frac{\mathbf{v}^a}{S^a \rho^a} \\ & \nabla [n(1-S^w)\rho^a] + (1-n)\nabla \bullet \mathbf{v}^s + \mathbf{v}^s \bullet \nabla (1-n) = 0 \end{aligned} \tag{10}$$

Where ρ^s = const. is assumed. Please note that this does <u>not</u> imply the assumption of an undeformable soil matrix, as only the grain density is assumed to be constant, not the matrix density of the solid. Supposing that the degree of saturation S^w is a function of the capillary pressure $p^c = p^w - p^a$ only and that ρ^π depends exclusively on p^π, one finds for the water – solid mass balance

$$\begin{aligned} & \left(\frac{n}{S^w}\frac{\partial S^w}{\partial p^c} + \frac{n}{\rho^w}\frac{\partial \rho^w}{\partial p^w}\right)\frac{\partial p^w}{\partial t} - \frac{n}{S^w}\frac{\partial S^w}{\partial p^c}\frac{\partial p^a}{\partial t} + \\ & \nabla \bullet \mathbf{v}^s + n\nabla \bullet [\mathbf{v}^w - \mathbf{v}^s] + \\ & \mathbf{v}^w \bullet \nabla n + \frac{n}{\rho^w}\mathbf{v}^w \bullet \nabla \rho^w + \frac{n}{S^w}\mathbf{v}^w \bullet \nabla S^w - \mathbf{v}^s \bullet \nabla n = 0 \end{aligned} \tag{11}$$

Rearranging (11) and using the relative velocity $v^{ws} = S^w n (v^w - v^s)$ and Darcy's law, the water-solid mass balance reads

$$\begin{aligned} & \left(\frac{n}{S^w}\frac{\partial S^w}{\partial p^c} + \frac{n}{\rho^w}\frac{\partial \rho^w}{\partial p^w}\right)\frac{\partial p^w}{\partial t} - \frac{n}{S^w}\frac{\partial S^w}{\partial p^c}\frac{\partial p^a}{\partial t} \\ & + \nabla \bullet \mathbf{v}^s + \frac{1}{S^w}\frac{k_r^w K}{\mu}\nabla \bullet \nabla p^w + \\ & + \frac{1}{S^w}\frac{k_r^w K}{\mu}\left(\frac{1}{\rho^w}\nabla \rho^w + \frac{1}{S^w}\nabla S^w\right)\nabla p^w + \\ & \frac{1}{S^w}\left(\frac{n}{\rho^w}\nabla \rho^w + \frac{n}{S^w}\nabla S^w\right)\mathbf{v}^s = \\ & \frac{k_r^w K}{\mu}\nabla \bullet [\rho \mathbf{g}] + \frac{k_r^w K}{\mu}\frac{1}{S^w}\left(\frac{1}{\rho^w}\nabla \rho^w + \frac{1}{S^w}\nabla S^w\right)\rho \mathbf{g} \end{aligned} \tag{12}$$

A corresponding reasoning holds for the air-solid mass balance.

The relative permeability k_{rw} of water is assumed to be a function of the degree of water saturation and porosity, the relative permeability of air k_{ra} being a function of the degree of saturation only. K is the intrinsic permeability and μ^π the dynamic viscosity of the fluid in question. v^s is the (Lagrangian) velocity of the solid skeleton. p^c denotes the matrix suction ($p^c = p^a - p^w$). The non-linear capillary pressure - saturation relation after Seker (1983) is used

$$S^w = \frac{1}{\sqrt[\Psi_1]{\frac{1}{\Psi_0}\log\left(\frac{100\frac{1}{[m]}p^c}{\rho^w\|g\|_2}\right)+1}} \tag{13}$$

with the parameters Ψ_0 and Ψ_1.

3. EXPERIMENTAL SETUP

3.1 Experimental procedure

The ongoing Gas Migration Test, initiated by the Japanese Radioactive Waste Management Funding and Research Center RWMC, is located at the Grimsel test site in Switzerland. It studies the gas migration through a sand-bentonite engineered barrier system under conditions that are close to those expected in a real repository for radioactive waste in a crystalline host rock.

For this purpose, a real scale Engineered barrier system (EBS) has been emplaced in a purpose-build cavern in the granite host rock (Ando et al., 2001). A cavern was excavated which was composed of an access tunnel and a vertical silo part allowing the implacement of a sand-filled concrete box and a surrounding sand-bentonite buffer. The access tunnel is backfilled with gravel and sealed with a concrete plug. The granitic rock around the cavern is intersected by a shear zone (including mylonitic porous fault gauges). This shear zone is the main permeable element within the granitic rock matrix.

The first part of the GMT experiment after the completion of the experimental set-up was devoted to the natural and artificial resaturation of the engineered barrier system. This part of the experiment has been completed by end of 2002.

The current experimental phase of the the experiment is the performance of the gas injection into the concrete box placed inside in the centre of the EBS and to study the effects of the induced gas and water flow on the system.

In this paper we focus on the resaturation phase of the EBS. During construction, the cavern was subject to atmospheric pressure and the materials of the EBS where emplaced with various degrees of saturation, all of them being less than 1. Consequently, the overall system is unsaturated, but as the gas injection test was to be performed under saturated conditions of the EBS, a resaturation phase of about 1.5 years was scheduled, during which natural flow of water was allowed as well as injections into the lower gravel layer (3) and the backfill (4) at different rates (cf. Fig. 1).

3.2 Geometry

The experimental setup consists of a concrete container with gas vent surrounded by an engineered barrier system made of a 20/80 % bentonite/sand mixture, placed in layers, a granular backfill of the upper cavern and a concrete plug. In total, 12 different materials are considered in the numerical model (cf. Fig.1). As the surrounding rock matrix is very impervious, only the relatively high-permeable shear zone is considered for flow outside of the EBS. Most important material parameters are given below. These parameters are obtained from independent laboratory test for the sand/bentonite (materials 6 and 8), from the literature for the other materials: there is no back-estimation from the calculation results or calibration involved.

3.3 Initial and boundary conditions

Initial condition for the resaturation calculations correspond to mechanical equilibrium under gravity and a lithostatic vertical stress at the top of the considered shear zone of 3 MPa.

Table 1: Simulation phases and injected specific rates

	Start [s]	**End [s**	**q [m^3/s m^3]**
stage0	0.0	$2.7\ 10^{6}$	-
stage1	$2.7\ 10^{6}$	$4.7\ 10^{6}$	$4\ 10^{-8}$
stage2	$4.7\ 10^{6}$	$6.7\ 10^{6}$	$1\ 10^{-8}$
stage3a	$6.7\ 10^{6}$	$1.2\ 10^{7}$	$6\ 10^{-8}$
stage3b	$1.2\ 10^{7}$	$1.55\ 10^{7}$	$2\ 10^{-8}$
stage4	$1.55\ 10^{7}$	$4.87\ 10^{7}$	$1\ 10^{-7}$

At the top of the shear zone a hydraulic pressure corresponding to 100 m water is imposed, according to field measurements. The domain around the cavern is depressurised due to an imposed atmospheric pressure in the cavern for half a year (accounting for the desaturation during construction) and for the EBS initial saturation of 16 % (backfill), and 66% (sand/bentonite) are assumed. An atmospheric pressure boundary condition for gas and water is imposed at the concrete plug closing the cavern and at one side of the shear zone, at about the heights of the cavern, where the zone is known to intersect a tunnel. Gas will be injected in the container and it's flow through the barrier system studied. A large number of sensors has been placed all over the set-up. The FE-mesh used consists of 41000 3D tetrahedral elements.

Table 2: Material parameters

Material	Mat no	Permeability [m^2]	ψ_0 [-] (Seker)	ψ_1 [-] (Seker)	Density [kg/m^3]	Porosity [-]	Young modulus [Pa]	Poissons ratio
shear zone	1	$1\ 10^{-18}$	7.0	0.30	2700	0.01	$1.5\ 10^{10}$	0.3
EDZ	2	$5\ 10^{-14}$	2.8	0.12	2700	0.01	$1.0\ 10^{10}$	0.3
backfill	3	$1\ 10^{-13}$	2.8	0.12	2650	0.3	$5.0\ 10^{9}$	0.3
gravel	4	$5\ 10^{-14}$	2.8	0.12	2650	0.3	$5.0\ 10^{9}$	0.3
sand	5	$1\ 10^{-15}$	2.8	0.12	2650	0.4	$5.0\ 10^{9}$	0.3
sand/bentonite 1	6	$1\ 10^{-19}$	4.5	0.60	1930	0.3	$1.0\ 10^{9}$	0.3
sand/bentonite 2	7	$1\ 10^{-16}$	4.5	0.60	1930	0.3	$1.0\ 10^{9}$	0.3
sand/bentonite 3	8	$1\ 10^{-18}$	4.5	0.60	1930	0.3	$1.0\ 10^{9}$	0.3
concrete (container)	9	$5\ 10^{-19}$	4-0	0.40	2390	0.2	$2.5\ 10^{9}$	0.3
fill (container)	10	$5\ 10^{-14}$	2.8	0.12	2500	0.3	$2.5\ 10^{9}$	0.3
vent	11	$5\ 10^{-14}$	2.8	0.12	2500	0.3	$2.5\ 10^{9}$	0.3
plug	12	$5\ 10^{-19}$	4.0	0.40	2700	0.2	$5.0\ 10^{9}$	0.3

4. RESULTS

The following figures show the pore water pressure for several sensor locations, comparing measured and calculated values. The coordinates of the respective points are given in the legends, the plots are ordered by heights. Sensor nomenclature codes the location by height level/azimut/radius, higher numbers corresponding to a greater height or radius. The sensors used are not to record suction pressures, only zero or positive pressures are measured (i.e. piezometer but no tensiometer). Experimental results are the dotted lines. Results show a good agreement between measurements and calculations. In particular, the layer behaviour and transient effect are reproduced. Injection occurs mainly in the backfill. A steep rise in the pressure is observed at around t=1.5e7s. At this time, the cavern seems to be saturated. In the calculation, this happens slightly earlier than in the experiment, most likely due to a too high initial saturation. Water pressure in the gravel (point 1_9_4 and 1_9_3) rises drastically, the pressure level being influenced among others by the EDZ (material 2), that form a high-permeable connection between the cavern and the gravel.

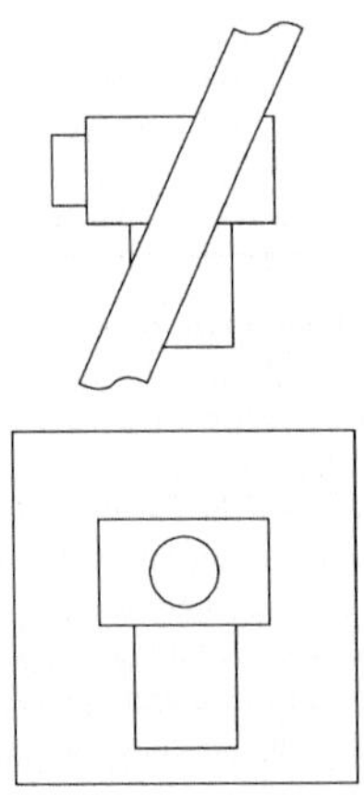

Figure 1: Sketch of GMT experimental setup (overview), showing shear zone and plug

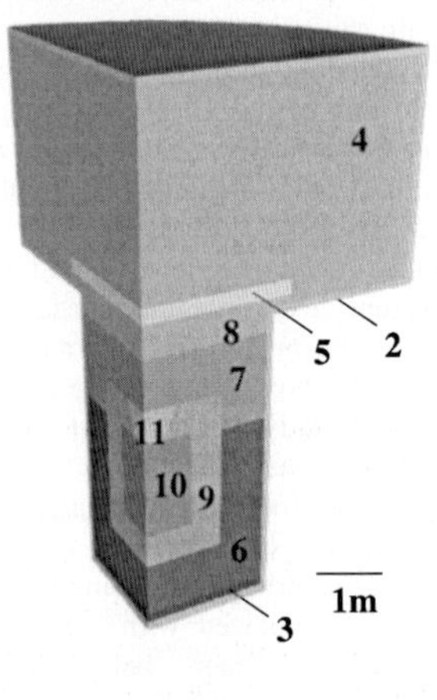

Figure 2: Sketch of GMT experimental setup (detail)

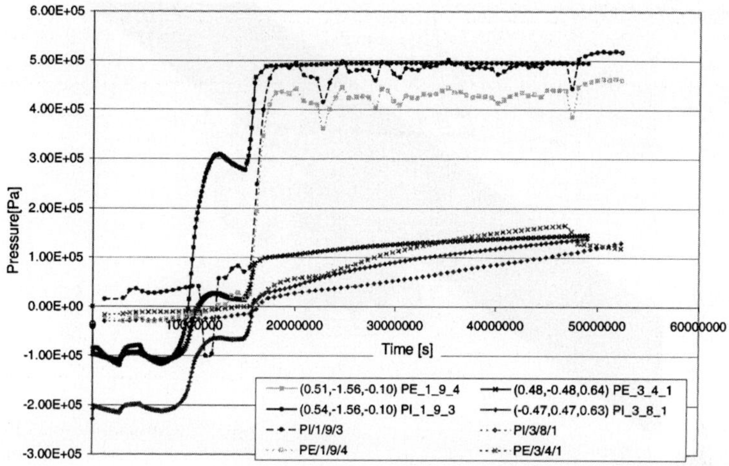

Figure 3: Experimental (dashed line) and simulated (continuous line) water pressure for pressure sensors in the gravel and the lower bentonite

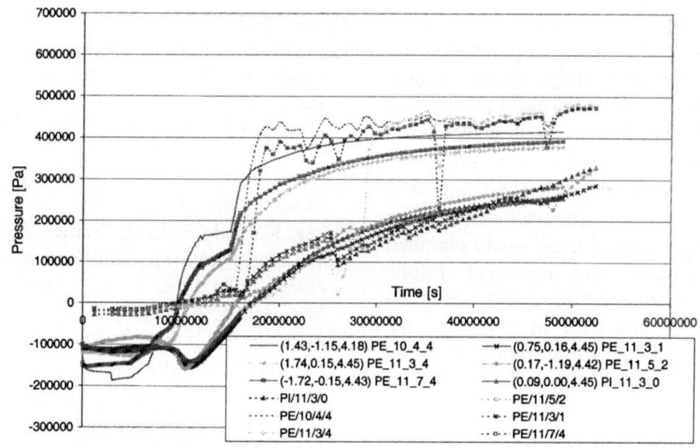

Figure 4: Experimental (dashed line) and c simulated (continuous line) water pressure for pressure sensors in the midle and upper bentonite

The sand/bentonite layers follow much slower, the pressure in the very tight materials 6 and 8 remaining rather low and the pressure in the more permeable layer 7 rising faster. In the experiment, a clear radial flow can be observed (points 11_7_4 or 11_7_3, more towards the outside and other points). This feature is reproduced by the simulation, however the detailed pressure evolution is poorly matched. This may indicate two phase flow properties of the materials other than assumed for the calculations, in particular a steeper suction-saturation relation.

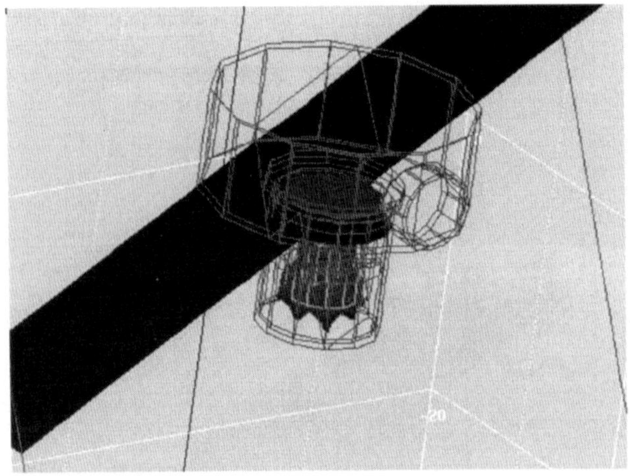

Figure 5: Calculated saturation in the EBS at time 3.9 10^7 s, saturation-isosurface S^w=0.7and horizontal cut, darker color corresponding to S^w=1

5. CONCLUSIONS

A complex 3D rock labortatory configuration has been modelled using the three phase hydro-mechanical coupled code MHERLIN. The structure involved 11 different materials with large differences in hydraulic and mechanical properties. A resaturation scenario was modelled and compared to measurement. Given that no parameter fitting was performed and that parameter used where obtained independently, calculated and measured results coincide well. All important features observed in the experiment could be modelled, in particular the transient behaviour of the resaturation and the differences of performance of the various materials could be reproduced. The modelling work of the gas injection phase is in progress.

Applicability of the model to the experiment and feasibility of three-phase calculations in such rather complex configurations has therefore been shown, both qualitatively and quantitatively. Numerical modelling of hydro-mechanical three phase coupling can, as in the presented example, yield useful and probably critical support for the design and the safety assessment in the context of e.g. nuclear waste disposal.

ACKNOWLEDGMENT

The modelling work presented in this paper was funded by the French National Radioactive Waste Management Agency, which is greatly acknowledged.

REFERENCES

Ando, K., Fujiwara, A., Tokuyama, S., Adachi, T., Saeki, T., Vomvoris, S., Fukudome, K. & Shimmura, A. *Total System evaluation of Gas Generation and Migration in the Radioactive Waste Repository*, in Proccedings of the Global 01 conference, Paris, 2001 .

Hutter, K., Laloui, L. & Vulliet, L. 1999. *Mixture models of saturated und unsaturated soils - A clarification*, Int. J. of Cohesive - Frictional Materials 4: 295 - 338

Klubertanz, G. 1999. *Zur hydromechanischen Kopplung in dreiphasigen porösen Medien: Modellbildung und Anwendung auf die Auslösung von Murgängen*, Ph. D. thesis, Soil Mechanics Laboratory, Swiss Federal Institute of Technology, Lausanne

Klubertanz G., Gachet, Ph., Laloui, L. & Vulliet, L. 1999. *Hydro-mechanical coupling in unsaturated porous media : an experimental and numerical approach*, NAFEMS world congress 1999, Newport, USA

Lewis, R.W. & Schrefler, B.A. 1999. *The finite element method in the static and dynamic deformation and consolidation of porous media*, Chichester, Wiley

Seker, E. 1983. *Etude de la déformation d'un massiv de sol non saturé.* Thèse no 183, LMS/DGC, Lausanne, Swiss Federal Institute of Technology

T-H-M MODELLING OF THE PROTOTYPE EXPERIMENT AT ÄSPÖ HRL (SWEDEN)

Alberto Ledesma & Guangjing Chen
Dept. Geotechnical Engng. and Geosciences
Tech. University of Catalonia (UPC), Barcelona, Spain

Abstract: The Prototype Repository Project is a full scale test of the KBS-3 Swedish concept for high level radioactive waste, being conducted at Äspö Hard Rock Laboratory and managed by SKB (Swedish Agency for radioactive waste disposal). The project includes the collaboration of different groups from other National Agencies, Companies and Universities, supported by the European Community. The field test involved the excavation of a tunnel and 6 vertical deposition holes in a crystalline rock environment. Each hole includes a canister with heaters surrounded by compacted bentonite blocks. The paper presents the thermo-hydro-mechanical simulation of the experiment performed by one of the groups involved in the project, and includes comparisons with some field measurements. A prediction for the future evolution of some typical variables (temperature, and degree of saturation) is also included.

1. INTRODUCTION

Prototype Repository is the short name for "Full-scale test of the KBS-3 method for deep geological disposal of spent nuclear fuel in crystalline rock", an international EC supported experiment being performed at the Äspö Hard Rock Laboratory (Sweden). The KBS-3 method was initially developed in Sweden and refers to the following main characteristics:

- Copper canister with cast steel insert
- Buffer of bentonite clay surrounding the canisters
- Emplacement in vertical bore holes
- Repository depth of 400-700 m in crystalline rock
- Backfilling of deposition tunnels with bentonite-based materials

The test site is a 65 m long TBM-bored tunnel including six 1.75 m diameter deposition holes of 8 m depth. The outer 25 m long part has two holes and it is separated from the inner 40 m section including 4 holes by means of a tight plug. Details of the work plan may be found in Dahlström (1998), Svemar & Pusch (2000) and Persson & Broman (2000).

The canisters were cylinders of 1.05 m diameter and 4.83 m height, installed in the deposition holes and surrounded by compacted bentonite rings 0.5 m height and 1.65 m of outer diameter. The gap between blocks and the rock was filled up with bentonite pellets. A small 1 cm gap was left between canister and bentonite blocks as a tolerance for installation purposes. The initial density of the bentonite blocks was 1.78 g/cm^3, reaching 2 g/cm^3 after saturation. The bentonite used was MX80 sodium bentonite from Wyoming.

After the installation of each heater, the bentonite blocks, the instrumentation and backfilling of the surrounding area, the corresponding system was switched on. That is, heaters were turned on according to the installation time schedule. First heater started last September 2001 (Goudarzi & Börgesson, 2003). Since that date, temperature, relative humidity, total pressure and water pressure from the instrumentation have been collected.

This paper presents part of the geo-mechanical modelling work performed by one of the groups involved in the Project. An "in-house" program, CODE_BRIGHT, has been used for that purpose. The main features of the code are described in next section, but basically, it is a finite element program that solves the energy balance equation, the water and air balance equations and the equilibrium equation in a porous deformable media.

Previous to the general simulation of the experiment, a comparison between the results of a Thermo-mechanical modelling of different geometries is presented. This is particularly important, as boundary conditions play a fundamental role in the thermal problem. Small changes on the geometry may lead to very different predictions on heater temperature. Finally, the prediction of temperature and degree of saturation in a mid plane for hole nr 1 is presented, and results are compared with the field data available. A list of the parameters and physical laws involved in the simulations has been included as well, in order to illustrate the basic data used in the analyses.

2. DESCRIPTION OF THE CODE

CODE_BRIGHT is a finite element code for the simulation of THM problems in geological media. It was initially developed for the analysis of those problems in saline media (Olivella et al, 1996), but it has been extended to cope with THM behaviour of other materials. In fact that code has been used in recent years to analyse THM problems in the context of radioactive waste disposal (Gens et al, 1995; Gens et al, 1998; Gens & Olivella, 2000). A brief description of the main features of the code is included here for consistency.

A porous medium composed by solid grains, water and gas is considered. Thermal, hydraulic and mechanical aspects are taken into account, including coupling between them in all possible directions. The problem is formulated in a multiphase and multispecies approach. The three phases are solid phase (*s*), liquid phase (*l*, water + air dissolved) and gas phase (*g*, mixture of dry air and water vapour). The three species are solid (-), water (*w*, as liquid or evaporated in the gas phase) and air (*a*, dry air, as gas or dissolved in the liquid phase).

The following assumptions are considered in the formulation of the problem:

- Dry air is considered a single species and it is the main component of the gaseous phase. Henry's law is used to express equilibrium of dissolved air.
- Thermal equilibrium between phases is assumed. This means that the three phases are at the same temperature
- Vapour concentration is in equilibrium with the liquid phase, the psychrometric law expresses its concentration.
- State variables (also called unknowns) are: solid displacements, $\boldsymbol{u}$ (three spatial directions); liquid pressure, P_l; gas pressure, P_g; and temperature, T.
- Balance of momentum for the medium as a whole is reduced to the equation of stress equilibrium together with a mechanical constitutive model to relate stresses with strains. Strains are defined in terms of displacements.
- Small strains and small strain rates are assumed for solid deformation. Advective terms due to solid displacement are neglected after the formulation is transformed in terms of material derivatives (in fact, material derivatives are approximated as eulerian time derivatives). In this way, volumetric strain is properly considered.
- Balance of momentum for dissolved species and for fluid phases are reduced to constitutive equations (Fick's law and Darcy's law).
- Physical parameters in constitutive laws are function of pressure and temperature. For example: concentration of vapour under planar surface (in psychrometric law), surface tension (in retention curve), dynamic viscosity (in Darcy's law), are strongly dependent on temperature.

The governing equations that CODE_BRIGHT solves are:

- Mass balance of solid
- Mass balance of water
- Mass balance of air
- Momentum balance for the medium
- Internal energy balance for the medium

The resulting system of Partial Differential Equations is solved numerically. Finite element method is used for the spatial discretization while finite differences are used for the temporal discretization. The discretization in time is linear and the implicit scheme uses two intermediate points, $t^{k+\varepsilon}$ and $t^{k+\theta}$ between the initial t^k and final t^{k+1} times. Finally, since the problems are non-linear, the Newton-Raphson method has been adopted following an iterative scheme.

3. PRELIMINARY ANALYSIS OF DIFFERENT MODEL GEOMETRIES

When analysing a large field test, one of the preliminary decisions refers to the geometry and boundary conditions adopted for the finite element mesh. This is apparently an easy task, but it involves many assumptions that may have an important effect on the final results.

For the Prototype Repository Project, six different geometries were considered in a preliminary thermo-mechanical analysis. Figure 1 shows the basic characteristics of those models. The case (6) corresponds to the actual geometry, assuming only a plane of symmetry. The rest of the geometries involve additional symmetries that are not strictly correct. The external boundaries were defined far away from the test, as usual in finite element computations. The effect of boundaries was particularly important for the thermal problem, the mechanical aspects being almost irrelevant in these cases.

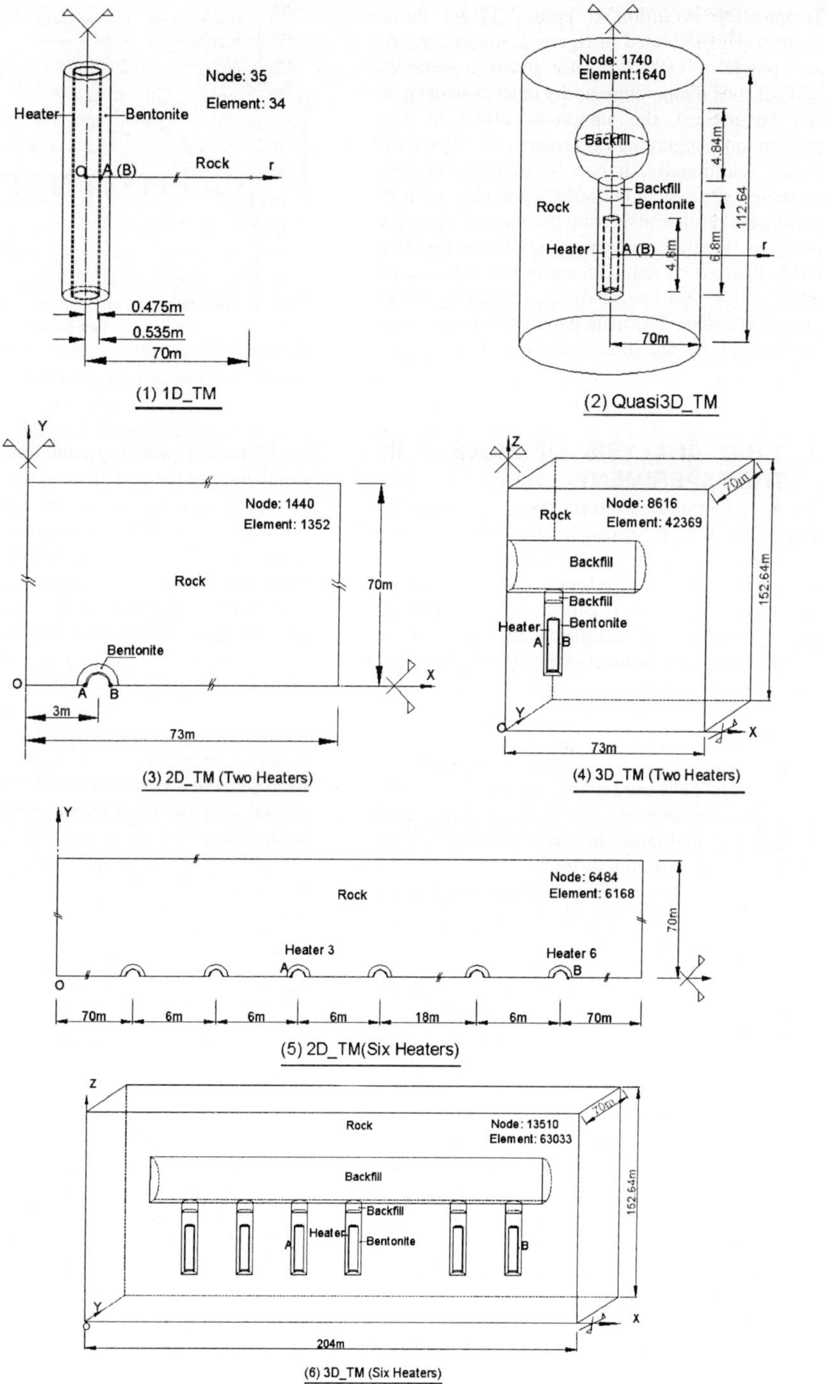

Figure 1. Geometries considered in the preliminary T-M analyses of the Prototype experiment

Temperature evolution at point "A" for those geometries is depicted in figure 2. It corresponds to a power of 1800 W/heater and to a value of 12ºC of rock temperature at the outer boundary. It can be noticed the important effect of the assumptions regarding geometry on the final result. Additionally, it can be pointed out that geometry (2), or "Quasi-3D" provides with a reasonable result considering the number of nodes involved in the corresponding finite element mesh. Because of that, Geometry (2) was used in most of the further analyses, including those presented below. The main parameters involved in the simulations are described in next section as well.

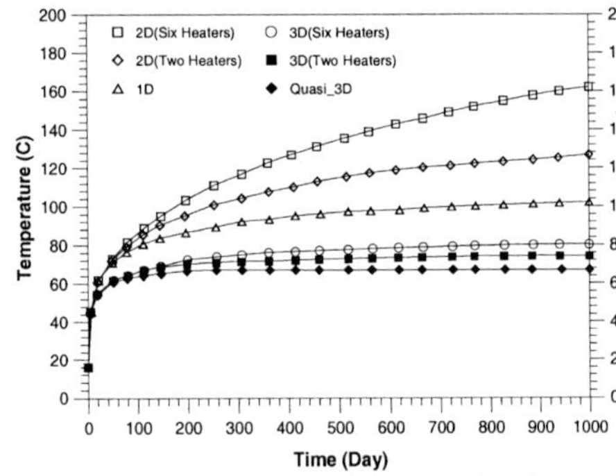

Figure 2. Temperature evolution at point A heater-bentonite for the geometries considered.

4. T-H-M ANALYSIS OF HOLE 1 IN THE EXPERIMENT

A thermo-hydro-mechanical predictive simulation was performed for the time evolution of hole 1. The Quasi-3D geometry (2D axisymmetric – case 2 in figure 1) was considered in these analyses. As it can be seen in figure 2, this geometry underestimates the temperature because only one heater is considered. However, due to the time delay for switching on the rest of the heaters, the predictions are considered to be realistic for the first months of the experiment.

The main parameters are presented in table 1. The physical laws used in the analyses correspond to those available in the code, and are also indicated in that table. In this simulation, due to the fact that maximum temperature is below 100ºC and the system is not closed, gas balance equation was not considered, in order to speed up the computational process.

In this particular case, a special attention was devoted to the coupling between the thermal and the hydraulic problems. In fact, a simple linear elastic model was assumed for the mechanical behaviour, although a coupling between intrinsic permeability and void ratio (and therefore volumetric deformation) has been considered.

On the other hand, for the thermal problem a dependence of the thermal conductivity coefficient on the degree of saturation was taken into account. The specific heat, however, was assumed constant. Most of the parameters involved were obtained from previous works related to bentonite and rock characterization (i.e. Börgesson, 2001), although information from other similar projects was also used (FEBEX, 2000).

Figures 3 and 4 present the results of the simulation for the mid plane of the deposition hole 1, regarding temperature and saturation degree. These results correspond to the location of some sensors for which some measurements are already available. Degree of saturation measurements have been obtained from relative humidity transducers by means of the psychrometric law.

It can be observed that the agreement is, in general, good. Temperature records may be eventually influenced by the installation of the rest of heaters. However, the trends and the temperature values seem to be consistent with measurements. For radius = 0.685 m, that is in the central area of the bentonite ring, there are two measurements available that show some scatter.

Saturation degree evolution has been reproduced reasonably. It should be pointed out that near the heater a wetting-drying cycle is observed, which is typical of coupled processes. Indeed, that point first increases its saturation degree, due to the water vaporisation that moves from the heater to the outer bentonite areas, and after a few days it dries because of this effect takes place eventually on that point. Finally water from the rock saturates the bentonite and the degree of saturation increases in a continuous manner. This coupling between temperature and water flow (in liquid or gas form) gives this cyclic response in the bentonite zone closer to the heater. This effect depends partly on the amount of water available for vaporisation. In this case, it was assumed that the slot between heater and bentonite blocks (1 cm) was filled initially with water, because of the initial emplacement conditions of the hole.

Table 1. Main parameters and physical laws used in the simulations of the experiment. "T" refers to thermal problem, "H" to hydraulic and "M" to mechanical information. "I" indicates initial conditions. In TM analyses, at r = 70 m , T=12ºC, and in THM analyses, at r = 70 m, T =16ºC; water pressure = 4 MPa at outer boundary. Heater power was 1800 W/heater in all cases.

		Unit	Steel	Rock	Bentonite	Backfill	Pellets	Slot
T	λ_{dry}	W/mK	50.16	2.5	0.3	1.5	0.1	0.1
	λ_{sat}	W/mK	50.16	2.5	1.3	1.5	1.0	0.6
	c	J/kgK	460	750	1091	1200	1091	1091
H	P_0	MPa	1000	4.0	60.0	0.12	0.4	0.001
	β		0.36	0.56	0.3	0.18	0.4	0.36
	k_0	m^2	10^{-30}	5×10^{-21}	0.5×10^{-20}	5×10^{-18}	4×10^{-19}	10^{-11}
	ϕ_0		0.001	0.003	0.366	0.4	0.706	0.99
M	b_s	$^\circ C^{-1}$	1.2×10^{-5}	7.8×10^{-6}	10^{-5}	10^{-5}		
	G	MPa	8750	20000	10	11.65		
	ν		0.3	0.25	0.3	0.3		
I	Temperature	$^\circ C$	12 (T and TM analyses); 16 (THM analyses)					
	Stress	MPa	0.0					
	Suction	MPa	0.0	0.0	55.0	1.0	2.5	0.0

T	λ_{dry}	thermal conductivity of the dry and water saturated porous medium	$\lambda = \lambda_{dry}^{1-S_r} \cdot \lambda_{sat}^{S_r}$
	λ_{sat}		For T and TM analysis $\lambda = \lambda_{sat}$ is assumed
	c	Specific heat	
H	P_0	Air entry value	Retention curve: $S_e = \dfrac{S_l - S_{rl}}{S_{ls} - S_{rl}} = \left[1 + \left(\dfrac{P_g - P_l}{P_0}\right)^{\frac{1}{1-\beta}}\right]^{-\beta}$
	β	Shape function	
	k_0	Intrinsic permeability	
	ϕ_0	Initial void ratio	Intrinsic permeability: $k = k_0 \dfrac{\phi^3}{(1-\phi)^2}\dfrac{(1-\phi_0)^2}{\phi_0^3}$ Liquid relative permeability: $k_{rl} = S_e^3$
M	b_s	Thermal expansion coefficient	$\Delta\varepsilon_v = 3b_s\Delta T$
	G	Shear modulus	
	ν	Poisson ratio	

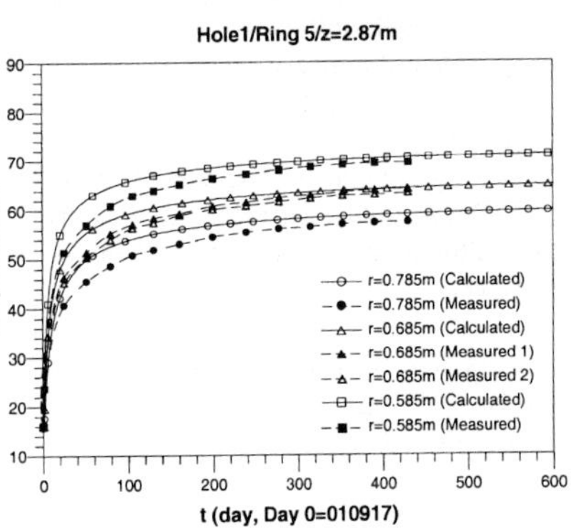

Figure 3. Temperature evolution for hole 1

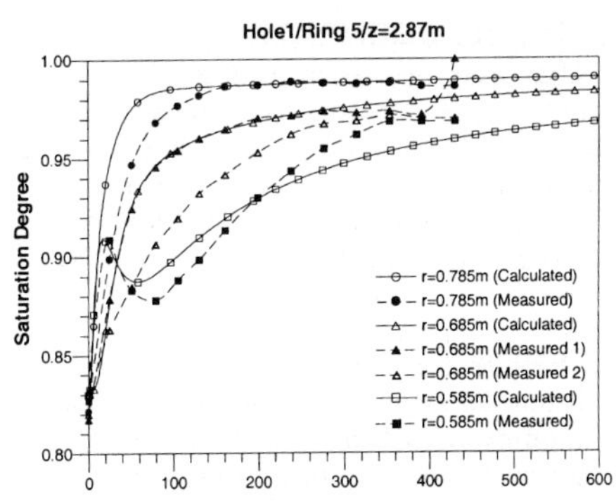

Figure 4. Degree of sat. evolution for hole 1

5. CONCLUSIONS

The paper presents some of the THM simulations carried out in the context of the Prototype Repository Project, a large scale field test being performed at Äspö Hard Rock Laboratory, Sweden. Only a few examples have been presented here. In particular, the analysis of different model geometries has been considered when studying the thermal problem. It has been shown that model geometry and boundary conditions may play a fundamental role in the final results, even for outer boundaries far away from the problem area. The "quasi-3D" geometry (2D – axisymmetric) has proven to give reasonable results in this case. When analysing temperatures, pore water pressures or displacements, the results are very sensitive to the model geometry assumptions and the evaluation of these effects are strongly recommended before selecting a final geometry. However, stresses are less sensitive to those changes.

In addition to these analyses, a comparison of measured and computed values of temperature and degree of saturation for some selected points of the mid plane of hole number 1 have been presented. The results show that despite the large amount of parameters and the difficulties in characterising those materials, the simulation can reproduce the main aspects of the processes involved, as, for instance, the wetting-drying cycles of the zone close to the heater. This effect has been simulated thanks to the consideration of the vapour flow in the system. This result is also consistent with other THM simulations performed in the context of radioactive waste disposal problems.

ACKNOWLEDGEMENTS

This work has been performed with the support of the European Union through Research Project FIKW-CT-2000-00055. Moreover, the support from ENRESA (Spanish Agency for radioactive waste disposal) and SKB is also gratefully acknowledged.

REFERENCES

Dahlström, L-O. 1998.*Test Plan for the Prototype Repository.* NCC Teknik, SKB, Sweden.

Börgesson, L. 2001.*Compilation of laboratory data for buffer and backfill materials in the Prototype Repository.* Äspö Hard Rock Laboratory, International Progress Report IPR-01-34, SKB, Sweden.

FEBEX. 2000. FEBEX Project. *Full-scale engineered barriers experiment for a deep geological repository for high level radioactive waste in crystalline host rock. Final Report.* ENRESA Technical Publication 1/2000. Madrid, Spain

Gens, A., García-Molina, A.J., Olivella, S., Alonso, E.E. & Huertas, F. 1998. *Analysis of a full scale "in situ" test simulating repository conditions.* Int. Journal Numerical and Analytical methods in Geomechanics, 22, p. 515-548.

Gens, A. & Olivella, S. 2000. *Non-isothermal multiphase flow in deformable porous media. Coupled formulation and application to nuclear waste disposal.* In Developments in Theoretical Geomechanics. Smith & Carter eds., Balkema, Rotterdam, p. 619-640.

Gens, A., Vaunat, J. & Ledesma, A. 1995. *Analysis of hydration of an engineered barrier in a radioactive waste repository scheme using an elastoplastic model.* Proc. 1st Int. Conf. on Unsaturated Soils, Alonso&Délage eds., Balkema, p. 1065-1073.

Goudarzi, R.& Börgesson, L. 2003. *Prototype Repository. Sensors Data Report (Period 010917-030301). Report nº 5.* Äspö Hard Rock Laboratory, International Progress Report IPR-03-23, SKB Sweden.

Olivella, S., Gens, A., Carrera, J., Alonso, E.E. 1996. *Numerical formulation for a simulator (CODE_BRIGHT) for the coupled analysis of saline media.* Engineering Computations, 13, n. 7, p. 87-112

Persson, G. & Broman, O. 2000. *Prototype Repository. Project Plan FIKW-CT-2000-00055.* Äspö Hard Rock Laboratory. International Progress Report, IPR-00-31. SKB, Sweden.

Svemar, Ch. & Pusch, R. 2000. *Prototype Repository. Project description FIKW-CT-2000-00055.* Äspö Hard Rock Laboratory. International Progress Report, IPR-00-31. SKB, Sweden.

INTERPRETATION OF SOME IN-SITU TRACER EXPERIMENTS IN FRACTURED CRYSTALLINE ROCK AT ÄSPÖ HARD ROCK LABORATORY

Ivars Neretnieks
Department of Chemical Engineering and Technology, Royal Institute of Technology
SE-100 44-Stockholm, Sweden

Abstract: Some recent, converging flow, tracer tests at the Äspö Hard rock laboratory in Sweden using non-sorbing as well as sorbing tracers in fractured crystalline rocks were analysed. The hydraulic and transport properties of the fracture networks were interpreted from hydraulic tests and borehole observations. A network model for flow was devised based on these observations. Laboratory data on diffusion and sorption properties were used to predict the Residence Time Distribution, RTD, of the sorbing tracers. Breakthrough curves for three strongly sorbing tracers were predicted using only laboratory data, the transmissivity data from borehole tests and the flow rate of the collected water. For the 5 m distance tests predictions were surprisingly accurate considering that no adjustments of any parameter values have been made.

For the 14-33 m distance tests the pure predictions were not as good but the model is deemed to capture the main interaction mechanisms. The results show that what at first seemed to be prominent rather isolated fractures belong to a complex network of intersecting fractures that strongly influence the flow distribution in the rock.

1. INTRODUCTION

A series of tracer tests were performed at the Äspö hard rock laboratory in Sweden. The tests were performed in a well-characterised rock volume. In this paper two sets of converging flow experiments are discussed. One set is over a distance of about five meters, the TRUE-1 experiments. The other is over 14-33 m in the TRUE-block experiments. The site is in a drift at about 400 m below the ground (Byegård et al. 1998 and Andersson et al. 2001).

In the TRUE-1 tracer tests the mixtures of non-sorbing and sorbing tracers were injected in packed off sections as short pulses over about four hours. The collection took place in another borehole about 5 m distant. The injected flow rate with tracers was only a small fraction of the water collected in the withdrawal hole. Two different injection points were used.

In the TRUE-block tests three injection points were used at geometrical distances between 14 and 33 m. In both locations the tests were deemed to be made in prominent fractures that could be identified in several other bore holes. The main flow was expected to take place in these fractures.

In the TRUE –1 test it was expected that most flow would be in what was called Feature A.

In the TRUE-block experiments one path was deemed to be in one fracture whereas the other paths could be in 2 or 3 major fractures.

Tracers were collected over several thousand hours and breakthrough curves analysed by different flow and transport models. The tracer breakthrough curves for the non-sorbing tracers can be used to obtain information on the mean residence time and on the spread in residence time. The latter is commonly called dispersion. The mean residence time can be used to assess the volume of flowpath if one assumes that the flow rate of all the collected water has the same residence time as that obtained by the tracer. Further, assuming that all the water has flown through one fracture, Feature A, the aperture of Feature A, can be determined. This was found to be about 3 mm. This is much larger than expected or reasonable. Typical fracture apertures are on the order of 0.1 mm.

It was also found in all cases that for the strongly interacting tracers it was not possible to predict the tracer behaviour using only laboratory data on diffusion and sorption properties. The tests seem to indicate that matrix diffusion and sorption properties must be more than 1000 times larger in the field than laboratory data indicate if it is assumed that the flow essentially takes place in one two-dimensional features (Byegård et al. 1998, Winberg et al. 2000, Elert and Svensson 2001, Anderson et al. 2001, Neretnieks 2002).

Another possible reason for the discrepancy could be that the collected flow has not arrived only through one 2 dimensional feature but that the collected flow travels through a network of many fractures. That would mean that the collected water has not only flown through one fracture but a large number of fractures. The flow volume would be divided over may fractures and each fracture could then have a smaller aperture. The same would apply for the Flow Wetted Surface, FWS. The collected stream of water on average would be in contact with a much larger FWS. This hypothesis is born out by the observed Conducting Fracture Frequency, CFF, in the rock surrounding the experimental site. We base this on the detailed analysis of the inflow data in the five boreholes that were used to explore the rock volume that surrounds Feature A (Neretnieks and Moreno 2002). For this case when it is assumed that the flow is through a multitude of fractures in a 3-dimensional network it is found that very good predictions can be made using observed fracture frequency data in the field and laboratory data for the sorption and diffusion properties. The reason for this is that the observed CFF will give approximately 30 times more fractures through which the pumped flow rate has passed and thus a 30 times larger FWS. This is just the needed value to have an agreement between observations and predictions both for the residence time of the non-sorbing and the sorbing tracers. (Neretnieks and Moreno 2003).

A similar analysis of the True-block experiments is also presented in this paper together with discussion of some basic mechanisms for solute transport in fractured rocks.

2. MODELLING OF FLOW AND SOLUTE TRANSPORT

We envisage that the fluid flow takes place in a 3-dimensional network of fractures (or channels) with a stochastic distribution of conductances (Moreno and Neretnieks 1993). In each channel the stream of water will be in contact with the rock surfaces. Intuitively it is conceived that the larger the contact surface there is for a given stream of water the stronger will be the interaction between flowing water and rock. This is the key issue in the paper. Several models have been proposed that account for the matrix diffusion effects. In all, the ratio of FWS to flowrate q enters as a key entity. In this paper we use a 3-dimensional fracture network model that is simplified by letting each fracture have constant properties. We call the fracture a channel.

For visualisation purposes we map the individual channel members on a cubic grid. Every member in the network has its own hydraulic properties chosen from a distribution determined from field observations. The dominating mechanism for solute interaction is the matrix diffusion and sorption on the interior micropore surfaces. This way of modelling flow and solute transport obviates the need to specifically model "hydrodynamic dispersion". "Dispersion" is caused by the combined residence time distributions in individual channels.

At the outlet of a single channel the concentration of a solute can be determined by the following simple expression. It is shown here for a Dirac input. The mass flux $\dot{m}$ in Equation (1) divided by the flow rate q_{coll} into which it is diluted, gives the concentration c (Neretnieks 1980).

$$\dot{m} = cq_{coll} = \frac{MG}{q\sqrt{\pi}} \frac{e^{-\frac{G^2}{q^2(t-t_w)}}}{(t - R_a t_w)^{3/2}} \tag{1}$$

The group G includes the Flow Wetted Surface, FWS, that is $2LW$ for a straight channel. G also includes the matrix property data, matrix porosity ε_p, pore diffusivity D_p and retardation factor R. q is the flow rate in the pathway. q_{coll} is the total collected flow rate, which also includes water from other paths arriving at the collection point. M is the tracer mass injected at time zero. We need to make this distinction for systems where there are several channels each getting their share of the injected mass proportioned to the flow rate in each channel. t_w is the water residence time, L the channel length and W its width. If the width and the matrix properties vary along the flow path then G may be obtained by integrating these entities along the flow path (Lee et al 1990).

$$G = LW\varepsilon_p\sqrt{D_p R} \tag{2}$$

Note that G always enters the equation as G/q.

The retardation factor in the matrix R is

$$R = 1 + \frac{1-\varepsilon_p}{\varepsilon_p} K_d \rho \tag{3}$$

The surface retardation factor is

$$R_a = 1 + a_w K_a \tag{4}$$

K_a is the surface sorption coefficient and a_w the FWS per volume of water in the fracture. The latter is related to the mean fracture aperture δ by

$$a_w = \frac{2}{\delta} \quad (5)$$

Equation (1) is convenient to use in cases where the tracer mass input varies over time. It can then be convoluted with the input mass flow that may vary in time. The effluent from a channel is mixed with the effluents from the other channels meeting at a channel intersection. The fluid is then redistributed to the channels leaving the intersection.

It should be noted that the entity G contains the magnitude of the flow wetted surface of the channel, $2LW$. It may be further noted that for large G/q, i.e. when matrix diffusion effects are large, the residence time of the water, t_w, will have a negligible influence on the Residence Time Distribution, RTD, of the sorbing tracer. This will be used when simulating the RTD of the sorbing tracers using only independent data.

This implies that we will not need any information on the RTD of the water to predict the RTD of the strongly sorbing tracers.

From the above it is seen that the more fractures there are that a given flow rate passes, the stronger will be the exchange of solute between flowing water and matrix.

The ratio FWS/q can be reformulated to be better suited for use in other models such as e.g. the Advection- Dispersion- Matrix diffusion model. In that model, which can be formulated as a stream tube model in a porous medium the fractures are not modelled explicitly. Instead one may consider the rock as containing a number of fractures per m^3 of rock. Each fracture has a FWS twice its size because both sides of the fracture are in contact with the water. An entity a_R can be defined that states the magnitude of FWS per m^3 of rock. A given flowrate of water q m^3/s flowing through a cross section of rock A_{ST} over a distance L will then be in contact with a FWS that is $a_R*A_{ST}*L$ That is the same as WL in equation (2). Thus a direct translation between the models is possible.

3. SIMULATION OF TRUE –1 EXPERIMENTS

An example of the predicted and observed breaktrough curve is shown in Figure 1. Because the model used is a stochastic model, results from several different realisations are shown. The agreement is surprisingly good considering that no adjustments of parameters were made. Laboratory date on matrix sorption and diffusion were used and data on the CFF from the experimental site.

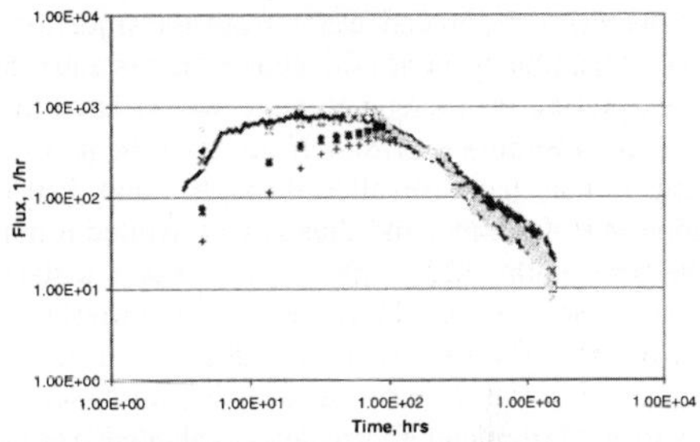

Figure 1. Breakthrough curves for Ba. Observed (full line) and predicted curves (various symbols)

It may be noted that also other models give similar results, e.g. the commonly used Advection Disperison-Matrix diffusion model Elert et al. (2001) and the Multi-Channel Neretnieks (2002) model, provided the FWS/q ratio is the same. In the two latter models information on dispersion is needed in addition to the other data. However, this cannot be obtained without using information from tests with non-sorbing tracers in the same pathway. This is not needed with the channel network model.

The observation that the influence of the entity, FWS/q, dominates the BTC in all models makes it simple to use the interpretations in Andersson et al (2001). They have evaluated the water residence time t_w, the matrix interaction parameter G and the surface retardation parameter R_a that also, and independently, can be used to assess the FWS of the flow paths. This information can be translated to a_R for any choice of the matrix property parameters. If it is assumed that the laboratory determined data are representative of the field then a_R can be obtained and compared to the observed CFF.

4. TRUE BLOCK EXPERIMENTS

For the TRUE-block experiments, another location at Äspö was chosen. Three different injection locations were used at geometrical distances of 14, 17 and 33 m from the collection point. There are a number of large fractures or minor fracture zones in the rock mass where the experiments were performed. The locations of the injection points and the common collection point suggest that the traced waters would move along 1, 2 and 3 of the major fractures to reach the collection point. The results show that if it is assumed that the flow is through one major fracture that carries the traced water from the injection to the collection point via the geometric distance then the aperture of a single fracture would have to be tens to several tens of mm. In some of the paths the tracers may have travelled through 2 and 3 of the identified fractures and thus have travelled a much longer path than the geometric distance. Accounting for this effect the apertures would have to be from 0.6 to 8 mm. Also these apertures are much larger than what is deemed possible. The transport parameter G was also evaluated from the field experiments assuming flow through one feature and over the geometric distance. This is compared to what is obtained using laboratory data in the transport group G. The results are that laboratory data give one to two orders of magnitude to small G to explain the RTD of the solutes. The ratios of G from field to laboratory data are shown in Table 1.

In the same way the ratios of the surface sorption coefficient K_a from field and laboratory data are compared. Also here, a difference of one to two orders of magnitude were found.

Table 1 Ratios of transport group G and K_a field to lab assuming 2D flow pattern and ratio FWS 3D to 2D.

	Distance m	Transport group G	Ka	Ratio FWS
		Ratio Field/lab	Ratio Field/lab	3D to 2D
Ba	14	49	36.7	7.5
Ca	14	94	26,7	7.5
Ca PathC2	33	51	93.3	17,6
Cs	14	73	3.6	7,5
Na Path C3	17	20	280	9.1
Na	14	34	40	7,5
Rb	14	56	8.6	7,5
Sr PathC3	17	12	440	9.1

5. EXPLORATION OF SOME REASONS FOR THE DIFFERENCES

There have been several suggestions for the differences between the field and the laboratory data both for the TRUE-1 and TRUE-block experiments. The main are:

1. The flow takes place through a 3-dimensional network of fractures instead of one or a few features.
2. The rock matrix nearest to the fracture surfaces is much more porous
3. The rock matrix nearest to the fracture surface has a much higher sorption capacity, K_a and K_d than the rock samples measured in the laboratory.
4. Fracture infill material such as secondary minerals and clays can have higher sorption coefficients than the original rock minerals

Cause 1. The CFF in True-1 was found to be such that a_R is about 8 m^2/m^3. The mean of the transmissivity of all fractures was about 5 times smaller than that of Feature A. However the fracture density is such that about 30 fractures would carry the collected flowrate in addition to Feature A. This would give a considerably larger FWS for the collected flow rate.

In the TRUE-block site the CFF is such that a_R is about 0.8 m^2/m^3. This also leads to considerably more fractures contributing to the FWS. The fifth column in Table 1 shows the increase in FWS if all fractures are equally contributing to the flow paths.

Cause 2. Some rock samples were found to have a higher porosity nearest the fracture surface extending some mm in depth. The increase in porosity could be from 0.4 % at a depth of a few mm to 2-4 % at the surface for some samples. Solutions of the diffusion equation for a varying porosity accounting also for the impact of the porosity on the diffusivity by Archies "law" shows that for the non-sorbing tracers an increase of G by less than a factor 2 is expected whereas for the strongly sorbing tracers and a surface porosity of 4% a increase of a factor of 6 could be possible. Table 2 summarises the enhancement on G due to an increased porosity for different contact times and for different sorption coefficients. Figure 2 show an example of the normalised concentration profile for a sorbing tracer for a low and a high porosity matrix and for a case of decreasing porosity with penetration depth.

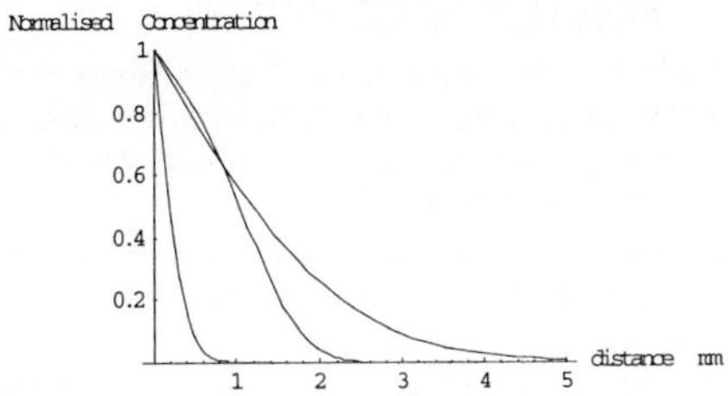

Figure 2. *Concentration profiles for 100 h contact time for a sorbing species with $K_d\rho = 1$, Left hand curve for a constant 0.4 % porosity, Right hand curve for a 4 % constant porosity and the middle curve for varying porosity.*

Table 2. Ratio of uptake in the matrix with porosity profile to that in the low porosity matrix. 2% /4% porosity at surface

Time h	Kdρ=0	1	100	1000
100	1.9/2.6	3.0/4.6	3.1/6.0	3.3/6.1
1000	1.3/1.7	2.4/3.2	3.3/5.5	3.1/6.0
10 000	1.0/1.05	1.8/2.0	3.0/3.4	3.3/4.0

Cause 3. There are indications that chemical alterations of e.g, feldspars in the rock matrix to secondary minerals could give rise to increases in sorption. This might further increase *G*.

Cause 4. facture infill minerals have been found in some fractures. This can contribute to the surface retardation effect; Equation (4)

6. DISCUSSION AND CONCLUSIONS

There are several possible causes for the very large differences between the field and the laboratory results. There are indications that infill materiel in the fractures and alteration of the rock nearest to the fracture surface can increase the retardation of the sorbing tracers more than what unaltered rock does. However, these effects alone are not sufficient to explain the more than 30 times larger retardation found in the field.

They also cannot explain the much to large fracture apertures that would have to be invoked to explain the residence times of the non-sorbing tracers.

The observations that there are many more fractures than the prominent ones that may contribute to the number of flow paths can to a large degree explain the differences for the non-sorbing as well as for the sorbing tracers.

The results of the analysis in this paper suggests that all the fractures in the rock mass contribute strongly to the FWS that the flow can access.

If this were generally the case, it is good news for the retardation of solutes such as radionuclides that can be considerably retarded during their transport from a repository to the biosphere. They would be given much more time to decay. The effect can be very strong because an increase in the FWS by a factor x gives the factor x^2 influence on the mean travel time of a species that diffuses in the matrix.

7. REFERENCES

Andersson P., Byegård J., Holmqvist M., Skålberg M., Wass E., Widestrand H. *True block scale project- tracer test stage,* SKB, IPR-01-33, October 2001.

Byegård J., Johansson H., Skålberg M. *The interaction of sorbing and non-sorbing tracers with different Äspö rock types. Sorption and diffusion experiments in the laboratory scale.* SKB Technical Report TR 98-18, Nov 1998.

Elert M., Svensson H. *Evaluation of modelling of the TRUE-1 radially converging tests with sorbing tracers,* SKB Technical Report TR-01-12, 2001.

Lee, H.S., L. Moreno, and I. Neretnieks, *Some properties of a channel network model,* Paper presented at Geoval 1990, Stockholm, May, 1990. Proceedings p 226-133.

Moreno L., Neretnieks I. *Fluid flow and solute transport in a network of fractures.* J. Contaminant Hydrology, 14, p 163-192, 1993.

Neretnieks, I., *Diffusion in the rock matrix: An important factor in radionuclide retardation?* J. Geophys. Res. 85, p 4379-4397, 1980.

Neretnieks I. *A stochastic multi-channel model for solute transport- Analysis of tracer transport in fractured rock,* J. Contaminant Hydrology, 55, p 175-211, 2002.

Neretnieks I, Moreno L. *Analysis of the TRUE experiments considering a 3D flow-pattern.* Paper presented at the international Groundwater symposium, Berkeley, California March 25-28, 2002, Proceedings. Ed. Findikakis A. N.

Neretnieks I., Moreno L. *Prediction of some in situ tracer tests with sorbing tracers using*

independent data, 61, p 351-360, 2003.
Winberg A., Andersson P., Hermanson J., Byegård J., Cvetkovic V., Birgersson L. *Final report of the first stage of the tracer retention understanding experiments*, Technical Report, TR 00-07 , Äspö Hard Rock Laboratory, March 2000.

THE ON-GOING PILLAR STABILITY EXPERIMENT AT THE ÄSPÖ HARD ROCK LABORATORY, SWEDEN

Christer Andersson[1], Mikael Rinne[2], Isabelle Staub[3], Toivo Wanne[4]
[1])SKB Sweden [2])FRACOM Finland [3])Golder Associates Sweden [4])Saanio&Riekkola Finland

Abstract: A Pillar Stability Experiment, is currently planned for SKB's Äspö Hard Rock Laboratory. The objectives of this experiment are: demonstrate the capability to predict brittle failure (spalling) in a fractured rock mass, demonstrate the effect of backfill (confining pressure) on the brittle failure response, and compare 2D and 3D mechanical and thermal numerical predicting capabilities. The 1 m thick pillar will be located between two large vertical boreholes. Preliminary modelling of the experiment has been carried out with Examine3D, FLAC3D, FRACOD and JobFem. The 2D and 3D numerical results are similar and indicate that a temperature increase of approximately 40-50 degrees C is sufficient to initiate spalling. The field part of the experiment is scheduled for completion by May 2004.

1. INTRODUCTION

The Swedish Nuclear Fuel and Waste Management Company (SKB) is responsible for the disposal of spent nuclear fuel in Sweden. The fuel is to be placed in copper canisters that will be deposited in vertical 8-m-deep 1.8-m-diameter boreholes at 300-700 m depth in crystalline rock. This will result in the formation of approximately 4500 rock mass pillars surrounding the emplacement boreholes. The stability of pillars in the mining industry is traditionally carried out using empirical methods. It is unknown if these methods are suitable for the design of a borehole emplacement pillar. Hence, SKB is conducting the Äspö Pillar Stability Experiment (APSE) to: 1) demonstrate our current capability to predict brittle failure (spalling) in a fractured rock mass, 2) demonstrate the effect of backfill (confining pressure) on the brittle failure response, and 3) compare the 2D and 3D mechanical and thermal predicting capabilities of existing numerical models. In this paper we present a brief overview of the experiment and some modelling results.

2. EXPERIMENT LAYOUT & DESIGN

A tunnel will be excavated at the 450-m level of Äspö Hard Rock Laboratory parallel to σ_2, The pillar for the experiment will be created by boring two large holes, Ø1.8m by 6m deep, in the floor of the tunnel such that a 1 m thick pillar remains between the hole walls (Fig. 1). It is anticipated that the pillar at this stage will respond elastically. To increase the pillar stresses, to induce brittle failure, the pillar will be heated by electrical heaters in small diameter boreholes. The resulting thermal expansion will elevate the pillar stresses under controlled conditions. In addition one of the boreholes that will form the pillar will be pressurized by 1 MPa fluid pressure contained in a specially designed bladder. This will provide the opportunity to measure the effect of confinement on the pillar response.

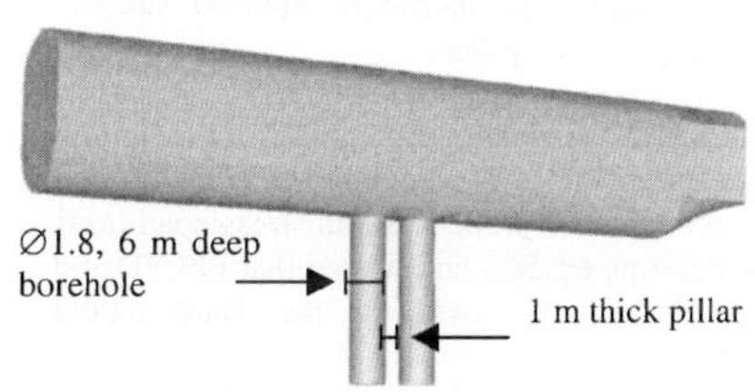

Figure 1. General layout of the experiment with a 1 m thick pillar between two 1.8 m diameter and 6 m deep boreholes. The tunnel is oriented perpendicular to the maximum principal in-situ stress, σ_1.

The experiment will be carried out in Äspö diorite which is the typical rock type in the HRL. The unconfined strength of the rock is 219 MPa, the crack damage stress (CDS) 190 MPa, the crack initiation stress (CIS) 118 MPa, Nordlund (1999) and the tensile strength 14.8 MPa, Staub (2003). The laboratory deformation properties of the intact diorite are: Young's modulus 68 GPa and Poisson's ratio, 0.24, (E=47 GPa was used in the sensitivity analysis). The tangential stresses required to initiate spalling are estimated to be approximately 150 MPa. The modelling results presented here have used the properties described above and an in-situ stress state of σ_1=30, σ_2=15 and σ_3=10 MPa. A complete characterization program of the actual pillar location and material properties will be carried out and used in the final model predictions

Because of the coupling between the mechanically-induced and thermally induced stresses the design of the experiment required several iterations. This was accomplished using the following numerical.tools.

- Examined3D (three dimensional boundary element) and Phase2 (two dimensional finite element) was used to scope the tunnel geometry and the dimensions of the boreholes and pillar (Andersson, 2003).
- JobFem (two dimensional finite element) was used to determine the size of the locations of the heaters. It was concluded that the heaters should have an effect of 200 W/m during 120 days of heating to elevate the temperature approx. 40 degrees C to ensure the required stress level.are reached
- FRACOD is a two dimensional fracture mechanics code used to analyze the growth and initiation of stress-induced fractures at discrete locations in the pillar.
- FLAC3D (three dimensional finitie difference) was used to estimate the coupled mechanical and thermally induced stresses throughout the pillar.

Initially all analysis were carried out assuming elastic parameters. These analyses were refined once the material properties and stress conditions were determined. It is anticipated that PFC2D and 3D will also be used in the final model predictions,

To be able to properly evaluate the results from the numerical models measurement of the rock mass response will be carried out during the experiment. At present three main monitoring methods will be used: (1)Acoustic Emission (AE) and Ultrasonic arrays will monitor the ongoing fracturing in the pillar. (2) Displacement measurements will monitor the pillar response during the spalling process. (3) Thermocouples will be used to compare the actual and predicted temperatures at certain points in the rock mass. The proposed geometry of the heaters and thermocouples in relation to the acoustic emission sensors is presented in Figure 2. In addition to the monitoring a complete characterization of the pillar before and after the experiment will also be carried out.

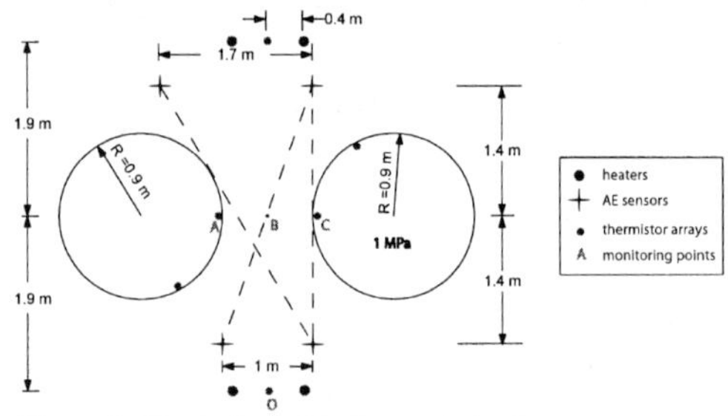

Figure 2. The design of the heater holes in relation to the holes for the thermocouples and the AE-sensors.

3. 2D TM MODELLING

The 2D TM modelling was carried out with the finite element numerical code JobFem (Stille, 1982). The aim of the modelling was to predict the thermo-induced stresses and the resulting total stresses in the pillar under the period of heating. The in-situ stresses were extracted from Examine3D for two horizontal sections in the pillar located 0.5 and 1.5 m below the tunnel floor. Heat flow and temperature induced stresses have been calculated under plane strain conditions.

The temperature increase and associated thermo-induced stresses are modelled at different stages of the experiment: 1) after excavation of the holes and before heating, 2) after applying the confining pressure in one of the holes, and 3) after 30, 60, 90, and 120 days of heating. The total stresses at each stage are obtained by adding for each node of the element mesh the excavation-induced stresses to the thermo-induced stresses.

The determination of input parameters for JobFem is presented in detail in Staub (2003). In order to better estimate the range of stresses that can be reached in the pillar the variability of principal stress, σ_1, Young's modulus and thermal conductivity have been accounted for by defining a range for "probable" values. The different combinations that have been modelled are presented in Table 1.

The complete results from the various can be found in Fredriksson (2003). The results presented in this article focus on the influence of mechanical and thermal properties of the rock mass for the horizontal section located 1.5 m below the tunnel floor, and assuming the mean value for the major principal stress, σ_1=30MPa.

Table 1. Parameters combinations used for the sensitivity analysis

σ_1, MPa	Section Depth (m)	E (GPa)	Thermal Conductivity λ (W/m, K)	Number of combinations
25	0.5 / 1.5	68	2.83	2
30	0.5 / 1.5	47 & 68	2.4 & 2.83	8
35	0.5 / 1.5	68	2.83	2

The temperature increase during the heating period of 120 days is illustrated in Figure 3 and shows that regardless of the variability in the thermal conductivity and Young's Modulus (Table 1) the expected temperature change is quite similar. The total major principal stress increase, monitored at the border of the confined hole (point C, Figure 2), is illustrated in Figure 4. Figure 4 shows that spalling will initiates between 60 and 120 days of heating assuming, that the stress magnitude required for spalling is 150 MPa. The range in stress magnitudes during the heating period in Figure 4 is a function of the rock mass modulus. For these analysis E=68 GPa represents the upper bound, equivalent to the intact rock while, E=47 GPa represents the lower bound, equivalent to the rock mass modulus.

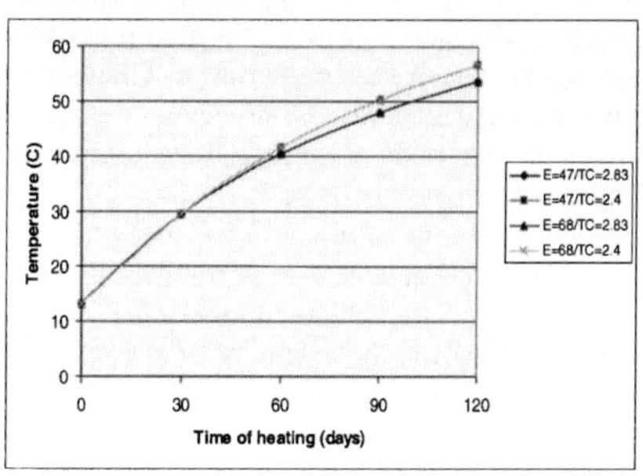

Figure 3. Temperature increase during heating, measured at the border of the confined hole, section 1.5 m below the tunnel floor. E: Young's modulus; TC: thermal conductivity (JobFem results).

Clearly the results are sensitive to the assumed E and highlights the need to accurately determine rock mass modulus.

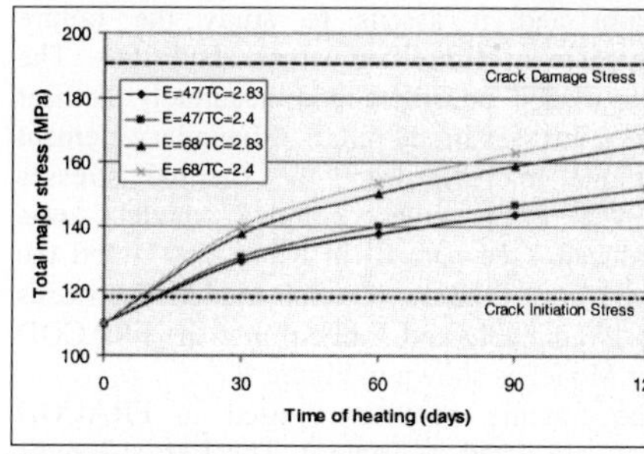

Figure 4. Major total stress increase during heating, measured at the border of the confined hole, section 1.5 m below the tunnel floor. E: Young's modulus; TC: thermal conductivity. The lines are an extrapolation between the five monitoring timepoints (JobFem results).

4. FRACTURE MODELLING USING FRACOD

FRACOD is a two dimensional BEM/DDM code for fracturing analysis in rocks, (see Part II in Rinne, 2003). The code is based on the principles of linear elastic fracture mechanics and has been developed to track both fracture initiation and propagation. The initiation of fracturing can be specified using any criterion, but once initiated fracture growth is controlled by the fracture toughness.

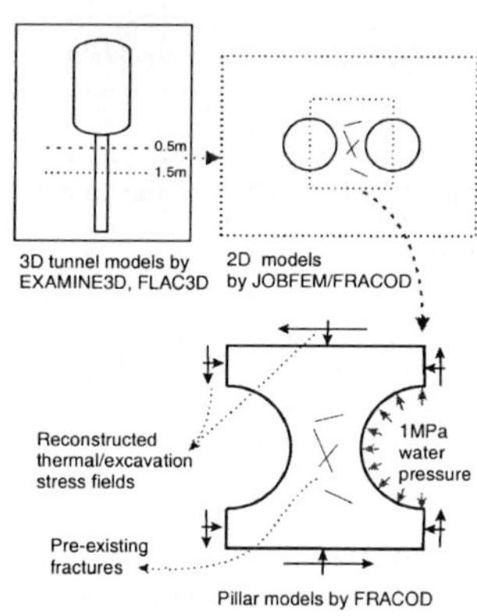

Figure 5. Numerical methods and the Pillar model used in APSE (FRACOD results).

Two FRACOD models have been established for the APSE configuration. The "Far-field model" simulates excavation in an infinite medium and it assists to study the failure mechanism in simplified stress conditions. The "Pillar model" considers more accurately the heat induced stresses in the pillar. A boundary element technique to reconstruct the thermal stresses obtained from other coupled models was introduced, (see part III in Rinne 2003) and the method was used to transfer modelled stresses from Examine3D and JobFem into the FRACOD Pillar Model as shown in Figure 5.

The fracture parameters used in FRACOD models are listed in Table 2. The Barton-Bandis model has been applied to estimate the fracture properties for newly initiated fractures.

Table 2. Input data for FRACOD models.

Fracture parameters		*Value / unit*
All fractures	*Mode I toughness*	$2.54 MPa \cdot m^{1/2}$
	Mode II toughness	$6.35 MPa \cdot m^{1/2}$
Pre-existing fractures	*Normal stiffness*	*61.5 GPa/m*
	Shear stiffness	*35.5 GPa/m*
	Friction angle	*31°*
	Cohesion	*1 MPa*
New initiated fractures	*Normal stiffness (S)*	*4643 GPa/m*
	Normal stiffness (T)	*320 GPa/m*
S (shear)	*Dilation angle (S)*	*2.3 °*
T (tensile)	*Dilation angle (T)*	*12.1 °*
	Friction angle	*49.0 °*
	Cohesion	*31 MPa*

The fracture evolution during the sequences of excavation and heating at 0.5 m below the tunnel floor is presented in Figure 6. The excavation induced stresses cause minor fracture initiation and propagation at borehole boundaries. Fracturing along borehole walls occur continuously with stress increase induced by heating. After 30 days of heating a major fracture starts to develop at the unconfined side of the pillar. After 90 days of heating major fractures are propagated deep into the pillar, and several fractures are initiated inside pillar region. After 120 days of heating fracturing in both shear and tensile mode occurs. Fracture growth at the right pillar side seems to be limited compared to left side due to the 1 MPa confinement.

After excavation, 30, 60, 90 and 120 days the maximum displacements are 1.8, 2., 2.2, 2.3 and 2.4 mm respectively. It must be noted that there are large uncertainties in predicting deformations associated with stress-induced fracturing.

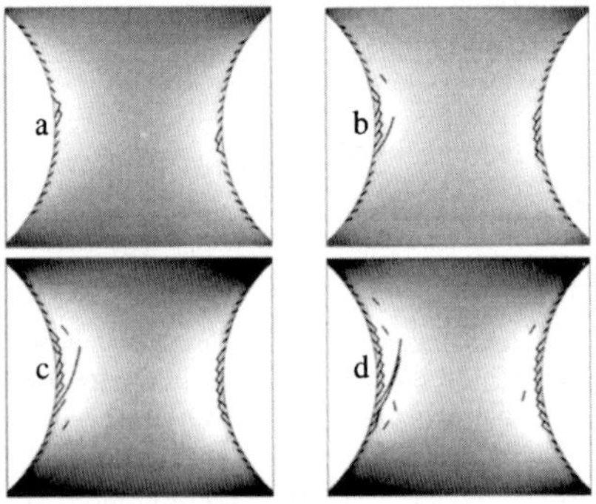

Figure 6. Fracturing at the borehole boundary after (a) excavation (b)-(d) after 30, 60 and 120 days of heating (FRACOD results).

5. 3D FLAC MODELLING

FLAC3D is a commercial finite difference code available from Itasca Consulting Group, Inc Itasca 2002). The code was specifically designed for geotechnical engineering analysis. FLAC3D provides the capability to conduct coupled thermal and mechanical analysis. The thermal model simulates the transient flux of heat in materials and the subsequent development of thermally induced stresses. The thermal-mechanical coupling is one-way; temperature change may induce a mechanical stress, however, mechanical changes in the body do not induce temperature change.

The initial temperature of 15°C rises about 50 degrees in the pillar area during the heating period of 120 days. The rising temperature induced stresses in the pillar area after 120 days of heating have increased the stresses about 40%, Wanne (2003).

After 120 days of heating the tangential stresses at the hole wall at the level of 1.5 m below the floor of the tunnel are about 175 MPa, which is close to the crack damage strength of Äspö diorite. The maximum stress at the top of the pillar is about 220 MPa being close to the uniaxial peak strength of Äspö diorite. Figure 7 shows the increase in stress at 0.5 and 1.5m during 120 days of heating. Figure 8 shows the maximum stress contours on a vertical section of the model. The crack initiation strength (σ_{ci} ~118 MPa) is exceeded after mechanical excavation and before heating, at the upper part of the pillar in each modelling case.

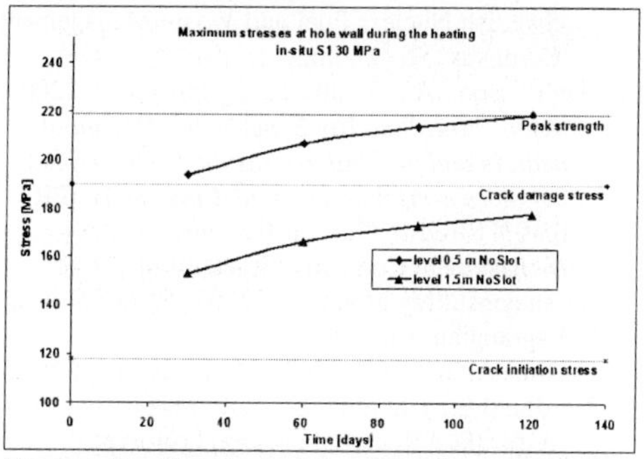

Figure 7. FLAC model with thermal load. Maximum principal stresses at the pillar area.

It should be noted that the strength values presented here correspond to the uniaxial stress state which is valid for the open borehole.

In the confined borehole where a triaxial situation will exist the strength values will be higher e.g. the peak strength of Äspö diorite is about 340 MPa with confinement of 20 MPa, Nordlund (1999).

For the determination of the influence of the thermal conductivity and Young's modulus on the outcome of the modelling results, several simulations were performed with alteration of these parameters. The sensitivity analyses showed that the resulting induced-stresses are dependent on the parameter values assumed; e.g., as the conductivity increased by 15% the stresses decreased by 4%, while when the modulus value was increased by 45% the stresses increased 13%, for the same time periods

The modelling results showed the areas where the uniaxial crack initiation strength of the rock is exceeded on the boundary of the open borehole which are also the areas where the AE activities would likely occur during the heating process. The tensile stresses in addition to high compressive stresses may also induce some AE activity from the very beginning of the test and are thus potential areas where AE activities will be detected.

6. MODEL EVALUATION STRATEGY

The Pillar experiment is specifically designed to monitor the brittle failure spalling process. This is a particularly difficult processes to simulate in numerical models as it requires the complete simulation of a rock mass transforming from a continuum to a discontinuum. The scoping

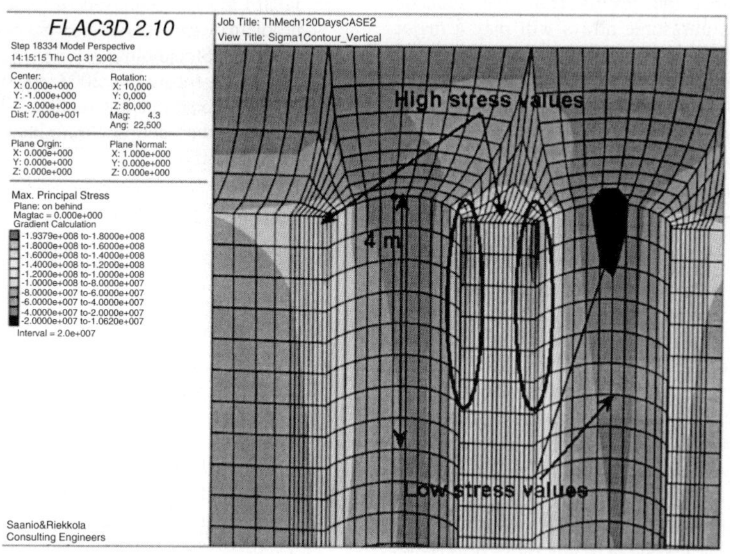

Figure 8. FLAC model with thermal load. Maximum principal stresses at the pillar area. Circled areas indicate possible exceeding of the crack initiation strength at the hole walls (in uniaxial compression.

calculations were restricted to continuum and fracture mechanics codes. However, during the final numerical simulations, the discrete element code PFC2D and possibly 3D (also available from Itasca Consulting Group) will also be used. By choosing these various numerical approaches it is anticipated that a more complete understanding of the failure process can be obtained.

One of the challenges faced by the experiment team is the comparison of numerical results with measurements and observations. Each team will be required to provide output at specific locations where point measurements will be made. All results will be presented and compared to actual measurements in TecPlot.

The output provided by the teams shall be reported before the excavation of the shafts and hence build on input data from lab tests on cores. This will give a measure of how good predictions that can be made without performing large scale calibrations of the models.

Once the experiment heating is completed a complete characterization of the pillar will be carried out. This will included visual observations, core sampling from the pillar, micro-crack density, and p-wave and s-wave velocity variation.

In addition to the discrete measurements it is anticipated that the results from the acoustic emission and ultrasonic arrays will play a major role in facilitating the comparison between the numerical model predictions and the measured response as these tools will capture the 3D temporal and spatial response of the pillar to the increasing temperature.

7. CONCLUSIONS

The results from the various numerical codes used for the scoping calculations give similar results. This is not too surprising given that all the codes are continuum codes using the same input parameters. The modelling results indicates that the excavation–induced stresses resulting from the proposed geometry for the experiment combined with the thermally-induced stresses are sufficient to induce spalling in the pillar. However, it should be noted that this type of experiment has not been completed under in-situ conditions and hence `surprises' must be anticipated. The sensitivity analysis briefly presented in this paper stresses the importance to properly assess the in situ conditions and rock mass parameters.

REFERENCES

Andersson, C. 2003. *Äspö Pillar Stability Experiment. Feasibility Study*. SKB, IPR-03-01. Swedish Nuclear Fuel and Waste Management Company, Stockholm.

Fredriksson, A., Staub, I., & Janson, T. 2003. *Äspö Pillar Stability Experiment. Design of heaters and preliminary results from coupled 2D thermo-mechanical modelling*. SKB, IPR-03-03. Swedish Nuclear Fuel and Waste Management Company, Stockholm.

Itasca consulting group. Inc. 2002. *FLAC3D* (Fast Lagrangian Analysis of Continua in 3 Dimensions). Version 2.1. Minneapolis, USA.

Nordlund,E., Chunlin, B. & Carlsson, B. 1999. Äspö Hard Rock Laboratory, Prototype repository. Mechanical properties of the diorite in the prototype repository at Äspö HRL – Laboratory tests. SKB, IPR-99-25.

Rinne, M. Shen, B. & Lee, H-S. 2003. Äspö Pillar Stability Experiment. Modelling of fracture stability by FRACOD. SKB, IPR-03-05.

Staub, I., Janson, T., & Fredriksson, A. 2003. Äspö Pillar Stability Experiment. Geology and properties of the rock mass around the experiment volume. SKB, IPR-03-02.

Stille, H., Groth, T., & Fredriksson, A. 1982. FEM-analys av bergmekaniska problem med JOBFEM / FEM-analysis of rock mechanical problems with JOBFEM. BeFo och KTH, Nr 307:1/82, Stockholm.

Wanne, T. & Johansson. 2003. Äspö Pillar Stability Experiment. Coupled 3D thermo-mechanical modeling. Preliminary results. SKB, IPR-03-04.

ALGORITHMS FOR PARALLEL FEM MODELLING OF THERMO-MECHANICAL PHENOMENA ARISING FROM THE DISPOSAL OF THE SPENT NUCLEAR FUEL

R. Blaheta, P. Byczanski, R. Kohut and J. Starý
Institute of Geonics AS CR, Ostrava, Czech Republic

Abstract: The analysis of stability and safety of underground repositories of the spent nuclear fuel requires the use of mathematical modelling of coupled T-H-M phenomena. The realization of reliable numerical simulations is a difficult task from many points of view including the aspect of high computational requirements concentrated mainly in the necessity of a repeated solution of large linear systems.

In our contribution, we address this aspect and describe numerical methods based on the use of efficient iterative solvers, which exploit the conjugate gradient (CG) method, its generalization and the space decomposition preconditioners. The efficiency of these solvers will be illustrated by the solution of elasticity and thermo-elasticity problems arising from the finite element analysis of selected benchmarks with computations performed on a PC cluster. The introduced ideas could be useful also for the solution of more complicated coupled problems.

1. INTRODUCTION

The solution of many geotechnical problems requires the use of mathematical modelling for understanding of natural and induced phenomena in the geological environment. This is the case of thermal, hydrological, mechanical (T-H-M) and other phenomena induced by mining, building of underground constructions, deep geological deposition of the spent nuclear fuel and other geotechnical activities. In some cases, the T-H-M phenomena can be investigated separately, neglecting their interactions. In other cases, more accurate coupled analysis is required. Such coupled analysis was shown to be important for the investigation of deep geological repositories of the nuclear waste, see Stephansson (1996, 2001).

The mathematical modelling of the T-H-M phenomena uses initial – boundary value problems for differential or variational equations involving the physical principles. We shall assume that these problems are discretized by the finite element or similar methods. What we want to point out is that the numerical solution can be computationally very expensive due to

- solution of problems in large 3D domains,
- requirements for an accurate analysis in some parts of the computational domains,
- possibly large time-scale of the investigated phenomena,
- complexity of the material behaviour and requirements of the coupled analysis.

The large computational demands indicate that a special care must be given to searching for accurate, robust and efficient numerical methods, which are capable to exploit the power of high-performance (parallel) computers.

In this paper, we shall touch the development of such numerical methods intended for the solution of the coupled evolution problems as e.g. thermo-elasticity, which is described in Section 2. Here we also discuss the discretization of the evolution problems. As the computational demands are concentrated mainly in the solution of the arising linear systems, we shall focus on the application of suitable, efficient and parallelizable iterative solvers for these linear systems. Section 3 deals with some general techniques enhancing the efficiency of the iterative solution of discrete evolution problems. Section 4 is devoted to a short discussion of the numerical results. In Section 5, we shall describe solvers, which exploit the domain decomposition and parallel computations. Here we also mention another division techniques as displacement decomposition or composite grid methods.

The performance of the introduced numerical methods will be illustrated by solving a simple 2D model extracted from the DECOVALEX project and large-scale real-life 3D model arising from a computation of stress changes induced by uranium ore mining. In a near future, we would like to test the introduced methods also by solving 3D thermo-elasticity problems relevant to the assessment of underground repositories of the spent nuclear fuel.

2. THERMO-ELASTICITY

In this section, we describe problems of thermo-elasticity and focus on their robust numerical solution. The considered problems are important for the investigation of nuclear waste repositories, as is illustrated by the simple model problem shown in Figure 1.

This problem represents a 2D cut through a repository tunnel located at the depth of 500 m. The nuclear waste, which is the source of the heat, is disposed in a borehole below the tunnel. We are interested in the temperature and stress changes in the near field. This model problem, which arose by simplification of the BMT3 benchmark formulated in the DECOVALEX project, see Jing et al. (1994), will be used for an illustration of the behaviour of some of the numerical methods presented later. For this, we shall use the discretization by linear triangular finite elements, see Figure 2.

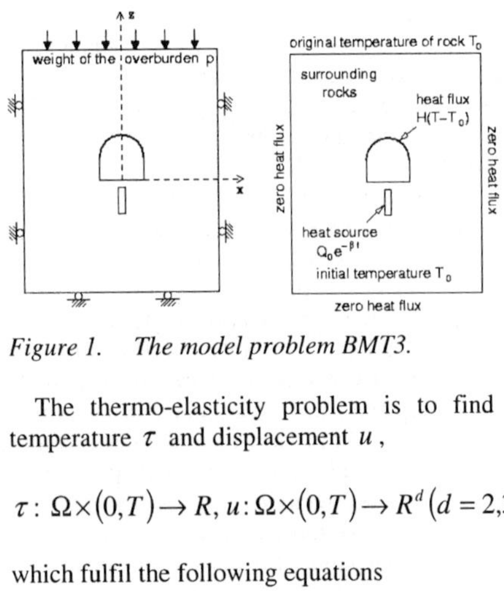

Figure 1. The model problem BMT3.

The thermo-elasticity problem is to find the temperature τ and displacement u,

$$\tau: \Omega\times(0,T)\rightarrow R,\ u: \Omega\times(0,T)\rightarrow R^d\,(d=2,3),$$

which fulfil the following equations

$$\kappa\rho\frac{\partial\tau}{\partial t}-\sum_{ij}\frac{\partial}{\partial x_i}\left(k_{ij}\frac{\partial\tau}{\partial x_j}\right)=q \quad in\ \Omega\times(0,T),$$

$$-\sum_j\frac{\partial\sigma_{ij}}{\partial x_j}=f_i \quad in\ \Omega\times(0,T),$$

$$\sigma_{ij}=\sum_{kl}c_{ijkl}\left[\varepsilon_{kl}(u)-\alpha_{kl}(\tau-\tau_R)\right] \quad in\,\Omega\times(0,T),$$

$$\varepsilon_{kl}(u)=\frac{1}{2}\left(\frac{\partial u_k}{\partial x_l}+\frac{\partial u_l}{\partial x_k}\right) \quad in\ \Omega\times(0,T)$$

together with the initial and boundary conditions.

Above, $\Omega\subset R^d$ represents the considered domain, $(0,T)$ is the time interval, q is the density of the heat source, f is the density of the volume forces, κ is the specific heat, ρ is the density of material, k_{ij} are coefficients of the heat conductivity, c_{ijkl} are the elastic moduli and α_{ij} are the coefficients of the heat expansion.

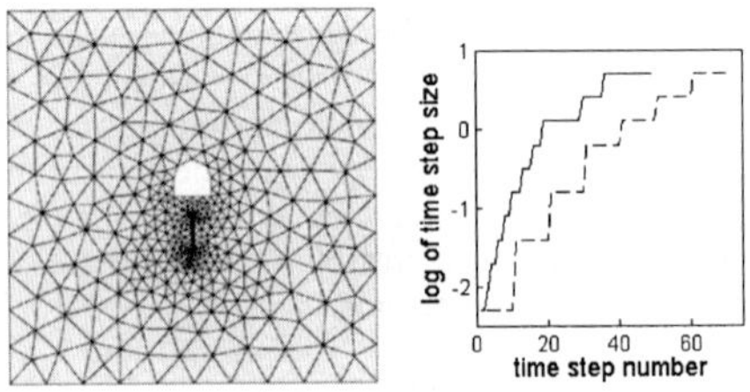

Figure 2. The FE mesh and time step sizes determined by the adaptive algorithm.

The initial-boundary value problem of thermo-elasticity can be discretized by the finite elements in space and the finite differences in time. Using the linear finite elements and the simplest time discretization, we come to the computation of vectors $\underline{\tau}^j$, $\underline{u}^j$ of nodal temperatures and displacements at the time levels t_j, $j=1,\ldots,N$, with the time steps $\Delta t_j=t_j-t_{j-1}$. This leads to the following algorithm:

$$find\ \underline{\tau}^0: M_h\underline{\tau}^0=\underline{\tau}_0,\quad \underline{u}^0: A_h\underline{u}^0=b^0=b_h\left(\underline{\tau}^0\right),$$

for $j=1,\ldots,N$:

$$find\ \underline{\tau}^j:\ B_h^{(j)}\underline{\tau}^j=\left[M_h+\theta\Delta t_jK_h\right]\underline{\tau}^j=c^j,$$

$$find\ \underline{u}^j: A_h\underline{u}^j=b^j.$$

Above, M_h is the capacitance matrix, K_h is the conductivity matrix, A_h is the stiffness matrix, $\theta\in\langle 0,1\rangle$ is a parameter,

$$c^j=\left[M_h-(1-\theta)\Delta t_jK_h\right]\underline{\tau}^{j-1}+\theta q_h^j+(1-\theta)q_h^{j-1}$$

$b^j=b_h\left(\underline{\tau}^j\right)$ and $\underline{\tau}_0$ is from the initial condition.

Note that q_h represents the heat sources and b_h represents volume and surface forces including the thermal expansion term.

We aim at the development of fully robust, stable methods and therefore we restrict our attention to implicit methods with $\theta \in \langle 1/2, 1 \rangle$. Particularly, we shall consider two cases with $\theta = 1/2$ and $\theta = 1$, which correspond to the Crank – Nicolson (CN) and backward Euler (BE) method, respectively. More details can be found e.g. in Quarteroni & Valli (1994). Finally, we chose the BE method for its higher stability (the CN scheme can show some local oscillations for large time steps).

The use of the BE method allows also to use adaptive time stepping scheme based on a local comparison of the BE and CN steps. To make this comparison computationally cheap, we shall use only the residual corresponding to the substitution of the BE time step $\underline{\tau}^j$ into the CN equation,

$$r^j = \left(M_h + 0.5\Delta t_j K_h\right)\underline{\tau}^j - \left(M_h - 0.5\Delta t_j K_h\right)\underline{\tau}^{j-1} - 0.5 q_h^j - 0.5 q_h^{j-1}.$$

Then the CN step can be approximated by the value $\underline{\tau}_{CN}^j = \underline{\tau}^j - r^j$ and the time steps can be controlled with the aid of the ratio $\eta = \| r^j \| / \| \underline{\tau}^j \|$ according to the following algorithm (k counts time step adaptation cycles):

for $k=1,2,\ldots$ **until** *stop* **do**

- **compute** $\underline{\tau}^{j,k}$, $r^{j,k}$ and evaluate η_k,
- **if** $\eta_k < \varepsilon_{\min}$ **then** $\Delta t_j \leftarrow 2 * \Delta t_j$
- **if** $\eta_k > \varepsilon_{\max}$ **then** $\Delta t_j \leftarrow \Delta t_j / 2$
- **if** $\eta_k \in \langle \varepsilon_{\min}, \varepsilon_{\max} \rangle \mid \eta_k < \varepsilon_{\min} \,\&\, \eta_{k-1} > \varepsilon_{\max}$
  then $\Delta t_{j+1} = \Delta t_j$, $\underline{\tau}^j = \underline{\tau}^{j,k}$, *stop*
- **if** $\eta_k > \varepsilon_{\max} \,\&\, \eta_{k-1} < \varepsilon_{\min}$
  then $\Delta t_{j+1} = \Delta t_j / 2$, $\underline{\tau}^j = \underline{\tau}^{j,k-1}$, *stop*

end

As a change in the step-size induces the change of the matrix $B_h^{(j)}$, it may be advantageous to do the step-size adaptation only in each k-th time step. In Figure 2 (right), we can see the step size adaptation in each time step (solid line) and each 10th time step (dashed line) for parameters $\varepsilon_{\min}$, $\varepsilon_{\max}$ equal to $10^{-3}, 10^{-2}$, respectively.

3. THE USE OF ITERATIVE SOLVERS

The computational requirements of the time stepping algorithm are concentrated in the solution of linear systems with the matrices $B_h^{(j)}$ and A_h. For large-scale and accurate models, the dimension of these matrices will be very large. Therefore, it will be advantageous to solve them iteratively. Then we have to answer the following questions

- which iterative method is suitable to use,
- what is the right accuracy of the iterations,
- what are suitable initial guesses for the solution of systems arising in the time steps.

As the matrices $B_h^{(j)}$ and A_h are symmetric and positive definite in our case, it seems that a good choice of the solution technique is the conjugate gradient (CG) method with a suitable preconditioning. This question will be further discussed in the next section.

The accuracy of solving the systems with matrices $B_h^{(j)}$ should be in correspondence with the accuracy of the time stepping scheme. We performed numerical tests with the adaptively controlled time stepping schemes $\varepsilon_{\min}$, $\varepsilon_{\max}$ equal to $10^{-3}, 10^{-2}$ or $10^{-4}, 10^{-3}$. These tests show that the accuracy of computed results is not deteriorated when we use a lower accuracy $\varepsilon = 0.01$ for stopping the iterations,

$$\left\| c^j - B_h^{(j)} \underline{\tau}^j \right\| \le \varepsilon \left\| c^j \right\| .$$

Surprisingly, even $\varepsilon = 0.1$ works still well with the less accurate adaptive scheme. The adaptive scheme only somewhat increased the number of time steps in this case. But with the more accurate adaptive scheme such low accuracy of iterations caused infinite cycling.

The efficiency of the iterative solution can be also increased by using a suitable initial guesses. There are several possibilities:

(i) using the previous time step (PS) values $\underline{\tau}^j, \underline{u}^j$ as initial guesses for $\underline{\tau}^{j+1}, \underline{u}^{j+1}$,

(ii) using the initial guesses given by the linear extrapolation (LE), i.e. for $\Delta t_j = \Delta t_{j-1}$, we get $\underline{\tau}^{j+1} = 2\underline{\tau}^j - \underline{\tau}^{j-1}$, $\underline{u}^{j+1} = 2\underline{u}^j - \underline{u}^{j-1}$,

(iii) improvement of the initial guesses by exploiting the conjugate directions (CD) from application of the CG method to the solution of the systems in some previous time step.

In the CD technique, we assume that $B_h^{(j)} = B_h$ does not change in a selected sequence of time steps $j = j_0, \ldots, j_1$. Then the conjugate directions $\{v_i\}$, $i = 1, \ldots, m$ from solving the system in the step j_0 can be used for improving the initial guesses in the subsequent steps where we take

$$\underline{\tau}^{j,0} = \underline{\tau}_{ini} + \sum_{i=1}^{m} \eta_i v_i, \quad \eta_i = \frac{\langle r_{ini}, v_i \rangle}{\langle A v_i, v_i \rangle}.$$

Here $\langle , \rangle$ denotes the inner product and $\langle A v_i, v_i \rangle$ are known from the previous application of the CG method. Then computation of $\underline{\tau}^{j,0}$ requires only computation of the residual $r_{ini} = c^j - B_h \underline{\tau}_{ini}$ and m inner products $\langle r_{ini}, v_i \rangle$. The same technique is applicable also for solving elasticity systems.

Let us focus again on the systems $B_h \underline{\tau}^j = c^j$. The slow change of the right hand sides (RHS) c^j with changing j motivates introduction of another technique, which consists in the gradual construction of an orthonormal basis for the representation or approximation of the next RHS vectors. This technique requires the solution of the system with the matrix B_h only in the case that the existing basis is not sufficient for the representation of the next RHS. In this case, the base is extended.

To be more specific, let us consider the j-th time step. Let $\{e_k\}$, $k \in K_j$, be the existing basis and let θ_k be (approximate) solutions of the systems $B_h \theta_k = e_k$. Then c^j can be represented as

$$c^j = \sum_{k \in K_j} \lambda_k e_k + \nu e_\perp,$$

where $e_\perp$ is orthogonal to all $e_k, k \in K_j$, $\|e_\perp\| = 1$, $\lambda_k = \langle c^j, e_k \rangle$.

If $|\nu| \le \varepsilon \|c^j\|$ then we take the solution $\underline{\tau}^j$ as

$$\underline{\tau}^j = \sum\nolimits_{k \in K_j} \lambda_k \theta_k,$$

otherwise we solve the system $B_h \theta_\perp = e_\perp$ and take

$$\tau^j = \sum\nolimits_{k \in K_j} \lambda_k \theta_k + \nu \theta_\perp.$$

If some of λ_k is small (typically $\lambda_k \le \varepsilon / 10$), then the new pair $e_j = e_\perp$, $\theta_j = \theta_\perp$ can replace e_k, λ_k. Otherwise the new pair extends the existing base. Due to the rounding errors, e_j should be reortho-gonalized to the previous basis vectors e_k.

Note that this technique, which can be denoted as the combination of the previous solutions (CPS), does not need the symmetry of B_h and is applicable also to the elasticity systems.

The efficiency of the described ideas is illustrated by the solution of the benchmark problem BMT3. The results are summarized in Table 1. The systems are reformulated in unknown increments of temperatures and displacements.

Table 1. Reduction of the total numbers of iterations for 400 time steps. Numbers of exploited conjugate directions (CD) and maximal previous solutions (CPS) can be seen in the subscripts.

Initial	τ: $\Delta t_j = 0.25$		u: $\Delta t_j = 0.25$	
gues:	$\varepsilon = 10^{-2}$	$= 10^{-3}$	$= 10^{-2}$	$= 10^{-3}$
Zero	2420	3632	8362	11159
PS	590	1457	810	4526
PS + CD	399_8	899_{10}	210_{22}	1126_{30}
LE	689	697	501	840
LE + CD	423_8	459_{10}	207_{22}	311_{30}
CPS	101_{18}	181_{21}	199_9	351_{12}

In the above table, ε denotes the relative accuracy of the iterative solution or the parameter in the CPS technique. The LE helps in the case of smooth results, which are not the case in column 2 due to a low accuracy ε. To emphasize the effect of the initial guess, we present here the results for constant $\Delta t_j = 0.25$ but the described ideas are useful also for some adaptive schemes.

4. BMT3 RESULTS

We use the BMT3 benchmark for an illustration of the efficiency of the described ideas concerning the acceleration of the iterative solution. As concerns the computed results, the temperature field is in a good agreement with the results reported by Jing et al (1994), see also Fig.3. The comparison of computed stresses and displacement is more complicated, because the original BMT3 assumed fractured rocks, which require to use homogenized properties. There is also a bigger variance in the corresponding results presented in Jing et al (1994).

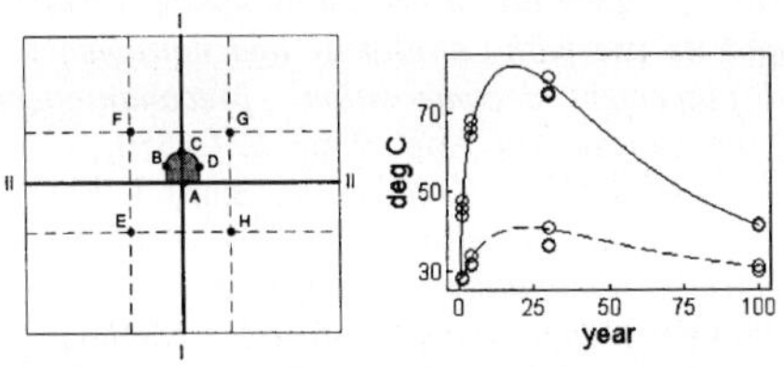

Figure 3. The comparison of the computed temperature behaviour in points E (solid line) and F (dashed line) with previously reported results (circles).

The BMT3 problem was solved by means of a simple code implemented in MATLAB. In near future, the described ideas will be also implemented and tested within a scientific FEM system GEM (Institute of Geonics AS CR). This system aims at the solution of large 3D problems, for which the iterative solvers are crucial. In the next section, we try to give an idea about the expected efficiency of a large 3D modelling with parallel computations.

5. SPACE DECOMPOSITION SOLVERS AND LARGE 3D MODELLING

In Section 3, we discussed an acceleration of iterative solvers, now we shall touch the question of the choice of a suitable iterative methods and show that efficient methods can be found in the class of space decomposition - subspace correction methods, see Blaheta et. al. (2003). Remember that we are interested in the solution of linear systems with symmetric positive definite matrices B_h and A_h for finding the unknown temperature and displacement increments, respectively.

The well-known domain decomposition (DD) methods, see e.g. Chan & Mathew (1994), use data partition induced by a decomposition of the computational domain. This decomposition can be used for two purposes: Firstly for parallel implementation of vector updates, inner products and matrix-vector multiplication, i.e. for parallel implementation of the CG method. Secondly, for a construction of efficient preconditioners.

The preconditioner C is an approximation of the FEM matrix A, which can be used for computations of pseudoresiduals $g=C^{-1}r$ and search directions, which are closer to errors $e=A^{-1}r$. Of course, the operation $C^{-1}r$ should be sufficiently computationally cheap. The basic preconditioner induced by a domain decomposition is the block diagonal one and its efficiency can be increased by extending the blocks or correspondingly extending the subdomains into overlapping ones, see Figure 4 (right). But even more important is to enable a global exchange of information by adding a small (coarse) but global subproblem. This defines so-called two-level preconditioner. The global subproblem can be created by discretization of the solved problem on a coarser grid or e.g. by the use of a simpler construction with aggregation of unknowns, see Blaheta et. al. (2003a, 2003b). The subproblems can be solved exactly by a direct method or approximately by inner iterations with accuracy ε_0. The latter approach can be more efficient and correctly used within a generalized CG method, see Blaheta (2002).

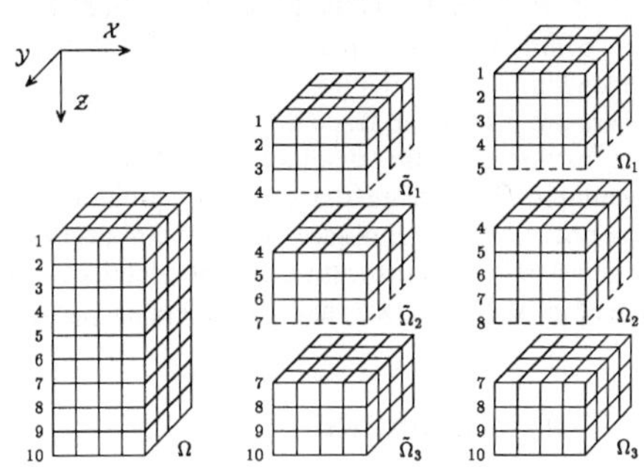

Figure 4. The decomposition of the domain.

To demonstrate the efficiency of the described methods, we show in Table 2 the numbers of iterations and computing times corresponding to

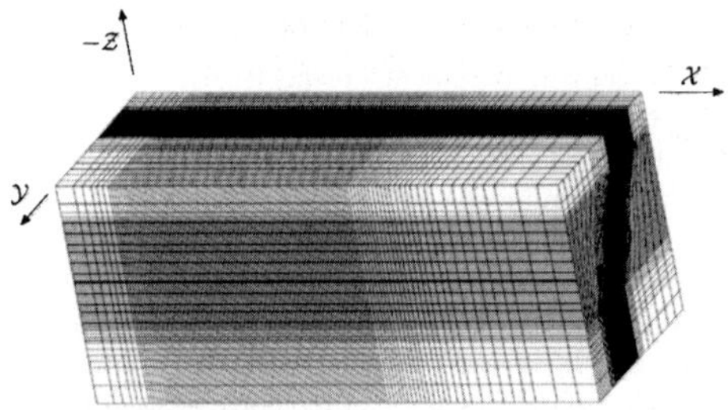

Figure 5. The FE mesh for the benchmark DR, about 4 million DOF.

the solution of linear system of about 4 million unknowns arising from the FE solution of elasticity problem with the mesh shown in Figure 5. This elasticity problem represents the computation of stresses induced by mining at the uranium ore deposit Rožná (DR) in the Czech Republic, for more details see Blaheta et al. (2002a, 2002b).

The computations were done in parallel on a cheap Beowulf cluster consisting of 8 PCs with the processors AMD Athlon 1.4 GHz, 768 MB RAM. The PCs are interconnected by Fast Ethernet LAN. The results in Table 2 correspond to computations on 3 and 8 PC's from the cluster.

Table 2. Solving the problem DR on a PC cluster.

method	# proc's	# iter's	time [s]
CG-DD-one-level	3	102	283
CG-DD–two-level	3	56	242
CG-DD- one-level	8	128	161
CG-DD- two-level	8	72	111

Note that similar methods can be applied also to the solution of systems arising from the time-dependent heat conduction – convection problems. There are also other space decomposition methods. Let us mention the displacement decomposition technique for solving the elasticity problems and composite grid technique, for solving problems, which need a local resolution. More details can be found in Blaheta (2002) and Blaheta et al. (2002b).

6. FINAL REMARKS

The aim of this paper was to show numerical techniques suitable for efficient modelling of large-scale coupled problems, which are typical for the assessment of repositories of the spent nuclear fuel. The use of iterative solvers and parallel computing is the essential part of these methods.

Note that the efficiency of the considered numerical methods is important not only for solving large–scale and complicated coupled problems but also for performing a further analysis of the obtained results for the assessment of sensitivity to changes of input parameters, influence of uncertainty in the model data, etc.

Acknowledgement

The work is supported by the Grant Agency of the Czech Republic under the project GAČR 105/02/0492.

REFERENCES

Blaheta, R. 2002.
GPCG - generalized preconditioned CG method and its use with nonlinear and nonsymmetric displacement decomposition preconditioners. Num. Linear. Alg. Appl. 9, pp. 527 - 550.

Blaheta, R., Jakl, O., Krečmer, K., Starý, J. 2002a.
Large-scale modelling in geomechanics with parallel computing on clusters of PC's. In Philipe Mestat (ed.), Numerical Methods in Geotechnical Engineering; Presses de l'Ecole Nationale des Ponts et Chausses, pp. 315-320.

Blaheta, R., Byczanski, P., Jakl, O., Kohut, R., Kolcun, A., Krečmer, K., Starý, J. 2002b.
Application of large-scale FEM computations for design of new extracting technologies in mining. HIPERGEOS Project report. See also http://www.ugn.cas.cz/math/kit/kit.htm.

Blaheta, R., Byczanski, P., Jakl, O., Starý, J. 2003.
Space decomposition preconditioners and their application in geomechanics. Mathematics and Computers in Simulation 61, pp. 409 - 420.

Chan, T.F. & Mathew, T.P. 1994.
Domain Decomposition Algorithms. Acta Numerica, Cambridge University Press.

Jing, L., Rutqvist, J., Stephansson, O., Tsang, C.F., Kautsky, F. 1994.
DECOVALEX Mathematical Models of Coupled T-H-M Processes for Nuclear Waste Repositories. SKI Report 94:16, Stockholm.

Quarteroni, A. & Valli, A. 1994.
Numerical Approximation of Partial Differential Equations. Springer-Verlag, Berlin.

Stephansson, 0., Jing, L., Tsang, C.F. (eds.) 1996.
Coupled Thermo-Hydro-Mechanical Processes of Fractured Media. Elsevier, Amsterdam.

Stephansson, O. (ed.) 2001. *DECOVALEX II.* Special issue of Int. J. Rock. Mechanics and Mining Sciences 38(1).

THERMO-MECHANICAL MODELING OF A SUBSURFACE INTERIM NUCLEAR WASTE STORAGE: BEHAVIOR IN WORKING CONDITIONS

Thouvenin G.[1] & Millard A.[2]
[1]) CEA Cadarache, DED/SAMRA/LGEM, 13108 St Paul lez Durance Cedex, France
[2]) CEA Saclay, DM2S/SEMT/LM2S, 91191 Gif-sur-Yvette Cedex, France

Abstract: Among the different concepts for shallow subsurface storage of nuclear fuel or waste considered by the French Atomic Commission, one consists in setting them in underground wells ventilated by natural convection. This paper deals with the estimation of the impact of an increase of temperature in the near field of the wells on the mechanical behavior of the surrounding rock. The aim of this study is to predict stresses induced by thermo-mechanical loads, in a qualitative way (traction or compression) as well as quantitative (maximal values), since they may be liable to damage the rock mass and the waste containers. In many disposal-oriented studies, heat power needs to be limited for chemical or hydraulical reasons; differently in this particular case of storage, the limit is determined by the mechanical strength of the rock and the damage of the rock is studied through the overstepping of a failure criterion.

1. INTRODUCTION

According to a law adopted by the French Parliament in 1991, the French Atomic Energy Commission (CEA) is in charge of the feasibility study of shallow subsurface storage of nuclear fuel or waste. Different concepts have been considered since this decision. One of them, called CERTI, consists in setting waste containers in underground wells ventilated by natural convection, a drift connected to the lower part of the well allowing the external cold air flow which is then collected by a higher drift after circulation in the well. Such structures of interim subsurface storage, under consideration in hard rocks such as limestone or granite, must be able to ensure the integrity of the containers for a later recovery. Contrary to high level waste disposal, they only need to be designed for one to three centuries, and besides they don't play any role of chemical and hydraulical barriers.

The problems raised by these structures are then related to the fact that the containers release a high heat flux, and a more or less significant part goes into the host rock. This heat supply increases the temperature in the near field of the wells and causes stresses, induced by thermo-mechanical (TM) coupling effects. The additional question is then to know if these stresses are likely to damage the rock, which could lead to a deterioration of the wells as well as of the containers, thus preventing their recovery. From a technical point of view, this paper is based on a thermo-elastic modeling of a 3-D elementary pattern of the repetitive shallow waste disposal structure, using the finite element code CAST3M, developed by CEA. A large range of TM properties is taken into account to contribute to the design of such a disposal in hard rocks such as limestone and granite. From the temperature fields calculated in the rock mass, the induced stresses are formulated in term of principal stresses and evaluated by means of a failure criterion.

2. DESCRIPTION OF THE CERTI CONCEPT

2.1 Geometry of the interim storage structure

The CERTI concept corresponds to a hillside storage in a geological medium consisting of hard rock (Motuelle, 2000). A series of wells are filled with waste containers, around which a flow of air is induced by natural convection to cool the containers (see fig.1).

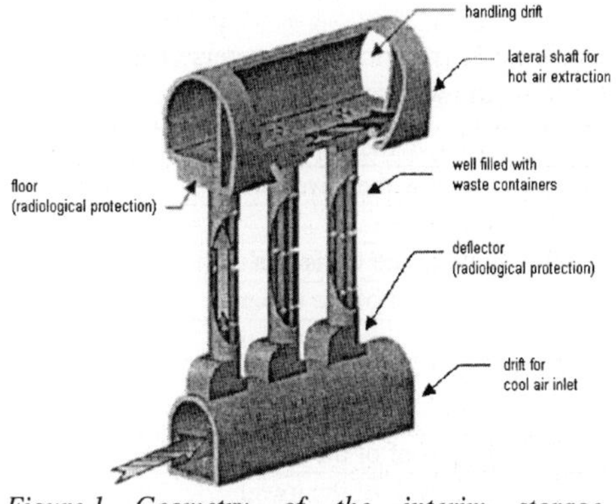

Figure 1. Geometry of the interim storage structure (from Motuelle, 2000).

The geometrical model represents in fact a quarter of the well and a section of the two lower and upper galleries with vertical planes limit characterized by symmetrical boundary conditions (see fig.2). Indeed, this pattern can be considered as repeated in x- and y-directions, because of the periodic structure of such a storage, and the approximation thus made will be even more justified considering a zone located at the center of storage and a large number of patterns (strictly speaking, it should be infinite in the two horizontal directions). All the geometrical parameters are given in table 1.

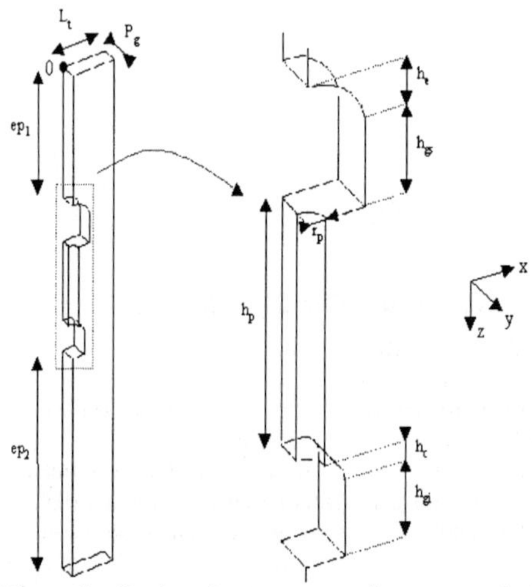

Figure 2. Designed pattern for numerical modeling with CAST3M.

Table 1. Geometrical parameters characterizing the well and the two drifts.

Thickness above the upper drift	ep_1	30 m
Thickness below the lower drift	ep_2	50 m
Radius of the well	r_p	1.75 m
Half-spacing of the wells in a drift	p_g	4.375
Height of the roof of the upper	h_e	4 m
Height of the wall of the upper	h_{gs}	6 m
Width of the upper drift	L_{gs}	5.15 m
Height of the well	h_p	18 m
Height of the roof of the lower	h_c	1.5 m
Height of the wall of the lower	h_{gi}	5.5 m
Width of the lower drift	L_{gi}	3.5 m
Section of the lower drift	S_{gi}	40 m^2
Half-spacing of the drifts	L_t	12.5 m

2.2 Thermo-mechanical equations, properties and loading

The layout of the CERTI concept being considered in a hard geological medium, a simple thermo-elastic law has been chosen to describe the behavior of the rock. Together with the mechanical equilibrium equation, it leads to a relation between the displacement vector $\vec{u}$ and the temperature T ; besides, deformations and stresses tensors can be derived from the classical following equations :

$$\Delta\vec{u}+\frac{1}{1-2\nu}\overrightarrow{\text{grad}}\,\text{div}\,\vec{u}=\frac{2(1+\nu)}{1-2\nu}\overrightarrow{\text{grad}}\,\alpha T \quad (1)$$

$$\varepsilon_{ij}=\frac{1}{2}\left(\frac{\partial u_i}{\partial x_j}+\frac{\partial u_j}{\partial x_i}\right) \quad (2)$$

$$\sigma_{ij}=E\left[\frac{1}{1+\nu}\varepsilon_{ij}+\frac{\nu}{(1-2\nu)(1+\nu)}\text{tr}\bar{\bar{\varepsilon}}.\delta_{ij}-\frac{1}{1-2\nu}\alpha T\delta_{ij}\right] \quad (3)$$

TM properties of the host rock were selected from the literature (Michaud & Tauveron, 2001), and presented in table 2.

Table 2. TM properties of the rock

Thermal conductivity	λ_T	2 W.m^{-1}.K^{-1}
Heat capacity	C_p	1000 J.kg^{-1}.K^{-1}
Density	ρ	2800 kg.m^{-3}
Thermal expansion coefficient	α	10^{-5} K^{-1}
Young's modulus	E	20 GPa
Poisson's ratio	ν	0.3

The loading of the structure is mechanically very simple since it consists in an initial isotropic stress field related to the dead weight. Concerning the mechanical boundary conditions, a zero normal stress is prescribed on the free boundaries of the pattern (wall of the drifts and well, and ground surface), and the symmetry planes are characterized by a zero normal displacement. Before excavation, the rock mass is supposed to be in a compressive stress state, and the principal minor stress σ_3, indicator of maximum compression, is equal to σ_{zz} (and $\sigma_1=\sigma_2=\sigma_{xx}=\sigma_{yy}$). This stress increases (in absolute value) with depth, from -1.1 MPa at the top of the wells to -1.6 MPa at its base, and to -3.1 MPa in lower limit of the model. The excavation of drifts and wells causes a disturbance of this initial stress field (see fig.3). It is noticed that, apart from the relatively high values related to the right angles of the model (5 MPa), σ_3 has increased on the wall of well to reach a maximum value of 2.6 MPa, located near the bottom of the well. Thus, the compressive stresses are relatively low in the whole structure since they do not exceed 2.6 MPa close to the well, and moreover, the top of gallery which will be more intensively loaded by the thermal stresses, is almost unloaded after excavation.

The cumulated effect of the radiation (with a constant heat flow released by the waste containers equal to 86400 W) and of the natural air convection can be approximated by a constant temperature gradient along the well, with a temperature T_b=45°C at the base and T_h=95°C at the top. This boundary condition is established from the assumption that the heat flow penetrating in the rock mass by conduction is negligible in front of the heat flow convected by the drifts.

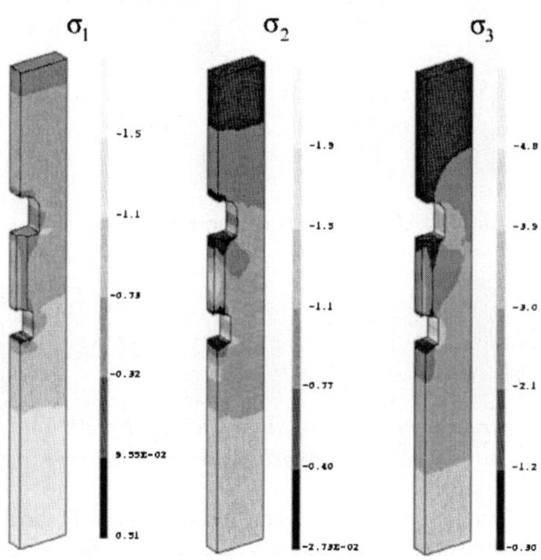

Figure 3. Initial stress field (in MPa) after excavation.

Lastly, the walls of the drifts are subjected to a convection condition with air circulating at a temperature T_2 of 20°C in the lower drift and T_1 of 60°C in the upper. From the geometry of each drift and air flow circulating in each well, the exchange coefficients h_1 and h_2 are respectively about 11 W/m^2/K for the upper drift and 8 W/m^2/K for the lower. Using these boundary conditions and a diffusion model taking of account the conduction of heat in the rock mass, the temperature field has been calculated by CAST3M (see fig.4).

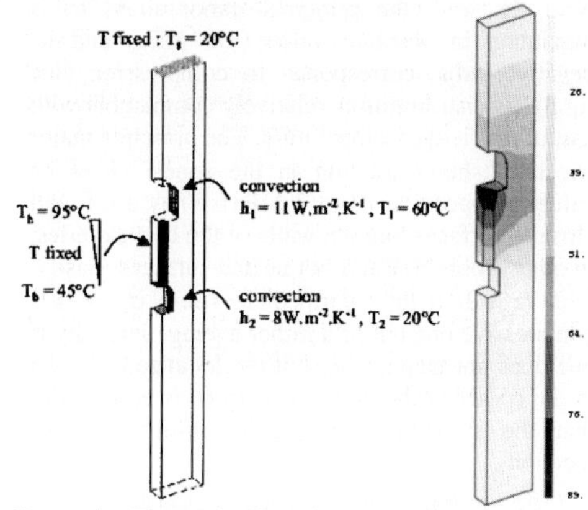

Figure 4. Thermal boundary conditions and computed temperature (in °C).

3. BEHAVIOR IN OPERATING CONDITIONS

3.1 Impact of thermal loading on stresses distribution

Generally, in a TM problem, an increase in temperature causes an expansion of the material which results in compressive stresses. The presence of free surfaces often make it possible to reduce compressive stresses in the normal direction on the surface and cause tensile stresses in the orthoradial.

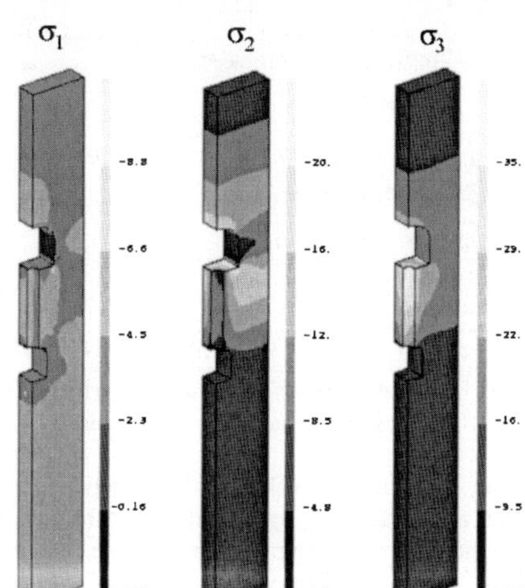

Figure 5. Distribution of stresses for steady state.

Near the well, the principal minor stress σ_3 is maximum in absolute value (-38 MPa) and its negative value corresponds to compression (see fig.5). Its distribution is relatively comparable with that of the field of temperature. The principal major stress σ_1 shows traction in the zones where its values are positive (0 to 3 MPa). They are found close to surfaces like the walls of the two galleries, in other words free and not heated surfaces. Lastly, with regard to the intermediate stress σ_2, it is a compressive one but of a rather average intensity if one does not take account of the localized stresses in the corner of the higher gallery (it is pointed out that the stress is theoretically infinite at this location).

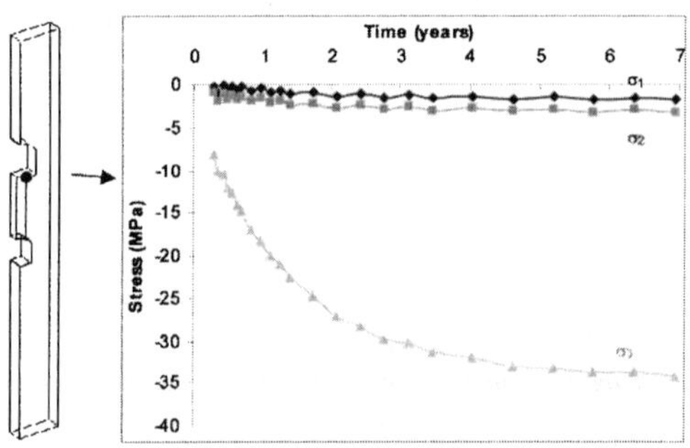

Figure 6. Time dependent evolution of principal stresses on the wall of the well

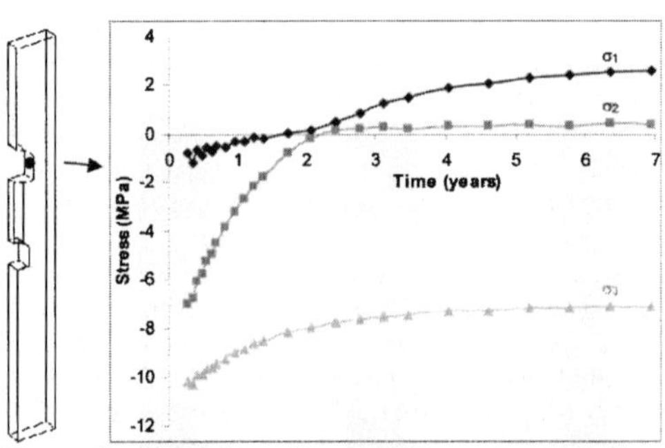

Figure 7. Time dependent evolution of principal stresses on a wall of the upper gallery.

On the wall of the well, the time dependent evolution of stresses (see fig.6) shows that the rock mass undergoes a compression in the three principal directions (maximum at the point illustrated in black) which increases gradually with time, with a principal stress σ_3 much higher than both σ_1 and σ_2. This situation is similar to a triaxial test with a low confining pressure ($\sigma_1 \approx \sigma_2$) and a load σ_3 applied in the orthoradial direction. On a vertical wall of the upper gallery, the stresses evolution is definitely different (see fig.7). The rock mass is initially in compression like previously but after approximately two years, tensile stresses appear in the principal directions 2 (in the horizontal plane according to the normal to the wall) and especially 1 (close to vertical direction z). Direction 3 (in the horizontal plane according to the axis of the gallery) is always subjected to a lower compression.

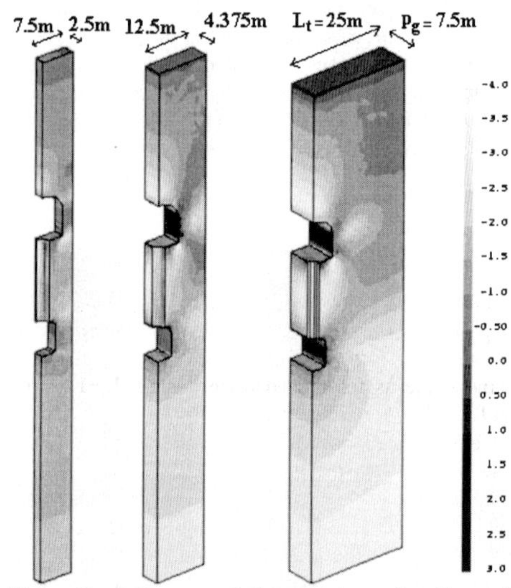

Figure 8. Influence of the spacing of wells and galleries on principal stress σ_1.

In the configuration of very close well and galleries (L_t=7.5 m and p_g=2.5 m), the temperature gradient is smaller and the temperature of the volume of rock in the vicinity of the well is higher. Moreover, the displacement of the rock mass is prevented on the two vertical symmetry levels in the normal direction, and these two plans are closer in this configuration. These two effects are then at the origin of higher compressive stresses (-47 MPa, i.e. approximately 20% more compared to the initial configuration) and lower tensile stresses. It may be noticed however that the estimation of stresses is erroneous because of an artefact of modeling. Indeed, only one pattern has been considered, leading by symmetry to only possible vertical displacements of symmetry planes. In reality, the storage would have a finite extension.

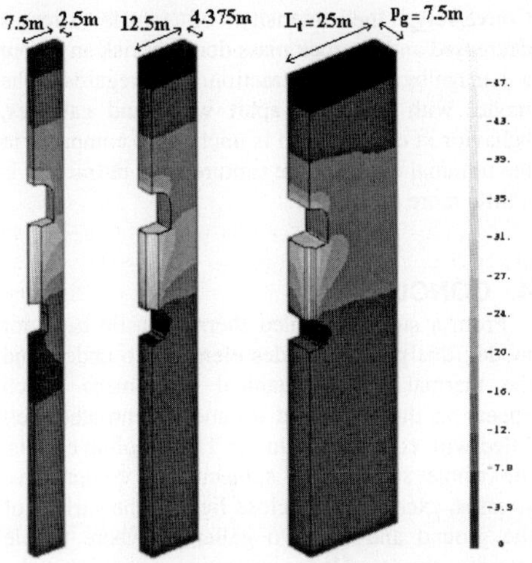

Figure 9. Influence of the spacing of wells and galleries on principal stress σ_3.

At the opposite, when the wells and galleries are wide apart (L_t=25 m and p_g=7.5 m), compressive stresses are close to -38 MPa, in other words the same value as for intermediate design (L_t=12.5 m and p_g=4.375 m). Consequently, for intermediate design between values (L_t=7.5 m and p_g=2.5 m) and (L_t=12.5 m and p_g=4.375 m), the behavior in the vicinity of the well is nearly the one of an insulated well. The level of compressive stress in the wall of the well can be written as follows, where ΔT is a specified value fixed by the boundary condition at the top of the wall and the initial temperature of the rock mass (i.e. 75°C) :

$$\sigma_3 = -\frac{E\alpha\Delta T}{1-2\nu} = -38\text{MPa} \qquad (4)$$

3.2 Evaluation of damage caused by thermal stresses

From the stress fields, the estimate of the damage of the interim storage structure is carried out by using a rupture criterion as post-processing. Two criteria were taken into account to study the influence of the three-dimensional form of the pattern: a parabolic 2-PS criterion based on only major and minor principal stresses, and a 3-PS one taking also account of the intermediate principal stress (where σ_c and σ_t are the uniaxial compressive and tensile strengths of the material):

$$(\sigma_1-\sigma_3)^2-(\sigma_1+\sigma_3)(\sigma_c+\sigma_t)+\sigma_c\sigma_t=0 \qquad \text{(2-PS)} \qquad (5)$$

$$(\sigma_1-\sigma_3)^2+(\sigma_2-\sigma_1)^2+(\sigma_3-\sigma_2)^2 -2(\sigma_1+\sigma_2+\sigma_3)(\sigma_c+\sigma_t)+2\sigma_c\sigma_t=0 \qquad \text{(3-PS)} \qquad (6)$$

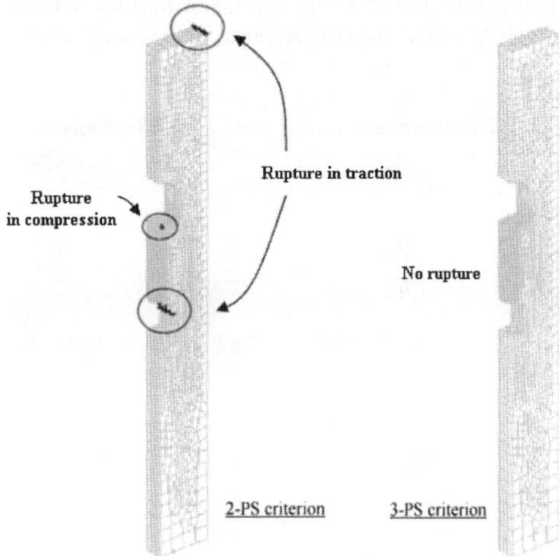

Figure 10. Localization of positive values of 2-PS and 3-PS criterion showing rupture.

For a hard rock mass, the compressive and tensile strengths measured on samples in laboratory are about several tens of MPa for σ_c (80 to 100 MPa) and some MPa for σ_t. If one considers for example σ_c=-90 MPa and σ_t=3 MPa, the rupture criterion, whether it is 2-PS or 3-PS, is never reached: there is thus no risk of damage of the rock under nominal thermal loading for such a rock mass. However, the values of E and σ_c which were used for these calculations are representative of a non fractured rock. The effect of fracturing on both parameters is more sensitive on the compressive strength than on the Young's modulus. Using Hoek & Brown approach (1997), for a standard rock of average quality (GSI=50), one can estimate that E reduces from 20 to 10 GPa (factor 2) whereas σ_c decreases from 80 to 15 MPa (factor 6), which corresponds to a reduction three times more important for σ_c than for E. To take into account the fractures, the stresses calculated for the intact rock must then be compared not to σ_c, but to β σ_c, with β=1/3. Figure 10 shows in black the points of the mesh grid for which the criterion gives positive values with σ_c=-30 MPa and σ_t=1 MPa. With the 3-PS criterion, there is always no damage. On the other hand, the rupture due to compressive stresses is reached just at the limit with the 2-PS criterion, in one point where the temperature is maximum. These results underline the importance to choose an appropriated criterion, since the confinement in a point (role played by σ_2 and taken into account in the 3-PS criterion) does strengthen the rock.

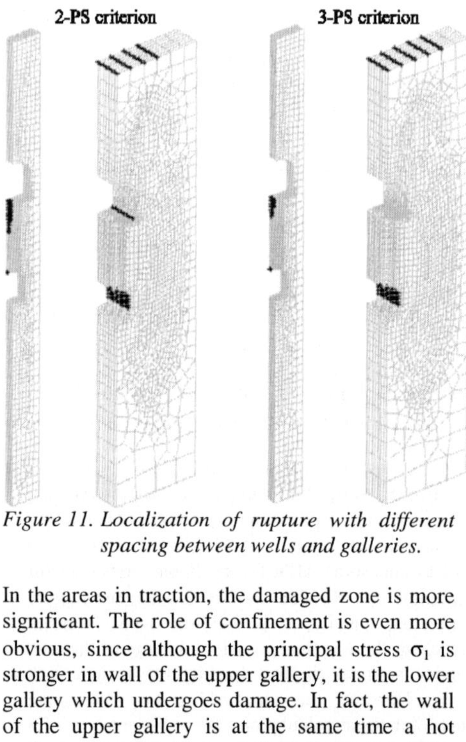

Figure 11. Localization of rupture with different spacing between wells and galleries.

In the areas in traction, the damaged zone is more significant. The role of confinement is even more obvious, since although the principal stress σ_1 is stronger in wall of the upper gallery, it is the lower gallery which undergoes damage. In fact, the wall of the upper gallery is at the same time a hot surface and a free face, so there is traction in its principal direction 1 but compression in directions 2 and 3. At the opposite, the wall of the lower gallery is a free but cold face, only subjected to traction according to direction 1 ($\sigma_2 \approx \sigma_3 \approx 0 << \sigma_1$). The same post-processing is presented on figure 11 with other spacing between wells and galleries. It is first noted that both 2-PS and 3-PS criteria do not give the same extent of the zone subjected to rupture in compression: the principal intermediate stress σ_2 plays a major confining role. Moreover, it had been noticed that the compressive stress σ_3 was larger for small values of L_t and p_g, resulting in a relatively large rupture area around the well. Conversely, the intensity of tensile stresses decreased and the rock mass does not risk any more a damage by rupture in traction. With regards to the model with the wide apart wells and galleries, behavior in compression is unchanged compared to the nominal case and the rupture zone in traction is a little more developed.

4. CONCLUSIONS

From a simple coupled thermo-elastic behavior model, this paper provides elements to understand the thermal and mechanical phenomena which appear in the near field around a ventilated well filled with containers. Due to TM coupling effects, the whole structure is subjected to compressive stresses, except for the close field of the surface of the ground and the two galleries, where tensile stresses appear. Under operating conditions, the consequence of the increase of temperature in the rock mass over the mechanical behavior of the work does not lead to a rupture of the rock. For the three designs considered, whatever the traditional formulations of rupture criteria used (2-PS conservative criterion or more realistic 3-PS criterion taking into account the confinement by the intermediate principal stress), a damage in compression is not expected compared to the maximal temperatures reached (95°C) even if the material is initially fairly fractured (Hoek & Brown GSI index equal to 50). This assertion does not remain systematically valid if a safety factor is taken into account, which decreases by as much the strength of the rock. Strengths in traction being relatively low, the rupture in traction is likely to appear close to surface of the ground and in the wall of the lower gallery, but seems to be not very problematic as for the behavior of the structure.

REFERENCES

Hoek, E. & Brown, E.T. 1997. *Practical estimates of rock mass strength*, Int. J. Rock Mech. Min. Sci., vol. 34, no. 8, pp1165-1186.

Michaud, V. & Tauveron, N. 2001. *Synthèse bibliographique sur les propriétés thermo-hydro-mécaniques des géomatériaux*, CEA Technical note, DED/SAMRA n°01-015.

Motuelle, C. 2000. Projet Entreposage de très Longue Durée - Description technique de la variante CERTI-Géo, CEA Technical note, DRRV/SSP/BCE n°00-010.

EFFECT OF COUPLING BEHAVIOR OF THE NEAR FIELD ON GROUNDWATER FLOW OF THE FAR FIELD FOR GEOLOGICAL DISPOSAL OF HIGH-LEVEL RADIOACTIVE WASTE

Hiroshi Kurikami[1], Akira Kobayashi[2], Masakazu Chijimatsu[3], Yutaka Sugita[1] and Yuzo Ohnishi[2]

[1)] Japan Nuclear Cycle Development Institute, Japan

[2)] Kyoto University, Japan

[3)] Hazama Corporation, Japan

Abstract: It is very important in performance assessment to estimate the impact of constructing high-level radioactive waste repositories deep underground on the regional groundwater flow system. This paper proposes a method of estimating the impact of coupled thermal, hydraulic and mechanical behavior near the repository. Since the disposal pits will be arrayed at regular intervals in the disposal panel that is a unit of the repository, the homogenization theory is applied to derive the representative parameters of the panel, such as permeability, according to the coupled analysis in the near field. Moreover, by using the homogenized permeability and the thermal conductivity, the coupled thermal and hydraulic analysis in the far field is conducted. This series of simulations is useful in estimating the regional groundwater flow system.

1. INTRODUCTION

Estimating the impact of constructing a high-level radioactive waste repository deep underground in the regional groundwater flow system is one of the most important aspects of performance and safety assessment. Excavation of tunnels for the repository is the largest factor affecting the groundwater flow during the construction of the repository. However, the influence of coupled behavior near the repository that occurs after closure of the repository has also to be considered because heat from the waste may cause heat convection of groundwater.

Coupled thermal, hydraulic and mechanical processes take place near the disposal site (the near field) after the placement of waste canisters. For instance, infiltration of groundwater into bentonite buffer material causes it to swell, and heat flux from the waste canisters generates thermal stress. On the other hand, in the surrounding rocks, the water pressure that has decreased due to excavation of the tunnels begins to recover. If the permeability of the rocks is low, however, groundwater in the rocks is drawn by the high suction of the buffer, and the rocks may dehydrate. The dehydration of the surrounding rock decreases disposal panel permeability and consequently affects the regional groundwater flow system (the far field). Moreover, heat conduction from the near field to the far field is considerable because it may result in heat convection of the groundwater flow system.

This paper, to evaluate these effects, proposes a method of connecting a near field model and a far field model. When a regional groundwater flow system is modeled, it is necessary to give the properties of the disposal site such as permeability, thermal conductivity, etc. To determine these properties, the homogenization theory is introduced. Homogenization is a method of determining the representative parameters of the media with microscopically periodic structures (Ene & Polisevski (1987), Kyoya & Terada (2001)). Figure 1 shows a disposal layout concept shown in JNC (1999). Disposal pits are arrayed at regular intervals (periodic) and make up the disposal panel. Therefore, disposal panel parameters can be obtained through homogenization.

Moreover, coupled thermal and hydraulic analysis in the far field is conducted using the homogenized parameters and the heat flux from the near field. This series of simulations shows that it is necessary to consider the effects of coupled processes in the near field on the regional groundwater flow system.

2. COUPLED PROCESSES IN NEAR FIELD

This chapter shows the coupled thermal, hydraulic and mechanical behavior in the near field by using THAMES that is the finite element simulator originally developed by Ohnishi et al (1985). It was applied to the near field mesh shown as Figure 2. The model domain is corresponding to the shaded area shown in Figure 1. It is assumed that tunnel interval is 10 m and pit interval is 4.4 m following the Japan Nuclear Cycle Development Institute H12 report (JNC(1999)).

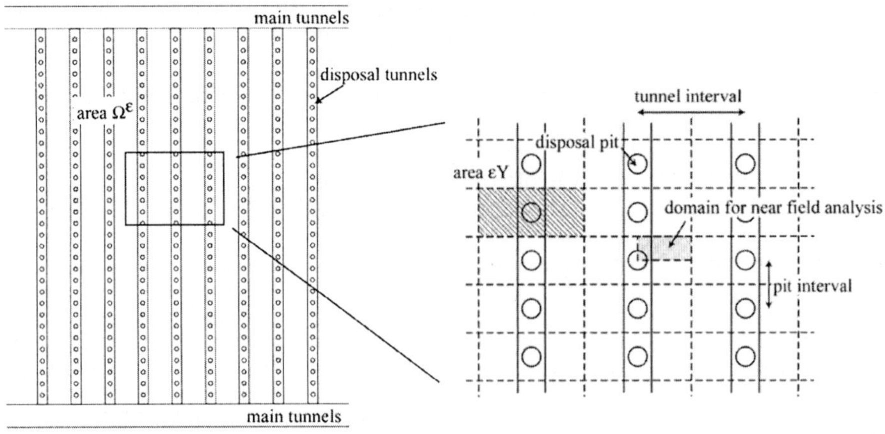

Figure 1. Macro-scale view and micro-scale view of disposal panel

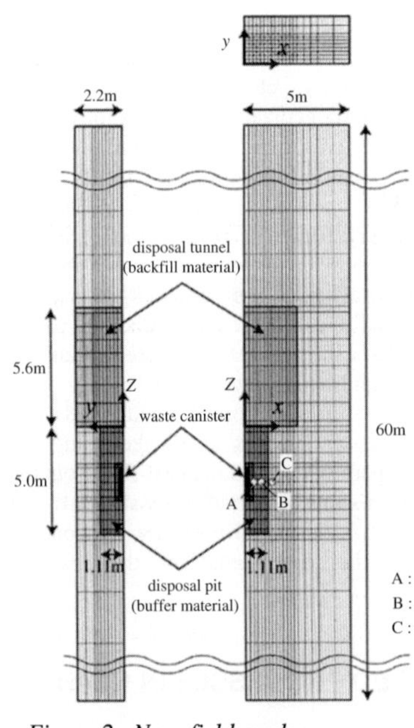

Figure 2. Near field mesh

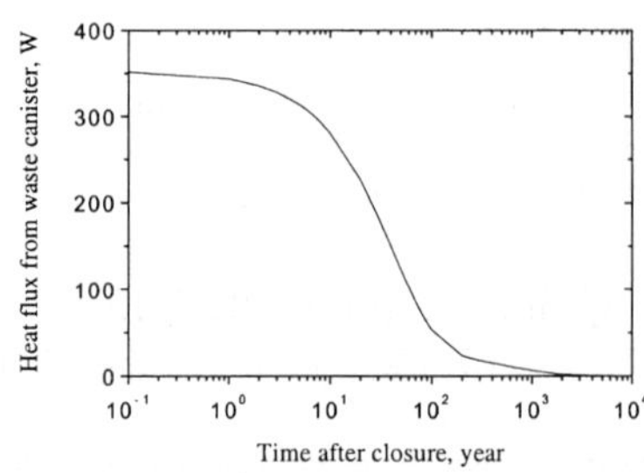

Figure 3. Time history of heat flux from waste

The height of the domain is 60 m so that the coupled behaviors in the buffer are not influenced by the boundaries. The surrounding rock is assumed linear, elastic, isotropic and homogeneous and the buffer is assumed the mixture of sand and bentonite. Material properties of the rock and the buffer are shown in Table 1. The same properties as buffer are given to the backfill. The waste canister is assumed rigid and impermeable.

Boundary conditions are as follows: Temperature and hydraulic head are fixed on the top and the bottom boundary and no flow boundaries for water and heat are specified on the other boundaries. And slide boundaries are given to all the boundaries. These boundary conditions satisfy the situation that disposal pits are arrayed at regular intervals. Initial state is assumed the time that the repository is closed: as initial conditions, hydraulic head in rock mass is 500 m without the effect of excavation, water content of the buffer is 14 %, water content of the backfill is 23% and temperature is 30 °C in all the domain. Heat flux from the waste canister is given as Figure 3. The points A to C in Figure 2 are the monitoring points for temperature and saturation degree. A is close to the waste canister, B is located at the center of the buffer and C is located at the rock close to the buffer.

Figure 4 and Figure 5 show the time histories of temperature and saturation degree at the monitoring

Table 1.Material properties

		Buffer	Rock mass
Dry density [kg/m^3]		1.6×10^3	2.67×10^3
Young's modulus [MPa]		58.74-1.87 w	3.7×10^4
Poisson ratio [-]		0.3	0.25
Intrinsic permeability K [m^2]		4.00×10^{-20}	1.0×10^{-18}
Van Genuchten parameters	θ_s [-]	0.403	5.0×10^{-3}
	θ_r [-]	0.000	0.000
	α[1/m]	8.0×10^{-3}	8.0×10^{-3}
	n [-]	1.6	1.6
Relative permeability r_k [-]		-	$S_e^{1/2}\left(1-\left(1-S_e^{1/m}\right)^m\right)^2$
Water diffusivity [cm^2/s] $D_\theta = \frac{a_1(\theta-\theta_s)}{(\theta-b_1)(b_1-\theta_s)} + \frac{a_2\theta}{b_2(\theta-b_2)}$	a_1	$2.99\times10^{-8}T$-3.78×10^{-7}	-
	a_2	$-1.50\times10^{-8}T$+1.49×10^{-7}	
	b_1	-2.49×10^{-3}	
	b_2	$5.59\times10^{-4}T$+3.93×10^{-1}	
Thermal water diffusivity D_T [m^2/s/K]		7.0×10^{-12}	-
Thermal conductivity λ [W/m/K]		4.44×10^{-1}+$1.38\times10^{-2}w$ +$6.14\times10^{-3}w^2$-$1.69\times10^{-4}w^3$	2.6
Specific heat capacity c [kJ/kg/K]		(34.1+4.18w)/(100+w)	9.0×10^2

w : water content, T : temperature [°C], S_e : effective saturation ratio, θ: volumetric water content

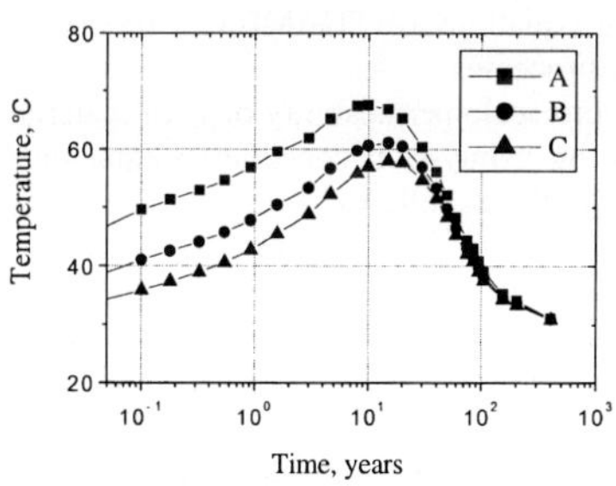

Figure 4. Time history of temperature

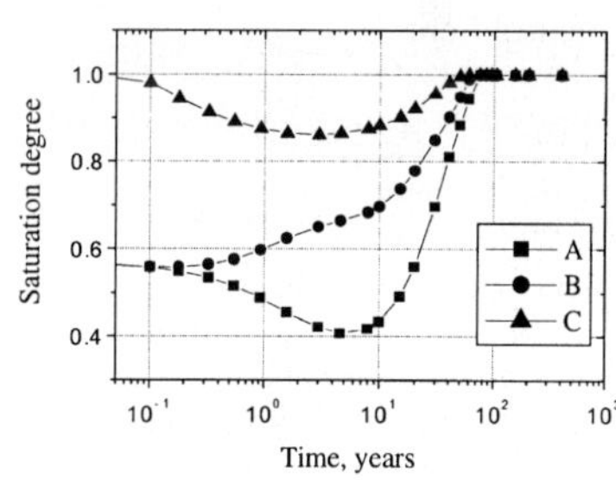

Figure 5. Time history of saturation degree

points, respectively. Temperature increases due to the heat flux from the waste canister in the earlier stages and decreases with decline of the heat flux. At the same time, groundwater infiltrates the buffer. The degree of saturation at the point A close to the waste decreases once due to the evaporation and gradually increases and terminates in saturation, while the dehydration at the point C in the rock is caused by the high suction of the buffer.

3. COUPLED PROCESSES IN FAR FIELD

On the other hand, the coupled processes in the far field are described in this chapter based on the results of the near field analysis. Assuming that mechanical behavior and unsaturated flow are negligible for regional groundwater flow system, only coupled thermal and hydraulic processes in saturated zone are considered. Governing equation of seepage flow considering density change is as follows.

$$\frac{\partial}{\partial x_i}\left[k_{ij}\left\{\frac{\partial h}{\partial x_j}-\left(1-\frac{\rho_w}{\rho_{w0}}\right)\frac{\partial z}{\partial x_j}\right\}\right]=0 \qquad (1)$$

where h is hydraulic head, ρ_w is water density and ρ_{w0} is basic water density. k_{ij} is permeability tensor, which can be written as equation (2) by using intrinsic permeability tensor K_{ij}.

$$k_{ij}=\frac{\rho_w g K_{ij}}{\mu_w} \qquad (2)$$

where g is gravity acceleration and μ_w is water viscosity.

On the other hand, energy conservation equation can be written as

$$\frac{\partial\{(\rho c)_m T\}}{\partial t}+\frac{\partial}{\partial x_i}\left(\rho_w c_w v_i T-\lambda_{ij}\frac{\partial T}{\partial x_j}\right)+Q_w=0 \qquad (3)$$

where T is temperature, $(\rho c)_m$ is average specific heat, t is time, c_w is specific heat capacity of water, v_i is Darcy's velocity vector, λ_{ij} is thermal conductivity and Q_w is heat source. Darcy velocity is determined from

$$v_i=k_{ij}\left\{\frac{\partial h}{\partial x_j}-\left(1-\frac{\rho_w}{\rho_{w0}}\right)\frac{\partial z}{\partial x_j}\right\} \qquad (4)$$

A finite element simulator derived from the Equation (1) and (3) is used. In the equations the parameters of disposal panel such as permeability tensor and thermal conductivity tensor have to be decided. To specify the parameters, the homogenization theory (Ene& Polisevski (1987)), Kyoya & Terada(2001)) is applied.

The homogenization is the method to determine the macroscopic parameters of the media with microscopically periodic structure. The homogenized permeability tensor $k^H{}_{ij}$ resulted from the application of the theory to the Equation (1) is

$$k_{ij}^H=\frac{1}{|Y|}\int_Y k_{ik}^\varepsilon\left(I_{kj}-\frac{\partial\chi^j}{\partial y_k}\right)dY \qquad (5)$$

where Y is the volume of the domain, I_{ij} is unit tensor, y_i is the local coordination system compared with x_i which is the global coordination system and superscript ε means the parameter depends on the micro structure. And the particular solution χ_i is defined from the following FEM-like equations.

$$\int_Y \frac{\partial v^1}{\partial y_i}k_{ij}^\varepsilon\left(I_{jk}-\frac{\partial\chi^k}{\partial y_j}\right)dY=0 \qquad (6)$$

where v^1 is a test function. The Equation (6) can be calculated using the numerical techniques of the FEM.

As the same approach, the homogenized thermal conductivity tensor $\lambda^H{}_{ij}$ can be written as follows.

$$\lambda_{ij}^H=\frac{1}{|Y|}\int_Y \lambda_{ik}^\varepsilon\left(I_{kj}-\frac{\partial\chi^j}{\partial y_k}\right)dY \qquad (7)$$

where the particular solution χ_i is gained from

$$\int_Y \frac{\partial v^1}{\partial y_i}\lambda_{ij}^\varepsilon\left(I_{jk}-\frac{\partial\chi^k}{\partial y_j}\right)dY=0 \qquad (8)$$

Strictly speaking, the governing equations for the near field are those of THAMES. However, the assumption can be permitted because the Equation (1) and the Equation (3). are almost same as the governing equations of THAMES in thermal and hydraulic processes.

The micro scale permeability of each material is described by using each constitutive equation for the permeability as follows.

$$k_{ij}^\varepsilon=\frac{\rho_w g K_{ij}}{\mu_w} \quad \text{in saturated rock and buffer}$$

$$k_{ij}^\varepsilon=\left(D_\theta\right)_{ij}\frac{\partial\theta}{\partial\psi} \quad \text{in unsaturated buffer}$$

$$k_{ij}^\varepsilon=\frac{r_k\rho_w g K_{ij}}{\mu_w} \quad \text{in unsaturated rock}$$

where D_θ is water diffusivity, θ is volumetric water content and ψ is suction.

The disposal pits are arrayed periodically in the horizontal direction, but not periodically in the vertical direction. Therefore, the height of the mesh domain in homogenization process is harmonized with that of the panel element for the far field analysis.

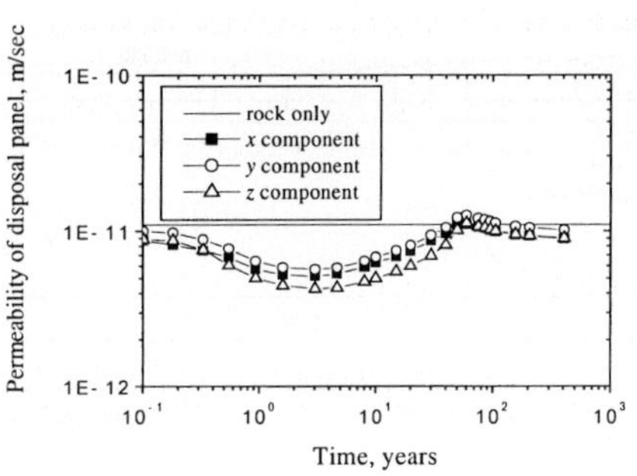

Figure 6. Time history of homogenized permeability

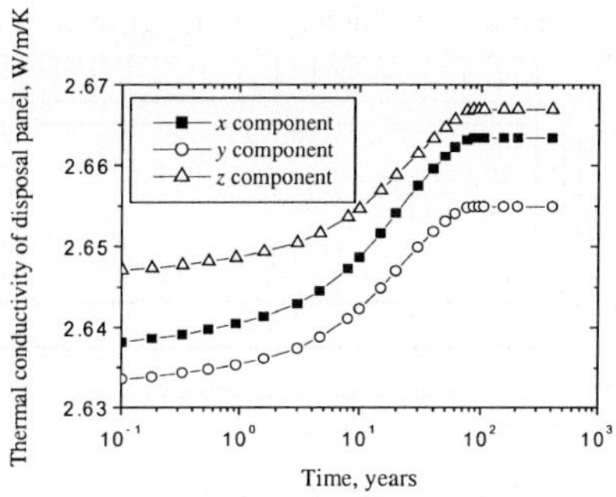

Figure 7. Time history of thermal conductivity

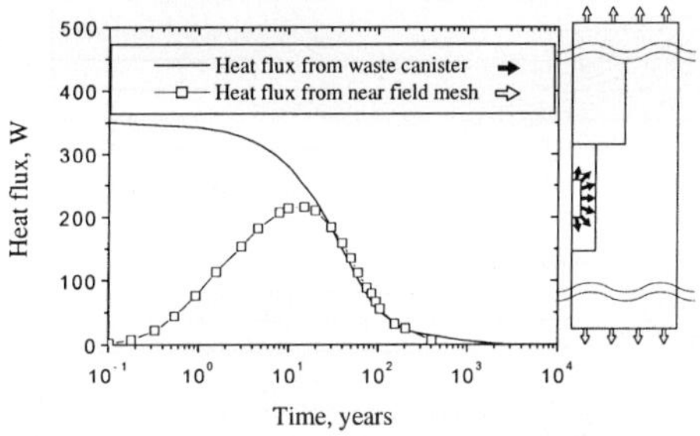

Figure 8. Heat flux from near field

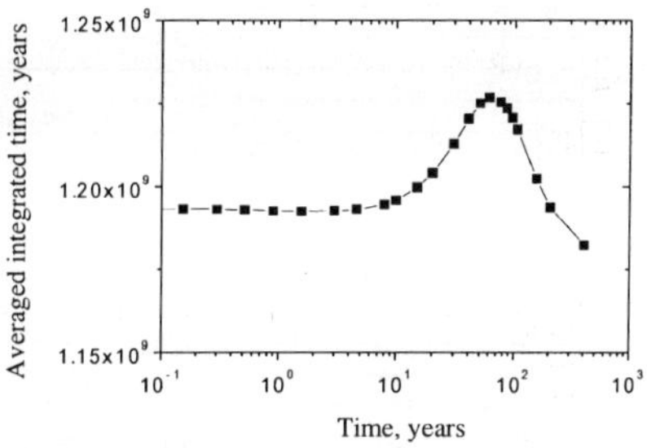

Figure 11. Integrated time through streamline

Using the results of the near field analysis described in the previous chapter, the homogenized tensors are derived. Figure 6 and Figure 7 show the time histories of the homogenized permeability and the homogenized thermal conductivity, respectively. The homogenized permeability decreases once, increases after that, decreases again and finally terminates to the steady state. The reason of the first decrease is the dehydration of rock near the pit. The subsequent increase is caused by the combination of the re-saturation of rock and temperature increment, and the final decrease of permeability is caused by temperature decrease. On the other hand, the homogenized thermal conductivity steadily increases as time. But the magnitude of increment is inconsiderable.

By the way, the heat flux from the panel is also very important for the global groundwater flow system. Figure 8 shows the time history of the heat flux from the near field mesh, compared with that from the waste canister. It can be seen that peak of the heat flux from near field is delayed and loosened compared with the value from the waste canister.

The far field model is shown in Figure 9. This model was made referring to that of a bench mark test of DECOVALEX (Jing et al(1999)) Phase III. The disposal panel is located at 500m in depth and the length is 2km. It is assumed that the left hand side is a watershed and the right hand side is the sea. Hydraulic head is specified to the groundwater surface shown as the figure and the other boundaries are impermeable. Heat flux from the bottom boundary is $0.078 W/m^2$ that is the average value in Japan and temperature is fixed on the top boundary. As the initial condition, the distributions of hydraulic head and temperature calculated under the boundary conditions mentioned above are given. The homogenized parameters and the heat flux shown in Figure 6 to Figure 8 are applied to the disposal panel elements.

Figure 10 shows the streamlines through the disposal panel under initial condition and those after 150 years. Under the initial condition, temperature gradient with depth is almost constant and the streamlines are smooth. On the other hand, after 150 years, temperature increase near the disposal panel causes thermal convection and streamlines are warped. Figure 11 shows the averaged time integrated through the streamlines. The integrated time is dependent of the time after the closure, which means that the groundwater migration path and the time also depend on coupled

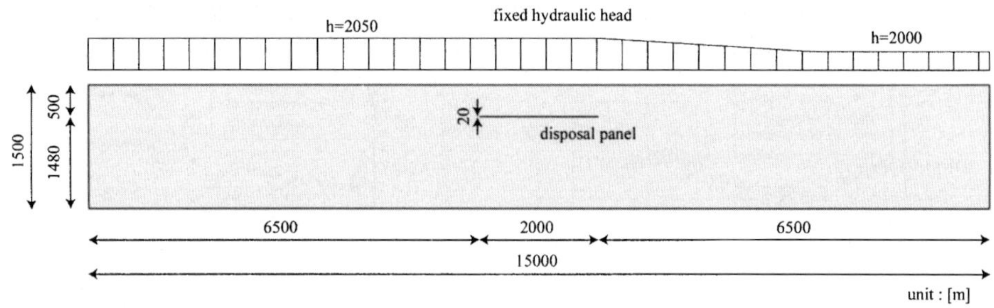

Figure 9. Model for groundwater flow system (far field)

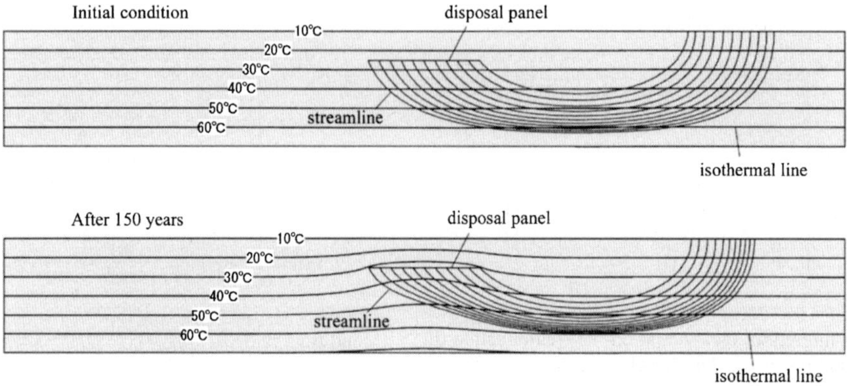

Figure 10. Streamlines through the disposal panel

phenomena in the near field. Although this study dealt with a simplified generic site, an actual site is also considered to be affected with coupled processes. However, the magnitude of the effect is site-specific, and the modeling should be taken into account corresponding to the site characteristics.

4. CONCLUSIONS

This paper proposes a method of estimating the interaction between coupled processes in the near field and those in the far field. Because disposal pits are arrayed periodically, representative parameters of the disposal panel could be obtained through the homogenization theory and applied to the far field analysis with the heat flux. It was clarified from the results that the homogenized parameters, especially permeability, are significantly influenced by the coupled processes in the near field, and that the regional groundwater flow system also depends on it. The proposed method is useful for combining the near field and the far field behavior.

REFERENCES

Ene, H. I. & Polisevski, D. 1987. *Thermal Flow in Porous Media.* D. Reidel.

Jing, L., Stephansson, O., Borgesson, L., Chijimatsu, M., Kautsky, F. and Tsang, C.-F. 1999. *DECOVALEX II project Technical report – Task 2C.* SKI report. 99/23

Japan Nuclear Cycle Development Institute. 1999. *H12 Project to Establish the Scientific and Technical Basis for HLW Disposal in Japan.* JNC TN1400 99-020

Kyoya, T. & Terada, K. 2001. *A Multiscale Structural Analysis and its Application to a Strength Evaluation for Fissured Media (in Japanese).* JSCE 680/III-55: pp.49-64.

Ohnishi, Y., Shibata, H. & Kobayashi, A. 1985. *Development of Finite Element Code for the Analysis of Coupled Thermo-Hydro-Mechanical Behaviors of a Saturated-Unsaturated Medium.* Proc. Int. Symp. on Coupled Process Affecting the Preformance of a Nuclear Waste Repository. Berkeley: pp.263-268.

IMPACT OF TEMPERATURE INCREASE ON NUCLIDE TRANSPORT IN CRYSTALLINE ROCK ON THE NEAR FIELD SCALE

Hua Cheng, Vladimir Cvetkovic
LWR, Royal Institute of Technology, Stockholm

Abstract: The TRUE ("Tracer Retention Understanding Experiments") programme at the Äspö Hard Rock Laboratory (Winberg et al., 2000, 2002) has since 1995 generated a unique database for quantifying retention of radionuclides in crystalline rock on the 5-30m scales. Temperature conditions in the TRUE analysis were about 15°C. In this study, we consider the effect of temperature increase to ca 60°C consistent with conditions after say 1000 years in the rock volume surrounding a KBS-3 type of repository, anticipated to persist over a relatively long time. Temperature elevation will decrease the mean aperture of a conducting fracture by approximately 30%, which in turn will enhance retention. Furthermore, diffusion in the rock matrix will increase at elevated temperatures by approximately factor 4, further enhancing retention. Sorption coefficients are assumed to be unchanged since there is still considerable uncertainty regarding sorption and its change with increasing temperature. We show that elevated temperature on the near field scale enhances nuclide retention, thereby providing an additional safety margin.

1. INTRODUCTION

Tracer Retention Understanding Experiments (TRUE) (Winberg et al., 2000, 2002) was performed at Äspö Hard Rock Laboratory (HRL) by the Swedish Nuclear Fuel and Waste Management Company (SKB). One of the objectives of the TRUE program is to understand the migration and retention of radionuclide in crystalline rock. The TRUE was performed in a series of experiments with successively increasing spatial scale and complexity. The first TRUE stage (TRUE-1) was aimed at performing tracer tests in a detailed spatial scale (0-10 m) and in a single fracture (Winberg et al., 2000; Cvetkovic et al., 2000). The TRUE Block Scale program (Winberg et al., 2002) was aimed at performing tracer tests in a block scale (10-100 m) and in a network of fractures. The Task Force program is currently performed to carry out performance assessment modelling using site characterisation data. The recently performed Task 6 (Cheng and Cvetkovic, in prep.) in the Task Force program is aimed at bridging between site characterisation (SC) and performance assessment (PA) approaches of solute transport and retention in fractured rocks. Five radioactive tracers (I, Sr, Co, Tc, Am) have been studied in Task 6.

In Sweden, a repository design of KBS-3 system has been developed (SKB, 1999). The KBS-3 is a multibarrier system to isolate the spent nuclear fuel. The spent nuclear fuel is placed in corrosion-resistant 5-m long copper canisters. Each of the canisters is surrounded by an engineered barrier system (EBS) of bentonite clay in separate deposition holes excavated along tunnels in crystalline rock at a depth of about 500 meters. The temperature in the geosphere surrounding the canisters is about 15°C initially at deposition. The temperature will increase to about 60°C after tens of years to hundreds of years (SKB, 1999). The temperature condition in the Task 6 tests has been assumed to be similar to the initial temperature condition at deposition, i.e., about 15°C, which is the same as that in the TRUE-1 tests.

The base cases considered in this paper are the cases of Task 6A and Task 6B at T = 15°C (Selroos and Elert, 2001). Tasks 6A and 6B were modelled on Feature A at Äspö Hard Rock Laboratory in south-eastern Sweden (Winberg et al., 2000). Task 6A is modelled on the same flow path as in the STT-1B test in TRUE-1, i.e., the path between the injection borehole KXTT1 and the pumping borehole KXTT3, and using the same pumping flow rate 400 ml/min as in STT-1B (Figure 1). Task 6B is modelled on the same flow path as in Task 6A, but with a flow rate 1000 times lower in order to match the performance assessment time scale.

In this study, we investigate the effect of increased temperature, from about 15°C to 60°C on solute transport and retention in rock fractures.

2. MODEL DESCRIPTION

The Lagrangian Stochastic Advective Reaction (LaSAR) framework (Cvetkovic et al., 1999) has been used in the prediction and evaluation of the TRUE program (Cvetkovic et al., 2000) and in the Task Force project modelling (Cheng and Cvetkovic, in prep). The same framework is employed in this study.

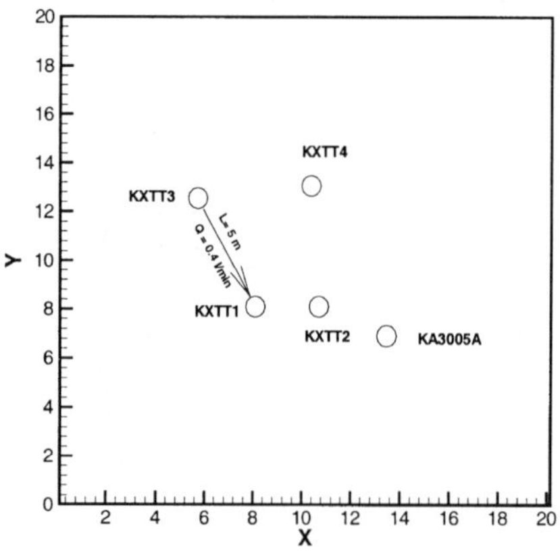

Figure 1. Test configuration for the sorbing tracer tests STT-1b. Q is the pumping rate. L is the distance between injection and pumping boreholes

2.1 Geometrical description

Feature A is conceptualised as a single planar heterogeneous fracture with variable spatial apertures. The distribution of the apertures is approximated to be lognormal with an exponential correlation structure. The porosity is assumed to be constant.

2.2 Processes considered

When a tracer is injected into the fracture, it will be advected and dispersed. The tracer will also be subject to mass transfer processes. In the modelling, the following mass transfer processes are considered: sorption on the fracture surface, diffusion into the rock matrix and sorption in the inner surface of the rock matrix. The following assumptions are made concerning the transport:

- All mass transfer processes are linear;
- The relationship between the water residence time τ and the parameter β is assumed to be linear (Cvetkovic et al., 1999, 2000);
- The tracers are transported into the rock matrix only by diffusion. The diffusion into the rock matrix is one-dimensional (the diffusive flux is perpendicular to the fracture plane);
- The dispersion in the fracture is longitudinal (parallel to the fracture plane).

2.3 Mathematical model

Based on the above assumptions, the governing equations for transport and retention become

$$\frac{\partial q}{\partial t}+\frac{\partial q}{\partial \tau}=\frac{D\theta}{b(\tau)}\frac{\partial N}{\partial z}-\frac{K_a}{b(\tau)}\frac{\partial q}{\partial t} \tag{1}$$

$$\frac{\partial N}{\partial t}=-K_d\frac{\partial N}{\partial t}+D\frac{\partial^2 N}{\partial z^2} \tag{2}$$

where q [$ML^{-2}T^{-1}$] is tracer discharge in the fracture, N [$ML^{-2}\,T^{-1}$] is immobilized q, z [L] is the coordinate orthogonal to the fracture plane, θ [-] is the matrix porosity, D [L^2T^{-1}] is the diffusivity in the matrix (both θ and D are assumed to be spatially uniform), $b(\tau)$ [L] is the Lagrangian half-aperture of the fracture obtained as $b(\tau)\equiv b[X(\tau)]$ with $X(\tau)$ being the advection trajectory (e.g., Dagan, 1984), τ [T] is the water residence time, and K_a [L] is the partition/distribution coefficient for sorption on the fracture surface. Linear equilibrium sorption in the rock matrix is assumed with K_d [-] being the distribution coefficient.

The solution for (1) and (2) along a streamline for a pulse injection is (Selroos and Cvetkovic, 1996; Cvetkovic et al., 2000)

$$\gamma(t;\tau,\beta)=\frac{H(t-\tau-\beta K_a)\beta\kappa}{2\sqrt{\pi}(t-\tau-\beta K_a)^{3/2}}\exp\left[\frac{-\beta^2\kappa^2}{4(t-\tau-\beta K_a)}\right] \tag{3}$$

where $\beta(\tau)=\int_0^\tau\frac{d\tau'}{b(\tau')}$ and $\kappa=\theta\sqrt{D(1+K_d^m)}$

By taking into account the dispersion effect and assuming that the joint distribution of τ and β at cross-section x is $g(\tau,\beta;x)$, the tracer breakthrough can be calculated by

$$q(x,t)=M\int_0^\infty\int_0^\infty\gamma(t;\tau,\beta)g(\tau,\beta;x)d\tau d\beta \tag{4}$$

For a continuous injection of mass M with rate ϕ(t), the breakthrough is obtained by

$$Q(x,t)=M\int_0^t\phi(t-t')dt'\int_0^\infty\int_0^\infty\gamma(t;\tau,\beta)g(\tau,\beta;x)d\tau d\beta$$
(5)

3. KEY PARAMETERS AND TEMPERATURE EFFECT

The two key parameters that control sorption and diffusion shown in (3) are K_a for surface sorption, and the parameter group κ for diffusion/sorption in the rock matrix. The parameter K_a and the parameter group κ have been determined in the laboratory for all tracers at a temperature of about 15°C. In Table 1 the laboratory values of K_a and the parameters determining κ are summarized. The quantities τ (water residence time) and β are dependent on the flow conditions in the field by definition and cannot be determined in the laboratory. The distribution of τ is inferred from the data of the breakthrough curves (BTCs) of HTO (tritiated water), and β is assumed to be perfectly correlated to τ with a linear relationship, $\beta = k\tau$.

Table 1. Summary of sorption and diffusion parameters at T = 15°C and T = 60°C

Tracer	$Kd^{*(1)}$ (m^3/kg)	Ka (m)	De×10^8 (m^2/h)	
			T=15°C	T=60°C
I-131	0	0	5.4	21.6
Sr-85	4.7×10^{-5}	8.0×10^{-6}	2.6	10.4
Co-58	0.008	0.008	1.9	7.6
Tc-99m	2.0	0.2	2.6	10.4
Am-241	5.0	0.5	2.6	10.4

(1) K_d^* is related to K_d in (2) by $K_d = K_d^* \rho(1-\theta)/\theta$.

For flow between two parallel plates, β may be calculated by τ/b where b is the half aperture between the two plates. In this case, $k = 1/b$. In a case of varying aperture, we assume $k = 1/b_e$ where b_e is the effective aperture related to β. The change of aperture affects the slope k.

By using $\beta = k\tau$, the parameter group that controls matrix diffusion/sorption can be written as $k\theta\sqrt{D(1+K_d^m)}$. The parameter group for surface sorption can be written as $K_a k$.

All the parameters and parameter groups discussed above may vary due to temperature variation. Carbol et al. (2002) have studied the temperature dependence of sorption parameters for cesium. However, the definite trend of variation of sorption parameters when temperature changes is still unclear. In this study, we therefore do not consider the variation of the sorption parameters with temperature. The porosity is also assumed to be unchanged by temperature. We focus our study mainly on the influences of temperature on the width of the effective aperture and the parameter group κ (mainly through the variation of the diffusivity D).

In the following, the temperature profiles from the SR 97 studies in the near field geosphere will be summarized (SKB, 1999). Three hypothetical sites, Aberg, Beberg and Ceberg, have been selected from different parts of Sweden as the model areas in the SR-97 studies. The initial temperatures of the three sites are between 11°C and 16°C. According to the modelling results, the temperature at the deposition boundary will reach a maximum of about 60°C, 50°C, 45°C, for Aberg, Beberg and Ceberg respectively, after about 20 years. The temperatures of the three sites are expected to remain at around 40°C to 60°C for up to hundreds of years and decrease thereafter.

In this paper we consider two scenarios for comparison. In the first one the temperature is at 15°C, corresponding to the initial stage of the repository. In the second scenario the temperature is at 60°C, corresponding to the maximum temperature after the closure of the repository.

Meanwhile, Temperature impact on two modelling tasks (Task 6A and Task 6B) is studied.

The sorption and diffusion parameters at T = 15°C are chosen to be the same as in the previous TRUE-1 (Cvetkovic et al., 2000) and Task 6 studies (Cheng and Cvetkovic, in prep.).

3.1 Initial parameters for Task 6A

The values for surface sorption coefficient K_a are obtained from the modelling input data set (MIDS) (Byegård et al., 1998) (for Sr-85). The data for other tracers (Co-58, Tc-99m and Am-241) provided by Tasks 6A and 6B modelling task specification (Selroos and Elert, 2001) are used.

The porosity is assumed to be θ = 0.02, based on the lower end of the TRUE-1 evaluation values (θ = 0.02 - 0.024) (Cvetkovic et al., 2000), which can be viewed as the best estimate. A porosity of 0.02 is also consistent with the range provided in Tasks 6A and 6B Specification (Selroos and Elert, 2001).

The values of sorption coefficient K_d for different tracers are based on the batch data (fraction 1 - 2 mm) (Byegård et al., 1998), which are about 10 times of the MIDS value. These values are also consistent with the TRUE-1 evaluation. In the modelling, K_d values are either chosen to be 10 times of the MIDS values or 10 times to the values specified in Task Specification. The values for K_a (the surface sorption coefficient) are chosen to be the same as the MIDS values since K_a has relatively less effect on retention.

In the empirical Archie's law for the formation factor, $F = \theta^m$, $m = 1.2$ is used in accordance with the results of the Task 6A studies. The diffusivity D is then calculated from $D = \frac{FD_w}{\theta}$. The effective diffusivity D_e is defined as $D_e = \theta D = FD_w$.

In the assumed inverse-gaussian distribution for the water residence time τ, the temporal moments from the TRUE-1 evaluation for the STT-1B tests are used, i.e., the mean residence time is 5 h, and the variance is 1.5 h^2. The linear relationship between β and τ is taken as $\beta = 3000\tau$, i.e., k = 3000 1/m (Cvetkovic et al., 2000).

3.2 Initial parameters for Task 6B

The same values of K_d, K_a, and the power in Archie's law are used for Task 6B as those for Task 6A.

The values of temporal moments for the inverse-Gaussian distribution of the water residence time τ are chosen based on simple reasoning of scaling. If the flow rate is 1000 time lower, then the mean travel time will be about 1000 time longer. Therefore for Task 6B, a mean residence time of 5000 h is used. The variance for Task 6B is chosen so that the coefficient of variation is the same as that in Task 6A. The variance becomes therefore 1.5×10^6 h^2. The proportionality of the linear relationship between β and τ is assumed to be same, i.e., $\beta = k\tau$, k = 3000 1/m.

3.3 Parameters at T = 60°C

The aperture may decrease 30% if the temperature increases from about 15°C to around 60°C (personal communication with Lanru Jing, KTH, Stockholm). While the diffusivity increases about 4 times. Table 1 lists the parameters at T = 15°C and T = 60°C. The parameter values at T = 60°C are obtained in the following way:

According to Nernst-Einstein equation (e.g., Daniels and Alberty, 1955), the diffusivity is proportional to temperature for diffusion in dilute solution. The diffusivity at T = 60°C are about 4 times of the values at T = 15°C. The decrease of the effect aperture by 30% implies that the proportionality k in the linear β–τ relation is increased by 30%. We then have $\beta = 3900\tau$ for 60°C. Four times increase of D implies only two times of increase of κ since $\kappa \sim \sqrt{D}$

4. RESULTS

The simulation results for Tasks 6A and 6B with pulse injection at T = 15°C and T = 60°C are presented below.

4.1 Breakthrough curves for Task 6A

Figure 2 shows the breakthrough curves (BTCs) for the pulse injection of Task 6A. The solid lines are the BTCs for tracer I-131, the dashed lines for Sr-85, the dashdot lines for Co-58, the longdash lines for Tc-99m, and the dashdotdot lines for Am-241. This line representation of the tracers will also be used in Figure 3. The thick lines are the BTCs at T = 15°C, and the thin lines are those at T = 60°C.

The BTCs in Figure 2 can be divided into three groups: a group for the conservative to weakly sorbing tracers, I-131 and Sr-85, a group for the moderately sorbing tracer Co-58, and a group for the strongly sorbing tracers, Tc-99m and Am-241. The increase of temperature apparently enhances the retention of the all tracers.

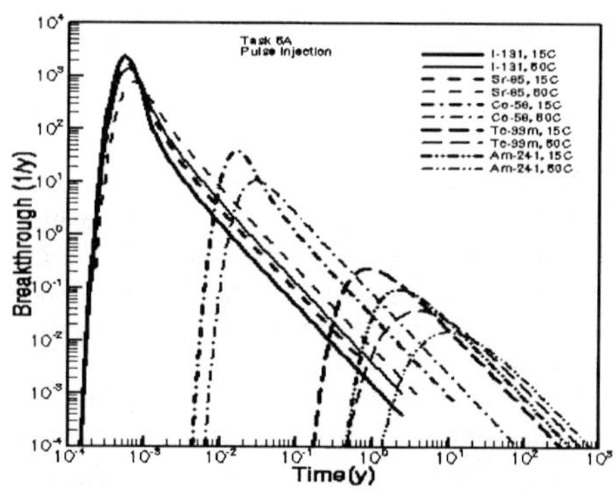

Figure 2. Breakthrough curves of Task 6A for pulse injection at T = 15°C (thick lines) and T = 60°C (thin lines).

For the first group of the tracers (the conservative to weakly sorbing tracers I-131 and Sr-85), the enhanced retention due to the increase of temperature is discernible, as the thin lines are slightly shifted to the right in the figures (the solid and dashed lines). While for the second group of the tracer (the moderately sorbing tracer Co-58), the effect of temperature increase is stronger, as can be seen from the difference between the thick dashdot line and the thin dashdot line. For the third group of the tracers (strongly sorbing tracers Tc-99m and Am-241), the temperature increase has the strongest retention effect among the tracers studied in this work. This can be observed from the longdash lines and the dashdotdot lines where the thin lines are shifted considerably to the right

Table 2 summarises the breakthrough times for 5%, 50% and 95% mass recovery for Task 6A. The quantitative effect of the enhanced retention due to the temperature increase is observed. Generally, the temperature increases have prolonged the recovery times. The impact of the increase of temperature becomes stronger for the more strongly sorbing tracers. The impact is also stronger on the breakthrough times for high percentage mass recovery.

The temperature effects for the tracers of the first group are as follows. The breakthrough times for 5% mass recovery for I-131 increases from 3.4 days to 3.8 days. For Sr-85 it increases from 3.7 days to 4.5 days. For the 50% mass recovery, the impact of temperature increase is stronger. For Sr-85, it changes from 6.2 days to 11.4 days. While for the breakthrough times of 95% mass recovery, for Sr-85, from 56 days to 341 days. For more strongly sorbing tracers, the impact is even stronger. For Am-241, for example, the breakthrough times for 50% mass recovery increases from 10 years to about 63 years.

Table 2. Breakthrough times for recovery of 5, 50 and 95% of the injected mass for Task 6A (years).

Tracer		T_5 (y)	T_{50} (y)	T_{95} (y)
I-131	15°C	3.9×10^{-4}	6.1×10^{-4}	2.0×10^{-3}
	60°C	4.3×10^{-4}	7.9×10^{-4}	1.2×10^{-2}
Sr-85	15°C	4.2×10^{-4}	7.1×10^{-4}	6.4×10^{-3}
	60°C	5.1×10^{-4}	1.3×10^{-3}	3.9×10^{-2}
Co-58	15°C	1.2×10^{-2}	2.5×10^{-2}	0.6
	60°C	2.2×10^{-2}	8.9×10^{-2}	4.2
Tc-99m	15°C	0.7	4.0	215
	60°C	2.9	25.1	1452
Am-241	15°C	1.6	10.0	540
	60°C	7.2	62.9	3643

4.2 Breakthrough curves for Task 6B

The BTCs for Dirac pulse injection for Task 6B are shown in Figure 3. Two BTCs are presented for each tracer. The thick lines are for the case when temperature T = 15°C, and the thin lines are for T = 60°C. Clearly the higher the temperature is, the stronger the retention. The tracers can also be divided into three groups as discussed in Task 6A. The temperature impact is even more profound than in Task 6A.

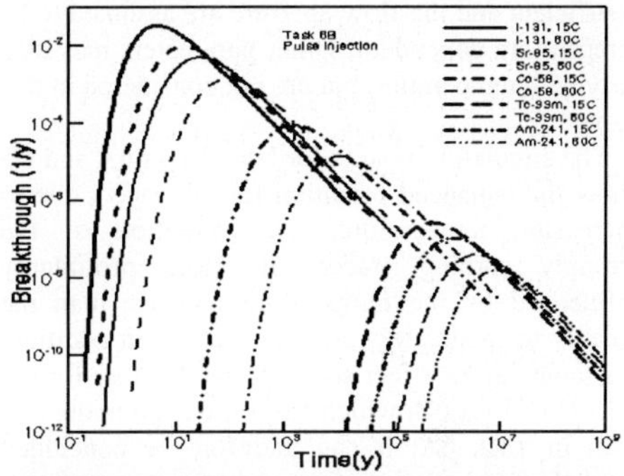

Figure 3. Breakthrough curves of Task 6B for pulse injection at T=15℃ (thick lines) and T=60℃ (thin lines).

Table 3. Breakthrough times for recovery of 5, 50 and 95% of the Dirac pulse injection for Task 6B.

Tracer		T_5 (y)	T_{50} (y)	T_{95} (y)
I-131	15°C	3.3	28.1	1624
	60°C	16.6	186.7	10970
Sr-85	15°C	10.4	97.8	5738
	60°C	67.9	658.3	45930
Co-58	15°C	1073	10360	610300
	60°C	7208	70150	4.2E+6
Tc-99m	15°C	3.7E+5	3.6E6	2.2E+8
	60°C	2.5E6	2.5E7	1.5E9
Am-241	15°C	9.4e+5	9.1E6	5.4E8
	60°C	6.3E6	6.2E7	3.6E9

Table 3 shows the times for 5%, 50% and 95% mass recovery for Dirac pulse injection in Task 6B for both T = 15°C and T = 60°C. The retention times for the different tracers follow the same trend of changes with temperature as in Task 6A, but are several orders of magnitude longer. The temperature impact starts to be apparent even for the conservative to weakly sorbing tracers (Figure 3). For example, the breakthrough times of 5% mass recovery for I-131 increases from 3.3 to 16.6 years as the temperature increases from 15°C to 60°C. For the more sorbing tracers, the temperature impact is very strong. The breakthrough time of 50% mass recovery for Tc-99m increases from 3.6 million years to about 25 million years.

5. DISCUSSIONS AND CONCLUSIONS

The influences of temperature on solute transport and retention have been studied. The diffusion coefficient and the flow aperture are assumed to be temperature dependent. Other parameters may also vary with temperature but are not considered in this paper.

The simulation results for both Tasks 6A and 6B show the enhanced retention for all tracers due to increasing temperature. The retention of the strongly sorbing tracers is more profoundly influenced by the temperature increase than the weakly to moderately sorbing tracers. In addition, the temperature effect seems to be stronger for the tracers in Task 6B, which has a lower pumping rate than in Task 6A. It can therefore be concluded from this study that the increased temperature may possibly provide additional safety margin for the retention of radionuclides, where strongly sorbing tracers acquire larger safety margins than weakly sorbing tracers.

Acknowledgements

This work was supported by the Swedish Nuclear Fuel and Waste Management Co. (SKB), Stockholm.

REFERENCES

Byegård, J., Johansson, H., Skålberg, M., and Tullborg, E.-L. 1998. The interaction of sorbing and nonsorbing tracers with different Äspö rock types: Sorption and diffusion experiments in the laboratory scale. TR-98-18, Swedish Nuclear Fuel and Waste Management Co. (SKB).

Carbol, P., Koth, L., Betti, M., Albinsson, Y., and Tullborg, E.-L., 2002. *Isotopic exchange of cesium with fracture filling minerals from Äspö HRL*, TRUE Bock Scale international seminar, Äspö, Sweden.

Cheng, H., and Cvetkovic, V., (in prep). *Modelling of sorbing tracer breakthrough for Task 6A, 6B and 6B2,* Swedish Nuclear Fuel and Waste Management Co. (SKB).

Cvetkovic, V., Selroos, J.-O., and Cheng, H., 1999. *Transport of reactive tracers in rock fractures*. J. Fluid Mech., 378, 335-356.

Cvetkovic, V., Cheng, H., and Selroos, J.-O., 2000. *Evaluation of Tracer Retention Understanding Experiments (first stage) at Äspö.* ICR-00-01, Swedish Nuclear Fuel and Waste Management Co. (SKB).

Dagan, G., 1984. *Solute transport in heterogeneous formations*. J. Fluid Mech., 145, 151-177.

Daniels, F., and Alberty, R. A., 1955. *Physical Chemistry*. Wiley, New York.

SKB, 1999. *Deep repository for spent nuclear fuel, SR 97 – Post-closure safety*. TR-99-06, Swedish Nuclear Fuel and Waste management Co. (SKB).

Selroos, J. O., and Cvetkovic, V., 1996. *On the characterization of retention mechanisms in rock fractures*. TR 96-20, Swedish Nuclear Fuel and Waste Management Co. (SKB).

Selroos, J.-O., and Elert, M., 2001. *Task 6A & 6B modelling task specification*, Swedish Nuclear Fuel and Waste Management Co. (SKB).

Winberg, A., Andersson, P., Hermanson, J., Byegård, J., Cvetkovic V., and Birgersson, L., 2000. Äspö Hard Rock Laboratory, *Final report of the first stage of the tracer retention understanding experiments*. TR-00-07, Swedish Nuclear Fuel and Waste Management Co.(SKB). ISSN 1404-0344.

Winberg, A., Andersson, P., Byegård, J., Poteri, A., Cvetkovic, V., Dershowitz, B., Doe, T., Hermanson, J., Gómez-Hernández, J-J., Hautojärvi, A., Billaux, D., Tullborg, E-L., Holton, D., Meier, P. and Medina, A., 2002. *Final Report of the TRUE Block Scale Project 4. Synthesis of flow, transport and retention in the block scale*. TR-02-16, Swedish Nuclear Fuel and Waste Management Co. (SKB).

DESICCATION AND REHUMIDIFICATION EFFECTS ON THE THERMOHYDROMECHANICAL BEHAVIOUR OF THE CALLOVO-OXFORDIAN ARGILLACEOUS ROCK

Su K., Hoteit N., Ozanam O.
Geosphere Department, French National Radioactive Waste Management Agency (Andra), France

Abstract: This paper presents experimental results concerning the impact of changes in relative humidity or water content on the mechanical properties of an argillaceous rock. The rock under investigation, known as Callovo-Oxfordian argillites, is situated at a depth of 500 m in Eastern France. An underground research laboratory is being built since 2000 by Andra for a feasibility study on deep geological disposal of high-level long-lived radioactive waste. Two contradictory desiccation effects are observed: i) hardening of the rock, and ii) damage due to micro-cracking associated with desiccation. The amplitude of these two phenomena in the argillaceous rock depends on the changes in relative humidity. For example, a decrease from 100% to 60% in relative humidity may multiply the Young's modulus of the rock by five. Meanwhile, the uniaxial compression strength of the rock becomes three to four times higher due to the humidity decrease, strain (shrinking) induced by the change of relative humidity may reach 0.5%

1. CONTEXT AND MINERALOGICAL COMPOSITION OF THE ROCK UNDER INVESTIGATION

Andra, the French National Radioactive Waste Management Agency, is carrying a multidisplinary programme on an argillaceous formation known as Callovo-Oxfordian argillites, with a view to studying the feasibility of deep radioactive waste disposal in France (Hoteit 1999, Ozanam 2000, Su 2000).

At a depth of 500 m and with a thickness of 130 m, the argillites under study have an initial permeability less then 10^{-13} m/s, with overlying and underlying limestone formations remaining relatively impermeable. The porosity of the argillites varies from 13 to 17%.

The following mineral species are identified in the Callovo-Oxfordian argillites (Hoteit 2002): clay minerals, calcite, quartz, dolomite, feldspar, pyrite and some trace of minor phases. Calcite and quartz were present in all samples, comprising 50 to 60% of the rock, with the clay matrix and the minor phase making up the rest. Three iron-bearing minerals have been identified: pyrite, hematite and some traces of siderite. Iron sulphide and oxide were encountered more or less continuously throughout the layer. Over a 130-m thickness, the argillites contain an average of 40 to 45% clay minerals. The clay matrix comprises five types of minerals: illite, regular mixed R0 illite-smectite, irregular mixed-layer R0 illite-smectite, chlorite and kaolinite. The mixed illite-smectite is the most abundant. Two types may be distinguished, type R1 richer in illite than in smectite (60 to 75% illite) in the lower part of formation, and type R0 (40 to 60% smectite) in the upper part (Figure 1).

From the geomechanical point of view, the host layer is divided into five geomechanical zones (Figure 1). That setup is based on the results from various well logs (gamma-ray, density, porosity, sonic velocity) as well as physical and mechanical characteristics obtained on core samples (density, porosity, uniaxial strength, Young's modulus). The five zones are also consistent with three geological sedimentary sequences: two stiff (higher carbonate) zones surround a central less stiff (more clayey) zone. However, the transition from one zone to the other is progressive. Within each zone, a finer breakdown may be proposed.

2. DESICCATION EFFECTS ON THE SHORT-TERM MECHANICAL BEHAVIOUR OF ARGILLITES

Typical stress-strain curves of the argillites under investigation and obtained by conventional triaxial tests are shown in Figure 2. Loading and unloading cycles are carried out in the pre-failure phase in order to calculate the Young's modulus and to analyse inelastic deformations. Experimental curves shows that the rock has a typical plastic-damage behaviour. Loading and unloading paths are affected by plasticity, hardening, damage and viscosity.

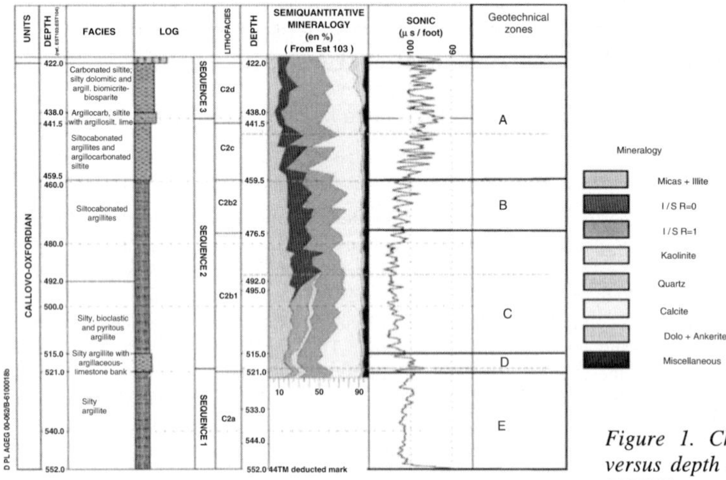

Figure 1. Change in mineralogy versus depth observed in borehole EST103

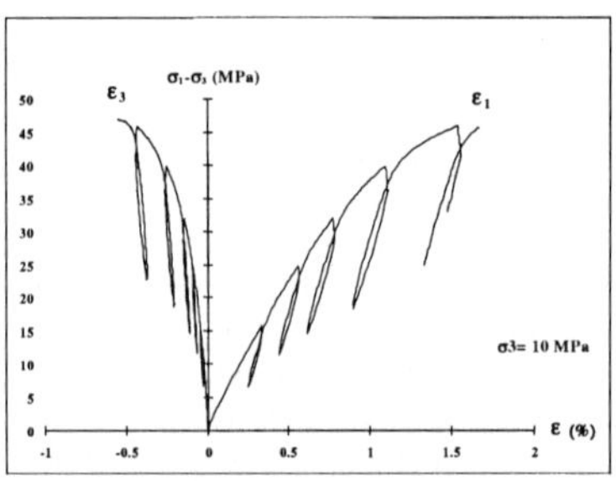

Figure 2a.

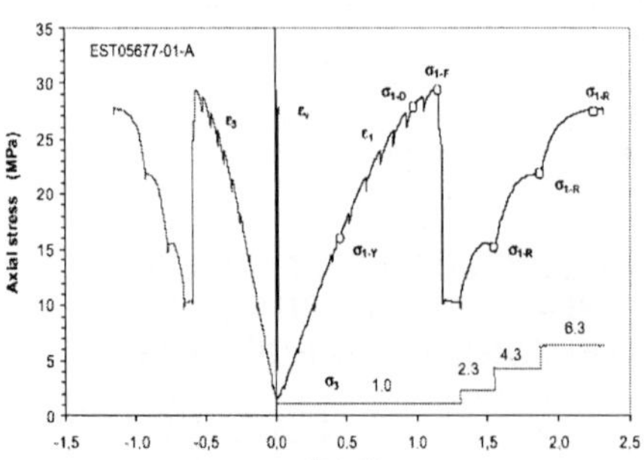

Figure 2b. (Zhang 2002)

Figure 2. Argillites' mechanical behaviour under classic triaxial compression tests, 2a) Triaxial test with unloading paths for determining damage evolution, 2b) Triaxial test with increasing step by step the confining pressure on the post-failure phase.

In order to study the effects of desiccation on the mechanical behaviour of argillites, several types of laboratory hydromechanical coupling tests were performed: uniaxial and triaxial tests on the samples with different water content, desaturation under isotropic confining pressure with monitoring associated displacement. In most cases, the samples were previously submitted to a cure under a given humidity level and at constant temperature.

Desiccation effects on the mechanical behaviour of the rock are characterised by the changes in the Young's modulus, uniaxial strength and associated strains.

2.1 Variations in the Young's modulus versus varations in water content

Desiccation changes significantly the elastic behaviour of the argillites, as shown in Figure 3, which presents strain-stress curves of three triaxial tests with different levels of water content measured after cure: The more desaturated the rock sample is, the deeper in axial and transversal directions the slope of stress-strains curves is. The Young's modulus is multiplied by 1.5 between a sample almost completely saturated (w=7.55%) and a sample with about 4.6% water content. Meanwhile, the strength of the rock increases also as the degree of saturation decreases and the onset of loss of linearity of stress-strain curves becomes higher.

From a set of uniaxial compression tests, the change in the Young's modulus as a function of relative humidity of the cure undergone by the samples before testing is presented in Figures 4a and 4b. They show that the Young's modulus increases from 2~5 GPa to 10~25 GPa as relative humidity decreases from 90-100% to 50%. The results reported on other clayey rocks (Ramambasoa 2001) confirm also that the Young's modulus of argillaceous rock is multiplied by 3 to 4 when the rock is desaturated. In addition, we note that when the samples are severely desaturated (water content near 30~0%), the Young's modulus starts to decrease.

The increase of the Young's modulus with desaturation may be interpreted by the increase of suction pressure inside the samples with desiccation, that plays the role of confining pressure to close pre-existing micro-cracks and to enhance the elastic properties of the material.

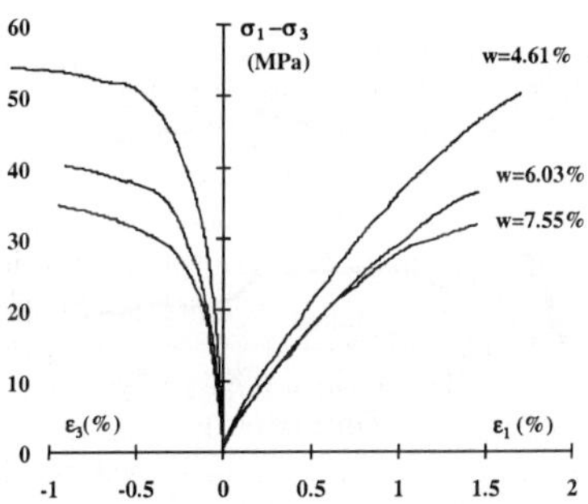

Figure 3. Triaxial tests on argillites samples with different water content under 10 MPa of confining pressure

The suction pressure is not uniform because of the heterogeneity in pore size and the humidity gradient existing inside the rock samples during the cure; consequently, the differential strains thus created may provoke micro-cracks and decrease the elastic properties of the material. There is a therefore competition between the role of the confining pressure and the creation of micro-cracks due to desiccation.

2.2 Variations in uniaxial compression strength versus variations in water content

As in the case of the Young's modulus, uniaxial compression strength increases as the degree of saturation decreases (figure 5a, figure 5b). Three phenomena are involved in the evolution of the strength: i) the release of pore water improves the cohesion of the interfaces of the rock grains, ii) the suction pressure improves the contact of interface and enhances their shear strength, iii) heterogeneity of induced strains and suction gradient within the rock may provoke micro-cracking and reduce the strength of the material. The result of those three competing phenomena governs the tendency of in the evolution of the strength. Overall, for the argillites under investigation, between 100% and 0% degree of saturation, the uniaxial strength is multiplied by three to six (Figure 5a). Such an evolution is also observed during triaxial compression tests (Figure 5b). the same phenomena are also observed on the Opalinus clay of Mont-Terri (Blümling, 2000).

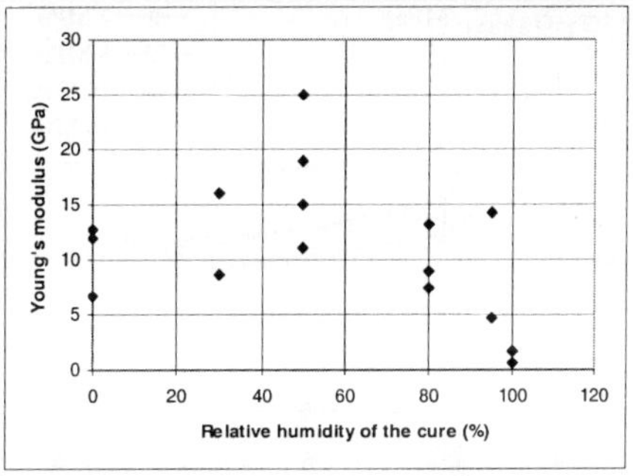

Figure 4a

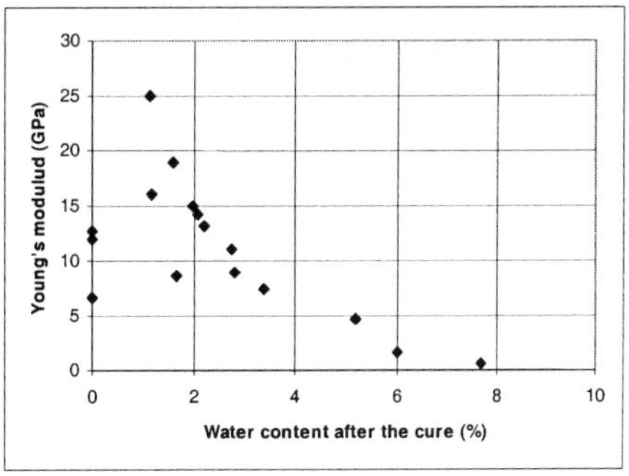

Figure 4b

Figure 4. Desiccation effects on the elastic properties of the argillites 4a). Evolution of the Young's modulus versus the relative humidity of the cure, 4b) Evolution of the Young's modulus

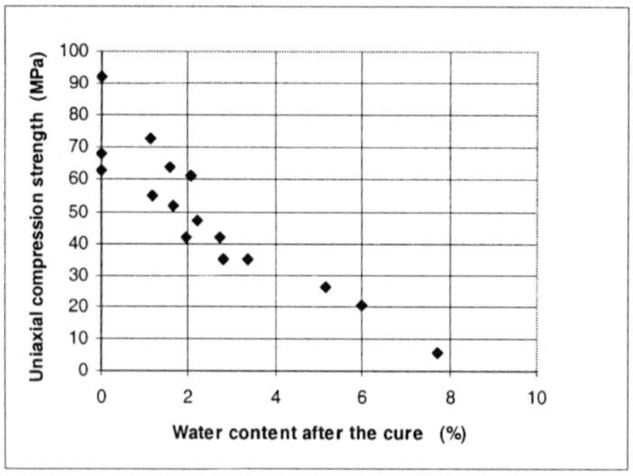

Figure 5a

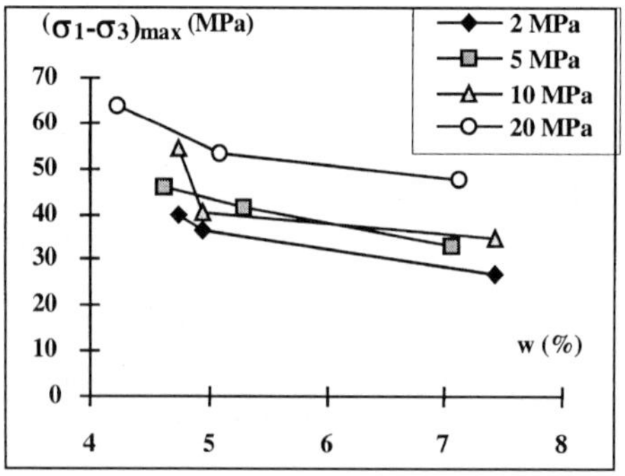

Figure 5b

Figure 5. Desiccation effects on the elastic properties of the argillites, 5a). Evolution of the uniaxial compression strength versus water content, 5b).). Evolution of the maximum strength of triaxial compression tests versus water content

2.3 Desiccation-induced strains

Submitted to the isotropic confining pressure, the deformation of the argillites is monitored during desiccation and rehumidification cycles. A special experimental set-up is developed allowing air to circulate with a given humidity level around the side of the samples.

Strain measurements show important shrinkage associated with desiccation (Figure 6). While relative humidity drops from 98% to 60%, associated strain reaches 0.5%, which is about the same magnitude of strain due to mechanical unloading at a gallery wall at a depth of 500 m in the argillites. That strain is also comparable to the strain corresponding to a 500°C-cooling in the argillites. In other words, the strain associated with desiccation is significantly greater than the strain due to temperature changes in a deep waste repository.

Under certain mechanical boundary conditions, the desiccation-induced strain may provoke enough tensile stress to create macro-fissures and then to increase the damage in the vicinity of the opening.

During the rehumidification phase, swelling is observed. However, the strain amplitude is four to five times less than the strain associated with shrinkage. That means that the desiccation-rehumidification cycle generates irreversible strain.

The swelling phenomenon in the rehumidification path should be distinguished from that observed under a perfect saturation case. The previous one is mainly due to variations in suction pressure, and the second one is due to the reorganisation of clay minerals resulting from the complex physical-chemical interactions (Ghoreychi 1997).

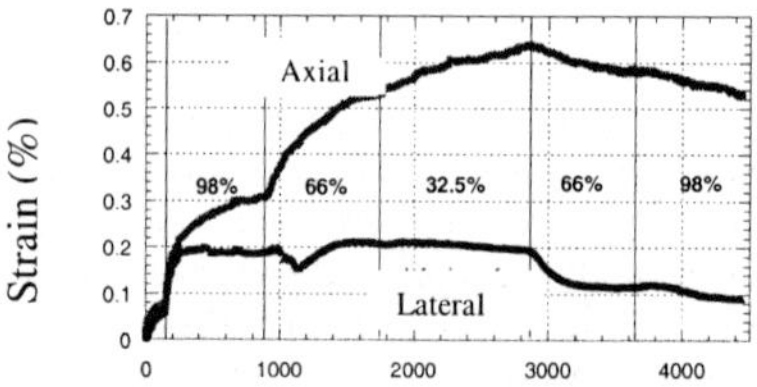

Figure 6. Evolution of the axial and lateral strains of the argillites under constant isotropic pressure (15 MPa) and variable relative humidity

3. DESICCATION EFFECTS ON THE LONG-TERM BEHAVIOUR OF ARGILLITES

The long-term mechanical behaviour of argillites impacts the reversibility and the performance of the repository through the following phenomena controlled by: i) the evolution of the damaged zone, ii) the evolution of cracks (opening/closure) and of permeability, iii) interaction with the engineered barrier.

Several successive sets of creep tests were carried out between 1996 and 2002. The tests results show that the argillites has a time-dependent behaviour even in undrained conditions. That observation confirms that the time-dependent behaviour of the argilites is thus dependent on both the viscoplasticity of the material and compaction phenomena of.

3.1 Creep-test results under saturated conditions

The amplitude of the differed deformations during creep tests is weak. The creep rate under undrained triaxial conditions is about 10^{-6} to 10^{-7}/h (~0.1 to 1% per year or 2.10^{-11} to 2.10^{-10} per second) according to the loading velocity and to the mode of calculation of those rates (Figure 7). The comparison between axial and lateral differed deformations allows to ignore volumetric deformations. The anisotropy of the differed parallel and perpendicular strains to the stratification is negligible. The strain rate is independent of the sample size, which may mean that the scale effects may not be significant for the rock under investigation in relation to its time-dependent behaviour.

As shown in Figures 7a and 7b, the differed strain increases as the deviatoric stress increases. The differed strain remains measurable whatever the deviatoric stress level. It was not possible, until now, to determine a deviatoric-stress threshold below which there would be no differed strain.

With regard to the influence of temperature, it is shown that the time-dependent behaviour may be enhanced by an increase in temperature (Figure 7a). At low deviatoric stress, temperature effects are discreet for an rise from 20°C to 50 °C in temperature. However, from 50°C to 120°C, an acceleration of the viscoplastic strain is observed obviously, whatever the level of deviatoric stress might be.

3.2 Desiccation effects on time-dependent behaviour

In the last creep phase of a creep test at 15 MPa, two specimens were exposed to ambiant air with a relative humidity of 24.0±3.4 % in order to examine the effect of moisture on the time-dependent behaviour, as shown in Figure 8. Since a direct measurement of the desaturation of the creeping specimens was impossible, an accompanying measurement of water-content change of the witness specimen was conducted under the same conditions as in the creep tests. As the specimens were rapidly desaturated during the first days, the deformations responded simultaneously with a large magnitude of about 0.3% shrinkage, irrespective of the loading direction. The fluctuations of the deformations corresponded to the moisture fluctuations. After sealing the specimens against the air conditions in the room, the fluctuations of the deformations were minimised. Over the last 40 days, under a stress of 15 MPa, the air-dried specimens deformed steadily backwards with negative rates of -10^{-11} per second (swelling).

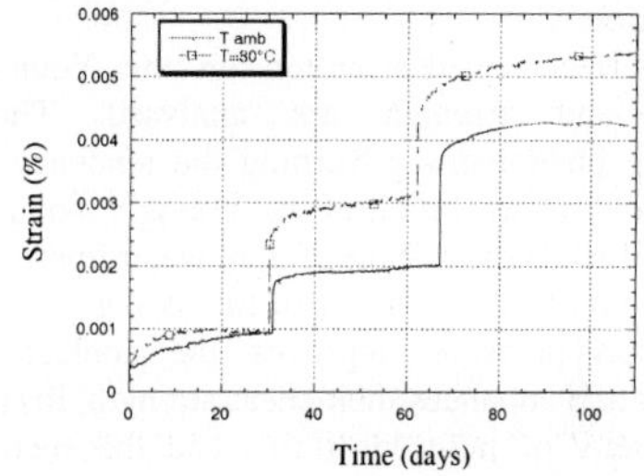

Figure 7a

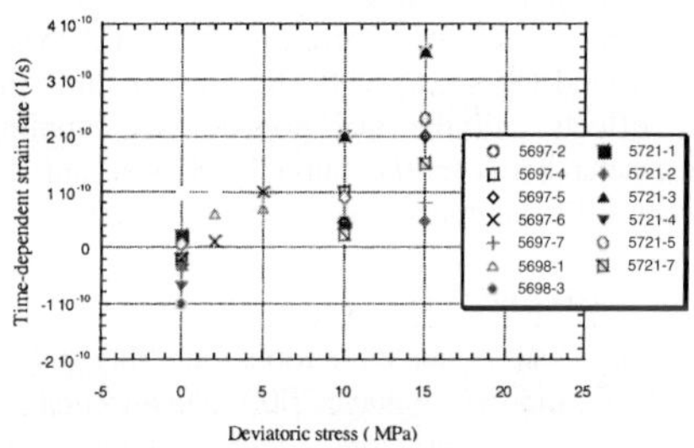

Figure 7b

Figure 7. Creep tests on the argillites. 7a) Effect of temperature and deviatoric stress, 7b) Strain rate versus deviatoric stress

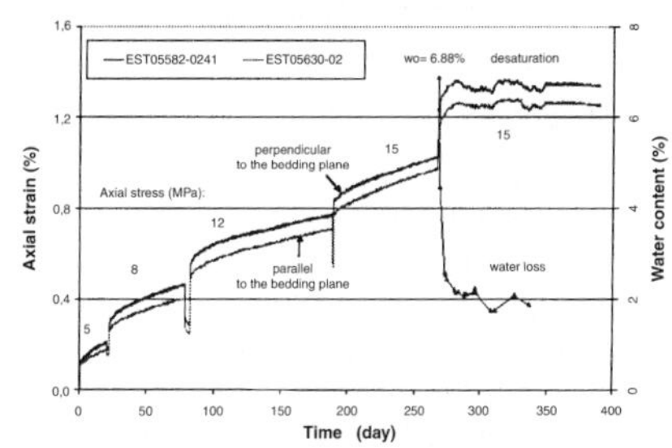

Figure 8. Creep tests on the samples parallel and perpendicular to the stratification and study of the effect of desiccation on the last step of loading (Zhang 2002).

4. CONCLUSIONS

The mineralogical composition of Callovo-Oxfordian argillites and of its clay minerals has been analysed. The high content of the illite/smectite mixed layer confirms the strong coupling potential between the water content and the mechanical behaviour of the formation under investigation.

The effects of desiccation on the Young's modulus and strength are analysed. Three competing phenomena governing the tendency of strength and the evolution of the Young's modulus are quantified. i) the release of pore water improves the cohesion of the interfaces of the rock grains, ii) the suction pressure improves the contact of interfaces and enhances their shear strength, iii) the heterogeneity of induced strains and the suction gradient within the rock may provoke micro-cracking and reduce the strength of the material.

In-situ geomechanical experiments programme (Su, 2000) scheduled to be carried out at Andra's underground laboratory will allow us to study the scale effects and the real potential of coupling phenomena between the humidity levels and the mechanical behaviour observed on the samples.

5. REFERENCES

Blümling, P. C. Bauer-Plaindoux, J.C. Mayor, H.J. Alheid and M. Fukaya.2000. *Geomechanical investigation at the underground rock laboratory Mont-Terri,* Proceeding of International Workshop on Geomechanics Hydromechanical and Thermohydromechanical Behavior of Deep Argillaceous Rocks, pp 275-283, 10/2000 – ANDRA/École des mines de Paris.

Ghoreychi M. 1997. *Comportement rhéologique et couplages thermo-hydro-mécaniques dans les argilites de l'Est : expérience macroscopiques et analyse microscopiques* – Actes des Journées scientifiques CNRS/ANDRA, pp. 119-126, Bar-le-Duc, 20-21/10/1997.

Hoteit N., Ozanam O. and Su K. 1999. *Geomechanical investigation of an argillaceous formation in the east of France* – International Workshop on the rock mechanics of nuclear waste repositories, June 5-6,Vail, Colorado, USA, 1999.

Hoteit N., Su K. and Masrouri F. 2002. *Swelling in the Callovo-Oxfordian Argillites – Characterisation and design parameters.* Narm's 2002, July 2002, pp 1171-1178, Toronto.

Ozanam O. Su K. and Hoteit N. 2000. *First in situ experiment in the French underground laboratory: vertical mine-by-test in the main shaft,* Waste Management (WM'00) conference, Tucson, 27 February – 3 March 2000.

Ramambasoa N. 2001. *Étude du comportement hydromécaniquedes argilites – Application au site de Tournemire* – Thèse de l'École polytechnique – janvier 2001.

Su K., Hoteit N. and Ozanam O. 2000. *Geomechanical investigation plan in an underground research laboratory for feasibility study of radioactive waste repository in deep argillaceous rock* – International Conference on Tunnels and Underground Structure (ICTUS), Singapour, 26-29 November 2000.

Su. K, Hoteit N. and Ozanam O. 2000. *Geomechanical investigation programme in the Meuse/Haute-Marne underground research Laboratory* – International Workshop on Hydromechanical and thermohydromechanical behaviour of deep argillaceous Rock, pp.229-238 Paris, October 2000.

Zhang Cl., Dittrich J., Muller J., Rothfuchs T. 2000. *Experimental study of the hydromechanical behaviour of the Callovo-Oxfordian argillites* – Deliverable N°4 of 5thEURATOM MODEX-REP project – FIKW-CT-2000-00029, 2002.

THERMO-MECHANICAL SIMULATIONS OF PILLAR SPALLING IN SKB APSE TEST BY FRACOD

Mikael Rinne[1], Baotang Shen[1], Hee-Suk Lee[1], Lanru Jing[1]
[1]Fracom Ltd. Farfarsbacken 14, FIN-02400, Kyrkslätt, Finland

Abstract: This paper summarises a study on pillar spalling using fracture propagation code FRACOD. The rock mass response in a heated pillar between two large boreholes in the planned Äspö Pillar Stability Experiment by SKB is modelled. To model all the planned loading phases, the code has been improved in many ways. Today it predicts the explicit fracturing process including fracture initiation. Barton-Bandis model has been applied to estimate the fracture properties for newly initiated fractures. It also simulates the development of Acoustic Emission (AE) events. A stress reconstruction technique has been developed to transfer excavation and heat induced stresses from other models into FRACOD. The modelling results suggest that the excavation induced stresses will cause slight fracturing in the pillar walls. Fracture propagation driven by thermal loading may lead to minor spalling.

1. INTRODUCTION

The Swedish Nuclear Fuel and Waste Management Company (SKB) is investigating the stability of a pillar between two closely located deposition holes at Äspö Hard Rock Laboratory (AHRL). This in-situ experiment is called Äspö Pillar Stability Experiment (APSE) and it is described more detailed elsewhere in this proceeding; see APSE by Andersson et al. (2003).

For this experiment a tunnel will be excavated at the 450 m depth. Two large holes with a diameter of 1.8 m and a depth of 6 m will be drilled in the tunnel floor to form a pillar with the width of 1 m. The loading configuration is designed to induce stresses in the pillar to a level close to spalling.

Electric heaters will be used to induce additional thermal stresses that aim to force the rock in the pillar walls to spall. To simulate the effect of confinement from backfill, a water pressure of 1 MPa will be applied in one of the holes. All experimental stages will be monitored by an Acoustic Emission (AE) and micro-seismic system. Convergence and strain measurements will be made to monitor the deformation.

To model all the planned loading phases, the code has been improved in many ways. This paper summarizes the recent code development and presents the central results of the APSE fracture models.

2. THEORETICAL BACKGROUND OF FRACOD

FRACOD utilizes the Displacement Discontinuity Method (DDM) principles. It was designed to simulate fracture propagation and interaction of multiple fractures in rock masses. Today it predicts the explicit fracturing process including, fracture initiation and fracture propagation, fracture sliding and opening. A brief description of formulation of fracture initiation and propagation is given below. More details can be found in Shen (2002) and in Rinne (2003).

2.1 Modelling fracture propagation

Both tensile and shear failure are common in rock masses. Therefore, to effectively predict rock fracture propagation, Shen and Stephansson (1993) suggested a fracture propagation criterion for both mode I and mode II fracture propagation, namely the F-criterion. According to the F-criterion, in an arbitrary direction (θ) at a fracture tip there exists a F-value, which is calculated by

$$F(\theta) = \frac{G_I(\theta)}{G_{Ic}} + \frac{G_{II}(\theta)}{G_{IIc}} \tag{1}$$

where G_{Ic} and G_{IIc} are the critical strain energy release rates for mode I and mode II fracture propagation; $G_I(\theta)$ and $G_{II}(\theta)$ are strain energy release rates due to the potential mode I and mode II fracture growth of a unit length.

The direction of fracture propagation will be the direction where F reaches the maximum value. If the maximum F reaches 1.0, fracture propagation will occur.

2.2 Modelling fracture initiation

Based on the laboratory test results and AE interpretation (e.g. Li, 1993), damage (fracture initiation) may start at a low stress level and increases with stress. Due to rock anisotropy, the chance of failure at a give location increases with stress until the strength is reached.

FRACOD uses a probabilistic approach to simulate fracture initiation. It is assumed that, at a candidate location, the probability of a fracture initiation depends upon the stress/strength ratio ($\sigma/\sigma_{strength}$):

$$p = 0; \qquad if(\sigma/\sigma_{strength}) < 0.5$$

$$p = 4\cdot(\sigma/\sigma_{strength} - 0.5)^2; \qquad if\,0.5 < (\sigma/\sigma_{strength}) < 1.0$$

$$p = 1.0; \qquad if(\sigma/\sigma_{strength} > 1.0)$$

where p is the probability of fracture initiation; $\sigma/\sigma_{strength}$ is the ratio of the stress to strength. Note that here we refer to both the tensile and compressive stress and strength. In FRACOD the Mohr-Coulomb intact rock strength criteria are used with a tension cut-off at the tensile strength.

The probability of fracture propagation is plotted in Figure 1.The probability curve has a shape similar to that of a typical AE account curve obtained during uniaxial compression tests of rock samples from Äspö (Li, 1993).

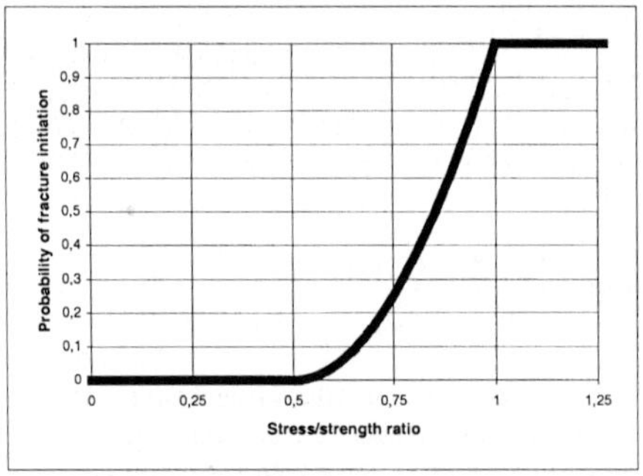

Figure 1. Probability of fracture initiation as a function of the stress/strength ratio.

Fracture initiation can occur at boundary or inside the rock mass. Both the excavation boundaries (e.g. borehole wall) and the grid points in the intact rock are frequently checked to determine any possible fracture initiation.

2.3 Prediction of Acoustic Emission (AE)

Acoustic Emission (AE) is a low energy seismic event. A seismic event can be considered as sudden inelastic deformation within a given volume of rock that radiates detectable seismic waves. A seismic event may be induced by fracture initiation, fracture propagation, or a sudden movement of existing fractures.

In FRACOD, the existing fractures and the grid points in the intact rock are checked in every step of the calculation to determine the possible fracture surface movement (including elastic, sliding or opening movement), and fracture propagation and initiation. If a fracture element is found to undergo a movement or there are new fracture initiations and propagations in intact rock in any calculation step, an AE event is predicted to occur in the vicinity of the failed fracture element or the new fracture element. Strictly speaking, the location (epicenter) of this AE event should be located at the center of the element. To avoid overlapping of many AE events from the same element in the process of failure, we assume each individual AE event occurs randomly in a circular region as shown in Figure 2.

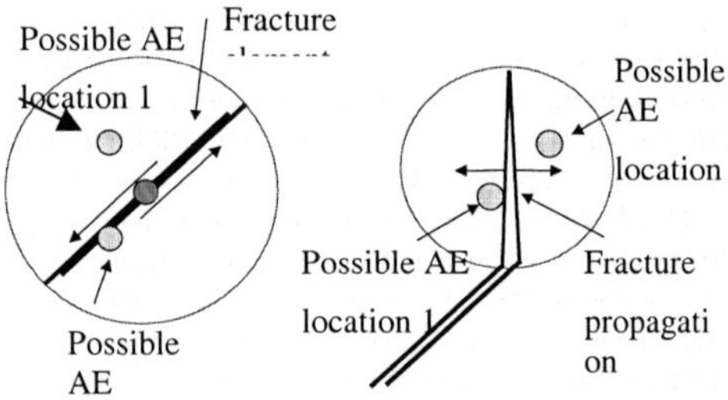

Figure 2. Predicted AE locations due to fracture sliding and fracture growth.

The AE magnitude is measured using the concept of moment sensor (Mendecki, 1997). For an idealised planar source, the seismic moments are defined as:

$$M = G\cdot\Delta D_s\cdot A \text{ (shear)} \qquad (2)$$

$$M = E\cdot\Delta D_n\cdot A \text{ (open)} \qquad (3)$$

where ΔD_s and ΔD_n are the change of displacement discontinuity (slip or open) of the source area A. G is the shear modulus and E is the Young's modulus.

2.4 Barton-Bandis model for newly initiated fractures

There are limited test data available for the properties of newly created fresh fractures. To provide an estimation of the fracture stiffness for the APSE project, the following factors have been considered: Values have to be theoretically sound and realistic. They also have to be feasible to implement in FRACOD. Barton-Bandis joint model has been used to estimate the normal stiffness of the new fractures. The shear stiffness obtained from the Barton-Bandis model was found to be too low for modelling random fracture initiation in FRACOD due to numerical reasons. Therefore the elastic shear deformation has been restricted and only fracture sliding causes shear deformation. Restricting shear and/or normal elastic deformation is common in solid mechanics when dealing with interface or contact problem (Zang, 1990).

The newly created fractures are also assumed to have the same friction angle and cohesion as the intact rock. Once sliding occurs, however, the cohesion will become zero. The dilation of the new fresh fracture is also estimated using the Barton-Bandis model. It is found that the dilation has major effect on the failure process.

The fracture properties are estimated separately for tensile fracture and shear fractures. Due to the difference in surface roughness, tensile fractures and shear fractures have different stiffness and dilation values. The estimated fracture properties for the APSE project see Andersson et al. (2003) in this proceeding.

3.0 INVERSE STRESS RECONSTRUCTION FOR EXCAVATION AND HEATING EFFECT

To model all experimental stages, thermal loading must be considered. FRACOD cannot directly simulate the effect of thermal loading and an alternative way to transfer thermal and initial stresses from other coupled numerical models has been sought. Here an inverse boundary element technique to reconstruct stress field and transfer it to FRACOD, is presented.

3.1 Principle of inverse BEM

There are two kinds of inverse problems in the solid mechanics. One is the source reconstruction problem and the other is the parameter identification problem (Bezerra & Saigal, 1995). For our application only source reconstruction technique is used. In the case of source reconstruction, usually unknown boundary conditions are determined using observed field quantities inside the domain of interest or over-prescribed boundary conditions. If on some portion of the boundaries both displacements and tractions are unknown, it cannot be solved directly in a forward boundary value problem senses because the number of unknowns is larger than number of equations. However, if some quantities, say stresses for our problem at hand, are known at some points inside the domain, the number of the equations can be increased to obtain a solution. Usually the solution of an inverse problem does not always satisfy stability and uniqueness requirements in itself, so it is generally an ill-posed problem (Martin and Lesnic, 2002). Therefore, special iterative or non-iterative numerical techniques should be applied to get a meaningful solution.

Based on the above concepts, stress distributions from other numerical modelling can be reconstructed after solving unknown boundary tractions with known stress distribution at some points inside the model. The reconstructed tractions along unknown boundaries can be used again to calculate stresses at all other points inside the domain using a conventional BEM code with the calculated boundary tractions directly used as prescribed boundary conditions in a BEM model, without the BEM code to have special functions for domain integral evaluations. An inverse BEM code using truncated singular value decomposition was developed to perform just such inversion tasks for this project. Details of the theory and verifications can be found in Rinne et al. (2003).

3.2 Reconstruction of stress field

The developed inverse stress reconstruction technique is applied to transfer the heat induced stresses from a 3D model (Examine3D) and a heat flow model (JobFem2D) into FRACOD. Since most stress concentration is anticipated inside pillar region, only pillar region is considered to reduce numerical efforts. Two 2D sections have been modelled, 1.5 m and 0.5 m below the tunnel floor. The equivalent boundary tractions of a 2D BEM model was calculated using the stress inversion technique, based on the stress fields produced by both a 3D BEM and a 2D FEM after excavation and heating stages (30, 60, 90 and 120 days).

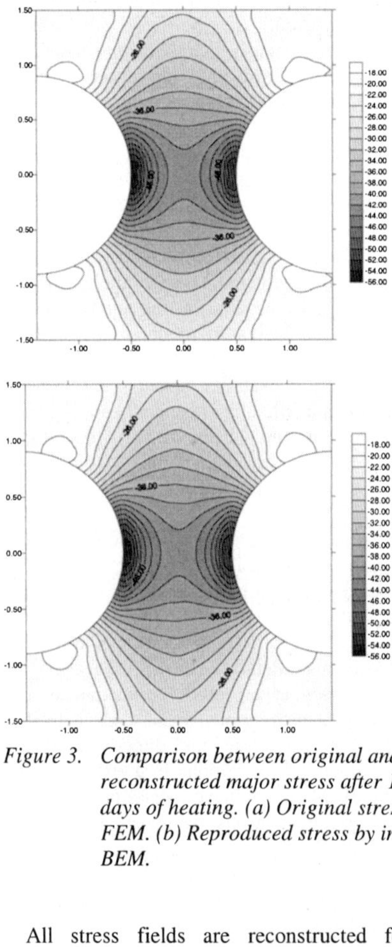

Figure 3. Comparison between original and reconstructed major stress after 120 days of heating. (a) Original stress from FEM. (b) Reproduced stress by inverse BEM.

All stress fields are reconstructed for the simulation of fracture process. The equivalent boundary tractions were directly inputted in FRACOD models to consider effect of excavation and heating in the model.

Figure 3. shows the comparison between original and the reconstructed major stress field after 120 days of heating. The reconstructed stress distribution shows very close agreements for all stages of stresses. Table 1. lists relative errors at some selected points inside the pillar. In most cases, errors for the major principal stress show below 1%. Errors are quite small especially in the case of thermal stresses compared to the case of excavation induced stresses. This is because thermal stress field were originally obtained from 2D model. 3D nature of excavation stress field would induce larger error. Even though errors seem to be big at some points, real differences are small because errors are relative and original values at those points are small.

Table 1. Comparison of relative errors at some points (%)

Location (x,y)		Excavation		t=30 day		t=120 day	
		σ_1	σ_3	σ_1	σ_3	σ_1	σ_3
0.00	0.00	-0.46	-4.76	0.96	1.09	0.47	0.21
0.31	0.00	-1.07	-7.51	0.62	-0.25	0.37	-0.36
0.19	0.00	-0.46	-4.76	-0.04	-1.18	0.08	-0.72
0.44	0.00	-0.94	-17.87	-0.86	-1.57	-0.20	-1.09
0.00	0.35	0.46	-4.57	0.47	-3.78	-0.02	-0.74
0.00	0.70	0.24	-7.75	-0.19	-5.57	-0.76	-0.53
0.00	1.23	4.40	-3.73	-1.27	6.15	-1.45	2.75

4. MODELLING FRACTURING PROCESS BY FRACOD

Major objectives of the Äspö Pillar Stability Experiment are to demonstrate the capability to predict spalling in a fractured rock mass and to investigate the effect of the backfill (confining pressure) on the rock mass response. The FRACOD modelling aims to predict properly the fracturing process and the development of AE events during the planned sequences of excavation-confinement-heating.

The geological conditions, parameters and stress conditions have been presented elsewhere in this proceeding, see Andersson et. al (2003). Two kinds of models have been established to simulate the fracturing process (See Figure 4.). The "Far-field model" simulates the effect of excavations in infinite medium without external boundaries. It is used to investigate parameter sensitivity and the effect of pre-existing fractures. The "Pillar model" is used to simulate real heating process and 3D excavation effects, with the aid of the stress reconstruction technique described above.

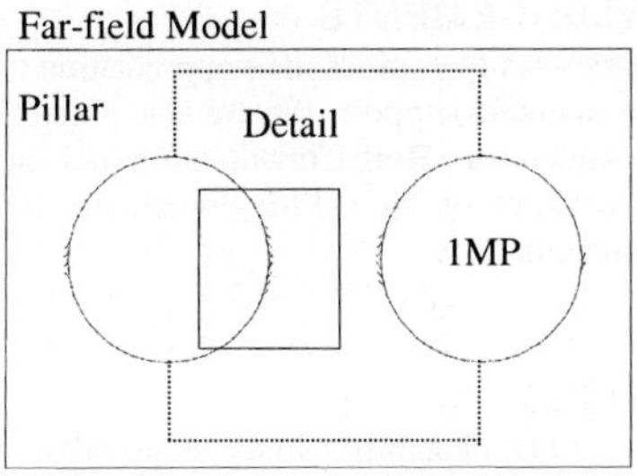

Figure 4. Farfield and Pillar Model.

4.1 Fracture initiation, propagation and AE events

According to the model, the maximum induced stress 0.5 m below the tunnel floor in the pillar boundary is about 137 MPa after excavation. This stress will cause minor fracture initiation and propagation at the borehole boundaries. Fracturing along the central borehole walls occur continuously as stresses increase, Figure 5. Fractures initiates also in the intact rock.

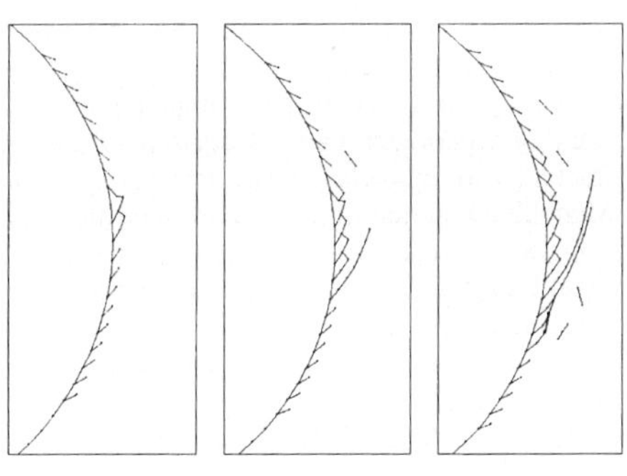

Figure 5. Fracture initiation, propagation and coalescence due to (a) excavation induced stresses (b) heating of 30 days (c) heating of 120 days.

As the stress increase, major fractures develop at the left side (unconfined), but the mid region of the pillar remains stable. The pattern of possible spalling from successive loading becomes more apparent though the pillar remains stable. In spite of the extensive fracturing at left side of the pillar, overall stability seems to be maintained after 120 days of heating.

Acoustic Emission evolution was modelled for each loading step. Most of the AE events occur in the vicinity of the central pillar walls, Figure 6. The propagation of major fractures creates intense AE events and they are active at the unconfined borehole wall for all loading stages as inferred from fracturing pattern, see detail in Figure 6.

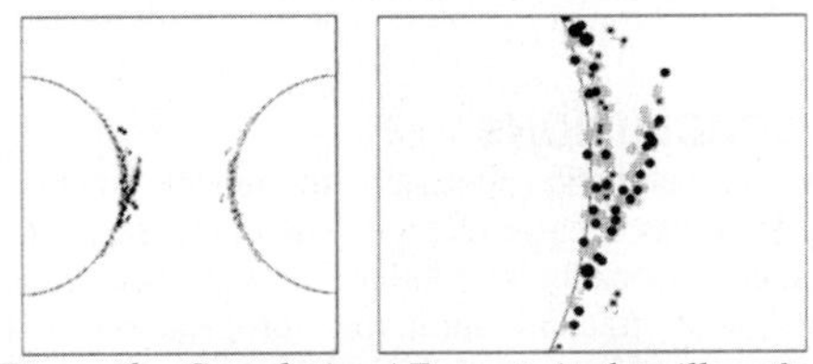

Figure 6. Cumulative AE events in the pillar after heating the pillar for 120 days (detail at right). The size of the spot corresponds to the magnitude of the AE event.

4.2 Effect of pre-existing fractures

The effect of pre-existing fractures was preliminary studied by using FRACOD models. Large fractures may reduce the stress concentration in the critical pillar region because they allow larger deformation compared to intact rock. Fracturing is reduced compared to the intact case due to lower stress level at the borehole wall.

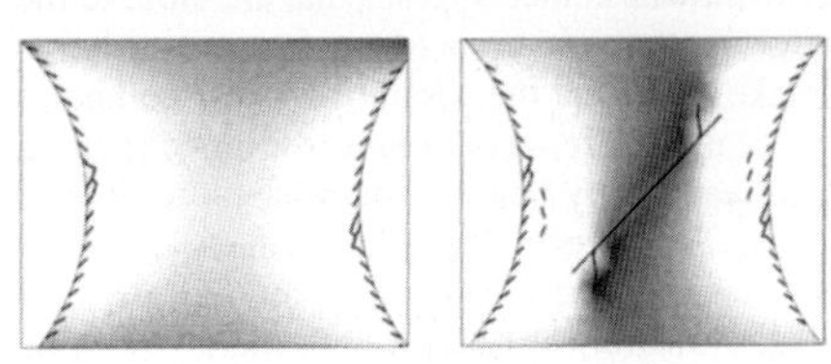

Figure 7. Stress distribution and fracturing of a model without pre-exiting fracture. Right, with an inclined pre-existing fracture.

However, instability may occur due to fracture initiation and propagation when pre-existing fractures are located inside the pillar in a way that they cause high stress concentration near the borehole wall (See Figure 7.).

4.3 Effect of fracture properties

Fracture properties seem to have a strong effect on the fracturing process. Most of the fracture parameters are sensitive for the actual loading configuration. For example, the effect of fracture toughness in mode I (K_{IC}) is more pronounced in uniaxial loading than in confined loading, because tensile failure will dominate under uniaxial loading conditions. Therefore, conclusions of fracture parameter sensitivity on rock failure cannot be given without a site specific fracture pattern and loading configuration.

5. CONCLUSIONS

A probabilistic approach to model fracture initiation has been developed. The model seems to predict reasonably the failure of the rock mass throughout fracture initiation, propagation and coalescence of the new initiated fractures.

The AE function implemented in the code is for idealised planar sources. It predicts the locations of AE events and relative magnitude of the AE events through the fracturing process. It is, however, not recommended to rely too much on the absolute value of the predicted AE magnitude because it is sensitive to the element size.

Stress field inside a solid or rock structure with combined mechanical and thermal loadings has been successfully reconstructed by the inverse boundary element technique.

Calculations with preliminary field data suggest that the planned loading geometry in the APSE experiment will induce stresses that are close to the limit for fracture initiation. Minor fracture initiation may take place in the vicinity of the boreholes before heating. Stress conditions leading to spalling will be reached by heating the pillar according to the plan. The core of the pillar remains intact in stress conditions corresponding to 120 days of heating. However, disturbances in the stress field, for example caused by unsymmetrical heat flow, might lead to some extent instable fracture propagation.

Confining pressure of 1.0 MPa at the right borehole slightly decreases the maximum stresses of the confined borehole wall, but the effect seems to be limited.

Properties of newly created fractures are found to be important parameters controlling the degree of rock failure. Many uncertainties still exist on these properties. Additional study on fracture stiffness and the fracture dilation angle of new fractures is needed to gain more reliable models of rock failure.

ACKNOWLEDGEMENTS

The authors wish to express their appreciation to SKB for the financial support. We are also grateful to Christer Andersson, Rolf Christiansson and the Steering Committee of the APSE experiment for valuable comments.

REFERENCES

Andersson C, 2003. Feasibility study: Äspö pillar stability experiment. International Progress Report, IPR-03-01. Swedish Nuclear Fuel and Waste Management Company, Stockholm.

Bezerra L, Saigal S, 1995. Inverse Boundary Traction Reconstruction with the BEM. Int J Solids Struct;32(10):1417-1431.

Li C, 1993. Deformation and Failure of Brittle Rocks under Compression. Doctoral Thesis. Lulea University of Technology, Sweden. 1993:118D, ISSN 0348-8373.

Martin L, Lesnic D, 2002. Boundary element solution for the Cauchy problem in linear elasticity using singular value decomposition. Comput Methods Appl Mech Engrg. 2002;191:3257-3270.

Mendecki AJ, 1997. Seismic Monitoring in Mines. Chapman & Hall (publisher).

Rinne M, Shen B, Lee H-S, 2003. Äspö pillar stability experiment. Part I: *Modelling of* fracture stability by FRACOD. Part II: Modelling Fracture Initiation and Acoustic Emission. Part III: Reconstruction of Stress Field Using an Inverse Technique. International Progress Report, IPR-03-05. Swedish Nuclear Fuel and Waste Management Company, Stockholm, 2003.

Shen B, 2002. FRACOD Ver. 1.1, *User's manual.* Fracom Ltd.

Shen, B. and Stephansson, O., 1993. Modification of the G-criterion of crack propagation in compression. Int. J. of Engineering Fracture Mechanics. 47(2), 177-189.

Zang W, 1990. Crack Kinks and Crack Closure Modelled by Integral Equation Methods. Doctoral thesis Royal Institute of Technology. Stockholm, Sweden.

Theme 2
Fundamentals of Modelling Coupled T-H-M-C Processes of Geosystems

T-H-M-C MODELLING OF ROCK MASS BEHAVIOUR – 1: THE PURPOSES, THE PROCEDURES AND THE PRODUCTS

J. Andersson[1] and J. A. Hudson[2]
1) JA Streamflow AB, Sweden
2) Rock Engineering Consultants, UK

Abstract: Coupled thermal, hydraulic, mechanical and chemical processes control the behaviour of the geosphere, but when modelling the geosphere all existing coupled mechanisms need not necessarily be represented numerically. Modelling is conducted for specific purposes to support performance measures and the required confidence level should be considered. Only some of the geosphere responses will significantly affect a particular project's function. For example, geological evolutionary modelling on the origin of mineral deposits may need to consider the coupling between groundwater flow and dissolution/precipitation processes. Geotechnical projects may need to consider the water pressure impact on stability. For geothermal reservoirs, the interactions between the in-situ stress, the rock fractures, the rock and water temperatures, and the water flow needs to be considered. Issues in geological radioactive waste disposal where coupled processes may be important include density dependent flow, saturation of the near-field and the development of an Excavated Disturbed Zone.

1. INTRODUCTION

Being able to model the behaviour of rock masses is of crucial interest for many applications. For geological characterisation, such modelling is needed to understand the long term evolutionary processes which have resulted in the current rock mass form. The emphasis is on the interpretation of past processes, although predictions, particularly related to natural hazards can be one of the objectives. For engineering purposes, however, such as an underground excavation projects, oil and gas production, geothermal energy extraction or nuclear waste repository engineering, the modelling emphasis is on providing the predictive capability which is essential for engineering design, construction and operation. The geological and engineering emphases are highlighted in Fig. 1.

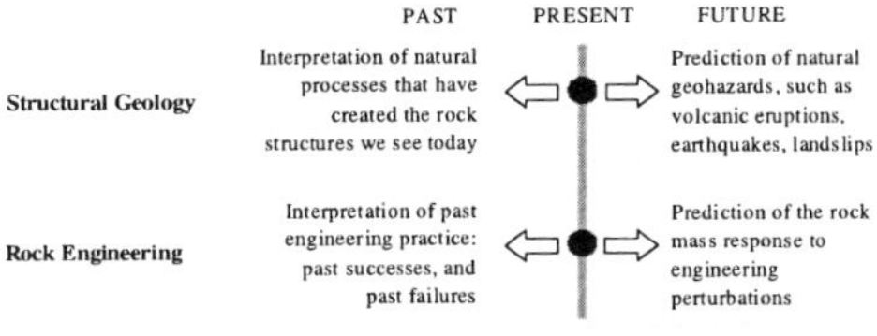

Figure 1. The purpose of modelling THMC processes in rock masses is to be able to interpret previous behaviour and to predict future behaviour.

There are two main challenges in predicting the behaviour of the rock mass:

- its evolution and response to disturbances is controlled by coupled thermal, hydraulic, mechanical and chemical processes; and
- the properties of the rock vary in space in a complex and not fully predictable manner.

A complete understanding, characterisation and modelling of the geosphere is thus unattainable, but complete coverage of all the geometrical details and mechanical processes is never needed in practice. How comprehensively the coupled mechanisms have to be modelled should be related to the objectives of the modelling. For many applications, existing understanding and computer codes are sufficient; whereas, for some other applications, development of more sophisticated tools is required.

In this paper, we discuss the philosophy of approach to THMC coupled modelling for engineering purposes, concentrating on the purposes, procedures and products. In our companion Part 2 second paper to this conference (Hudson and Andersson, 2003), we focus on the spatial variability problem.

2. THE THMC COUPLINGS

When modelling the behaviour of a rock mass it is convenient to consider the six main aspects of the modelling problem.

Geological: site geometry, lithology, fractures
Thermal: heat loads, heat flow
Hydrological: water pressures, water flow
Mechanical: rock stress, stiffness, strength
Chemical: water chemistry, swelling rocks
Engineering: effects of excavation

In a THMC model, the geology is taken into account via the host rock geometry and properties. The engineering is taken into account by the perturbations applied to the model and mainly relates to the Excavation Disturbed Zone. The construction of an underground excavation will lead to changes in the rocks surrounding the excavation, resulting in localized mechanical deformation, alteration in the stress distribution and changes in the water flow and hydraulic properties of the surrounding rock volume. The zone of altered properties is termed the Excavation Disturbed Zone (EDZ). Thus, all six aspects of the GTHMCE model will be taken into account in a THMC model.

If we consider that there are individual binary interactions between these six aspects, there are (6x6 - 6) = 30 separate interactions. The concept of such interactions and the use of an interaction matrix within rock engineering systems have been discussed by Hudson (1992); and the structuring of the features, events and processes related to radioactive waste disposal within an interaction matrix has been used by SKB in Sweden (SKB, 1999).

However, it is not necessary to develop an all-encompassing model. It is only necessary to include those interactions that will significantly affect the predictive capability, given the objective of the modelling.

3. CURRENT MODELLING METHODOLOGIES

A review of techniques, advances and outstanding issues in numerical modelling for rock mechanics and rock engineering has recently been written by Jing (2003). Two text passages in the conclusions of that paper are especially important in the current context.

- "The most important step in numerical modelling is not running the calculations, but the earlier 'conceptualization' of the problem regarding the dominant processes, properties, parameters and perturbations, and their mathematical presentations. The associated modelling component of addressing the uncertainties and estimating their relations to the results is similarly important. The operator should not 'dive in' and just use specific approaches, codes and numerical models, but first consider the specific codes and models to evaluate the harmony between the nature of the problem and the nature of the codes, plus studying the main uncertainties and their potential effects on the results."
- "Success in numerical modelling for rock mechanics and rock engineering depends almost entirely on the quality of the characterization of the fracture system geometry, physical behaviour of the individual fractures and the interaction between intersecting fractures."..."Today's numerical modelling capability can almost handle very large scale and complex equations systems, but the quantitative representation of the physics of fractured rocks remains generally unsatisfactory, although much progress has been made in this direction."

In fact, the full development of T-H-M-C modelling is still at an early stage. For example, we do not have satisfactory answers to the following questions. Do current codes provide the information that is required? How can the codes be validated? How can the codes or the approaches in applying them, be modified to fit the ambition of the modelling? What kind of quality control instruments need to be introduced?

These issues are discussed within the 'purposes, procedures and products' theme of our paper, and within the separate scientific and engineering themes of the conference. Further attention is then drawn to the current outstanding issues. We conclude with a list of highlighted issues and we make recommendations and predictions for the future of T-H-M-C modelling.

4. THE OBJECTIVE, UNCERTAINTIES AND CONFIDENCE

4.1 Objectives

Although the geosphere is a system of fully coupled processes, as we have said this does not directly imply that all existing coupled mechanisms must be represented numerically. Modelling is conducted for specific purposes and the required confidence level should be considered. It is necessary to match the confidence level with the modelling objective. For a given underground construction, only some geosphere responses directly affect its function, and minor mistakes made in design may be correctable during the construction phase. In other cases, knowledge of the long term evolution of the geosphere may be required and the associated prediction may require consideration of a wide range of couplings. The modelling style and content depends on the objectives — which are to be judged, inter alia, against performance measures and the confidence required, given the available capabilities in terms of codes and data.

4.2 Emphasis - Performance measures

In engineering and applied science it is appropriate, not to judge a model by whether it is absolutely true or not, but to what extent it is relevant to the issues being explored. In order to quantify this, performance measures should be identified, making it possible to explore how assumptions, uncertainties and confidence affect the issues of concern.

Useful performance measures should be sought in the context of model predictions that may affect decisions. The decisions may concern a wide range of issues e.g. licensing a nuclear waste repository, selecting a proper grouting scheme or deciding whether a scientific investigation is good enough for publication. A common performance measure for a nuclear waste repository is the yearly risk of death to an individual due to the repository evolution and potential radionuclide releases. Such overall performance measures are usually to be found in regulations (e.g. SSI, 1998). For a tunnel project, an overall performance measure may be the total cost of construction and operation. However, useful performance measures can also be related to construction components, such as the expected volume of failed key blocks in a tunnel, or the expected water inflow. Table 1 in Section 5 provides more examples of issues of concern in different applications. All such issues are good candidates for developing performance measures.

4.3 Uncertainties

Coupled THMC modelling has to incorporate uncertainties. These uncertainties mainly concern uncertainties in the conceptual model and uncertainty in data, where the latter is related to fact that geologic media usually show strong spatial variability.

Model (or conceptual) uncertainty concerns the uncertainty originating from an incomplete or oversimplified understanding of the structure of the analysed systems and the constituent interacting processes (see the first quotation in Section 3). Although, the fundamental THMC couplings, see Section 2, are always present, they need not always be included in the modelling. This would make modelling cumbersome, or even impossible, and would certainly imply non-optimal modelling and data collection efforts. The need to include any specific interaction needs to be weighted against its impact on the performance measures.

On the other hand, overlooking interactions may have severe consequences in modelling utility. For example, in hydropower tunnels hydraulic fracturing may be an issue (if the water pressures are high in relation to the rock stress), but the failure consequence would be overlooked unless this process were to be included in the description of the mechanical system (Stille et al., 2003).

In modelling a system, procedures are needed to handle this type of uncertainty. Clearly, the experience and knowledge of the involved experts is the key to identifying the proper mechanisms, but the identification process is significantly enhanced by using system identification tools, such as the interaction matrices referred to in Section 2 and further described in Harrison and Hudson (2000).

Another aspect of model uncertainty in geologic media concerns the basic principles for describing the host rock mass, the G of the GTHMCE modelling described in Section 2. Among geologists, such principles are often called the conceptual model of the rock mass. Examples of different conceptual models of the rock may be 'typical Swedish low fractured crystalline rock', 'complex deformation zones', 'sedimentary rock', volcanic regions', 'high stress rock' etc. Clearly, the conceptual model affects how data are interpreted and the overall confidence in the system description.

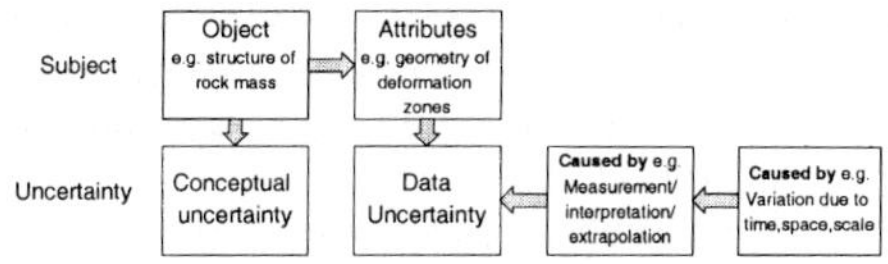

Figure 2. Conceptual uncertainty, data uncertainty and spatial variability may all be related.

Another main source of uncertainty is the uncertainty in input data. Data uncertainties may be caused by, for example, measurement errors, interpretation errors, or the uncertainty involved in extrapolation when the parameter varies in space or in time. Conceptual uncertainty, data uncertainty and spatial variability may all be related (Figure 2). Our second Part 2 paper (Hudson and Andersson, 2003) will elaborate further on data uncertainty and spatial variability. As with conceptual uncertainty, the judgement whether the data uncertainty needs to be reduced, must be weighted against its impact on performance.

Assessing data uncertainty is also important when judging the need to model coupled processes. For example, in the upscaling Benchmark Test of DECOVALEX III, participating teams found the means of interpreting fracture information to have a much more profound impact on calculated flow than the extent to which HM couplings were considered or not (Andersson and Knight, 2003).

4.4 Confidence

Confidence in a model is the total assembly of motives, indications, and arguments in support of the model. However, high confidence is not synonymous with low uncertainty. If the uncertainty description is well founded, the confidence can be high in the model. Conversely, if a model with low uncertainty has a poor foundation, the confidence in the model should be low.

Confidence is essentially a qualitative entity, but judgement of confidence can be based on several specific analyses (Andersson, 2003) as follows.

- A similar and unbiased treatment of all different data and interpretations that explains several different observations enhances confidence.
- Consistency (i.e. no conflicts) between different discipline model interpretations and with the geological evolution model enhances confidence.
- Data and data sources need to be quality assured according to predefined protocols.
- Small estimated uncertainties and inability to produce many different alternative interpretations from the same database are indications of confidence — although not strict proof.
- An indication of confidence is to what extent measurement results from late stages of an investigation compare well with previous predictions.

When evaluating the confidence in THMC predictions, it is necessary to match this level of confidence with an understanding of how much confidence is needed. Will uncertainties and lack of confidence really affect performance measures and to what extent are these significant in relation to other uncertainties?

4.5 Validation

Validation essentially concerns establishing under which circumstances a given model is adequate, i.e. will the model predict the performance measures with sufficient certainty and confidence. A computer model validation may be split into verifying that the code solves the mathematical equations satisfactorily (code verification) and to explore to what extent the mathematical models embedded in the code actually represent reality.

Verifying coupled, and thus generally non-linear, THMC codes is a complex undertaking, which will not be further elaborated on here. There are more or less established procedures for code verification.

Comparing THMC models with real data is a fundamentally more difficult problem — since each experiment or measurement series on a rock mass is unique. This makes it impossible to carry out genuinely repeated experiments. This does not mean that validation is an impossible task. For example, the DECOVALEX project (e.g. Stephansson, 2001) demonstrates how in-situ experiments can be analysed through different approaches and partially different conceptual models, which in the end leads to a better understanding of ongoing processes and interactions. Engineering experience, although biased, provides a large source of information and demonstrates that predictions and design for the underground is possible. Furthermore, design sometimes fails and mishaps occur, providing yet another important source of information for improving understanding — thus further contributing to the validation of models.

5. CAPABILITIES AND APPLICATIONS

5.1 Subject-specific emphases

The emphasis on the need for THMC modelling differs among disciplines, but all applications concerned with the rock mass need to at least consider the THMC couplings. Table 1 lists important examples of THMC emphases in different subjects, based on the authors' judgement.

Table 1. THMC emphases for different applications

Subject	Emphases / Performance required
Geological processes	Understanding past natural rock mass mechanisms (see Fig. 1) Basis for conceptual model Geophysics interpretation
Geotechnical projects	Optimal engineering: hazard and environmental impact mitigation, service life predictions, cost optimisation
Oil and gas production	Optimal flow. Long term managment of resource
Geothermal reservoirs	Optimal extraction of heat and energy
Radioactive waste disposal	Long term function of isolating barriers. Radionuclide migration 'Performance Confirmation' and monitoring

5.2 Geological processes

Geological modelling essentially concerns developing a description of a rock volume in terms of rock types, brittle/ductile deformations and structures. Even if a key input to such descriptions are observations on site and from boreholes, geological modelling would be impossible without a basic conceptual model (see Section 4.3). The conceptual model in turn builds on hypotheses on the (long term) geological evolutionary processes.

Geological modelling thus needs to consider various coupled processes. This concerns understanding the deformation history and the past and current stress fields, where thermal and hydraulic processes affect the large scale rock mechanics evolution. The evolution of the flow field, which in turn is affected by the fracturing of the rock will also determine the occurrence of mineral deposits affected by dissolution and precipitation processes. Furthermore, a substantial part of the geologic data is obtained from geophysics, the interpretation of which usually builds on assessing the coupled processes mobilised by the geophysical perturbation being invoked.

5.3 Geotechnical projects

The overall aim of a geotechnical project is to design, construct and maintain a surface or underground facility for a particular function. In the design and construction process, there is a need to optimise with regard to e.g. construction costs, service life costs, overall costs, hazard and environmental impacts.

Depending on the type of project and facility, there are numerous examples where coupled processes need to be considered in geotechnical projects. The water pressure affects rock mass strength and deformation modulus. Effective grouting to mitigate ingress of water needs to consider how the grouting pressure affects fractures to be grouted. Hydraulic fracturing and jacking may occur in pressurised hydropower tunnels if stress levels are too low.

5.3 Oil and gas production

The main emphases in oil and gas production are how to explore sites and how to operate resources for optimal flow, and long term resource management. The rock structure, rock stress field, primary and secondary rock permeabilities, wellbore locations and orientations, and fluid flow are all coupled and hence there are many interactions that require modelling.

Significant progress has been made in petroleum engineering modelling over the last 15 years or so, not only in the development of the coupled models and the incorporation of the necessary interactions, but especially in the use of computer projected models so that the understanding and perception of the modelling output is enhanced.

5.4 Geothermal reservoirs

The main emphasis for developing and managing geothermal reservoirs is to devise schemes for optimal extraction of heat and energy. In the hot dry rock projects, cold water is injected in one borehole. The water travels down the borehole, passes through the hot rock mass, and travels back up another borehole. Thus, intrinsic couplings in the modelling of this process relate to the interactions between the in-situ stress, the rock fractures (geometry and mechanical-hydrogeological properties), the rock and water temperatures, and the water flow.

5.5 Radioactive waste disposal

The overall emphasis of geological radioactive waste disposal is to site and construct the repository facility and to demonstrate its safety. There are several couplings affecting repository performance, and identifying processes and interactions is an essential part of repository performance assessment (see e.g. SKB, 1999). However, as with other applications, all couplings are not important.

As regards chemical couplings, it is clear that the water composition to a large extent depends on mixing processes, i.e. is coupled to flow (e.g. Laaksoharju et al., 1999). The water composition, in turn, affects groundwater flow through density effects (see e.g. Voss and Andersson, 1993). However, the mechanical impact on chemistry and vice versa appears to be quite weak.

In assessing the repository performance implications of the THM couplings explored in the international DECOVALEX III project, Andersson (ed, 2003) suggest that modelling the saturation of the buffer between the heat producing waste container and the near field rock needs to include the interaction between heat, vapour and flowing water transport. Also in understanding the permeability field and its anisotropy, coupled effects may be crucial; processes like fracture dilatation caused by past stress changes may have a large impact. Other coupled THM effects may be of a less concern.

Coupled processes need also be considered in relation to short term monitoring and 'performance confirmation', Arguably, it will be necessary to explain and predict monitored changes at a repository site, even if these changes may not have implications for repository safety. Confidence in the safety predictions may be in jeopardy in case observed changes in the short term are not reasonably understood. Here, more coupled phenomena, including chemical couplings, may potentially be significant. For example, changes in the water inflow to underground openings are likely to be a function of chemical effects and flow (e.g. precipitation of calcite from water migrating to the drifts), but possibly also for mechanical interactions.

6. CONCLUSIONS

The full development of T-H-M-C modelling is still at an early stage and it is not evident whether current codes provide the information that is required. However, although the geosphere is a system of fully coupled processes, this does not directly imply that all existing coupled mechanisms must be represented numerically. Modelling is conducted for specific purposes and the required confidence level should be considered. It is necessary to match modelling content and the confidence level with the modelling objective.

REFERENCES

Andersson, J., 2003. *Site Descriptive Modelling - Strategy for Integrated Evaluation.* R-03-05, Swedish Nuclear Fuel and Waste Management Co., Stockholm.

Andersson , J., ed, 2003. *T-H-M in Safety Assessments. Findings of DECOVALEX III*, SKI Report (in progress).

Andersson, J. & Knight, L., 2003. *Approaches to Upscaling THM Processes in a Fractured Rock Mass and its Significance for Large-Scale Repository Performance Assessment. Summary of Findings in BMT2 and WP3 of DECOVALEX/BENCPAR,* SKI Report (in progress).

Harrison, J.P. & Hudson, J.A., 2000. *Engineering Rock Mechanics, Part 2: Illustrative Worked Examples*, Elsevier, Oxford, U.K. 502p.

Hudson J. A., 1992. *Rock Engineering Systems: Theory and Practice.* Ellis Horwood, London, 185p.

Hudson, J. A. & Andersson, J., 2003. T-H-M-C modelling of rock mass behaviour: 2 – the input data and rock mass partitioning. *Proceedings of the GEOPROC2003 conference*, Stockholm.

Jing, L., 2003, A review of techniques, advances and outstanding issues in numerical modelling for rock mechanics and rock engineering. *Int. J. Rock Mech. Min. Sci,* (in press).

Laaksoharju, M., Tullborg, E.-L., Wikberg, .P, Wallin, B., Smellie J., 1999, Hydrogeochemical conditions and evaluation at the Äspö HRL, Sweden. *Applied Geochemistry* 14, (7) pp 835-859.

SKB, 1999, *Deep repository for spent nuclear fuel, SR 97 – Post-closure safety.* TR-99-06, Swedish Nuclear Fuel and Waste Management Co., Stockholm.

SSI, 1998, SSI:FS 1998:1.

Stephansson, O. (Ed.), 2001. Decovalex II Special Issue. *Int. J. Rock Mech. Min. Sci.*, 38(1): 161p.

Stille, H., Andersson, J., Olsson, L. 2003. *Information Based Design in Rock Engineering.* SveBeFo Report 61, Stockholm, Sweden.

Voss, C. I., Andersson, J., 1993. Regional flow in the Baltic Shield during Holocene coastal Rregression. *Ground Water,* 31(6).

T-H-M-C MODELLING OF ROCK MASS BEHAVIOUR: 2 – THE INPUT DATA AND ROCK MASS PARTITIONING

J. A. Hudson[1] and J. Andersson[2]
1) Rock Engineering Consultants, UK
2) JA Streamflow AB, Sweden

Abstract: This paper follows on from the immediately preceding Part 1 paper by Andersson and Hudson on approaches to coupled T-H-M-C modelling. Here we focus on the geometrical and mechanical heterogeneity of rock masses and the need to cope with this problem in order to provide input to the modelling. The T-H-M-C modelling is intended to adequately capture the rock mass behaviour commensurate with the modelling requirements and so due consideration has to be given to the discontinuous, inhomogeneous, anisotropic and not-elastic (DIANE) characteristics. A particularly effective technique is to partition the rock mass into domains, within which the properties can be considered as essentially constant, or at least statistically homogeneous. We discuss the mapping function associated with domain partitioning and how this can be used to generate the required input for numerical modelling. A case example is presented to illustrate the methodology as it can be applied to borehole data.

1. INTRODUCTION

In the conclusions of the companion Part I paper, we noted that, although numerical codes have become one of the main tools for modelling rock mass behaviour, the full development of T-H-M-C modelling is still at an early stage. Consequently, it is not self-evident whether current codes can provide the information that is required for rock engineering design, especially when coupled mechanisms are a significant feature for the particular engineering project. On the one hand, the modelling should not be unnecessarily complicated; on the other hand, the model should be an adequate representation of reality within the context of the engineering objective.

Hence, the main issues are ensuring that

1. the necessary physical variables and associated mechanisms are included in the model,
2. the appropriate input data are available, and
3. the modelling is quality assured.

We discussed the first issue and its ramifications in the Part I paper. In this Part II paper we will discuss the second issue. The third issue is currently being studied in the DECOVALEX/ BENCHPAR research programme and will be reported on in due course.

2. THE INPUT DATA PROBLEM

In order to model rock mass behaviour with a numerical code, it is necessary to input values for the relevant parameters. In a relatively simple code, e.g. one that calculates the stresses and displacements in an elastic rock around openings with a simple geometry, it is necessary to input values for the far-field stresses and the elastic constants (Young's modulus and Poisson's ratio). But even in this case, problems associated with the input data immediately arise. What are the far-field stresses? Are the local stresses significantly perturbed, so that the far-field stress is not representative of the local stress around the engineering structure? Is the Young's modulus the same in all directions and locations? Is the rock mass, elastically isotropic, transversely isotropic, orthotropic or fully anisotropic? What is the Poisson's ratio of a fractured rock mass?

These questions are answered by considering whether the modelling is adequate given the engineering objective. In a fractured rock mass, the Young's modulus and Poisson's ratio will vary with direction and location—but is this significant in terms of the information required from the modelling? Considerable benefit has been derived from modelling in which the rock is assumed to be a CHILE material (Continuous, Homogeneous, Isoptropic and Linearly Elastic), despite the fact that we know rock masses are DIANE (Discontinuous, Inhomogeneous, Anisotropic and Not Elastic). It is necessary to harmonize the complexity of the modelling with the information required for the engineering design within the constraints of the capabilities of current codes.

Similarly, for the various applications and aspects of the THMC modelling that we have highlighted in Part I, the modeller and engineer have to find a method of benefiting from the modelling, despite knowing that not all the features can be included.

There are additional problems, such as those related to time and scale. Many civil engineering structures are designed to have a life of around 120 years: yet nowhere in the modelling or design calculations is the figure of 120 years explicitly input. This is because we do not currently have sufficient knowledge of time-dependent processes in fractured rock masses. Nor do we have an adequate methodology in place to account for scale effects and the fact that the properties of large rock masses may be represented by the results of tests on small laboratory sized specimens.

We have reached the following stages with the separate DIANE features.

- *Discontinuities.* It is now possible to model the fractures, the main problem being to characterize their 3-D geometry and to characterize the geometry and mechanical properties of large fractures.
- *Inhomogeneity.* This is still a major problem, both in knowing the inhomogeneity characteristics and then modelling them.
- *Anisotropy.* This is a more tractable subject because anisotropy is incorporated into some variables and parameters, e.g. stress, strain, stiffness, hydraulic conductivity, and can be more easily determined than inhomogeneity from a site investigation.
- *Non-Elasticity.* Considerable progress has been made in incorporating plastic, viscous, and failure behaviour into the codes, but representing the long-term time-dependent behaviour is still an outstanding requirement.

Given the inhomogeneity problem and the structure of the currently available codes, it is helpful to be able to establish zones of the rock mass where the rock properties are essentially statistically homogeneous—again within the context of the modelling objectives. We will concentrate on this subject in the next two sections.

3. ROCK MASS DOMAINS

The idea is to partition the rock mass into domains, as illustrated in Fig. 1, so that the domains can be considered as having constant properties for the purposes of modelling.

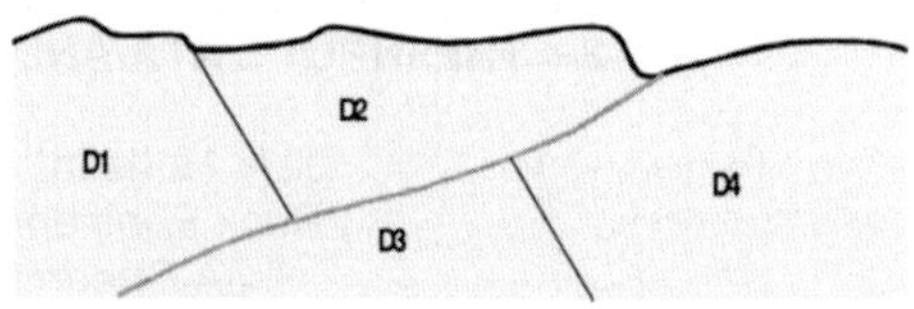

Figure 1. Example partitioning of a rock mass into domains.

The use of such domains for structural geology and engineering geology studies has been noted by Hudson and Harrison (2002).

3.1 Rock Mass Domain Definition

For engineering purposes, we define a rock domain as a rock volume within which a certain rock property or function of rock properties is statistically homogeneous. An immediate corollary of this definition is that the rock mass could well be partitioned into different domains for different rock properties or different functions of rock properties.

3.2 Methodology for Establishing the Rock Mass Domains

In Hudson and Harrison (2002), a methodology is proposed for establishing rock mass domains, as shown in Table 1.

Table 1. The four steps to be followed in establishing rock mass domains.

A: Establish purpose of the domain partitioning ⇓	There can be a variety of rock engineering purposes: e.g. for selection of excavation method, for support design, for choice of and values of input parameters to numerical models, presentation of representative and extreme values.
B: Choose governing parameters ⇓	Depending on the purpose established in (A), different suites of parameters should be chosen to define the domains. For example, fracture characteristics for input to discrete fracture network modeling.
C: Select and apply the mapping function ⇓	The mapping function is used to combine the parameter values into one or more Domain Indices that characterize the domain. Any mathematical function of the selected parameters can be used to form a Domain Index.

D: Partition rock mass into domains	Using the Domain Index values, the rock mass is partitioned into domains according to the requirements established in (A), and the results presented in an appropriate format (e.g. a map, tables of representative numerical values, via computer-based visualization tools).

3.3 Types of Rock Mass Domain

In the general rock engineering context, the mapping function could be a classification index value, such as RMR or Q. In the THMC modelling context, we should like the modelling to occur within one overall rock domain in which all the rock properties were statistically homogeneous. However, because of the different natures of the T, H, M and C components, they are not likely to have the same domains, as illustrated by the example content of Table 2.

Table 2. Example of the different possible THMC domains in Fig. 1

Thermal	D1 + D2 + D3 + D4 as one domain. It could well be that for the required analysis the thermal properties of each of the sub-domains can be considered as the same.
Hydro-geological	Given the through going fault, it is possible that D1 + D2 are one domain, and D3 + D4 are a different domain.
Mechanical	If the four domains are in separated geological formations, then the four separate D1 through to D4 domains may be required.
Chemical	The chemical domains, involving dissolution and/or precipitation, might also require the four separate domains, D1 through D4.

This example, with its generic extrapolation, means that in the numerical modelling consideration must be given to the procedure for dealing with the separate components of the input data. This may involve alternative and sequential rock mass domain partitioning, as described in Hudson and Harrison (2002).

These are the main issues relating to the partitioning itself. In the next Section, we discuss how a methodology can be developed — even in the face of uncertain information.

4. IMPLEMENTATION OF THE ROCK MASS DOMAIN METHODOLOGY

4.1 Strategy and the role of the mapping function

We have emphasized that the modelling and the associated rock mass characterization should be determined by the rock engineering objective. This means that different types of information will be required for THMC modelling for the different types of application, as illustrated by the themes of this conference and the associated papers. However, site investigation has limited methods of approach, e.g. surface outcrop studies, borehole and borehole core measurements, and geophysics, which means that we will not have as much information as we would like and concessions have to be made on both the site investigation and modelling sides.

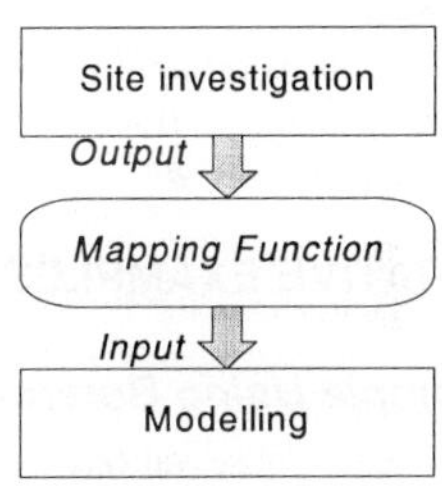

Figure 2. The mapping function serves as a link between the site investigation and modelling, ensuring appropriate site investigation output, appropriate modelling input and suitable rock mass domains.

The mapping function described in Section 3.2 can provide this intermediary role, serving to define a component of the site investigation output and to ensure that regions of the rock mass are characterized in a suitable way for modelling input.

4.2 Dealing with uncertainty

However, we recognize that we will be working with incomplete knowledge and it is necessary to distinguish between lack of knowledge, spatial variability and 'random variation'. So, how do we incorporate uncertainty/confidence into the methodology?

At an early stage of an investigation it is probable better to use few rock domains and instead accept rather high uncertainty within these few domains. Division into more distinct domains can only be considered when confidence increases. There will always be a trade-off between defining the domains – and the confidence in their description and geometrical extent.

4.3 Dealing with discipline integration

Another aspect is ensuring that the geological, thermal, hydrogeological, mechanical, chemical and engineering aspects are all integrated. No one person will be an expert in all these disciplines and therefore it is necessary to ensure that the disciplines are harmonized and integrated. The main features in this integration have been outlined by Andersson (2003) and are listed below.

- Consistency in evaluation of primary data
- Consistency in 3-D modelling results
- Consistency in being aware of inter-disciplinary processes
- Consistency in uncertainty and confidence assessments
- Consistency in reporting/documentation
- Consistency in application of procedures/protocols/QA

5. ILLUSTRATIVE EXAMPLES

5.1 An Example Using Borehole Data

Geological data observations cover many different aspects such as observations of lithology, fracturing, alteration, mineralisation, tectonic indications, and hydraulic indications from flow etc. The tools for observing these characteristics are both by direct visual inspection and using indirect tools, such as various types of geophysical logging tools, packer tests, down-hole cameras, radar and seismics.

When constructing a geological model of the rock mass to be used for purposes such as design of tunnels, hydraulic modelling and safety performance studies, it is useful to simplify the geological information so that it can be contained within a few classes that describe certain characteristics of the rock. Such a simplification was made in a site descriptive modelling exercise conducted by SKB (Andersson et al., 2002a) using a single borehole interpretation.

The aim of a geological single hole interpretation is to identify rock segments ('sections') along the borehole having similar geological characteristics. These rock segments are then to be used in the 3D geometric modelling of rock mass domains, i.e. it is a way to simplify and reduce the complex geological information to a few alternative interpretations that can be used in the 3D geometrical modelling, and subsequent selection of rock mass domains.

In the SKB study, the following borehole data were considered most valuable for the construction of the geological model:

- Lithology (rock types).
- Alteration (oxidization, chloritization, epidotization, weathering, tectonization).
- Brittle deformation (i.e. observations of fracturing, core loss and crushed zones).

Table 3 shows eight different types of combinations that can occur when classifying lithology, alteration and fracturing along the borehole. The classification is based on dividing the lithology into segments with either one dominating rock type (single) or with a mix of rock types (mixed). Alteration and brittle deformation are divided into sections of high or low proportion of alteration and fracture frequency.

Table 3: Approch to borehole rock segment division (from Andersson el., 2002a)

Lithology	Alteration	Fracturing	Rock Segments
Single	Low	Low	R1
Mixed	Low	Low	R2
Single	High	Low	R3
Single	Low	High	R4
Mixed	High	Low	R5
Mixed	Low	High	R6
Single	High	High	R7
Mixed	High	High	R8

5.2 A Characterization Example Conducted at the Äspö HRL, Sweden

A series of site investigations are planned as part of the Swedish nuclear waste disposal programme. In order to provide the necessary rock mechanics support for these site investigations, a strategy has been developed (Andersson et al. 2002b) which includes the following items:

- quality control of rock mechanics data, including techniques to reduce uncertainties;

- a descriptive rock mechanics methodology (which has been verified by a Test Case using protocols and an auditing methodology) enabling the crystalline rock mass to be characterized in terms of constructability and performance assessment; and
- the use of the rock mechanics information for designing the repository.

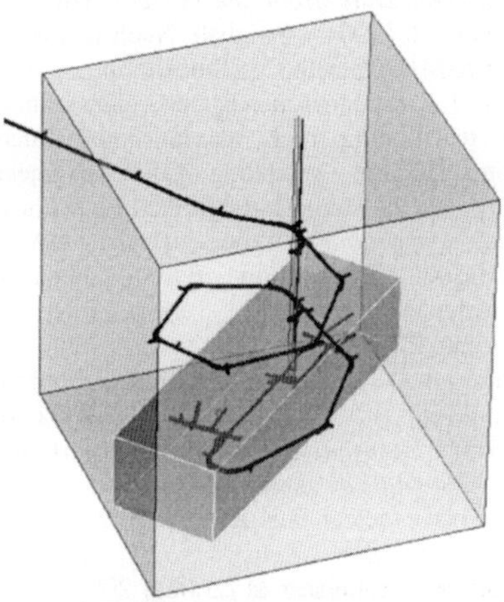

Figure 3. The rock volume(darkly shaded,600 m x 180 m x 120 m) used for the Test Case at the Äspö HRL in Sweden with the access tunnel and shaft, from Staub et al. (2002).

The approach to the Rock Mechanics Site Descriptive Model was tested by predicting the rock mechanics properties of a 600 m x 180 m x 120 m rock volume at the Äspö Hard Rock Laboratory (HRL) in Sweden using limited borehole data of the type typically obtained during a site investigation (Hudson, 2002). These predicted properties were then compared with 'best estimate' properties obtained from a study of the test rock volume using additional tunnel and tunnel borehole data. This exercise was known as the Test Case, and directly indicates how some of the input required for modelling can be specified and tested.

Three modelling techniques were used to predict the rock properties:

- the 'empirical approach' — the rock properties were estimated using rock mass classification schemes and empirical correlation formulae;
- the 'theoretical approach' — the rock properties were estimated using numerical modelling techniques; and
- the 'stress approach' — the rock stress state was estimated using primary data and numerical modelling.

These approaches are described separately and respectively in Röshoff et al.(2002), Staub et al. (2002) and Hakami et al. (2002). The work describing the 'best estimates' of the actual rock conditions is described in Makurat et al. (2002).

The Test Case exercise was invaluable in forcing the generic approach to the Rock Mechanics Site Descriptive Strategy to become a specific approach for the Äspö HRL rock volume being considered. The predictions of the rock mass modulus between fracture zones made by the empirical and theoretical teams were similar and close to the 'best estimate' values. The predictions of the rock mass modulus in fracture zones were more varied, because few data were available and hence there was greater choice in the assumptions necessary to make the predictions. The empirical and theoretical teams' predictions of the rock strength were also more varied, mainly because few data were available and different interpretations of rock mass strength were used.

The rock stress predictions, based on database information and numerical modelling, were satisfactory and helped to focus attention on the key factors involved. In particular, the use of numerical modelling assists in evaluating potential stress changes in the vicinity of fracture zones. Methods of dealing with conceptual uncertainty and spatial variability of stress were successfully introduced into the rock stress characterization and predictive methodology.

The information developed from the Test Case exercise enables the input parameters to modelling to be chosen more coherently and accurately, and support the methodology for establishing rock domains.

6. CONCLUSIONS AND RECOMMENDATIONS

- In this Part II paper, we have concentrated on the input data for T-H-M-C modelling and the associated problem of partitioning the rock mass into domains within each one of which the rock properties are assumed to be constant or at least statistically homogeneous. We have emphasized that this should be done within the context of the project/modelling objective and with some uncertainty.

- The separate DIANE features of discontinuousness, inhomogeneity, anisotropy and non-elasticity are rock mass characteristics that should also be considered.
- A four-step procedure for establishing suitable rock mass domains was described, involving
  A. Establish the purpose of domain partitioning
  B. Choose governing parameters
  C. Select and apply mapping function
  D. Partition rock mass into domains.
  The mapping function can serve as an intermediary between the site investigation and the modelling.
- Dealing with uncertainty and discipline integration are important aspects of the modelling.
- We presented two illustrative examples of obtaining input information. The first concerned partitioning of a borehole into domains (rock segments lengths with similar characterisitcs). The second concerned a Test Case in which a 600 m x 180 m x 120 granitic rock mass was studied to establish how additional information would influence the estimated rock properties and their variation.
- We concluded the Part I paper stating that the full development of T-H-M-C modelling is still at an early stage and it is not evident whether current codes provide the information that is required. This also applies to the input data and rock mass partitioning Part II conclusion: we have practical techniques for obtaining the input information for simple codes, but not for the new generation of T-H-M-C codes — which require a more systematic and overall approach to the problems described herein.

REFERENCES

Andersson J, Berglund J., Follin S., Hakami E., Halvarson J., Hermanson J., Laaksoharju M. Rhén I. & Wahlgren C.H., 2002a. *Testing the Methodology for Site Descriptive Modelling. Application for the Laxemar area.* SKB TR 02-19, Swedish Nuclear Fuel and Waste Management Co., Stockholm.

Andersson, J., Christiansson, R. & Hudson, J.A., 2002b. *Site Investigations Strategy for Development of a Rock Mechanics Site Descriptive Model.* SKB TR 02-01, Swedish Nuclear Fuel and Waste Management Co., Stockholm.

Hakami, E., Hakami H. & Cosgrove J., 2002. Strategy for a Rock Mechanics Site Descriptive Model: Development and testing of an approach to modelling the state of stress. SKB report R 02-03, Swedish Nuclear Fuel and Waste Management Co., Stockholm.

Hudson J. A. (ed.), 2002. *Strategy for a Rock Mechanics Site Descriptive Model: A Test Case based on Data from the AEspoe HRL.* SKB Report R 02-04 Swedish Nuclear Fuel and Waste Management Co., Stockholm.

Hudson J A & Harrison J P, 2002. The principles of partitioning rock masses into structural domains for modeling and engineering purposes. In *Mining and Tunnelling Innovation and Opportunity.* Proc. of the NARMS-TAC2002 Symposium, Eds. Hammah R., Bawden W., Curran J., Telesnicki M. Univ. Toronto Press, 623-628.

Makurat, A., Løset, F., Hagen, A.W., Tunbridge, L., Kveldsvik V.& Grimstad, E. 2002. *Äspö HRL: A Descriptive Rock Mechanics Model for the 380-500 m level.* SKB report R 02-11, Swedish Nuclear Fuel and Waste Management Co., Stockholm.

Röshoff, K., F. Lanaro & L. Jing, 2002. *Strategy for a Rock Mechanics Site Descriptive Model: Development and testing of the Empirical Approach.* SKB report R 02-01, Swedish Nuclear Fuel and Waste Management Co., Stockholm.

Staub, I., A. Fredriksson & N. Outters, 2002. *Strategy for a Rock Mechanics Site Descriptive Model: Development and testing of the Theoretical Approach.* SKB report R 02-02, Swedish Nuclear Fuel and Waste Management Co., Stockholm.

SIMULATION OF CONSOLIDATION AND TRANSPORT PROCESSES IN CLAYEY ROCKS

Maxim G. Khramchenkov, Kazan State University, Russia

Abstract: A united mathematical model for the rheological and transport properties of saturated clays is proposed. The foundation of the model is the unification of filtration's consolidation theory and the theory of the stability of lyophobic colloids, which is based on the conception of 'disjoining' pressure as a surplus in relation to hydraulic pressure. This pressure is caused by surface capacities and exists in water films between clay particles. In this work it is shown that the problem of the shrinkage of a clay layer can be reduced to the well known problem. We obtained the approximate solution for pressing the water out of a clay layer. The solution that we obtained requires introduction of a concept for the limit shear stress for clays. We investigated the model, and explained some characteristic features of transfer processes in clays (the existence of anomalous high pressures in clays, the flocculation at diffusion in clays, etc.). It is shown that solutions which we received are in harmony with results of experiments.

1. INTRODUCTION

Properties of clays and clayey rocks, and also the processes in them depend on a number of factors. Then the mathematical simulation of the properties and processes, as one of the methods of their examination, is a rather difficult problem. Physically it is clear that the specific properties of clay rocks (low permeability, plasticity in moist condition) are caused by the existence of clay minerals in their composition, and these properties are a manifestation of surface capacities, which exist between particles of the clay minerals, which are included in the composition of clays. The most useful conception of the activity of surface capacities is the conception of disjoining pressure between colloid particles, Mitchell (1976). In this work we provide a description of the physical and mechanical clay properties and transport processes in them. The description is based on methods of theory of filtration consolidation, Nikolaevskiy (1996), and also on the theory of stability of lyophobic colloids (theory of Derjaguin-Landau-Verwey-Overbeeck, or DLVO theory), which uses the conception of disjoining pressure.

2. MECHANICS OF CLAYS AND CLAYEY ROCKS

The specific properties of clay rocks (low permeability, plasticity in moist condition) are caused by the existence of clay minerals in their composition. Clay minerals are hydroalumosilicates of calcium, magnesium, potassium and natrium and they are structurally arranged as thin sheet particles (about 50- 100 nm in diameter and a few nm in height). Because of the isomorphic replacement in the crystal matrix these particles usually carry an electrical charge (usually negative), which is compensated by cations. Cations are adsorbed on to the surface of particles and at hydration they are able to dissociate (totally or partly) and make the double diffusion layer, Mitchell (1976). Before we proceed to theoretical examination of clay properties, we describe the structure of clay rocks. We will consider clay rocks as a solid porous skeleton (not essentially coherent), which pores are not fully filled by clay particles. In that case any process, connected with the deformation of clay rocks, could be divided into two stages: 1) free porosity (part of pore volume, which is not filled by clay particles and their interlayer intervals) is not equal zero; 2) free porosity is equal to zero, it means that all pores are filled by clay particles and interlayer films of water. We begin examination from the first stage. We will consider that the filtration is going only in the free part of the pores. Now we can write mass balance equations for liquid and solid phases:

$$\frac{\partial(m_{p,c}\rho_f)}{\partial t}+\mathrm{div}(\rho_f m_{p,c}\vec{V_f})\pm j=0, \qquad (1)$$

$$\frac{\partial\left[(1-m_p-m_c)\rho_s\right]}{\partial t}+\mathrm{div}[\rho_s(1-m_p-m_c)\vec{V_s}]=0 \qquad (2)$$

In these equations ρ_f is the density of fluid (water), ρ_s - density of solid phase, $\vec{V_f}$ - fluid velocity in transport pores, $\vec{V_c}$ – fluid velocity between clay particles, $\vec{V_s}$ - solid phase velocity,

$m_p = m$ - free (transport) porosity, m_c - part of film interlayer water in volume (porosity of clay minerals), j – exchanging flow in the system «transport pores – water between clay particles». We consider that ρ_f and ρ_s are constants. Then we receive the equation

$$\dot{\theta} + \mathrm{div}\vec{q} = 0, \tag{3}$$

where $\mathrm{div}\vec{V_s} = \dot{\theta}$, $\mathrm{div}[m(\vec{V_f} - \vec{V_s})] = \vec{q}$ and $\vec{V_c} = \vec{V_s}$ for clay particles (meaning that particles of clay are able to move only together with a solid skeleton). Here θ is common shrinkage, $\vec{q}$ - filtration velocity. Now we write conditions of mechanical balance. We suppose that an external load G is exerted to clay. Then the distribution of load between the rock components gives an equation

$$G_{ij} = (1 - m - m_c)\sigma_{ij} - m_c(p + \Pi)\delta_{ij} - mp\delta_{ij}, \tag{4}$$

which after the designation

$$(1 - m - m_c)(\sigma_{ij} + p\delta_{ij}) = \sigma_{ij}^f \tag{5}$$

can be written in the form

$$G_{ij} = \sigma_{ij}^f - m_c\Pi\delta_{ij} - p\delta_{ij}. \tag{6}$$

In these equations p – pressure in liquid, Π - disjoining (additional to hydraulic) pressure in water interlayer between clay particles, the aim of introduction of the σ_{ij}^f will be explained below. Now proceed to thermodynamic balance. Denote the free energy of the solid phase through macroparameters σ^f, m, m_c, θ, Π. For this purpose we will write the first and second law of thermodynamics for the solid phase

$$\mathrm{dU_s} = \delta Q^{(e)} + \delta A^{(i)}, \quad T\,\mathrm{dS_s} = \delta Q^{(e)} + \delta Q'. \tag{7}$$

Here U_s is an internal energy of the skeleton, S_s - entropy of solid phase, $\delta Q^{(e)}$ - external stream of heat, $\delta Q' \geq 0$ - uncompensated heat, $\delta A^{(i)}$ - elementary work of the internal surface strength.

For isothermal conditions

$$\delta A^{(i)} = \mathrm{d}U_s - T\,\mathrm{d}S_s + \delta Q' = \mathrm{d}F_s + \delta Q'. \tag{8}$$

Then in conditions of the absence of mass forces we have the equation

$$\frac{\delta A^{(i)}}{\mathrm{dt}} = \int_{V_s} \sigma_{ij}^s \frac{\partial u_i}{\partial x_j}\mathrm{dV} = \int_{V_s} \frac{\partial \sigma_{ij}^s u_i}{\partial x_i}\mathrm{dV}, \tag{9}$$

where u_i is a component of the vector of transference velocity of skeleton's particles, σ_{ij}^s is the skeleton's stresses. After transformation of the volume integral into a surface integral, we obtain:

$$\frac{\delta A^{(i)}}{\mathrm{dt}} = \int_{V_0} \frac{\partial(\sigma_{ij}^f \langle u_i \rangle)}{\partial x_i}\mathrm{dV_0} + (-p)\int_{\Sigma_S} u_i n_i\,\mathrm{ds} + (-\Pi)\int_{S_{S-C}} u_i n_i\,\mathrm{ds}, \tag{10}$$

where $\sigma_{ij}^f = (1 - m - m_c)(\langle i_{ij}^s \rangle - \langle \sigma - p \rangle \delta_{ij})$ – is a tensor of effective tensions, corresponding to (4), $\sigma = -(1/3)\sigma_{ii}^s$ – is the pressure in the solid phase, i_{ij}^s – is a deviator of the σ_{ij}^s. It is clear that $\int_{\Sigma_S} u_i n_i\,\mathrm{dS} = \dot{V_s}$ - the velocity of changing of the solid phase volume. During the first stage of deformation, until the full disappearing of transport pores, water in the interlayer between the clay particles doesn't perceive the effective tensions of skeleton and it is in thermodynamics balance with the water in the transport pores, so it satisfies the equation Π=0. So during the first stage from the condition $\rho_s V_s = M_s$ = const we have

$$\frac{1}{V_0}(\dot{F_s} + \frac{\delta Q'}{\mathrm{dt}}) = \sigma_{ij}^f e_{ij} + \frac{(1 - m - m_c)}{\rho_s}\dot{\rho_s} p, \tag{11}$$

where e_{ij} is a tensor of the deformation's velocities. Let us examine the elastic skeleton, when $\delta Q'/dt=0$. Then from the last equation we find that

$$\dot{F_s} = T_{ij} e'_{ij} - \sigma^f \dot{\theta} + (1 - m - m_c)p\dot{\rho_s}/\rho_s. \tag{12}$$

In this equation e_{ij}' is a deviator of the tensor of the deformation's velocities. So far as the disjoining pressure Π is a function of the thickness of the water interlayer h between clay particles, that is why condition Π=0 is equivalent to the condition h=const. This means that the volume which is filled with clay particles and interlayer water V_c is preserved. Let us write it as $m_c V_0 = V_c$, V_c=const. Then from the last equation we have $\dot{m_c}/m_c = -\dot{V_0}/V_0 = \dot{\theta}$, so

$$m_c = m_c^{(0)} \exp(-\theta). \tag{13}$$

Because of $(1 - m - m_c)\rho_s V_0 = M_s$=const, then

$$\frac{\mathrm{d}[\ln V_0 + \ln \rho_s + \ln(1 - m - m_c)]}{\mathrm{dt}} = 0, \tag{14}$$

and we receive one equation for three variables. In the most simple case, when ρ_s=const, it can be written in such a form

$$\frac{1-m-m_c}{1-m^{(0)}-m_c^{(0)}}=1+\theta \cdot \quad (15)$$

Then for the free energy we have a dependence $F_S = F(J_2', \theta, \rho_S)$, where J_2' – is the second invariant of the tensor of deformations, then $\dot{F}_s = \frac{\partial F}{\partial J_2'}\varepsilon_{ij}' e_{ij}' + \frac{\partial F}{\partial \theta}\dot{\theta} + \frac{\partial F}{\partial \rho_s}\dot{\rho}_s$. Comparing the right parts of the equations for the $\dot{F}_s$ gives for a concrete view of the F for the elastic skeleton $F = \frac{\lambda}{2}\theta^2 + \mu J_2' - \upsilon\theta\sqrt{J_2'}$, Nikolaevskiy (1996), where λ, μ, ν = const, next correlation

$$-\sigma^f = \lambda\theta - \nu\sqrt{J_2'}, \quad T_{ij} = (2\mu - \nu\theta / \sqrt{J_2'})\varepsilon_{ij}' \cdot \quad (16)$$

The last equations play the roles of rheological correlations. They give us an ability to obtain a closed system of equations for filtration consolidation of clays at the first stage. At the second stage m=0, we have to specify the law of the filtration in such a system, because we have ignored the filtration of water between clay particles until this moment (at the first stage we consider that for the $\vec{q}$ Darcy's Law is operating). Let us use the correlation for case m=0:

$$m_c = \beta - (1-\beta)\theta, \quad \beta = m_c^{(0)} + m^{(0)}, \quad (17)$$

so $\dot{m}_C = -(1-\beta)\dot{\theta}$. In this case we must take into account the last term of (10). This item describes the pressure of water between clay particles

$$\Pi \int_{S_{S-C}} u_i n_i \, dS / V_0 = -\Pi \dot{V}_c / V_0 = -\Pi(\dot{m}_c + \dot{\theta} m_c) \cdot (18)$$

Then for the elastic pore environment we have analogically in this case

$$-\sigma^f - [m_c - (1-\beta)]\Pi = \lambda\theta - \nu\sqrt{J_2'}, \quad T_{ij} = (2\mu - \nu\theta / \sqrt{J_2'})\varepsilon_{ij}' \quad (19)$$

As far as in the first equation there is a term which contains the disjoining pressure which itself depends on the thickness of interlayer film of water h, it is necessary to determine the form of function $h(\theta)$, to connect microparameter h with the macrovariables of the process. From one side we have the correlation (15) for m_c and from another side $m_c = S_c h / V_0, S_c = \text{const}$. Here S_c is a half-square of the surface of clay particles which are situated in the volume V_0. Let us differentiate the left and right part of the last correlation with respect to time, so we receive

$$h = \frac{h^{(1)}}{m^{(1)}}[\beta - (1-\beta)\theta]\exp(\theta) \cdot \quad (20)$$

Here $h^{(1)}, m_c^{(1)}$ - constants, which provides a connection of the first and second stages. The last correlation relates to the shrinkage θ<0. So the process of consolidation of clay rocks at the second stage determines the value h and, consequently the disjoining pressure as a function of the macrovariable θ. At the moment the disjoining pressure is able to give an addition to hydraulic pressure, in terms of the value of θ. This fact can explain the nature of anomalous high seam pressures in deep-laying clay-containing reservoirs.

3. RHEOLOGICAL PROPERTIES OF CLAYS

We examine the rheology of clays on the example of the one-dimensional problem of clay layer's shrinkage under a constant load. From previously obtained equations and from the law of Darcy for the first stage of water pressing out we receive:

$$\sigma^f - p = G, \quad \frac{\partial q}{\partial z} + \dot{\theta} = 0, \qquad \sigma^f = \lambda\theta, \quad q = -\frac{k}{\eta}\frac{\partial p}{\partial z} \quad (21)$$

System (21) can be easily led to the equation

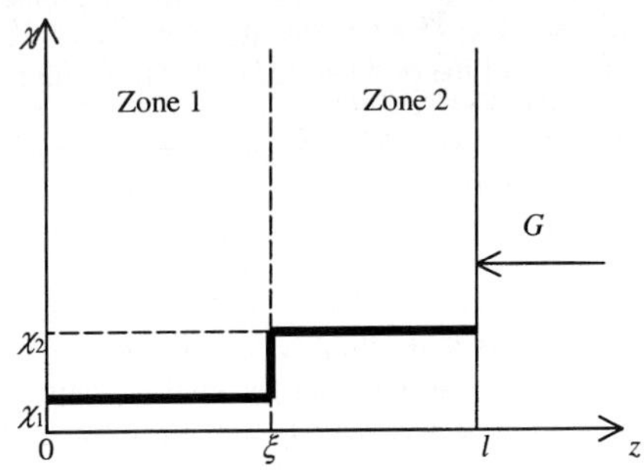

Figure 1. Formation of two shrinkage zones with different values of piezoconductivity and the mobile border

$\frac{\partial \theta}{\partial t} = \frac{\lambda}{\eta}\frac{\partial}{\partial z}(k\frac{\partial \theta}{\partial z})$. To solve this equation we have to introduce a function $k(\theta)$. In a simplest case we can decide that the process of clay's shrinkage and water 'wringing' exists in two zones - «fast» and «slow», with different coefficients χ_1 and χ_2, where the parameter $\chi=k/(\lambda\eta)$ can be considered as an effective coefficient of piezoconductivity. When the shrinkage in the first zone reaches the value of θ_0 (which is determined from condition $m=0$), the coefficient χ_2 of the «fast» zone immediately becomes equal to χ_1 in the «slow» zone, so the border between the two zones is mobile (fig. 1).

Let us write this problem in more detail. The process of shrinkage is described by the last equation, which is written for two zones. The motion of the border between the two zones is determined from the condition of equality of the filtration expenses. Then for two zones we have instead of (4) the next system of equations:

$$\frac{\partial\theta_1}{\partial t}=\chi_1\frac{\partial^2\theta_1}{\partial z^2},\quad 0\le z<\xi,$$
$$\frac{\partial\theta_2}{\partial t}=\chi_2\frac{\partial^2\theta_2}{\partial z^2},\quad \xi<z\le l, \tag{22}$$

where θ_1, θ_2 – are shrinkages correspondingly in the first and second zones,

$$\theta_1(z=0)=\theta_{\max},\quad \theta_2(z=l)=0. \tag{23}$$

On the border of the contact of two zones we have the conditions

$$\theta_1(\xi)=\theta_2(\xi)=\theta_0,\quad \theta_0=\text{const},$$
$$\chi_1\frac{\partial\theta_1}{\partial z}(\xi)=\chi_2\frac{\partial\theta_2}{\partial z}(\xi). \tag{24}$$

This problem is the N.N. Verigin's problem, Nikolaevskiy, 1996. We do not have exact solutions of the problem (22) – (24), because we can speak only about approximate solutions. Let us turn to more "rude" approximations of the process to provide an opportunity to reveal and examine some physically important peculiarities of the process. We ignore the filtration between clay particles, after the shrinkage has become equal θ_0. So θ_0 is a limit value for the shrinkage. This situation also is described by the system (21), but with one reservation. The pressure of the water in clay doesn't become equal to zero when $\theta=\theta_0$. It becomes equal to some constant value p_0. We can regard it as a limit value of the flow's tension with a reversed sign. Value θ_0 is determined independently and the first equation of (21) takes a form $\lambda\theta_0-p_0=G$. It contains two empirical constants λ и p_0, so we can use it to determine one of them by the known other. In this positing the problem (22) – (24), accounting for the reservation about p_0, it is a classical one-dimension problem of filtration consolidation of soil, and its solutions are well-examined, Nikolaevskiy (1996). The solution for the shrinkage of the soil layer-thickness l is

$$\theta\approx\theta_{fin}\left(1-\frac{8}{\pi^2}e^{-N}\right),\quad N=\frac{\pi\chi}{4l^2t}, \tag{25}$$

where θ_{fin} – is a final shrinkage. In our case $\theta_{fin}=\theta_0$. Let us differentiate the last correlation with respect to t. Then we obtain

$$\dot\theta=\frac{2\chi\theta_0}{\pi l^2}=\frac{2\chi}{\pi l^2\lambda}(G-G_0). \tag{26}$$

Here we introduced a designation $G_0=-p_0$, which corresponds to agreement about the sign of tension in the continuous mechanics. The last law is a law of flow with a limit tension of the shift (Bingham's body). This law as a limit case describes the rheology of clays, which are examined in experiments, Mitchell (1976). So, following this model we can receive the theory of the rheological properties of clays and this theory is in conformity with experimental results.

4. DIFFUSION IN CLAYS

Let us examine the process of diffusion in clays at the first stage of their consolidation, before the moment when $m=0$. We will suppose that the diffusion solution contains cations of the same type, as cations of the exchanging complex of clay; meaning that there is no ion exchange reaction. We will also suppose that in every point of environment the equilibrium between the solution in transport pores and the solution between clay particles is established immediately (its parameters we will be marked by the overline). Like this we will successively build the model of diffusion in clays in a local equilibrium approach. The conditions of equilibrium of cations (index 1) and anions (index 2) of two solutions are the equality of the chemical potentials $\overline{\mu_1}=\mu_1$, $\overline{\mu_2}=\mu_2$, where

$$\mu_i=\mu_i^0+RT\ln C_i+ez_i\varphi. \tag{27}$$

Here the index «0» indicates standard values of chemical potentials of cations and anions, z_1 and z_2 are valencies of cations and anions correspondingly, e is charge of the electron, φ is an electrical potential of the solution. We suppose that in the solution which is situated in transport pores $\varphi=0$ and in the solution between clay particles $\varphi\neq 0$ because it's determined by exchange cations of clay. Also suppose that $z_1=1$, $z_2=-1$ (binary 1-1 electrolyte). So, adding the first equation with the second we obtain $\overline{C_1 C_2} = C_1 C_2$, (28)

This equation the condition of Donnan's equilibrium. We have to supplement it with conditions of electrical neutrality in the solution, situated in transport pores $C_1 = C_2 = C$ and in the solution between clay particles

$\overline{C_1} - \overline{C_2} = 2q/h$. (29)

Here q is a surface concentration of exchanging cations in clay (supposed that they are fully dissociated). From Norrish (1954), we have for h that

$h/d_0 \approx 1/\sqrt{C}$, (30)

where d_0 is the thickness of clay particle ($\approx$ 1nm), C is measured in mole/liter. Then for pore clays we have

$m_c = hS_c/2V_0 = \alpha h S_0/2V_0$. (31)

Here S_c is a square of surface of clay particles in the representative volume V_0, α is a part of clay in rock, $S_c = \alpha S_0$. Taking for $S_0 V_0^{-1} \approx 2 \cdot 10^9$ м $^{-1}$, Mitchell (1976), we receive in the defined coordinate system that

$m_c \approx \alpha/\sqrt{C}$. (32)

Solving chemical potential equations for concentrations in the solution between clay particles we receive

$\overline{C_1} = \sqrt{C} + \sqrt{C + C^2}, \overline{C_2} = -\sqrt{C} + \sqrt{C + C^2}$. (33)

For a diffusion in transport pores and in space between a clay particles we have an equations

$$\frac{\partial (mC_i)}{\partial t} + J_i = \frac{\partial}{\partial x}(mD\frac{\partial C_i}{\partial x}),$$

$$\frac{\partial (m_c \overline{C_i})}{\partial t} = J_i + \frac{\partial}{\partial x}[m_c D(\frac{\partial \overline{C_i}}{\partial x} + \frac{z_i \overline{C_i} F}{RT}\frac{\partial \varphi}{\partial x})],\ i=1, 2;$$

where J_i is an exchanging member "pore-clay". The right parts of the last equations have an unusual form because of the Nernst – Einstein law for a diffusion in electrically active environment

$$j_i = -D_i(\frac{\partial \overline{C_i}}{\partial x} + \frac{z_i \overline{C_i} F}{RT}\frac{\partial \varphi}{\partial x}),\ i=1, 2, \quad (34)$$

and additionally there are two more conditions

$$m_c \sum_i z_i \overline{C_i} = \alpha q S_0,\ \sum_i z_i j_i = 0,\ i=1, 2. \quad (35)$$

Because we suppose that $D_1 = D_2 = D$, $z_1 = +1$, $z_2 = -1$ we obtain an equation for a diffusion in clays

$$\frac{\partial}{\partial t}(mc + m_c\sqrt{C + C^2}) = D\frac{\partial}{\partial x}[(m + m_c\frac{\sqrt{C}}{\sqrt{C+1}})\frac{\partial C}{\partial x}] \quad (36)$$

Now let us find the law, which describes the behavior of m_c and m in dependence on a load and a concentration. Let's express the porosity m_c through the concentration C from (31):

$V_0 m_c \sqrt{C} = \alpha$. (37)

So far as shrinkage's definition $\dot{\theta} = \mathrm{d}(\ln V_0)/\mathrm{d}t$ leads to $V_0 = V_0^{(0)} \exp\theta$ then for the finished process of consolidation at the first stage the maximum shrinkage is $\theta_{\max} = G/\lambda$ then from the last but one equation for m_c we receive

$$m_c = \frac{\alpha}{V_0^{(0)}\sqrt{C}\exp(G/\lambda)} \approx \frac{\alpha}{V_0^{(0)}\sqrt{C}(1 + G/\lambda)}, \quad (38)$$

and m and m_c are connected by the early obtained correlation. Then

$m = \beta - m_c - \theta(1-\beta)$. (39)

After denoting

$$\beta - \theta(1-\beta) = \beta_0,\ \frac{\alpha}{V_0^{(0)}\exp(G/\lambda)} = \alpha_0, \quad (40)$$

we have

$$\frac{\partial}{\partial x}[D(\beta_0 - \frac{\alpha_0}{\sqrt{C}} + \frac{\alpha_0}{\sqrt{1+C}})\frac{\partial C}{\partial x}] = 0 \quad (41)$$

(This equation is correct for steady state diffusion process in clay). The solution of the equation (41) is

$$D(\beta_0 C - 2\alpha_0\sqrt{C} + 2\alpha_0\sqrt{1+C} = Qx + Q_0 \quad (42)$$

where Q, Q_0 = const.

Suppose that $0 \le x \le 1$. Then from the boundary conditions we obtain $Q_0>0$, $Q<0$. Let us find the form of *C(x)* function(or *x(C)*). After estimating the extreme points of the solution we obtain the form of the function *C(x)* as illustrated on fig.2. The point *A* responds to the jump of *C* from the one stable solution's branch to the second, and is a point of flocculation. The point x_0 responds to *C=0*. It's interesting that the graph of *C(x)* when $x \to x_0$ thrusts to the point x_0 under zero angle because the solution, for $C \to 0$, tends to infinity.

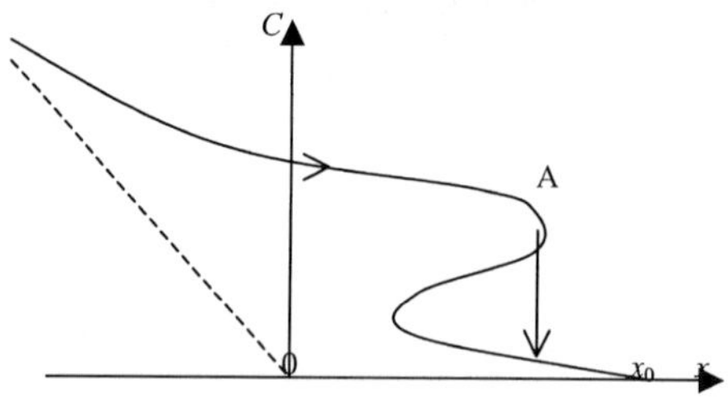

Figure 2. The curve of the concentration's alternation at diffusion in clay with a leap in the point A (point of the floculation)

It is an interesting fact that the parameter α_0 depends on load in conformity with its definition. If in (40) we take $G \to \infty$ then equation (36) can be written in the form

$$\partial C / \partial t = D \partial^2 C / \partial x^2 . \quad (43)$$

This equation is an ordinary equation of diffusion. The last fact is important for theoretical mineralogy. Physically such a transition responds to mineral transformation of swelling clays (smectites) into non-swelling clays under the exposure of rising pressure, because the transition of equation (36) into the equation of the diffusion opens possibilities for models of complex geological processes.

5. FILTRATION IN CLAYS

Correlations we have received in the previous part are useful in the analysis of the influence of different factors (concentration of solutions, pressure) on the penetrability of clay rocks. Let us use the formula of Archie in the form of

$$k = A \exp(n \ln 10 / 0{,}03), \quad A = \text{const}. \quad (44)$$

This formula contains the porosity of rocks. So far as we discern the porosity of transport pores and the porosity of interlayer films of water, then it is clear that $n = m_c + m$. So using $n = \beta - \theta(1-\beta) = \beta_0$ we see that if we want to receive the dependence of clay rocks on the concentration, we have to examine the character of flow in transport pores and pores, which are created with interlayer space of clay particles, in more detail. It was done in Khramchenkov (2000). The new thing we need to add to the results of this work is the ability to account for the influence of the reduction's load on curves of permeability's dependence upon concentration through the value of n according to the last formula. Usually for clay rocks we can consider that $m \ll m_c$, then from the last formula we receive

$$n \approx n_0 \exp(-G/\lambda), \quad n_0 = \alpha / V_0^{(0)} \sqrt{C} . \quad (45)$$

It is clear that the load G can be expressed through the depth H and density γ_u of rocks, which lay above: $G = \gamma_u H$. Then n_0 has a physical meaning of porosity of this type of clays at the zero depth (at the surface) and the last correlation for n takes the form

$$n \approx n_0 \exp(-\gamma H), \quad \gamma = \gamma_u \lambda^{-1}, \quad (46)$$

which fully coincides with the recommended dependence of clay rock at their depth.

6. CONCLUSIONS

It was important for us to offer a model of the physical and mechanical properties of clay which doesn't use a priori suppositions about its rheology, but to use only known facts about the properties of components of clayey rocks. Such a model was created by the unification of the theory of filtration consolidation and the theory of DLVO. On the basis of this model we solved a simple problem of clay's shrinkage under load and we showed that the model gives an ability to describe the rheology of clays, which was obtained experimentally. We created the mathematical model of diffusion in clay. It is shown that the solution of the final equation has a peculiarity (point of flocculation). We modified the model for filtration properties of clayey rocks, because we included the weight of rocks and clay layer shrinkage in the model as active factors. The results are in agreement with experiments. The developments are a physical substantiation of well-known hydrogeological schemes for the calculation of yield clay water into an aquifer.

7. ACKNOWLEDGEMENT

The author thanks INTAS, proj. num. 99-1810.

REFERENCES

Khramchenkov, M.G. 2000. *Mathematical simulation of clayey rock's transport and storage properties*. Environmental Geoscience 4: pp. 401 – 405 (in Russian).

Mitchell, J.K. 1976. *Fundamentals of Soil Behavior*. John Wiley & Sons, Inc. New York.

Nikolaevskiy, N.V. 1996. *Geomechanics and Fluidodynamics*. Kluwer Academic Publishers. Dordecht.

Norrish, K. 1954. *Swelling of montmorillonite*. Disc. Faraday Soc. 18: pp. 120 – 134.

VERIFICATION AND VALIDATION OF A THREE-DIMENSIONAL FINITE-ELEMENT CODE FOR COUPLED THERMO-HYDRO-MECHANICAL AND SALINITY (T-H-M-C) MODELLING IN FRACTURED ROCK MASSES

Tin Chan[1], Varut Guvanasen[2] and Frank W. Stanchell[1]
[1]Atomic Energy of Canada Limited (AECL)
[2]HydroGeoLogic Consulting, U.S.A.

Abstract: MOTIF is a three-dimensional finite-element code developed to simulate groundwater flow, heat transfer and solute transport in deformable fractured porous media. The code has been subjected to an extensive verification and updating programme since the onset of its development. In this paper, additional verification and validation works with an emphasis on thermo-hydro-mechanical processes are presented. The verification results are based on cases designed to verify thermo-hydro-mechanical coupling terms, and isothermal and non-isothermal consolidations. A number of validation case studies have been conducted on the code. Example results are reported in this paper.

1. INTRODUCTION

Over the past two decades AECL has developed a three-dimensional code, MOTIF (Model Of Transport In Fractured/porous media), for detailed modelling of groundwater flow, heat transport, mechanical equilibrium and solute transport in a fractured rock mass. The initial development was completed in 1985 (Guvanasen 1985). Since then the code has undergone extensive updating, verification – comparison with known analytical or numerical solutions – and validation – comparison with experiments – (Chan et al. 2000). In the latter document sixteen test cases were reported to verify the code for groundwater flow, heat transfer and solute transport in fractured or porous rock. In this paper, additional verification and validation studies with an emphasis on thermo-hydro-mechanical (T-H-M) processes are presented.

2. DESCRIPTION OF THE MOTIF CODE

The MOTIF code is a three-dimensional finite-element code capable of simulating steady state or transient coupled/uncoupled variable-density, variable- saturation fluid flow, heat transport, and conservative or non-conservative solute (including single species radionuclide) transport in deformable fractured/ porous media. In the code, the porous medium component is represented by hexahedral elements, triangular prism elements, tetrahedral elements, quadrilateral planar elements, and lineal elements. Discrete fractures are represented by biplanar quadrilateral elements (for the equilibrium equation), and monoplanar quadrilateral elements (for flow and transport equations).

The code solves four main governing equations: the equilibrium equation, the flow equation, the heat transport equation, and the solute transport equation. These four equations for deformable porous media are presented below. The treatment of discrete fractures is presented in Guvanasen and Chan (2000).

Generalized equations describing isothermal hydroelastic equilibrium in poroelastic materials were given by Biot (1955). These equations were subsequently extended to describe thermohydroelastic phenomena by assuming that: 1) the fluid and solid are in thermodynamic equilibrium, and 2) a change in fluid density has no effect on the solid strain. Based on these assumptions, the following equations can be written tensorially (Guvanasen and Chan 2000):

$$e_{ij} = C^{-1}_{ijkl}\left(\tau_{kl} - \tau^{o}_{kl}\right) + \Delta p U_{ij} + \Delta T W_{ij} \qquad (1)$$

$$e_{ij} = \tfrac{1}{2}\left(\frac{\partial u_i}{\partial x_j} + \frac{\partial u_j}{\partial x_i}\right) \qquad (2)$$

$$\tau_{kl} = C_{ijkl} e_{kl} - \Delta p U^{*}_{ij} - \Delta T W^{*}_{ij} + \tau^{o}_{ij} \qquad (3)$$

$$U^{*}_{ij} = C_{ijkl} U_{ij};\ W^{*}_{ij} = C_{ijkl} W_{ij} \qquad (4)$$

where e_{ij} is the strain tensor, u_i the deformation in the i-th direction, C_{ijkl} the elastic constant tensor, τ_{ij} the stress tensor, τ_{ij}^{o} the initial stress tensor, U_{ij} the isothermal hydroelastic constant tensor, W_{ij} the isothermal thermoelastic constant tensor, x_i the i-th direction coordinate, p the fluid pressure, p_{ref} the reference fluid pressure, $\Delta p = p - p_{ref}$, T the temperature, T_{ref} the reference temperature, $\Delta T = T - T_{ref}$.

The stress expression in Equation (3) is used in the following quasi-static equilibrium equation assuming the acceleration term to be negligible.

$$\frac{\partial \tau_{ij}}{\partial x_j} + F_i = 0 \qquad (5)$$

where F_i is the body force term in the i-th direction.

The constitutive relationships for discrete fractures are based on those of Barton et al. (1985). Finite-element implementation is detailed in Guvanasen and Chan (2000).

Based on the principle of mass conservation and Darcy's law, a modified form of Biot's (1955) equation for flow through deformable porous media is written tensorially as (Guvanasen and Chan 2000):

$$\frac{\partial \rho_f \zeta}{\partial t}+\frac{\partial}{\partial x_i}\frac{\rho_f k_{ij}\rho_o g}{\mu_f}\frac{\partial}{\partial x_j}(h+\delta x_3)-\rho_f Q=0 \tag{6}$$

$$\zeta=U^*_{ij}e_{ij}+\frac{1}{M}\Delta p-\left(U^*_{ij}e_{ij}+\theta_o\beta_f\right)\Delta T \tag{7}$$

$$\frac{1}{M}=\frac{1}{R}-U^*_{ij}U\approx\frac{\theta_o}{\rho_f}\frac{\partial \rho_f}{\partial p}=\theta_o C_f \tag{8}$$

$$\delta=\frac{\rho_f-\rho_o}{\rho_o};\ h=\frac{p}{\rho_o g}+x_3 \tag{9}$$

where: M is the bulk hydroelastic modulus for fluid, R Biot's isothermal hydroelastic constant, θ_o the initial porosity, C_f the fluid compressibility, k_{ij} the permeability tensor, t the time, ζ the change in volumetric fluid content per unit volume, ρ_f the fluid density, ρ_o the reference fluid density, β_f the fluid volumetric thermal expansion coefficient, g the gravitational acceleration, μ_f the fluid dynamic viscosity, h the reference hydraulic head, and Q the specific volumetric fluid generation rate.

Assuming negligible conversion between hydromechanical and thermal energies and thermodynamic equilibrium between the fluid and solid phases, the equation for heat transport in porous media can be written (Combarnous and Bories 1975):

$$\frac{\partial}{\partial t}\left[\theta\rho_f c_f+(1-\theta)\rho_s c_s\right]T+\frac{\partial}{\partial x_i}q_i\rho_f c_f T$$
$$-\frac{\partial}{\partial x_i}H_{ij}\frac{\partial T}{\partial x_j}-Q_H=0 \tag{10}$$

$$H_{ij}=c_f\rho_f a_{ijkl}\frac{q_k q_l}{\sqrt{q_m q_m}}+\lambda^T_{ij} \tag{11}$$

where: c_f is the specific heat capacity for fluid, c_s the specific heat capacity for solid, ρ_s the solid skeleton density, H_{ij} the thermal hydrodynamic dispersion tensor, q_i the Darcy velocity, a_{ijkl} the dispersivity tensor, $\lambda^T{}_{ij}$ the thermal conductivity tensor, and Q_H the specific enthalpy production rate.

Based on the principle of mass conservation, the equation for non-conservative solute transport in porous media can be written (Bear 1972):

$$\frac{\partial}{\partial t}\left[\theta+(1-\theta)K_d\rho_s\right]C+\frac{\partial}{\partial x_i}q_i C \tag{12}$$
$$-\frac{\partial}{\partial x_i}D_{ij}\frac{\partial C}{\partial x_j}+\lambda\left[\theta+(1-\theta)K_d\rho_s\right]C-Q_C=0$$

$$D_{ij}=a_{ijkl}\frac{q_k q_l}{\sqrt{q_m q_m}}+\theta T_{ij}D_d \tag{13}$$

where: C is the solute concentration, K_d the distribution coefficient, D_{ij} the hydrodynamic dispersion tensor, T_{ij} the tortuosity tensor, D_d the molecular diffusion coefficient in free water, λ the degradation rate constant, and Q_C the specific solute production rate. The distribution coefficient, K_d, may be of Freundlich or Langmuir type or a general function of solute concentration.

3. VERIFICATION CASES

In this section we present a number of hitherto unpublished verification tests and a summary of some validation examples. The presentation is divided into three levels: single-element verification, multiple-element verification and validation case studies.

3.1 Level 1 - Verification of thermohydromechnical coupling

A series of one-element tests was conducted to verify the T-H-M coupling in Equation (7) (the flow equation). The tests were conducted only for the case of isotropic media in the transient mode.

The test problem was initially simplified by assuming no lateral flow so that only the storage terms with all the coupling parameters may be isolated from the rest of the equation. For an isotropic medium with no flow, integrating Equation (6) using Equations (7) and (8) over a unit time step with an assumption that the change in fluid density is negligible leads to:

$$\zeta=\alpha\varepsilon_V+\frac{1}{M}\Delta p-(3\alpha\beta_s+\theta_0\beta_f)\Delta T=Q \tag{14}$$

$$\alpha=\frac{E}{3(1-2\nu)H} \tag{15}$$

where α is Biot's coupling coefficient, E Young's modulus, H the hydroelastic modulus, ε_V the volumetric strain, and β_s the linear thermal expansion coefficient of the solid.

In the event that mechanical stresses are equal in all directions, the principal mechanical stresses may be determined using the following expression:

$$\tau_{11} = \tau_{22} = \tau_{33} = \alpha(H\varepsilon_V - \Delta p - 3\beta_s H\Delta T) \qquad (16)$$

To test the coupling in the storage terms of the flow equation, five test cases were used for verification purposes: 1) Isothermal Undrained Test (UIT), 2) Isothermal Drained Test (DIT), 3) Isothermal Injection Test (INJ), 4) Non-isothermal Undrained Test (UNT) and (5) Non-isothermal drained test (DNT).

The tests were performed using a single element with dimensions 1 m x 1 m x 1 m. Parameters used in the tests are: E = 35 GPa, ν = 0.25, H = 29.2 GPa, α = 0.8, c_f = 4.5 x 10^{-10} Pa^{-1}, θ = 0.005, β_s = 10^{-6} °C^{-1}, β_f = 2.4 x 10^{-6} °C^{-1} and ΔT = 10 °C.

In the UIT, an incremental confining pressure was applied to the unit element. At the same time, no fluid was allowed to leave the element. The applied confining pressure was defined as a pressure that resulted in a displacement of -2.5 × 10^{-7} m in all directions.

In the DIT, an incremental confining pressure of 1 MPa was applied to the unit element. At the same time, the fluid was maintained at 0 m on the boundary.

In the INJ, the confining pressure was kept constant. At the same time, the fluid was elevated to 100 m on the boundary.

In the UNT, an incremental confining pressure was applied to the unit element. At the same time, no fluid was allowed to leave the element. The applied confining pressure was defined as a pressure that resulted in a displacement of -2.5 × 10^{-7} m in all directions. A temperature elevation of 10 °C was imposed on the unit element.

In the DNT, an incremental confining pressure of 1 MPa was applied to the unit element. At the same time, the fluid was maintained at 0 m on the boundary. A temperature elevation of 10 °C was imposed on the unit element.

As shown in Table 1, principal mechanical stresses, change in specific volumetric water content, and fluid pressure were accurately determined by MOTIF. The analytical values in the table were calculated using Equations (14) and 16).

Table 1. Summary of results for Level 1 verification

Test		Principal stress, τ_{11}, τ_{22}, and τ_{33} (Pa)	Pressure head (m)	Change in volumetric water content (d'less)
UIT	A	-2.31×10^5	27.26	0
	M	-2.31×10^5	27.18	0.7×10^{-17}
DIT	A	-1×10^6	0	-3.43×10^{-5}
	M	-1×10^6	0	-3.44×10^{-5}
INJ	A	0	100	-2.90×10^{-5}
	M	-6×10^{-11}	100	-2.92×10^{-5}
UNT	A	-1.37×10^7	1663.5	0
	M	-1.37×10^7	1652.3	-5×10^{-13}
DNT	A	-1×10^6	0	-4.63×10^{-5}
	M	-1×10^6	0	-4.64×10^{-5}

Notes: A – Analytical, M – MOTIF.

3.2 Level 2 – Verification of isothermal and non-isothermal consolidation

Two transient simulations were conducted in addition to the one-element verification cases. The two simulations comprise: one-dimensional isothermal consolidation simulation; and one-dimensional non-isothermal consolidation simulation. These two simulations were designed to examine the coupling between two dominant modules: the equilibrium module and the flow module. The heat transport module is weakly coupled to the flow module via the velocity terms; and unilaterally coupled with the equilibrium module via the thermal expansion terms.

One-dimensional isothermal consolidation simulation

This is a particular case of a column of soil supporting a load (per unit area) of p_o and confined laterally in a sheath so that no lateral expansion can occur. It is also assumed that no water can escape laterally or through the bottom, while it is free to escape at the upper surface by applying the load through a very porous slab.

This case was simulated by assuming that the column was oriented horizontally so that the column was confined in the x_2 (y) and x_3 (z) directions. The upper surface of the column corresponds to x_1 = 0, and the bottom of the column is located at x_1 = h.

An analytical solution of this scenario was presented by Biot (1941) and is reproduced (with some modifications) below:

$$p = -4\frac{\sigma_o}{\pi}\sum_{n=1}^{\infty}\frac{(-1)^n}{(2n-1)}\exp\left(\frac{-(2n-1)^2\pi^2 ct}{4h^2}\right)\bullet$$

$$\bullet\cos\left(\frac{(2n-1)(h-x)\pi}{2h}\right) \qquad (17)$$

where:

$\sigma_o = \dfrac{a-a_i}{\alpha a}p_o$, $a = \dfrac{1-2\nu}{2G(1-\nu)}$, and

$$a_i = \frac{a}{1+\dfrac{\alpha^2 a}{\theta_o c_f}}, \quad \frac{1}{c} = \alpha^2 a \frac{\mu}{k} + \theta_o c_f \frac{\mu}{k}$$

and p is the water pressure.

The total deformation, u, at $x_1 = 0$ is given by:

$$u = ap_o h - \frac{8}{\pi^2}(a - a_i)hp_o \sum_{n=1}^{\infty} \frac{1}{(2n-1)^2} \exp\left(\frac{-(2n-1)^2 \pi^2 ct}{4h^2} \right) \quad (18)$$

In the interest of brevity, the reader is referred to Guvanasen and Chan (2000) for the list and values of parameters used in the simulation. Ten uniform 0.1 m×0.1 m×0.1 m eight-noded brick elements (along the x_1 direction) were used in the simulation. A spatial distribution of water pressure heads (in m of water) at 15 s is presented in Figure 1, in which it can also be seen that the MOTIF-calculated pressure profile is in excellent agreement with the analytical solution. Figure 2 shows a plot of soil column displacement immediately adjacent to the porous slab (at $x_1 = 0$) versus time. The MOTIF solution agrees remarkably well with the analytical solution.

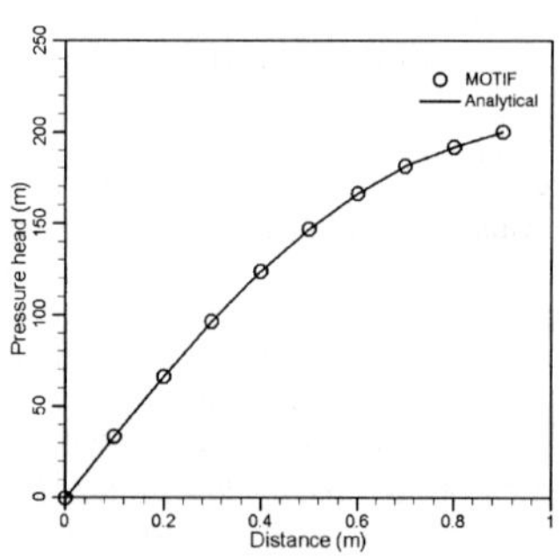

Figure 1. Isothermal consolidation verification: head vs distance at 15 s.

One-dimensional non-isothermal consolidation simulation

This case is identical to the above isothermal case; however, the soil column is heated uniformly by an internal, uniformly distributed heat source with an intensity of 100 kW/m^3. All the boundaries surrounding the column are assumed to be adiabatic. An analytical solution of this scenario was derived using Laplace transformation. Details of this verification case may be found in Guvanasen and Chan (2000).

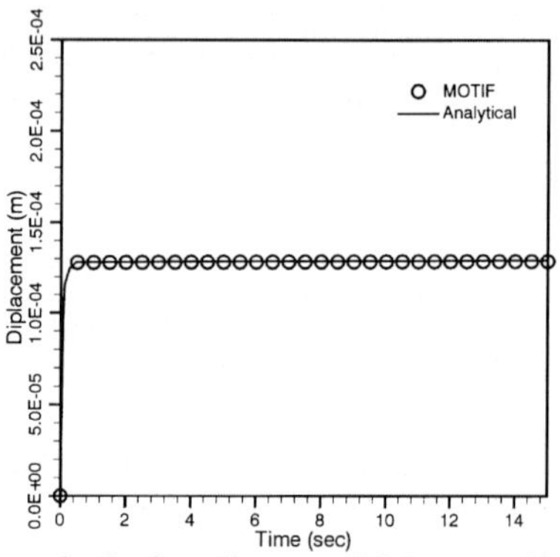

Figure 2. Isothermal consolidation verification: diplacement vs time at $x_1 = 0$.

3.3 Level 3 - Validation case studies

This section deals with the comparison of MOTIF modelling results with three field experiments and a large bench-scale experiment. While the inherent experimental uncertainties in fractured rock environments are such that an absolute validation is difficult to achieve, the comparison provides some confidence in the suitability of the code to model various subsurface processes in fractured or porous rock.

In the AECL URL (Underground Research Laboratories) Drawdown Experiment Program, MOTIF was used to predict the hydrogeologic disturbances caused by the excavation of the URL shaft to a depth of 255 m (Davison and Guvanasen 1985). Planar elements were used to represent the fracture zones and 3D continuum elements were used to represent the background rock mass. In general, the predicted drawdown histories agreed well with measurements, Figure 3. The overall trend of observed seepage into the shaft also agreed well with the predictions. However, the magnitude of the seepage rate was generally over-predicted by a factor of three. Possible causes for the various discrepancies may include the difference between the assumed and actual excavation schedule, the existence of vertical fractures connecting the shaft with a major fracture zone, and reduction of the hydraulic conductivity near the excavation due to stress concentration (Chan et. al. 1987).

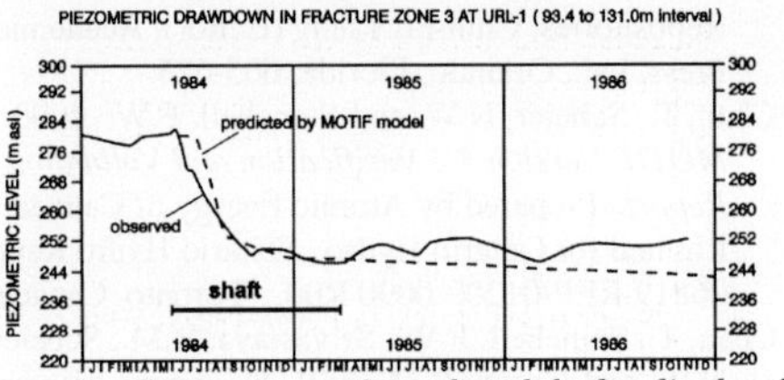

Figure 3. Comparison of predicted hydraulic head history with observation.

The MOTIF code is also being used in modelling groundwater tracer tests in the Moderately Fractured Rock (MFR) Experiment Program. These tests were conducted in a block of moderately fractured rock located between two major low-dipping fracture zones at the URL site. Tracer Test 2 (TT2) involved an injection of conservative tracers into a single borehole interval and withdrawal from intervals in three boreholes. The distances between injection and withdrawal intervals ranged from 34 m to 58 m. Figure 4 shows a comparison between the predicted normalized tracer breakthrough curves (deterministic and one stochastic realization, Chan et al. 2001) and measurement.

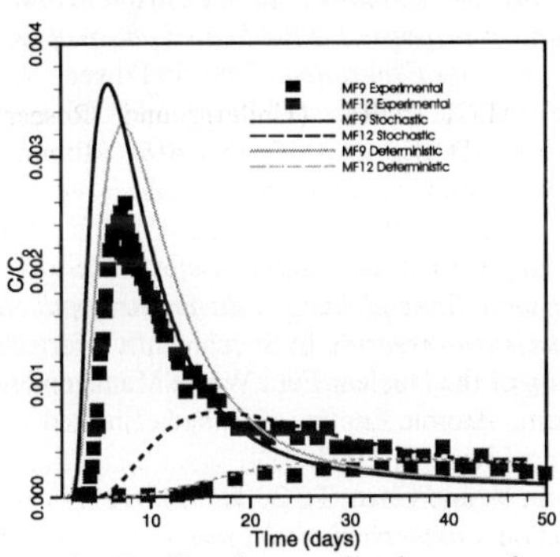

Figure 4. Predicted normalized tracer breakthrough curves (deterministic and stochastic realization #70) compared with measurement.

Another example is the participation in (validation) Test Case 6 (TC6) (Rutqvist et al. 1996) of the international DECOVALEX I project.

Hydraulic tests were conducted at 356-m depth on a single fracture intersecting a 500-m deep borehole in crystalline rock at the Rock Mechanics Laboratory of Lulea University of Technology in Sweden. Three types of hydraulic tests were included: (i) pulse test, (ii) step-pressure (hydraulic jacking) test and (iii) constant pressure test. MOTIF was used to simulate tests (ii) and (iii). Figure 5 shows the general agreement between the results of MOTIF modeling and field test of hydraulic jacking. The modeling results of the constant pressure injection test also agree well with the field test (not illustrated). In particular, the model properly simulated the increase of well flow with time, due to fracture opening. This would not be possible for a model that does not simulate fracture deformation.

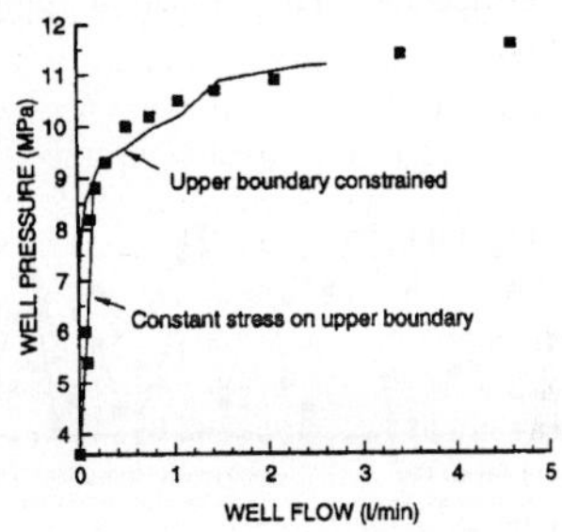

Figure 5. Comparison of Modelling and Field Test Results of Hydraulic Jacking

A recent example is coupled T-H-M modelling of the Tunnel Sealing Experiment (TSX) in URL (Guo et al. 2002). To provide data for preliminary validation a surface laboratory experiment known as the Thermal Evaluation of Material Test (TEMT) was conducted. A steel vessel, 1.47 m long with an interior diameter of 0.74 m, was filled with a medium-grained sand and heated by circulating hot water. An array of thermistors monitored the evolution of temperature. Comparison between the physical test results and MOTIF simulation results, Figure 6, shows that MOTIF can be used to simulate convection dominated heat transfer in a porous medium.

4. CONCLUSIONS

The six verification tests presented in this paper, along with other verification tests presented in the cited literature, have established the ability of the MOTIF finite-element code to accurately solve the equations of groundwater flow, solute and heat transport, and coupled thermo-hydro-mechanical phenomena in porous media, or in fractured media that can be adequately characterized by equivalent porous media (EPM) elements or a combination of EPM and discrete fracture elements.

MOTIF has been applied to several field experiments in fractured rock settings. These tests provide some confidence in the suitability of the model to fractured rock. However, the inherent experimental uncertainties in fractured rock environments are such that an absolute validation is difficult to achieve. In the tests noted here, at least one model realization of the geology was found that was in reasonable agreement with the known characterization and the measured results. Further efforts to improve the validation of the code against experiments, as well as further participation in international verification and validation projects, would be beneficial.

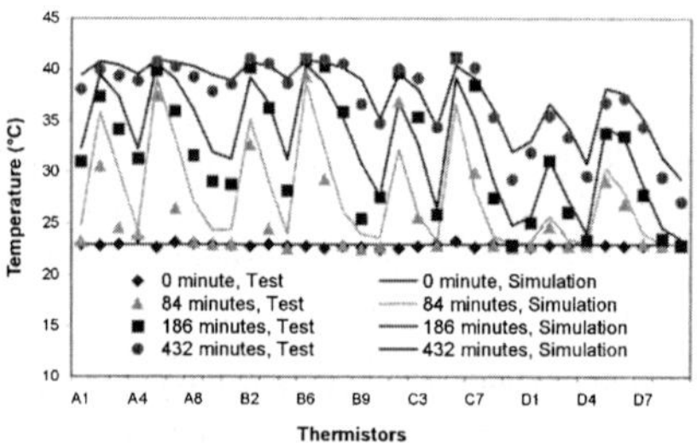

Figure 6. *Comparison of simulated and measured temperatures for 33 thermistors in the TEMT*

ACKNOWLEDGEMENT

The authors are grateful to Ontario Power Generation for funding support.

REFERENCES

Barton, N.R., Bandis, S.C., and Bakhtar, K., 1985. *Strength, deformation, and conductivity of rock joints*, Int. J. Rock Mech. Min. Sci. and Geomech. Abstr., 22(3): 249-268.

Bear, J., 1972. *Dynamics of fluids in porous media.* Elsevier, New York, NY, 764 pp.

Biot, M.A. 1941. *General theory of three-dimensional consolidation.* Journal of Applied Physics 12: pp. 155-64.

Biot, M.A., 1955. *Theory of elasticity and consolidation for a porous anisotropic solid*, J. Applied Physics, Vol. 26, pp.182-185.

Chan, T., Reid, J.A.K. and Guvanasen, V. 1987. *Numerical modelling of coupled fluid, heat, and solute transport in deformable fractured rock.* In Coupled Processes Associated with Nuclear Waste Repositories, Chin-Fu Tsang (Editor), Academic Press, Inc., Orlando, Florida, 605-625.

Chan, T., Scheier, N.W. and Stanchell, F.W. 2000. *MOTIF Version 3.2 Verification and Validation Report.* Prepared by Atomic Energy of Canada Limited for Ontario Hydro. Ontario Hydro Report 06819-REP-01200-0090 R00. Toronto, Canada.

Chan, T., Stanchell, F.W., Srivastava, R.M., Scheier, N.W. and Jensen, R.M. 2001. *Stochastic flow and transport simulations of a three-dimensional tracer test in moderately fractured plutonic rock.* In Monestiez, P., Allard, D. & Froidevaux, R. (ed.). GeoENV III – Geostatistics for Environmental Applications, pp. 205-216. Dordrecht: Kluwer Academic.

Combarnous, M.A., and Bories, S.A., 1975. *Hydrothermal convection in saturated porous media.* Advances in Hydroscience, Vol. 10, edited by V.T. Chow, Academic Press, New York, NY, pp.231-307.

Davison, C.C. and Guvanasen, V. 1985. *Hydrogeological characterization, modeling and monitoring of the site of Canada's Underground Research Laboratory.* Atomic Energy of Canada Limited report, AECL-8676.

Guo, R., Chandler, N.A. & Dixon, D.A. 2002. *Modelling the thermally induced hydraulic and mechanical response for the heated phase of the Tunnel Sealing Experiment.* Ontario Power Generation Nuclear Waste Management Division Report 06819-REP-01200-10095-R00. Atomic Energy of Canada Limited for Ontario Power Generation, Toronto.

Guvanasen, V. 1985. *Development of a three-dimensional finite-element code and its application to geoscience research.* In Seventeenth Information Meeting of the Nuclear Fuel Waste Management Program. Atomic Energy of Canada Limited Technical Record, TR-299.

Guvanasen, V. and Chan, T., 2000. *A three-dimensional numerical model for thermohydromechnical deformation with hysteresis in a fractured rock mass*, Int. J. Rock Mech. and Mining Sc., 37: 89-106.

Rutqvist, J., Follin, S., Khair, K., Nguyen, S. and Wilcock, P. 1996. *Experimental investigation and mathematical simulation of a borehole injection test in deformable rocks.* In Coupled Thermo-Hydro-Mechanical Processes of Fractured Media, O. Stephansson, L. Jing and C.-F. Tsang (Editors), Elsevier, 425-448.

WATER FLOW AND DIFFUSION PROBLEM IN BENTONITE: MOLECULAR SIMULATION AND HOMOGENIZATION ANALYSIS

Y. Ichikawa[1], S. Prayongphan[1], K. Kawamura[2], K. Kitayama[3]
[1])Nagoya University, Nagoya 464-8603, Japan
[2]) Tokyo Institute of Technology, Tokyo 152-8551, Japan
[3]) Nuclear Waste Management Organization of Japan, Tokyo 108-0014, Japan

Abstract: We here treat a diffusion problem coupled with water flow in bentonite. The remarkable behavior originates from molecular characteristics of its constituent clay mineral, namely montmorillonite, and we show the behavior based on a unified simulation procedure starting with the molecular dynamic (MD) method and extending the obtained local characteristics to a macroscale behavior by the multiscale homogenization analysis (HA: Sanchez-Palencia, 1980). Not only the macroscale effective diffusion property but also the adsorption behavior is well defined based on this method.

1. INTRODUCTION

Saturated bentonite consists of hydrated smectite minerals, mainly montmorillonite, and quartz particles, called the macro-grain. See a schematic diagram of the microstructure given in Figure 1.

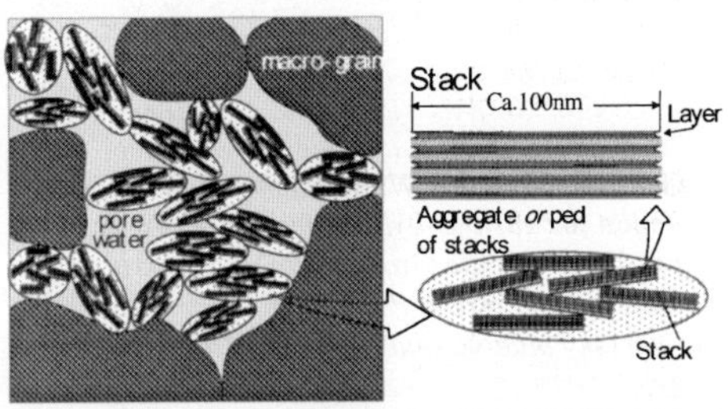

Figure 1. Microstructure of bentonite.

The bentonite shows peculiar characteristics of water permeability and chemical transport of many cations and anions. These are caused by the surface characteristics of smectite minerals. One mineral is a nanometer scale of platelet, called the layer, with the thickness ca.1nm and the width ca.100x100nm. Commonly the surface is negatively charged. Some layers form a stack, and a group of stacks constitutes an aggregate or ped.

The water flow and diffusion behavior of such a finely microinhomogeneous material can be understood through the following procedures; 1) a characterization of the microstructure, 2) an identification of the local material properties based on experiment or simulation, and 3) an extension of the microscale characteristics to the macroscale behavior.

For characterizing the microstructure we use a confocal laser scanning microscope (CLSM). By CLSM we can specify a 3-D configuration under atmospheric condition. Smectite minerals are extremely fine and poorly crystallized, so it is difficult to determine the properties by experiment. We inquire into the physicochemical properties by a molecular dynamics (MD) simulation method. Then, we develop a multiscale homogenization analysis (HA) method to extend the microscopic characteristics to the macroscopic behavior. We show numerical examples of a coupled water-flow and diffusion problem.

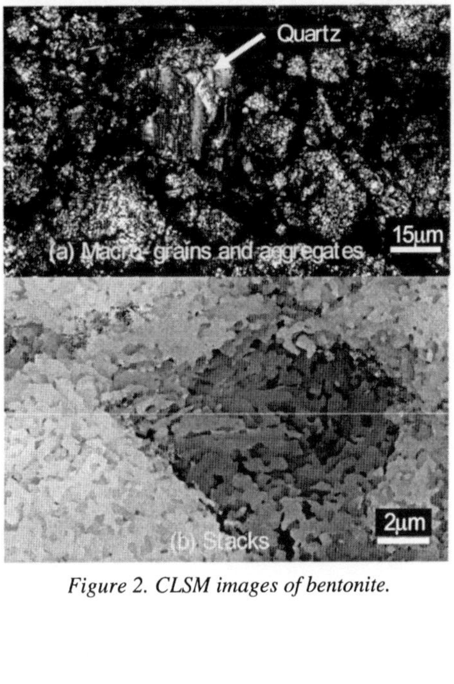

Figure 2. CLSM images of bentonite.

2. MD SIMULATION OF HYDRATED SMECTITE

In a MD system several thousands of atoms or molecules, called particles, are provided in a unit cell under 3-D periodic boundary conditions. We solve the equations of motion of all particles in the system. Interactions between atoms are given by atom-atom potential models including Coulomb, van der Waarls, short range repulsion and radial covalent terms:

$$u_{ij}(r_{ij}) = z_i z_j e_2 / 4\pi\varepsilon_0 r_{ij} \qquad \text{[Coulomb term]}$$
$$- c_i c_j / r^6 \qquad \text{[van der Waarls term]}$$
$$+ f_0 (b_i + b_j) \exp\{(a_i + a_j - r_{ij})/(b_i + b_j)\} \qquad \text{[short range repulsion term]}$$
$$+ D_{1ij} \exp(-\beta_{1ij} r_{ij}) + D_{2ij} \exp(-\beta_{2ij} r_{ij}) + D_{3ij} \exp[-\beta_{3ij}(r_{ij} - r_{3ij})^2] \qquad \text{[radial covalent term]}$$

A 3-body term is added to H-O-H interaction both for intra-atomic and hydrogen bonds because of the sp3 hybrid orbital:

$$u_{jik}(\theta_{jik}, r_{ij}, r_{ik}) = -f_a[\cos\{2(\theta_{jik} - \theta_0)\}-1]\,(k_1 k_2)^{1/2}$$
$$k_i(r_{ij}) = 1 / [\exp\{g_r(r_{ij} - r_m)\}+1]$$

Here r_{ij} is the distance of i-j atoms, and θ_{jik} is the angle of i-j-k atoms. Parameters {z, a, b, c} for each atom, $\{D_1, D_2, D_3, \beta_1, \beta_2, \beta_3, r_3\}$ for each atom pair, and $\{f_k, \theta_0, g_r, r_m\}$ for a 3-body term, are determined to reproduce structural and physical properties of several oxide crystals (ice-Ih, ice-II, ice-IX, quartz, corundum, gibbsite, feldspars, muscovite, etc.) and water. In details, see Kawamura & Ichikawa (2001).

(NPT)-ensemble MD calculations are carried out at temperature 273K and 0.1MPa pressure (1atm). Pressure and temperature are controlled by scaling of basic cell parameters and scaling of atom velocities, respectively. The differential equation of motion is solved by a finite difference method of Verlet algorithm. A time increment for the difference equation is chosen as 0.4fs. The long-range Coulomb interaction is calculated by Ewald method.

We here use a beidellite system as the smectite mineral instead of montmorillonite because in our target bentonite both are included and have similar properties. The molecular formulation with *n*-mole water is denoted as $Na_{1/3} Al_2 [Si_{11/3} Al_{1/3}] O_{10} (OH)_2 \cdot nH_2O$ (*n* is called the hydration number). For this hydrated beidellite we investigate the structure of interlayer water molecules and exchangeable cations, diffusive motions of interlayer molecules and ions, swelling properties, viscosity of water, properties of mixed system of water and clay, etc. Some snap shots are given in Figure 3 for n=0, 0.333, 1, 3 and 7. Then we calculate distributions of water viscosity and diffusion coefficient near the clay surface by using n=5000 system (Figure 4). We observe a high viscosity near the surface.

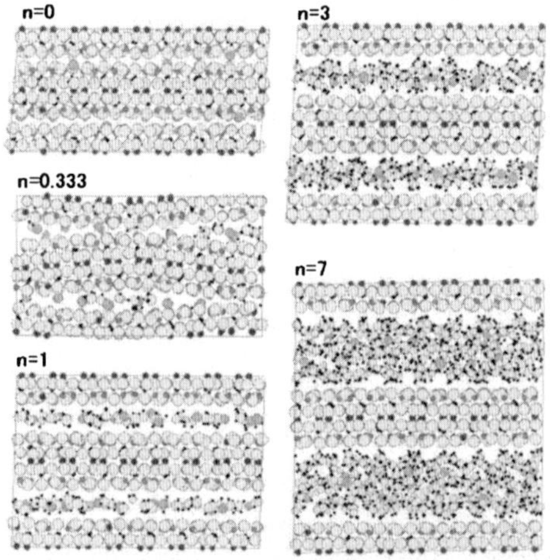

Figure 3. Structural snap shots of hydrated beidellite for various hydration numbers n at 293K and 0.1MPa. Circles are hydrogen (dark blue •), silicon (yellow green ◦), aluminium (pink ○), sodium (sky blue ○), and oxygen (yellow ○) atoms.

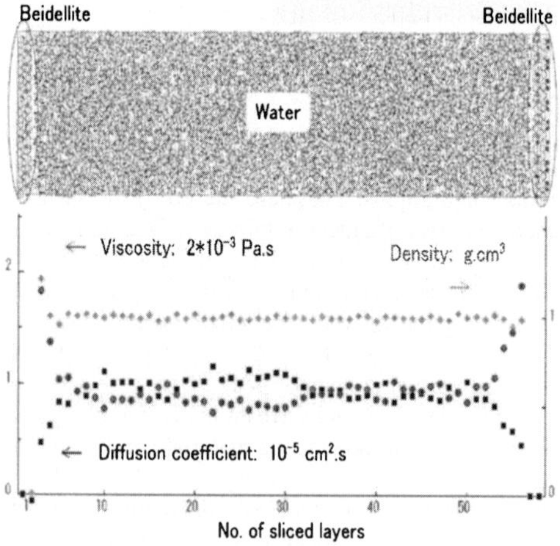

Figure 4. Snap shot (upper) and distributions of viscosity, diffusion coefficient and density.

3. MULTISCALE HOMOGENIZATION IN FLOW PROBLEM

The diffusion in a porous media is affected by the flow field, so we treat here the seepage flow problem in the bentonite starting with the Stokes' equation and applying the multiscale HA method.

We assume mesoscale and microscale domains are periodic (Figure 5). Let $\boldsymbol{x}^0$, $\boldsymbol{x}^1$, and $\boldsymbol{x}^2$ be coordinate systems defined in macro-domain Ω_0, meso-domain Ω_1 and micro-domain Ω_2. If we perform a laboratory seepage/consolidation test on a typical Japanese bentonite, Kunigel V1, the macroscale specimen is of scale 10^{-1}m, the mesoscale quartz grains are of 10^{-4}m, and the microscale clay minerals are of 10^{-8}m. So, we estimate the scale factor $\varepsilon=10^{-4}$, and introduce the relations $\boldsymbol{x}^1=\boldsymbol{x}^0/\varepsilon$, $\boldsymbol{x}^2=\boldsymbol{x}^1/\varepsilon$.

Since the flow velocity is very slow in bentonite, the flow field in the fluid domain Ω_f is given by the following incompressible Stokes' equation:

$$-\frac{\partial P^\varepsilon}{\partial x_i}+\frac{\partial}{\partial x_j}\left(\eta\frac{\partial V_i^\varepsilon}{\partial x_j}\right)+F_i=0,\qquad \frac{\partial V_i^\varepsilon}{\partial x_i}=0, \tag{1}$$

where V_i^ε is the velocity vector with the viscosity η whose distribution in the microscale domain is determined by MD, P^ε the pressure, and F_i the body force.

Since we have multiscale systems $\boldsymbol{x}^0$, $\boldsymbol{x}^1$, and $\boldsymbol{x}^2$, the differentiation in (1) changes to

$$\frac{\partial}{\partial x_i}=\frac{\partial}{\partial x_i^0}+\frac{1}{\varepsilon}\frac{\partial}{\partial x_i^1}+\frac{1}{\varepsilon^2}\frac{\partial}{\partial x_i^2}, \tag{2}$$

Then we introduce an asymptotic expansion

$$\begin{aligned}\boldsymbol{V}^\varepsilon(\boldsymbol{x})&=\varepsilon^4\boldsymbol{V}^0(\boldsymbol{x}^0,\boldsymbol{x}^1,\boldsymbol{x}^2)+\varepsilon^5\boldsymbol{V}^1(\boldsymbol{x}^0,\boldsymbol{x}^1,\boldsymbol{x}^2)+..,\\ P^\varepsilon(\boldsymbol{x})&=P^0(\boldsymbol{x}^0)+\varepsilon^1P^1(\boldsymbol{x}^0,\boldsymbol{x}^1,\boldsymbol{x}^2)+..,\end{aligned} \tag{3}$$

where $\boldsymbol{V}^\alpha(\boldsymbol{x}^0,\boldsymbol{x}^1,\boldsymbol{x}^2)$ and $P^\alpha(\boldsymbol{x}^0,\boldsymbol{x}^1,\boldsymbol{x}^2)$ $(\alpha=0,1,2,\cdots)$ are periodic functions.

Substituting (2) and (3) into $(1)_1$ yields

$$\begin{aligned}&\varepsilon^{-1}\text{-term:}\ \frac{\partial P^0}{\partial x_i^1}+\frac{\partial P^1}{\partial x_i^2}=0,\\ &\varepsilon^0\text{-term:}\ \frac{\partial}{\partial x_j^2}\left(\eta\frac{\partial V_i^0}{\partial x_j^2}\right)-\frac{\partial P^2}{\partial x_i^2}=\frac{\partial P^0}{\partial x_i^0}+\frac{\partial P^1}{\partial x_i^1}-F\end{aligned} \tag{4}$$

because P^0 is a function of only $\boldsymbol{x}^0$. $(4)_1$ implies P^1 is a function of $\boldsymbol{x}^0$ and $\boldsymbol{x}^1$. Then we introduce characteristic functions $v^k_i(\boldsymbol{x}^2)$ and $p^k(\boldsymbol{x}^2)$ by

$$\begin{aligned}V_i^0&=-\left(\frac{\partial P^0}{\partial x_k^0}+\frac{\partial P^1}{\partial x_k^1}-F_k\right)v_i^k(\boldsymbol{x}^2),\\ P^2&=-\left(\frac{\partial P^0}{\partial x_k^0}+\frac{\partial P^1}{\partial x_k^1}-F_k\right)p^k(\boldsymbol{x}^2)\end{aligned} \tag{5}$$

Substituting (5) into $(4)_2$ yields the following *microscale Stokes' equation*:

$$\frac{\partial}{\partial x_j^2}\left(\eta\frac{\partial v_i^k}{\partial x_j^2}\right)-\frac{\partial p_2^k}{\partial x_i^2}+\delta_{ki}=0\quad\text{in }\Omega_2 \tag{6}$$

Similarly substituting (2) and (3) into the incompressible condition $(1)_2$ yields

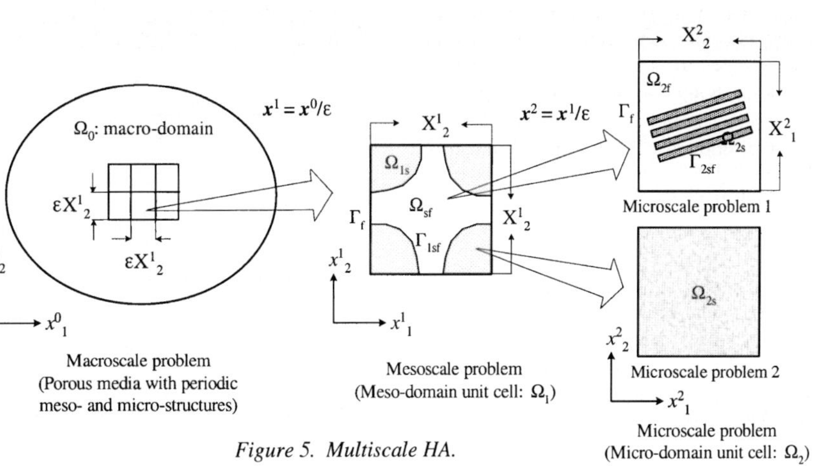

Figure 5. Multiscale HA.

$$\varepsilon^2\text{-term:}\quad \frac{\partial V_i^0}{\partial x_i^2}=0,$$
$$\varepsilon^3\text{-term:}\quad \frac{\partial V_i^0}{\partial x_i^1}+\frac{\partial V_i^1}{\partial x_i^2}=0, \qquad (7)$$
$$\varepsilon^4\text{-term:}\quad \frac{\partial V_i^0}{\partial x_i^0}+\frac{\partial V_i^1}{\partial x_i^1}+\frac{\partial V_i^2}{\partial x_i^2}=0$$

$(7)_1$ and $(5)_1$ give the following *microscale incompressible equation*:

$$\frac{\partial v_i^k}{\partial x_j^2}=0 \quad \text{in } \Omega_2 \qquad (8)$$

We solve (6) and (8) under periodic boundary conditions, and determine $v^k_i(\boldsymbol{x}^2)$ and $p^k(\boldsymbol{x}^2)$.

Next let us introduce average operations for meso-domain Ω_1 and micro-domain Ω_2 by

$$\langle \bullet \rangle_1=\frac{1}{|\Omega_1|}\int_{\Omega_1}\bullet\, dx^1, \qquad \langle \bullet \rangle_2=\frac{1}{|\Omega_2|}\int_{\Omega_2}\bullet\, dx^2$$

Averaging $(7)_3$ and applying periodic conditions yield the following *multiscale Darcy's equation*

$$-\frac{\partial}{\partial x_j^0}\left\{K_{ji}\left(\frac{\partial P^0}{\partial x_j^0}-F_j\right)\right\}=0 \quad \text{in } \Omega_0, \qquad (9)$$

$$K_{ji}=\langle K_{ji}^2\rangle_1,$$
$$K_{ji}^2=\langle v_i^j\rangle_2, \ \langle V_i^0\rangle_2=-K_{ji}^2\left(\frac{\partial P^0}{\partial x_j^0}-F_j\right) \qquad (10)$$

Comparing the homogenized K_{ji} with the conventional Darcy's coefficient K^*_{ji} we have the relation $K^*_{ji}=\varepsilon^4\gamma_w K_{ji}$ where γ_w is the unit mass weight of water. Note that the first order approximations of velocity and pressure which will be introduced the following diffusion problem can be written as

$$V_i^\varepsilon(\boldsymbol{x}) \simeq \varepsilon^4 V_i^0(\boldsymbol{x}^0,\boldsymbol{x}^1,\boldsymbol{x}^2), \quad P^\varepsilon(\boldsymbol{x}) \simeq P^0(\boldsymbol{x}^0). \qquad (11)$$

The target bentonite, Kunigel V1, consists of about 50% of montmorillonite in weight and the rest of macro-grains, mainly quartz particles. The mass density of a crystal is 2.7g/cm^3, both for montmorillonite and for quartz, so one third (1/3) of the bentonite with its dry density 1.8g/cm^3 is void (i.e., the void ratio e=0.5). Based on a measurement of the sizes of macro-grains, we understand sizes vary widely from 5 to 100μm in diameter. Considering the peak value of this grain-size curve and taking into account the fact that permeability is affected by smaller particles, we set the size of macro-grains as 15μm and the size of a meso-domain as 45μm (3 times of the grain size). On the other hand, the size of one montmorillonite crystal is about 100x100x1 nm, and several parallel platelets together form a stack. Here, we assume that a stack consists of six platelets with an interlayer distance 0.56nm which corresponds to two-layers of hydrated water molecules. We set this as one of the microscale domains. Since the width of a platelet is very larger than the thickness, the domain in local x^2_1 direction is set to be infinite. The inter-group distance X of Figure 6 is specified by the given data of saturated density. If the inter-group distance X becomes smaller than 0.56nm, the inter-lamellae distance is set to be the same as X. As mentioned above, the local value of viscosity is calculated by MD (Figure 4). Note that after calculating K^2_1 of (10) using this pseudo-one dimensional model, we set that the homogeneous K^2 is equal to $K^2=K^2_{11}/3$, because of random distribution of lamellae groups. Another micro-domain is of a quartz grain, which is completely impermeable. Furthermore, due to MD calculation of quartz-water system, we consider no singular distribution of water viscosity in vicinity of quartz.

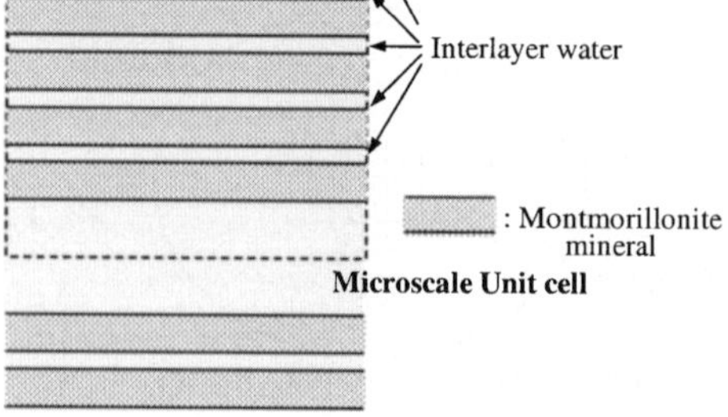

Figure 6. Microscale unit cell for hydrated smectite (corresponding to the microscale problem 1 in Figure 5).

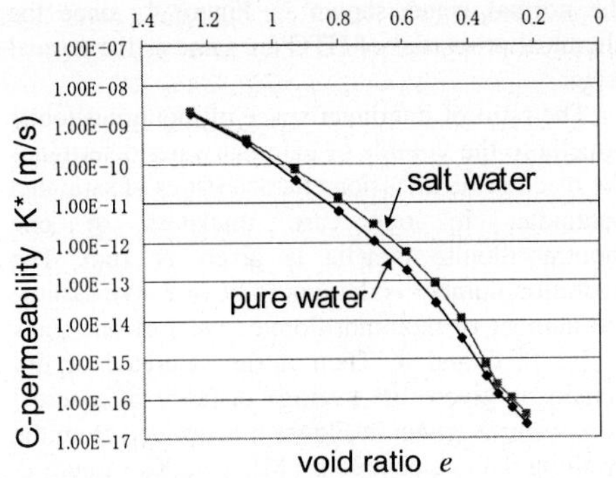

Figure 7. Permeability of bentonite due to change of void ratio calculated by MD and HA. Here the salt water case is calculated similarly starting with MD.

4. MULTISCALE HOMOGENIZATION IN DIFFUSION PROBLEM

Transport of materials in bentonite is controlled by the interlayer structure of montmorillonite lamellae and by interlayer cations. The governing equation of transport is given by

$$\frac{\partial c^\varepsilon}{\partial t} + v_j^\varepsilon \frac{\partial c^\varepsilon}{\partial x_j} - \frac{\partial}{\partial x_i}\left(D_{ij}^\varepsilon \frac{\partial c^\varepsilon}{\partial x_j} \right) + f^\varepsilon = 0 \quad \text{in } \Omega_f \quad (12)$$

where c^ε is the concentration, $\boldsymbol{v}^\varepsilon$ the velocity vector, D^ε_{ij} the diffusion coefficient, f^ε the source term and Ω_f the flow region. The boundary conditions (BC) and the initial condition (IC) are given by

$$\text{(Dirichlet BC);} \quad c^\varepsilon(\boldsymbol{x},t) = \hat{c}^\varepsilon(t) \quad \text{on } \partial\Omega_c \quad (13)$$

$$\text{(Neumann BC);} \quad -D_{ij}^\varepsilon \frac{\partial c^\varepsilon}{\partial x_j} n_i = \hat{q}^\varepsilon(t) \quad \text{on } \partial\Omega_c \quad (14)$$

$$\text{(IC);} \quad c^\varepsilon(\boldsymbol{x},t) = \hat{c}^\varepsilon(\boldsymbol{x}) \quad \text{at } t = t_0 \quad (15)$$

4.1 Multiscale HA

As mentioned in the flow problem, we introduce the multiscale HA. The perturbation is introduced for the concentration $c^\varepsilon(\boldsymbol{x};t)$ by

$$c^\varepsilon(\boldsymbol{x};t) = c^0(\boldsymbol{x}^0;t) + \varepsilon c^1(\boldsymbol{x}^0,\boldsymbol{x}^1,\boldsymbol{x}^2;t) + \cdots \quad (16)$$

where c^α's (α=0,1,2,...) are periodic functions

Substituting (16) and (2) into (12) and taking $\varepsilon \to 0$, we get the following partial differential equations corresponding to each ε-term.

$$\varepsilon^{-4}\text{-term:} \quad \frac{\partial}{\partial x_i^2}\left(D_{ij}^\varepsilon \frac{\partial c^0}{\partial x_j^2} \right) = 0$$

Since c^0 is a function only of $\boldsymbol{x}^0$, this is automatically satisfied.

$$\varepsilon^{-3}\text{-term:} \quad \frac{\partial}{\partial x_i^2}\left[D_{ij}^\varepsilon \frac{\partial c^1}{\partial x_j^2} \right] = 0$$

This suggests that c^1 is independent from $\boldsymbol{x}^2$, so we can define the *first characteristic function* $N^k{}_1(\boldsymbol{x}^1)$:

$$c^1(\boldsymbol{x}^0,\boldsymbol{x}^1,\boldsymbol{x}^2;t) = -N_1^k(\boldsymbol{x}^1) \frac{\partial c^0(\boldsymbol{x}^0;t)}{\partial x_k^0} + \bar{c}^1 \quad (17)$$

where $\bar{c}^1(\boldsymbol{x}^0;t)$ is an integration constant.

$$\varepsilon^{-2}\text{-term:} \quad \frac{\partial}{\partial x_i^2}\left[D_{ij}^\varepsilon \left(\frac{\partial c^0}{\partial x_j^0} + \frac{\partial c^1}{\partial x_j^1} + \frac{\partial c^2}{\partial x_j^2} \right) \right] = 0 \quad (18)$$

Since the term $\partial c^0 / \partial x_j^0 + \partial c^1 / \partial x_j^1$ is a function only of $\boldsymbol{x}^0$ and $\boldsymbol{x}^1$, we can introduce the *second characteristic function* $N^k{}_2(\boldsymbol{x}^2)$ by

$$c^2(\boldsymbol{x}^0,\boldsymbol{x}^1,\boldsymbol{x}^2;t) = -N_2^k(\boldsymbol{x}^2)\left(\delta_{kl} - \frac{\partial N_1^l}{\partial x_k^1} \right) \frac{\partial c^0}{\partial x_l^0} + \bar{c}^2 \quad (19)$$

Here (17) is used. Substituting this into (18) yields the following *microscale equation* [MiSE], together with X^2-periodic boundary condition for the microscale domain Ω_2:

$$\frac{\partial}{\partial x_i^2}\left[D_{ij}^\varepsilon \left(\delta_{jk} - \frac{\partial N_2^k}{\partial x_j^2} \right) \right] = 0 \quad \text{in } \Omega_2 \quad (20)$$

$$\varepsilon^{-1}\text{-term:} \quad \frac{\partial}{\partial x_i^1}\left[D_{ij}^\varepsilon \left(\frac{\partial c^0}{\partial x_j^0} + \frac{\partial c^1}{\partial x_j^1} + \frac{\partial c^2}{\partial x_j^2} \right) \right] + \frac{\partial}{\partial x_i^2}\left[D_{ij}^\varepsilon \left(\frac{\partial c^1}{\partial x_j^0} + \frac{\partial c^2}{\partial x_j^1} + \frac{\partial c^3}{\partial x_j^2} \right) \right] = 0 \quad (21)$$

We introduce the volume average for the microscale domain Ω_2, then the last term vanishes because of X^2-periodicity. Substituting Eqn (11) into this yields the following *mesoscale equation* [MeSE] together with X^1-periodic boundary condition for the mesoscale domain Ω_1:

$$\frac{\partial}{\partial x_i^1}\left[D_{ik}^{H_2}\left(\delta_{kl}-\frac{\partial N_1^l}{\partial x_k^1}\right)\right]=0 \quad \text{in } \Omega_1 \tag{22}$$

$$D_{ik}^{H_2}=\left\langle D_{ij}^{\varepsilon}\left(\delta_{jk}-\frac{\partial N_2^k}{\partial x_j^2}\right)\right\rangle_2 \tag{23}$$

$$\begin{aligned}\varepsilon^0\text{-term: } & \frac{\partial c^0}{\partial t}+v_j^{\varepsilon}\left(\frac{\partial c^0}{\partial x_j^0}+\frac{\partial c^1}{\partial x_j^1}+\frac{\partial c^2}{\partial x_j^2}\right)\\ & -\frac{\partial}{\partial x_i^0}\left[D_{ij}^{\varepsilon}\left(\frac{\partial c^0}{\partial x_j^0}+\frac{\partial c^1}{\partial x_j^1}+\frac{\partial c^2}{\partial x_j^2}\right)\right]\\ & -\frac{\partial}{\partial x_i^1}\left[D_{ij}^{\varepsilon}\left(\frac{\partial c^1}{\partial x_j^0}+\frac{\partial c^2}{\partial x_j^1}+\frac{\partial c^3}{\partial x_j^2}\right)\right]\\ & -\frac{\partial}{\partial x_i^2}\left[D_{ij}^{\varepsilon}\left(\frac{\partial c^2}{\partial x_j^0}+\frac{\partial c^3}{\partial x_j^1}+\frac{\partial c^4}{\partial x_j^2}\right)\right]+f^{\varepsilon}=0\end{aligned} \tag{24}$$

We make the average for the microscale domain Ω_2 and for the mesoscale domain Ω_1, then we have the following *macroscale equation* [MaSE]:

$$\frac{\partial c^0}{\partial t}+v_l^H\frac{\partial c^0}{\partial x_l^0}-\frac{\partial}{\partial x_i^0}\left(D_{il}^H\frac{\partial c^0}{\partial x_j^0}\right)+f^H=0 \text{ in } \Omega_0 \tag{25}$$

$$\begin{aligned} v_l^H &=\left\langle v_k^{H_2}\left(\delta_{kl}-\frac{\partial N_1^l}{\partial x_k^1}\right)\right\rangle_1,\\ D_{il}^H &=\left\langle D_{ik}^{H_2}\left(\delta_{kl}-\frac{\partial N_1^l}{\partial x_k^1}\right)\right\rangle_1, \quad f^H=\left\langle f^{H_2}\right\rangle_1\end{aligned} \tag{26}$$

The macroscale equation [MaSE] can be solved together with the boundary conditions (13) and (14) and the initial condition (15).

4.2 Diffusion of tritium HTO: Verifying the effective diffusion coefficient

Tritium water ^3H_2O, denoted as HTO, is non-adsorptive, so it is appropriate to check the applicability of our analysis. The diffusion coefficient of HTO in free water is reported as 2.44×10^{-5} cm^2/s=769.48cm/year (Klitzsche *et al.*, 1976).

We assume that in the neighbourhood of montmorillonite surface the microscale distribution of diffusivity of HTO varies in the same profile as the normal water shown in Figure 4, since the chemical properties of HTO are same as the normal water.

The ratio of interlayer space of montmorillonite lamella to the volume of external water determines the macroscale diffusion characteristics of saturated bentonite. In our case thickness of one montmorillonite lamella is given as 1nm. The hydration number is thought to be n=2. We assume the number of montmorillonite layers of one stack is 10, 8, 6 and 4. Then if the saturated or dry density is given, the average distance of external pore water X given in Figure 6 is decided from the swelling data calculated by MD (see Kawamura & Ichikawa, 2001).

The microscale model Figure 6 is pseudo one-dimensional. Because of random distribution of stacks we assume that the macroscale effective diffusion coefficient is given as $D^H=D^H_{11}/3$ where D^H_{11} is the value directly calculated by (23) using the model of Figure 6.

Results for the cases of 10, 8, 6 and 4 layers are given Table 1. The calculated data are verified well by the experimental one. Note that it is conceived that the average number of lamellae of one stack is about 8 by experiment (Fujii & Nakano, 1984).

Table 4: Effective diffusion coefficient $D^H=D^e$ of HTO in bentonite (Kunigel V1)

		$D^H=D^e$ [$\times10^{-11}$ m^2/s]
No. of layers	4	5.81
	6	7.38
	8	9.77
	10	11.40
Experimental data (Sato et al., 1994)		4.2
		5.1

4.3 Diffusion of cesium

We simulate the through-diffusion experiment of cesium (Cs) for Kunipia F by using a two-scale model (that is, excluding the mesoscale domain). The dry desity of bentonite 1.4 Mg/m^3. The hydration number n is 2.5.

The macroscale and microscale models are sown in Figure 8 and 9 with its finite element meshes. The diffusion coefficient in the external pore water is 2.0×10^{-9} m^2/s, and one in the interlayer space 2.62 $\times10^{-10}$ m^2/s, which is calculated by MD. The upper stream boundary concentration given at AB in Figure 8 is given as 1.0×10^{-3} mol/l.

The microscale model involves adsorption edge parts where the volume is determined as follows: the total edge surface is estimated as 35 m^2/g by the BET method (Baeyens & Bradbury, 1997) and the half is effective, so the total amount of Cs which can be adsorbed on edge surface is m_{lim}=6.51x10^{-5} mol/g. Here the ion radius r of Cs is assumed as 0.334 nm, and the shape of one clay mineral as 100x100x1 nm. The adsorption volume is set at each edge site in the microscale model with the thickness 0.1 nm where the above amount of adsorption occurs. That is, in the edge volume we have the condition $c \leq c_{lim}$ where the limit value c_{lim} is calculated by using m_{lim}.

The calculated results (i.e., the time-dependent change of concentration at point a, b, c and d in Figure 8) are given in Figure 10 compared with the cases without adsorption.

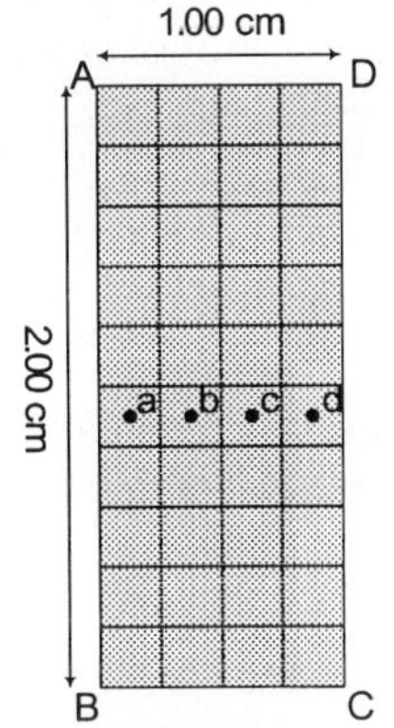

Figure 8. The macroscale model of through-diffusion experiment.

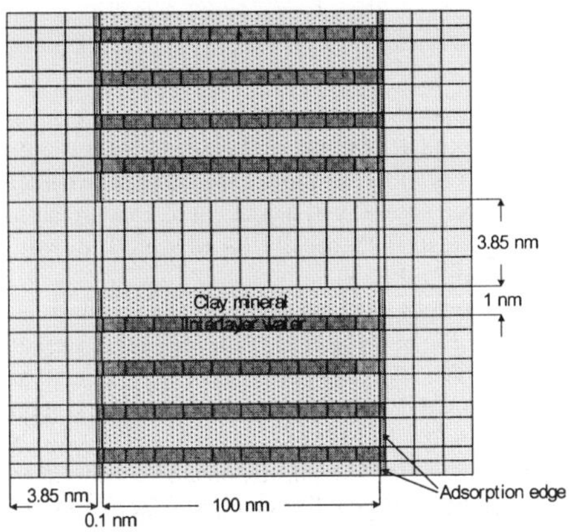

Figure 9. The microscale model.

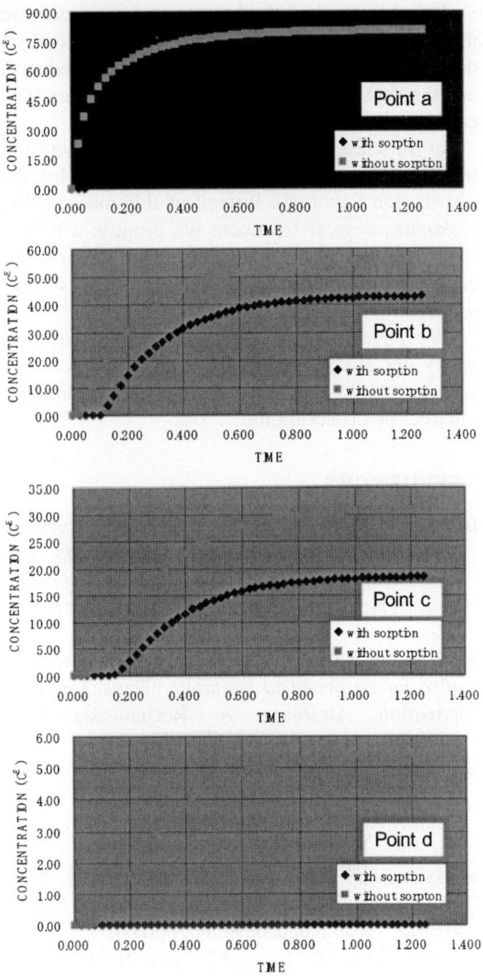

Figure 10. Calculated results: Time-dependent change of concentration.

5. CONCLUSIONS

We here showed that for bentonite clay, we can determine the nano-scale material properties such as diffusion coefficient and viscosity by molecular dynamics (MD) simulation and extend the micro-scale characteristics to the macroscale behavior by the multiscale homogenization analysis (HA) method. A seepage flow and diffusion problem is treated. The micro/macro problem can be simulated well by this procedure if we know the microscale geometrical characteristics.

By the results of the seepage analysis we understand that the permeability coefficient is drastically changed as the void space is reduced, and this fact affects the long-term deformation (i.e., consolidation) behavior strongly.

In the classical diffusion theory the adsorption term of bentonite is commonly treated by the distribution factor K_d. Instead of this macroscopic phenomenological treatment we propose a micro-scale HA procedure which exactly represents the edge adsorption characteristics at the edges of clay minerals.

We have planned a series of experiments to verify our analysis method. In particular it is our keen interest to specify the microscale geometry by microscope observations.

REFERENCES

Baeyens, B., Bradbury, M. 1997. *A mechanistic description of Ni and Zn sorption on Na-montmorillonite Part I: Titration and sorption measurement,* J. Contaminant Hydr., 27, pp.199-222.

Fujii, K., & Nakano, M. 1984. *Chemical potential of water adsorbed to bentonite,* Trans. Jap. Soc. Irrigation, Drainage & Reclamation Eng. (JSIDRE), 112, pp.43-53, (in Japanese).

Kawamura, K., & Ichikawa, Y. 2001. *Physical properties of clay minerals and water -By means of molecular dynamics simulations-,* Bull. Earthq. Res. Inst., Univ. Tokyo, 76, pp311-320.

Klitzsche, C.S. Weistroffer, K., & Elshasly, E.M. 1976. *Grundwasser der Zentralsahara: Fossile Vorrate,* Geol. Rundschau, 65, pp.264.

Sanchez-Palencia, E. 1980. *Non-Homogeneous Media and Vibration Theory,* Lec. Notes in Phy. 127, Springer-Verlag, Berlin.

Sato, H., & Shibuya, T. 1994. *Study on adsorption and diffusion mechanism of nuclides in buffer material and geosphere,* JNC Technical Review, 91(5), pp.71-89 (in Japanese).

ANALYSIS OF THE HYDRAULIC INTERACTION BETWEEN CLAY BUFFER AND HOST ROCK IN A LARGE SCALE TEST

H.R. Thomas[1], P.J. Cleall[1], N.Chandler[2], D. Dixon[2] and H.P.Mitchell[1]
[1]) Geoenvironmental Research Centre, Cardiff School of Engineering, Cardiff University
[2]) Atomic Energy of Canada Ltd., Pinawa, Manitoba.

Abstract: This paper examines the hydraulic interaction between emplaced engineered clay buffers and host rock formations. This has been achieved via the analysis of the re-saturation behaviour of a large scale in-situ experiment. Experimental results from a testing programme performed by Atomic Energy of Canada Limited, (AECL), are examined via both direct interpretation of the experimental results and numerical simulations of the test. Analysis of seepage rates into an open borehole indicate the presence of an EDZ prior to the emplacement and some 'healing' of the EDZ after buffer placement. It is also found that correct representation of moisture infiltration into the buffer is essential to capture the variations in pore water pressure in the host rock.

1. INTRODUCTION

The development of models for flow and deformation behaviour of porous materials has had significant attention in recent years. In particular several decades of research have been dedicated to high level nuclear waste disposal. Currently the preferred option of a number of disposal authorities is use of a deep geological repository, (Backblom and Martin, 1999). Although the host rock is generally highly impervious to water flow the relatively high potentials for flow coupled with possible excavation damage zones mean that the rock alone cannot be relied upon as a barrier. Therefore the interaction between an emplaced clay barrier and the host rock is a focus for research.

Large scale experiments in underground research laboratories, (SKB, 1993, Huertas et al., 2000), have played an important role in demonstrating full scale behaviour. Such experiments have become a key component of a number of extensive programmes of research investigating various aspects of behaviour. Additionally they serve as major sources of data for use in numerical model validation and development.

One such programme of research has been undertaken by Atomic Energy of Canada Limited, (AECL). As part of this programme a series of large scale in-situ tests have been performed including the Buffer/Container Experiment and the Isothermal Test.

The Buffer/Container Experiment was designed to investigate performance of a single borehole emplacement, with a full size heater representing a container of nuclear fuel. In particular, thermal, hydraulic and mechanical interactions exhibited within a sand-bentonite buffer and the granite host rock were examined.

The experiment involved placing a 0.6m diameter 2.2m long heater in a 1.24m diameter and 5m deep borehole at the 240 m Level of AECL's Underground Research Laboratory. The heater was surrounded by a 50:50 bentonite:sand buffer. The experiment was operated by AECL over the period 1991 – 1994, (Graham et al., 1997).

The Isothermal Test was designed in conjunction with the Buffer/Container Experiment and removed the heating phase to investigate the hydraulic interaction between the rock and the buffer in the absence of thermal gradients, (Chandler, 2000). AECL operated the "Isothermal Test" over the period 1992 – 1999.

The experiment, as for the Buffer/Container Experiment, was based on a 1.24 m diameter 5 m deep borehole at the 240 m Level of AECL's Underground Research Laboratory. A 50:50 bentonite:sand buffer was compacted to a depth of 2m into the base of the borehole, (Chandler, 2000, Dixon et al., 2002). A 1.25 m thick concrete plug was then installed above the buffer to act as a vertical resistance to any swelling that might occur. The granite rock around the experiment could be considered to be un-fractured and nearly saturated, (Chandler, 2000). The layout of the experiment is shown in Figure 1.

In this paper the interaction between the buffer and the host rock in these experiments is investigated. In particular the resaturation of the Isothermal Test is analysed via both direct interpretation of the experimental results and numerical simulations of the test.

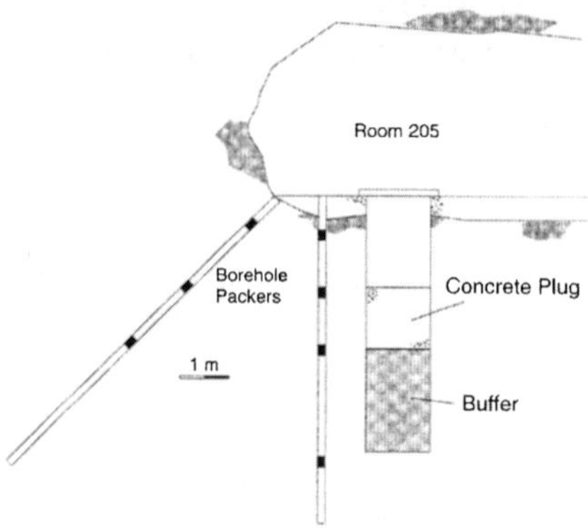

Figure 1. Configuration of the Isothermal Test and adjacent packer systems (After, Chandler 2000).

2. EXPERIMENTAL RESULTS

For both experiments extensive instrumentation allowed for monitoring of the hydraulic and mechanical performance of the system before, during and after the test. In the Buffer/Container experiment the thermal response of the system was also measured.

The Isothermal Test can be considered as being made up of three main phases; firstly, a pre buffer placement stage with an open borehole; secondly, the buffer-rock interaction phase after the buffer has been placed and finally a decommissioning phase after the buffer has been excavated.

The pre-placement stage started immediately after the borehole had been drilled. During this stage, instrumentation was installed in the borehole to measure the rate of seepage from the saturated rock into the empty borehole.

After placement of the buffer, hydraulic packers and piezometers installed in the rock monitored the hydraulic gradients towards the borehole. Also thermocouple psychrometers were installed within the buffer to measure transient water contents.

During the decommissioning stage, end-of-test moisture contents were measured, for use as a comparison against those measured by the thermocouple psychrometers.

Instrumentation to measure seepage into the empty borehole was re-installed shortly after the buffer removal was completed. This allowed comparison against the pre-placement seepage measurements.

2.1 Seepage from host granite into empty borehole

A series of collection rings were employed to measure inflow, from the host granite, into the empty borehole both before and after the experiment. The total seepage into the borehole was divided into four zones, SR1, SR2, SR3 and SR4. Zone SR1, collected water seeping into the borehole from the tunnel, and from the damaged rock between the concrete tunnel floor and the rock. Zone SR2, which corresponds to the position of the concrete plug, was installed to isolate the bottom two metres of the borehole. Zones SR3 and SR4 collected seepage that would be flowing towards the buffer, with zone SR4 also measuring seepage into the base of the borehole.

Seepage results before and after the experiment are shown in Table 1. It can be seen that the measured inflow into the lower two zones is 35% less after the test was completed, while the upper two zones recorded an increase in inflow. During the inflow measurements the borehole was sealed creating a near 100% humidity local environment. Possible reasons for these variations are discussed in section 4.

2.2 Hydraulic gradients in the rock

A series of pneumatic piezometers (IRP1-10) and hydraulic piezometers (IRH1 and 2) were installed within 0.9 m of the outer diameter of the borehole. Two boreholes were installed with packer systems to measure pore water pressures in the rock further from the borehole (IRW).

Results from these instruments are shown in Figure 2 at two discrete intervals in time, namely 50 and 1000 days into the test. It can be seen that the pore water pressure close to the borehole gradually increased over this period of time, following emplacement of the buffer. This recovery of pore water pressure, relative to the hydrostatic pressures, is due to the reduced seepage into the borehole after the placement of the buffer. The coupling of moisture movement into the buffer and recovery of pore pressure in the host rock is investigated further in section 4.

2.3 Water content in the buffer

Transient moisture contents were measured by psychrometers within the buffer. These measurements have been converted into a water uptake calculation, Chandler (2000), based on a simplifying assumption of constant dry density. The results obtained are presented in Figure 3.

Table 1: Comparison of Measured and Modelled Inflow Rates

Seepage Collection Zone	Measured inflow (pre-placement) x 10-6 litres/min	Measured inflow (post-decommissioning) x 10-6 litres/min	COMPASS calculated inflow x 10-6 litres/min	COMPASS calculated inflow With EDZ x 10-6 litres/min
SR1	1.6	3.4	3.7	4.4
SR2	2.7	9.4	9.8	10.7
SR3	10.0	6.5	8.6	9.7
SR4	34.0	22.0	16.4	29.4
Total	48.3	41.3	38.5	54.2

Also during the decommissioning stage, thorough sampling of the water content of the buffer was conducted (Dixon and Chandler 2000). It was observed that the centre of the buffer had a much lower moisture content than the external edge. With the average buffer water content at the centre of the borehole being between 0.5 and 2.0% lower than the average buffer pore water content adjacent to the rock. It was also found that the buffer tended to absorb more water in the lower layers. This is probably due to the bottom of the buffer being in direct contact with the saturated rock, while the top of the buffer was in contact with the concrete plug.

3. THEORY

To investigate the behaviour observed in both the Isothermal Test and Buffer Container Experiment numerical simulations have been undertaken. In particular coupled hydraulic-mechanical analyses of the Isothermal test and thermal-hydraulic-mechanical analyses of the Buffer Container Experiment have been undertaken. These analyses have been performed via a finite element computer code, COMPASS (Thomas and He, 1997 & 1998).

The theoretical formulation employed in COMPASS is based on considering an unsaturated soil as a three-phase porous medium consisting of solid, liquid and gas. A set of coupled governing differential equations can be developed to describe heat and moisture flow within a deformable porous media. The primary variables of the model are pore water pressure, u_l, pore air pressure, u_a, temperature, T and displacement, $\boldsymbol{u}$.

The numerical solution of the governing equations is achieved by a combination of the finite element method for the spatial discretisation and a finite difference time stepping scheme for temporal discretisation.

The Galerkin weighted residual method is employed to formulate the finite element discretisation. An implicit mid-interval backward difference algorithm is implemented to achieve temporal discretisation. With appropriate initial and boundary conditions the set of non-linear coupled governing differential equations can be solved.

4. NUMERICAL ANALYSIS RESULTS

A series of numerical simulations of each of the phases of the Isothermal Test have been performed. For these analyses a two-dimensional, axisymmetric geometry has been used as a high degree of radial symmetry was observed in the experimental results.

The problem under consideration consists of three materials; the bentonite/sand buffer, the granite rock, and the concrete barrier. Since the main focus of attention is the hydraulic interaction between the buffer, and the rock, attention has been concentrated here on these two materials. The concrete barrier was assumed, for the purposes of the analyses presented here, to be both impermeable and rigid. The material parameters used are given in detail in Thomas et al., (2003).

For the analysis of the buffer-rock interaction phase (second phase) of the experiment two hydraulic conductivity relationships have been considered. Firstly a 'conventional' approach has been used where the unsaturated hydraulic conductivity increases with saturation. Secondly an 'effective' conductivity relationship has been used. In this case reduction of the material's hydraulic conductivity due to expansion of the microstructure of the buffer, as the material saturates, is taken into account. This microstructural expansion tends to reduce the void spaces in the macrostructure, since the material, overall, is constrained from swelling. Further details of this approach can be found in Thomas et al., (2003).

4.1 Seepage into borehole

Initially the first phase of the experiment was simulated so that the seepage into the rock prior to emplacement of the buffer material could be analysed. The simulation was carried out for a period of three years to represent the time when the borehole was open before buffer emplacement. At the end of this phase the model estimated that steady state pore water pressure profiles had been achieved within the rock. Inflow from the rock into the empty borehole was then calculated.

A reasonable comparison can be seen in Table 1 between the results of this analysis to the experimentally measured results. It can be seen that these model results compared more favourably with the decommissioning data. For the pre-placement phase it was observed in the experiment that there was a larger amount of seepage into the bottom two zones (SR3 and SR4), and less into the upper two zones (SR1 and SR2), than was calculated in the numerical simulation.

The drilling of the emplacement borehole could have caused damage to the rock. The precise changes to material parameters are unknown. However, some estimates have been made with an increased hydraulic conductivity and porosity and a revised moisture retention relationship. As a first assessment of the problem, an EDZ extending 30 cm from the edge of the emplacement borehole and Room 205, surrounding the borehole and cavern, was incorporated into the numerical analysis.

Results from these analyses are also presented in Table 1. It can be seen that including an EDZ in the analysis increased the total seepage into the buffer-filled portions of the borehole (SR3 and SR4). The numerical results correlated more closely to the pre-placement experimental results than the results produced without an EDZ, particularly for zones SR3 and SR4. The experimentally measured results for zone SR2 were very low for the pre-placement stage. However a good match was found between the decommissioning seepage for zone SR2 and the numerically modelled results.

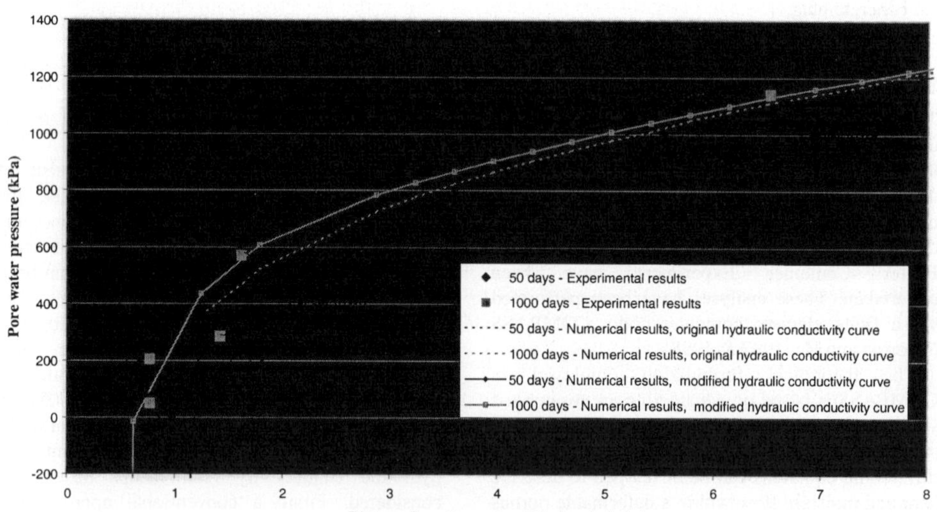

Figure 2. Pore water pressures in the rock at the mid-height of the buffer following emplacement: exponential relationship accounting for swelling in the buffer.

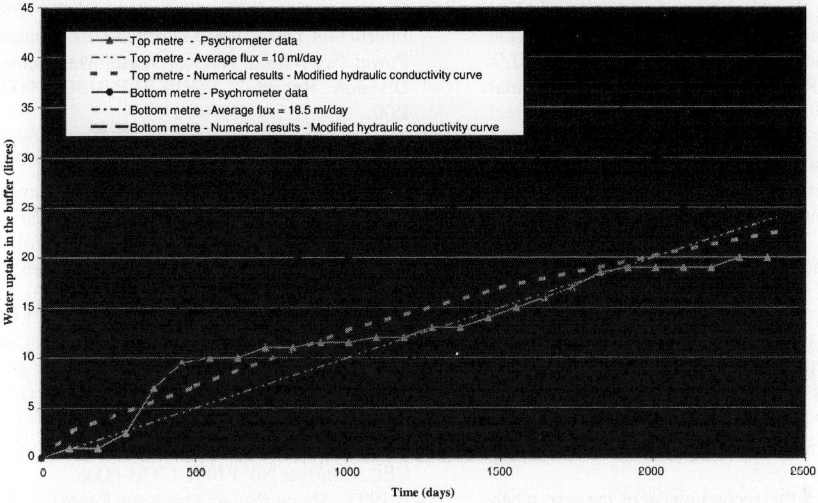

Figure 3. Pore water pressures in the rock at the mid-height of the buffer following emplacement: accounting for swelling in the buffer

It is postulated here, that prior to the emplacement of the buffer material flow towards the borehole was increased due to the presence of an EDZ. During the experiment the buffer swelled, and to some degree plugged the cracks formed in the EDZ, thereby returning the hydraulic properties at the rock-buffer boundary back towards those of unfractured rock. The numerical results appear to support this theory.

4.2 Seepage into buffer

The buffer-rock interaction phase, with 7 years of water uptake, was then analysed. Comparison of simulated and experimental values within the buffer, not shown here, found that the numerical analysis significantly over-predicted the resaturation of the buffer. As shown in Figure 2 little difference between the numerical values of the simulated results for 50 days, and those for 1000 days for pore water pressure within the rock is found. In fact an overall trend of slight decrease in the pore water pressures within the rock is simulated. However the experimental results demonstrate a recharge of pore water pressures close to the buffer/rock interface, with the 1000 day pore water pressures being much higher than the 50 days pore water pressures.

Clearly it can be concluded that these results do not match the experimentally observed values. The hydraulic regime observed in the rock adjacent to the buffer is not correctly described. Also neither the pattern of moisture distribution nor the rate of infiltration has been replicated within the buffer. This phase was then analysed incorporating an 'effective' hydraulic conductivity relationship for the buffer as described previously. A much improved correlation with the experimental data is found for the buffer resaturation. Figure 2 shows the pore water pressure distribution in the rock. It can be seen that the pore water pressure has increased in the rock, corresponding with the experimental results. Within the buffer an improved correlation is also found.

To further test the value of the 'effective' approach, numerical simulation results of transient water content are compared against the experimental results in Figure 3. A reasonable match between the numerical and the experimental results is obtained.

5. CONCLUSIONS

This paper has examined the hydraulic interaction between emplaced engineered clay buffers and host rock formations. This has been achieved via the analysis of the re-saturation behaviour of a large scale in-situ experiment, the Isothermal test, carried out at AECL's Underground Research Laboratory. Experimental results have been examined via both direct interpretation of the experimental results and numerical simulations of the test.

Comparisons of numerical simulations, achieved using the computer code COMPASS, have been presented. Various simulations have been presented of each phase of experiment.

Analysis of seepage rates of the first and last phases of the experiment, when the borehole is open, tend to indicate that prior to the emplacement of the buffer material flow towards the borehole was increased due to the presence of an EDZ. Also, seepage after the buffer had been removed was reduced indicating some 'healing' of the EDZ had occurred.

Simulation of the second phase of the test, when the buffer is in-situ, showed that correctly representing moisture infiltration into the buffer was essential to capture the variations in pore water pressure in the host rock. In particular when an 'effective' conductivity for the buffer was used, in comparison to a 'conventional' relationship, an improved correlation with experimental in-flow within the buffer was found. Correspondingly the pore water pressure variations in the rock correctly followed the trends observed in the experimental results.

REFERENCES

Backblom, G., & Martin, C.D. 1999 *Recent experiments in hard rocks to study the excavation response: Implications for the performance of a nuclear waste geological repository.* Tunnelling and underground space technology 14 377-394.

Chandler, N.A. 2000. *Water inflow calculations for the Isothermal Buffer-Rock-Concrete Plug Interaction Test.* Ontario Power Generation, Nuclear Waste Management Division Report, 06819-REP-01200-10046-R00.

Dixon, D.A., Chandler, N. Graham, J. & Gray, M.N. 2002. *Two large-scale sealing tests conducted at Atomic Energy of Canada's underground research laboratory: the buffer-container experiment and the isothermal test.* Canadian Geotechnical Journal 39(3) pp. 503-518

Dixon, D.A. & Chandler, N.A.. 2000. *Physical properties of the materials recovered from the Isothermal Buffer-Rock-Concrete Plug Interaction Test conducted at the URL.* Ontario Power Generation, Nuclear Waste Management Division Report, 06819-REP-01300-10003-R00.

Graham, J., Chandler, N.A. Dixon, D..A., Roach, P.J. To, T. & Wan, A.W.L.. 1997. *The Buffer/Container Experiment: Results, synthesis, issues.* Atomic Energy Canada Limited Report, AECL-11746, COG-97-46-I.

Huertas, F., Fuentes-Cantillana, J.L., Jullien, F. Rivas, P., Linares, J., Farina, P. Ghoreychi, M., Jockwer, N., Kivkmaier, W., Martinez, M.A. Samper, J., Alonso, E. & Elorza, F.J.. 2000. *Full-scale engineered barriers experiment for a deep geological repository for high-level radioactive waste in crystalline host rock (FEBEX project).* End of Contract Report on CEC, Contract No. F14W-CT95-0006.

SKB 1993. *Stripa Project Overview Report. Vol. I: Executive summary.* ISBN 91-971906-2-4.

Thomas, H.R. & He, Y.1997. *A coupled heat-moisture transfer theory for deformable unsaturated soil and its algorithmic implementation.* Int J for Num Meth in Engng, 40, 3421-3441.

Thomas, H.R. & He, Y. 1998. *Modelling the behaviour of unsaturated soil using an elasto plastic constitutive relationship.* Geotechnique, 48(5): 589-603.

Thomas, H.R., Cleall, P.J., Chandler, N., Dixon, D. & Mitchell, H.P. 2003. *Water infiltration into a large-scale in-situ experiment in an underground research laboratory.* Geotechnique, 53(2): 207-224.

MODELLING GROUNDWATER PRESSURE AND THERMAL LOADING IN THREE-DIMENSIONAL DISCONTINUOUS DEFORMATION ANALYSIS

Q.H. Jiang[1], M.R. Yeung[2], N. Sun[2]

[1]) College of Water Conservancy and Hydropower Engineering, Wuhan University, Hubei, China

[2]) Department of Civil Engineering, The University of Hong Kong, Hong Kong, China

Abstract: The Discontinuous Deformation Analysis (DDA) is a powerful numerical method for the analysis of discontinuous rock mass behaviour. In this paper, three-dimensional (3-D) DDA is extended to allow the consideration of groundwater pressure and thermal loading. A general model of water pressure distribution on the surfaces of rock blocks is established. The water pressure submatrices and thermal loading submatrices are derived by minimizing the total potential energy and added to the global matrices. These extensions to 3-D DDA have been implemented into a 3-D DDA computer program. Three test cases given show that the 3-D DDA results are reasonable.

1. INTRODUCTION

The Discontinuous Deformation Analysis (DDA) is a numerical method for blocky system (Shi, 1988). Due to its rigorous scheme of block kinematics and equilibrium conditions achieved by minimizing the total potential energy, DDA has attracted the attention of researchers and geotechnical engineers worldwide. With many people contributing to its development and application, the original two-dimensional (2-D) DDA is well developed in terms of both theory and computer coding (e.g. Lin et al., 1996; Chen et al., 1997). However, 2-D DDA is applicable mainly to two-dimensional plane strain or plane stress problems. There is an obvious need for a three-dimensional (3-D) model because of the highly directional nature of jointed rock mass behaviour, making the application of 2-D DDA to many practical problems inappropriate. While various researchers are working on 3-D DDA, only preliminary work on its development has been published (Shi, 2001; Yeung et al., 2003; and Jang and Lee, 2002). Methods for analysing groundwater pressure and thermal loading in 3-D DDA have not been reported.

Coupled thermal-hydro-mechanical (T-H-M) behaviour of fractured rocks is an important consideration in the design and safety assessment of many civil and environmental engineering works, such as slopes, underground storage caverns, geothermal energy extraction facilities and nuclear waste repositories. Assuming that the block element has uniform strain and is impervious, Kim et al. (1999) reported a method to couple fluid flow in fractures with stress and deformation of blocks using 2-D DDA. Because the constant strain mode cannot describe the deformability of blocks of complex shapes, Jing et al. (2001) developed a numerical model for coupled hydro-mechanical processes in fractured hard rocks using 2-D DDA with finite element meshes, and the strain-block geometry incompatibility in some DDA formulas was overcome by proper internal discretization of deformable blocks using triangular or quadrilateral elements. Yeung and Blair (1999) used 2-D DDA to analyse the thermal-mechanical behaviour of the in-situ rock block tested in the Large Block Test (LBT) of the Yucca Mountain Project. In this paper, the authors attempt to model water-block interaction and thermal-mechanical response using 3-D DDA and present the relevant algorithms, equations and examples. In this first attempt, the emphasis is on the contributions of groundwater pressure and thermal loading to the global stiffness matrices and equations of motion. The effects of fluid and heat flows on the coupled T-H-M analysis will need further research.

2. BASIC PRINCIPLES OF DDA

2.1 Displacement and deformation of a rock block

In DDA, time steps are used to drive the computation. Large displacement and large deformation result from the accumulation of small displacements and small deformations within the individual time steps. Given that each time step satisfies the condition of infinitesimal displacement and deformation and assuming that each block has uniform stress and strain, the displacement and deformation of a block are determined by 12 independent deformation variables:

$$[D_i]=[u_0,v_0,w_0,\alpha_0,\beta_0,\gamma_0,\varepsilon_x,\varepsilon_y,\varepsilon_z,\gamma_{xy},\gamma_{yz},\gamma_{zx}]^T \quad (1)$$

where (u_0,v_0,w_0) are the rigid body translations in the x-, y- and z-directions, respectively, of the block's centre of mass (x_0, y_0, z_0); (α_0, β_0, γ_0) are rotations of the block around the x-, y- and z-axes, respectively; and (ε_x, ε_y, ε_z, γ_{xy}, γ_{yz}, γ_{zx}) are the three normal strains and three shearing strains, respectively. The displacement of an arbitrary point in the block can be expressed in terms of the deformation variables of [D_i]:

$$\begin{bmatrix} u \\ v \\ w \end{bmatrix}=[T_i][D_i] \quad (2)$$

where

$$[T_i]=\begin{bmatrix} 1 & 0 & 0 & 0 & (z-z_0) & -(y-y_0) & (x-x_0) & 0 & 0 & 0 & (y-y_0) & 0 \\ 0 & 1 & 0 & -(z-z_0) & 0 & (x-x_0) & 0 & (y-y_0)/2 & 0 & (z-z_0)/2 \\ 0 & 0 & 1 & (y-y_0) & -(x-x_0) & 0 & 0 & (y-y_0)/2 & (x-x_0)/2 \end{bmatrix} \quad (3)$$

$$[T_i]=\begin{bmatrix} 1 & 0 & 0 & 0 & (z-z_0) & -(y-y_0) & (x-x_0) & 0 & 0 & 0 & (y-y_0)/2 & 0 & (z-z_0)/2 \\ 0 & 1 & 0 & -(z-z_0) & 0 & (x-x_0) & 0 & (y-y_0) & 0 & (x-x_0)/2 & (z-z_0)/2 & 0 \\ 0 & 0 & 1 & (y-y_0) & -(x-x_0) & 0 & 0 & 0 & (z-z_0) & 0 & (y-y_0)/2 & (x-x_0)/2 \end{bmatrix}$$

[T_i] is called the displacement conversion matrix of the block.

2.2 General equilibrium equations

A number of individual blocks can be connected to form a block system satisfying the constraints between blocks. For a system of n blocks, its behaviour is governed by the following system of simultaneous equations:

$$\begin{bmatrix} K_{11} & K_{12} & K_{13} & \cdots & K_{1n} \\ K_{21} & K_{22} & K_{23} & \cdots & K_{2n} \\ K_{31} & K_{32} & K_{33} & \cdots & K_{3n} \\ \vdots & \vdots & \vdots & \ddots & \vdots \\ K_{n1} & K_{n2} & K_{n3} & \cdots & K_{nn} \end{bmatrix}\begin{bmatrix} D_1 \\ D_2 \\ D_3 \\ \vdots \\ D_n \end{bmatrix}=\begin{bmatrix} F_1 \\ F_2 \\ F_3 \\ \vdots \\ F_n \end{bmatrix}$$

(4)

Each 3-D block has 12 degrees of freedom, so in the coefficient matrix of Equation 4, each element [K_{ij}] is a 12×12 submatrix. [D_i] and [F_i] are 12×1 submatrices, with [D_i] containing the kinematic variables of block i and [F_i] containing the loads on block i distributed to the 12 kinematic variables. The submatrix [K_{ii}] depends on the material properties of block i, and [K_{ij}] ($i \neq j$) is determined by the contact condition between block i and block j.

For the given constant-strain mode, we can obtain the global equilibrium equation by minimising the general potential energy Π_p under the actions of stresses and external forces including the inertial force and damping force. The elements k_{ij} of [K_{ij}] and f_i of [F_i] can be obtained from Eqs. (5) and (6), respectively:

$$k_{ij}=\frac{\partial^2 \Pi_P}{\partial d_{ri}\partial d_{sj}} \qquad r,\ s=1-12 \quad (5)$$

$$F_i=-\left.\frac{\partial \Pi_P}{\partial d_{ri}}\right|_{[D_i]=0} \qquad r=1-12 \quad (6)$$

3. MODELLING OF GROUNDWATER PRESSURE

3.1 Fitting function for piezometric surface

In geological modelling, various geo-information can be expressed as functions of spatial variables. From groundwater observation data, we can construct a three-dimensional piezometric surface using a relevant fitting function. Let (x, y, z) be a point on the piezometric surface with the x-y plane being horizontal. The height of the piezometric surface above the x-y plane, z, can be given by a fitting function (Chen et al., 1998):

$$z(x,y)=\sum_{i=1}^{N} E_i\left(\tfrac{r_i^2}{R^2}\ln\tfrac{r_i^2}{R^2}+1-\tfrac{r_i^2}{R^2}\right) \quad (7)$$

where

$$r_i^2=(x-x_i)^2+(y-y_i)^2 \quad (8)$$

and R is the "influence radius", excluding the N observed points on the piezometric surface $(x_i, y_i, z_i), i=1,2,...,N$, which are given by the following N equations:

$$z_j(x_j,y_j)=E_j+\sum_{i=1,i\neq j}^{N} E_i\left(\tfrac{r_{ji}^2}{R^2}\ln\tfrac{r_{ji}^2}{R^2}+1-\tfrac{r_{ji}^2}{R^2}\right),$$

$$j=1-N \quad (9)$$

where

$$r_{ji}^2=(x_j-x_i)^2+(y_j-y_i)^2 \quad (10)$$

The coefficients $E_i\,(i=1-N)$ are obtained by solving the above N equations (Equation 9).

3.2 A general water pressure distribution model

The piezometric surface described by Equation 7 can be discretised into a series of triangular facets as shown in Figure 1. The triangular facet *E'F'G'* can be described by the equation

$$A_1x + B_1y + C_1z + D_1 = 0 \quad (11)$$

The equation of the block surface *EFG* acted on by this piece of the piezometric surface is

$$A_2x + B_2y + C_2z + D_2 = 0 \quad (12)$$

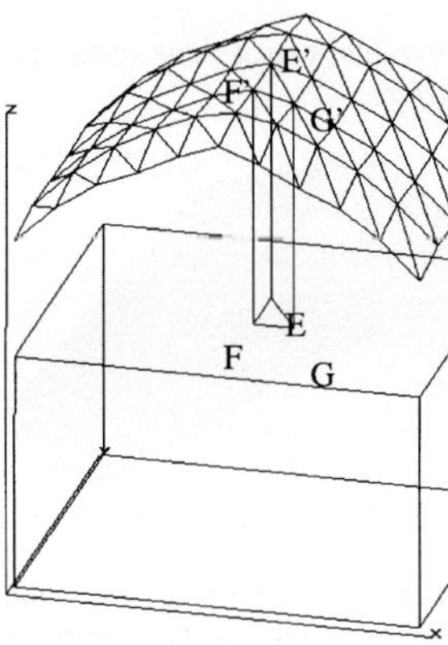

Figure 1. A general water pressure distribution model.

Let $\vec{n} = [n_x \quad n_y \quad n_z]$ be the unit outward normal of the block face containing *EFG* and γ be the unit weight of water. Assuming that the pressure head at a point is approximated by the height of the piezometric surface above the point, the water pressure on EFG is given by

$$[F_x \quad F_y \quad F_z]^T = (ax + by + c)[n_x \quad n_y \quad n_z]^T \quad (13)$$

where

$a = \gamma(A_1/C_1 - A_2/C_2)$;

$b = \gamma(B_1/C_1 - B_2/C_2)$; and

$c = \gamma(D_1/C_1 - D_2/C_2)$.

The potential energy of the water force is given by

$$\Pi_w = -\iint_s (F_x u + F_y v + F_z w)ds$$

$$= -\frac{1}{|n_z|}[D_i]^T \iint_{D_{xy}} [T_i]^T \begin{bmatrix} F_x \\ F_y \\ F_z \end{bmatrix} dxdy \quad (14)$$

where S is the loaded area *EFG* and D_{xy} is the projection of S on the x-y plane. The case of *EFG* being vertical, i.e. $n_z = 0$, is a special case that would require special treatment but is not included in this paper.

The derivatives of Π_w

$$f_{ri} = -\frac{\partial \Pi_w(0)}{\partial d_{ri}} \qquad r = 1-12 \quad (15)$$

form a 12×1 submatrix which is added to the submatrix $[F_i]$ in Equation 4:

$$\frac{1}{|n_z|}\iint_{D_{xy}} [T_i]^T \begin{bmatrix} F_x \\ F_y \\ F_z \end{bmatrix} dxdy \to [F_i] \quad (16)$$

To simplify the analytical solutions required, the following basic integrations over the projected area D_{xy} are introduced:

$$S = \iint_{D_{xy}} 1dxdy \quad S_x = \iint_{D_{xy}} x\,dxdy \quad S_y = \iint_{D_{xy}} y\,dxdy$$

$$S_{xx} = \iint_{D_{xy}} x^2dxdy \quad S_{yy} = \iint_{D_{xy}} y^2dxdy \quad S_{xy} = \iint_{D_{xy}} xy\,dxdy \quad (17)$$

The analytical solutions for these integrations can be found in Shi (1988).

Now, let

$[a_1] = [0 \quad 0 \quad n_x]^T[a \quad b \quad c]$

$[a_2] = [0 \quad 0 \quad n_y]^T[a \quad b \quad c]$

$[a_3] = [0 \quad 0 \quad n_z]^T[a \quad b \quad c]$

$[a_4] = (-n_y[Z_0] + n_z[Y_0])[a \quad b \quad c]$

$[a_5] = (n_x[Z_0] - n_z[X_0])[a \quad b \quad c]$

$[a_6] = (-n_x[Y_0] + n_y[X_0])[a \quad b \quad c]$

$[a_7] = n_x[X_0][a \quad b \quad c]$

$[a_8] = n_y[Y_0][a \quad b \quad c]$

$$[a_9] = n_z[Z_0][a \quad b \quad c]$$

$$[a_{10}] = (\frac{1}{2}n_x[Y_0] + \frac{1}{2}n_y[X_0])[a \quad b \quad c]$$

$$[a_{11}] = (\frac{1}{2}n_y[Z_0] + \frac{1}{2}n_z[Y_0])[a \quad b \quad c]$$

$$[a_{12}] = (\frac{1}{2}n_x[Z_0] + \frac{1}{2}n_z[X_0])[a \quad b \quad c]$$

where

$$[X_0] = [1 \quad 0 \quad -x_0]^T, [Y_0] = [0 \quad 1 \quad -y_0]^T$$

$$[Z_0] = [-A_2/C_2 \quad -B_2/C_2 \quad -(D_2/C_2 + z_0)]^T$$

Furthermore, let

$$[a_r] = \begin{bmatrix} a_{11}^r & a_{12}^r & a_{13}^r \\ a_{21}^r & a_{22}^r & a_{23}^r \\ a_{31}^r & a_{32}^r & a_{33}^r \end{bmatrix} \qquad (18)$$

The analytical solution for the matrix integration in Equation 15 is obtained:

$$f_{ri} = [a_{11}^r S_{xx} + a_{22}^r S_{yy} + (a_{12}^r + a_{21}^r) S_{xy} + (a_{13}^r + a_{31}^r) S_x + (a_{23}^r + a_{32}^r) S_y + a_{33}^r S] / |n_z| \qquad (19)$$

3.3 Effect of groundwater on stability of jointed rock slopes

3.3.1 Sliding on a planar surface

To illustrate the effect of groundwater on the stability of jointed rock slopes, a three-block rock slope system as shown in Figure 2a is considered. The bottom and left boundaries of the slope shown are fixed in their respective normal directions, and the other boundaries are free. The x-y plane is horizontal and the dip angle of the sliding surface is 21°. The cohesion and friction angle of all joints are 0 and 25°, respectively. The material constants for all blocks are: Young's modulus E = 1 GPa, Poisson's ratio ν = 0.3, and unit weight $\gamma = 27 kN/m^3$. Two analyses are performed. In the first analysis, the slope is subjected to its self-weight only, and Figure 2b shows the results from 3-D DDA at the end of time step 500. It can be seen that no obvious movement occurs and the slope is stable. In the second analysis, the slope is subjected to its self-weight and groundwater pressures from the piezometric surface shown in Figure 2a. Figure 2c shows the results from 3-D DDA at the end of time step 500. The 3-D DDA results show clearly that groundwater pressures cause the sliding in this case.

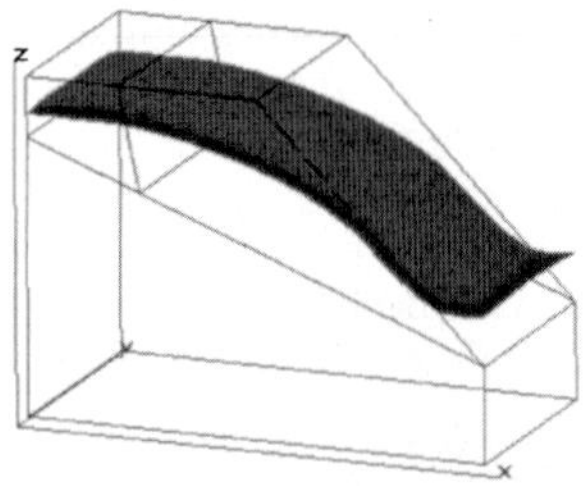

Figure 2a. Three-block rock slope system and piezometric surface.

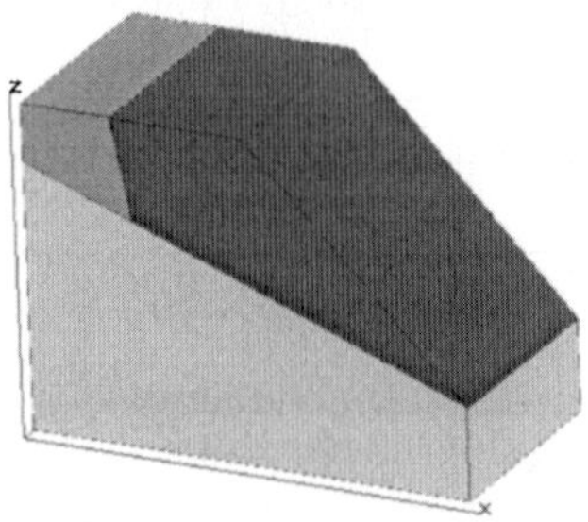

Figure 2b. 3-D DDA results for three-block slope without piezometric surface.

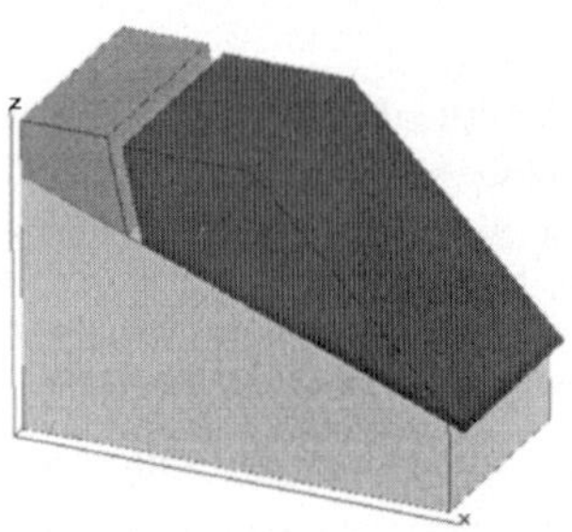

Figure 2c. 3-D DDA results for three-block slope with piezometric surface.

3.3.2 Sliding on a circular surface

Another case of sliding on a circular surface is considered. The block system in this case is shown in Figure 3a along with the piezometric surface. The bottom and left boundaries of the slope shown are fixed in their respective normal directions, and the other boundaries are free. The material constants for all the blocks are: Young's modulus E

= 1 *GPa*, Poisson's ratio ν = 0.3, and unit weight $\gamma = 14 kN/m^3$. For the joints, the cohesion and friction angle are 0 and 60°, respectively. The slope is subjected to its self-weight and the groundwater pressures from the piezometric surface. Figure 3b shows the results from 3D DDA at the end of time step 1000. The 3-D DDA results show that blocks slide along the pre-existing circular surface in a complicated manner.

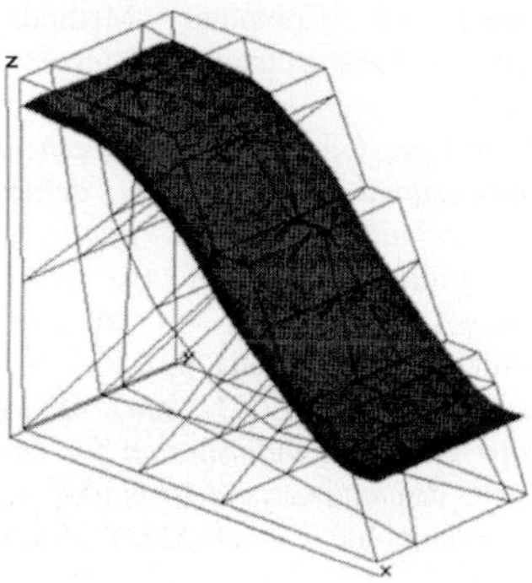

Figure 3a. Rock slope model with circular sliding surface and piezometric surface.

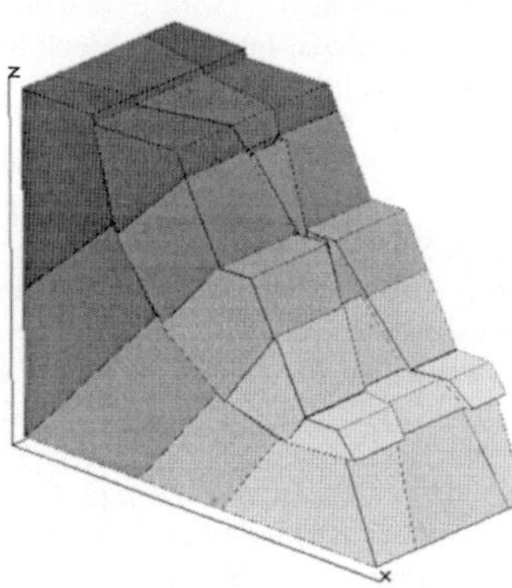

Figure 3b. 3-D DDA results for rock slope model with circular sliding surface and piezometric surface.

4. MODELLING OF THERMAL LOADING

4.1 Thermal loading submatrices

In a heated body, the strains at a point may be considered as consisting of two parts. One part is from free thermal expansion, and the other part is from the initial stress state of the body. Since the first part is uniform in all directions at a given point in an isotropic body, no shear strains result from the first part. If we assume that the block expands uniformly, the strains from the free thermal expansion of the block are

$$\Delta\varepsilon_x = \Delta\varepsilon_y = \Delta\varepsilon_z = \alpha\Delta T$$

$$\Delta\gamma_{xy} = \Delta\gamma_{yz} = \Delta\gamma_{zx} = 0 \tag{20}$$

where the temperature change is ΔT and the thermal expansion coefficient is α. The thermal stresses acting on the block can then be expressed as

$$\Delta\sigma_x = \Delta\sigma_y = \Delta\sigma_z = \frac{E\alpha\Delta T}{1-2\nu}$$

$$\Delta\tau_{xy} = \Delta\tau_{yz} = \Delta\tau_{zx} = 0 \tag{21}$$

where E is the Young's modulus and υ is the Poisson's ratio.

Denoting the volume of the block as V_i, the potential energy associated with the temperature change is given by

$$\Pi_t = -\iiint(\varepsilon_x\Delta\sigma_x + \varepsilon_y\Delta\sigma_y + \varepsilon_z\Delta\sigma_z)dxdydz$$
$$= V_i[D_i]^T[\Delta\sigma] \tag{22}$$

where

$$[\Delta\sigma] = [0 \quad 0 \quad 0 \quad 0 \quad 0 \quad 0 \quad E\alpha\Delta T/(1-2\upsilon)$$
$$E\alpha\Delta T/(1-2\upsilon) \quad E\alpha\Delta T/(1-2\upsilon) \quad 0 \quad 0 \quad 0]$$

The derivatives of Π_t

$$f_{ri} = -\frac{\partial\Pi_t(0)}{\partial d_{ri}} \qquad r = 1-12 \tag{23}$$

form a 12×1 submatrix which is added to the submatrix $[F_i]$ in Equation 4:

$$V_i[\Delta\sigma] \rightarrow [F_i] \tag{24}$$

4.2 Thermal-mechanical response of a fractured rock block

To demonstrate the capability of 3-D DDA to deal with thermal-mechanical behaviour of fractured rock masses, an example is given of a fractured rock block acted on by thermal loading, as shown in Figure 4a. The block is 5 m × 5 m × 2.5 m in size. Its bottom face shown in the figure is fixed, and the other faces are free. The block is cut by three sets of fractures, one horizontal and two steeply inclined. The friction angle and cohesion of all fracture surfaces are zero. The material constants for the block are: Young's modulus E = 10 MPa, Poisson's ratio υ = 0.3, and thermal expansion coefficient $\alpha = 0.00002K^{-1}$. Figure 4b shows the 3-D DDA results for the case when the temperature of the block is increased linearly from 10°C to 80°C. It can be seen that the thermal loading causes fractures to open and blocks to slide relative the each other.

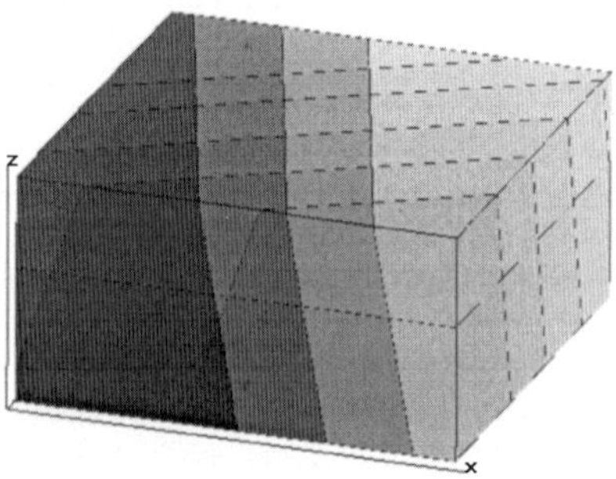

Figure 4a. A fractured rock block.

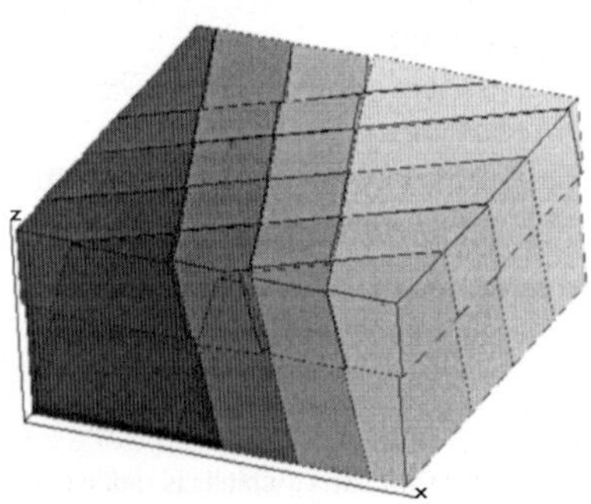

Figure 4b. 3-D DDA results of a thermally-loaded fractured rock block.

5. CONCLUSIONS

This paper describes the basic principles of 3-D DDA and puts forward models to account for groundwater pressure and thermal loading in 3-D DDA. The water force and thermal loading submatrices are derived by minimizing the total potential energy and added to the global matrices. These extensions to 3-D DDA have been implemented into a 3-D DDA computer program, and numerical results from three test cases are reasonable.

REFERENCES

Chen, C.Y., Zhang, J.M., Du, Y.L. & Wang, S.J. 1998. *3-D visualization of geoinformation in slope engineering and its application to the permanent shiplock slope in the Three Gorge Project* Chinese Journal of Geotechnical Engineering 20(4): pp.1-6.

Chen G. Q., Miki S., and Ohnishi Y. 1997. *Development of the interactive visualization system for DDA*. Proc. of the 9th Intern. Conference on Computer Methods and Advances in Geomechanics, Wuhan, China, pp.495-500.

Jang, H.I. & Lee, C.I. 2002. *Development of a three-dimensional discontinuous deformation analysis technique and its application to toppling failure*. Proc. of the 5th intern. conference on analysis of discontinuous deformation, Wuhan, China, pp.225-229.

Jing, L.R., Ma, Y. & Fang, Z.L. 2001. *Modeling of fluid flow and solid deformation for fractured rocks with discontinuous deformation analysis (DDA) method.* Int. J Rock Mech Min Sci 38: pp.343-355.

Kim, Y.I., Amadei, B. & Pan, E. 1999. *Modeling the effect of water, excavation sequence and rock reinforcement with discontinuous deformation analysis*. Int J Rock Mech Min Sci 36: pp.949-970.

Lin, C.T., Amadei, B., Jung, J. & Dwyer, J. 1996. *Extensions of discontinuous deformation analysis for jointed rock masses*. Int J Rock Mech Min Sci 33: pp.671-694.

Shi, G.H. 1988. *Discontinuous deformation analysis-a new numerical model for the statics and dynamics of block system.* Ph.D. Thesis, Dept. of Civil Engineering, University of California, Berkeley.

Shi, G.H. 2001. *Theory and examples of three dimensional discontinuous deformation analyses*. Proc. of the 2nd Asian Rock Mechanics Symposium, Beijing, China, pp.27-32.

Yeung, M.R., & S.C. Blair 1999. Analysis of large block test data using the DDA *method.* Proceedings of the Third International Conference on Analysis of Discontinuous Deformation, Vail, Colorado, USA, pp.141-150.

Yeung, M.R., Jiang, Q.H. & Sun, N. 2003. *Validation of block theory and three-dimensional discontinuous deformation analysis as wedge stability analysis methods.* International Journal of Rock Mechanics & Mining Sciences 40(2): pp.265-275.

PENETRATION-INDUCED PORE PRESSURE MAGNITUDES – METHODS TO DETERMINE TRANSPORT PARAMETERS FROM TERRESTRIAL AND MARINE PENETROMETER TESTING

D, Elsworth[1], D.S. Lee[1], H. Long[1], and P.B. Flemings[2]
[1]Pennsylvania State Univeristy / Department of Energy and Geo-Environmental Engineering
[2] Pennsylvania State University / Department of Geosciences

Abstract: New approaches are developed to use peak penetration-induced pore pressure magnitudes to index profiles of in situ permeability. These methods potentially enable permeability profiles to be determined from continuous sounding by cone penetrometers, and from freefalling marine penetrometers that self-embed into seafloor sediments. Solutions are developed for the unsteady and partially-drained, fluid pressure field that develops around a volumetric dislocation that represents a penetrometer either advanced at constant rate, or decelerating as it self-embeds after freefall. Inertial effects are ignored. These solutions provide a consistent framework to view the penetration process, and enable penetration-induced pore pressures to be represented in non-dimensional form, and related to conventional sounding indices inclusive of tip resistance, Q_t, friction factor, F_r, and pore pressure ratio, B_q, among others. These relationships are used to define permeability magnitudes in the intermediate range where response is partially drained.

1. INTRODUCTION

Penetrometers have been used as a minimally invasive, rapid, and consequently inexpensive method of determining soil properties, in situ (e.g. Mitchell and Brandon, 1998). Originally developed to correlate measurements of penetration resistance with magnitudes of soil strength, techniques have been extended to incorporate both measurement and analysis of penetration-generated pore fluid pressures. The generation and subsequent dissipation of these fluid pressures are related to the transport properties of the penetrated medium, and this response may be used to determine permeability, directly. Specifically, the peak magnitude of pore fluid pressure, measured at the tip, and its rate of dissipation, may be correlated directly with permeability magnitudes. Pore fluid pressures are generated by the mechanical displacement of the soil surrounding the penetrometer tip, represented in its simplest form as hydromechanical coupling (Elsworth, 1993; Elsworth, 1998). Measured pore pressure magnitudes may be correlated with empirical (Robertson et al., 1986) and process-based (Levadoux and Baligh, 1986) models for behavior, either incorporating models of failure in the tip-process zone (Teh and Houlsby, 1991, Song et al., 1999), or accommodating pseudo-linearized behavior (Elsworth, 1993). The pore pressure response may be used to determine permeability magnitudes around the tips of terrestrial cone penetrometers, advanced either from the surface (Elsworth and Lee, 2003) or beyond the base of evolving drillholes (Long et al., 2003), or from marine lances, freefalling through the water column and self-embedding into soft sediments on the seafloor (Lee and Elsworth, 2003). Similar approaches can be applied to each, the specifics of which are documented in the following.

2. ANALYSIS OF PORE PRESSURE RESPONSE

Relations may be developed to represent the steady pore fluid pressures that develop around a penetrometer under steady penetration, at penetration rate, U. These models necessarily employ simple linearized constitutive relations, but incorporate the important influence of a porous medium migrating past the penetrometer tip, albeit in a simplified form. Dislocation models (Elsworth, 1991; 1993) may be applied to represent a penetrometer of infinitesimal-radius, but suffer the disadvantage that penetration-induced pressures become singular at the assumed penetrometer tip. The approximate solution for a finite radius penetrometer avoids this shortcoming, as explored in the following.

2.1 Infinitesimal- radius penetrometer

Dislocation models allow undrained to drained conditions to be represented during penetration and following arrest for penetrometers (Elsworth, 1991). The same parameters are used to represent undrained and partially drained behaviour, including estimation of cavity expansion stresses that represent tip resistance, q_t (Elsworth, 1993). Pore pressures generated behind and ahead of the tip are defined by representing the penetrometer as a series of volumetric dislocations, arranged along the trajectory of the penetrometer, and activated sequentially as the tip advances (Elsworth, 1993).

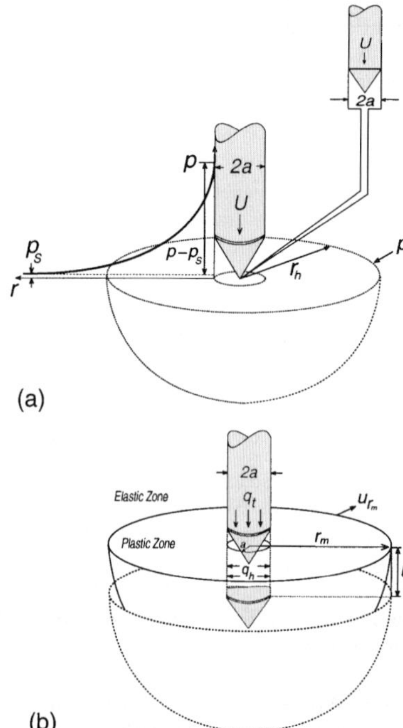

Figure 1 Geometry of the process zone surrounding an advancing penetrometer. (a) Hydraulic behaviour defined with pressure, p, induced at the tip, and remote pressure, p_s, at radius, r_h. (b) Mechanical behaviour defined by mechanical process zone of radius, r_m, with tip stress of magnitude, q_t.

As this linear distribution is successively "inflated," and retained in this inflated state, pore pressures that are induced around the tip and shaft may be straightforwardly determined as (Elsworth, 1991)

$$p - p_s = \frac{Ua\gamma_w}{4K}\frac{a}{x}e^{-\frac{U(\sqrt{x^2}-x)}{c_v}} \qquad (1)$$

where c_v is coefficient of consolidation, a is the radius of the penetrometer, and x is the location where the excess pore pressure, p-p_s, is measured, relative to the most recently activated dislocation, that concurrently represents the penetrometer tip (x positive behind the tip). Importantly, along the penetrometer shaft, the steady pore pressures decay with the reciprocal of length from the tip, and are defined as $4K(p-p_s)/U\gamma_w a$, enabling permeability to be, in principle, recovered from the steady pressure distribution at the tip and along the shaft. The main problem in applying this method is that the soil is assumed linearly elastic and induced pressures are singular at the penetrometer tip. Hence, an alternative representation is sought.

2.2 Finite radius penetrometer

Similar to the point moving dislocation, approximate solutions may be developed for the fluid pressure field that develops around a migrating dislocation of finite size. The approach is similar to that for a point dislocation and considers penetration as analogous to injecting a fluid volume equivalent to that introduced, per unit time, by the insertion of the penetrometer column (Figure 1), of diameter $2a$, and distributed over the spherical contour representing the penetrometer tip, that moves at a penetration rate (or synonymous insertion velocity), U. This fluid volume, equivalent to $dV = \pi a^2 U$, per unit time, is injected into a static location, and the resulting difference between this and the point dislocation analysis is that the dynamic flow system is assumed as illustrated in Figure 1(a). The principal requirement is for the system to be in a dynamic steady state, and the contour of injection is now of finite radius, not infinitesimal. The resulting approximate solution to this spherically symmetric geometry is obtained by applying continuity and Darcy's law to yield

$$p - p_s = \frac{\gamma_w}{4\pi K a}\left[1-\frac{a}{r_h}\right]dV = \frac{Ua\gamma_w}{4K}\left[1-\frac{a}{r_h}\right] \qquad (2)$$

where p_s is the static pore fluid pressure at radius r_h from the penetrometer, relative to the pressure measured at the penetrometer face, p. This simply represents the solution where fluid flux is injected at the interior surface of the spherical shell (Figure 1(a)) at a rate equivalent to the insertion volume, per unit time, of the penetrometer. Where dilation occurs at constant rate $dV \cong U\pi a^2 A$, where A represents the average volume change in the process zone and is related to mean volume strain in the process zone, ε_v, $\varepsilon_v = \Delta V/V = a^2 A/r_m^2$. Where the soil is contractile, A is positive. Where no void volume change results, A is null, and dilation is represented by negative values. Most importantly, A may be recovered from routinely measured cone metrics, enabling the induced excess pore pressure at the tip to be evaluated by substituting $dV \cong U\pi a^2 A$ into Equation 2, where $a/r_h << 1$, as

$$p - p_s = \frac{U\gamma_w aA}{4K}\left[1 - \frac{a}{r_h}\right] \approx \frac{U\gamma_w aA}{4K} \qquad (3)$$

Notably, this solution is similar to that recovered for a moving point dislocation (Equation 1), but includes three important improvements: the effect of failure in a mechanical process zone is incorporated, as are the influence of both dilation in the process zone and effect of a finite size penetrometer tip. Despite these additions, the approximate solution is not able to discriminate between pore pressures measured on the tip, shoulder, or shaft. Therefore although unable to differentiate between the response resulting from different tip geometries and locations, the most crucial components regulating the response are included.

2.3 Piezocone indices

A variety of normalized indices are available to define the end-bearing, frictional, and pore pressure response recorded around the tip of an advancing piezocone. The dimensionless magnitudes of tip resistance, Q_t, and sleeve friction, F_r, may be defined as

$$Q_t = \frac{q_t - \sigma_{vo}}{\sigma'_{vo}}; \quad F_r = \frac{f_s}{q_t - \sigma_{vo}} \qquad (4)$$

where q_t is the corrected tip resistance, f_s is the magnitude of sleeve friction, defined in units of stress, σ_{vo} is the initial magnitude of in situ vertical stress, and the prime denotes effective stress. In addition to these mechanical parameters, pore pressure ratio, B_q is defined as

$$B_q = \frac{p - p_s}{q_t - \sigma_{vo}} = \frac{p - p_s}{\sigma'_{vo}}\frac{1}{Q_t} \qquad (5)$$

where the second part of Equation 5 results from the substitution of Equation 4. Notably, Equation 5 links B_q and Q_t. Normalized sleeve friction, F_r, may be related to pore pressure, though knowledge of strength parameters resulting in,

$$F_r = \frac{c}{(q_t - \sigma_{vo})} + \left[M - B_q + \frac{M}{Q_t} + \frac{\sigma'_h + Mp_s}{q_t - \sigma_{vo}}\right]\tan\phi \qquad (6)$$

where it is implicitly assumed that the coefficient of friction at the sleeve-soil interface is $\tan\phi$ and cohesion is c. This enables the dimensionless sleeve friction to be determined from fundamental material characteristics. With these quantities determined, an evaluation of the permeability of the penetrated soil may be defined.

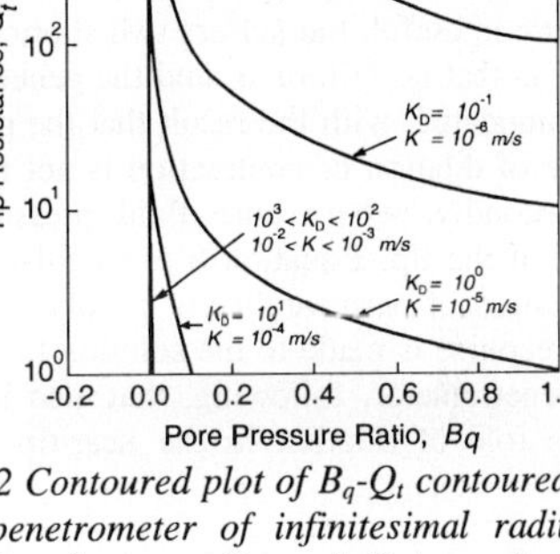

Figure 2 Contoured plot of B_q-Q_t contoured for K_D for a penetrometer of infinitesimal radius. The conversion between K_D and K is performed for effective stresses at 10 m depth, for penetration of a standard cone at 2 cm/s and for A=1.

2.3.1 Infinitesimal-radius penetrometer

With magnitudes of penetration induced pore pressure, p-p_s, available in Equation 1, substituting this relation directly into Equation 5 enables plots of Q_t versus B_q to be contoured for permeability, for the case of an infinitesimal radius penetrometer. The resulting relationship is

$$B_q = \frac{1}{\sigma'_{vo}}\frac{1}{Q_t}\frac{Ua\gamma_w}{4K}\frac{a}{x}e^{-\frac{U(\sqrt{x^2}-x)}{c_v}} = \frac{1}{K_D Q_t}\frac{a}{x}e^{-\frac{U(\sqrt{x^2}-x)}{c_v}} \qquad (7)$$

where $K_D = (4K\sigma'_{vo})/(Ua\gamma_w A)$, with $A=1$, and is a non-dimensional index, directly proportional to coefficient of permeability, K. Where measurements of pressure are taken on the shaft, only, Equation 7 reduces to a simpler form as x is constrained to be positive, behind the tip. In this instance, the role of the non-dimensional penetration rate, $U_D = Ua/2c_v$ (Elsworth, 1991), is eliminated in controlling the form of the shaft pressure distribution during steady penetration, where it varies as $(p\text{-}p_s) \propto 1/x$. The magnitude of peak pressure is modulated by a different non-dimensional parameter, K_D, as Equation 7 becomes

$$B_q = \frac{1}{K_D Q_t}\frac{a}{x} \qquad (8)$$

where the factor a/x is defined by the location of the pore pressure transducer, and is sensibly of the order of 1 to 10, on the shaft. Plots of B_q-Q_t are shown in Figure 2 for $a/x=1$, contoured for values of K_D, and permeability, where σ'_{vo} is taken as $100\ kPa$. This figure notes the variation of K_D, which is directly proportional to permeability, K, enabling permeability to be determined, at least in theory. This figure is useful, but suffers two shortcomings. The first is that no failure around the penetrometer is accommodated, with the result that the potential influence of dilation or contraction is not incorporated. Secondly, where pore fluid pressures are recorded at the tip, Equation 8 is singular and no determination of permeability is possible. For this reason, recourse is made to the solution for a finite radius penetrometer, following, that also incorporates the role of dilation in the near-tip process zone.

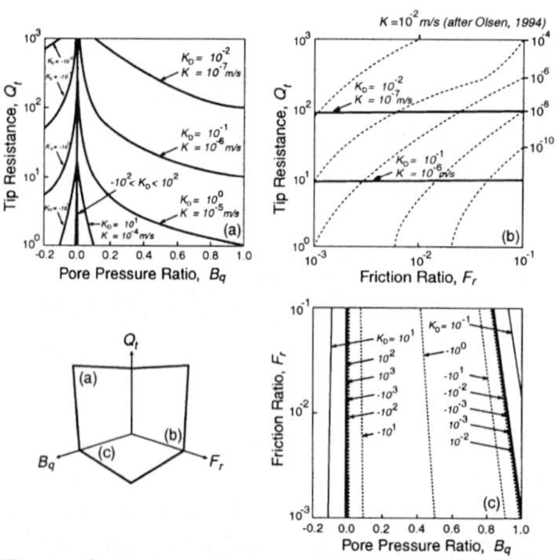

Figure 3 Contoured plots of (a) B_q-Q_t, (b) F_r-Q_t, and (c) B_q-F_r contoured for K_D for a penetrometer of finite radius. Values of K are shown for A=1 and σ'_{vo}=100 kPa. The empirical results of Olsen (1994) are shown dashed in Figure (b).

2.3.2 Finite-radius penetrometer

The problem of singular pressures at the tip of a moving dislocation is circumvented if the approximate solution for steady pressure distribution around a migrating cavity (defined in Equation 3) is used. Substituting Equation 3 into Equation 5 enables the pore pressure ratio, B_q, to be defined as

$$B_q \approx \frac{Ua\gamma_w A}{4K}\frac{1}{(q_t - \sigma_{vo})} = \frac{Ua\gamma_w A}{4K\sigma'_{vo}}\frac{1}{Q_t} = \frac{1}{K_D Q_t} \qquad (9)$$

This defines a relation between B_q-Q_t that is plotted in Figure 3(a) for a variety of magnitudes of the non-dimensional parameter, K_D, directly proportional to permeability. This is similar to Figure 2, but is representative of a finite-radius penetrometer tip, where dilation or compaction of the tip process zone is accommodated.

Noting from Equation 9 that $B_q \approx 1/K_D Q_t$ enables Equation 6 to be transformed to a relation between F_r-Q_t with $F_r = f(c,\phi,K_D,M,Q_t)$ and plotted in Figure 3(b) for various K_D, where $c \to 0$ and $M \approx 1$.

3. PORE PRESSURE REPONSE

Field data correlating cone sounding data with field-measured permeabilities are necessary to confirm the utility of the models. To investigate the veracity of the proposed models, two applications are possible. The first involves attempting to determine magnitudes of K_D, by sequentially using alternate pairs of the penetrometer indices (F_r, B_q, and Q_t) to determine which twin-pairs give consistent predictions of K_D for terrestrial penetrometer. The second involves proving independent corroboration permeabilities, and also soil strength parameters for marine penetrometer, including the Pop-Up Pore Pressure Instrument (PUPPI). Both applications are documented in the following.

3.1 Penetrometer advanced at constant rates

The data used here are for normally consolidated varved clays at the National Geotechnical Experimentation Site in Amherst, Massachusetts (DeGroot and Lutenegger, 1994). The penetration is "partially drained" making this analysis appropriate, and correlating permeability data are available from laboratory tests on recovered samples. A full suite of the three parameters B_q-Q_t-F_r is available and these data are used in a triplet of pairs, B_q-Q_t, F_r-Q_t, and B_q-F_r, to evaluate K_D. With K_D determined, magnitudes of permeability are recovered by determing the magnitude of the dilation parameter, A, directly for the sleeve friction data (F_r) and plotted with depth in Figure 4.

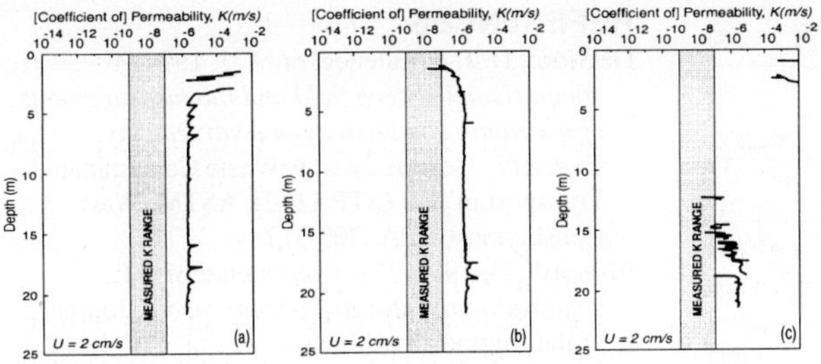

Figure 4 Plots of predicted permeability with depth at the Amherst Test Site, CPTU-1. Permeability profiles are determined from the data pairs (a) Bq-Qt, (b) Fr-Qt, and (c) Bq-Fr to recover magnitudes of K_D.

Permeability measurements, completed both on laboratory samples and in a predrilled well, are in the range 10^{-7} to 10^{-9} *m/s*, with the field measured magnitudes populating the higher portion of the range, circa 10^{-7} *m/s*. Estimates of permeability using B_q-Q_t and F_r-Q_t data are consistent for both soundings, are uniform with depth, and are of the order of 10^{-6} *m/s*. These only marginally overestimate the field measured magnitudes and are taken as reasonable predictions of the permeability. The consistency between estimates recovered independently from B_q-Q_t, and F_r-Q_t data is taken as particularly encouraging. The use of B_q-F_r data also produces reasonable estimates with both field measured permeabilities and those predicted from the other cone indices. Data missing from the profile are where predictions of permeability are negative. Since for these data, induced pore pressures are negative, they will only result in admissible permeabilities (positive) if the dilation parameter is negative, a factor not recovered in these particular data.

3.2 Self-embedding penetrometer

A solution is developed for the build-up, steady and post-arrest dissipative pore fluid pressure fields that develop around a penetrometer that self-embeds from freefall into the seabed. Arrest from freefall considers deceleration under undrained conditions in a purely cohesive soil, with constant shear strength with depth. Consider a lance falling through the water column that has reached terminal velocity, U_0, and subsequently impacts the soft sediments of the seabed, as illustrated in Figure 5. The non-dimensional pressure, P_D, may be used to define the build-up of pressure following the impact of the penetrometer with the surface of the seabed. The penetrometer impacts the seabed at velocity U_0, represented in dimensionless magnitude as U_D, and decelerates to arrest. Non-dimensional pressures are plotted as the product $P_D x_D$, since it is known that the peak pressures, shown in this format asymptote to unity, as $P_D x_D \rightarrow 1$. In Figure 6, the influence of tip taper is prescribed for various taper angles, penetration rates, monitoring locations, and material strengths on the penetrometer shaft for build-up and dissipation, respectively.

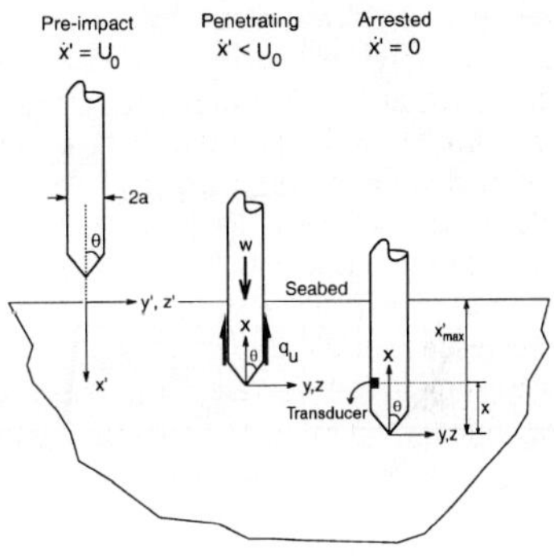

Figure 5 Schematic of a lance falling at terminal velocity, U_0, and impacting the seabed. For the evolving partially-drained analysis of embedment-generated pore pressures, the coordinate system is fixed to the penetrometer tip.

Since the non-dimensional pressure P_D, includes the magnitude of permeability, k, the peak generated pressure may be used to evaluate transport parameters of both permeability and consolidation coefficient. From the pressure, P_D, permeability may be determined for peak insertion pressure, p-p_s, as

$$\frac{k}{\mu} = \frac{U_0 a}{4(p - p_s)x_D} \tag{11}$$

provided behaviour has asymptoted to $P_D x_D = 1$. These practical solutions provide feasible mechanisms to evaluate seabed strength and fluid transport properties of the penetrated seabed sediments.

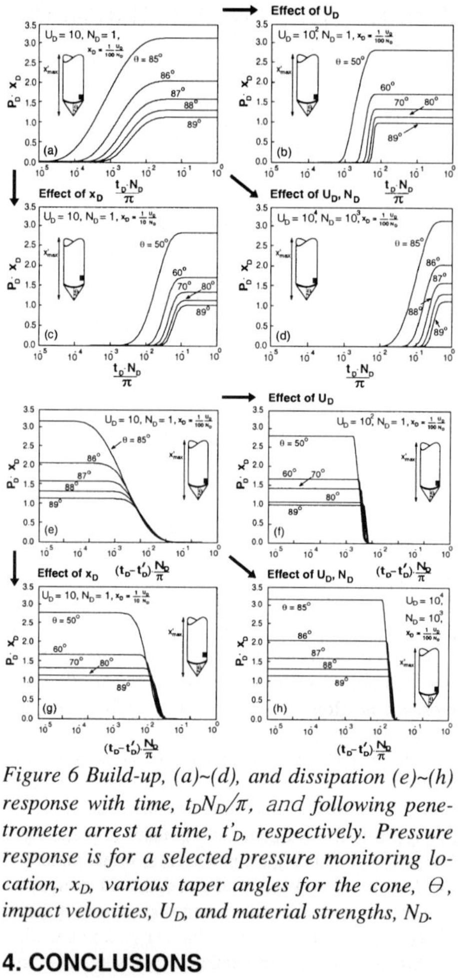

Figure 6 Build-up, (a)~(d), and dissipation (e)~(h) response with time, $t_D N_D/\pi$, and following penetrometer arrest at time, t'_D, respectively. Pressure response is for a selected pressure monitoring location, x_D, various taper angles for the cone, Θ, impact velocities, U_D, and material strengths, N_D.

4. CONCLUSIONS

Solutions are developed for the pore pressure fields that develop around penetrometers either advanced at constant or decelerating rate and under partially-drained flow conditions. The maximum pore pressure, recorded at the penetrometer is suggested as indicative of the permeability of the surrounding soil, since the peak-pressure is conditioned by the ability for fluids to escape. This full suite of solutions promotes a consistent framework to understand the processes operating during penetrometer penetration and arrest, and to enable the conventional sounding indices inclusive of tip resistance, Q_t, friction factor, F_r, and pore pressure ratio, B_q to be related to penetration-induced pore pressures, and therefore to permeability.

REFERENCES

DeGroot, D. J. & Lutenegger, A. J. 1994. *A comparison between field and lab measurements of hydraulic conductivity in a varved clay.* Hydraulic Conductivity & Waste Contaminant Transport in Soil (STP 1142), ASTM, West Conshohocken/PA, 300-317.

Elsworth, D. 1991. *Dislocation analysis of penetration in saturated porous media.* Journal of the Engineering Mechanics. Vol. 117. No. 2. 391-408.

Elsworth, D. 1993. *Analysis of piezocone dissipation data using dislocation methods.* Journal of Geotechnical Engineering. Vol. 119. No.10. 1,601-1,623.

Elsworth, D. & Lee, D.S. 2003. *Permeability determination from On-the-fly piezocone sounding.* (under review). Journal of Geotechnical and GeoEnvironmental engineering.

Lee, D.S. & Elsworth, D. 2003. *Indentation of a free-falling sharp penetrometer into a poroelastic seabed.* (under review). International Journal for Numerical and Analytical Methods in Geomechanics.

Long. H., Elsworth, D. & Flemings, P.B. 2003. *Composite dislocation analysis of piezoprobe penetration.* (in preparation). J. Geophys. Res.

Mitchell, J.K. & Brandon, T.L. 1998. *Analysis and use of CPT in earthquake and environmental engineering.* Geotechnical Site Characterization, Robertson & Mayne (eds), Balkema, Rotterdam.

Robertson, P. K., Campanella, R. G., Gillespie, D., & Greig, J. 1986. *Ues of piezometer cone data.* Proc. ASCE Spec. Conf. In Situ '86. Use of In Sit Tests in Geotechnical Engineering, Blacksburg, 1,263-1,280, ASCE.

Levadoux, J. N. & Baligh, M. M. 1986. *Consolidation after undrained piezocone penetration. I:* Prediction. Journal Geotechnical .Engineering. Vol. 112. No. 7. 707-726.

Song, C. R., Voyiadjis G. Z., & Tumay, M. T. 1999. *Determination of permeability of soils using the multiple piezo-element penetrometer.* International Journal for Numerical and Analytical Methods in Geomechanics. Vol. 23, No. 13. 1,609-1,629.

Teh, C. & Houlsby, G. 1991. *An analytical study of the cone penetration test in clay.* Geotechnique. 41. 17-34.

A DOUBLE-POROSITY POROELASTIC MODEL TO RELATE P-WAVE ATTENUATION TO FLUID FLOW IN VUGGY CARBONATE ROCK

Jorge Parra[1] *and Chris Hackert,*[1] *Steven Pride*[2]
[1]) *Southwest Research Institute, San Antonio, Texas, USA*
[2]) *Geosciences Rennes, University de Rennes 1, Rennes Cedex, FRANCE*

Abstract: A pressure source in a fluid-filled porous formation (a South Florida aquifer) excites P waves that compress the pore volume. This results in increasing pore pressure, which induces fluid flow in the carbonate rock. Fluid flow is modeled by assuming that the correlation length based on the heterogeneous pore structure corresponds to a double porosity medium. We determine pore structure using computer tomography. We also use well logs to estimate fluid and poroelastic properties, and we construct a double-porosity model to predict attenuation and spectral fluid pressure responses in the frequency range of 0.01-100 kHz. We find that the high fluid pressure is controlled by permeability, and the wave motion can excite slow waves observed as oscillations in the low frequency part of the spectrum. The results demonstrate how rapidly the slow waves vanish in low permeability regions, and that the fluid pressure drop is greater for fluid transport between the large vugs and the matrix.

1. INTRODUCTION

When compressional, or P, waves excited by a seismic source propagate in an aquifer that is characterized by heterogeneities of different dimensions, such as variations in lithology and vuggy porosity, different regions respond with different fluid pressures. The associated fluid-pressure diffusion attenuates the wave energy. Three types of waves can be identified – a fast and a slow P wave and a shear wave. For the fast P wave, the pore fluid and porous matrix are compressed simultaneously; for the slow P wave, the porous matrix relaxes when the pore fluid is compressed.

In general, in a homogeneous fluid-filled porous medium, when the passing wave induces small fluid pressure gradients between regions of compression and extension, these gradients cause fluid flow relative to the solid. The fluid flow incurs viscous loss, resulting in a small attenuation of the passing wave. However, in real rocks this phenomenon is overwhelmed by local flow effects such those observed by Mavko and Nur (1975). These effects are caused by the pore fluid between regions of different compliances under the compression (or extension) induced by the passing wave. In an application given by Gurevich et al. (1997) for a medium composed by thin alternating layers of two poroelastic materials with different compliances, propagation of a P-wave will squeeze the fluid from more compliant into the less compliant layers. This local flow of the pore fluid is accompanied by the viscous loss and results in the attenuation of the passing wave. On the other hand, in an aquifer when the P-waves compress an element containing mesoscopic heterogeneity, the different porosity types should respond with different changes in their fluid pressure. An internal equilibrium then takes place with fluid flowing from the more compliant high-pressure regions to the relatively stiff low-pressure regions. Such mesoscopic flow attenuates energy and affects the wave amplitude.

In this paper we investigate the attenuation and fluid pressure due to mesoscopic flow in a double-porosity medium. The medium we analyze is an aquifer characterized by pore structure formed by a matrix and vuggy porosity. Attenuation curves of double porosity models are calculated using the theory developed by Pride and Berryman (2003). The Green function for a volume source in a double porosity effective medium is developed for calculating fluid pressure spectral responses and seismic signatures based on typical vuggy carbonate rock properties from South Florida.

2. PROBLEM FORMULATION

To calculate the pore pressure response due to a volume source we use the Green's function based on the effective Biot theory. We write the coupled system of equations directly from the constitutive relations given by Biot (1962). These are the total stress of the isotropic porous medium, the stress in the porous fluid, the momentum balance equation for total stress, and the generalized Darcy's law. Following Parra (1991) and Boutin et al. (1987), the coupled system of differential equations in the frequency domain for a volume source (V) and a force (F) is given by:

$$\mu\nabla^2 u + (\lambda+\mu)\nabla\nabla \bullet u + \omega^2\bar{\rho} - \bar{\alpha}\nabla p = -F \qquad (1a)$$

$$\theta\nabla^2 p - \beta p - \bar{\alpha}\nabla \bullet u = -V\,, \qquad (1b)$$

where u is the particle vector displacement of the solid, p is the fluid pressure, and

$$\bar{\rho} = \rho + \rho_f^2\,\omega^2\theta$$
$$\bar{\alpha} = \alpha + \rho_f\,\omega^2\theta$$
$$\lambda = K_d - 2\mu/3$$

$$\theta = \frac{jk(\omega)}{\eta\omega}$$

where ρ is effective bulk density, and 8, ∀ and ∃ are effective constants related directly to the complex Biot constants H, C and M by:

$$\beta = \frac{1}{M} \qquad (2a)$$

$$\alpha = \frac{C}{M} \qquad (2b)$$

$$\lambda + \mu = H - G - C^2/M\,. \qquad (2c)$$

We also have the effective shear modulus relation, : = G, and the effective parameter, θ, which includes the effective fluid inertia in the relative motion parameter, $\tilde{\rho}$, i.e.,

$$\theta = -\frac{1}{\omega^2\tilde{\rho}}\,, \qquad (3)$$

where $\tilde{\rho} = -\eta/[j\omega\,k(\omega)]$, $k(\omega)$ is the complex frequency-dependent dynamic permeability, and η is the fluid viscosity.

The effective complex Biot constants H, C, and M are defined in terms of the effective drained modulus of the composite K_d, the effective undrained bulk-modulus K_u, and the effective Skempton's coefficient B as

$$H = K_u + 4G/3$$
$$C = BK_u \qquad (4)$$
$$M = \frac{B^2}{1 - K_d/K_u}K_u$$

where the effective complex functions K_d, K_u, and B are given for double porosity materials by Pride and Berryman (2002).

The next step is to find the Green's function from Eq. (1). Parra (1991) solved this equation and developed a solution for the pressure source (or volume) using the Sommerfeld integrals. The fluid pressure in the frequency domain is given by

$$p = \frac{V_o(\omega)}{4\pi\omega^2\bar{\rho}}\left[\frac{\delta_1}{\alpha_3}\left(\frac{e^{j\lambda_1 R}}{R}\right) - \frac{\delta_2}{\alpha_3}\left(\frac{e^{j\lambda_2 R}}{R}\right)\right] \qquad (5)$$

where R is the distance between the source and receiver, and the parameters λ_1, λ_2 are the wave numbers associated with fast and slow P waves travelling in an effective porous medium. The first and second terms in Eq. (5) are the fast and slow P waves, respectively. Wave numbers that control the phase velocity and attenuation of the wave travelling in the effective porous medium are summarized as follows:

$$\lambda_1 = [(b+\Delta)/2]^{1/2} \qquad (6)$$

$$\lambda_2 = [(b-\Delta)/2]^{1/2}$$

where $b = k_p^2 - \bar{\alpha}^2/(\lambda+2\mu)6 - \beta/6$

and $\Delta = [b^2 + 4k_p^2\beta/6]^{1/2}$

in which $k_p^2 = \omega^2\bar{\rho}/(\lambda+2\mu)$.

In Eq. (5), the factors of the terms containing the exponentials are obtained from the relations given by

$$\delta_m = \frac{\lambda_m^2 - k_p^2}{2\theta(\lambda_1^2 - \lambda_2^2)};$$

for $m = 1, 2$ and $\alpha_3 = 1/2\bar{\rho}\omega^2$.

3. RESULTS

We use cores and well logs to construct and verify a double porosity model. Core data provide images of pore structure, such as matrix porosity and range of vug dimensions and physical properties (Figure 1). Velocity, porosity and density logs provide poroelastic properties. Permeability is derived from NMR well logs. These data are given in Parra and Hackert (2002) and Parra et al. (2002). Based on the solution given by Pride and Berryman (2002), we assume a spherical vug of radius a surrounded by a homogeneous porous matrix.

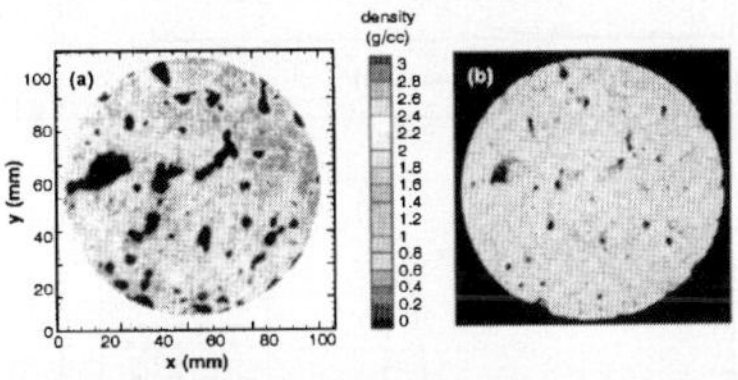

Figure 1. (a) Computed density from x-ray CT data, and (b) photograph of the end of the core.

The bulk modulus K_I of the rock matrix in terms of the grain density, ρ_s, the P- and S-wave velocities of the dry skeleton (v_p, v_s), and the porosity ϕ_I, allow us to constrain the double porosity model by use of the well log data and the equation,

$$K_I = (1-\phi_I)\rho_s(v_p^2 - \frac{4}{3}v_s^2). \tag{7}$$

From this data, we determine the following fluid and carbonate properties to construct the model:

$$K_s = 72.2GP \qquad \eta = 0.001P \bullet s$$
$$G_s = 38.6GP \qquad v_p = 2276m/s$$
$$\rho_s = 2650kg/m3 \qquad v_s = 1036m/s$$
$$v_f = 1500m/s \qquad \phi_1 = 35\%$$
$$\rho_s = 1000m/s \qquad \phi_2 = 45\%$$

Alternatively, to simulate the bulk modulus of the vuggy structure we consider that soft materials are embedded in the interconnected pore matrix containing vugs. In this case we use the relationship (Pride et al., 2002),

$$K_2 = K_s \frac{(1-\phi_2)}{(1+c\phi_2)}, \tag{8}$$

Where K_s is the modulus of the solid grain, ϕ_2 is porosity, and c is a consolidation parameter that represents the sediments and fluid inside the vugs (unconsolidated material). We use c = 200. The effective compressional modulus (harmonic average of the K_I and K_2 drained moduli) and the shear bulk modulus, G, of the double porosity medium can be obtained, respectively, from

$$K = \frac{K_I K_2}{d_2 K_I + d_I K_2}, \tag{9a}$$

$$G = (1-\phi)\rho_s v_s^2, \tag{9b}$$

Where ϕ is the effective porosity of the double porosity medium given by

$$\phi = \phi_I d_I + \phi_2 d_2. \tag{10}$$

The values of d_I =0.963 and d_2 = 0.037 are for a little sphere embedded in a stiff porous matrix with interconnected pore structure, as given in Pride and Berryman (2003).

Vug size varies in the pore matrix, with typical lengths from 1-6 cm. In the core in Figure 1, vugs are 15-20 mm in length. Based on NMR logs, we find that permeability in the region of interest varies from 1-4 Darcies. Using these data we produce attenuation curves for vugs of dimensions a = 1, 1.5, 2 and 3 cm, in an effective medium having a permeability of 3 Darcies, as shown in Figure 2a. In this case the fluid flows easily from pore matrix to vugs and vice versa. Attenuation peaks are shifted to low frequency range as vug dimension increases. The mesoscopic heterogeneity (vug) of characteristic length L = 1.92a equilibrates the pressure at a frequency

$$f = D/L^2 \tag{11}$$

where D is the pore-pressure diffusivity given by

$$D = \frac{k}{\eta}\frac{BK}{\alpha} \tag{12}$$

and where k is permeability, B is Skempton's coefficient, K is drained bulk modulus, and α is the Biot-Willis coefficient of the medium. For given fluid and rock physical properties, frequency is inversely proportional to the characteristic length,

L. This makes attenuation peaks shift downward with increasing vug size. The calculated equilibrium frequencies (f_{meso}) for heterogeneities with radii of 1.0 cm, 1.5 cm, 2.0 cm, and 3.0 cm are 4446 Hz, 1976 Hz, 1112 Hz, and 494 Hz, respectively.

To relate attenuation to fluid dynamics, we calculate the fluid pressure response due to a pressure source (or volume source) located 100 m from a pressure detector. The pressure response is calculated in MPa as a function of frequency, as shown in Figure 2b, for a = 1 to 3 cm. Each curve shows an increase in pressure to a maximum and then a sharp decrease to a minimum to values near the mesoscopic frequency (f_{meso}). The plots show that as vug dimension increases pore pressure decreases in the frequency range of 3-1000 Hz. The pressure peaks are shifted to the lower frequencies as vug size increases. An internal equilibrium takes place as the pulse deforms the pore space, and fluid will tend to flow from compliant high-pressure regions to relatively stiff low-pressure regions (less compliant regions). Such mesoscopic flow reduces wave amplitude. In fact, the maximum pressure for each correlation length (vug dimension) occurs at a frequency less than the corresponding mesoscopic frequency and at lower attenuations. Lower frequency corresponds to long periods for a travelling pressure pulse. Long periods (low frequencies) provide sufficient time for the flow to occur during one-half period of the wave cycle. Thus, the fluid pressure will equilibrate and the rock will be in a relaxed state.

At high frequencies (short periods) there is not enough time for flow to occur, and the rock is in an unrelaxed state, thus, the rock is stiffer at high frequencies and has higher wave velocities. At intermediate frequencies, interactions between fluid and solid is at a maximum, and a maximum attenuation occurs. In this case, the wave amplitude is reduced practically to zero.

In Figure 2b we observe that in the frequency range between 3-750 Hz, the spectra exhibit rapid oscillations. This part of the spectra captures the slow P-wave motion due to propagation of a pulse in the fluid (water) of the aquifer when the matrix is relaxed (or in state of equilibrium). Oscillations in the spectrum arise from phase differences in the slow and fast waves. As seen in Figure 2b, these rapid oscillations occur for each of the pressure peaks. This suggests that the slow wave can be excited at frequencies equal to or less than the mesocopic frequency.

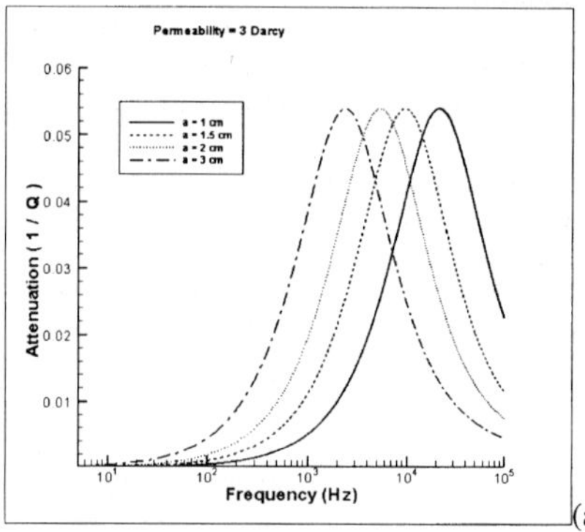

(a)

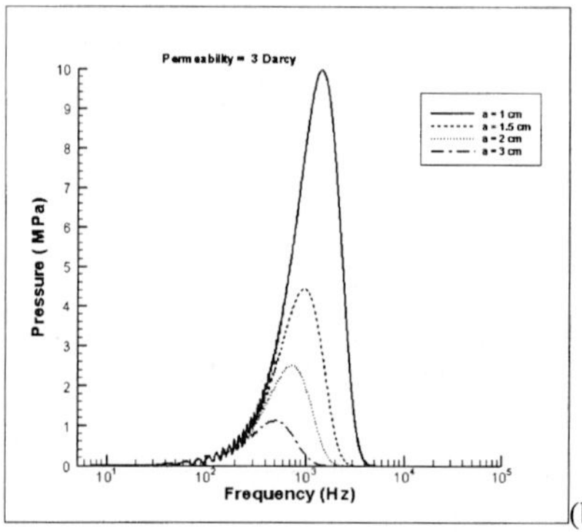

(b)

Figure 2. (a) Attenuation curves and (b) pressure spectral response illustrating the length scale effect on attenuation and fluid pressure.

To analyze this effect in more detail, we calculate the spectrum of slow P waves only by using the second term of Eq. (5). Figure 3 shows the spectra of slow P waves for a = 1-3 cm. The curves show pressure sensitivity to vugs in the pore matrix; the fluid pressure is sensitive to vugs at frequencies greater than 100 Hz. For frequencies less than approximately 100 Hz, the fluid pressure amplitude should remain unchanged.

The high permeability of the carbonate aquifer (easy flow) provides the right conditions for part of the P-wave energy to compress the fluid and excite slow waves. This phenomenon can also be seen in the time domain. A pressure source having a peak source frequency of 300 Hz excites fast and slow P waves that are recorded by a pressure detector at a horizontal distance of 100 m. The waveforms are calculated for a = 1, 1.5, 2, and 3 cm vugs, in an effective medium of permeability = 3 Darcies (see Figure 4). In the traces, the fast P wave arrives with a velocity of 2450 m/s, which agrees with the P-wave velocity log. The second event observed in each trace is the slow P wave that travels with a velocity of 1250 m/s. As expected when vug size increases, the amplitude of the fast P wave is more attenuated, and the amplitude of the slow P wave remains approximately the same. It is not affected by vug size as long as permeability does not vary.

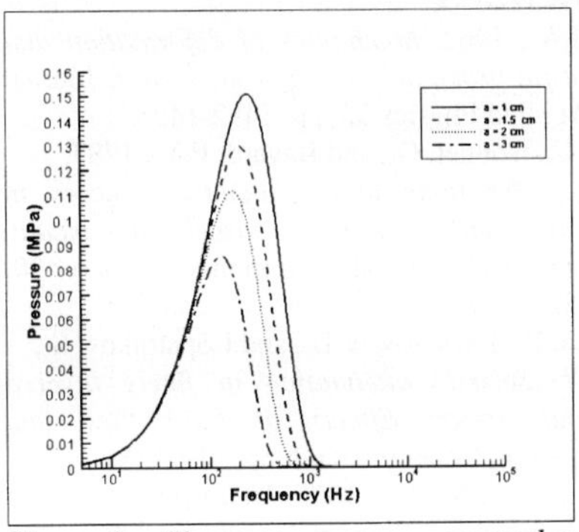

Figure 3. Slow-wave pressure spectral responses illustrate the length scale effect on fluid pressure.

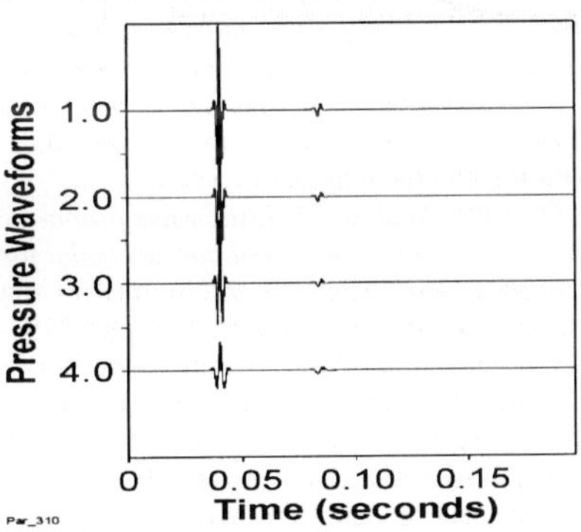

Figure 4. Pressure waveforms illustrating the length scale effect on fast and slow P-wave events. Waveforms correspond to vug sizes of 1, 1.5, 2, and 3 cm, in a matrix of 3 Darcy permeability.

In the second example given in Figure 5, we examine how permeability affects interconnected vugs. We assume that fluid will flow from the vugs to the pore matrix with the same permeability for a given length scale. We vary permeability from 1-3 Darcies for a vug size of 2 cm. As in the previous example, the mesoscopic heterogeneity has a characteristic length (a = 2 cm), and the fluid pressure equilibrates when the frequency is $f = D / L^2$. According to Eq. (12), pore pressure diffusivity D is proportional to permeability, which causes the mesoscopic attenuation peaks to shift upward to higher frequency with increasing permeability for a given heterogeneous medium, as shown in Figure 5a. For low frequencies, fluid pressure will attenuate more as permeability decreases (Figure 5b). This means that at lower frequency (long periods), fluid motion (from vugs to pore matrix) becomes unrelaxed in low permeability environments, and the presence of slow waves is diminished (i.e., less fluid pressure decreases the generation of slow waves). In a more permeable medium (e.g., 3 Darcies), fluid pressure increases and the slow wave can be excited and will be less attenuated.

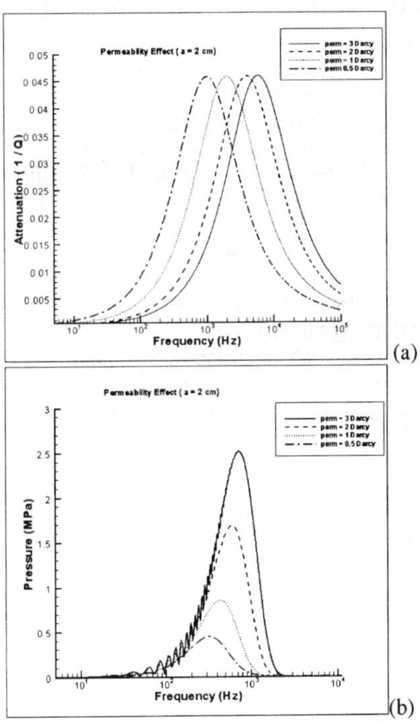

Figure 5. (a) Attenuation curves and (b) pressure spectral response illustrate the effect of permeability effects on fluid pressure attenuation.

Finally, we calculate synthetic seismograms for this model to show the effect of fast and slow P waves as a function of permeability. Each trace of Figure 6 represents different permeability. The waveforms were produced for a peak source frequency of 300 Hz. At this frequency, both waves are sensitive to permeability. For frequencies ≤100 Hz (see Figure 5), as permeability decreases, the amplitude of the fast P wave remains about the same, but slow wave amplitude decreases as permeability decreases.

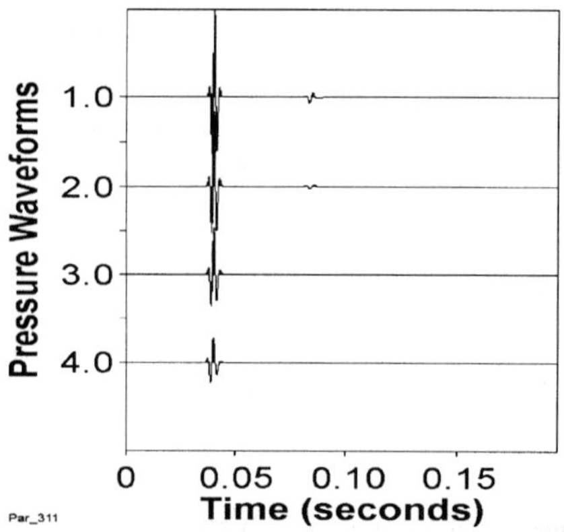

Figure 6. Pressure waveforms illustrating permeability effects on fast and slow P-wave events. Waveforms are for 0.5, 1, 2, and 3 Darcy permeability, with a vug of radius a=2 cm.

4. CONCLUSIONS

The effective double-porosity model was successfully adapted to the dynamic poroelastic wave equation to simulate P-wave and slow wave attenuation and spectral responses due to poroelastic fluid flow in vuggy carbonate rocks. We demonstrated that P-wave attenuation is sensitive to the pore structure of the carbonate rock formation. We found that the slow wave has a characteristic spectral signature that can be detected with seismic techniques in formations having interconnected vuggy porosity. This suggests that the hydraulic properties of vuggy zones can be inferred from P-wave attenuation measurements. The present solution allows us to relate the amplitude of fluid pressure oscillations to the attenuation of the wave motion, thus enhancing our understanding of coupling between the poroelastic fluid-flow and the vuggy porous matrix when conducting monitoring surveys.

ACKNOWLEDGMENTS

Program work was supported by contract DE-FC26-02NT15343 from the U.S. Department of Energy, National Petroleum Technology Office. The assistance of P. Halder (DOE) is gratefully acknowledged.

REFERENCES

Biot, M.A., 1962. *Mechanics of deformation and acoustic propagation in porous media*, Journal of Applied Physics 33: pp. 1482-1498.

Boutin, C., Bonnet, G., and Bayard, P.Y., 1987. *Green functions and associated sources in infinite and stratified poroelastic media*, Geophysical Journal Astronomic Society 90: pp. 521-550.

Gurevich, B., Zyrianov, V.B., and Lopatnikov, S., 1997. *Seismic attenuation in finely layered porous rocks: Effects of fluid flow and scattering*, Geophysics 62: pp. 319-324.

Mavko, G., and Nur, A., 1975. *Melt squirt in aesthenosphere*, Journal Geophysical Research 80: pp. 1444-1448.

Parra, J.O., Hackert, C.L., Collier, H., and Bennett, M., 2001. *A methodology to integrate magnetic resonance and acoustic measurements for reservoir characterization*, Report DOE/BC/ 15203-3, National Petroleum Technology Office, Department of Energy, Tulsa, OK. (www.reservoirgeophysics.swri.edu).

Parra, J.O., 1991. *Analysis of elastic wave propagation in stratified fluid-filled porous media for interwell systemic applications*, Journal of the Acoustic Society of America 90: pp. 2557-2575.

Parra, J.O., and Hackert, C.L. 2002. *Wave attenuation attributes as flow unit indicators*, The Leading Edge, June 2002: pp. 564-572.

Pride, S., Tromeur, E., and Berryman, J., 2002, *Biot slow-wave effects in stratified rocks*, Geophysics 67: pp. 271-281.

Pride, S., and Berryman, J., 2002. *Attenuation of P waves by wave-induced fluid flow*, Proceedings of the second Biot Conference on Poromechanics, Auriault et al. (eds.), Grenoble, France, pp. 739-744,.

Pride, S., and Berryman, J., 2003. *Linear dynamics of double-porosity and dual-permeability materials*, submitted to Geophysical Letters.

THERMO-MECHANICAL YIELDING OF A CLAY

Cane Cekerevac and Lyesse Laloui

Soil Mechanics Laboratory, LMS
Swiss Federal Institute of Technology at Lausanne, EPFL
CH-1015 Lausanne, Switzerland
E-mail: Lyesse.Laloui@epfl.ch

Abstract: The paper presents results of an experimental study of thermal effects on the mechanical behaviour of a saturated clay, with emphasis on the determination of the onset of yielding. The study was performed on CM clay (Kaolin) using a temperature-controlled triaxial apparatus. Applied temperatures were between 22 °C and 90 °C. Various methods are used to identify the yield points (pseudo-elastic limit) and to define the shape of the yield surface in the invariant stress space p' (effective mean pressure)- q (deviatoric stress). Yield surface obtained at 90 °C is compared with results at ambient temperature. Based on this comparison, thermo-mechanical yielding is discussed and yield limit evolution with temperature is presented.

1. INTRODUCTION

Research interest in the thermo-mechanical behaviour of soils is growing as a result of an increasing number of geomechanical problems involving thermal effects. These problems with non-isothermal situations are mainly encountered in the field of environmental geomechanics (Vulliet et al. 2002). The thermal effect is dominant in some applications such as geothermal structures where heat exchanger piles are used (Laloui et al. 2003). In spite of the practical relevance of the thermo-mechanical applications, the effect of high temperatures on soil behaviour is not yet completely understood. As an example, very few experimental results are available for the determination of the shear yield surfaces for different temperatures. This is due to the complex influence of temperature on the behaviour of soils, and the fact that thermo-mechanical testing of soils is much more complex than conventional isothermal testing.

This paper is a contribution to the subject of thermal effects on the mechanical behaviour of clays. We concentrate here on the presentation of experimental results obtained on a Kaolin clay under triaxial conditions at two temperatures (22 °C and 90 °C), in view of determining the evolution of the shear yield surface with temperature. A new temperature-controlled triaxial apparatus developed by the authors was used (Cekerevac et al. 2003). Temperature effects were identified by comparing the results of tests performed at high temperature (90 °C) with the results of the same type of tests carried out at ambient temperature (22 °C).

2. MATERIAL AND EXPERIMENTAL PROGRAM

2.1 Tested material

This study was performed on CM clay, commercially known as Kaolin. Samples were obtained by mixing de-aired distilled water with clay powder at a water content two times the liquid limit of the soil. After preparation, isotropic consolidation of the slurry was carried out under a confining pressure of 100 kPa. The soil samples, 55mm diameter and 110 mm in height, were then obtained by trimming the clay cakes. The grain size distribution revealed that about 97% by weight was smaller than 0.06 mm and 45% was smaller than 0.002 mm. The unit weight of the solid particles was $\gamma_S = 25.8\,\text{kN/m}^3$. The liquid limit was w_L=45% and the plasticity limit was w_P= 24%. The mineralogical content of the material obtained from X-ray diffraction analysis yielded 67% kaolinite, 2% illite or mica and 31% quartz.

2.2 Experimental program

Figure 1 shows the loading paths in three-dimensional space (mean effective stress-shear stress-temperature). The procedure for carrying out the tests in triaxial conditions was as follows:

i) Isotropic consolidation (path 0 → 1), followed by an unloading in some cases (path 1 → 1'),

ii) Drained heating (path 1 or 1'→ 2),

iii) Drained shearing (path 2 → 3, and 1 or 1'→ 4 for tests at ambient temperature), by applying a standard triaxial shearing path ($\partial q/\partial p' = 3$) or by applying a constant p' path. q and p' are the shear and the effective mean stresses, respectively.

To ensure saturation of the samples, a constant back pressure of 100 kPa was applied during isotropic consolidation and kept constant during the subsequent mechanical and/or thermal loadings. Back pressure also reduces evaporation of water in the drainage system. The value of 100 kPa was chosen after tests with different back pressure values (100 kPa to 300 kPa) to ensure that the obtained results were independent of the back pressure value. Based on preliminary tests, the heating rate of 10 °C per 3 hrs was chosen to keep avoid excess pore pressure generation during heating. This was confirmed by the measurement of pore water pressure during heating. The consolidation step took about 2 days and heating from 22 °C to 90 °C about 4 days. Shear tests were carried out with a strain rate of 0.3 %/hrs in drained conditions.

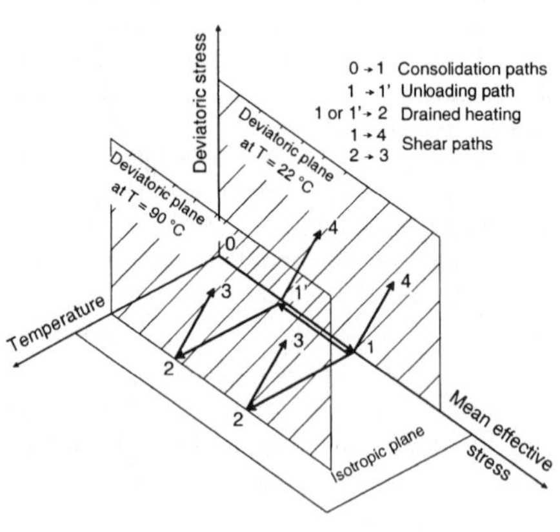

Figure 1. Thermo-mechanical testing paths.

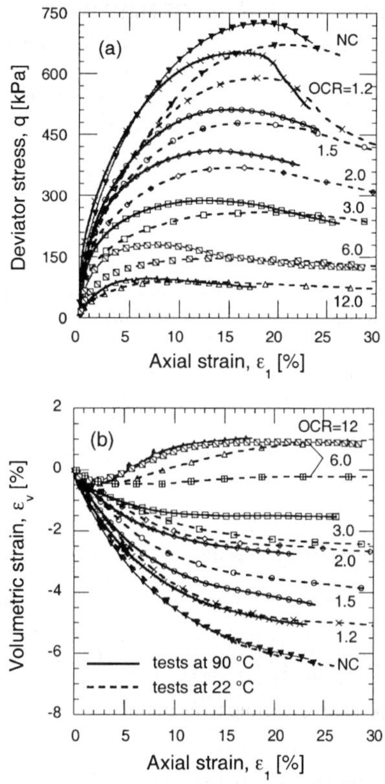

Figure 2. Drained triaxial tests at ambient (22°C) and high temperatures; consolidation pressure 600 kPa, 22 °C – dashed lines, 90 °C – solid lines.

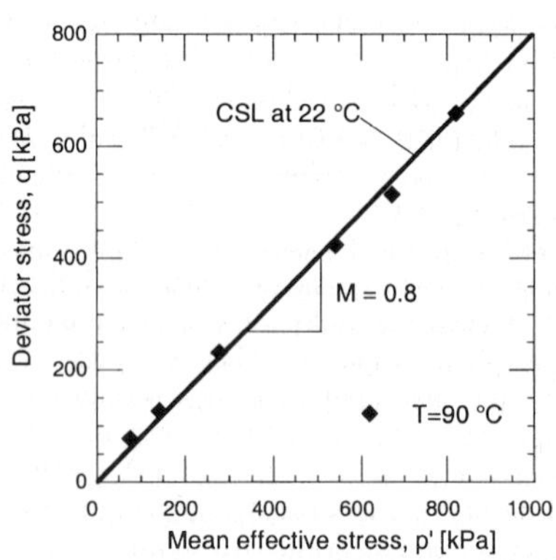

Figure 3. Friction angle at Critical State for temperatures of 22 °C and 90 °C.

3. EXPERIMENTAL RESULTS

The results of the drained triaxial compression shear tests with the same initial states and different temperatures (22 °C and 90 °C) are compared for several overconsolidation ratios (OCR) (Figure 2). The initial confining pressure was 600 kPa.

The results indicate that specimens tested at high temperature show higher shear strength (Figure 2a). However, at large strains where ($\partial q/\partial \varepsilon_s \approx \partial \varepsilon_v/\partial \varepsilon_s \approx 0$) the shear stresses obtained for high temperature tend to the same critical state as samples tested at ambient temperature (ε_s is the deviatoric strain). No clear trend can be seen on the volume variation for samples sheared at different temperatures (Figure 2b).

It may be seen that initial secant modulus increases with increasing in temperature for all of the samples.

The strength envelope at critical state of Kaolin clay at ambient temperature shows a straight line, with a slope M = 0.8 that corresponds to a friction angle of $\phi' \approx 21°$. Figure 3 combines the results of drained shear tests at critical state obtained at ambient and high temperatures (90°C). It is clearly shown that the strength envelope at critical state is independent of testing temperature.

4. THERMAL YIELDING

Thermal yielding represents the evolution of the yield limit induced by heating. The evolution for the hydrostatic limit is presented elsewhere (Laloui and Cekerevac, 2003). We will focus here on the shear yield limit. Starting with isotropic consolidation up to 600 kPa, an isotropic unloading to different values of confining pressures was applied. The defined yield limit then corresponded to a mean effective stress of 600 kPa and a specific void ratio. The unloading steps moved the stress state inside this yield surface. Since the stress point was within the yield surface, initial strains due to a shear stress would be mainly elastic, until the stress point reached the yield surface, and after that it became plastic (Figure 4). The combination of pre-yield and post-yield linearity permits bilinear extrapolation techniques for identifying the yield point (Graham et al. 1982). Yield points were determined from a stress-strain criterion by using various procedures, and some of which are shown in Figure 5.

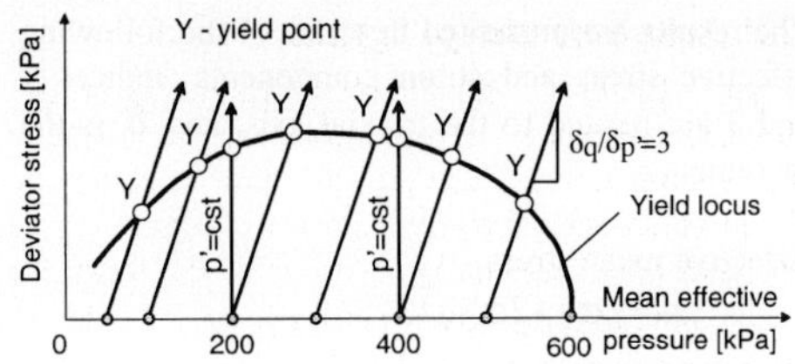

Figure 4. Testing paths used for shearing at 22 °C and 90 °C.

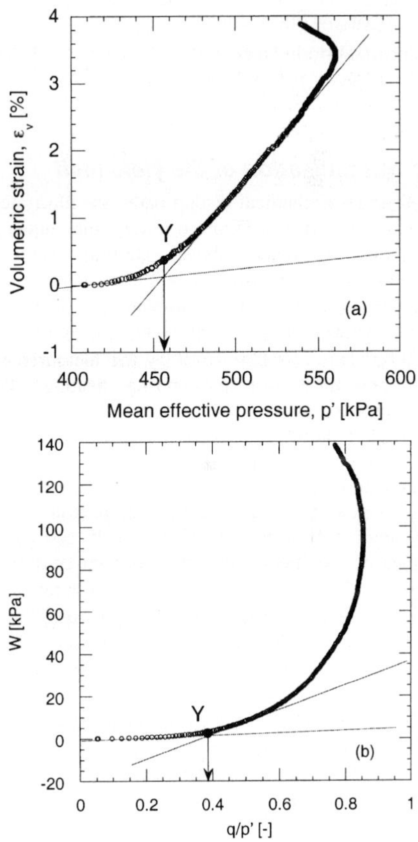

Figure 5. Identification of the yield points from different stress-strain criterion; a) volumetric strain versus mean effective pressure and b) dissipated stress energy versus ratio q/p'.

The results are presented in terms of the following effective stress and strain components (indices 1 and 3 are related to the triaxial axis, and δ is the increment):

Effective mean stress,

$$p' = (\sigma'_1 + 2\sigma'_3)/3$$

Deviator stress,

$$q = \sigma'_1 - \sigma'_3$$

Dissipated strain energy,

$$W = \sum(\sigma'_1 \delta\varepsilon_1 + 2\sigma'_3\, \delta\varepsilon_3)$$

Volumetric strain increment,

$$\delta\varepsilon_v = \delta\varepsilon_1 + 2\delta\varepsilon_3$$

Distortional strain increment,

$$\delta\varepsilon_s = 2/3\,(\delta\varepsilon_1 - \delta\varepsilon_3)$$

4.1 Normalisation of the yield limit

Thermo-mechanical testing paths are illustrated in $q : v : p'$ space in Figure 6 (*v: specific volume).* As was mentioned above, starting from a preconsolidation pressure of 600 kPa (point 0), all samples were isotropically unloaded to different effective mean pressures (for example to point A or B) before being sheared. Isotropic unloading allows changes of the specific volume (v = 1+e) along the swelling line with an elastic strain. Thus, the initial specific volume in isothermal conditions is different for the different samples depending on the values of isotropic stress (for ex. v_a versus v_b). Moreover, samples tested at high temperature were submitted to drained heating which caused a change of specific volume at a constant confining pressure (B $\rightarrow$ C), starting from different v = constant sections (either point A or B). Thus, samples at the same confining pressures and different temperatures also had different initial void ratios (for ex. v_b versus v_c). During drained shearing (path C $\rightarrow$ D), v changes and so the paths will move through an infinity of constant v sections of the Roscoe surface (Schofield and Wroth, 1968). Because of that, obtained yield points will be situated in different v = constant planes.

If we suppose that each v = constant section is of the same shape but different in size, we can scale them down to the same form. The scaling factor depends on specific volume and could be defined as *the equivalent pressure on the normal consolidation line at a given specific volume*, p'_e. The value p'_e could be obtained from the normal consolidation line by using specific volume at yielding, v_y:

$$p'_e = \exp\frac{N - v_y}{\lambda} \qquad (1)$$

where N is the specific volume of the isotropic normally consolidated line at p' = 1 kPa and λ is the slope of the NCL in the v - $\ln p'$ plane. The procedure is illustrated in Figure 7. If we suppose that the sample yields at point Y (p_y, q_y, v_y), the equivalent pressure is obtained by tracing a line *v=constant* to the NCL and then reading the corresponding value of the mean effective stress (Atkinson and Bransby, 1978).

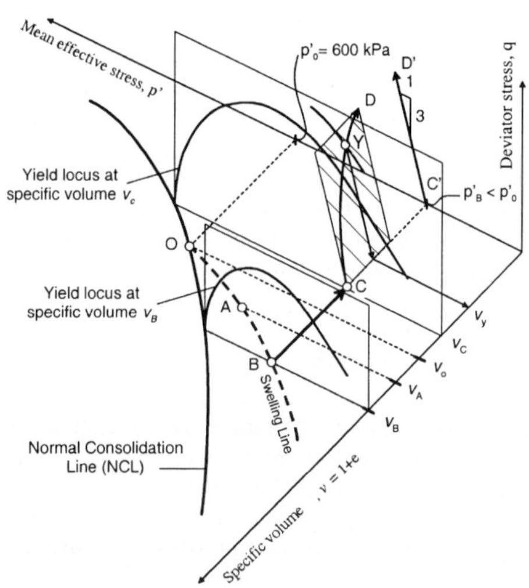

Figure 6. Thermo-mechanical testing paths illustrated in $q : v : p'$ space.

4.2 Yield locus at 22 °C and 90 °C

The yield stresses were identified by using the method explained above. All tests were with an initial confining pressure of 600 kPa.

Specific volume at yielding was observed from initial moisture contents and volume change during shearing. Identified yield stresses were normalised with respect to the equivalent pressure, which were calculated using specific volume at yielding (Equation 1). Normalised yield envelopes, interpolated through these yield points, are shown in Figure 8.

Based on this result, it may be concluded that the pseudo-elastic yield limit shrinks with increase in temperature. It is clear that this decrease of the yield surface is not homothetic and depends on stress history. In other words, drained heating produces thermal softening. In order to sustain constant effective stress during drained heating, thermal softening must be compensated by strain hardening in NC or lightly OC samples. As a result, the strain hardening stress-strain curves obtained at high temperature are different from those obtained at ambient temperature (Figure 2).

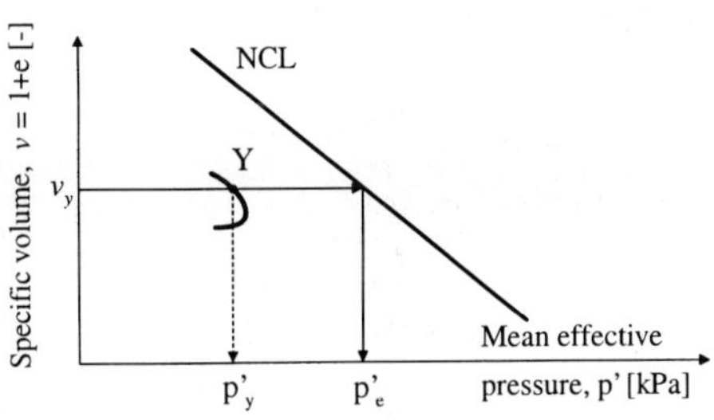

Figure 7. Procedure of obtaining the equivalent pressure, p'_e corresponding to the yield point Y (p_y, q_y, v_y).

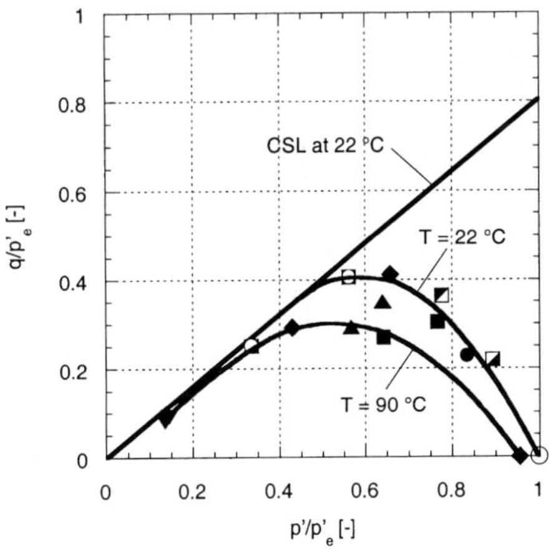

Figure 8. Normalised yield envelope of Kaolin clay at T=22 °C and T=90 °C; CSL is the critical state line.

5. CONCLUSIONS

Some results of a comprehensive laboratory study using a temperature controlled triaxial cell has been presented. Thermal effects on the mechanical behaviour of a Kaolin clay were analysed by comparing tests at ambient (22 °C) and high (90 °C) temperatures. Several thermo-mechanical loading paths were used to gain insight into the effect of temperature on the behaviour of clays. The main conclusions drawn from this study are as follows:

- High temperatures induce a more ductile mechanical behaviour. The heated specimen exhibited a higher shear strength. This effect is slightly more important for samples initially normally consolidated than for overconsolidated ones. At large strain when the critical state is reached the deviator stress drops and has almost the same value as for samples tested at ambient temperature.
- The yield limit shrinks with increase in temperature.
- Friction angle at critical state (CSL) and the normally consolidated line (NCL) seem to be independent of temperature.

REFERENCES

Atkinson J.H, Bransby P.L. 1978. *The Mechanics of Soils - An Introduction to Critical State Soil Mechanics*. McGraw-Hill London.

Cekerevac C, Laloui L, Vulliet L. 2003. *Thermo-mechanical testing of soils: equipment, procedures and experimental results.* Submitted for publication.

Graham J, Pinkney R.B, Lew K.V, Trainor P.G.S. 1982. *Curve fitting and laboratory data.* Canadian Geotechnical Journal, 1982, 19: 201-205.

Laloui L, M Moreni, L Vulliet. 2003. *Comportement d'un pieu bi-fonction, fondation et échangeur de chaleur.* Canadian Geotechnical Journal, 2003, Vol 40(2): 388-402.

Laloui L, Cekerevac C. 2003. *Thermo-plasticity of clays: an isotropic yield mechanism.* Submitted for publication.

Schofield A, Wroth P. 1968. *Critical State Soil Mechanics.* McGRAW-HILL, London.

Vulliet L, Laloui L, Harding R. 2002. *Environmental geomechanics: an introduction.* In Environmental Geomechanics, L. Vulliet, L. Laloui & B. Schrefler (Eds.). EPFL-Press, 3-12.

AN ELASTOPLASTIC DAMAGE MODEL FOR UNSATURATED ARGILLITES

Shao J.F., Jia Y. and Kondo D.
LML, UMR 8107 CNRS, University of Lille, France

Abstract: We present a new elastoplastic damage model for unsaturated hard sedimentary rocks like argillites. The thermodynamic potential for dry material is formulated from relevant results of micromechanics by assuming an isotropic damage. Elastic properties are affected by induced damage. We account for the coupling between damage evolution and plastic flow. For unsaturated material, elastic and plastic behaviours are influenced by capillary pressure. The model's parameters are determined from triaxial compression tests. We present finally comparison between numerical simulations and experimental data for various loading paths and saturation conditions.

1. INTRODUCTION

A number of geomaterials (rocks and concrete) exhibit coupled elastoplastic damage behaviour. The plasticity is related to sliding process while damage is attributed to growth of microcracks. In many engineering constructions, these materials are submitted to drying and wetting processes. The change of saturation condition may significantly modify their mechanical properties. Thus, constitutive models for geomaterials should include these basic features. During the recent years, many advances have been performed for unsaturated soils and clay. On the other side, different plastic and damage models have also been proposed for rocks and concrete in dry condition. However, only a little number of works is performed for plastic and damage modelling of unsaturated hard rocks and concrete. This paper presents a new elastoplastic damage model for one class of semi-ductile geomaterials in saturated and unsaturated conditions. The model is formulated in the framework of irreversible thermodynamics of open systems. Relevant results from micromechanics analysis will be used for a rigorous formulation of thermodynamic potential of cracked material. Applied to unsaturated condition, effects of capillary pressure on plastic flow are taken into account. The proposed model is applied to a class of sedimentary rocks, argillites. These materials have been largely studied in the framework of experiments in the underground research laboratory conducted by the French Agence National de Gestion des Déchets Radioactifs (ANDRA).

2. PRESENTATION OF MODEL

In this section, we first present the formulation of the model for dry material, and then the extension to unsaturated porous material. By adopting the assumption of small strains, the total strain tensor is decomposed into an elastic part and plastic part:

$$\varepsilon_{ij} = \varepsilon_{ij}^e + \varepsilon_{ij}^p \;, \quad d\varepsilon_{ij} = d\varepsilon_{ij}^e + d\varepsilon_{ij}^p \qquad (1)$$

2.1 Constitutive equations of dry material

We consider an elastic solid weakened by a set of microcracks. The elastic free energy, used as thermodynamic potential, can be estimated by using micromechanics approaches (Krajcinovic 1989, Pensee and Kondo 2001). In this work, we assume an isotropic distribution of microcracks. We limit the present study to the case of fully open microcracks. However we account for an energy coupling between damage evolution and plastic flow. Therefore, the thermodynamic potential for dry material is obtained:

$$\Psi = \frac{1}{2}\left[K_0(1-\alpha d)(\varepsilon_v^e)^2 + 2\mu_0(1-\beta d)e_{ij}^e e_{ij}^e\right] + (1-\chi d)\Psi_p^0(\gamma_p) \qquad (2)$$

Where ε_v^e is the elastic volumetric strain and e_{ij}^e components of the elastic deviatoric strain. K_0 and μ_0 are respectively the initial bulk modulus and shear modulus of undamaged material. The scalar variable d characterises the isotropic damage. The scalar valued function Ψ_p^0 represents the locked plastic energy of undamaged material. Two parameters α and β define deterioration of elastic properties due to damage and the parameter χ accounts for the damage effect on plastic hardening function. The values of α and β can be determined from micromechanical analysis. In the case of opened microcracks, we have:

$$\alpha = \alpha^{open} = \frac{4(\lambda_0 + 2\mu_0)(9\lambda_0 + 2\mu_0)}{9(\lambda_0 + \mu_0)\mu_0} \tag{3a}$$

$$\beta = \beta^{open} = \frac{32(\lambda_0 + 2\mu_0)(9\lambda_0 + 10\mu_0)}{90(\lambda_0 + \mu_0)(3\lambda_0 + 4\mu_0)} \tag{3b}$$

where λ_0 and μ_0 are initial Lame constants of undamaged materials. The standard derivation of the thermodynamic potential leads to the state equation:

$$\sigma = \frac{\partial \Psi}{\partial \varepsilon^e} = \boldsymbol{C}(d) : \varepsilon^e \tag{4}$$

The effective elastic stiffness tensor of damaged material $\boldsymbol{C}(d)$ is given by:

$$\boldsymbol{C}(d) = 2\mu^0(1 - \beta d)\boldsymbol{J} + 3k^0(1 - \alpha d)\boldsymbol{K} \tag{5}$$

$$K_{ijkl} = \frac{1}{3}\delta_{ij}\delta_{kl}, I_{ijkl} = \frac{1}{2}(\delta_{ik}\delta_{jl} + \delta_{il}\delta_{jk}),$$
$$J_{ijkl} = I_{ijkl} - K_{ijkl}$$

The thermodynamic force associated with the damage variable can be determined:

$$Y_d = \frac{1}{2}\left[\alpha K^0 (\varepsilon_v^e)^2 + 2\beta\mu^0 e_{ij}^e e_{ij}^e\right] + \chi\Psi_p^0(\gamma_p) \tag{6}$$

The incremental form of the constitutive equation (4) can also be obtained:

$$d\sigma = \boldsymbol{C}(d) : d\varepsilon^e + \boldsymbol{C}' : \varepsilon^e d(d) \tag{7}$$

The fourth order tensor $\boldsymbol{C}'$ is the derivative of the elastic tensor with respect to damage variable:

$$\boldsymbol{C}' = \frac{\partial \boldsymbol{C}(d)}{\partial d} = -2\mu^0\beta\boldsymbol{J} - 3k^0\alpha\boldsymbol{K} \tag{8}$$

2.2 Extension to unsaturated media

We present now the extension of the constitutive equation (7) to partially saturated porous media. The material is assumed to be saturated by a liquid phase (noted by index w) and a gas mixture (noted by index g). The gas mixture is a perfect mixture of dry air (noted by index da) and vapour (noted by index va). Based on most experimental data of unsaturated rocks and soils (Fredlund and Rahardjo 1993), and on the theoretical background of micromechanical analysis (Château and Dormieux 1998), the poroelastic behaviour of unsaturated material should be non-linear and depends on the water saturation degree. We consider here the particular case of spherical pores which are dried or wetted under a capillary pressure equal to the superficial tension on the air-solid interface. By adapting the macroscopic non-linear poroelastic model proposed by Coussy & al. (1998) to unsaturated damaged porous media, the incremental constitutive equations in isothermal conditions are expressed as follows:

$$\begin{aligned} d\sigma_{ij} &= C_{ijkl}(d)d\varepsilon_{kl}^e + C'_{ijkl}\,\varepsilon_{kl}^e\, d(d) \\ &- b(d\,p_g - S_w dp_c)\delta_{ij} \end{aligned} \tag{9}$$

$$\begin{aligned} dp_g &= dp_{va} + dp_{da} = \frac{p_g}{\phi_g}\left[-b\,d\varepsilon_v^e + \left(\frac{dm}{\rho}\right)_w^e\right] \\ &+ \frac{p_{va}}{\phi_g}\left(\frac{dm}{\rho}\right)_{va}^e + \frac{p_{da}}{\phi_g}\left(\frac{dm}{\rho}\right)_{da}^e \end{aligned} \tag{10}$$

$$\begin{aligned} dp_c &= dp_g - dp_w \\ &= -M_c\left[-b\,S_w\,d\varepsilon_v^e + \left(\frac{dm}{\rho}\right)_w^e\right] \end{aligned} \tag{11}$$

The parameter b denotes the Biot's coefficient of porous medium and S_w the water saturation degree, and ϕ_g the porosity occupied by gas mixture. The scalar variable $(dm/\rho)_k^e$ represents the elastic variation of mass of the kth fluid phase per unit initial volume of porous medium. In

Equation (11), the variable p_c defines the capillary pressure. The parameter M_c is Biot's modulus related to capillary pressure, which is a function of water saturation degree through the water retention curve $p_c(S_w)$:

$$M_c = -\frac{1}{\phi_0 + \varepsilon_v}\left(\frac{dp_c}{dS_w}\right) \quad (12)$$

2.3 Plastic characterisation

In partially saturated materials, most experimental data have shown that the plastic deformation is controlled by two independent force variables: the net stress tensor, $\sigma' = \sigma + bp_g\delta$, and capillary pressure p_c (Alonso et al. 1990; Fredlund and Rahardjo 1993). However, for hard rocks, plastic deformation due to pure change of capillary pressure can be neglected. Based on experimental data from argillites (Chiarelli 2000), a quadratic function is used to describe plastic yielding surface:

$$f_p = q'^2 + A\,h(\theta')\,\alpha_p\,(p'-C)\,p_r = 0 \quad (13)$$

where p' is the mean net stress, q' deviatoric stress and θ' Lode's angle associated to the net stress tensor. The function $h(\theta')$ defines the dependency of yield function on the Lode angle. For the reason of simplicity, we have here taken $h(\theta') = 1$. The parameter C denotes the material cohesion for a given saturated condition. The fixed parameter $p_r = 1MPa$ is used to obtain the failure parameter A without dimension.

The plastic hardening law of material is derived from the definition of plastic energy in thermodynamic potential. We have chosen the generalised plastic shear strain γ_p as the plastic hardening variable. Based on the experimental data obtained on argillites, the following specific function of plastic energy is proposed (Bourgeois 2002):

$$\Psi_p^0 = \left[\alpha_p^0 + (1-\alpha_p^0)\right]\gamma_p - B(1-\alpha_p^0)\ln\frac{B+\gamma_p}{B} \quad (14)$$

The plastic hardening law is then obtained:

$$\alpha_p = \frac{\partial \Psi}{\partial \gamma_p} = (1-\chi d)\left[\alpha_p^0 + (1-\alpha_p^0)\frac{\gamma_p}{B+\gamma_p}\right] \quad (15)$$

We can see that the induced damage results in a deterioration of plastic hardening (reduction of frictional angle) of material.

We use a non-associated plastic flow rule in order to better describe the transition from plastic contractance to dilatance. Based on the previous work by Pietruszczak et al. (1988), the following function is adopted:

$$g = q' + A(1-\chi d)\,\eta\,h(\theta')(p'-C)\ln\left(\frac{p'-C}{I_0}\right) = 0 \quad (16)$$

The boundary between contractance and dilatance is defined by the condition $\partial g/\partial p' = 0$. In this model, we assume the following linear function:

$$\bar{f}_{cd} = q' - A(1-\chi d)\,\eta\,h(\theta')(p'-C) = 0 \quad (17)$$

The parameter η is the slope of the boundary line between contractance – dilatance zones. The plastic strain rates are given by:

$$\dot{\varepsilon}_{ij}^p = \dot{\lambda}_p \frac{\partial g(\sigma', d)}{\partial \sigma'_{ij}} \quad (18)$$

2.4 Damage characterisation

For the damage evolution law, a standard normal dissipation schema is employed. In the case of time independent dissipation, the damage evolution law is derived from the damage criterion which is a scalar valued function of the thermodynamic force associated to damage variable. We propose here a simple linear function for damage criterion:

$$f_d(Y_d, d) = Y_d - (Y_d^0 + m\,d) = 0 \quad (19)$$

where the parameter Y_d^0 is the initial threshold of damage surface and m defines the kinetics of damage evolution. The damage evolution rate is then given by:

$$\dot{d} = \dot{\lambda}_d \frac{\partial f_d(Y_d, d)}{\partial Y_d} \tag{20}$$

The plastic and damage multipliers $\dot{\lambda}_p$ and $\dot{\lambda}_d$ should be determined from the coupled consistence conditions $\dot{f}_p = 0$ and $\dot{f}_d = 0$.

2.5 Influence of capillary pressure

In most unsaturated rocks and soils, elastic properties and plastic flow depend on the capillary pressure which is related to water saturation degree through water retention curve (Fredlund and Rahardjo 1993). In this work, for simplicity, we neglect the variation of elastic constants with capillary pressure. However, we intend to account for the influence of capillary pressure on plastic behaviour of argillites. Only a small number of triaxial compression tests with different water saturation degrees are available. We can only provide a first approximation of such a influence. We consider that the failure parameter A (see Equation 13) linearly increases with capillary pressure:

$$A = A_s + 0.95\left(\frac{p_c}{p_r}\right) \tag{21}$$

where A_s is the failure parameter for saturated material. However, we have noticed that the variation of material cohesion C is not significant and then neglected in this work.

2.6 Identification of model's parameters

The proposed model contains 10 parameters: two elastic constants for undamaged material, five parameters for plastic behaviour and three parameters for damage characterisation. These parameters can be identified from a series of triaxial compression tests with different confining pressures. The initial elastic parameters are generally determined from the linear part of stress-strain curve before damage and plastic initiation. For most rock materials, the elastic parameters vary with confining pressure. This is not considered in this work as the emphasis is put on the plasticity and damage. We use average values of elastic parameters obtained in various triaxial compression tests. The damage parameters are related to the variation of elastic properties and they can be estimated from unloading cycles during triaxial compression tests. The plastic parameters are identified from general responses in triaxial compression tests, namely the initial plastic yield stresses, failure stresses (peak points of stress strain curves) and plastic hardening function. For the argillites tested (Chiearelli 2000), the following representative values of parameters have been obtained and are given in Table 1. Notice that the two parameters determining the effective elastic tensor are calculated by Equation 3 from micromechanics, we have obtained $\alpha = 2.42$ and $\beta = 1.59$.

Table 1: Representative values of parameters for argillites

Elastic parameters	plastic parameters	Damage parameters
$E_0 = 5800$ (MPa) $\nu = 0.14$	$\alpha_p^0 = 0.017$ $A_s = 40$ $C = 7MPa$ $B = 0.0019$ $\eta = -0.0242$	$Y_d^0 = 2.5.10^{-5}$ (MPa) $m = 3.0$ $\chi = 1.0$

3. NUMERICAL SIMULATIONS

In this section, we present some numerical simulations of triaxial compression tests performed on argillites in natural saturation condition and in different water saturation degree.

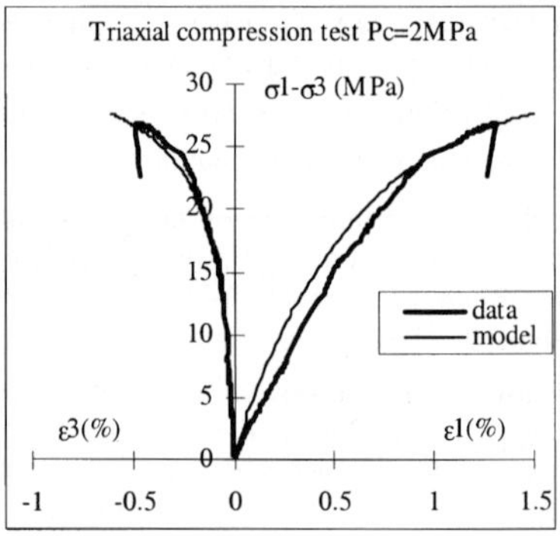

Figure 1a: Simulation of a triaxial compression test with 2MPa confining pressure (thin lines are model's simulations)

On Figures 1a and 1b, we present simulations of two triaxial compression tests performed on saturated samples respectively with two different confining pressures. We can see a general good agreement between model's simulations and tests data. On Figure 2, we show the variations of elastic modulus and Poisson's ratio as functions of lateral strain, for a triaxial compression tests with a $10MPa$ confining pressure. There is a qualitatively good agreement between numerical predictions and experimental data. We can see that the elastic modulus decreases while Poisson's ratio increases during damage evolution. This represents an advantage of the proposed model with respect to classic isotropic damage models where only the elastic modulus is affected by damage and Poisson's ratio remains constant. However, some quantitative scatter exist, we need further investigations on the damage evolution law and parameter's identification.

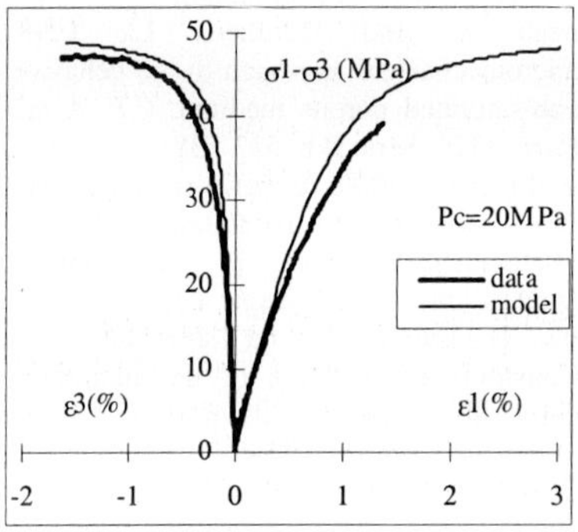

Figure 1b: Simulation of a triaxial compression test with $20MPa$ *confining pressure (thin lines are model's simulations)*

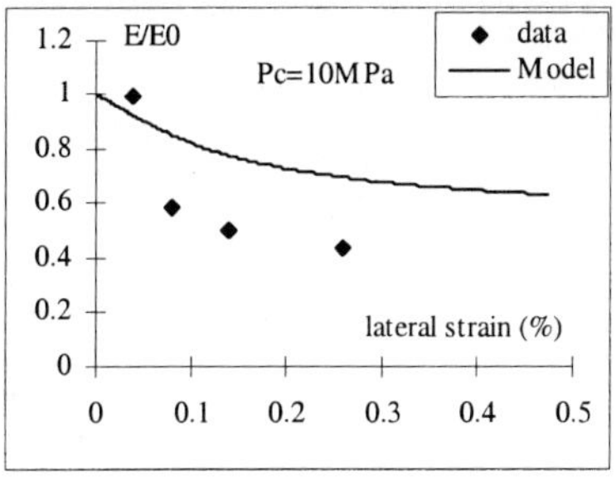

Figure 2a: Variation of elastic modulus as a function of lateral strain

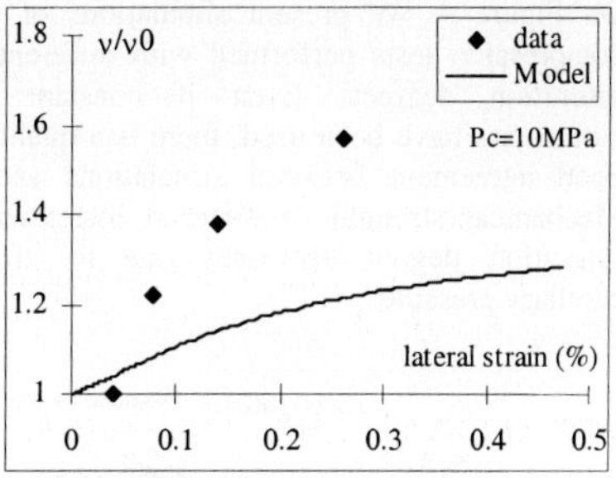

Figure 2b: Variation of Poisson's ratio as a function of lateral strain

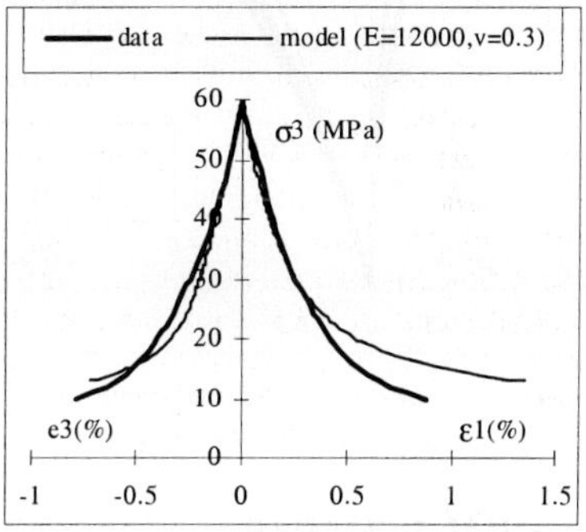

Figure 3: Simulation of a lateral extension test with an initial confinig pressure of $60MPa$

On Figure 3, a lateral extension test is simulated on saturated sample. In this test, the sample is first submitted to a hydrostatic stress to a given value of confining pressure. Then the lateral stress is decreased while the axial stress is hold constant. We should point out that in this work, for simplicity, we have chosen constant initial elastic parameters which are the average values from triaxial compression tests. The scatter among the experimental values is generally important. It has been seen that the elastic modulus observed in this lateral extension test is significantly larger than the average values used. Such a difference may be partially attributed to the initial water saturation degree of sample tested. If this sample is not fully saturated, we could obtain higher values of elastic modulus than those in saturated samples. Thus, in the simulation of this test, we have used a higher value of elastic modulus to better fit the elastic response. We obtain a good agreement between simulations and data, in particular the inelastic behaviour.

On Figure 4, we present simulations of triaxial compression tests performed with different water saturation degrees. Even if constant elastic parameters have been used, there is a qualitatively good agreement between simulations and data. Mechanical strength of material increases when saturation degree decreases due to effect of capillary pressure.

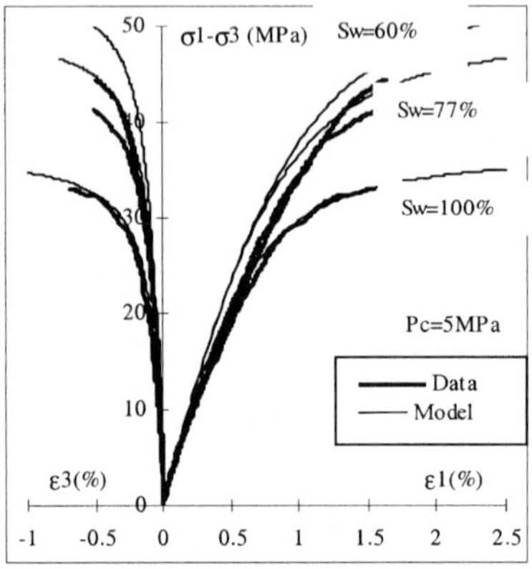

Figure 4: Simulation of triaxial compression tests with different water saturation degrees

4. FINAL REMARKS

We have presented a new coupled elastoplastic damage model for the class of semi-ductile geomaterials like rocks and concrete. An isotropic damage is used for simplicity. The effective elastic tensor is obtained from micromechanical analysis. The elastic modulus and Poisson's ratio are independently affected by damage. There is an energy coupling between plastic flow and damage evolution. The constitutive model is extended to partially saturated rocks. Plastic behaviour of material depends on the net stress tensor and capillary pressure. A non-associated flow rule is used to describe transition from plastic contractance to dilatance. A first application of the model to argillites is presented. Comparisons between simulations and experimental data are shown for various loading conditions. The proposed model seems to be able to describe the main feature observed for this class of material in saturated and unsaturated conditions. However, more experimental data will be necessary to carry out a further validation of the proposed model. The proposed model is easy to be implemented into a FEM code, and then could be used for numerical modelling of various engineering problems, for instance, feasibility studies of underground storage of nuclear waste.

ACKNOWLEDGEMENTS

The authors are grateful to the French Agence National de Gestion des Déchets Radioactifs (ANDRA) for the financial support to this work.

REFERENCES

Alonso,E. E., Gens, A. & Josa, A. 1990, A constitutive model for partially saturated soils, *Géotechnique* 40 (3), 405-430

Bourgeois F. 2002. Development of constitutive models for partially saturated rocks and concrete with microcracks, *doctoral thesis, University of Lille*, (in French).

Chateau X. and Dormieux L., 1998. A micromechanical approach to the behavior of a non-saturated porous medium, *C.R. Acad. Sci. Paris,* t326, Série II b, 533-538.

Chiarelli A.S., 2000. Experimental investigation and constitutive modelling of coupled elastoplastic damage in hard claystones, *Doctoral thesis (in French), University of Lille .*

Coussy O., Eymard R. and Lassabatère T. 1998. Constitutive modelling of unsaturated drying deformable materials, *Journal of Engineering Mechanics*, 124 (6), 658-667

Fredlund D.G. and Rahardjo H. 1993. *Soil mechanics for unsaturated soils*, John Wiley & Sons, New York

Krajcinovic D. 1989. Damage mechanics, *Mechanics of materials* 8, 117-197

Pensee V. and Kondo D. 2001. A 3-D micromechanical analysis of damage by mesocracking, *C.R. Acad. Sci.* Paris, t329, Série II b, 271-276

Pietruszczak S., Jiang J. and Mirza F.A.,1988. An elastoplastic constitutive model for concrete, *Int J. Solids & Structures* 24 (7), 705-722

STUDY ON THE TIME-TEMPERATURE EQUIVALENT PRINCIPLE FOR ROCKS

Quansheng Liu[1], Chongge Wang[2], Tsutomu Yamaguchi[3]
[1]) Institute of Rock and Soil Mechanics, CAS, Wuhan, Hubei, China
[2]) Shandong University of Science and Technology, Taian, Shandong, China
[3]) Institute for Geo-Resources and Environment, AIST, Tsukuba West, Ibaraki, Japan

Abstract: Based on the theory of irreversible process thermodynamics, non-linear stress-strain-temperature equations are derived, together with an expression for time-temperature equivalence. In addition, an equation of shift factor for time-temperature equivalence is also obtained. The parameters in the equations are experimentally determined and the main curves for creep compliance and cohesion of TGP granite are obtained by a series of creep tests. As a result, it is proved that both deformation and strength of the TGP granite follow the time-temperature equivalent principle.

1. INTRODUCTION

It is of a great significance to understand how the mechanical behaviours and properties of rock masses change with temperature, such as for nuclear waste repositories and deep mining at certain temperatures. The key to this problem is how to make predictions to long-term response of rocks based on mechanical models and test results within a short time of experiments. It is put forward in this paper that the problem can be resolved by means of time-temperature equivalent principle for rocks.

Study on time-temperature equivalence for polymers can be traced back to the mid 1970s when Williams, Landel and Ferry gave the shift factor equation of relaxation modulus curve for polymers near glazy and visco-elastic states (Aklonis & Macknight, 1983). Ferry (1980) and Christensen (1982) summarized the research achievements in this field in their works. Although large amount of study has been carried out and many papers have been published on independent temperature and time effects of rocks, so far little study has been made on combined effects of them. Whereas, publication of study on time equivalence to temperature has hardly been reported, except for Paterson (1978) and Wang (1988) for the concept of replacing time by temperature.

Comparing time effect with temperature effect on rock, a very interesting phenomenon appears: the effect of raising temperature on rock properties is equivalent to that of prolonging time or reducing strain rate. For instance, with temperature increasing, both rock strength and deformation modulus decrease while deformation and toughness increase. If the strain rate is reduced, the same result can be obtained. The key is to prove the existence of such equivalence. On the other hand, properties of rock are much more complex than polymers because of its inhomogeneity, anisotropy and non-linearity.

In this paper time-temperature equivalence for rocks is investigated for non-linear behaviours of rocks, based upon inner-variable theory of irreversible process and creep tests.

2. THEORY OF NONLINEAR THERMO-VISCO- ELASTICITY

2.1 Evolution and constitutive equations

Biot (1954) obtained the linear motion equation for n state variables q_i in a closed system subject to generalized forces Q_i and absolute temperature θ, by introducing inner variables and applying Onsager's principle, given as follows:

$$\frac{\partial \psi}{\partial q_i} + b'_{ij}\dot{q}_j = Q_i \qquad (i = 1, 2, \cdots, n) \tag{1}$$

where, ψ is the Helmholtz's free energy, b'_{ij} are linear coefficients in Onsager's principle considered as material constants, q_j are state variables in the system which may include strain components, temperatures and non-observable inner variables, and Q_i are generalized forces. The ones corresponding to strains are stresses, and the ones corresponding to non-observable state variables are equal to zero. Equation (1) can be generalized for solving non-linear problems of irreversible process of rocks. Given that among n state variables in a closed system, there are m observable and ($n-m$) non-observable ones respectively. By making expansion of Taylor's series on Helmholtz free energy ψ near the equilibrium state with terms of second orders (or higher) derivatives omitted in eq. (1), we can derive the evolution equations (2a) and (2b) corresponding to the observable variables and inner variables, respectively. (The readers can refer to authors' another paper (Liu and Wang, 2002) for detailed procedures of derivation)

$$\frac{\partial \psi_e}{\partial q_k} + b'_{kj}\frac{dq_j}{dt} = Q_k + d'_k\Theta \quad (k=1, 2, \ldots, m) \tag{2a}$$

$$a'_{rs}\, q_s + b'_{rj}\frac{dq_j}{dt} = d'_r\,\Theta \quad (r, s=m+1, m+2, \ldots, n) \tag{2b}$$

Letting $b'_{ij} = B_d(q_k, \theta)b_{ij}$, $a'_{rs} = A_f(q_k)a_{rs}$, $d'_i = D_t(q_k)d_i$ in eq.(2), here B_d, A_f, D_t being non-linear coefficients and b_{ij}, a_{rs}, d_i being the coefficients in the linear equations, we have the following non-linear evolution equations of n generalized coordinates q_i under the influence of generalized forces and temperature:

$$a_{ij}\, q_j + b_{ij}\frac{dq_j}{d\gamma} = \tilde{Q}_i + \frac{D_t}{A_f}\Theta \tag{3}$$

where the generalized force $\tilde{Q}_i$ is given by

$$\tilde{Q}_i = \frac{1}{A_f}\left(Q_i + A_f\, a_{il}\, q_l - \frac{\partial \psi_e}{\partial q_i}\right) \tag{4}$$

and γ is the *conversion time* expressed by eq. (5)

$$\gamma = \int_0^t \frac{dt}{a_{\varepsilon\theta}[q_k, \theta]} \tag{5}$$

where $a_{\varepsilon\theta}[q_k, \theta] = B_d / A_f > 0$ and $a_{\varepsilon\theta}$ is a thermodynamic shift factor.

By adopting similar procedures used for linear equations by Biot (1954) and in consideration of generalized forces corresponding to the inner variables being zero, we have the non-linear constitutive equation of thermo-visco-elasticity (see Liu et al, 2001 for details) written for i=1, 2, 3 as

$$\begin{aligned} Q_i = {} & \frac{\partial \psi_e}{\partial q_i} - 2A_f\int_0^t [G(\gamma-\gamma') - G_e]\frac{dq_i}{d\tau}d\tau \\ & + A_f\int_0^t [K(\gamma-\gamma') - K_e]\frac{de}{d\tau}d\tau \\ & - A_f\int_0^t \beta(\gamma-\gamma')\frac{d}{d\tau}\left(\frac{D_t}{A_f}\Theta\right)d\tau \end{aligned} \tag{6a}$$

and for i=4, 5, 6 as

$$Q_i = \frac{\partial \psi_e}{\partial q_i} + A_f\int_0^t [G(\gamma-\gamma') - G_e]\frac{dq_i}{d\tau}d\tau \tag{6b}$$

where G_e and K_e are Lamé's elastic coefficients, $e = \varepsilon_{11} + \varepsilon_{22} + \varepsilon_{33}$, and

$$\gamma = \int_0^t \frac{d\tau}{a_{\varepsilon\theta}[\varepsilon, \theta(t)]}, \quad \gamma' = \int_0^\tau \frac{d\tau'}{a_{\varepsilon\theta}[\varepsilon, \theta(t)]} \tag{7a}$$

$$G(\gamma) = G_e + \sum_\alpha G^{(\alpha)} e^{-\lambda_\alpha \gamma} \tag{7b}$$

$$K(\gamma) = K_e + \sum_\alpha K^{(\alpha)} e^{-\lambda_\alpha \gamma} \tag{7c}$$

$$\beta(\gamma) = \beta_e + \sum_\alpha \beta^{(\alpha)} e^{-\lambda_\alpha \gamma} \tag{7d}$$

where $G(\gamma)$ and $K(\gamma)$ are thermo-visco-elastic relaxation modulus, $\beta(\gamma)$ is the thermo-relaxation function related to thermal expansion.

2.2 One-Dimensional Stress Strain Temperature Equation

In the case of one-dimensional problems, there is only one observable state variable q, i.e. the axial strain ε and the generalized force Q corresponds to the axial stress σ. From eq. (6) we have a one-dimensional non-linear thermo-visco-elastic equation (8) as follows:

$$\begin{aligned} Q = {} & \frac{d\psi_e}{dq} + A_f\int_{0^-}^t [E(\gamma-\gamma') - E_e]\frac{dq}{d\tau}d\tau - \\ & A_f\int_{0^-}^t \beta(\gamma-\gamma')\frac{d}{d\tau}\left(\frac{D_t}{A_f}\Theta\right)d\tau \end{aligned} \tag{8}$$

The unit work done by the external force is then

$$\delta W = \sigma\delta\varepsilon = Q\delta q = Q\frac{dq}{d\varepsilon}\delta\varepsilon \tag{9}$$

From eq. (8) and eq. (9), we can derive the expression for σ

$$\sigma = Q\frac{dq}{d\varepsilon} = \frac{d\psi_e}{d\varepsilon} + \frac{A_f}{2}E_e\frac{dq^2}{d\varepsilon} + A_f\frac{dq}{d\varepsilon}\int_{0^-}^{t} E(\gamma-\gamma')\frac{d}{d\tau}\left(q - \frac{D_t}{A_f}\alpha_e\Theta\right)d\tau \tag{10}$$

where $\beta(\gamma) = \alpha_e E(\gamma)$, and α_a being the thermal expansion coefficient.

2.3 One-dimensional Expressions for the Time-Temperature Equivalence

Let $\Delta E(\gamma-\gamma') = E(\gamma-\gamma') - E_e$ in eq. (10). Considering also that ψ_e is the free energy near equilibrium state and is a function of strain ε, we can assume

$$\frac{d\psi_e}{d\varepsilon} = A_e E_e \varepsilon \tag{11}$$

From eq.(10) and (11), we have

$$\sigma = A_e E_e \varepsilon + A_f \int_{0^-}^{t} \Delta E(\gamma-\gamma')\frac{d\varepsilon}{d\tau}d\tau - A_f\int_{0^-}^{t}\beta(\gamma-\gamma')\frac{d}{d\tau}\left(\frac{D_t}{A_f}\Theta\right)d\tau \tag{12}$$

where $\gamma = \gamma(t), \gamma' = \gamma'(\tau)$ is conversion time, $\beta(\gamma)$ is the thermal relax function, and A_e, A_f and D_t are functions of strain ε.

In the case of one-dimensional relaxation, strain ε remains constant and temperature variation Θ_R also becomes a constant in a homothermous state. With condition of Θ=0 when τ=0, the third term on the right side of eq. (12) vanishes and $\frac{d\varepsilon}{d\tau}$ is equal to zero except for $\tau = 0$. Then eq. (12) becomes

$$\sigma = A_e E_e \varepsilon + A_f \Delta E\left(\frac{t}{a_\varepsilon[\varepsilon,\theta]}\right)\varepsilon \tag{13}$$

The expression of relax modulus E_r is written

$$E_r = \frac{\sigma}{\varepsilon} = A_e E_e + A_f \Delta E\left(\frac{t}{a_\varepsilon[\varepsilon,\theta]}\right) \tag{14}$$

In the case of $a_\varepsilon[\varepsilon,\theta] = a_\varepsilon[o,\theta] \equiv a_\theta$ is no longer the function of strain ε, the eq. (14) becomes

$$E(t,\theta) = [A_e(\theta) - A_f(\theta)]E_e + A_f(\theta)E\left(\frac{t}{a_\theta}\right) \tag{15a}$$

When $A_e = A_f$ this equation becomes

$$E(t,\theta) = A_f(\theta)E\left(\frac{t}{a_\theta}\right) \tag{15b}$$

where A_f and a_θ are vertical and horizontal shift factors, respectively. It can be observed from eq. (15) that the stress relaxation process at a certain temperature can be represented by a main curve at a reference temperature from relaxation curves of modulus vs. time, superposed at different temperatures. In this way, the longer-term mechanical response of a rock at lower temperatures can be described using the results from shorter-term relaxation tests at higher temperatures. The expression for creep modulus can also be obtained using the method similar to the above, as given by

$$J(t,\theta) = B_f(\theta)J\left(\frac{t}{a_\theta}\right) \tag{16}$$

where B_f and a_θ are the same vertical and horizontal shift factors respectively. Equation (16) has the same form as equation (15).

3. MAIN CURVES AND SHIFT FACTORS

3.1 Equation of shift factors

The creep compliance and/or relaxation modulus of a rock can be measured either as the function of temperature within a certain time or as the function of time at a constant temperature. In principle, a complete curve of compliance vs. time can be measured at any temperature. Such a complete curve is called main curve of relaxation modulus at that temperature.

It is known from eq. (16) that the time-temperature equivalence of the creep compliance can be reflected in the following form

$$J(\theta,t) = B_f(\theta)\,J(\theta_g, t_{ref}) \tag{17}$$

here B_f is the vertical shift factor and t_{ref} is the reference time of the main curves at reference temperature θ_g, defined as

$$t_{ref} = t / a_\theta \tag{18}$$

where a_θ is the time shift factor that plays an important role in time-temperature equivalent effect. It is only a function of θ_g and θ.

Concerning the value of the creep compliance $J(\theta_g, t_{ref})$ at the reference temperature, there is a corresponding point on the main curve where the value of creep compliance $J(6,t)$ is the same as that of $J(\theta_g, t_{ref})$, namely,

$$J(6,t) = J(6_g, t_{ref}) \tag{19}$$

The variation of density of rock resulting from temperature variation is very small. Therefore it can be considered that the intrinsic variation of creep compliance itself is also small. Consequently the effect of vertical shift factor in eq. (17) can be neglected and only the horizontal shift factor plays a dominant role.

Because the viscosity of rock is related to the temperature, the following relation becomes tenable

$$\eta(\theta) = \int_0^{+\infty} \frac{dt}{J(\theta,t)} \tag{20}$$

here $\eta(6)$ is the viscosity of rock.

In addition, because a_θ is a function of only temperature, by substituting eq. (18) and (19) into eq. (20), we have

$$\eta(\theta) = \int_0^{+\infty} \frac{a_\theta}{J(\theta_g, t_{ref})} dt_{ref} \tag{21}$$

where a_θ can be transposed outside the integral since it is a constant. By comparing eq. (20) and (21), we have

$$a_\theta = \frac{\eta(6)}{\eta(\theta_g)} \tag{22}$$

Using the semi-empirical Doolittle's formula (Aklonis and Macknight, 1983), we have

$$\ln \eta = \ln D_1 + D_2\left(\frac{1}{h} - 1\right) \tag{23}$$

where D_1 and D_2 are two constants of material properties and h is called the free-volume fraction ($h = V_f / V$, here V stands for the whole volume of rock and V_f is the free volume after subtracting the micro voids volume between grains). The Eq. (23) indicates that the viscosity is related to the flowing nature of rock, which, in turn, is closely related to the free volume. The viscosity decreases with the increase of free volume fraction.

It is further assumed that the above free-volume fraction has the following quadratic polynomial form

$$h = h_g + \alpha_f(\theta - \theta_g) + \beta_f(\theta - \theta_g)^2 \tag{24}$$

where both α_f and β_f are thermal expansion coefficients of the free-volume fraction with temperature variation relative to the reference temperature θ_g. Substituting eq. (24) into eq. (23), and then into eq. (22), in the meanwhile, letting $\theta = \theta_g$ in the denominator, we can obtain

$$\log a_\theta = \frac{C_a(\theta - \theta_g) + C_b(\theta - \theta_g)^2}{C_c + (\theta - \theta_g) + C_d(\theta - \theta_g)^2} \tag{25}$$

where

$$C_a = -\frac{D_2}{2.303 h_g}\ ;\quad C_b = -\frac{D_2 \beta_f}{h_g \alpha_f}\ ;\quad C_c = \frac{h_g}{\alpha_f}\ ;\quad C_d = \frac{\beta_f}{\alpha_f}.$$

There is one more term in both numerator and denominator compared with the WLF equation derived by Williams, Landel and Ferry (Aklonis and Macknight, 1983). The coefficients in eq. (25) are related to temperature and have different meanings than ones in WLF equation, in which these coefficients are treated as constants. The Eq. (25) is the shift factor equation of time coordinate and the expression of time-temperature equivalence of rocks.

3.2 Shift Factor Parameters

Uniaxial compressive creep tests were carried out for granite sampled from Three Gorges Project (TGP) site in China at seven temperatures from 20°C to 300°C in order to determine the parameters C_a, C_b, C_c, and C_d in eq. (25). The curves of creep compliance vs. time in logarithmic coordinates were obtained from the test results, as shown in Figure 1. The coefficients in eq. (25) can be fitted from the above curves. These four coefficients have different values with different reference temperatures. Therefore, three situations were chosen as reference temperatures, 20°C, 100°C and 150 °C, respectively, for explaining the problem.

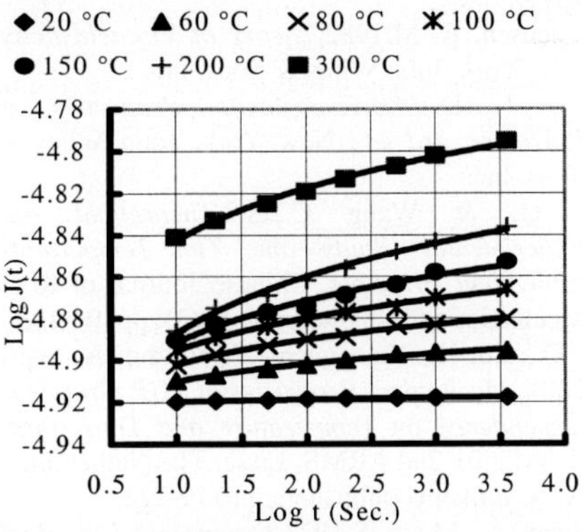

Figure 1. Logarithmic curves of creep compliance vs. time for TGP Granite

The long-term creep curve at 20 °C is just the main curve of creep compliance vs. time at the reference temperature of θ_g =20 °C. The parameters in shift factor can be determined according to the main curve and creep curves at other temperatures. By parallel moving the curves at temperatures 60°C, 80°C, 100°C and 150°C in positive direction of abscissa t and finding out the corresponding points on the main curve where the $J(t)$ values are the same as those before moving curves, we could find out the corresponding values of t_{ref} at the above reference temperatures and the four other temperatures. With eq. (25) and the four additional simultaneous equations the parameters C_a, C_b, C_c and C_d, were solved as listed in Table 1.

The values of the parameters at reference temperatures of 100 °C and 150 °C can also be obtained in the same way, also listed in Table 1.

Table 1. The values of shift factor parameters

θ_g	C_a	C_b	C_c	C_d
20	7.516	-0.0147	15.924	-0.0025
100	4.448	0.0176	144.878	-0.0009
150	1.807	0.0260	257.095	0.00230

Now we can consider the inverse problem for how a main curve be formed at a reference temperature. This is done by applying eq. (25) to the logarithmic curves of creep compliance vs. time at temperatures of 200 °C, 300 °C, etc. and moving the curves along the positive direction of time axis according to the calculated horizontal shift factor a_θ. By doing so, the main curve at the referential temperature 20°C can be formed through superposition as shown in Figure 2.

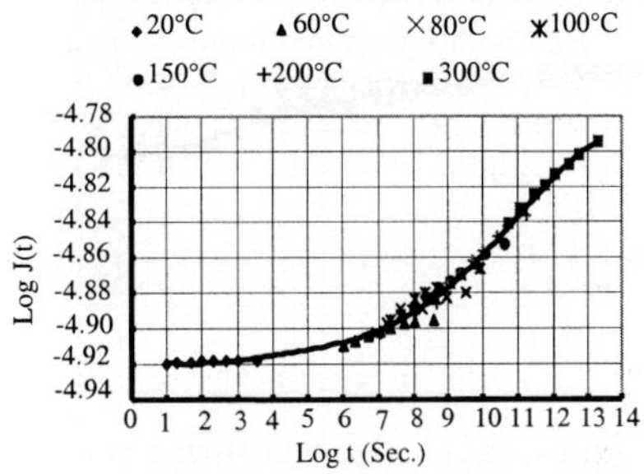

Figure 2. Major curve of TGP Granite at referential temperature θ_g=20°C

3.3 Time-Temperature Equivalence of Cohesion

Similarly, the group of logarithmic curves of cohesion vs. time of the rock are obtained shown in Figure 3, by a series of triaxial compressive creep tests at seven temperatures from 20°C to 300°C. In the same way as mentioned above we can also obtain the main curve (Figure 4) at the reference temperature at 20°C as well as the corresponding shift factor parameters (Table 2).

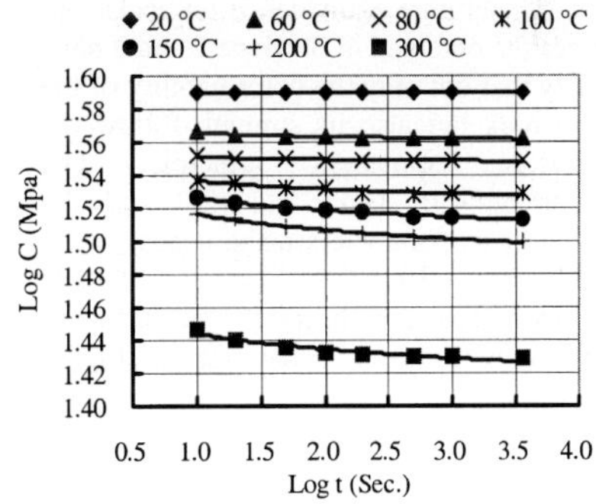

Figure 3. Logarithmic curves of cohesion vs. time at different temperatures

Table 2. Value of shift factor Parameters at the referential temperature of 20°C

θ_g(°C)	C_a	C_b	C_c	C_d
20	227.64	-0.613	101.31	-0.00704

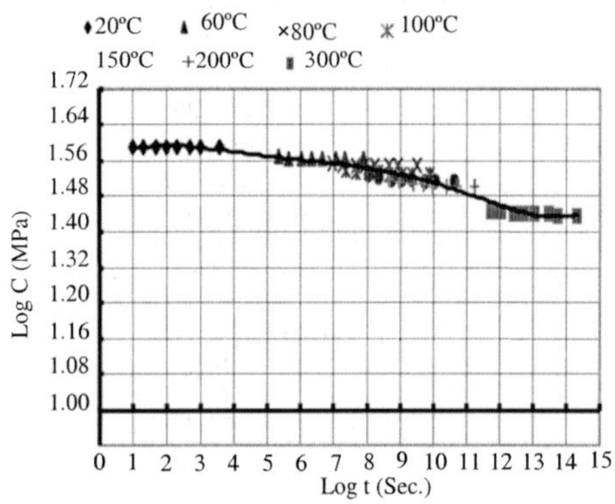

Figure 4. Main curve of cohesion of TGP granite at referential temperature of 20°C

It is shown that cohesion of the TGP granite also follows time-temperature equivalent principle.

4. CONCLUSIONS

The creep compliance of rocks reflects their deformability whereas the cohesion reflects their strength. In Section 1, the expressions are derived for the time-temperature equivalence for rocks and the testing results obtained in Sections 2 and 3 have shown clearly that not only the deformability of the TGP granite rock but also its strength follow the time-temperature equivalent principle. The establishment of shift factor eq. (25) and the determination of its parameters make it possible to predict correctly the long-term mechanical response of rock at lower temperatures according to its short-term mechanical behaviour at higher temperatures.

ACKNOWLEDGEMENT

The works performed in this paper were supported by The Chinese natural Science Foundation (NSFC) Project No. 40172098 and Innovation Foundation Project of Institute of Rock and Soil Mechanics, CAS, No Q110206.

REFERENCES

Aklonis,J. J. & Macknight, W. J. 1983. *Introduction to Polymeter Viscoelasticity* (2nd *Ed.*), New York: John Wiley and Sons.

Biot, M. A., 1954.*Theory of Stress-strain Relations in Anisotropic Viscoelasticity and Relaxation Phenomena*, J. Appl. Phys., 25(11): pp1385-1391.

Christensen, R. M.1982.*Theory of Viscoelasticity*, New York: John Wiley & Sons, Inc.

Ferry, J. D.1980.*Viscoelastic Properties of Polymers, 3rd ed.*, New York: John Wiley & Sons, Inc.

Liu, Q. & Wang C.2002.*Theoretical and Experimental Study on Time-Temperature Equivalent Principle*, Chinese Journal of Rock Mechanics and Engineering, 21(2): pp193-198.

Liu, Q., Xu, X., Yamaguchi, T. & Cho, A. Sept., 2001. *Mechanical Properties of TGP Granite in Dependence on Temperature and Time*, Proc. IRSM2001-2nd ARMS, Lisse, The Netherlands: A. A. Balkema publishers: pp119-122.

Paterson, M. S.1978.*Experimental Rock Deformation-The Brittle Field*, Berlin Heidelberg, New York: Springer-verlag.

Roger, I. & Tanner.1985.*Engineering Rheology*, UK: Oxford Science publication, Oxford University Press.

Wang, S. 1988. *Some Problems in the Development of Rock Mechanics at High Temperature and High Pressure*, Papers of the First Workshop on High Temperature and High Pressure Rock Mechanics, CSRME, China Academic Periodical Press: pp1-7 (in Chinese)

ON THE SIGNIFICANCE OF HYDRODYNAMIC CONTROL FOR RADIONUCLIDE RETENTION IN FRACTURED POROUS MEDIA

V. Cvetkovic
LWR, Royal Institute of Technology, Stockholm

Abstract: It is demonstrated that the approximate means of quantifying hydrodynamic control of retention is reasonably accurate for low values of the transport resistance on the 100m and 1000m scales; for high values, the approximate expression may significantly underestimate retention. Our results emphasize the need for further development of practical methodologies for quantifying statistical distributions of transport resistance by effectively combining field measurements, numerical simulations and theoretical/analytical considerations.

1. INTRODUCTION

The efficiency of a rock formation as a transport barrier depends on fluid flow and on radionuclide retention in the rock due to a variety of physical and chemical processes. Open fissures or fractures in the rock provide pathways through which water and radionuclides may travel. Although most radionuclides have a strong tendency to sorb to mineral grains in the rock, tracers first have to diffuse from fractures into the rock matrix in order to access the extensive pool of sorption sites (Neretnieks, 1980). Diffusion in turn depends on mass transfer properties of the rock matrix and on the hydrodynamics of fracture networks, emphasizing the interaction between water flow, advective transport and retention processes. Although models for reactive transport in discrete fracture networks have been around for some time, it is only recently that a theoretical framework is available for systematic studies of the hydrodynamic impact on retention (e.g., Cvetkovic et al., 1999, 2002).

In this study, we address the hydrodynamic control of retention in fractured porous media. The hydrodynamic control of retention in fracture networks can be reduced to the distribution of a single parameter β, referred to as ''transport resistance". Two specific objectives of the study are: (i) to summarize two main modelling approaches (continuum and discrete) and conditions for their equivalence, where from linearization of β is deduced, and (ii) to test the applicability of the linearization of β for 100m and 1000m scales using results from site-specific simulations (Outters and Shuttle, 2000; Outters, 2002).

2. CONTINUUM AND DISCRETE APPROACHES

The problem of tracer transport and retention is formulated by partitioning the tracer into a *mobile* component (with concentration C) and an *immobile* component with concentration C^*. In our following discussion, we shall neglect radioactive decay and surface sorption for simplicity.

2.1 Continuum (Eulerian) formulation

Let J [M/T] denote the tracer discharge from a rock volume. The conventional modelling approach is based on a one-dimensional formulation and has the solution in the Laplace domain (Shapiro, 2001)

$$\hat{J}(s) = M\hat{\phi}\exp\left\{\frac{UL}{2D_f}\left[1-\sqrt{1+\frac{4D_f}{U^2}\left(s+\frac{2S_f}{n_f}\kappa\sqrt{s}\right)}\right]\right\}$$

(1)

where q [L/T] is the volumetric flow rate per unit cross-sectional area, n_f [-] is the porosity of fractures, $U = q/n_f$ is the mean velocity, S_f [L^2/L^3] is the specific surface of fractures, D_f[L^2/T] is the longitudinal coefficient of dispersion in fractures, D_e is "effective diffusivity" of the "rock matrix"(immobile zone), θ is the porosity of rock matrix, R is the retardation coefficient for rock matrix, s is a Laplace transform variable, $\phi(t)$ denote the tracer injection rate at the upstream boundary and M[M] the total tracer mass injected in the system.

2.2 Discrete (Lagrangian) formulation

A discrete view of the transport considers either random *trajectories* of discrete particles, or random *streamtubes* of tracer ``parcels" carried by a constant flow rate through a network of fractures. For pulse injection of mass M, we compute the tracer discharge as $J(t) = M\gamma(t,\tau)$, the Laplace transform of γ is $\hat{\gamma}(s) = \exp(-s\tau - B\sqrt{s})$ (Cvetkovic et al., 1999) with s being the Laplace transform variable, τ[T] is the water residence time, and B[T/L] is referred to as "total transport resistance"; both τ and B can be expressed as sums of random increments. If the material (retention) properties quantified by θ, D_e and R are assumed uniform in the entire domain, we have

$$B = \beta\sqrt{\theta D_e R}\,;\; \beta = \sum_{i=1}^{N}\frac{l_i}{V_i b_i}\,;\; \tau = \sum_{i=1}^{N}\frac{l_i}{V_i} \qquad (2)$$

where i denotes a fracture/segment, and N is the total number of fractures/segments along a trajectory. β is referred to as "transport resistance" and quantifies the hydrodynamic control of retention.

To compute tracer discharge J, we recognize that γ is a conditional solution of transport since τ and β are random variables. Let $f(\tau,\beta)$ denote the joint probability density function for τ and β. Then the Laplace solution for J at a discharge surface is

$$\hat{J}(s) = M\hat{\phi}\int_0^{\infty}\int_0^{\infty}\hat{\gamma}(s/\tau,\beta)f(\tau,\beta)d\tau d\beta \qquad (3)$$

The key to computing statistics of radionuclide discharge is therefore the joint pdf $f(\tau,\beta)$. It can be shown that for strongly sorbing tracers, the distribution of τ has little impact and we may approximate $f(\tau,\beta) = \delta(\tau - \tau_0)f(\beta)$, where τ_0 is a characteristic water residence time for the fracture network. If there is no retention, then $\gamma = \delta(t - \tau)$, and $J = M\int \phi(t-\tau)f(\tau)d\tau$.

If flow and transport is assumed to take place in uniform fractures, we can write

$$\beta = 2\sum_{i=1}^{N}\frac{W_i l_i}{Q_i}; \quad \tau = \sum_{i=1}^{N}\frac{b_i W_i l_i}{Q_i} \qquad (4)$$

where W_i is the width and Q_i the volumetric flow rate of a streamtube in fracture segment i. Note that if transport takes place along a single streamtube, $Q_i = Q = const.$, $2\sum W_i l_i$ defines the total area over which the tracer is in contact with the rock matrix; hence it has been referred to as "flow-wetted surface" (Moreno and Neretnieks, 1993).

2.3 Conditions of equivalence

If β can be approximated as a linear function of τ, then $\beta = k\tau$ where k [1/L] is a constant coefficient. Recognizing that $\hat{\gamma} = \exp(-s\tau - \beta\kappa\sqrt{s})$, and accounting for time-dependent injection, we get

$$\hat{J}(s) = M\hat{\phi}(s)\int_0^{\infty}\exp\left[-\tau(s + k\kappa\sqrt{s})\right]f(\tau)d\tau \quad (5)$$

The integral is the Laplace transform of $f(\tau)$ if Laplace transform variable is replaced by $s + k\kappa\sqrt{s}$. In other words, we have $\hat{J}(s) = \hat{f}[H(s)]$ and $H(s) \equiv s + k\kappa s^{1/2}$, where $\hat{f}(s)$ is the Laplace transform of $f(\tau)$.

Linearization $\beta = k\tau$ implies $f(\tau,\beta) = f(\tau)\delta(\beta - k\tau)$. For semi-infinite domain and pulse injection, and transport governed by the ADE, $f(\tau)$ is an inverse-gaussian density with Laplace transform $\hat{f}(s) = \exp(\sqrt{a} - c\sqrt{a+s})$, where $a^2 = U^2/4D_f$ and $c = L/\sqrt{D_f}$.

Finally, we get

$$\hat{J}(s) = \exp\left\{\frac{UL}{2D_f}\left[1 - \sqrt{1 + \frac{4D_f}{U^2}\left(s + k\kappa\sqrt{s}\right)}\right]\right\} \qquad (6)$$

Comparing (6) with (1), we find that the conditions of equivalence between the (one-dimensional) continuum approach and the discrete approach are:

- Linearization of β as $\beta = k\tau$ where k is a constant implying $f(\tau,\beta) = f(\tau)\delta(\beta - k\tau)$;
- $k \equiv 2S_f/n_f$
- $f(\tau)$ defined consistent with the advection-dispersion process (most appropriately inverse-gaussian);

3. EXAMPLES FOR ÄSPÖ

Transport resistance as quantified by β ((2) or (4)) is dependent on the structure of the fracture network and prevailing boundary conditions for flow. The key issue for PA applications is how to infer the distribution of β based on site characterization data. The full description requires the joint density $f(\tau,\beta)$, however, for more strongly tracers the marginal density $f(\beta)$ is sufficient for most practical purposes.

If S_f and n_f can be estimated in the field, then $\beta = k\tau$ with $k \equiv 2S_f/n_f$ [1/L] is a linearized model for hydrodynamic control of retention in a fractured porous medium. This approach has been used for interpreting field observations (e.g., Shapiro, 2001), as well as for PA/SA (Selroos et al., 2002).

In the following, we shall use "discrete fracture network" (DFN) simulations (Outters and Shuttle, 2000; Outters, 2002) to exemplify the estimate of β using $k = 2S_f/n_f$.

The DFN simulation cases were constructed to emulate the hydraulic behaviour of a typical Swedish granitic rock volume. In an earlier study (Outters et al., 2000), the geometric and hydraulic properties of the DFN model were based on the prototype repository and the TRUE Block Scale site at the Äspö Hard Rock Laboratory, Sweden. Since then, the DFN for the prototype repository was updated. Stigsson et al. (2001) carried out an extensive analysis of pump tests which has led to an improved understanding of the fracture set definition and their hydraulic properties; their model was used in Outters (2002).

3.1 100m scale

The size of a simulation rock volume was selected based on safety assessment and computation time considerations. The shape of the domain is a 100 m ×100 m ×100 m cube (Outters, 2002). Since the DFN model characteristics are stochastic, the modelling is carried out with Monte Carlo simulations using 20 equi-probable realizations. The DFN realizations were generated by the FracMan(c) software. A total of about 20000 three dimensional disc-shaped fractures were identified in each realization. Following each trajectory, τ and β were computed using (2) or (4). Details of the simulations are given in Outters (2002) and further applications in Cvetkovic et al. (2002).

3.2 1000m scale

A series of DFN simulation data was obtained by Outters and Shuttle (2000) based on the FracMan code. A large number of disk shaped fractures with pre-defined radius and orientation distribution are placed randomly throughout the simulation region in addition to deterministic fracture zones. Pressure boundary conditions are applied to the simulation domain and flow in the equivalent pipe network is solved numerically by assuming laminar flow in each pipe.

Once the flow network is constructed, pathways through the network are identified using the code PAWorks (Dershowitz et al., 1999) that is based on a group theory search algorithm. For the particular simulations described here, a large number of starting positions (waste canisters) are placed in a spatial pattern consistent with current repository design. Canisters intersecting individual fractures are identified, as is the dominant pathway connecting each to the discharge plane. The possibility of downstream bifurcation in the transport pathway was neglected in these simulations. Once each particle trajectory is identified, the length, aperture, and velocity for each segment is recorded. β and τ were computed using (2) or (4). Further interpretation of the simulation results was given in Painter et al. (2002).

4. RESULTS

In Figure 1, we compare the complementary cumulative distribution functions (CCDF) and cumulative distribution function (CDF) for β obtained directly from simulations using (2) or (4), with the approximate value of β using the linear expression $\beta = k\tau$. For the 100m scale, k is estimated using fracture density and flow porosity as $k = 2S_f / n_f$, whereas for 1000m scale k is inferred directly from simulations, as the correlation between β and τ.

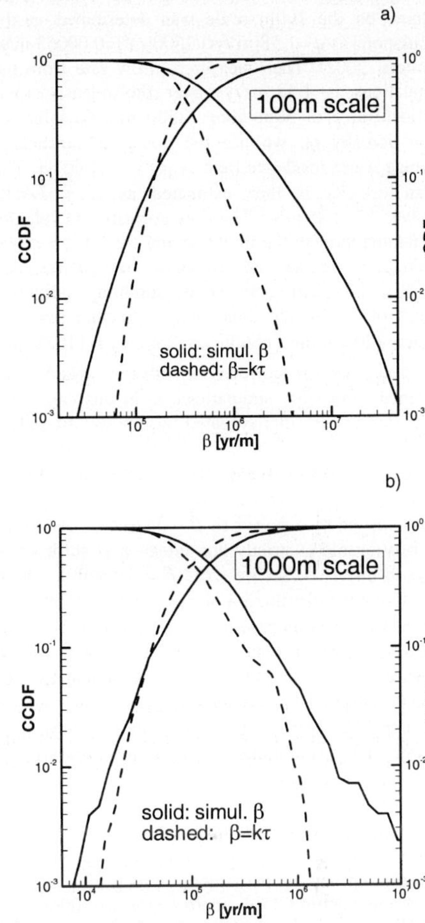

Figure 1. Cumulative distribution function (CDF) and complementary cumulative distribution function (CCDF) for β on: (a) 100 m scale, and (b) 1000 m scale. The solid curves have been obtained from DFN simulations based on site-specific data, whereas the dashed curves were obtained from simulated distributions of water residence time τ, and assuming $\beta = k\tau$ applicable. For the 100 m, k was estimated from field and simulated data as described in the text as 11628 1/m, whereas on the 1000 m scale, k was obtained assuming linear correlation between τ and β as 3241 1/m.

The average flow rate through the system rock volume on the 100m scale was determined in the simulations as q =1.58[m^3/yr]/2500[m^2]=0.00063 m/yr (Outters, 2002). The "measured" flow rate from the simulations is 1.58 m^3/yr over the release cross-sectional area of 50m×50m= 2500 m^2. To infer the flow porosity n_f, we use the computed arithmetic average water residence time as $<\tau>$ = 35.05 yr. The mean velocity is then estimated as $U = L/<\tau>$ =100/35.07=2.85 m/yr. The flow porosity and velocity are related through the flow rate and we get $n_f = q/U$ = 0.00022. To estimate an "effective" half-aperture, we use S_f = 1.29 1/m (obtained by summing individual values of S_f for the three sets of fracture given in Outters (2002)), and obtain $k = 2S_f/n_f = 11628\,1/\mathrm{m}$. We note that the correlation between τ and β as obtained from DFN simulations, is in this case non-linear and is best approximated by a power fit as $\beta = 3241\tau^{1.5}$.

For the 1000m scale, the correlation between simulated τ and β (Outters and Shuttle, 2000) is well approximated by a power fit $\beta = 3416\tau^{1.103}$. Although the relationship is strictly non-linear, it is sufficiently close to linear so that we write $\beta = k\tau$ with k =3416 1/m. Using the distribution of τ as obtained from DFN simulations, we compare in Figure 1b the CCDF and CDF of β directly obtained from simulation with β approximated by $\beta = 3416\tau$, where CCDF and CDF of τ are those from DFN simulations.

Comparison between Figures 1a and 1b shows that for considered conditions, the linearized expression $\beta = k\tau$ is a better approximation for the 1000m scale than for the 100m. In both cases, the approximate expression $\beta = k\tau$ underestimates dispersion; the water residence time τ varies less than transport resistance β. The discrepency is particularly strong in the high range (tail part) of CCDF, more so for the 100m scale (Figure 1a) than the 1000m scale (Figure 1b). In this high range, the approximation $\beta \sim \tau$ is conservative. In the lower range (reflected by the tail of the CDF), the approximation $\beta = \tau$ is non-conservative, since the simulated values of β are lower. However, the discrepency is relatively small and the linearization $\beta = k\tau$ may provide a reasonable approximation for conditions where field data on water residence time statistics is available.

5. CONCLUSIONS

Using the linearized model $\beta = k\tau$ implies that $f(\beta) \sim f(\tau)$. The accuracy of $f(\beta)$ estimation thus depends on the accuracy of the k estimate, and on the accuracy of the estimated water residence time density $f(\tau)$.

The approximate means of quantifying β as $\beta = k\tau$ is reasonably accurate for low values of β on the 100m and 1000m scale. The approximate approach yields a comparatively poor representation for high values of β on the 100m and 1000m scales, underestimating retention considerably.

For PA applications, we need to infer β statistics based on site characterization data. More specifically, we need to infer the marginal density $f(\beta)$. Several approaches for estimating $f(\beta)$ are possible, however, at this time there is no general consensus on which one is most preferable. In view of the heterogeneity and complexity of fractured porous media, and uncertainties involved, it is most likely that several complementary approaches for estimating $f(\beta)$ from field data need to be used, rather than a single approach.

6. ACKNOWLEDGEMENTS

This work was supported by the Swedish Nuclear Fuel and Waste Management Co. (SKB), Stockholm.

REFERENCES

Cvetkovic, V., Painter, S., and Selroos, J. 2002. *Comparative measures for radionuclide containment in the crystalline geosphere.* Nuclear Science and Engineering, 142: 292-304.

Cvetkovic, V., Selroos, J.-O., and Cheng, H., 1999. *Transport of reactive tracers in rock fractures.* Journal of Fluid Mechanics, 378, 335-356.

Dershowitz, B., Eiben, T., Follinf, S., and Andersson, J., 1999. *SR-97 – alternative models project. Discrete fracture network modelling for performance assessment of Aberg.* R-99-43, Swedish Nuclear Fuel and Waste Management Co. (SKB).

Moreno, L., and Neretnieks, I., 1993. *Flow and nuclide transport in fractured media: The importance of the flow-wetted surface.* Journal of Contaminant Hydrology, 13 : 49-71.

Neretnieks, I., 1980. *Diffusion in the rock matrix: An important factor in radionuclide retention.* Journal of Geophysical Research, 85: 4379-4397.

Outters, N., 2003. *A generic study of discrete fracture network transport properties using fracman/mafic.* R-03-XX, Swedish Nuclear Fuel and Waste Management Co. (SKB).

Outters, N. and Shuttle, D., 2000. *Sensitivity analysis of a discrete fracture network model for performance assessment of aberg*. R-00-48, Swedish Nuclear Fuel and Waste Management Co. (SKB).

Painter, S., Cvetkovic, V., and Selroos, J., 2002. *Power-law velocity distributions in fracture networks : Numerical evidence and implications for tracer transport*. Geophysical Research Letters, 29(14): 10.1029/2002GL014960.

Selroos, J.-.O, Walker, D.D., Ström, A., Gylling, B., and Follin, S., 2002. *Comparison of alternative modelling approaches for groundwater flow in fractured rock*. Journal of Hydrology, 257: 174-188.

Shapiro, A., 2001, *Effective matrix diffusion in kilometer-scale transport in fractured crystalline rock*. Water Resources Research, 37(3), 507-522.

Stigsson, M., Otters, N., and Hermanson, J., 2001. *Äspö Hard Rock Laboratory, Prototype Repository Hydraulic DFN model no:2*. IPR-01-39, Swedish Nuclear Fuel and Waste Management Co. (SKB).

FUNDAMENTAL THERMODYNAMIC REQUIREMENTS FOR POROUS MEDIA DESCRIPTION

T.J.T. (Tim) Spanos[1,2], Maurice B. Dusseault[3], Norman Udey[1]
1) PRISM Production Technologies Inc., Edmonton Alberta
2) University of Alberta, Edmonton Alberta
3) University of Waterloo, Waterloo Ontario

Abstract: In general, existing porous media descriptions are thermodynamically incomplete or improperly constituted. This incompleteness arises because of two widely accepted but incorrect assumptions:

- Porosity is a scalar invariant (rather than a thermodynamic state variable)
- The energy state for a representative elementary volume can be described by a unique energy function

These two assumptions lead to flawed thermodynamical statements for porous media containing multiphase fluids, with a number of consequences that hinder theoretical developments in thermomechanics, and impair development of practical applications. The article seeks to fully express the equilibrium thermodynamics relationships for porous media, based on a first-principles derivation.

1. INTRODUCTION

The existence of important coupled diffusion-dynamic processes in porous media is well-documented empirically. The Russian literature contains many descriptions of earthquake-induced altered (generally enhanced) production from reservoirs of modest porosity (i.e.: elastic conditions with no potential for compaction). Anecdotal evidence for increased well levels, release of fine-grained material into wells, and enhanced stream flow after earthquakes (Manga et al. 2003) has been reported. Triggering of sympathetic secondary earthquakes at a distance is well-known, and it is also known that the time of propagation for the triggering energy is far slower than the velocity of compressional, shear, or other common waves that are sufficiently conservative to travel hundreds of kilometres.

Based on theoretical developments (e.g. de la Cruz et al. 1993, Hickey et al. 1995, Spanos, 2001), experiments on dynamic excitation of liquid flow rate in laboratory cells under constant pressure head were carried out. Low-frequency excitation was applied externally (by pressure impulses or by impacting on the solid body of the cell), and increased flow rates of 40-100% were observed without alteration of the external heads or pumping (Davidson et al. 1999, Dusseault, 1999). Findings were applied in several heavy oil fields in east-central Alberta (Spanos et al. 1999, 2003, Dusseault et al. 1999, Dusseault et al. 2002) with reasonable success. These results could not be predicted by any of the standard theories based on Darcy and Biot-Gassmann formalisms.

In this article, we will present the theoretical equilibrium thermodynamics basis that has allowed these developments to take place. Much of what is presented here can be found in a broader treatment of porous media thermomechanics (Spanos, 2001), but there are several subtleties and ideas herein that may be considered new.

2. INTERNAL ENERGY

Based on a fundamental revisitation of equilibrium thermodynamics applied to multiphasic porous media, the internal energy for the fluid and solid phases are given by (de la Cruz et al, 1993; Spanos, 2001):

$$dU_f = -\eta p_f du_{kk}^{'f} + T_f dS_f \qquad (1)$$

$$dU_s = \sigma_{ik}^{s} du_{ik}^{'s} + T_s dS_s \qquad (2)$$

where:

$$u_{kk}^{'f} \equiv \vec{\nabla} \bullet \vec{u}_f + \frac{\eta - \eta_o}{\eta_o} \qquad (3)$$

$$u_{ik}^{'s} \equiv u_{ik}^{s} + \frac{1}{3}\frac{\phi - \phi_o}{\phi_o}\delta_{ik} \qquad (4)$$

and $\phi \equiv 1 - \eta$. According to (1) and (2),

$$U_f = U_f(S_f, u'^f_{kk}), \quad U_s = U_s(S_s, u'^s_{ik}) \tag{5}$$

and the stresses are given by

$$\eta p_f = \frac{\partial U_f}{\partial u'^f_{kk}}, \quad \sigma^s_{ik} = \frac{\partial U_s}{\partial u'^s_{ik}} \tag{6}$$

3. THERMODYNAMIC DEFORMATION: SUMMARY OF EQUATIONS

3.1 Solid properties

Here, we present in summary form the equations that describe solid strain, solid stress, and solid internal energy of the porous medium. First, the internal energy statement:

$$dU_s = \sigma^s_{ik} du'^s_{ik} + T_s dS_s \tag{7}$$

For the solid stress

$$\sigma^s_{ik} = -\phi_o K_s \alpha_s (T_s - T_o)\delta_{ik} + 2\mu_M (u'^s_{ik} - \tfrac{1}{3}\delta_{ik} u'^s_{jj}) + \phi_o K_s \delta_{ik} u^s_{jj} \tag{8}$$

For the solid strain

$$u'^s_{ik} = u^s_{ik} + \tfrac{1}{3}[(\phi - \phi_o)/\phi_o]\delta_{ik} \tag{9}$$

And, the total solid internal energy associated with pressure and stresses is:

$$U_s(S_s, u'^s_{ik}) = \phi_o U_{os}(S_s) - \frac{T_o K_s \alpha_s}{c_V}(S_s - S_o)u'^s_{jj} + \mu_M (u'^s_{ik} - \tfrac{1}{3}\delta_{ik} u'^s_{jj})^2 + \tfrac{1}{2}\phi_o K_{s(ad)} u'^{s\,2}_{jj} \tag{10}$$

3.2 Entropy

For molecular mixtures, entropy is the Boltzmann constant times the logarithm of the number of ways in which the components of a system may be rearranged. If we now allow for macroscopic components to be rearranged and go up in scale by orders of magnitude, the fashion in which the system may be rearranged now has new degrees of freedom. For example, in the case of a granular medium, we may rearrange the grains from highly ordered packing to looser packing, thus increasing the entropy (or "mixed-uppedness", to paraphrase Gibbs). Thus, the entropy statement must be sufficiently general and complete to describe how the energy of the system is arranged at all scales. In the case of equilibrium thermodynamics one may write for entropy:

$$S_{os}(T_s) = -\phi_o \frac{dF_{os}(T_s)}{dT_s} \tag{11}$$

Thermostatic entropy balance of the solid component is given by

$$S_s - S_{os} = \phi_o c_V (T_s - T_o)/T_o + \phi_o K_s \alpha_s u'^s_{jj} \tag{12}$$

and the heat capacity by

$$c_V = -T_o \frac{d^2 F_{os}(T_o)}{dT_o^2} \tag{13}$$

In this formulation, the Helmholtz Free Energy "F", in general, represents the total energy necessary to create the system, minus the heat obtained from an environment at temperature T. If one also allows for irreversible granular motions, then changes in porosity will also cause changes in the entropy. Such a system would involve irreversible deformations (non-equilibrium thermodynamics formalism), but dealing with such a system is beyond the goal of this article. However, for reversible motions of the elastic matrix involving only mechanical energy (strain energy) and thermal energy (heat), the Helmholtz Free Energy for the solid may be written as

$$dF_s = \sigma^s_{ik} du'^s_{ik} - S_s dT_s \tag{14}$$

$$F_s(T_s, u'^s_{ik}) = \phi_o F_{os}(T_s) - \phi_o K_s \alpha_s (T_s - T_o) u'^s_{jj} + \mu_M (u'^s_{ik} - \tfrac{1}{3}\delta_{ik} u'^s_{jj})^2 + \tfrac{1}{2}\phi_o K_s u'^{s\,2}_{jj} \tag{15}$$

The enthalpy represents the internal energy of the system plus the work required to make room for it in an environment with constant pressure.

$$dH_s = u'^s_{ik} d\sigma^s_{ik} + T_s dS_s \tag{16}$$

$$H_s(S_s, u'^s_{ik}) = \phi_o U_{os}(S_s) - \frac{T_o K_s \alpha_s}{c_V}(S_s - S_o)u'^s_{jj} + 3\mu_M (u'^s_{ik} - \tfrac{1}{3}\delta_{ik} u'^s_{jj})^2 + \phi_o(\tfrac{1}{2}K_{s(ad)} + K_s)u'^{s\,2}_{jj} - \phi_o(p_s - p_o)u'^{s\,2}_{jj} - \phi_o K_s u'^{s\,2}_{jj} - K_s(\phi - \phi_o) \tag{17}$$

The Gibbs Free Energy represents the internal energy of the system plus the work required to make room for it in an environment with constant pressure, minus the heat obtained from an environment at temperature T.

$$dG_s = u'^s_{ik} d\sigma^s_{ik} - S_s dT_s \tag{18}$$

$$G_s(T_s, u_{ik}^{'s}) = \phi_o F_{os}(T_s) - 2\phi_o K_s \alpha_s (T_s - T_o) u_{jj}^{'s} + 3\mu_M (u_{ik}^{'s\,2} - \tfrac{1}{3}\delta_{ik} u_{jj}^{'s\,2}) + \tfrac{3}{2}\phi_o K_s u_{jj}^{'s\,2} \tag{19}$$

3.3 Fluid properties

Now we specify the fluid equilibrium thermodynamic properties. The internal energy is

$$dU_f = -\eta p_f du_{ik}^{'f} + T_f dS_f \tag{20}$$

$$U_f(S_f, u_{ik}^{'f}) = \eta_o U_{of}(S_f) - (T_o \alpha_f / c_p)(S_f - S_o) p_f + \tfrac{1}{2}\eta_o K_{ad}^f u_{jj}^{'f\,2} \tag{21}$$

The fluid compressive strain is given by:

$$u_{kk}^{'f} = \nabla \bullet u_f + (\eta - \eta_o)/\eta_o \tag{22}$$

The ambient temperature is

$$T_o = \eta_o \frac{dU_{of}(S_o)}{dS_o} \tag{23}$$

The Helmholtz Free Energy is

$$dF_f = -\eta p_f du_{jj}^{'f} - S_f dT_f \tag{24}$$

$$F_f(T_f, u_{ik}^{'f}) = \eta_o F_{of}(T_f) + \tfrac{3}{2}\eta_o K_f u_{jj}^{'f\,2} \tag{25}$$

The thermostatic fluid entropy is

$$S_f - S_{of} = \eta_o c_p^f (T_f - T_o)/T_o + \eta_o \alpha_f p_f \tag{26}$$

which may be summed with equation (12) to obtain the entropy of the system.

The enthalpy is

$$dH_f = -u_{jj}^{'f} d(\eta p_f) + T_f dS_f \tag{27}$$

$$H_f(S_f, u_{jj}^{'f}) = \eta_o U_{of}(S_f) - \frac{T_o \alpha_f}{c_p^f}(S_f - S_o) p_f + \tfrac{1}{2}\eta_o K_{ad}^f u_{jj}^{'f\,2} - \eta_o p_f u_{jj}^{'f} \tag{28}$$

The Gibbs Free Energy is

$$dG_f = -u_{jj}^{'f} d(\eta p_f) - S_f dT_f \tag{29}$$

$$G_f(T_f, \sigma_{ik}^f) = \eta_o F_{of}(T_f) - 2\eta_o \alpha_f (T_f - T_o) p_f - \eta_o p_f u_{jj}^{'f} \tag{30}$$

Our exposition is limited to equilibrium thermodynamics and reversible elasticity in the solid phase. The porosity is clearly shown to be a variable on the same footing as the strain tensor. This new dynamic variable allows for additional degrees of freedom in the equations of motion. This results in new physical processes that remain to be predicted by the equations of motion. Two of these new processes will be studied in the discussion presented in sections 6 and 7

4. EFFECT OF HETEROGENEITY

When porosity is heterogeneous and therefore varies with position, the strain associated with the stress tensor, defined as

$$u_{ik}^{'s} \equiv u_{ik}^s + \tfrac{1}{3}\frac{\phi - \phi_o}{\phi_o}\delta_{ik}, \tag{31}$$

must be generalized to (Sahay et al, 2001)

$$\bar{u}_{ij}^{s\,''} = \bar{u}_{ij}^s + \tfrac{1}{3}\frac{(\phi - \phi_o + \bar{u}_k^s \partial_k \phi_o)}{\phi_o}\delta_{ij} \tag{32}$$

in order to retain the same relations with the stress tensor being a function of the single quantity $\bar{u}_{ij}^{s\,''}$. (The subscript k is the summation convention subscript.) In other words, there now exists an intrinsic porosity gradient ($\partial\phi_o$) that must be accounted for; as a result, the free energy functional of the porous matrix must now be written as:

$$F_s(\bar{T}, \bar{u}_{ik}^{s\,''}) = \phi_o F_o^s(\bar{T}) - \phi_o K_s \alpha_s (\bar{T} - T_o)\bar{u}_{jj}^{s\,''} + \mu_M (\bar{u}_{ik}^{s\,''} - \tfrac{1}{3}\delta_{ik}\bar{u}_{jj}^{s\,''})^2 + \tfrac{1}{2}\phi_o K_s \bar{u}_{jj}^{s\,''2} \tag{33}$$

The entropy of the solid matrix is now given by

$$S^s - S_o^s = \phi_o c_V^s (\bar{T} - T_o)/T_o + \phi_o K_s \alpha_s \bar{u}_{jj}^{s\,''} \tag{34}$$

and the internal energy is given by

$$U^s(S^s, \bar{u}_{ik}^{s\,''}) = \phi_o U_o^s(S) - \frac{T_o K_s \alpha_s}{c_V}(S^s - S_o^s)\bar{u}_{jj}^{s\,''} + \mu_M (\bar{u}_{ik}^{s\,''} - \tfrac{1}{3}\delta_{ik}\bar{u}_{jj}^{s\,''})^2 + \tfrac{1}{2}\phi_o K_{ad}^s \bar{u}_{jj}^{s\,''2} \tag{35}$$

These and related issues are developed in more detail elsewhere (e.g. Sahay et al, 2001, Spanos, 2001).

5. EQUATIONS OF WAVE MOTION

Now that the thermodynamic statements have been derived more completely by the treatment of porosity as a thermodynamic state variable for a porous medium, it is possible to determine the effects that this new dynamic variable has on the equations of motion. In the following section, it is shown that this new thermodynamics variable causes two new dynamic processes to be predicted, porosity diffusion and porosity wave propagation.

The equation of dilational wave motion for the solid phase is written as:

$$\frac{\rho_{\eta s}}{\eta_o}\left(\frac{\partial^2 \eta}{\partial t^2} + 2a_{\eta s}\frac{\partial \eta}{\partial t} - v_{\eta s}^2 \nabla^2 \eta\right) = \frac{\rho_{ps}}{K_f}\left(\frac{\partial^2 p_f}{\partial t^2} + 2a_{ps}\frac{\partial p_f}{\partial t} - v_{ps}^2 \nabla^2 p_f\right) \quad (36)$$

And for the liquid phase it is

$$\frac{\rho_{\eta f}}{\eta_o}\left(\frac{\partial^2 \eta}{\partial t^2} + 2a_{\eta f}\frac{\partial \eta}{\partial t} - v_{\eta f}^2 \nabla^2 \eta\right) = \frac{\rho_{pf}}{K_f}\left(\frac{\partial^2 p_f}{\partial t^2} + 2a_{pf}\frac{\partial p_f}{\partial t} - 2b_{pf}\nabla^2 \frac{\partial P_f}{\partial t} - v_{pf}^2 \nabla^2 P_f\right) \quad (37)$$

The terms in these equations are now defined. Moduli of dilation are given in (38) and (39):

$$\sigma_M = \mu_f (1-\eta_o)\left(\left(\frac{\mu_M}{\mu_s(1-\eta_o)} - 1\right)\right) \quad (38)$$

$$\alpha_1 = \frac{\eta_o - \delta_f}{\delta_s}, \; \alpha_2 = \frac{\delta_f}{\delta_s} \quad (39a)$$

where

$$\delta_f < \eta_o, \; \delta_s < 1 - \eta_o \quad (39b)$$

The shear moduli are μ_f for the fluid shear viscosity, μ_s for the solid shear modulus, and μ_M for the macroscopic (matrix) shear modulus. Under the conditions of porosity and pressure variations, the effective densities of the liquid phase are:

$$\rho_{\eta f} = \rho_f - \frac{\rho_{12}}{\eta_o}(\alpha_1 + 1)$$
$$\rho_{pf} = \rho_f - \frac{\rho_{12}}{\eta_o}(\alpha_2 - 1) \quad (40)$$

Reference to Landau and Lifshits (1975) will clarify the meaning of these relationships. The fluid attenuation and bulk attenuation associated with the porosity are:

$$a_{\eta f} = \frac{1}{2}\frac{\mu_f}{K}\frac{\eta_o(\alpha_1 + 1)}{\rho_{\eta f}}$$
$$b_{\eta f} = \frac{2}{3}\frac{(\mu_f + \alpha_1 \sigma_M)}{\rho_{\eta f}} \quad (41)$$

The fluid attenuation (gain), fluid bulk attenuation and fluid wave speed associated with the pressure are:

$$a_{pf} = -\frac{\mu_f}{2\kappa}\frac{\eta_o(\alpha_2 - 1)}{\rho_{pf}}$$
$$b_{pf} = \frac{1}{2\rho_{\eta f}}\left(\xi_f + \frac{4}{3}(\mu_f - \alpha_2\sigma_M)\right) \quad (42)$$
$$v_{pf}^2 = \frac{K_f}{\rho_{pf}}$$

The effective densities of the solid phase under conditions of porosity and pressure variations are:

$$\rho_{\eta s} = \alpha_1 \rho_s - \frac{\rho_{12}(\alpha_1 + 1)}{(1-\eta_o)}$$
$$\rho_{ps} = \alpha_2 \rho_s - \frac{\rho_{12}(\alpha_2 - 1)}{(1-\eta_o)} \quad (43)$$

The solid attenuation and solid wave speed associated with the porosity, then with the pressure, are given in these equations (Eqs. 44, 45):

$$a_{\eta s} = \frac{1}{2}\frac{\mu_f}{\kappa}\frac{\eta_o^2(\alpha_1 + 1)}{\rho_{\eta s}\alpha_1(1-\eta_o)}, \; and \; v_{\eta s}^2 = \frac{K_\eta}{\rho_{\eta s}} \quad (44)$$

$$a_{ps} = \frac{1}{2}\frac{\mu_f}{K}\frac{\eta_o^2(\alpha_2 - 1)}{\rho_{ps}(1-\eta_o)}, \; and \; v_{ps}^2 = \frac{K_M}{\rho_{ps}} \quad (45)$$

The bulk modulus associated with the porosity (inverse of the pore compressibility) and the macroscopic bulk modulus of the medium are given by:

$$K_\eta = \alpha_1 K_M - K_s \frac{\eta_o}{(1-\eta_o)}, \; and \; K_M = K_s + \frac{4}{3}\frac{\mu_M}{(1-\eta_o)} \quad (46)$$

Notice in the previous equations that the effect of fluid flow through the pores, governed by the permeability κ and the fluid viscosity μ_f, enters into the equations of motion only in the attenuation term. In other words, wave attenuation arises because of liquids moving through pore throats (i.e.: displacement, not strain). Also note that the ratio of the two is also known as the mobility of the fluid, or the inverse of the hydraulic conductivity.

Finally, it is important to note that at all times the following statement of porosity diffusion is satisfied, and that this is a statement of conservation independent of Newton's Second Law of Motion, therefore it is necessary to achieve physical completeness of the problem statement.

$$\frac{\partial \eta}{\partial t} = \delta_s \nabla \bullet v_s - \delta_f \nabla \bullet v_f \tag{47}$$

6. POROSITY DIFFUSION

It was first observed by Geilikman et al. (1993) that a porosity diffusion process is predicted by the equations of motion associated with this thermodynamic description in the limit when the inertial terms and bulk attenuation are assumed small. Under these assumptions, the equations of motion may be written as follows (Spanos, 2001); first, for the solid, including a source term:

$$\frac{\partial \eta}{\partial t} - D_s \nabla^2 \eta = -cM_{ik} \frac{\partial^2 \delta(\vec{r})}{\partial r_i \partial r_k} \theta(t) + B \frac{\partial p_f}{\partial t} \tag{48}$$

Then, the equation of motion for the fluid phase is:

$$\frac{\partial p_f}{\partial t} - D_f \nabla^2 \eta = C \frac{\partial \eta}{\partial t} \tag{49}$$

Here, M_{ik} is the seismic moment, $\delta(r)$ is the Dirac delta function and $\theta(t)$ is the step function in time to account for the sudden impulse. The various coefficients in Eqs (48) and (49) are now specified:

$$D_s = \frac{v_{\eta M}^2}{2a_{\eta M}} \quad and \quad D_f = \frac{v_{pf}^2}{2a_{pf}} \tag{50}$$

$$B = 2\frac{\alpha_2 \rho_{pM} \eta_o a_{pM}}{\alpha_1 \rho_{\eta M} K_f} \quad and \quad C = \frac{\rho_{\eta f} K_f a_{\eta f}}{\alpha_1 \rho_{pf} a_{pf}} \tag{51}$$

$$c = \frac{\eta_o}{2\alpha_1 \rho_{\eta M} a_{\eta M}} \quad and \quad v_{\eta M}^2 = \frac{K_\eta}{\rho_{\eta M}} \tag{52}$$

$$\rho_{\eta M} = \rho_{\eta s} + \frac{\alpha_2 K_M}{\alpha_1 K_f} \quad and \quad \rho_{pM} = \rho_{ps} + \rho_{pf} \frac{K_M}{K_f} \tag{53}$$

$$a_{\eta M} = a_{\eta s} \frac{\rho_{\eta s}}{\rho_{\eta M}} + a_{\eta f} \frac{\rho_{\eta f}}{\rho_{\eta M}} \frac{\alpha_2 K_M}{\alpha_1 K_f}$$
$$a_{pM} = a_{ps} \frac{\rho_{ps}}{\rho_{pM}} + a_{pf} \frac{\rho_{pf}}{\rho_{pM}} \frac{\alpha_2 K_M}{\alpha_1 K_f} \tag{54}$$

The moment tensor has the following general form (Geilikman et al. 1993):

$$M_{ik} = (K_s - \tfrac{2}{3}\mu_M)\delta_{ik} \int_s dS' b_j(r') n_j(r') + \mu_M \int_s dS'(b_i(r') n_k(r') + b_k(r') n_i(r')) \tag{55}$$

where n_i is the normal to the plane of the fault, b_i is the fracture displacement vector in the plane of the fault in the direction r' (analogous to the Burger's vector in crystallographic distortions), and the integration extends over the area of the source S'.

7. POROSITY-PRESSURE WAVE PROPAGATION

The equations of motion (36) and (37) may be solved as an eigenvalue problem, yielding the various solutions for P waves and porosity waves (Spanos, 2001).

The motions of a plane wave propagating through a porous medium may be described as follows:

$$\Omega_\eta = |\Omega_{o\eta}| e^{i\phi_\eta} e^{i(k_r x - \omega t)} e^{-k_i x} \tag{56}$$

$$\Omega_p = |\Omega_{op}| e^{i\phi_p} e^{i(k_r x - \omega t)} e^{-k_i x} \tag{57}$$

Here, k_i is attenuation, ω/k_r is the phase velocity, $\partial\omega/\partial k_r$ is the group velocity, $|\Omega_{o\eta}|/|\Omega_{op}|$ is the relative magnitude of the two modes, and $\phi_\eta - \phi_p$ is the phase angle. Upon substituting these two solutions into the equations of motion, one obtains two algebraic equations of the form

$$a_1 \omega^2 + b_1 \omega + c_1 = 0 \tag{58}$$

$$a_2 \omega^2 + b_2 \omega + c_2 = 0 \tag{59}$$

With the following defined terms:

$$a_1 = \frac{\rho_{\eta f}}{\eta_o} e^{i\phi_\eta} + \frac{\rho_{pf}}{K_f} e^{i\phi_p} \tag{60}$$

$$b_1 = \frac{2\rho_{\eta f}}{\eta_o}(a_{\eta f} + b_{\eta f}k^2)e^{i\phi_\eta} + \frac{2\rho_{pf}}{K_f}(a_{pf} - b_{pf}k^2)e^{i\phi_p} \quad (61)$$

$$c_1 = \frac{\rho_{pf}}{K_f}v_{pf}^2 k^2 e^{i\phi_p} \quad (62)$$

$$a_2 = \frac{\rho_{\eta s}}{\eta_o}e^{i\phi_\eta} + \frac{\rho_{ps}}{K_f}e^{i\phi_p} \quad (63)$$

$$b_2 = \frac{2\rho_{\eta s}}{\eta_o}a_{\eta s}e^{i\phi_\eta} + \frac{2\rho_{ps}}{K_f}a_{ps}e^{i\phi_p} \quad (64)$$

$$c_2 = \frac{\rho_{\eta s}}{\eta_o}v_{\eta s}^2 k^2 e^{i\phi_\eta} - \frac{\rho_{ps}}{K_f}v_{ps}^2 k^2 e^{i\phi_p} \quad (65)$$

The two second-order equations (58) and (59) yield four solutions for ω. The first solution is a seismic wave in which the fluid and solid compress almost in phase. However, since the fluid is normally an order of magnitude more compressible than the solid, fluid flow is induced and the shear and bulk viscosities give rise to attenuation. This attenuation is minor for low frequency waves and much greater for high frequency waves.

The second solution is a seismic wave in which the fluid and solid compress almost completely out of phase. This solution induces so much fluid flow and thus attenuation that in general it is unable to propagate very far from the source.

The third solution is porosity diffusion. In this solution the fluid is unable to compress because the motions are too slow, therefore it releases the stresses imparted on it by the matrix through fluid motions observed as pressure and porosity diffusion. Apparently, the compressible-incompressible transition comes at about a frequency of strain excitation of 10^{-2} to 10^{0} Hz for the common range of liquids and phase compressibilities. Above this transition frequency, the fluid will compress (strain) in response to excitation.

The fourth and last solution is a pressure-porosity wave in which fluid flow at the incompressible limit of fluid motions is coupled to elastic deformations of the matrix. The attenuation of this wave is highly frequency dependent, but it is quite conservative at low frequencies. Unlike the previous three solutions which each exist as an independent process, this solution is always coupled to porosity diffusion. It can leave behind an increase in pressure as it propagates, converting some of the inertial energy associated with the propagating wave to a static energy associated with a higher pressure.

8. PRACTICAL IMPLICATIONS

The existence of a slow displacement wave transiting a liquid-saturated porous medium and leading to internal pressure increases has many implications. The principle is now used in practice to enhance fluid recovery in oil reservoirs (Spanos et al. 2003) and to help mobilize trapped immiscible contaminating phases in shallow aquifers (Gray, 2002). However, a practical limit on the exploitation of this displacement wave is the presence of a compressible phase in the fluid. Free gas attenuates this wave through its compression, therefore the pore fluid cannot behave incompressibly, and attenuation is extremely rapid. Once a continuous gas phase exists in a porous medium, that is, once the gas phase occupies ~20% of the pore volume, it is unlikely that the displacement wave can be generated. In the case where there are small amounts of free gas present as disseminated discrete bubbles, the wave can transit, but its amplitude decays rapidly with distance because of the compression of the gas. In practical cases involving free gas in the pore fluid around an excitation well, other steps are taken, such as co-injection of fluid at higher pressures, to drive the gas into solution so that the fluid can again behave incompressibly.

The basic equations of motion contain coefficients of shear viscosity for the fluids. This gives rise to a dependency of porosity diffusion on the rate of fluid transmission through the pore space. In low viscosity cases such as crustal fluids in deep interconnected fracture networks, transmission is relatively rapid; in reservoirs largely saturated with viscous oil, transmission of pressure effects can be quite slow. For a 10,000 cP oil filling 88% of the pore space of a 30% porosity 2-3 Darcy sand, it took about 5 weeks to see a substantial response at distances of 300 m from the excitation well which was being aggressively pulsed at the right frequency (Spanos et al., 2003).

The process of pressure-porosity diffusion, the fourth solution mentioned in Section 7, requires solid-fluid coupling to exist. If the pore liquid cannot flow, as is likely in the case of ductile smectitic shales (Dusseault 2003), or if the liquid is present as a fully occluded phase so that it is in the form of liquid "bubbles" in the solid mineral, as in a vesicular basalt, the wave cannot exist. This also means that the energy is confined by impermeable boundaries such as low-porosity shales, which is beneficial to practical applications because energy loss through geometric spreading is reduced. The impermeable barriers act as reflectors, confining the dynamic energy to the bounded stratum.

The existence of this wave helps account in a more rational and physically consistent manner for many natural phenomena such as those mentioned in Section 1. Geilikman et al. (1993) hypothesized that the substantial delays that are observed in sympathetic earthquake triggering are explained by the slow velocity (100-150 m/s) of this wave in stiff porous media. In other words, delayed triggering is not likely to be the result of the easily-detected and relatively short wavelength seismic waves such as the P-wave, S-wave and Stoneley wave; rather, it is more likely that the conservative and long-wavelength porosity diffusion (or porosity dilation) wave, with its effect on pressure, is the trigger for sympathetic earthquakes. In earthquake coda, the porosity dilation wave will be evidenced at transit times about 20 times greater than the transit time for P- and S-waves, and they will be of long wavelength.

Capillary blockage in oil-water systems can be reduced because of the effect of porosity dilation that takes place during the transit of the displacement wave. Fluid must move into and out of the pores in a short time, therefore a small mass in the pore throats is accelerated rapidly, generating a force acting on the fluid at the pore scale ($\vec{F} = m\vec{a}$). Dividing both sides by the effective area of the pore throat, we observe that a directional dynamic pressure is brought to bear at the pore-throat scale, which helps to overcome capillary blockages arising from the surface tension between immiscible liquids. Of course, in the absence of a dilating porous solid, such an effect is impossible, and this is the essence of the concept of solid-liquid inertial coupling.

These and other aspects of potential exploitation of the existence of these liquid-saturated porous media displacement waves are discussed in greater detail in another paper in these Proceedings (Dusseault 2003), along with process diagrams and additional references. However, it is important to emphasize that a great deal more development is needed to fully exploit the more complete description of inertial and diffusive fluid flow in liquid-saturated porous media. What has been done in this article is partly a re-statement of developments made in the recent past, with a more complete exposition based on better understanding of the micromechanics. Many other applications will be shown to exist. Seismological theory should be revisited so that dispersion, attenuation, and other issues that are currently treated empirically can be analyzed more fundamentally and in greater detail, allowing us to extract more useful information about the physical systems that are being probed seismically. For example, with the more rigorous thermodynamics presented here, it is possible to make a first-order (i.e. elastic) prediction of the attenuation of seismic waves as they pass through porous media. Because all phases undergo compression and rarefaction as a porosity dilation wave passes, a small amount of the energy must be converted into heat, in a manner similar to compression-expansion cycles in a gas. Although the quantities are small compared to a gas, they nonetheless exist, and can now be accounted for quantitatively from first principles, rather than being treated uniquely empirically through measurements.

We envision other applications arising from this more complete understanding of diffusive-inertial coupling in porous media. In the applications we have briefly mentioned, it has been convenient to apply the dynamic energy in the form of a sharp impulse to the liquid phase (i.e. a pressure pulse), but other possibilities exist as well, such as a mechanical source. The key factor in the excitation is that the impulse must be tailored so that the energy is applied at the right frequency for the medium so that the liquid phase is excited at the incompressible limit. Higher frequencies will generate seismic waves, lower frequencies will generate pure diffusive (Darcian) flow. However, in the field, porous media are heterogeneous with respect to the liquid (oil and water) saturations, to the porosity, to the permeability, and to parameters such as the bulk matrix compressibility. It would appear that excitation over a narrow band of frequencies would be preferred to excitation at a single frequency.

In closing, a few comments are made for general clarification. First, the porosity diffusion wave (or porosity dilation wave) is not to be confused with the Biot "slow wave". The displacement wave is inertially coupled to the incompressible excitation frequency limit of the pore liquid, not to a single or lumped phase. Second, the excitation associated with the generation of a displacement wave is not the same as that associated with seismic wave excitation: the latter are strain waves generated at higher frequencies. Third, although differing claims have been made, it does not appear that large-amplitude, high-frequency, short wave-length seismic waves can have applications in dislodging fine-grained minerals (except at short distances) or in substantially enhancing fluid flow. An additional reason for this (in addition to the displacement wave context) is that it is impractical to generate large-amplitude vibrations at high frequencies. In contrast, for low frequencies and long wavelengths of the porosity diffusion wave, large amplitudes can be generated by aggressive pressure pulsing or some form of asymmetrical mechanical vibration.

SYMBOLS USED

Most of the terms are rigorously defined in a mathematical manner in the text, or are mentioned close to the first time that they occur. Most are presented again here for ease of reference.

The equations of motion, Eqs. (36) and (37) in Section 5, contain many coefficients (a_i, b_i, c_i, etc.) that are defined mathematically in that section. Similarly, in Section 6, when a particular solution is written, the coefficients (B, c, C, D_i) are defined explicitly in terms of physical quantities. In Section 7, where the plane wave solution for the porosity diffusion wave is given, all the symbols and coefficients are defined immediately after they are written down, or have been previously defined.

Barred values (e.g. $\bar{u}$) are values averaged at the appropriate scale.

Roman characters

F	Helmholtz Free Energy
G	Gibbs Free Energy
H	enthalpy
K	bulk modulus (K_M – macroscopic)
M_{ik}	seismic source displacement tensor
S	entropy
T	temperature (absolute)
U	internal energy (strain energy)
c_p	fluid heat capacity
c_V	solid heat capacity
p	pressure
u_{kk}, u_{ik}	strain (tensorial notation)
v	velocity (v_{pf} – fluid wave speed)

Greek characters

α_s, α_f	coefficients of thermal expansion
α_1, α_2	dilation moduli, Eq 39
δ_{ik}	Kroenecker delta (tensorial notation)
ρ	phase density (ρ_f, ρ_s...)
η	porosity
ϕ	1 - η (1 – porosity)
κ	permeability (K & k are used elsewhere)
σ_{ik}	stress (tensorial notation)
σ_M	macroscopic modulus of dilation
μ_M	macroscopic (matrix) shear modulus
μ_f	fluid shear viscosity
μ_s	solid shear modulus

Subscripts and superscripts

ad	adiabatic, as in K_{ad}
f	the fluid phase, as in $u_{kk}^{'f}$ or μ_f
s	the solid phase, as in U_s, S_s
ik, kk	Einsteinian tensorial notation, as in σ_{ik}
o	initial or ground state, as in ϕ_o, T_o
M	a macroscopic measure, as in μ_M
η	referring to porosity, as in $\rho_{\eta f}$
p	referring to pressure, as in ρ_{pf}
′	for total strain, u′defined in Eqs 3 & 4

REFERENCES

Davidson, B.C., Spanos, T.J.T. & Dusseault, M.B. 1999. *Laboratory experiments on pressure pulse flow enhancement in porous media.* Proc, CIM Can. Pet. Society, Regina Technical Meeting.

de la Cruz, V., Sahay, P.N. & Spanos, T.J.T. 1993. *Thermodynamics of porous media.* Proc. of the Royal Society of London A, 433, 247-255.

Dusseault, M.B. 1999. *Petroleum Geomechanics – excursions into coupled behavior.* Journal of Canadian Petroleum Technology, Distinguished Author Series, December, 6 pages.

Dusseault, M.B. 2003. *Coupled processes and petroleum geomechanics.* Proceedings GeoProc 2003, Stockholm, Sweden, 14 p

Dusseault, M.B., Davidson, B.C., & Spanos, T.J.T. 1999, *A new workover tool for CHOP wells.* Proceedings of the CSPG and Canadian Petroleum Society Annual Technical Mtg., Calgary, paper 99-77, 9 pp.

Dusseault, M.B., Shand, D., Meling, T., Spanos, T.J.T., & Davidson, B.C. 2002. *Field applications of pressure pulsing in porous media.* Proc. 2nd Biot Conf. on Poromechanics, Grenoble, Balkema, Rotterdam, 639-645.

Geilikman, M.B, Spanos, T.J.T. & Nyland, E. 1993. *Porosity diffusion in fluid –saturated media.* Tectonophysics, 217, 111-115.

Gray, Peter 2002. Personal communication. (Private geol. Eng. consultant, Kitchener ON)

Hickey, C.J., Spanos, T.J.T. & de la Cruz, V. 1995. *Deformation parameters of permeable media.* Geophysical Jour. Internat., 121(2), 359-370.

Landau, L. and Lifshits, E.F. 1975, *Fluid Mechanics*, Toronto, Pergamon.

Manga, M., Brodsky, E.E. & Boone, M. 2003. *Response of streamflow to multiple earthquakes.* Geophysical Research Letters, 30(5), 1214 ff.

Sahay, P.N., Spanos, T.J.T. & de la Cruz, V., 2001, *Seismic wave propagation in inhomogeneous and anisotropic porous media.* Geophysics Journal International 144, 1-20.

Spanos, T.J.T. 2001, *The Thermophysics of Porous Media.* Chapman & Hall/CRC Press, Monographs and Surveys in Pure and Applied Mathematics series. 212 pp.

Spanos, T.J.T., Davidson, B.C., Dusseault, M.B., Shand, D., & Samaroo, M. 2003. *Pressure pulsing at the reservoir scale – a new IOR approach.* Journal of Canadian Petroleum Technology, 42(2), 16-28.

Spanos, T.J.T., Dusseault, M.B. & Davidson, B.C. 1999, *Pressure pulsing at the reservoir scale: a new IOR approach.* Proceedings of the CSPG and Canadian Petroleum Society Annual Technical Mtg., Calgary, paper 99-11, 11 pp.

MODELLING CONTAMINATION OF CLAYS

N. Boukpeti[1], R. Charlier[1], T. Hueckel[2]
[1]) Department GeomaC, University of Liege, B – 4000 Liege, Belgium
[2]) Department of Civil and Environmental Engineering, Duke University
Durham, NC 27708, U.S. A.

Abstract: A constitutive model for the behaviour of clays in contact with organic liquids is presented. The model was originally proposed by Hueckel (1997), based on experimental data reported by Fernandez and Quigley (1985,1991). It is recognized that the physico-chemical phenomena taking place at the scale of the clay particles are responsible for changes in permeability and deformation in the clay. Two types of chemical strains are identified: an elastic expansive strain at stress level lower than the plastic yield limit, and a plastic compaction strain at higher stress level. The plastic yield limit is assumed to decrease as the contaminant mass concentration increases. This is referred to as chemical softening. The chemo-plastic behaviour of a cylindrical clay layer subjected to vertical flow of contaminant is simulated.

1. INTRODUCTION

The chemical sensitivity of clays plays a role in a number of geotechnical fields, such as boreholes drilling, slope stability, radioactive waste disposal, and clay barriers in landfills. When a clay is in contact with a liquid, which is not in chemical equilibrium with the water present in its microstructure, physicochemical phenomena or chemical reactions may occur, affecting the clay microstructure, and producing changes in the mechanical and hydraulic behaviour of the clay. Motivated by the problem of integrity of clay barriers in landfills, Fernandez and Quigley (1985,1991) conducted permeation experiments on clay specimens with several organic liquids having a dielectric constant lower than that of water. Their results indicate large increases in permeability up to four orders of magnitude. These variations are attributed to a decrease in the volume of adsorbed water as the dielectric constant of the liquid decreases, which implies a larger volume of liquid free to move. In addition, flocculation (reorientation of the particles) was observed, with formation of particles clusters leaving large macro pores between them. In their latter series of experiments, Fernandez and Quigley (1991) showed that contamination at a relatively modest constant compressive effective stress, produced a visible plastic compressive strain healing the damage, and nearly no change in permeability.

The plastic strain induced during contamination at a constant effective stress can be simulated by using a chemo-plasticity model, Hueckel (1997), in which chemical softening is introduced. In this paper we consider a version of this model in axisymmetric stress state, where only a small change in the shape of the yield curve is introduced. The model is used, within the finite elements program LAGAMINE, to simulate the chemo-plastic behaviour of a cylindrical clay layer subjected to vertical flow of contaminant. The data used to calibrate the model are those from the tests by Fernandez and Quigley (1991).

2. CHEMO-ELASTO-PLASTIC MODEL

The constitutive equations are expressed in terms of the rates of the following variables: the effective isotropic stress $p' = \sigma'_{ii}/3$; the deviator stress $q = \sqrt{(3/2 s_{ij}s_{ij})}$; the volumetric strain $\varepsilon_v = \varepsilon_{ii}$; and the deviator strain $\varepsilon_q = \sqrt{(2/3 e_{ij}e_{ij})}$, where s_{ij} and e_{ij} refer to the stress and strain deviator, respectively. Compression is taken positive. The material is assumed to be isotropic and its deformation can be decomposed into elastic and plastic parts. The chemical effects on the material behaviour are described in terms of the contaminant mass concentration c, the ratio of contaminant mass to total fluid mass.

2.1 Elastic behaviour

The elastic law is modified from the classical relation for soils, by adding a term representing the elastic volumetric expansion generated in the clay when in contact with organic contaminant at concentration c. The chemo-elastic law can be expressed in incremental form as

$$\dot{\varepsilon}_v^e = \frac{\kappa}{v}\frac{\dot{p}'}{p'} - \beta\dot{c}, \quad \dot{\varepsilon}_q^e = \frac{1}{3G}\dot{q} \tag{1}$$

where v=1+e is the specific volume (e is the void ratio), G is the elastic shear modulus, κ is the slope of the elastic response in the (v,ln p'/p'_{ref})-plot (p'_{ref} is a reference pressure). For the elastic chemical expansion coefficient β, Hueckel (1997) proposed the following form

$$\beta = -F_0\beta_0 \exp\left[\beta_0\left(1-c+\ln c\right)\right]\left(\frac{1}{c}-1\right) \tag{2}$$

where constants F_0 and β_0 are soil and contaminant dependent. The constant F_0 (F_0<0) represents the maximal elastic volumetric expansion reached at a concentration of 100% (c=1), whereas β_0 defines the slope of the strain-concentration curve at the point $F_0/2$. The elastic strain-concentration relation is given by

$$\varepsilon_v^e = F_0 \exp\left[\beta_0\left(1-c+\ln c\right)\right] \tag{3}$$

and is plotted in Fig. 1.

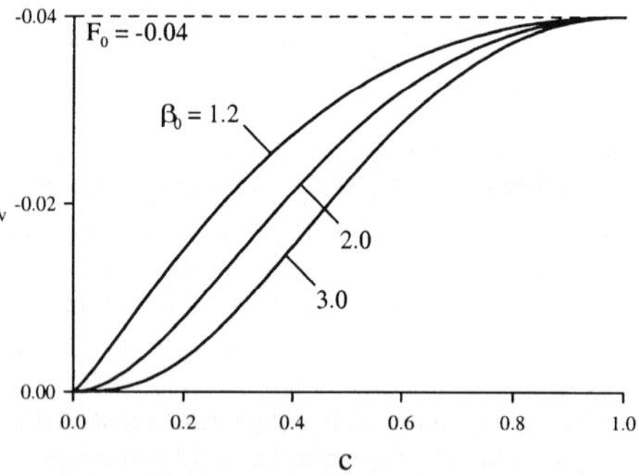

Figure 1. Chemo-elastic volumetric deformation ε_v due to an increase in concentration c.

2.2 Isotropic behaviour

During isotropic loading, plastic deformation takes place when the isotropic stress p' reaches the preconsolidation pressure p_c'. The pressure p_c' is a measure of the size of the yield surface on the isotropic axis and can be viewed as an hardening/softening parameter (the specific shape of the yield surface is described in the next section). An essential feature of the proposed model is the decrease of p_c' with respect to an increase in contaminant concentration. This can be expressed as

$$p_c' = p_c'^{*} S(c) \tag{4}$$

where $p_c'^{*}$ represents the preconsolidation pressure in non-contaminated conditions (i.e., c=0), and $S(c)$ is the chemical softening function. Hueckel (1997) proposed the form

$$S(c) = \exp(-ac) \tag{5}$$

The constant a, which depends on the soil and the contaminant, represents the strength of chemical softening. Figure 2 shows the chemical softening curves in the (c,p')-plane for three values of the constant a.

The variable $p_c'^{*}$ evolves as a function of the plastic volumetric strain in the same manner as in a Cam clay-type model

$$\frac{\dot{p}_c'^{*}}{p_c'^{*}} = \frac{v}{\lambda-\kappa}\dot{\varepsilon}_v^p \tag{6}$$

The constant λ represents the slope of the isotropic mechanical response plotted in the (v,ln p'/p'_{ref})-plane. By differentiating (4) and using (6) one can deduce the incremental hardening/softening law

$$\frac{\dot{p}_c'}{p_c'} = \frac{v}{\lambda-\kappa}\dot{\varepsilon}_v^p + \frac{\frac{\partial S}{\partial c}}{S(c)}\dot{c} \tag{7}$$

or for the specific form (5) of $S(c)$

$$\frac{\dot{p}_c'}{p_c'} = \frac{v}{\lambda-\kappa}\dot{\varepsilon}_v^p - a\dot{c} \tag{8}$$

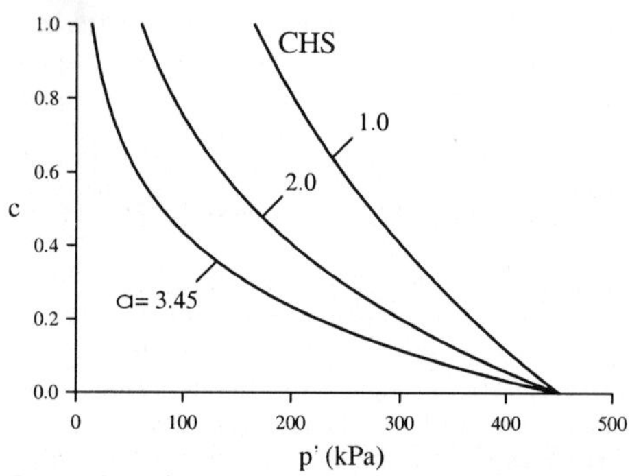

Figure 2. Chemical softening (CHS) curves for various values of the constant a, in the diagram concentration vs effective isotropic stress.

This relation contains two competing terms; the first term represents plastic hardening as a function of the volumetric part of plastic strain, the second term describes chemical softening due to an increase in contaminant concentration. Let us consider the plastic response to an increase in contaminant concentration at constant isotropic stress. The condition $p'=p_c'=0$ in equation (7) implies

$$\dot{\varepsilon}_v^p = -\frac{\lambda-\kappa}{v}\frac{\frac{\partial S}{\partial c}}{S(c)}\dot{c} = \frac{\lambda-\kappa}{v}a\dot{c} \tag{9}$$

Therefore, the plastic volumetric strain varies linearly with concentration (neglecting the small variations of v).

2.3 Plasticity model

The plastic behaviour is represented by a cap model, where the yield curve is composed of two parts, a frictional line and a volumetric cap. This is a modification from the original model of Hueckel (1997), where an elliptical yield curve was used. The cap of the yield curve is defined by

$$f_1 = q^2 + M^2 p'(p' - p_c') = 0 \tag{10}$$

where M is the critical state constant and represents the slope of the frictional line. The yield function f_1 is similar to the one of the Cam clay model and is represented in the (q,p')-plane by a quarter of an ellipse (see Fig. 3). The corresponding plastic deformations obey an associated flow rule

$$\dot{\varepsilon}_{ij} = \dot{\chi}\frac{\partial f_1}{\partial \sigma_{ij}} \tag{11}$$

The hardening/softening law is given by the evolution equation for p_c' (7). Alternatively, equations (6) and (4) can be used to determine the size of the yield curve.

The equation of the frictional line reads

$$f_2 = q - Mp' = 0 \tag{12}$$

For this part of the model the flow rule is non-associated

$$\dot{\varepsilon}_{ij} = \dot{\chi}\frac{\partial g_2}{\partial \sigma_{ij}} \tag{13}$$

and the plastic potential has the form

$$g_2 = q - q_0 = 0 \tag{14}$$

This implies that the volumetric component of the plastic strain rate is zero. There is no hardening/softening for this part of the model.

2.4 Strain-stress relation

The incremental strain-stress relation is obtained following the standard procedure in which the consistency condition, the flow rule and the hardening rule are used. This relation reads

$$\begin{bmatrix}\dot{\varepsilon}_v \\ \dot{\varepsilon}_q\end{bmatrix} = \begin{bmatrix} \frac{\kappa}{vp'} + \frac{1}{H}\frac{\partial f_1}{\partial p'}\frac{\partial f_1}{\partial p'} & \frac{1}{H}\frac{\partial f_1}{\partial p'}\frac{\partial f_1}{\partial q} \\ \frac{1}{H}\frac{\partial f_1}{\partial p'}\frac{\partial f_1}{\partial q} & \frac{1}{3G} + \frac{1}{H}\frac{\partial f_1}{\partial q}\frac{\partial f_1}{\partial q} \end{bmatrix}\begin{bmatrix}\dot{p}' \\ \dot{q}\end{bmatrix} + \begin{bmatrix} \frac{1}{H}\frac{\partial f_1}{\partial c}\frac{\partial f_1}{\partial p'} - \beta \\ \frac{1}{H}\frac{\partial f_1}{\partial c}\frac{\partial f_1}{\partial q} \end{bmatrix}\dot{c} \tag{15}$$

with

$$H = -p_c'\frac{v}{\lambda-\kappa}\frac{\partial f_1}{\partial p_c'}\frac{\partial f_1}{\partial p'} \tag{16}$$

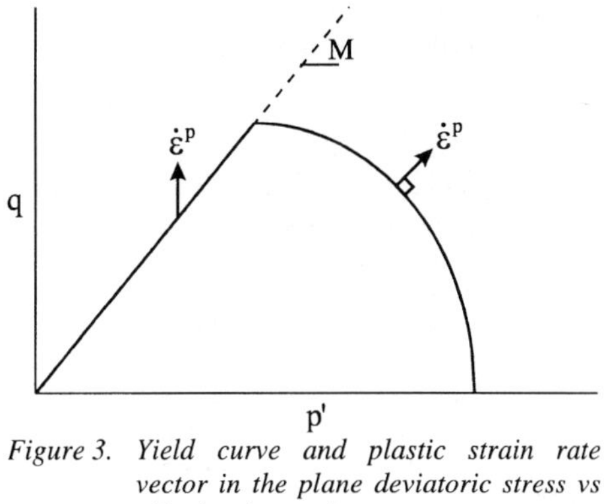

Figure 3. Yield curve and plastic strain rate vector in the plane deviatoric stress vs effective isotropic stress.

3. MODEL CALIBRATION

The chemo-elasto-plastic model described above requires the determination of eight constants: five mechanical constants and three constants describing the influence of contaminant concentration on the mechanical behaviour. The mechanical constants are identical to those used in the Cam clay model; they can be determined from a series of isotropic compression tests and triaxial tests.

Ideally, the two chemo-elastic constants, F_0 and β_0, and the chemo-plastic constant, a, would be easily determined from tests where the concentration of contaminant increases while the stresses remain constant. Considering first a low increase of concentration at low stresses, such that the yield criterion is not met, allows the evaluation of F_0 and β_0. Then, the constant a can be determined from tests with higher values of concentration and stresses, such that the chemical plastic compaction is observed. In practice, the increase of contaminant concentration is applied either by mean of a flow of contaminant (transport by advection and dispersion) or by considering the process of molecular diffusion. In the first case, changes of the pore pressure distribution in the clay specimen, due to chemical effects on hydraulic conductivity, result in changes of effective stresses. In the case of molecular diffusion, because the clay has a very low permeability, the volumetric changes produced by the chemical effects result in pore pressure development and in turn affect the effective stresses. Identification of the constants describing the chemical effects on clay requires some efforts to determine the evolution of effective stresses associated with an increase in contaminant concentration, in order to be able to isolate the part of the deformation due to the chemical effects.

4. DETERMINATION OF CHEMICAL CONSOLIDATION IN PERMEATION EXPERIMENTS WITH ETHANOL

The purpose of this section is to compare the predictions of the chemo-elasto-plastic model with experimental data showing the chemical effects of organic liquids on clay. For this exercise, the permeation experiments conducted by Fernandez and Quigley (1985,1991) on specimens of Sarnia clay are considered, with the focus on the results obtained for permeation of ethanol. A simplified representation of these experiments can only be considered at this time, as the data available are not sufficient for a complete and reliable characterization of the model.

The permeation experiments were carried out on 2 cm thick clay specimens placed into an aluminium cylinder of 5.38 cm diameter. The testing equipment allows the generation of a constant flow rate using a piston moving at a constant velocity. The chamber of the piston is connected to the fluid chamber in the permeameter cell located above the clay specimen, which is toped by a porous stone. A pressure transducer is connected to the hydraulic circuit. The fluid outlet in the lower aluminium plate is kept at atmospheric pressure. An external vertical load is applied to the specimen by means of a spring-loaded cap prior to permeation. Vertical settlements are monitored using a dial gauge and a settlement rod mounted on top of the specimen.

This series of experiments aimed at measuring the hydraulic conductivity of the clay when permeated with water, pure organic liquids (e.g., ethanol, dioxane, benzene), or mixtures of organic liquid and leachate at various concentrations. The hydraulic conductivity was determined with the help of Darcy's law using the steady state value of the measured fluid pressure. The fluid pressure applied on the upper boundary of the specimen is equilibrated by the vertical effective stress at the base of the clay specimen, where the pore pressure vanishes. For the steady state condition, the distribution of vertical effective stress along specimen height is assumed as linear. Using fluid pressure measurements, Fernandez and Quigley (1991) evaluated the values of maximal vertical effective stress acting at the base of the specimen for the steady state condition. These values are relatively high, reaching up to 700 kPa. Therefore, a significant part of the measured settlement is due to the increase in effective stress, the remaining part being attributed to the chemical effects. The measured settlements are for chemically stationary states, after several pore volumes have passed through the specimen and the effluent mass concentration has reached the influent value.

A simple model of the permeation experiment with ethanol is now presented, where the distribution of effective stresses and contaminant concentration is uniform. Let us consider an axisymmetric specimen with the same dimensions as described above, subjected to uniaxial deformation, with zero vertical displacement at its base. The specimen is subjected initially to a vertical load applied on the upper boundary, and the pore pressure in the specimen is zero. The cases of the external stress 160 and 80 kPa are considered, which correspond to values greater than the preconsolidation pressure of the specimen prior to the initiation of any flow. Therefore, the conditions are those of a normally consolidated material. The horizontal effective stress (not measured) is assumed to be constant and half of the value of the vertical stress (K_0=0.5). The specimen is subjected to an increase in vertical effective stress σ_v', equal to the evaluated value of the seepage stress reached in the middle of the clay specimen at the steady state. In addition, the contaminant mass concentration increases from 0 to the value in the influent permeant.

Using the chemo-elasto-plastic model implemented in the finite elements program LAGAMINE, and the material parameters listed in Table 1, the volumetric strains due to changes in effective stress and concentration are predicted. The value of λ was estimated from the volumetric response measured by Fernandez and Quigley (1991) for the application of external vertical effective stress, with the assumption that K_0=0.5. The constant κ was taken as $\lambda/5$, which is typical for clays. Values typical for clays were selected for the elastic shear modulus G and the critical state constant M.

For the parameters representing the chemical effects, we first considered the set of values estimated by Hueckel (1997) using an interpretation of the permeation tests of Fernandez and Quigley (1991) (i.e., F_0=-0.029, β_0=1.32, a=3.45). With these parameters, the chemical consolidation for the tests at 160 kPa was greatly overestimated.

Therefore, we decided to adjust the two parameters F_0 and β_0 until a reasonable match between the model response and the experimental data was obtained. The results shown in Fig. 4b for various concentrations, c=0, 0.4, 0.8, and 1.0, were obtained with F_0=-0.07 and β_0=1.32. Figure 4c depicts the mechanical and the chemical parts of the volumetric strain. The mechanical part is computed using the model and the measured variations in effective stresses (Fig. 4a). The part of the strains due to chemical effects is deduced from measurements by subtracting the predicted mechanical part of the strain.

In Figure 5, the model prediction is compared with experimental data for the case of 80 kPa external stress. A reasonable agreement is obtained. The simulations differ from the test data within the experimental accuracy range.

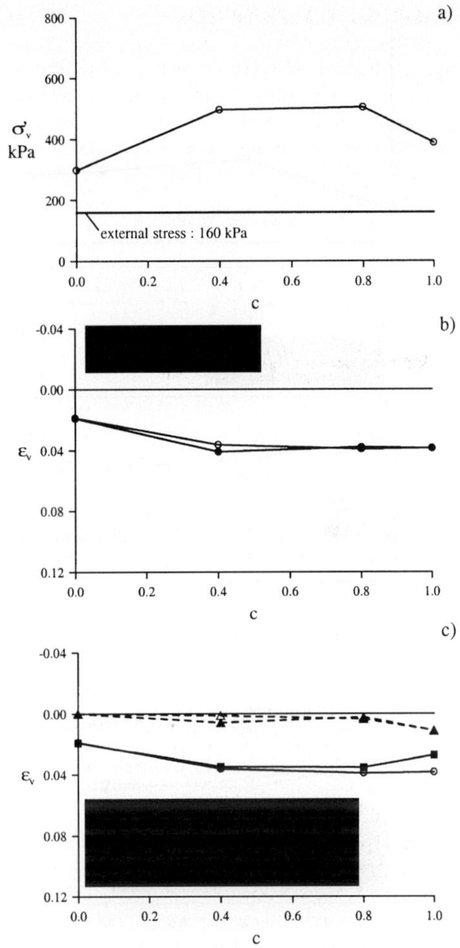

Figure 4. Experimental data and model prediction for the volumetric response of Sarnia clay permeated with ethanol under 160 kPa; a) effective vertical stress vs concentration, b) and c) volumetric deformation vs concentration.

Table 1. Constitutive constants for Sarnia clay and ethanol.

κ	λ	G(kPa)	M	F_0	β_0	a
0.0064	0.032	10000	1.2	-0.07	2.5	3.45

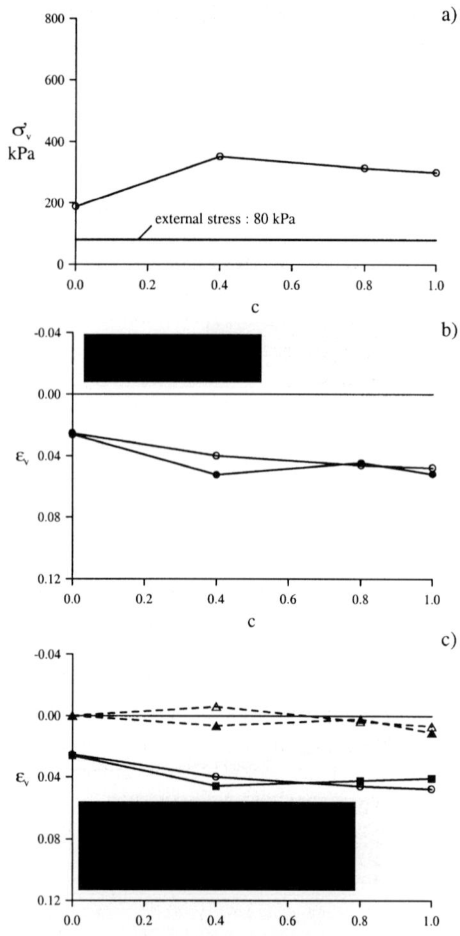

Figure 5. Experimental data and model prediction for the volumetric response of Sarnia clay permeated with ethanol under 80 kPa. a) effective vertical stress vs concentration, b) and c) volumetric deformation vs concentration.

5. CONCLUDING REMARKS

A chemo-elasto-plastic model for the behaviour of clays in contact with a single contaminant has been presented. The model has been implemented in the finite elements program LAGAMINE providing an efficient tool for solving problems involving clay-contaminant interaction. Comparison of model predictions with available experimental data shows reasonable agreement within the experimental accuracy range. One important lesson from this exercise is the realization of the type of experimental data needed for a proper and reliable characterization of the model. In particular, adequate evaluation of the evolution of effective stresses in the vertical and horizontal directions during permeation is required, in order to represent accurately the mechanical loading taking place. The vartiations of contaminant concentration in space and time is also needed.

REFERENCES

Fernandez, F. & Quigley R. M. 1985. *Hydraulic conductivity of natural clays permeated with simple liquid hydrocarbons.* Can. Geotech. J. 22: pp. 205-214.

Fernandez, F. & Quigley R. M. 1991. *Controlling the destructive effect of clay-organic liquid interactions by application of effective stresses.* Can. Geotech. J. 28: pp. 388-398.

Hueckel, T. 1997. *Chemo-plasticity of clays subjected to stress and flow of a single contaminant.* Int. J. Num. Anal. Meth. Geom. 21: pp. 43-72.

DEPENDENCE OF SUBCRITICAL CRACK GROWTH IN ROCKS ON WATER VAPOR PRESSURE

Yoshitaka NARA and Katsuhiko KANEKO
Graduate School of Engineering, Hokkaido Univ., JAPAN

Abstract: It is considered that subcritical crack growth is caused by stress corrosion. In the case of rocks, stress corrosion is a weakening process of a strong siloxane bond structure, and the corrosive agent is water. In this study, the dependence of subcritical crack growth on water vapor pressure was investigated by using Double-Torsion test. Rock studied was Kumamoto andesite. During experiments, the partial pressure of water was kept constant by controlling the temperature and the humidity. When the temperature and humidity were higher, the crack growth rate became higher than that under the lower temperature and humidity. Crack growth rate in andesite was approximately proportional to the partial pressure of water vapor. The activation energy for subcritical crack growth in andesite was evaluated as 110 kJ/mol, and the crack velocity under various water vapor pressures could be estimated.

1. INTRODUCTION

In fracture mechanics, crack growth takes place when the stress intensity factor reaches a critical value, that is, fracture toughness. However, crack grows even if the stress intensity factor is smaller than the fracture toughness. This phenomenon is called as subcritical crack growth, which is one of the key factors to cause the time-dependent behaviors of rocks (Atkinson, 1984). Hence, the knowledge of subcritical crack growth in rocks is very important for the evaluation of the long-term stability of structures in rock mass.

Subcritical crack growth is considered to be caused by stress corrosion, which is a weakening process of a bond structure in a solid by a chemical reaction (Anderson and Grew, 1977). In the case of the silicate materials, stress corrosion is a weakening process of a strong siloxane bond structure, and the corrosive agent is water. The crack growth behavior should depend on the activity of water in environment. It has been reported that the crack growth rate in water saturated rocks was 2~3 orders of magnitude greater than that in the dry rocks (Waza et al., 1980). However, under air condition, the dependence of the crack growth rate in rocks on the activity of water, namely, the partial pressure of water vapor, has not been made clear yet.

In this study, the dependence of subcritical crack growth behaviors in rocks on water vapor pressure is investigated by conducting fracture mechanics tests under several water vapor pressures.

2. STRESS CORROSION CRACKING

In the case of silicate materials, stress corrosion is presented as following chemical reaction (Michalske and Freiman, 1982):

$$\equiv Si - O - Si \equiv + H_2O \rightarrow 2[\equiv Si - OH]$$

When the crack growth behavior is controlled by stress corrosion under air condition, the crack velocity, v, is related to the stress intensity factor, K_I, as follows (Sano, 1977):

$$v = v_0 P(H_2O)^{n_w} \exp\left(\frac{-E^* + bK_I}{RT}\right) \quad (1)$$

where $P(H_2O)$ is the water vapor pressure, E^* is the activation energy, R is the gas constant, T is the absolute temperature, and v_0, n_w and b are constants determined experimentally. The constant n_w means the influence of the water vapor pressure on the crack propagation. In the case of glass, n_w is about unity (Wiederhorn, 1967). However, the value of n_w has not been made clear for rocks. Therefore, it is very important to determine the value of n_w for rocks.

In this study, eExperimental data was analized by transforming Eq.(1) into the next form:

$$\ln v = A + \frac{b}{RT} K_I \quad (2)$$

$$(\because A = n_w \cdot \ln P(H_2O) + \ln v_0 - \frac{E^*}{RT})$$

3. DOUBLE-TORSION TEST

The load-relaxation method of Double-Torsion (DT) test is adopted in this study (Evans, 1972). A schematic illustration of a DT specimen is shown in *Figure 1*, where the notations of the specimen are noted. In this study, the guide groove was set upward.

This method gives a wide range of the data in the K_I-v diagram, generally 10^{-1}~10^{-10} m/s of the crack velocity, from only a single experimental run. Therefore, the load-relaxation method has been used in many studies on subcritical crack growth.

In the load-relaxation method, K_I and v are expressed as follows (Williams and Evans, 1973):

$$K_I = Pw_m \sqrt{\frac{3(1+p)}{Wd_n d^3}} \quad (3)$$

$$v = -f \frac{S_0 P_0}{BP^2} \frac{dP}{dt} \quad (4)$$

$$(\because B = \frac{3w_m}{Wd^3 G})$$

where P is the load applied to the specimen, w_m is the moment arm, p is Poisson's ratio, W is the width of the specimen, d is the thickness of the specimen, d_n is the thickness on the crack plane, f is the constants depend on the shape of the crack front, S_0 is the initial compliance of the specimen, P_0 is the initial load, dP/dt is the load-relaxation rate, and G is shear modulus. Atkinson (1979a) indicated that f is about 0.2 for quartz. Then, f was set as 0.2 in this study. As shown in Eq.(3), because the stress intensity factor is independent of the crack length, DT test is convenient for opaque materials such as rocks.

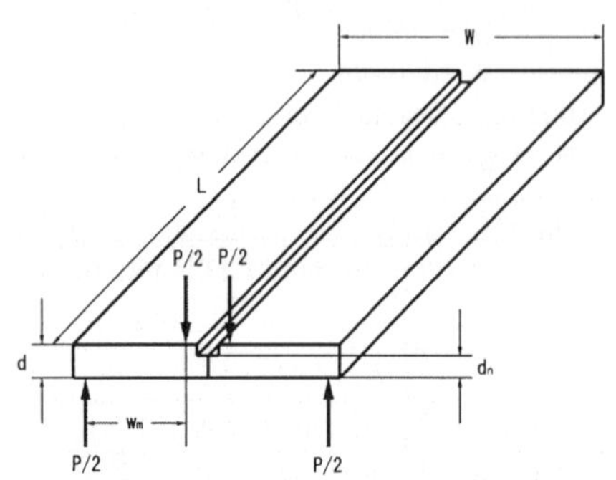

Figure 1. An illustration of DT specimen.

There are some conditions about the size of the specimen and the crack length, *a*. Atkinson (1979b) experimentally indicated that *W* should be greater than 12*d*. Pletka et al. (1979) suggested that *L* should be greater than 2*W* from measurements of fracture toughness. Hence, the following condition is necessary for the size of the DT specimen:

$$12d \le W \le L/2 \quad (5)$$

Trantina (1977) showed the following condition on the crack length by finite element analysis:

$$0.55W < a < L - 0.65W \quad (6)$$

In this study, the size of the specimen is decided taking above conditions into account. The size is that $W = 45$, $L = 120$, $d = 3$, $d_n = 2$, and the width of the guide groove is 2 mm. The value of w_m in Eq.(3) is 18 [mm].

4. TESTING APPARATUS

Figure 2 shows the schematic drawing of the apparatus used in this study. The apparatus is set in a room in which the temperature and humidity can be kept constant within 0.1 K in the range of 283~333 K and within 1 % in the range of 40~75 %, respectively. Hence, it is possible to conduct experiments under various water vapor pressures. For the calculation of the value of the water vapor pressure, next equations were used (Plate, 1982):

$$\left.\begin{array}{l} P(H_2O) = e_s \times h \\ e_s = 6.11 \times 10^{7.5t/(237.3+t)} \times 10^5 \\ t = T - 273.15 \end{array}\right\} \quad (7)$$

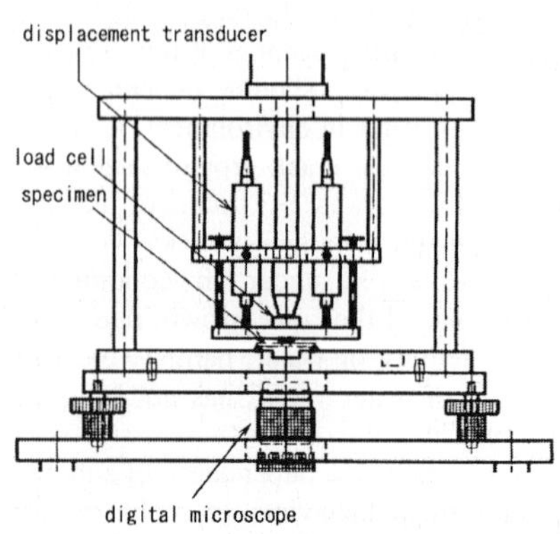

Figure 2. DT testing apparatus.

where e_s is the saturation vapor pressure of water, h is the relative humidity, and t is the temperature in degree centigrade.

5. ROCK STUDIED

Rock studied is Kumamoto andesite, JAPAN. The rock-forming minerals of this rock are plagioclase, pyroxene, hornblende and so on. Some phenocrysts of about 3 mm length are distributed in groundmass.

P-wave velocities in orthogonal three directions in this rock were 4.84, 4.89, and 4.91 km/s. Therefore, this rock can be assumed isotropic.

Elastic modulus and Poisson's ratio were determined by uniaxial compression test. Young's modulus was 31.9 GPa, Poisson's ratio was 0.27, and shear modulus was 12.6 GPa.

6. EXPERIMENTAL PROCEDURE

Specimens were dried at 470 K for about 10 days in order to remove water in rock specimens before conducting DT tests. Then, specimens were kept in a deciccator with silica gel.

After above drying process, precracking was done. In precracking, a specimen was loaded slowly, and the displacement at the loading point was fixed occasionally and a constant load was maintained. When the crack started to advance, the load relaxation was detected. In order to confirm the crack growth and the condition expressed in Eq.(5), all of specimens were observed by the digital microscope set in the DT apparatus during precracking.

After precracking, specimens were kept in environment of DT tests for about 20 hours, and then DT tests were carried out. In this study, in order to avoid hysteresis of the load-relaxation method (Sano, 1988), all of experiments were conducted under the same loading condition. The loading condition was that pre-load was applied to the specimen about 15 N, and the loading point was then shifted about 0.27 mm within 0.4 seconds and kept for about 2 hours in all experiments.

7. EXPERIMENTAL RESULT

K_I-v relations observed in this study are shown in *Figure 3*, where results obtained under two different water vapor pressures were shown. It can be seen that the reproducibility of testing results of Kumamoto andesite under the same environmental condition is very high from this figure. In addition, it can be seen that the crack growth behaviors are facilitated under high water vapor pressure. This is in harmony with the theory of stress corrosion cracking in rocks. The obtained results are summarized in Table 1, where K_I^* means K_I at $v = 10^{-5}$ [m/s], and v^* means v at $K_I = 1.6$ [MN/m$^{3/2}$]. K_I^* and v^* were determined in order to compare the stress intensity factor

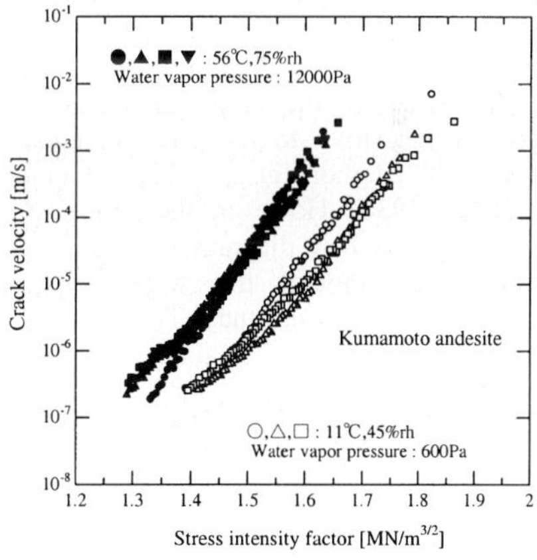

Figure 3. K_I-v relations of andesite.

Table 1. The summary of experimental results.

condition	A	b [m$^{5/2}$/mol]	K_I^* [MN/m$^{3/2}$]	v^* [m/s]	$P(H_2O)$ [Pa]
low temperature & low humidity	-52.6	0.060	1.62	$6.59*10^{-6}$	574 (284.1K,44%rh)
	-50.0	0.055	1.67	$2.99*10^{-6}$	574 (284.1K,44%rh)
	-46.2	0.051	1.66	$2.29*10^{-6}$	561 (284.1K,43%rh)
high temperature & high humidity	-58.9	0.086	1.51	$1.67*10^{-4}$	12126 (329.9K,71%rh)
	-48.9	0.068	1.52	$1.02*10^{-4}$	12239 (329.8K,72%rh)
	-49.5	0.069	1.52	$9.99*10^{-5}$	12283 (329.3K,74%rh)
	-48.0	0.066	1.52	$8.24*10^{-5}$	12410 (329.8K,73%rh)

and the crack velocity of each environmental condition, respectively. It can be recognized that the values of v^* are larger and the values of K_I^* are smaller under high water vapor pressure than under low water vapor pressure. Therefore, it can be suggested that water acts as the corrosive agent of stress corrosion cracking in andesite. On the other hand, with respect to A and b, the differences were small. Hence, these can be assumed as constants.

8. DISCUSSION

K_I-v relations obtained in this study show that the crack growth behavior in andesite is promoted by water vapor under air condition. These are consistent with the results of uniaxial compression tests and Brazilian tests by Jeong et al. (2003), in which the strengths of Kumamoto andesite decreased with increasing water vapor pressure. Therefore, stress corrosion can be considered as a primary process of subcritical crack growth in andesite, and it is important to consider the effect of water vapor on the crack growth behavior.

For glass and rocks, it has been reported that the value of the slope of the K_I-v diagram is sensitive to the environmental condition (Wiederhorn et al., 1980 ; Lajtai and Bielus, 1986). However, the values of the slope of the K_I-v diagram were almost constant even though the water vapor pressure changed in this study. The value of the slope is considered to be almost constant in environmental conditions for this study. The mean value of b was determined as 0.065.

Water vapor pressure obviously affected the crack velocity at the same stress intensity factor as shown in *Figure 3*. In order to estimate quantitatively the influence of water vapor on subcritical crack growth, the relation between water vapor pressure and the crack velocity at a given stress intensity factor is investigated. When K_I and T in Eq.(1) can be assumed as constants, Eq.(1) can be transformed into the next form:

$$ln\, v = n_w \cdot ln\, P(H_2O) + C$$

(8)

$$(\because C = ln\, v_0 + \frac{-E^* + bK_I}{RT})$$

In this study, v^* was expressed as a function of $P(H_2O)$, and the value of n_w in Eq.(1) was evaluated. The relation between $P(H_2O)$ and v^* is shown in *Figure 4*. The value of n_w was 1.11 for andesite. Therefore, the crack growth rate can be considered to be proportional approximately to water vapor pressure. This agrees with the case of glass (Wiederhorn, 1967).

On the basis of above results, the activation energy can be calculated, and the theoretical values of the crack growth rate under various temperatures and water vapor pressures can be evaluated. Because the crack velocity at a given stress intensity factor is proportional to water vapor pressure, Eq.(1) becomes:

$$v = v_0 P(H_2O) exp\left(\frac{-E^* + bK_I}{RT}\right)$$

(9)

This equation can be transformed into:

$$\left.\begin{aligned} ln\, v &= X + ln\, P(H_2O) + \frac{Y}{T} \\ X &= ln\, v_0 \\ Y &= \frac{-E^* + bK_I}{R} \end{aligned}\right\}$$

(10)

$P(H_2O)$, T and b are known, and v can be determined from experimental results. Thus, unknown quantities in Eqs.(10) are the activation energy, E^*, and v_0. Substituting v for v^* (= 10^{-5} [m/s]), X and Y were calculated

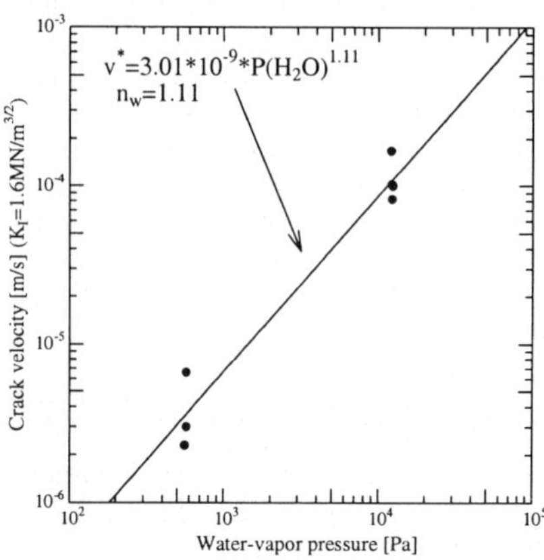

Figure 4. The relation between water vapor pressure and crack velocity.

as -16.3 and -724.0, respectively. From the value of X, v_0 was determined as $8.01*10^{-8}$.

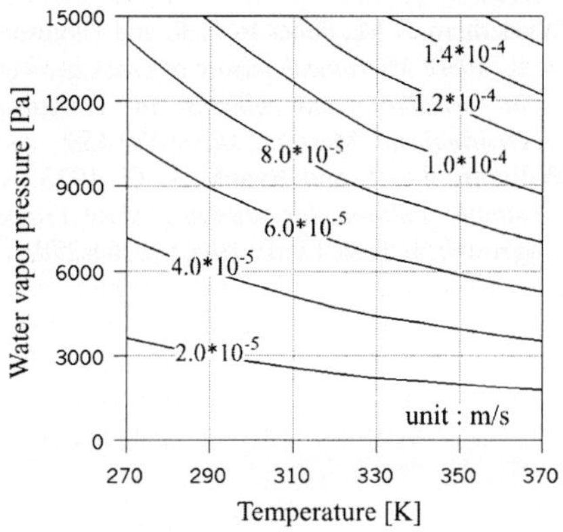

Figure 5. The distribution of the crack velocity under various temperature and water vapor pressure.

When $K_I = K_I^*$ (= $1.6*10^6$ [N/m$^{3/2}$]) and b = 0.065, the activation energy, E^*, was estimated as $E^* \approx 110.0$ [kJ/mol]

For quartz, Scholz (1972) determined the activation energy as about 100 kJ/mol by static fatigue tests. For Westerly granite, Carter et al. (1981) determined as 100 ~ 130 kJ/mol by creep tests under high temperature (743 ~ 1038 K). The value of the activation energy obtained in this study agrees well with above values.

Substituting the values of E^* and v_0 into Eq.(9), the theoretical values of the crack velocity can be calculated. In *Figure 5*, the distribution of the theoretical crack velocity at $K_I = K_I^*$ under various temperature and water vapor pressure is shown. The relation between the observed values, v^{ob}, and the theoretical values, v^{th}, of the crack velocity at $K_I = K_I^*$ is shown in *Figure 6*, where the straight line indicates the line of $v^{th} = v^{ob}$. It can be shown that the theoretical crack velocity agrees well with the observed crack velocity. Therefore, It can be made sure that the crack velocity at a given stress intensity factor under various water vapor pressure can be estimated.

It is concluded that the stress corrosion controls the crack growth behaviour in rock, and the water vapor plays a major role of it. It is a noteworthy result to estimate the activation energy. In order to estimate

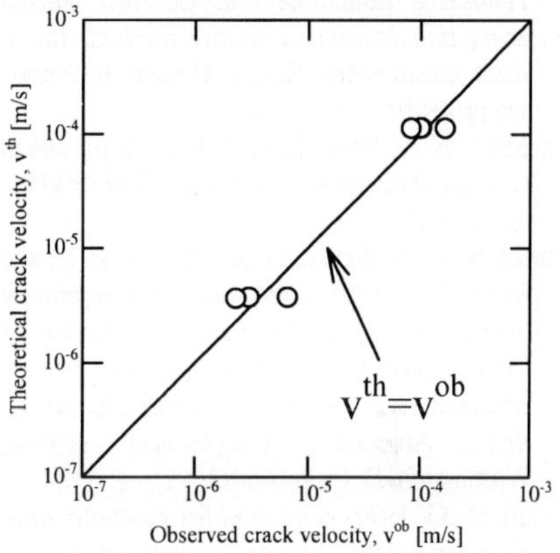

Figure 6. The relationship between the theoretical and observed crack velocity.

precisely the crack growth behaviour in rocks, it is essential to clarify the effect of water vapor and the value of activation energy of rocks.

9. CONCLUSION

The dependence of subcritical crack growth behaviour on water vapor pressure was investigated by using Kumamoto andesite. It was made clear that the crack velocity at a given stress intensity factor was proportional to water vapor pressure. The activation energy for subcritical crack growth in andesite was evaluated as 110 kJ/mol. The theoretical value of the crack velocity under various water vapor pressure can be estimated. It can be concluded that subcritical crack growth in rock is controlled by the stress corrosion and the water vapor affects on the crack growth behaviour strongly, and that the crack growth behavior can be estimated theoretically.

REFERENCES

Anderson, O. L. and Grew, P. C. 1977. *Stress corrosion theory of crack propagation with applications to geophysics*, Rev. Geophys. Space Phys., 15(1), pp.77-104.

Atkinson, B. K. 1979a. *A fracture mechanics study of subcritical tensile cracking of quartz in wet environment*, PAGEOPH, 117, pp.1011-1024.

Atkinson, B. K. 1979b. *Fracture toughness of Tennessee sandstone and Carrara marble using double torsion testing method*, Int. J. Rock Mech. Min. Sci. & Geomech. Abstr., 16, pp.49-53.

Atkinson, B. K. 1984. *Subcritical crack growth in geological materials*, J. Geophys. Res., 89(B6), pp.4077-4114.

Carter, N. L., Anderson, D. A., Hansen, F. D. and Kranz, R. L. 1981. *Creep and creep rupture of granitic rocks*, in Mechanical Behavior of Crustal Rocks (Edited by Carter, N. L., Friedman, M., Logan, J. M. and Stearns, D. W.), American Geophysical Union, Washington, D. C., pp.61-82.

Evans, A. G. 1972. *A method for evaluating the time-dependent failure characteristics of brittle materials – and its application to polycrystalline alumina*, J. Mater. Sci., 7, pp.1137-1146.

Jeong, H. S., Obara, Y. and Sugawara, K. 2003. *The strength of rock under water vapor pressure*, J. Mining and Materal Processing Institute of Japan, 119, pp.9-16. (in Japanese with English abstract)

Lajtai, E. Z. and Bielus, L. P. 1986. *Stress corrosion cracking of Lac du Bonnet granite in tension and compression*, Rock Mech. Rock Eng., 19, pp.71-87.

Michalske, T. A. and Freiman, S. W. 1982. *A molecular interpretation of stress corrosion in silica*, Nature, 295, pp.511-512.

Plate, E. J. 1982. Engineering meteorology, Elsevier Scientific Publishing Company, Amsterdam, 742 pp.

Pletka, B. J., Fuller Jr., E. R. and Koepke, B. G. 1979. *An evaluation of Double-Torsion testing-Experimental*, ASTM STP 678, pp.19-37.

Sano, O. 1977. *Fundamental study on the mechanism of brittle fracture of rocks*, Ph. D. Thesis, Kyoto Univ.

Sano, O. 1988. *A revision of the double-torsion technique for brittle materials*, J. Mater. Sci., 23, pp.2505-2511.

Scholz, C. H. 1972. *Static fatigue of quartz*, J. Geophys. Res., 77(11), pp.2104-2114.

Trantina, G. G. 1977. *Stress analysis of the Double-Torsion specimen*, J. Am. Ceram. Soc., 60, 338-341.

Waza, T., Kurita, K. and Mizutani, H. 1980. *The effect of water on the subcritical crack growth in silicate rocks*, Tectonophys., 67, pp.25-34.

Wiederhorn, S. M. 1967. *Influence of water vapor on crack propagation in soda-lime glass*, J. Am. Ceram. Soc., 50(8), pp.407-414.

Wiederhorn, S. M., Fuller Jr., E. R. and Thomson R. 1980. *Micromechanisms of crack growth in ceramics and glasses in corrosive environments*, Met. Sci., 14, pp.450-458.

Williams, D. P. and Evans, A. G. 1973. *A simple method for studying slow crack growth*, J. Test. Eval., 1(4), pp.264-270.

THERMOMECHANICAL MODELING OF THE MICROSTRUCTURED POROUS MEDIA WITH INCLUSIONS

Pasquale Giovine
Dipartimento di Meccanica e Materiali, Università Mediterranea,
Reggio Calabria, Italy

Abstract: Dynamic and thermodynamic balance principles for porous materials consisting of three components are deduced by using, as a logical tool, only three fundamental principles proposed by Truesdell for classical mixtures; the principles are adapted to the context of soils mechanics where the mixture is considered immiscible and the porous solid matrix is modeled as a continuum with ellipsoidal microstructure.

1. INTRODUCTION

In Giovine's paper (1996) porous solids with large vacuous interstices were considered as continua with ellipsoidal structure for which the microdeformation is a pure strain: it means that each void has different microdeformation along principal axes, but rotates locally with the matrix of the material. That model generalizes previous 'void theories' of, *e.g.*, Nunziato & Cowin (1979), Capriz & Podio Guidugli (1981) or Wilmanski (1996), which does not predict size effects in torsion, as observed by Lakes (1986).

When fluids can seep through pores, interacting mechanically with the solid skeleton, the material is composed of more than one constituent; thus we need to use a mixture theory in which we could clearly make out each part filled by different constituents on a scale which is rather large in comparison with molecular dimensions: so we put forward a new continuum theory of an immiscible mixture consisting both of a continuum with ellipsoidal microstructure (the porous elastic solid) and of two classical media (see, also, the conservative case examined by Giovine (2000)). In accordance with Biot (1956), we consider virtual mass effects due to diffusion; we also introduce the microinertia associated with the rates of change of the constituents' local densities, as well as the one due to the deformation of the pores close to their boundaries.

Truesdell's principles, he refers to as metaphysical, were stated in his lecture on thermomechanic theory for classical mixtures (see lecture 5 in Truesdell (1984)):

1. All properties of the mixture must be mathematical consequences of properties of the constituents.

2. So as to describe the motion of a constituent, we may in imagination isolate it from the rest of the mixture, provided we allow properly for the actions of the other constituents upon it.

3. The motion of the mixture is governed by the same equations as is a single body.

Thus the third principle simply asserts that, in its motion as a whole, a body does not know whether it is a mixture or not; but in this paper the skeleton of the medium consists of a continuum with microstructure, as defined by Capriz (1989), and therefore it must satisfy balance equations there proposed. The first and the second principles affirm that the whole is no more than the sum of its parts and that the mixture's constituents, imagined as splitted in geometry, must be considered united in physics by suitable forces or energies, respectively. We should also notice how, unlike balance equations, constitutive proposals for dependent fields are usually affected by microstructural independent variables in addition to gross ones.

The proposed model focuses on the case of the contaminant's transport in soils and is applied to the study of the coupling of adsorption and diffusion in porous materials with large voids by Giovine (2003) and to the investigation of the hyperbolicity's condition of the subsystem describing the liquid phase by Giovine & Palumbo (2003).

2. NOTATION AND KINEMATICS

Let us consider a three-phase system composed by an elastic body with large pores where a mixture of two immiscible fluids (or a suspension of solid particles in a fluid or a fluid component carrying an adsorbate) flows through.

The constituents of the system are supposed to be superimposed, so that they contemporaneously occupy a given region B_τ of the three-dimensional Euclidean space E, at a certain time τ in an interval $[\tau_0, \tau_1]$ during which the motion is observed.

Therefore every place x in the body is simultaneously occupied by a material particle of each constituent whose independent motion is described by the velocity field $v_i := v_i(x, \tau)$, for $i = f, a, s$, where the indices f, a and s are used to specify the fluid, adsorbate or skeleton component, respectively.

If ρ_i denotes the bulk mass density of ith-constituent, then

$$\rho := \rho_f + \rho_a \quad \text{and} \quad \rho_t := \rho_f + \rho_a + \rho_s \tag{1}$$

are the density of the fluid mixture and of the total system, respectively, while

$$c := \frac{\rho_a}{\rho} \tag{2}$$

is the mass concentration of the adsorbate in the fluid inclusion.

Velocities of the fluid mixture v and of the total system v_t are defined by

$$v := (1-c)v_f + cv_a \qquad \text{and} \tag{3}$$

$$\rho_t v_t := \rho_f v_f + \rho_a v_a + \rho_s v_s \tag{4}$$

while

$$j := c(v_a - v) \tag{5}$$

is the flux of molecular diffusion, which is null if the fluid and the adsorbate components have the same velocity field.

A kinematic condition of compatibility applies for velocities on the boundary of the body:

$$v_i \cdot n = 0, \tag{6}$$

for $i = f, a, s$, with n the outer normal of ∂B_τ, therefore we have also

$$v \cdot n = 0, \quad v_t \cdot n = 0 \quad \text{and} \quad j \cdot n = 0. \tag{7}$$

The material time derivative of a quantity (•) following the motion of the ith-constituent is

$$(\bullet)'^{i} := \frac{\partial(\bullet)}{\partial\tau} + [\text{grad}(\bullet)]v_i\,, \tag{8}$$

where grad(•) := $\partial(\bullet)/\partial x$, while the material time derivative following the motion of the fluid mixture is

$$(\bullet)\acute{Y} := \frac{\partial(\bullet)}{\partial\tau} + [\text{grad}(\bullet)]v\,. \tag{9}$$

Thus the peculiar accelerations and the mean fluid one are defined by

$$a_i := v_i'^{i}, \quad \text{for } i = f, a, s, \quad \text{and} \quad a := v\acute{Y}, \tag{10}$$

respectively.

The hypothesis according to which the solid phase of the mixture, that is the porous skeleton, has an ellipsoidal microstructure is made esplicit by introducing a second order symmetric tensor field with positive determinant U ($\in \text{Sym}^+$), namely a microstretch, which only takes into account contractions or expansions of the large pores in the material.

The portion of space occupied by the ith-constituent of the mixture in each spatial point is the volume fraction β_i of it, which is related to its bulk density ρ_i through the true mass density γ_i by the relation $\rho_i = \gamma_i \beta_i$.

The sum of volume fractions is less or equal to one: $\beta_f + \beta_a + \beta_s = 1 - \beta_v \leq 1$, where β_v is the volume fraction of the bare sites of matter in pores. The solid matrix is called unsaturated if $\beta_v > 0$, and saturated if $\beta_v = 0$: in this case pores are completely filled by the inclusion and the volume fraction β_s, closely related to the determinant of the microstretch U, is equal to $(1 - \beta_f - \beta_a)$, which represents an explicit constraint on contractions or expansions of pores. Here, we do not impose the saturation constraint and so $\beta_v \in (0, 1)$.

Finally, for the kinetic energy we assume a density per unit volume, which generalizes the one considered by Biot (1956),

$$k := \frac{1}{2}\sum_{j=f,a} \rho_j\left[v_j^2 + \mu_j(\beta_j)\left(\beta_j'^{j}\right)^2\right] + \frac{1}{2}\rho_s\left[v_s^2 + \mu_s(U)U'^s \cdot U'^s\right]. \tag{11}$$

The first three terms in the square brackets are classical and represent the translational kinetic energy of the gross motion; the other three terms introduce the inertia due to the local microvariations of the volume of inclusions, as well as the one related to the admissible expansional motion of pores' boundaries; some particular expressions of non-negative functions μ_f, μ_a and μ_s, which also depend on geometrical features of pores, are given by Giovine (1990, 1999, 2000).

On the contrary, in this theory the effects of the translational virtual inertia, representing the local nonuniformities in the flow of solid and fluid constituents when they move with respect to one another, and usually inserted by Biot (1956) in the expression of the kinetic energy, have to be considered in the constitutive choices.

3. BALANCE OF MASS AND OF MICROINERTIA

The equation of mass balance for each constituent in local form is obtained by applying Truesdell's principles. If we assume that the fluid component does not exchange the mass with other components, the second principle requires that

$$\begin{aligned}
&\frac{\partial\rho_f}{\partial\tau} + \operatorname{div}(\rho_f v_f) = 0 \\
&\frac{\partial\rho_a}{\partial\tau} + \operatorname{div}(\rho_a v_a) = \rho\alpha, \qquad (12)\\
&\frac{\partial\rho_s}{\partial\tau} + \operatorname{div}(\rho_s v_s) = -\rho\alpha
\end{aligned}$$

where α stands for the mass growth of adsorbate; whereas the third principle states the conservation of mass for the whole mixture:

$$\frac{\partial\rho_t}{\partial\tau} + \operatorname{div}(\rho_t v_t) = 0. \qquad (13)$$

In addition we write other two equations in terms of mean and removal quantities, instead of the peculiar ones, which are, together with equation (13), perfectly equivalent to those postulated in equations (12):

$$\begin{aligned}
&\frac{\partial c}{\partial\tau} + v\cdot\operatorname{grad} c + \rho^{-1}\operatorname{div}(\rho j) = (1-c)\alpha, \\
&\frac{\partial\rho}{\partial\tau} + \operatorname{div}(\rho v) = \rho\alpha.
\end{aligned} \qquad (14)$$

Let us now consider balances of micromomentum; from the expression (11) of the kinetic energy, we assume local equations of balance of micromomentum in accordance with Truesdell's second principles:

$$\begin{aligned}
&\rho_f\mu_f(\beta_f)\beta_f''^{f} + \frac{1}{2}\rho_f\frac{d\mu_f}{d\beta_f}(\beta_f)\left(\beta_f'^{f}\right)^2 = \\
&\qquad = \rho_f\gamma_f - \zeta_f + \operatorname{div}s_f + \phi_f, \\
&\rho_a\mu_a(\beta_a)\beta_a''^{a} + \frac{1}{2}\rho_a\frac{d\mu_a}{d\beta_a}(\beta_a)\left(\beta_a'^{a}\right)^2 = \\
&\qquad = \rho_a\gamma_a - \zeta_a + \operatorname{div}s_a + \phi_a, \qquad (15)\\
&\rho_s\mu_s(U)U''^{s} + \frac{1}{2}\rho_s\frac{d\mu_s}{dU}(U)U'^{s}\cdot U'^{s} = \\
&\qquad = \rho_s C - Z + \operatorname{div}\Omega + \Phi
\end{aligned}$$

(see also equations (7) and (8) obtained by Giovine (2000) in the conservative case).

On the left-hand side of equations (15) it is possible to notice the full Lagrangian derivative of the density of kinetic coenergy related to the fluid, adsorbate or solid component, respectively. In particular they are the Lagrangian derivatives of respective homogeneous parts of second degree in $\beta_j'^{j}$ or U'^{s}, being the homogeneous part of first degree of constitutive character and included in the supply force $-\zeta_j$ and $-Z$, respectively; the kinetic coenergy is easily obtained from the kinetic energy by means of a Legendre transformation (see Capriz & Giovine (1997)).

On the right-hand side of the same equations we recognize usual quantities introduced in (microstructured) immiscible mixtures (see Passman & Batra (1984) and Giovine (1990, 2000)): $\rho\gamma_j$ and $\rho_s C$ are the densities per unit volume of microstructural body forces for the fluid, or adsorbate, and solid phase, respectively; s_j is the microstress vector and Ω the third-order microstress tensor, which is symmetric in the first two indices and normally related to the boundary microtractions, even if, in some cases, it could express weakly non-local internal effects; ϕ_j and Φ represent the growth of micromomenta which permits the interchange between phases.

Tensor fields C, Z and Φ are symmetric: the first one can be interpreted as an externally controlled pore pressure; the second one, which does not necessarily sum to zero as internal forces do, includes interactive forces between the gross and fine structures as well as internal dissipative contributions due to the stir of the pores' surface; for the third one, one should notice that micromomentum is not necessarily conserved for the mixture.

Besides, by summing the first two equations in (15) and by using relations (8) and (9) frequently, we obtain the balance of micromomentum for inclusions:

$$\rho\frac{\partial(\mu\dot{\beta})}{\partial\tau} + \rho v\cdot\operatorname{grad}(\mu\dot{\beta}) = \rho\gamma - \zeta + \operatorname{div}s + \phi, \qquad (16)$$

where $\beta := \beta_f + \beta_a$ is the volume fraction of the fluid mixture, while we introduced the following quantities:

$$\begin{aligned}
&\mu\dot{\beta} := (1-c)\mu_f(\beta_f)\beta_f'^{f} + c\mu_a(\beta_a)\beta_a'^{a}, \\
&\gamma := (1-c)\gamma_f + c\gamma_a, \\
&s := \sum_{j=f,a}\left[s_j - \rho_j\mu_j(\beta_j)\beta_j'^{j}(v_j - v)\right], \qquad (17)\\
&\zeta := \sum_{j=f,a}\left[\zeta_j - \frac{1}{2}\rho_j\frac{d\mu_j}{d\beta_j}(\beta_j)\left(\beta_j'^{j}\right)^2\right], \\
&\phi := \phi_f + \phi_a + \rho_f\alpha\left[\mu_f(\beta_f)\beta_f'^{f} - \mu_a(\beta_a)\beta_a'^{a}\right]
\end{aligned}$$

4. BALANCE OF MOMENTUM AND OF MOMENT OF MOMENTUM

The balance laws of linear momentum in local form for constituents are

$$\rho_f a_f = \rho_f b_f + \operatorname{div} T_f - \rho m_f,$$
$$\rho_a a_a = \rho_a b_a + \operatorname{div} T_a + \rho m_a - \rho\alpha v_a, \tag{18}$$
$$\rho_s a_s = \rho_s b_s + \operatorname{div} T_s - \rho m + \rho\alpha v_s,$$

where b_i and T_i, for $i = f, a, s$, are the density per unit mass of body forces and the stress tensor of indicated phase, respectively, while $-\rho m_f$ and ρm_a are the density per unit volume of interaction forces between phases for the fluid and adsorbate components, respectively, and $m := m_a - m_f$; in equation $(18)_3$ we applied the conservation law of linear momentum in order to obtain the interaction force for the solid.

Moreover the balance laws of moment of momentum are, by applying the conservation law of rotational momentum,

$$\operatorname{skw} T_f = -\rho M_f, \quad \operatorname{skw} T_a = \rho M_a,$$
$$\operatorname{skw}\left(T_s - UZ^T - \operatorname{grad} U \varnothing \Omega\right) = -\rho M, \tag{19}$$

where 'skw' indicates the skew part of a tensor, that is skw(•) := 1/2 [(•) – (•)^T], $-M_f$ and M_a are the skew tensors of growth of rotational momentum for the fluid and adsorbate components, respectively, $-M := M_a - M_f$ and tensor product '$\varnothing$' reads as: $(\operatorname{grad} U \varnothing \Omega)_{ij} := U_{ih,k}\, \Omega_{jhk}$.

Therefore none of stress tensors is 'a priori' symmetric, while the law $(19)_3$ takes into account the particular structure of the porous solid (see equation (6) in Giovine (1999)).

If we denote by T_t Cauchy's stress tensor of our material and by b_t the density of body forces, then by Truesdell's third principle the balances of linear momentum and of moment of momentum for the whole mixture in local form turn out to be

$$\frac{\partial \rho_t v_t}{\partial \tau} + \operatorname{div}\left(\rho_t v_t \otimes v_t\right) = \rho_t b_t + \operatorname{div} T_t,$$
$$\operatorname{skw}\left(T_t - UZ^T - \operatorname{grad} U \varnothing \Omega\right) = 0, \tag{20}$$

where $(u \otimes v)_{ij} := u_i v_j$ (see Capriz (1989)).

But the first principle implies that these relations must be consequences of equations (18) and (19) too: by summing those equations separately and by considering definitions (4) and (8) and mass balances (12), we exactly obtain equations (20) provided that

$$\rho_t b_t := \rho_f b_f + \rho_a b_a + \rho_s b_s,$$
$$T_t := \sum_{i=f,a,s} \left[T_i - \rho_i\left(v_i - v_t\right) \otimes \left(v_i - v_t\right)\right]. \tag{21}$$

As already done for mass equations (14), we here write the other balances of linear and angular momentum in terms of mean and removal quantities, once again obtained from (18) and (19), respectively:

$$\frac{\partial(\rho v)}{\partial \tau} + \operatorname{div}(\rho v \otimes v) = \rho(b + m) + \operatorname{div} T,$$
$$\frac{\partial(\rho j)}{\partial \tau} + \operatorname{div}\left(\frac{\rho}{c(1-c)}\, j \otimes j\right) = \rho(d + n) +$$
$$+ \operatorname{div}\left[S + \rho j \otimes \left(\frac{2}{1-c}\, j - v\right)\right] + T \operatorname{grad} c - \tag{22}$$
$$- \rho\left[(\operatorname{grad} v) j + \alpha(1-c) v\right],$$
$$\operatorname{skw} T = \rho M, \quad \operatorname{skw} S = \rho N, \tag{23}$$

where

$$b := (1-c) b_f + c b_a, \quad d := c\left(b_a - b\right),$$
$$n := (1-c) m_f + c m_a,$$
$$T := T_f + T_a - \frac{\rho}{c(1-c)}\, j \otimes j, \tag{24}$$
$$S := (1-c) T_a - c T_f,$$
$$N := (1-c) M_a - c M_f.$$

5. BALANCE OF ENERGY AND ENTROPY INEQUALITY

In this section we begin by writing relations which give peculiar balances of internal energy by also taking into account the power density of internal and external actions due to microstructure. If we introduce the peculiar specific internal energy ε_i, the heat flux vector q_i, the rate of heat generation λ_i, and finally the growth of energy ε_i^+, due to interactions, we have the following equations in local form:

$$\rho_f \varepsilon_f'^{\,f} = T_f \cdot \operatorname{grad} v_f + s_f \cdot \operatorname{grad} \beta_f'^{\,f} - \operatorname{div} q_f +$$
$$+ \rho_f \lambda_f + \rho m_f \cdot v_f + \left(\zeta_f - \phi_f\right) \beta_f'^{\,f} - \rho \varepsilon_f^+,$$
$$\rho_a \varepsilon_a'^{\,a} = T_a \cdot \operatorname{grad} v_a + s_a \cdot \operatorname{grad} \beta_a'^{\,a} - \operatorname{div} q_a \tag{25}$$
$$+ \rho_a \lambda_a - \rho m_a \cdot v_a + \left(\zeta_a - \phi_a\right) \beta_a'^{\,a} + \rho \varepsilon_a^+ -$$
$$- \rho\alpha\left[\varepsilon_a - \frac{1}{2} v_a^2 + \frac{1}{2} \mu_a(\beta_a)\left(\beta_a'^{\,a}\right)^2\right],$$

$$\rho_s \dot{\varepsilon}_s^s = T_s \cdot \text{grad}v_s + \Omega \cdot \text{grad}U'^s - \text{div}q_s +$$
$$+ \rho_s \lambda_s + \rho m \cdot v_s + (Z - \Phi) \cdot U'^s - \rho\left(\varepsilon_a^+ - \varepsilon_f^+\right) +$$
$$+ \rho\alpha\left[\varepsilon_s - \frac{1}{2} v_s^2 + \frac{1}{2}\mu_s(U)\left(U'^s\right)^2\right],$$

where the conservation of energy growths was used and account is taken of the work done by the respective terms in the balance of micromomentum.

Truesdell's third principle instead requires the conservation of energy for the whole mixture:

$$\rho_t\left(\frac{\partial \varepsilon_t}{\partial \tau} + v_t \cdot \text{grad}\,\varepsilon_t\right) = \text{sym}T_t \cdot \text{sym}(\text{grad}v_t) +$$
$$+ \text{skw}\left(UZ^T + \text{grad}U \varnothing \Omega\right) \cdot \text{skw}\left(\text{grad}v_t\right) + \qquad (26)$$
$$+ s \cdot \text{grad}\dot{\beta} - \text{div}q_t + \rho_t \lambda_t + (\zeta - \phi)\dot{\beta} +$$
$$+ (Z - \Phi) \cdot U'^s + \Omega \cdot \text{grad}U'^s,$$

where we used definitions (17), the balances of linear and angular momentum (20) and we indicated with 'sym' the symmetric part of a tensor, that is sym(•) := 1/2 [(•) + (•)^T].

But again the first principle states that this equation must be a consequence of the previous ones, therefore, by summing them, by considering definitions (1)-(4), by using relations (8) and (9) and mass balances (12)-(14), we exactly obtain equations (26) provided that the internal energy ε_t, the heat flux vector q_t, and the rate of heat generation λ_t, for the total mixture depend on peculiar quantities through the following relations:

$$\varepsilon_t := \sum_{i=f,a,s} \frac{\rho_i}{\rho_t}\left[\varepsilon_i + \frac{1}{2}(v_i - v_t)^2\right] +$$
$$+ \frac{1}{2\rho_t}\left[\sum_{j=f,a} \rho_j \mu_j \left(\beta_j^j\right)^2 + \rho_s \mu_s \left(U'^s\right)^2 - \rho\mu\dot{\beta}^2\right],$$

$$q_t := \sum_{i=f,a,s}\left[q_i - T_i^T v_i\right] - \sum_{j=f,a} \beta_j^{\,j} s_j -$$
$$- \Omega U'^s + T_t^T v_t + \beta\dot{s} + \frac{1}{2}\rho\mu\dot{\beta}^2 v - \qquad (27)$$
$$- \sum_{j=f,a} \rho_j\left[\varepsilon_j + \frac{1}{2} v_j^2 + \frac{1}{2}\mu_j\left(\beta_j^j\right)^2\right] v_j +$$
$$+ \rho_t\left[\varepsilon_t + \frac{1}{2} v_t^2\right] v_t - \rho_s\left[\varepsilon_s + \frac{1}{2} v_s^2\right] v_s,$$

$$\lambda_t := \sum_{i=f,a,s} \frac{\rho_i}{\rho}\left(\lambda_i + b_i \cdot v_i\right) + \sum_{j=f,a} \gamma_j \beta_j^j -$$
$$- b_t \cdot v_t - \gamma\dot{\beta}.$$

We can besides obtain the balances of mean internal energy ε for inclusions from equation (25):

$$\rho\dot{\varepsilon} = T \cdot \text{grad}v + s \cdot \text{grad}\dot{\beta} - \text{div}q + \rho\lambda + (\zeta - \phi)\dot{\beta} +$$
$$+ \rho\left(\varepsilon_a^+ - \varepsilon_f^+\right) - \rho m \cdot v - \rho\alpha\left[\varepsilon - \frac{1}{2}v^2 + \frac{1}{2}\mu\dot{\beta}^2\right], \qquad (28)$$

where the internal energy ε, the heat flux vector q and the rate of heat generation λ of inclusions are defined by:

$$\varepsilon := (1-c)\varepsilon_f + c\varepsilon_a + \frac{j \cdot j}{2c(1-c)} +$$
$$+ \frac{1}{2}\left[(1-c)\mu_f\left(\beta_f^f\right)^2 + c\mu_a\left(\beta_a^a\right)^2 - \mu\dot{\beta}^2\right],$$

$$q := \sum_{j=f,a}\left[q_j - T_j^T v_j - \beta_j^j s_j\right] +$$
$$+ T^T v + \beta\dot{s} + \rho\left[\varepsilon + \frac{1}{2}\left(v^2 + \mu\dot{\beta}^2\right)\right] v - \qquad (29)$$
$$- \sum_{j=f,a} \rho_j\left[\varepsilon_j + \frac{1}{2} v_j^2 + \frac{1}{2}\mu_j\left(\beta_j^j\right)^2\right] v_j,$$

$$\lambda := \sum_{j=f,a}\left[\frac{\rho_j}{\rho}\left(\lambda_j + b_j \cdot v_j\right) + \gamma_j \beta_j^j\right] -$$
$$- b \cdot v - \gamma\dot{\beta}.$$

Now we introduce peculiar specific entropy η_i, and growth of entropy due to interactions η_i^+, along with temperature θ and entropy η_t for the total body.

The second law of thermodynamics for each constituent affirms that each entropy production is always non-negative; it means that, in local form, the following inequality has to hold:

$$\rho_i\left(\dot{\eta}_i^i - \frac{\lambda_i}{\theta}\right) + \text{div}\left(\frac{q_i}{\theta}\right) \geq \eta_i^+ - \rho\alpha_i\eta_i\,, \qquad (30)$$

where $\alpha_f = 0$ and $\alpha_a = -\alpha_s = \alpha$.

By applying the requirement according to which also total growth of entropy vanishes, the entropy inequality for the mixture reads

$$\rho_t\left(\frac{\partial \eta_t}{\partial \tau} + v_t \cdot \text{grad}\,\eta_t - \frac{\lambda_e}{\theta}\right) + \text{div}\left(\frac{q_e}{\theta}\right) \geq 0\,, \qquad (31)$$

where the effective heat flux vector q_e and the effective heat supply λ_e have to satisfy relations

$$q_e := \sum_{i=f,a,s}\left[q_i + \rho_i \eta_i \theta_i \left(v_j - v_t\right)\right],$$
$$\qquad (32)$$
$$\lambda_e := \sum_{i=f,a,s} \frac{\rho_i}{\rho_t}\lambda_i$$

because of the first Truesdell's principle.

6. CONCLUSIONS

In this work we propose an immiscible mixtures consisting of a continuum with ellipsoidal microstructure (the porous elastic solid) and of two classical fluids (a fluid carrier and an adsorbate) to characterize, for example, the transport of a contaminant with rainwater through the soil (see Giovine (2003)).

Our model about the porous matrix includes, as a particular case, previous 'void theories' which does not anticipate any size effects in torsion; on the other hand the virtual inertia considered in our immiscible mixture allows the presence of an inviscid drag term depending on changes of the radius of the material elements, other than on the relative velocities, a term which is usually absent in previous theories (see Giovine (1990)).

General and peculiar balance equations, as well as relations between quantities here introduced, are obtained in all generality by applying Truesdell's metaphysical principles for classical mixtures appropriately suited to our immiscible model.

In particular, we mean to focus on the fact that our system of equations differs from those of the classical mixtures as there is an additional balance equation (15) for each component which models the microstructural force systems, besides it takes into account the changes of constituent volume fractions and of lacunae's volume, which occur without affecting the gross motion.

In addition, there are new terms in the energy equations corresponding to work done by respective terms in the balance equations of micromomentum.

Finally, once constitutive proposals for dependent fields are given, the dependence on all microstructural variables, along with all the gross ones, should be taken into consideration, as we made in the papers Giovine (2003) and Giovine & Palumbo (2003), where we studied the coupling of adsorption and diffusion in porous materials with large voids and the condition of hyperbolicity in the subsystem describing the liquid phase, respectively.

ACKNOWLEDGEMENTS

This research has been supported by the Italian M.U.R.S.T. through the project "Cofin2002 - Mathematical Models for the Material Science".

REFERENCES

Biot, M.A. 1956. *Theory of propagation of elastic waves in a fluid-saturated porous solid. I. Low-frequency range.* J. Acoust. Soc. Am. 28: pp.168-178.

Capriz, G. 1989. *Continua with microstructure.* Springer Tracts in Natural Philosophy 35, New York: Springer-Verlag.

Capriz, G. & Giovine, P. 1997. *On microstructural inertia.* Math. Model Meth. Appl. Sc. 7: pp.211-216.

Capriz, G. & Podio Guidugli, P. 1981. *Materials with Spherical Structure*, Arch. Rat. Mech. Analysis 75: pp.269-279.

Giovine, P. 1990. *Sulla dinamica di una miscela di due fluidi comprimibili e non miscibili.* Rend. Mat. Acc. Lincei 1(9): pp.377-385.

Giovine, P. 1996. *Porous solids as materials with ellipsoidal structure.* In R.C. Batra & M.F. Beatty (eds.), Contemporary Research in the Mechanics and Mathematics of Materials, CIMNE, Barcelona: pp.335-342.

Giovine, P. 1999. *A linear theory of porous elastic solids.* Trans. Porous Media 34: pp.305-318.

Giovine, P. 2000. *A mixture theory for microstructured porous media.* Zeit. angew. Math. Mech. 28: pp.153-156.

Giovine, P. 2003. *On adsorption and diffusion in microstructured porous media*, Proc. of IUTAM 2003 on "Physicochemical and Electromechanical Interactions in Porous Media", Maastricht, 18-23 May 2003, in press.

Giovine, P. & Palumbo, A. 2003. *Modeling contaminant transport in soils*, Proceed. of Mediterranean Conference on "Modelling and Simulation '03", Reggio Calabria, 25-27 June 2003, Session on Modeling, Technical Note n.178 on the CD rom, Reggio Calabria: NeuroLab.

Lakes, R.S. 1986. *Experimental microelasticity of two porous solids.* Int. J. Solids Structures 22: pp.55-63.

Nunziato, J.W. & Cowin, S.C. 1979. *A nonlinear theory of elastic materials with voids.* Arch. Rat. Mech. Analysis 72: pp.175-201.

Passman, S.L. & Batra, R.C. 1984. *A thermomechanical theory for a porous anisotropic elastic solid with inclusions.* Arch. Rat. Mech. Analysis 87: pp.11-33.

Truesdell, C. 1984. *Thermodynamics of diffusion.* In Rational thermodynamics, 2nd-edition, Springer-Verlag, New York: pp.218-236.

Wilmanski, K. 1996. *Porous media at finite strains. The new model with the balance equation for porosity.* Arch. Mech. 48: pp.591-628.

ANALYSIS OF MECHANICAL AND HYDRAULIC PROPERTIES OF CRACKED STRUCTURE BY THE RATIO OF CRACK OPENING DEPENDENCY (RCOD)

Dr.Akira Sato[1], Dr.Yoshiaki Hirakawa[2] and Dr.Katsuhiko Sugawara[3]
1)Faculty of Engineering, Kumamoto University, Japan
2) NEWJEC Inc., Japan
3)Graduate School of Science and Technology, Kumamoto University, Japan

Abstract: The Ratio of Crack Opening Dependency (RCOD) was proposed as a new index to simultaneously evaluate the mechanical and hydraulic properties of cracked structure which contains many open cracks within. The RCOD is an index which is defined as the ratio of total amount of crack opening and the total deformation of cracked structure. The efficiency of RCOD as an index to evaluate the coupled problem was numerically analyzed by applying S̃CDDM (stress compensation-displacement discontinuity method) to the two dimensional multi crack model. It was shown that RCOD is an useful index which can directly evaluate the deformability and the permeability of the cracked structure without any parameters such as stresses or crack size. The possibility of application of RCOD to the field measurements was also discussed.

1. INTRODUCTION

The rock mass properties such as deformability and permeability are highly dependent on the existence of cracks (Eshelby (1920), Bieniawski (1979), Kinoshita (1992), Yoshida (1999)). Especially, open cracks give crucial effects to these factors. Moreover, the deformability and the permeability of rock mass are strongly related to each other, and the coupling analysis between these factors is inevitable for the rock mass evaluation.

In this study, the Ratio of Crack Opening Dependency (RCOD) is proposed as a new index to simultaneously evaluate the mechanical and hydraulic property of cracked structure which contains many open cracks within. Generally, the relation between the water flow and the crack opening distribution is expressed by the cubic law (Elsworth (1992)), the RCOD gives a new approach to the evaluation of permeability of cracked structure. The RCOD is an index which is defined as the ratio of total amount of crack opening and the total deformation of cracked structure. The efficiency of RCOD as an index to evaluate the coupled problem is numerically analyzed by applying S̃CDDM (stress compensation-displacement discontinuity method) (Hirakawa (2000), Sato (2001)) to the two dimensional multi crack model. In order to estimate the mechanical interaction of multi cracks, the homogenization method is applied in the analysis. Here, the relation between RCOD and effective compliance of the model is discussed. Then, the permeability of the model is also estimated under the assumption that the water permeates through elliptic crack, and the relation between RCOD and the permeability is discussed.

2. RCOD

In the case that cracked structure is subjected to the uniform tensile stress σ as shown in Figure 1, the total strain of the structure ε consists of two components. One is the strain due to the elastic deformation of the matrix ε_0 and another is the one due to the crack opening ε_c.

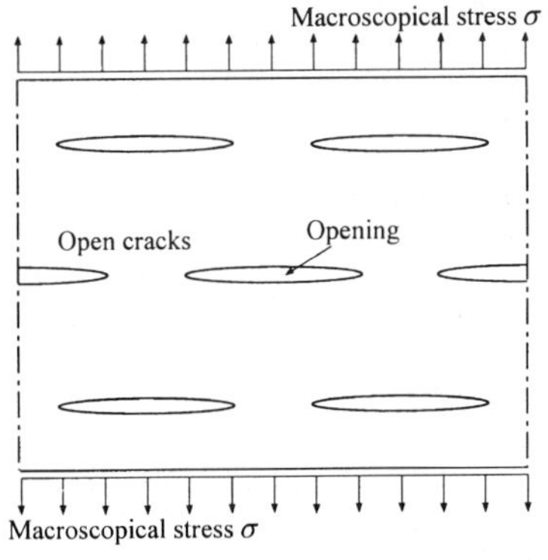

Figure 1. Cracked structure subjected to the macroscopical tensile stress σ.

$$\varepsilon = \varepsilon_0 + \varepsilon_c \quad (1)$$

Let us introduce the RCOD (ratio of crack opening dependency) R as a new index to evaluate the cracked structure. The RCOD is defined by the ratio between the total strain and crack opening strain, and given by the following equation.

$$R = \frac{\varepsilon_c}{\varepsilon} \quad (2)$$

That is, the RCOD R reaches to zero when the elastic deformation is dominant in the total deformation. On the other hand, R reaches to one when the crack deformation is dominant. By substituting equation (2) into equation (1), the following relation is obtained.

$$\frac{\lambda}{\lambda_0} = \frac{1}{1-R} \quad (3)$$

where λ_0 and λ are the compliance of the matrix and the one of whole cracked structure, and equation (3) represents that the effective compliance λ / λ_0 is simply expressed as a function of R.

3. CRACK HOMOGENIZATION MODEL

In order to analyze the interaction between cracks, the two dimensional homogenized crack distribution is taken into consideration. As shown in Figure 2, crack distribution inside the rectangular unit cell ABCD is prescribed, as this cell represents the corresponding cracked structure. Various kinds of crack arrangements in the unit cell are possible. Here the single open joint problem in which an open joint with the length of $2a$ is located at the center of the unit cell (Figure 2) is assumed. Then, the whole mass of the cracked structure is expressed as the infinite numbers of the unit cells as shown in Figure 3. Here-in-after, this infinite model is named as crack homogenization model. Characteristic points of the model are that the crack arrangement around a representative crack is same at everywhere in the model, and that all unit cells and cracks deform uniformly. Various kinds of unit cell arrangements are also possible. In the analysis, alternative model shown in Figure 3 is taken into consideration.

4. ANALYSIS OF CRACK HOMOGENIZATION MODEL

4.1 Compliance

In the case that the macroscopical uniform stress $\{\sigma^0_{xx}, \sigma^0_{yy}, \sigma^0_{xy}\}$ is acting at far field of the crack homogenization model, the macroscopical uniform strain $\{\varepsilon^0_{xx}, \varepsilon^0_{yy}, \varepsilon^0_{xy}\}$ is generated. The relation between the strain and the stress is expressed by using compliance $\{\lambda_{ij\,kl}\}$ as

$$\begin{Bmatrix} \varepsilon^0_{xx} \\ \varepsilon^0_{yy} \\ \varepsilon^0_{xy} \end{Bmatrix} = \begin{bmatrix} \lambda_{xx\,xx} & \lambda_{xx\,yy} & \lambda_{xx\,xy} \\ \lambda_{yy\,xx} & \lambda_{yy\,yy} & \lambda_{yy\,xy} \\ \lambda_{xy\,xx} & \lambda_{xy\,yy} & \lambda_{xy\,xy} \end{bmatrix} \begin{Bmatrix} \sigma^0_{xx} \\ \sigma^0_{yy} \\ \sigma^0_{xy} \end{Bmatrix} \quad (4)$$

In order to evaluate the compliance $\{\lambda_{ij\,kl}\}$, SC-DDM (Hirakawa (2000), Sato (2001)) is applied in this study.

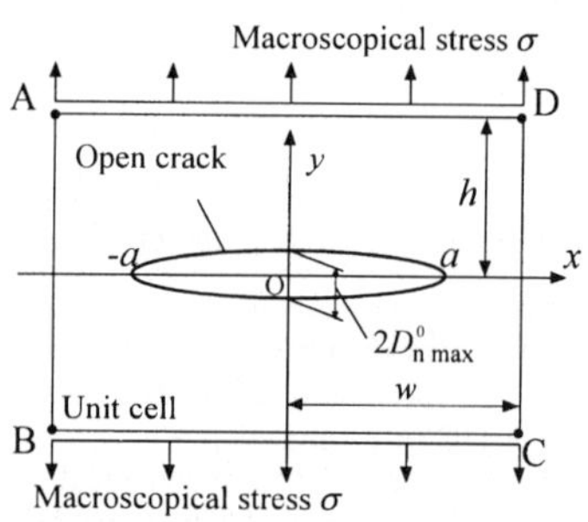

Figure 2. A unit cell for single crack set.

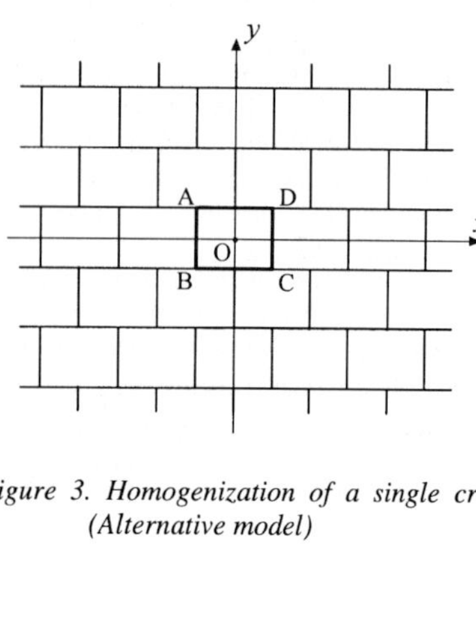

Figure 3. Homogenization of a single crack set. (Alternative model)

4.2 Crack opening and deformation of unit cell

The displacements of each crack and unit cell can be also expressed as a function of uniform stress { σ_{ij}^0 }. Here, the share and normal components of displacement discontinuities at each point on the crack are denoted by { D_s^0, D_n^0 }, and given by

$$\begin{Bmatrix} D_s^0 \\ D_n^0 \end{Bmatrix} = \begin{bmatrix} \phi_{xx} & \phi_{yy} & \phi_{xy} \\ \chi_{xx} & \chi_{yy} & \chi_{xy} \end{bmatrix} \begin{Bmatrix} \sigma_{xx}^0 \\ \sigma_{yy}^0 \\ \sigma_{xy}^0 \end{Bmatrix} \quad (5)$$

where $\{\phi_{ij}\}$ and $\{\chi_{ij}\}$ are the dimensionless coefficient determined at each point on the crack.

The shear and normal components of displacement discontinuities on the side of a unit cell are denoted by { V_s^0, V_n^0 }, and given by

$$\begin{Bmatrix} V_s^0 \\ V_n^0 \end{Bmatrix} = \begin{bmatrix} \xi_{xx} & \xi_{yy} & \xi_{xy} \\ \zeta_{xx} & \zeta_{yy} & \zeta_{xy} \end{bmatrix} \begin{Bmatrix} \sigma_{xx}^0 \\ \sigma_{yy}^0 \\ \sigma_{xy}^0 \end{Bmatrix} \quad (6)$$

where $\{\xi_{ij}\}$ and $\{\zeta_{ij}\}$ are the dimensionless coefficient determined at each point on the. The coefficients $\{\phi_{ij}\}$, $\{\chi_{ij}\}$, $\{\xi_{ij}\}$ and $\{\zeta_{ij}\}$ are also evaluated by SC-DDM.

5. COEFFICIENT OF PERMEABILITY

The permeability of the cracked structure is highly dependent on the existence of open cracks, and open cracks can be major permeation paths. Here, shape of open cracks is approximated by elliptic, and one dimensional water permeation in the elliptic tube is assumed (Figure 2). In the case of single open crack problem, one open crack exists in the unit cell. The flow rate of water of a unit cell Δu is given as a function of maximum crack opening $D_{n\,max}^0$ by

$$\Delta u = -\frac{1}{\mu} \cdot \frac{a^2}{4} \cdot \frac{\eta^2}{1+\eta^2} \cdot \frac{dp}{dz} \quad (7)$$

where η is the aspect ratio of the crack defined by $D_{n\,max}^0/a$, dp/dz is the hydraulic pressure gradient in the z-direction (flow direction), and μ is the viscosity of water. Since the crack density of the homogenization model shown in Figure 3 is defined by h/w, the flow rate of water, u, in the unit area $2w$ x $2w$ is given by

$$u = \frac{w}{h} \cdot \frac{1}{(2w)^2} \Delta u \quad (8)$$

By substituting equation (7) into equation (8), the following relation is obtained.

$$u = -\frac{1}{\mu} \cdot \frac{1}{16} \cdot \left(\frac{w}{h}\right) \cdot \left(\frac{a}{w}\right)^2 \cdot \frac{\eta^2}{1+\eta^2} \cdot \frac{dp}{dz} \quad (9)$$

Therefore, the coefficient of permeability of the crack homogenization model, k, is defined by

$$k = \frac{a^2}{16} \cdot \left(\frac{w}{h}\right) \cdot \left(\frac{a}{w}\right)^2 \cdot \frac{\eta^2}{1+\eta^2} \quad (10)$$

Since the relation between crack length a and the maximum crack opening $D_{n\,max}^0$ is numerically analyzed by SC-DDM, the coefficient of permeability of the model is obtained.

6. RESULTS

6.1 Crack opening and unit cell deformation

An example of unit cell deformation is shown in Figure 4. Here the distribution of deformation on the side AD is drawn. The magnitude is normalized by the differences between the maximum deformation $V_{n\,max}^0$ and minimum deformation $V_{n\,min}^0$ in order to clarify the distribution of deformation. A part of crack homogenization model is also drawn in Figure 4 in order to confirm the correspondence to the arrangement of unit cells. It is found that the distribution of deformation is almost similar when the crack distance is h/w=0.5. The total strain ε (equation (2)) is determined as the mean value of unit cell deformation. As well as the deformation of unit cells, the distribution of crack opening differs with the unit cell interval h/w and crack length a/w. An example of the results is shown in Figure 5. Here, crack opening displacement D_n^0 is normalized by the maximum value of crack opening displacement $D_{n\,max}^0$ in order to clarify the distribution of crack opening. The horizontal axis is also normalized by the each crack length. For the sake of comparison, distribution of single crack, which exists on the infinite elastic body and subjected to uniaxial tensile stress σ, is also drawn by dotted line in Figure 5. It is found that each profile is relatively similar to the distribution of single crack and that crack shape of each case can be approximated by ellipse. The strain ε_c in equation (2) is determined from the crack opening distribution.

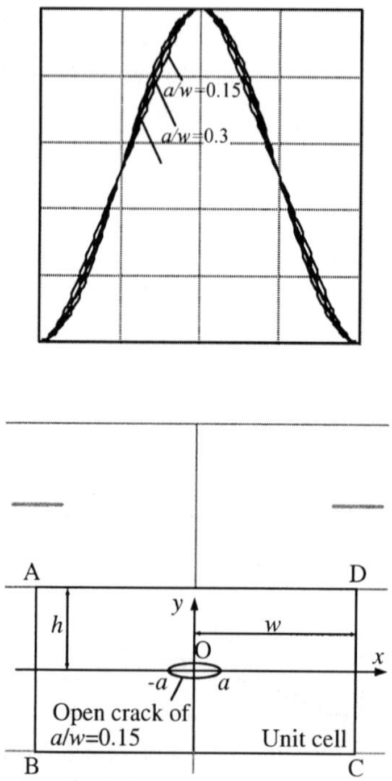

Figure 4. Normalized displacement on the side AD of unit cell.

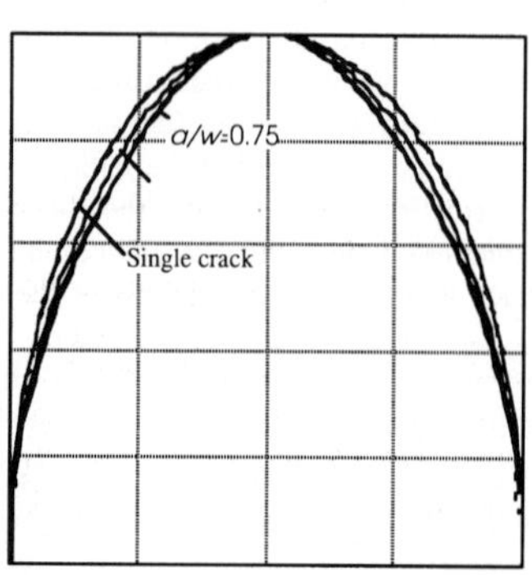

Figure 5. Normalized crack opening displacement.

6.2 Deformability of the model

The relation between effective compliance λ/λ_0 and crack length a/w is shown in Figure 6 for four cases of unit cell intervals h/w=0.125, 0.25, 0.5 and 1. It is obvious that the compliance becomes larger when the crack interval h/w becomes smaller, and that the crack interval is a crucial factor to the deformability of cracked structure.

The relation between effective compliance λ/λ_0 and the RCOD (equation (3)) is shown in Figure 7. As this figure shows, all results of the effective compliance are plotted on the curve given by equation (3) and the theoretical relation is proved in the case of crack homogenization model. That is, effective compliance can be directly evaluated from the RCOD.

6.3 Coefficient of permeability

The relation between coefficient of permeability k and crack length a/w is shown in Figure 8 for four cases of unit cell intervals h/w=0.125, 0.25, 0.5 and 1. The difference of coefficient of permeability becomes larger when the crack length become larger.

The relation between coefficient of permeability k and the RCOD is shown in Figure 9. As this figure shows, almost all results for four cases of unit cell intervals h/w=0.125, 0.25, 0.5 and 1 are plotted on the same curve and coefficient of permeability can be directly evaluated from the RCOD.

6.4 Relation between deformability and permeability

The relation between effective compliance λ/λ_0 and coefficient of permeability k is shown in Figure 10. As this figure shows, almost all results are also plotted on the almost same curve, and the characteristic point of the result is that the value of k is almost proportional to the value of λ/λ_0 on the log-log plots in the range of $\lambda/\lambda_0 > 2$.

6.5 Applicability to field measurements

Generally, two parameters, such as deformation and stress, are necessary for the evaluation of the effective compliance, and it is almost impossible to determine the stress state if the materials contains many open cracks within. However, the total deformation of the cracked structure is easily measured. The crack opening displacements can be

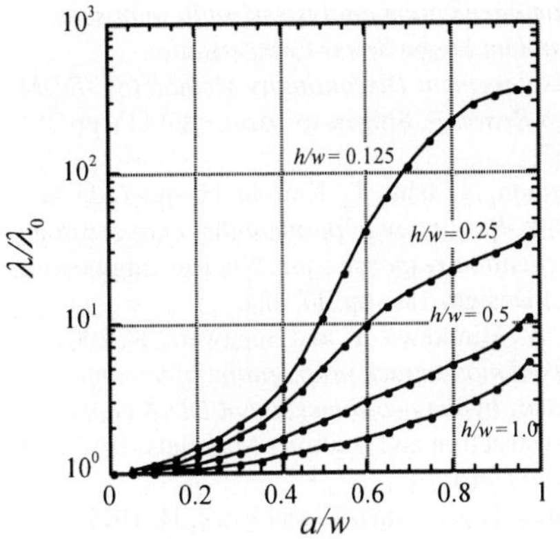

Figure 6 Relation between compliance and crack length.

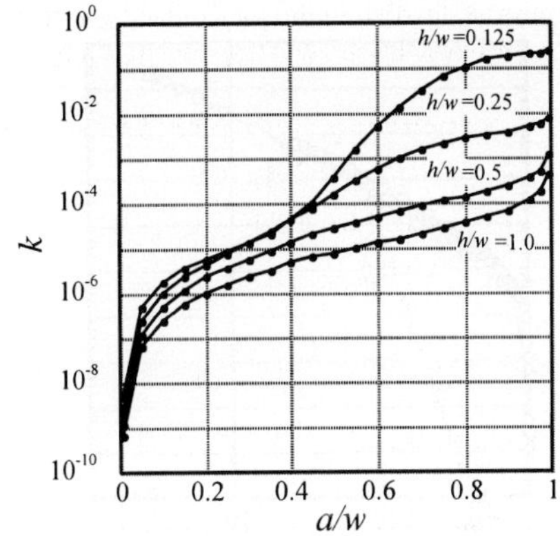

Figure 8. Relation between coefficient of permeability and crack length.

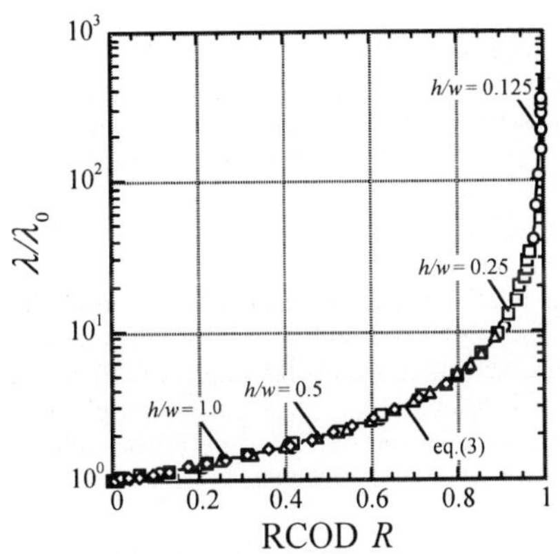

Figure 7. Relation between compliance and the RCOD.

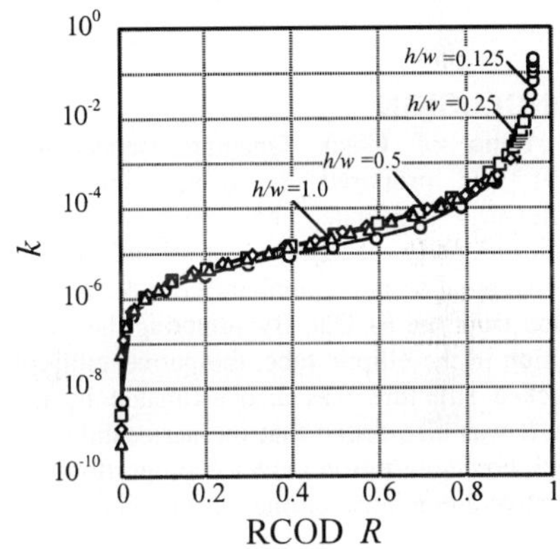

Figure 9. Relation between coefficient of permeability and the RCOD.

also easily measured from the observation of boreholesby borehole televiewers and other instruments. Therefore, it is possible to determine the RCOD from the measurements of these deformation, and it is the useful point of the RCOD that the deformability of the cracked structure can be evaluated only from the measurements of deformation.

In the case of the evaluation of the permeability, many parameters, such as state of stress, density and geometry of crack and heterogeneity of rock mass, are necessary (Yoshida (1999)). The evaluation of these parameters in the field measurements are almost impossible. However, the RCOD is the index which directly indicate the degree of crack opening and the total amount of permeation path. Even though the analysis was conducted under the assumption that the shape of water permeation path is ellipse, there is a possibility that the RCOD could be an useful parameter to estimate the permeability of the cracked structure.

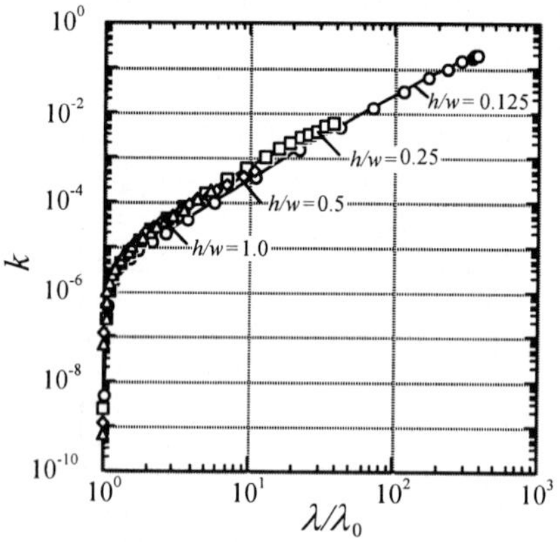

Figure 10. Relation between coefficient of permeability and the effective compliance.

7. CONCLUSION

The Ratio of Crack Opening Dependency (RCOD) was proposed as a new index to simultaneously evaluate the mechanical and hydraulic property of cracked structure. It was shown that the effective compliance can be directly evaluated from the RCOD. By assuming the water permeation in the elliptic tube, the permeability of the cracked structure was also estimated by the RCOD. It was also found that the permeability of the crack homogenization model is proportional to the effective compliance on the log-log plots in the case that the effective compliance is more than two.

REFERENCES

Bieniawski, Z.T. 1979. *The geomechanics classification in rock engineering applications*, Proc. of 4th Cong. of ISRM, 2: pp.41-47.

Crouch, S. L. and Starfield A. M. 1983. *Boundary Element Methods in Solid Mechanics*, George Allen & Unwin, London, pp.79-109.

Elsworth, D. and Mase, C.R. 1992. *Groundwater in rock engineering, Comprehensive rock engineering: Principles, practice and projects*, Chapter8, 1, Hudson, J.A., pp.201-226.

Eshelby,J.D. 1920. *The determination of the elastic field of an ellipsoidal inclusion and related problem* , Proc. Roy. Soc. London, A(241), pp.376-396.

Hirakawa, Y., Sugawara, K., Sato, A. and Eto, S. 2001.*Analysis of Stress Intensity Factor – Homogenization analysis of multi joint set problem by the Stress Compensation-Displacement Discontinuity Method (SC-DDM) (1st Report)-*, Shigen-to-sozai, 117(11), pp.251-257.

Kinoshita, N., Ishii, T., Kuroda, H. and Tada, H. 1992. *Prediction of permeability changes in an excavation response zone*, Nuclear Engineering and Design, 188, pp217-224.

Sato, A., Hirakawa, Y. and Sugawara, K. 2001. *Mixed mode crack propagation of homogenized cracks by the two-dimensional DDM analysis*, Construction and Building Materials, 15(5-6), pp.247-261.

Yoshida, H., Itoyama, T. and Horii, H. 1999. *Coupling analysis of deformation and flow in jointed rock mass during cavern excavation, Applied Mechanics*, J. of Applied Mechanics, JSCE, 2, pp.325-334.

CHARACTERIZING IN THE LABORATORY PERMEABILITY CHANGES INDUCED BY DEVIATORIC STRESS IN CLAYEY ROCKS

C. Coll, J. Desrues, P. Bésuelle, G. Viggiani
Laboratoire 3S, CNRS-UJF-INPG, Grenoble, France

Abstract: Hydromechanical coupling is a major issue when dealing with fracturation problems in stiff overconsolidated clays in civil, environmental and geology engineering. Clays and clayey rocks are mainly characterized by a strong hydromechanical coupling and very low permeability leading to numerous difficulties when trying to study them experimentally. For that purpose, experimental set up and special procedures to investigate the evolution of such material permeability with stress-strain behaviour in specimens undergoing triaxial tests up to failure are presented. The experimental device is composed of a high pressure triaxial cell (confining pressure, pore pressure up to 60 MPa, deviatoric stress up to 270 MPa) in order to perform permeability measurement under triaxial compression using the pulse decay method. In order to characterize precisely the strains during a triaxial path, we use a local strain measurement device composed of 7 LVDTs. It allows us to measure strains in 4 radial directions (90° from each other) and in 3 axial directions (120° from each other) in order to detect a possible loss of homogeneity in the strain field e.g. strain localization in shear bands.

1. INTRODUCTION

The study of the feasibility of storing radioactive wastes in deep geological formation includes the assessment of the efficiency of the host rock. Stress relief, due to excavations, may have consequences on the hydromechanical properties of the host geomaterial. Development and propagation of microcracks and fractures weakens its mechanical properties. Moreover these cracks can change the fluid flow in the rock. SELFRAC project comes within this framework. This project, founded by the European Community, is coordinated by EIG EURICE GIE (European Underground Research Infrastructure for Disposal of Nuclear Waste in Clay Environment, Belgium) and involves several european laboratories. It deals with experimental and numerical studies of fractures and self-healing within an excavation disturbed zone in stiff clayey geomaterials. Within this project, we are developing special techniques and procedures in order to understand and quantify the increase of the permeability, related to crack proliferation around excavations in clayey rocks. This paper is a description of our experimental set up and of procedures that we are implementing. The main objective is to perform drained triaxial tests, up to failure, and to measure during the deviatoric load, at several stage, permeability, using a transient method (pulse technique) in order to characterize the evolution of the permeability with deviatoric stress and hence with damaging. The permeability of the studied clays is very low (less than 10^{-18} m^2), therefore difficult to be measured. However, it might significantly increase if strain localization and/or fracturing occur in the coarse of deviatoric loading. Such change of permeability is likely to be abrupt and could represent an indicator of strain localization.

We describe the high pressure triaxial cell in which tests will be performed. Then we present the strain local measurements developed to identify and characterize the different mechanical stages that can be correlated to permeability evolution and in particular strain localization. The experimental procedure focused for materials having low permeability and hydromechanical coupling (i.e. effective stress has an important role) is described. Finally we present two materials involved in the SELFRAC project and with which we have to improve our experimental developments.

2. EXPERIMENTAL OBJECTIVES

In order to characterize hydromechanical coupling in stiff clays we want to observe the evolution of the permeability during several stages of the mechanical behaviour namely the elastic stage, the elasto-plastic stage, then failure and softening if they exist. In porous media, permeability is mainly linked to the porous structure (diameter, pores shape, distribution of interconnected pores) and linked to cracks (size, shape, interconnection). Along a triaxial loading path, porous structure and cracks are developing allowing permeability evolution. In this framework a lot of work has been done to study hydro-mechanical coupling in argillaceous rocks and in rocks. They highlighted permeability evolution with confining pressure, pore pressure and effective stress (Zoback and Byerlee (1975); Zhang *et al.* (2000)). Others have correlated volumetric strains and permeability on natural clay soils (Leroueil *et al.* (1990); Nagaraj *et al.* (1994)). However concerning permeability evolution with deviatoric stress on stiff clays, literature is poor. Wang J.-A. and Park (2002), Azeemuddin *et al.* (1995), Keaney *et al.* (1998), Rhett and Teufel (1992), Zhu and Wong (1997) describe some experimental results on sedimentary rocks and sandstone. Coste *et al.* (2000) present tests on MHM Argilite; they damaged specimens by increasing deviatoric stress under a high effective stress in order to avoid failure. Damaging i.e. a decrease of undrained and drained bulk modulus is accompanied by an increase of the permeability (2 orders of magnitude) and an increase of the Biot coefficient. There is no study of the permeability evolution of MHM Argilite as a function of progressing damage.

3. EXPERIMENTAL SET UP

To study hydromechanical coupling in deep stiff clays we need to supply mean stress relatively high linked to the in situ stress state. Thus, to perform triaxial tests on stiff clays with permeability measurement we use a special triaxial apparatus (high pressure cell) with two independent drainage line. We have developed also a system for local strain measurements composed of 7 LVDTs which are suitable for tests in oil (confining fluid) under high level of stress. Tests are performed on cylindrical specimens (up to 80 mm in height and 40 mm in diameter).

3.1 Triaxial apparatus

Numbers in brackets refer to figure 1. Consolidated drained tests will be performed in a high pressure cell (confining pressure up to 60 MPa) connected to two independent drainage lines (4 & 5). The first one provides a pore pressure up to 60 MPa at the top of the specimen, the second one is linked to the bottom of the specimen. Hence we can control pore pressure in an independent way at the top and at the bottom of the specimen at the same time, as to impose an hydraulic gradient. For the axial load, the cell has a vertical self-contained piston (1) which can provide a deviatoric stress up to 270 MPa. It can be controlled in displacement or in pressure. An axial force transducer (2) is located at the end of the axial piston in order to know the value of the deviatoric force without perturbations due to the friction of the piston during loading.

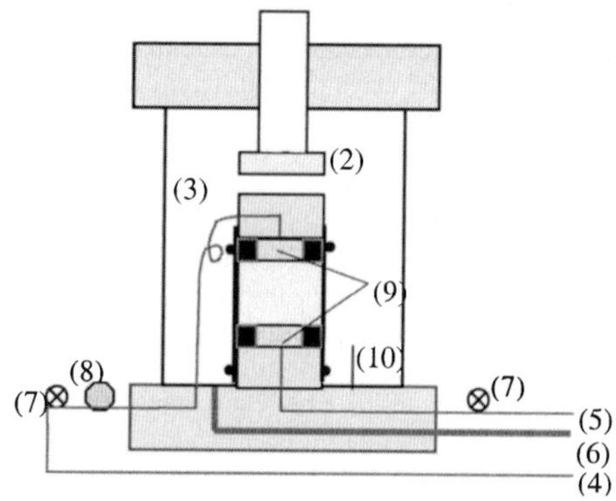

Figure 1. Schematic of the triaxial apparatus used to characterize hydromechanical coupling

The different pressures (two interstitial pressure, cell pressure and deviatoric pressure) are supplied by four identical but independent high pressure generators. A generators is, schematically, a piston moving in a pressure chamber and controlled by a motor. A pressure transducer (8) is located on the top drainage line before a valve (7) in order to measure pore pressure during undrained stages. Specimens are isolated from the confining fluid (3) using a neoprene jacket. Porous ceramics high entry air value (0.5 MPa) are used on both sides of the specimen (9) in order to avoid desaturation of the porous discs when the specimen comes in contact with it. If the specimen has an initial estimated negative pore water pressure when it is placed over the saturated ceramic disc at atmospheric pressure it will immediately tend to extract water up from the ceramic disc. The porous discs are sealed in a steel hollow and enlarged platen. The diameter of the porous disc (16 mm) is smaller than the specimen one (40 mm) to reduce friction at the specimen ends. They must be saturated under high pressure. Ceramic disc permeability has to be higher than the specimen one. Before each test the permeability of the porous ceramic is checked because this permeability could be affected by a previous test (chemical or mechanical effects). Pore fluid is prepared in the laboratory in order to have the same chemical composition than the in situ one. This is to prevent chemical reactions (swelling) between the specimen and the fluid. In addition all the experimental set up (triaxial cell, jacks) are in an enclosure in which temperature is regulated (+/- 0.1°C). A thermocouple (10) is located in the cell chamber.

3.2 Local strains measurement system

The current way to characterize the mechanical behaviour of soils or rocks consists in the study of the experimental load-displacement curve (F-d) or of the deviatoric stress-axial strain one (q-ε). Local internal measurement (in the confining cell) of specimen axial displacement have been performed in order to exclude bedding errors at both ends of the specimen in contact with porous discs. In addition we can also use an external axial measurement to measure the piston axial displacement. In order to have a better precision axial displacement are measured using three miniature LVDT transducers, mounted at 120° from each other. The transducers are glued on the neoprene membrane in order to measure axial displacement in the central part of the specimen (figure 2).

In addition, four miniature LVDTs have been performed and used (two diametrically opposite pairs) for direct measurement of local radial deformations. These transducers are in contact with the neoprene membrane at middle height of the specimen (figure 2). They are capable of measuring large strains in particular if barrelling of the specimen occurs. Average axial and radial strains are computed from these measurements.

Measuring radial and axial strains in several points of the specimen surface allows to detect with larger precision a possible loss of homogeneity of strains during loading. This is a way to detect a strain localization point on the q-ε curve and to locate shear bands or fractures in the specimen.

This technique has been previously used in our laboratory for tests on sandstone specimens (Bésuelle *et al.* (2000)). A symmetrical response of the local transducers was observed until stain localization. Figure 3 indicates why the responses of several transducers diverge from each other when a shear band appears. Among several LVDTs axial measurements one or more become predominant. This is the same principle for radial measurements; there is a divergence in the transducers response when deformations are not more homogeneous.

Moreover using this system allows to measure specimen volumetric strains and to characterize the evolution of the specimen volumetric shape.

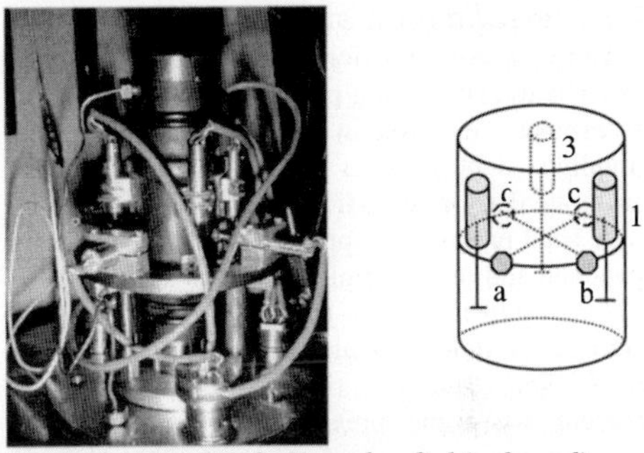

Figure 2. Axial(1, 2, 3) and radial (a, b, c, d) miniature LVDTs for local internal strain measurement.

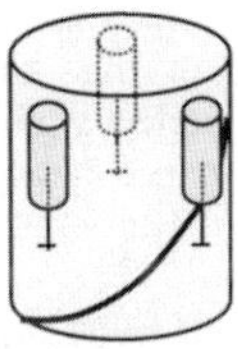

Figure 3. Shear band detection principle with local transducers. The response of LVDT n°1 is affected by the shear band crossing the specimen, whereas this is not the case for the others.

4. EXPERIMENTAL PROCEDURES

4.1 Specimens preparation

The quality of results of laboratory tests, especially for stiff clays, depends on the alteration that may have occurred during the preparation of laboratory specimens (coring, transport to the laboratory, storage, core handling...). Theoretically we want to perform tests on intact specimens, that is not so realistic in practice. Natural, saturated specimens, tested in the laboratory, undergo several mechanical disturbances. The effects of these disturbances are important for argillaceous rocks especially when the material is sampled at great depth (Tanaka *et al.* (2002)). In order to avoid desaturation during storage blocks of material to be tested are covered with wax and are wrapped in a waterproof liner. To minimise storage disturbances it is better to store them in a regulated temperature and humidity room. Cylindrical specimens can be trimmed using a whee or cored mechanically depending on the stiffness of the material.

4.2 Consolidation stage

Isotropic consolidation is the first stage of the test. The total confining pressure is increased up to the estimated effective in situ stress (existing at the sampling depth). This stage allows also to resaturate the specimen if it is slightly desaturated.

A back pressure is applied at the bottom of the specimen and the top drainage line is closed (valve (7) in figure1) while total isotropic stress is increased in order to maintain a constant effective stress. The saturation time corresponds to the moment when the pressure at the top of the specimen equals the back pressure. This time is physically the time for the pressure to diffuse it through the specimen. Theoretically, if the system is saturated there is no flow during this stage. Practically we ensured saturation measuring the Skempton coefficient which has to be near to 1. It can be noted that, in this kind of materials, saturation is never perfectly ensured. Moreover saturation phase can induce damaging and it would be of interest to quantify this effect.

4.3 Deviatoric load with permeability measurement

Specimens are loaded under drained conditions. Pore pressure is measured in order to check possible excess pore water pressure. Several times before failure, loading is stopped. Axial force is kept constant and when pressure equilibrium is installed through the specimen, permeability measurement is performed using a transient method: the pulse technique. This method has been introduced by Brace *et al.* (1968) to measure very low permeability (less than 10^{-17} m^2). The more classical steady state method is inadequate because of the very low flow induced. In addition the time to reach steady state can be very large, indicating stability problems in time (long term behaviour of the specimen, chemical effects). Theoretical data analysis of pulse method is based on the diffusion equation of pressure wave. The specimen is surrounded by two reservoirs and the system is under equilibrium (at the same water pressure). A pulse of pressure is created at one end of the specimen and the propagating wave induced is recorded at the other end. Pore pressure then reaches a new equilibrium. The shape of the wave (figure 4) allows to determine the specific storage of the material and the time to reach the equilibrium indicates the permeability (Hsieh *et al.* (1981)). Thus we can obtain the evolution of the permeability with axial strain, volumetric strain and axial stress. This procedure has been tested on concrete specimens having a low permeability. Figure 4 represents pore pressure evolution at the bottom of the specimen during permeability measurement with the pulse technique. Pressure is normalised with respect to the initial volume. Each curve refers to a different axial load. These data clearly indicate that permeability increases with deviatoric loading.

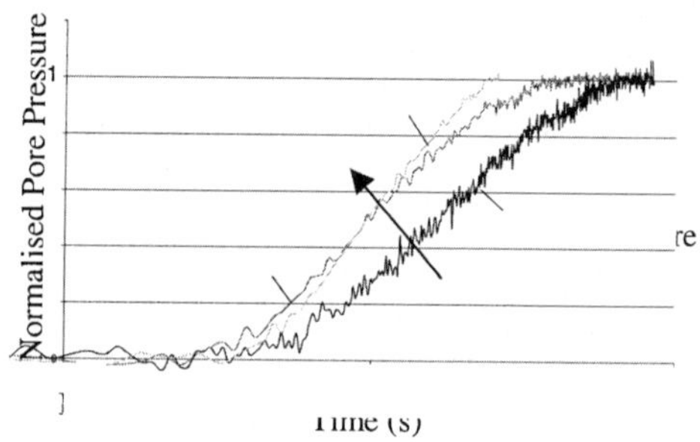

Figure 4. Pulse performed on concrete at different stages of loading.

5. MATERIALS TO BE TESTED

Two tertiary marine overconsolidated stiff clays have been selected for testing in SELFRAC project: Boom Clay (Rupelian age) and Opalinus Clay (Aalenien age). Boom Clay and Opalinus Clay both are expected to have a good potential to serve as a host rock for a repository of high radioactive waste because of their stability for a long time, of their very low permeability and because of the self-healing mechanism that can occur in these clays. Table 1 indicates the hydromechanical properties of these two materials.

5.1 Boom Clay

Boom Clay formation, which extends over a large part of North Europe, has been extensively investigated (Beer *et al.* (1977), Horseman *et al.* (1987), Baldi *et al.* (1991), Sultan (1997), Romero Morales (1999)). Specimens in this study come from Mol in Belgium where underground facilities exist. They have been taken at 223 m depth in a 100 m average thickness formation. The in situ stress state is slightly anisotropic (K_0 = 0.9).

According to the Unified Soil Classification, Boom Clay is an inorganic clay of high plasticity, a fat clay with a plastic limit w_P = 23-29 %, a liquid limit w_L = 55-80% and a plasticity index PI = 32-51%. These data have been obtained in the laboratory. The soil consists of clay minerals (60% of weight of the total solid) with kaolinite/illite (20-30% of weight clay minerals) and smectite (10-20% of weight clay minerals). The non-clayey fraction is dominated by quartz (20% of total weight) and by some other components as pyrite (<5%) and carbonates (<5%). Note in table 1 the very low permeability contrasting with a high porosity. Properties heterogeneity with depth is observed for this clay(Baldi *et al.* (1991); Romero Morales (1999)).

Table 1. Hydro-Mechanical Properties of ***Boom Clay*** (Horseman *et al.* (1987); Bernier and Van Cauteren (1998)) and ***Opalinus Clay*** (Thury *et al.* (2000); Harrington *et al.* (2001)).

HM Properties	*Boom Clay*	*Opalinus Clay*
Compressive Uniaxial Strength [MPa]	2	8
OCR	2.4	?
Natural water content [%]	19-25	8
Porosity [%]	39	5-10
Permeability (vertical) [m/s]	2.10^{-12}	8.10^{-14}
Permeability (horiz.) [m/s]	$4,5.10^{-12}$	1.10^{-13}

5.2 Opalinus Clay

Specimens of Opalinus Clay are coming from the geological formation located at the Mont-Terri facility (Switzerland) at an average depth of 500 m. This formation has been formed 180 millions years ago. The in situ stress state is strongly anisotropic (K_0 = 1-4). Opalinus clay is a clayey silt of low to medium plasticity with a plastic limit w_P = 23 %, a liquid limit w_L = 39% and a plasticity index PI = 16% measured in the laboratory. It is mainly composed of silt (60%) and clay fraction represents about 30%. Mineralogy of Opalinus Clays at Mont Terri consists of illite/kaolinite for clay minerals and quartz and calcite represent the non clayey fraction (Thury *et al.* (2000)). Opalinus clay mechanical behaviour is strongly dependent on the water content (Kharchafi and Descoeudres (1995)) and the permeability is very low. Boom Clay and Opalinus Clay will be studied in a similar way but performing tests on Opalinus clay presents larger difficulties because it is stiffer and it has a lower permeability.

6. CONCLUSIONS

An experimental set up (high pressure triaxial cell and local strain measurement system) and an experimental procedure have been developed in order to quantify the permeability evolution with deviatoric stress on clays of very low permeability. Developed techniques have been improved on concrete and sandstone. We are able to detect with precision strain localization and we are able to measure with the pulse method, permeability, at different stages of loading. A testing program on both Boom Clay and Opalinus is in progress.

REFERENCES

Azeemuddin M., Roegiers J. C., Suri P., Zaman M. and Kukreti A. R. (1995). Stress-dependant permeability measurements of rocks in a triaxial cell. 35th U.S. symp. Rock Mech.

Baldi G., Hueckel T., Peano A. and Pellegrini R. (1991). Further check and development of the thermo-hydro-geomechanical model for Boom Clay and a clay based buffer material with an emphasis on thermo-mechanics of water solid system. Bergamo, Italy, ISMES.

Bernier F. and Van Cauteren L. (1998). Instrumentation programme near the surface of an advancing tunnel in Boom Clay. 2nd Int. Symposium on Hard Soils-Soft Rocks, Naples-Italy.

Bésuelle P., Desrues J. and Raynaud S. (2000). "Experimental characterisation of the localisation phenomenon inside a Vosges sandstone in a triaxial cell." Int. J. Rock Mech. Min Sci. & Geomech. Abstr. **37**: 1223-1237.

Brace W. F., Walsh J. B. and Frangos W. T. (1968). "Permeability of granite under high pressure." Journal of Geophysical Research **73**(6): 2225.

Coste F., Bounenni A., Chanchole S. and Su K. (2000). A method for measuring mechanical, hydraulic and hydromechanical properties during damaging in materials with low permeability. Hydromechanical and thermohydromechanical behaviour of deep argillaceous rock, Paris.

Harrington J. F., Horseman S. T. and Noy D. J. (2001). Swelling and osmotic flow in a potential host rock. 6th International Workshop on Key Issues in Waste Isolation Research, Paris-France.

Horseman S. T., Winter M. G. and Entwistle D. C. (1987). Geotechnical characterization of boom clay in relation to the disposal of radiactive waste, Commission Européenne.

Hsieh P. A., Tracy J. V., Neuzil C. E., Bredehoeft J. D. and Silliman S. E. (1981). "A transient laboratory method for determining the hydraulic properties of tight rocks - I. Theory." INT. J. Rock. Mech. Min. Sci. & Geomech. Abstr. **18**(245-252).

Keaney G. M. J., Meredith P. G. and Murrel S. A. F. (1998). Laboratory study of permeability evolution in a 'tight' sandstone under non-hydrostatic stress conditions. Eurock'98, Trondheim, Norway.

Kharchafi M. and Descoeudres F. (1995). "Comportement différé des roches marneuses encaissant les tunnels." Communication du Laboratoire de Mécanique des Roches de Lauzanne **n°158**: 1.2.58-1.2.67.

Leroueil S., Bouclin G., Tavenas F., Bergeron L. and La Rochelle P. (1990). "Permeability anisotropy of natural clays as a function of strain." Canadian Geotechnical Journal **27**: 568-579.

Nagaraj T. S., Pandian N. S. and Narasimha Raju P. S. R. (1994). "Stress-state-permeability relations for overconsolidated clays." Geotechnique **44**(2): 349-352.

Rhett D. W. and Teufel L. W. (1992). Stress path dependence of matrix permeability of north sea sandstone reservoir rock. 33rd U.S. symp. Rock Mech.

Romero Morales E. (1999). Characterization and thermo-hydro-mechanical behaviour of unsaturated Boom Clay : an experimental study. Barcelone, Université polytechnique de Barcelonne.

Sultan N. (1997). Etude du comportement thermo-mécanique de l'argile de Boom: expériences et modélisation. Paris, Ecole Nationale des ponts et chaussées.

Tanaka H., Ritoh F. and Omukai N. (2002). "Quality of samples retrieved from great depth and its influence on consolidation properties." Canadian Geotechnical Journal **39**: 1288-1301.

Thury M., Boisson J. Y. and Bossar P. (2000). "Le laboratoire souterrain du Mont Terri (Suisse) et premiers résultats des études hydrogéologiques d'une formation argileuse." Hydrogeologie **2**: 13-22.

Wang J.-A. and Park H. D. (2002). "Fluid permeability of sedimentary rocks in a complete stress-strain process." Engineering geology **63**: 291-300.

Zhang M., Takahashi M., Morin R. H. and Esaki T. (2000). "Evaluation and application of the transient-pulse technique for determining the hydraulic properties of low permeability rocks - Part 1 : Theoretical Evaluation." Geotechnical Testing Journal **23**(1): 83-90.

Zhu W. and Wong T.-F. (1997). "The transition from brittle faulting to cataclastic flow: permeability evolution." J. Geophys. Res. **102**(B2): 3027-3041.

Zoback M. D. and Byerlee J. D. (1975). "The effect of microcrack dilatancy on the permeability of Westerly granite." Journal of Geophysical Research **80**(5): 752-755.

COUPLED THERMAL, HYDRAULIC AND MECHANICAL SIMULATION WITH A THEORETICAL MODEL FOR SWELLING CHARACTERISTICS

Hideo Komine[1], Hiroshi Kurikami[2], Masakazu Chijimatsu[3], Akira Kobayashi[4], Yutaka Sugita[2] and Yuzo Ohnishi[4]
[1]) Ibaraki University, Japan
[2]) Japan Nuclear Cycle Development Institute, Japan
[3]) Hazama Corporation, Japan
[4]) Kyoto University, Japan

Abstract: Buffer material as a part of an engineered barrier system for high-level waste disposal is expected to seal the initial gaps between the buffer material and the surrounding rock during the resaturation process. Since the self-sealing ability represents the swelling characteristics of the buffer material, it is very important to estimate the swelling behavior during the resaturation process. This paper expands the theoretical model for evaluating the swelling characteristics of saturated buffer material proposed by Komine to deal with unsaturated media, and it is applied to the coupled thermal, hydraulic and mechanical (THM) model used in Japan. The new model is validated by comparison with a series of laboratory experiments quoted from the literature.

1. INTRODUCTION

Buffer material for high-level waste disposal is expected as a barrier with groundwater preventive ability, self-sealing ability, adsorption ability, etc. As a buffer material, bentonite has attracted attention because its swelling characteristics are useful for sealing the initial cracks within the buffer material or the initial gaps between the buffer material and the surrounding rock.

Komine & Ogawa (1996, 2003) have proposed a theoretical model for evaluating the swelling characteristics of saturated compacted bentonite-based buffer material. This model is based on the Gouy-Chapman diffuse double layer theory and the van der Waals force. Specifying the physical and chemical properties of the buffer material, such as the density of montmorillonite, exchangeable ion capacity, etc, the swelling pressure or swelling strain of the saturated buffer material can be calculated. However, the buffer material is unsaturated during the resaturation process when the swelling takes place. Therefore, evaluation of swelling characteristics in an unsaturated condition is also important.

During the resaturation process, coupled THM behavior occurs. Although analytical and experimental studies on the coupled processes have been performed for a few decades, the swelling characteristics have not been strictly modeled.

This paper proposes a new equation to predict the swelling pressure of the buffer material during the resaturation process based on Komine's model, and introduces it to the coupled THM model proposed by Chijimatsu et al. (2000) to build a more reliable model. In this model, the swelling pressure is assumed to be a function of the degree of saturation by combining the swelling potential to the free energy of the unsaturated buffer material. The proposed model is validated by comparison with a series of laboratory experiments conducted by Chijimatsu et al. (1999) and Suzuki et al (1999).

2. SWELLING CHARACTERISTICS

2.1 Evaluation of maximum swelling pressure in saturated media

Swelling characteristics of compacted bentonite is dominated by the repulsive and the attractive forces between two parallel clay layers. Komine has introduced new ideas to the previous model to deal with the compositions and conditions of compacted bentonite (Komine & Ogata(1996, 2003)). In the new equation, it is assumed that the swelling clay minerals such as montmorillonite in bentonite dominate the swelling characteristics of the buffer material independently of the nonswelling minerals and that each exchangeable cation (Na^+, Ca^{2+}, K^+, Mg^{2+}) contributes the forces. Using the new model, not only the compositions of bentonite but also the conditions of environment such as groundwater composition, temperature, dry density can be taken into account. Table 1 shows the constants and parameters required to the model. If the coupled THM processes are considered, temperature and dry density in the environmental properties are most important because they are variables in the coupled processes. If the effects of

Table 1. Constants required

Physical constans	
Electonic charge	1.602×10^{-19} C
Boltzmann constant	1.38×10^{-23} J/K
Hamaker constant	2.2×10^{-20} J
Abogadro's number	6.023×10^{23}
Bentonite properties (bentonite Kunigel V1)	
Density of montmollironite	2.77×10^{3} kg/m^3
Density of the other minerals	2.81×10^{3} kg/m^3
Density of sand	2.66×10^{3} kg/m^3
Specific surface of montmorillonite	810
Specific surface of the other minerals	0
Content of montmorillonite	48 %
exchange capacity of Na^+	0.405 mequiv./g
exchange capacity of Ca^{2+}	0.287 mequiv./g
exchange capacity of K^+	0.009 mequiv./g
exchange capacity of Mg^{2+}	0.030 mequiv./g
Nonhydrated radius of Na^+	1
Nonhydrated radius of Ca^{2+}	2
Nonhydrated radius of K^+	1
Nonhydrated radius of Mg^{2+}	2
Thickness of montmorillonite layers	9.60×10^{-10} m
Environmental parameters	
Static permittivity of pore water, Temperature, Ion consentration, Dry density, Content of bentonite	

chemical reactions or salinity in groundwater are considered, this model is useful to build coupled thermal, hydraulic, mechanical and chemical (THMC) model.

2.2 Evaluation of process of swelling pressure in unsaturated media

During the resaturation process of the buffer material, the swelling pressure in unsaturated condition is important, though the model explained in the previous section is limited to the saturated condition. In this section, a new equation is proposed to treat swelling pressure in the unsaturated condition.

Free energy change, which consists of matric potential, osmotic potential and gravity potential, is equal to potential swelling pressure (Nakano et al.(1984)). In the case of buffer material, it is approximately equal to the suction, because the

Table 2. Definition of F against saturation degree

	Function F	Swelling Pressure
(A)	$F = \frac{1}{2}\sigma_{sw\max} S_r^{-\frac{1}{2}} \frac{\partial S_r}{\partial \psi}$	$\sigma_{sw} = \sigma_{sw\max}\sqrt{S_r}$
(B)	$F = \sigma_{sw\max} \frac{\partial S_r}{\partial \psi}$	$\sigma_{sw} = \sigma_{sw\max} S_r$
(C)	$F = 2\sigma_{sw\max} S_r \frac{\partial S_r}{\partial \psi}$	$\sigma_{sw} = \sigma_{sw\max} S_r^{2}$
(D)	$F = 3\sigma_{sw\max} S_r^{2} \frac{\partial S_r}{\partial \psi}$	$\sigma_{sw} = \sigma_{sw\max} S_r^{3}$

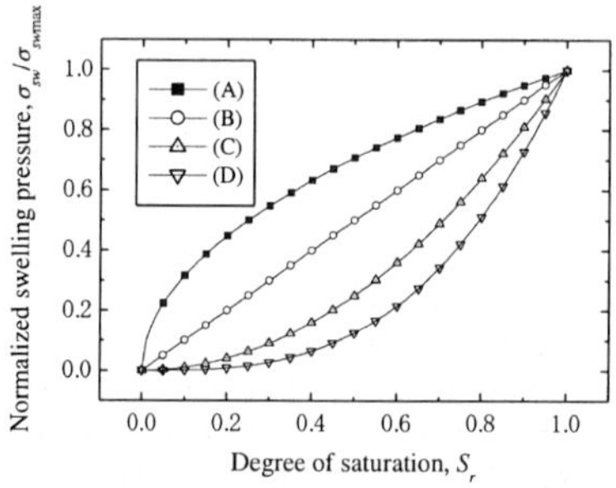

Figure 1. Swelling pressure with degree of saturation

gravity potential is negligible by comparison with the osmotic and the matric potential, where the suction is sum of the osmotic and the matric potential. Therefore, the swelling pressure can be expressed by the suction. Although the unsaturated buffer material has large suction, it is consumed during the resaturation process not only by the swelling but also by the soil structure changes. We define the function F that illustrates the ratio of the suction contributing to the swelling pressure.

$$\Delta\sigma_{sw} = -F\Delta\psi \tag{2}$$

where ψ is suction and σ_{sw} is swelling pressure. The function F must satisfy the following equation.

$$\sigma_{sw\max} = \int_0^{\infty} F d\psi \tag{3}$$

where $\sigma_{sw\max}$ is the maximum swelling pressure from Komine's model. The prospective functions for F are shown in Table 2. Where F is assumed to depend only on degree of saturation S_r. Figure 1 shows the relationship between the swelling

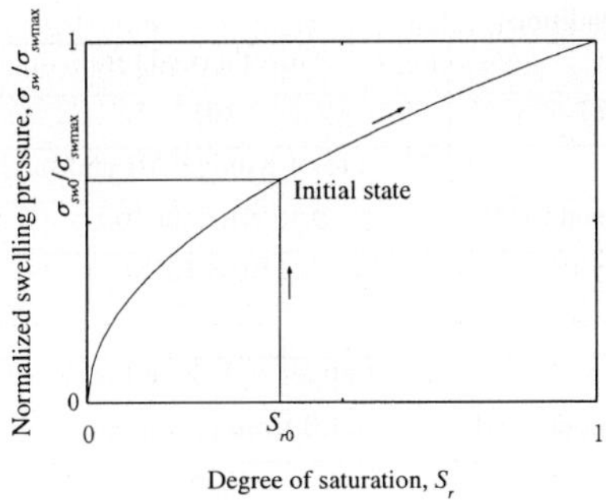

Figure 2. Swelling pressure against degree of saturation

pressure and the degree of saturation for each function given in Table 2. The model (A) represents the condition under that swelling pressure rises in the early stages of the resaturation process, while the model (D) represents the condition under that the swelling pressure cannot be expected for the self-sealing ability in the early stages. Figure 2 shows the resaturation process where initial degree of saturation is S_{r0}. When the resaturation begins, the swelling pressure rises to the initial value σ_{sw0} immediately and continues to rise with change of degree of saturation gradually.

3. COUPLED T-H-M PROCESSES IN BUFFER

To introduce the new model for the swelling characteristics to the fully coupled THM processes in the resaturation stages, the model explained above is applied to THAMES that is the finite element simulator treating the fully coupled THM process which was improved by Chijimatsu et al(2000). This chapter describes the governing equations of the coupled processes.

3.1 Conservation laws

3.1.1 Law of momentum conservation

Conservation law of momentum represents the equilibrium equations as

$$\Delta\sigma_{ij,j} = 0 \tag{4}$$

where $\Delta\sigma_{ij}$ is tensor of total stress increment.

3.1.2 Law of fluid mass conservation

Law of fluid mass conservation can be written as

$$\frac{\partial\theta\rho_w}{\partial t} + q_{i,i} + Q = 0 \tag{5}$$

where θ is volumetric water content, ρ_w is water density, t is time, q_i is groundwater flux vector and Q is sink/source.

3.1.3 Law of heat energy conservation

Law of heat energy conservation can be described as

$$\frac{\partial\{(\rho c)_m T\}}{\partial t} + q^h_{i,i} + Q^h = 0 \tag{6}$$

where $(\rho c)_m$ is specific heat of mixture of gas, liquid, and solid, T is temperature, q^h_i is heat flux vector and Q^h is heat sink/source.

3.2 Constitutive equations for bentonite

This paper explains only constitutive equations in the unsaturated bentonite because the other constrains are almost general. Assuming the elasticity of bentonite, the increment of the total stress in unsaturated media is written as follows.

$$\Delta\sigma_{ij} = \frac{1}{2}C_{ijkl}\left(\Delta u_{k,l} + \Delta u_{l,k}\right) - \Delta\sigma_{sw}\delta_{ij} - \beta\Delta T\delta_{ij} \tag{7}$$

where C_{ijkl} is elastic modulus tensor, u_i is displacement, β is the parameter regarding to thermal stress and δ_{ij} is Kronecker's delta. The second term which represents swelling pressure is calculated by the Equation (2).

Groundwater flux is expressed by

$$q_i = -\rho_w (D_\theta)_{ij}\theta_{,j} - \rho_w (D_T)_{ij} T_{,j} \tag{8}$$

where $(D_\theta)_{ij}$ is water diffusivity and $(D_T)_{ij}$ is thermal water diffusivity. D_θ is dependent of both degree of saturation and temperature and shows concave up when it is plotted against the volumetric water content. It represents that the moisture movement in the unsaturated media occurs both in liquid and gas phase, where the former is dominant in high saturation, while the latter is dominant in law saturation. D_θ of the bentonite in Japan is given by the following experimental equation, which is the combination of the moisture movement in gas and liquid phase.

$$(D_\theta)_{ij} = \delta_{ij}(D_{\theta l} + D_{\theta v}) \tag{9}$$

Table 3. Experimental conditions

Case	Isothermal tests		Anisothermal test
	(a), (b)	(c)	(d)
Material	Bentonite Kunigel V1		Mixture of Kunigel V1 and sand
Specimen size [mm]	ϕ 20 × height 20	ϕ 50 × height 100	ϕ 50 × height 100
Dry density [kg/m^3]	1.8×10^3	1.8×10^3	1.6×10^3
Initial water content [%]	0.5, 9.0	10.0	7.0
Temperature [℃]	25	25	30 at upper end, 40 at lower end
Infiltration pressure [mH_2O]	0.0 from lower end	1.0 from upper end	1.0 from upper end
Max swelling pressure [MPa]	3.7	3.7	0.35

Table 4. Material properties

		Bentonite Kunigel V1	Mixture
Dry density [kg/m^3]		1.8×10^3	1.6×10^3
Young's modulus [MPa]		1140.0-55.64w	58.74-1.87 w
Poisson ratio [-]		0.3	0.3
Intrinsic permeability K [m^2]		4.00×10^{-21}	4.00×10^{-20}
Van Genuchten parameters	θ_s [-]	0.333	0.403
	θ_r [-]	0.000	0.000
	α [1/m]	6.0×10^{-4}	8.0×10^{-3}
	n [-]	2.5	1.6
Water diffusivity [cm^2/s] $D_{\theta v} = \frac{a_1(\theta-\theta_s)}{(\theta-b_1)(b_1-\theta_s)}, D_{\theta l} = \frac{a_2\theta}{b_2(\theta-b_2)}$	a_1	$1.76 \times 10^{-8}T - 3.04 \times 10^{-7}$	$2.99 \times 10^{-8}T - 3.78 \times 10^{-7}$
	a_2	$-1.48 \times 10^{-7}T + 2.98 \times 10^{-6}$	$-1.50 \times 10^{-8}T + 1.49 \times 10^{-7}$
	b_1	-3.68×10^{-3}	-2.49×10^{-3}
	b_2	$5.22 \times 10^{-3}T + 2.68 \times 10^{-1}$	$5.59 \times 10^{-4}T + 3.93 \times 10^{-1}$
Thermal water diffusivity D_T [m^2/s/K]		2.0×10^{-12}	7.0×10^{-12}
Thermal conductivity λ_m [W/m/K]		$5.58 \times 10^{-1} + 6.17 \times 10^{-2}w - 5.28 \times 10^{-4}w^2 - 9.67 \times 10^{-6}w^3$	$4.44 \times 10^{-1} + 1.38 \times 10^{-2}w + 6.14 \times 10^{-3}w^2 - 1.69 \times 10^{-4}w^3$
Specific heat capasity c [kJ/kg/K]		(32.3+4.18w)/(100+w)	(34.1+4.18w)/(100+w)

w : water content, T : temperature [°C]

where subscript l means liquid and v means vapor, or gas phase. And the isotropy is assumed. On the other side, D_T is assumed a constant because of its difficulty to define.

Water retention curve is described by van Genuchten model shown as bellow.

$$S_r = \frac{\theta - \theta_r}{\theta_s - \theta_r} = \{1 + |\alpha\psi|^n\}^{-m} \tag{10}$$

$$m = 1 - \frac{1}{n}$$

where θ_r and θ_s is residual and maximum volumetric water content, respectively. And α, n and m are van Genuchten parameters.

Heat flux, q^h, consists of the advective term, q^a, and the dispersive term, q^d, as

$$q_i^h = q_i^a + q_i^d \tag{11}$$

where

$$q_i^a = S_r q_i T \tag{12}$$

$$q_i^d = -\lambda_m T_{,i} \tag{13}$$

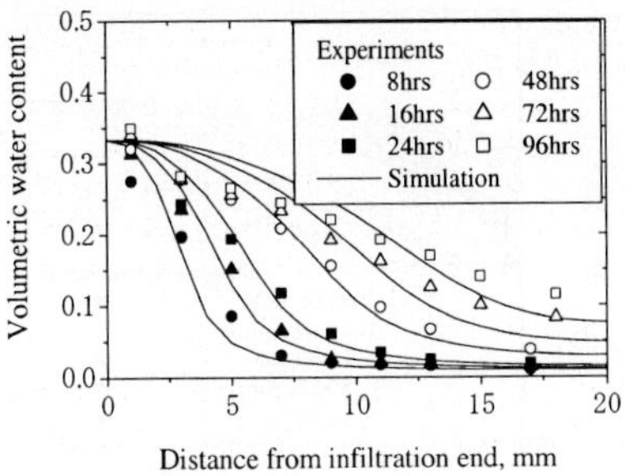

Figure 3. Distribution of volumetric water content in simulation and experiments of the test (a)

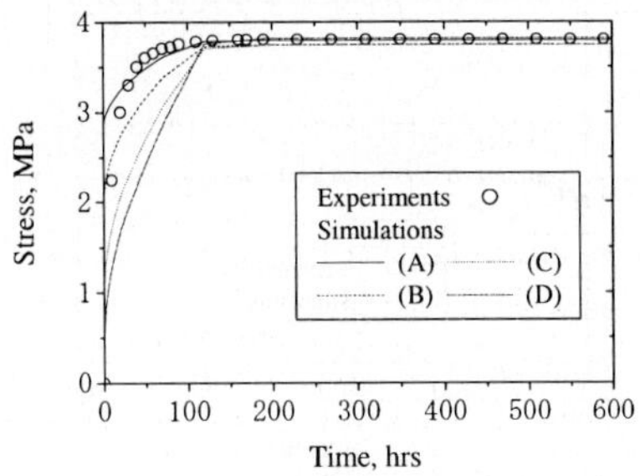

Figure 4. Time histories of stress in the test (b)

where λ_m is thermal conductivity of the mixture.

4. VALIDATION

4.1 Experimental conditions

To examine the applicability of the new model, it was applied to a series of laboratory swelling experiments conducted by Chijimatsu et al.(1999) and Suzuki et al(1999). These experiments were performed to estimate the hydraulic and the mechanical characteristics of the buffer during the resaturation process.

Table 3 and Table 4 show the experimental conditions and the material properties, respectively. The parameters shown in Table 1 are used for Komine's model. There were three tests (a), (b) and (c) in which bentonite Kunigel V1 was used on isothermal condition and one test (d) in which mixture of Kunigel V1 and sand (sand 30wt%) was used under temperature gradient condition. The object of the test (a) was the evaluation of the infiltration behavior. The distribution of the water content at each time was measured by using several specimens under the same condition, where the specimen was divided into every 2mm height and the water content of each piece was measured. On the other side, the test (b) and (c) were performed to examine the swelling pressure. The experimental conditions of the test (b) were the same as those of the test (a), though the initial water contents were different. The size of the specimen of the test (c) is larger than that of the test (a) and (b). The object of the test (d) was to estimate the swelling pressure under temperature gradient condition during the resaturation process. In the test, horizontal pressure at the upper, the middle and the lower parts and the vertical pressure at the top surface of the specimen were measured as time. Comparing this series of tests with the numerical results comprehensively, the numerical model can be validated.

4.2 Results

The height of each finite element was 1mm, which resulted in 20 elements for the test (a) and (b) and in 100 elements for the test (c) and (d). The boundary conditions are as follows. Slide boundaries were applied to all the boundaries. The constant hydraulic pressure and constant temperature were specified to the upper and/or the lower boundary following the test conditions shown in Table 3, while the other boundaries were impermeable and adiabatic. As the initial conditions, 25 °C and 30 °C were applied to the isothermal and the anisothermal tests, respectively. Initial water content was also shown in Table 3. To examine the effect of the function F proposed in this paper, the models (A) to (D) in Table 2 were applied to the simulation of the test (b). In the other tests, only the model (A) was used.

Figure 3 shows the comparison of the volumetric water content distribution of the test (a) between the experiment and the simulation. Good agreement gives the validity of the moisture movement in the unsaturated bentonite.

Figure 4 shows the time histories of the vertical stress in the test (b) corresponding to each F with the experimental results. The resaturation time of all the simulations was about 150 hours and was almost same as that of the experiment. The figure shows that the model (A) is the most suitable for F compared with the other models, though the initial stress is slightly large. It suggests that swelling pressure rises mainly in the early stages of the infiltration with respect to the compacted bentonite. That is the reason why the model (A) was applied to the other tests.

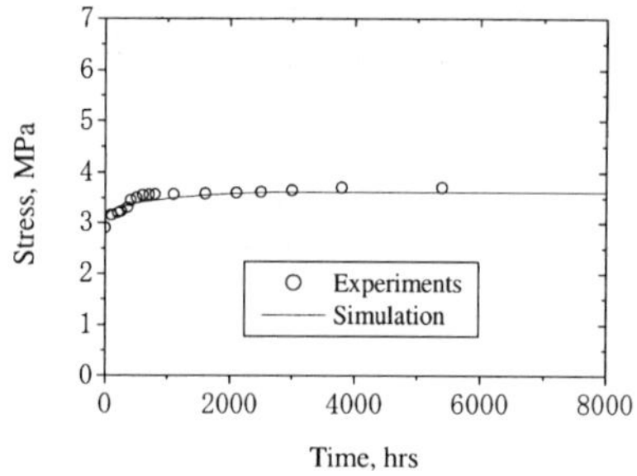

Figure 5. Time histories of stress in the test (c)

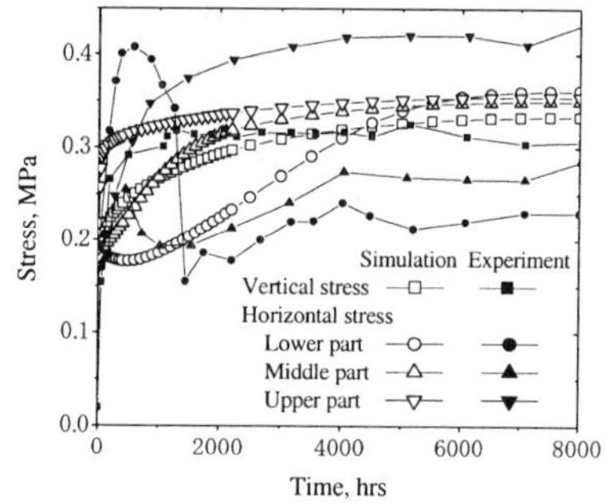

Figure 6. Time history of stress in the test (d)

Figure 5 shows the time history of the vertical stress in the test (c). As well as the test (b), good agreement between the experiment and the simulation was obtained. It means that the model proposed is useful for every scale.

Figure 6 shows the time histories of the vertical and the horizontal stress with the experimental result for the test (d). Simulated vertical stress agrees with the monitored. On the other hand, the horizontal stress was slightly different. One reason is the initial disturbance of the experiment because of the complicatedness of the apparatus. However, the most influential reason is the anisotropy of the swelling pressure. The direction of compaction of the buffer material plays an important role for the swelling behavior. The further study on the anisotropy has to be continued.

5. CONCLUSIONS

This paper proposes a new model for evaluating the swelling pressure during the resaturation process, and introduces it to the coupled thermal, hydraulic and mechanical model. Compared with the series of laboratory experiments, the following conclusions were drawn.

- Assuming that the swelling pressure is a function of the degree of saturation, good agreement between the simulation and the experiment was obtained. This indicates a one-to-one correspondence between the swelling pressure and the degree of saturation.
- Swelling pressure is likely to occur in the earlier stages of the resaturation process, and the value is strongly related to the initial degree of saturation.
- Although the proposed model cannot strictly simulate the anisotropy of swelling pressure, the results are useful.

Since the model involves the Komine's theoretical model for evaluating swelling pressure in saturation, it has the potential to include the effects of chemical change in bentonite and pore water. Using this model, fully coupled thermal, hydraulic, mechanical and chemical analysis can be conducted in the future.

REFERENCES

Chijimatsu, M. & Taniguchi, W. 1999. *Coupled thermal, hydraulic and mechanical analysis in the near field for geological disposal of high-level radioactive waste (in Japanese).* JNC TN8400 99-014.

Chijimatsu, M., Fujita, T., Kobayashi, A. & Nakano, M. 2000. *Experiment and validatioin of numerical simulation of coupled thermal, hydraulic and mechanical behaviour in the engineered buffer materials.* Int. J. Num. Anal. Mech. Geomech. 24: pp.403-424.

Komine, H. & Ogata, N. 1996. *Prediction for swelling characteristics of compacted bentonite.* Can. Geotech. J. 33: pp.11-22.

Komine, H. & Ogata, N. 2003. *New equations for swelling characteristics of bentonite-based buffer materials*, Can. Geotech. J. 40(2): pp.460-475

Nakano, M., Amemiya, Y., Fujii, K. & Ishida, A. 1984. *Water Penetration and Swelling Pressure of Restrained Unsaturated Clay(in Japanese).* JSIDRE 112: pp.55-66.

Suzuki, H., Chijimatsu, M. & Fujita, T. 1999. *Measurement of moisture movement and swelling pressure under temperature gradient (in Japanese).* JNC TN8400 99-020.

ON THE CONSTITUTIVE MODELLING OF THERMO-HYDRO-MECHANICAL COUPLING IN ELASTIC MEDIA WITH DOUBLE POROSITY

N. Khalili[1], A.P.S. Selvadurai[2]
[1]) School of Civil and Environmental Engineering,
The University of New South Wales, Sydney 2052, Australia
[2]) Department of Civil Engineering and Applied Mechanics,
McGill University, Montreal, QC, Canada, H3A 2K6,

Abstract: The partial differential equations governing non-isothermal flow and deformation in an elastic medium with double porosity are presented. The governing equations satisfy an effective stress concept, Darcy's law, Fourier' law and the equations of Hookean thermoelasticity. The equations are derived using a systematic macroscopic approach that satisfies conservation laws applicable to balance of linear momentum, mass and energy. The Thermo-Hydro-Mechanical coupling take into account processes associated with: thermal expansion, thermal convection by moving fluid, fluid flux due to temperature gradients, heat flux due to pressure gradients, fluid and heat exchange between the two pore system, and the heat of phase compression.

1. INTRODUCTION

The thermo-hydro-mechanical (THM) coupling, influencing heat and mass transport in deformable porous media has received increasing attention in various branches of science and engineering, primarily due to its importance to processes such as secondary enhancement of oil production, storage of thermal energy in the form of sensible and latent heat, remediation of contaminated sites, isolation of hazardous waste, etc. THM processes are also of interest to applications related to insulation materials, heat emitting buried cable systems, tumour treatment, and in the cooling of microelectronic components. Constitutive models have been developed over the past two decades in relation to topics such as coupled heat and moisture transfer in agricultural engineering and in the study of THM processes in natural and engineered barriers for the storage of heat emitting nuclear fuel waste. In spite of their importance in the fields of petroleum engineering, reservoir engineering, and nuclear waste management, THM coupling effects in media with double porosity have rarely been investigated (Bai & Roegiers 1994).

In this paper, the governing equations for THM processes in an elastic medium with double porosity are derived using a systematic macroscopic approach, that satisfies appropriate constitutive and conservation laws. The comprehensive array of THM processes taken into account, include: thermo-mechanical deformations, fluid transport, heat transfer, fluid flux due to temperature gradients, heat flux due to pressure gradients, fluid and heat exchange between the double pore system, heat transfer due to fluid movement, and the presence of heat of phase compression. Three phases (solid, pore fluid and fissure fluid) and two constituents (solid and liquid) are identified. Each phase is viewed as an independent continuum, endowed with its own kinematics, mass, momentum and energy. Both solid and fluid constituents are assumed slightly compressible, but if required, either can be made incompressible. Matrix displacement vector, pore fluid pressure, fissure fluid pressure, solid temperature, pore fluid temperature and fissure fluid temperature are introduced as the primary variables.

2. THE CONCEPTUAL MODEL

The formulation presented consists of three separate, yet inter-connected, models involving 1) deformation, 2) flow model, and 3) heat transfer. The deformation model is based on the classical theory of thermoelasticity, the effective stress concept and the equations of equilibrium. The fluid flow through the medium is modelled using the double porosity concept. Two interacting continua are identified: one representing the porous blocks and the other the fissure network. The two flow regions are connected through a leakage term, describing transfer of fluid from the porous blocks into the fissure network, and vice versa. The rate of fluid transfer at any point depends upon the fluid pressure difference between the porous blocks and the fissure network, as well as the permeability of the porous blocks and the fissure spacing. The coupling between the flow and the deformation fields is established through the effective stress parameters, ζ_1 and ζ_2, which relate pore and fissure fluid pressures to matrix deformation vector. The heat transfer model, is based on the assumption of thermal non-equilibrium between the phases. At any material point, three temperatures are introduced; one each for the solid, the pore fluid and the fissure fluid, phases. The modes of heat transfer are assumed to be: conduction, advection, and heat exchange. The thermal interaction between the phases is established through heat exchange terms. The coupling between the heat transfer model and the flow and the deformation models is established through thermal expansion coefficients and the heat of phase compression. In the case of the flow model, an additional coupling is also established through the thermal energy transfer due to the leakage term.

3. THE DEFORMATION MODEL

To formulate the deformation model, we define the effective stress in the solid skeleton of a doubly porous medium, σ'_{ij}, as (Khalili & Valliappan 1996),

$$\sigma_{ij} = \sigma'_{ij} - \zeta_1 p_1 \delta_{ij} - \zeta_2 p_2 \delta_{ij} \; ; \quad i, j = 1,2,3 \tag{1}$$

where σ_{ij} is the total stress and p_α is the fluid pressure (positive for compressive pressure and negative for tensile or suction), with subscripts $\alpha = 1$ and $\alpha = 2$ referring to the porous blocks and fissure network, respectively, and, δ_{ij} is the Kronecker's delta function. The effective stress parameters, ζ_1 and ζ_2, are defined as (Khalili & Valliappan 1996; Loret & Rizzi 1999; Callari & Federico 2000; Pao & Lewis 2002),

$$\zeta_1 = (c_p - c_s)/c \quad \text{and} \quad \zeta_2 = (c - c_p)/c \tag{2}$$

where c_s is the elastic compressibility of solid constituent, c is the underlying elastic drained compressibility of the fissured porous media, and c_p is the elastic drained compressibility of the porous blocks. Assuming, the compressibility of solid grains to be negligible, Equation 2 yields $\zeta_1 + \zeta_2 = 1$. Notice that in the limiting case of zero fissure volume we have $c_p = c$ thus, $\zeta_1 = 1$ and $\zeta_2 = 0$, and Equation 2 reduces to that proposed by Terzaghi for saturated elastic medium with a single porosity. The Duhamel-Neumann extension to Hooke's law gives the thermoelastic constitutive relationship as,

$$\sigma'_{ij} = 2G\varepsilon_{ij} + (\lambda\varepsilon_{kk} - \gamma T_s)\delta_{ij} \tag{3}$$

where G and λ are the Lamé constants, T_s is the temperature of the solid constituent, and $\gamma\delta_{ij} = -(\partial\sigma'_{ij}/\partial T_s)\big|_{\varepsilon_{kk}}$ and γ is related to the volumetric thermal expansion of the fissured porous medium, c_T, through $\gamma = (c_T/c)$, with $c = (2G/3 + \lambda)^{-1}$. The strain ε_{ij} is the related to the displacement components, u_i, through

$$\varepsilon_{ij} = \frac{1}{2}\left(\frac{\partial u_i}{\partial x_j} + \frac{\partial u_j}{\partial x_i}\right) \tag{4}$$

Considering the equations of equilibrium,

$$\frac{\partial \sigma_{ij}}{\partial x_j} + F_i = 0 \tag{5}$$

and combining it with Equations 1 to 4 yields,

$$\frac{\partial}{\partial x_j}[G\frac{\partial u_i}{\partial x_j} + (\lambda + G)\frac{\partial u_j}{\partial x_i} - (\zeta_1 p_1 + \zeta_2 p_2 + \gamma T_s)\delta_{ij}] + F_i = 0 \tag{6}$$

in which F_i is i^{th} component of the body force per unit volume. The set of partial differential Equations 6, for govern the deformation phenomenon in a thermoelastic double porous medium.

4. THE FLOW MODEL

The flow model is obtained by combining the equations of linear momentum and mass balance for the liquid phase. Neglecting inertial effects, the equation of linear momentum balance can be written as,

$$v^r_{\alpha i} = -\frac{k_{\alpha ij}}{\mu_\alpha}\left(\frac{\partial p_\alpha}{\partial x_j} + \rho_\alpha g_j\right) - k_{\alpha Tij}\frac{\partial T_\alpha}{\partial x_j} \tag{7}$$

where $k_{\alpha ij}$ is the intrinsic permeability and $k_{\alpha Tij}$ is the coefficient of thermal coupling for fluid flux, analogous to the Soret effect in molecular diffusion, in which mass flux is caused due to both a concentration gradient and a temperature gradient. Also, μ_α and ρ_α are the dynamic viscosity and the density of fluid respectively, g_i is the ith component of gravitational acceleration, and $v^r_{\alpha i}$ is the relative velocity of the fluid defined as,

$$v^r_{\alpha i} = n_\alpha (v_{\alpha i} - v_{si}) \tag{8}$$

where n_α is the porosity, $v_{\alpha i}$ is the absolute fluid velocity and v_{si} is the solid skeleton velocity,

$$v_{\alpha i} = \frac{\partial u_{\alpha i}}{\partial t} \quad \text{and} \quad v_{si} = \frac{\partial u_i}{\partial t} \tag{9}$$

where $u_{\alpha i}$ and u_i are the components of displacements of fluid and solid, respectively.

For conservation of fluid mass, we require

$$-\frac{\partial}{\partial x_i}(\rho_\alpha n_\alpha v_{\alpha i}) = \frac{\partial}{\partial t}(n_\alpha \rho_\alpha) + (-1)^{\alpha+1} \rho_\alpha \Gamma \tag{10}$$

where Γ is the leakage term controlling the rate of fluid transfer between the fissure network and the porous blocks, defined as,

$$\Gamma = \eta (p_1 - p_2) \tag{11}$$

in which η is the leakage parameter. Substituting Equation 8 into 10, we obtain

$$-\frac{\partial}{\partial x_i}(\rho_\alpha v^r_{\alpha i}) - \frac{\partial}{\partial x_i}(\rho_\alpha n_\alpha v_{si}) = \frac{\partial}{\partial t}(\rho_\alpha n_\alpha) + (-1)^{\alpha+1} \rho_\alpha \Gamma \tag{12}$$

and introducing the Lagrangian total derivative concept with respect to the moving solids, $d_s(.)/dt = \partial(.)/\partial t + v_{si}\,\partial(.)/\partial x_i$, and the moving fluid, $d_l(.)/dt = \partial(.)/\partial t + v_{li}\,\partial(.)/\partial x_i$, Equation 12 can be rearranged to,

$$-\rho_\alpha \frac{\partial}{\partial x_i}(v^r_{\alpha i}) = n_\alpha \frac{d_\alpha \rho_\alpha}{dt} + \rho_\alpha \frac{d_s n_\alpha}{dt} + \rho_\alpha n_\alpha \frac{\partial v_{si}}{\partial x_i} + (-1)^{\alpha+1} \rho_\alpha \Gamma \tag{13}$$

Considering a barotropic fluid $\rho_\alpha = \rho_\alpha(p_\alpha, T_\alpha)$,

$$\frac{d_\alpha \rho_\alpha}{dt} = \rho_\alpha c_\alpha \frac{d_\alpha p_\alpha}{dt} - \rho_\alpha c_{\alpha T} \frac{d_\alpha T_\alpha}{dt} \tag{14}$$

in which $c_\alpha = \frac{1}{\rho_\alpha}\frac{\partial \rho_\alpha}{\partial p_\alpha}\Big|_{T_\alpha}$ and $c_{\alpha T} = -\frac{1}{\rho_\alpha}\frac{\partial \rho_\alpha}{\partial T_\alpha}\Big|_{p_\alpha}$ are the coefficients of compressibility and thermal expansion for fluid, respectively. Noting $n_\alpha = \frac{V_\alpha}{V}$,

$$\frac{d_s n_\alpha}{dt} = \frac{1}{V}\left(\frac{d_s V_\alpha}{dt} - n_\alpha \frac{d_s V}{dt}\right) \tag{15}$$

and assigning $\alpha = 1,2$, Equations 7, 13, 14 and 15 can be combined to give

$$\frac{\partial}{\partial x_i}\left[\frac{k_{1ij}}{\mu_1}\left(\frac{\partial p_1}{\partial x_j} + \rho_1 g_j\right) + k_{1Tij}\frac{\partial T_1}{\partial x_j}\right] = n_1 c_1 \frac{d_1 p_1}{dt} + \frac{1}{V}\frac{d_s V_1}{dt} - n_1 c_{1T}\frac{d_1 T_1}{dt} + \Gamma \tag{16a}$$

$$\frac{\partial}{\partial x_i}\left[\frac{k_{2ij}}{\mu_2}\left(\frac{\partial p_2}{\partial x_j} + \rho_2 g_j\right) + k_{2Tij}\frac{\partial T_2}{\partial x_j}\right] = n_2 c_2 \frac{d_2 p_2}{dt} + \frac{1}{V}\frac{d_s V_2}{dt} - n_2 c_{2T}\frac{d_2 T_2}{dt} - \Gamma \tag{16b}$$

In Equations 16, $\frac{d_s V_1}{V}$ and $\frac{d_s V_2}{V}$ represent the change in the pore and fissure volume over the current volume of the porous medium with respect to the moving solid. For isothermal conditions, $\frac{d_s V_1}{V}$ and $\frac{d_s V_2}{V}$ can be calculated from (Khalili and Valliappan 1996),

$$\frac{d_s V_1}{V} = (c_p - c_s) d_s \frac{\sigma_{ii}}{3} + [(1 - n_2)c_p - (1 + n_1 - n_2)c_s] d_s p_1 + n_2 (c_p - c_s) d_s p_2 \tag{17a}$$

$$\frac{d_s V_2}{V} = (c - c_p) d_s \frac{\sigma_{ii}}{3} + n_2 (c_p - c_s) d_s p_1 + [c - (1 + n_2) c_p] d_s p_2 \tag{17b}$$

Extending Equations 17 to include thermal effects, we have,

$$\frac{d_s V_1}{V} = (c_p - c_s) d_s \frac{\sigma_{ii}}{3} + [(1 - n_2) c_p - [(1 + n_1 - n_2) c_s] d_s p_1 + n_2 (c_p - c_s) d_s p_2 + n_1 c_T d_s T_s \tag{18a}$$

$$\frac{d_s V_2}{V} = (c - c_p) d_s \frac{\sigma_{ii}}{3} + n_2 (c_p - c_s) d_s p_1 + [c - (1 + n_2) c_p] d_s p_2 + n_2 c_T d_s T_s \tag{18b}$$

The last terms on the left hand side of 18a and 18b represent the change in the pore and fissure volume due to a change in the volume of the fissured porous medium with solid temperature, respectively.

Now, using Equations 1 to 3, Equation 18 can be rearranged to,

$$\frac{d_s V_1}{V} = \zeta_1 d_s \varepsilon_{ii} + [(\zeta_2 - n_2)(c_p - c_s) + (\zeta_1 - n_1) c_s] d_s p_1 - (\zeta_2 - n_2)(c_p - c_s) d_s p_2 - (\zeta_1 - n_1) c_T d_s T_s \tag{19a}$$

$$\frac{d_s V_2}{V} = \zeta_2 d_s \varepsilon_{ii} + (\zeta_2 - n_2) c_p d_s p_2 - (\zeta_2 - n_2)(c_p - c_s) d_s p_1 - (\zeta_2 - n_2) c_T d_s T_s \tag{19b}$$

Substituting (19) into (16), noting that $\frac{1}{V}\frac{d_s V}{dt} = \frac{\partial v_{si}}{\partial x_i}$, and introducing the approximations $v_{\alpha i}\frac{\partial p_\alpha}{\partial x_i} << \frac{\partial p_\alpha}{\partial t}$, $v_{si}\frac{\partial (.)}{\partial x_i} << \frac{\partial (.)}{\partial t}$ or $\frac{d_\alpha p_\alpha}{dt} \approx \frac{\partial p_\alpha}{\partial t}$, $\frac{d_s (.)}{dt} \approx \frac{\partial (.)}{\partial t}$, yield the differential equations governing non-isothermal fluid flow through fissured porous media,

$$\frac{\partial}{\partial x_i}\left[\frac{k_{1ij}}{\mu_1}\left(\frac{\partial p_1}{\partial x_j} + \rho_1 g_j\right) + k_{1Tij}\frac{\partial T_1}{\partial x_j}\right] = [n_1 c_1 + (\zeta_2 - n_2)(c_p - c_s) + (\zeta_1 - n_1) c_s]\frac{\partial p_1}{\partial t} - (\zeta_2 - n_2)(c_p - c_s)\frac{\partial p_2}{\partial t} + \zeta_1 \frac{\partial^2 u_i}{\partial t \partial x_i} - n_1 c_{1T}\frac{\partial T_1}{\partial t} - n_1 c_{1T} v_{1i}\frac{\partial T_1}{\partial x_i} - (\zeta_1 - n_1) c_T \frac{\partial T_s}{\partial t} + \Gamma \tag{20a}$$

$$\frac{\partial}{\partial x_i}\left[\frac{k_{2ij}}{\mu_2}\left(\frac{\partial p_2}{\partial x_j} + \rho_2 g_j\right) + k_{2Tij}\frac{\partial T_2}{\partial x_j}\right] = [n_2 c_2 + (\zeta_2 - n_2) c_p]\frac{\partial p_2}{\partial t} - (\zeta_2 - n_2)(c_p - c_s)\frac{\partial p_1}{\partial t} + \zeta_2 \frac{\partial^2 u_i}{\partial t \partial x_i} - n_2 c_{2T}\frac{\partial T_2}{\partial t} - n_2 c_{2T} v_{1i}\frac{\partial T_2}{\partial x_i} - (\zeta_2 - n_2) c_T \frac{\partial T_s}{\partial t} - \Gamma \tag{20b}$$

5. THE HEAT TRANSFER MODEL

Assuming the pore-fluid, fissure-fluid, and solid temperatures are not in local equilibrium, the energy balance equation for the solid phase can be written as,

$$\frac{\partial}{\partial x_i}(n_s \lambda_{sij}\frac{\partial T_s}{\partial x_j}) = n_s \rho_s C_s (\frac{\partial T_s}{\partial t} + v_{si}\frac{\partial T_s}{\partial x_i}) - T_s(\zeta_1 - n_1) c_T \frac{\partial p_1}{\partial t} - T_s(\zeta_2 - n_2) c_T \frac{\partial p_2}{\partial t} + T_s \gamma \frac{\partial^2 u_i}{\partial t \partial x_i} + \kappa_{s1}(T_s - T_1) + \kappa_{s2}(T_s - T_2) \tag{21}$$

in which λ_{sij} is the thermal conductivity of solid phase, C_s is the heat capacity of the solid, κ_{s1} is the coefficient of heat exchange between the solid and the pore-fluid, and κ_{s2} is the coefficient of heat exchange between the solid and the fissure-fluid. The Equation (21) accounts for the flux of thermal energy due conduction and convection, the change in the internal energy of the solid phase due to a change in pore and fluid pressures, change in the internal energy due to volume change of the solid skeleton, and the exchange of energy between the solid phase and the pore-fluid and between the solid phase and the fissure-fluid. The energy balance equation for the pore-fluid is in turn written as,

$$\frac{\partial}{\partial x_i}(n_1 \lambda_{1ij}\frac{\partial T_1}{\partial x_i}) = C_1(\frac{\partial n_1 \rho_1 T_1}{\partial t} + \frac{\partial v_{1i} n_1 \rho_1 T_1}{\partial x_i}) + n_1 T_1 \beta_1 \left.\frac{d_1 \varepsilon_1}{dt}\right|_{T_1} + C_1 T_1 \rho_1 < \Gamma > - C_2 T_2 \rho_2 < -\Gamma > + \kappa_{1s}(T_1 - T_s) + \kappa_{12}(T_1 - T_2) \tag{22}$$

with

$$\left.\frac{d_1\varepsilon_1}{dt}\right|_{T_1} = -\frac{1}{\rho_1}\left.\frac{d_1\rho_1}{dt}\right|_{T_1} = -c_1\frac{d_1 p_1}{dt} \quad (23)$$

in which λ_{1ij} and C_1 are the thermal conductivity and the heat capacity of the pore-fluid, respectively. κ_{12} is the coefficient of heat exchange between the pore-fluid and the fissure-fluid, and $\beta_1 = \partial p_1/\partial T_1|_{\varepsilon_1} = c_{1T}/c_1$. In Equation 22, < > refers to the Clebsch-Macaulay singularity function. The third term on the right hand side represents the change in the internal energy due to change in the volume of pore-fluid, the fourth term represents the convective heat transfer due to fluid exchange between the porous blocks and the fissure network. Noting that

$$\frac{\partial n_1\rho_1 T_1}{\partial t} + \frac{\partial v_{1i} n_1\rho_1 T_1}{\partial x_i} = T_1\Big(\frac{\partial n_1\rho_1}{\partial t} + \frac{\partial v_{1i} n_1 \rho_1}{\partial x_i}\Big) + n_1\rho_1\Big(\frac{\partial T_1}{\partial t} + v_{1i}\frac{\partial T_1}{\partial x_i}\Big) \quad (24)$$

and using Equation 23, Equation 22 can be written as,

$$\frac{\partial}{\partial x_i}\Big(n_1\lambda_{1ij}\frac{\partial T_1}{\partial x_i}\Big) = n_1 C_1\rho_1\frac{\partial T_1}{\partial t} - n_1 c_{1T} T_1\frac{\partial p_1}{\partial t} + C_1 T_1\rho_1 <\Gamma> - C_2 T_2\rho_2 <-\Gamma> + \kappa_{s1}(T_1 - T_s) + \kappa_{12}(T_1 - T_2) + n_1 C_1\rho_1 v_{1i}\frac{\partial T_1}{\partial x_i} \quad (25)$$

Similarly, the energy balance equation for the fissure-fluid is written as,

$$\frac{\partial}{\partial x_i}\Big(n_2\lambda_{2ij}\frac{\partial T_2}{\partial x_i}\Big) = n_2 C_2\rho_2\frac{\partial T_2}{\partial t} - n_2 c_{2T} T_2\frac{\partial p_2}{\partial t} - C_1 T_1\rho_1 <\Gamma> + C_2 T_2\rho_2 <-\Gamma> + \kappa_{s2}(T_2 - T_s) + \kappa_{12}(T_2 - T_1) + n_2 C_2\rho_2 v_{2i}\frac{\partial T_2}{\partial x_i} \quad (26)$$

in which λ_{2ij} and C_2 are the thermal conductivity and the heat capacity of the fissure-fluid, respectively. Notice that in Equations 23 to 26 the term $C_m T_m$ could be replaced by the enthalpy, h_m.

6. A SUMMARY OF THE GOVERNING EQUATIONS

In summary, the partial differential equations describing the coupled THM processes in an elastic medium with double porosity can be written as,

Deformation

$$G\frac{\partial^2 u_i}{\partial x_j\partial x_j} + (\lambda + G)\frac{\partial^2 u_j}{\partial x_i\partial x_j} - \zeta_1\frac{\partial p_1}{\partial x_i} - \zeta_2\frac{\partial p_2}{\partial x_i} - a_{\varepsilon T_s}\frac{\partial T_s}{\partial x_i} + F_i = 0 \quad (27a)$$

Fluid Flow

$$\frac{\partial}{\partial x_i}\frac{k_{1ij}}{\mu_1}\left(\frac{\partial p_1}{\partial x_j} + \rho_1 g_j\right) + \frac{\partial}{\partial x_i}k_{1Tij}\frac{\partial T_1}{\partial x_j} = a_{11}\frac{\partial p_1}{\partial t} - a_{12}\frac{\partial p_2}{\partial t} + \zeta_1\frac{\partial^2 u_i}{\partial t\partial x_i} - a_{1T_s}\frac{\partial T_s}{\partial t} - a_{1T_1}\frac{\partial T_1}{\partial t} - n_1 c_{1T} v_{1i}\frac{\partial T_1}{\partial x_i} + \Gamma \quad (27b)$$

$$\frac{k_{2ij}}{\mu_2}\left(\frac{\partial p_2}{\partial x_j} + \rho_2 g_j\right) + \frac{\partial}{\partial x_i}k_{2Tij}\frac{\partial T_2}{\partial x_j} = -a_{21}\frac{\partial p_1}{\partial t} + a_{22}\frac{\partial p_2}{\partial t} + \zeta_2\frac{\partial^2 u_i}{\partial t\partial x_i} - a_{2T_s}\frac{\partial T_s}{\partial t} - a_{2T_2}\frac{\partial T_2}{\partial t} - n_2 c_{2T} v_{1i}\frac{\partial T_2}{\partial x_i} - \Gamma \quad (27c)$$

Heat Transport

$$\frac{1}{T_s}\frac{\partial}{\partial x_i}\Big(n_s\lambda_{sij}\frac{\partial T_s}{\partial x_j}\Big) = a_{T_sT_s}\frac{\partial T_s}{dt} - a_{T_s1}\frac{\partial p_1}{\partial t} - a_{T_s2}\frac{\partial p_2}{\partial t} + a_{T_s\varepsilon}\frac{\partial^2 u_i}{\partial t\partial x_i} + \frac{\kappa_{s1}}{T_s}(T_s - T_1) + \frac{\kappa_{s2}}{T_s}(T_s - T_2) + a_{T_sT_s} v_{si}\frac{\partial T_s}{\partial x_i} \quad (27d)$$

$$\frac{1}{T_1}\frac{\partial}{\partial x_i}\Big(n_1\lambda_{1ij}\frac{\partial T_1}{\partial x_i}\Big) = a_{T_1T_1}\frac{\partial T_1}{\partial t} - a_{T_11}\frac{\partial p_1}{\partial t} + C_1 T_1\rho_1 <\Gamma> - C_2 T_2\rho_2 <-\Gamma> + \frac{\kappa_{1s}}{T_1}(T_1 - T_s) + \frac{\kappa_{12}}{T_1}(T_1 - T_2) + a_{T_1T_1} v_{1i}\frac{\partial T_1}{\partial x_i} \quad (27e)$$

$$\frac{1}{T_2}\frac{\partial}{\partial x_i}(n_2\lambda_{2ij}\frac{\partial T_2}{\partial x_i}) = a_{T_2T_2}\frac{\partial T_2}{\partial t} - a_{T_22}\frac{\partial p_2}{\partial t} -$$

$$C_1T_1\rho_1 < \Gamma > + C_2T_2\rho_2 < -\Gamma > + \tag{27f}$$

$$\frac{\kappa_{2s}}{T_2}(T_2 - T_s) + \frac{\kappa_{21}}{T_2}(T_2 - T_1) + a_{T_2T_2}v_{2i}\frac{\partial T_1}{\partial x_i}$$

subject to the constitutive constraints,

$$\frac{d\sigma_{ii}}{3} = c^{-1}d\varepsilon_{kk} - \zeta_1 dp_1 - \zeta_2 dp_2 - a_{\varepsilon T_s}dT_s$$

$$\frac{dm_1}{\rho_{1o}} = \zeta_1 d\varepsilon_{ii} + a_{11}dp_1 - a_{12}dp_2 - a_{1T_s}dT_s - a_{1T_1}dT_1$$

$$\frac{dm_2}{\rho_{2o}} = \zeta_2 d\varepsilon_{ii} - a_{21}dp_1 + a_{22}dp_2 - a_{2T_s}dT_s - a_{2T_2}dT_2$$

$$ds_s = a_{T_s\varepsilon}d\varepsilon_{ii} - a_{T_s1}dp_1 - a_{T_s2}dp_2 + a_{T_sT_s}dT_s \tag{28}$$

$$ds_1 = -a_{T_11}dp_1 + a_{T_1T_1}dT_1$$

$$ds_2 = -a_{T_22}dp_2 + a_{T_2T_2}dT_2\ ;\ \Gamma = \eta(p_1 - p_2)$$

with

$$\zeta_1 = \frac{c_p}{c} - \frac{c_s}{c}, \qquad \zeta_2 = 1 - \frac{c_p}{c}$$

$$a_{11} = n_1c_1 + a_{12} + (\zeta_1 - n_1)c_s,$$

$$a_{22} = n_2c_2 + a_{12} + (\zeta_2 - n_2)c_s,$$

$$a_{12} = a_{21} = (\zeta_2 - n_2)(c_p - c_s),$$

$$a_{\varepsilon T_s} = a_{T_s\varepsilon} = \gamma,\ a_{1T_1} = a_{T_11} = n_1c_{1T}, \tag{29}$$

$$a_{2T_2} = a_{T_22} = n_2c_{2T},\ a_{1T_s} = a_{T_s1} = (\zeta_1 - n_1)c_T,$$

$$a_{2T_s} = a_{T_s2} = (\zeta_2 - n_2)c_T,\quad a_{T_sT_s} = \frac{n_s\rho_sC_s}{T_s},$$

$$a_{T_1T_{1l}} = \frac{n_1\rho_1C_1}{T_1},\quad a_{T_2T_2} = \frac{n_2C_2\rho_2}{T_2}$$

where m_1 and m_2 are the mass pore-fluid and mass fissure fluid per reference volume, respectively. ρ_{1o} and ρ_{2o} are the densities of the pore and fissure fluid at the reference configuration, respectively. Similarly, s_s, s_1, and s_2 are the entropies of solid, pore fluid and fissure fluid, respectively. The constitutive equations in (28) enjoy the major symmetry due to the reversibility of the physical processes taken into account. This result can also be derived using Biot's free energy approach, within the general framework of the mixture theory.

7. CONCLUSIONS

It is shown that the development of the equations governing THM processes in elastic media with double porosity can be approached in a systematic manner where all the constitutive equations governing deformability, fluid flow and heat transfer are combined with the relevant conservation laws. The double porosity nature of the medium requires the introduction of dependent variables applicable to the deformable solid, and the fluid phases in the two void spaces. The governing partial differential equations are linear in view of the linearized forms of the constitutive assumptions invoked in the formulations. The linearity of these governing equations makes them amenable to solution through conventional mathematical techniques applicable to the study of initial boundary value problems in mathematical physics (Selvadurai, 2000). Such solutions should serve as benchmarks for appropriate computational developments.

REFERENCES

Bai, M. and Roegiers, J.C. 1994. *Fluid flow and heat flow in deformable fractured porous media.* Int. J. Engng Sci. 32(10): pp. 1615-1633.

Callari C. & Federico F. 2000. *FEM validation of a double porosity elastic model for consolidation of structurally complex clayey soils.* Int. J. Numer. Analy. Meth. 24 (4): pp. 367-402.

Khalili, N., and Valliappan, S. 1996. *Unified theory of flow and deformation in double porous media.* European Journal of Mechanics, A/Solids. 15(2): pp. 321-336.

Loret, B. and Rizzi, E. 1999. *Strain localisation in fluid saturated anisotropic elastic plastic porous media with double porosity.* J. Mech. Phys. Solids. 47: pp. 503-530.

Pao W.K.S. & Lewis R.W. 2002. *Three-dimensional finite element simulation of three-phase flow in a deforming fissured reservoir.* Comput. Methods. Appl. M. 191 (23-24): pp. 2631-2659.

Selvadurai, A.P.S. 2000. Partial Differential Equations in Mechanics, Vols. 1 and 2, Springer-Verlag, Berlin.

SIMULATION OF COUPLED FLUID FLOW AND SOLUTE TRANSPORT IN A ROUGH FRACTURE

Wang Jinguo, Zhou Zhifang
College of Civil Engineering, Hohai University, Nanjing, P. R. China, 210098

Abstract: Several studies have been carried out to analyze the applicability of fractal geometry to fracture surface. At present, the self-affine model seems to be the most suitable fractal model for fracture rough surface. In this paper, the Weierstrass-Mandelbrot function method is used to generate a fractal fracture surface. On the basis of modelling fracture surface and aperture, the transport of coupled fluid flow and solute in the fractured media is modeled. The result shows that the solute transport law more accords with actual circumstance when the rough degree of fracture surface is considered. Compared this with the solute transport in the smooth fracture with average aperture, the concentration front in the rough fracture will lag it in the smooth fracture. Furthermore, the solute transport in the rough fractured media is of non-uniformity and anisotropy.

1. INTRODUCTION

In the research of resources and environment, such as geological storage of nuclear wastes, landfill of refuse and intrusion of saline water, transport of pollutant along with groundwater in fractured media has become one of the hotspots in recent years. This research becomes very difficult due to complexity of fractured media themselves. Solute transport in single fracture system is the basis of studying the solute transport mechanism in fractured media, and can help people to better understand the convection – diffusion mechanism of solute in fractures.

Starting from the 1960s, many scholars made indoor and field test and research on the law of transport of solute and radioactive elements in single fracture systems (Moreno, 1982; Novakowski, 1994; Hadamman, 1991); and there have been large number of numerical simulation research results (Noorishad and Mehran, 1982; Christopher, 1995) considering influential factors such as diffusion of solute from fracture to matrix, the absorption action of fracture surface, variation of water flow speed and dispersivity and radioactivity decaying. However, all simulations hitherto made assume that tracer transports in straight and smooth fractures. Considering the changes in surface shape and status of fractures, Wendl and Himmelsbach (2002) simulated the solute transport in a rough fractures by assuming that fracture aperture follows the logarithmic normal distribution, made comparison with indoor experiment data, and obtained fairly good effect. The roughness of fracture discontinuity is a natural fractal phenomenon. In resent years, Carr et al. (1987), Lee et al. (1990), Yu et al. (1991) and Zhou and Xiong (1996) studied the roughness characteristics of discontinuities using fractals, and obtained the correlation formula of discontinuity roughness with dimensions D. In this article, rough fracture faces are simulated using fractal theory, and characteristic FEM is used to simulate the concentration distribution of tracer solute transport in fractal fractures.

2. SIMULATION OF GEOMETRIC SHAPES OF FRACTURE

2.1 Simulation of fracture surface

Fracture surface are rough, with certain fractal characteristics. A fracture itself can be seen as congruence of the upper and lower fracture surfaces, therefore the fracture surface and distribution of fracture apertures can be simulated on the basis of fractals theory. The fractal method is to obtain the fracture aperture distribution by using successive accumulation method or random Brown function on the basis of fractal dimensions of fracture, fracture surface dislocation value and area contact rate on the fracture surface.

It is difficult to describe the complex forms and status of natural fracture surfaces with ordinary functions. The fractal theory can be used to describe extremely irregular geometric patterns, and many researches have indicated that it is reasonable to simulate rough fracture surfaces with fractals theory. At present, the self-affine fractal model is regarded as the best fractal model to simulate fracture surface. In the following, random Brown function method (Weierstrass- Mandelbrot function) is used to simulate a fractal fracture surface.

It is known from fractals theory that, if dimensions of a fracture surface are D, a random Brown function with the exponent of H can produce a fracture surface with fractal characteristics by using the following formula.

$$z(x,y)=\sum_{k=1}^{\infty} C_n \lambda^{-Hk} \sin[\lambda^k (x\cos B_k + y\sin B_k) + A_k] \tag{1}$$

where C_n are mutually independent random numbers according to standard normal distribution; A_n and B_n are mutually independent random numbers according to homogeneous distribution over $[0,2\pi]$; H=3-D, and D is the number of fractal dimensions of the fracture surfaces.

According to a certain D, the curved surface height z of an arbitary point can be obtained if λ is given (commonly $1.0<\lambda<1.5$). A fracture surface with fractal characteristics can be obtained by overall distribution (x,y). Fig. 1 shows a fracture surface generated by simulation using random Brown function method, the fractal dimensions is 2.3 and the parameter $\lambda=1.25$.

2.2 Simulation of fracture aperture distribution

As the aperture at different positions of the fracture surface varies, it would be more reasonable to use the position-dependent method to simulate the fracture aperture.

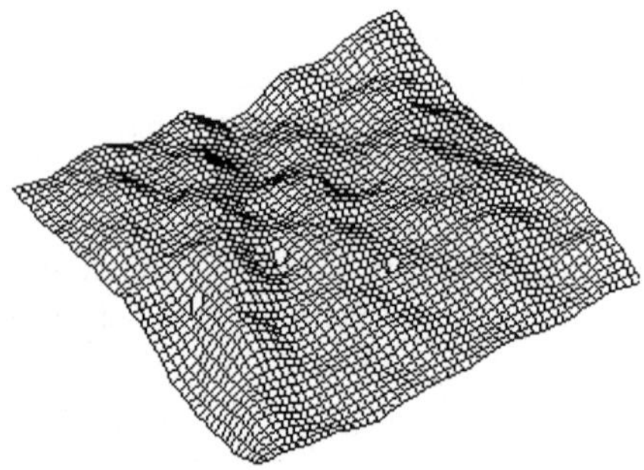

Figure 1. Generation of fractal fracture face (number of fractal dimensions being 2.3).

According to the property of Brown curved surface, if the dislocation values of the upper and lower fracture surfaces are $(\Delta x, \Delta y)$, then

$$\Delta z = z(x+\Delta x, y+\Delta y) - z(x,y) \tag{2}$$

would follow the standard normal distribution with zero as the average value and $(\Delta x^2+\Delta y^2)^H$ as the variance. In the Δz generated using the above formula, there are two series, one positive and the other negative. Assuming the absolute value of the negative number with the maximum absolute value in the negative series is Δz^*, the fracture aperture $b(x,y)$ can be estimated using the following formula:

$$b(x,y)=\Delta z(x,y)+r\Delta z^* \tag{3}$$

Adjust the value of r, so that $b(x,y)$ meets the requirement of area contact rate on the fracture face (it is deemed as contact when the aperture is less than or equal to zero), now the condition of fracture aperture distribution can be generated. When r=1, there is only one contact point between the two walls of the fracture faces, i.e., the contact area is 0. In fact, the area contact rate of a natural fracture is normally greater than zero. Therefore there should be $0<r<1$. Take $b(x, y)$ on the left of the above formula as the statistic average of apertures, with given Δx and Δy, r can be determined by optimization.

Assuming the dislocation value $\Delta x = 0.15$ mm, $\Delta y = 0.15$ mm and area contact rate as 12.24%, now the parameter r=0.34 and the average aperture is 0.142mm. The fracture aperture distribution obtained by simulation with the above method is as shown in Fig. 2.

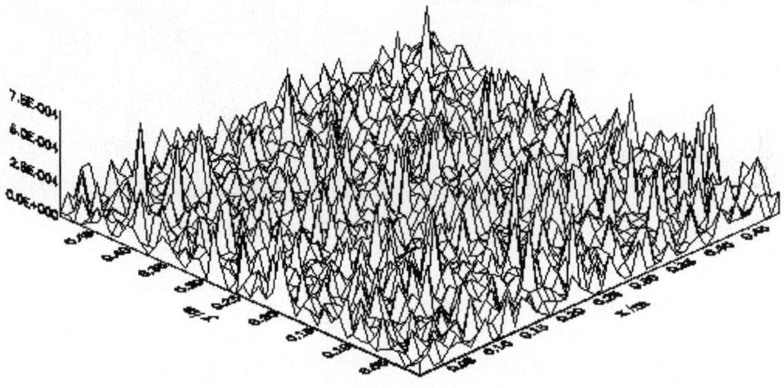

Figure 2. Fracture aperture distribution generated using fractals method

3 SIMULATION OF TRACER SOLUTE TRANSPORT

3.1 Tracer solute transport model

The fracture surface is discreted into many square grids of equal size, and the aperture on each square grid is deemed as constant. As each unit is very small, its seepage coefficient can be determined by the law of cubicity.

$$k_i = \frac{g}{12\mu} b^2(x_i, y_i) \tag{4}$$

where $b(x_i, y_i)$ is the aperture of the *ith* unit at the center of form; g the gravitational acceleration; μ the kinematic viscosity coefficient of the fluid. When the solute is tracer, at 15℃, $\mu = 0.0114 cm^2/s$, and when the solute concentration is high, μ is a variable changing with concentration.

The water flow movement in the fracture can be seen as the stable flow movement of confined water between two water stop sheets, so the governing equation of water flow can be described as:

$$\frac{\partial}{\partial x}(k_i b_i \frac{\partial H}{\partial x}) + \frac{\partial}{\partial y}(k_i b_i \frac{\partial H}{\partial y}) = 0 \tag{5}$$

The governing equation of trace solute transport:

$$\frac{\partial}{\partial x}(D_{xx}\frac{\partial C}{\partial x} + D_{xy}\frac{\partial C}{\partial y}) + \frac{\partial}{\partial x}(D_{yx}\frac{\partial C}{\partial x} + D_{yy}\frac{\partial C}{\partial y})$$

$$-\frac{\partial(v_x C)}{\partial x} - \frac{\partial(v_y C)}{\partial y} = \frac{\partial C}{\partial t} \tag{6}$$

The equation of motion in subsurface flow:

$$v_x = -k_i \frac{\partial H}{\partial x}, \quad v_y = -k_i \frac{\partial H}{\partial y} \tag{7}$$

where: C is the solute concentration; u_x, u_y the velocities in subsurface flow ; R_d coefficient of retardation; D_{xx}, D_{yy}, D_{xy} and D_{yx} are the components of coefficient of hydrodynamic dispersion tensors, which can be expressed as:

$$D_{xx} = \frac{1}{\overline{u}}(\alpha_L u_x^2 + \alpha_T u_y^2) + D^*$$

$$D_{xy} = D_{yx} = \frac{1}{\overline{u}}(\alpha_L - \alpha_T) u_x u_y + D^* \tag{8}$$

$$D_{yy} = \frac{1}{\overline{u}}(\alpha_T u_x^2 + \alpha_L u_y^2) + D^*$$

where, α_L and α_T are the longitudinal and transversal dispersivity, $\overline{u}$ the average velocity of water flow in the unit, and D^* the molecular diffusion coefficient.

3.2 Numerical simulation and interpretation of results

On the basis of analyzing the fracture surface form and aperture, the water flow and solute transport in the fracture are simulated using the FEM and Characteristic FEM (Huyakorn, 1983) methods. Assuming the number of fractal dimensions of the fracture surface is 2.3 and the area contact rate between the two fracture faces is 12.24%, the area of 50×50 cm is splitted into 400 square units. In the calculation, the radial and transversal dispersivity is taken respectively as

$$\alpha_L = 0.5\,\text{cm}, \quad \alpha_T = 0.2\,\text{cm}$$

Two boundary conditions are considered.

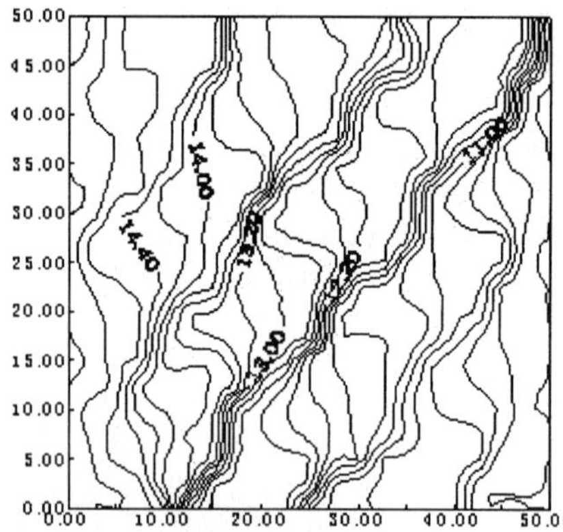

Figure3. Boundary condition (1) isoline of fracture seepage (cm)

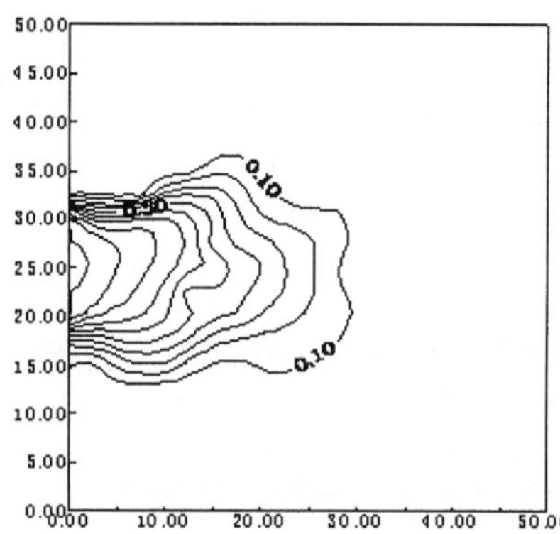

Figure 4. Boundary condition (1) concentration isoline (cm)

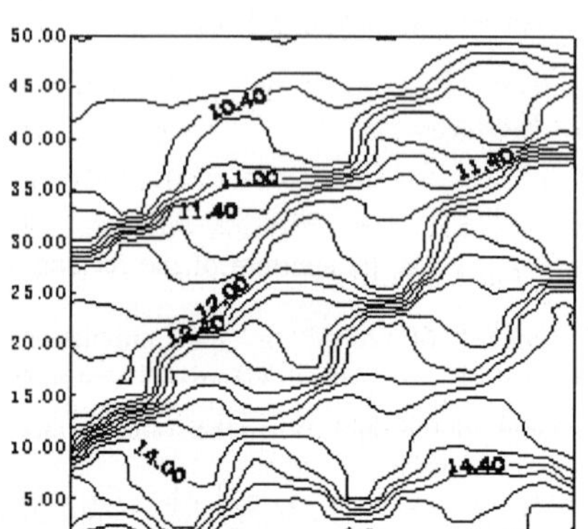

Figure 5. Boundary condition (2) isoline of fracture seepage (cm)

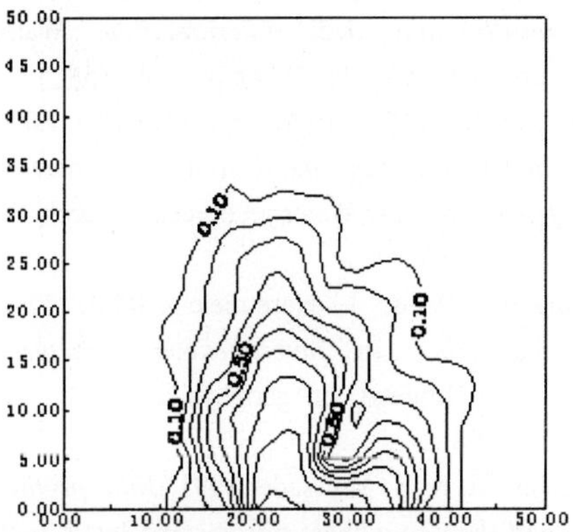

Figure 6. Boundary condition (2) concentration isoline (cm)

Boundary condition (1). Water flow boundary condition: the side parallel with x axis is a water stop boundary, the side parallel with y axis is the boundary with constant water head. Concentration boundary: within 20~30cm of the left boundary is Class I boundary, with C/C_0=1.0, and the area of 0~20cm, 30~50cm is the Class II boundary condition with solute flux as zero, and remaining boundary as Class I boundary with concentration as zero.

Boundary condition (2). Water flow boundary condition: the side parallel with y axis is a water stop boundary, the side parallel with x axis is the boundary with constant water head. Concentration boundary: within 20~30cm of the lower boundary is Class I boundary, with C/C_0=1.0, and the area of 0~20cm, 30~50cm is the Class II boundary condition with solute flux as zero, and remaining boundary as Class I boundary with concentration as zero.

In Fig. 3 ~ Fig. 6, the water flow and concentration isolines at the time of 200s are given respectively.

It can be seen from water flow simulation result (Fig. 3 and Fig. 5) that, there is obvious evidence of channeling flow in the seepage through rough fractures. This is accordant to the research result of Tsang C. F. et al. It can be seen from Fig. 4 and Fig. 6 that, under the above-mentioned hydrodynamic conditions, the solute show obvious unhomogeneity in its transport in the fracture because of the unhomogeneity and anisotropy of the seepage, and this is determined by the uneven distribution characteristics of the fracture apertures.

4 CONCLUSIONS

On the basis of fractal theory, the model of solute transport in fractured rock mass, which is in vie of the fractal characteristics of the fracture surface, has been established, and numerical simulation has been performed on the solute transport through a rough fracture. Results show that there is obvious unhomogeneity and channeling flow in the seepage in rough fracture, and there is also unhomogeneity in solute transport. When compared with the case in a smooth fracture surface with the corresponding average aperture, the solute transport front through a rough fracture is more lagging behind, with the characteristics of stage wise. It can be seen from the simulation results that it better agrees with the actual condition to simulate the solute transport in it by considering the fractal characteristics of the fracture surface. When we are definitely know the statistic characteristics (aperture, occurrence, fracture face roughness, etc.) of fractures in the area under study, the above-mentioned model can be used to simulate the solute transport through rough fractures.

REFERENCES

Carr. J. R. 1989. *Fractal characterization of joint surface roughness in welded tuff at Yucca Mountain, Nevada.* 30th U. S. Symp. On rock mech, Rotterdam Balkema Publisher.

Christopher A. Kennedy & William C. Lennox. 1995. *A control volume of solute transport in a single fracture.* Water resources research. 31(2): pp. 313~322

Edson Wendland &Thomas Himmelsbach. 2002. *Transport simulation with stochastic aperture for a single fracture – comparison with a laboratory experiment.* Advances in water resources, 25(1): pp. 19-23

Haldemman W. R., Y. Chuang, T. C. Rasmussen & D. D. Evans. 1991. *Laboratory Analysis of flow and solute transport through a fracture embedded in porous tuff.* Water resources research. 27(1): pp. 53~65

Huyakorn P. S., B. S. Lester and J. W. Mercer. 1983. *An efficient finite element technique for modeling transport in fractured porous media.* Water resources research. 19(3): pp. 841~854

Jahan Noorishad & Mohsen Mehran. 1982. *An upstream finite element method for solution of transient transport equation in fractured porous media.* Water resources research. 18(3): pp. 588~596

Luis Moreno & Ivars Neretnieks. 1985. *Analysis of some laboratory trace transport in natural fracture surfaces.* Water resources research. 21(7): pp. 951~958

Neretnieks, I., T. Eriksen & P. Tahtinen, 1982. *Tracer movement in a single fracture surface in granitic rock: Some experimental results and their interpretation.* Water resources research. 18(4): pp. 849~858

Novakowski K. S. & P. A. Lapcevic. 1994. *Field measurement of radial solute transport in fractured rock.* Water resources research. 30 (1): pp. 37~43

Tsang C. F. & I. Neretnieks. 1998. *Flow channeling in heterogeneous fractured tocks.* Reviews of geophysics, 36(2): pp. 275-298

Xiaobin Yu & B. Vayssade. 1991. *Joint profiles and their roughness parameters.* Int. J. rock mech Min. Sci. 28(4): pp. 333-336

Y. H. Lee. 1990. *Fractal dimension as a measure of the roughness of rock discontinuity profiles.* Int. J. rock mech Min. Sci. 27 (6): 453-464

Zhou Chuangbing & Xiong Wenlin. 1996. *Relation between joint roughness coefficient and fractal dimension.* J. Wuhan Univ. of Hydr. & Elec. Eng. 29(5): pp. 1-5

Theme 3
Coupled T-H-M-C processes for Oil/Gas Reservoir Engineering

COUPLED THERMO-MECHANO-CHEMICAL PROCESSES IN SHALES: THE PETROLEUM BOREHOLE

Maurice B. Dusseault
Earth Sciences, University of Waterloo, Waterloo Ontario

Abstract: Shales present particular problems to drilling activities. Swelling, shrinkage and strength degradation may occur from pressure, chemical and thermal diffusion processes, leading to hole control problems that may lead to excessive times, hole loss, even blowouts. The article will describe qualitatively the major physical processes, field conditions, and explain how various mud systems address these issues.

1. INTRODUCTION

Shale accounts for ~75% of rocks in basins and presents problems to the drilling industry. Borehole instability is related to in situ state, geological history, shale mechanical and transport properties, and drilling and mud practices.

Shale is an indurated, fine-grained, fissile rock with >33% clay minerals. Fissility is the tendency to split easily along compaction-enhanced bedding planes. Clayey rocks without fissility, generally of porosity >15%, are called mudstones, mud rocks, clay shales or claystones. In petroleum practice, "shale" generically describes all clayey fine-grained rock. Clay content varies from 15% to >90%; the high surface area of phyllosilicates, combined with surface charge distribution, gives rise to mechanical and geochemical properties that lead to swelling or shrinking, slaking, softening and weakening. Chemo-mechanical coupling is of great importance, in contrast to other rock types.

In shales, the presence of adsorbed water and the volume changes that this water experiences because of geochemical, stress and thermal effects is fundamental, not only to drilling, but to all engineering activities in shales.

This article will try to clarify relevant physical processes in shale stability in drilling. However, a quantitative framework for coupling stress-strain-strength behavior and transport processes in shale must be based on induced volume changes, and this issue is not explicitly discussed.

2. SHALE BEHAVIOR

Clay groups (phyllosilicates) in shale are characterized by different basal spacings in X-ray diffraction tests using glycerol-treated oriented clay specimens: K=Kaolinite (7.2 Å), I=Illite (9.9 Å), C=Chlorite (14 Å), V=Vermiculite (14.8 Å), and S=Smectite (17 Å) are the common families. Vermiculite is mildly swelling, but smectites with no confining stresses freely swell in fresh water.

Clays bind polar water molecules to free surfaces by electrostatic attraction from mineral lattice charge deficiencies. Charge density specifies the degree of water bonding (number and strength of bonded mono-layers); the surface area available dictates total adsorption potential. For example, K has more activity per surface area than S, but because of the huge difference in grain size (K ~10-20 m^2/g, S ~400-800 m^2/g), S adsorbs far greater amounts of water. Even quartz (SiO_2) adsorbs up to 9-11 monolayers of water (Dolinšek et al. 1998), but because of low surface area (<2 m^2/g), the effect is negligible overall.

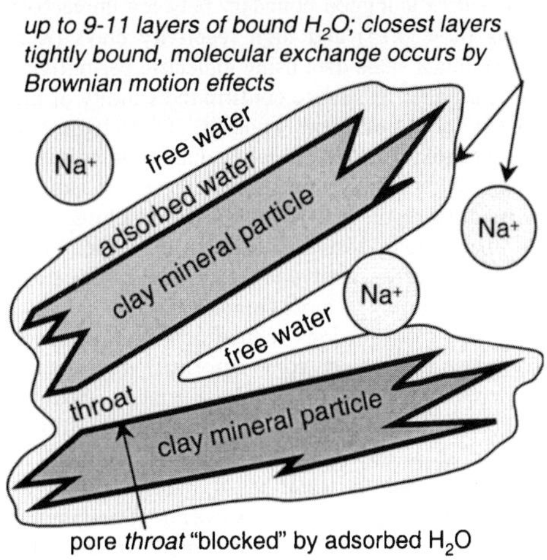

Figure 1. Clays, cations and adsorbed water

The second major behavioral factor is pore fluid geochemistry. Aqueous cations lead to competition between ions and clays for water. Ions are point sources; charges are strong but drop off by $1/r^2$; clay surface charges are weaker but near planar surface do not decay much with distance. High cation concentrations thus suppress the adsorbtive capacity of clays, reducing swelling tendencies. Incorporating pore fluid geochemistry is basic to any attempt to correctly describe shale behavior.

Other important factors include cation type (K^+, Na^+, Ca^{++}...), clay specie charge deficiency, temperature, and so on. For example, increased kinetic energy at high temperatures overcomes bonding forces and reduces adsorbed layer thickness, leading to additional shale compaction under high T (thermal consolidation).

To capture combined effects of surface area and pore water ionic state, the Reactivity Coefficient λ has been defined and used for shale (Fam et al. 2000). It is a dimensionless number, the product of specific surface, S_a, mineral gravity, G_s, water density, ρ_w, and adsorbed layer thickness, ϑ:

$$\lambda = S_a G_s \rho_w \vartheta. \quad (1)$$

S_a captures the clay effect, i.e., the amount of H_2O physically absorbed by the clays. S_a is measured by glycerol adsorption, from sedimentation tests, or from liquid limit tests (Fam and Dusseault 1999). The symbol ϑ represents double layer thickness calculated theoretically (and consistently) or from measured pore fluid ionic activity.

$\lambda \sim 0.15$ is a good boundary between unreactive clays ($\lambda < 0.15$) and high reactivity clays ($\lambda > 0.15$). Intact shale data using dielectric permittivity and magnetic resonance confirm the validity of the Reactivity Coefficient concept. Dielectric data permit rapid shale reactivity evaluation in practical cases. Along with other simple tests, it can be carried out on the drill rig using drilling chips.

Smectites, by virtue of their large surface area, are particularly sensitive to the exchangeable cation type. In nature, the most common cations are Ca^{++} and Na^+. Calcium ions reduce bound water layer thickness and provide some deformation resistance by cross-linking clay platelet surfaces. Sodium smectite, on the other hand, has the highest affinity for water of any common phyllosilicate, and is therefore used in water-based muds (WBM) to build viscosity and to suspend fine-grained materials used to increase the mud weight. The most geochemically sensitive shales are almost invariably smectitic with saline pore fluids.

Confining stress impedes swelling forces. In the borehole wall, radial effective stress, σ'_r, has been greatly reduced, allowing expansion into the hole. Conversely, processes that increase σ'_r tend to have stabilizing effects; combined with geochemistry, this forms the basis of much shale stabilization efforts during drilling (i.e. higher mud densities).

3. THE PETROLEUM BOREHOLE

Borehole stress is well-treated in the literature, so only a few relevant issues are discussed here.

Large shear stresses in the borehole wall, $(\sigma'_\theta - \sigma'_r)/2$, generate forces that weaken and yield shale, leading to sloughing. Borehole breakouts are evidence of yield, and likely all boreholes in shales are somewhat yielded, yet drilling proceeds. Also, yield is accompanied by microfissuring, which increases permeability by many orders of magnitude. Mud density (p_w) is used to increase σ'_r to support the wall. If $p_w > p_o$, the in situ pressure, hydrostatic wall support suppresses deterioration. If pressures communicate, support at the wall is destroyed, even if p_w is large. In aqueous muds, slow Darcian diffusion leads to delayed sloughing.

A scale effect exists in borehole stability: smaller diameter holes are more stable. Often, this is forgotten, and small diameter borehole strategies can be extremely successful in rapid deep drilling (advance rate is a function of borehole area).

Shale fissility is important; drilling ~normal to fissility leads to fewer problems than drilling ~parallel to fissility (Fig 2).

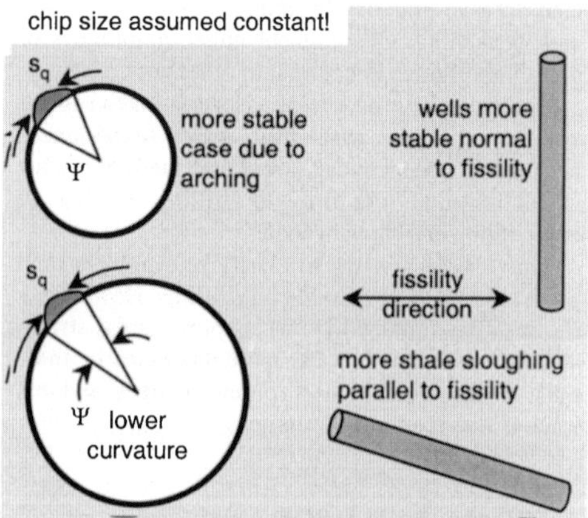

Figure 2. Size & orientation affect shale stability: Scale factors and fissility effects

Temperature affects stability in several ways. First, shale shrinks thermoelastically if cooled, and expands if heated (β_T ~12-15 με/°C, MacGillivray 1996). Heating increases shear stress, cooling decreases it. Cooling also reduces the rate of geochemical alterations and Fickian diffusion processes. For stability, cooling is generally positive, heating is negative.

Finally, high porosity and reactive shales are strongly anisotropic, and are likely to be neither linear elastic nor with confining stress- independent stiffness. These factors are affect analysis and interpretation, yet they remain ill understood.

4. PROCESSES AND COUPLING

Clearly, mechanical, diffusive, thermal and chemical processes all act together. This section attempts to clarify some of the processes. Darcian, Fickian and Fourier diffusion processes take place simultaneously, and interact intimately. They are affected by geochemically- and mechanically-induced changes in shale properties (Sønstebø & Horsrud 1996, Fam and Dusseault 1998, and many others).

In general, intrinsic loss of strength in shale solely due to temperature changes is not important, as these changes are rarely more than 30°C, and they can be neglected as compared to the effects that temperature has on stresses and diffusive processes.

4.1 Undrained behavior

Sudden unloading during drilling generates large pore pressure reduction in the immediate borehole wall because of the Skempton effect. This contributes to stability, but the effects decays through diffusive equilibration as a function of the shale permeability and the driving pressure arising because of $p_w - p_o$ in WBM. Also, this effect does not take place in fractured shales, and is more rapidly destroyed due to mechanical damage in the wall. Anything that leads to reduction of the amount of adsorbed water (e.g. a highly saline WBM, increased temperature) accelerates the disappearance of undrained behavior. Delayed sloughing of shale in boreholes may be linked closely to the rate of decay of the undrained effect.

4.2 Diffusive process coupling

Darcy diffusion in shale is a complex issue because water exists in all states from tightly bound, to water of hydration of ions, to fully free water. Pore throats are blocked by adsorbed water, therefore in intact smectitic shales the "permeability" is very low and is a function of local gradient (non-newtonian flow) because of varying "viscosities" of bound water. For example, magnetic resonance tests showed only ~5% free water in a smectitic Pierre II shale of 15% porosity, but ~50% free water in a 6% porosity Q-I shale from Ontario.

Restricted pore throats (Fig 1) impede Fickian diffusion, in part because hydrated ions have a larger effective diameter. However, as the salinity of the pore fluid increases as with salt WBM, Fickian, as well as Darcian diffusion rates increase. The chemical coupling to these diffusion processes is clear, and it is also clear that Fickian diffusion affects Darcy flow by altering the proportion of immobilized water. Also, it is worth mentioning that the water that is called "bound water" on clays and "water of hydration" on ions is not fully immobile: slow exchange between the bound water and the free-flowing water can take place, and this can be confirmed by magnetic resonance tests using deuterated water (DHO) replacing H_2O.

Whenever Fickian ionic transfer takes place in aqueous solutions, water of hydration accompanies the ions. Thus, a mass flux of water of some magnitude must occur, even in the absence of an externally applied pressure gradient.

Heat flow is purely by conduction in all but intensely fractured shales, therefore a pure thermal problem can usually be treated as uncoupled. However, a change in temperature directly affects all other transport properties. Increased temperature reduces water viscosity (Darcian), increases Browian motion (Fickian), increases stresses (microfissuring with k increases), reduces water layers (increased kinetic energy), and accelerates all chemical reaction processes.

Osmotic processes arise wherever there is a strong salinity difference between fluids in contact (the need for a semi-permeable membrane is usually fulfilled by the shale, albeit imperfectly). The direction of flow is in the direction of the stronger ionic solution, so that a salt mud in the borehole, for example, will tend to osmotically dehydrate surrounding shales. Osmotic flow can be counteracted by high pressure gradients.

4.3 Chemical-mechanical coupling

Chemical changes in shales affect their mechanical properties. For example, cation exchange involving replacement of Na^+ by Ca^{++} decreases the porosity (at constant σ'), thereby increasing the stiffness and perhaps increasing the strength (see below).

Geilikman and Dusseault (1996) approached this problem in a coupled formulation and showed that there exists an optimum mud weight that is a function of material properties and rate of diffusion. For complex systems, finding this optimum *a priori* is difficult. Nevertheless, understanding the nature and direction of the possible physical changes can go a great distance toward finding the optimum conditions quickly.

4.4 Local strain and weakening

A phenomenon widely ignored in geomechanics is that of rock weakening because of local strains associated with diffusive processes, a form of mechanical-diffusive coupling.

The propagation of a diffusion front is accompanied by small-scale volume changes in shales. Near the diffusion front, differential strains are large enough to cause loss of cementation bonds, deteriorating the shale cohesive strength. This is particularly important in shales because cohesive strength generally is far more important than frictional strength (the friction angle of phyllosilicates is much lower than other typical rock-forming minerals). This weakening is more important the sharper the front, therefore it is more important in convective processes.

Even if a volume reduction process takes place (e.g. K^+ or Ca^{++} diffusion), a process that is generally assumed to be a strengthening one because the porosity is reduced, the shearing resistance is degraded by the local weakening that follows the propagating front. This explains laboratory data on shales (e.g. Horsrud et al. 1998) that show weakening accompanying diffusive processes.

5. IMPROVING SHALE STABILITY

There are many mud systems and techniques available to increase borehole stability. In this discussion, all immiscible systems are considered to be oil-based muds; others, air excepted, are water-based. Special chemicals such as liquid silicates that lead to precipitation of cementitious materials in the borehole wall are not dealt with here: the mechanism appears straightforward, but success depends on diffusion as well. In most cases, the concentrations of the active agents must be controlled to remain above the levels at which beneficial effects ensue. High mud weights are not the solution to many problems, it can exacerbate instability in fractured systems and even in some diffusion cases if it increases the flux of a damaging specie into the shale.

Proper mud system control is therefore an optimisation problem, and the number of variables is so large and the uncertainties so severe that much optimisation is carried out in practice. It is not unusual to start in a new area with a conservative approach and gradually seek the economically optimum case by trial-and-error. This paper seeks to provide an understanding that can accelerate this process, while avoiding catastrophic hole loss.

5.1 Mud cooling

In a long open hole, the drilling mud cools the hole bottom, but heats the formations higher up in the hole, and the last casing shoe is the most affected location. Cooling the mud temperature at surface as much as possible increases cooling of the borehole wall at all points in the open hole. This reduces σ'_θ, thereby reducing destabilizing forces; it also slows down all diffusion processes, retarding fluxes that can lead to changes in system properties. Clearly, mud cooling is more effective in long open-hole sections drilled in regions with high geothermal gradients, but it should be considered whenever sloughing at the casing shoe is identified as a problem. Because of the modest thermal conductivity of shales, cooling effects last long enough that even long trips benefit from cooling. Cooling also improves bit performance and reduces the rate of thermal drilling mud degradation.

5.2 Oil-based muds (OBM)

OBM support mechanisms are almost purely mechanical, though the highly saline emulsified aqueous phase also helps. The immiscible oil forms a capillary fringe at the wall, and because the throats in shale are small, oil entry pressures are high (>5-10 MPa). Thus, $p_w - p_o$ acts right at the wall, almost like a tunnel liner, greatly improving stability in ductile shales. The saline aqueous phase tends to cause osmotic flow toward the hole, but the magnitude of this force is small, no more than 1 MPa). Any osmotic dewatering tends to slightly shrink the shale as well, reducing σ'_θ.

OBM alone is ineffective in coping with fractured shales because the fracture aperture is sufficient to allow the oil to penetrate the wall, equilibrating the pressures and destroying the support effect.

5.3 Saline muds

WBM salinity (and also in the aqueous phase of OBM) higher than the formation pore water has several beneficial effects. Osmotic flow causes shale shrinkage, as does the increased salinity, and this shrinkage reduces shear stresses. Swelling of smectites is suppressed. The most effective system is a NaCl-saturated mud; this mud also tends to be of higher density (>1.3), but drilling speeds are reduced, exposing shales for longer periods. It is generally only used in drilling salt sequences.

5.3 Calcium muds

Ca^{++} reduces the thickness of the adsorbed water layer and this has a large beneficial value in smectitic shales through the shrinking effect. Calcium smectite is far less reactive than sodium smectite, and the divalent cations help to bond clay particles together somewhat. Two types are used: low pH gypsum mud and high pH lime muds. Polymeric additives (e.g carboxymethylcellulose) help to block microfissures and sustain good wall sealing. Drill chips tend to be firm and intact, easing problems of mud properties control (similar effects exist for saline muds as well).

5.4 KCl-glycol muds

Hydrated K^+ ions tend to fit better in the interplatelet structure of shales and thus reduce swelling and even induce some shrinking in smectitic shales. In these muds, high molecular weight glycols of limited water solubility are used in sufficient concentrations to maintain an appreciable amount of colloidally dispersed particles. These particles block microfissures and thus act to sustain a low permeability in the wall, increasing the effect of the support pressure. Apparently, there is also some effect of glycol adsorption onto the clay surfaces, reducing water layer adsorption.

5.5 Coping with fractured shales

Fractured shales are usually quartz-illite shales; it is quite rare to have a smectitic fractured shale for good geological and mineralogical reasons. Smectite is diagenetically altered with depth, time and temperature to illite, coliberating SiO_2 in solution to precipitate as quartz. This process involves a large reduction in volume, which gives rise to the fracturing (analogous to cooling fractures in basalt sills). The fracture network allows rapid pressure equilibration with the borehole, and therefore no support effect. The typical small size of the shale fragments allows them to easily "fall" into the borehole.

A first possible solution is to lower p_w to reduce flux of whole mud into the fractures. Note that fissure permeability is strongly aperture dependent, and increasing the pressure in the rock also increases the aperture of the fractures.

The next approach is to somehow block the fissures, which are too large to be plugged by common mud solids (clay particles, polymer molecules...). A good approach is to add material to the mud with a designed range of sizes so that fracture bridging ensues, and then the finer-grained materials can bridge and further reduce flow.

The naturally occurring asphaltic material gilsonite has proven effective in plugging fractures. It is introduced in a finely ground form, and behaves in a ductile manner in situ,

5.6 Better drilling practices

There are many routine practices in drilling that can improve borehole stability. For example, aggressive running into the hole can cause hydraulic fracturing and perhaps accelerate geochemical and Darcian transport processes. Aggressive tripping out of the hole can create sharp pressure declines, momentarily reducing support pressures so that hole caving takes place.

More rapid drilling (slim holes, optimised bit choice, good hydraulics and mud design) reduces shale exposure time. Since all of the diffusive processes are time-dependent, instability is less commonly observed, and slim holes are inherently easier to drill. However, the hydraulic system must be studies to insure adequate hole cleaning without excessive flow pressure effects (called equivalent circulating density effects in the oil industry).

5.7 Underbalanced drilling

In regions where the shales are relatively strong (uplifted, eroded and old regions in general), underbalanced drilling has proved highly effective. Using gas or foams and special practices, p_w is kept lower than p_o, eliminating pressure gradients that introduce detrimental species that can lead to diffusive degradation of properties. Drilling underbalanced also means penetration rates that are usually twice as fast as mud drilling, reducing exposure time. Just before casing placement, the hole is filled with a mud to control pressures, so the time of exposure to detrimental conditions is small.

However, because $p_w < p_o$, there is a destabilizing hydrodynamic effect (seepage force)

acting in the borehole direction, and the shear stresses are larger than for balanced drilling. The shale must be able to resist these additional forces without excessive sloughing.

6. EXPERIMENTS AND ANALYSIS

In principle, all the processes described above can be expressed in the form of fully-coupled differential equations and solved mathematically. It is also possible to achieve coupling though hierarchally nested, iterative and convergent mathematical processes. For a simple Darcy diffusion process, the procedure could be:

- Given initial and boundary conditions, solve nodal pressures $\{p\}$ for a time increment Δt.
- Use $\delta\{p\}$ in a σ-ε model to calculate volume changes $\{\delta V\}$ arising from σ' changes.
- Use $\{\delta V\}$ as a series of source/sink terms in the flow model to recalculate $\{p\}$.
- Iterate until numerical convergence.
- Assess the mechanical state to see if yield or microfissuring has occurred.
- Modify constitutive parameters in the σ-ε model, perhaps also the flow model (Δk).
- Take another time step, and continue the process.

This method converges rapidly if just the $\{p\}$ and $\{\Delta V\}$ method is used (first four steps), but for each additional diffusion process and for large changes in properties, the complexity rises by a large factor, and convergence is affected.

Should we pursue comprehensive testing to delineate the range of possible diffusivities and constitutive parameters of particular shales for predictive numerical modeling? The writer believes that true predictions will never be a practical goal:

- Even a single shale type exhibits wide variations in properties such as clay content, cation state, permeability, and so on.
- Inherent stratigraphic variability means that an open borehole passes through many "different" shale beds of different properties, all exposed to the same wellbore fluids.
- Obtaining coupled diffusivity parameters is extremely challenging, and a wide variety of conditions (p, T, C) must be explored.
- In a borehole, large T gradients are set up; these can be predicted relatively easily, but the coupled effects are complicated.
- Shale is strongly anisotropic in diffusivity and σ-ε as well as strength parameters. Deviated wells encounter anisotropy in any orientation.
- In situ stress conditions vary among and within sedimentary basins, even at the scale of a few hundred metres (around a salt dome or near a fault for instance).
- There is fundamental uncertainty as to the pore water pressure in shale strata, and in situ p_o measurements are rarely possible (low k).
- Even mechanical constitutive laws for shale are ill-known at present. Sampling damage cannot be avoided in taking shale core from depth; quarry sample use remains contentious.
- As the drill bit penetrates, microfissuring occurs, plasticity processes develop, and the intrinsic properties of the shale change.
- Microfissuring and yield lead to permeability and volume changes, thus all coupled processes are highly non-linear in nature.

Given the extreme environment in deep boreholes, the chances of collecting sufficient field data to calibrate models seem remote.

Complex behavior, uncertainty, many coupled processes, and massive non-linearities lead to difficulty in formulating a reasonable model. Functionalities for non-linear coupled processes must be based on tests, and the experimental effort that this requires, even for one shale type, is enormous. All aspects of shale behavior are not fully understood at present; one example is the possible large effect of capillarity on test results (Schmidt et al. 1994). Even with extensive data, fundamental uncertainties exist, particularly with shales, as compared with other rocks.

What is to be done? The proper approach is the following:

- Based on physics, delineate the relevant diffusion and stress-strain processes.
- Separate these into first-order (>10% effect), second-order (1-10% effect), and third-order (<1% effect) processes.
- Constitute mathematical models using only the first-order effects.
- Bound the functionalities and properties empirically by using existing data, field observations, and physical considerations.
- Focus experimental work on the most ill defined parameters associated with first-order effects.
- If economically warranted, carry out large-scale physical simulations using special drilling benches and specimens obtained from quarries.
- Apply the results, using numerical simulation, to improving the efficiency and economics of drilling though shale in difficult areas.

- Collect data through monitoring drill cuttings and cavings shape and quantity, geophysical logging data, and simple well-site tests.
- Apply the knowledge in an integrated geomechanics strategy to improve practices.

This is essentially the "observational approach" in geomechanics. The economic rewards are large: the cost of borehole instability worldwide is on the order of $3-4 billion, and if costs of blowouts that are in part facilitated because of borehole stability are included, the figure is far higher.

7. CONCLUSIONS AND DISCUSSION

Some general comments on borehole stability can now be made. Some of these comments can be generalized to the behavior of shales in other geomechanical conditions, such as in tunnel walls, dam foundations, and in weathering slopes.

It is best to avoid instability problems proactively, rather than reacting only when problems occur. Once problems are noticed, shales have been weakened, properties altered, and it may be too late to rectify them easily. Instability problems also tend to compound because drilling slows (longer exposure times), swabbing pressures on trips increase, and so on.

Underbalanced drilling for the shallow, upper part of the hole should be considered whenever it appears that shale strength is adequate. Models to aid such decisions are rudimentary; assessment must be made with various screening criteria such as geological age, density, wave velocities, and so on, combined with experience in the region. Highly fractured shales or very weak ductile shales are poor candidates for underbalanced drilling.

Ductile unfractured shale (i.e. no natural fractures) can be drilled effectively using OBM. It is prohibited in some areas for environmental reasons and is far more expensive than WBM, so land-based drilling tends to use WBM.

If WBM must be used, and reactive (smectitic) unfractured shale is known to cause problems, a WBM should be formulated to eliminate osmotic flow into the borehole wall, while keeping $p_w - p_o$ small, so as to keep gradients modest yet still sufficient to provide support. Osmotic flow elimination means balancing the ionic activity of the drilling mud with that of the shale encountered. Because shale pore water becomes more saline with depth, in long open-hole sections the ionic strength of the drilling mud will have to be increased gradually with depth, and this leads to a favorable osmotic gradient in the upper part, combined with a neutral gradient in the lower part. Activity control can be achieved with many soluble salts, but all of them also tend to reduce the effectiveness of the bentonite clay that is used to control viscosity and suspend weighting material (usually barite, $BaSO_4$). NaCl-saturated mud, used to drill long halite sections, is formulated with an extremely fine-grained acicular mineral, attapulgite, to build viscosity; this mineral is inactive, and the viscosity effect is mechanical, not geochemical. If a choice exists, the cation used should be one that reduces the smectite activity, reducing swelling tendencies; examples are Ca^{++} and K^+, the latter commonly used with glycols.

One should not approach drilling with high mud weights in the expectation that this will preclude problems. This can accelerate detrimental fluxes, and also, if a high weight is used initially, it gives less capability to react in the future. High mud weight slows drilling, increasing exposure time.

The role of stress-induced microfissures is important, as they allow pressure equalization across the wall, facilitating chip detachment. The use of glycols in over-saturated solutions was mentioned, and to a lesser degree, gilsonite gives some positive effects. As with most of the processes discussed herein, the effects are twofold: maintenance of a better pressure support, as well as retardation of the flux of destabilizing species. The concentration of these agents must be maintained at some level to guarantee effectiveness, and this level has been determined empirically in many cases. There are also other, generally polymeric substances that have a similar beneficial effect, but gilsonite seems to be particularly effective in WBM. (A chemically treated version of gilsonite of reduced solubility in oil is available for OBM.)

Most WBM will benefit from the addition of 5-10% oil, preferably a more viscous oil such as 30-80 cP viscosity lease oil. The oil is emulsified by the clays and shearing, and colloidal oil droplets of various sizes help block the microfissures and large pores. Low viscosity oils are less effective.

Because pressure gradients tend to flatten with time, it is important to keep mud weight increases in reserve if troubles are encountered. A sudden mud weight increase usually has the effect of stabilizing the hole, but if the mud weight is maintained too high from the onset, little room is left for manoeuvre if problems arise. Many strategies have evolved to exploit low mud weight benefits during drilling. For example, during active drilling the mud weight may be reduced to speed advance, but just before the pipe is withdrawn to change the bit, a higher mud weight is used to assure stability during the long period of time when the mud is not circulating.

If long open-hole sections are drilled in areas of high geothermal gradient, thermal destabilization may occur. A diagnostic criterion is if the source of large-scale sloughing is from the shales at the shoe rather than from further downhole.

Drilling fractured shale presents its own challenges. OBM is rarely effective because it cannot easily block the natural fissures, and industry attempts to develop additives for OBM such as graded lost circulation materials and oil-resistant gilsonites have not met with complete success. WBM is far more effective in drilling fractured shale, as the clays and chemicals used in WBM plug the natural fissures more effectively, reducing the magnitude of Darcian flux (fractured shale is rarely reactive). Also, it is easier to design a graded particulate additive to enhance the plugging capabilities of the mud.

To assess whether fractured shale is responsible for excessive cavings, careful visual examination of the borehole returns is advised. It is usually easy to immediately identify two types of fragments: the drill cuttings and the sloughed cavings. Pressure control requirements permitting, in fractured shale it may be most expedient and effective to reduce the mud weight, reducing fracture dilation and enhanced flux into the borehole wall.

Other practices such as drilling the borehole in the best orientation to minimize shear stresses and to encounter shales at a large angle to bedding (>45°) must be considered when possible, particularly where there are stress-induced problems or fissility issues.

The writer does not wish to imply that physics and testing are unimportant because of complexities. In fact, the more we understand the nature of the physical processes, the better we can anticipate and thereby avoid problems, or arrive at rational explanations once problems are noted. Though it may not be feasible to do explicit predictive modelling, parametric modelling is vital to assess the impact and direction of changes induced by coupled processes.

Finally, although this article discussed shale behavior in the context of elevated temperatures and large gradients, the lessons learned with respect to coupled processes are of interest in other applications, and helps us to understand many deep processes involved in rock formation through diagenesis.

REFERENCES

Dolinšek, J., Bharatam, J., Dusseault, M.B. & Pintar, M.M. 1998 *Two-dimensional NMR study of surface water dynamics in hydrated silica spheres.* Physical Review B, 58(11), 7340-7346.

Fam, M.A. & Dusseault, M.B. 1998. *Borehole stability in shales: a physico-chemical perspective.* Proc. SPE/ISRM Eurock '98, Balkema, Rotterdam, 461-470.

Fam, M. & Dusseault, M.B. 1999. *Determination of the reactivity of clay-fluid systems using liquid limit data.* Canadian Geotechnical Journal 36(1), 161-165.

Fam, M., Dusseault, M.B., & Fooks, J. 2003. *Drilling in mudrocks: rock behavior issues.* J. Petroleum Science and Eng., at the press.

Geilikman, M.B. & Dusseault, M.B. 1996. *Mechano-chemical corrosion of shales.* Proc. 2nd NARMS '96, Montreal, Canada, 959-964.

Horsrud, P., Bostrøm, B., Sønstebø, E.F. & Holt, R.M. 1998. *Interaction between shale and water-based drilling fluids: Laboratory exposure tests give new insight into mechanisms and field consequences of KCl contents.* SPE 48986. SPE Annual Tech. Conf. & Exhibition, New Orleans, 27-30 September.

Horsrud, P., Holt, R.M., Sønstebø, E.F., Svanø, G. & Bostrøm, B. 1994. *Time dependent borehole stability: Laboratory studies and numerical simulation of different mechanisms in shale.* SPE/ISRM 28060, EUROCK'94, Delft, NL, Balkema, Rotterdam. 8pp.

MacGillivray, D., Davidson, B. & Dusseault, M.B. 1996. *One-dimensional thermal conductivity measurements in quartz-illite and smectitic shales.* Proc. Eurock '96, ISRM Int. Symp., Italy, Balkema, 107-113.

Schmitt, L., Forsans, T. & Santarelli, F.J. 1994. *Shale testing and capillary phenomena.* Int. J. Rock Mechanics Mining Sciences & Geomechanics Abstracts, 31, 411-427.

Sønstebø, E.F. & Horsrud, P. 1996. *Effects of brines on mechanical properties of shales under different test conditions.* Proc. Eurock '96, Barla (ed.), Balkema, Rotterdam.

A COUPLED MECHANICAL-THERMAL-PHYSICO-CHEMICAL MODEL FOR THE STUDY OF TIME-DEPENDENT WELLBORE STABILITY IN SHALES

Choi, S.K., Tan, C.P. and Freij-Ayoub, R
CSIRO Petroleum, Australia

Abstract: The analysis of wellbore stability in shales needs to take into account the interaction of the drilling fluid and formation rock. Depending on the borehole trajectory, formation rock properties, in-situ stress, mud pressure, type of drilling fluid and dissolved ions, and temperature of drilling fluid and formation, the interaction may involve invasion of mud filtrate into the formation, single- or two-phase flow of formation pore fluid, transport and diffusion of salt ions (solutes), mechanical deformation caused by anisotropic swelling of shales, thermal stress, poroelastic effects, and possibly failure of the formation rock. Such failure can be almost instantaneous or time-dependent. To solve the coupled partial differential equations describing the complex interaction of the various processes and mechanisms, a finite element program has been developed to specifically model the mechanical-thermal-physico-chemical interactions associated with wellbore stability in shales. In the model, the shale is treated as a transversely isotropic medium. A brief description of the models and the underlying theories are presented in this paper. Some model results are presented and discussed.

1. INTRODUCTION

Wellbore stability in shales is strongly influenced by the evolution of the effective stress field around the wellbore as a result of the interaction of a number of factors and processes (See for example Tan et al., 1998; Choi and Tan, 1998). The important factors include orientation of the wellbore, magnitude and orientation of the in-situ principal stresses, properties of the bedding planes and other major geological structures, type of drilling fluid and dissolved ions, mud pressure, formation pore pressure, temperature of drilling fluid and formation rock, and the fluid transport, thermal and geomechanical properties of the shale, and thermoporoelastic effects. These factors can have an important influence on processes such as pore fluid flow and solute transport induced by the difference in pressure and solute concentration between the drilling fluid and formation, invasion of formation by mud filtrate, swelling of shale due to change in water potential, thermal diffusion caused by difference in temperature between drilling fluid and formation, and induced thermal expansion of pore fluid and shale. These processes and their interaction can have significant influence on the evolution of the effective stress field around the wellbore. Some of the processes are strongly coupled and must be treated accordingly. A coupled finite element program has been developed to model these processes and the effects of their interaction on time-dependent wellbore stability.

2. MAJOR PROCESSES AND MECHANISMS

The processes can be considered as three initially independent processes. These include the undrained load-deformation-failure process and the fluid and heat flow processes. The fluid flow process may include solute transport. Other mechanisms include swelling of shale caused by change in water potential resulting from the other processes. The main coupling parameters are stress, pore pressure and temperature.

2.1 Load-deformation-failure process

The borehole is assumed to have a long straight section, which can be vertical or inclined, and the borehole axis may not align with any of the principal in-situ stresses or with the direction normal to the bedding planes. Under some conditions, cross-sectional planes which are initially perpendicular to the borehole axis before drilling may no longer be planar after creation of the borehole and a 3-dimensional model is required.

To reduce the computational effort, especially for coupled analyses, a generalised plane strain (Leknitskii, 1981; Aoki, 1994) instead of a fully 3-dimensional modelling approach is adopted.

Due to the relatively high porosity and low permeability, fluid saturated pores and the existence of bedding planes in most shales, it is necessary to take into account the effects of material anisotropy and poroelastic effects (Biot, 1956) on wellbore stability in such formations. The constitutive equation for a transversely isotropic poroelastic material can be found, for example, in Cheng (1997).

2.2 Fluid flow process

Loss of confinement resulting from the creation of a borehole can lead to wellbore failure. The weight of the drilling fluid provides some of the support (for the wellbore) which was originally provided by the drilled out material. However, when drilling under an overbalance condition in shales without an effective flow barrier present at the wellbore wall, invasion of the mud filtrate into the formation may occur. Due to the saturation and low permeability of shales, a small volume of mud filtrate penetrating the formation will result in a considerable increase in pore pressure near the wellbore wall. The increase in pore pressure reduces the effective mud support and can lead to a less stable wellbore condition.

The penetration of mud filtrate will depend on the wettability of the drilling fluid and pore fluid to the shale, miscibility of the drilling fluid and pore fluid, and geometrical properties of the inter-connected pores within the formation. If the drilling fluid is a non-wetting fluid, invasion of the mud filtrate into the formation will only occur if the mud pressure exceeds the formation pressure by at least a threshold value referred to as breakthrough pressure (Tan et al., 1996a) given by:

$$P_{bt} = P_c = P_m - P_f = \frac{4}{\delta}\left[(S_t \cos\theta)_{pf} - (S_t \cos\theta)_{mf}\right] \quad (1)$$

where P_{bt} is breakthrough pressure, P_c is capillary pressure, P_m is mud pressure, P_f is formation pressure, S_t is surface tension, θ is contact angle (fluid-air contact), δ is pore diameter and subscripts mf and pf refer to mud filtrate and pore fluid respectively.

If the fluids are immiscible, a pressure difference equal to the capillary pressure of the two fluids exists permanently across the mud filtrate-pore fluid interface when flow occurs. Depending on the size of the pores, the mud filtrate may not be able to penetrate the smaller pores.

The apparent permeability, a function of the proportion of pores penetrated by the mud filtrate, can be determined from the pore size distribution curve, absolute permeability and porosity of the rock material (Tan et al., 1996a). The methodology is based on a simple capillaric model which represents a porous medium by a bundle of parallel capillaries (Scheideggar, 1974). From the law of Hagen-Poiseuille, the absolute permeability of such a model is given by:

$$k = \frac{\phi \bar{\delta}^2}{T^2} \quad (2)$$

where ϕ is porosity, $\bar{\delta}$ is pore diameter parameter and T is parameter similar to tortuosity.

The parameter $\bar{\delta}$ in the above equation can be determined from the differential pore size distribution, $\alpha(\delta)$:

$$\bar{\delta}^2 = \int_0^\infty \delta^2 \alpha(\delta) d\delta \quad (3)$$

The differential pore size distribution is defined as:

$$\alpha(\delta) = \frac{df(\delta)}{d\delta} \quad (4)$$

where $f(\delta)$ is cumulative pore size distribution.

Using Equation 2, the parameter T for a rock material can be determined from the measured absolute permeability and porosity, and the parameter calculated from the cumulative pore size distribution of the rock material.

The modified form of Equation 2 for the apparent permeability, k_{app} is given by:

$$k_{app} = \frac{f(\delta)\phi\bar{\delta}^2_{app}}{T^2} \quad (5)$$

where $\bar{\delta}_{app}$ is an apparent pore diameter parameter.

For a water-based drilling fluid, the front of the filtrate which penetrated into the pores of the rock would gradually mix with the pore fluid. As a consequence, the threshold breakthrough pressure will decrease with time as a result of the reduction in interfacial tension between the two fluids.

2.3 Solute and fluid flow induced by osmosis – chemical potential mechanism

The fine pores and negative charge of clay on pore surfaces make shales exhibit membrane behaviour (Whitworth and Fritz, 1994). Hence, the flow of water out of (or into) shales due to the chemical potential mechanism is somewhat similar to the flow of water through a semi-permeable membrane (Mody and Hale, 1993; van Oort et al., 1995; Tan et al., 1996b). The driving force involved in the water transportation (for zero overbalance condition) is the chemical potential gradient across the membrane that is generally related to the difference in solute (salt) concentration (water activity) (Tan et al., 1996b):

$$\frac{\partial \Phi}{\partial t} = C_c[\nabla(K \cdot \nabla\Phi)] \quad (6)$$

where Φ is rock aqueous potential, t is time, C_c is $\frac{\partial \Phi}{\partial \theta_w} = 1/(M_v + \phi/K_w)$, θ_w is volumetric water content, M_v is coefficient of volume change of rock $(1/(K_r + 4/3G_r))$, K_r is bulk modulus of rock, G_r is shear modulus of rock, ϕ is rock porosity, K_w is bulk modulus of water, K is hydraulic conductivity of rock = k/μ, k is absolute permeability, μ is fluid viscosity which is a function of salt concentration, salt type and temperature, and ∇ is mathematical operator gradient $(\delta/\delta x, \delta/\delta y, \delta/\delta z)$.

The solute flow process includes advection (or convection), dispersion and diffusion. The equation for the process can be written as

$$\frac{\partial \phi C}{\partial t} = -\nabla(C \cdot q - \theta_w D_s \cdot \nabla C - \theta_w D \cdot \nabla C)(1 - \alpha_r) \quad (7)$$

where ϕ is rock porosity, q is water flux, t is time, C is solute concentration of pore fluid, D_s is coefficient of dispersion, D is coefficient of molecular diffusion, α_r is reflection coefficient which describes the solute leakage through the membrane system, and θ_w is volumetric water content.

The osmotic pressure for a non-ideal semi-permeable membrane is given by

$$\varphi = \alpha_p \frac{RT}{V} \ell n(A_w)$$

$$= \alpha_p \frac{RT}{V} \frac{M_w}{M_s} C V_a C_o \quad (8)$$

where φ is osmotic pressure, α_p is osmotic pressure coefficient, R is gas constant, T is absolute temperature (°K), V is partial molar volume of water, A_w is water activity, M_w is molecular weight of solvent, M_s is molecular weight of salt, V_a is valency of salt and C_o is osmotic coefficient of salt in the solution.

For an ideal semi-permeable membrane, all solutes are reflected by the membrane and only water molecules can pass through the membrane. However, shale exhibits a non-ideal semi-permeable ('leaky') membrane behaviour to water-based solutions because it has a range of pore sizes including wide pore throats that result in significant permeability to solutes. The wide throats reduce the solute interaction with the pore surfaces which increases the permeability of the membrane to solutes.

The reflection coefficient (α_r) and osmotic pressure coefficient (α_p) would only be the same for either a fully permeable membrane or an ideal semi-permeable membrane. The reflection coefficient (α_r) can be determined from either empirical relationships with osmotic pressure coefficient based on back-analyses of the experimental results obtained for a given drilling fluid-shale system or theoretical equations based on thermodynamics (van Oort et al., 1995; Zeynaly-Andabily et al., 1996).

2.4 Thermoporoelasticity mechanism

The drilling fluid often has a temperature different from the formation as a result of thermal gradient down the borehole. The temperature difference between the drilling fluid and the formation will result in heat transfer between the two media leading to change in pore pressure due to thermoporoelastic effects (Choi and Tan, 1998).

Due to the low permeability of shales, the coefficient of thermal diffusivity is at least a few orders of magnitude greater than the coefficient of fluid diffusivity. Hence, heat transfer in the formation will be dominated by diffusion, and convective transfer by fluid flow may be ignored. Since the coefficient of thermal expansion of pore fluid is much larger (in the order of 100 times) than the coefficient of rock solid, temperature change will result in a change in pore pressure.

The governing equations for conductive heat transfer are given by:

$$\nabla \cdot (c_o \nabla T) = \dot{T} \quad (9)$$

$$c_o = \frac{\lambda}{\rho s}$$

where c_o is thermal diffusivity, λ is thermal conductivity, ρ is mass density, s is specific heat (or thermal capacity), T is temperature and $\dot{T} = \frac{\partial T}{\partial t}$.

The effects of temperature change on pore fluid pressure (Wang et al., 1996) is given by:

$$c\nabla^2 p = \dot{p} - c'\dot{T} \quad (10)$$

$$c = \frac{2\kappa G B^2 (1+\nu_u)^2 (1-\nu)}{9(1-\nu_u)(\nu_u-\nu)}$$

$$c' = \frac{c}{\kappa}\left[\frac{2\alpha_s(\nu_u-\nu)}{B(1-\nu)(1+\nu_u)} + \phi(\alpha_f - \alpha_s)\right]$$

where c is generalised consolidation coefficient, κ is $\frac{k}{\mu}$, k is intrinsic permeability, μ is viscosity of pore fluid, G is shear modulus, ν is drained Poisson's ratio, ν_u is undrained Poisson's ratio, B is Skempton's pore pressure coefficient, α_s is coefficient of thermal expansion of solid matrix, α_f is coefficient of thermal expansion of pore fluid and ϕ is porosity.

The pore fluid flow is described by:

$$\frac{\partial \Phi}{\partial t} = c[\nabla(\kappa \cdot \nabla \Phi)] \qquad (11)$$

where Φ is rock water potential.

The pore pressure change will result in a change in effective stress due to poroelastic effects. In addition, the formation material away from the wellbore wall will be constrained from expansion/contraction due to temperature change. Consequently, thermal stress will be generated which will change the effective stress. The total changes in effective stresses as a result of mechanical deformation and changes in pore pressure and temperature (Wang et al., 1996) are given by:

$$\Delta\sigma'_{ij} = 2G\Delta\varepsilon_{ij} + \frac{2G\nu}{1-2\nu}\Delta\varepsilon_{kk}\delta_{ij} - \alpha\Delta p\delta_{ij}$$

$$- \frac{2G(1+\nu)}{3(1-2\nu)}\alpha_s \Delta T \delta_{ij} \qquad (12)$$

$$\alpha = 1 - \frac{K}{K_s}$$

where δ_{ij} is Kronecker delta, K is drained bulk modulus of rock framework, and K_s is bulk modulus of mineral constituents.

If a transient temperature field exists in the saturated rock, the difference in coefficient of thermal expansion of the pore fluid and the rock matrix can induce changes in pore pressure and stresses. This can lead to thermal-induced pore fluid flow and deformation of the rock around the borehole.

2.5 Swelling and hydrational stress mechanism

The total aqueous potential (pressure and chemical potential) of the pore fluid increases with the increase in pore pressure and/or chemical potential (decrease in salt concentration). When the total aqueous potential of the pore fluid increases, water will be adsorbed into the clay platelets. The water adsorption will result in either the platelets moving further apart i.e., swelling if they are free to expand, or the generation of hydrational stress if swelling is constrained. The process can be represented by (Tan et al., 1997):

$$\Delta\varepsilon_{oi} = Q_i \Delta \log \Phi \qquad (13)$$

where ε_{oi} is strain due to change in rock water potential in direction i, Q_i is volume change parameter at a given stress in direction i, and Φ is rock water potential.

It is assumed that rock water potential will affect the rock matrix (effective) stress, and therefore the change in swelling strain can also be expressed as:

$$\Delta\varepsilon_{oi} = C_i \Delta\sigma_i' \qquad (14)$$

where C_i is volume change coefficient at a given stress in direction i, and σ_i' is effective stress in direction i.

The volume change coefficient can be obtained from the following equation

$$C_i = \frac{d(\varepsilon_{oi})}{d(\log \sigma_i')} \qquad (15)$$

The generation of hydrational stress due to the rock material constrained from swelling is given by

$$\Delta\sigma_{oi} = [D]\Delta\varepsilon_{oi} \qquad (16)$$

where σ_{oi} is hydrational stress in direction i; and [D] is stiffness matrix.

3. INTEGRATION OF THE PROCESSES AND MECHANISMS

The model takes into account changes in formation pore pressure arising from fluid flow and induced by deformation of the formation rock, and changes in effective stresses as a result of change in the formation pore pressure.

In the mud pressure penetration model, the geometry of the pore network was represented as a series of parallel cylindrical tubes. Although the model may not represent the actual flow processes in the pore network at microscopic scale, the approach enables certain properties of the porous medium (or equivalent continuum) for certain flow process at the macroscopic scale to be derived. The main effect of the mud pressure penetration process is in modifying the effective permeability.

Chemical potential difference will affect the flow rate of dissolved solute ions and solvent (or pore fluid - the mud filtrate is considered to be pore fluid once it has infiltrated the pores of the shale). The main consideration of dissolved ions is their contribution to osmotic pressure. This is achieved by keeping track of the concentration of the dissolved ions (Equation 7). The effect of osmotic pressure on the flow of pore fluid is taken into account by including the osmotic pressure in the rock water potential, and use the spatial gradient of rock water potential (Equation 6) as the driving force for pore fluid flow.

The effects of temperature can be prevalent for some formation rocks and in-situ conditions. Temperature-induced changes in pore pressure and rock matrix stress (thermo-poro-elasticity model) can be modelled using Equations 10 to 12. In addition, to take into account the chemical potential effects, gradient of rock water potential instead of pore pressure gradient is used in Equation 11.

As the effects of (rock matrix) swelling are mainly on effective stress, they can be taken into account by keeping track of changes in rock water potential or effective stresses, and the strain induced (swelling and hydrational stress model, Equations 14 to 16). The hydrational stress can be considered as additional internal stress in the equilibrium equations.

For all the processes and mechanisms, the major coupling parameters are stress, pore fluid pressure and temperature.

4. MODEL RESULTS AND DISCUSSIONS

Analyses have been conducted to study the effects of the different processes and their interaction on wellbore stability. These include coupling of some or all of the processes and mechanisms. Examples of the model results are shown in Figures 1 and 2. Figure 1 shows the change in pore pressure at different distance (normalized with respect to wellbore radius) from the wellbore center with time when drilling in a formation which is 30°C hotter than the drilling fluid, taking into account the effects of borehole deformation, mud filtrate invasion, thermal diffusion and thermoporoelastic effects. Figure 2 shows the model results which take into account borehole deformation, mud filtrate invasion, fluid flow, solute transport, chemical potential mechanism, and poroelastic effects. The model results show that stability can be time-dependent because some of the processes such as mud filtrate invasion, fluid flow and thermal diffusion are time-dependent. The results also show that thermoporoelasticity effects can induce a pore pressure change of about 0.7 MPa close to the wellbore wall for every 1°C temperature difference between the drilling fluid and formation. The model results also show that the chemical potential mechanism can improve wellbore stability by lowering the pore pressure below the initial value in the formation near to the wellbore.

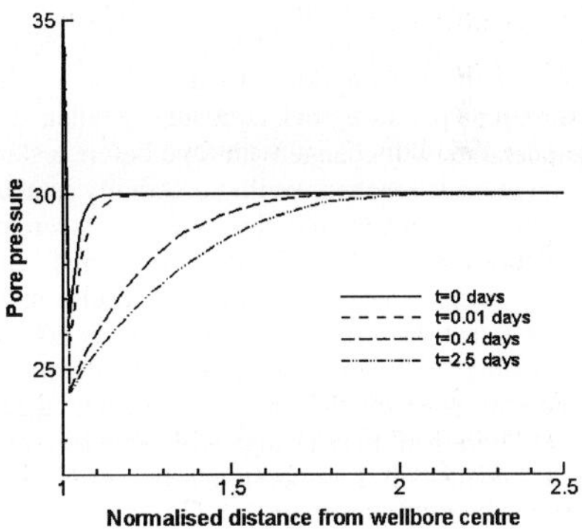

Figure 1. The effects of coupled interaction of mechanical deformation, fluid flow & a colder drilling fluid on pore pressure.

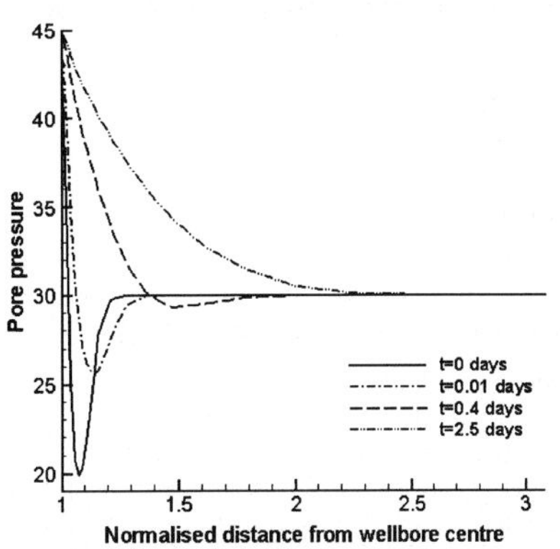

Figure 2. The effects of coupled interaction of mechanical deformation, fluid flow & a lower water activity drilling fluid on pore pressure.

5. CONCLUSION

A finite element program has been developed which can model the effects of the interaction of a number of important factors and processes on wellbore stability in shales. The factors and processes which have been incorporated into the model include the orientation of the wellbore, magnitude and orientation of the in-situ principal stresses, properties of the bedding planes, properties of drilling fluid and dissolved ions, mud pressure, formation pore pressure, temperature of drilling fluid and formation rock, and the fluid transport, thermal, geomechanical and properties of shale, swelling and hydrational stress mechanism, and thermoporoelastic effects. The results of some preliminary modelling studies showed that, as some of the important field variables such as formation pore pressure, rock aqueous potential, stress and temperature will change with time before a steady state condition is reached, wellbore stability in shales can be time-dependent and this must be taken into consideration in drilling fluid design and shale instability management. The models results in this paper show that the temperature or chemical potential of the drilling fluid can have significant influence on how the pore pressure field in the formation adjacent to the wellbore wall may change with time as a result of heat diffusion, thermoporoelastic effect, and chemical potential mechanism.

Results of the modelling study conducted to date indicate that, in order to predict both short and long term stabilities of wellbore in shales, it is important to take into account the interaction of the processes and mechanisms described in this paper. The coupled model provides a tool for studying the relative importance of the various factors and processes on wellbore stability in shales, and how stability may change with time.

REFERENCES

Aoki, T. (1994). *Stress-strain-pore pressure interaction for anisotropic porous rock*, Taisei Technical Research Report, 27, 359-366.

Biot, M.A. (1956). *General solutions of the equations of elasticity and consolidation for a porous material*. J. Appl. Mech., Trans. Am. Soc. Mech. Engrs., Vol. 78, pp. 91-96.

Cheng, A.H.-D. (1997). *Material coefficients of anisotropic poroelasticity*. Int. J. Rock Mech. Min. Sci., Vol. 34, No. 2, pp. 199-205.

Choi, S.K. and Tan, C.P. (1998). *Modelling of effects of drilling fluid temperature on wellbore stability*. SPE 47304, Proc. SPE/ISRM EUROCK' 98, Trondheim, Norway, 8-10 July 1998, Vol. 1, pp. 471-477.

Leknitskii, S.G. (1981). *Theory of Elasticity of an Anisotropic Body*. Mir Publ.

Mody, F.K. and Hale, A.H. (1993). *A borehole stability model to couple the mechanics and chemistry of drilling fluid interaction*, Proc. SPE/IADC Drilling Conf., Amsterdam, The Netherlands, 473-490

Scheidegger, A.E. (1974). *The Physics of Flow Through Porous Media*, 1974, Univ of Toronto Press, Britain.

Tan, C.P., S.S. Rahman, B.G. Richards and F.K. Mody (1998) *Integrated Approach to Drilling Fluid Optimization for Efficient Shale Instability Management*. Proc. SPE Int. Oil and Gas Conf. and Exhibition in China, Beijing, China.

Tan, C.P., Richards, C.G., Rahman, S.S. and Andika, R. (1997). *Effects of swelling and hydrational stress in shales on wellbore stability*. Asia Pacific Oil and Gas Conference and Exhibition, Kuala Lumpur, Malaysia, 345-349.

Tan, C.P., E.M. Zeynaly-Andabily and S.S. Rahman (1996a). *A Novel Method of Screening Drilling Muds Against Mud Pressure Penetration for Effective Borehole Wall Support*. IADC/SPE Asia Pacific Drilling Technology Conf., Kuala Lumpur, Malaysia.

Tan, C.P., B.G Richards and S.S. Rahman.(1996b). *Managing Physico-Chemical Wellbore Instability in Shales with the Chemical Potential Mechanism*. SPE Asia Pacific Oil and Gas Conf., Adelaide, Australia. (1996b).

van Oort, E., Hale, A.H. and Mody, F.K. (1995). *Manipulation of coupled osmotic flows for stabilisation of shales exposed to water-based drilling fluids*. Proc. 70th SPE Annual Technical Conference and Exhibition, Dallas, USA, 497-509.

Wang, Y., Papamichos, E. and Dusseault, M.B. (1996). *Thermal Stresses and Borehole Stability in Rock Mechanics*. Proc. 2nd NARMS, Rock Mechanics Tools and Techniques, pp. 1121-1126.

Whitworth, T.M. and Fritz, S.J. (1994). *Electrolyte-induced solute permeability effects in compacted smectite membranes*. Applied Geochemistry, 533-546.

Zeynaly-Andabily, E.M., Chen, H., Rahman, S.S. and Tan, C.P. (1996). SPE 36396 *Management of wellbore instability in shales by controlling the physical-chemical properties of muds*. SPE Asia Pacific Drilling Technology Conf., Kuala Lumpur, Malaysia, pp. 253-261.

MECHANICAL BEHAVIOUR OF CHALK RESERVOIRS: NUMERICAL MODELLING OF WATER SENSITIVITY EFFECTS

F. Collin[1], Y.J. Cui[2], Ch. Schroeder[1] and R. Charlier[1]
[1])Department GeomaC, University of Liège, Belgium
[2]) CERMES, Ecole Nationale des Ponts et Chaussées, France

Abstract: Chalk is the constituent material of numerous oil reservoirs in North Sea. The mechanical behaviour of a saturated chalk has been largely studied. However, different aspects of its behaviour are not yet well understood: material characteristics depend on the saturating fluids and chalk response is time-dependent. This paper proposes the PASACHALK numerical model: an elasto-plastic constitutive law is presented, which reproduces the different plastic mechanisms of the chalk (pore collapse and shear failure) and the influence of pore fluids. The water sensitivity of this soft rock is explained by the existence of suction effects in chalk. Finally, a simulation of a hypothetical reservoir is proposed to show the response of the elasto-plastic model during depletion phase and water injection phase.

1. INTRODUCTION

Compaction of chalky reservoirs during oil extraction and other important problems like the "casing collapse" or the "chalk production" are related to the mechanical properties of chalk. Controlling compaction is very important because reservoir deformations imply seabed subsidence that endangers the offshore stations.

The most remarkable behaviour of chalk is its water-sensitivity: water injection in the reservoir induces additional subsidence. Though many studies have been already performed on chalks, the basis mechanism of the water sensitivity was not defined. The ongoing EC Research Program Pasachalk aim is to give a better understanding of the chalk behaviour. The origin of the research comes from the comparison between experimental results obtained on Lixhe chalk and on Jossigny silt which showed that the influence of water on pure high porosity chalk is similar to that on partially saturated soils (Delage *et al.* 1996). Another extensive experimental analysis of the influence of the saturating fluid on Lixhe chalk behaviour concluded that the water-weakening effect might be suction related. Hence the idea appeared to apply the knowledge, the approach, and the tools of the partially saturated soil mechanics to the understanding, description, and modelling of chalk behaviour during changes in saturation fluids, such as when water flooding.

This paper presents the developed constitutive model, which is a cap type plasticity model coupled with the Barcelona one (Alonso *et al.* 1990) for taking the suction effect into account.

The model parameters are calibrated based on the experimental results. The validation of the model is performed on a water flooding experiment. We show that the model is able to reproduce qualitatively and quantitatively the observed basic phenomena.

Unsaturated flow formulation is necessary here as far as suction has to be known. The flow model used is based on works in relation with the problem of nuclear waste disposal (Collin *et al.* 2002a). For each fluid (Water and oil), balance equations and state equations are written. In partial saturation conditions, the permeability and the storage law have to be modified: a generalised Darcy's law defines the fluid motion (Bear 1972). Numerous couplings existing between mechanics and flows are considered.

2. ELASTO-PLASTIC MODEL

Experiments performed on chalk samples have shown two plastic mechanisms: the pore collapse for high mean stresses (contractant behaviour) and the frictional failure for low mean stresses. The pore collapse could be caused by the breakdown of physico-chemical bonds between the grains inducing some grain-to-grain slip (Monjoie *et al.* 1990). The frictional failure corresponds to a plastic distortion inducing an increase of porosity.

The two-evidenced plastic mechanisms are modelled by two yield surfaces combined within a cap model: the modified Cam-Clay model is used for pore collapse whereas an internal friction model for friction failure (Collin *et al.* 2002b). Experimental results show that the chalk strength under extension can be overestimated using an internal friction model, a third yield surface is then adopted to limit traction stresses.

Obviously, the so-defined yield curve is not continuously derivable at the intersections, leading to numerical difficulties. However, recent publications provide an elegant way to solve this problem (Simo & Hughes 1998).

As far as the suction effect is concerned, the model adopts the approach developed in the Barcelona Basic Model (Alonso *et al.* 1990) where the suction is considered as an independent variable. Suction modifies yield surfaces and produces reversible and irreversible deformations.

The mechanical model is expressed in terms of the following stress invariants and suction:

$$I_\sigma = \sigma_{ii} \quad (1)$$

$$II_{\hat{\sigma}} = \sqrt{\frac{1}{2}\hat{\sigma}_{ij}\hat{\sigma}_{ij}}, \hat{\sigma}_{ij} = \sigma_{ij} - \frac{I_\sigma}{3}\delta_{ij} \quad (2)$$

$$III_{\hat{\sigma}} = \frac{1}{3}\hat{\sigma}_{ij}\hat{\sigma}_{jk}\hat{\sigma}_{ki} \quad (3)$$

$$\beta = -\frac{1}{3}\sin^{-1}\left(\frac{3\sqrt{3}}{2}\frac{III_{\hat{\sigma}}}{II_{\hat{\sigma}}^{3}}\right) \quad (4)$$

$$s = p_o - p_w \quad (5)$$

Where β is the Lode angle, s is the suction, p_o and p_w are the oil and water pressures.

2.1 General formulation

The general elasto-plastic relations are formulated in rate form. The strain rate is composed of a mechanical part (superscript m) and of suction one (superscript s). Each contribution is partitioned in an elastic (superscript e) and a plastic component (superscript p):

$$\dot{\varepsilon}_{ij} = \dot{\varepsilon}_{ij}^{m,e} + \dot{\varepsilon}_{ij}^{s,e} + \dot{\varepsilon}_{ij}^{m,p} + \dot{\varepsilon}_{ij}^{s,p} \quad (6)$$

The mechanical elastic part is related to the Jaumann objective stress rate $\tilde{\sigma}_{kl}$ through Hooke's law. For the plastic parts, a general framework of non-associated plasticity is adopted in order to limit dilatancy. In that case, the plastic flow rate is derived from a plastic potential g_α:

$$\dot{\varepsilon}_{ij}^{m,p} = \dot{\lambda}^p \frac{\partial g_\alpha}{\partial \sigma_{ij}}, \quad (7)$$

where λ^p is a scalar multiplier and g_α is the plastic potential related to the plastic mechanism α.

Elastic and plastic deformations related to suction changes are defined following expressions given in Barcelona Basic Model. Irreversible deformations are induced when the suction becomes higher than a suction level s_0.

$$\dot{\varepsilon}_{ij}^{s,e} = \frac{\kappa_s}{(1+e)}\frac{\dot{s}}{(s+p_{at})}\delta_{ij} = h_{ij}^e\dot{s} \quad (8)$$

$$\dot{\varepsilon}_{ij}^{s,p} = \frac{\lambda_s - \kappa_s}{(1+e)}\frac{\dot{s}}{(s+p_{at})}\delta_{ij} = h_{ij}^p\dot{s} \quad (9)$$

Where e is the void ratio, p_{at} is the atmospheric pressure, κ_s and λ_s are elastic and plastic coefficients.

Combining equations (6) to (9), one obtains the following expression:

$$\tilde{\sigma}_{kl} = C_{klij}^e\left(\dot{\varepsilon}_{ij} - h_{ij}^e\dot{s} - \dot{\lambda}^p\frac{\partial g_\alpha}{\partial \sigma_{ij}} - h_{ij}^p\dot{s}\right) \quad (10)$$

Considering a general hardening/softening plastic law depending on the internal variable ζ, the consistency condition and equation (10) give the expression of multiplier λ^p and the stress rate can be computed:

$$\tilde{\sigma}_{kl} = (C_{klij}^e - C_{klij}^p)\dot{\varepsilon}_{ij} - M_{kl}\dot{s} \quad (11)$$

The first term of the right part is the classical expression of an elasto-plastic formulation. The second term is related to the suction.

2.2 CamClay pore collapse model

The Modified Cam-Clay yield surface is defined by the following expression:

$$f_1 \equiv II_{\hat{\sigma}}^2 + m^2\left(I_\sigma + \frac{3c}{\tan\phi_C}\right)(I_\sigma - 3p_0) = 0 \quad (12)$$

Where c is the cohesion, ϕ_C is the friction angle in compression path, p_0 is the pre-consolidation pressure, which defines the size of the yield surface, and m is a coefficient introduced to take into account the effect of the third stress invariant.

The plastic flow is supposed to be associated and the internal variable is the pre-consolidation pressure p_0, which is related to the volumetric plastic deformations $d\varepsilon_v^p$ following the cinematic equation:

$$dp_0 = \frac{1+e}{\lambda - \kappa} p_0 d\varepsilon_v^p \qquad (13)$$

where λ is the compression coefficient and κ is the elastic coefficient.

2.3 Internal friction model

A more sophisticated model can be built from the Drucker-Prager's cone by introducing a dependence on the Lode's angle β, in order to match more closely the Mohr-Coulomb criterion. It consists of a smoothed Mohr-Coulomb plasticity surface. The formulation based on the idea of Van Eekelen (1980) is used. It can be written in a very similar way to the Drucker-Prager's criterion:

$$f_2 \equiv II_{\hat{\sigma}} - m\left(I_\sigma + \frac{3c}{\tan\phi_C}\right) = 0 \qquad (14)$$

A non-associated plasticity is considered here using a plastic potential definition similar to Eq. 14 where the dilatancy angle ψ is used instead of the frictional angle.

The internal variables of the model are the frictional angles ϕ_C (for compression paths), ϕ_E (for extension paths) and the cohesion c, which are related to the equivalent plastic strain.

2.4 Suction effect on yield surface

Several phenomena are usually evidenced for unsaturated soils:

1. The pre-consolidation pressure p_O and the material stiffness increase with suction. This is described by the LC concept of the Barcelona model (Alonso *et al*, 1990):

$$p_0(s) = p_c\left(\frac{p_0^*}{p_c}\right)^{\frac{\lambda(0)-\kappa}{\lambda(s)-\kappa}} \qquad (15)$$

where p_O^* is the pre-consolidation pressure for null suction, p_c is a reference pressure, $\lambda(0)$ is the compression coefficient at zero suction, $\lambda(s)$ is the compression coefficient at suction s.

However, chalk experiment does not show an increase of plastic stiffness with suction i.e. water saturated chalk and oil saturated chalk present the same plastic compressibility (Schroeder *et al.* 2000). Another expression of LC curve has been adopted:

$$p_0(s) = p_0(0) + \Delta p_0 \frac{s}{s+s^*} \qquad (16)$$

Where Δp_O is the pre-consolidation pressure difference between water and oil saturated sample, and s^* is a numerical parameter.

2. Cohesion increases with suction, this is modelled using Eq. 17. The influence of suction on friction angle depends on the material studied. Experiment on chalk shows that friction angle is independent of the saturating fluid.

$$c(s) = c(0) + k\,s \qquad (17)$$

where k is a material constant, $c(0)$ is the cohesion at saturated state.

3. Suction changes may create irreversible strains. In the Barcelona model, this is modelled thanks a yield surface, the SI "Suction Increase" curve. When suction becomes higher than a suction level s_O, plastic strains are created. This yield criterion is introduced in our constitutive law:

$$f_4 \equiv s - s_0 = 0 \qquad (18)$$

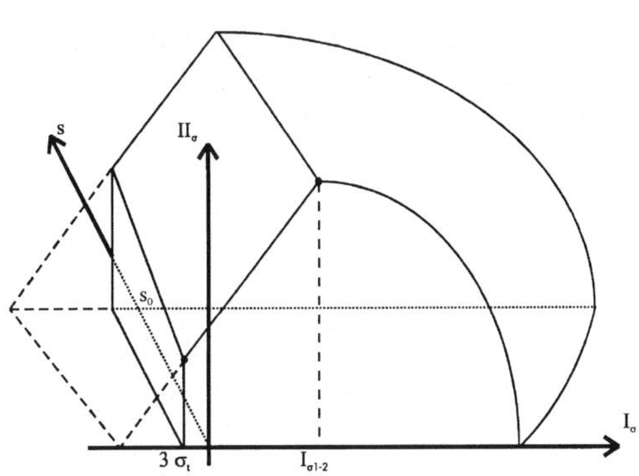

Figure 1. Cap model in the stress space

Figure 1 presents all the yield surfaces in the stress space.

2.5 Waterflooding experiment

The water-flooding test performed on Lixhe chalk initially saturated by SoltrolTM (Schroeder & al. 1998) allows validation of the developed model.

The sample with an initial porosity n = 40.55 % and permeability k_{int} = 1mDarcy has dimensions of 25 mm of diameter and 50 mm of height. Four strain gauges are glued on the sample (located respectively from the injection side at a distance of 4, 12, 22 and 30 mm), which aims to monitor the evolution of the axial deformation with the waterfront. In addition, an axial LVDT records the global axial deformation.

The initial stress state is isotropic at a level of 18 MPa, just below the expected pore collapse for 'oil-like' plug. The injection water pressure is equal to 0.9 MPa. Just before the injection front, a small swelling is measured by the strain gauges but a brutal and quasi-instantaneous compaction appears at the waterfront. The final amplitude of the compaction is around 2%-3%.

Experiment results of triaxial tests on saturated samples allow us to define the yield surface of oil-like and water-like plugs. The transition between these two cases is characterized by results of suction controlled oedometer tests.

The parameters of the mechanical model are presented in Table 1.

Table 1. Parameters of the Cap model

	Parameter	Value	Unit
Elasticity			
	K_{water}	612	MPa
	G_{water}	500	MPa
	K_{oil}	726	MPa
	G_{oil}	700	MPa
Frictional mechanism			
	ϕ_{water}	25	°
	ϕ_{oil}	25	°
	c_{water}	1.5	MPa
	c_{oil}	2.0	MPa
	k	0.167	-
CamClay + suction LC			
	$p_{0,water}$	10	MPa
	$p_{0,oil}$	22	MPa
Hardening rule			
	λ	0.195	-
	s^*	0.2	MPa

The comparison of the numerical results with the experimental data is shown in Figure 2. The injected water volume evolution is similar. After 3500 sec, the waterfront reaches the top; no more oil is driven out of the sample and water is produced at the top. The computed axial strains at the four gauges present a small swelling followed by a brutal collapse of around 2.5% (Figure 3).

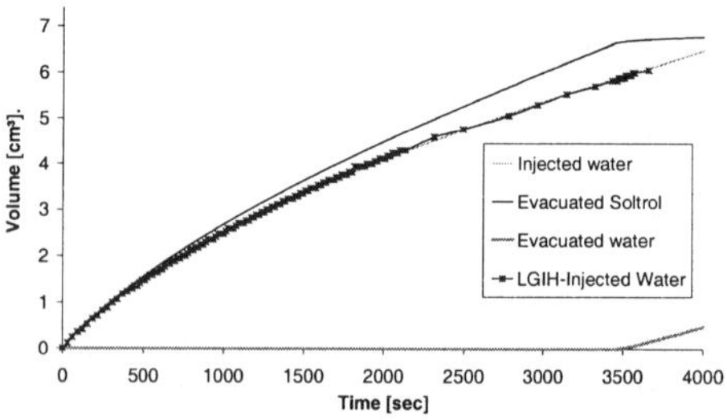

Figure 2. Fluid exchange during water flooding

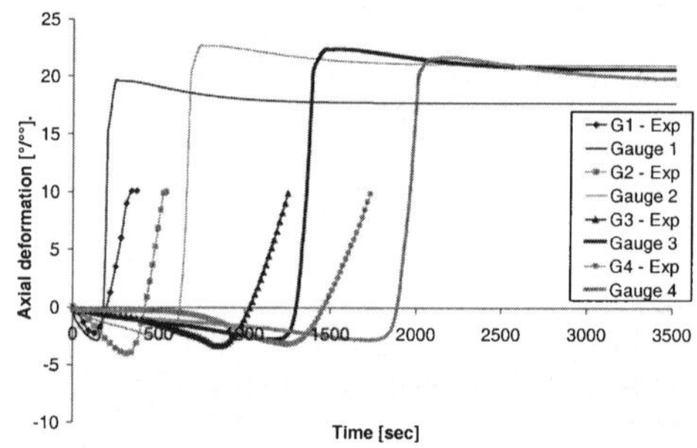

Figure 3. Axial strains at the four gauges

The experiment value of strain stops at 1%, which corresponds to the gauge failure.

The good consistence obtained between the numerical simulation and the experimental responses shows the validity of the developed model for describing the chalk behaviour during water injection.

3. COUPLED RESERVOIR MODELLING

The reservoir model used here is similar to the hypothetical one proposed by Homand (2000): it is 2.5 km long and 300 m thick. The sideburden and the underburden are assumed to be rigid; a dead load corresponding to the weight of the overburden acts on the upper edge of the reservoir. The initial total stress conditions are those of Ekofisk field i.e. 62 MPa vertical stress and 55 MPa horizontal stress (Teufel *et al.* 1990).

All the reservoir edges are supposed to be impervious: the fluids inflows or outflows come only through the wells. In this computations well pressures are controlled in order to reproduce a depletion and an injection phase. The production scheme is the one proposed by Homand (2000) and presented in Figure 4.

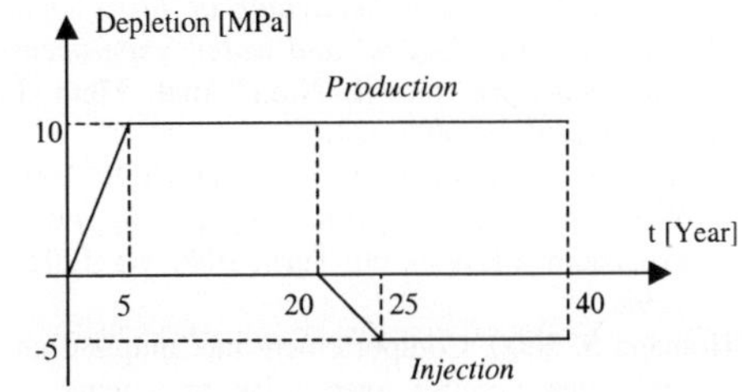

Figure 4. Production scheme

In this model, the injection well is located on the left edge and the distance between the injection and the production well is 2.25km. The initial pressure condition supposed a uniform suction value of 300 kPa (neglecting gravity effect on the reservoir thickness) corresponding to an oil pressure of 49 MPa.

The computation is performed using the elasto-plastic model and the flow model, with the same parameter used in the water flooding simulation. Except the permeability was modified, related to the scale change between the laboratory sample and the reservoir field.

In the production phase, the oil pressure decrease at well to reach a depletion of 10 MPa after 5 years. Due to hydro-mechanical couplings, this pressure decrease leads to a compaction of the reservoir (increase of the effective stress). Figure 5 shows that compaction is low after 1 year (40 cm). However, up to 5 years, irreversible deformations appear and compaction reaches a value of 13 meters at the production well after 20 years.

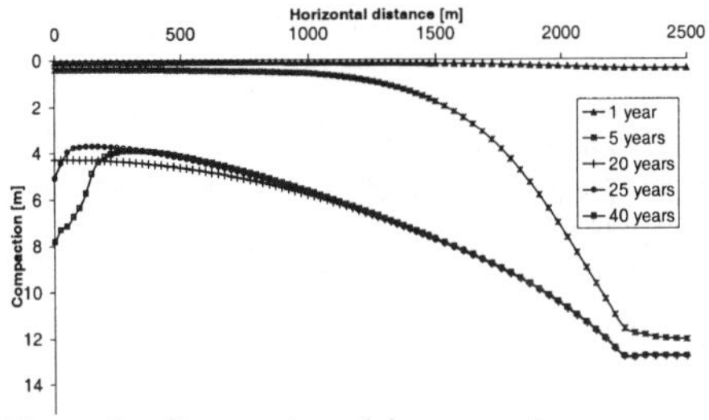

Figure 5. Compaction of the reservoir

When injection phase begins, the water-re-saturation compaction is well predicted by the code. Additional deformations appear and are related to decrease of suction at the injection well.

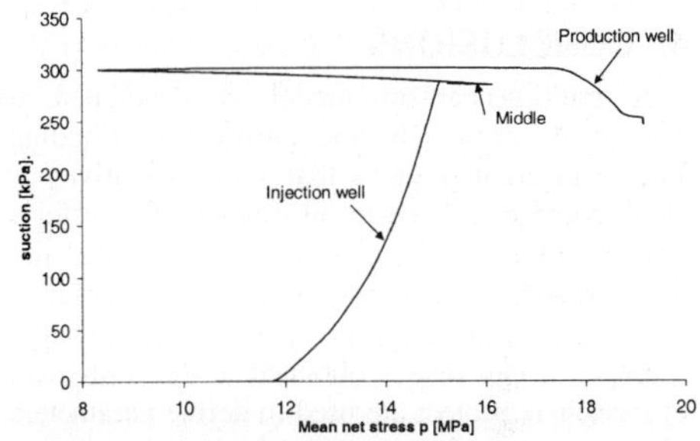

Figure 6. Stress path in the reservoir

Figure 6 presents stress paths (p-s plane) followed in finite elements located near the production well, the injection well and between them. During production phase, stress path in the three points is nearly horizontal: this means that suction does not vary a lot during production phase. When injection phase begins, one observes a brutal decrease of suction at the injection well. This leads to the additional compaction when the stress path reaches the LC curve. However, at the production well and at the midpoint, suction does not vary so much during the time scale considered in this simulation because the water front does not yet reach them.

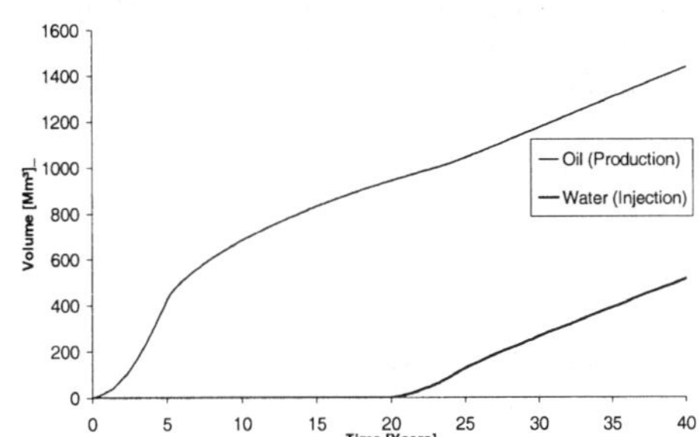

Figure 7. Oil production and water injection volumes

Oil production is clearly affected by the water injection (Figure 7). The oil production rate increases when the injection begins. Indeed the water flows out oil present in the reservoir.

The results show that the constitutive relationship developed within Pasachalk project is able to reproduce the reservoir compaction related to water injection.

4. CONCLUSIONS

A multimechanism model is developed to describe the chalk behaviour during water flooding. The main assumption is that water sensitivity of chalk could be related to oil-water suction effects. The Barcelona model concepts are integrated into the PASACHALK constitutive law in order to reproduce suction dependency behaviour of chalk.

Experimental results obtained in the framework of Pasachalk project are used to define parameters. The validation of the constitutive law is achieved thanks to the modelling of water flooding experiment. Comparisons between experimental and numerical results show that our finite element code is able to reproduce the fluid mass transfer and the mechanical water sensitive behaviour of chalk.

This numerical tool may be used for simulations of oil reservoir chalk in order to reproduce the compaction during water injection. The computations presented here correspond to a hypothetical reservoir. The results show that the model is able to reproduce the compaction during production phase. These deformations are related to an increase of 'effective' stress. During injection phase, an additional compaction is predicted. Even if the reservoir pressure increases, the suction decrease leads to compressive deformations. Suction is a pertinent mechanism to explain the mechanical behaviour of chalk.

Another important aspect of chalk is its time dependent behaviour. We are currently developing an elasto-visco-plastic model including suction effects. This constitutive law will be able to deal with the main issues related to chalk.

ACKNOWLEDGMENTS

The authors are grateful to the EC–Joule programme, the FNRS and the Communauté Française de Belgique for their supports.

REFERENCES

Alonso, E.E. Gens, A. & Josa, A. 1990. *A constitutive model for partially saturated soils.* Géotechnique 40(3): pp. 405-430.

Bear, J. 1972. *Dynamics of fluids in porous media.* American Elsevier Environmental science series 1972.

Collin, F. Li, X.L. Radu, J.P. & Charlier, R. 2002a. *Thermo-hydro-mechanical couplings in clay barriers.* Engineering Geology (64): pp. 179-193.

Collin, F. Cui, Y.J. Schroeder, C. & Charlier, R. 2002b. *Mechanical behaviour of Lixhe chalk partly saturated by oil and water: experiments and modelling.* Int. Jl. Num. Anal. Meth. In Geom. (26): pp. 897-924.

Delage, P. Schroeder, C. & Cui, Y.J. 1996. *Subsidence and capillary effects in chalk.* Proceedings Eurock 96, Turin 1996: pp. 1291 – 1298.

Homand S. 2000. Comportement mécanique d'une craie très poreuse avec prise en compte de l'effet de l'eau : de l'expérience à la simulation. Thèse de doctorat Université de Lille I.

Monjoie, A. Schroeder, C. Prignon, P. Yernaux, C. da Silva, F. & Debande, G. 1990. *Establishment of constitutive laws of chalk and long term tests.* Proc. 3rd Sea Chalk Symposium, Copenhagen June 1990.

Schroeder, C. Bois, A.P. Maury, V. & Hallé, G. 1998. Water/chalk (or collapsible soil) interaction: Part II. Results of tests performed in laboratory on Lixhe chalk to calibrate water/chalk models. SPE/ISRM Eurock'98 Trondheim (SPE 47587) 1998.

Schroeder C., Illing P., Charlier R., Collin F., Delage P., Cui Y.J., De Gennaro V., De Lebeek A., Keül P., Bois A.P. – EC PROJECT PASACHALK. Mechanical behaviour of partially and multiphase saturated chalks Fluid-skeleton interaction : Main factor of chalk oil reservoirs compaction and related subsidence. Final report Contract N°JOF3CT970033 of Non Nuclear Energy Programme JOULE III, 2000.

Simo, J.C. & Hughes, T.J.R. 1998. *Computational Inelasticity.* Interdisciplinary applied mathematics, 7:198-218.

Teufel L.W. & Farrel H.E. 1990. *In situ stress and natural fracture distribution in the Ekofisk field, North Sea.* Proc. Third North Sea Chalk Symposium, Copenhagen.

Van Eekelen, H.A.M. 1980. *Isotropic yield surfaces in three dimensions for use in soil mechanics.* International Journal for Numerical and Analytical Methods in Geomechanics 4:98-101.

COUPLED ANALYSIS OF SAND STABILITY IN PETROLEUM WELLBORES

Gang Han[1], Maurice B. Dusseault[2]
[1]) Geomechanics Consultant
[2]) University of Waterloo, Waterloo, Ontario, Canada

Abstract: There are several mechanisms that may be responsible for sand destabilization in petroleum wellbores, including strength decrease by capillarity and chemical reactions and the variation of pore pressure because of oil/water relative permeability changes. A micromechanical model that quantitatively captures the changes in capillary strength with water saturation in unconsolidated or weakly consolidated sand is proposed. The effect of chemical reactions on rock strength is considered through an explicit approach, which can be easily integrated with the capillarity model to generate a coupled elastoplastic model that describes stress redistribution around a wellbore after water breakthrough. Coupling between stress and pore pressure in the geomechanical model allows direct analysis of the effect of fluid relative permeabilities. The conclusions illuminate the relative importance of each mechanism, and can be used as a foundation to evaluate sand production risk in multiphase fluid environments.

1. INTRODUCTION

It is estimated that, on average, oil companies today produce three barrels of water for each barrel of oil (Bailey, 2000), seventy percent of which comes from weakly consolidated or unconsolidated sandstone. The intrusion of formation water entering water-wetted but oil-saturated sand, which is the usual case in oil fields, may make the rock weaker and easier to fail (Skjærstein et al., 1997; Han et al., 2002).

Experiments have been carried out to study the effect of water saturation (moisture content, humidity, etc.) on the stability of various rock types including shale (Colback & Wild, 1965), chalk (Papamichos et al., 1997), and sandstone (Bruno et al., 1996; Hawkins & McConnell, 1992). The main possible reasons for strength loss are:

- Chemical reactions between water and solids and the dissolution of cementing minerals may weaken the rock;
- Decreases in the surface tension and capillary force that lower the cohesive strength;
- Fluid pressure changes because the relative permeability of oil decreases with an increase in water saturation; and,
- The particles plucked out of the rock skeleton by fluid flow and the swelling of clay materials may block pore throats, locally increase the pressure gradient, and thus increase the destabilizing force.

It is exceedingly difficult to quantitatively describe all these possible physical and chemical changes (Han and Dusseault, 2002). This is not surprising: the influence of water saturation on stability has been known for a long time, but few attempts have been made to predict them, in comparison with the efforts placed on "dry" sand production models (i.e. no capillarity, just Terzaghi's principle).

In this paper, based on a verified microscopic capillary model, a semi-empirical strength model, and a coupled elastoplastic geomechanics model, the effects of capillarity, chemical reactions, and pore pressure are considered and incorporated into stress calculations. The results reveal which processes play major roles in destabilizing sand in oil/water fluid environments at different stages of water breakthrough.

2. CAPILLARITY MODEL

Biphasic fluid conditions in porous media (i.e. oil-water, oil-gas, gas-water) generate liquid menisci among particles, whence weak capillary forces arise to cohesively bond particles. Some doubt the significance of capillarity on rock stability (Lord et al., 1998); others believe it plays an important role in sand production after water breakthrough into an oil well (Papamichos et al., 1997; Bianco and Halleck, 2001).

Consider two identical particles contacting tangentially (Figure 1). Assuming the shape of the liquid bridge is a toroid, the capillary pressure in the water/oil menisci can be written as (Nomenclature is at the end)

$$\Delta P = \gamma\left(\frac{1}{x_p} - \frac{1}{r}\right) \quad (1)$$

where x_p is the radius of curvature at the point p in the horizontal plane, and r is the radius of

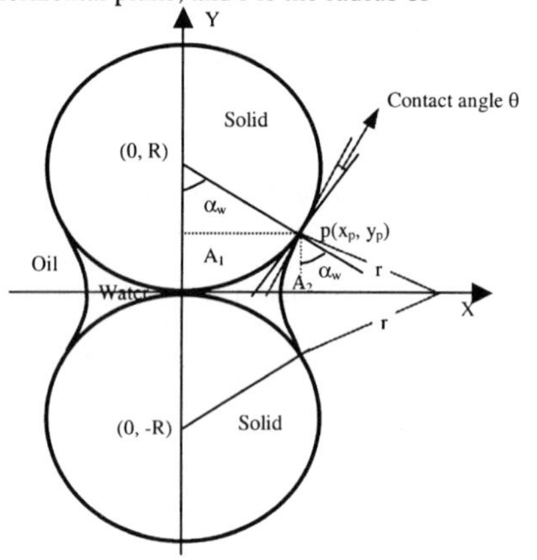

Figure 1. Model for capillarity calculations

curvature in the vertical plane going through the axis of symmetry. The Pressure Difference Method is used to calculate capillary cohesive force resulting from capillary pressure, since it has been demonstrated to be more reasonable (Han and Dusseault, 2002):

$$F_c = \pi x_p^2 \Delta P \tag{2}$$

In particulate mechanics, based on statistical-geometrical considerations, rock tensile strength σ_{T_C} can be related to the cohesive force F_c of a single bond (Schubert, 1984):

$$\sigma_{T\_C} = \frac{1-\phi}{\phi}\frac{F_c}{4R^2} \tag{3}$$

where R is radius of sand spheres and ϕ is rock porosity. Applying a Mohr-Coulomb (M-C) strength criteria to Eq. 3, the unconfined compressive strength can be approximated:

$$\sigma_{UCS} = 2\frac{1-\phi}{\phi}\frac{\sin\varphi}{1-\sin\varphi}\frac{F_c}{R^2} \tag{4}$$

where φ is the graphical M-C friction angle. Thus, for unconsolidated sand, the rock strength is related to rock porosity, friction angle, particle size and capillary force. Notice that in the equation the only variable changing with water saturation is F_c.

The geometrical relationships among these variables can be written down:

$$x_p = R\sin\alpha_w \;;\; r = \frac{R - R\cos\alpha_w}{\cos(\alpha_w + \theta)} \tag{5}$$

$$\eta\phi S_w = -\frac{\alpha_w}{2} + \frac{1}{2}\tan\alpha_w + (\frac{\alpha_w}{2} - \frac{\pi}{4})(\frac{1-\cos\alpha_w}{\cos\alpha_w})^2 \tag{6}$$

Hence, capillary strength can be linked with the variation of water saturation through Eq. 6.

3. EFFECT OF CHEMICAL REACTIONS

If formation water, a powerful chemical agent, moves into a new environment, chemical reactions and physical changes have to take place in order to reach a new equilibrium because of different compositions, density, or concentrations of active ions such as sodium, potassium, carbonate, etc. Water-weakening reactions include: quartz hydrolysis which decreases rock strength through reducing silica-to-hydrogen bond energy; CO_3^{--} cement dissolution that physically changes the shape and size of cementitious deposits; new ferruginous mineral deposition in the rock pore system; and, clay swelling that locally alters driving forces and therefore rock stability. These reactions can continue even after the rock is fully saturated, instead of quickly diminishing with water saturation as in the case of capillarity. Figure 2 shows an experimental relationship for shear strength with time for fully saturated sand and shale: the main strength reduction occurs within the first 20 hours, then the strength remains almost unchanged for many days.

However, as discussed earlier (Han and Dusseault, 2002), chemical reactions are too difficult to be rigorously quantified with respect to sand stability. An empirical approach using a time-exponential relationship is recommended for incorporating chemically reduced strength into stress calculations:

$$C_{o\_ch} = a\exp(-bt) \tag{7}$$

where a and b are coefficients determined by curve-fitting, and t is time.

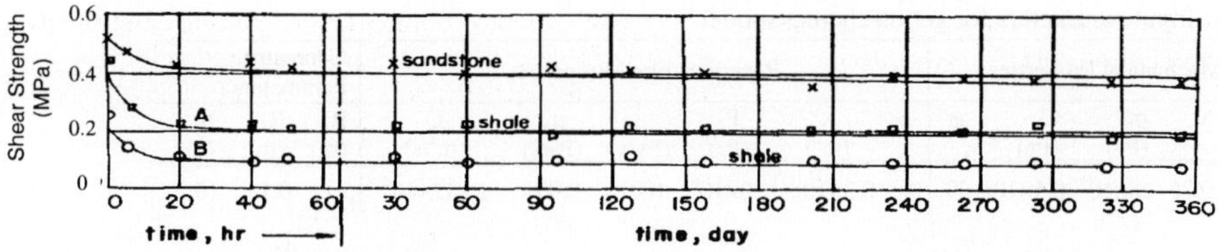

Figure 2. Evolution of water-related shear strength with time (after Tamada, 1970)

4. COUPLED ELASTOPLASTIC GEOMECHANICS MODEL

After water breakthrough, two pressures exist: water pressure and oil pressure. Assuming steady-state fluid flow in an infinite reservoir, they can be calculated by

$$P_w(r) = P_2 - \frac{Q_w \mu}{2\pi k k_{rw} h} \ln\left(\frac{R_2}{r}\right) \tag{8}$$

$$P_o(r) = P_2 - \frac{Q_o \mu}{2\pi k k_{ro} h} \ln\left(\frac{R_2}{r}\right) \tag{9}$$

where P_w, P_o are water and oil pressure at distance "r" from the wellbore, respectively; P_2 is far-field reservoir pressure; k is absolute permeability; k_{rw} is the relative permeability to water, and k_{ro} is the relative permeability to oil. The difference of the two pressures is equal to the capillary pressure:

$$P_w - P_o = -P_c \tag{10}$$

Following the development of effective stress theory by Bishop and Blight (1963), the pore pressure in water-oil environments can be

$$P(r) = P_w(r) S_w - P_o(r)(1 - S_w) \tag{11}$$

For an elastic isotropic formation with a Biot coefficient of α, stress equilibrium around a wellbore in a one-dimensional cylindrical system can be expressed as

$$\frac{\partial \sigma'_r}{\partial r} + \frac{\sigma'_r - \sigma'_\theta}{r} = \alpha \frac{\partial P}{\partial r} \tag{12}$$

where α is the negative Biot constant defined in $\sigma' = \sigma + \alpha P$ and varying from ϕ (porosity) to 1, depending on rock lithology and state of consolidation. Meanwhile, the stresses inside the plastic zone must fulfill the M-C failure criterion:

$$\sigma'_\theta = 2C_o \tan\beta + \sigma'_r \tan^2\beta \tag{13}$$

Because capillary strength resulting from fluid menisci mainly prevents particle separation rather than rotation, it acts more as a part of cohesive strength. Therefore, the cohesive shear strength C_o in Eq. 13 can be expressed as:

$$C_o(S_w) = C_{o\_init} - C_{o\_ch} - \sigma_{T\_C}(S_w) \tan\varphi \tag{14}$$

where C_{o_init} is the initial cohesive shear strength (before water breakthrough); C_{o_ch} and σ_{T_C} are the reduced strengths that arise because of chemical reactions and capillarity.

5. CALCULATIONS AND DISCUSSIONS

Before any further calculations, the water saturation in the microscopic model developed above should be calibrated to experimentally determined values. The saturation discrepancy between the model and reality results mainly from two sources that the microscopic model cannot address: one is connate water saturation (S_{wc}) and immobile oil saturation (S_{oi}); the other is the wettability effect of irregular particle surfaces. If we assume that water saturation remains as a constant (S_{wc}) until water breakthrough occurs, the calibration can be carried out as

$$S'_w = S_{wc} + S_w * (1 - S_{wc} - S_{oi}) / S_{w0} \tag{15}$$

where S_{w0} is the saturation at which capillary pressure becomes zero.

Tables 1, 2, and 3 list the inputs for the geomechanics model and the capillarity model, and include relative permeability data.

Table 1: Input parameters for geomechanics model

Rock Mechanical Properties					Reservoir Flow Properties						Geometry Parameters		
E (Pa)	ν	σ_h (Pa)	C_o (Pa)	φ	ϕ	k (m^2)	P_2 (Pa)	μ (Pa/s)	μ_w (Pa/s)	Q (m^3/s)	R_2 (m)	R_1 (m)	h (m)
3×10^9	0.25	28×10^6	0.5×10^6	30	0.3	0.3×10^{-12}	10×10^6	0.01	0.001	1.157×10^{-3}	50	0.1	10

Table 2: Input parameters for capillarity model

R (m)	θ(°)	γ (N/m)
1×10^{-4}	0	0.036

Table 3: Relative permeabilities vs. saturation

S_w	k_{rw}	k_{ro}
0.32	0	1
0.375	0.003	0.653
0.415	0.008	0.436
0.4555	0.017	0.311
0.495	0.028	0.214
0.535	0.057	0.14
0.575	0.091	0.089
0.615	0.134	0.049
0.655	0.184	0.019
0.694	0.242	0.001
0.734	0.301	0

5.1 Effect of o/w relative permeabilities

Figure 3 shows the calculated pressure variations with water saturation at different distances from the wellbore (r = 0.2 m, 1.0 m, 2.55 m). Interestingly, pore pressure first decreases with saturation until some critical saturation (S_w = 0.45), and the decrease in magnitude can be as high as several MPa; then, it increases continuously to a value (when S_w = 0.734) even higher than initially (when S_w = 0.32). Furthermore, the pressure decrease becomes more significant closer to the wellbore (i.e.: @ r = 0.2m). Physically, because water is a less viscous and more mobile fluid than oil, less energy (i.e.: a lower pressure drawdown) is needed to drive it into the wellbore; consequently, the increase of water relative permeability raises the pore pressure whereas that of oil relative permeability lowers it. The synthesis of both effects indicates that pore pressure in a water-dominant fluid system is relatively higher than in an oil-dominant fluid system.

Figure 4 describes the redistribution of effective stresses around a wellbore producing oil and water simultaneously. Corresponding to the changes of pore pressure, the effective normal stresses first shift significantly away from the wellbore, along with an increase of the shear stress (i.e. more rock more likely experiences shear failure in higher water saturation areas), then move back towards the opening because of the increased pressure at later stages of water invasion. The stress increase reaches several MPa magnitude.

5.2 Effect of capillarity

Compared to pore pressure, the variation of capillary strengths with water saturation is much smaller (see Figure 5): only on the order of kPa. It

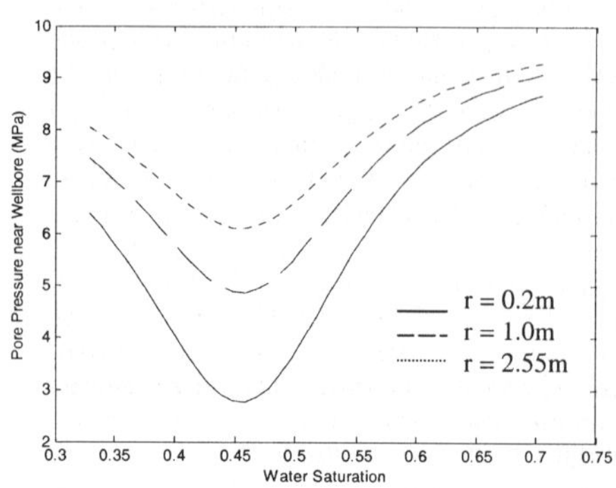

Figure 3. Variations of pore pressure with saturation

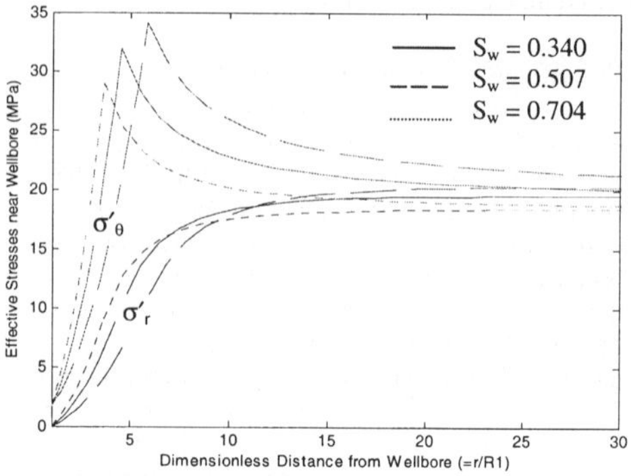

Figure 4. Variations of effective stresses with saturation

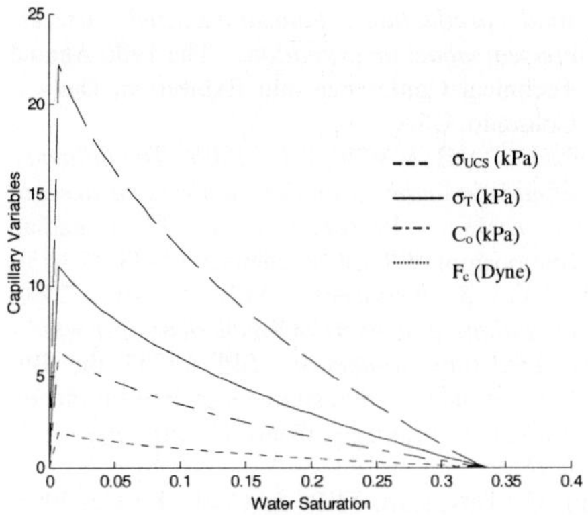

Figure 5. Variations of capillary strengths with saturation (without the effect of connate water saturation)

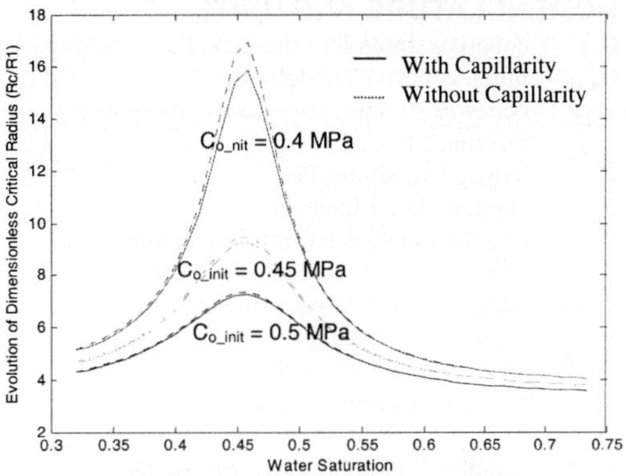

Figure 6. Effect of capillary strength on plastic yield front (with the effect of connate water saturation)

should be noted that the effect of connate water saturation is not considered in the capillarity model until it is applied into the geomechanics model. The solid lines in Figure 6 describe the relationship between dimensionless critical radius and saturation when capillary strength changes are taken into account. Comparing to the dashed lines that treat rock strength as a constant, i.e. no capillary strength so rock instability results only from pore pressure variations by virtue of relative fluid permeability changes, the capillary effect is far less significant than the effect of relative permeabilities unless the initial rock strength is relatively low (e.g. C_{o_init} = 0.4 MPa). This defines when capillary strength plays an important role in stabilizing sand: i.e., after the rock experiences shear yield and most of its initial strength has been destroyed. Only at this time can the effect of capillarity strength variations become a significant destabilizing factor.

5.3 Effect of chemical reactions

Even though it is hard to rigorously quantify the effects of chemical reactions, one qualitative way is to consider a reduced initial cohesive shear strength, as shown in Figure 6. When the rock initial strength is decreased from 0.5 to 0.4 MPa because of the chemical reactions discussed above, the dimensionless critical radius (r/R_1) can increase around 2.5 times at the same saturation (e.g. S_w = 0.45). This indicates that strength-weakening related chemical reactions can significantly affect the amount of failed rock around a wellbore.

6. CONCLUSIONS

It is widely accepted that water breakthrough can lead to sand destabilization, with consequent influx of sand. Several mechanisms may be responsible, including decrease of capillary strength from water-oil menisci, weakening chemical reactions such as quartz hydrolysis and carbonate dissolution, decrease of pore pressure due to oil/water relative permeability changes with water saturation, blockage of rock pore throats by fragments or particles plucked out of the rock skeleton by fluid flow, and effects associated with shale swelling. In this paper, the first three mechanisms are quantified and compared within the context of rock stability.

Calculations indicate that for consolidated and weakly consolidated strata oil/water relative permeability changes and strength loss through chemical reactions may be largely blamed for rock instability after an oil well starts to produce free water. Both can lead to significant changes of the plastic (shear yielded) radius and stress alterations on the order of MPa, compared to capillary strength typically on the order of kPa. However when initial rock cohesive strength is low or becomes low, such as when chemical reactions take place, the contribution of capillarity to stabilize the rock through water-oil menisci is magnified and should not be overlooked. When water saturation reaches some level (i.e. at late stages of water invasion), only chemical reactions can substantively destabilize sand (the other two mechanisms diminish).

NOMENCLATURE AND UNITS

C_o cohesive strength of the rock, Pa
C_{o_init} initial cohesive strength, Pa
C_{o_ch} cohesive strength affected by chemical reactions, Pa
E Young's modulus, Pa
F_c capillary bond force, N
F_f destabilizing force from fluid gradient, Pa
k absolute permeability of reservoir, m^2
k_{rw}, k_{ro} water and oil relative permeabilities, m^2
P_c capillary pressure, Pa
P_w, P_o water and oil pressures, Pa
P_2 far-field reservoir pressure, Pa
Q fluid production, m^3/s
Q_w, Q_o water and oil production rates, m^3/s
r radius of liquid bridge between particles, m distance from wellbore, m
R radius of the particles, m
S_w water saturation, dimensionless
S_{wc} connate water saturation, dimensionless
S_{oi} immobile oil saturation, dimensionless
S_{w0} the saturation at which capillary pressure becomes zero, dimensionless
x_p, y_p the coordinates of point p, m
α Biot coefficient, dimensionless
α_w volume angle of wetting fluid, radian
β failure angle ($= \pi/4 + \varphi/2$), radian
γ surface tension between two fluids, N/m
φ internal friction angle defined in Mohr-Coulomb, degree
ϕ rock porosity, dimensionless
η the coefficient defined to balance porosity difference between the model and reality, dimensionless
σ'_r, σ'_θ effective radial and tangential stresses, Pa
σ_{T_C} capillary tensile strength of the rock, Pa
σ_{UCS} Uniaxial Compressive Strength, Pa

REFERENCES

Bailey, B., Crabtree, M., Tyrie, J., Kuchuk, F., Romano, C. & Roodhart, L. 2000 Spring. *Water Control*. Oil Field Review: 30-50.

Bianco, L.C.B. & Halleck, P.M. May 2001. *Mechanisms of arch instability and sand production in two-phase saturated poorly consolidated sandstones.* SPE 68932, SPE European Formation Damage conference, Hague, Netherlands.

Bishop, A.W. & Blight, G.E. 1963. *Some aspects of effective stress in saturated and partly saturated soils.* Géotechnique 13(3): 177-197.

Bruno, M.S., Bovberg, C.A. & Meyer, R.F. 1996. *Some influences of saturation and fluid flow on sand production: laboratory and discrete element model investigations.* The 1996 Annual Technical Conference and Exhibition, Denver, Colorado, USA.

Colback, P.S.B. & Wild, B.L. 1965. *The influence of moisture content on the compressive strength of rocks.* Proceedings of 3rd Canadian Symposium of Rock Mechanics: 65-83.

Han, G. & Dusseault, M.B. Feb. 2002. *Quantitative analysis of mechanisms for water-related sand production.* SPE 73737, the SPE International Symposium and Exhibition on Formation Damage Control, Lafayette, LA, USA.

Han, G., Dusseault, M.B. & Cook, J. Oct.2002. *Quantifying rock capillary strength behavior in unconsolidated sandstones*, SPE 78170, the SPE/ISRM Rock Mechanics Conference, Irving, TX, USA.

Han, G. & Dusseault, M.B. May 2003. *Description of fluid flow around a wellbore with stress-dependent porosity and permeability.* J. Petro. Sci. & Eng. In press.

Han, G. 2003. *Rock Stability under Different Fluid Flow Conditions.* PhD thesis. University of Waterloo, Waterloo, ON, Canada.

Hawkins, A.B. & McConnell, B.J. 1992. *Sensitivity of sandstone strength and deformability to changes in moisture content.* Quarterly Journal of Engineering Geology (25): 115-130.

Lord, C.J., Johlman, C.L. & Rhett, D.W. Jul. 1998. *Is capillary suction a viable cohesive mechanism in chalk?* SPE 47310, the SPE/ISRM Eurock 98, Trondheim, Norway.

Papmichos, E., Brignoli, M. & Santarelli, F.J. 1997. *An experimental and theoretical study of a partially saturated collapsed rock.* Mechanics of Cohesive-Frictional Materials (2): 251-278.

Schubert, H. 1984. *Capillary forces – modelling and application in particulate technology*, Powder Tech. (47): 105-116.

Skjærstein A., Tronvoll, J., Santarelli, F.J., & Jøranson, H. Oct. 1997. *Effect of water breakthrough on sand production: experimental and field evidence.* SPE 38806, the SPE Annual Technical Conference and Exhibition, San Antonio, TX, USA.

Tamada, B. 1970. *Studies on the collapse phenomenon of sedimentary rock by absorption of water at landslide area.* Rock Mechanics in Japan (1): 68-70.

COUPLED ANALYSIS OF DAMAGE FORMATION AROUND WELLBORES

L. do N. Guimaraes[1], A. Gens[2], S. Olivella[2]

[1]) Federal University of Pernambuco. Av. Acadêmico Hélio Ramos, s/n, Centro de Tecnologia e Geociências, Departamento de Engenharia Civil, Cidade Universitária, CEP: 50740-530, Recife – PE, Brazil

[2]) Geotechnical Engineering Department, Technical University of Catalunya, c/ Jordi Girona 1-3, Edificio D-2, 08034 Barcelona, Spain.

Abstract: In offshore operations, sulfate scale may form when seawater is injected into the formation during waterflooding operations. When seawater (rich in sulfate) mixes with a formation porewater containing cations Sr^{2+} and Ba^{2+}, sulfate minerals will precipitate resulting in reservoir matrix damage and scale in production well. A computer code developed to perform coupled THMC analysis was used to investigate this problem. The mathematical formulation is based on the simultaneous solution of the mass balance equations (of the various species considered), energy balance equation and momentum balance equation (equilibrium). Besides the matrix damage, the model is also able to predict the composition of water produced in the well, that frequently causes scale in production tubing. In the simulation, a multiphase porous medium was considered containing water and oil. A stratified formation with large variations of permeability was assumed. The analysis was able to reproduce qualitatively major features of the phenomena involved.

1. INTRODUCTION

The paper presents a study of scale formation around boreholes when stimulating oil production by waterflooding. When seawater (rich in sulfate and poor in Ca^{2+}, Sr^{2+} and Ba^{2+}) mixes with a formation porewater containing cations Sr^{2+} and Ba^{2+}, sulfate minerals will precipitate resulting in reservoir matrix damage and scale in production well. This phenomenon can have very serious consequences on oil production as it reduces formation porosity and permeability, it may block flow by clogging perforations or forming a thick lining in production tubing or it may coat and damage downhole completion equipment.

Reactive transport formulations can be very useful in helping to prevent the problems caused by this phenomenon. Through numerical modeling of the injection process it is possible to estimate the damage zones evolution in the reservoir and the composition of water produced in the well. Indeed, identifying the location and composition of the scale deposit is the first step in designing a cost-effective remediation program (Crabtree et al., 1999).

The numerical code used is the finite element program CODE_BRIGHT. It is based on a mathematical formulation that combines an already existent thermo-hydro-mechanical formulation (Olivella et al., 1995) with the reactive transport equations in a fully coupled way (Guimarães, 2002).

To illustrate the matrix damage problem in the reservoir and scale formation in production wells an application using the numerical code is presented. In this example, sulfate scaling occurs by mixing of injected water (seawater: rich in sulfate and poor in Ca^{2+}, Sr^{2+} and Ba^{2+}) with a connate water rich in Sr^{2+} and Ba^{2+} cations. From this mixing strontium sulfate salt ($SrSO_4$) and barite ($BaSO_4$) precipitations are expected. This kind of scaling is typical from the North Sea (Osborne et al., 1994) and also common in Campos Basin, Brazil (Silva et al., 2001).

2. CODE DESCRIPTION

In the finite element program CODE_BRIGHT, Olivella el al. (1995) implemented the governing equations for the thermo-hydro-mechanical (THM) problem in a deformable and multiphase porous media. Recently, Guimarães (2002) implemented the reactive transport equations in a fully coupled way with the THM equations. From this coupling a THMC (thermo-hydro-mechanical and geochemical) formulation had resulted that has been used in a number of environmental and geotechnical engineering problems (Gens et al., 2002). As a general formulation, however, it can be extended to other fields such as the petroleum engineering application presented in the paper. In the next sections the CODE_BRIGHT THM and geochemical formulations will be briefly described.

In the code, all coupled equations, in discrete form, are solved simultaneously using the Newton-Rapshon method.

2.1 THM formulation

The governing equations for the THM problem are balance equations and constitutive equations. Mass conservation equations apply to water and oil. When the porous medium is deformable, the momentum balance equation (mechanical equilibrium) is also taken into account. In non-isothermal problems, the internal energy balance for the total porous medium must be considered.

In the example of application presented, some THM formulation simplifications were adopted: the porous medium is not deformable (therefore the mechanical problem is not solved) and it is assumed that the injected water is at the same temperature as the reservoir (the thermal problem is not resolved). Hence, from the full THM formulation only the hydraulic problem equations will be solved, i.e. the water (w) and oil (o) mass balance equations:

$$\frac{\partial}{\partial t}(\rho_\alpha S_\alpha \phi) + \nabla \cdot (\rho_\alpha \mathbf{q}^\alpha) = f^\alpha \quad ; \quad \alpha = w, o \qquad (1)$$

were ϕ is the porosity, ρ_α the density of α phase, S_α the degree of saturation and $\mathbf{q}^\alpha$ the volumetric flux, given by Darcy's Law. f^α is a source/sink term, which in the absence of chemical reactions is zero. The unknowns of these equations are the water (P_w) and oil (P_o) pressures. Darcy flux is given by $\mathbf{q}^\alpha = -(\mathbf{k}\ k_{r\alpha}/\mu_\alpha) \cdot (\nabla P_\alpha - \rho_\alpha \boldsymbol{g})$, where $\boldsymbol{g}$ is the gravity vector, $\mathbf{k}$ is the intrinsic permeability tensor, $k_{r\alpha}$ is the relative permeability (depends on the α fluid) and μ_α is the α fluid viscosity. Another important hydraulic constitutive equation is the retention curve (or capillarity curve) that relates the degree of water saturation with the capillarity pressure ($S_w = f(P_o - P_w)$, with $S_w + S_o = 1$).

2.2 Reactive transport formulation

The reactive transport formulation has two main components: transport equations and the chemical equilibrium model. The transport equations describe the mass continuity of each chemical species in the porous medium. In the chemical equilibrium problem, a relationship between the species concentrations is established that allows a reduction in the degrees of freedom of the reactive transport problem (Guimarães, 2002).

Reactive transport equations: In the formulation proposed by Guimarães (2002), several dissolved chemical species (solutes) are transported in the water phase through the porous medium and, simultaneously, they react among themselves (homogeneous reactions) and, also, with species present in the solid phase (heterogeneous reactions). The homogeneous reactions include: aqueous complex formation, acid/base and oxidation/reduction. The relevant heterogeneous reaction for the case presented in this paper refers to dissolution and precipitation of minerals.

Local equilibrium is assumed for all the chemical reactions except for dissolution/precipitation of minerals where kinetics can also be assumed. The use of one or the other of those two hypotheses must be based on the comparison between rates (or characteristic times) of concentration changes due to reaction and due to transport. Local equilibrium is the limiting case for kinetics when transport characteristic times are much higher then those of the chemical reactions.

The solute transport mechanisms considered in the formulation are: advection (due to water phase movement), molecular diffusion and mechanical dispersion.

The resultant transport equation and the variables used in its development are summarized in Table 1. Transport equations unknowns are the total concentrations U_j that includes precipitated minerals in local equilibrium. They have the convenient property of not depending on the equilibrium reactions. Dependent variables in transport equation (as the total aqueous concentration Ua_j) are related with total concentrations by the chemical equilibrium model described below.

Chemical equilibrium model: Most reactive transport formulations use the mass action law to solve the chemical equilibrium equations. In this formulation an alternative (though thermodynamically equivalent) approach is used, based on the minimization of Gibbs Free Energy. This approach has a wider application range extending to highly non-ideal brine systems.

To compute the concentrations of the species in equilibrium, a Newton-Raphson algorithm is applied to the direct minimization of Gibbs Free Energy. Lagrange multipliers are used to incorporate the restrictions of the problem. Its implementation is adapted to problems with high dynamic minerals appearance/disappearance as in the case presented in next section.

Table 1. Summary of reactive transport equations

Transport equations:	$\frac{\partial}{\partial t}(\phi S_w \rho_w U_j) + \nabla \cdot (\rho_w U a_j \mathbf{q}_w + \mathbf{D}_w \nabla U a_j) = R_j$
THM variables:	ϕ = porosity S_w = water saturation ρ_w = water density
Geochemical variables:	U_j = total concentration Ua_j = total aqueous concentration R_j = (source term due to mineral dissolution/preciptation (kinetic condition))
Flow mechanisms:	$\mathbf{q}_w$ = Darcy's flux (advection) $\mathbf{D}_w$ = hydrodynamic dispersion tensor (diffusion + mechanical dispersion)

3. EXAMPLE OF APPLICATION

In this example of application, a stratified formation with high permeability variation between the layers was assumed. Stratification was considered due to its importance regarding the mixing of the two types of water (injected and connate). The mixing may occur in the reservoir and also in the production well (Sorbie e Mackay, 2000). Figure 1 shows the 2D problem geometry together with the horizontal intrinsic permeabilities ($\mathbf{K}_h$) values for each layer. In this figure, the presence of an aquifer in the lower part of the formation can be observed. The wells are completed between layers 1 and 5. Vertical permeabilities ($\mathbf{K}_v$) were assumed as 10% of horizontal permeabilities. A uniform porosity of 20% was assumed for all the foemation. For the transport of solutes, the longitudinal and transversal dispersivities are 1m and 0.1m respectively.

Porous medium is multiphase with two fluid phases: water and oil. Water transports dissolved species (cations and anions) present in seawater (injected water) and formation (connate) water. The geochemical model proposed by Yuan and Todd (1991) was adopted to model the mixture of high saline water types. This model gives the activity coefficients for the Na-Ca-Ba-Sr-Mg-SO_4-Cl system extrapolated to typical reservoir temperatures and pressures. It is based on the Pitzer models for brine solutions and was implemented in the reactive transport module of CODE_BRIGHT.

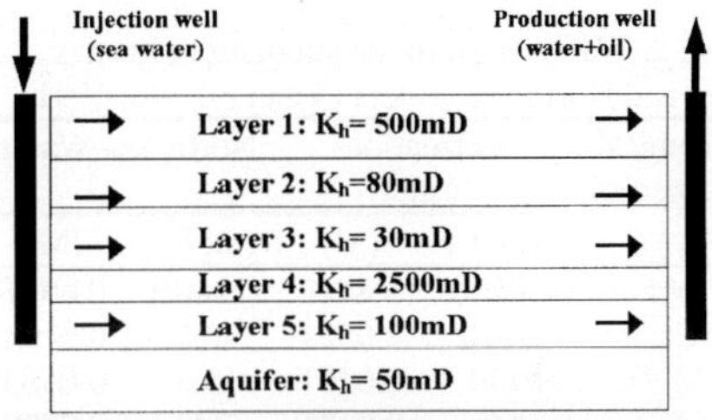

Distance between wells: 400m

Layer	Thickness (m)	Horizontal Permeability (Kh)
1	2	500mD
2	1.5	80mD
3	2	30mD
4	1	2500mD
5	1.5	100mD
Aquifer	2	50mD

Total thickness: 10m. $K_v = 0.1\ K_h$.

Figure 1. Problem geometry and permeabilities of layers.

In this numerical simulation, seawater (rich in SO_4^{2-}) is injected in a formation with porous water containing Ba^{2+} e Sr^{2+} cations. The mixture of waters results in $BaSO_4$ and $SrSO_4$ minerals precipitation (Figure 2). Formation water is from the Forties field at North Sea. Seawater from North Sea and formation water chemical compositions are shown in Table 2. Among other authors, Osborne et al. (1994) relates that these sulfate scaling are typical from North Sea. It is also very common in Brazilian reservoirs such as Campos Basin (Silva et al., 2001). In the geochemical modeling, local equilibrium was assumed for reactions describing the dissolution/precipitation of minerals

The initial water pressure is 30MPa and the degree of saturation is S_w = 0.31 for the reservoir and S_w = 0.77 for the aquifer. It was considered that seawater is injected at a pressure of 38MPa whereas a water pressure of 30MPa was assumed in the production well. Formation and injected water temperatures are constant and equal to 30°C.

The finite element grid adopted has 8000 elements (200 x 40 rectangular elements). To solve the linear systems resulting from the iterative Newton-Raphson scheme, a parallelized conjugated gradient algorithm for multiple processor machines was used. Rectangular elements dimension are: 2m x 0.25m, that guarantees a Peclet number in the dominant flux direction (horizontal) less than 2 for the transport problem.

Table 2. Waters used in the simulation. Forties Field and North Sea waters (Yuan e Todd, 1991).

Chemical species	Formation Water		North Sea Water	
	(mg/L)	(M)	(mg/L)	(M)
Na^+	30200	1.4412	11000	0.4961
Ca^{2+}	3110	0.0851	403	0.0104
Mg^{2+}	480	0.0217	1320	0.0563
K^+	430	0.0121	340	0.0090
Ba^{2+}	250	0.0020	0	0
Sr^{2+}	660	0.0083	0	0
Cl^-	53000	1.6401	19800	0.5790
SO_4^{2-}	0	0	2480	0.0268
HCO_3^-	360	0.0065	135	0.0023

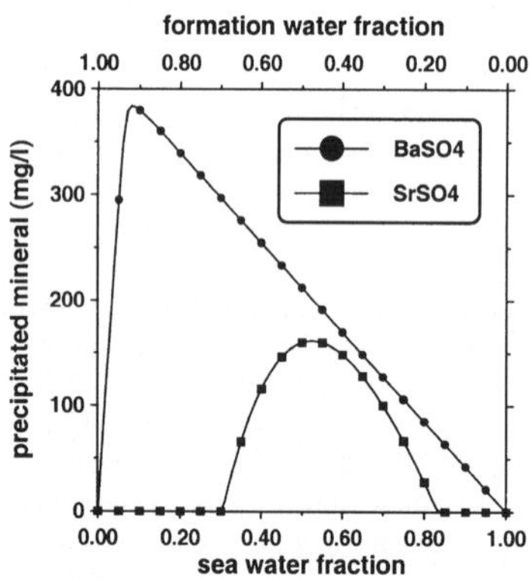

Figure 2. Precipitation of SrSO4 and BaSO4 minerals when mixing the formation and connate waters (Table 2) at 25oC and 30MPa.

Figure 3 shows the distribution of degree of saturation in the formation at two different times, illustrating oil displacement by the injected water. In this figure it is possible to observe that oil displacement by waterflooding is faster at layer 4 (the most permeable) than in the other layers. Thus, seawater injected at layer 4 will reach the production well before the layers with lower permeability. This means that, while the other layers are producing oil and formation water, layer 4 can already be producing seawater causing mixture of the two waters in the producing well.

In Figures 4 and 5 it can be observed how $SrSO_4$ and $BaSO_4$ minerals precipitate more near layer 4 then the rest of the reservoir zones. Seawater injected at layer 4 flows faster than in the adjacent layers 3 and 5 and so diffuses and mixes with formation water of these layers. It is observed that regions of maximum minerals concentration do not coincide for the two minerals. This is explained by the fact that both minerals compete for the same SO_4^{2-} in the seawater. Despite the large extent precipitation in the formation (especially $BaSO_4$) their concentrations are small and insufficient to promote significant porosity and permeability changes in the porous medium. Hence, for this problem it can be seen that the model does not predict matrix damage in reservoir due to mineral precipitation. It is important to note, however, that mineral precipitation can alter totally concentrations of species that take part in chemical reactions in the reservoir and in the water produced in the well.

The chemical composition of produced water was obtained through the simulation of the flow of water and solutes through the reservoir. With this data it was possible to estimate how much mineral can potentially precipitate at the production well (Figure 6). As comparison, a simulation considering conservative transport (without chemical reactions) was also performed.

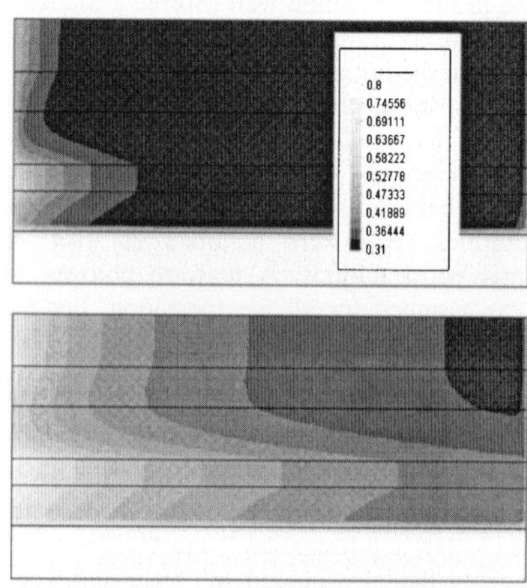

(a)

(b)

Figure 3. Distribution of water degree of saturation at: (a) 28 days and (b) 228 days.

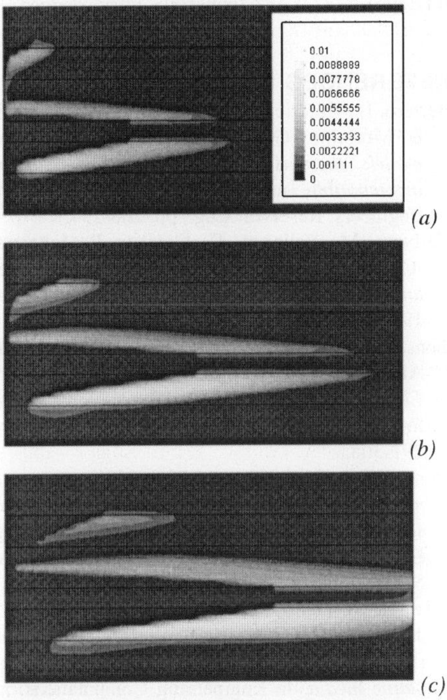

Figure 4. Distribution of SrSO4 concentration at: (a) 100, (b) 178, and (c) 278 days

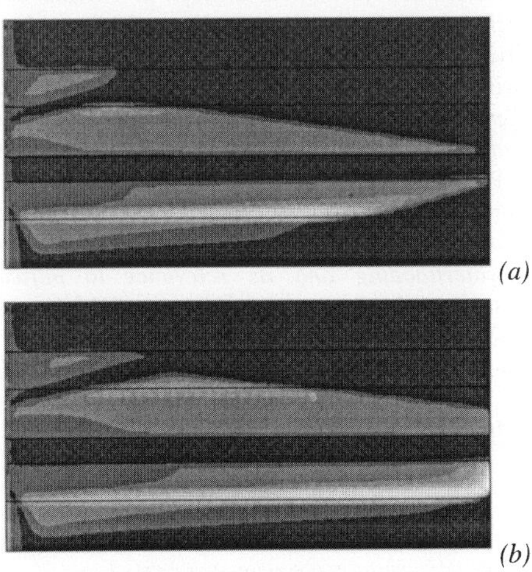

Figure 5. Distribution of BaSO4 concentration at: (a) 128 and (b) 178 days.

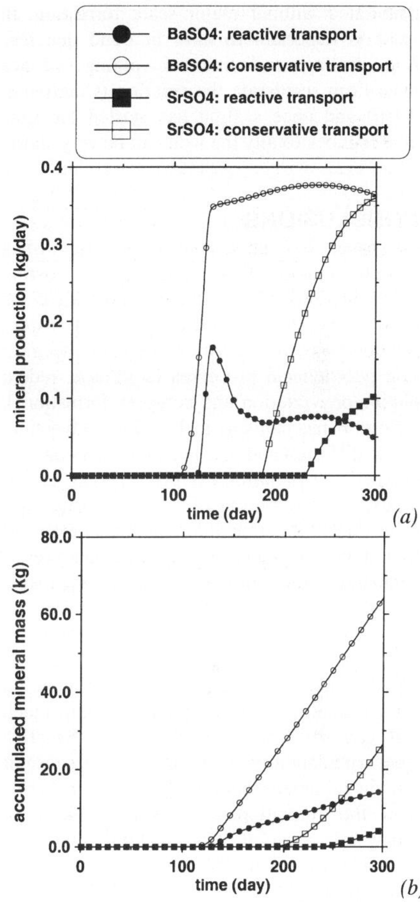

Figure 6. Minerals that potentially precipitate at production well as result of incompatible water mixing: (a) produced mineral per day and (b) accumulated total mass

Reactive transport simulation is more realistic because of the consideration of $SrSO_4$ and $BaSO_4$ precipitation in reservoir (Bertero et al., 1988), predicting a lower scaling in the production well than the conservative transport simulation (Figure 6). This can be justified by the fact that in a reactive transport simulation part of the mineral that could cause scale in the production well has already precipitated in the reservoir. For kinetic reasons, they are called *potentially* precipitated minerals because the local equilibrium assumption at the production well is not necessarily valid. Produced water can be $SrSO_4$ and $BaSO_4$

supersaturated without major scale formation. In this case is important to have in mind that pre-existing scaling and internal piping surface roughness can accelerate the reaction (Crabtree et al., 1999) and once scaling has started the time required to block totally the well can be very short.

4. CONCLUSIONS

This paper has presented a summary of a coupled formulation that combines an existing THM formulation with reactive transport equations in a fully coupled way. The reactive transport formulation takes into account some of the most relevant geochemical processes (acid/base, redox, dissolution/precipitation and complex formations). The transport mechanisms included are advection, molecular diffusion and mechanical dispersion.

A numerical simulation exercise was carried out to evaluate the possibility of matrix damage in a reservoir due to incompatible chemical mixing of different types of water. In the case analyzed a large mineral precipitation in the formation capable to damage it was not identified. However, the numerical simulation indicated the possibility of scaling formation at the production well. It was also shown that a conservative transport simulation has significant limitations for predicting how much mineral can scale at production well. Reactive transport simulation is more realistic because it allows mineral precipitation at the reservoir. In this problem the formation heterogeneity plays an important role since it increases the mixing of incompatible waters in the reservoir and in the producing well.

Future research in this area involves the detailed study of auto sedimentation caused by changes in reservoir and producing well temperature and pressure as well as reservoir 3D modeling. In addition, it is also envisaged to study other types of scaling such as carbonated mineral scaling and other mechanisms that cause matrix damage such as particle transportation and deposition. The flexibility of the formulation incorporated in CODE_BRIGHT allows the tackling this variety of petroleum engineering problems.

ACKNOWLEDGEMENTS

The authors are grateful for the financial support given by CNPq (Conselho Nacional de Desenvolvimento Cientíco e Tecnológico) and ANP (Agência Nacional do Petróleo) in Brazil. The assistance of the Ministerio de Ciencia y Tecnología of Spain through research grant BTE2001-2227 is also thankfully acknowledged.

REFERENCES

Bertero, L., Chierici, G.L., Gottardi, G., Mesini, E. & Mormino, G. 1988. *Chemical equilibrium models: their use in simulating the injection of incompatible waters.* In Society Petroleum Engineers. Reservoir Eng., pp. 288-294.

Crabtree, M., Eslinger, T., Fletcher, P., Johnson, A., King, G. 1999. *Fighting Scale – Removal and Prevention.* In Oilfield Review, Autumn 1999: pp. 30-45.

Gens, A., Guimarães, L. N., García-Molina, A. J. & Alonso, E. E. 2002. *Factors Controlling Rock-Clay Buffer Interaction in a Radioactive Waste Repository.* In Engineering Geology, v. 64, pp. 297-308.

Guimarães L. N. 2002. *Análisis multi-componente no isotermo en medio poroso deformable no saturado.* Phd Thesis, Geotechnical Engineering Department, Technical University of Catalunya, Spain.

Olivella, S., Gens, A., Carrera, J., Alonso, E. E. 1995. *Numerical formulation for a simulator (CODE_BRIGHT) for the coupled analysis of saline media.* In Engineering Computations, n. 13, pp. 87-112.

Osborne, C. G., Ravenscroft, P. D., McCracken, I. R. 1994. *Predicted water chemistry trends in North Sea formation brines.* In Marine and Petroleum Geology, n. 11 (1), pp. 20-23.

Silva, F. L., Carageorgos, T., Bredikovetsky, P. Rosário, F. 2001. *Modelagem da cinética de precipitação da barita em projetos de injeção de água.* In 1o Congresso Brasileiro de P&D em Petróleo e Gás, Natal - Brazil, pp.161.

Sorbie, K. S., Mackay, E. J. 1997. *Mixing of injected, connate and aquifer brines in waterflooding and its relevance to oilfield scaling.* In Journal of Petroleum Science & Engineering, 27, pp. 85-106.

Yuan, M. D., Todd, A. C. 1991. *Prediction of sulfate scaling tendency in oilfield operations.* In SPE Production Engineering, n. 6, pp. 63-72.

DEVELOPMENT OF 3D FEM SOFTWARE FOR TWO-PHASE FLOW AND ITS APPLICATION TO HORONOBE NATURAL GAS SIMULATION

H. Li[1], M. Sato[1], T. Sakai[2]
[1])Geoscience Research Laboratory (GRL), Yamato, Kanagawa, Japan
[2]) Mitsui Mining Engineering Co., LTD, Tokyo, Japan

Abstract: A newly developed numerical simulator of two-phase flow using three-dimensional finite element method is presented in this paper. It is described that the fundamental simultaneous equations, the deduction to implicit pressure explicit saturation formulation and their finite element discretization method. Furthermore, its practical application to the numerical simulation project of predicting *Horonobe* natural gas product is also introduced.

1. INTRODUCTION

In recent years, the necessity of evaluating multiphase flow engineering problems is becoming more and more outstanding over the world. It rises from diverse newly blooming fields such as compressed air energy storage underground, Nakagawa (1990), Nakata (2002), contaminated soil and groundwater remediation, Helmig (1997), geothermal energy extraction, Fujinawa (1991), Fujii (2002), dissoluble natural gas extraction and carbon dioxide sequestration in coal seam, Gunter (2002), where it is deficient in customized but directed software to facilitate the complicated numerical simulations. Although a few applications are available, most of them are limited to finite difference method without mature pre-post processors or for other traditional industries, e.g., TOUGH, Pruess (1987) and COMET2, Pekot (2001). The former has no satisfactory interfaces yet and the latter is slightly revised from the primitive application to petroleum reservoir simulation. Against this background, Geoscience Research laboratory is extending its three-dimensional finite element method simulators on unsaturated subsurface water or mine gas flow (MGF) from single to the multi-phase flow. The updated software is gradually improved through various practical applications and expected to grow to become one helpful simulator.

2. FUNDAMENTAL THEORY ON TWO-PHASE FLOW

2.1 Gas and water two-phase flow dominating simultaneous equations

$$\nabla \cdot [b_w M_w (\nabla p_w + \gamma_w \nabla Z)] + q_w = \frac{d}{dt}(\Phi b_w S_w) \quad (1)$$

$$\nabla \cdot [b_g M_g (\nabla p_g + \gamma_g \nabla Z)] + R_{sw} \nabla \cdot [b_w M_w (\nabla p_w + \gamma_w \nabla Z)] = \frac{d}{dt}(\Phi b_g S_g + R_{sw} \Phi b_w S_w) \quad (2)$$

where $\nabla\cdot$ is divergence operator and ∇ is gradient operator. b_n (n = g or w, i.e., gas or water) is flow shrinkage factor (-); $b = 1/\beta$, where β is formation volume factor (-). Formation volume factor is the ratio of the volume (m^3) at a certain state V to that at the standard state V_s, i.e., $\beta = V/V_s$. Since V is the function of pressure, $\beta_n = \beta_n(p_n)$. So is b. $M_n = k_n K/\mu_n$ is phase mobility; $k_n \in [0,1]$ is phase relative permeability (-), which is the function of phase saturation; K is abstract permeability tensor (m^2); μ_n is phase viscosity (kg/m/sec). The capillary pressure (Pa) $P_c = p_g - p_w \approx 0$ is regarded in deep underground. $\gamma_n = \rho_n \cdot g$, where ρ_n is phase mass density (kg/m^3) and g is gravitational acceleration (9.8 m/sec^2). Z is elevation (m). $R_{sw} = R_{sw}(p_w)$ is gas solubility (-) in water. Φ is porosity; if necessary $\Phi = \Phi(p_w)$. The phase saturation (-) must of course satisfy $S_g + S_w = 1$.

2.2 The deduction of Right Hand Side

The time derivative in R.H.S. of the dominating equations can be deduced and expressed as the function of $\dot{p}_w$ and $\dot{S}_w$.

$$\frac{d}{dt}(\Phi b_w S_w) = S_w{}^n(\Phi^n b_w{}' + \Phi' b_w{}^{n+1}) \cdot \dot{p}_w + \Phi^{n+1} b_w{}^{n+1} \cdot \dot{S}_w \quad (3)$$

$$\frac{d}{dt}(\Phi b_g S_g) = S_g{}^n(\Phi^n b_g{}' + \Phi' b_g{}^{n+1}) \cdot \dot{p}_w - \Phi^{n+1} b_g{}^{n+1} \cdot \dot{S}_w \quad (4)$$

$$\frac{d}{dt}(R_{sw}\Phi b_w S_w) = S_w{}^n(\Phi^n b_w{}' R_{sw}{}' + \Phi' b_w{}^{n+1} R_{sw}{}^{n+1}) \cdot \dot{p}_w + \Phi^{n+1} b_w{}^{n+1} R_{sw}{}^{n+1} \cdot \dot{S}_w \quad (5)$$

where the superscript n and $n+1$ indicate the value of variables at time t and $t+\Delta t$ respectively. Thus, the dominating simultaneous equation (1) and (2) can be rewritten as

$$\nabla \cdot [b_w M_w (\nabla p_w + \gamma_w \nabla Z)] = d_{11}\dot{p}_w + d_{12}\dot{S}_w \quad (6)$$

$$\nabla \cdot [b_g M_g (\nabla p_w + \gamma_g \nabla Z) + R_{sw} \cdot b_w M_w (\nabla p_w + \gamma_w \nabla Z)] = d_{21}\dot{p}_w + d_{22}\dot{S}_w \quad (7)$$

where

$$d_{11} = S_w{}^n(\Phi^n b_w{}' + \Phi' b_w{}^{n+1})$$
$$d_{12} = \Phi^{n+1} b_w{}^{n+1}$$
$$d_{21} = S_g{}^n(\Phi^n b_g{}' + \Phi' b_g{}^{n+1}) + S_w{}^n(\Phi^n b_n{}' R_{sw}{}' + \Phi' b_w{}^{n+1} R_{sw}{}^{n+1})$$
$$d_{22} = \Phi^{n+1}(-b_g{}^{n+1} + b_w{}^{n+1} R_{sw}{}^{n+1})$$

2.3 Pressure-saturation formulation

Eliminating $\dot{S}_w$ from (6) and (7) yields

$$d_{22}\nabla \cdot [b_w M_w (\nabla p_w + \gamma_w \nabla Z)] - d_{12}\nabla \cdot [b_g M_g (\nabla p_w + \gamma_g \nabla Z) + R_{sw} b_w M_w (\nabla p_w + \gamma_w \nabla Z)] = (d_{11} \times d_{22} - d_{21} \times d_{12})\dot{p}_w \quad (8)$$

From (6)

$$\dot{S}_w = \nabla \cdot [b_w M_w (\nabla p_w + \gamma_w \nabla Z) - d_{11}\dot{p}_w] / d_{12} \quad (9)$$

If the initial values of p_w and S_w are $p_w{}^n$ and $S_w{}^n$, from (8) $p_w{}^v$ can be sought implicitly. From (9), $S_w{}^v$ can also be sought explicitly. When $\left|p_w{}^{v+1} - p_w{}^v\right| \le \varepsilon$ is satisfied, $p_w{}^{v+1} \to p_w{}^{n+1}$.

3. NUMERICAL DISCRETIZATION

3.1 finite element in space

Make $\lambda_w = (d_{22} - d_{12}R_{sw})b_w M_w$, $\lambda_{wg} = \gamma_w \lambda_w$, $\lambda_g = -d_{12} b_g M_g$, $\lambda_{gg} = \gamma_g \lambda_g$, $\lambda_p = \lambda_w + \lambda_g$, $\lambda_Z = \lambda_{wg} + \lambda_{gg}$ $d_\tau = d_{11} \times d_{22} - d_{21} \times d_{12}$, then (8) becomes

$$\nabla \cdot [\lambda_p \nabla p_w + \lambda_Z \nabla Z)] = d_\tau \cdot (\partial p_w / \partial t) \quad (10)$$

Unknown variable p_w is discretized as follows

$$p_w = \Psi = N_k \Psi_k = \sum_{k=1}^{n_0} N_k \Psi_k \quad (11)$$

where N_k is interpolation shape function; Ψ_k is pore pressure at gridpoint; n_0 is the number of gridpoints. Taking advantage of the *Galerkin* method yields

$$\sum_{e=1}^{N} \int_{V_e} \left\{ \frac{\partial}{\partial x_i}[\lambda_p (\frac{\partial N_k}{\partial x_j}\Psi_k) + \lambda_Z (\frac{\partial X_3}{\partial x_j})] - d_\tau (\frac{\partial \Psi}{\partial t}) \right\} N_m dV_e = 0 \quad (12)$$

where N is the number of elements; V_e is the volume of e th element. Applying partial integral and *Gauss* theorem to (12) gets

$$[A_{mk}]\{\Psi_k\} + [F_{mk}]\{\frac{\delta_t \Psi_k}{\Delta t}\} = \{Q_m\} - \{G_m\} \quad (13)$$

where

$$A_{mk} = \sum_{e=1}^{N} \int_{V_e} \lambda_p \cdot \frac{\partial N_m}{\partial x_i} K_{ij} \frac{\partial N_k}{\partial x_j} dV_e \, ;$$

$$F_{mk} = \sum_{e=1}^{N} \int_{V_e} (d_\tau N_k) \cdot N_m dV_e \, ;$$

$$Q_m = \sum_{e=1}^{N} \int_{\Gamma_e} [(d_{22} - d_{12} R_{sw}) b_w \bar{u}(w) - d_{12} b_g \bar{u}(g)] N_m d\Gamma_e \, ;$$

where

$$\bar{u}(w) = M_w \left(\frac{\partial \Psi}{\partial x_j} + \gamma_w \frac{\partial X_3}{\partial x_j}\right) n_i$$

$$\bar{u}(g) = M_g \left(\frac{\partial \Psi}{\partial x_j} + \gamma_g \frac{\partial X_3}{\partial x_j}\right) n_i$$

$$G_m = \sum_{e=1}^{N} \int_{V_e} \lambda_Z \frac{\partial N_m}{\partial x_i} K_{i3} dV_e \, .$$

3.2 Finite difference in time

Time derivative of the variable are approximated as follows,

$$\frac{\partial \Psi}{\partial t} = \frac{\Psi(t+\Delta t) - \Psi(t)}{\Delta t} = \frac{\Psi^{n+1} - \Psi^n}{\Delta t} \tag{14}$$

$$\Psi = (1-\omega)\Psi^n + \omega \cdot \Psi^{n+1} \tag{15}$$

where $\omega = 1$ for backward difference; $\omega = 1/2$ for central difference. Accordingly, after some deduction (13) can be changed into

$$\left([A_{mk}]^{n+1/2} + \frac{1}{\omega} \frac{[F_{mk}]^{n+1/2}}{\Delta t} \right) \{\Psi_k\}^{n+1} = \frac{1}{\omega} \left(\{Q_m - G_m\}^{n+1/2} + \frac{[F_{mk}]^{n+1/2}}{\Delta t} \{\Psi_k\}^n \right) - (2 - 2\omega)[A_{mk}]^{n+1/2} \{\Psi_k\}^n \tag{16}$$

3.3 To seek new saturation

The same discretization method is applied to renew the value of saturation accordingly.

$$\left\{ (S_w^m)^{n+1} \right\} = \left\{ (S_w^m)^n \right\} - \left[F_S^m \right]^{-1} F_p^m (\Psi^{n+1} - \Psi^n) + \left[F_S^m \right]^{-1} \left\{ Q_w^m - A_w^m \Psi_k - G_w^m \right\} \cdot \Delta t \tag{17}$$

where

$$F_S^m = \sum_{e=1}^{N} \int_{V_e} (d_{12} N_k N_m) dV_e \, ;$$

$$F_p^m = \sum_{e=1}^{N} \int_{V_e} (d_{11} N_k N_m) dV_e \, ;$$

$$Q_w^m = \sum_{e=1}^{N} \int_{\Gamma_e} b_w \bar{u}(w) d\Gamma_e \, ;$$

$$A_w^m = \sum_{e=1}^{N} \int_{V_e} \frac{\partial N_m}{\partial x_i} b_w M_w \frac{\partial N_k}{\partial x_j} dV_e \, ;$$

$$G_w^m = \sum_{e=1}^{N} \int_{V_e} \frac{\partial N_m}{\partial x_i} \gamma_w b_w M_w \frac{\partial X_3}{\partial x_j} dV_e \, .$$

4. HORONOBE PROJECT

The numerical simulation project in the survey of natural gas productivity in Horonobe town, Tensio County, Hokkaido Prefecture was supported by Japan Nuclear Cycle Development Institute (JNC) in the end of 2002. There, from up to down, the underground strata are mainly divided into three layers, i.e., Upper Koetou (0 ~ -48m) , Lower Koetou (-48 ~ -435m) and Wakanai (-435m ~) . The simulation consists of 250 and 500 m deep single vertical well in case of plain and lining, single horizontal tunnel at depth of 125, 250, 375, 500 m respectively, double neighbouring well models and some parametric study.

The physical mechanical parameters are shown as Table 1.

Table 1. Physical mechanical parameters

Parameter	Unit	Value
absolute permeability	m^2	1.e-15 *
		1.e-17 **
		1.e-18***
relative permeability	-	= saturation
water viscosity	kg/m/s	1.e-3
gas viscosity		1.1e-5
water mass density	kg/m^3	1.e+3
gas mass density		7.17e-1
initial porosity	-	0.6 *
		0.6-0.36 **
initial saturation	-	1.000-0.986 *
		0.977 **

* for Koetou ** for Wakanai *** for Lining

Additionally, the elastic compression of porosity is expressed as follows,

$$\Phi = \Phi_0 (1 + c_p \times \Delta p_w) \tag{18}$$

where $c_p = 1.e^{-9}$ Pa .

The volume compression of gas

$$b_g = p_w \times \frac{273.15}{273.15 + T} \tag{19}$$

where the unit of porous pressure is Atmospheric pressure and temperature is Celsius degree.

The solubility of gas into water is shown as

$$R_{sw} = 0.033 + 0.3 \times 10^{-6} p_w \tag{20}$$

where the unit of pressure is Pascal.

The vertical model (only one-fourth is adopted because of its symmetry) is illustrated as Figure 1 and its enlargement for the vicinity of well bottom as Figure 2.

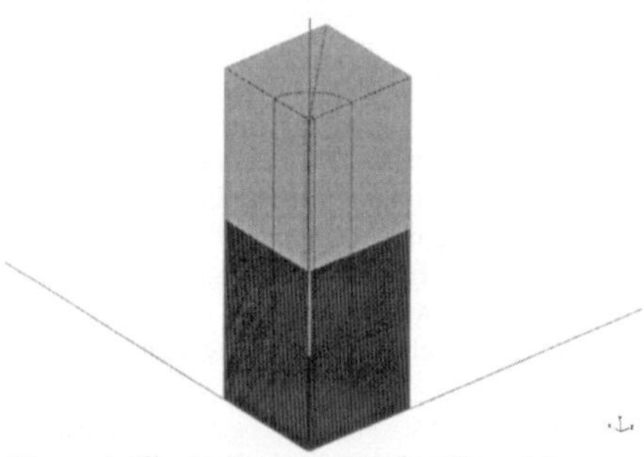

Figure 1. Illustration of vertical well model

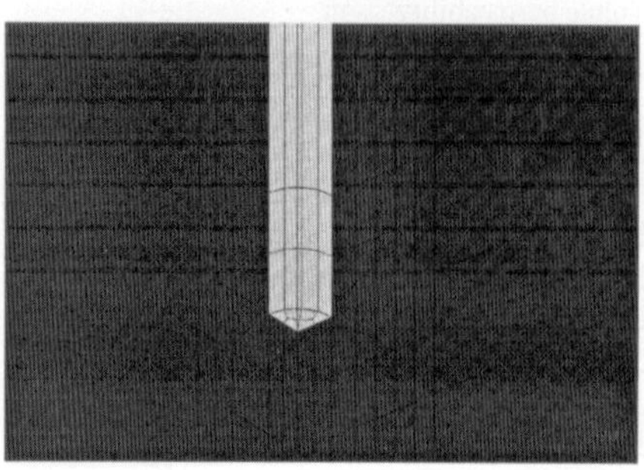

Figure 2. Local enlargement of Fig. 1

The horizontal tunnel model is illustrated as Figure 3; its enlargement as Figure 4.

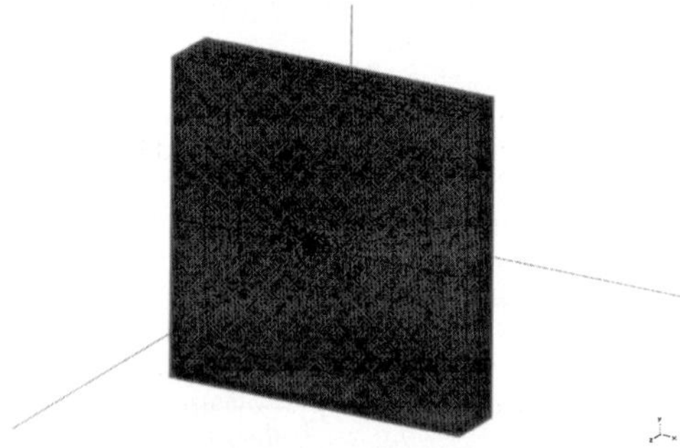

Figure 3. Illustration of horizontal tunnel model

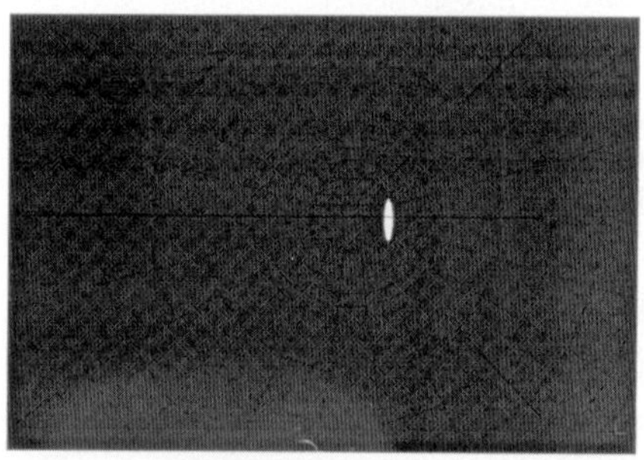

Figure 4. Enlargement of Fig. 3

The geometry of models is shown in Table 2

Table 2. Model Geometry (unit: m)

radius of vertical well	3.25
radius of horizontal tunnel	4.0 (about)
total length of tunnel at –125m	150.6
total length of tunnel at –250m	934.1
total length of tunnel at –375m	150.6
total length of tunnel at –500m	871.3
range of vertical well model	500×500×700
range of horizontal tunnel model	100×100×20

The initial pressure condition is set according to

$$p(Z) = \gamma_w \cdot H(Z) \tag{21}$$

where Z is the coordinate (the origin is at the intersection between well axis and –700m surface) in vertical direction (m); $H(Z) = 700 - Z$ is the height of water head (m).

The side surfaces of vertical well model are set proof for flow (Exterior boundaries are wide enough to avoid any visible influence. Flow rate there approximates zero accordingly) and saturation is fixed as initial values. The pressure of top boundary is fixed as zero (It in actual corresponds to one atm.); that of bottom is fixed as 7Mpa. The exterior boundaries of horizontal tunnel model are set pressure fixed as the corresponding subsurface water pressure distribution. The well and tunnel walls are set pressure fixed as zero.

As a preliminary estimate, stepwise excavation is not functioned in the program temporarily. The simulation results consist of pore pressure, saturation, solubility distribution change, total gas and water well or tunnel output and portion of vicinity of well bottom (one diameter height) after 24 hour and 2 year. The following figures are some of examples.

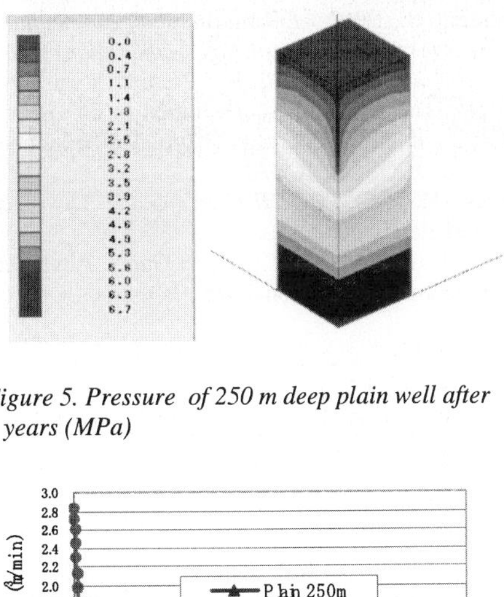

Figure 5. Pressure of 250 m deep plain well after 2 years (MPa)

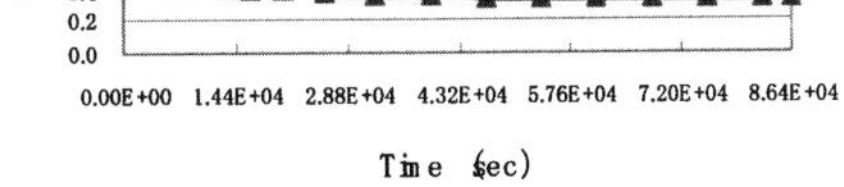

Figure 6. Gas gush rate from plain well

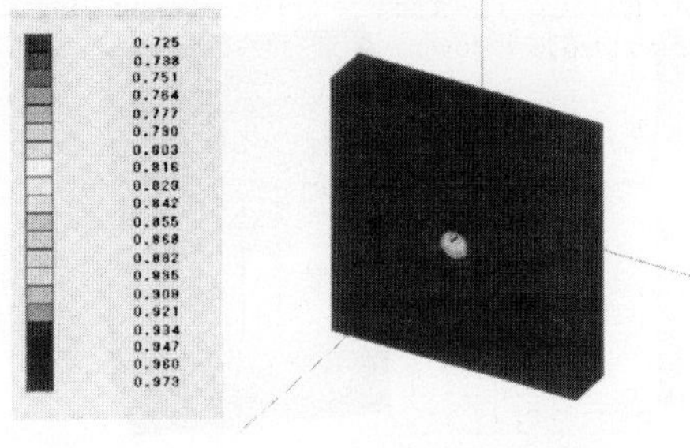

Figure 7. Saturation of tunnel GL-250 after 2 years

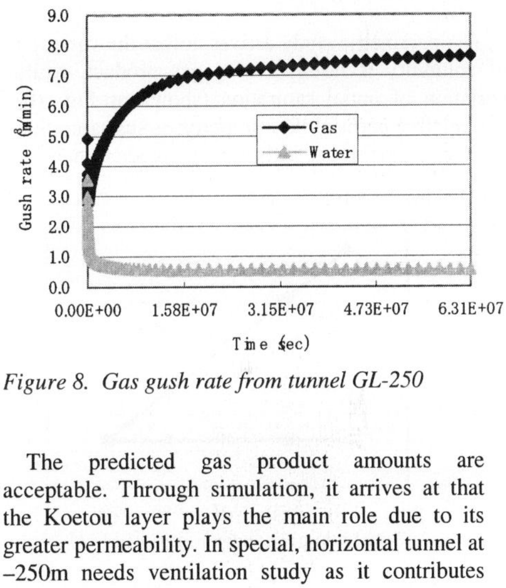

Figure 8. Gas gush rate from tunnel GL-250

The predicted gas product amounts are acceptable. Through simulation, it arrives at that the Koetou layer plays the main role due to its greater permeability. In special, horizontal tunnel at –250m needs ventilation study as it contributes most to product. Unlike vertical well, gas accumulation tendency around horizontal tunnel keeps going on because gas doesn't rise up like that around vertical well according to their different pressure gradient. This makes the permeability of gas in the vicinity of tunnel are gradually improved along time going by.

As one of example of double well model results, Fig.9 reflects the interference between neighbouring wells. Indeed, consideration on the interaction among wells and intersecting tunnels needs complicated overall model afterwards.

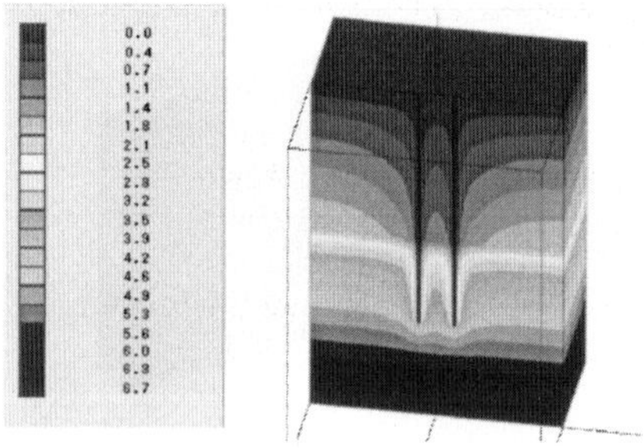

Figure 9. Pressure of double-well after 2 years (MPa)

The parametric study arrives at that the variation of solubility is insensitive to gas product, while variation of initial saturation (shown as Fig. 10) and relative permeability property versus saturation are rather sensitive to influence the prediction.

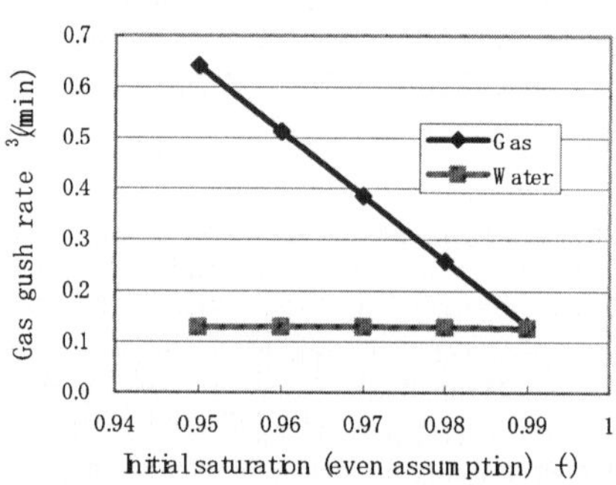

Figure 10. Initial saturation parametric study in case of 250 deep plain well after 2 years

5. CONCLUSIONS

The 3D FEM two-phase flow program with friendly user interface is preliminary established. Finite element method makes the software to be able to flexibly dispose more complicated model geometry. Its functions and efficiency have also begun to be examined through practical application. This work offers a console to lift a series of our existing single flow software to the advanced version of multiphase flow.

REFERENCES

Nakagawa, K. 1990. *A study on numerical simulation of porous flow coupled behaviour and its application to geotechnical problems*. Ph.D Dissertation of Central Research Institute of Electric Power Industry (In Japanese): pp. 60 - 135.

Nakata, M. 2002. *A study on the Compressed Air Energy Storage in hard rock cavern*. Ph.D Dissertation and Special report of Technical Research Institute of Mitsui Construction CO., Ltd (15) (In Japanese).

Helmig, R. 1997. *Multiphase flow and transport processes in the subsurface*. Springer-Verlag.

Fujinawa, K. 1991. *Theoretical studies on two-phase heat conduction in saturated porous media using the finite element method – studies on heat transfer in porous media (I)*. Trans. JSIDRE 152: pp.83-90.

Fujii, H. 2002. *A study on completion and configuration of heat exchange wells in Ground-Coupled Heat Pump systems with groundwater flow*. J. Geotherm. Res. Soc. Japan 24(2): pp.191-205.

Gunter, B. 2002. *Enhanced coalbed methane recovery and CO_2 storage: simulation issues and model comparisons*. Present status and perspective of CO_2 sequestration in coal seam; Proc. Intern. Workshop, Tokyo, 5-6 September 2002. JCOSC & JCOAL.

Pruess, K. 1987. *TOUGH User's Guide*. Lawrence Berkeley Laboratory.

Pekot, L. 1998. *COMET2 User's Guide*. Advanced Resources International, Inc.

A COUPLED FLOW-TRANSPORT-DEFORMATION MODEL FOR UNDERGROUND COAL GASIFICATION

Jishan Liu[1], Cliff Mallett[2], Andrew Beath[2], Derek Elsworth[3] and Barry Brady[4]
[1]) School of Oil & Gas Engineering, The University of Western Australia
39 Stirling Highway, Crawley WA 6009, Australia, Email: jishan@cyllene.uwa.edu.au
[2]) CSIRO Exploration & Mining
[3]) Energy Institute and Department of Energy and Geo-Environmental Engineering
Pennsylvania State University, University Park, PA16802, USA
[4]) Faculty of Engineering, Computing and Mathematics, The University of Western Australia

ABSTRACT: A coupled flow-transport-deformation model is developed to simulate underground coal gasification (UCG) though which coal is converted to useful gases in-situ without the need for mining. During UCG the carbon in a coal seam is burnt while high heat energy is released. As a result of burning, new cavities are created in the coal seam. These cavities will modify the permeability field dramatically. The modified permeability field will in turn affect the gasification. It is apparent that UCG is a coupled phenomenon involving hydrological, chemical, thermal and mechanical processes. In this study, UCG is simulated as a selective mining method through introducing a cavity evolution equation. The equation defines the selectivity as a function of porosity, and can be understood as the influence of the residual carbon on the gasification rate. When the porosity is equal to the original value (the initiation of gasifying), the gasification rate reaches its maximum; when the porosity is equal to the final value (the completion of gasifying), the gasification rate is reduced to zero. The introduction of this equation into classical flow and transport equations constrains the whole gasification process from initiation to completion. Numerical results of a typical five-spot gasification layout have confirmed the validity of this approach.

1. INTRODUCTION

Underground coal gasification (UCG) is a process by which coal is converted to useful gases in situ without the need for mining. The gases can subsequently be used to produce heat, generate power or synthesize a variety of chemical products. UCG is realized through circulating gasification agents (air or oxygen and water) into a coal seam. The product gases are transported to the surface for processing and utilization. This conceptually simple technology eliminates mining, coal transport, post-mining reclamation, and management of combustion residues with minimum start-up capital requirements, operating costs and construction time. UCG has had an established operational history in the former USSR since 1930s and a number of field trials have been conducted all over the world: UK Newman Spinney trial (1950s); American trials (1970s and 1980s) (Britten and Thorsness, 1989; Cena et al, 1988; Hill et al., 1984); Belgium trials (1982-1987); Spain trials (1992-1999); and China trials (1990s). The Spanish trial was completed successfully and a feasibility study has now been initiated by the UK (1999-2003) into whether UCG can enhance the value of the coal reserves both onshore and offshore. In Australia, feasibility studies of UCG have been conducted since 1999. A trial burn was initiated in 1999 in Queensland, and the process is still in operation (Walker et al., 2001; Blinderman and Jones, 2002). Some of these trials may have led to a technology ready for commercialization but UCG has not become a mature method for in-situ extraction of coal in the western world.

Practical application of UCG is restricted partly by lack of satisfactory quantitative descriptions of the processes involved. During gasification a cavity is formed in the coal seam and high heat energy is released. Both the formation of the reactor cavity and the high temperature resulting from the coal burning affect the overburden deformation. If the cavity is too large, it may result in collapse of the overburden, either piecewise or catastrophically. The host rock deformation modifies the permeability field in the UCG zone, and the change in permeability affects the migration of the potential contaminants resulting from UCG. It is apparent that UCG is a coupled phenomenon involving hydrological, chemical, thermal and mechanical processes. Quantification of these processes and their interactions has been the research target of recent studies. A model of UCG was developed to study the impact of changing parameters on the recession rate of a coal face reacting with hot gases

(Beath et al., 2000a, 2000b, 2001). The model is strictly one-dimensional, representing a slab of infinite dimensions with gas reacting on one face only. A double-diffusive natural convection model was developed to simulate an idealized situation in UCG (Van Der Eyden et al., 1998). The model is capable of producing most of the essential features of the process. In several studies (Trent and Langland, 1983; Stephenson et al, 1983; Hettema et al., 1998), overburden deformation was investigated. Flow and chemical processes in a deep underground gasifier were simulated by using a channel-type model (Batenburg et al., 1994), a cavity-type model (Britten and Thorsness, 1989; Coeme et al., 1993; Pirlot et al., 1998), and boundary element method (Mathy et al., 1994). In these previous studies, the coupling among flow, transport and gasification has not been investigated. In this study, UCG is simulated as a selective mining method through introducing a cavity evolution equation. The equation defines the selectivity as a function of porosity, and can be understood as the influence of the residual carbon on the gasification rate. When the porosity is equal to the original value (the initiation of gasifying), the gasification rate reaches the maximum; when the porosity is equal to the final value (the completion of gasifying), the gasification rate is reduced to zero. The introduction of this equation into classical flow and transport equations constrains the whole gasification process from initiation to completion.

2. GOVERNING EQUATIONS

2.1 DERIVATION OF GASIFICATION EQUATION

Assuming coal is packed with a number of spherical particles and the surface of a particle is itself receding during gasification, the rate of gasification may be expressed as

$$\frac{dn}{dt} = -\frac{4\pi r^2 k_0}{s} C_1 \tag{1}$$

Where n represents the number of moles remaining in the ungasified core, s is the cross-sectional area of gasifying mineral (carbon) on the solid sphere, r is the radius of the sphere, k_0 is a rate constant, t is time, and C_1 is the concentration of a gasification agent (such as oxygen). Assuming V is the molar volume, the number of moles, n, is defined as

$$n = \frac{4\pi r^3}{V} \tag{2}$$

Differentiating equation (2) with respect to time, t, and equating to equation (1) yields

$$\frac{dr}{dt} = -\frac{Vk_0 C_1}{s} \tag{3}$$

Equation (3) represents constant velocity movement of the gasification interface. If r_0 is the initial radius of the particle, α is the fraction gasified, it may readily shown that

$$\alpha = 1 - \frac{r^3}{r_0^3} \tag{4}$$

Where upon differentiation with respect to time, t, becomes

$$\frac{d\alpha}{dt} = -\frac{3r^3}{r_0^3}\frac{dr}{dt} \tag{5}$$

Combining equations (3), (4) and (5) results in the following equation

$$\frac{d\alpha}{dt} = \frac{3Vk_0}{r_0^3 s}(1-\alpha)^{2/3} C_1 \tag{6}$$

Assuming G is the mass fraction of a total gasifiable mineral (carbon) per unit medium mass, the relation between gasification-modified porosity, ϕ, and the gasification fraction, α, can be defined as

$$\phi - \phi_0 = G\alpha \tag{7}$$

Where ϕ_0 is the original porosity. If all of the gasifiable mineral is gasified, the porosity should reach the maximum value, ϕ_f, defined as

$$\phi_f = \phi_0 + G \tag{8}$$

Combining equations (6), (7) and (8) gives

$$\frac{d\phi}{dt} = k(\phi_f - \phi)^{2/3} C_1 \tag{9}$$

where k is defined as

$$k = \frac{3k_0 V G^{1/3}}{r_0^3 s} \quad (10)$$

The concentration of the product gas, C_2, can be defined as

$$C_2 = \frac{\rho_c}{\rho_w \phi_0} \frac{d\phi}{dt} \quad (11)$$

where ρ_c and ρ_w are coal density and water density, respectively. Substituting equation (9) into (11) and differentiating with respect to *t* yield

$$\frac{dC_2}{dt} = \frac{\rho_c k}{\rho_w \phi_0} (\phi_f - \phi)^{2/3} C_1 \quad (12)$$

In heterogeneous systems, the importance of sample geometry cannot be over emphasized. This is particularly true for underground coal gasification in which the surface of a solid particle is itself receding during the course of gasification. Equation (12) can be modified to incorporate the effect of sample geometry as

$$\frac{dC_2}{dt} = \frac{\rho_c k}{\rho_w \phi_0} (\phi_f - \phi)^A C_1 \quad (13)$$

where *A* is defined as a geometric factor (*A* is equal to 0.667 for the case of uniform spheres). The term, $(\phi_f - \phi)^A$, can be understood as the influence of the residual carbon on the gasification rate. When $\phi = \phi_0$, the gasification rate is equal to the maximum value; when $\phi = \phi_f$, the gasification rate is reduced to zero. Equations (9), (12) and (13) will be used to define the source/sink terms in classic flow and transport equations, as discussed in the following section.

2.2 COUPLED EQUATIONS

It is assumed that carbon is distributed uniformly in a coal seam. Under this treatment, underground coal gasification can be treated as reactive flow and transport in geological media. Once injected into the coal seam, a reagent (oxygen) will burn the coal. This will in turn create cavities in the coal seam. These cavities will modify the permeability field dramatically. The modified permeability field will then affect the gasification. The coupling of these processes can be quantified through the following set of equations, similar to those for in-situ leaching (Liu and Brady, 2003): (1) Flow Equation; (2) Gasification Reagent Transport Equation; (3) Product Gases Transport Equation; (4) Cavity Evolution Equation; and (5) Cavity-Dependent Permeability Equation. We have implemented these equations into and solved by FASTFLO, a powerful commercial PDE solver for multiphysics. These equations are defined as

$$\phi_{,t} = (\phi K(\phi) h_{,i})_{,i} \quad (14)$$

$$V_{,i} = -K(\phi) h_{,i} \quad (15)$$

$$(\phi C_1)_{,t} = -(\phi C_1 V_i)_{,i} + (\phi D_{1ij} C_{1,j})_{,i} - \beta \frac{G\rho_c}{\rho_w} \phi_{,t} \quad (16)$$

$$(\phi C_2)_{,t} = -(\phi C_2 V_i)_{,i} + (\phi D_{2ij} C_{2,j})_{,i} + \frac{G\rho_c}{\rho_w} \phi_{,t} \quad (17)$$

where

$$K(\phi) = K_0 \left(\frac{\phi}{\phi_0} \right)^3 \quad (18)$$

and the porosity evolution, $\phi_{,t}$, is defined by equation (12) or (13). Where ϕ is porosity; $K(\phi)$ is the porosity-dependent hydraulic conductivity, *h* is total hydraulic head; V_i is the ith velocity component; x_i is coordinates; C_1 is the concentration of gasification agent (oxygen); D_{ij} are components of the dispersion tensor; ρ_c is the mass density of the coal; ρ_w is the mass density of water; β is the stoichiometric coefficient; C_2 is the concentration of the product gas, *k* is a rate constant; *G* is the burnable coal fraction; K_0 is the initial hydraulic conductivity; ϕ_f is the final porosity; and ϕ_0 is the initial porosity. Equations (14) and (15) are fluid continuous equation and Darcy's law, respectively. Equations (16) and (17) are transport equations of the gasification reagent and the product gas, respectively. Equation (18) describes the evolution of the hydraulic conductivity during the course of gasification. Equation (13) defines the evolution of cavity. These equations are derived based on the following assumptions:

- The coal can be treated as a continuous medium or an equivalent continuous medium;
- Gasification increases the porosity and permeability;
- Mass transfer rates are assumed to be large enough such that gasificationss are kinetically controlled.
- The influence of thermal effects is ignored.

3. EXAMPLE OF SOLUTIONS

In following simulations, underground coal gasification is treated as a selective mining technique. In a conventional mine, miners excavate a specific amount of ore and ship it to a mill for processing. UCG is quite different. Instead of sending miners to excavate the ore, "packets" of small unit volumes of gasification solution, all with a particular concentration of the reagent (oxygen) designed to convert a specific mineral (carbon) into useful gases, are sent into the coal seam through injection wells. These packets are called "in-situ miners" in this study. Each in-situ miner, along with many others, rapidly moves into the coal seam through the injection well, as illustrated in Figure 1(a). From this point on, the in-situ miner moves only on one particular path (pathway), created by hydraulic gradients. All the in-situ miners that enter the coal seam simultaneously at different points travel at slightly different velocities. Those with a straight pathway to the production well move the fastest, while those entering the ore directly opposite the production well move the slowest. At the entry within the wellbore, the frontal area is very small and each in-situ miner is squeezed very thinly and moves very fast. As the in-situ miner travels on its pathway, its frontal cross-sectional area starts to expand and the movement of the in-situ miner begins to slow down. During all the time, the in-situ miner is transporting through the coal seam, its chemical reagents are reacting with a specific mineral. When the in-situ miner moves slowly there is more time for gasifying a certain weight of a target mineral contained within its unit volume. As more and more product gases are added to the "in-situ miner", they tend to interfere with the gasification process itself. Some product gases and reagents are retarded for various reasons and they are left behind. In some instances, the reagents are depleted before the in-situ miner has reached a production well. These conceptual explanations are illustrated numerically in the following section.

The model is applied to a two-dimensional gasification system consisting of one central production well and four surrounding injection wells. The distance between wells is assumed as $25m$. The original porosity and the final porosity are assumed as 0.20 and 0.45, respectively. A zero pressure condition is applied for the production well, and a unit pressure for the injection wells. A unit concentration of the injected gasification agent is maintained throughout the simulations. Results after 10 iteration steps are illustrated in Figure 1. As shown in Figure 1(a), an ideal gasification process, under which the gasification pressure in the gasifier is always slightly less than hydrostatic pressure of fluid in coal seam and surrounding strata, has been created under the applied boundary conditions. This creates pressure gradient directed towards the gasifier. As a result, no flow from gasifier into surroundings is allowed, thus preventing the loss of valuable product as well as contamination of underground environment. The contours of the injected gasification reagent are illustrated in Figure 1(b). The injected agent (oxygen) is consumed while it flows from the injection wells towards the production well. Contours of the product gas concentration are illustrated in Figure 1(c). The product gas is produced initially near the wellbore entries, and then transported towards the production well. As the gasification front advances, the gasification region expands and less product gas remains in the gasified region. The distribution of gasified regions is illustrated in Figure 1(d) where numerical values represent the gasification percentage. The gasification has completed in the darkest area while only a fraction of coal has burnt in other areas (60-100% burnt in the less dark area and 20-60% in the white area).

4. CONCLUSIONS

During underground coal gasification the carbon in the coal seam is converted into gases while high heat energy is released. As a result of burning, new cavities are created in the coal seam. These cavities will modify the porosity and permeability field dramatically. The modified permeability field will then affect the gasification. It is apparent that UCG is a coupled phenomenon involving hydrological, chemical, thermal and mechanical processes. Through this study, a preliminary computer model has been developed to simulate UCG as a selective mining method. A cavity evolution equation is developed to define the selectivity as a function of porosity. The equation can be understood as the influence of the residual carbon on the gasification rate. When the porosity is equal to the original value (the initiation of gasifying), the gasification rate reaches the maximum; when the porosity is equal to the final value (the completion of gasifying), the gasification rate is reduced to zero. The introduction of this equation into classical flow and transport equations constrains the whole gasification process from initiation to completion. Numerical results of a typical five-spot gasification layout have confirmed the validity of this approach. Building on this success, the numerical model will be improved through incorporating thermal effects, overburden deformation, and more complex gasification reactions. These studies will be conducted in the near future.

ACKNOWLEDGEMENTS

The authors wish to acknowledge the financial support from Australia Research Council through the discovery project DP0342446.

REFERENCES

Batenburg, D. W., Biezen, E. N. J., Bruining, J., 1994, *A new channel model for underground gasification of thin, deep coal seams*, In Situ, 18(4), 419-451.

Beath A, Wendt M and Mallett C, 2000a. *Optimization of Underground Coal Gasification for Improved Performance and Reduced Environmental Impact*, Proceedings, Australian Institute of Energy, 9th Australian Coal Science Conference, November 26-29 2000, Brisbane.

Beath AC, Wendt MN and Mallett CW, 2001. *A Generalized Mathematical Model of Underground Coal Gasification*, Eighteenth Annual International Pittsburgh Coal Conference Proceedings, December 4-7, Newcastle, Australia.

Beath, A. C., Wendt, M. and Mallet, C. 2000b. *Underground Coal Gasification as a method for reducing the greenhouse impact of coal utilization*, Proceedings of 5th International Conference on Greenhouse Gas Control Technologies, August 13-16, Cairns, Australia.

Blinderman, M. S., and Jones, R. M., 2002, *The Chinchilla IGCC project to date: underground coal gasification and environment*, Gasification Technologies Conference, San Francisco, October 27-30.

Britten, J. A., and Thorsness, C. B. 1989. *A model for cavity growth and resource recovery during underground coal gasification*. In Situ, 13:1-53.

Coeme, A., Pirard, J. P., Mostade, M., 1993, *Modeling of the chemical processes in a longwall face gasifier at great depth*, In Situ, 17(1), 83-104.

Hettema, M. H. H., Wole, K-H. A. A. and Pater, C. J., 1998. *The influence of steam pressure on thermal spalling of sedimentary rock: theory and experiments*, Int. J. Rock Mech. Min. Sci. 35(1), 3-15.

Liu, J. and Brady, B. H., 2003, *Simulations of A Coupled Hydro-Chemo-Mechanical System in Rocks*, Geotechnical and Geological Engineering, in press (accepted on 11/1/03).

Mathy, B., Pirlot, P., Pirard, J. P., Coeme, A., Mastade, M., Ro, P. D., 1994, *Flow modeling in un underground gasifier at great depth by the boundary element method*, In Situ, 18(4), 399-418.

Pirlot, P., Pirard, J. P., Coeme, A., Mostade, M., 1998, *A coupling of chemical processes and flow in view of the cavity growth simulation of an underground coal gasifier at great depth*, In Situ, 22(2), 141-156.

Stephenson, D. E., Dass, S. T., and Shaw, D. E., 1983, *Numerical modeling of subsidence induced by underground coal gasification, including thermal effects*, In Situ, 7(1), 27-51.

Trent, B. C. and Langland, R. T., 1983. *Subsidence modeling for underground coal gasification*, In Situ, 7(1), 53-58.

Van Der Eyden, J. T., Van Der Meer, T. H., Hanjalic, K., Biezen, E., and Bruining, J., 1998. *Double-diffusive natural convection in trapezoidal enclosures*, Int. J. Heat Mass Transfer. Vol. 41, No. 33, pp. 1885-1898.

Walker, L. K., Blinderman, M. S., Brun, K., 2001, *An IGCC project at Chinchilla, Australia based on underground coal gasification*, Gasification Technologies Conference, San Francisco, October 8-10.

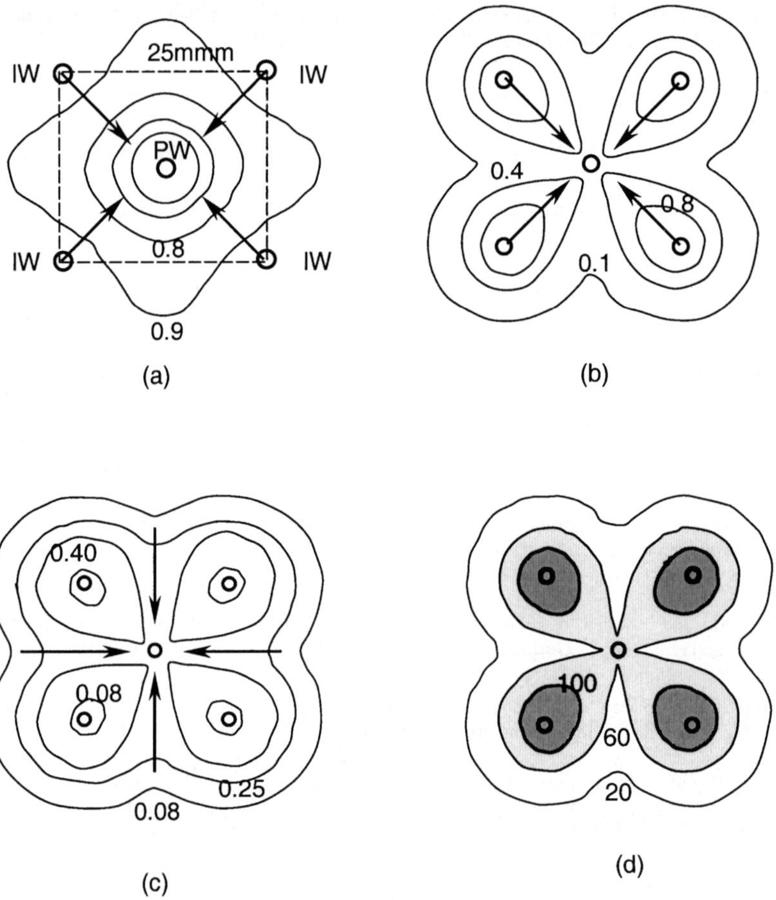

Figure 1. Numerical results for a typical five-spot underground coal gasification layout. (a). Contours of hydraulic head. Arrows represent the flow directions of fluid. IW and PW represent injection well and production well, respectively. It is shown that the gasification pressure in the gasification region is lower than hydrostatic pressure in the coal seam and surrounding strata. (b). Contours of gasification agent concentration. The injected agent (oxygen) is consumed while it flows from the injection wells towards the production well. Arrows represent the flow directions of the gasification agent. (c). Contours of product gas concentration. The product gas is produced initially near the wellbore entries, and then transported towards the production well. As the gasification front advances, the gasification region expands and less product gas remains in the gasified region. Arrows represents the flow directions of the product gas. (d). Distribution of gasified regions where numerical values represent the gasification percentage. The gasification has completed in the darkest area while only a fraction of coal has burnt in other areas (60-100% burnt in the less dark area and 20-60% in the white area).

INVESTIGATING THE RELATIONSHIP BETWEEN FAULT PERMEABILITY AND EFFECTIVE STRESS USING CONSTRAINTS FROM RESERVOIR INDUCED SEISMICITY (RIS)

Rebecca J Lunn[1], Aderson F do Nascimento[2], Patience Cowie[3]
[1]) School of the Built Environment, Heriot-Watt University, Edinburgh
[2]) Universidade Federal do Rio Grande do Norte, Programa de Pós-Graduação em Geodinâmica e Geofísica
[3]) Department of Geology and Geophysics, University of Edinburgh

Abstract: This research uses observations of RIS beneath Açu Reservoir, NE Brazil, to investigate the damage zone permeability of geological faults. High-resolution digital seismic monitoring of the reservoir has provided detailed information on the locations of seismic events. The temporal distribution of these events shows them to be directly related to annual fluctuations in the reservoir level. Model simulations, using a decoupled hydromechanical formulation (i.e. a static permeability field decreasing exponentially over depth due to increased confining pressure), show that for pressure-diffusion to be a hydrogeologically consistent mechanism for RIS, preferential flow must occur within 2D fault planes embedded in a 3D low permeability matrix. Further, the observed spatial and temporal variability of earthquakes indicate that these faults must have heterogeneous permeability fields with significant spatial structure; pockets of high and low permeability of the order of 0.5 - 1.5 km in diameter. Predictions of the maximum pressure change in these faults at hypocentral depths indicate <0.05 kPa is required to trigger seismic events.

1. INTRODUCTION

This research uses observations of reservoir induced seismicity (RIS) at Açu reservoir, NE Brazil, to investigate the spatial and temporal evolution of effective stress in the region and its relationship to fault permeability. Açu reservoir was constructed in 1983 and has a capacity of 2.4×10^9 m^3 maintained by a 34 m high earth-filled dam constructed on Precambrian shield. Annual reservoir variation is 3-6 m which results in annual seismic activity due to a proposed mechanism of pore pressure diffusion (Ferreira *et al.* (1995), do Nascimento *et al.* (2003a)). Digital data at Açu have revealed the seismic activity in remarkable detail (do Nascimento *et al.* 2003a). The majority of earthquake activity is clustered within several well-defined zones and individual zones are active over discrete periods of time. In this paper we present the main features of the seismicity data and propose that the observed spatial and temporal evolution of seismicity at Açu is defined by pressure diffusion through local fault planes with heterogeneous permeability structures.

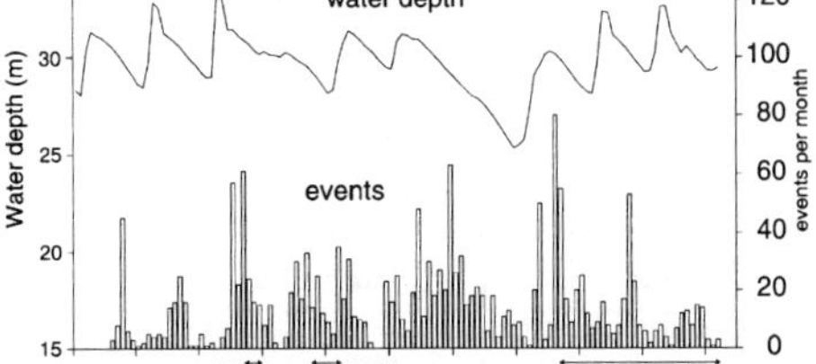

Figure 1. Monthly variation of reservoir water depth (top line) and monthly seismic activity (histograms) at Açu reservoir.

Table 1. Activation periods of clusters recorded by the digital seismic network and permanent station BA1.

Period of activation	Active seismic cluster
Nov. 1994 - 15/Aug./1995	a, BA1 cluster
15/Aug./1995 - 21/Oct./1995	a
1/Oct./1996 - 22/Nov./1996	b
25/Nov./1996 - 30/Dec/1996	c,b
30/Dec./1997 - May/1997	a,b

2. SEISMIC ACTIVITY AT AÇU

Seismic activity in the Açu area has been monitored over a ten year period from 1987-1997. However, accurate hypocentral information is only available from 1994 to 1997, when a network of three-component digital seismographs were operational which have provided a very accurate assessment of hypocentral with errors of ≈100m do Nascimento *et al.* 2003a.

With such well located events provided by the digital monitoring it has been possible (do Nascimento *et al.* 2003a) to examine the relationship between the migration of events and the water level variation (Figure 2 and Figure 3). Figure 2 shows three spatially identifiable seismic clusters, **a**, **b** and **c** that are separated both temporally and spatially (vertical sections from the projection lines shown on Figure 2 showed the clusters to be on separate fault planes). Table 1 summarises the temporal evolution of each cluster. A more detailed picture of their temporal evolution is shown in Figure 3. Here, the evolution of each of the clusters is shown simultaneously by plotting the depth of each event against the time delay calculated from the previous reservoir level peak. Three observations are apparent from Figures 1-3 and Table 1 that are obviously consistent with a mechanism for RIS of pressure diffusion:

(i) The onset of peak seismicity is delayed with increasing depth of cluster (Table 1, Figure 3)
(ii) The duration of seismic activity reduces with depth of cluster: this is consistent with the amplitude of the pressure wave decreasing with depth, causing a reduction in the time that the pressure exceeds a given value (Figure 3).
(iii) There is an apparent cessation in seismic activity during the 1993 drought (Figure 1).

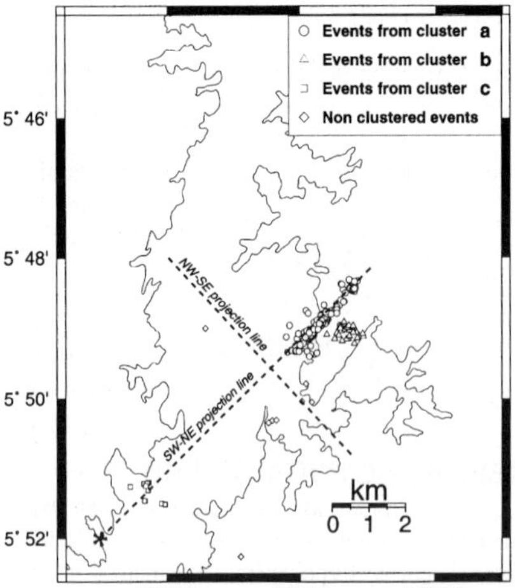

Figure 2. The epicentres located by the digital network in map view.

There are also, however, three further observations that are not immediately consistent with this mechanism:

(iv) The average speed of propagation of the pressure wave differs between clusters; for clusters **a** and **b** it is ≈ 15 m/day and for cluster **c** ≈ 21 m/day(calculated from Table 1).
(v) Each cluster shows a rapid onset to peak seismicity followed by a long tail with gradually diminishing event frequency (Figure 3).
(vi) There is no apparent relationship between temporal *within-cluster* evolution and depth: for pressure diffusion in a homogeneous medium within-cluster events should get gradually deeper over time (Figure 3).

In the following sections we demonstrate that pressure-diffusion is a hydrogeologically valid mechanism for RIS at Açu as supported by observations (i)-(iii), and that the seemingly inconsistent observations of seismicity (iv)-(vi) can be explained by flow within faults with spatially structured heterogeneous permeability fields.

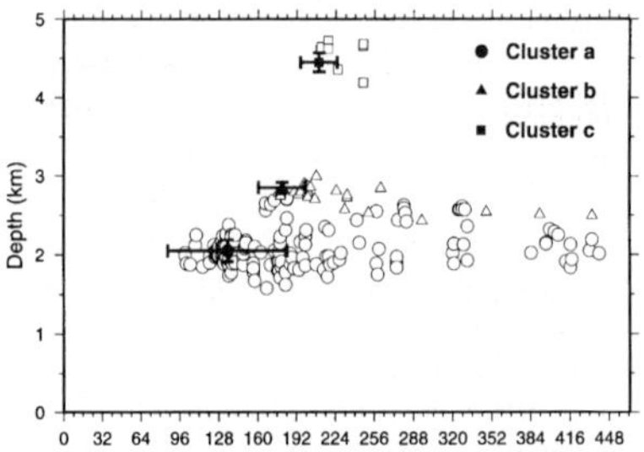

*Figure 3. Time delay from the previous reservoir peak plotted against depth for clusters **a**, **b**, and **c**. The filled symbols denote the modal average delay time, error bars show the interquartile range.*

3. IS PRESSURE-DIFFUSION CONSISTENT WITH THE HYDROGEOLOGY AT AÇU?

To assess the hydrogeological feasibility of pressure diffusion as a mechanism for RIS at Açu, it is first necessary to construct a numerical model of induced seismicity. Several previous authors including Roeloffs (1988) and Kalpna (2000) have modelled RIS employing a range of different modelling assumptions including: the level of hydromechanical coupling; the representation of rock properties; and the model geometry. Açu reservoir is an unusual case of RIS in that the region is relatively flat with only a 34m high reservoir dam. During the year, however, the spatial extent of the reservoir varies by around 50% and any flow model must be able to incorporate this. Kalpna (2000) is the only previous author to model RIS in 3D. He uses a Green's function approach that requires a homogeneous subsurface permeability field. At Açu, we specifically wish to investigate the effect of heterogeneous permeability fields, so a new model is required. Since the data show no obvious undrained mechanism for RIS (Figure 1), we have adopted the simplest available approach of assuming an exponential decrease in permeability due to crack closure in response to increased confining pressure at depth.

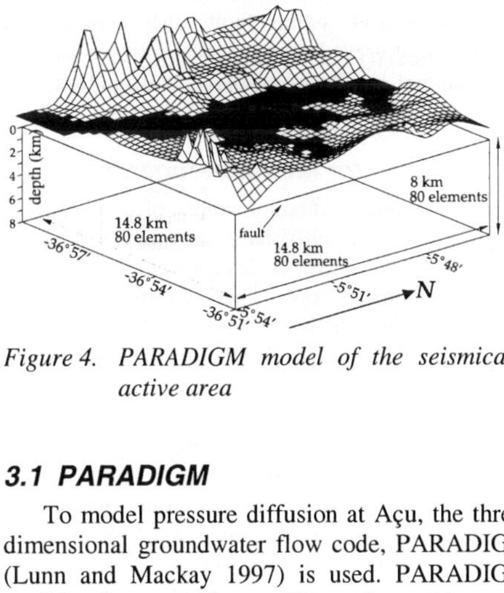

Figure 4. PARADIGM model of the seismically active area

3.1 PARADIGM

To model pressure diffusion at Açu, the three-dimensional groundwater flow code, PARADIGM (Lunn and Mackay 1997) is used. PARADIGM models flow within a 3D rock matrix with embedded 2D planar faults. Faults have a separate 2D planar mesh and can be at any location and orientation: interaction with the 3D matrix mesh is provided by a fully implicit flow solution that allows leakage between fault nodes and the nearest matrix nodes. Further details of PARADIGM and its application at Açu are presented in Nascimento et al. (2003b).

3.2 Modelling groundwater flow at Açu

Development of a realistic conceptual model of groundwater flow beneath Açu requires accurate information on boundary and initial conditions for the seismically active area. Very few hydrological data sets are available for the Açu region, however, two key points can be accurately determined:

- There is zero groundwater recharge away from surface water features (inferred from xerophilous regional vegetation, precipitation and potential evaporation data).
- Surface water features are topographically higher than the surrounding water table (inferred from stream gauging data and xerophilous vegetation below dam).

In other words, there is no groundwater recharge at the upper boundary and flow from surface water features is outwards (leaking into groundwater). The upper boundary in the model is represented as a sinusoidally oscillating reservoir level whereby nodes within the reservoir at any given time are assigned a fixed head. The reservoir boundary itself varies according to the topographic data as the water level rises and falls. Lateral boundary conditions for the seismically active region are not possible to assess from data, therefore, an approach of using a regional scale model to provide boundary conditions for a finer scale local model has been adopted. A detailed description of the modelling approach taken is given in Nascimento et al (2002b).

3.3 Model calibration and results

Two conceptual models of groundwater flow beneath Açu have been simulated using PARADIGM. In the first case, the *matrix model*, pressure diffusion is within a 3D homogeneous matrix and the diffusion coefficient, *D*, is calibrated such that the peak pressure reaches 2 km depth after 4.5 months. This time delay matches the observed delay in peak seismicity for cluster **a**, which has the most events. The matrix mesh for the local scale model is that shown in Figure 4, without the fault plane.

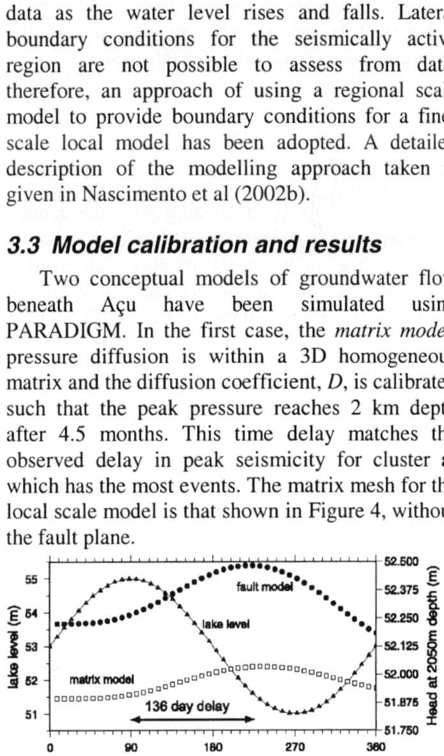

Figure 5. Calibration of the matrix and fault model.

In the second case, the *fault model*, the same matrix mesh is employed, but with the addition of a single 2-D embedded planar fault. The orientation and dip of the fault have been estimated from the seismic location data for cluster **a**. The matrix permeability is held constant at a typical value for intact crystalline rock and the fault diffusivity is calibrated to match the observed 4.5 month time delay of the pressure wave to 2 km depth.

The results for both models are shown in Figure 5, in each case the calibrated diffusivity is $\approx 5\times10^3$ m^2/day. In the case of the matrix model, $D=K/S_S$, where K is matrix hydraulic conductivity (m/day) and S_S is specific storage (m^{-1}). If we select a value of order 10^{-8} for specific storage for intact rock (e.g. Roeloffs (1988)) then this corresponds to a value of K of order 10^{-5}, which is at the upper limit of values reported for crystalline rock (Singhal and Gupta 1999) and would imply heavy fracturing and fissures. This is inconsistent with our original choice for S_S and since any increase in S_S would necessarily increase K (in order to keep D constant) it is not possible to find physically consistent values of K and S_S for this model.

In the case of the fault model, the calibrated diffusion coefficient $D=T_f/S_f$ where T_f is the fault transmissivity (m^2/day) and S_f is the fault storativity (-). Taking a value for T_f of 2×10^{-4}m^2/day from Rutquist (1998) leads to a corresponding storativity S_f of 4×10^{-8}. This is in keeping with the actual measured value from Rutquist (1998) for $S_f = 2\times10^{-8}$. Hence, the fault model produces physically consistent hydraulic parameters and demonstrates that pressure-diffusion through a permeable fault plane embedded in a low permeability matrix is a hydrogeologically valid model for RIS at Açu.

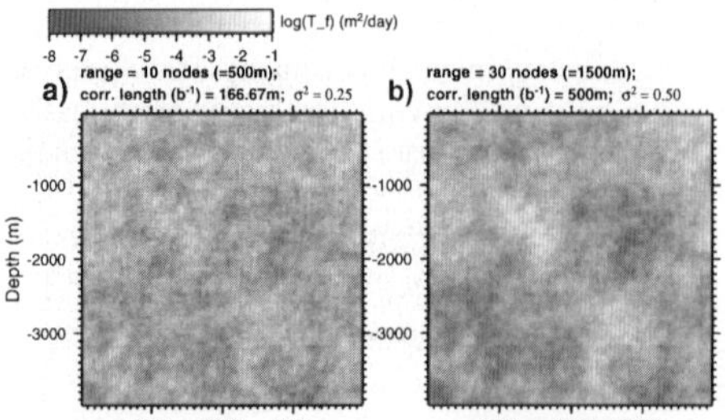

Figure 6. Fault transmissivity fields for two different values of the correlation length and variance

4. EXPLAINING THE SPATIAL AND TEMPORAL VARIABILITY OF EARTHQUAKES

Three observations described in Section 2 are not immediately consistent with the above model of pressure-diffusion at Açu. In each case, we will show that these can be explained by the presence of heterogeneous subsurface permeability structures. The first of these observations (iv – Section 2) is that the average speed of propagation of the pressure wave differs between clusters. Analysis of the data in Figure 2 and other vertical projections clearly show that the clusters are on different fault planes. Therefore, the differing speeds of propagation must be caused by different mean transmissivity values on the individual fault planes.

The remaining two observations (v) and (vi) relate to the spatial and temporal variability of earthquake evolution within individual clusters. In the following sections, we investigate the hypothesis that this variability is consistent with pressure diffusion within faults that contain a spatially structured heterogeneous transmissivity field. Detailed results from these investigations at Açu are presented in Nascimento et al (2003c).

4.4 Evidence for fault heterogeneity

Field and microstructural studies of fault zones (e.g. Caine *et al.* (1996)) have shown them to be composed of two distinct compartments: a fault core and a damage zone. This two compartment structure results in heterogeneity of hydraulic conductivity in the plane, due to both the existence of the low permeability core and the heterogeneous nature of the fractured damage zone. Caine (1996) also presented a conceptual scheme for fault-related fluid flow. According to the proportion of damage zone to the proportion of fault core, they defined four end-member architectural styles, each associated with a characteristic hydraulic conductivity structure: (1) localised conduits, (2) distributed conduits, (3) localised barriers and (4) combined conduit-barriers. Field observations of the faults in the Açu region show them to be typically characterised by zones of intense fracturing that demonstrate both opening and minor shear offsets. In terms of fluid flow, they are best described as distributed conduits in which the permeability is dominated by a fractured damage zone. These fractures may vary in orientation, frequency and aperture to produce a spatially variable heterogeneous fault permeability structure.

4.5 Incorporating fault heterogeneity in the pressure diffusion model

To incorporate heterogeneity in the pressure diffusion model a spatially varying transmissivity field is introduced within the fault plane. Fault transmissivity is log-normally distributed and is described by a spatially correlated random field with an exponential covariance structure. Mean fault permeability is significantly greater than that within the rock matrix to maintain the conceptual model of preferential flow within a 2D fault plane embedded in a 3D rock matrix. The effect of increasing compressive stress with depth is incorporated in both the matrix and fault by applying an exponential reduction in permeability with depth. In the case of the fault, this is achieved by first generating the log-normal spatially correlated field and then applying the exponential transformation. Figure 6 shows the resulting transmissivity fields within the fault plane for two different statistical parameter sets: a correlation length of 167 m with a variance for $\log T_f$ of 0.25; a correlation length of 500 m with a variance for $\log T_f$ of 0.5. As expected, the right hand field demonstrates more spatial structure with a greater range of log transmissivity values; patches of low and high permeability are 1-1.5 km across. The reduction in transmissivity due to increasing compressive stress is clearly visible in both cases.

4.6 Simulating earthquake evolution

Pressure diffusion was simulated for a number of statistical parameter sets using 5 realisations in each case. The number of realisations was constrained by the high CPU time, however, it is still possible to see how changes in statistics are influencing earthquake evolution (but not to derive accurate population statistics).

To compare simulation results with the observation data, it was assumed that earthquakes occur at the time of maximum pressure increase at a given depth. The simulations were compared using 3 statistics derived from both the observation data and the model predictions: the median time delay from reservoir peak; the interquartile range (IQR); and the range. The results of these simulations in the 1800-2200m depth range are given in Table 2; for the model predictions, values are ensemble statistics for the 5 realisations. More detailed simulation results, including additional depth ranges are presented in Nascimento *et al.* (2002c).

Table 2. The correlaion length, C_l, and variance of $\log T_f$, σ^2, derived from the observation data and model predictions.

Delay time: 1800-2200m	**Median (days)**	**IQR (days)**	**Range (days)**
Observation data	149	68	359
C_l=0 m σ^2=0.5	136	16	56
C_l=167 m σ^2=0.25	120	32	96
C_l=167 m σ^2=0.5	120	48	120
C_l=500 m σ^2=0.25	120	40	160
C_l=500 m σ^2=0.5	120	48	224

Table 2, column 2, shows that the median time delay does not vary significantly between transmissivity fields. This is explained by the fact that the mean transmissivity profile (exponentially decaying over depth) has been calibrated, prior to the introduction of any random variation, to achieve the observed 4.5 month delay for cluster **a** at 2 km depth.

Columns 3 and 4 of Table 2 show the IQR and range, in each case the observation data are higher than those derived from the ensemble statistics. However, large variations are apparent across the different statistical parameter sets. In general, the IQR and range are increasing with increases in correlation length and variance of log transmissivity. In addition, the observation data themselves represent a single statistical realisation and an analysis of the individual realisations for a correlation length of 500m and a variance of 0.5 shows that even with only 5 realisations the IQR varies between 28 to 56. Hence, it seems likely that further simulations could reproduce the observed value of 68.

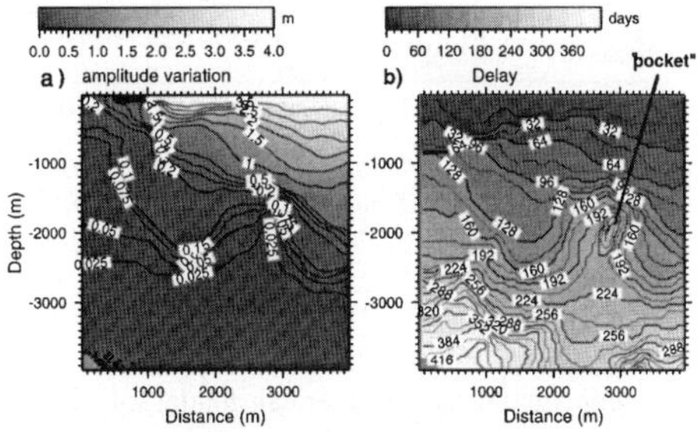

Figure 7. Contours of maximum (a) pressure change and (b) time delay from reservoir peak for the transmissivity field in Figure 6(b).

Since the observation data represent pressure-diffusion in a single realisation of fault transmissivity, it is interesting to compare in more detail the observation data and simulation results from an individual transmissivity field. Figure 7 shows contour plots of the maximum pressure change within the fault plane and the time delay from reservoir peak for an individual realisation from the final parameter set in Table 2. Figure 7(a) shows that seismic events are predicted to occur due to a maximum pressure change of 0.5 kPa (equivalent to < 0.05m of head), hence the fault must be very close to failure. Analysis of Figure 7(b) shows that the high correlation length is leading to the presence of 'pockets' of long time delay; these are observed in all 5 five realisations and are not present in predictions from any of the other parameter sets. Figure 8 shows the predicted event frequency histogram corresponding to the plots in Figure 7 alongside that derived from the observation data for cluster **a**. In each realisation, the pockets in Figure 7(b) lead to a bimodal distribution (Figure 8) of seismic event frequency in the individual histograms. The observation data also display this bimodal behaviour.

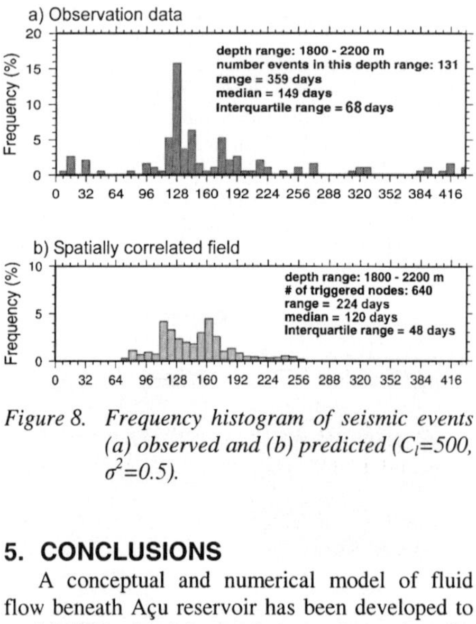

Figure 8. Frequency histogram of seismic events (a) observed and (b) predicted (C_l=500, σ^2=0.5).

5. CONCLUSIONS

A conceptual and numerical model of fluid flow beneath Açu reservoir has been developed to model RIS. Model simulations show that for pressure-diffusion to be a hydrogeologically consistent mechanism for RIS, preferential flow must occur within 2D fault planes embedded in a 3D low permeability matrix. Further, the observed spatial and temporal variability of earthquakes at Açu indicate that these faults must have heterogeneous permeability fields with significant spatial structure; pockets of high and low permeability of the order of 0.5 - 1.5 km in diameter. Predictions of the maximum pressure change in the faults at hypocentral depths, indicate that only small changes in the effective stress field <0.05 kPa are required to trigger seismic events.

REFERENCES

do Nascimento, A.F., Cowie, P.A. Lunn, R.J. & Pearce R.G. 2003a. *Spatio-temporal evolution of induced seismicity at Açu reservoir, NE Brazil.* Geophys. J. Int. (in review).

do Nascimento, A.F., Lunn, R.J. & Cowie, P.A. 2003b. *Numerical modelling of pore pressure diffusion in a reservoir-induced seismicity site in NE Brazil.* Geophys. J. Int. (in review).

do Nascimento, A.F., Lunn, R.J. & Cowie, P.A. 2003c. *Modelling the heterogeneous hydraulic properties of faults using constraints from Reservoir Induced Seismicity.* Geophys. J. Int. (in review).

Ferreira, J.M., Olivera, R.T., Assumpção, M., Moreira, J.A.M., Pearce, R.G. & Takeya, M.K. 1995. *Correlation of seismicity and water level – an example from northeastern Brazil.* Bull. Seismol. Soc. Am. 85: pp1483-1489.

Roeloffs E.A. 1988. *Fault stability changes induced beneath a reservoir with cyclical variations in water level.* J. Geophys. Res. 93(B3): pp2107-2124.

Kalpna R.C. 2000. *Green's function based stress diffusion solutions in the porous elastic half space for time varying finite reservoir loads.* Phys. Earth Planet. Int. 120: pp93-101.

Caine, J.S., Evans, J.P., & Forster, C.B. 1996. *Fault zone architecture and permeability structur.* Geology 24(11): pp1025-1028.

Rutqvist, J. Noorishad, J., Tsang, C. & Stephansson, O. (1998). *Determination of fracture storativity in hard rocks using high-pressure injection testing.* Water Res. Res. 34(10): pp 2551-2560.

Singhal, B.B.S. & Gupta, R.P. 1999. *Applied hydrogeology of fractured rocks.* Kluwer Academic Publishers, The Netherlands.

ACKNOWLEDGEMENTS

AFDN was supported by CNPq, ANP, CTPETRO and FINEP (Brazil). PC was supported by a University Research Fellowship from the Royal Society of London.

VISUAL NUMERICAL SIMULATION OF COUPLED GAS LEAK FLOW AND COAL-ROCK DEFORMATION IN PARALLEL COAL SEAMS

Peide Sun

Department of Environmental Engineering,Hangzhou University of Commerce, Hangzhou, 310035, China

Abstract: Underground coal seams are mostly distributed in adjacent multi-layers in most countries of the world, especially in China. Therefore, the interaction between the solid coal-rock and gas must be taken seriously into account for simulating the action of methane gas leak flow. In this study numerical solutions using the SIP (Strong Implicit Procedure) method to the coupled mathematical model for double parallel coal seams are also developed. Visual numerical simulations for the prediction of the safety range using protection layer mining are performed with experimental data from a mine with potential danger of coal and gas outbursts. Analyses show that the numerical simulation results are consistent with the measured data on the spot. The coupled model shows a positive future for applications in a wide range of the gas-leak-flow-related problems in gas drainage engineering and mining safety engineering.

1. INTRODUCTION

Underground coal seams are mostly distributed in adjacent multi-layers in most countries of the world, especially in China. Many problems in mining engineering and rock engineering remain unsolved such as prediction and control of methane gas emission from the coalface during mining, safety mining area prediction issues related to coal and gas outbursts in protective layer mining, prediction of methane gas drainage rate near to the mining seam and from original coal seam, as well as prediction of gas seepage rate in multi-coal-seam. These problems, in essence, are linked with rock mass deformation and gas leak flow by Sun (1998, 1999, 2000, 2002).

Under the action of a pore gas pressure gradient in the multi-coal-seam system, gas flow emission occurs near to the mining seam, which in turn leaks through the layer of intercalation with lower permeability, and flows into the mining coal faces or gas drainage bores. This is referred to as gas leak flow by Sun (1998). As gas flows during coal mining and gas drainage, the effective stress of the solid skeleton of coal/rock mass changes, and the pore pressure in the coal seam or gas drainage bores also changes, which lead to a gas flow through a network made of micro-pores and micro-cracks. Moreover, the gas flowing through coal/rock cracks also alters with the variance of the gas permeability in Sommerton (1975), Sun (1990) and Zhao (1994). Therefore, the interaction between the coal/rock mass and gas must be taken seriously into account for simulating the action of methane gas flow.

It is a well-established fact that the description of the interaction between coal/rock mass and gas in the gas leak flow system is difficult. Nonetheless, the problem of gas leak flow must be studied for the sake of coal mining safety and gas drainage in the multi-coal-seam. In this study the gas leak flow systems for coupled coal/rock mass deformation and gas seepage in double coal seams are studied with the principle of rock mechanics and coal gas seepage.

2. THE SOLID-GAS COUPLED MATHEMATICAL MODEL

2.1 Equations for gas leak flow

According to the continuity equations for the gas leak flow by Sun (1998), we have Eq.1, which is expressed in the tensor notation in the Cartesian coordinate system.

$$(T_{mi}P_m^2,_i)_{,i} \pm \frac{T_3}{b_m^{\bullet} b_3^{\bullet}}(P_1^2 - P_2^2) = Sun(P_m,t)\frac{\partial P_m^2}{\partial t} - 2P_m\frac{\partial e_m}{\partial t}$$

(m=1,2; i,j=1,2,3)

(1)

$$Sun(P_m,t) = \frac{n_m}{P_m} + \frac{a_m b_m (1 - e^{-\lambda_m^* \sqrt{t}})}{P_m (1 + b_m P_m)^2} \tag{2}$$

$$\lambda_m^* = \frac{12}{d_m} \sqrt{\frac{D_n P_n}{\pi b_m P_m}} \tag{3}$$

where, $Sun(P,t)$ is named as Sun-function which is a state one describing the methane gas flow. In general, the gas flow grows fast as the value of the Sun function increases, and vice versa. The full deduction detail of the Sun function is given by Sun (1998). Where, the subscript m denotes the coal seam or intercalation level with m=1 representing the bottom coal seam, m=2 standing for the top one and m=3 symbolizing the intercalation level; T_m is the gas permeability; The P_m is the pore gas pressure; b_m^* is the average thickness; n_m is the porosity; a_m is the maximum gas adsorption coefficient; b_m is the gas adsorption coefficient; D_m is the gas diffusion coefficient; e_m is the deformation rate; d_m is the average coal micro-grain size; the P_n equals one atmosphere; and the D_n is the gas diffusion coefficient when the pore pressure equals P_n.

2.2 *Equations for coal/rock deformation*

The equations for the coal/rock mass deformation are obtained by Sun (1998) and using the equilibrium Eq.4 for steady-state-equation of motion with the total stress in Bear (1972), Jaeger and Cook (1979).

$$\sigma_{ij,j}^m + F_{mi} = 0 \qquad (m = 1,2) \tag{4}$$

Taking into account the action of the pore gas pressure, we have the Eq.5 by Sun (1998).

$$U_{mj,ji}(\lambda_m + G_m) + U_{mi,jj} G_m + F_{mi} + (\alpha_m P_m)_{,i} = 0 \tag{5}$$

Where, the $U_{mx,xx}$ is the displacement function; λ_m is the Lame-constant, δ_{ij} is the Kronecker function, G_m is the shear modulus and the F_{mi} is the free term for the stress.

2.3 *Solid-gas coupled model for the gas leak flow system*

With the continuity equations for gas leak flow and the equilibrium equations for coal/rock mass deformation provided, the coupled mathematical model for the interaction between coal/rock mass and gas in the gas leak flow system can be described with the boundary and initial conditions (Eq.6) by Sun (1998).

$$\left.\begin{aligned}
&(T_{1i} P_{1,i}^2)_{,i} + T_3 \frac{P_1^2 - P_2^2}{b_1^* b_3^*} = Sun(P_1,t)\frac{\partial P_1^2}{\partial t} - 2P_1 \frac{\partial e_1}{\partial t} \\
&(T_{2i} P_{2,i}^2)_{,i} - T_3 \frac{P_1^2 - P_2^2}{b_2^* b_3^*} = Sun(P_2,t)\frac{\partial P_2^2}{\partial t} - 2P_2 \frac{\partial e_2}{\partial t} \\
&U_{mj,ji}(\lambda_m + G_m) + U_{mi,jj} G_m + F_{mi} + (\alpha_m P_m)_{,i} = 0 \\
&e_m = U_{mi,i}, \\
&P_m(x,y,z,0) = P_{mo}(x,y,z); \\
&P_m(x_o,y_o,z_o,t)\big|_{t\in(0,+\infty)} = P_a; \\
&T_{mi}\partial P_m / \partial \vec{n}\big|_{t\in(0,+\infty)} = \vec{Q}_m(x,y,z,t); \\
&e_m(x,y,z,t)\big|_{t=0} = e_{mo}; \\
&U_{mi}(x,y,z,t)\big|_{t=0} = 0; \\
&U_{mi}(x,y,z,t)\big|_{t\in(0,+\infty)} = U_{mi}.
\end{aligned}\right\} \tag{6}$$

where m=1,2; j=1,2,3; P_{m0} is the initial pore gas pressure; P_a is the pressure in free space; e_{m0} is the initial deformation rate; U_{mi} is the boundary value for displacement; and Q_m is the quality vector for gas seepage rate from boundary coal-mass surface.

3. VISUAL SIMULATIONS OF THE COUPLED MODEL

3.1 *Implementations*

The SIP methods by Stone (1986) for the gas leak flow equations and the rock mass deformation equations for double parallel coal seams have been developed by Sun (1998,2002) with Microsoft Visual C++ 6.0 under Windows2000 on a Pentium IV PC. The program is suitable for isotropic heterogeneous coal seams with irregular shapes as well as anisotropic heterogeneous coal seams. The systems with the first or the second boundary conditions as well as the hybrid boundary conditions can all be solved. The overview of the numerical implementations of the SIP methods for the solid-gas coupled mathematical model for double parallel coal seams is described as follows.

The equations for the gas leak flow and the equations for the coal/rock mass deformation are regarded as two separate, yet coupled systems to numerically solve them. Firstly, the solid coal/rock mass deformation rate e, the effective total stress Θ', the initial pore gas pressure values at both the upper and lower coal seams P_2 and P_1 at the time $t = t_0$, are substituted into the gas leak flow equation for the upper coal seam to calculate the pore pressure values (at each grid point) of the upper coal seam at the time $t_1 = t_0 + \Delta t$ (denoted as $P_2(t_1)$). Secondly, the pore gas pressure values of the lower coal seam at the time t_1 (denoted as $P_1(t_1)$) are calculated by coupling the numerical results of the first step. Thirdly, the coal/rock mass deformation rate at the time t_1 (denoted as $e(t_1)$) is calculated by substituting $P_1(t_1)$ and $\alpha_1(t_1)$ into the equations for coal/rock mass deformation. Consequently, the effective total stress at the time t_1 (denoted as $\Theta'(t_1)$) is also calculated. Finally, $e(t_1)$ and $\Theta'(t_1)$ are substituted into the equations for gas leak flow to get the pore gas pressure values at the time $t_2 = t_1 + \Delta t$ (denoted as $P_2(t_2)$ and $P_1(t_2)$).

The above steps are performed iteratively along the time axis to simulate the interaction between the gas leak flow and the coal/rock mass deformations in double parallel coal seams. The spatial precision (i.e. the sizes of grids for describing the coal seam planes) as well as the time precision (i.e. the time increment value) for numerical simulations is determined by practical engineering requirements.

3.2 Testing case

The proposed solid-gas coupled mathematical model and its numerical implementations have been tested with experimental data measured from the coalface N1709 in the Datong mine 2, Songzao mining district, China. As shown in Figure 1, the coal seams 7 and 8 are mostly horizontal and the average distance between the two seams is 7.2 meters. The coalface N1709 in coal seam 7 has a length of 120 meters. The details of the mechanical and physical properties of the two coal seams are given respectively in Tables 1 and 2.

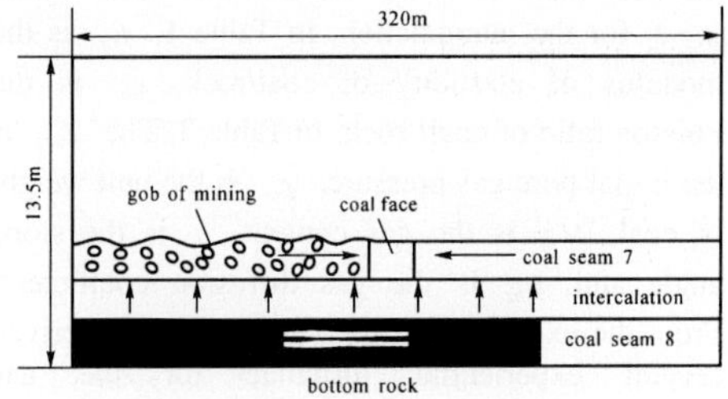

Figure 1a. A model for gas leak flow

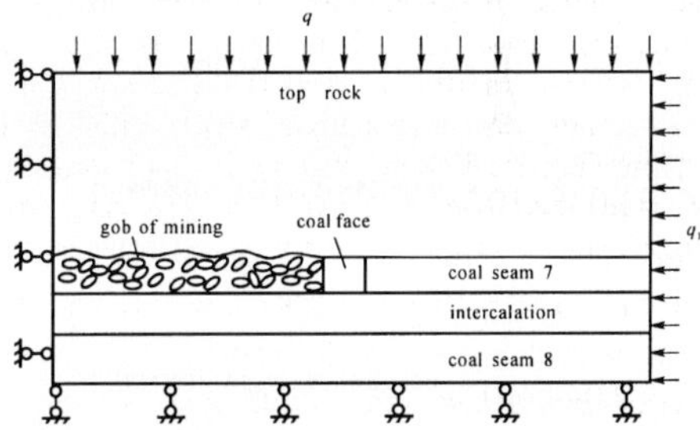

Figure 1b. A model for solid deformation
Figure 1. Illustration of the testing environment

Table 1. Mechanical properties for coal seams 7 and 8 of Datong mine 2.

Coal seam	E_m (GPa)	v_m
Coal seam 7	0.65	0.294
Coal seam 8	0.41	0.292

Table 2. Physical properties for coal seams 7 and 8 of Datong mine 2.

Parameters	**Coal seam 7**	**Coal seam 8**
P_{m0}, MPa	0.80	1.20
b_m^*, m	0.90	2.60
γ_m, g/cm^3	1.45	1.45
n_m, %	1.58	1.56
a_m, cm^3/g	34.48	34.83
b_m, $(kg/cm^2)^{-1}$	0.12	0.13
M_m, cm^3/g	18.20	21.19
α (°)	3-5	3-5
D_m, cm^2/s	5.59×10^{-8}	5.59×10^{-8}

In Table 1 and 2, the subscript m represents the corresponding coal seam with $m=1$ standing for the coal seam 8, $m=2$ for the coal seam 7 and $m=3$ for the intercalation. In Table 1, E_m is the modulus of elasticity of coal/rock, v_m is the Poisson ratio of coal/ rock. In Table 2, The P_{m0} is the initial pore gas pressure, γ_m is the unit weight of coal, M_m is the gas content, α is the slope angle and D_m is the gas diffusion coefficient. From the experimental data, we have also derived several experiential formulae for the gas permeability and the effective stress coefficient of pore pressure by Sun (1998).

$$T_1 = 2.044\times10^{-4} e^{0.050286\Theta_1^{'}+0.22911P_1^2-0.005669\Theta_1^{'}P_1}$$

(for coal seam 8) (7)

$$T_2 = 1.038\times10^{-4} e^{0.049739\Theta_2^{'}+0.28838P_2^2-0.006494\Theta_2^{'}P_2}$$

(for coal seam 7) (8)

$$T_3 = 2.044\times10^{-6} e^{0.050286\Theta_1^{'}+0.22911P_1^2-0.005669\Theta_1^{'}P_1}$$

(for intercalation) (9)

$$\alpha_1 = \frac{0.0543 - 0.00032\Theta_1^{'} + 0.09748P_1 - 0.00058\Theta_1^{'}P_1}{1 + 0.00096P_1 + 0.001734P_1^2}$$

(for coal seam 8) (10)

$$\alpha_2 = \frac{0.0233 - 0.00011\Theta_2^{'} + 0.06686P_2 - 0.00032\Theta_2^{'}P_2}{1 + 0.00033P_2 + 0.000969P_2^2}$$

(for coal seam 7) (11)

3.3 Prediction of effective protective range

The Datong mine 2, Songzao mining district of China is a mine with lots of coal and gas outbursts. Protective layer mining techniques have been used in this mine for many years. On the basis of *in-situ* parameters and study, we analyze the effective protective range of the down-proximate layer (the coal seam 8) after coal mining the coal seam 7 as up-protective layer. Taking the coalface N1709 of the coal seam 7 as the firstly-exploiting up-protective layer and accounting for a uniform coal-mining speed at 1.6 meters per day, we analyze the redistribution and changing of the pore gas pressure in the down-proximate layer (the coal seam 8) caused by the methane gas gushing up from the down-proximate layer when the coal face advances till it reaches 310 meters (corresponding to the 194th day). In practice, the couple coal seams are on average of 330 meters in the bury depth with an overloaded stress q_o of 8.25 Mpa. The interaction process of the coal/rock mass-elastic deformation and the gas leak flow caused by the action of mining is reflected in Eq.6. With a little abstraction from practice and under the condition of the particular mining scene, we mainly study the distribution law of the pore pressure in the down-proximity layer. Figure 2 is the plane boundary for the gas leak flow.

The numerical solution for the testing case is performed with the SIP method. Figure 3 shows some running visual results with our current implementation on a Pentium IV PC. In Figure

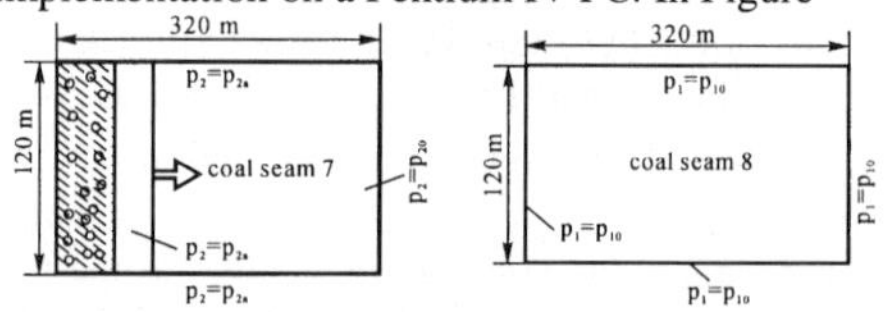

Figure 2. A plane boundary for gas leak flow

3a~3e, 3D contour for distributions of the pore gas pressure in the coal seam 8 when the upper-protect seam (the coal seam 7) is being exploited along the direction of the run of the coal seam at respectively the 100th, the 120th, the 140th, the 160th, and the 194th day. In the figures, the horizontal axis shows the length along the direction of the coalface and the value is $2.4 * Number$ in meter, and the vertical axis shows the length of the coal seam along the direction of the run of the coal seam and the value is $4.0 * Number$ in meter. The contour line is distribution of the pore gas pressure and the unit is MPa.

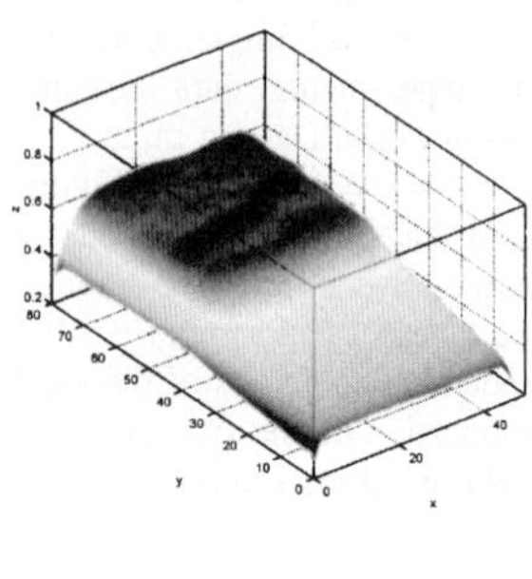

Figure 3a. 100th day

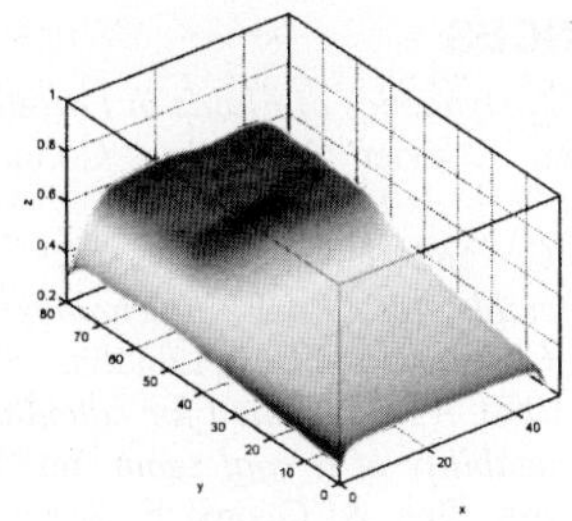

Figure 3b. 120th day

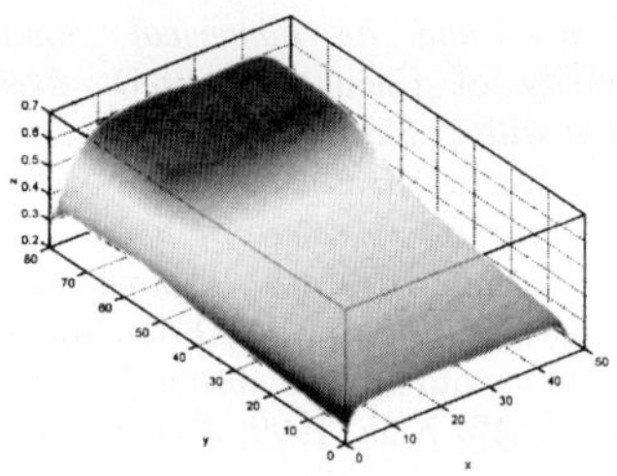

Figure 3c. 140th day

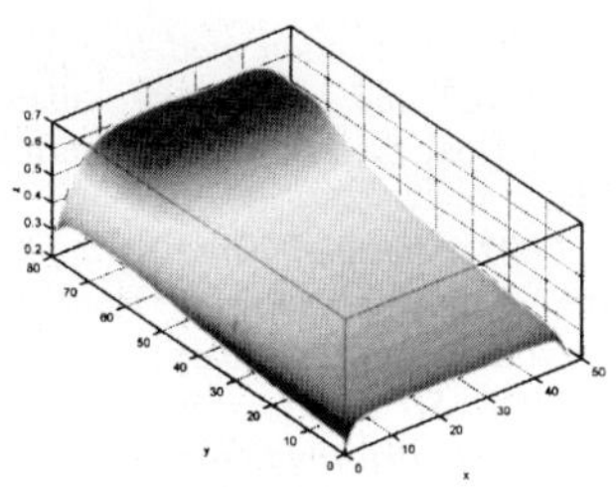

Figure 3d. 160th day

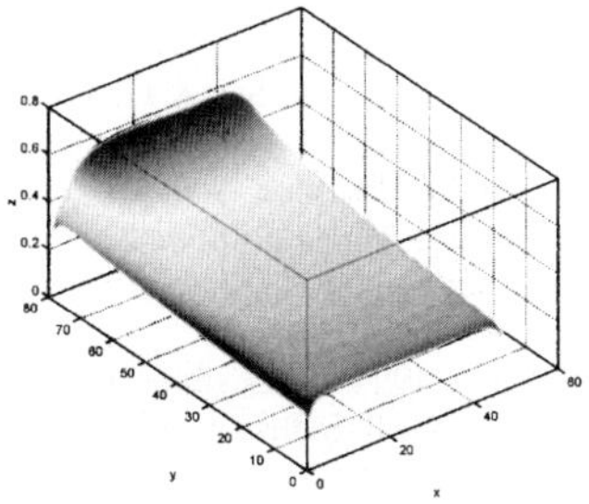

Figure 3e. 194th day

Figure3. 3D contour line for distributions of pore pressure in the coal seam 8.

The 3-demensional contour line in Figure 3 clearly shows that the pore pressure distributions in the coal seam 8 are time-dependent with the coal seam 7 (as the protective-seam) exploited firstly. If the minimum value of pore gas pressure were chosen as 0.5 MPa, the zone with the pore pressure less than 0.5 MPa in the seam 8 in Figure 3 would be the safety protective one. That is to say, it would be safe to work in this zone with the pore gas pressure less than 0.5 MPa in the coal seam 8, and coal/gas outbursts would not appear. The above deduction has been proved to be right by practice on the spot.

Again, if the minimum value of the pore gas pressure were chosen as 0.5 MPa in the coal seam 8, we could get a curve for pressure down angle along the run of the coal seam (Figure 4). Of course, the minimum value of the pore pressure could be decided by observation and study on the spot. In Figure 4, we show both the simulated curve (with α the pressure down angle) and the measured curve (with β the pressure down angle). From Figure 4 we can know that the distribution of the simulation values of the pore pressure is consistent with the distribution of the practical one. More importantly, the pressure down angle in the simulated curve ($\alpha = 46^0$) is less than the pressure down angle in practice ($\beta = 54^0$), which means that the pore gas pressure distribution by numerical analysis inclines to more safety.

It can also be seen from Figure 4 that the pore pressure of methane gas drops fast in the direction of gob within 100 meters from the coalface in the coal seam 8. The pore pressures in the coal seam 8 drops dramatically from 0.9Mpa to 0.4 MPa in the direction of gob within 50 meters to 100 meters away from the coalface.

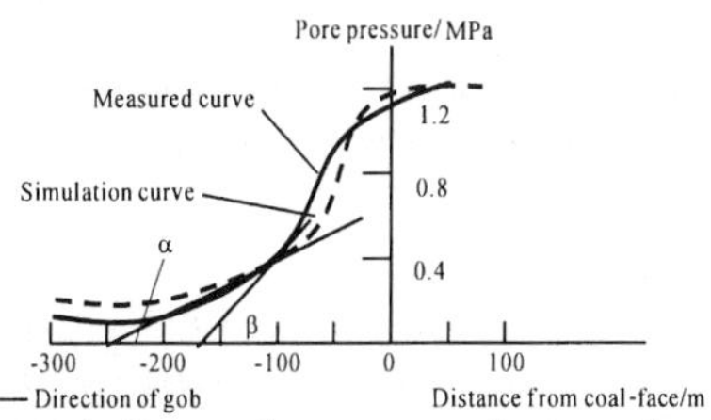

Figure 4. Curves for pressure down angle along the run of the coal seam 8.

4. CONCLUSIONS

In this paper, a mathematical model for coupled coal/rock mass deformations and gas leak flow in parallel coal seams is presented from a new viewpoint of solid-gas interaction. It consists of equations for the gas leak flow, equations for coal/rock mass deformation and the boundary/initial conditions. The model is characterized by: (1) The main consolidation of gas leak flow through the coal/rock mass due to gas drainage or coal mining is described by the linear-elastic deformation in the coal/rock mass; (2) The mechanism for gas leak flow with both seepage and diffusion is given by the equations of gas leak flow, and the gas desorption due to the pore pressure decreasing is also given by equations for the gas leak flow; and (3) The interaction between the gas leak flow and coal/rock mass elastic-deformation is also described by the coupled model.

The interaction between the elastic deformation of coal/rock mass and gas leak flow has been thoroughly studied with the built equations. Numerical simulations have been performed with the SIP method with robustness and efficiency. Furthermore, the simulation results have shown a promising consistency with the measured data on the spot.

The solid-gas coupled model can be applied to a wide range of gas-leak-flow-related problems in mining engineering, gas drainage engineering and mining safety engineering. For instance, we can analyse with the numerical simulation results for the effective protective-range at the down-proximity seam quantitatively rather than qualitatively. We can predict the change of the pore gas pressure along the run of the coal seams and the trend distribution in the down-proximate seam at any time when the upper-protective layer is being exploited. This quantitative analysis can guide safety exploitation for miner and protective measure for coal and gas outbursts, etc.

ACKNOWLEDGEMENTS

This study was supported by the Visiting Scholar Foundation of Key Laboratory in University of China. The author would like to thank to Prof. Xuefu Xian with Chongqing University and to Dr. Huagen Wan with Zhejiang University for their great help.

REFERENCES

Bear, J. 1972. *Dynamics of Fluids in Porous Media.* American Elsevier Publishing Co.Inc. New York.

Jaeger J. C. and Cook N. G. W. 1979. *Fundamentals of Rock Mechanics.* Third Edition, Chapman and Hall, London.

Peide Sun. 1990. *A new method for calculating the gas permeability of a coal seam.* Int. J. Rock Mech. Min. Sci. & Geomech. Abstr. 27(4): pp.325-327.

Peide Sun. 1998. *A study on the interaction mechanics for coal seam deformation and gas leak flow and its numerical simulations.* University of Chongqing, Ph.D. thesis, (in Chinese with English abstract).

Peide Sun, Xuefu Xian. 1998. *Solid-gas coupled analysis for safety range of upper protective layer mining.* Chinese Journal of Rock Mechanics & Engineering. 17(supp): 932-936 (in Chinese with English abstract).

Peide Sun, Xuefu Xian. 1999. *A study on coupled models for coal seam deformation, gas leak flow and its applications.* Journal of China Coal Society. 24(1): pp.60-64 (in Chinese with English abstract).

Peide Sun. 2000. *Interaction modeling for mining safety range and numerical simulations.* Journal of Coal Science & Engineering. 6(2): pp.41-46.

Peide Sun. 2002. *Sun model and its applications.* Zhejiang University Press, Hangzhou (in Chinese with English summary).

Peide Sun. 2002. *SIP analysis on coupled models for coal seam deformation and gas leakage flow.* Journal of China Coal Society. 27(5): pp.276-280 (in Chinese with English abstract).

Sommerton, W. J., Soylemezoglu, I. M., Dudley, R. C. 1975. *Effect of stress on permeability of coal.* Int. J. Rock Mech. Min. Sci. & Geomech. Abstr. 12(2): pp.129-145.

Stone H.L. 1986. *Iteration solution of implicit approximations of multi-dimensional partial differential equations.* J. Soc. Ind. Appl. Math. 5(6): pp.530-558.

Yangsheng Zhao.1994. *Rock fluid mechanics in mine.* China Coal Industry Press, Beijing (in Chinese).

Yangsheng Zhao, Zhongming Jin, Jun Sun.1994. *Mathematical model for coupled solid deformation and methane flow in coal seams.* Appl. Math. Modelling. 18(4): pp.328-333.

A COUPLED GEOMECHANICAL-RESERVOIR MODEL FOR THE MODELLING OF COAL AND GAS OUTBURSTS

Choi, S.K. and Wold, M.B.
CSIRO Petroleum, Melbourne, Australia

Abstract: Coal mining under conditions of sufficiently high pore gas pressure can produce the hazard of dynamic failure of coal known as coal and gas outbursts. During an outburst, coal fragmentation occurs and the fragmented coal is projected into the mine opening, accompanied by a rapid release of gas. A dual porosity multi-component coalbed methane reservoir simulator has been coupled with a coupled geomechanical – fluid flow analysis code, to investigate gas outburst initiation and evolution during roadway development. The model is three-dimensional and can take into account factors such as rate of mining (or face advance), mine geometry, magnitude and orientations of in-situ principal stresses, strength of coal, large deformation, reservoir pressure, mixed gases sorption properties of coal, stress dependent permeability and relative permeabilities for two-phase (gas and fluid) flow in a fractured porous medium, and poroelastic effects. Reasonably good agreement between model predictions on outburst initiation and field data has been obtained. Some results on the modelling of the post-initiation deformation and fragmentation of the ejected coal are presented and discussed.

1. INTRODUCTION

During mining or roadway development, a change in stress and pore pressure around the new opening will occur. At some stage, an outburst may be initiated which can be related to the size of the opening and/or the proximity to some geological structures. When outbursting occurs, the rock/coal/gas system transforms from a stable to an unstable state with the release of a significant volume of gas over the duration of the outburst. Some mechanisms must exist that enable the system to overcome the energy barrier provided by the strength of the coal. They can be related to major geologic structure, natural heterogeneity, and/or the state of the system. During an outburst, the system may go through several meta-stable states before reaching the final equilibrium state. As coal is a soft rock, the amount of strain energy stored as a result of elastic deformation can be quite limited. In order for coal to fail in a violent manner, another form of energy needs to be available. This is supplied by the adsorbed gas in the micropores within the coal matrix and the compressed gas in the cleats and other natural fractures. Pre-mining gas drainage is effective in preventing outbursts by removing this source of potential energy.

After an outburst has been initiated, the outburst coal starts to deform at a high strain rate, and the fragmented coal will tend to behave as a particulate material. Depending on the gas pressure and the availability of new sources of free gas, the gas can cause further failure and fragmentation of the coal, and the expanding gas will provide a drag force to propel the fragmented coal further into the mine opening. During an outburst, fluid flow in the outburst coal will transform from flow in a fractured porous medium to dense particle flow, and then to dilute particle flow as the ejected coal fragments and particles move further away from each other.

2. NUMERICAL MODELLING OF OUTBURSTS

Numerical modelling of outbursts requires the quantification of a number of processes and factors. These include gas adsorption and desorption, coal deformation and failure, coal fragmentation, gas dynamics and transport of outburst coal (particle flow).

2.1 Gas adsorption and desorption

The adsorption properties of coal are normally represented by the adsorption isotherm which shows the amount of gas adsorbed at a certain temperature and partial pressure. For single component gas, the Langmuir adsorption isotherm (Langmuir, 1918) is usually used for coal and is given by

$$\frac{V}{V_L} = \frac{p_g}{P_L + p_g} \qquad (1)$$

where V is the volume of gas adsorbed per unit mass of coal at pressure p_g, V_L is Langmuir volume, which is the volume of gas needed for monolayer coverage of the surfaces of the micropores per unit mass of coal, and P_L is Langmuir pressure, defined as the pressure at which the volume of gas adsorbed is half the Langmuir volume.

The extended Langmuir model (ELM) is the most common approach used to represent multicomponent gas adsorption in coal. This is based on direct extension of the single component isotherm to a multicomponent system giving the analogous expression

$$\frac{V_i}{V_{Li}} = \frac{p_{gi}}{P_{Li} + p_{gi}} \qquad (2)$$

where subscript i represents the ith gas component.

After initiation, the way an outburst evolves can be strongly influenced by the amount of free gas in the outburst coal. It is therefore important also to understand the kinetics of the adsorption and desorption processes.

In the Langmuir treatment of the kinetics of gas adsorption and desorption, the rate of adsorption is equated to the rate of desorption as follows:

$$\frac{P}{\sqrt{2\pi mkT}}(1-\theta) = \frac{e^{-Q/kT}}{\tau_o a}\Theta \qquad (3)$$

where P is gas pressure, m is the mass of the gas molecule, k is Boltzmann constant, T is absolute temperature, θ is fraction of available sites occupied $(0 \le \theta \le 1)$, Θ is total surface area of adsorption sites, Q is energy of adsorption, a is area of an adsorption site, and τ_o is average resident time.

It was observed by Groszek (1997) that, at very low pressure, very little energy is required for the adsorbed molecules to move away from the interface, and the rate of desorption can be very high.

In the conventional approach, the rate of mass transport between adsorbed gas in the matrix and free gas in the cleats and other fractures is assumed to involve a two-step process corresponding to desorption/diffusion within the coal matrix, followed by long-range flow in the cleat system. Desorption of gas from the coal surface occurs at a much faster rate than the diffusion of the gas through the coal matrix. The rate of diffusion is often assumed to be sufficiently represented by Fickian diffusion which is usually the rate-limiting step in the desorption process.

If the coal matrix can be assumed to be made up of spheres with a characteristic radius a (Ancell et al., 1979, 1980), the rate of mass transport of gas between the primary and secondary porosity system is given by

$$\frac{c_d}{r^2}\frac{\partial}{\partial r}\left(r^2\frac{\partial c}{\partial r}\right) = \frac{\partial c}{\partial t} \qquad (4)$$

where c_d is the micropore diffusion coefficient, c is gas concentration, and r is radial distance from the centre of the sphere. The rate of mass transfer therefore depends on the characteristic radius of the spheres, the micropore diffusivity, the partial pressure of the free gas in the secondary porosity system, and the gas content (or concentration of gas in the matrix spheres).

However, considering that small fragments and fine particles can be generated very rapidly during an outburst with loss of confinement and rapid drop in pressure around the particles, rapid gas desorption will occur. At the time when the particles have just been formed, it is assumed that the gas content and pressure within the particle is uniform. An amount of free gas resulting from the formation of the new surfaces will be available, together with the gas in the bulk phase existing in the pores that are connected to the new surfaces. The adsorbed gas in the very small particles may transform into free gas in probably a very short time. For the larger particles, a high pressure gradient will exist within the particles, and viscous flow, in addition to Fickian diffusion, will become an important mass transport mechanism within the matrix. Within the particles, it is assumed that the viscous flow can be adequately represented by Darcy's law :

$$u_i = -\frac{k}{\mu}\frac{\partial p}{\partial x_i} \qquad (5)$$

where u_i is apparent velocity, k is permeability, μ is viscosity, x_i is spatial coordinate.

In 3-dimensional Cartesian space, taking into account desorbed gas as a source for free gas, and assuming that the change in porosity and gas density over a small time step can be ignored, the continuity equation is given by

$$\frac{\partial V}{\partial t} = \frac{1}{\rho_s}\nabla \bullet u_i = \left(\frac{V_L}{P_L + p_g}\right)\left(\frac{p_g}{P_L + p_g}\right)\frac{\partial p_g}{\partial t} \qquad (6)$$

where ρ_s is the density of coal.

Equation (6) can be used as the source term in the mass balance and momentum equations to calculate gas flux and drag force acting on the coal fragments.

2.2 Coal deformation and failure – continuum state

During mining of a roadway, stress concentration will occur in certain regions around the mine opening. Under sufficiently high stress, initiation and growth of microcracks can occur. Initially, this will tend to occur in a smooth and gradual manner, and the differences between the states at each stage of progressive failure can be very small, with a smooth transition from intact to failed state. Failure at this stage may manifest itself as plastic deformation resulting from distributed damage evolution, strain localisation, subcritical crack growth and coalescence, and degradation of mechanical properties including strength and bulk and shear moduli. The degree of "plastic" damage just prior to outburst occurrence can have important influence on the fragmentation process during post-initiation outburst evolution. However, the plastic damage is rate-independent while the post-initiation damage (or fragmentation) is strongly rate-dependent. It is therefore important for the two distinct stages of failure to be modelled using different approaches.

The Mohr-Coulomb criterion is used to represent the shear failure of coal. In the modelling studies, after yielding has occurred, the total strain is assumed to consist of three components: the elastic, plastic and damage components.

$$\varepsilon = \varepsilon^e + \varepsilon^p + \varepsilon^d \qquad (7)$$

Damage is expressed as a function of total plastic strain.

$$D = 1 - e^{-\alpha\varepsilon_p} \qquad (0 \le D \le 1) \qquad (8)$$

The elastic properties of the coal are functions of the degree of plastic damage.

$$c_d = c_o(1 - \chi D) \qquad (9)$$

$$\phi_d = \phi_o(1 - \beta D) \qquad (10)$$

$$E_d = E_o(1 - \gamma D) \qquad (11)$$

$$\upsilon_d = \upsilon_o(1 - \kappa D) \qquad (12)$$

where c is cohesion, ϕ is angle of internal friction, subscript o represents the initial (undamaged) material property, subscript d represents the damaged material property, and the values for α, χ, β, γ, and κ have to be determined in the laboratory under triaxial compression to beyond the peak strength, under strain-controlled conditions.

2.3 Coal fragmentation

When an outburst has been initiated, macroscopic continuum damage mechanics is adopted to model the creation of new cracks, and the propagation and coalescence of new and existing cracks. Although it is less rigorous than the micro- and meso-damage approaches, it provides a computationally feasible alternative for some complex problems such as the modelling of rock fragmentation. The constitutive models are based mainly on experimental observations at the macroscopic scale (Grady and Hollenbach, 1979; Kipp et al., 1980). It is assumed that the fragmenting coal can be treated as an isotropic, continuous and homogeneous material with pre-existing microcracks. The outburst coal is considered to deform at a high strain rate, and confinement will be lost after the coal has detached from the face. The coal will continue to deform under the influence of the free gas in the pore space, and is subjected mainly to tensile stress induced by the compressed gas. The propagation of the pre-existing microcracks occurs mainly under mode I (opening mode), and damage is isotropic. Based on the above assumptions, damage can be sufficiently represented by a scalar variable. It is also assumed that a critical strain criterion can be used to predict the onset of fragmentation. The assumption of isotropy enables the use of a model which is tractable and requires only a few model parameters that are physically meaningful. Most importantly, their values can be measured in the laboratory.

In the model, the degree of damage of the outburst coal is a function of strain rate. The coal is assumed to be completely destroyed (or lose its strength and stiffness) when the crack density has exceeded a certain critical value for a particular type of coal. This is reflected by the values of effective bulk modulus and Poisson's ratio both approaching zero. The size of the fragments is assumed to be directly related to the maximum strain rate experienced during the loading history.

Based on the above assumptions, the model can therefore be represented by the following equations to provide initial estimate of particle size. Subsequent size evolution may depend on other factors such as collision among particles, but they are not considered here.

$$F = 1 - e^{-\alpha C_d} \qquad (13)$$

$$\upsilon_e = \upsilon(1 - \beta F) \qquad (14)$$

$$\text{where } \lim_{C_d \to \infty} \beta = 1 \qquad (15)$$

$$K_e = (1 - \gamma F)K \qquad (16)$$

$$r = \omega\left(\frac{K_{IC}}{E_{eff} R}\right)^{2/3} \qquad (17)$$

$$c_d = f(\varepsilon_{dp})\left(\frac{K_{IC}}{E_{eff} R_{max}}\right)^2 \qquad (18)$$

C_d is crack density, F is regularised damage parameter, υ is Poisson's ratio, K is bulk modulus, K_{IC} is mode I fracture toughness, r is radius of fragment, R is strain rate, R_{max} is maximum strain rate during loading history, ε_{dp} is cumulative irreversible strain due to plastic deformation and damage, and subscript e refers to effective material property. α, β, γ, ω, and E_{eff} are empirical constants and material properties that need to be determined through laboratory measurements.

2.4 Gas dynamics and transport of outburst coal

Particle flow refers to the two-phase (particles and fluid) flow where solid particles are immersed in a continuous carrier fluid (liquid or gas). These correspond to the fragmented coal and free gas during an outburst in which the fragmented coal is expelled into the mine opening by the gas. In modelling particle flow, two conditions have to be distinguished – dilute flow (the solid particles are highly dispersed) or dense flow. For dilute flow, the solid and fluid mixture may be treated as single-phase fluid with modified rheological properties. However, for dense flow, which is more related to the early stages of coal ejection from the face, the mixture has to be treated as a two-phase medium, taking into consideration the interaction between the coal fragments and the continuous fluid phase.

However, the number and size of the particles (or coal fragments) are not constant, but constantly evolving, and due to the very large number of fragments, it is almost impossible to model them explicitly. Instead, the coal fragments and the carrier fluid are treated as two super-imposing continua. For such a case, the particle flow can be represented by the Navier-Stokes equations for compressible gas, and the interaction between the coal fragments and the carrier fluid can be represented by suitable interfacial exchange terms in the momentum equations

$$-2\mu_s \nabla \bullet (\phi_s \varepsilon(u_s)) + \rho_s \phi_s (u_s \nabla) u_s = -\phi_s \nabla p_f + M + \phi_s F_b \qquad (19)$$

$$-2\mu_f \nabla \bullet (\phi_f \varepsilon(u_f)) + \rho_f \phi_f (u_f \nabla) u_f = -\phi_f \nabla p_f + M + \phi_f F_b \qquad (20)$$

The drag force (Ergun, 1952) of the gas on the coal fragments (assumed to be spherical) is given by

$$F_d = f(n, Re)\frac{\pi}{4} d_p^2 \frac{\rho_g u_r^2}{2} \qquad (21)$$

where

$$Re = \frac{u_r d_p \rho_g}{\mu_f} \qquad (22)$$

ϕ is volume fraction, u is velocity, μ is viscosity, ρ is density, ε is stretch tensor, p is pressure, M is interfacial momentum transfer, F_b is body force, n is local porosity, Re is Reynold's number, d_p is average particle diameter, u_r is relative velocity, subscript f is fluid phase, and subscript s is solid phase.

To model the violence of outbursts, it is necessary to calculate the initial momentum and the initial drag force of the expanding free gas on the motion of the coal fragments. After expulsion from the face, the coal fragments are travelling at reasonably high speed and desorbing gas at the same time. Further fluid-particle interaction is not tracked. It is considered that the initial momentum gives a good indication of the violence of the event.

3. MODEL RESULTS AND DISCUSSIONS

Modelling studies have been conducted to simulate the mining of a 5 m wide roadway at a depth of 500 m. The input values for some of the model parameters are shown in Tables 1 and 2. Some of the model results are shown in Figures 1 to 4. Figures 1 and 2 show the deformation of coal near the face during an outburst when mining through weak and very strong coals.

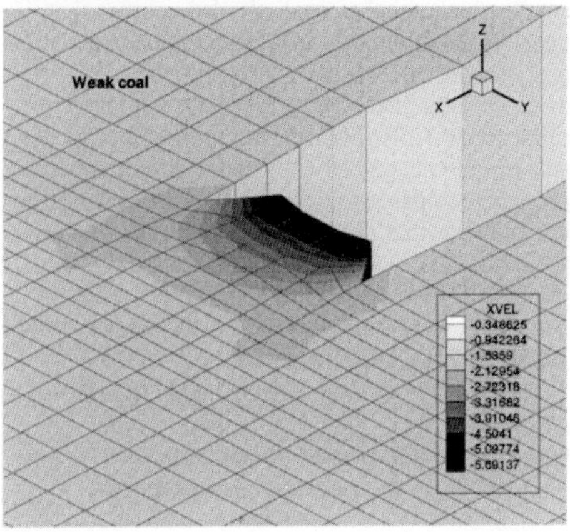

Figure 1. Velocity (m/s) contours during model outburst in weak coal.

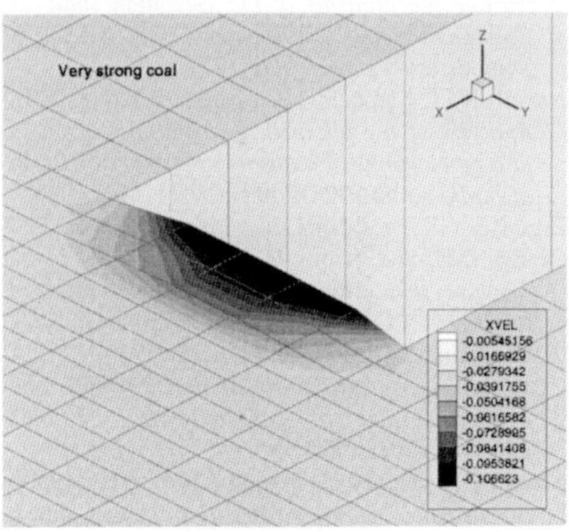

Figure 2. Velocity (m/s) contours during model outburst in very strong coal.

The model results show that, under the model conditions, the outburst process can be suppressed if the coal is strong enough. The model results (see Figure 3) also show that, when the mine face approaches some major geological structures like faults and dykes, an outburst is likely to be more violent compared to mining through uniform coal.

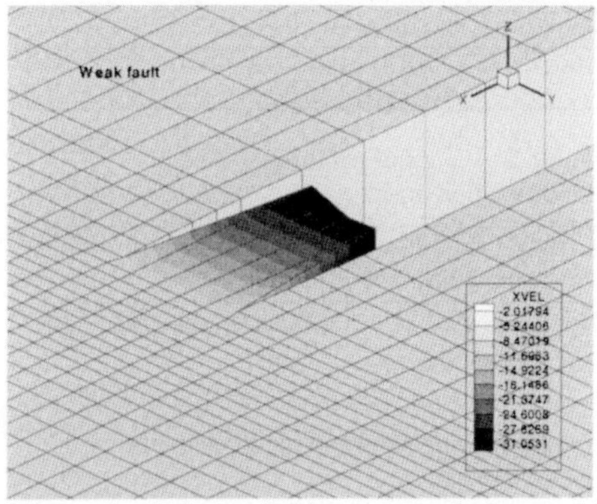

Figure 3. Velocity (m/s) contours during model outburst when mining through weak fault

Figure 4 shows the geometry of the cavity that can be created when mining through weak coal.

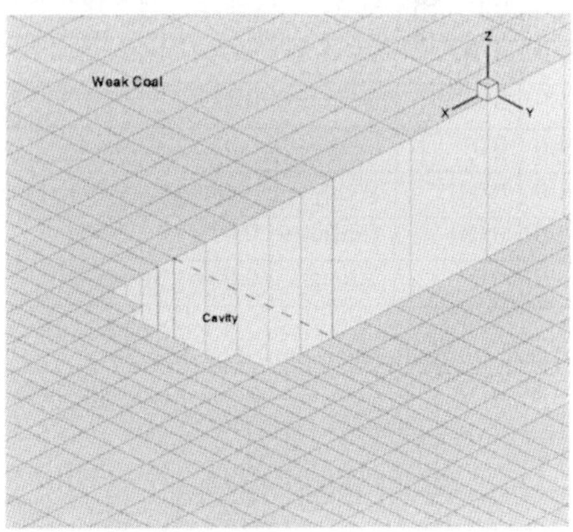

Figure 4. Model prediction of cavity size in weak coal under model conditions.

The predicted particle size distribution, based on the values of the input parameters shown in Tables 1 and 2, are shown in Tables 3 to 4. Table 3 shows the predicted range of fragment size as a function of coal strength. Table 4 shows the model computed drag force acting on the coal at different relative velocity between the outburst gas and the coal fragments. The size of fragments appears to be smaller when mining through weaker coal.

Table 1. Values of parameters used in strain softening model

α	χ	β	γ	κ
100.0	1.0	0.0	0.0	0.0

Table 2. Values of parameters used in fragmentation model

Critical strain (fragmentation)	α	β	γ	f (ε_{dp})
0.3	1.0	0.0	1.0	ε_{dp}

Table 3. Values of parameters used in fragmentation model and predicted range of particle sizes

ω ($s^{-2/3}$)	(K_{IC}/E_{eff}) ($\sqrt{m}$)	Coal type	Smallest particles (m)	Largest fragments (m)
1.0	0.1	Weak	0.093	1.0
1.0	0.1	Uniform	0.115	1.91
1.0	0.1	Strong	0.171	2.18
1.0	0.01	Weak	0.02	0.22
1.0	0.01	Uniform	0.025	0.41
1.0	0.01	Strong	0.037	0.47
1.0	0.001	Weak	0.0043	0.046
1.0	0.001	Uniform	0.005	0.088
1.0	0.001	Strong	0.008	0.101

Table 4. Acceleration caused by drag force

Radius of Particle (mm)	Relative velocity (ms^{-1})	Drag Force (N) * 10^3	Acceleration (ms^{-2})
5	15	7.22	9.84
5	10	4.81	6.56
5	5	2.40	3.28

4. CONCLUSIONS

A model has been developed which takes into account the major processes and mechanisms that can influence both outburst-proneness and post-initiation outburst behaviour. The model has been applied to simulate the effects of gas diffusion mechanisms in the coal matrix, coal strength, coal damage, geological structures and gas composition on outbursts. The model is able to provide estimates of the amount of outburst coal and released gas if laboratory data is available for the model input parameters.

Form the model results, the likelihood of outbursts appears to be less in stronger coal. Also the degree of violence decreases with an increase in coal strength, and the size of the coal fragments is smaller for a weaker coal. It also appears that, if the strength of the coal is high enough, the outburst process can be subdued. During the study, the model results also suggest that CO_2 outbursts tend to be more violent mainly because of the greater adsorption capacity of CO_2 under the same partial pressure compared to CH_4. The results however do not indicate that, under the same initial reservoir pressure, coal seams rich in CO_2 are more outburst prone than seams rich in CH_4.

ACKNOWLEDGEMENTS

The authors would like to thank ACARP and CSIRO for their financial support for this work. They also thank the Australian coal mining community for their continuing support.

REFERENCE

Ancell, K.L., Price, H.S. and Ford, W.K. (1979). *An investigation of the gas producing and storage mechanisms of the Devonian shale at Cottageville field.* SPE paper 7938, AIMMPE Symposium, Denver, May 1979.

Ancell, K.L., Lambert, S.L., Johnson, F.S. (1980). *Analysis of the coalbed degasification process at a seventeen well pattern in the Warrior Basin of Alabama.* SPE/DOE paper 8971, May 1980.

Ergun, S. (1952). Chem. Eng. Prog., Vol. 48.

Grady, D.E. and Kipp, M.E. (1979). *Oil shale fracture and fragmentation at high rates of loading*. Proc. 20th U.S. Symposium Rock Mechanics, Austin, TX, June 3-6, 1979.

Groszek, A.J. (1982). *Prediction method and countermeasure of gas outburst.* Uchida Rokakuho Press, Tokyo.

Kipp, M.E., Grady, D.E. and Chen, E.P. (1980). *Strain-rate dependent fracture initiation.* Int. J. Fracture, Vol. 16, No. 5, pp. 471-478.

Langmuir, I. (1918). *The adsorption of gases on plane surfaces of glass, mica and platinum.* J. Am. Chem. Soc., Vol. 40, pp. 163-172.

APPLICATION OF FLUID-SOLID COUPLING THEORY IN OIL FIELD CASING DAMAGE FORECAST

Jianjun Liu[1,2], Xiating Feng[1]
[1]) Institute of Rock and Soil Mechanics / Chinese Academy of Sciences, Wuhan China
[2]) Institute of Mechanics of Porous Media / Wuhan Polytechnic University, Wuhan China

Abstract: Oil/water well casing damage is a challenging difficulty to oil reservoir engineering. This paper discussed different mechanisms of casing damage and presented numerical simulation results based on a fluid-solid coupling theory. Using a 3-D finite element method, the authors simulated the rock deformation patters causing casing damages in Daqing oil field, China, during water-flooding operations. According to the simulation results, some technical measures were advanced to forecast and prevent casing damage.

1. INTRODUCTION

Casing damage has become a very serious concern in China since 1970s. According to statistic data, the number of casing damage well has exceeded 14000, and distributed in many oil fields. Along with adoption of high-pressure water injection and high pressure hydraulic fracturing in oil fields, casing damage become a challenging technical difficulty in the Chinese and international petroleum industry. Many research have been devoted to solve this problems and had identified important factors affecting casing damage, such as geologic structure, development operation and drilling program. Although stress concentration around well bore is the direct reason of casing damage, it is the flow field change that induced geostress redistribution. Therefore, it is necessary to study the fluid-solid coupling during oil development to identify casing damage mechanisms. This paper presents mathematical models for casing damage forecast based on fluid-solid coupling theory. Technical measures and critical injection pressure to prevent casing damage are given in the end for Daqing oil field. The works performed are based on extensive previous research, by the author and their colleagues, such as performed by Hu (1998), Liu et al. (2001a,b), Yin (2000), Wang (1994), Zhang and Liu (2000), Liu et al. (2002), Zhou (1993), Zhang (2001), Lian et al. (1998) and Chou (1996).

2. MECHANISM OF CASING DAMAGE INDUCED BY SEEPAGE

According to the casing damage characters, casing damages in Daqing oil field can be classified into three kinds: shear-sliding damage, tension damage, and squeeze damage.

2.1 Shear-sliding casing damage

Field investigation shows that many casing damages occurred in shallow layers, not in deep layers that have larger geo-stress. By theoretical analyses, we found that strata slippage is the naturally reason of casing damage. Shear damage is induced by geologic fault slippage because water or oil flow into fault, and intenerates the mechanical parameters, such as cohesion C and interior friction angle Φ . When porous pressure increases to critical value, fault will slippage, and cause casing collapse.

As we know, the determinative factor of rock mass stability is faults in rock mass, which directly restrict the procedure of rock mass deformation and failure. During geostress redistribution after load action, deformation and failure of rock mass often occurs at the most frail sites, that is not always the sites bearing the largest load. From geological data of oil field development, underground rock layer has more or less weak formation. If the injection pressure arrives at a certain value, water will be absorbed into weak formation through fractures. Once the formation is filled with water, the physical and mechanical property of formation will change. By experimental studies, it is found that cohesion C and interior friction angle Φ decline with water content ratio rising. According to Mohr-Coulomb criterion, when applied shear stress equals rock's intrinsic shear stress (that is $\tau = \tau_f$), rock mass is in a critical equilibrium status. When $\tau > \tau_f$, rock mass will destabilize. When $\tau < \tau_f$, rock mass is in a stability status.

If maximum major principal stress σ_1 and minimum main stress σ_3 are known, then normal and shear stress can be calculated by the following equations

$$\sigma_n = \sigma_1 \cos^2 \beta + \sigma_3 \sin^2 \beta \tag{1}$$

$$\tau_n = \frac{\sigma_1 - \sigma_3}{2} \sin 2\beta \tag{2}$$

where σ_n is normal stress, τ_n is tangential stress, β is the incidence of fault.

According to the Mohr-Coulomb Criterion and Terzaghi's effective stress principle, the maximum equilibrium pore pressure $p_{\max}$ can be derived as

$$\begin{aligned} p_{\max} &= [\sigma_1 \cos\beta(\cos\beta tg\phi - \sin\beta) \\ &+ \sigma_3 \sin\beta(\cos\beta + \sin\beta tg\phi) + C]/tg\phi \end{aligned} \tag{3}$$

Because C, ϕ are the functions of the water content ratio, $p_{\max}$ is also related to the strata incidence and water content ratio. Before developing an oil reservoir, in-situ stresses should be obtained firstly. Using the in-situ stress values and strata incidence, the maximum pore pressure can be calculated through numerical simulation. In this way, reasonable water injection pressure values can be decided.

2.2 Tensile damage

Tensile damage is induced by vertical strata displacement. When large volume water is injected into reservoir, porous pressure rises, and leads to strata elevated. Because the casing, cement shealth and strata are cemented together, the vertical displacement of strata can result in tensile stress in casing. If the tensile stress exceeds the casing's tensile strength, casing pipe will be destabilized. Daqing oil field is one of the earliest oil fields that adopt water injection development. Because of injection water, formation pressure is not uniform. Land deformation and casing damage happen frequently in Daqing oil field. Most of oil field surface is uplifted 80 mm. In south-I zone of Daqing oil field, an area of 1.7 sq. m, the uplift has reached 0.4 meter. The maximum uplift value is 2 meter, and more than 88% water or oil well casing are damaged in this area. The study on the characteristic of vertical deformation of solid matrix under high-pressure water injection becomes therefore important for casing well damage prevention. Since 1990, Daqing petroleum Co. erected many surveying points to monitor the land surface deformation. Analysis of the observation data showed that the vertical deformation of surface and pressure difference had a linear correlation.

Supposing that the cement shealth is strong bonding and the vertical deformation of strata induced casing being tensile. Supposing also that the elongation of casing equals to the vertical deformation of the ground. According to Hook's law, tensile stress in casing can be calculated by

$$\sigma = \varepsilon E = \frac{\Delta H}{H} E \tag{4}$$

where ε denotes the strain of the casing, E is the elastic modulus, H is the formation thickness and ΔH is the vertical deformation of reservoir formation, respectively.

2.3 Squeezing casing damage

Squeezing casing damage is induced by large pressure gradient around the well bore as a pressure releasing effect in the water injection wells. The pressure gradient makes casing undergo squeezing stresses. The larger pressure gradient is, the larger extrusion-stress the casing bears. When the release discharge is too large, the casing may be squeezed to produce radial deformations. During pressure releasing in well injection, total stress is un-varied and the declining pore pressure will cause the increasing of the effective stress of rock matrix.

3. MATHEMATIC MODEL OF CASING DAMAGE FORECAST AND ITS NUMERICAL SOLUTION

The mathematical description of fluid-solid coupling in deformable reservoir is a problem of great practical importance to subsurface hydrology, including applications to underground waste containment, enhanced recovery of petroleum, aquifer remediation, and seismic phenomena in geological formations and unconsolidated geological materials. In order to characterize this intricate porous framework-fluid system, a variety of methods in continuum physics have been developed. The theories developed thus far have not been fully assimilated into numerical simulations for field applications due to their mathematical complexity. The combination of casing, cement shealth and porous reservoir is a complex coupled system. Based on the principles of rock mechanics and fluid mechanics in porous media, a mathematical model to forecast casing damage, with the seepage effect included, is developed in this paper. It includes three parts: flow equation, solid deformation equations and the coupling relations.

3.1 Governing equations of fluid flow through deformable media

Supposing that the fluid flow in an oil reservoir is a linear Darcian flow and the fluid phases are immiscible. Based on the mass conservation law and the Darcy's law, the momentum conservation equations of fluid phases are given as

$$\nabla\cdot\left[\frac{K_{ij}k_{r\alpha}}{B_\alpha\mu_\alpha}(\nabla p_\alpha-\rho_\alpha g\nabla D)\right]+\phi\frac{\partial}{\partial t}\left(\frac{S_\alpha}{B_\alpha}\right)+\frac{S_\alpha(1-\varphi_0)}{B_\alpha(1+\varepsilon_V)^2}\frac{\partial\varepsilon_V}{\partial t}+\frac{q_\alpha}{\rho_{\alpha s}}=0 \quad (5)$$

where α signifies the three immiscible phases of water, oil and gas, K_{ij} the absolute permeability tensor, D the elevation of the reservoir, ϕ the porosity, g is the gravity acceleration, ε_V the volume strain of the formation rock, and φ_0 the initial porosity of reservoir, respectively. For fluid of phase α, $k_{r\alpha}$ denotes the relative permeability, p_α the fluid phase pressure, ρ_α the fluid density, S_α the saturation, B_α the compressibility coefficient, and $\rho_{\alpha s}$ the density of fluid in standard conditions, respectively. The term q_α denotes the source or converge item of fluid and it equals mass discharge of fluid.

In order to solve the multi-phase flow equations, supplemented equations are added to consider the saturation restriction and capillary pressure, as given by

$$S_o+S_g+S_w=1 \quad (6)$$

$$p_{cgo}=p_g-p_o,\ p_{cow}=p_o-p_w \quad (7)$$

where p_o, p_g and p_w are the oil, gas and water pressure, p_{cgo} is the capillary pressure between gas and oil, p_{cow} is the capillary pressure between oil and water.

3.2 Equations of solid deformation

Supposing that the solid matrix has an elasto-plastic behavior, the theory of plasticity can then be used to describe the matrix deformation in oil reservoir. Equations of the solid deformation include the following three groups: equilibrium equations, geometry equations and constitutive equations.

According to the revised Terzaghi's effective stress principle, the equilibrium equation of the reservoir matrix can be written

$$\sigma'_{ij,j}+(\alpha p\delta_{ij})_{,j}+f_i=0 \quad (8)$$

where σ'_{ij} is the tensor of effective stress; α the Biot coefficient, δ_{ij} the Kronecker function and f_i the body force term, respectively.

The geometry equations is given by

$$\varepsilon_{ij}=\frac{1}{2}(u_{i,j}+u_{j,i}) \quad (9)$$

where u_i is the displacement vector and ε_{ij} is the strain tensor, respectively.

The constitutive equation for elasto-plastic materials is given by

$$\{d\sigma'_{ij}\}=[D_{ep}]\{d\varepsilon_{ij}\} \quad (10)$$

where $d\sigma'_{ij}$ is the increment of effective stress, $[D_{ep}]$ is the elasto-plastic matrix, and $d\varepsilon_{ij}$ is the increment of strain, respectively. The Druck-Prager criterion and associated flow law were used in the elasto-plastic model. The above equations established the mathematics model of fluid-solid coupling for flow through a nonlinearly deformable porous reservoir. Site-specific boundary conditions must be applied to the above equations for solving practical problems.

3.3 Coupling relation between seepage field and deformation field

Coupling relations between seepage field and deformation field include relations among porosity and permeability change versus stress and pore pressure, as expressed in a exponential function in equation (11) and the effects of flow on mechanical parameters of rock matrix as given in equation (12-13), obtained through experimental studies.

$$\begin{cases} \varphi = \varphi_0 e^{-c_p(\sigma_v - \alpha p)} \\ k = ae^{-b\sigma_v + cp} \end{cases} \tag{11}$$

$$\sigma_c = \sigma_{c0} e^{-a_1 p}, E = E_0 e^{-a_3 p} \tag{12}$$

$$c = c_0 - A.S_w, \phi = \phi_0 - B.S_w \tag{13}$$

where c_p is the pore volume compressibility, σ_v the volume stress, φ is porosity, k is permeability, σ_c the peak strength of rock matrix, c_0 and ϕ_0 the cohesion and interior friction angle of the dry rock, and S_w the water content ratio, respectively. A, B, a, b, c are the fitting coefficients obtained by experiments.

3.4 Numerical solution techniques

The numerical solution method for the above fluid-solid coupling model is an iterative computation process. To reduce the computational complexity, the solid deformation and fluid flow are regarded as two coupled equation systems, solved by FEM. The equilibrium in solid matrix is solved using Eq.(6) with an added coupling item $\alpha p \delta_{ij}$ and the pore pressure is treated as an equivalent initial stress term. The flow equation (5) is solved with an added term of volume strain ε_v, reflecting the effect of solid deformation on fluid flow. It can be treated as a source or converge. In each iterative loop, the solid matrix deformation is solved firstly. The stress and strain results are then taken as inputs for the flow calculation with modified hydraulic parameters. After flow model is solved, the pore pressure values are transferred into solid matrix deformation model and begins next iterative loop. In this way, the flow and deformation of oil reservoir can be simulated.

4. FIELD APPLICATIONS

Above models were applied to Daqing Petrolrum Co. for simulating causes for casing damages and obtained some useful results. The formation rock is mudstone.

4.1 Ground uplift

We simulated the ground uplift that is the main cause for tensile casing damage in south-II pool of Daqing Petroleum Co. by a finite element program. The oil field we simulated is 16 m² in area, defined by faults around its boundaries. Based on the site geology, model is divided into five rock seams with 8568 elements and 10400 nodes. The faults are treated as special joint elements. The ground surface is treated as a free surface. The bottom of the FEM model is at a depth of 1300m and is treated as fixed in displacement. The restraining barriers are supposed to be impermeable, and the pore pressure is zero. Figure1 shows the simulated results of vertical displacement of the ground in South-II pool in 1994.

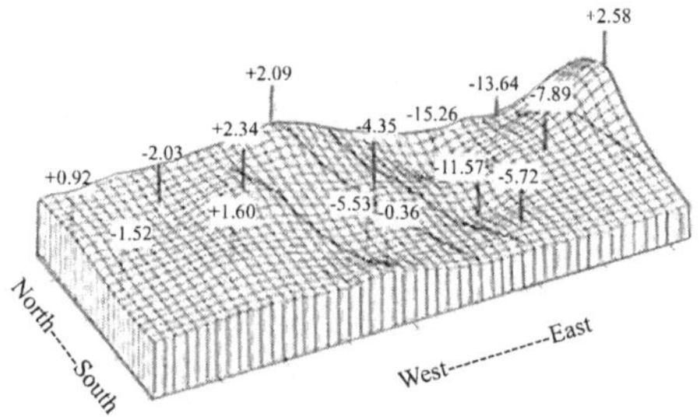

Figure 1 Vertical deformation in south-II pool

4.2 Injection vs. shear deformation

In order to simulate causes for shear casing damage induced by joint slippage, we define the stability coefficient K according to tectonic mechanics theory, given as

$$K = \sum R / \sum T \tag{14}$$

where $\sum R$ is the total anti-slip force and $\sum T$ is total slip force. K > 1 means that the block is stable and K = 1 indicate a critical stability status. During oil reservoir development, the porosity, permeability, cohesion, and interior friction angle vary because of seepage changing. In the simulation, we adopted Goodman's joint element method. From numerical simulation results, we observed increasing frequency of micro-seismicity and vertical displacement with increasing injection pressure, shear deformation along inclined strata layers causing shearing casing damage, reduced formation strength with increasing water saturation, and variations of in-situ stress with injection.

4.2 Releasing injection pressure

Large amount of field data have shown that the squeezing casing damages often occur at the moment of releasing injection pressure which causes sharp increase of the effective stress in the formation rocks and is the main cause for squeezing casing damage. The simulated region is a cube with a well bore in the center (Fig. 2). Length, width, and height of the simulation region are 100m, 100m and 55m, respectively. Oil seam is in the middle, with a thickness of 5m. The distance from top of the model to the ground surface is 1000m. The region was divided into 13482 elements with 21948 nodes. Displacements in x, y, z directions were constrained at the bottom, surface forces acting on the surrounding and top surfaces were decided by measured in-situ stresses, with a vertical stress of 20 MPa and horizontal stress of 30MPa, respectively.

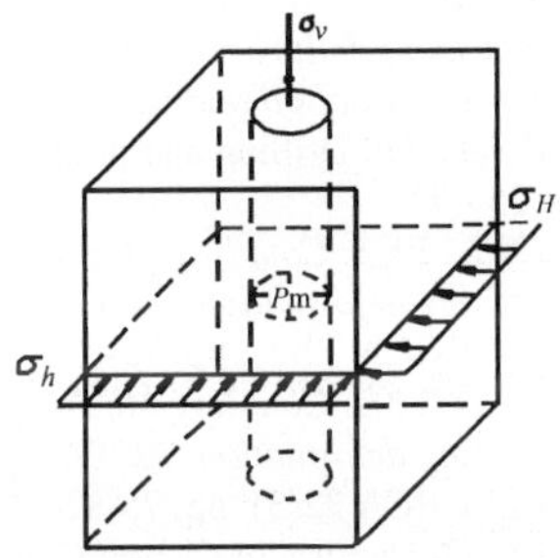

Figure 2 Simulation model for squeezing casing damage causes

We simulated fluid pressure distribution according to a given pressure difference. Figure 3 shows the changing pressure releasing in the injection well. From the figure, it can be found that fluid pressure declines soon on the 1st day. During the overflow procedure, the initial instantaneous discharge is very large, but with the time going the discharge diminishes rapidly (Fig. 4).

We found that at the same pressure difference, the lower permeability is, the larger the squeezing stress for casing will be, when subjected to the same open flow. With the known casing squeezing strength, a diagram (Fig. 5) between flow discharge and squeezing stress will indicate the maximum open flow value for preventing squeezing casing damage. For example, if the squeeze strength of casing is 26MPa, the pressure difference is 12 MPa, and the permeability is 50mD, the maximum discharge without causing casing damage is 14m^3/d.

During injection well pressure releasing, the open flow discharge must be controlled under the value in order to prevent casing damage.

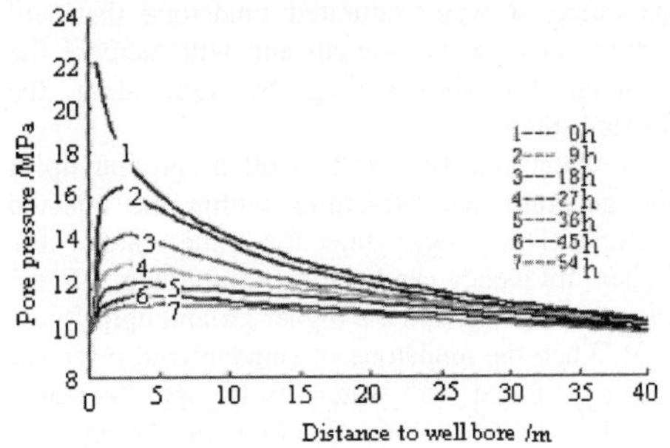

Figure 3 Pressure change during water overflow

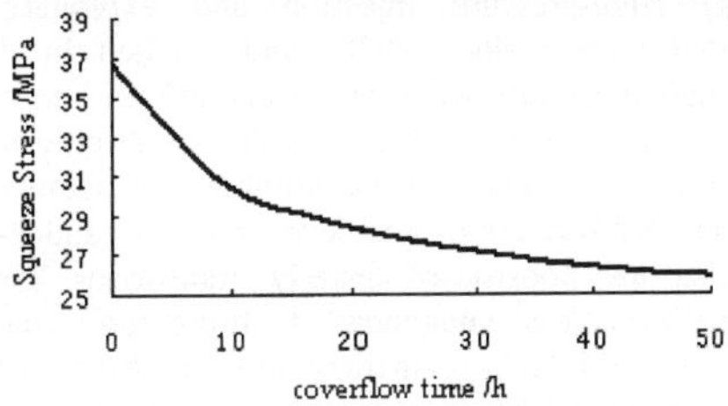

Figure 4 Squeeze pressure changes during pressure releasing

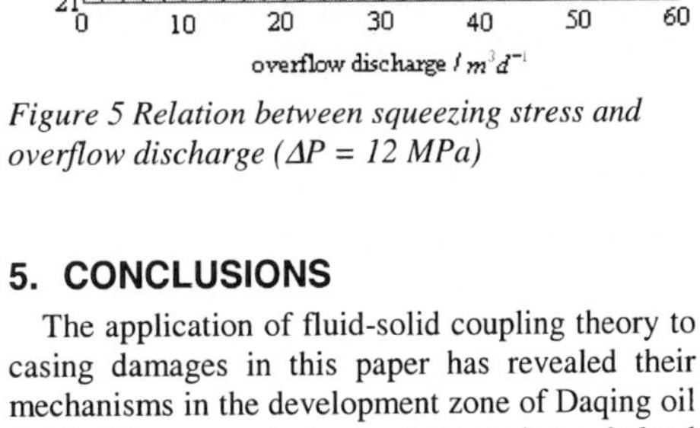

Figure 5 Relation between squeezing stress and overflow discharge (ΔP = 12 MPa)

5. CONCLUSIONS

The application of fluid-solid coupling theory to casing damages in this paper has revealed their mechanisms in the development zone of Daqing oil field. The numerical simulations have helped establish practical measures for protection of casings in the area.

For predominantly laterally shear-sliding casing damages the results are:

1) In present geological conditions the increase in differential pressures is most likely to enlarge the range of water-saturated mudstone that will further alleviate its strength and will increase the potential for shear-sliding by lubricating the fracture surfaces.

2) Water injection will result in ground uplift and generate micro-fractures within the injected stratum. The longer injection time causes the higher frequency and larger amplitude of the micro-seismicity, and the higher ground uplift.

3) When the mudstone of standardized reservoir bed is saturated with water, its strength decreases and plays a role of lubrication in block sliding.

For predominantly vertically extensional casing damages the results are:

1) High-pressure injection and exploitation started therein since 1970s and resulted in the formation of abnormal high pressure in the S0 formation. There are three factors for forming the abnormal pressures: a) the lithology between the S0 and S-I formations and between the S-I and S-II formations consist of mainly mudstones (and sometimes less siltstones), high-pressure water injected in the S-II formation would probably seep through under-developed barriers into the S0 and S-I formations; b) water seepage around the casing as a result of bad quality of well cementing; and c) if damaged wells were not closed instantaneously, high-pressure water might flow through subsurface fractures that result from high fluid pressure causing damage. Furthermore, exploitation in both the S0 and SI formations up to now helped to form the high pressure in the S0 formation. If the pressure is high enough, it will uplift the above rock and snap the casing.

2) The best way of protection against casing damage is to effuse high pressure in the S0 formation in order to keep stratum pressure at a low level. The measures taken include keeping the pressure of injected water as low as possible (no more than lithostatic pressure), closing damaged wells instantaneously and no injection of water both in aborted damaged wells and in replacement wells. In addition, 3D geodetic leveling is performed in the area twice a year to monitor geodetic deformation for prediction of casing damages.

For predominantly radial squeezing casing damages, through numerical simulation the authors draw the following conclusions:

1) Pressure difference between well bore and reservoir is a key factor causing squeezing casing damage. During pressure releasing in injection wells, the large pressure gradient and concentrated rock stress induce squeezing stresses to the casings. Squeezing casing damage often occur at the moment of releasing pressure.

2) The lower the permeability of reservoir is, the larger pressure difference between well bore and reservoir becomes and the higher squeezing stresses will be produced to the casings.

ACKNOWLEDGEMENTS

The financial supports from The Special Funds for Major State Basic Research Project under Grant no. 2002CB412708 and National Nature Science Foundation of China under Grant no. 50179034 are gratefully acknowledged.

REFERENCES

Hu Z.B. 1998. *Casing failure mechanisms and protection measures of oil and water well in Daqing oil field.* Oil Drilling and production 20(5): pp. 95 - 99.

Liu J.Z., Liu X.L. 2001. *Coupling Mechanics Models of casing deformation.* ACTA PETROLEI SINICA 22(3): pp. 75-82

Liu J.Z., Liu X.L. 2001. *Coupling Mechanics Models of casing deformation.* ACTA PETROLEI SINICA 22(3): pp. 75-82

Yin Z.M. 2000. *Mechanism of Casing Damage in Saertu South Pool of Daqing Oil Field.* Thesis for doctoral degreee of Chinese Academy of Sciences.

Wang Z.M. *Casing damage of Oil and Water well and controlling method.* Beijing:Oil Industrial Press.1994

Zhang G.D., Liu C.L. 2000. *Effect of suroundingrock rheology on oil well casing damage.* Chinese Journal of Rock Mechanics and Engineering. 19(1): pp. 971-975

Liu J.J., Liu X.G., Hu Y.R. 2002. *Study on Fluid Flow through Low permeability reservoir.* Chinese Journal of Rock Mechanics and Engineering. 21(1): pp. 48-54

Zhou W.Y. 1993. *Advanced Rock Mechanics.* Chinese Hydropower Press.

Zhang X.Y.2001. *Some Problem on Simulation of Casing damage by Finite Element Method.* Natural Gas Industry. 21(1): pp.62-65

Lian Z.H., Zhao G.Z., Zhang X.P. 1998. *A Study on Computer Emulation Software for Casing Failure.* Natural Gas Industry. 18(2): pp.35-38

Theme 4
Coupled T-H-M-C Processes for Geothermal Energy Engineering

PERMEABILITY IN LAYERED RESERVOIRS: FIELD EXAMPLES AND MODELS ON THE EFFECTS OF HYDROFRACTURE PROPAGATION

Sonja L. Brenner[1], Agust Gudmundsson[2]

[1]) Department of Earth Science, University of Bergen, Norway. Sonja.Brenner@geo.uib.no
[2]) Geoscience Centre, University of Göttingen, Germany. Agust.Gudmundsson@gwdg.de

Abstract: Calcite veins in a limestone-shale succession in SW England were formed by geothermal water and injected along normal faults. Vein length/thickness ratios indicate that their fluid overpressure was 20 MPa, suggesting a depth to the fluid sources beneath the present exposures of several hundred metres. Many subvertical veins are restricted to limestone layers, reducing the temporary permeability. When hydrofracture fluid overpressure is the only loading, soft layers tend to arrest hydrofractures, whereas stiff layers favour their propagation. Veins dissecting many layers form subvertical extension fractures in the limestone layers but thinner inclined shear fractures in the shale layers. Vein aperture changes between layers suggest that flow channelling may be important in layered reservoirs.

1. INTRODUCTION

All fractures generated by internal fluid overpressure are here referred to as hydrofractures. The fracture-generating fluid may be oil, gas, magma, groundwater, or geothermal water. Hydrofractures include dykes, inclined sheets, mineral veins, many joints, and the man-made hydraulic fractures that are used in the petroleum industry to increase the permeability of reservoir rocks. Hydrofractures are primarily extension fractures (Gudmundsson et al. 2001). The difference between the total fluid pressure in a hydrofracture and the normal stress, which for extension fractures is the minimum compressive principal stress, σ_3, is referred to as the fluid overpressure.

The formation and development of hydrofractures is, together with the formation of shear fractures (faults), presumably one of the main mechanisms for the generation and maintenance of permeability in fluid-filled heterogeneous (e.g. layered, fractured) reservoirs such as those for groundwater, geothermal water and petroleum. The permeability development in fractured reservoirs depends largely on fluid overpressure and transport in hydrofractures (Nelson 1985, Singhal & Gupta 1999). Fluid reservoirs are commonly composed of heterogeneous and anisotropic, especially layered, rocks, in which the mechanical properties vary considerably. For example, the stiffnesses of common rock types, expressed by their Young's moduli, span more than two orders of a magnitude (Bell 2000).

The hydrological processes of fluid flow and transport through rock fractures and fluid pressure changes are coupled with mechanical processes such as fracturing (Tsang 1991; Neuzil 2003) in that the mechanical properties of the host rock and the fluid overpressure in hydrofractures determine the propagation and aperture variation of hydrofractures. The size and shape of hydrofractures, in turn, influence their fluid overpressure and ability to transport fluids. The permeability evolution of a fluid reservoir does thus depend on hydromechanically coupled processes.

In this paper we present field examples of hydrofractures as mineral veins in layered sedimentary rocks and compare these observations with numerical models. We investigate the propagation of hydrofractures and then use analytical models based on field measurements to estimate the fluid overpressure and depth to the fluid sources of the veins. We also explore the effects of layering on aperture variations of such hydrofractures.

2. HYDROFRACTURE PROPAGATION

Hydrofractures propagate by advancing their tips when the associated tensile stresses exceed the tensile strength of the host rock. The propagation of a hydrofracture depends on its fluid pressure in comparison with tectonic stresses, as well as on the mechanical properties of the host rock.

Field studies of calcite veins formed by geothermal water were carried out in the Lower Jurassic "Blue-Lias-Formation", a limestone-shale succession cut by many E-W trending normal faults, at the Somerset Coast, Southwest England (cf. Whittaker & Green 1983). Most veins trend subparallel to, and occur in the damage zones and the cores of, the normal faults (Figs 1 and 2). Also, there are observations of veins being injected into the limestone layers of the damage zone directly from the fault plane. Thus, the normal faults, active mainly during Late Jurassic to Early Jurassic (Nemcok et al. 1995), transported geothermal water from the sources at deeper levels in the basin up to and into the present exposures.

Figure 1. A dense network of calcite veins runs subparallel to a normal fault plane, particularly in the hanging (right) wall, in a limestone-shale succession at Kilve, Somerset Coast, SW England. The veins extend only for a few metres into the damage zone. View west; the person provides a scale.

Most mineral veins in the damage zones, however, are restricted to the limestone layers, that is, are stratabound extension fractures (Fig. 2). This means that the permeability was less than it would have been in case the veins propagated through many layers.

Theoretical tip tensile stresses of hydrofractures in homogeneous, isotropic rocks are normally so high that continuous and buoyant hydrofractures should reach the surface. Many hydrofractures, however, become arrested at contacts in heterogeneous, anisotropic rocks, especially when the host rock consists of alternating soft and stiff layers such as in limestone-shale successions (Gudmundsson & Brenner 2001).

Figure 2. In the fault core of this normal fault at Kilve, SW England, there is a dense network of white calcite veins. In the damage zone, most of these are restricted to grey limestone layers and end abruptly at the contacts with the darker shale layers. View east; the measuring steel tape is 50 cm.

We use the boundary-element program BEASY (1991) to explore why many hydrofractures, such as the observed veins, are stratabound; why, for example, the calcite veins in the limestone layers became arrested on meeting with the shale layers. For each model the stress field around the hydrofracture tip is calculated. Tip propagation is simulated in a semi-static way using models that vary slightly in boundary conditions between runs.

In the models (Figs 3 and 4), the tip of a vertical hydrofracture propagates through a succession of ten layers of equal thicknesses but extremely different stiffnesses. The fluid overpressure in the hydrofracture is the only loading and has a constant value of 6 MPa. It is applied only in the lowermost layer so that the tip propagates ahead of a stationary fluid front. The lowermost layer J is very stiff (E = 100 GPa), layer I is very soft (E = 1 GPa), and layer H is moderately stiff (E = 10 GPa). This succession of 3 layers (stiff, soft, and moderate) is repeated to the top layer A with E = 100 GPa. With these stiffnesses we cover the stiffness range of common solid rocks, while all layers have a Poisson's ratio of 0.25, because its variation between layers is normally much smaller than that of Young's modulus (Bell 2000). The hydrofracture tip is allowed to propagate upwards along vertical internal springs, each with a stiffness of 6 MPa m^{-1}, representing a discontinuity with a soft, elastic infill.

In the first model (Fig. 3), three vertical internal springs have been added to layers I, H and G, so that the hydrofracture tip can propagate to the bottom of the soft layer F (E = 1 GPa). The soft layer F makes the tip wide and blunt, dissipates most of the fracture-tip tensile stress, and may arrest the hydrofracture. In the second model (Fig. 4), internal springs have been added to each of the layers I through B, allowing the hydrofracture tip to propagate to the bottom of the stiff surface layer A (E = 100 GPa). The sharp and narrow tip gives rise to a tensile stress concentration in layer A which is similar to the in situ tensile strength of many low-strength rocks (Haimson & Rummel 1982; Amadei & Stephansson 1997) so that the tip may continue its propagation through layer A and to the surface.

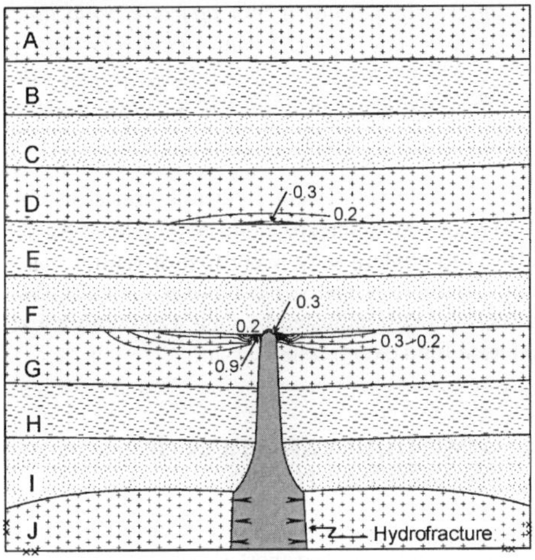

Figure 3. Boundary-element model of a vertical hydrofracture propagating through ten layers of different stiffnesses. Horizontal arrows indicate the hydrofracture's fluid overpressure of 6 MPa in the lowermost layer J. Crosses indicate the fastening of the model in the lower corners. The opening displacement is exaggerated. Three vertical internal springs (through layers I, H and G) allow the hydrofracture tip to propagate up to the bottom of the soft layer F, which makes the tip wide and blunt, dissipates most of the fracture-tip tensile stress (contours of σ_3 in megapascals), and may arrest the hydrofracture.

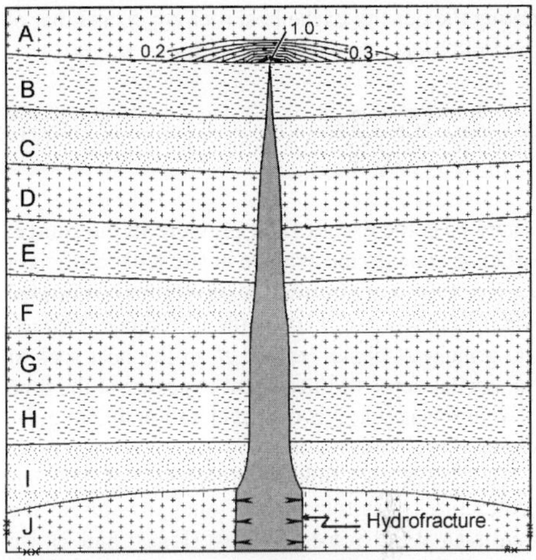

Figure 4. Same model as in Fig. 3, except that the hydrofracture tip has been allowed to propagate to the bottom of the stiff surface layer A. The sharp and narrow tip gives rise to relatively high tensile stress concentration in layer A, and the tip may continue its propagation through this layer.

The results indicate that, when the fluid overpressure is the only loading, the stresses of a hydrofracture tip become suppressed on meeting with soft layers such as shale, resulting in its arrest. By contrast, high tensile stresses at sharp tips would make it likely for a hydrofracture to propagate through stiff layers such as limestone.

3. FLUID OVERPRESSURE IN HYDROFRACTURES

The static fluid overpressure p_o of a fluid-filled extension fracture, modelled as a through crack with the outcrop length L as the controlling dimension, is (Gudmundsson et al. 2001):

$$p_o = \frac{b_{\max} E}{2L\left(1-\nu^2\right)} \tag{1}$$

where $b_{\max}$ is the maximum aperture (thickness) of the fracture, E is Young's modulus of the host rock and ν its Poisson's ratio.

Using this fluid overpressure, the depth h of the source of the fluid that formed the hydrofractures below the present exposure can be estimated. Denoting the density of the host rock by ρ_r, and that of the fluid by ρ_f, then (Gudmundsson et al. 2001):

$$h = \frac{p_o - (p_e + \sigma_d)}{(\rho_r - \rho_f)g} \tag{2}$$

where p_e is the excess fluid pressure in the source, which is equal to the average in situ tensile strength T_0 (because a hydrofracture is normally initiated when p_e reaches T_0); σ_d is the difference between the principal stresses σ_1 and σ_3 in the host rock at the crustal depth of the current exposure at the time of vein formation; g is the acceleration due to gravity.

To obtain the fluid overpressure of the hydrofractures that formed the mineral veins in the field area at the Somerset Coast (Figs 1 and 2), we measured the length and aperture (thickness) of 239 calcite veins exposed in subhorizontal limestone layers. We used only veins that have both ends visible in the outcrop, are continuous, and are non-restrictive (that is not dissecting other veins or other discontinuities). All the measured veins are extension fractures that show no evidence of subsequent shear deformation.

There is a reasonable correlation between the length and aperture of these veins, the typical aspect (length/aperture) ratio being 400 (Fig. 5). Using this aspect ratio, a reasonable Young's modulus of 15 GPa for limestone, and a Poisson's ratio of 0.25 (Bell 2000), equation (1) yields a fluid overpressure at the time of vein formation of 20 MPa.

In equation (2), we use limestone density 2400 kg m^{-3} (Bell 2000) for ρ_r and water density at 90°C as 965 kg m^{-3} (Smits 2000) for ρ_f. σ_d cannot be greater than $4T_0$, because of frictional equilibrium on pre-existing, optimally oriented small-scale faults subject to high fluid pressure (Gudmundsson et al. 2001). The in situ tensile strength of many solid rocks is 0.5 to 6 MPa, most commonly 2-3 MPa (Haimson & Rummel 1982; Amadei & Stephansson 1997). Using these values, equation (2) gives a depth of 355 to 710 m. With very extreme parameter variation the maximum depth is 1200 m. Isotopic results by Davison (1995) indicate origin of the water 1000 m below the present exposures, in good fit with these results. This suggests that the geothermal water forming the veins was of rather local origin, that is from sources several hundred metres below the current exposures in the sedimentary basin.

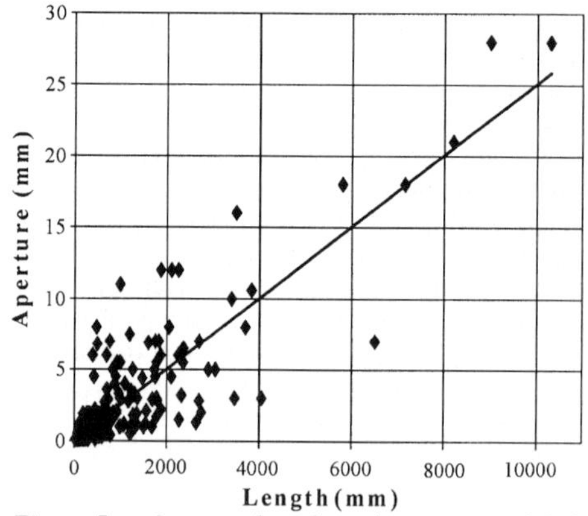

Figure 5. Outcrop length and aperture of 239 calcite veins, measured in limestone exposures at Kilve, SW England. Linear regression constrained through the origin (so that a fracture of zero length has a zero aperture, solid line) gives an aperture of 1 mm as typical for a vein of 400 mm length (R^2 = 0.740, significant with 99% confidence according to a Fisher t-test).

The calcite veins were formed at crustal depths of at least 1 km (Nemcok et al. 1995). Therefore, so they are unlikely to have formed as tension fractures (with σ_3 negative) but rather as hydrofractures, as is also indicated by their aspect ratios and inferred fluid overpressure.

4. APERTURE VARIATION OF HYDROFRACTURES

The aspect ratio and shape of a hydrofracture and hence its ability to transport fluids, depend on its fluid overpressure and the mechanical properties of the host rock. Hydrofracture aperture variation may result in channelling of the fluid flow along the widest parts of the fracture (Tsang & Neretnieks 1998).

Analytical models of the aperture variation of a two-dimensional extension fracture subject to a constant overpressure in a homogeneous, isotropic rock show that the fracture opens into a flat ellipse (Sneddon & Lowengrub 1969; Valko & Economides 1995). For layered rocks, numerical models of the aperture variation of vertical hydrofractures with fluid overpressure as the only loading indicate that the aperture of hydrofractures tends to be smaller in stiff layers than in soft layers (Brenner & Gudmundsson 2002).

In the study area most veins are restricted to the limestone layers (Fig. 2). Comparatively few veins dissect the shale layers and those that do mostly follow inclined shear fractures in the soft shale layers. Thus, during vein formation the shale had no tensile strength and failed in shear rather than in extension. The veins occupying inclined shear fractures are thinner than the same veins in the limestone layers, where they are vertical extension fractures. This is because where the veins are inclined they are no longer perpendicular to the minimum principal compressive stress, σ_3, but normal to a higher stress σ_n (Fig. 6).

Figure 6. Aperture variation of a calcite vein at Kilve, SW England. In the grey limestone layer, the vein is a vertical extension fracture, with σ_3 as its normal stress. In the shale layer, the vein is an inclined shear fracture, subject to a higher normal stress σ_n and therefore thinner than the vertical part. View east; the limestone layer in the upper half of the picture is 25 cm thick.

We now explore the aperture variation of such a hydrofracture in a layered crust dissecting many layers using the boundary-element program BEASY (1991). In the model (Fig. 7) the layers are in a succession of moderately stiff layers (A, C, E etc.) of Young's modulus E = 15 GPa as for limestone, alternating with soft layers (B, D etc.), E = 1 GPa as for shale (Bell 2000). Poisson's ratio of all the layers is 0.25. In all the stiff layers the hydrofracture is a vertical extension fracture with a fluid overpressure of 20 MPa (cf. section 3). In all the soft layers, the hydrofracture is inclined at 45° and the overpressure is reduced to 14 MPa. For this reduction we assume that the minimum principal compressive stress, σ_3, is horizontal and has a magnitude of 12 MPa less than the vertical stress σ_1 (because $\sigma_d = 4T_0$) so that σ_n on a 45°-inclined fracture is 6 MPa higher than on a vertical fracture. The results show that for the parts of the hydrofracture where the fluid overpressure is reduced due to its inclination the aperture is smaller than for the vertical parts. This is so even if the inclined parts occur inside the soft layers, where the aperture would be expected to be greatest if the fracture was vertical throughout.

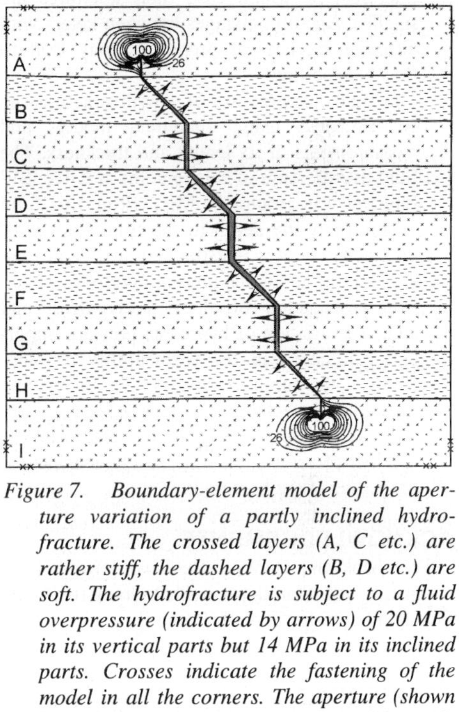

Figure 7. Boundary-element model of the aperture variation of a partly inclined hydrofracture. The crossed layers (A, C etc.) are rather stiff, the dashed layers (B, D etc.) are soft. The hydrofracture is subject to a fluid overpressure (indicated by arrows) of 20 MPa in its vertical parts but 14 MPa in its inclined parts. Crosses indicate the fastening of the model in all the corners. The aperture (shown without exaggeration) is slightly smaller inside the soft layers.

The aperture of a hydrofracture is clearly a function of its fluid overpressure and the mechanical properties of the host rock and can therefore change considerably in a fluid reservoir consisting of layers with different stiffnesses. Aperture changes between layers suggest that there may be preferential flow (flow channelling) in layers with certain mechanical properties in a fluid reservoir.

5. CONCLUSIONS

1. Field examples and numerical models show that mechanical layering is important for the

propagation of hydrofractures. While soft layers suppress tensile stresses and blunt hydrofracture tips, thereby favouring their arrest, stiff layers concentrate tensile stresses, make hydrofracture tips sharper and ease their propagation.

2. Mechanical layering also affects aperture variations of hydrofractures. The aperture of a vertical (extension) vein is commonly greater in soft layers than in stiff layers. In soft layers, however, the veins are often inclined (shear fractures) in which case their apertures are smaller. Aperture changes between layers with different mechanical properties may lead to flow channelling.
3. The fluid overpressure of a hydrofracture can be estimated using analytical models and measurements of its length and thickness. The fluid overpressure of calcite veins at Kilve, SW England was around 20 MPa. From this fluid overpressure the depth to the fluid sources of the hydrofractures is estimated at several hundred metres below the present outcrops.
4. Both propagation (or, alternatively, arrest) and aperture variation of hydrofractures are important parameters in the permeability development of fractured reservoirs. To understand these hydromechanically coupled processes, one has to consider the fluid overpressure of the hydrofractures as well as the mechanical properties of the host rock.

ACKNOWLEDGEMENTS

We thank the GeoProc-Reviewers for helpful comments and Statoil for a PhD-Grant (to Agust Gudmundsson) for Sonja L Brenner.

REFERENCES

Amadei, B. & Stephansson, O. 1997. *Rock Stress and its Measurement*. Chapman & Hall, London.

BEASY. 1991. *The Boundary Element Analysis System User Guide*. Computational Mechanics, Boston.

Bell, F.G. 2000. *Engineering Properties of Soils and Rocks*. Blackwell, Oxford.

Brenner, S.L. & Gudmundsson, A. 2001. *Permeability development during hydrofracture propagation in layered reservoirs*. Norges Geologiske Undersøkelse Bulletin 439: pp. 71-77.

Davison, I. 1995. *Fault slip evolution determined from crack-seal veins in pull-aparts and their implications for general slip models*. Journal of Structural Geology 17: pp. 1025-1034.

Gudmundsson, A. & Brenner, S.L. 2001. *How hydrofractures become arrested.* Terra Nova 13: pp. 456-462.

Gudmundsson, A., Berg, S.S., Lyslo, K.B. & Skurtveit, E. 2001. *Fracture networks and fluid transport in active fault zones*. Journal of Structural Geology 23: pp. 343-353.

Haimson, B.C. & Rummel, F. 1982. *Hydrofracturing stress measurements in the Iceland research drilling project drill hole at Reydarfjordur, Iceland.* Journal of Geophysical Research 87: pp. 6631-6649.

Nelson, R.A. 1985. *Geologic Analysis of Naturally Fractured Reservoirs.* Gulf Publishing, Houston, Texas.

Nemcok, M., Gayer, R., & Miliorozos, M. 1995. *Structural analysis of the inverted Bristol Channel Basin: implications for the geometry and timing of fracture porosity.* In: Buchanan, J.G. & Buchanan, P.G. (eds.). *Basin Inversion.* Geological Society Special Publication 88: pp. 355-392.

Neuzil, C.E. 2003. *Hydromechanical coupling in geologic processes.* Hydrogeology Journal: pp. 41-83.

Singhal, B.B.S. & Gupta, R.P. 1999. *Applied Hydrogeology of Fractured Rocks.* Kluwer, London.

Smits, A.J. 2000. *A Physical Introduction to Fluid Mechanics.* Wiley, New York.

Sneddon, I.N. & Lowengrub, M. 1969. *Crack Problems in the Classical Theory of Elasticity. Wiley, New York.*

Tsang, C.F. 1991. *Coupled hydromechanical-thermochemical processes in rock fractures.* Reviews of Geophysics 29: pp. 537-551.

Tsang, C.F. & Neretnieks, I. 1998. *Flow channeling in heterogeneous fractured rocks.* Reviews of Geophysics 36: pp. 275-298.

Valko, P. & Economides, M.J. 1995. *Hydraulic Fracture Mechanics.* Wiley, New York.

Whittaker, A. & Green, G.W. 1983. *Geology of the country around Weston-super-Mare.* Geological Survey of Great Britain, London.

THE EFFECT OF THERMAL, CHEMICAL, HYDROLOGICAL, AND MECHANICAL FACTORS ON WATER/ROCK INTERACTION IN HDR GEOTHERMAL SYSTEMS

S. Fomin, Z. Jing and T. Hashida
Fracture Research Institute, Tohoku University, Sendai 980-8579, Japan

Abstract: The numerical model FRACSIM developed by the research group in Tohoku University for the hot-dry-rock (HDR) geothermal reservoir is proved to be an appropriate approximate model capable to address simultaneously the problems associated with hydraulic stimulation, fluid circulation and heat extraction. The structure of the fractured rock is approximated with the network models of "fractal geometry". The adopted procedure makes it possible to characterize geothermal reservoirs by parameters measured from the field data. As is proved both theoretically and experimentally for the fluid flows in the fractured and porous media with relatively high fluid velocities, the values of the effective thermal conductivities in the longitudinal and transverse to the flow directions are directly proportional to the values of the fluid velocity. The water/rock chemical interaction (WRCI) is the other factor that inevitably exerts an influence on the permeability of the high temperature deep HDR circulating system. In the present study the mathematical model, which accounts for the effects of thermal dispersion and WRCI (in terms rock chemical dissolution and precipitation), is proposed. The numerical analysis based on the 3D model of the reacting fluid flow within the fractured media proves the importance of the thermal dispersion and WRCI factors for the assessment of the overall reservoir thermal and hydraulic performance.

1. INTRODUCTION

Many numerical models have been proposed for simulating the HDR systems (Kohl & Hopkirk, 1995; Watanabe and Takahashi, 1995; Jing et al., 2000; etc). The numerical model FRACSIM-3D developed by the research group in Tohoku University is proved to be an appropriate approximate model capable to address simultaneously the problems associated with hydraulic stimulation, fluid circulation and heat extraction and can quantitatively predict the 3D reservoir growth behaviour. One of the aspects of the general problem of reservoir modelling is the proper approximation of the fracture distribution within the rock. The latter is of crucial importance for simulating the hydraulic fracturing or stimulation and for predicting the reservoir long-term performance. In fractured rocks, groundwater flow occurs predominantly through the connected network of discrete fractures. A series of geophysical investigations has confirmed that subsurface fracture networks can commonly be described by fractal geometry. The model presented here is based on the relationship between the fracture length and a number of fractures, as suggested in Watanabe & Takahashi (1995). It incorporates the elements of the approximate model discussed in Jing et al. (2000), where the fracture shear displacements and openings, variation of the shape of the stimulating rock volume, pressure compliant fracture apertures are taken into account. The natural rock fracture surfaces, which are proved to be of fractal geometry as well (Brown, 1995), are modelled on the basis of the spectral synthesis method, which accounts for the fractal nature of the fracture surfaces. This model is used for numerical simulation of the fracture dilation due to the offset of the fracture surfaces and accounts for the experimentally acquired data of the mating rough fracture surfaces profiles. Since the fluid velocity within the geothermal reservoirs created in the fractured rock can be relatively high and the characteristic sizes of the solid blocks of the rock, which constitute the fissured reservoir media, are relatively large, the high mean values of heat flow rates are typically attained within the stimulated region. In this situation the effect of thermal dispersion is an important factor, which is included into the mathematical model of the heat transport within the natural or artificially developed fissured geothermal reservoirs. Mathematical model also accounts for the effects of rock chemical dissolution and precipitation.

2. DESCRIPTION OF THE MODEL

In the present 3D model the fracture network, which consists of the randomly distributed penny-shaped subsurface fractures, is mapped on a regular cubical grid of the discretized reservoir domain. It is assumed that the flow properties of a stochastic fracture network depend on the fluid pressure.

Fractures are generated stochastically within a fracture generation volume of $(L+2r_{max})^3$, where L is the characteristic linear dimension of the model volume and r_{max} is the radius of the largest fracture. The fracture centers are uniformly-randomly distributed. In 3D case the radius of the penny-shaped fracture is used as a characteristic fracture dimension (instead of fracture length in 2D-model) and will be quoted as a fractal fracture length in the further discussion. The random fractures are generated as follows. In a fractal fracture length distribution of fractal dimension D, the number of fractures N_r whose characteristic length (in terms of fracture radius r) is greater than r, is given by $N_r=Cr^{-D}$. Hence, the number of fractures between the specified upper and lower limits is given by $N_{r_{min}}^{r_{max}} = C(r_{min}^{-D} - r_{max}^{-D})$, where r_{min} and r_{max} are the lower and upper fracture radius limits, respectively. Consider some fraction, α, of this total number counting from r_{min} upward and the corresponding size r_α of the largest object in that fraction, then $N_{r_{min}}^{r_\alpha} = \alpha C(r_{min}^{-D} - r_{max}^{-D}) = C(r_{min}^{-D} - r_\alpha^{-D})$. The latter yields the fractal fracture length distribution

$$r_\alpha = [(1-\alpha)r_{min}^{-D} + \alpha r_{max}^{-D}]^{-1/D}, \tag{1}$$

where α is a random parameter in the interval [0,1]. Thus, for the purpose of simulation, it is necessary to specify the length of the smallest and largest fractures, as well as fractal dimension D. Using equation (1), fracture r_α can be generated by simply changing the α value. For any generated fracture, the initial (i.e. undisturbed) fracture aperture, a_0, when evaluated at zero effective stress, is assumed to be proportional to the fracture radius, $a_{i0} = \beta \cdot r_i$, where i=1, 2,..., N_f and r_i is the radius of each fracture from the whole set of fractures N_f; β is a constant of proportionality, which is chosen to allow the undisturbed fracture network to match (at least approximately) the *in situ* measured permeability. Fracture orientation distribution is based on the fracture orientations that obtained (usually from well logs) by field observations at the HDR geothermal reservoir. Fractures are generated until the fracture density (total area of fracture surfaces per rock volume) reaches the experimentally observed level.

2.1 Reservoir dynamics

Fracture apertures are affected by the effective normal stress at the fracture surface and by shear displacement that determines the fit of the opposing rough surfaces. Shear stability is expressed by a simple friction law, when slip is taking place if $\tau > (\sigma_n - P)\tan(\phi_i + \phi_{dil}^{eff})$, where τ is a shear stress, σ_n is the rock stress normal to the fracture surface, P is a fluid pressure in fracture, and ϕ_{dil}^{eff} is an effective shear dilation angle at a given effective normal stress. The basic friction angle ϕ_i is a material property of the fracture walls. The effective shear dilation angle ϕ_{dil}^{eff} is a property of both the fracture wall asperities and effective normal stress

$$\tan(\phi_{dil}^{eff}) = \tan(\phi_{dil})/[1+9(\sigma_n - P)/\sigma_{nref}], \tag{2}$$

where σ_{nref} is the effective normal stress applied to cause a 90% reduction in the compliant aperture and ϕ_{dil} is the laboratory measured shear dilation angle at very low effective stress. The amount of shear displacement depends on the fracture shear stiffness and on the amount of "excess" shear stress available. Referring to the theory of elasticity, the shear displacement of a fracture U can be expressed as:

$$U = [\tau - (\sigma_n - P)\tan(\phi_i + \phi_{dil}^{eff})]/H_s, \tag{3}$$

where H_s is a shear stiffness of the fracture. The change in aperture is a product of the displacement and the tangent of the effective shear dilation angle $a_s = U\tan(\phi_{dil}^{eff})$. An expression for the aperture a of a sheared fracture with unabrading asperities in contact is

$$a = (a_0 + U\tan(\phi_{dil}))/[1+9(\sigma_n - P)/\sigma_{nref}]. \tag{4}$$

The above equations provide a simple set of approximate correlations used in a model that contains many thousands of fractures each of which might undergo a small displacement. Since the flow rate in the reservoir is relatively high (especially during reservoir hydraulic stimulation) and, therefore, form drag due to solid obstacles may be comparable with the surface drag due to friction. Moreover, in some numerical models, for instance the 3D Finite Element Program FRACTure (Kohl & Hopkirk; 1995), the terms that govern the surface drag is ignored initially, referring to the fact of high flow rate, and only Forchheimer drag is incorporated. To this end, it would be quite useful to obtain the reliable estimates (at least to the order of magnitude) of the forces acting in the fractured media and validate the model for momentum equation before proceeding with complex numerical simulations. This can be made on the basis of scale analysis of the momentum equation that accounts for the both effects mentioned above. This equation can be presented as: $\nabla P = -\frac{\mu q}{2\pi K r_w h_w}(\tilde{\mathbf{u}} + c_F \operatorname{Re}|\tilde{\mathbf{u}}|\tilde{\mathbf{u}})$, where $c_F \approx 0.55$ is a dimensionless form-drag (Forchheimer) constant, q is a flow rate in the borehole, $\tilde{\mathbf{u}} = \mathbf{u}/u_0$ is a non-dimensional value of filtration velocity vector $\mathbf{u}(u_x, u_y, u_z)$ which is scaled with $u_0 = q/(2\pi r_w h_w)$, $\operatorname{Re} = \sqrt{K} u_0 \rho_L / \mu$ is Reynolds number, r_w is borehole radius, h_w is characteristic vertical sizes of the model area; K permeability of the media, μ and ρ_L fluid viscosity and density, respectively. For the typical conditions of the geothermal reservoir exploitation, h_w=100m, ρ_L=1000kg/m^3, q=200·10^{-3}m^3/s, μ=3·10^{-4}s·N/m^2, r_w=0.1m and K=10^{-14}m^2, we have Re≈10^{-3}. Hence, flow is laminar and the Forchheimer drag force can be neglected in the model. In this case the linear Darcy momentum equations can be employed, namely, $u_f = -(K_f/\mu)(\partial P/\partial f)$, where f=x, y ,z.

At this point it should be noted that in FRACSIM model the fractured reservoir is considered as an orthotropic media with symmetric permeability tensor and a diagonal matrix is formed whenever the coordinate axes are parallel with the principal axes along which the coordinates of the permeability tensor are K_x, K_y, K_z . Accounting for the mass conservation equation, the Darcy flow model leads to the following equation for pressure distribution

$$\frac{\partial}{\partial x}\left(K_x \frac{\partial P}{\partial x}\right) + \frac{\partial}{\partial y}\left(K_y \frac{\partial P}{\partial y}\right) + \frac{\partial}{\partial z}\left(K_z \frac{\partial P}{\partial z}\right) = 0. \qquad (5)$$

The permeability in the x, y and z directions, is strongly affected by the fracture distribution and fracture dimensions. Since the fracture apertures depend on pressure in the fluid and normal and shear stresses (see equations (3) and (4)), the permeability of the reservoir (defined by its components K_x, K_y, K_z), should be treated as a functions of spatial coordinates (x, y, z) and pressure P. As a result the problem is formulated as strongly non-linear, with variable permeability and, therefore, numerically it can be solved only by iterations.

2.2 Fracture surface model

As it was already mentioned, the above model contains the unknown values of shear dilation angles ϕ_{dil} , which can be predicted for the specified rock by analysing the effect of shears displacement of the syntactic fracture on the fracture dilation. This can be made with analytic-experimental method (it couples the analytical algorithm for fracture mathematical modelling and experimental data for the natural fracture topography). The natural rock fracture surfaces, which are proved to be of fractal geometry (Brown, 1995), are modelled on the basis of the spectral synthesis method, which accounts for the fractal nature of the fracture surfaces. For describing the fracture profile as a function of spatial frequency v=2π/λ, which is inversely proportional to wavelength λ, the power spectral density $S(v)$ is introduced in the spectral synthesis method. The topography of the rough fracture surface, measured by the laser profilometer, is approximated by Fourier components of various amplitudes and varies frequencies. The sum of these makes up the fracture surface, and defines the spatial correlation between the heights of any two points on the surface. The power spectrum is the plot of the squared amplitude of each of the Fourier components as a function of its spatial frequency; it's usually plotted in a logarithmic space. The power spectrum of a 2D rough fracture surface generally takes the form of a straight line. The gradient of this line, β, depends upon the fractal dimension D_S of the surface. Generally if a fracture profile $h(x)$ has a spectral density $S(v)$ which scales proportional to $1/v^{\beta}$, then if β is in a range $1 \le \beta \le 3$, the profile $h(x)$ has the fractal dimension described by the relationships $D_S = (5-\beta)/2$. If function $h^j(x)$ denotes the fracture surface profile recorded by the profilometer along the jth scan-line on the interval (0,Z), then the discrete Fourier transform of this function, H^j, can be presented as $H^j(v, Z) = \Delta x \sum_{n=0}^{N-1} h_n^j \exp[-2\pi i v n \Delta x]$, where i is the imaginary unity and h^j_n=$h^j(x_n)$=$h^j(n\Delta x)$ for n=0,1,2,...,N-1. The usual selection of discrete frequency value for computation of $H^j(v,Z)$ is v=v_k=k/Z=$k/N\Delta x$, where k=0,1,2,...,N-1. Power spectral density for each individual profile is defined as $S^j(v_k) = |H(v_k)|^2 / Z$. Since fracture surface can be approximated by the finite set of m scanned profiles the ensemble average power spectral density can be readily computed as

$S(\nu_k)=\frac{1}{m}\sum_{j=1}^{m}S^j(\nu_k)$. In discrete case, obviously $S(\nu_k)$ is a function of the frequency $\nu_k=k/N$, where k=0, 1, 2,..., N-1. The discrete inverse Fourier transform can be presented in its sinusoidal form as

$$h^j(x_n)=\sum_{k=0}^{N-1}\left(A_k^j\cos\varphi\frac{2\pi kn}{N}+B_k^j\sin\frac{2\pi kn}{N}\right) \qquad (6)$$

where A_k and B_k are the amplitudes (geometrically represent the amplitudes of the fracture profile fluctuations) of the real and imaginary parts of the Fourier components as a function of ν_k and n=0,1,2,..., N-1. Hence, along the specified scanned line from the set of m scanned lines on the fracture surface the power spectral density can be determined as $S^j(\nu_k)=\left[(A_k^j)^2+(B_k^j)^2\right]/Z$, where k=0, 1, 2,..., N-1. To obtain a spectral density that scales proportional to $1/\nu^\beta$, and hence to obtain a fractal profile, the following must be true:

$$(A_k^2+B_k^2)\propto 1/\nu_k^\beta, \qquad (7)$$

where k=0,1,2,..., N-1, and $1\leq\beta\leq3$. The method to produce a fractal profile is simply to choose those components A_k and B_k that obey equation (7) and subject them to an inverse Fourier transform represented in discrete terms by equation (6) from which the fractal profile can be generated.

2.3 Heat and mass transfer process

Since the fluid velocity within the geothermal reservoirs created in the fractured rock can be relatively high and the characteristic sizes of the solid blocks of the rock, which constitute the fissured reservoir media, are relatively large, the high mean values of heat flow rates are typically attained within the stimulated region. In this situation the effect of thermal dispersion is an important factor that should be included into the mathematical model of the heat transport within the natural or artificially developed fissured geothermal reservoirs. Employing different theoretical methods such as, volume averaging, moment-matching, homogenization, moment-difference expansion, and others (Hsu and Cheng, 1990; Koplik et al., 1988), and also experimental approaches it was proved that the approximate macro-scale model of heat transport within the fractured media in whole provides a quite sufficient level of accuracy and can be readily used instead of the exact model on the micro-scale level. The length L, which characterizes the whole volume of the fractured media, is assumed to be significantly larger than the spatial representative scale l for the unit cell (solid block), so that the small parameter can be introduced as $\varepsilon=l/L<<1$. For convenience, further we will assume that z-axis of the Cartesian coordinates coincides with the axis of symmetry of the borehole and the dominant flow in the reservoir is orthogonal to z- axis. In the publications cited above it was rigorously proved that with the accuracy of $O(\varepsilon)$ the macro-scale temperature equation for this flow can be formulated as:

$$c\rho\frac{\partial T}{\partial t}+c_L\rho_L\mathbf{u}\cdot\nabla T= \\ =\frac{\partial}{\partial x}\left(k_\parallel\frac{\partial T}{\partial x}\right)+\frac{\partial}{\partial y}\left(k_\parallel\frac{\partial T}{\partial y}\right)+\frac{\partial}{\partial z}\left(k_\perp\frac{\partial T}{\partial z}\right), \qquad (8)$$

where $k=k_Lm+k_S(1-m)$, $c=k_Lm+k_S(1-m)$, $\rho=\rho_Lm+\rho_S(1-m)$ are stagnant effective thermal conductivity, effective specific heat and effective density, respectively; indexes L and S indicate liquid and solid phase; m is porosity; t is time; $k_\parallel$ and $k_\perp$ are the longitudinal (parallel to the dominant flow) and the orthogonal (transverse to the main flow) effective thermal conductivities that account for the stagnant effective conductivity k and thermal dispersion which occurs due to hydrodynamic mixing of the interstitial fluid at the solid block scale. The effect of thermal dispersion, results in the apparent increase of thermal conductivity of the porous media. It is well documented that $k_\parallel, k_\perp$ depend on the fluid velocity and geometry of the media. For the relatively high velocities, (the typical Peclet numbers: $Pe=u_0l/d_s\sim10^2$ - 10^3) the components of effective conductivity tensor for the fractured and porous media of different structure and geometry can be presented as $k_\parallel=k+A(c_L\rho_L|\mathbf{u}|l)^\gamma$ and $k_\perp=k+B(c_L\rho_L|\mathbf{u}|l)$, where $|\mathbf{u}|=(u_x^2+u_y^2+u_z^2)^{1/2}$, parameters B, A and γ depend on the geometry of the fracture, fracture distribution and physical situation. For instance, if the medium is such that mixing effect dominates, then γ=2. For the medium where dispersion takes place mostly owing to the velocity variation inside the network (Koplik et al., 1988), which apparently is our case, γ=1. The above equations with the corresponding boundary conditions constitute the basic mathematical description of the heat transport in the geothermal reservoir.

Operation of the HDR geothermal system is expected to proceed for a long time. Water/rock chemical interaction inevitable exerts an influence on the permeability of these high temperature and pressure circulating systems due to dissolution and deposition processes. The model for the water/rock chemical interaction adopted in FRACSIM numerical code is based on a set of simple equations proposed and discussed in detail by Jing et al (2002). For a single fracture, variation of dissolved rock concentration C and fracture aperture a are described by the following equations

$$\frac{\partial C(\xi,t)}{\partial \xi} = \frac{2lR(T(\xi,t))}{\rho_L u_\xi}[C_\infty(T(\xi,t)) - C(\xi,t)], \quad (9)$$

$$\frac{\partial a(\xi,t)}{\partial t} = \frac{2R(T(\xi,t))}{\rho_S}[C_\infty(T(\xi,t)) - C(\xi,t)], \quad (10)$$

where $R(T)$ and $C_\infty(T)$ are the temperature-dependent reaction rate and the saturation concentration; ξ is coordinate along the fracture, l is fracture length, u_ξ is flow velocity along the fracture. These equations, coupled with temperature equation, are used for simulating the fracture aperture variation due to dissolution/deposition within the reservoir fracture network during the production stage of the reservoir exploitation.

2.4 The numerical algorithm

The numerical simulation of the fractured reservoir behaviour can be decomposed into two main parts. Namely, hydraulic fracturing part (here at first the values of the fracture shear dilation angles should be obtained for the typical rock in reservoir), and, after that, the model of the stimulated reservoir is employed for evaluating the heat and mass transfer processes during the production stage of reservoir exploitation.

The algorithm of natural fracture surface simulation includes: (i) experimental measurement of the fracture surface topography; (ii) computation of the power spectral density for this fracture surface as a function of spatial frequency; (iii) plotting this function in logarithmic coordinates and computing the plot's gradient β and surface fractal dimension D_S ; (iv) application of these parameters for constructing synthetic fractal surfaces that share the same physical properties as the natural fractures surfaces. If there is no interpenetration, increasing lateral displacement of the synthetic surfaces causes aperture to increase. Fracture dilation versus fracture shear displacement is plotted for different rock samples. The dilation is more pronounced for the fractal surfaces with lower fractal dimension (greater β). The shear dilation angle ϕ_{dil} can be computed as a mean angle between the plot tangent and horizontal axis. This value is also higher for the surfaces with lower fractal dimension.

Initially, a 3D stochastic network of many thousands of fractures (which length satisfy equation (1)) is generated within a cubic domain $L \times L \times L$. Equations that govern the reservoir dynamics and heat and mass transfer processes in the fractured media are solved numerically by the finite difference method, when the computational domain is composed from $l \times l \times l$ of square grid elements (blocks) which typical scale $l \ll L$, as discussed in Jing et al. (2002). For each block, some fractures will intersect with its faces. The intersected fractures on each face are assumed to be independent of each other and are approximated statistically to cuboid fractures, each of intersecting width l_j, aperture a_j and interposing fracture length l (equal to block length). Fluid is inferred to flow through every intersected fracture, since the fracture network is assumed to be well connected. The quantity of a fluid flow from block to block is controlled by Darcy law with the permeability contribution from each fracture governed by the sum of the products of the fracture apertures to the 3rd power and length of the intersection of the fracture with the block face. Summing the contribution of each fracture, the effective permeability connecting the blocks of the grid is used to perform the flow computations $K_f = \sum_j (a_j^3 l_j)/(12l^2)$, where f=x,y,z.

Stresses in the upper crust are typically inhomogeneous, usually with one principal stress approximately normal to the Earth's surface and the other two principal stresses sub horizontal. The present model uses the vertical stress and two horizontal principal stresses, which are taken to be the maximum and minimum principal stresses. Stimulation, by increasing fluid pressure in a low permeability fractured rock volume, causes shear of some of the pressurized existing fractures, what in turn leads to a permanent increase in aperture via shear dilation. The stimulated region is characterized by high permeability, which is surrounded by significantly lower undisturbed permeability region. In order to predict the maximum reservoir size, the present stimulation code calculates the flow rate and pressure distribution during stimulation within the field; afterwards the fracture shear displacements are obtained. The discretized steady-state flow equation (5) is solved by the Gauss-Seidel method. The fluid loss through the boundary is estimated by using a semi-analytic model of compliant fractures with a spherical "fluid dumping" region (with the model volume embedded at its central part) which is bounded externally by fixed pressures. The numerical algorithm accounts iteratively for the fluid flow that affects the fracture aperture and pressure distribution. The changes in pressure and fracture aperture induce a shear displacement that consequently leads to a new change in pressure distribution. Provided the spatial distribution of permeability is given, the pressure is calculated by iteration; the aperture value responds to the deviation of the local pressures according to equations (3) and (4). The latter yields the change of transmissibility of the rock. Then a new calculation (equation (5)) for the pressure distribution is required. These steps are repeated until the flow rate and the transmissibility converge. Using this method, the maximum reservoir size is determined. After the fluid velocities are determined, equation (8) for the temperature field within the reservoir is solved numerically by the Alternative Direction Implicit method while the constant temperature boundary conditions are assumed. On the next step of reservoir numerical simulation at production stage, variation of the fracture apertures due to chemical dissolution/precipitation is calculated (equations (9), (10)). The latter affects the permeability of the rock. For the changed permeability, new calculations for the pressure and temperature distributions are required. Hence, similarly to the stimulation part, at this stage the iterative algorithm is employed.

3. RESULTS AND CONCLUSIONS

Conclusions drawn on the basis of a large number of computations are following.

Fracture dilation is more pronounced for the fractal surfaces with lower fractal dimension D_S. As soon as the variation of the fracture dilation (which is measured geometrically) due to shear offset is computed, the shear dilation angle ϕ_{dil} can be obtained as a mean angle between the shear dilation plot tangent and horizontal axis. This value is higher for the surfaces with lower fractal dimension.

The thermal output of the reservoir is greatly affected by the size of the stimulated reservoir volume. Therefore the mathematical model for the pressure distribution in the reservoir (5) should be coupled with the basic shear stress–fracture displacement relationships (3) and (4) which are used for defining the size of stimulated volume.

Ignoring the effect of thermal dispersion at the high flow rates, which are typical for the fractured reservoirs, leads to the underestimation of the computed output temperature up to 20 percent, which constitutes a significant fraction of total heat obtained during the reservoir exploitation. Hence, thermal dispersion is an important factor that affects the assessment of the fractured reservoir productivity.

Results show that for the high temperatures, high flow rates, and bigger HDR geothermal systems, the WRCI may cause a permeability enhancement that lead to the increase of the flow rate.

ACKNOWLEDGEMENTS

A part of this work was supported by the Grant-in-Aid for Scientific Research (A) (No. 14205149), The Japan Society for the Promotion of Science.

REFERENCES

Brown, S.R., 1995. *Simple Mathematical Model of rough Fracture.* Journal of Geophysical Research 100: pp. 5941-5952

Hsu, C.T. & Cheng, P. 1990. *Thermal dispersion in porous medium.* International Journal of Heat and Mass Transfer 33: pp. 1587-1587.

Jing, Zh., Watanabe, K., Willis-Richards, J. & Hashida, T. 2002. *A 3-D water/rock chemical interaction model for predicting of HDR/HWR geothermal Reservoir performance.* Geothermics 31: pp.1-28.

Jing. Z., Willis-Richards, J., Watanabe K. & Hashida, T. 2000. *A Three-Dimensional Stochastic Rock Mechanics Model Of Engineering Geothermal Systems in Fractured Crystalline Rock*, Journal of Geophysical Research 105(B10): pp. 23663-23679.

Kohl, T. & Hopkirk, R.J. 1995. *"FRACTure" – a simulation code for forced fluid flow and transport in fractured, porous rock.* Geothermics 24(3): pp.333-343.

Koplik, J., Redner, S. & Wilkinson, D. 1988. *Transport and dispersion in random networks with percolation disorder.* Physical Review A 37: pp. 2619-2636.

Watanabe, K. and Takahashi, T., 1995. *Fractal Geometry Characterization of Geothermal Reservoir Fracture Networks*, Journal of Geophysical Research 100(B1): pp.521-528.

COUPLED T (THERMAL) - H (HYDROLOGICAL) - C (CHEMICAL) PROCESS OF GEOTHERMAL ALTERATION, BASED ON EXPERIMENTAL AND KINETIC CONSIDERATIONS

Junko Hara and Noriyoshi Tsuchiya
Graduate School of Environmental Studies / Tohoku University, Japan

Abstract: The style and intensity of hydrothermal alteration in active geothermal systems reflect the thermal regime and chemical composition of waters resulting from hydrothermal fluid-mineral interactions. This study uses an experimental approach to deduce dynamic mineral formation mechanisms in the active geothermal setting, to understand the interaction processes that produced Na-rich hydrothermal alteration zones in the East Hachimantai Geothermal Field (Japan). Hydrothermal experiments were conducted under saturated vapour pressure conditions, at 75 – 250 °C, using pyroclastic rock samples. The hydrothermal (laboratory) alteration of Tamagawa Welded Tuff did not proceed according to expected stoichiometric constraints, with respect to observed mineral formation and whole rock chemical compositions, and there is a great disparity in mineral dissolution rates and incorporation of elements into new-formed minerals. Mass transport coefficients and enrichment of certain major element in the reacted solutions resulting from hydrothermal fluid/rock interaction also varied, which lead us to interpret a new diffusion – advection – reaction rate equation that enables kinetic and mass transport parameters to be quantitatively estimated.

1. INTRODUCTION

Alteration mineral assemblages characterize the many varied hydrothermal-chemical reactions that occur in active geothermal systems. The effect of chemical and temperature conditions on the occurrence of secondary minerals is well known, based on field observation and mineral synthesis laboratory experiments. It is less clear, however, how specific chemical species become enriched in one part of an active hydrothermal system.

The East Hachimantai geothermal area is located within a caldera structure, and constitutes one of the most famous active geothermal regions in NE Japan. In the past, thick hydrothermally altered pyroclastic sequences accumulated in and around the caldera structure.

A zonal distribution of clay minerals occurs in the East Hachimantai thermal area, characterized by smectite – illite/smectite – illite/chlorite, with the latter tending to occur in the vicinity of hot ‘upflow’ zones. The smectite is characteristically Ca-smectite. The occurrence of Ca-bearing zeolite minerals, such as laumontite and wairakite correspond to the presence of hot hydrothermal fluids near in the center of the geothermal resource. In contrast, Na-smectite and Na-zeolite (e.g. clinoptilolite – mordenite – analcime) in marine sediments and pyroclastic sequences tend to envelope the main thermal area. Inoue *et al.* (2001) and Hara *et al.* (2001) have described the style and distribution of alteration in the Hachimantai area. The Na-enriched alteration zones contain higher Na concentrations than marine sediments, and tend to correspond to areas affected by meteoric water infiltration.

Here, we have examined water/rock interaction processes in the geothermal systems, from the perspective of kinetic/rate reaction processes and mass transport phenomena. By careful attention to the dynamic function on water/rock interaction processes, this study applies experimental analysis and development of a geochemical simulator to answer questions about reaction mechanisms in the active hydrothermal setting.

2. MATERIALS AND METHODS

2.1 Materials

Starting materials in our experiments comprised Tamagawa Welded Tuff, which is widely and thickly distributed in the Hachimantai volcanic area. Rock samples were collected from the Kowasegawa Welded Tuff Member. These pyroclastic rocks contain characteristic hydrothermal alteration assemblages from the active geothermal area. The rock has a hyalophilic texture, with altered plagioclase phenocrysts in a fine-grained matrix (Table 1).

2.2 Methodology

Batch type experiments were conducted to understand chemical functions and reaction mechanisms related to hydro-geochemical alteration processes. Experiments were conducted for up to 26 days, over a wide temperature range (of 150 – 300 °C), which is appropriate for most geothermal environments, and saturated vapor pressure. Rock sample weight was 0.5 g, which was reacted with 4 ml pure water in 8 ml pressure vessel. Rock samples were crushed and sieved to obtain a ϕ = 0.25 – 0.5 mm size fraction, with a specific surface area of 5.4 m^2/g by BET method (Yasa Ionics, NOVA1200).

The effect of mass transport phenomena on the reaction process was inferred using a series of (tube) flow type experiments, at saturated vapor pressure, over a temperature range of 75 – 250 °C, although the temperature was homogeneous in each experiment. The reaction vessel used was 1 m in length, with an inner radius of 1.25 cm. It has three outlets at 33 cm intervals within the pressure vessel. The volume of the vessel is 491 ml. Rock grains of ϕ = 1.0 – 1.5 cm filled the vessel, to achieve an estimated porosity in the tube of 56 %. The geometric method indicated a specific surface area of the rock grains to be ~3.47 m^2/g. Fluid velocity in the experiments was 1.5 - 9.9 ml/sec, which is considered to be high compared to estimated velocities in subsurface water reservoirs (Veld, 1995). We set reaction time by consistently using 56.16 litres of water for all experiments.

The compositions of resultant solutions from the batch- and flow-tube experiments were analyzed by Inductively Coupled Plasma spectrometry (Hitachi P-4000). Rock samples were analyzed as follows: secondary minerals were identified by XRD (Mac Science Co., Ltd., M21X) and SEM-EDX (Hitachi S-4700T); sample surfaces were examined using SEM, SIMS and XPS (Perkin Elrner, PHII6600).

3. INCONGRUENT DISSOLUTION

3.1 Reaction rate

Pyroclastic rocks dissolve incongruently and dissolution rate is different for their constituent matrix and crystal components. Here, dissolution rate was estimated by the "*initial dissolution method*". Dissolution rate for each element was calculated following the conceptual kinetic equation, using initial dissolution weights. The dissolution rate equation is:

$$\frac{\partial C}{\partial t} = k_d \cdot S \cdot (Cs - C) \qquad (1)$$

where C denotes the concentration of bulk solution, k_d dissolution rate constant, Cs solubility, and S surface area.

A plot of ln (Cs-C) versus time lines for major elements at 250 °C is shown in Fig. 1, with values for reaction order approximating 1. The slope of the ln (Cs-C) versus time lines are indicative of respective dissolution rate constants, and concur with inferred reaction rate equations. Incongruent dissolution is strongly affected by temperature change, with the major rock formation elements having the following relationship with respect to rate of dissolution: Si > Al > Na >> K > Ca.

Dissolution rate constants for major elements are summarized as a function of temperature in Fig. 2. Apparent activation energies have the same relationship as dissolution rate. Alkali metal and alkali earth metal ion have ~20 to 40 kJ/mol apparent activation energy, due to a diffusion effect from the mineral surface. The existing reaction condition does not correspond to the critical reaction rate, but this condition is more applicable for natural water-rock interactions, because the nature of incongruent dissolution on the mineral surface controls fluid chemistry and metastable reaction processes.

3.2 Observation of reacted surface of minerals

The chemical compositions of mineral surfaces were analyzed using SIMS. We found that a surface layer of a few micron thickness forms on the mineral surface and controls dissolution to the fluid system. Alteration intensity versus mineral depth (i.e. distance from the surface of the mineral) was assessed for K, Al, Ca and Na relative to Si content, with a steady decrease in K/Si and Ca/Si values indicating a zone of depletion extends ~0.3μm into the plagioclase crystal. On the other hand, Na/Si ratio (i.e. Na content) increased with depth during 3 hour, 9 and 18 day experiments, with a zone of enrichment being ~0.7μm thick. In our experiments Na-montmorillonite forms on the surface of other minerals with a greater Na concentration than in marine sediments from the study area. The heterogeneous layer on mineral surfaces formed by leaching and enrichment of selected elements.

3.3 Reaction mechanisms

Based on our experimental results, we infer that a chemical reaction process develops from a "reaction determining process" to a "diffusion determining process", when a heterogeneous layer is formed on the mineral surface. The inference means that corresponding reaction equations must combine reaction and diffusion determining steps. The following fundamental equation is proposed:

$$\frac{dC}{dt} = \frac{k_d S}{1 + k_d \delta / D}(Cs - C) - k_p \quad (2)$$

where δ denotes the thickness of the reaction layer, D is diffusion coefficient, k_p precipitation rate, and t is time – other symbols are the same as in (1). In the case of $k_d \delta / D$>>1, (2) becomes a diffusion determining step equation, and in the case of $k_d \delta / D$<<1 (2) becomes a reaction determining step equation.

4. MASS TRANSPORT COEFFIECINT

4.1 Reaction determining step

Mass transport coefficients were based on results from our flow system experiments, which require initial clarification of the rate determining reaction step.

The reaction rate was inferred by assessing material balance between input/output materials and material remaining inside the system, assuming density and volume do not change in the system. The differential equation is written as:

$$\frac{\partial C_A}{\partial X} = -\frac{k_d}{v} \cdot (Cs - C_A)^n + \frac{k_p}{v} \cdot (C_A - Cs^*)^n \quad (3)$$

Reaction order (n) may approximate unity and input concentration is 0 at X=0, so

$$C_A = (Cs + Cs^*) \cdot \{1 - \exp(-(k_d + k_p) \cdot t^*)\} \quad (4)$$

with retention time t^* in the system defined as x/v. According to (4), the slope of the approximation line on a plot of ln{(Cs+Cs*)-C} versus t^* will indicate the apparent rate constant, which can be estimated using the steady state concentration, with the solution chemistry consistent with the first order rate equation (Fig. 3).

The effect of temperature and flow rate on the overall reaction rate constant is evaluated in Figs. 4 and 5. Figure 4 shows Arrhrenius plots of Si and Na, which have the lowest and highest apparent rate constants, respectively. Activation energy of Si is ca. 16 [kJ/mol] and Na ranges from 1.7 to 7.4 [kJ/mol]. Activation energy for Si is little affected by flow rate, but the activation energy for Na is increased with lower flow rate. By considering inferred values of activation energy, we infer that the system is likely to be controlled by mass transport phenomena. It is likely that the apparent rate constant reflects the mass transport coefficient. If the apparent rate constant is represented by mass transport coefficient, then the Sherwood number, Sh, can be assessed. The Sh number is the ratio between diffusion coefficient and mass transport coefficient, and is written as :

$$Sh = \frac{k_c d_p}{D} \quad (5)$$

kc denotes mass transport coefficient, and *dp* particle diameter. Dependence on the Reynolds number for a given apparent rate constant can be demonstrated using the Sherwood number. Figure 5 shows the relationship between Sh and Re number, which has a linear relationship on a log scale. In contrast to values for activation energy, the slope of network formation elements (e.g. Si) is less than for network modifier elements (e.g. Na) in the log Sh – log Re diagram. Network modifier elements have higher velocity dependence than network formation elements. Slopes for network modifier elements are scattered compared to network formation elements, due to temperature effects. for apparent rate constant may be considered using the Sherwood number. Figure 5 shows the relationship between Sh number and Re number, which have a linear relationship on a log scale. In contrast to values for activation energy, the slope of network formation element (Si) is less than for network modifier element (Na) in the log Sh – log Re diagram. In other words, network modifier elements have higher velocity dependence than network formation elements. Slopes for network modifier elements are scattered compared to network formation elements, due to temperature effects.

4.2 Mass transport coefficients of major elements

Apparent reaction rate constants indicating mass transport factors, and mass transport coefficients indicating diffusion processes from mineral surface to bulk solution, are introduced into an advection – diffusion – reaction model.

The coefficient is set as $k_c = D/\delta$. Incorporating these considerations into a reaction term, and consideration of reaction data from the batch-type experiments, leads to a mathematical water/rock interaction model:

$$\frac{\partial C}{\partial t} = D\frac{\partial^2 C}{\partial x^2} - v\frac{\partial C}{\partial x} + \frac{k_d S}{1+\frac{k_d}{k_c}}(Cs - C) - k_P(C - Cs^*) \quad (6)$$

The steady state concentration was fitted by (6) using the thickness of the reaction layer (δ) as a fitting parameter. δincreases with reaction length and differs for the respective elements, as shown in Fig. 6. An outstanding factors, for determination of the mass transport coefficient is change of temperature, flow velocity condition and the distance from water (fluid) input. Specifically, mass transport coefficients depend on physical condition. δNa is greater at lower temperature conditions. On the other hand, δCa is greater at higher temperature conditions. The relationship between major elements depends on temperature conditions, but δ values most strongly depend on flow rate condition.

5. NUMERICAL MODELLING OF GEOTHERMAL ALTERATION

5.1 Method of calculation

Based on the previously described reaction mechanism, we were able to use a numerical water/rock interaction model to quantitatively evaluate data from our experimental system.

Multi-component reactions within a Si – Al – Na – K – Ca – H_2O system were considered, involving 16 minerals secondary minerals: quartz, chalcedony, gibbsite, kaolinite, K-feldspar, Na-montmorillonite, illite, clinoptilolite, mordenite, analcime, albite, Ca-montmorillonite, laumontite, wairakite, prehnite and clinozoisite, which occur in geothermal systems. Dissolution rate constant, solubility and thickness of surface layer were based on experimental results. Surface free energy, mineral density and thermodynamic data for each mineral are taken from published sources (e.g. Aene and Shnel, 1971; Arrnorson, *et al.*, 1982). If the solvent attains equilibrium state with secondary minerals they are stoichiometrically precipitated, according to nucleation theory. Numerical calculation was performed using the Runge-Kutta method.

The flow system was modeled using (6). Calculation conditions are: porosity 0.56, water/rock ratio 96, and fluid velocity 10^{-4} [m/s], which are based on experimental condition (except for fluid velocity). No secondary minerals appeared when using the experimental velocity, so a relatively low velocity was adopted in the model.

5.2 Secondary minerals

Numerical calculations were undertaken for systems with a water/rock ratio of 8, which was typical for our batch type experiments. If temperature conditions are set at 250°C, then the first precipitated mineral was Na-montmorillonite, then chalcedony and illite, which is consistent with our experimental results. Na-montmorillonite was observed in batch experiments at the same conditions, which agrees with modeling and demonstrates how Na-montmorillonite can be formed from non-marine sediment/meteoric water interaction.

Figs. 7-A and 7-B shows examples of calculated results from our flow system at 75 °C, and 250 °C respectively. Physical parameters, such as porosity, water/rock ratio and fluid velocity are the same in both cases. Pure water infiltrated from the 0 cm point. Ca-montmorillonite was precipitated below 20 cm depth point, with the amount precipitated material increasing with reaction time. Laumontite precipitated later, and was distributed in lower part. Illite appeared at an early stage, and was abundant in the shallow part, as it reflects K content and has a high reaction rate at high water/rock ratio conditions.

Figure 7-B shows calculated result at 250 °C, with, Na-montmorillonite and chalcedony precipitating at an early stage (compared to the low temperature condition). Clinoptilolite precipitated at a late stage. The results are consistent with secondary minerals occurrence in the experimental system.

According to calculated results shown in Fig. 7, assemblage of secondary minerals was strongly temperature dependent. Ca-bearing minerals preferentially precipitated at lower temperature conditions, and Na-bearing minerals were precipitated at higher temperature, which was an effect of element behaviour due to incongruent dissolution and transport phenomena.

6. DISCUSSION

Taking account chemical reaction and mass transport differences between elements, then we can infer the distribution of secondary minerals by considering thermodynamic constraints, as well as kinetic and transport phenomena. Solubility (Shikazono, 1988) and mass transport coefficient of Ca is large at low temperature, and therefore Ca is readily dissolved in shallow zones and transported to deeper levels. As a result, Ca-bearing alteration minerals, such as Ca-montmorillonite, occur widely in a geothermal field. Solubility and mass transport coefficient of Ca decreases with increasing temperature, so Ca-zeolites precipitate in the active zones.

On the other hand, dissolution rate and mass transport coefficient of Na is large at high temperature conditions. Plagioclase is a stable mineral at high temperature but zeolite is a metastable phase. Na-bearing minerals formed as a result of plagioclase dissolution are considered to be metastable at high temperature, so Na-minerals such as Na-smectite and Na-zeolite form at the cool margin and/or outside the main thermal area.

7. CONCLUSIONS

Geothermal systems are characterized by incongruent dissolution reactions in host rocks, which lead to leaching or enrichment of mineral surfaces. The effect of surface layer reactions must be considered in developing multi-component equilibrium or water/rock interaction models to take into account experimentally determined kinetic and mass transport effects, as well as field observations. The origin of Na in the East Hachimantai geothermal area is much debated, since the hydrothermal fluids are remarkably Na-rich, and it may be that the source of the Na can be directly attributed to seawater or an altered seawater fluid source. This study shows that the elemental source of Na, now fixed in the Na-rich secondary minerals, is more likely to come from land-deposited sediments, and that it is possible to 're-construct' the chemistry of the high-Na hydrothermal fluid responsible for geothermal alteration in pyroclastic rocks.

REFERENCES

Aene, E. N. and Shnel, O. 1971. *Interfacial tensions electrolyte crystal – aqueous solution, from nucleation data.* Journal of Crystal Growth, vol.11, pp. 233 – 242.

Arnorson, S., Sigurdsson, E. and Savarsson, H. 1982. *The chemistry of geothermal waters in Iceland. I. Calculations of aqueous speciation from 0 to 370°C.* Geochim. Cosmochim. Acta, vol.64, pp. 1513 – 1532.

Hara, J., Tsuchiya, N. and Inoue, A. 2001. *Mineralogical alteration zoning in active geothermal systems of the Hachimantai volcanic region, northeast Japan.* 10th WRI proceedings, pp. 847 – 850.

Inoue, A., Hara, J. and Imai, A. 2001. *Genesis of Na-series rock alteration widespread in the southeasten area of Hachimantai geothermal field: Water-rock interactions driven by descending groundwater and fossil seawater.* The Society of Resource Geology, vol. 51, No.2, pp. 101 – 120.

Velde, B. 1995. *Origin and Mineralogy of Clays.* Springer – Verlag press., pp.334.

Shikazono, N. 1988. *Hydrothermal alteraration associated with epithermal vein-type deposits in Japan; a review.* Mining Geology Special Issue, No.12, pp. 47 – 55.

Table 1. Model composition of starting materials

Mineral name	%
Phenocryst	55.8
Plagiclase(oligoclase)	25.6
Tridynamite	18.2
Pyroxene (enstatic)	4.2
Illmenite	0.6
Pyrite	2.2
Ca - montmorillonite	5.0
Matrix	44.2
Needle like plagiclase	31.2
Glass	13.0

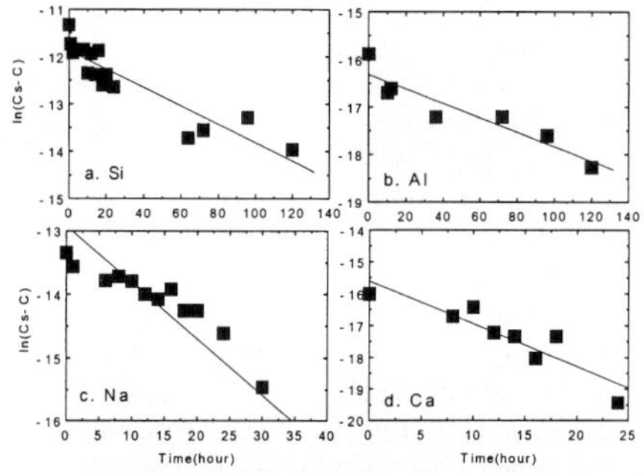

Figure 1. Ln(Cs – C) – time diagram for major elements at 250 °C under non-stirring condition. Reaction order is approximated to unity.

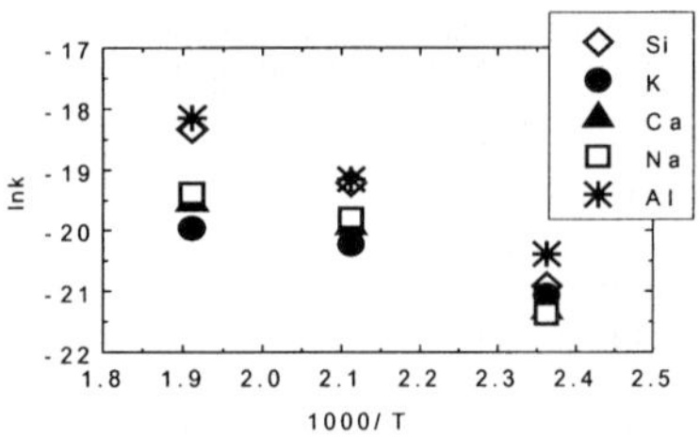

Figure 2. Arrehnius plot for major rock formation elements. Ln k for non-stirred experimental conditions. Each activation energy is estimated as follow: $Ea_{,Si}$ = 57 [kJ/mol], $Ea_{,Al}$ = 41[kJ/mol], $Ea_{,Na}$ = 38 [kJ/mol], $Ea_{,Ca}$ = 33[kJ/mol], $Ea_{,K}$ = 21 [kJ/mol].

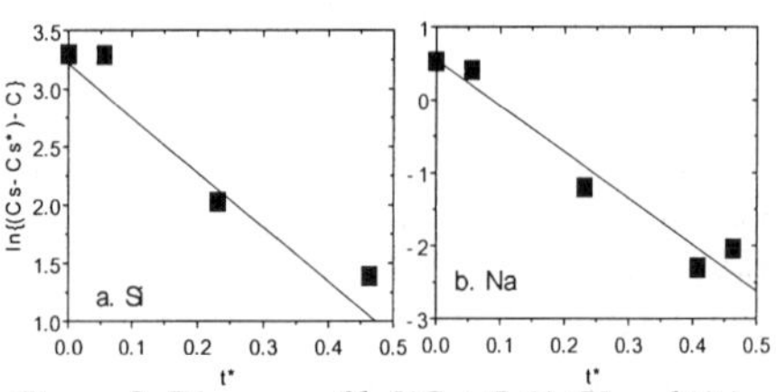

Figure 3. Diagram of ln[(Cs+Cs)-C] and t* in the case of Si and Na, for 150 ℃, and 9.9 [ml/min] experiment. Reaction order is unity for them.*

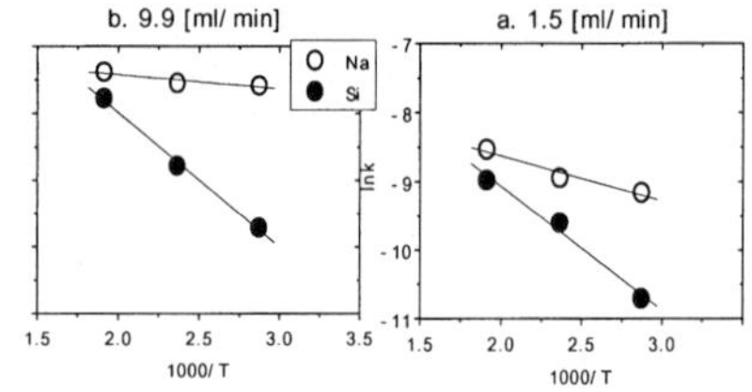

Figure 4. Arrhrenius plots for Si and Na at flow rate of 1.5 (a) and 9.9 (b) [ml/min].

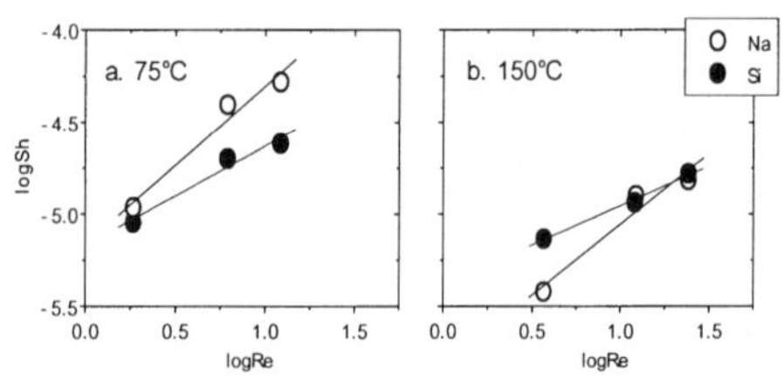

Figure 5. log Sh – log Re diagram for Si and Na at 75 (a) and 150 (b) [℃].

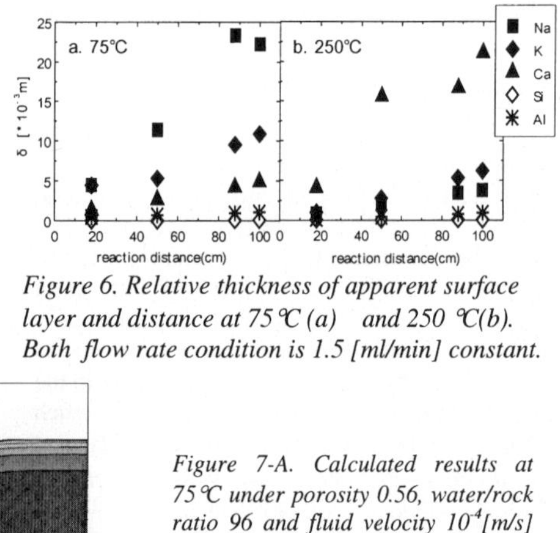

Figure 6. Relative thickness of apparent surface layer and distance at 75℃ (a) and 250 ℃(b). Both flow rate condition is 1.5 [ml/min] constant.

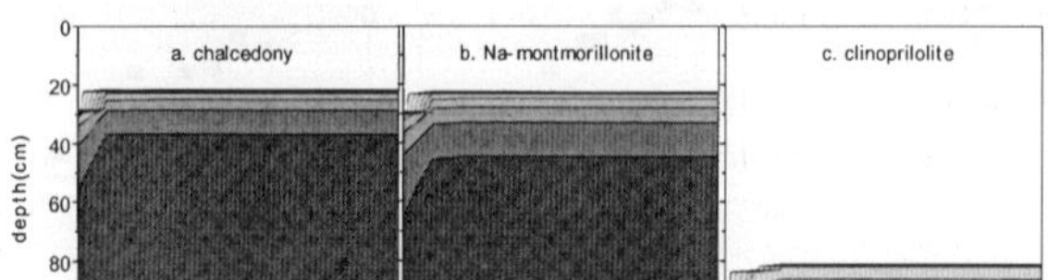

Figure 7-A. Calculated results at 75℃ under porosity 0.56, water/rock ratio 96 and fluid velocity 10^{-4}[m/s] conditions. Darker colure part is higher precipitation value region.

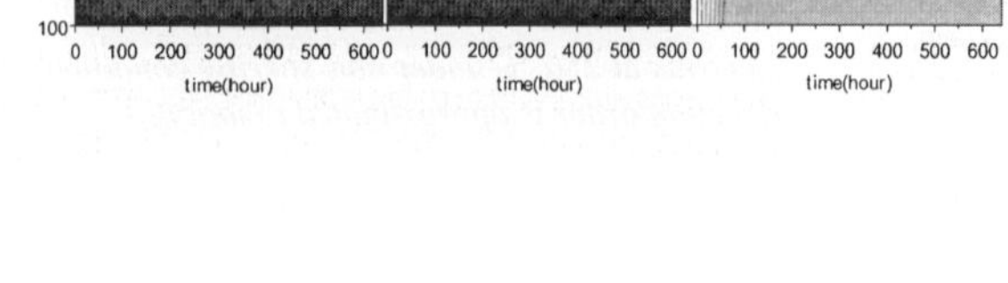

Figure 7-B. Calculated results at 250 ℃ under the same conditions at 75 ℃.

SUPERCRITICAL WATER/ROCK INTERACTIONS AND GENERATION OF ARTIFICIAL GEOTHERMAL RESERVOIRS IN DEEP-SEATED HIGH TEMPERATURE ROCK MASSES

Toshiyuki Hashida and Toru Takahashi

Fracture Research Institute, Tohoku University

01 Aza-Aoba, Aramaki, Aoba-ku, Sendai, 980-8579 Japan

Abstract: Laboratory results from simulated hydraulic stimulation tests on granite are presented where fracture and fluid flow behavior were examined under temperatures up to 600 °C and confining pressures up to 100 MPa at various injection rates. The results show that under high temperatures, fluid flow dominant behavior occurs for relatively lower injection rates, whereas macroscopic fracture tends to takes place at higher injection rates. Continuous injection tests under supercritical water conditions reveal that the fluid flow dominant behavior is due to the generation of a grain-scale extensive microcracks induced by the interaction between supercritical water and rock-forming minerals. It is proposed to utilize the discovered supercritical water-induced microcracking phenomenon in order to create an artificial reservoir in deep-seated rock masses whose temperature and pressure conditions exceed the critical point of water.

1. INTRODUCTION

Most Hot Dry/Wet Rock (HDR/HWR) development programs, and all geothermal projects presently completed for commercial energy utilization, have focused on water/vapor reservoir(s) of limited size and temperatures below the critical temperature of water.

To date, high-temperature rock masses have been identified by deep exploration drilling at Kakkonda, Japan (Ikeuchi et al., 1996); Larderello, Italy (Capetti et al., 1985); and Nesjavellir, Iceland (Steingrimsson et al., 1990). As exemplified by the field survey at the Kakkonda geothermal field in Japan, a rock mass of extremely high temperatures above the critical point of water may be found just below existing geothermal reservoirs. Thus, it may be one of the useful options to exploit the very high temperature rock mass underneath existing geothermal reservoirs in order to enhance the geothermal energy extraction. In this paper, such high temperature rock masses whose temperature and pressure conditions exceed the critical point of water are referred to as a supercritical rock mass. It is also suggested that no significant natural fracture system is expected to exist in such supercritical rock masses for a volcanic arc system. This may call for the creation of an artificial reservoir and water circulation system in order to exploit supercritical rock masses.

In this study, simulated hydraulic stimulation tests were undertaken under supercritical water environments in order to examine the feasibility of creating water pathway in deep-seated supercritical rock masses. The present experimental results suggest that the utilization of the interaction between supercritical water and rocks provides the possibility of generating a micro-fracture network for the thermal extraction from supercritical rock masses.

2. EXPERIMENTAL PROCEDURE

The rock used in this study was Iidate Granite, which was taken from a quarry in Fukushima Prefecture, Japan. Iidate Granite is a coarse-grained rock of average grain size of 1.5mm and is composed of 37.1% quartz, 34.0% K-feldspar, 6.3% plagioclase and other minerals.

Simulated hydraulic stimulation tests were conducted on the granite using thick-walled cylindrical specimens. The outer and inner diameter of the cylindrical specimens were 45mm and 5mm, respectively, and the specimen height was 90mm. A schematic illustration of the hydraulic stimulation test apparatus used is given in Fig.1. The maximum temperature and confining pressure achievable with this apparatus are 600°C and 150MPa, respectively. The maximum axial load is 100ton and the corresponding maximum axial stress is approximately 625MPa for the above-mentioned specimen dimensions. The cylindrical specimen and metallic spacers were inserted into a 0.3mm thick silver jacket to form a specimen assembly, as shown in Fig. 1. The specimen assembly was then placed between the upper and lower piston. Aluminium disks of 1 mm thickness were placed between the specimen and metallic spacer in order to prevent the leakage during the simulated hydraulic stimulation tests. After applying confining pressure onto the cylindrical specimen, the temperature was elevated at a constant rate of 3 °C/min up to a predetermined value. The heating of the specimen was undertaken using an internal heater. The temperature control was conducted using a thermocouple placed in the middle height of the specimen, and the temperature distribution was monitored at the upper and lower positions. The borehole temperature was also measured using a thermocouple during the hydraulic stimulation tests. After the temperature reached the predetermined level, water was then injected into the borehole at a constant flow rate. The flow rates used in this study were in the range of 0.5 mm^3/sec -300 mm^3/sec.

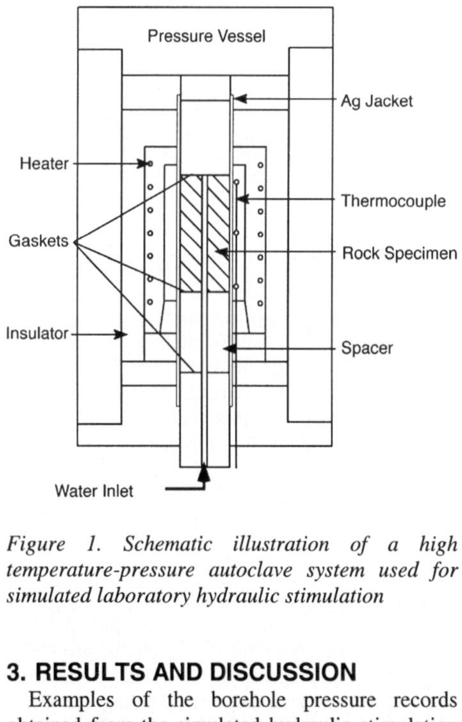

Figure 1. Schematic illustration of a high temperature-pressure autoclave system used for simulated laboratory hydraulic stimulation

3. RESULTS AND DISCUSSION

Examples of the borehole pressure records obtained from the simulated hydraulic stimulation tests are shown in Fig. 2. Figure 2(a) shows the variation of the borehole pressure with respect to time at a room temperature (25°C), under the confining pressure of 100MPa and the injection rate of 5.0 mm^3/sec. Figure 3 (a) shows the fracture profile observed on the cross section perpendicular to the specimen axis. The specimen has been impregnated with a red dye penetrant for ten minutes. As exemplified by the example, a macroscopic planar fracture was induced in the subcritical temperature regime at the injection rate. Figure 3 (a) illustrates the tensile type fracture observed for relatively low differential stresses. Figure 2(b) shows the pressure change for 600°C. The injection rate (5.0 mm^3/sec) is the same as for the room temperature. At this high temperature condition (600°C), no macroscopic fracture took place and no significant difference between the borehole and confining pressure was observed up to the maximum flow rates used in this study (max. 300 mm^3/sec), indicating that flow dominant behavior was occurring. However, at 500°C a macroscopic fracture was induced for the injection rates greater than 200 mm^3/sec, accompanied by a breakdown behavior. Figure 3(b) shows the fracture pattern induced at 500°C. It is noted that an extensive 'cloud-like' region of grain-scale microcracks is formed around the macroscopic fracture in the supercritical temperature regime. The generation of a volume-like microcrack region may be beneficial for increasing the heat-exchanging surface, in contrast with a simple planar fracture observed in the subcritical temperature regime. For higher differential stresses, shear type fracture has been shown to occur. The above-mentioned observation indicates that the fracture behavior of the granite is brittle at the supercritical temperature regime for the strain rates employed in the hydraulic stimulation tests.

The results of the simulated hydraulic stimulation tests are summarized in Fig. 4. The observed stimulation behaviors can be classified into three types: tensile fracture, shear fracture, and no macroscopic fracture. Their occurrence conditions are plotted in terms of temperature, differential stress and injection rate in Fig. 4. The differential stress in the diagram is the difference between the axial stress and the confining pressure. Figure 4 also gives the approximate boundaries between the region of the three stimulation behaviors, which have been predicted using a simple fracture and fluid flow analysis. The tensile fracture is assumed to initiate when the hoop stress at the borehole wall exceeds the fracture strength of the rock. The tensile strength determined from hydraulic injection tests is plotted in Fig. A1, as a function of the testing temperature. The borehole pressure attainable under a given injection rate are estimated from the permeability at the temperature and pressure, using the following equation:

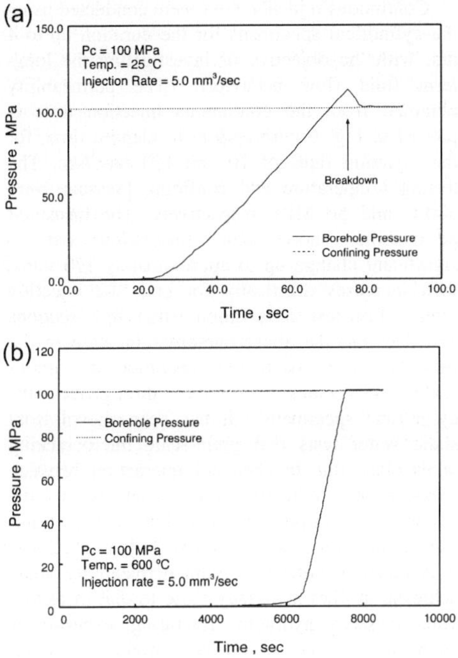

Figure 2. Representative curves of borehole pressure versus time for simulated hydraulic stimulation tests conducted at (a) a room temperature (25°C) and (b) 600°C. The confining pressure is 100 MPa, and the injection rate is 5 mm3/sec.

$$\Delta P = P_b - P_c = \frac{Q}{2\pi L(k/\mu)} \ln \frac{r_o}{r_i}, \quad (1)$$

where ΔP is the difference between the borehole pressure P_b and confining pressure P_c, Q is the constant flow rate, L is the height of the specimen, r_o and r_i is the outer and inner radius of the specimen, μ is the water viscosity at the temperature and pressure conditions. The value of water viscosity was taken from literature values for pure water. The permeability data were taken from our previous report (Takahashi et al., 2003), where hydraulic injection tests have been conducted on the same granite under various temperatures and confining pressures. Then, the hoop stress at the borehole wall is calculated using the predicted borehole pressure and confining pressure, and compared with the fracture strength to estimate the occurrence condition for the tensile fracture under the given temperature and injection rate.

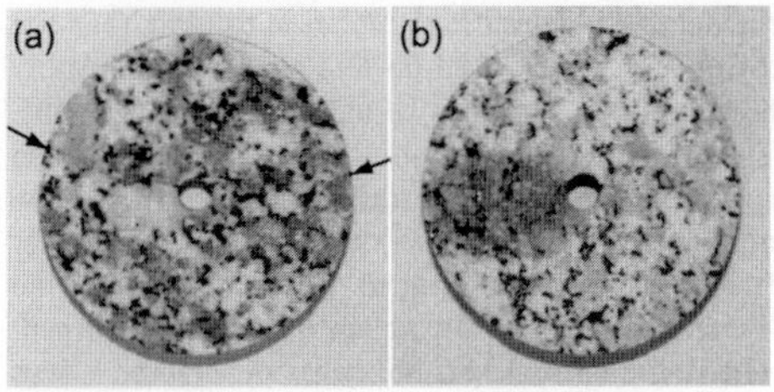

Figure 3. Fracture patterns induced at (a) room temperature (25°C) and (b) 500°C. (a) 25°C, 5 mm^3/sec (b) 500°C, 300 mm^3/sec

The occurrence condition of shear fracture is examined on the basis of the Coulomb criterion. The averaged shear stress across the fracture plane in the simulated hydraulic stimulation tests is plotted in Fig. 5, as a function of the effective normal stress across the fracture plane. The steady-state pore pressure distribution given from Equation (1) is averaged over the fracture plane and is used to compute the effective normal stress. Triaxial compression tests have been performed on the granite using the same apparatus shown in Fig. 1 (Takahashi, 2000). The peak shear stresses obtained from the triaxial compression tests are also plotted in Fig. A2. In the plot of Fig. 5, there is general agreement between the two types of the experimental results. Thus, it is thought that the occurrence of the shear fracture in the simulated hydraulic stimulation tests can be approximately predicted by the Coulomb criterion. Based on the comparison, the critical condition for the shear fracture due to hydraulic stimulation was estimated using the experimental results of the triaxial compression tests, as given in Fig. A2, and the averaged value of pore pressure. The detailed discussion of the triaxial compression tests can be found elsewhere (Takahashi, 2000.).

Even though the experimental results are still limited, it appears that the predicted occurrence conditions provide the general trend compatible with the experimental results. Thus, it may be possible to examine the fracture process to be expected for hydraulic stimulation in supercritical rock masses on the basis of the fracture mechanism map given in Fig. 4. The ratio of the minimum horizontal tectonic stress to the vertical tectonic stress usually falls in the range of 0.5-1.0 for brittle rock masses (Brace & Kohlstedt, 1980.). The actual difference of the tectonic stresses may be significantly smaller in supercritical rock masses due to the high temperature creep deformation (Fournier, 1999; Muraoka et al., 2000). If we limit the depth of hydraulic stimulation to 6km, then the differential stress may be less than approximately 80MPa for the majority of tectonic stress gradients. The fracture mechanism map of Fig. 4 then suggests that either tensile fracture or fluid flow dominant behavior may occur rather than shear fracture. A more detailed FEM analysis of the hydraulically induced fracture has been performed, where the growth of mixed-mode fractures was examined taking into account the formation of fracture process zone. The numerical analysis has also shown that the most likely mode of fracture is the extension of tensile fracture (Itaoka et al., 2002).

It is particularly noted that the region of fluid flow dominant behavior expands significantly with respect to the tensile fracture region when the temperature increases. The flow velocity at the borehole wall is typically in the range of 0.05-0.5 mm^3/sec for HDR/HWR field hydraulic stimulation experiments conducted to data. The injection rates used in this study correspond to the flow velocity of 70-700 mm^3/sec. This observation may suggest that either tensile fracture or fluid flow dominant behavior take place when the injection rate of that range is employed. The exploitation of the fluid flow dominant behavior may be useful to enhance the heat-exchanging surface for the thermal extraction from supercritical rock masses. It is judged to be worthwhile to investigate the flow dominant behavior observed at the high temperature in more detail.

Continuous injection tests were conducted using the cylindrical specimens for the duration up to 4 hrs, with the objective of investigating the long-term fluid flow behavior. The permeability obtained from the continuous injection tests is plotted in Fig. 6 with respect to elapsed time, for the injection rates of 10 and 150 mm^3/sec. The testing temperature and confining pressure were 500°C and 50 MPa, respectively. The measured permeability shows some fluctuations but no significant change up to approximately 170 mins, then increases drastically for both the injection rates. Post-test observation using thin sections revealed that the about one-order increase in the permeability is due to the generation of a grain-scale microcrack network throughout the cylindrical specimens. It has been shown from static water tests that grain-scale microcracking takes place due to chemical interaction between rock-forming minerals and water when the temperature and pressure conditions of pore water exceed the critical point of water (Fujii et al., 2001). It is noted that the maximum permeability value achieved at 4hrs is already close to that measured after massive hydraulic fracturing experiments conducted to date, which have produced a water circulation system with fractured type reservoirs under subcritical temperature regime. Thus, it may be useful to utilize the phenomenon of "supercritical water induced microcracking" in order to increase the permeability due to grain-

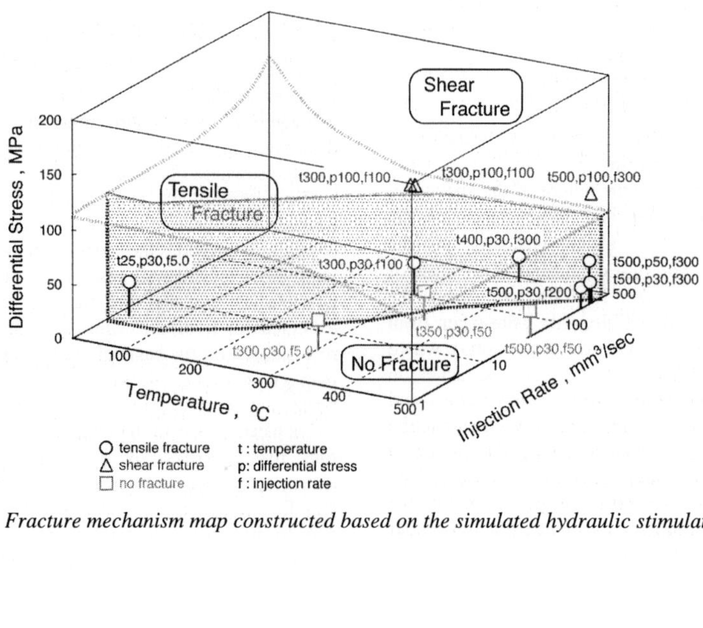

Figure 4. Fracture mechanism map constructed based on the simulated hydraulic stimulation tests.

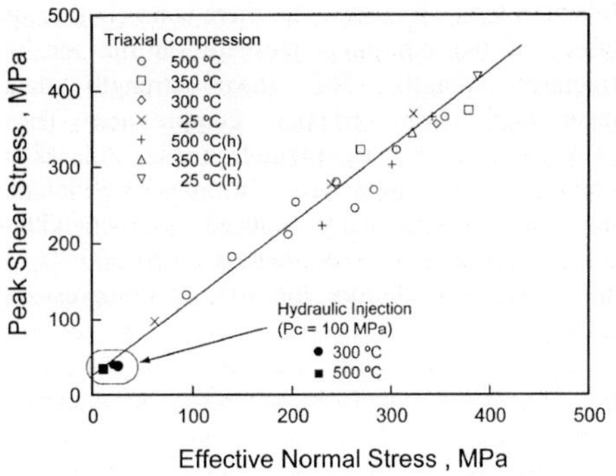

Figure 5. Comparison of the shear strength data obtained from simulated hydraulic stimulation tests and triaxial compression tests. Those specimens which were pre-heated up to 500°C are designated (h).

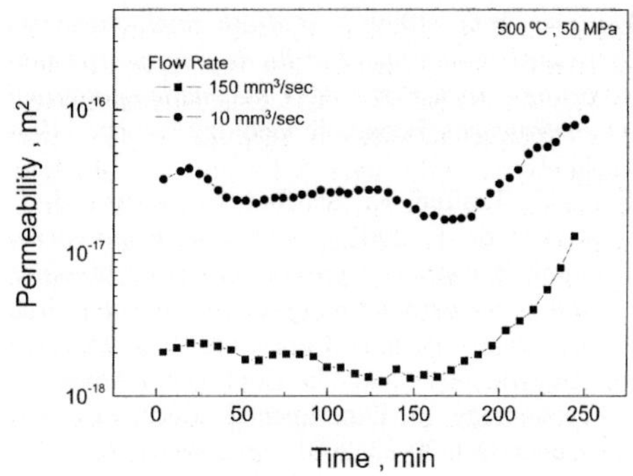

Figure 6. Permeability versus elapsed time for continuous injection tests

scale microcrack network and to create an artificial thermal extraction system in supercritical rock masses.

It should be mentioned here that there is the significant difference in the initial permeability values for the two different injection rates, as shown in Fig. 6. The borehole temperature measured by a thermocouple was observed to decrease due to the hydraulic injections although the stainless steel piping leading to the test section was preheated up to 300°C with a band type heater. The temperature at the outer specimen surface was, on the other hand, kept constant throughout the injection tests. The temperature drop at the borehole was approximately 5°C for 10 mm^3/sec, and 70°C for 150 mm^3/sec, respectively. The aperture of pre-existing microcracks is thought to increase due to the thermal stresses induced by the temperature gradient. The thermal stresses may be responsible for the higher initial permeability for the injection rate of 150 mm^3/sec compared with that for 10 mm^3/sec. Further study should be undertaken to examine quantitatively the effect of thermal stresses on the measured permeability.

4. CONCLUSIONS

Simulated hydraulic stimulation tests were performed using thick-walled cylindrical specimens of Iidate Granite at temperatures up to 600°C and confining pressures up to 100 MPa. The objective of the laboratory experiment was to examine the feasibility of creating an artificial reservoir for the thermal extraction from supercritical rock masses whose temperature and pressure conditions exceed the critical point of water. Either macroscopic fracture or fluid flow dominant behavior was observed to take place in the simulated hydraulic stimulation tests, depending on the injection rate used. The fluid flow dominant behavior was due to the generation of grain-scale extensive microcracks, induced by interaction between supercritical water and rock-forming minerals. The supercritical water-induced microcracking phenomenon may offer the feasibility of creating a volume-like reservoir in supercritical rock masses.

ACKNOWLEDGMENTS

A part of this work was supported by the Grant-in-Aid for Scientific Research (A) (No. 14205149), and the Research for the Future Program (JSPS-RFTF 97P00901) , The Japan Society for the Promotion of Science.

REFERENCES

Brace W.F. & Kohlstedt D.L. 1980. *Limits on lithosperic stress imposed by laboratory experiments.* Journal of Geophysical Research 85(B11): pp. 6248-6252.

Cappetti G., Romano C., Cigni U., Squarci P., Stefani G. & Taffi L. 1985. *Development of deep exploration in the geothermal areas of Tuscany, Italy.* GRC, Int. Sympo. on Geothermal Energy: pp. 303-309.

Fournier R.O. 1999. *Hydrothermal processes related to movement of fluids from plastic into brittle rock in the magmatic-epithermal environment.* Economic Geology 94: pp. 1193-1211.

Fujii, Y., Tanifuji, K., Stafford, C., Takahashi, T. & Hashida, T. 2001. *Crack healing and micro-crack formation in granite under the presence of high temperature and pressurized water.* Proc. of Asian Pacific Conf. on Fracture and Strength '01 and Int.Conf. on Advanced Technology in Experimental Mechanics '01; Sendai, Oct. 20-22, 2001: pp. 1041-1046.

Ikeuchi K., Komatsu R., Doi N., Sakagawa Y., Sasaki M., Kamenosono H. & Uchida T. 1996. *Bottom of hydrothermal convection found by temperature measurements above 500 °C and fluid inclusion study of WD-1 in Kakkonda geothermal field, Japan.* Geothermal Resources Council Transactions 20: pp. 609-616.

Itaoka M., Sato K. & Hashida T. 2002. *Numerical simulation of geothermal reservoir formation induced by hydraulic fracturing at great depth.* Geothermal Resources Council Transactions 26: pp. 297-302.

Muraoka H., Yasukawa K. & Kimbara K. 2000. *Current state of development of deep geothermal resources in the world and implications to the future.* Proc. of World Geothermal Congress 2000; Kyushu-Tohoku, Japan, May 28-June 10, 2000: pp. 1479-1484.

Steingrimsson B., Gudmundsson A., Franzson H., & Gunnlaugsson E. 1990. *Evidence of a supercritical fluid at depth in the Nesjavellir field.* Proc. 15th Workshop on Geothermal Reservoir Engineering, Stanford University, CA; SGP-TR-130: pp. 81-88.

Takahashi T. 2000. *Fracture and fluid flow characteristics and hydraulic stimulation behavior in granite under supercritical water environments.* Doctoral dissertation, Tohoku University.

Takahashi T., Tanifuji K., Stafford C. & Hashida T. 2003. *Permeability enhancement by microfracturing in granite under supercritical water conditions.* JSME International Journal, Series A 46(1): pp. 24-29.

APPENDIX

The fundamental fracture properties of Iidate Granite are given in this Appendix. Figure A1 gives the temperature dependence of the tensile fracture strength obtained from the hydraulic injection tests. It is seen that there is no significant effect of the confining pressure on the tensile fracture strength. The shear strength data determined from triaxial compression tests (Takahashi, 2000) are plotted in Fig. A2, as a function of the temperature. In order to eliminate the effect of thermally induced microcracking, some specimens were pre-heated up to 500 °C in air environment before the triaxial compression tests.

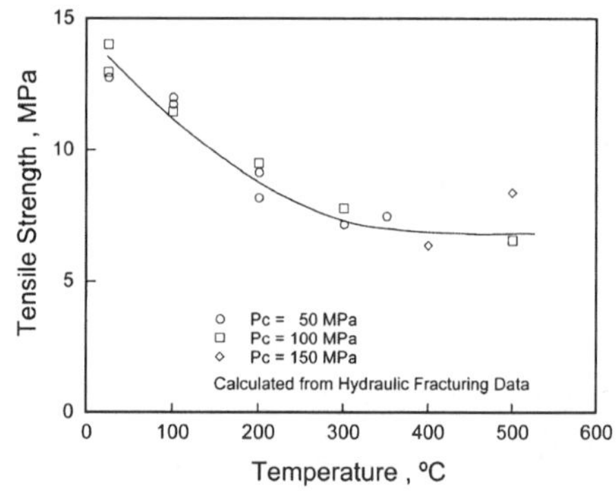

Figure A1. Temperature dependence of tensile fracture strength determined by simulated hydraulic stimulation tests

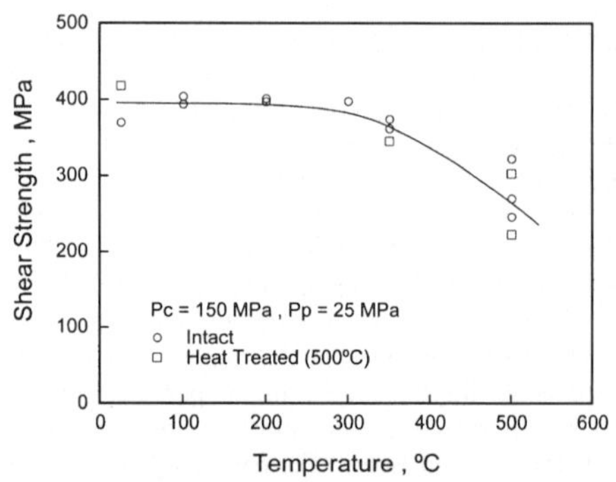

Figure A2. Temperature dependence of shear fracture strength determined by triaxial compression tests

COUPLED THM MODELING OF THE STIMULATED PERMEABLE FRACTURES IN THE NEAR WELL AT THE SOULTZ-SOUS-FORÊTS SITE (FRANCE)

A. Hosni[1], S. Gentier[1], A. Genter[1], J. Riss[2], D. Billaux[3] and F. Dedecker[3]
[1])BRGM, 3 avenue Claude Guillemin, B.P. 6009, 45060 Orléans cedex 2, France
[2]) CDGA, University Bordeaux 1, Av. des Facultés, 33405 Talence, France
[3]) Itasca Consultants S.A., 64 chemin des Mouilles, 69130 Ecully, France

Abstract: The stimulation of the deep wells drilled in the Soultz-sous-Forêts geothermal site (France) by the injection of cold and pressurised water demonstrates that the access zones of the fluid to the fracture network is limited to a few fractures. In the near well at the scale of some meters, a decrease of the impedancy is outlined via hydraulic vibration tests and is presently not well understood. As a consequence, the thermo-hydro-mechanical responses in the nearest of the stimulated fractures must be investigated. A numerical approach, based on the representation of the rock mass by circular particles, is developed taking into account the mineral phases of the different type of granite present in the vicinity of the permeable fractures. The model results show that the THM load can lead to a degradation which depends on the type of granite, creating the possibility of fissure propagation leading to an increase of porosity in the near well.

1. INTRODUCTION

This work is a part of a European research program initiated at the Soultz-sous-Forêts site (France) in 1987 which aims to develop a pilot plant for extracting heat accumulated in the deep fractured granitic basement. Heat extraction must be carried out by forced circulating of a fluid between three wells drilled at 5 km depth.

The experiments conducted so far at the site in range of depths between 3000 and 3800 m show that hydraulic stimulation techniques can considerably increase the permeability in a rock volume that can extend up to several hundred meters around the well, thereby serving to establish connections between the wells. The major impact of the hydraulic stimulation could be explained by hydromechanical mechanisms (ruptures in fracture planes). However in the near well at the scale of some meters, a decrease of the impedancy is outlined via hydraulic vibration tests and is presently not well understood. As at this scale, the hydromechanical effect cannot be dissociated from the thermal effect due to the injection of a cold water in a hot media, the thermo-hydro-mechanical responses of the stimulated fractures must be investigated.

Then, a numerical approach, based on a representation of the modelled environment by discrete circular particles that interact via contacts points is developed to study the thermo-hydro-mechanical (THM) behaviour of the access zones with the Particle Flow Code PFC^{2D}, Itasca (1995). Indeed, modelling with PFC^{2D} offers the possibility to study the THM behaviour in a transient state and to take into account the real mineral structure for the environment of the most stimulated fractures which are in a limited number. Effectively, investigations conducted during stimulation and injection tests in the open hole sections of first two wells by different techniques (UBI, flow logs, temperature logs) showed that the fluid has access to the heat exchanger along only three main fractured zones which were identified in each well, Evans (1996), Tenzer (1997). The three main fractured zones which support flow were characterised from a petrographic, structural and geometric standpoints from borehole imaging and well log, Genter (2000). Five main petrographic facies have been identified corresponding for four of them to more or less altered and deformed granites and for the fifth to quartz vein.

The main objective is to quantify the possible effects of the THM behaviour of one of the stimulated fracture in relation with its petrographic, structural and physical environment. This modelling procedure takes into account the properties of the main rock facies surrounding the permeable zone.

We will present in this paper some first results obtained for one of these stimulated fractures located at a depth of 3224 m in one well.

2. CHARACTERIZATION OF THE MINERAL PHASES

The stimulated fracture chosen for the THM analysis is at the interface between a hydrothermalized granite and quartz vein. At the meter scale of the model, the particle modelling approach requires to quantify the granulometry and the distribution of each mineral phase to define at best the distribution of particle radii.

An algorithm for an automatic identification of each mineral phases has been developed to process images of sections in hydrothermalized granite in order to quantify the granulometry of each phase (Riss et al., 2001). Percentages for each mineral phase are presented in the Table 1. For the quartz vein, having no specific information, it has been considered as completely monomineral with a heterogeneous radii distribution.

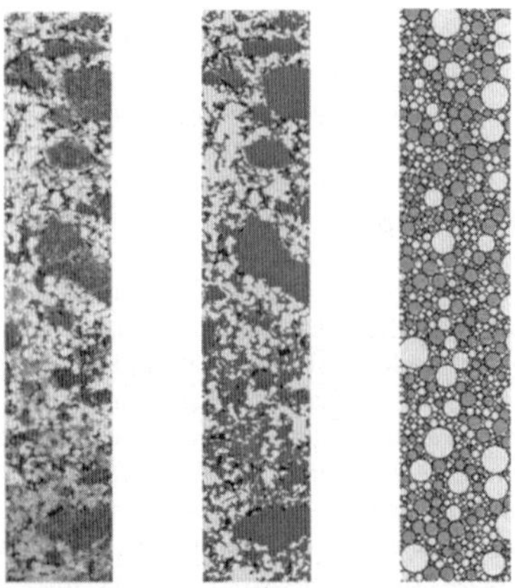

Figure 1. *Quantitative analysis of the hydrothermalized granite. a: original sample ;b: sample reconstructed after automatic identification of each mineral phases ; c: sample reconstructed for PFC model from statistical analysis of original samples*

To complete the model, physical, mechanical and thermal properties must be attributed to each particle. The parameters used for the model are presented in Table 2, Gentier (2001). The parameters of altered plagioclases have been assimilated to those of some clay minerals because of the alteration process of plagioclases. In this case, we have assume a transformation of all the plagioclases (100%).

Table 1. Percentage and particle radii of the main mineral phases introduced in the model : hydrothermalized granite (HYDR) and Quartz vein (VEIN).

Facies and radius	K Feldspars.	Altered Plagio-clases	Quartz	Black minerals (Biotite mainly)
HYDR	22%	39%	32%	7%
R_{min} (mm)	3	2	1,5	1
R_{max} (mm)	12	6	3,5	2
R_{mean} (mm)	7,5	4	2,5	1,5
VEIN	0%	0%	100%	0%

Table 2. Thermo-mechanical properties for the main mineral phases.

	K Feldspars.	Altered Plagio-clases	Quartz	Black minerals (Biotite mainly)
λ	1,612	1,612	5,2	1,976
α	0,385	1,4	3,5	1,4
C	679	707	700	700
d	2560	2329	2653	3051
E	38445	21076	57900	62185
ν	0,28	0,28	0,08	0,21
R_c	173	103	207	200

λ: thermal conductivity (W/m°C); α: linear thermal expansion coefficient (10^{-5}/°C); C: particle specific heat at constant volume; d: density (kg/m^3); E: Young's modulus (MPa), ν: Poisson's ratio and R_c: compressive strength (MPa).

3. NUMERICAL APPROACH – NEAR WELL

3.1 PFC^{2D} modelling

In the model, the mineral phases are represented by circular particles obeying to a statistical law based on the data presented in the table 1 and by respecting the spatial distribution for each phase.

In Figure 2, we present a simplified two dimensional geometrical model obtained for the permeable fracture identified between a quartz vein and hydrothermalized granite and located at 3224 m depth. The dimension of the modelled zone

is equal to 70 cm and the measured dip for the fracture is 81°.
From the mechanical standpoint, the particles have a density and are interlinked by contacts which are defined by their normal and tangential stiffnesses, a friction coefficient and a microscopic cohesion. A "cohesive" behaviour model serves to define a fracture threshold that determines the normal and tangential forces that can be withstood by the inter particle links. Hence, a micro crack appear when the contact between two adjacent particles is broken. When a connection between several micro cracks takes place, it can generate a new fracture in the model.

The first difficulty encountered in a discrete element approach is to correctly define the microscopic input parameters, which help to reconstruct the macroscopic behaviour determined from the measurements for each facies.
This step is carried out in several stages, the first being to fit the macroscopic elastic parameters (Young's modulus and Poisson's ratio), followed by fitting the parameters of plastic behaviour (cohesion and friction angle for the Mohr-Coulomb criterion used in the present paper).

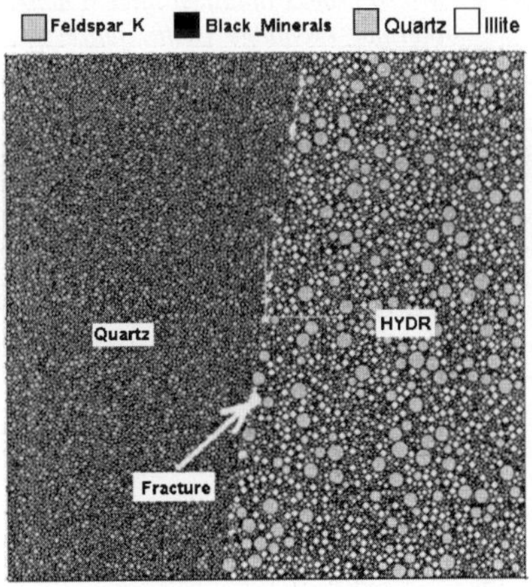

Figure 2. Vertical cross section within the simplified geometrical model

In this study, the macroscopic parameters considered for the two facies are presented in the Table 3.

From the thermal standpoint, each particle is considered to be a heat reservoir, which exchange heat with its neighbours by conduction in transient state through conduits represented by cemented bonds. Heat exchange disappears once the contact between the two particles is broken thus allowing to simulate a decrease of the equivalent thermal conductivity according to the number of the broken contacts in the model.

Table 3: Mechanical parameters considered for the hydrothermalized granite (HYDR) and quartz vein (VEIN).

	Elastic parameters	Plastic parameters
HYDR	E = 33,5 MPa ν = 0,26	Φ = 37,5° C = 35,8 MPa
VEIN	E = 57,9 MPa ν = 0,26	Φ = 36,5° C = 50,5 MPa

In this model, we assume that the temperature variations due to mechanical deformation of the particles are negligible. We also ignored heat transfer by convection.

Deformations and particle radius variations induced by thermal variations are calculated in the model by the followed equations:

$$\Delta\varepsilon_{ij} = \alpha\,\Delta T\,\delta_{ij} \qquad (1)$$

$$\Delta R = \alpha\,R\,\Delta T \qquad (2)$$

where α represent the linear expansion coefficient and R the particle radius.

The hydraulic pressures are simulated in the model by imposing a constant force on the open portions of the fracture.

3.2 Boundaries and Initial THM conditions

The natural stress field were determined by hydraulic fracturing, Klee (1993), in combination with the determination of the maximum stress direction based on borehole images. The values obtained for the principle stresses at a depth of 3224 m are respectively equals to –81 MPa for the vertical stress (σ_v), -83 MPa for the maximal horizontal stress (σ_H) which is oriented N170° from the east and –42 MPa in the perpendicular direction (σ_h).

In situ measurements also indicate the presence of a hydrostatic pressure field which is done by the expression above:
$P = \rho\, g\, z$, where ρ represent the fluid density assumed to be equal to a value of 1000kg/m^3 in the model and z the depth.

Initially, we assume a constant temperature of 150°C for the rock mass and a temperature of 20°C for the injected fluid which is maintained constant during the calculations.

We also assume an adiabatic conditions and zero displacements on the boundaries of the model.

The hydraulic solicitations (pressures) are applied in the model in several successive phases which are defined from the data obtained during the hydraulic tests in the well. In a first stage, the water pressure applied in the fracture was increased gradually from 0 to a value of 41 MPa which is closed to the minimal horizontal stress (σ_h). During the following stages, the water pressure was increased by 1 MPa, 1,5 MPa, 2 MPa, 2,5 MPa and 3 MPa.

4. RESULTS AND DISCUSSION

After the first stage of hydraulic pressure, the contact forces between particles are primarily in compression and no contact was broken. When the heat exchange is applied in the model, the distribution of forces reveals greater tensional forces (black lines) in the quartz vein in comparison with the hydrothermalized granite. Also, the compressive forces (light grey lines) are much less developed in the quartz vein compared with the hydrothermalized granite (Figure 3).

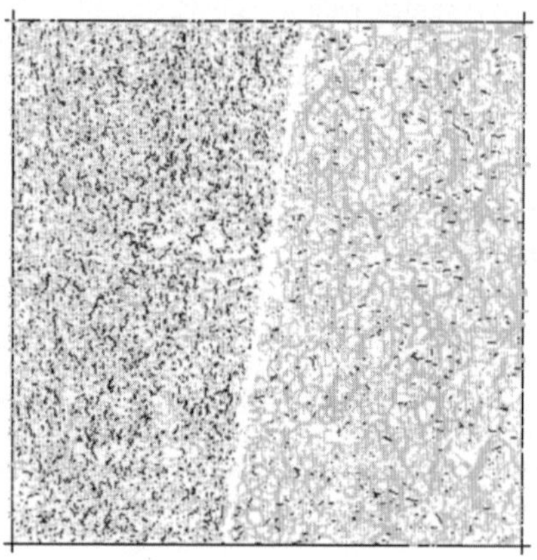

Figure 3. Distribution of forces after 6 hours of heat exchange.

A lot of micro cracks occur within the quartz vein and the hydrothermalized granite. However, the two facies behave differently with a very high micro crack development within the quartz vein (Figure 4).

Also, the number of micro cracks is ten times more significant in quartz vein than in the hydrothermalized facies and the main part of micro cracks appears during the first four hours of heat exchange (Figure 5).

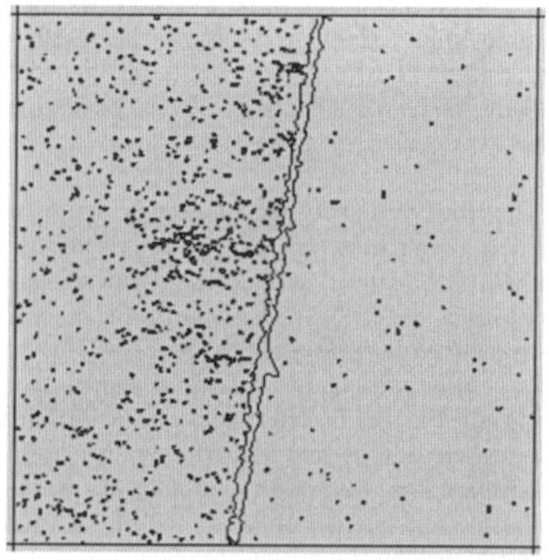

Figure 4. *Micro cracks obtained after 6 hours of thermal exchange (Left: quartz vein , right: hydrothermalized granite)*

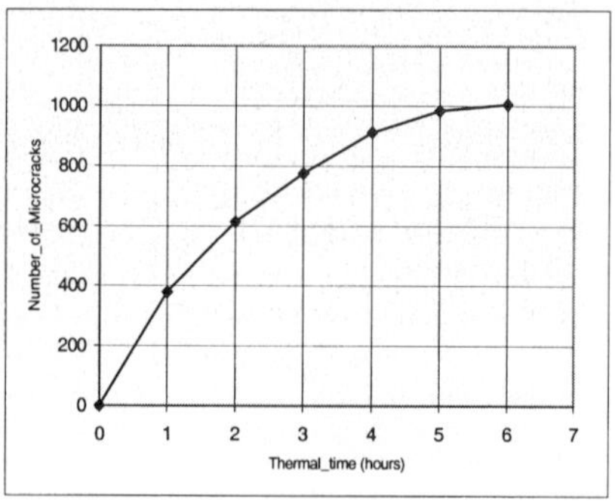

Figure 5. Evolution of the total number of micro cracks versus time during the thermal exchange

The percentage of contact bonds broken in the quartz vein after six hours of heat exchange is

4,25% whereas it is only 0,55% in the hydrothermalized granite.
Difference in the behaviour between the two facies is primarily of thermal origin. In fact, the heterogeneity of the mineralogical content of the hydrothermalized granite gives it thermal properties that are relatively lower than those of the facies composed exclusively of quartz.

We also studied the influence of the particle sizes of the quartz on the behaviour presented below and no significant difference was noted since the same percentage of micro cracks was obtained, i.e. 4,25%, by multiplying in the model the radius of the quartz particles by two.

In general, the fracturing of the inter-particle bonds are mostly in tension (only 38 micro cracks are from shear mechanism) and are of thermal origin particularly in the quartz vein.

In the model, the micro cracks of thermal origin obtained in the quartz vein can coalesce initiating small new fractures (Figure 6), and also contribute to increase porosity in the near wall.

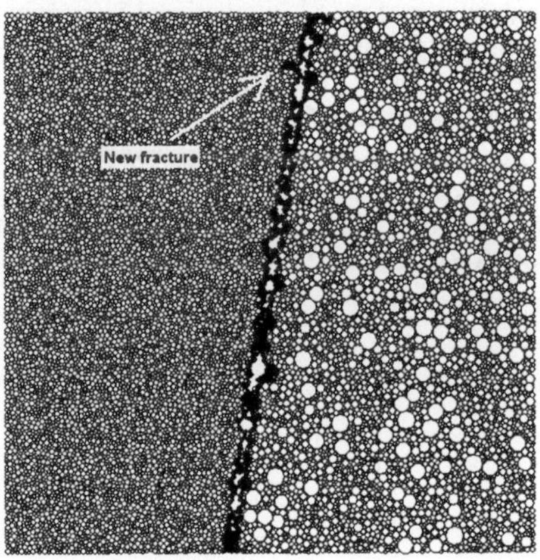

Figure 6. Propagation of a new fracture in the quartz facie

At this local scale (metric scale), the modelling shows that the thermal balance is reached after 48 hours of heat exchange, i.e. temperature of all particles is equal to 20°C (Figure 7).

If we assume an initial temperature for the injected fluid equal to 70°C instead of 20°C currently used in the previous model, the total number of micro cracks after six hours of heat exchange is almost 40% less significant than in the reference model and the micro cracks are mainly located in the quartz facies (Figure 8).

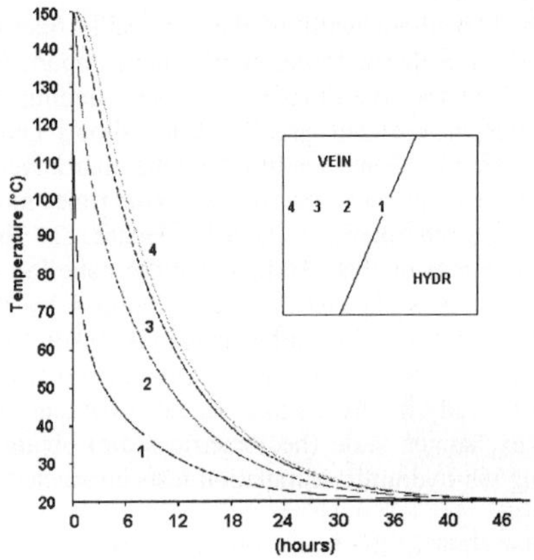

Figure 7. Temperature versus time calculated in the quartz vein

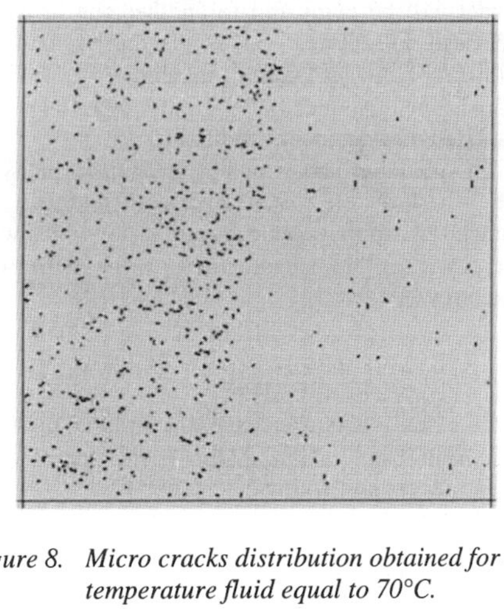

Figure 8. Micro cracks distribution obtained for a temperature fluid equal to 70°C.

A second case study, in which the fracture is assumed to be located exclusively in a hydrothermalized facies, revealed better resistance to thermal loading than the previous case. In fact, the percentage of contact bonds broken is closed to 0,25% of the total contacts in the model.

5. CONCLUSION AND PERSPECTIVE

The main results obtained by the numerical approach developed above to study the thermo-hydro-mechanical behaviour of the stimulated fractures are:

- the immediate walls of the permeable fracture can undergo thermal alterations whose importance depends on the type of rocks in the surrounding.
- hydrothermalized granite facies shows better resistance to thermal loading than the quartz facies which develops more micro cracks (ten times more than hydrothermalized facies). Hence, a first interpretation of this result is that the cohesion of the fracture walls can be much reduced by the thermal effect. This information is taken into account in a model presently in preparation which aims to study the hydro mechanical behaviour at a heat exchanger scale (hectometric scale) obtained during the hydraulic stimulation tests conducted in the site.
- alterations of the facies are mainly from thermal origin and could lead to initiate new fractures if the facies thermal properties surrounding the permeable fracture are high.

In the future, the results obtained by the current model will be compared to another one in which stresses are applied as a boundary conditions allowing us to better quantify the shear displacements induced along the fracture than with the fixed displacement conditions.

The model will be implemented in the study of the others access zones and will integrate the cataclazed and brecciated granite facies which were identified in the surrounding of the permeable fractures.

ACKNOWLEDGEMENT

This work was carried out with funding from European Community, ADEME (French Agency for Energy) and BRGM.

REFERENCES

Evans, K., Kohl T., Hopkirk, R., Rybach, L. 1996. *Studies of the nature of non-linear impedance to flow within the fractured granitic reservoir at the European Hot Dry Rock Project site at Soultz-sous-Forêts, Franc*". ETH Zurich Final Report July 1996, 144p.

Genter, A., Traineau, H., Bourgine, B., Ledesert, B., Gentier, S. 2000. *Over 10 years of geological investigations within the European Soultz HDR project, France*. Proceedings of the World Geothermal Congress 2000, Kyushu-Tohoku, Japan, May 28 - June 10, 2000, Full length paper on Cd-Rom, Editors E. Iglesias, D. Blackwell, T. Hunt, J. Lund, S. Tamanyu, 3707-3712.

Gentier, S., et al.. 2001. *Modélisation de l'interface puits-échangeur (Soultz-sous-Forêts) :acquisition des données de base et développements préliminaires.* rapport BRGM RP-51764-FR, 349 p

Gérard. A., Baumgaertner. J., Baria R., Jung R., Gentier S., Genter A. 2002. *Elements for a conceptual model of the underground heat exchanger at Soultz-sous-Forêts (France)-situation beginning 1998.* Geologisches JahrBuch Sonderhest, Procedding of the international Conference for HDR forum, Strasburg France 1998, 281-290.

Itasca, 1995. *PFC^{2D}, Particle Flow Code in 2 Dimensions. User's Manual.* Itasca Consulting Group, Minneapolis, MN.

Klee, G., Rummel, F. 1993. *Hydrofrac stress data for the European HDR Research Test Site Soultz-sous-Forêts.* Int. J. Rock Mech. Min. Sci & Geomech. Abstr., 30(7) : 973-976.

Riss, J., Gentier, S., Genter, A. 2001. *Granitic core cross section : numerical modelling.* Proceedings of the 8th International Symposium for Stereology and Image Analysis.

Tenzer, H., Schanz, U., 1997. *Development and characterization of HDR heat exchanger. Reprocessing of fracture orientation in drill hole GPK2 and evaluation of production logging data in drill holes GPK1 and GPK2.* Periodic report from SWBU, 01/02/96-31/01/1997.

EFFECT OF THERMAL DEFORMATION ON FRACTURE PERMEABILITY IN STRESSED ROCK MASSES

Takatoshi ITO[1], Daniel SWENSON[2], Kazuo HAYASHI[1]
[1]) Institute of Fluid Science, Tohoku University, Japan
[2]) Mechanical Engineering Department, Kansas State University, US

Abstract: We analyse the effect of thermal contraction of rock on fracture permeability. The analysis is carried out by using a 2D FEM code which can treat the coupled problem of fluid flow in fractures, elastic and thermal deformation of rock and heat transfer. In the analysis, we assume high-temperature rock with a uniformly-distributed fracture network. The rock is subjected to in-situ confining stresses. Under the conditions, low-temperature fluid is injected into the fracture network. Our results show that even under confining environment, the considerable increase in fracture permeability appears due to thermal deformation of rock, which is caused by the difference in temperature of rock and injected fluid. However, for the increase of fracture permeability, the temperature difference is necessary to be larger than a critical value, ΔT_c, which is given as a function of in-situ stresses, pore pressure and elastic properties of rock.

1. INTRODUCTION

Let us consider fluid flow through fracture network in rock formations. When the fluid temperature is lower than that of rock, the fluid flow removes heat of rock and the rock cools down. The cooling will cause thermal contraction of rock, and as a result fractures are likely to open and to be more permeable. In reality, the test results of fluid injection into rock formations at depths of about 1000 m, show that cold-fluid injection is effective to increase permeability of the rock formations (e.g., Horne et al., 1988, Ariki and Hatakeyama, 1997, Clotworthy, 2000).

However, it should be considered that rock masses are subjected to remote compressive stresses to cause fracture closure. If the confining effect is too significant, the cooling associated with the fluid flow may not increase fracture permeability so much. Thus it is not clear whether the cooling is always effective to enhance fracture permeability or not. Ariki and Akibayashi (2001) has tried to explain the permeability-increasing phenomena resulted from the cold-fluid injection by taking into account of microscopic change in rock porosity due to thermal deformation of the rock, however, roles of fracture and remote compressive stresses were not considered in their analysis.

In this research, taking into account of the confining effect due to remote compressive stresses, we examine how the cooling will change fracture permeability under various conditions. Such examination will contribute to geothermal energy development and to clarification of various geophysical phenomena associated with fluid flow through fractures.

2. NUMERICAL SIMULATION

2.1 Problem definition

We considered here a jointed rock, i.e. a formation of rock blocks separated by thin joints / fractures. Fluid flows through the fractures, while fluid permeation into the rock blocks is neglected. The rock formation is at high temperature initially, and is subjected to in-situ confining stresses. Under the conditions, low-temperature fluid is injected from a point into the fracture network. For such a case, we examined how the fluid injection changes the fracture network permeability assuming 2-dimensional cases for simplicity.

2.2 Modeling

All calculations were performed using the Geocrack2D finite element code, which was developed to solve coupled structure / fluid / thermal problems. Summary and basic equations of the code are given in Swenson et al. (1995). A simulation consists of arbitrarily shaped rock blocks with linear / non-linear contact and discrete fluid paths between the blocks. Heat transfer occurs by conduction in the rock blocks and conduction and transport in the fluid.

Figure 1 shows the configuration of the model used for the present analysis. Small squares filling the model represent the rock blocks, and lines between the squares represent the fractures. The blocks and fractures are divided into the finite element mesh for numerical calculation, but the mesh is not displayed in Figure 1. The whole size of the model is 3.5 x 3.5 m^2, and the size of the square rock block is 0.1 x 0.1 m^2. The formation of rock blocks and fractures are subjected to the biaxial compressive stresses S_x and S_y. Due to the

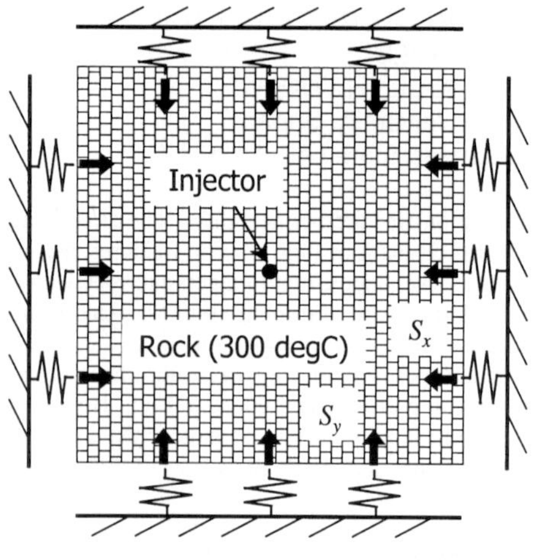

Figure 1. Configuration of the model used for the present analysis.

compression, the fractures are mechanically closed initially, however, it is assumed that there is a mismatch of the fracture surfaces and the pore space is available for fluid flowing. The residual fracture aperture at the initial condition is assumed 20 μm. The mismatch of fractures produces the stiffness for the fracture opening / fracture normal stress behavior. In the present model, the fracture stiffness is represented by a spring which connects opposite sides of the fractures. The stiffness of the spring is assumed 6000 GPa/m (= k_n), and is assumed to be vanished, i.e. $k_n = 0$, at fracture opening, since the mismatch of fractures then disappears. The stiffness for the fracture sliding / fracture shear stress behaviour is characterized also by a spring with the stiffness of 250 GPa/m, although it is assumed constant independently of fracture opening and sliding. Young's modulus, Poisson's ratio and linear expansion coefficient of rock are assumed to be 60 GPa, 0.2 and 7.5×10^{-6} $(\text{degC})^{-1}$, respectively. In the initial stage, the temperature of rock blocks and fluid, T_r, is 300 degC, and the fluid pressure is zero. From the time $t = 0$, low temperature fluid is injected into the fracture network from a point at the center of the model. The fluid has a constant temperature of T_f at the inlet, and it flows through the fracture network to the outlets at the model boundary while cooling the rock blocks. The fluid pressure at the single inlet and the multiple outlets are kept constant at 0.1 MPa and zero respectively. Fluid viscosity is assumed constant at 1.25×10^{-4} Pa/sec independently of temperature. Therefore we can evaluate the overall change in the fracture permeability from the variation of the injection flow rate. If the fracture opening becomes larger and the fractures get more permeability due to the cooling, the injection flow rate becomes larger. On the other hand, the special interface elements that have a function of spring, are set up around the model as shown in Figure 1. One side of the interface is attached to the mesh, and the other side is assumed fixed. The interface stiffness represents a far-field elastic stiffness so that we can analyze elastic problems for an in-finite media by using a relatively small model such as Figure 1.

2.3 Results

An example of the simulation is shown in Figure 2 for the case that the injection temperature T_f is 100 degC and the confining stresses are $S_x = S_y =$ 20 [MPa]. This figure shows the variation of the injection flow rate Q with the duration of injection, t. Q increases gradually with t, even though the compressive stresses are relatively large such as 20 MPa. Figures 3 and 4 show the distributions of temperature T and average stress σ_{avg} respectively, where compressive stresses are taken to be positive in this paper. Cooled region is developed in a concentric configuration around the injection point at the centre of each figure. According as the temperature decline of rock blocks, the average stress σ_{avg} decreases in the region which is

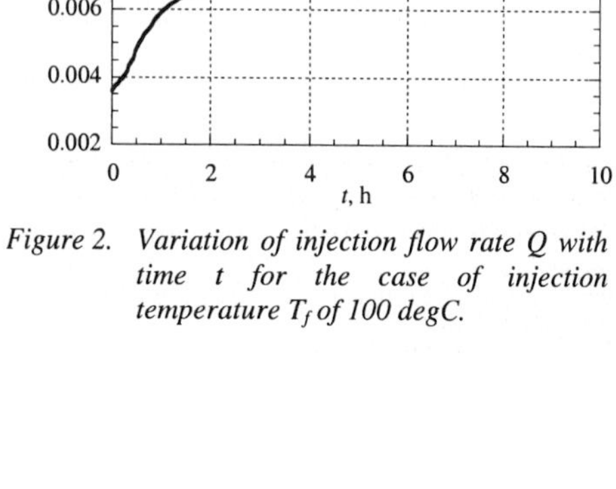

Figure 2. Variation of injection flow rate Q with time t for the case of injection temperature T_f of 100 degC.

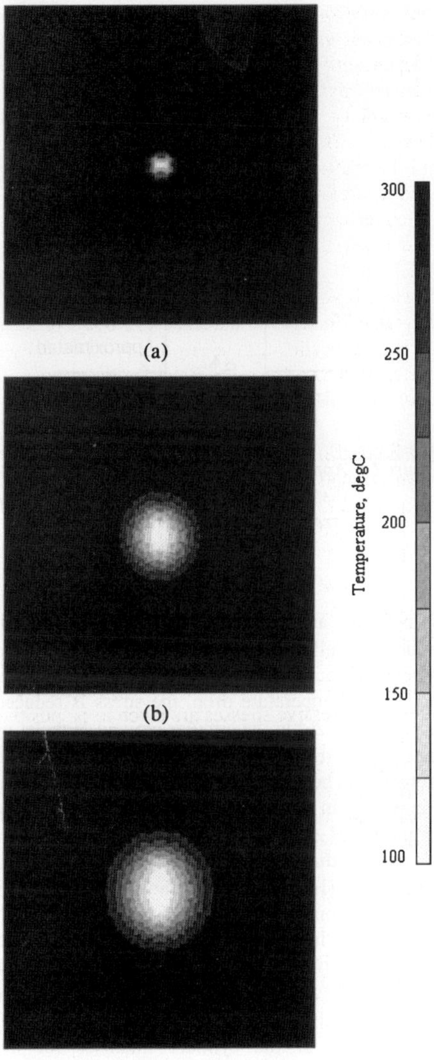

Figure 3. Temperature distribution at (a) t = 2 [h], (b) 6 [h] and (c) 10 [h] (T_f = 100 [degC]).

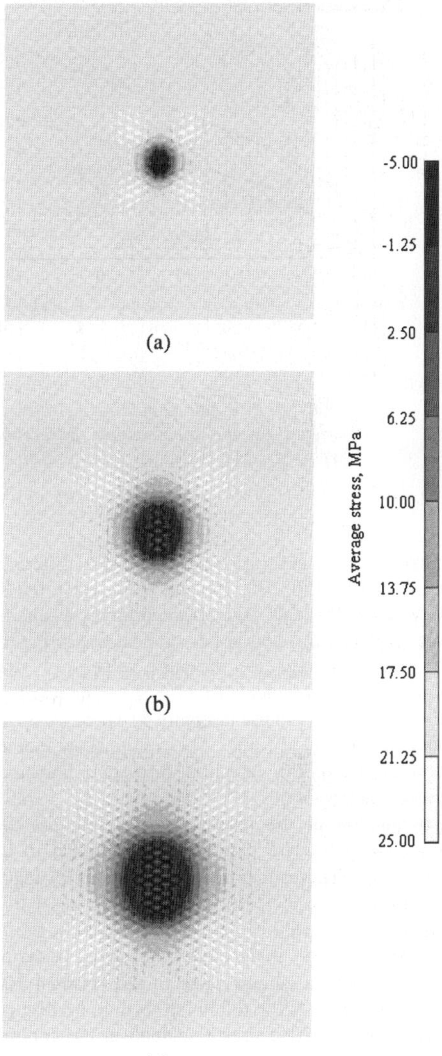

Figure 4. Averaged stress distribution at (a) t = 2 [h], (b) 6 [h] and (c) 10 [h] (T_f = 100 [degC]).

approximately overlapped with the cooled region. The stress decline is steeper at a region closer to the injection point, and there appears the region in which σ_{avg} is reduced to zero. These results suggest that the cooling associated with the cold fluid injection reduces the fracture normal stress around the injection point, and the stress drop causes fracture opening which results in the increase of fracture permeability.

Next we examined the effect of the injection temperature on the $Q - t$ relationship. In the similar way as described above, the simulation was

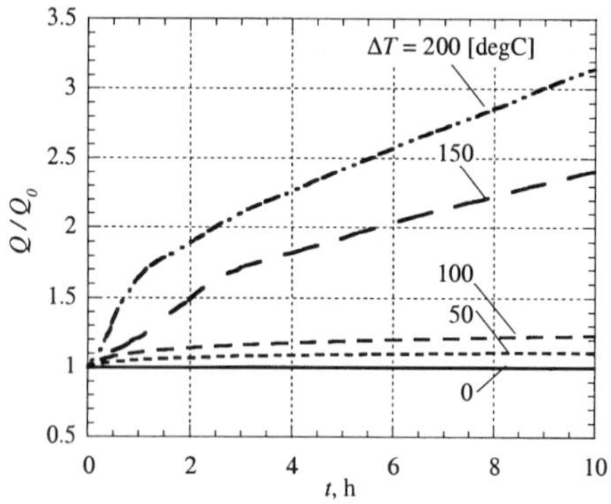

Figure 5. Change in the relationship between injection flow rate Q vs. time t, which is caused by the temperature difference ΔT (= $T_r - T_f$).

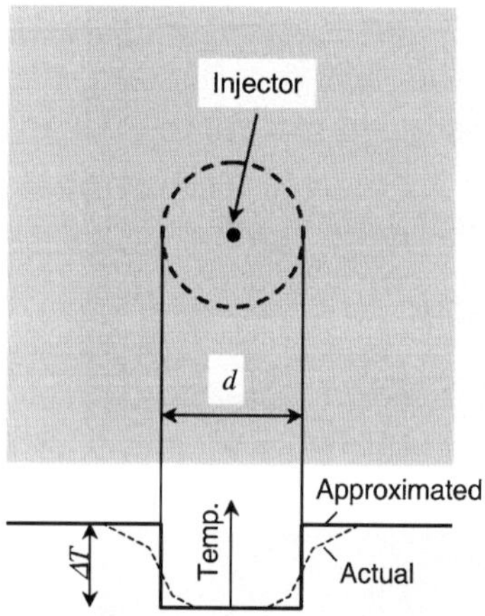

Figure 6. Approximation on temperature distribution in order to estimate the stress drop caused by cooling.

carried out for several cases where T_f is different, i.e. T_f = 100, 150, 200, 250 and 300 [degC], but T_r is fixed at 300 degC. We obtained finally a suite of the simulated $Q - t$ curves as summarized in Figure 5 corresponding to various values of the temperature difference ΔT (= $T_r - T_f$). The result of Figure 2 is also plotted in Figure 5, where Q_0 is the magnitude of Q at $t = 0$. The results show that as one expected, the injection flow rate increases more steeply with ΔT. However, by careful examination of the results, it is found that the increase of fracture permeability is limited to be small for the cases that ΔT is lower than 100 degC, but its apparent increase appears for the cases that ΔT is larger than 150 degC. This result suggests that the cooling / permeability increasing behavior is changed at a certain value of ΔT between 100 and 150 degC. In the following section, we discuss the reason why the drastic change in the cooling /permeability increasing behavior occurs depending on the value of ΔT.

3. DISCUSSION

The cold fluid injection will reduce temperature around the injection point, and it will develop a temperature distribution such as schematically shown by a dashed line in the lower part of Figure 6. For simplicity, we approximate here the non-linear temperature distribution as a step-like distribution shown by a solid line in that figure.

Namely we assume that the fluid injection will induce a uniform temperature drop, ΔT, in a circular region with diameter d. For such a case, due to the temperature drop, the stress is reduced uniformly by $\Delta\sigma$ in the circular region. Now let us consider the relationship between ΔT and $\Delta\sigma$. The problem can be divided into three parts as shown in Figure 7. In the problem (I), traction with the magnitude of $\Delta\sigma$ is applied on the wall of a circular cavity with diameter d in an infinite medium. In the problem (II), traction with the magnitude of $\Delta\sigma$ is applied on the surface of a cylinder with diameter d. In the problem (III), a cylinder with diameter d is shrunk by the uniform drop in temperature, ΔT. The increase in the cavity diameter for the problem (I) is denoted by Δd, and the decrease of the cylinder diameter for the problems (II) and (III) are denoted by Δd_1 and Δd_2, respectively. Those values can be evaluated by the theory of elasticity, and they result in

$$\Delta d = \frac{1+\nu}{E}\cdot\Delta\sigma\cdot d \tag{1}$$

$$\Delta d_1 = \frac{1-\nu}{E}\cdot\Delta\sigma\cdot d \tag{2}$$

$$\Delta d_2 = \alpha\cdot\Delta T\cdot d \tag{3}$$

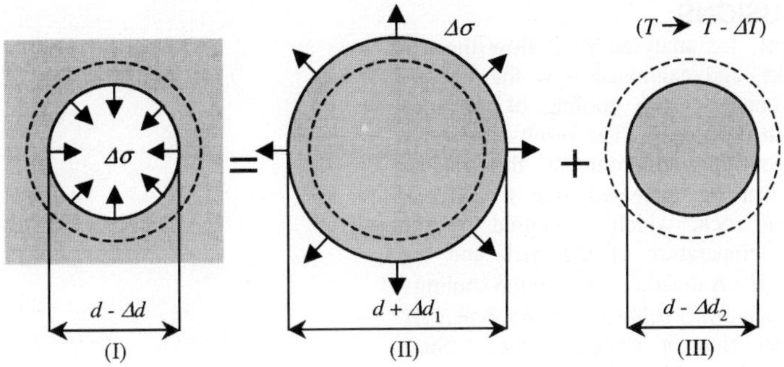

Figure 7. Three problems (I)-(III) and their relationship.

where E, ν and α are Young's modulus, Poisson's ratio and linear expansion coefficient of the rock block, respectively. Furthermore, the following equation must be satisfied.

$$\Delta d + \Delta d_1 - \Delta d_2 = 0 \tag{4}$$

Substituting Equations (1)-(3) into Equation (4), we obtain the relationship between ΔT and $\Delta\sigma$ as follows;

$$\Delta\sigma = \frac{E}{2} \cdot \alpha \cdot \Delta T \tag{5}$$

Note that this stress drop $\Delta\sigma$ occurs in hydrostatic manner in the circular region with radius d (see Figure 6). Namely the normal stress component in any direction decreases uniformly anywhere in that region by $\Delta\sigma$.

On the other hand, before the fluid injection is started, the stress state in the rock blocks is in equilibrium with the remote compressive stresses. Then the cooling associated with fluid injection decreases the stress by $\Delta\sigma$ in the circular region around the injection point. Therefore, the average stress in the circular region during the fluid injection, σ_{avg}, is given by

$$\sigma_{avg} = S_{avg} - \Delta\sigma \tag{6}$$

where S_{avg} is the average magnitude of the remote compressive stresses. Furthermore we make the following two assumptions, (i) σ_{avg} is equivalent to fracture normal stress on average, and (ii) when the fracture normal stress is less than fluid pressure in the fracture, p, the fracture opens and its permeability increases drastically. In this case, the condition required for the drastic increase of fracture permeability is given by

$$\sigma_{avg} - p = S_{avg} - \Delta\sigma - p = 0 \tag{7}$$

Substituting Equation (5) into Equation (7) finally yields

$$\Delta T = \frac{2}{\alpha \cdot E}\left(S_{avg} - p\right) \equiv \Delta T_c \tag{8}$$

Thus if the temperature difference ΔT becomes larger than ΔT_c defined by Equation (8), the fracture permeability will increase drastically. ΔT_c, is referred to the critical temperature difference. Note that ΔT_c is independent of the size of the cooled region, i.e. d.

For the numerical simulations presented in the previous section, Equation (8) predicts ΔT_c of 89 degC taking into account of $\alpha = 7.5\text{x}10^{-6}$ (degC)$^{-1}$, $E = 60$ [GPa], $S_{avg} = 20$ [MPa] and $p = 0$ [MPa]. The estimated value is reasonable, although it is slightly lower than the critical temperature difference which is expected in the range of 100 and 150 degC from the numerically-obtained results of Figure 5. Such discrepancy may be caused by the fact that the approximated temperature distribution as shown in Figure 7 by the solid line, overestimates the effect of cooling associated with fluid injection.

4. CONCLUSIONS

In this paper, we analysed fluid flow through fractured rocks, and examined how the fracture permeability changes with cooling of the rock associated with fluid flow. The results show that even under confining environment, the fracture permeability can be increased due to thermal deformation of rock, which is caused by the difference in temperature of the rock and the injected fluid, ΔT. A drastic change in the cooling / permeability increasing behavior occurs, when ΔT becomes larger than a critical value. Such phenomenon can be explained taking into account the fact that the cooling associated with fluid injection causes a localised stress drop around the injection point. When ΔT exceeds the critical value, the stress drop reduces fracture normal stress to be less than fluid pressure in fractures. As a result, the fractures open and their permeability increase drastically.

REFERENCES

Ariki, K. & Hatakeyama, K. 1997. *Effects of the injection temperature on injectivity of geothermal well.* J. Geoth. Res. Soc. Japan 19 (in Japanese): pp.197-208.

Ariki, K. & Akibayashi, S. 2001. *Effects in the injection temperature on injection-capacity of geothermal wells – Numerical simulation on the measured downhole pressure.* In 2001 Annual Meeting Geoth. Res. Soc. Japan Abstr. with Programs, Nagoya, 3-5 Dec. 2001 (in Japanese): p. A06.

Clotworthy, A. 2000. *Reinjection into low temperature loss zone.* In Proc. World Geoth. Cong. 2000, Kyushu-Tohoku, 28 May-10 June 2000: pp. 2063 - 2067.

Horne, R.N., Grant, M.A. & McCabe, W.J. 1988. *Injection into heterogeneous reservoirs: experience in New Zealand.*, In E. Okandan (ed), Geothermal Reservoir Engineering, Kluwer Academic Publ.: pp. 91-101.

Swenson, D., DuTeau, R & Sprecker, T. 1995. *Modeling flow in a jointed geothermal reservoir.* In Proc. World Geoth. Cong., Florence, 18-31 May 1995: pp. 2553 - 2558.

NUMERICAL FLOW AND HEAT TRANSFER MODEL OF THE POROUS-FRACTURING HYDROTHERMAL SYSTEM OF THE PARATOON THERMAL WATER FIELD

Igor I. Krashin[1], Lyudmila V. Semendyaeva[1], Alexander I.Zinin[2], Galina A. Zinina[2]

[1]) -VSEGINGEO, Moscow , Russia

[2]) -SSC IPPE, Obninsk, Russia

Abstract: Results of numerical study of filtration and heat transfer processes in the middle area of Paratoon field have been presented. Equation system of the model used includes continuity equation, a generalization of Darcy law for the case of variable density flow, thermal energy conservation equation, and closing relationships for calculating the porous solution density and viscosity. The area modeled is 11 km^2, its vertical size – 3 km. The model stratigraphy is represented by interbedding strata of various conductivity (tuffs, andesites, diorites). Vertical faults of high conductivity were included into the model area. Initially stationary temperature field was calculated, calibration calculations were fulfilled, to simulate the temperature field observed. Then, non-stationary calculation followed – pumping out via boreholes during 30 years. According to calculations, in total boreholes yield of 22000 m^3/day, temperature in the central productive area does not change considerably. The model developed can be used for prediction calculations.

INTRODUCTION

The Paratoon thermal water field is located within the Central Kamchatka volcanic area, 25 km apart from Petropavlovsk-Kamchatski. The field is confined to the structural depression of the Paratoonka river valley, which is a graben of complicated structure formed by block-wise shifts of crystalline foundation in quaternary period.

The field has a three-level structure. The lower stage is represented by volcanogenic-sedimentary, effusive-pyroclasticrocks of the Myocenic paratoon formation 2000–4000 m thick. Volcanogenic-sedimentary rocks of the alnoite series form the middle layer of 125–500 m thickness. The upper layer is formed by the early quaternary conglomerates, sandstones, aleurolites and argillites of 40–50m thickness. The profile has a "corona" of non-segmented quaternary sediments of various geneses with prevailing alluvial sandy-gravel-pebbled and aqueo-glacial deposits of total thickness ranging from 40 to 180 m.

Fractured and jointing volcanogenic-sedimentary effusive-pyroclastic rocks of Neogene, as well as the Early quaternary conglomerates filling the unevenness of the ancient relief serve as the main collector for thermal waters. The aleurolites and argillites of the same age bedding at depths of 140–180 m overlapping them represent a relative aquitard, separating the rocks of productive strata from loose quaternary sediments containing cold waters.

The presence of a concentrated heat source is a principal cause of natural circulation of water in the hydrothermal system. In this case, the temperature in deep bedding of rocks in the system reaches high values, over 100 °C. Pressure at 3000 m depth is 300 atm. With these ranges of temperature and pressure variation, properties of water appear to change considerably (viscosity, density, heat capacity), which is one more cause for the water circulation intensification.

The main source of replenishment of thermal water store is cold underground waters of unconfined alluvial water-bearing level formed by means of atmospheric precipitation and the Paratoonka river. The river's structure creates favorable conditions for the accumulation of considerable reserves of underground waters, because the surface run-off, deep one, and subsurface drainage are directed from flanges towards depression.

As it has been shown by the investigations, there are several sites within overall field area (340 – 400 km^2) typified by the presence of concentrated heat sources developed in the foundation rocks which are the basement for the volcanogenic-sedimentary strata forming the field (Kiryukhin, 1996). The Middle site of the Paratoon field (2.4 – 3.5 km) associated with the northern part of Paratoon hydrothermal system is one of the most perspective in the aspect of thermal waters; numerical modeling of groundwater flow and heat exchange processes has been carried out with the orientation to its hydrogeologic conditions.

Results of three-dimensional modeling of the Middle – Paratoon site have been presented.

1. SCHEMATIZATION OF THE AREA MODELED

The model domain boundary was chosen so that the model area is significantly larger than the volume of the heated zone, and it includes the cold water recharge zone as well.

Figures 1-2 show the model geometry in planar aspect and vertical cross-section. The Figure 1 shows lines of six faults (the 6–th one is situated over the northern part of the modeled area boundary). Besides, the figure shows the line of vertical section (x=1500 m), some results of calculations to be presented below. The modeled area is 11 km^2, its size in vertical direction – 3 km. The upper boundary is located on the topographic surface.

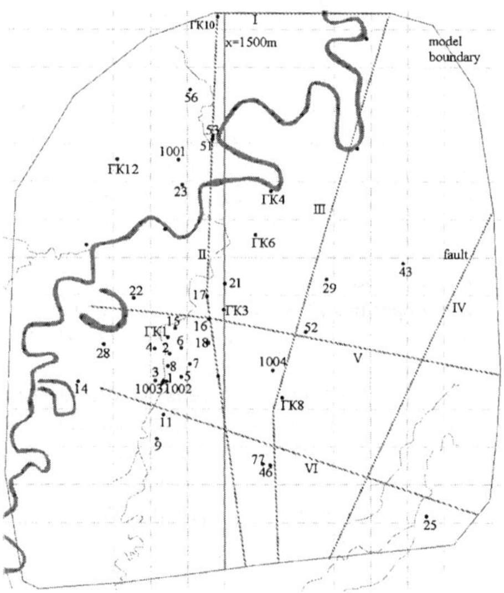

Figure 1. Area modeled in planar presentation..

The model is subdivided into layers of various thickness over its depth. The layer thickness cases were defined by averaging the data on hydrogeologic cross sections of the boreholes $\Gamma K-4$, 5, 8, $\Gamma K-1$, 22, $\Gamma K-3$, 11, 6, 51, 23, 21, 19, $\Gamma K-2$, $\Gamma K-6$, and $\Gamma K-8$.

The upper stratum of non-segmented loose sediments is represented by three layers:

layer 1 – sandy-pebbled sediments (average thickness 60 m);

layer 2 – aleurolites (average thickness 38 m);

layer 3 – conglomerates (average thickness 55 m).

The main productive series of fractured volcanogenic-sedimentary rocks of ~1270 m thickness was modeled as alternating beds of tuffs and interlayers of andesites, diorites, porphyrites. The lower series of ~1600 m – effusive-pyroclastic rocks of the Paratoon suite of Myocene. The values of hydrogeologic parameters of the layers identified are presented in Table 1. Thermoconductivity coefficient of 2.0 W/m·°K was specified for the rocks in three upper layers, and 2.5 W/m·°K in the rest ones, with heat capacity of 1750 J/kg·°K, density of rocks is 2500 kg/m^3.

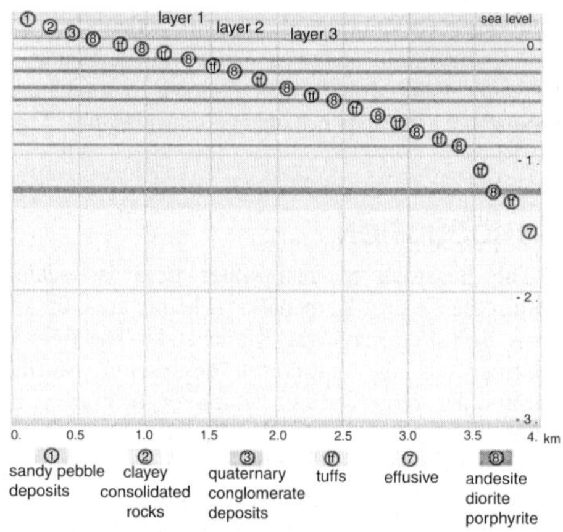

Figure 2. Vertical cross-section of the model (layers are numbered according to Table 1).

Table 1. Hydrogeologic properties of the rocks.

Designation	**N**	K_f^{xy} **(m/day)**	K_f^{z} **(m/day)**	**Porosity** Φ
layer 1	1	5	1	0.35
layer 2	2	0.005	0.005	0.3
layer 3	3	1	0.1	0.3
prod_h	4 (tf)	0.9	0.18	0.05
prod_m	5 (tf)	0.5	0.1	0.05
prod_l	6 (tf)	0.2	0.04	0.05
Effusive	7	0.05	0.05	0.05
Interlayer	8	0.025	0.025	0.01
Fault	9	0.5	0.5	0.05
Layer 2p	10	0.002	0.002	0.3
Fault	11	0.1	0.1	0.05

Space variation of hydraulic conductivity of the productive rock mass was modeled as planar zones

(Fig. 6); they are presented in Table 1 as materials 4, 5, 6 belonging to the tuff layers. The properties of effusive rock mass are defined via material 7, and properties of interlayers consisting of andesites, diorites, porphyrites are determined by material 8; those of faults – by materials 9, 11 (properties of faults I, II, V are determined by material 9, and the rest ones – by the 11-th material). Vertical faults penetrate the modeled area from the bottom boundary to the 3-rd layer foot. In this case, the finite-difference scheme plotting accounts for the fault thickness, which was assumed to be 5 m. The input data for numerical study were taken after a series of preliminary calculations, Krashin (2002).

The heat flux distribution at the bottom boundary was modeled as planar zones (fig.3), their boundaries being determined by isolines of temperature field evaluated as a result of experimental hydrogeologic works. In the calculations the heat flux values of Q1, Q2, Q3 were specified to be equal to 2.07, 1.91, 0.4 (W/m^2). The source's integral power of 7880 KW was used.

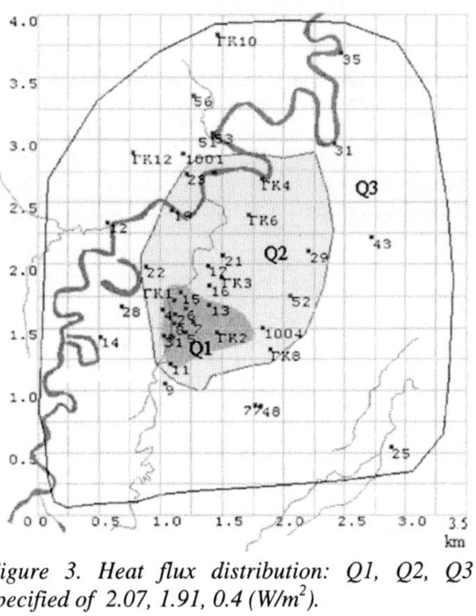

Figure 3. Heat flux distribution: Q1, Q2, Q3 specified of 2.07, 1.91, 0.4 (W/m^2).

The filter spacing intervals were chosen as follows. The information on the water inflow intervals was analysed, and a continuous spacing interval of filter was chosen for each borehole, including the water inflow zones with considerable stream yields. Total yield of the borehole was distributed over the filter interval in proportion to the hydraulic conductivity. Yields of water supply boreholes (1, 3, 4, 5, 6, 8, 11, 15, 21, 22, $\Gamma K-1$, $\Gamma K-2$, $\Gamma K-3$, $\Gamma K-4$, $\Gamma K-6$, 51, 53) vary in the range of 1262-1856 (m^3/day). Total yield of the boreholes is 11000 m^3/day. Calculations were also carried out with a doubled yield value – 22000 m^3/day.

2. MATHEMATICAL STATEMENT OF THE PROBLEM

Numerical studies have been carried out in three-dimensional (x, y, z) geometry. The system of equations of the mathematical model includes the continuity equation, a generalization of Darcy law for the case of variable density flow, equation of thermal energy conservation, and closing relationships for the calculation of the pore solution density and viscosity.

Pressure P and temperature of the pore solution T are used as the variables of state for the system studied. A local thermal equilibrium of the system is assumed, i.e., equality of the pore solution temperature and that of rocks at each point of the continuum studied.

The impermeability conditions are used at the lateral boundary for the groundwater flow and heat flux.

At the bottom boundary there is a source of heat q(x, y, t) – thermal flow from the bottom boundary and the impermeability conditions for the groundwater flow. At the upper boundary, the first type condition is specified for pressure P:

$$P(x, y, z_{topo}, t) = P_{atm}(x, y, t). \qquad (1)$$

The non-linear boundary condition is used for temperature T, which depends on the current direction of flow on the boundary surface. If

$$\vec{n}^T \vec{w} \leq 0 \qquad (2)$$

(input flow), then

$$T(x, y, z_{topo}) = T_0(x, y, t), \qquad (3)$$

otherwise (output flow)

$$\frac{\partial T}{\partial n} = 0 . \qquad (4)$$

Here $z = z_{topo}(x, y)$ – the level of the model roof, $\vec{w}$ - Darsy's velocity, $\vec{n}$ - external normal. It is a special feature of the boundary conditions that the choice of specific variant of conditions is possible only after solving the fluid flow problem and defining the velocity field. However, the latter depends, in its turn, on the solution of heat transfer equation, i.e., also on the boundary conditions of these equations. The value of $T_0(x, y, t)$ was assumed to be equal to 10°C.

The formulas for calculations of density, temperature, and viscosity in the range of temperature of 150 - 600°C and pressure from 5.0 to 18.0 MPa are taken from Reference book, Kirillov (1984).

3. TECHNIQUE FOR THE BOUNDARY PROBLEM APPROXIMATION.

The model uses an irregular mesh of trihedral prisms for the presentation of filtration properties and supplementary thereto - a mesh of hexagonal prisms for the approximation of filtration and heat transfer equations by the finite difference control volume method. Numerous foundations of hexagonal right angle prisms (additional planar mesh) are presented in Fig. 4. Nodes of the basic planar mesh are geometric centers of hexahedral cells of the additional planar mesh. Fig.4 shows that the boundaries of the area modeled and the mesh of faults is accurately approximated by planar mesh subdivision. The faults are assumed to be strictly vertical.

The 3D-model of the Paratoon field of thermal water contains about 104000 control volumes. Finite difference mesh step in the vertical direction is varied on zones in the range of 5-60 (m). Number of mesh intervals in the vertical direction is 92.

4. RESULTS OF NUMERICAL STUDIES.

4.1 Stationary solution

At the first step of modeling the problem was considered with the objective of reproducing the natural temperature field (the yield of all productive boreholes set to be equal to zero). Fig. 5 shows as an example the temperature field isolines in vertical x=1500 m, (the section line is shown in Fig.1). For the source's integral power of 7880 KW, maximum temperature for the variant under study amounts to 110°C. The temperatures in productive layer in the high conductivity zone vary in the range of 80 -90°C, which is in an agreement with the observation data.

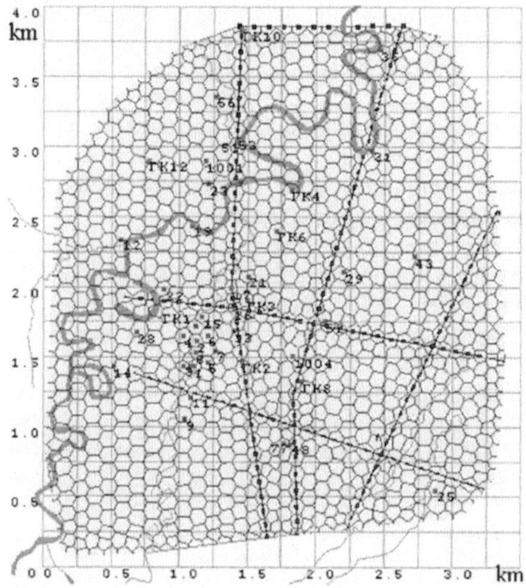

Figure 4. Calculation mesh in planar presentation.

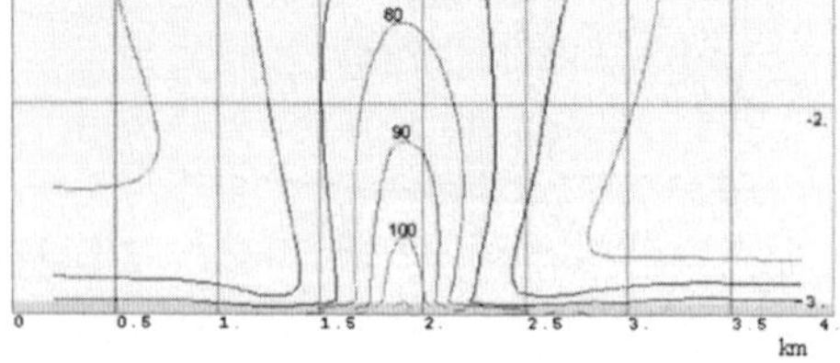

Figure 5. Temperature distribution in cross section x=1500 м

The rate of discharge Q^- on the ground surface for the stationary condition is 3150 m^3/day. In the stationary solution, water inflow Q^+ through the section of the upper boundary with Vz <0 is equal to flow Q^- upwards through the boundary section with Vz >0 (z-axis is directed upwards). The infiltration supply average for the model area is $Q^+/S=104$ mm/year, which amounts to 7.5% from precipitation (in annual precipitation of 1380 mm/year).

It can be seen by the temperature distribution (Fig. 5) that cold water from peripheral parts of the model are immersing with a simultaneous heating from the local thermal source. Hot water is discharged to a well-penetrable sub-fluvial alluvium of the Paratoonka river.

4.2 Non-stationary problem – 30-year operation of thermal water field

A non-stationary problem of 30-year operation of thermal water field using boreholes with pumping has been considered. The calculations have been carried out with two values of total well yield - 11000 m^3/day and 22000 m^3/day, for various arrangements of filters: in the top and bottom areas of the productive layer. The borehole yields were defined as constant values for the 30-year operation. The filter boundary intervals were evaluated using information on the water inflow intervals. For the variant with filter arrangement at the bottom of productive area, the values of filter boundaries were assumed to be similar for all boreholes, equal to 1100 m and 1500 m.

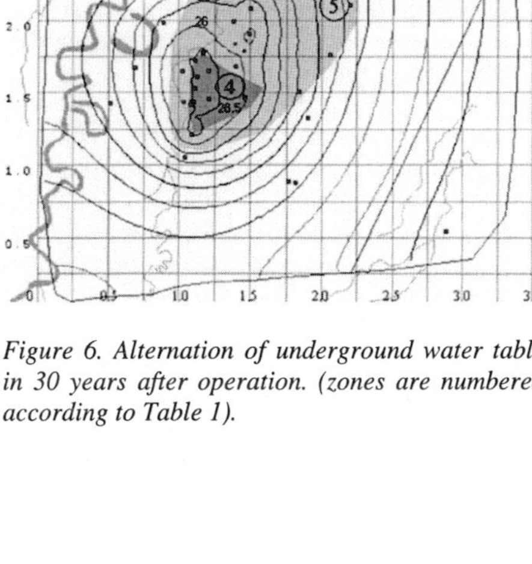

Figure 6. Alternation of underground water table in 30 years after operation. (zones are numbered according to Table 1).

Temperature field for natural conditions (the borehole yield equal to zero) was used as the initial condition for solving the non-stationary problem.

Fig. 6 shows isolines of underground water variation in 30 years of operation for total yield of 22000 m^3/day and for the borehole filters arrangement in the upper part of productive layer.Maximum variation of underground water table for the variant under study amounts to 26.5 m.

The thermal water discharge to the ground surface is terminated after 30-year operation. Calculated values of the input Q^+ and output Q^- flows are equal to 21600 m^3/day and 0.0 m^3/day, respectively, with total yield of the boreholes of -22000 m^3/day. The infiltration supply Q^+/S average in area is 708 mm/year (51.3% from the average annual precipitation).

Figures 7-8 show the distribution of vertical component of velocity in cross-section x = 1500 m after 30 years of operation. Isoline Vz=0 separates the areas of water upflow and downflow. Water withdrawal during the 30-year period resulted in a notable alteration of the field of vertical component of the rate only in the upper part of the model.

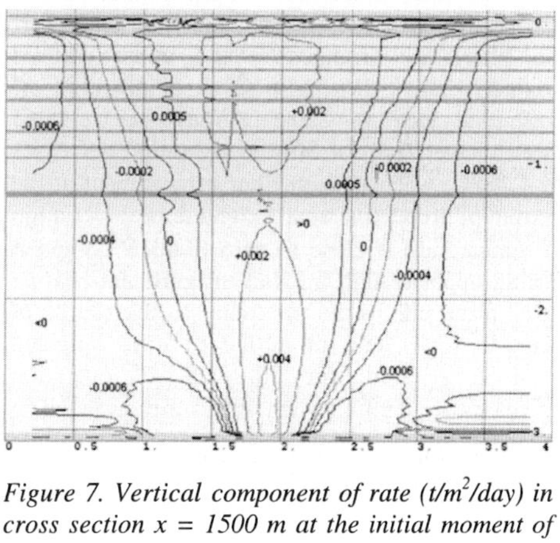

Figure 7. Vertical component of rate (t/m^2/day) in cross section x = 1500 m at the initial moment of time.

Calculations have shown that the arrangement of filters does not render any notable effect on water balance. At the same time, a certain difference in the thermal balance should be mentioned. With filter location at the upper part of the productive area, the quantity of heat removed by the uptake boreholes appears to be 7.6% higher than in that case when filters are arranged in the bottom part of productive area (70300 KW and 65300 KW, respectively) in total yield of boreholes of 22000 м^3/day and heat source at the bottom boundary equal to 7880 KW.

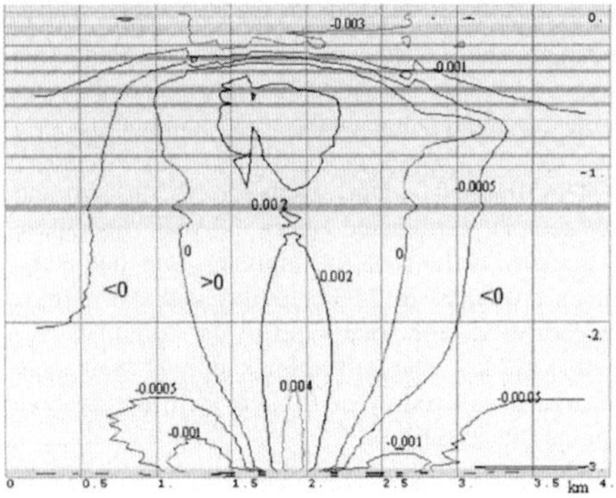

Figure 8. Vertical component of rate (t/m^2/day) in cross section x = 1500 m in 30 years after the beginning of operation..

In this case, temperature of the central part of productive layer does not change significantly. If the filters are arranged in the upper part of productive area, then, as a result of pumping off with total yield of 22000 m^3/day the weighted mean temperature for 30-year operation decreases by **1.88°C**, whereas with filter arrangement at the bottom of productive area weighted mean variation of temperature is close to zero **(0.0015°C)** and it has an opposite sign, i.e., an insignificant heating takes place due to the proximity to the source of heat. In total yield of 11000 m^3/day, weighted mean temperature for 30 years of operation decreases by **0.69°C** when filters are arranged in the upper part of productive area.

Weighed mean decrease/increase of temperature is estimated by the formula:

$$\Delta T_{mixed} = \frac{\Sigma Q_i \cdot \Delta T_i}{\Sigma Q_i}, \qquad (5)$$

where Q_i - the i-th borehole yield, ΔT_i - decrease/increase of temperature average for the length of filter of the i-th borehole after 30 years of operation.

CONCLUSIONS

Water exchange of the hydrothermal system can be determined totally by the deep heat source concentrated therein. In the natural water circulation, water exchange takes place characterized by the immersion of cold water with simultaneous heating thereof and hot water discharge into a well-penetrable sub-fluvial alluvial layer.

The temperature field obtained as a result of solving the equation systems in three-dimensional geometry satisfactorily simulates the temperature field of Paratoon field existing under natural conditions.

In the operation of thermal water field using boreholes with pumping, the discharge of thermal waters into the overlapping highly penetrable level (alluvium of the Paratoonka river) is discontinued, and a slow process of changing the temperature conditions begins. For the field operation during a 30-year period with total yield of 22000 m^3/day the weighted mean temperature decreases by **1.88°C**.

In case of total yield of 11000 м^3/day, the weighted mean temperature decrease obtained by calculation will amount to **0.69 °C** with filters located at the top of productive layer.

In case of the borehole filters arrangement in the upper part of productive area, the quantity of heat removed by the uptake boreholes amounts to 70300 KW in total yield of boreholes of 22000 м^3/day and heat source at the bottom boundary equal to 7880 KW.

REFERENCES

Kirillov P., Yuriev Yu., Bobkov V. 1984. *Reference book on thermal - hydraulics calculations (nuclear reactors, heat exchangers, steam generators).* M., Energoatomizdat Publishers, in Russian, p.296.

Kiryukhin A., 1996. *Modelling Studies: Dachny GeotermalReservoir, Kamchatka, Russia.* Geothermics, v.26, number 1 p. 63-90.

Krashin I., Semendyaeva L., Zinin A., Zinina G. 2002. *Numerical study of filtration and heat transfer in the porous-fracturing hydrothermal system of Paratoon field of thermal water.* In: Proceedings of Conference "Modern problems of hydrogeology and hydroecology". St-Petersburg, in Russian, p. 413 –421.

MICROCRACK FORMATION AND FRACTURE CHARACTERISTICS IN GRANITE UNDER SUPERCRITICAL WATER CONDITIONS

Toru Takahashi* and Toshiyuki Hashida*
*Fracture Research Institute, Tohoku University
01 Aramaki-Aza-Aoba, Aoba-ku, Sendai, 980-8579 Japan

Abstract : In this study, the triaxial compression experiments were conducted to investigate the temperature and pore pressure effects on fracture strength of a granite. The experiments were conducted at temperatures and pressures up to 600°C and 150MPa both under water (supercritical) and air-dry environment. Experimental results show that the strength of the granite under air-dry conditions is temperature insensitive within the temperature range of this study. The strength of the granite was also nearly constant up to 350°C under water-saturated conditions. In contrast, under temperatures above the critical point of water (374°C and 22MPa or more temperature and pressure), the strength decreases rapidly with increasing temperature. We observed the micro-structural change around fracture surface using 100μm thick thin section of granite samples. Optical observation results around the fracture surface revealed that the crack density gave a higher value in supercritical water condition than in sub-critical water conditions. It was shown that the strength decrease under supercritical water conditions was due to the formation of micro-fractures induced by the effect of supercritical water as well as the stress effect, which may be referred to as supercritical water enhanced stress corrosion cracking. These experimental results suggest that it may be possible to generate a volume-like (microcrack predominant) reservoir in supercritical rock masses utilizing the stress corrosion cracking phenomena under supercritical water conditions, which is more advantageous to extract the heat energy compared with a crack-like reservoir.

1. INTRODUCTION

Geothermal energy is one of "green" energy resources that is expected to play a significant role for a sustainable development. In order to increase the amount of geothermal energy available, new extraction systems which are called as HDR (Hot Dry Rock) and HWR (Hot Wet Rock) systems are being investigated (Abé *et al.*, 1999). Recent survey at the Kakkonda geothermal field in Japan has discovered a very hot granite rock just beneath the existing hydrothermal reservoir (Ikeuchi *et al.*, 1996, Kato *et al.*, 1996, Uchida *et al.*, 1996). In the Kakkonda geothermal field, the rock temperature has been shown to reach 500°C at a depth of approximately 3500 m. The temperature condition is not unique to the Kakkonda field. Hot rock masses having a temperature well above 350°C have been also found in Italy (Cappetti *et al.*, 1985), Iceland (Steingrimsson *et al.*, 1990) and other countries. If we could use the heat energy stored in this class of high temperature rock masses, the total amount of geothermal energy is expected to increase significantly.

As demonstrated by the field survey at the Kakkonda geothermal field, the temperature and pressure condition of the deep-seated high temperature rock mass exceeds the critical point of water (374°C and 22MPa or more temperature and pressure). It has also shown that no significant natural fracture network and no significant water is present in the rock mass. The rock mass underlying the existing geothermal reservoir is a neo-granitoid pluton. This observation at the Kakkonda field suggests that it is required to form an artificial water pathway, for example by using the hydraulic stimulation technology, in order to extract the heat energy from a deep-seated high temperature rock mass of volcanic geological setting. This calls for investigation regarding the mechanism of reservoir formation during hydraulic stimulations under supercritical water environment. However, the temperature and pressure condition of water in this geothermal field exceeds the critical point of water, so it is important to evaluate the fracture characteristics of rocks under supercritical water environment in order to predict the formation process of deep geothermal reservoirs.

In this paper, triaxial shear deformation tests were performed on a granite under high temperature and pressure water conditions up to 600°C and 150MPa in order to obtain the fundamental knowledge of fracture behaviour under supercritical water environment. The effects of temperature and fluid pressures were examined based on thin section observations of micro-structural change.

2. EXPERIMENTAL PROCEDURE

2.1 Triaxial deformation test

The rock sample used in this study is Iidate Granite, which was taken from Fukushima Prefecture, Japan. Iidate Granite is a coarse-grained rock sample of average grain size of 1.5mm, and is composed of 37.1% quartz, 21.8% K-feldspar, 34.0% plagioclase, 6.3% biotite and 0.6% other minerals.

The triaxial deformation tests were conducted on the granite using cylindrical specimens. The specifications of rock sample were 20mm diameter and 60mm length, and were cut with the vertical axis parallel to the rift plane. The rift plane is a plane of weakness in Iidate Granite. Some samples were initially pre-heated up to 500 °C at a rate of 3 °C per minute, at atmospheric pressure. Pre-heated samples are necessary to investigate the micro-structural effect of thermal expansion mismatching of different constituent minerals in granite. Part of samples are washed in distilled water and saturated at room temperature over 24 hours, and others are dried in a vacuum oven over 24 hours.

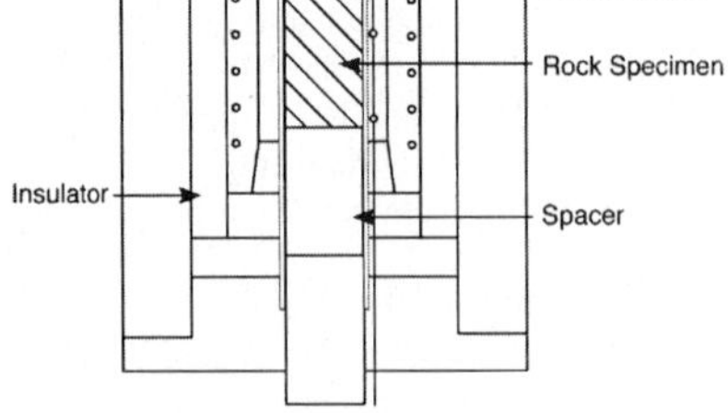

Figure 1. Schematic illustration of the triaxial high temperature and pressure autoclave vessel

The triaxial compression experiments were conducted to investigate the temperature and pore pressure effects on fracture strength of a granite. The experiments were conducted at temperatures and pressures up to 600 °C and 150MPa both under water (supercritical water) and air-dry environment. Schematic illustration of the triaxial deformation test apparatus is given in Fig.1. The maximum temperature and confining pressure achievable with this triaxial apparatus are 600°C and 150 MPa, respectively. The cylindrical specimen and silicon nitride (Si_3N_4) spacer were inserted into a 0.3mm thick silver jacket to form specimen assembly, as shown in Fig.1. The specimen assembly was then placed between the upper and lower piston. Sealing tapes were attached to the end region of the silver jacket in order to prevent the fluid flow between the jacket and pistons during the triaxial deformation tests. After applying confining pressure onto the cylindrical specimen, the temperature was elevated at a constant rate of 5 °C per minute up to a predetermined value. The heating of the specimen was undertaken using an internal heater. The temperature control was conducted using a thermocouple placed in the middle height of the specimen, and the temperature distribution was monitored at the upper and lower positions. The temperature difference was less than 3% with respective to the middle position through the specimen height. After the temperature reached the predetermined level, the samples were deformed at a constant deformation rate of 0.02mm per second.

2.2 Shear stress and normal stress calculation

The axial load and axial displacement are logged in PC during triaxial deformation test. Using these logged data, it is possible to derive the shear stress and normal stress acting on the slip surface by using the following equations,

$$\tau = \frac{1}{2}(\sigma_1 - \sigma_3) sin2\theta \qquad (1)$$

$$\sigma_n = \sigma_3 + \frac{(\sigma_1 - \sigma_3)}{2}(1 - cos2\theta) \qquad (2)$$

where τ is the shear stress, σ_n is the normal stress, σ_1 is the axial stress, σ_3 is the confining pressure and θ is the fracture angle. Effective normal stress,

σ^{eff}_n, can be calculated by subtracting pore pressure from the normal stress using following equation,

$$\sigma_n^{eff} = \sigma_n - P_p \qquad (3)$$

where P_p is the pore pressure.

3. RESULTS AND DISCUSSION

Figure 2 shows the relationship between axial displacement and differential stress. Temperatures were in the range of 25 to 600°C and confining pressure and pore pressure were 100MPa and 25MPa, respectively. The tests were performed at a constant ram displacement rate of 0.02 mm/min. It is seen that nonlinear deformation and softening behavior were observed and peak strength was reducing with increasing temperature. The fractured specimen was shown in Fig. 3. Thus, the brittle shear fracture was occurred in any case of temperature conditions. Using the fracture angle

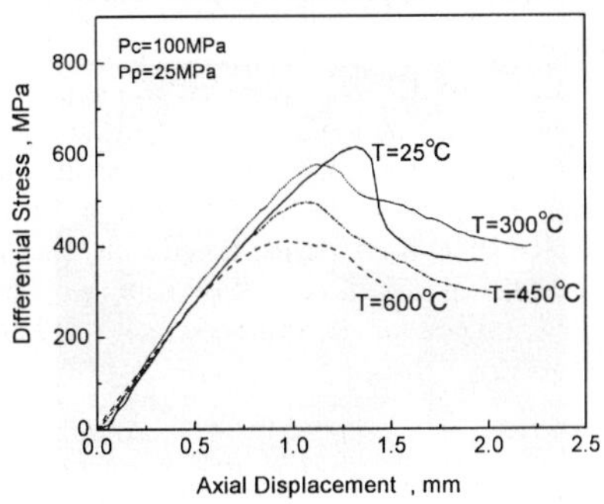

Figure 2. Stress-Strain curves obtained by triaxial compression test. Confining pressure is 100MPa and pore pressure is 25MPa.

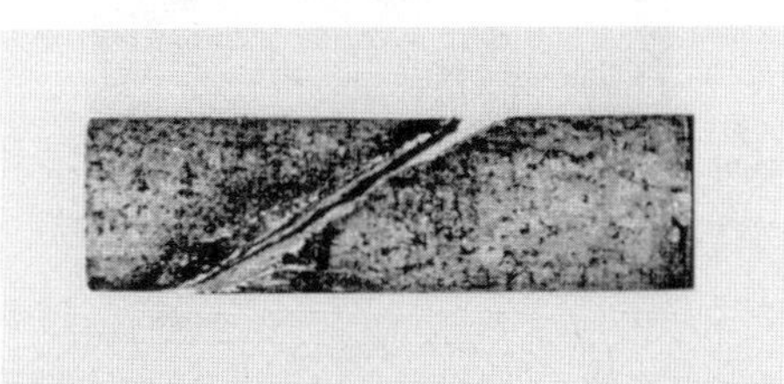

Figure 3. Brittle shear faulting in a compressive test of Iidate Granite (Temperature : 500°C, Confining Pressure : 100MPa, Pore Pressure : 25MPa).

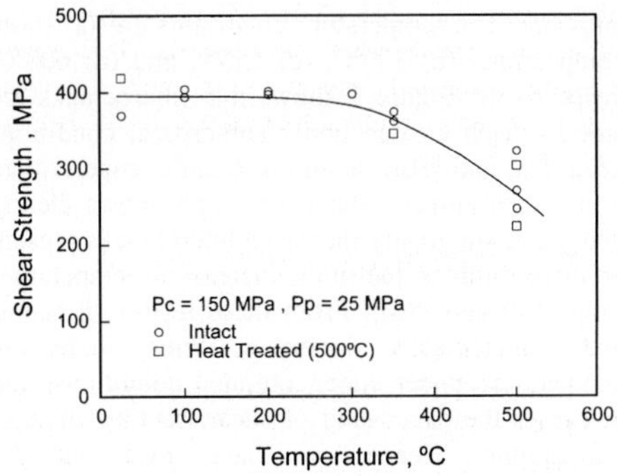

Figure 4. Temperature dependence of shear fracture strength obtained by triaxial compression test.

measured from the deformed specimen shown as Fig. 3, the shear stress and normal stress acting on the fracture surface is calculated by means of equation (1) – (3).

The temperature dependence of the shear strength is shown in Fig. 4 at various temperature conditions. Confining pressure and pore Pressure was constant at 150MPa and 25MPa, respectively. . Under sub-critical water conditions (<300°C), the shear strength does not change so much if temperature is rising. On the other hand, under supercritical environment (>350°C), the shear strength decreases rapidly with increasing temperature.

Figure 5 and 6 shows the relationship between effective normal stress and peak shear stress acting on the fracture surface. Figure 5 shows the all "dry" data, and Fig. 6 shows the "water-saturated" data at various temperature conditions. Figure 5 shows linear relationship between effective normal stress and peak shear stress. In contrast, under high temperature and water saturated (wet) conditions, the shear strength shows much lower trend than under dry conditions (Fig. 6). This result shows that the existence of water has affected the strength of rock and the effect has appeared more notably under supercritical water environment. In order to examine the factor to which the shear strength decrease, microscopic observation using the polarization microscope was conducted. Figure 7 shows the polarized micro-photograph of quartz grains near the shear fracture surface, which have undergone shear fracture test at conditions of 100MPa confining pressure and 25MPa pore pressure. The temperature conditions are (a) room temperature, (b) 300ºC, (c) 450ºC and (d) 600ºC, respectively. Figure 7 shows that microcracks are not so much created under sub-critical conditions (Fig. 7(a) and 7(b)). In contrast, under supercritical water environment, the photograph shows clearly that there are greatly increased microcracks present in these samples due to the increase in temperature (Fig. 7(c) and 7(d)). From these results, it seems that microcracks increases with increasing temperature under water-saturated conditions, and it causes the decreasing of shear fracture strength of granite. Hence, we have measured the microcrack densities of quartz grain in order to

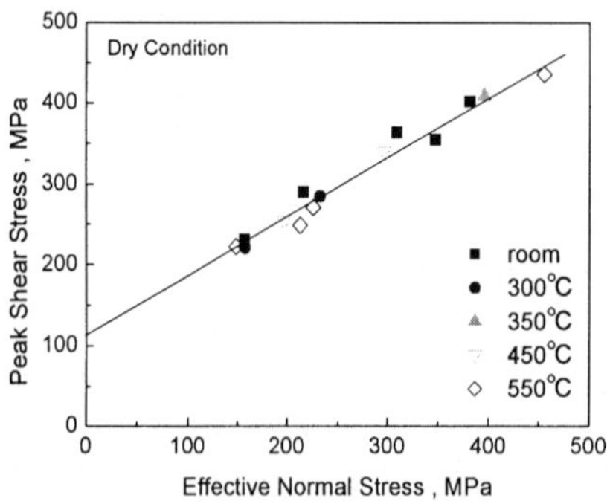

Figure 5. *Relation between peak shear stress and effective normal stress under "dry" condition.*

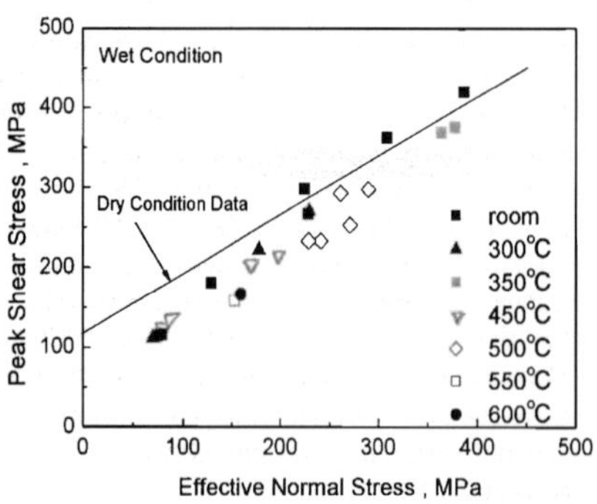

Figure 6. *Relation between peak shear stress and effective normal stress under "wet" condition*

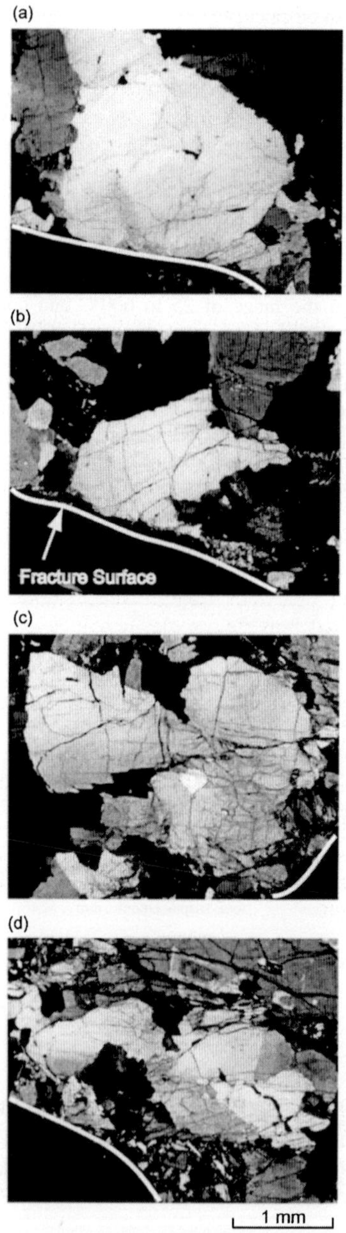

Figure 7. *Polarized micro-photograph of a quartz grain which has undergone triaxial compression test at condition of 100MPa confining pressure and (a) room temperature, (b) 300°C, (c) 450°C and (d) 600°C.*

determine the effect of water and temperature dependence. Microcrack densities are measured using thin section photograph of quartz grain. The quartz grain used for measurement is taken from both near and far from the fracture surface. The trends of microcrack densities in quartz grain are shown in Fig. 8. In Fig. 8, "Fractured Zone" data shows the density near the shear fracture surface, and the density far from the fracture surface is shown as "Far Field" data. Figure 8 shows that there is no difference between "wet" and "dry" condition in far field. The microcrack density is slightly increases with increasing temperature. In contrast, crack density in fractured zone is increases up to 300°C in a dry environment, but it doesn't increase so much even if the temperature reaches at 600°C. On the other hand, in a wet environment, microcrack density is increasing rapidly with increasing temperature up to 600°C. Optical observation results around the fracture surface revealed that the crack density gave a higher value in supercritical water condition than in sub-critical water conditions. It was shown that the strength decrease under supercritical water conditions was due to the formation of micro-fractures induced by the effect of supercritical water as well as the stress effect, which may be referred to as supercritical water enhanced stress corrosion cracking. These experimental results suggest that it may be possible to generate a volume-like (microcrack predominant) reservoir in supercritical rock masses utilizing the stress corrosion cracking phenomena under supercritical water conditions, which is more advantageous to extract the heat energy compared with a crack-like reservoir.

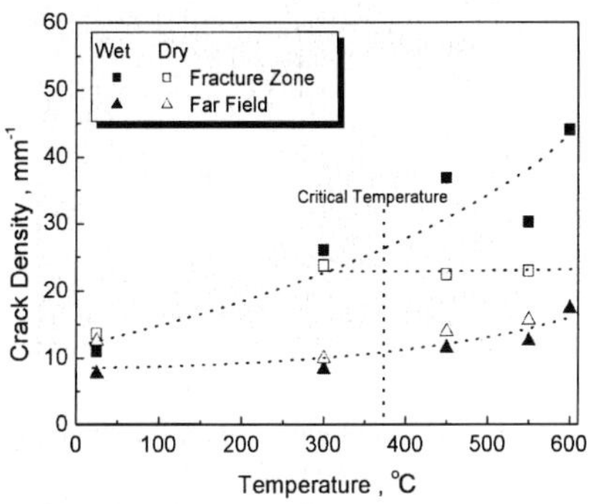

Figure 8. Temperature dependence of microcrack density in quartz grain.

4. CONCLUSIONS

Triaxial shear deformation tests were performed on Iidate Granite under supercritical water environment, with the objective of investigating the reservoir formation mechanism in supercritical rock masses. The testing temperature was up to 600°C and the confining pressures up to 150MPa. Triaxial compression test results show that, in wet environment, the shear strength of Iidate Granite reduces with increasing temperature. The decrease tendency is remarkable especially under supercritical water environment. From the micro-structural observation results, the strength decrease is caused by the formation of microcracks on fracture surface, which is induced by the effect of supercritical water enhances stress corrosion cracking. From these results, It can be said that it may be possible to create artificial pathway and generate a microcrack predominant reservoir in supercritical rock masses.

ACKNOWLEDGEMENT

The work in this report is supported through the Grant-in-Aid for Scientific Research (A) (No.14205149), and the Japan Society for the Promotion of Science under Grant-in-Aid for Research for the Future Program (JSPS-RFTF 97P00901).

REFERENCES

Abé, H., H. Niitsuma and R. Baria (ed.), 1999, Special Issue on Hot Dry Rock/Hot Wet Rock academic review, Geothermics, 28.

Cappetti, G., C. Romano, U. Cigni, P. Squarci, G. Stefani and L. Taffi, 1985, *Development of deep exploration in the geothermal area of Tuscany, Italy*, Geothermal Resources Council Int. Symp. on Geothermal Energy, pp. 303-309.

Ikeuchi, K., R. Komatsu, N. Doi, Y. Sakagawa, M. Sasaki, H. Kamenosono and T. Uchida, 1996, *Bottom of Hydrothermal Convection Found by Temperature Measurements above 500°C and Fluid Inclusion Study of WD-1 in Kakkonda Geothermal Field, Japan*, Geothermal Resources Council Transactions, Vol. 20, pp. 609-616.

Kato, O., N. Doi, K. Ikeuchi, T. Kondo, H. Kamenosono, M. Yagi and T. Uchida, 1996, *Characteristics of Temperature Curves and Fracture Systems in Quaternary Granite and Tertiary Pyroclastic Rocks of NEDO WD-1a in the Kakkonda Geothermal Field, Japan*, Proc.

8th Int. Symposium on the Observation of the Continental Crust through Drilling, pp. 241-246.

Staingrimsson, B., A. Gudmundsson, H. Franzon and E. Gunnlaugsson, 1990, *Evidence of a supercritical fluid at depth in the Nesjavallir field*, Proc. 15th Workshop on Geothermal Reservoir Engineering, Stanford University, California, SGP-TR-130, pp. 81-88.

Uchida, T., M. Yagi, M. Sasaki, H, Kamenosono, N. Doi and S. Miyazaki, 1996, *Investigation of Deep-Seated Geothermal Reservoir by NEDO's 4000m Well in Kakkonda, Japan*, Proc. 8th Int. Symposium on the Observation of the Continental Crust through Drilling, pp. 73-78.

ESTIMATION OF CRITICAL PORE PRESSURE FOR SHEAR SLIP OF FRACTURES AT THE SOULTZ HOT DRY ROCK GEOTHERMAL RESERVOIR USING MICROSEISMIC MULTIPLETS

H. Moriya[1], H. Niitsuma[1] and R. Baria[2]

[1]) Graduate School of Environmental Studies, Tohoku University, Japan
[2]) GEIE - Exploitation Minière de la Chaleur, France

Abstract: Pore-pressure of seismically activated fractures during hydraulic stimulation of geothermal reservoir at Soultz HDR field, France is estimated based on the high-resolution mapping method of induced microseismic events. The microseismic events with similar waveforms are analyzed, and the seismically activated fracture planes are estimated using the precise source locations determined by the multiplet-clustering analysis. The tectonic stress measured by hydraulic fracturing method in borehole is introduced into calculation of critical pore-pressure for frictional shear slip of fractures according to Coulomb's law. The estimated critical pore-pressure suggests that the fluid pressure in pre-existing fractures are higher value near the permeable fracture zones in the well, and declines as the fractures are away from the zones.

1. INTRODUCTION

Hydraulic fracturing is often conducted at geothermal field in order to enhance the reservoir for extracting thermal energy. The monitoring and understandings of fluid flow in the reservoir is indispensable for effective stimulation of pre-existing fractures (Pine and Batchelor 1984). The locations of seismically and maybe hydraulically activated fractures can be estimated using the induced seismic events associated with shear slip of pre-existing fractures.

The precise mapping method of induced microseismic events has succeeded in revealing the structures, which may indicate fluid flow zone in the reservoir (e.g. Fréchet,, 1989; Moriya et al., 1994; Niitsuma et al., 1999). The multiplet-clustering analysis is one of the high-resolution mapping techniques. The similar microseismic events are analyzed to determine precise source locations, and the seismic clusters can be derived from the relative source locations of multiplets, where the cross-spectrum analysis and the idea of the clustering analysis are introduced (Moriya et al., 2003). The multiplet-clustering analysis can estimate fine scale structures and larger structures interconnected with the fractures derived from each multiplet. In addition, the fracture orientations can be also estimated using the source locations of each multiplet, since the similar microseismic events within a multiplet are considered to be radiated on the same fracture plane.

In this paper, we suggest a method for estimation of pore-pressure during hydraulic stimulation of geothermal reservoir by using the induced microseismic multiplets. The results of the application to the microseismic events at Soultz field are demonstrated showing the feasibility of the method.

2. SOULTZ HDR FIELD

The European HDR project site at Soultz-sous-Forêts is located on a local horst structure in the Rhine Graben, where Hercynian age granites are covered by a 1.4 km thick sedimentary section (Baria et al., 1999). Well GPK-1 was drilled to 3,590 m depth (open hole below 2,850 m depth), and since 1987 it has been used for a number of

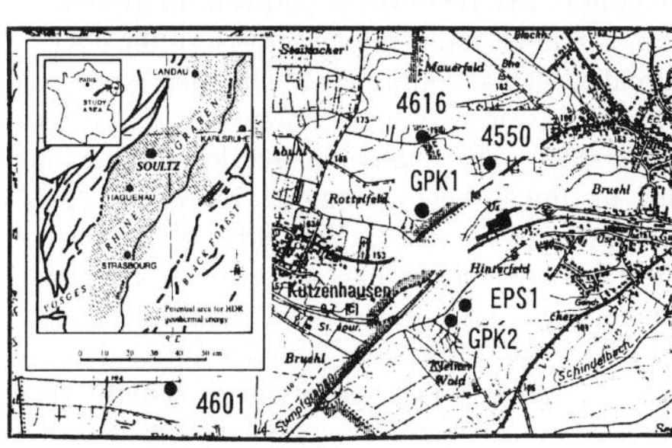

Figure 1. Deployment of wells in Soultz HDR field, France.

detailed experiments (Figure 1).

A major hydraulic fracturing experiment was performed in September and October 1993. In the September test, 25,000 m^3 of fresh water were injected between 2,850-3,350 m at progressively higher rates to 40 l/s and pressures of up to 10 MPa over a 17 day period. Through the test, it was demonstrated that the fracture network in the basement rock was well developed, with enhanced permeability and a substantial increase in transmissivity (Baria et al., 1999; Evans, 2000). The three 4-component seismic detectors and hydrophone were set at depths of 1,500 m, 1,420 m, 1,600 m and 2,850 m, respectively. Using these detectors, source locations for more than 10,182 events have been determined. Figures 2 shows source locations for all of the analysed events determined by JHD, where the map view and cross section looking toward N65°E are shown. As shown in this figure, the seismic cloud of shallow has approximate dimensions of 0.5 by 1.2 by 1.5 km, striking N30°W and dipping nearly vertically (Phillips, 2000).

3. ESTIMATION OF FRACTURE PLANES BASED ON MULTIPLET-CLUSTERING ANALYSIS

The multiplet-clustering analysis is a method to estimate relative positions of the multiplet clusters as well as the relative source locations within individual multiplets (Moriya et al., 2003). We have applied the multiplet-clustering analysis to the induced microseismic events as following procedures.

As the first step, the similar microseismic events were identified using the coherency function, and the similar events were automatically classified into groups. In case of the experiment in September 1993, a total of 5,490 events (58.3% of the events) were identified as multiplets (containing more than 3 similar waveforms), with coherency ranging from 0.80 to 0.99. The relative source locations within each multiplet were estimated by using cross-spectrum analysis, where the relative differences of P- and S-wave arrival times were used to determine relative locations among similar events within multiplet. Source locations, relative to a reference source location, were determined within each multiplet by assuming that the velocity structure was homogeneous along the wave propagation path. The RMS errors for the relocation were about 0.1 ms. In order to apply the multiplet-clustering

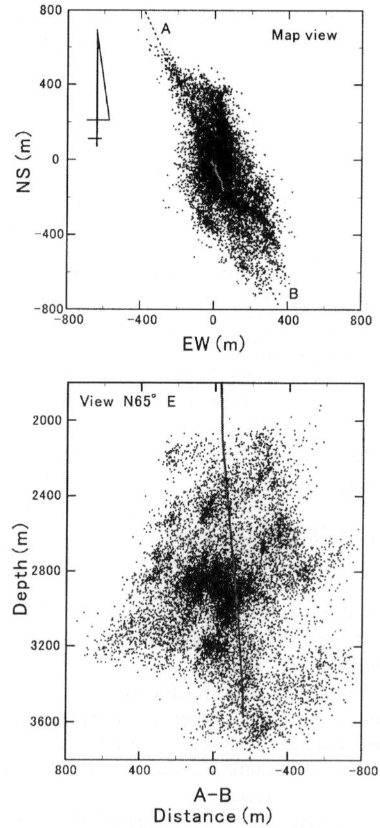

Figure 2. Source locations of induced microseismic events, where the events determined by JHD method. The cross section is the view toward N65 °E. Thick line denotes fracturing well GPK1.

analysis, we selected a total of 142 groups of multiplets. As a pre-processing, we applied a low pass filter with a cut-off frequency of 150 Hz to the raw signals. After checking the similarity of the waveforms, we made stacks of aligned waveforms for each multiplet, by a method similar to that of Rowe et al. (2002). Arrival time differences for the stacks were determined relative to the master event. Using the time differences the relative locations among the multiplet clusters were estimated, assuming a homogeneous velocity structure having P- and S-wave velocities of 5,850 m/s and 3,340 m/s, respectively (Moriya et al., 2003).

Figures 3(a) and 3(b) show source locations of multiplet events before and after multiplet-clustering analysis (Moriya et al., 2003). Following multiplet-clustering analysis, small planar clusters corresponding to each multiplet are identified, within 3 larger seismic clouds denoted by "A", "B" and "C." Two separate clouds, corresponding to "A" and "B," are identified at a depth interval of 2,750 to 3,000 m, as shown in Figure 3(b). Closed box denotes permeable fracture zones detected by well loggings. The 2 clouds are considered to be permeable fracture zones that intersect the borehole, which is consistent with the well-logging results. Because a multiplet would be induced on the same fracture plane, we have estimated the fracture planes as an approximated plane using the distribution of the source locations within the multiplet. A multiplet plane is defined as the plane best fitting the distribution of the source locations within the multiplet distribution (Moriya et al., 2002). Figure 4 shows the stereographic projection of poles of the estimated fracture planes.

4. CALCULATION OF CRITICAL PORE-PRESSURE FOR SHEAR SLIP OF FRACTURE

We estimate distribution of critical pore-pressure of fractures for shear slip during hydraulic stimulation. We introduce the fracture planes derived from the source locations of multiplets. The procedures for calculation of the critical pore-pressure are described as follows.

The stress directions and magnitudes had been measured by hydraulic fracturing method in borehole. The stress magnitudes are described as following equations (Klee and Rummel, 1993).

$$
\begin{aligned}
S_h &= -5.9 + 0.0149Z \\
S_H &= -25.3 + 0.0336Z \\
S_V &= -1.3 + 0.0255Z \\
P_h &= 0.9 + 0.0098Z, \qquad (1)
\end{aligned}
$$

where S_h, S_H, S_V and P_h denote minimum horizontal stress, maximum horizontal stress, vertical stress and hydrostatic pressure in MPa, respectively. Z denotes the depth in meters.

According to Coulomb's law of friction, the relationship among shear stress, normal stress, coefficient of friction and pore-pressure at the moment of shear slip (critical condition) due to increasing of pore-pressure are described as follow,

$$\tau - \mu_S(\sigma_N - P_P) = 0, \qquad (2)$$

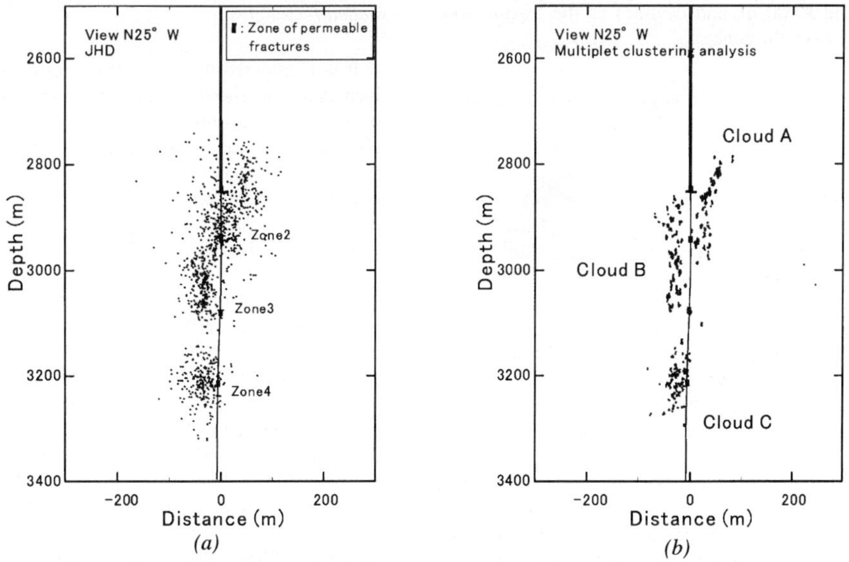

Figure.3. Source locations of multiplets determined by (a) JHD and (b) multiplet-clustering analysis. Closed boxes along well describe the permeable fracture zones detected by well loggings. Thick line denotes fracturing well GPK1. The casing shoe is at the depth of 2,850m.

where τ, σ_N, μ_S and P_P denote shear stress, normal stress, coefficient of friction and pore-pressure on a fracture surface, respectively. The cohesion is ignored in above mentioned formulation. The critical pore-pressure for shear slip P_C can be led as follow,

$$P_C = P_P - P_h$$
$$= \sigma_N - \frac{\tau}{\mu_S} - P_h . \quad (3)$$

P_C means the increase of fluid pressure to induced shear slip. When the orientation of a plane is given, the normal and shear stresses acting on the plane can be calculated. In our case, we have estimates of fracture orientations and their depths from the source distribution of individual multiplets, and the depth of fracture plane can be estimated from the center of gravity of source locations of each multiplet. Figure 5 shows the calculated P_C, where the coefficient of friction is assumed to be 0.8 because it is impossible to directly measure the in-situ friction coefficient. Zones in Figure 5 describe the depth intervals of permeable fracture zones in fracturing well GPK1, where the permeable fractures are confirmed using FMI and temperature loggings (Evans, 2000).

The estimated critical pore-pressure for shear slip suggests that the fluid pressure in fractures is higher value near the fractured zones 1 and 2 around 2,900 m, and declines as the fractures are away from the zones.

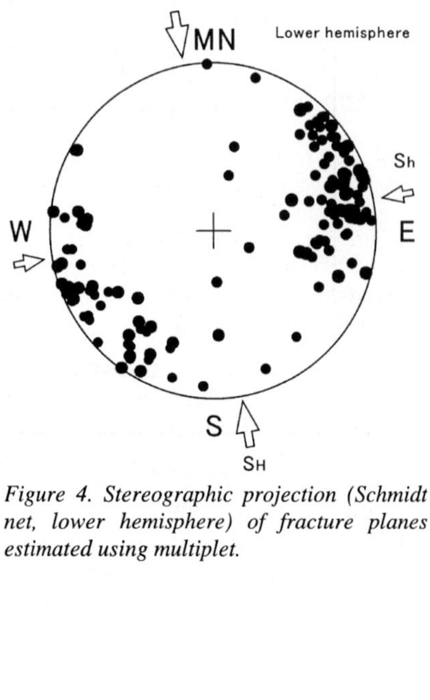

Figure 4. Stereographic projection (Schmidt net, lower hemisphere) of fracture planes estimated using multiplet.

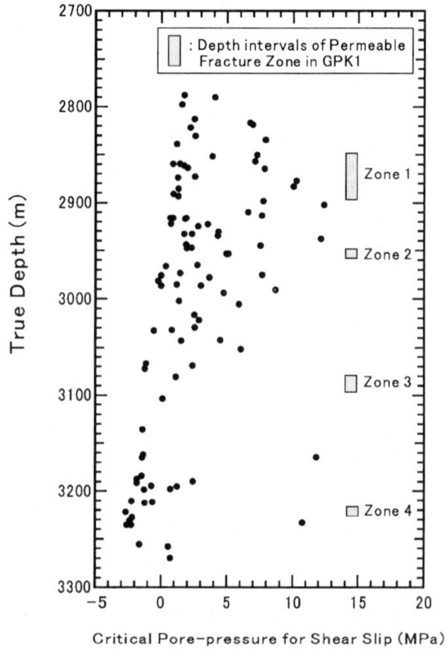

Figure 5. Calculated critical pore-pressure for shear slip of fractures versus depth of estimated fractures. The zone of permeable fracture, which are detected by well loggings are indicated by shaded rectangles.

It is considered that some fractures opened with increasing pore-pressure maybe after shear slip because the maximum wellhead pressure is about 10MPa. However, Mode I fractures can not radiate seismic events having enough energy, and we can not detect these seismic events. Therefore, we consider only shearing fractures for estimation of critical pore-pressure using induced microseismic events.

The result in Figure 5 implies that the fluid pressure was transmitted into fractures, and that the pore-pressure around zones 1 and 2 increased up to near maximum fluid pressure in fracturing well. It is reported that the fractures in zones 1 and 2 are permeable under low flow rate (Evans, 2000). Then, the distribution of critical pore-pressure in Figure 5 is reasonable.

On the other hand, the P_C becomes below zero around 3,200m. The same result can also obtained when the critical pore-pressure is calculated using the orientation of fractures detected by well logging. Then, the reason for lower value of P_C would be that the stress magnitudes are estimated lower than actual value.

5. CONCLUSION

Through observation of induced microseismic events and examination of well logging data, it has been understood that the permeable zone seems to be created by migration of shearing joints during hydraulic fracturing test, and that microseismic multiplets is considered to be seismic events associated with shearing of fractures which are favourably oriented to stress field.

Then, we have examined a method to estimate critical pore-pressure for shear slip of fractures during hydraulic stimulation of geothermal reservoir, and we have introduced the multiplet-clustering analysis, which is a precise mapping method of induced microseismic events. The multiplet-clustering analysis has revealed zones created by shearing fractures. The orientation of fractures has been estimated using source locations of multiplets, and critical pore-pressure has been calculated for each fracture plane, where the stress field estimated by hydraulic fracturing test is also used as well as the estimated fracture orientation. The distribution of estimated pore-pressure has suggested that fluid pressure in fractures is larger around hydraulically alternated fracture zones, and declines as the fractures are distant from the permeable zones. We would be able to say that the estimation method based on induced microseismic multiplets is effective to evaluate fluid pressure of subsurface fractures away from wells.

ACKNOWLEDGEMENT

We thank Dr. K. F. Evans (ETH, Switzerland) for detailed technical discussions and helpful comments. This work was carried out as a part of the MTC International Collaborative Project, and by a grant from the Industrial Technology Research Grant Program in 2000, supported by NEDO. We would also like to thank SOCOMINE (now GEIE) for providing the data from the European HDR site at Soultz-sous-Forêts, which is supported mainly by the European Commission, BMBF (Germany), and ADEME (France).

REFERENCES

Baria, R., Baumgärdner, J., & Gérard, A. 1999. *European HDR research programme at Soultz-sous-Forets (France) 1987-1996*. Geothermics, 28: pp.655-669.

Evans, K. F. 2000. T*he effect of the 1993 stimulations of well GPK1 at Soultz on the surrounding rock mass: Evidence for the existence of a connected network of permeable fractures*. Proc. World Geothermal Congress 2000, 6: pp.3695-3670.

Fréchet, J., Martel, L., Nikolla, L. & Poupinet, G. 1989. *Application of the cross-spectral moving-window technique (CSMWT) to the seismic monitoring of forced fluid migration in a rock mass*. Int. J. Rock Mech. Min. Sci. Geomech. Abstr., 26: pp.221-233.

Klee, G. & Rummel. F. 1993. *Hydraulic fracturing stress measurements in the geothermal borehole GPK1, Soultz-sous-Forets*. Report no. 04-93, MeSy, Bochum.

Moriya, H., Nagano, K. & Niitsuma, H. 1994. *Precise source location of AE doublets by spectral matrix analysis of triaxial hodogram*. Geophysics, 59: pp.36-45.

Moriya, H., Nakazato, K., Niitsuma, H. & Baria, R. 2002. *Detailed fracture system of the Soultz-sous-Forêts HDR field evaluated using microseismic multiplet analysis*. Pure Appl. Geophys., 159: pp.517-541.

Moriya, H., Niitsma, H & Baria, R. 2003. Multiplet-clustering Analysis Reveals Structural Details within Seismic Cloud at the Soultz Geothermal Field, France. *Bull. Seismol. Soc. Am.* (to be published)

Niitsuma H., Fehler, M., Jones, R., Wilson, S., Albright, J., Green, A., Baria, R., Hayashi, K., Kaieda, H., Tezuka,K., Jupe, A., Wallroth, T., Cornet, F., Asanuma, H., Moriya, H., Nagano, K., Phillips, W. S., Rutledge, J., House, L., Beauce, A., Alde, D. & Aster, R. 1999. *Current status of seismic and borehole measurements for HDR/HWR development*. Geothermics, 28: pp.475-490.

Phillips, W. S. 2000. *Precise microearthquake locations and fluid flow in the geothermal reservoir at Soultz-sous-Forêts, France*. Bull. Seism. Soc. Am., 90(1). Pp.212-228.

Pine, R. J., Batchelor, A. S. 1984. Downward migration of shearing in jointed rock during hydraulic injections, *Int. J. Rock Mech. Min. Sci.*, 21: 249-268.

Rowe, C. A., Aster, R. C., Phillips, W. S., Jones, R. H., Borchers, B. & Fehler, M. C. 2002. *Using Automated high-precision repicking to improve delineation of microseismic structures at the Soultz geothermal reservoir*. Pure Appl. Geophys, 159: pp.563-596.

Theme 5
Coupled T-H-M-C Processes in Geological Systems

MODELLING OF SEDIMENT COMPACTION DURING BURIAL IN SEDIMENTARY BASINS.

K Bjørlykke[1], F Chuhan[2], A Kjeldstad[3], E Gundersen[1], O Lauvrak[4] and K Høeg[1]
[1]) Department of Geology, University of Oslo
[2]) Esso Explor [3]) Shell Norway, [4])Norsk Hydro

Abstract: The reduction in porosity due to mechanical compaction and chemical processes are principally different processes and must be modelled separately even if they overlap in time. Experimental compaction of loose sand show that stresses exceeding 20-30 MPa cause significant grain crushing particularly in coarse-grained sand. Evidences of such grain fracturing are commonly observed in sandstone reservoirs. Quartz cementation in sandstones is mostly controlled by kinetics and therefore by temperature. Compaction of mudstones must be modelled as a function of effective stress and chemical reactions. Then sediments may compact chemically at low temperatures. The effective stress is then more important than temperature due to higher kinetic reaction rates. The rate of compaction controls water fluxes in the deeper parts of sedimentary basins. Modelling shows that these compaction-driven fluxes are too small to be significant in terms of transport of heat and solids. This represents a major constraint on advective mass transport during burial diagenesis. The physical properties of sediments and their seismic response also change during burial and depend very much on the initial composition.

1. INTRODUCTION

Compaction is the process causing reduced porosity during burial in sedimentary basins. It was suggested already by Athy (1930) that the loss of porosity (φ) with depth (z) could be formulated as an exponential decay process

$$\varphi = \varphi_0 e^{-cz} \qquad (1)$$

where φ_0 is the initial porosity near the surface (sea floor), and c is the rate at which compaction occurs with increasing dept. The problem is that the degree of compaction as a function of depth is a highly variable parameter dependant on the sediment composition, stress and temperature. Modeling compaction processes with the aim of predicting porosity at a certain depth in a sedimentary basin requires that all the processes are well understood. We also need to calibrate our models using data from wells in addition to experimental compaction in the laboratory.

The most important factor controlling sediment compaction is the sediment composition since both the mechanical and the chemical processes are so dependant on the lithology.

Each lithology must be described in terms of its mineralogy and grain texture treated separately. The present paper discusses the principles of mechanical and chemical compaction for different types of sediment.

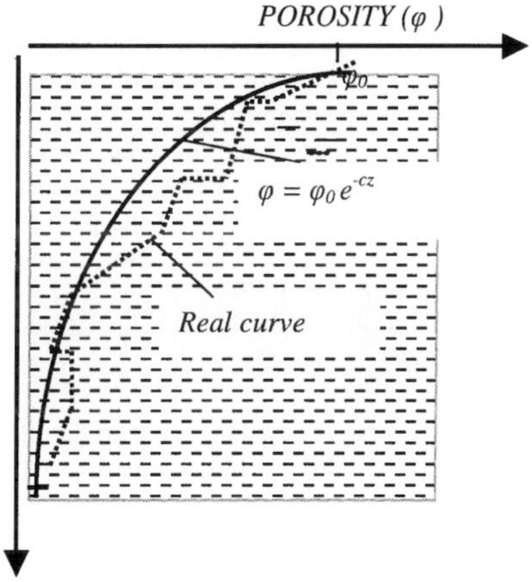

Figure 1.Porosity- depth curve following an exponential function (Athy1930) compared to a more realistic compaction curve reflecting variations in the compressibility of the different layers.

2. MODELLING OF SEDIMENT COMPACTION

Mechanical compaction of sediments includes the rearrangement and breakage of grains which is a function of the overburden stress. Sediment compaction and fluid flow are coupled because the reduction in porosity requires a removal of fluids.

Other processes can not be coupled because they are principally very different in terms of driving forces. A model attempting to include all the processes involved in sediment compaction would be far too complex, and practical predictions should be based on an identification of the quantitatively most important processes.

We must distinguish between uncertainty in modelling due to imperfect understanding of the processes and the uncertainty due to poorly constrained rock parameters. The main problem in basin modelling is in most cases the prediction of the rock composition in a sedimentary sequence. The composition of sediments at the time of deposition is a function of the source rock lithology, weathering and erosion, and subsequent transport and depositional environments. After deposition, sediment composition and textures may be significantly altered by dissolution of minerals due to freshwater (groundwater) flow. Sediments near the sea floor may also become cemented sourced by dissolving marine organisms or by chemical reactions between the pore water and the sea water. This is most commonly observed as carbonate and silica cements.

Mechanical compaction must be modelled as function of the vertical effective stresses (σ_v'), which is the difference between the pore pressure (u) and the total overburden stress (σ_v). (e.g. Lambe and Whitman 1979).

$$\sigma'_v = \sigma_v - u \qquad (2)$$

The ratio between an increment in the effective stress ($\Delta\sigma_v'$) and the resulting compressive strain is the compression modulus which varies with sediment type and effective stress level (i.e. overburden depth. In soil mechanics practice the one-dimensional compression modules (M) for sediments may be expressed as (Janbu,1963):

$$M = mp(\sigma'_v / p_a)^{1-a} \qquad (3)$$

Here m is a non dimensional modulus number, a = exponent; $0 < a < 1$, and p_a = reference stress at 100 kPa. This value of M is the value to be used for the next stress increment, σ'_v (i.e. a tangent modulus). It has been found that Eq. 3 is able to model one-dimensional compression behaviour of most sediments (normally and overconsolidated) by specifying appropriate values for the parameters m and a. For a = 1, M is independent of effective stress level and for a = 0, M is proportional to σ'_v.

For a sediment in the overconsolidated stress range, a = 0 may be appropriate while in the normally consolidated range the a-value is closer to 1. However Eq. 3 does not reflect the behaviour at high effective stresses when crushing of sand particles occur. This is discussed in the section below. From the above discussion we may conclude that the simplified relationship often used in basin modelling does not reflect realistic conditions in a sedimentary basin.

3. COMPACTION OF QUARTZ-RICH SAND

The mechanical compressibility of sand can be determined experimentally. The amount of compaction that can occur in sand by sliding and grain rotation is limited. Even at moderate overburden (2-4 MPa or 2-500 m depth) coarse-grained sand undergoes grain breakage (Fig 2) which is similar to that observed in naturally compacted reservoir sandstones (Fig 3). The brittle behaviour is probably not very time dependant and may be determined in laboratory experiments. In the case of hard grains like quartz, grain crushing is more important than in mode softer and more ductile grains. Sand grains coated with clay are less likely to fracture and therefore remain less compressible than quartz sand (Chuhan et al. 2001).

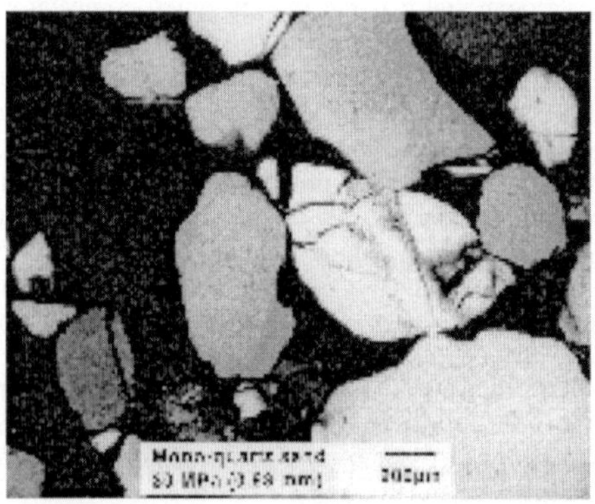

Fig 2. Grain crushing produced as a result of experimental compaction at 30 MPa corresponding to an overburden of 3-4 km at hydrostatic pore pressures.

Fig. 3 Cathodoluminescence picture of natural fractures in quartz grains. The fractures are healed with quartz cement which appears darker. From the Lavrans Field at 4800 m depth.

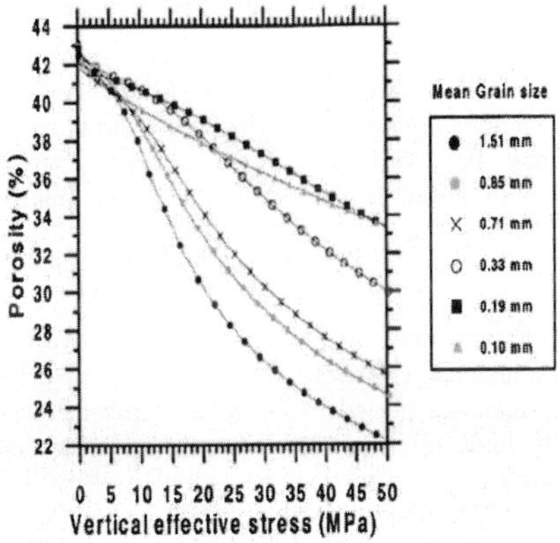

Fig. 4 Results from experimental compaction of sand from Chuhan et al. 2002. Note that fine grained well-sorted sand is less compressible than coarse-grained sand.

Experimental compaction shows that the loss of porosity as a function of increasing stress is a function of both grain composition and grain size (Pittman and Larese 1991, Chuhan et al. 2002) (Fig.4). The porosities observed in shallow reservoirs in the North Sea at about 2 km depth are about 30-35 % in the best sorted medium-grained sands (Giles et al. 1992, Ramm and Bjørlykke 1994) which correspond closely to those we measured during experimental compaction of similar sand in the laboratory (Chuhan et al. 2002).

Modelling the mechanical compaction of sand as a function of increasing overburden stress requires a set of material parameters defining experimental data the compressibility for sand with different mineralogical and textural composition. Chemical compaction of sediments involves porosity loss by dissolution and precipitation of minerals (Houseknecht 1987) and is principally different from mechanical compaction. These chemical processes must be modelled as a function of the thermodynamic stability of the minerals and the kinetic reaction rate. The driving force for compaction by dissolution of grains and precipitation of cements is the increased solubility at grain contacts due to stress and, in the case of soluble minerals, this process is a function of the effective stress and clay minerals at grain contacts (Renard et al 2001; Gundersen et al 2002 a,b). In the case of clay minerals and silicate minerals like quartz the kinetic reactions are so slow that chemical compaction can be ignored at temperatures below about 80°C, which in many sedimentary basins will correspond to about 2.0-2.5 km overburden and 20-25 MPa vertical effective stress.

When the rate of quartz cementation becomes significant at temperatures exceeding 80-100°C, the porosity loss is mainly due to quartz cementation. Calculations show that the advective transport of silica in the pore water can be ignored as the low flow rates due to compaction are so low and because of the low solubility and solubility gradients of silica. The rate of import or export of silica is a function of the water flux (F), the solubility of quartz as a function of temperature (α_T) and the geothermal gradient (∇T)

$$Vc = Ft \nabla T \alpha_T / \rho \quad (4)$$

Here F is the water flux ($cm^3/cm^2/s$) perpendicular to the isotherms, α_T is $1\text{-}3 \cdot 10^{-6}$ (Wood 1986), the geothermal gradient, ∇T = 30°C/km and ρ is the density of quartz (2.65 g/cm^3). From this we can calculate what it takes to precipitate 1% quartz cement (Vc=0.01). The fluid flux integrated over time must be $3 \cdot 10^9$ (cm^3/cm^2) (Bjørlykke 1994). This corresponds to a 30 km long column of water. If the porosity is 10%, 300 km of sediments must be compacted to nearly zero porosity to provide this amount of water. In most of the basin the pore water will be subsiding relative to the surface which implies that it will in most cases be heated and therefore dissolve minute amounts of quartz rather than precipitate quartz (Bjørlykke 1993). Advective transport of silica can therefore be ignored in sedimentary basins without hydrothermal activity. Silica transport by diffusion may be significant over shorter distances (<1m) but silica concentrations in the pore water are normally very close to quartz saturation and the concentration gradients in the pore water required to drive diffusion become very small. The increased solubility at grain contact provide, however, higher gradients on a very small scale (< 10 cm) (Walderhaug & Bjørkum 2003).

When the diagenesis is isochemical at a scale of 1-10m, the compaction and shortening (ΔZ) of a sedimentary sequence (Z) is equal to the porosity loss ($\Delta\ \varphi$) due to quartz cementation. The reduction in stratigraphic thickness is defined by:

$$\Delta Z/Z = (\varphi - \Delta\varphi)/\varphi \qquad (5)$$

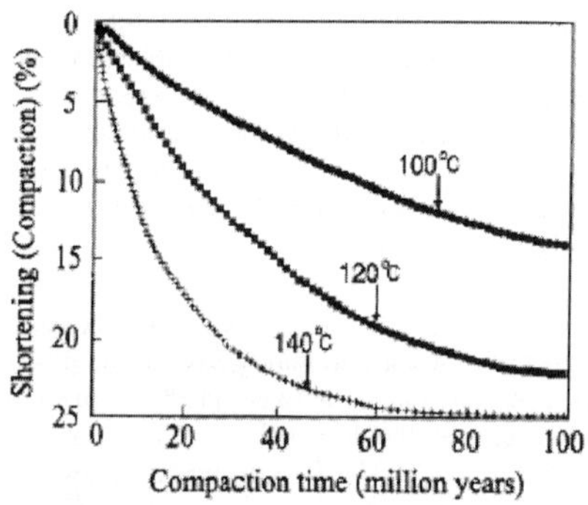

Fig. 5. Modelling of sandstone compaction as a function of time at constant temperatures. From Walderhaug et al. (2000)

Recently, significant progress has been made in the quantification of chemical compaction of sandstones (Giles 1997, Lander and Walderhaug 1997, Walderhaug 1996, Walderhaug et al. 2000, Walderhaug et al. 2001, Oelkers et al. 1996. Their models assume that temperature is the main rate limiting factor for quartz cementation and the resultant chemical compaction of sand sones Fig 5

Dissolution at grain contacts and precipitation of cement requires increased solubility caused by stress at grain contacts. This has been referred to as pressure solution. The stress at grain contacts is driving the process and quartz cementation can be modelled as a coupled process of solution and transport (Gundersen et al 2002 a,b , Renard et al 2001). In the case of quartz there is considerable evidence to suggest that the stress is not rate-limiting for this process at low temperatures (80-150°C) (Bjørkum 1996), as the rate of precipitation of quartz cement is mainly a function of temperature and the surfaces available for precipitation (Bjørkum and Nadeau 1998). The rate of quartz cementation (r) can then be expressed as:

$$r = a\ 10^{b(ct+d)} \qquad (6)$$

Here a is a constant related to the grain surface area available for quartz cementation, c is the heating rate as a function of time (t), and d is the initial temperature. b is a rate constant which needs to be calibrated. From this equation the volume of quartz cement precipitated can be calculated by integrating temperature over time during burial. Thus the rate of quartz cementation as a function of temperature controls compaction and the reduction in sediment thickness (Fig 5).Mechanically fractured grains similar to those produced experimentally are common in coarse- to medium-grained sandstone reservoirs (Chuhan 2002). The fractures are normally healed by quartz cement and may be overlooked in a normal petrographic examination. In cathodoluminescence microscopy healed fractures filled with quartz cement precipitated at low temperatures stand out from the detrital quartz (Chuhan et al. 2002). Healed fractures may often contain trapped hydrocarbon inclusions (Karlsen et al. 1993). Well-sorted sandstones buried to 2-2.8 km depth in the North Sea have relatively small amounts of quartz cement and the porosities may be as high as 35% (Giles et al. 1992, Bjørlykke et al. 1992 and 1994, Ramm and Bjørlykke 1994, Ramm 1992).

4. COMPACTION OF MUD AND MUDSTONES

Compaction of mudstones is, as for sand, mostly a function of effective stress at low temperatures (<80°C) (Aplin et al. 1999, Bjørlykke 1997). Carbonate cementation sourced by biogenic carbonate may, however, occur at rather shallow depth so that carbonate concretions or layers can be rendered almost incompressible at shallow depth. The precipitation of quartz cement in mudstones follows much the same principles as in sandstones and is primarily controlled by the temperature. Dissolution of biogenic silica has a strong effect as a source of silica on quartz cementation at temperatures exceeding about 80°C. When modelling the compaction of mudstones the carbonate content is critical. In mudstones with relatively low carbonate content the mechanical compaction continues to greater depth until the dissolution and precipitation of silicate minerals become significant, and this is mainly controlled by temperature. In the Northern North Sea there is a marked contrast between the compressibility of Pliocene and Pleistocene sediments and the underlying Eocene and Oligocene sediments resulting in a pronounced velocity inversion with depth (Fig. 6). Smectite becomes unstable between about 80°C and 100 °C and dissolves, being replaced by mixed-layered minerals and illite. Kaolinite becomes unstable at temperatures around 130°C, and when relatively coarse clastic kaolinite grains dissolve and fine-grained illite precipitates, a temperature-controlled compaction occurs. This reaction involves the dissolution of K-feldspar grains, which may cause further compaction of the grain framework. Massive quartz cementation will start at about 80 °C in mudstones with abundant biogenic silica, producing a siliceous mudstone with extremely low compressibility. In the absence of biogenic silica the precipitation of quartz cement filling the pores of the mudstones must be derived from silica released from clay mineral reactions such as smectite-illite transformation and kaolinite-illite reaction. In addition detrital quartz will, as in sandstones, be dissolved at grain contacts and against mica and clay mineral to provide a source of quartz cement.

5. COMPACTION OF CARBONATE SEDIMENTS

Carbonate sand may compact mechanically by grain crushing in the same manner as quartz sand. Sand from calcareous fossils (microporous carbonate) initially has a very high compressibility, but at stresses exceeding about 15-20 MPa they behave similarly to quartz sand (Fig. 7).

Recent ooids from Florida tested in the same way as described by Chuhan et al. (2001) show that they are less compressible at high stresses than coarse grained quartz sand. This may be due to the fact that carbonate grains are softer and more ductile thus distributing the force at grain contact over a larger area, reducing the stress resulting in less grain fracturing. Carbonate mud compacts mechanically in a way which is similar to siliceous mud and may have similar compressibility (Lind 1999). The mineral composition in particular the aragonite content is an important factor in the chemical compaction.

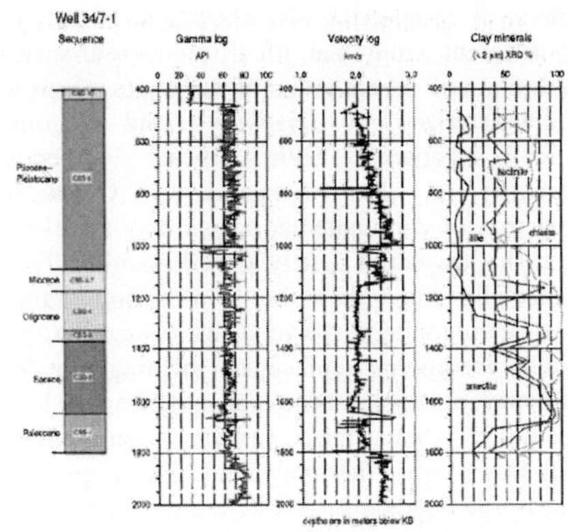

Figure 6. Well data from well 34/7-1 in the northern North Sea. Note from the log that the Tertiary sequence is generally muddy. The Pliocene and Pleistocene mud however compact much more readily than the underlying smectite-rich Oligocene and Eocene mudstones. This is clear from the velocity log which indicates nearly 3 km/s at the base of the Pliocene sequence but and less than 2 km/s in the underlying smectite rich mud. From Thyberg et al. (2000)

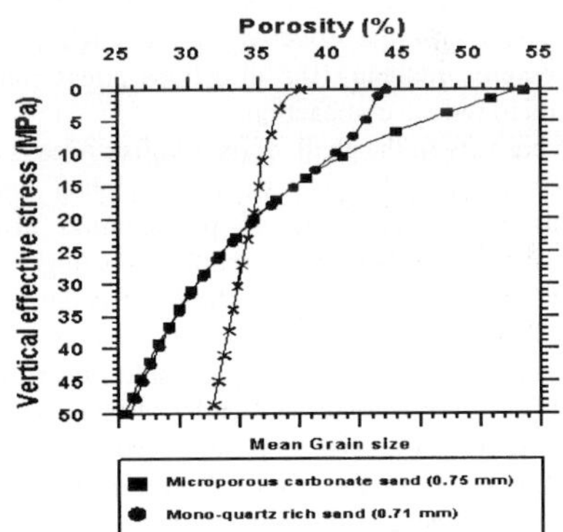

Fig 7. Compaction of carbonate and siliceous sand following the same procedure as in Chuhan et al. 2002. Microporous carbonate sand comprises carbonate fossil shells which initially compact readily. At stresses above 20 MPa, however, it is no more compressible than quartz sand. Recent carbonate ooids from Florida are less compressible than quartz at high stresses.

Chemical compaction of carbonate sediments is very different from that of siliceous sediments. Dissolution and precipitation of carbonate minerals are characterized by relatively rapid reaction kinetics compared with silicate minerals particularly at low temperatures. Chemical compaction of carbonate sediments may therefore start at shallow depth shortly after deposition. This is especially true when the load-bearing grains consist of aragonite, dissolve at shallow depth (low stresses). Neomorph replacement of aragonite by calcite may, however, occur at an early stage at low temperature hindering further compaction. Dissolution of thermodynamically unstable minerals like aragonite is often accompanied by precipitation of early calcite cement, which will stabilize grain structure and thus preserve porosity during burial. Aragonite fossils and ooids may dissolve leaving open moulds while the dissolved carbonate is precipitated as cement.

Chemical compaction is important both in carbonate sand and mud. The driving force here is also here the increased solubility due to stress at grain contacts or along stylolites and the dissolved ions must be transported by diffusion into the pore water and precipitated. Because of the high kinetic precipitation rate the dissolution of carbonate may be rate limiting and the chemical compaction will be a function of the effective stress. High fluid overpressures reducing the effectives stress thus will tend to reduce compaction.

The porosity in the chalk of the Ekofisk Area is a function of the effective stress (Campbell and Gravdal 1995). The rate of precipitation also depends on grain surface properties; coatings of clay or organic matter may inhibit or delay the precipitation of cement. Large calcite crystals, e.g. from crinoids, may provide very suitable sites (low solubility) for precipitation, while surfaces with very small carbonate crystals may retard precipitation of cement. This therefore influences the rate of compaction. Mainly due to high fluid pressures, the initial porosity in parts of the Ekofisk Chalk reservoir was high (up to 50%). When oil production started and the pore pressures were reduced by 5-10 MPa, the chalk structure (matrix) started to collapse. This caused much higher mechanical compaction and sea floor subsidence than originally anticipated and led to large additional field development costs (Gutuerrez et al. 2000).

6. MODELLING COMPACTION-DRIVEN FLUID FLOW.

Modelling of fluid flow is based on the Darcy Equation:

$$F = k \cdot \nabla P / \mu \qquad (7)$$

The viscosity μ is normally well constrained and the fluid flux (F) is a function of the loss of porosity in the underlying sediments during compaction. Dewatering of minerals, thermal expansion of water and generation of hydrocarbon may also add to the flux, Barker (1972), Daines (1982). If the porosity-depth curve and the subsidence rate are known, the compaction-driven flux can be modelled.

The compaction process is always a disequilibrium process in the sense that pressure gradients higher than hydrostatic are required for the fluids to escape even if the pressure gradients may be close to hydrostatic. The pressure gradients depend on the rate of fluid flux compared to permeability along the most permeable flow pathways. Layers of very low permeability (k) and also sealing faults can cause increased pressure gradients (∇P) so that the effective stress is reduced. This is a negative feedback preventing the build-up of high overpressures due to mechanical compaction.

As discussed above, in the deeper parts of sedimentary basins the compaction of siliceous sediments is mainly a function of the rate of heating during subsidence. In basins without hydrothermal input the rate of compaction will then change only slowly with time, and the pressure gradient will be mainly controlled by the permeability. Compaction will continue even as the pore pressure is increased and the effective stress is reduced. The main problem in modelling fluid flow is the prediction of the distribution of permeability. Sedimentary basins are heterogeneous with respect to primary sedimentary facies, diagenesis and tectonic development. Modelling based on idealised homogeneous basin properties may help in understanding the fundamental processes but fluid transport is very much controlled by the heterogeneities of the basin. The permeability of the rock matrix may vary from more than 1 Darcy in sandstones to much less than 1 nannoDarcy (10^{-9} Darcy) in shales. We can see from the Darcy equation that the pressure gradient (∇P) for a given flux is inversely proportional to permeability, and for vertical flow it is the harmonic average of the permeabilities of all the sedimentary layers which is critical. We have no method for predicting the permeability of shales accurately and a change from 1.10^{-9} Darcy to 3.10^{-9} Darcy would reduce the pressure gradient to 1/3.

The decrease in fluid (water) density with depth in sedimentary basins could potentially drive large scale thermal convection (Wood 1986). Modelling show however that the conditions required for Rayleigh convection is rarely met because even very thin clay or shale layers will provide effective flow separation so that the height required for Rayleigh convection is not reached, Bjørlykke et al. (1988)

Realistic modelling of fluid flow in a sedimentary basin in 3 D is "impossible" because the permeability distribution cannot be predicted. To determine the pressure in one part of the basin the effective permeability in all directions for long distances must be accurately determined.

The physical properties at any depth are a function of the effective stress, temperature and of the primary sediment composition which is very difficult to predict. This is because the sediment supply and deposition is difficult to model.

Consequently overpressue cannot be predicted reliably because the permeability distribution can not be sufficiently constrained. In the North Sea Basin the overpressured compartments are often separated from normally pressured areas by sealing faults (Buhrig 1989). In the deeper parts of sedimentary basins with siliceous sediments the fluid flux is the result of temperature controlled compaction and therefore independent of pore pressure and permeability. Even when modelling only vertical flow (1D), it is very difficult to determine the permeability of the different layers to calculate the overall effective permeability. The permeability parallel to bedding is in most cases several orders of magnitude higher than perpendicular to bedding, and the lateral drainage and offsets of sandstones by faults are usually decisive factors for build-ups of overpressure. Lateral drainage is frequently the main factor controlling the pressure.

At Haltenbanken, Offshore Norway, the Jurassic reservoirs in the Smørbukk and Lavrans fields is normally pressured, probably due to lateral flow up to the surface to the east (Olstad et al. 1997). To the west, separated by a main fault, the reservoirs (i.e. the Kristin Field) are overpressured close to fracture pressure because the lateral drainage path to the east is hindred by a fault which now appears to be sealing. Evidence suggests that when at shallower depth, this fault previously was more permeable and allowed petroleum migration (Olstad et al. 1997). The permeability of faults will change continuously during burial, and this is a very difficult parameter to determine for basin modelling. If the permeability is sufficiently low in overlying shales and also laterally, the fracture pressure will be reached, and fluid flow is then controlled by the permeability of the hydrofractures rather than by the rock matrix (Gundersen et al. 2002). In many overpressured basins the pore pressure is controlled by the fracture pressure of shales at the top of the structures, in which case the permeability becomes a dynamic variable adjusting itself to the flux by opening and closing fractures.

7. COMPACTION AND PHYSICAL PROPERTIES

The physical properties of sediments change continuously during burial as a function of the mechanical and chemical processes described above. This changes also the seismic velocities and attributes. Inverting the seismic signals to interpret sediment properties, like porosity and fluid content is very difficult because the solutions are rarely unique.

Experimental compaction of sediments and modelling of chemical compaction may help to predict the physical properties of different lithologies as function of the burial history. This will provide a better basis for the interpretation of the seismic data (Fig 8).

The most difficult part is to elucidate the initial sediment composition prior to the availability of drilling information. In basin modelling, however, we need to predict sediment properties as they vary from deeper to shallower burial depth (backstripping). After drilling, when the sediment composition is known, this modelling is clearly better constrained.

8. CONCLUSIONS

1) Realistic modelling of sediment compaction during burial is difficult because several very different processes are involved. Modelling of compaction in sedimentary basins is very dependent on the primary lithology because each layer may have very different mineralogy and textural characteristics controlling mechanical and chemical compaction.

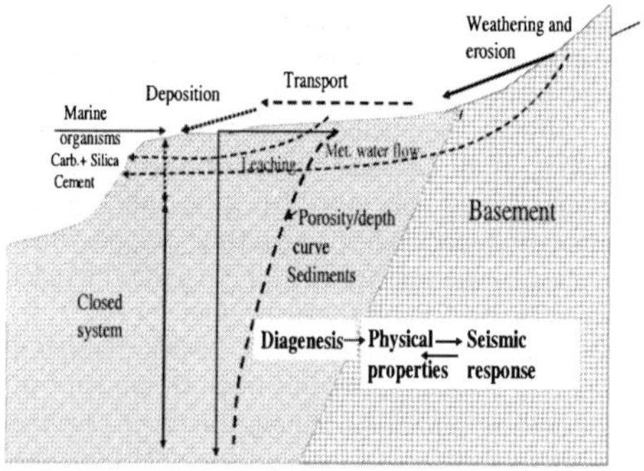

Fig 8 Simplified outline of geological processes controlling sediment supply, deposition, early diagenesis and late diagenesis.

2) Laboratory compression tests of sediments with different grain size/shape characteristics and mineralogy provide useful data for modelling mechanical compaction. The effect of breaking (fracturing) of sand particles on compressibility is particularly important.
3) Chemical compaction is controlled by the thermodynamics and kinetics of mineral reactions and can be modelled when calibrated against sediments of known subsidence history in terms of stress, temperature and time. The mineralogy and the texture must also be known because they are required as input for the chemical compaction models.
4) The compaction-driven fluid flux is a direct function of the porosity reductions in the underlying layers down to basement, of mineral dewatering, and of other phase changes.
5) A realistic prediction of the permeability distribution in three dimensions in sedimentary basins seems impossible given the wide ranges of values for different types of sediments and the heterogeneities of the basins. Pore pressures and fluid fluxes in three dimensions can not be modelled reliably. When the fracture pressure is reached at high overpressure, the fluid flow becomes decoupled from the permeability of the rock matrix and is mostly a function of the permeability of very thin hydrofractures. The permeability is then coupled to the fracture spacing and width which again is a function of the fluid flux and the rate of compaction.
6) Modelling of compaction is an important part of basin modelling because it is coupled with sediment density and basin subsidence. It is therefore also linked to sedimentary facies which control important input parameters.
7) The physical processes involved in sediment compaction and fluid flow are in principle well understood but more experimental data on compaction of different types of sediments are required. Great progress has been made in recent years in modelling of chemical compaction particularly for quartz-rich sandstones.

The limitations in terms of predicting compaction and fluid pressure seem at the present stage not to be the mathematical modelling but the poor constraints of the 3D distribution of sediment properties in the basin. The most difficult problem is to predict the distribution of sediment properties based on facies analyses, provenance, diagenesis and structural geology. We are faced with the problem that realistic basin modelling requires a fairly detailed input with respect to the distribution of different lithologies and their physical properties in a basin.

ACKNOWLEDGEMENTS

This research has been carried out with the support from the Research Council of Norway.

REFERENCES

Aplin, A.C., Fleet, A. J. & Macquaker, J. H. S. (eds.) 1999. *Mud and Mudstones: Physical and fluid flow properties*. Geological Soc. London, Spec. Publ. 158.

Athy, L.F. 1930. *Density, porosity and compaction of sedimentary rocks.* Bull. Am. Assoc. Petrol. Geol. 14: pp. 1-24.

Barker, C. 1972. *Aquathermal pressuring-role of temperature in development of abnormal pressure zones.* Bull. Am. Assoc. Petrol. Geol. 56: pp. 2068-2071.

Bjørkum, P.A. 1996. *How important is pressure solution in causing dissolution of quartz in sandstones.* Journ. Sedimentary Research 66: pp. 147-154.

Bjørkum, P.A. & Nadeau, P.H. 1998. *Temperature controlled porosity/permeability reduction; fluid migration and petroleum exploration in sedimentary basins.* APPEA Journal 38, 453-464.

Bjørlykke, K. 1993. *Fluid flow in sedimentary basins.* Sedimentary Geology 86: pp.137-158.

Bjørlykke, K. 1994. *Fluid flow processes and diagenesis in sedimentary basins. In Parnell, J.*

(ed.), Geofluid: Origin, Migration and Evolution of Fluids in Sedimentary Basins. Geol. Soc. London, Spec. Publ. 78: pp. 127-140.

Bjørlykke ,K 1999. *Principal Aspects of Comapaction and fluid flow in mudstones.* In Aplin, A. C. Fleet, A.J. & Macquaker, J.H.S. (eds) Muds and Mudstones: Physical and fluid flow properties. Geological Society of London. Pec. Publ. 158, 73-78.

Bjørlykke, K., Mo, A. & Palm, E. 1988. *Modelling of thermal convection in sedimentary basins and its relevance to diagentic reactions.* Marine and Petroleum Geology 5, 338 - 351.

Bjørlykke, K., Nedkvitne, T., Ramm, M. & Saigal, S. 1992. *Diagenetic processes in the Brent Group (Middle Jurassic) reservoirs of the North Sea: an overview.* In Haszeldine, A.C., Giles, R.S. & Brown, M.R. (eds.), Geology of the Brent Group. Geol. Soc. Spec. Publ. 61: pp. 263-287.

Buhrig, C. 1989. *Geopressured Jurassic reservoirs in the Viking Graben: modelling and geological significance.* Mar. Petrol. Geol. 6: pp. 31-48.

Campbell, S.J.D. & Gravdal, N. 1995. The prediction of high porosity chalks in the East Hod Field. Petroleum Geoscience 1: pp. 57-69.

Chuhan, F.A., Kjeldstad, A. Bjørlykke K. & Høeg, K. 2002. *Porosity loss in sand by grain crushing experimental evidence and relevance to reservoir quality.* Marine and Petroleum Geology 19, 39-53.

Chuhan, F.A., Kjeldstad, A., Bjørlykke, K. & Høeg, K. 2003. *Experimental compression of loose sands –simulating porosity reduction in petroleum reservoirs during burial.* Candian Geotechnical Journal (Accepted).

Daines, S.R. 1982. *Aquathermal pressuring and geopressure evaluation.* Am. Assoc. Petrol. Geol. XX: pp. 931-939.

Giles, M.R. 1997. *Diagenesis: A Quantitative Perspective. Implications for Basin Modelling and Rock Property Prediction.* Kluwer Academic Publisher. 526 p.

Giles, M.R., Stevenson, S., Martin, S.V., Cannon, J.C., Hamilton, P.J., Marshall, J.D. & Samways, G.H. 1992. *The reservoir properties and diagenesis of the Brent Group: a regional perspective.* In Hazeldine, A.C., Giles, R.S. & Brown, M.R. (eds.), Geology of the Brent Group. Geol. Soc. Spec. Publ. 61: pp. 289-327.

Gundersen, E., Renard, F., Dysthe, D.K., Bjørlykke, K. & Jamtveit, J. 2002a. *Coupling between pressure solution creep and diffusive mass transport in porous rocks.* Journal of Geophysical research 107(B11): p.2317, doi.10.1029/2001JB 000287.

Gundersen, E., Renard, F., Dysthe, D.K., Bjørlykke, K. & Jamtveit, J. 2002b. *Pressure solution in sandstones, rate limiting processes and the effect of clays.* In De Meer, S., Drury, M.R., D.E. Bresser, J.H.P. & Pennock, G.M. (eds.), Deformation Mechanisms, Rheology and Tectonics: Current Status and Future Perspectives. Geol. Soc. London, Spec. Publ. 200: pp. 41-56.

Guitierez,M. Oino,L.E. and Høy,K. 2002. *The effect off luid content on the mechanical behaviour of fractures in Chalk.* Int. Journal of Rock Mechanics and Rock Engineering. Vol 3.No2.93-117.

Houseknecht D.W. 1987. *Asessing the relative importance of compaction processes and cementation to reduction of porosity in sandstones.* Am. Ass. Petrol. Geol. Bull 71, 633-642.

Janbu, N. 1963. *Soil copresibility as determined by oedometer and triaxial tests.* Proc. European. Conf. on SMFE: Vol. 1 420-431.

Karlsen, D.A., Nedkvitne,T., Larter, S.R. & Bjørlykke, K. 1993. *Hydrocarbon composition of authigenic inclusions - application to elucidation of petroleum reservoir filling history.* Geochemica et Cosmochemica Acta 57: pp. 3641 -3659.

Kjeldstad, A., Langtangen, H.P., Skogseid, J. & Bjørlykke, K. . 2002. *Simulations of deformations, fluid flow and heat transfer in sedimentary basins.* In: Langtangen, H.P. & Tveit, A. (eds.), Advanced Topics in Computational Partial Differential Equations - Numerical Methods and Diffpack programming. Springer Verlag.

Lambe,T,W. and Whitman, R.V. 1979. *Soil Mechnics-SI version.* Wiley.

Lind.I.L.1993.*Laoding experiments on carbonate ooze and chalk from leg 130.* In Berger et al 1993. Proeceedings of the Ocean Drilling Program, Scientific Results.Vol 30. p673-686.

Lander, R.H. & Walderhaug, O. 1999. *Predicting porosity through simulating sandstone compaction and quartz cementation.* AAPG Bulletin 83(3): pp. 433-449.

Olstad, R., Bjørlykke, K. & Karlsen, D.K. 1997. *Pore water flow and petroleum migration in the Smørbukk Field area, offshore Norway.* In Møller-Pedersen, P. & Koestler, A.G. (eds.), Hydrocarbon Seals - Importance for Exploration and Production. Nor. Petrol. Soc. Elsevier 7: pp. 201 - 216.

Oelkers, E.H., Bjørkum, P.A. & Murphy, W.M. 1996. *A petrographic and computational investigation of quartz cement and porosity reduction in North Sea sandstones.* American Journal of Science 296: pp. 420-452.

Ramm, M. 1992. *Porosity-depth trends in reservoir sandstones, theoretical models related to Jurassic sandstones offshore Norway.* Marine and Petroleum Geology 9: pp. 553-567.

Ramm, M. & Bjørlykke, K. 1994. *Porosity/depth trends in reservoir sandstones: assessing the quantitative effectes of varying pore- pressure, temperature history and mineralogy, Norwegian Shelf data.* Clay Minerals 29: pp. 475-490.

Renard, F. Dysthe, D. Feder,J., Bjørlykke, K. and Jamtveit B. 2001. *Enhanced pressure solution creep rates induced by clay particles: Experimental evidence in salt aggregates.* Geophysical Research Letters. 28,1295-1298.

Thyberg, B.I., Jordt, H., Bjørlykke, K. & Faleide, J.I. 2000. *Relationships between sequence stratigraphy, mineralogy and geochemistry in Cenozoic sediments of the northern North Sea.* In Nøttvedt, A. et al. (eds.), Dynamics of the Norwegian Margin. Geol. Soc. London, Spec. Publ. 167: pp. 245-272.

Walderhaug, O. 1996. *Kinetic modeling of quartz cementation and porosity loss in deeply buried sandstone reservoirs.* Bull. Am. Assoc. Petrol. Geol. 80: pp. 731-745.

Walderhaug, O., Lander, R.H., Bjørkum, P.A., Oelkers, O.E., Bjørlykke, K. & Nadeau, P.H. 2000. *Modelling of quartz cementation and porosity in reservoir sandstones: examples from the Norwegian Continental Shelf.* Int. Ass. Sedimentologyists, Spec. Publ. 29: pp. 39-49.

Walderhaug, O., Bjørkum, P.A, Nadeau, P.H. & Langnes, O. 2001. *Quantitative modelling of basin subsidence caused by temperature-driven silica dissolution and reprecipitation.* Petroleum Geoscience 7: pp. 107-114.

Walderhaug. O. and Bjørkum,O.A.2003. *The effects of stylolite spacing on quartz cementation in the lower Jurassic Stø formation, Southerm Barents Sea.* Journ. Sedimentary Research 73,146-156.

Wood.J.R. 1986. *Thermal mass transfer in systems containing quartz and calcite.* In Gautier, D.L. (Ed) Roles of Organic matter in sediments diagenesis. SEPM Special Publication 38, 169-180.

VAPORIZATION-INDUCED OVERPRESSURES AS A TRIGGER FOR THE HAZARDOUS COLLAPSE OF LAVA DOMES

D. Elsworth[1], B. Voight[1], J. Simmons[1], S. Young[1], B. Winkler[1]

[1])College of Earth and Mineral Sciences, Pennsylvania State University, University Park, PA 16802

Abstract: Rainfall-triggered collapses of hot lava domes are described relative to the fluid overpressurization that accompanies the infiltration of rainfall into the carapace. Similar rainfall-triggered modes of failure are apparent for both hemispherical (endogenous) and lobate (exogenous) domes. Failure modes for these disparate geometries are linked by the necessary incursion of rainfall into the hot dome and the resulting impeded drainage of interior gas pressures – either self-generated by the vaporizing infiltration front, or the impeded drainage of interior pressurized gases. Limit equilibrium models are developed to assess the stability of idealized hemispherical or lobate domes, with driving forces generated by interior gas overpressures. For endogenous domes, shallow failures may develop in the absence of gas pressurization from the core, but observed deep-seated failures require that the free escape of conduit gases be impeded. For exogenous domes, failure is driven to the base of the lobe by the desire to maximize overpressures.

1. INTRODUCTION

The growth and ultimate collapse of lave domes represents an important and hazardous feature in the growth of silicic volcanoes [Miller, 1994]. Collapse of the highly gas-charged dome materials [Sparks, 1997; Denlinger and Hoblitt, 1999; Wylie et al., 1999] may spawn hazardous and highly mobile pyroclastic flows [Nakada and Fujii, 1993; Abdurachman et al., 2000], and may retreat retrogressively into the dome core, potentially unroofing the conduit and triggering the renewed effusion of deep magmatic products as Vulcanian [Voight et al., 2000] and fire fountain eruptions. Such collapses may potentially remove 90% of the dome structure, and mobilize tens of millions of cubic meters of incandescent (scoria and ash) materials.

In the quiescent stage of construction, lavas extrude from the conduit and may form a spectrum of structures, ranging from conduit-capping cupolas in highly gas-charged magmas (Soufrière and Peléan type), to pendant lobes (Merapi type) where the dome rocks are strong in relation to the magnitude of internal gas overpressures [Sato et al, 1992]. For this full spectrum of structures, but in particular for domes of the Soufrière and Peléan type (endogenous), near continuous raveling failure of the dome may be punctuated by less-frequent and more deeply-transecting failures, possibly triggered by the combined action of slope oversteepening [Fink and Griffiths, 1998; Sparks et al., 2000], of gas overpressurization of the dome interior [Voight and Elsworth, 2000; Elsworth and Voight, 2001], and by hydrothermal weakening of the dome or its substrate [Lopez and Williams, 1993; Voight and Elsworth, 1997; Crowley and Zimbleman, 1997; Watters et al., 2000]. At Soufrière Hills volcano (SHV), these complementary and potentially additive effects of dome oversteepening and interior gas-pressurization adequately honor observed near-dome tilt [Watson et al., 2000], RSAM [Miller et al., 1998], and gas discharge histories that span major dome collapse events triggered during periods of active dome growth (e.g. the collapse of June 25, 1997). However, collapses that occurred during periods of effusive quiescence, and in the absence of apparent seismic or intrusive triggers, cannot be explained by slope oversteepening, or in a traditional sense, by mechanisms of gas overpressurization.

Numerous lava dome collapses have coincided with intense rainfall during these times of low effusive growth, and absent the typical solid earth triggers [Voight et al., 2000; Matthews et al., 2002; Yamasato et al., 1998; Mastin, 1994]. The potential overpressurization of the dome by rainfall-induced fluid overpressures is a common process that links the observed rainfall-triggering of collapses at both endogenous (Soufrière Hills and Pelée) and exogenous (Merapi and Unzen) lava domes. At least 2 dome collapses on the endogenous dome of SHV [Matthews et al., 2000], at least 4 dome collapses on the exogenous dome of Merapi volcano [Voight et al., 2000], and upwards of 18 collapses on the exogenous dome of Unzen volcano [Yamasato et al., 1998] have been associated with heavy rainfall.

The following investigates the potential for hydraulic penetration of rainfall into the fractured surface of lava domes, its potential to both develop and seal-in gas overpressures, and evaluates the likely influence of these overpressures on instability of both endogenous and exogenous domes. Instability is described via a limit-equilibrium model incorporating rainfall-induced fluid overpressures. Fluid overpressures are modulated by rainfall intensity and duration, thermal controls on water infiltration, extent of vaporization in the hot interior, and the ability of a saturated rind to cap these pressures. The approximations of induced fluid pressures are used to evaluate the likelihood of collapse, provide order of magnitude estimates of destabilizing forces, and project likely timing and severity of failure.

2. STABILITY MODEL

Limit equilibrium models are applied to the specific geometries that represent both endogenous and exogenous domes, as represented in Figure 1. A force-balance may be completed on the free-body isolated at the toe, enabling the relative stability to be defined as a "factor of safety, F_S." The factor of safety is defined as the ratio of forces resisting failure to those promoting failure, resolved relative to the direction of potential failure. In general, the evolving instability may be considered in terms of a force balance where the disturbing influence of the downslope components of the water liquid, F_W, and vapor forces, F_V, and by the fluid uplift force (vapor or liquid water depending on liquid water penetration depth), F_U, applied to the basal plane are balanced against the restoring forces provided by the cohesive, c, and frictional strength, ϕ. Correspondingly, the factor of safety may be defined generically for either dome geometry as

$$F_S = \frac{cs + (w\cos(\alpha) - F_U)\tan(\phi)}{w\sin(\alpha) + F_W + F_V}, \quad (1)$$

for a disturbed block of weight, w , resting on a slope of inclination, α.

Rainwater will infiltrate the dome, principally along fractures, to a depth determined by both thermal controls and dome geometry. If sufficiently chilled by this, and prior quenching, the infiltrating liquid will act as a near-impervious "cap" trapping the underlying overpressures as the liquid at the infiltration front flashes to steam. The vapor overpressure is limited by the supernatant water pressure, with a small additional contribution from the air entry pressure, ignored in this treatment.

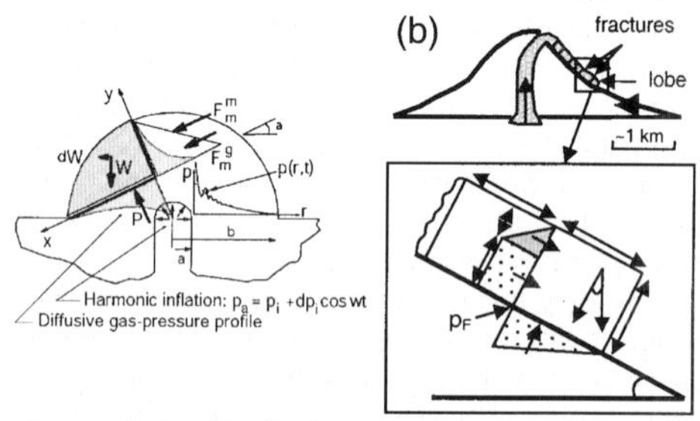

Figure 1. Free-body diagrams of (a) endogeonous and (b) exogenous dome.

3. THERMAL HYDROLOGY

The thermal hydrology is conditioned by the magnitudes of rock and fluid thermal conductivities, densities, and heat capacities, all defined within relatively narrow ranges. Hydraulic penetration to depth d into a ubiquitously fractured lava dome of fracture spacing s may be estimated by equating the quenching potential (thermal flux) of rainfall, Q_W, discharged from a tributary catchment area into a fracture, with the thermal flux conducted across the fracture wall, Q_T, as the penetrating water chills the margin to below boiling [Figure 2].

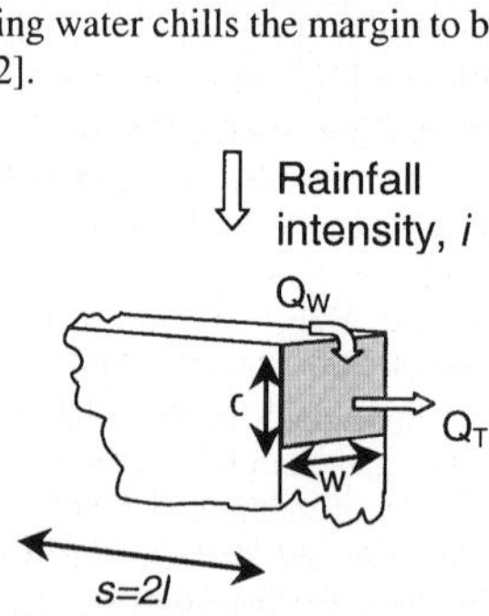

Figure 2. Geometry describing simplified thermal hydrology of a quenched surface containing vertical fracture pathways.

This energy balance results in a non-dimensional depth penetration of [Carslaw and Jaeger, 1959]

$$4\frac{\kappa_R d}{is^2} = \frac{\rho_W c_W}{\rho_R c_R}\frac{\Delta T_W}{\Delta T_R}[t_D + f(t_D)], \quad (2)$$

where ΔT_R is the temperature change of the rock due to quenching, ΔT_W is the required increase in temperature to boil the water (heat of vaporization is omitted), λ_R is the thermal conductivity of the rock and t_D is a diffusive time, defined both by the thermal diffusivity of the rock, κ_R, and the duration of the rainfall event, t, as $t_D=\kappa_R t/l^2$ where $l=s/2$, with rainfall of intensity, i, and water constants of density, ρ_W, and specific heat, c_W.

For anticipated dome temperatures in the range 200°C to 800°C, equation (2) is shown in Figure 3 for a rainfall event typical of collapse-triggering deluges at Montserrat, i.e. 75 mm in 3 hours, an event not atypical of Merapi [Lavigne et al., 2000]. For these parameters, $t_D \leq 10^{-2}$, which represents either shorter duration events (relative to thermal diffusivity) or widely spaced fractures. For fracture spacing typical (5 m to 100 m spaced fractures and faults) of dome environments, this behavior mode prevails and is represented in area II of Figure 3. This relationship represents the case where the rainfall event chills the fracture margin but does not significantly cool the interior. This is the behavior of interest in promoting collapse that develops concurrently with the rainfall event. If the collapse is significantly delayed, remnant heat will slowly remove the penetrating water.

Apparent from Figure 3, possible depth penetrations are of the order of 0.5 to 11 m for the hottest case of an 800°C dome, for fracture separations of 5 to 100 m. If the dome has been cooled by multiple prior events, depth penetration to 25 m is feasible at 400°C and 75 m at 200°C, each for the most widely spaced fractures (100 m). Importantly, the most hydraulically significant (i.e. most open) fractures may develop at progressively wider separations when subject to multiple episodes of quenching by prior storms [Keer et al., 1978]. This results from the positive feedback of the most open seed fracture, accepting quenching fluid that chills its margins and enlarges and deepens it, allowing it to preferentially accept more water from subsequent storms.

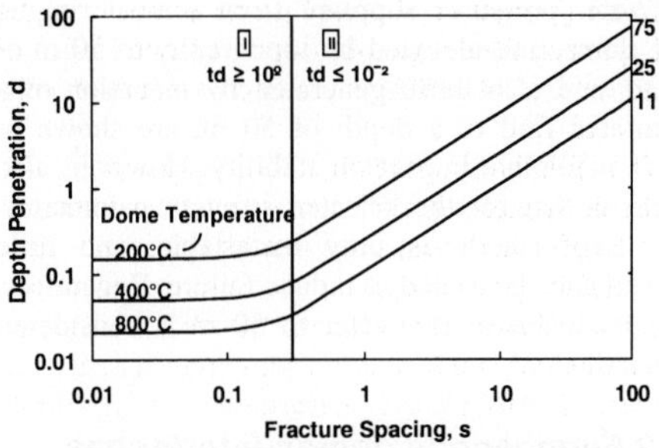

Figure 3. Quenching depth penetration for discrete storm discharging 75 mm of rainfall over 3 hours.

4. STABILITY EVALUATIONS

Anticipated collapse modes are examined for contrasting small (200 m) and large (350 m) endogenous dome geometries, for varied inclinations of the detachment plane, and under different conditions of interior water- and gas-pressurization, and for lobe failures resulting from only rainfall induced overpressures [Figure 1].

4.1 Endogenous domes-water and rind pressures only

Interior dome strengths compatible with the expulsion of spines ($c=0.5$ MPa and $\phi=25^{\circ}$) are selected for the remaining evaluations. The influence of uplift water pressures alone, in destabilizing both small and large domes is illustrated in Figure 4.

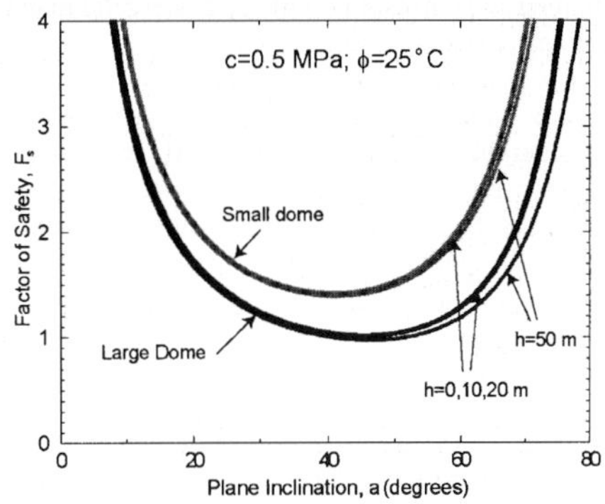

Figure 4. Variation of F_S with inclination of transecting failure plane, showing the influence of water pressures in the absence of gas pressurization. Pressure heads, h, shown in equivalent meters of water.

Water pressures supplied from a moat ringing the dome, and elevated by topography to 50 m of head, and also those generated by incursion of a saturated rind to a depth of 50 m, are shown to have negligible impact on stability. However, also clear is that for the selected strength parameters, the larger dome is only metastable, and little disturbance is needed to induce failure. Regardless, the incursion of rainwater to 50 m is considered improbable.

4.2 Endogenous domes-interior gas pressures

The large dome is only metastable when unpressurized, as illustrated in Figure 5a. The application of a relatively minor uniform interior pressure (0.2 MPa) is sufficient to bring the dome to failure. However, the application of a strong pressurization of the core (to 10 MPa) has less destabilizing influence than a much smaller pressure (0.2 MPa) that is uniformly applied. Regardless of the form of the applied interior pressure (uniform or radially diminishing), any resulting collapse will be at a high angle, and would be unlikely to retrogress to the dome core. This suggests that it is difficult to generate a dome-removing failure by any of the pressurization mechanisms previously discussed, in a "large" dome. This has some implications to the implied hazard posed by such domes, whether precursory pressurization signals (RSAM and tilt, etc.) may be minimal or absent, and where the relative collapse volume may be relatively high, but less gas-charged and correspondingly less energetic than the collapse of smaller domes.

The behavior of the small dome is illustrated in Figure 5b. Because effective stresses are considerably smaller in the 200 m dome, in comparison with the 350 m dome, interior gas pressures have a relatively greater influence on stability. Minor interior pressures (uniform pressures to 0.2 MPa) are capable of bringing the small dome close to instability, but must exceed 0.2 MPa to result in collapse. This may be too high a threshold. Conversely, radially diminishing core pressures exercise a more pronounced control. Achievable core pressures of 5 MPa are capable of approaching failure, slightly increased core pressures have the potential to develop low angle rupture.

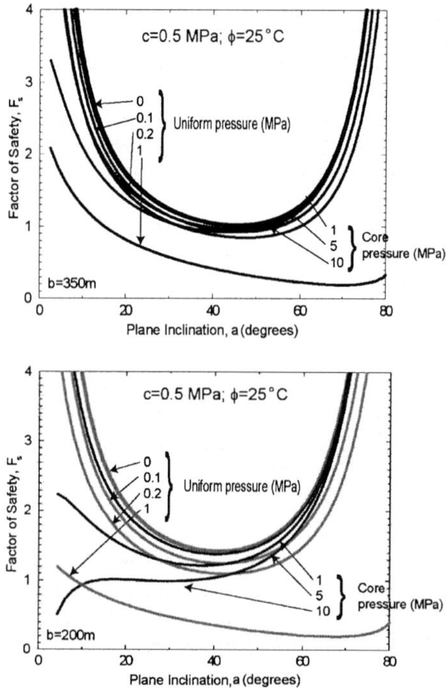

Figure 5. Variation in F_S for failure of (a) a large dome (radius=350m) and (b) a small dome (radius=200m) subject to uniform and core pressurization.

4.3 Endogenous domes-collapse styles

Collapse styles anticipated for different dome sizes, effusive styles, and resulting modes of pressurization are summarized in Figure 6. For the large dome, uniform interior pressures of 0.1 MPa are adequate to induce failure through one side at an inclination of about 50°. When mirrored on both sides, this failure would remove about 20% of the edifice, honoring the observed collapse volume, but only loosely approximating the style.

For the same strength parameters as previous, the unpressurized small dome is stable, but may be brought close to instability by steady core pressures of the order of 5 MPa. Minor augmentation by uniform rind-trapped pressures of the order of 0.2 MPa is sufficient to promote low-angle failure to as shallow as 40° and remove about 20% of the edifice on one side. Alternatively, if effusive gas discharge is flux, rather than overpressure-controlled, a three-order of magnitude (×500) reduction in permeability in a rind of 5 m will allow core pressures to double to 10 MPa. Following shut-in, failure is encouraged to pierce the evolving highly-pressured core, and develop on a low angle (10°-20°), dome-transecting trajectory. A transect inclined 10° would isolate the upper 75% of the dome and honor observations of the March 2000 collapse of SHV including the likely unroofing of the conduit.

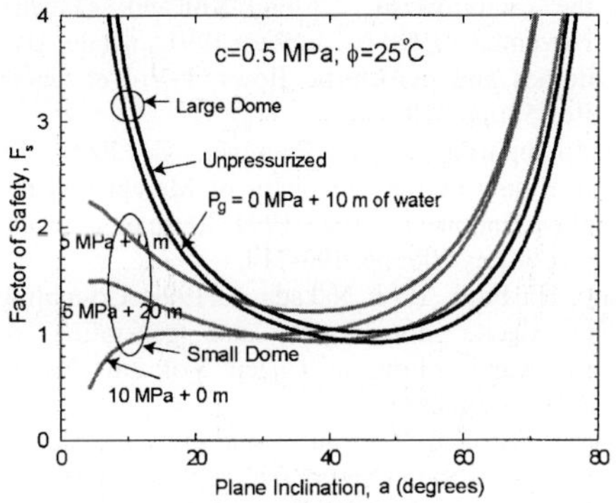

Figure 6. Summary of plausible collapse modes for small and large domes subject to applied uniform and radially diminishing core pressures.

4.4 Exogenous domes

At Merapi there have been at least seven occurrences of 100 mm/hr storm events in a twenty-year period (once every ~3 years), and smaller storm events of greater than 25 mm/hr occur more frequently, comprising ~5% of all 1990 storms [Lavigne et al., 2000]. The destabilizing influence of these infrequent storms is superimposed on a steady rainfall boundary condition corresponding to the high average annual rainfall at Merapi (4.5 m/yr) [Lavigne et al., 2000] in Figure 7. Selected, are infrequent storms of 75 mm in 3 hours and 200 mm in 2 hours, at respective frequencies of 4/yr and 0.3/yr. Cumulatively, these discrete events account for less than 10% of the distributed rainfall.

Figure 7 illustrates the deterioration to collapse of a 50 m thick lobe, fractured at 50 m intervals, over the decade following its emplacement. The lower locus of stability for the 50 m thick lobe is defined by the decrease in factor of safety caused by storm events. As evident in Figure 7, the average annual rainfall alone cannot induce failure in the 50 m thick lobe over the decade considered, nor can 75 mm of rain in 3 hour storm events. However, if a well-timed storm of 100 mm/hr, lasting 2 hours, occurs after about 6 years following lobe-emplacement, failure is likely. A storm of this same duration and intensity brings the lobe very close to failure at ~3.2 years. Also noteworthy from Figure 7 is that the average annual rainfall alone, not considering acute storm events, can bring a 100 m thick lobe to failure in less than one year following emplacement. This results from the inherent lower stability of thicker lobes, stemming from their amplification of downslope driving forces.

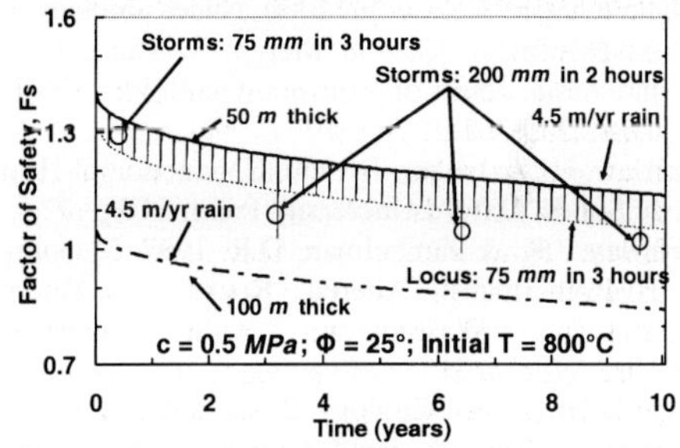

Figure 7. Variation in F_S over time for an exogenous lobe with 50m fracture spacing showing influence of rainfall induced overpressures.

5. CONCLUSIONS

The mechanistic models developed broadly replicate observed styles of rainfall-preceded collapses of the endogenous dome of Soufrière Hills volcano and the exogenous domes of Merapi and Unzen volcanoes. The high-angle collapse of an oversized metastable endogenous dome (SHV July 1998), during a period of volcanic repose, and absent obvious precursors other than a deluge of rain, is consistent with triggering by high-level gas pressurization within the dome, generated and capped by the rainfall itself. The near-complete removal of a much smaller dome, in a period of reinitiating effusive activity, again absent other precursors, is consistent with pressurization of the dome core, critically augmented by the partial sealing of the dome carapace.

Infiltration of rain and subsequent development of water and gas overpressures within an exogenous lobe, can be sufficiently large to trigger their collapse in the absence of solid earth triggers, thus reproducing observations at both Merapi and Unzen. Thicker lobes are intrinsically less stable due to their inherent amplification of downslope driving forces in overcoming cohesive resistance.

Importantly, the proposed mechanisms result in a monotonic decrease in stability with time rendering a once stable lobe metastable.

ACKNOWLEDGEMENTS

This work is a result of partial support from the National Science Foundation under grants CMS-9908590, EAR-9909673, and EAR-0116826. This support is gratefully acknowledged.

REFERENCES

Abdurachman, E.K., et al. 2000. Nuees ardentes of 22 November 1994 at Merapi Volcano, Java, Indonesia. Journ of Volc and Geoth Res 100(1-4): pp.345-361.

Carslaw, H., & Jaeger, J. 1959. Conduction of Heat in Solids. Oxford University Press: 510 p.

Crowley, J.K. & Zimbelman, D.R. 1997. Mapping Hydrothermally Altered Rocks on Mount Rainier, Washington, with Airborne Visible/Infrared Imaging Spectometer (AVIRIS) Data. Geology 25: pp. 559-562.

Denlinger, R.P. & Hoblitt, R.P. 1999. Cyclic eruptive behaviour of silicic volcanoes. Geology 25(5): pp. 459-462.

Elsworth, D. & Voight, B. 2001. The mechanics of harmonic gas-pressurization and failure of lava domes. Geophys. J. Int., of the R. Astronomical Soc. 145: pp. 187-198.

Elsworth, D., et al. 2003. Some Models of Rainfall-Triggered Collapse of Endogenous Lava Domes in Potentially Gas-Effusive Environments. To be published.

Fink, J.H. & Griffiths, R.W. 1998. Morphology, eruption rates, and rheology of lava domes: Insights from laboratory models. Journ of Geophys Res 103(B1): pp. 527-545.

Keer L, et al. 1978. Unstable growth of thermally induced interacting cracks in brittle solids: further results. Earthquake Res. Eng. Lab. Tech. Rept. No. 78-2-8, Northwestern University.

Lavigne F., et al. 2000. Lahars at Merapi volcano, Central Java: an overview. J of Volc and Geoth Res 100: pp. 423-456.

Lopez, D.L., & Williams, S.N. 1993. Catastrophic volcanic Collapse: Relation to Hydrothermal Processes. Science 260: pp. 1794-1796.

Mastin, L. 1994. Explosive tephra emissions at Mount St. Helens, 1989-1991 – The violent escape of magmatic gas following storms. Bull Geol Soc Amer 106: pp. 175-185.

Matthews, A. et al. 2002. Rainfall-induced volcanic activity on Montserrat. Geophys Res Lett 29(13): 4pp.

Miller, A.D., et al. 1998. Seismicity associated with dome growth and collapse at the Soufriere Hills volcano, Montserrat. Geophys Res Lett 25(18): pp. 3401-3404.

Miller, T.P. 1994. Dome growth and destruction during the 1989-1990 eruption of Redoubt Volc. J Vol & Geoth Res 26(1-4): pp.197-212.

Nakada, S. & Fujii, T. 1993. Preliminary report on the activity at Unzen Volcano (Japan). November 1990-November 1991: dacite lava domes and pyroclastic flows. J Vol & Geoth Res 54: pp.319-333.

Ratdomopurbo, A., & Poupinet, G. 2000. An overview of the seismicity of Merapi volcano (Java, Indonesia), 1983-1994. Journ of Volc and Geoth Res 100: pp. 193-214.

Sato, H., Fujii, T., & Nakada, S. 1992. Crumbling of dacite dome lava and generation of pyroclastic flows at Unzen Volcano. Nature 360: pp. 664-666.

Sparks, R.S.J. 1997. Causes and consequences of pressurization in lava dome eruptions. Earth and Planetary Sci Lett 150: pp. 177-189.

Sparks, R.S.J., et al. 2000. Control on the emplacement of the andesite lava dome of the Soufriere Hills volcano, Montserrat by degassing-induced crystallization, Terra Nova 12: pp. 14-20.

Voight, B. & Elsworth, D. 1997. Failure of volcano slopes. Geotechnique 47(1): pp. 1-31.

Voight, B. & Elsworth, D. 2000. Instability and collapse of lava domes. Geophys Res Lett 27(1): pp. 1-4.

Voight, B., et al. 2000. Deformation and seismic precursors to dome-collapse and fountain-collapse nuees ardentes at Merapi Volcano, Java, Indonesia; 1994-1998, Journ of Volc and Geoth Res 100(1-4): pp.261-287.

Watson, I.M., et al. 2000. The relationship between degassing and ground deformation at Soufriere Hills Volcano, Montserrat. Journ of Volc and Geoth Res 98(1-4): pp.117-126.

Watters, R.J., et al. 2000. Rock Mass Strength Assessment and Significance to Edifice Stability, Mount Rainier and Mount Hood, Cascade Range Volcanoes. Pure Appl Geophys 157: pp. 957-976.

Wylie J.J., et al. 1999. Instability of magma flow from volatile-dependent viscosity. Science 285 (5435): pp. 1883-1885.

Yamasato, H., et al. 1998. Effect of rainfall on dacitic lava dome collapse at Unzen volcano, Japan. Papers in Met and Geophy 48(3): pp. 73-78.

BENTONITES FROM ISHIRINI (LIBYA) AS NATURAL ANALOGUES OF LONG TERM THERMAL AND CHEMICAL EFFECTS: ISOTOPIC AND FLUID INCLUSION EVIDENCE

Irena Kolaříková, Radek Hanus
Charles University, Faculty of Science, Dept. of Mineralogy and Geochemistry, Prague, Czech Republic

Abstract: Smectites from the Ishirini bentonite deposit (Libya) were analyzed for their oxygen and hydrogen isotopic ratios. Based on published isotopic fractionation data the results argue for a main meteoric water origin of the fluids responsible for the transformation of pyroclastic rocks to bentonite at temperature range from 30° to 90 °C.

The thermal effect of later volcanic intrusions was further evaluated by detailed fluid inclusion study on quartz separated from silicified rings which arose by the thermal action of basalt dike. Five different groups of fluids were distinguished based on salinity and homogenization temperature.

15 cm wide disturbed zone was created in silicified rings and structural discontinuities enable the circulation of hot magmatic fluids towards the clay. Important mineralogical changes were observed in the close proximity of the bentonite/volcanic contact.

1. INTRODUCTION

The Ishirini deposit has been examined as a potential natural analogue of the bentonite clay barrier. The objective of this project is to qualify the observed effects of thermal loading in several bentonites of the Ishirini area.

Isotope studies are frequently carried out to investigate the conditions of smectite formation (Yeh and Savin, 1977). In this study $\delta^{18}O$ and δD isotopic compositions of smectite were used to characterize the nature of the fluids and the temperature during the transformation of pyroclastic rocks to a large bentonite deposit. It is supposed, that argilitization was related to meteoric water warmed up by hot volcanic rocks in the Ishirini area. However, bentonites could represent a multiphase alteration product of meteoric water as well as hydrothermal fluids.

2. GEOLOGICAL SETTING

The Ishirini area belongs to the NE part of Jabal al Hasawnah (fig. 1). Five geological units can be distinguished in the region - Hasawnah formation (Cambrian), thin sequence of Miocene sediments (Ahurfin formation), a volcanic region represented by the number of phonolite and andesite bodies (Miocene), Pliocene Hishar formation (bentonites) and Plio-Pleistocene basalts. The bentonite deposit is concentrated in SE part of Ishirini. In this area mainly pyroclastic rocks of andesitic composition and few lava flows have been altered to clay.

The bentonites are composed mainly of montmorillonite (determined by X-ray diffraction) with Ca^{2+} as dominant interlayer ion. The crystallochemical formula was calculated from the chemical analysis (Rieder, 1977): $(Na_{0,01}K_{0,02}Mg_{0,01}Ca_{0,27})[(Al_{1,42}Mg_{0,18}Fe^{3+}_{0,32}Ti_{0,08})(Si_{3,66}Al_{0,34})O_{10}(OH)_2]$. Quartz and kaolinite are subordinate.

Some of bentonite sequences were crosscut by sub-volcanic basaltic domes and small intrusions which injected the Hishar formation.

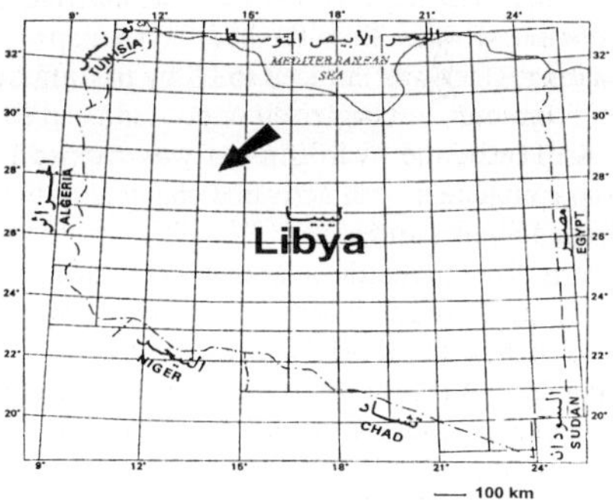

Figure 1. Location of the Ishirini deposit

These intrusions produced a large metamorphic aureole due to the high temperature of the intrusion. The intrusion temperature of basalt dike has been calculated using clinopyroxene thermometers (Lindsley, 1983). The temperature values are in the range of 830 - 900 °C.

3. SAMPLE PREPARATION AND ANALYTICAL METHODS

Most of samples from Ishirini contain smectite in excess of 90%. Isotope analysis was performed on the < 0.2 μm size-fraction of the bentonites, which exclusively consisted of smectites as determined by X-ray diffraction. Before separating the finest fraction agglomerating constituents such as carbonates and amorphous silica were removed (Jioneck, 1995). Carbonates were dissolved with acetic acid while amorphous silica was removed with sodium carbonate. Furthermore, the influence of present amorphous phases which can influence the isotope ratios was considered during measurements. The treatment of samples does not affect smectite's isotopic composition (Yeh and Savin, 1977).

The smectite samples were heated at 180 °C for eight hours under vacuum to eliminate interlayer water. The samples were reacted with fluorine gas using a modified Taylor and Epstein type fluorine line to liberate oxygen for isotope analysis (Taylor and Epstein, 1961). Fluorine gas was purified.

An experimental device was used for the analysis of hydrogen isotopes ratios. The samples were heated up to ~ 1300 °C with a high frequency induction coil under vacuum for the liberation of water from the hydroxyl groups. The resulting water was reduced to H_2 by passing over hot uranium (Bigleisen et al., 1952). Subsequently, the hydrogen gas was trapped in a sample container with activated charcoal at liquid nitrogen temperature.

Mass spectrometric analyses were carried out using a Finnigan MAT spectrometer. The responsibility of the total analytical procedure was in the range ± 0.2 ‰ for $\delta^{18}O$ and ± 1 ‰ for δD.

Microthermometric studies of fluid inclusions were performed on wafers using a Linkam THM SG 600 heat-freezing stage. The stage was calibrated with melting-point standards at T > 25 °C and natural and synthetic fluid inclusions at T < 0 °C. The rate of heating was monitored in order to obtain an accuracy of ± 0.2 °C during freezing, ± 1 °C when heating over the 25 - 500 °C range and ± 4 °C over the 500 - 600 °C range. Salinity, expressed as wt. % $NaCl_{eq.}$ was calculated from microthermometric data using equations from Bodnar (1993). In volatile-bearing fluid inclusions, CO_2 was identified by melting of a solid below -56.6 °C. Molar fractions of CO_2 and CH_4 were determined in individual fluid inclusions according to Thiéry et al. (1994).

4. RESULTS AND DISCUSSION

The $\delta^{18}O$ and δD values for smectites are presented in tables 1 and 2. Isotopic ratios of oxygen and hydrogen are reported relative to SMOW standard.

Table 1. Stable isotope composition of smectites from the Ishirini bentonite deposit

sample	$\delta^{18}O$ SMOW (‰)	δD SMOW (‰)
MK 3 I	15.8	-63
MK 3 II	14.6	-53
MK 3 III	16.4	-56
MK 3 IV	16.8	-59
MK 3 V	17.0	-57
MK 3 VI	19.7	-73

In Ishirini $\delta^{18}O$ ratios of the smectites vary from +14.6 to +19.7 and δD ratios range from -73 to -53.

During clay mineral formation ^{18}O and D are enriched in the newly formed mineral. The isotope fractionation between mineral and fluids in isotopic equilibrium is a function of the temperature. The isotopic composition of the parent rock is negligible (Yeh and Savin, 1977), especially when large water - rock ratios can be expected during the early stage of alteration.

The thermal effect of later volcanic intrusions is clearly demonstrated by $\delta^{18}O$ variation in smectites (fig. 2 and table 2).

Peculiar geochemical (fig. 2) and mineralogical data (fig. 3) were obtained from the bentonite in the close proximity (20 - 30 cm) of the dike.

Table 2. Evolution of the oxygen isotopic composition of smectites from the Ishirini bentonite deposit in respect to the distance from the bentonite/volcanic intrusion contact

sample	dist. from the contact (cm)	$\delta^{18}O$ SMOW (‰)	t (°C)
MK 1 I	20	12.4	200
MK 1 II	40	12.8	
MK 1 III	100	13.3	
MK 1 IV	150	13.6	
MK 1 V	200	13.8	
MK 1 VI	300	14.0	170

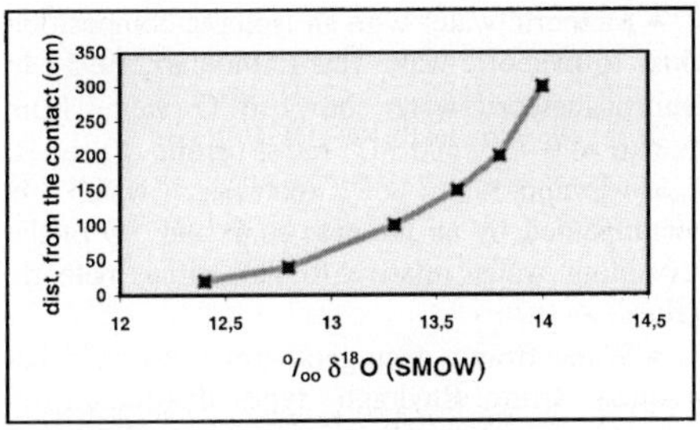

Figure 2. Evolution of the oxygen isotopic composition of smectites from the Ishirini bentonite deposit in respect to the distance from the bentonite/volcanic intrusion contact

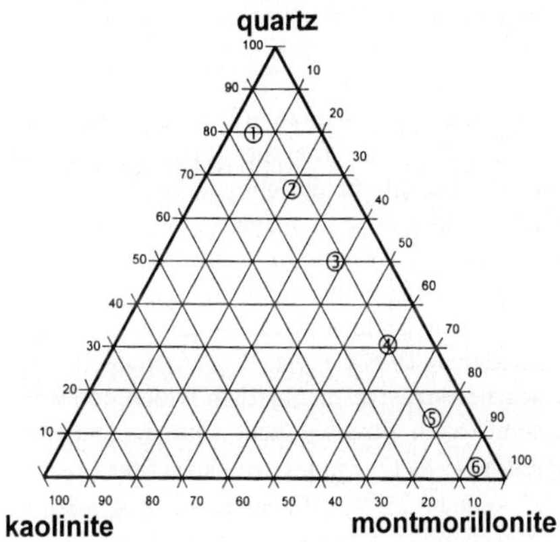

① 15 cm far from the contact
② 20 cm far from the contact
③ 30 cm far from the contact
④ 40 cm far from the contact
⑤ 50 cm far from the contact
⑥ 90 cm far from the contact

Figure 3. QKM diagrame showing the proportion of major components of the bentonite collected at increasing distance from the volcanic contact
(Semiquantitative mineralogy of selected samples was determined using recalculation program CQPA and XRD and chemical analyses data.)

These data cannot be explained by the thermal process alone and indicate possible water - rock interactions following the basaltic intrusion.

The succession of fluids after the thermal effect of later volcanic intrusion has been studied by examining relationships between fluid inclusions, their host mineral and the geometry of the host microstructures.

Fluid inclusions occur either as primary inclusions within the quartz of the silicified zone (I, II and III), or secondary inclusions in "milky" quartz of the silicified zone (I and II) 5 and 10 cm far from the intrusive contact.

Aqueous-carbonic fluid inclusions characterized by irregular shapes occur as primary fluid inclusions in zone I and II. These scarce inclusions exhibit variable size (10 - 20 μm). At room temperature, these inclusions contain liquid H_2O, liquid CO_2 and gaseous CO_2. The CH_4 content ranges from 30 mol % to 35 mol. % in zone I and < 10 mol. % in zone II. Minimum trapping temperatures have been recorded in the range of 430 - 470 °C in zone I, respectively 295 - 300 °C in zone II.

Hypersaline fluid inclusions were observed as isolated inclusions in zone I. These inclusions contain one halite cube. Eutectic temperatures are very low (< -48 °C) indicating the presence of divalent cations (Ca^{2+} and Mg^{2+}). The inclusions homogenise to the liguid between 270° and 300 °C. Th-tot values are in most cases lower than the halite melting temperatures, which occurs from 300° to 320° C. Salt content has been estimated using the triangular composition diagram of Vanko et al. (1988). The fluid composition is 30 - 40 % NaCl, 6 -10 % $CaCl_2$ and 50 - 64 % H_2O.

Low salinity aqueous fluids occur in primary inclusions (zone III) and secondary inclusions (zone II). Aqueous inclusions are characterised by Tm-ice ranging from -8.4 to -7.5 °C coresponding to salinites of 11.1 to 12.0 wt.%$NaCl_{eq.}$ (in zone II) and from -6.4 to -5.9 °C coresponding to salinites of 9.1 to 9.7 wt.% $NaCl_{eq.}$ (in zone III).

The study of aqueous fluid inclusions reveals relationships between the decrease in salinity of the fluids and decrease of minimum trapping temperature. The highest Th-tot and salinity were measured in inclusions of zone I, which are typically the oldest aqueous inclusions. The lowest Th-tot and Tm-ice pairs are found in zone III. The continuous trend in Tm-ice is interpreted as a result of mixing processes, especially a dilution of the early aqueous fluid type.

Table 3. Summary of the microthermometric data for the different groups of fluid inclusions

	primary FI (zone I)	secondary FI (zone I)	primary FI (zoneII)	second. FI (zoneII)	primary FI (zone III)
component	CO_2-H_2O-(CH_4)-NaCl	H_2O-NaCl-$CaCl_2$	CO_2-H_2O-NaCl	H_2O-NaCl	H_2O-NaCl
Tm-CO_2 (°C)	-60.8 to -58.0		-57.6 to -56.8		
Th-CO_2 (°C)	1.3 to 4.5		24.2 to 27.1		
Tm-cla (°C)	8.5 to 9.8		8.0 to 8.5		
Tm-ice (°C)		-34.5 to -30.9		-8.4 to -7.5	-6.4 to -5.9
Th-tot (°C)	430 to 470	270 to 300	295 to 300	140 to 173	120 to 150
salinity (wt.% $NaCl_{eq}$)	0.7 to 3.0	30.1 to 39.5	3.0 to 3.9	11.1 to 12.0	9.1 to 9.7

5. INTERPRETATION OF SMECTITE ISOTOPIC DATA

To gain information about alteration temperature and nature of formation water during argilitization from smectite isotopic data it is essential, that the minerals have retained the isotopic composition they acquired during their formation. In natural systems the exchange of structurally bound oxygen and hydrogen in clays is slow, in particular, at temperatures prevailing at the Earth's surface (Jioneck, 1995).

In the case of Ishirini the possibility of isotope exchange after argilitization is negligible. It can be assumed that the low hydraulic permeability of the bentonites prevented an extensive flow of water through the clay.

According to the geothermometric equations knowledge of the isotope composition of the smectites as well as the alteration fluids is required before the temperature of smectite formation can be calculated.

The isotopic composition of alteration fluids is unknown, but one can estimate the nature of fluids in a temperature range. To calculate the isotopic composition of water in equilibrium with the smectites at different temperatures, the relationship between oxygen fractionation and temperature and hydrogen fractionation and temperature in the water smectite system can be combined according to Savin and Lee (1988) and Capuano (1992).

Several end-member fluids can be distinguished in Ishirini:

- Meteoric water with an isotopic composition close to meteoric water line (Alhodiari, 1995). In Ishirini meteoric water shows $\delta^{18}O$ values from -6.3 to -4.9 ‰ and δD ratios from -40 to -30 ‰. Evaporation of meteoric water is accompanied by an increase in D and ^{18}O of the remaining water relative to the initial meteoric values.
- Water from a deep geothermal reservoir that resulted from Rayleigh type distillation of seawater accompanied by partial reequilibration with the lavas. The mean $\delta^{18}O$ and δD values of these fluids are +5 ‰ and 0 ‰. If water rises from the deep reservoir to the Earth's surface liquid - vapour phase separation can take place. In this case a residual solution is enriched in heavy isotopes.

The isotopic composition of fluids in equilibrium with the smectites are expected to be close to the meteoric water line at temperatures between 30° and 90 °C. Therefore, it can be assumed, that predominantely heated meteoric water is responsible for the formation of the bentonites. The slight enrichment of ^{18}O and D isotopes compared with meteoric water may be due to steam loss or mixing with both meteoric water and water from hydrothermal reservoir. The production of andesitic lavas, 20.2 Ma ago (Abdelkader, 1997), was accompanied by subvolcanic intrusive magmatism which could be considered as a possible heat source. At lower temperatures $\delta^{18}O$ values of smectites would almost exceed +25 ‰. If heated water of ~ 80 °C was responsible for bentonitization (Christidis et al., 1995) smectites should have $\delta^{18}O$ values of +19 ‰. Alteration by water similar to the fluids of the deep hydrothermal reservoir (Dietrich et al., 1992) is not likely as well. The isotopic composition of these fluids at temperatures above 200 °C should provide smectites with $\delta^{18}O$ value around +12 ‰.

In the interpretation of the smectite isotopic composition two restrictions have to be considered:

- According to Savin and Lee (1988) tetrahedral aluminium and iron in the smectite structure should influence isotopic fractionation. An Fe and $Al^{(IV)}$ free smectite compared to nontronite should provide a difference in $\delta^{18}O$ of nearly 5 ‰. The chemical variability of the Ishirini smectites is very small. Therefore, the chemical variability of smectites on isotope fractionation is assumed to be negligible. The influence of the chemical composition of smectites on D fractionation is not known.

• Temperature has a strong effect on the isotopic composition of rainwater. In rainwater a temperature drop of 1 °C causes a $\delta^{18}O$ decrease of 0.7 ‰ and δD decrease of 5.6 ‰ (Dansgaard, 1964). Therefore a mean temperature decrease of 4 °C would cause a shift in the area characterizing the fluids existing in Ishirini so that the smectite formation temperature must be considered 10 °C higher as suggested according present day conditions.

6. CONCLUSIONS

Isotopic composition of smectites in bentonites from Ishirini indicate an alteration of the pyroclastic rocks predominantely by heated meteoric water at temperatures ranging from 30 to 90 °C. Alteration by fluids similar to the present day hydrothermal solutions from the deep geothermal reservoir can be excluded. Geothermometric calculations based on $\delta^{18}O$ and δD ratios of the smectites can only provide a temperature range for the bentonite formation. In addition, several uncertainties have to be considered, e.g. the reability of the fractionation factors, deviations in the climatic conditions or isotopic exchange with fluids percolating through the newly formed bentonites.

Isotopic compositions of smectites from different Ishirini bentonites indicate similar meteoric origin of fluids responsible for argilitization. These fluids must have been a part of an extensive alteration system.

Reactive rocks, preferentially composed of volcanic glass present a prerequisite of bentonite formation. Furthermore, a sufficient hydraulic permeability is important which is assured in coarse pyroclastic rocks. On the other hand, there must be also a sufficient supply of water during a long time period.

Fluid inclusion studies evaluated the thermal effect of volcanic intrusion which injected the bentonite formation.

Carbonic-aqueous fluid inclusions (quartz zone I and II) corresponde to the abnormal heat flow.

The fluids from the zone I display a halogen signature similar to that obtained for magmatic fluids. Thus, a magmatic source for these fluids can be considered. In addition, the high salinity indicated by the presence of salt crystals in the fluid inclusions is in a good agreement with this hypothesis.

Low salinity aqueous fluids represent the cooling of hydrothermal system.

The results also show clear trend in mineralogical composition of the samples taken in a distance profile from the intrusion (fig. 3). There are also changes in the isotopic composition of montmorillonite (fig. 2).

• the thermal effect was strong enough to affect the bentonite

• the sampled region was close to the contact, that also low temperature thermal effects could be seen

• the cooling rate was very fast so there was not enough time to affect a large area of Hishar formation

• the temperature gradient was quite sharp, so that the intrusion heat has affected the bentonite, which was max. 3 m far from the contact

• presence of hypersaline fluids caused changes in the bentonite, small amount of montmorillonite was transformed into kaolinite

REFERENCES

Abdelkader V. A. 1997. *Radiometric age determination of the basalts and andesites in the Ishirini area.* Sebha, MS University Archiv.

Alhodiari A. 1995. *Geology and mineralogy of Ishirini deposit.* Sebha, MS University Archiv.

Bigleisen J., Perlman M. L. and Prosser H. C. 1952. *Conversion of hydrogenic materials to hydrogen for isotopic analysis.* Anal. Chem. 24: pp. 1356 - 1357.

Bodnar R. J. 1993. *Revised equation and table for determining the freezing point depression of H_2O - NaCl solutions.* Geoch. et Cosmoch. Acta 57: pp. 683 - 684.

Capuano R. M. 1992. *The temperature dependence of hydrogen isotope fractionation between water and clay minerals: Evidence from a geopressured system.* Geoch. Cosmoch. Acta 56: pp. 2547 - 2554.

Christidis G. E., Scott P. W. and Marcopoulos T. 1995. *Origin of the bentonite deposits of Eastern Milos, Greece: geological, mineralogical and geochemical evidence.* Clays Clay Miner. 43: pp. 63 - 77.

Dansgaard W. 1964. *Stable isotopes in precipitation.* Tellus 16: pp. 436 - 468.

Dietrich V., Nuesch R. and Balanti D. 1992. *The origin of the Milos bentonite deposits.* Abstr. 6th Congr. Geol. Soc. Greece: pp. 31 - 33.

Jioneck M. 1995. *Vorschriften für die Probenahme für die Isotopenanalyse.* MS Geological Survey, Marktredwitz.

Lindsley D. 1983. *Pyroxene thermometry.* Amer. Miner. 68: pp. 47-493.

Rieder M. 1977. *Micas: Calculation of crystallochemical formulas by a FORTRAN IV Comp. program.* Věst. ÚÚG 52: pp. 333 - 342.

Savin S. M. and Lee M. 1988. *Isotopic studies of phyllosilicates.* In S. W. Bailey (ed.), Hydrous phyllosilicates; Mineral. Soc. Am., Rev. Mineral.19: pp. 189 - 223.

Taylor H. P. and Epstein S. 1961. *Relation between oxygen isotope ratios in coexisting minerals of igneous and metamorphic rocks, I. Principles and experimental results.* Bull. Geol. Soc. Am. 37: pp. 461 - 480.

Thiéry R., Kerkhof A. M. and Dubessy J. 1994. *V-X properties of CH_4-CO_2 and CO_2-N_2 fluid inclusions:modelling for $T< 31\ ^oC$, $P < 400$ bars.* Europ. Journ. of Mineral. 6: pp. 753 - 771.

Vanko D. A., Bodnar R. J. and Sterner S. M. 1988. *Synthetic fluid inclusions: VIII.Vapour-saturated halite solubility in part of the system NaCl-$CaCl_2$-H_2O, with application to fluid inclusions from oceanic hydrothermal systems.* Geoch. et Cosmoch. Acta 52: pp. 2451 - 2456.

Yeh H. W. and Savin S. M. 1977. *Mechanism of burial metamorphism of argillaceous sediments.* Bull. Geol. Soc. Am. 88: pp. 1321 - 1330.

THE EVOLUTION OF PERMEABILITY IN NATURAL FRACTURES – THE COMPETING ROLES OF PRESSURE SOLUTION AND FREE-FACE DISSOLUTION

A. Polak[1], H. Yasuhara[1], D. Elsworth[1], J. Liu[2], A. Grader[1], P. Halleck[1]
[1]) Energy Institute and Department of Energy and Geo-Environmental Engineering, Pennsylvania State University, 403 Academic Activities University Park, PA, 16802 USA.
[2]) School of Oil & Gas Engineering, The University of Western Australia, 35 Stirling Highway, WA 6009, Australia.

Abstract: Fracture permeabilities are shown surprisingly sensitive to mineral dissolution at modest temperatures (c. 20°-80°C) and flow rates. Net dissolution may either increase or decrease permeability, depending on the prevailing ambient THMC conditions. These behaviours have important ramifications for constitutive laws for flow and transport. Flow-through tests are completed on a natural fracture in novaculite at temperatures of 20°C, 80°C, 120°C, and 150°C, and on an artificial fracture in limestone at 20°C. Measurements of fluid and dissolved mass fluxes, concurrent X-ray CT and imaging, and post-test sectioning and SEM are used to constrain the progress of mineral dissolution and its effect on transport properties. For the novaculite, under constant effective stress, fracture permeability decreased monotonically with an increase in temperature, with fracture permeability reducing by two-orders-of-magnitude over the 900 h test. For the limestone, an initial decrease in permeability over the first 935h of the test, switched to a net increase in permeability as distilled water was subsequently circulated for the final 500h of the test.

1. INTRODUCTION

A variety of experimental results illustrate the competing roles of processes that increase pore or flaw connectivity (including dilatant shear, microcracking, thermal cracking and focused dissolution), and those that destroy or counteract the evolution of pore interconnectivity (including shear and hydrostatic compaction, fracture healing, dislocation creep, and pressure solution prompted by water-film and free-face diffusion). The sense of permeability change is controlled by the processes that dominate under a given set of conditions; this is in turn largely controlled by the evolving conditions of effective stress and temperature in the sample, and species concentrations within the flowing fluid.

Data constraining these roles, particularly in fractures, where the effects are most apparent, are sparse. However, they are reported at elevated temperatures (>300ºC) in granite [Moore et al. 1994], and at lower temperatures (50ºC-150ºC) in tuff [Lin et al., 1997]. These are augmented by results for available composite aggregates of quartz [Elias and Hajash, 1992], halite [Gratier, 1993], calcite [Zhang, et al., 1994] and albite [Hajash et al., 1998], at moderate temperatures (23ºC-150ºC), and the same material suites at elevated temperatures and pressures [e.g. Zoback and Byerlee, 1975; Siddiqi, et al., 1997]. The limited studies on fractures [Moore et al., 1994; Lin et al., 1997; Durham et al., 2001] suggest an increased sensitivity of their transport properties to thermal, hydraulic, mechanical, and chemical processes, over porous medium flows. This is apparent even at temperatures as low as 100ºC, where the mobile dissolved species is silica, the test duration is of the order of a month [Elias and Hajash, 1992; Lin et al., 1997], and where permeability may be reduced by a factor of 104 [Lin et al., 1997].

In this work, we present results documenting surprising changes in permeability magnitudes, concurrent with flow-through tests in novaculite, and in limestone, at temperatures in the range 20ºC – 150ºC, and at confining stresses of 3.5 MPa. These tests are constrained by measurements of fluid flux, mineral mass flux, and of evolving porosity by X-ray CT.

2. EXPERIMENTAL METHOD

Flow-through tests are conducted on cylindrical specimens with a through-going axial fracture as illustrated in Figure 1.

Water is circulated through the specimen, confined within a Viton sleeve, and pressurized by the confining fluid within the core holder. Flow rates are prescribed,

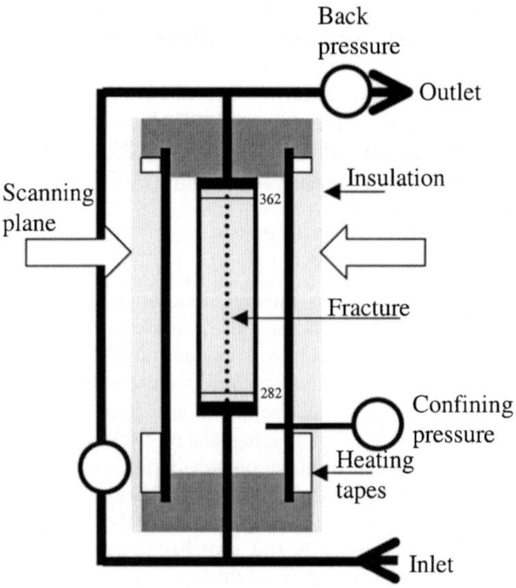

Figure 1. Schematic of flow-through experiment for the novaculite experiment. Scan section, lines and numbers are relative to core holder base (in mm).

and end-to-end pressure-drop is measured, to enable permeability to be evaluated directly. Heat tapes enable the entire core-holder and influent fluid to be raised to a prescribed temperature, in the range 20ºC – 150ºC, and water samples are collected through the duration of the test to assay for mineral mass fluxes. The final suite of data enable mass balances to be completed on fluid mass, mineral mass, and periodic scanning by X-ray CT enable dissolution-related changes in porosity to be similarly mapped in space, and in time – conditional on the resolution of the imaging system.

3. EXPERIMENTAL RESULTS

Experiments are reported on a natural fracture in novaculite, and an artificial fracture in limestone. Each material exhibited characteristically different behaviour.

3.1 Novaculite

The flow experiment was conducted on a cylindrical core (25 mm diameter × 90 mm length) of low porosity (<1%) Arkansas novaculite containing a single diametral natural fracture. Arkansas novaculite has a uniform grain size of the order of 1-6 μm, and quartz content of > 99.5% [Lee at al., 1991]. The test was run over 900h, and at staged temperatures of 20, 80, 120, and 150ºC.

3.1.1 Fluid and mass transport behavior

Changes in fracture transport characteristics that result from the removal or redistribution of mineral mass within a fracture may be constrained by calculating the evolution of permeability with the concurrent loss or redistribution of mineral mass. Mass loss is unambiguously recorded by the effluent flux, but redistribution may only be discerned by imaging. Independent measurements of fluid and mineral flux, concurrent with non-invasive imaging, are used to constrain the processes controlling mass redistribution within the fracture.

3.1.2 Observations

The evolving change in fracture aperture, under a constant effective stress of 3.5 MPa, is evaluated from the recorded flow-rates and differential pressures, as shown in Figure 2. The initial aperture of 12.3 μm at 20ºC, falls to 11.4 μm with the initiation of the test. This closure is interpreted as the minor crushing of asperities and interstitial propping grains, as the fracture seats. Heating the sample to 80ºC results in a sharp decline in effective hydraulic aperture that ultimately asymptotes to 8.3 μm at 75 hours.

Rapid heating of the fracture to 120ºC, under conditions of constant effective stress, sharply reduces the fracture aperture to 4.4 μm within 60 hours followed by a slow decrease of a few tenths of a micron spread over the following 100 hours (at constant temperature). The slow monotonic reduction in the fracture aperture is attributed to the redistribution of quartz within the fracture. This could result from the combined effects of dissolution of contacting fracture asperities or propping grains, and precipitation in flow-throats. Two sharp decreases in fracture aperture are apparent at 103 h and 115 h and are correlated with shutdown of the water injection pump for 2 min and 1.5 hours, respectively. A third resulted from pump failure. In all instances, changes are sufficiently rapid (visible instantly after the flow reinitiated) that asperity breakage under elevated effective stress is the likely cause. These stress-related reductions in aperture are irreversible. The

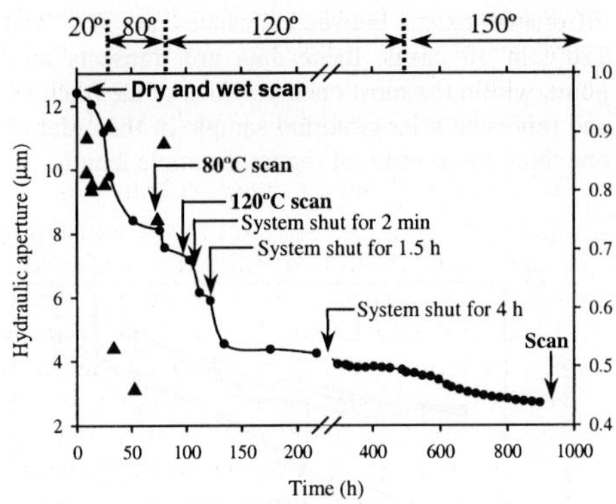

Figure 2. Change in fracture aperture with test duration. Hydraulic aperture evaluated from measured steady flux and axial pressure differential. Scan sequences, temperature transitions, brief test interruptions and aqueous concentrations of silicon.

core was finally heated to 150°C (430 h after the beginning of the experiment), resulting in a slow steady decrease in aperture over the following 500 hours at constant temperature (from 3.8 to 2.7 μm). The overall change in aperture over the 900 h test is of the order of 9.6 μm, representing an 80% reduction in the initial aperture, and slowing significantly with test progress.

The changes in Si concentration measured during the first one hundred hours of the test are shown in Figure 2. Of the 10 samples taken, the first 6 were at 20°C, 3 at 80°C and 1 at 120°C. The elevated concentration at 20°C, relative to the concentration at 80°C (0.85 vs. 0.58 ppm respectively), likely represents the rapid removal of crushed particles surrounding the contacting asperities. Concentrations measured at temperatures of 80°C to 120°C are in the range 0.75 to 0.88 ppm, with a mean of 0.77 ppm. A mean concentration of 0.77 ppm, distributed through the 0.0196 m^3 of water injected during the test, corresponds to 0.015g of Si removed, or 0.03g of quartz.

3.1.3 X-ray CT imaging

Non-invasive imaging may be used to further constrain the role of asperity dissolution in mediating changes in fracture apertures. Imaging gives a reliable estimation of the aperture, and at sufficiently high resolution, may discriminate between net dissolution and the combined effects of dissolution and precipitation that merely redistribute mineral mass.

A montage of scans is shown in Figure 3. Panels are for the core confined to 3.5 MPa and scanned first dry, and subsequently wet at 20°C, 80°C, and 120°C. The panel illustrates the evolution of the sub-horizontal aperture with the progress of the test, with some reduction in aperture qualitatively apparent. Quantitative changes in aperture may be recovered from the scan area where the rock wedge is absent. The aperture in this region is larger than the pixel size and movement of the fracture wall may be followed and measured in sequential scans.

Figure 4 shows a 4.2 mm wide section of the fracture, 47 mm from the inlet (location 322) as identified in the inset of Figure 3. Figure 4a identifies loss or gain of mass, by pixel, on adjacent walls of the sub-horizontal fracture, and Figure 4b the overall change in aperture. A threshold CT-number of 1700 defines the transition from rock to pore space. Apparent is a general reduction in aperture with progress of the experiment at constant temperature, and with increase in temperature. The most open area of the fracture is used as a proxy to represent closure of the tighter fracture interior, as displacements are observable at the resolution of imaging. Closure magnitudes vary with location, and in the range 70-200 μm, are up to a factor of four larger than measured changes in mean hydraulic aperture. This amplification likely results from rotations at the core periphery that amplify closure on the interior portion of the fracture.

At ambient temperature, the measured null change in hydraulic aperture is matched by a substantial imaged change of c. 86 μm. This may result from calibration errors in switching saturants from air to water, or from the dissolution loss of proppant infill that does not result in aperture change. This dissolution would be consistent with the anomalously high early time Si concentrations (Figure 2). Raising the temperature from 20°C to 80°C results in a recorded change in hydraulic aperture of c. 3.8 μm, and a comparable imaged adjustment of c. 0-50 μm. From 80°C to 120°C, the small 0.8 μm recorded reduction in hydraulic aperture is below the threshold resolution of the imaging, and the imaging records an anomalous increase in aperture. This mismatch may result from extrapolations at the limit of the resolution of the scanning technique, where no significant

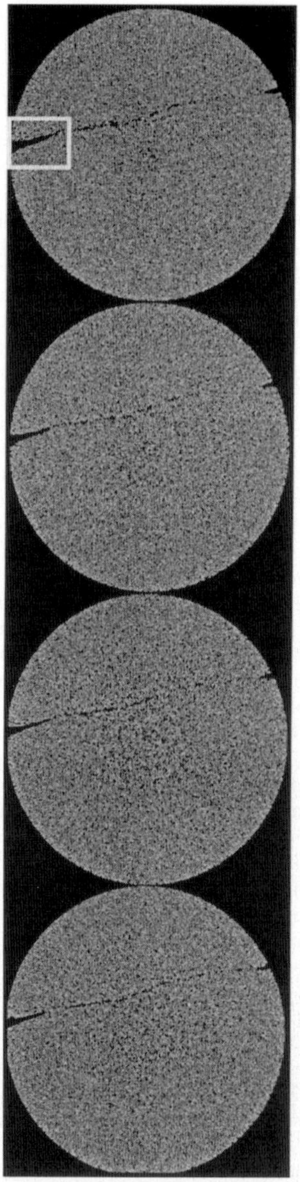

Figure 3. Scanning sequences for the core 47 mm downstream of the inlet of the 90 mm long core. The boxed region is processed in Figures 4 to define dissolved mass and relative displacement of fracture faces.

difference exists between surfaces at 80°C and 120°C.In all cases, these data are transects at a point, within the most open portions of the fracture, and represent a longitudinal sample of the order of one three-thousandth of the total sample length.

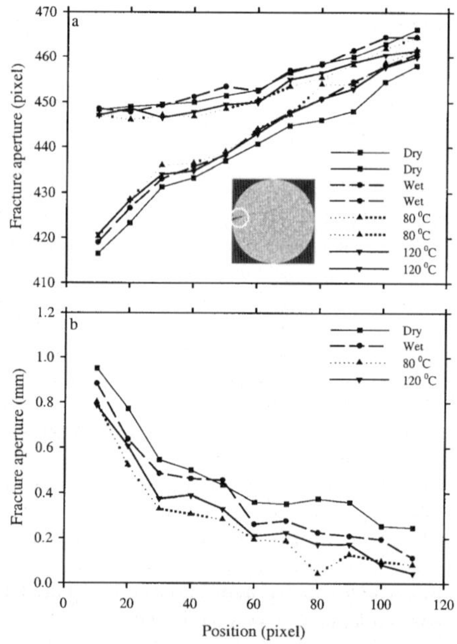

Figure 4. Distribution of fracture aperture, for (a) positions of the fracture walls, and for (b) fracture aperture.

3.2 Limestone Fracture

A circulation test within an artificially fractured sample of limestone was run for a total of 1500h, at a constant temperature of 20°C. During the test, pressure drop across the sample was monitored continuously, and aqueous samples were taken at frequent intervals. Sectional scans were completed to define changes in aperture.

Measurements of differential pressure are used to define changes in equivalent aperture of the fracture, with time. In the initial part of the experiment (0-932 h) the equivalent aperture reduced from 24 to 17 μm. The reduction rate is 0.2 μm per day for the first 21 days after which the aperture reaches steady state at 19.5 μm. In the second part of the experiment the equivalent aperture decreased to 12 μm over a period of 150h. The final part is characterized by a drop in the driving pressure along the fracture from 520 to 6 kPa in the final 402 h. An equivalent fracture aperture could be evaluated, although the geometry is not likely well represented by parallel fracture walls. Scanning results of the first scanning sequence and the fourth scanning sequence (0 and 1492 hours, respectively) at location 665, which is 85 mm from the core's inlet, is shown in Figure 5. The changes between the two sequences are visible at the area of the fracture where the wormhole was created. Two fractures are visible; the bottom fracture did not flow, as it was a local fracture. The flow took place at the upper fracture that with the current resolution of the CT system of 47μm is not showing any changes between the two sequences except at the wormhole. The enlargement on the left side shows that even in the first scanning sequence the fracture looks closed probably because the fracture's aperture is smaller then the resolution of the system. A profile crossing of the fracture, which is marked by the black line on the enlargement, shows a decrease in the CT numbers at the fracture (<2000) compared to the CT number of the matrix (~2200). If the fracture's aperture were larger, the difference in the CT numbers between the matrix and the center of the fracture would have been much higher. Because of the finite resolution, the pixellated measurements are a volume-weighted average CT number of the rock and water. The scanning image of the later sequence (1492 hours after the beginning of the experiment) shows the wormhole opening that was created inside the fracture. The profile crossing the fracture shows that the CT numbers in the fracture are much smaller than the CT numbers of the matrix (1300 vs. 1900). The fracture is marked by a peak CT number-trough of 600 at 1492 hours, versus just 200 at test initiation. This represents the true widening of the fracture as pixels are more fully filled with fracture void, rather than a volume-weighted mixture of void and solid. The differences in the CT numbers in the matrix and the fracture, between the two different sequences are due to the difference in the calibration of the CT system, and are not a change in any of the chemical or physical parameters of the sample.

4. CONCLUSIONS

The results indicate a general correspondence between the mineral mass removed from contacting asperities of a conductive fracture, the measured change in hydraulic aperture, and the non-destructively observed removal of mass from the fracture contact area. For the novaculite sample, measured changes in hydraulic aperture are high, amounting to an 80% reduction in fracture aperture over the 900 h duration test, and a corresponding hundredfold decrease in fracture permeability. The majority of this change in aperture occurs below a temperature of only 120°C. The measured change in aperture magnitude is consistent between hydraulic measurements and recorded mineral efflux, and agrees nominally with imaged observations of fracture aperture. Hydraulic measurements of aperture change are absolute, and require no assumptions other than the applicability

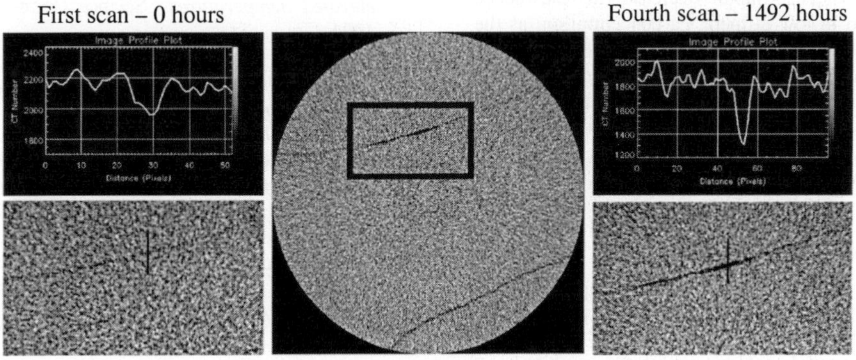

Figure 5. Scanning images for the core at longitudinal location 665 at 2 different scanning sequences, (a) 0 hours, and (b) 1492 hours.

of the parallel plate approximation to flow. The ensemble hydraulic and dissolution data constrain the proposed mechanistic model, and non-destructive imaging further elucidates behaviour. The hydraulically measured 10 μm change in fracture aperture is at the 35 μm resolution limit of the imaging technique. Despite being at this limit, the imaging corroborates the projected behaviour from the hydraulic and dissolution tests, confirming a net reduction in fracture aperture. Imaged changes in aperture are of the same order of those observed hydraulically, and estimated from the chemical data. Regardless, the changes in aperture are surprisingly large and rapid, even for modest test temperatures and flow rates, suggesting that benign changes in the thermal, stress, or chemical environment may exert profound changes in transport characteristics.

In the limestone study, continuous flow of water at ambient temperatures for 1450 h, measurement of differential pressure and chemical concentration of the inflow and outflow along with CT imaging enabled the understanding of the processes occurring inside the fracture. The flow of ground water in the first part of the experiment caused the impedance to increase monotonically as asperities were removed, and the fracture walls closed. The switch from ground water to distilled water caused an initial drop in the impedance as a result of a pressure change that occurred while switching the water composition and scanning the sample. The distilled water had a profound impact on the hydraulic impedence with a rapid increase that was due to fast removal of asperities from the contact areas and a rapid closure of the fracture aperture. This rapid decrease in impedence resulted in low effluent Ca concentration, as the amount of contact area was small or because most of the dissolved Ca precipitated along the fracture face and was not measured at the outflow. As the fracture aperture decreased, the flow localized into a single channel(s) and the impedance reduced significantly.

ACKNOWLEDGEMENTS

This work is a result of partial support from the U.S. Department of Energy under grants DOE-BES-DE-FG02-00ER15111, DOE-DE-PS26-01NT41048, and the Australian Research Council under grant DP0209425. This support is gratefully acknowledged. The authors also thank Mike Howard from the Arkansas Geological Commission for providing a bulk sample of Arkansas novaculite.

REFERENCES

Durham, W.B., W.L. Bourcier, and E.A. Burton, *Direct observation of reactive flow in a single fracture.* Water Resour. Res., 37, 1, 1-12, 2001.

Elias, B.P., and A. Hajash, *Changes in quartz solubility and porosity due to effective stress: an experimental investigation of pressure solution.* Geology 20:451-454, 1992.

Gratier, J.P., *Experimental pressure solution of halite by an indenter technique.* Geophys. Res. Letters 20(15):1647-1650, 1993.

Hajash, A., T.D. Carpenter, and T.A. Dewars, *Dissolution and time-dependent compaction of albite sand: experiments at 1000c and 1600C in pH-buffered organic acids and distilled water.* Tectonophisics, 295(1-2):92-115, 1998.

Lee, V.W., S.J. Mackwell, and S.L. Brantley, *The effect of fluid chemistry on wetting textures in novaculite.* J. Geophys. Res. 96(B6):10023-10037, 1991.

Lin, W., J. Roberts, W. Glassley, and D. Ruddle, *Fracture and matrix permeability at elevated temperatures. Workshop on significant issues and available data.* Near-field/Altered-zone coupled effects expert elicitation project, San Francisco, November 1997.

Moore, D.E., D.A. Lockner, and J.A. Byerlee, *Reduction of permeability in granite at elevated temperatures.* Science 1558-1560, 1994.

Siddiqi, G., B. Evans, G. Dresen, and D. Freund, *Effect of semibrittle deformation on transport properties of calcite rocks.* J. Geophys. Res. 102(B7):14765-14778, 1997.

Zhang, S., S. Cox, and M. *Paterson, the influence of room temperature deformation on porosity and permeability in calcite aggregates.* J. Geophys. Res. 99:761-775, 1994.

Zoback, M. and J. Byerlee, *The effect of microcrack dilatancy on the permeability of Westerly Granite.* J. Geophys. Res. 80:752-755, 1975.

EARTH CRUST STRUCTURE AS A RESULT OF ROCK FRACTURING AT HIGH PRESSURE AND TEMPERATURE CONDITIONS

Victor Nikolaevskiy, Igor Garagash
[1)] Institute of Earth Physics, Moscow, Russia

Abstract: The well-known features of the Earth's crust are interpreted as the result of rock fracturing under deep thermodynamic conditions. For this aim, triaxial failure data are scaled up to the crust taking into account temperatures and rock types. The critical depth of hydraulically permeable cracks coincides with the Mohorovicic boundary and that relates to the crust genesis. The annihilation of a crack system at this depth is in accordance with seismic velocity jump known from geophysical exploration. The features of crust floors as well as total thickness, fault inclination, etc, are explained by the suggested mechanical approach.

1. INTRODUCTION

In 1978 discussion about the limit depth of cracks inside the Earth's crust took place in our Institute. The participants were J.R. Rice, B.V. Kostrov, the authors of this presentation, et al. The answer was found later when the rock testing data were collected, Nikolaevskiy (1979) and it was simple: the Mohorovicic boundary is the limit depth for an open crack system. This conclusion contained, as one can see, also the genesis of the crust itself. The result of the following study is given below.

2. THE FISSURED EARTH'S CRUST

Failure of rocks depends on pressure and temperature conditions. The process includes microfracturing, dilatant opening of cracks, rupture localization into shear bands, etc. Triaxial standard test data permit one to find typical features, corresponding to definite PT-intervals. The adequate classification, initiated by Z. Bieniawski, J. Bernaux, H. Einstein et al., can be done by differences in the finite states of ruptured samples. The accounting for temperature effect is essential as is shown in Figure 1, published earlier, Nikolaevskiy (1996), and added here with the amphibole data.

If we compare the finite states of granite tested samples with deep seismic sounding of the Earth crust, the striking correspondence of seismic boundaries and the rupture state transitions under high PT - parameters become evident. The active tectonics, that is, shear stress action, provides the appearance of permeable faults, filled with debris of rocks that are cataclastically destroyed. The faults accelerate water (vapor) migration and this is the factor of genetic transformation of the crust rocks.

The crust consists of upper, middle and lower floors. In the first one, where the thermodynamic interval is $p \leq 0.2\ GPa$, $T \leq 200^0 C$, faults are narrow and the finite destruction is realized as vertical (I) or inclined (II) faults (Figure 1), whereas outside of them the rock mass remains in a slightly damaged state.

The fault thickness depends on the ratio of solid friction resistance (to slippage) and the strength of the intact rock material. At the transition to the middle crust they begin to have the same order, the breaking of faults edges takes place. The stick-slip phenomenon, Brace (1972), is an indicator of competition of fault asperity crushing and slippage. This happens in the PT zone (III) where the faults are widened due to dilatancy. Then low velocity zones (*LVZ*) as porous/fractured layers appear.

Because vertical rock stress is reducing with depth compared to the horizontal ones beneath the first 4 km, Bradshow & Zoback (1988), the faults plateau and *LVZ* are quasi-horizontal and compose wave-guides between the Forsch boundary (*F*) and the Conrad *(C)* boundary. They can be understood also as large scale localization of shear bands.

The thickness of *LVZ* (*waveguides*) can reach kilometers. Their low electric resistance is connected with filling of dilatant voids with brines or graphite or other sediment films.

For "normal" geothermal conditions, the interval (III)

$$0.2\,GPa \leq p \leq 0.5\,GPa, \quad 200^0 C \leq T \leq 500^0 C$$

containing a number of *LVZ*, corresponds to the middle crust.

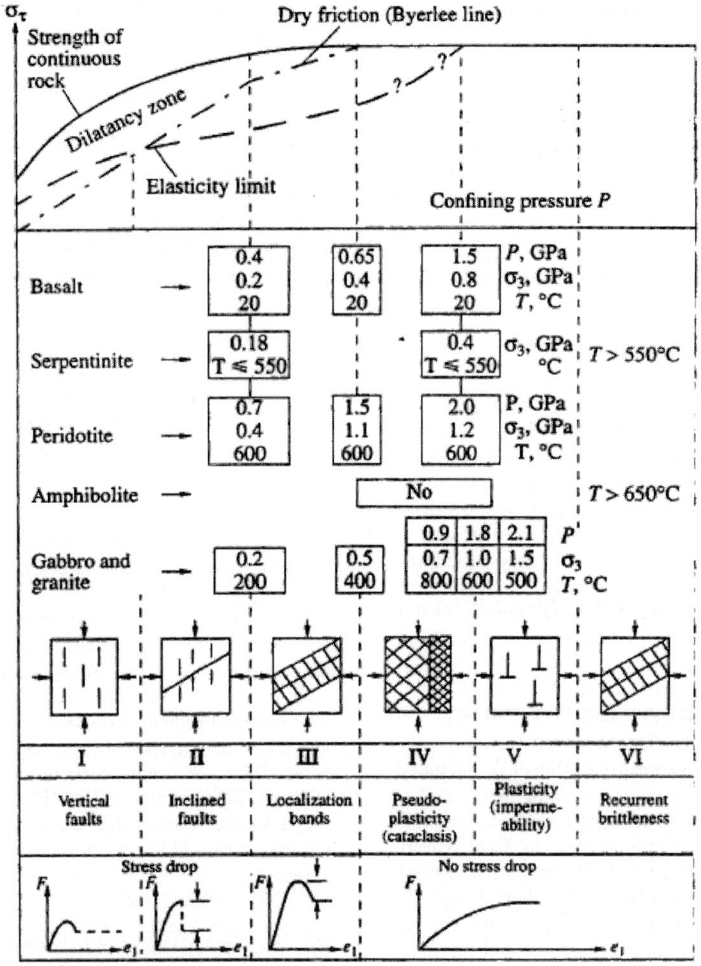

Figure 1. Classification of finite fractured states of rocks at high PT-conditions under pressure anisotropy

The *LVZ* zones can serve as feeding basins for reservoirs in the upper crust and its sedimentary massifs. The hypothesis of oil reserves connection with voids in crystalline bedrocks, Nikolaevskiy (1996), is supported by the experience of reservoirs Romashkino (Tatarstan), White Tiger (Vietnam), etc.

Transition to the lower crust corresponds to change of fracturing by faults to the global cataclastic rupture. The lower crust possesses the least resistance to the shear and the detachment of the crust floors becomes possible, see "two floor" plate tectonics suggested by Lobkovsky (1988).

The thermodynamics interval

$$0.5\ GPa \le p \le 1\ GPa, \qquad 400^0C \le T \le 600^0C$$

corresponds (Figure 1, interval IV) to half-brittle (cataclastic) type of failure that is realized in the lower crust. Here localization of damage has much more peculiar forms revealed by deep seismic sounding and it is not yet understood adequately.

3. THE MOHOROVICIC BOUNDARY

At depths of the continental Mohorovicic boundary (~ 35 km)

$$p = 1\ GPa, \qquad T = 600^0 C$$

the failure is realized at the expense of multiplication of defects and dislocation motion inside mineral crystal lattices. The "true", dislocation plastic screen appears, as follows from the triaxial tests with granite done by Tullis & Yund (1977).

In other words, at the Moho the crack system integrity is annihilated and the global hydraulic permeability of the crust is lost (Nikolaevskiy, 1979). Then the transition (phase or chemical) from the crust rock, saturated with water vapor, to "dry" rocks, (Bott, 1971), or with juvenile waters of the *upper mantle* becomes possible.

J. Tullis compared her data with the California region which has higher temperature than "normal" (Figure 3) and missed the coincidence with the Moho state.

4. STRESSES IN FISSURED CRUST

The stresses, shown in Figures 1 –3, correspond to the strength standards for *intact* geomaterials with regard to the weakening temperature factor. For example, the estimated shear stress at the Moho, required for the cataclastic fracture in the lower crust at a pressure of *1 GPa*, is *0.2 GPa* that is given in Figure 2.

After failure takes place, one has to use data on the *residual* strength of rocks. When applied to natural conditions, this means construction of a deep stress profile of the fractured crust with the use of the well-known Byerlee (1967) law,

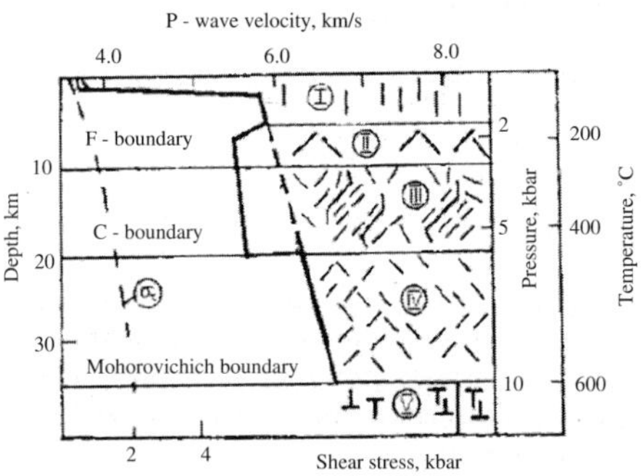

Figure 2. The fissured Earth crust and "standard", Bott (1971), seismic profiles of Europe

$$\tau = 0.85\,p \qquad (p < 0.2\ GPa), \tag{1}$$

$$\tau = 60 + 0.6\,p, \qquad (0.2 < p < 1.7\ GPa) \tag{2}$$

Here τ is the intensity of shear stresses and the pressure intervals (1) and (2) correspond to the upper and lower crust respectively. However, the upper limits in (2) exceed the values at the Moho. This is because of neglecting of temperature weakening effects in situ.

We emphasize that the Byerlee law describes the rheology of the upper (cool) crust but the lower crust has to be governed by the special theory of high-temperature creep of rocks in cataclastic states.

The data of Scott (1991) for brittle-ductile changes in sandstones are also done under room temperature but reveal the porosity factor at finite failure states. They can explain partly structures found by deep soundings in the lower crust.

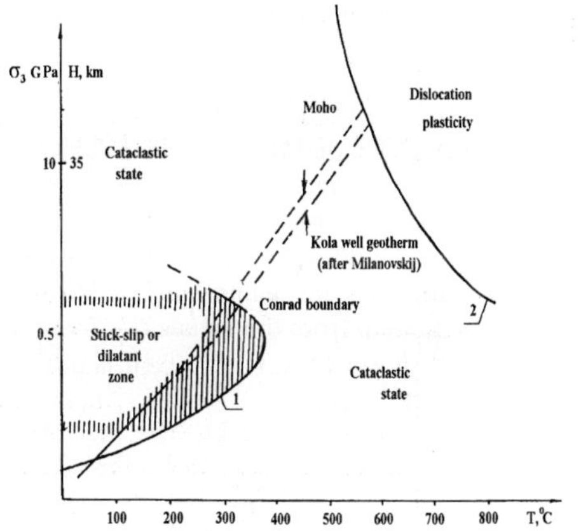

Figure 3. Geotherm and the Earth crust boundaries: (1) – stick-slip zone, (2) – the Moho states

There is one important question: why are tectonic stresses changing relatively slowly as we can see in the field? Definitely, there is a special mechanism of their redistribution typical for a wave energy transfer in a cataclastic medium. Its elements (blocks) can rotate. This adds the balance of the moment of momentum to the conventional impulse balance equations as well as spin (surplus) velocity of an individual block. The constitutive laws were suggested (Nikolaevskiy, 1996), that led to the Sin-Gordon equation with its soliton solution.

A more complicated equation system, that includes the conventional strains, was studied numerically (Mikhailov & Nikolaevskiy, 2000). The elastic shear has a form of the fast precursor wave as well as is present in a much slower rotation wave.

The rigid blocks rotate in a field that was described by Markis & Roussus (2000). The panorama of the city of Alger after earthquake shocks in May 2003 had shown that conserved houses attached to blocks had random residual inclination and this approved the rotation wave action at a ground surface.

The relative block rotation releases radon from mineral lattices at fresh wear of solid surfaces. Radon is observed in surface waters and plays the role of a tectonic events indicator, see Nikolaevskiy (1996, 1998). Subsequent radon appearance at large distances corresponds to a wave with the velocity ~ *10 km/day* measured at ground surface.

5. P-VELOCITY JUMP AT THE MOHO

The Moho was discovered by reflection of seismic waves that corresponds to the velocity of P-waves increase up to *8.2 km/s*. Let us show that the crack net annihilation leads to the necessary jump from the values *6 – 7 km/s* typical for the lower crust which itself is often *LVZ* huge layer.

There are two types of cracks. The first one when crack edges are touching each other. The second is when cracks are dilatant (open). Due to the cataclastic flow they can transfer from one state to another, Nikolaevskiy (1996).

We can estimate changes of elastic modulus of rock masses if there is isotropic net of N disc-like cracks of radius a in a unit volume, Garagash (1983), that is,

$$\lambda = \lambda_o + \frac{64}{15}\frac{1-\nu_o}{2-\nu_o}G_o\upsilon,$$

$$G = G_o - \frac{32}{15}\frac{1-\nu_o}{2-\nu_o}G_o\upsilon, \qquad (3)$$

$$\lambda_o = 2G_o\frac{(1-\nu_o)}{1-2\nu_o}$$

The Lame λ_0, G_0 and the Poisson ν_o coefficients of intact rock are used as well as crack concentration $\upsilon = N\langle a^3\rangle$. This appears at the integration of the displacement jump over the crack surface and is the ratio of the disturbed volume to the volume element per a crack.

Note that volume modulus K is the constant:

$$K = \frac{1}{3}(3\lambda + 2G) = \frac{1}{3}(3\lambda_o + 2G_0) \qquad (4)$$

These relations are valid for sufficiently small concentrations ($\upsilon \approx 0.1$ that corresponds to a distance between cracks that is twice bigger than their sizes) but some arguments can be mobilized to obtain the results for bigger concentrations. We have obtained the following differential equations:

$$\frac{d\lambda}{d\upsilon} = \frac{32}{45}\frac{3K-2\lambda}{2K-\lambda}(K-\lambda)$$

$$\frac{dG}{d\upsilon} = \frac{32}{45}\frac{3K+4G}{2K+2G}G \qquad (5)$$

If $\lambda = \frac{1}{3}(3K-2G) = \lambda_0$, $G = G_o$ at $\upsilon = 0$ then

$$G = [1 + \{1 + \frac{3K(3K+4G_o)}{4G_o^2} \times \exp(\frac{64}{15}\upsilon)\}^{1/2} \exp(-\frac{64}{45}\upsilon)] \times \qquad (6)$$

$$\frac{2G_o^2}{3K+4G_o}$$

In the case of dilatant cracks we have the two equations for the Young modulus $E = 2(1+\nu)G$ and the Poisson coefficient:

$$E = E_0[1 - \frac{16}{45}(10-3\nu_0)\frac{1-\nu_0^2}{2-\nu_0}\upsilon],$$

$$\nu = \nu_0[1 - \frac{16}{15}(3-\nu_0)\frac{1-\nu_0^2}{2-\nu_0}\upsilon];$$

$$\nu = \frac{3K-2G}{2(3K+G)}. \qquad (7)$$

Their evolution is described by the equations

$$\frac{d\ln E}{d\upsilon} = -\frac{16}{45}(10-3\nu)\frac{1-\nu^2}{2-\nu}$$

$$\frac{d\ln \nu}{d\upsilon} = -\frac{16}{15}(3-\nu)\frac{1-\nu^2}{2-\nu} \qquad (8)$$

In Figure 4 the graphs of shear modulus G and the Poisson coefficient ν are plotted for growing concentrations of cracks υ. Curve 1 shows the decrease of G with growth of υ. Curve 2 shows the practical constancy of the Poisson coefficient under changes of υ.

The density $\rho = 3100 kg/m^3$ is assumed as constant and all the changes of seismic velocity $V_p = \sqrt{(K + 4/3G)/\rho}$ happen through effective moduli. The lower crust has $G_o = 6 \times 10^{10}\,Pa$ and the crack net healing at its bottom is equivalent to decrease of υ from 0.8 value to zero. If $\nu_o = 0.3$, the curve, denoted by cross symbols, is showing that the minimum velocity in the lower crust at $\upsilon = 0.8$ is $V_p = 7400\ m/s$ and increases up to $V_p = 8230\ m/s$ at full healing of cracks ($\upsilon = 0$).

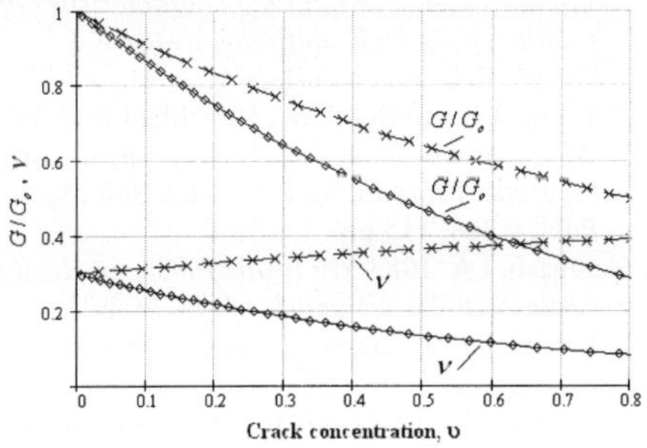

Figure 4. Ratio G/G_o and coefficient ν as functions of crack concentration υ (Cross symbol corresponds to closed cracks, diamonds - to open cracks)

In the case of dilatant cracks K and G are changed according to equations (13) and the effect is much stronger at much lower crack concentrations. This is shown by the curves denoted by diamond symbols (Figure 6). The minimum velocity in the lower crust at $\upsilon = 0.23$ equals to $V_p = 6200\ m/s$ ($\nu = 0.2$) and is growing to $V_p = 8230\ m/s$ ($\nu = \nu_o = 0.3$) at full healing of cracks ($\upsilon = 0$).

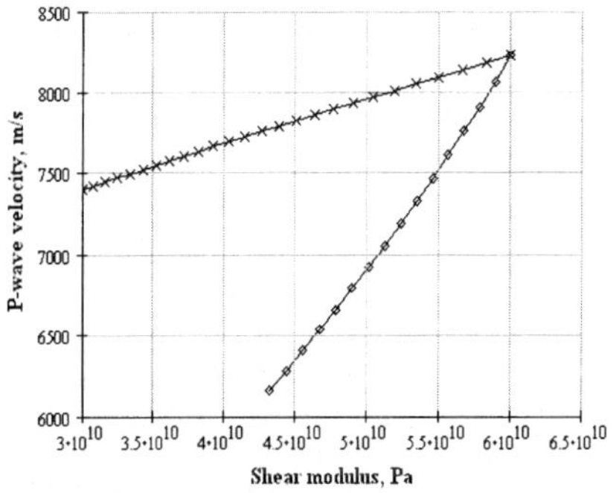

Figure 5. Increase of P-wave velocity with shear modulus (cross symbols correspond to closed cracks, diamonds - to open cracks).

6. PHASE TRANSITIONS AT MOHO

According to Bott (1971), the transformation of gabbro into granate - granulites and then into eclogite can explain the velocity increase at the Moho. Really, the rock phase or chemical changes adjoin the crack net annihilation beneath the Moho. This provides, (Nikolaevskiy, 1996), the necessary transition to "dry" rocks of the upper mantle. The petrophysical aspect of the water absence in the upper mantle is illustrated in Figure 6, where the Kennedy - Ito phase limits, (Bott, 1971), are given also. Here symbol *X* indicates the conditions on the Venus's surface where a basaltic plateau was observed.

The basalt – gabbro (the amphibole - in the case of water presence) transition to eclogite happens in upper mantle at the continental condition.

Sometimes faults are growing into the upper mantle. This happens if the lower crust is overheated as beneath the sedimentary basins. Then amphiboles are unstable (Figure 1) and the faults appear at the Moho. This point is tightly connected with the problem of hydrocarbons abiogen origin because there is assumption that excess methane is dissolved in the asthenosphere.

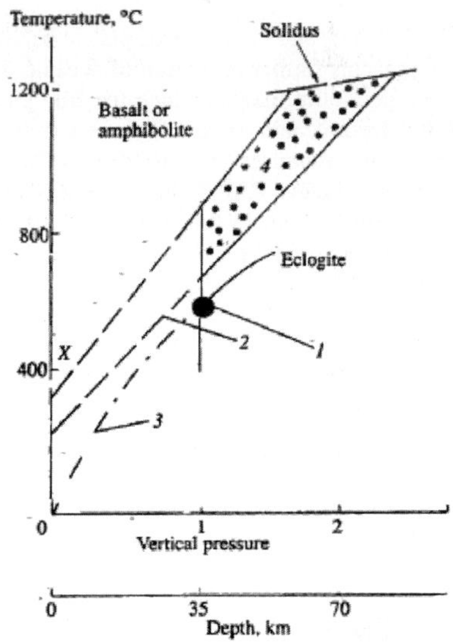

Figure 6. Penetration of water down to Mohorovichich boundary (1) prevents phase transition "eclogite - basalt" (2) according to the geoterm 3 in the Standard Earth Crust. Pressure is given in GPa. Granate –granulite is shown (4).

The meteor water and gases penetration through the crack system contributes to the genetic processes in the crust. Below the crust, the water content changes to a juvenile one.

The difference in the crust thickness is explained here by the dependency of the dislocation plasticity (Figure 3) on two parameters – pressure and temperature – and by corresponding changes of the crust rocks into a mantle type.

The conventional "isostasy" principle (Bott, 1971) of the crust thickness was motivated by capacity of mantle rock to flow, compensating the vertical pressure deviations from lithostatic level. We can see that Figure 3 yields the alternative explanation.

The Kirby & Kronenberg (1984) data showed that at temperatures above 600^0C the semi-brittle rocks were flowing, that is, became dependent on strain rates. Therefore some compensating flows in the upper mantle are possible and there is no accumulation of elastic energy (and no earthquake sources) in mantle out of subduction zones.

The transition zone in the oceanic crust is represented by serpentine, the product of the Hess reaction of olivine with water. The serpentine forms in the lower oceanic crust, because of water percolation from the ocean through brittle - fissured basalts, and exhibits true plastic behavior, which prevents further migration of oceanic water into the peridotite mantle. Since the true plastic, that is, impermeable state of the serpentine is reached at pressures of *0.2-0.4 GPa* and temperatures below *550°C*, the thickness of the oceanic crust is about 11 km, if we account for the ocean water weight, Nikolaevskiy (1979), Lobkovsky (1988).

7. CONCLUSION

The global structure of the Earth crust is in accordance with residual states of rocks in an active tectonic regime, that is, when the shear intensity is reaching the rock limit strength at high temperatures and pressures.

REFERENCES

Bernaix J. 1974. *Properties of rock and rock masses.* Adv. Rock Mech. 1, pt. A. Washington D.C.: 9 – 97.

Bott, M.H.P. 1971. *The Interior of the Earth.* London: Arnold.

Bieniawski Z.T. 1968. *Fracture dynamics of rock.* Int. J. Fracture Mech. 4:415 – 430.

Brace, W.F. 1972. *Laboratory studies of stick-slip and their applications to earthquakes.* Tectonophysics 14 (3/4): pp. 189 – 200.

Bradshaw, G.A. & Zoback, M.D. 1988. *Listric normal faulting, stress refraction and the state of stress in the Gulf Coast Basin.* Geology 16: pp. 271 – 274.

Byerlee, J.D. 1967. *Frictional characteristics of granite under high confining pressure.* J. Geoph. Res. 57, pp. 341–371.

Einstein H.H., Bruhy R.W., Hirschfeld R.C. 1970. *Mechanics of jointed rock. Experimental and theoretical studies.* MIT, R70-62, Soil Mech. Publ. # 268, 115 pp.

Garagash, I.A. 1983. *On brittle failure of elastic bodies with large number of cracks.* In: Mechanics of Tectonic Processes. Alma-Ata: Nauka.

Kirdy S.H., KronenbergA.K. 1984. *Deformation of clinopyroxenite: evidence for a transition in flow mechanics and semi-brittle behavior.* J. Geophys. Res. 89 (B5): pp.3177 – 3192.

Lobkovskiy, L.I. 1988. *Geodynamics of spreading zones, subduction and two - floor plate tectonics.* Moscow: Nauka.

Makris, N. & Roussus, Y.S. 2000. *Rocking response of rigid blocks under near-source ground motions.* Geotechnigues 50 (3): pp. 242 – 262.

Mikhailov, D.N. & Nikolaevskii, V.N. 2000. *Tectonic waves of the rotational type generating seismic signals.* Physics of Solid Earth 36 (11).

Nikolaevskiy, V.N. 1979. *The Moho as a limiting depth of brittle-dilatant state of rocks.* Proc. Acad. Sci. (DAN) 249(4).

Nikolaevskiy, V.N. 1996. *Geomechanics and Fluidodynamics.* Dordrecht: Kluwer.

Nikolaevskiy, V.N. 1998. *Tectonic stress migration as nonlinear wave process along earth crust faults.* In Adachi, T. & Oka, F. (eds.), Localization and Bifurcation for Solis and Rocks. Rotterdam: Balkema.

Scott T.E. & Nielsen K.C. 1991. *The effects of porosity on the brittle-ductile transition in sandstones.* J. Geophys. Res.96: 405.

Tullis, J. & Yund, R.A. 1977. *Experimental deformation of dry Westerly granite.* J. Geophys. Res. 82: 5705.

COMPACTION AND DIAGENESIS OF SANDSTONES – THE ROLE OF PRESSURE SOLUTION

Hideaki Yasuhara[1], Derek Elsworth[1], and Amir Polak[1]

[1])Department of Energy and Geo-Environmental Engineering and The Energy Institute, Pennsylvania State University, University Park, PA 16802, PA

Abstract: A model is presented for the compaction of granular aggregates that accommodates the serial processes of grain-contact dissolution, grain-boundary diffusion, and precipitation at the pore wall. Importantly, this treatment follows the progress of grain interpenetration as contact areas grow, mass transport lengths increase, and rate-limiting processes may switch with the progress of compaction. A simple repeating closed system incorporates two stressed grains in contact and enables the progress of compaction, and the evolution of the mass concentration of the pore fluids to be followed with time, for arbitrary mean stress, fluid pressure, and temperature conditions. Hindcast comparisons with experimental results for the compaction of quartz sand in a closed system (Elias and Hajash, 1992) show excellent agreement for rates of change of porosity, the asymptotic long-term porosity, and for the development of silica concentrations in the pore fluid with time.

1. INTRODUCTION

The compaction and concurrent cementation of granular aggregates is observed to be controlled by the rates of grain-contact crushing, grain-contact interpenetration, and the redistribution of mobilized mass from these contacts. Together, these processes control the rate of porosity loss, and evolution of strength with time. One of the main mechanisms of diagenetic compaction and deformation in sedimentary rocks is pressure solution, which involves three serial processes: dissolution at the stressed interface of grain-to-grain contacts, diffusive transport of dissolved matter from the interface to the pore space, and finally precipitation at the less-stressed surface of the grains. A variety of models have been proposed to describe deformation or compaction of porous aggregates mediated by pressure solution. These models can be classified to two main types. In the first set, a viscous law (e.g., Rutter, 1976; Raj, 1982; Dewers and Hajash, 1995) is utilized to match laboratory experiments, which are generally conducted over a relatively short period, on the order of hundreds of days. These models take into account the chemistry of the minerals and the transport of dissolved matter and are formulated from a thermodynamic standpoint. In the second set of models a purely plastic law is applied to interpret equilibrium compaction phenomena (Palciauskas and Domenico, 1989; Stephenson, 1992). Such models are developed from simple energy and volume balance considerations and quantitatively describe the compaction of aggregates in terms of their applied stress and temperature history.

In this work we follow the stress- and temperature-dependent controls on compaction where porosity loss is controlled by the evolving interpenetration of grains comprising aggregate. Interpenetration is mediated by rates of dissolution at grain-to-grain contacts, diffusion of aqueous solution along the interface of grain-to-grain contacts, and precipitation on the exposed pore wall. Importantly, resulting changes in grain geometry and pore fluid concentration are explicitly accommodated as interpenetration proceeds.

2. CONCEPTUAL MODEL

In this work a mechanistic model is presented to describe the compaction of porous aggregate that incorporates the important effects of grain interpenetration that accompany the redistribution of mass by pressure solution. The rate of compaction is controlled by the resulting mass redistribution from grain contacts to the pore walls; this is through the serial processes of dissolution at the grain interfaces, diffusive transport along the grain interfaces, and finally by precipitation on free walls, as schematically illustrated in Figure 1. Mass transfer from the grain contacts to pore walls irreversibly modifies the geometry of the grains. The reduction in contact stresses that accompanies the interpenetration of the grains provides a reinforcing feedback in slowing the rate of porosity reduction

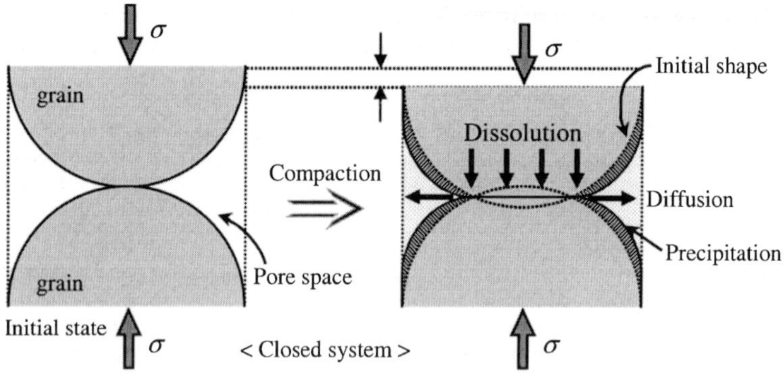

Figure 1. Schematic of pressure solution. Initially, two hemispheres are in contact. As compaction proceeds, the mineral at grain contacts dissolves due to high stress concentration. Solutes diffuse from the interface into the pore space. Finally, precipitation occurs as a result of oversaturation of solutes in the pore fluid. Grain geometries are no longer spherical.

by reducing dissolution rates and by concurrently increasing diffusion pathway lengths. The interaction of these mechanisms enables the decaying rate of porosity reduction to be followed from a mechanistic standpoint, and the evolution of quartz concentrations in the pore fluids to be followed.

2.1 Rate of compaction

Pressure solution incorporates the three serial processes; dissolution at grain contacts, diffusion along the interfacial water film, and precipitation on pore walls.

First, dissolution at the grain contacts provides a source of mass into the pore via the interface, and is defined in terms of the dissolution mass flux, dM_{diss}/dt, the rate of addition of dissolved mass into solution at the interface, given by Yasuhara et al. (2003):

$$\frac{dM_{diss}}{dt} = \frac{3\pi V_m^2 (\sigma_a - \sigma_c) k_+ \rho_g d_c^2}{4RT}, \tag{1}$$

where, V_m is molar volume of the solid (2.27×10^{-5} $m^3\ mol^{-1}$ for quartz), σ_a is the disjoining pressure (e.g., Heidug, 1995) equal to the amount by which the pressure acting at grain-to-grain contacts exceeds the hydrostatic pore pressure, k_+ is the dissolution rate constant of the solid, ρ_g is the grain density (2650 kg m^{-3} for quartz), d_c is the diameter of the grain-to-grain contacts, R is the gas constant, T is the temperature of the system, and σ_c is the critical stress, which defines stress state where the compaction of grain aggregate will effectively halt. This stress is determined considering the energy balance under applied stress and temperature conditions, given by Revil (1999) modified from Stephenson et al. (1992):

$$\sigma_c = \frac{E_m\left(1 - T/T_m\right)}{4V_m}, \tag{2}$$

where, E_m and T_m are the heat and temperature of fusion, respectively (E_m = 8.57 kJ mol^{-1}, T_m = 1883 K for quartz).

Secondly, based on Fick's first law, diffusion along interface of the grain contacts is defined in terms of the diffusive mass flux, dM_{diff}/dt, as (Yasuhara et al., 2003):

$$\frac{dM_{diff}}{dt} = \frac{2\pi\omega D_b}{\ln\left(d_c/2a\right)}\left(C_{int} - C_{pore}\right), \tag{3}$$

where, D_b is the diffusion coefficient, and $(C_{int})_{x=a}$ and $(C_{pore})_{x=d_c/2}$ are concentrations in the interface and pore space, respectively. ω is the thickness of the water film trapped at the interface, which is of the order of 10^{-9} m and dependent on applied stresses (e.g., Mullis, 1991; Heidug, 1995). However, this parameter remains ill-constrained with no magnitude established by either theoretical or empirical methods, and our calculations are made with an assumed thickness ω, of 4 nm (e.g. Revil, 2001).

Finally, precipitation of solute to the free faces of pore wall is defined in terms of the precipitation mass flux, dM_{prec}/dt, the rate of deposition of solute from the pore space onto the grain surfaces, is defined as (Yasuhara et al., 2003 modified from Canals and Meunier, 1995):

$$\frac{dM_{prec}}{dt} = V_p \frac{A}{M} k_- \left(C_{pore} - C_{eq}\right), \tag{4}$$

where, V_p is the volume of pore space, A is the relative grain surface area, M is the relative mass of the fluid, which are the dimensionless parameters defined by Rimstidt and Barnes (1980), k_- is the precipitation rate constant of solid, and C_{eq} is the solubility of quartz.

In the applied stress and temperature condition, concentration of the material strongly influences compaction process, and the concurrent calculation to obtain the concentration is significant. Concentrations may be defined at the interface between grain contacts and pore fluid for both the closed and open system. For the closed system (e.g Yasuhara et al., 2003):

$$\begin{Bmatrix} C_{int} \\ C_{pore} \end{Bmatrix}_{t+\Delta t} = \begin{bmatrix} D_1 + V_p/4\Delta t & -D_1 \\ -D_1 & D_1 + D_2 + V_p/2\Delta t \end{bmatrix}^{-1} \left[\begin{Bmatrix} dM_{diss}/dt \\ D_2 \cdot C_{eq} \end{Bmatrix}_{t+\Delta t} + \frac{1}{4\Delta t} \begin{bmatrix} V_p & 0 \\ 0 & 2V_p \end{bmatrix} \begin{Bmatrix} C_{int} \\ C_{pore} \end{Bmatrix}_t \right] \tag{5}$$

where,

$$D_1 = \frac{2\pi\omega D_b}{\ln(d_c/2a)}, \quad D_2 = V_p \frac{A}{M} k_-, \tag{6}$$

and Δt denotes the time step.

Diffusion coefficient D_b, dissolution rate constant k_+, and precipitation rate constant k_- of quartz in Eq. (1), (3), and (4) have all Arrhenius-type dependence with temperature, given by:

$$D_b = D_0 \exp(-E_D/RT), \tag{7}$$

$$k_+ = k_+^0 \exp(-E_{k+}/RT), \tag{8}$$

$$k_- = k_-^0 \exp(-E_{k-}/RT). \tag{9}$$

Appropriate magnitudes are selected for these constants defining the temperature dependence as $D_0 = 5.2 \times 10^{-8}$ m^{2}s^{-1} and E_D = 13.5 kJ mol^{-1} (Revil, 1999), k^0_+ = 1.59 mol m^{2}s^{-1} and E_{k+} = 71.3 kJ mol^{-1} (Dove and Crerar, 1990), and k^0_- = 0.196 s^{-1} and E_{k-} = 49.8 kJ mol^{-1} (Rimstidt and Barnes, 1980).

2.2 Overall computational procedure

The three processes of dissolution, diffusion, and precipitation, coupled with the associated change in geometry, are combined to define the progress of porosity reduction with time. The main points of the procedure are as follows.

Initially, the system is treated as a closed system, with the representative repetitive structure comprising two hemispherical grains in contact, as illustrated in Figure 1. The representative pore volume is assumed water filled and at an equilibrium concentration of quartz. Thus, concentrations of quartz in the interface and pore space are identical to the solubility of quartz, C_{eq}. Once stressed, during time step Δt, dissolution, diffusion, and precipitation mass are simultaneously calculated by Eqs. (1), (3), and (4), respectively. Physically, the dissolved mass evaluated from Eq. (1) is supplied to the interface, and domain shortening proceeds as this mass passes along the interface by diffusion. Some portion of the dissolved matter is removed from the interface as it exits into the pore fluid, as defined by Eq. (3). Depending upon the relative concentration differential between the pore fluid solution and the equilibrium concentration, a portion of the mass removed to the pore fluid is deposited to the grain surface (Eq. (4)) and the geometry of the grain is correspondingly modified. As a result, the porosity of the system is evaluated and updated. Simultaneously, concentrations in the interface and pore fluid are updated utilizing Eqs. (5) and (6). To conduct a consistent calculation until the system reaches an equilibrium state, an iterative procedure is utilized (e.g. Yasuhara et al. (2003)).

3. NUMERICAL SIMULATIONS

In the previous section, a mechanistic model was used to describe the compaction of the porous aggregate. To check its validity, we compare our model with existing experimental measurements (Elias and Hajash, 1992), and examine effects in porosity-loss due to applied stresses, temperatures, and grain sizes.

3.1 Comparison with experimental data

Compaction experiments have been completed on aggregate samples of quartz sand, of mean grain diameter 180-250 μm, at 150 °C (Elias and Hajash, 1992). Changes in porosity and changes in silica concentration in the pore fluid were measured through time to investigate the role of effective

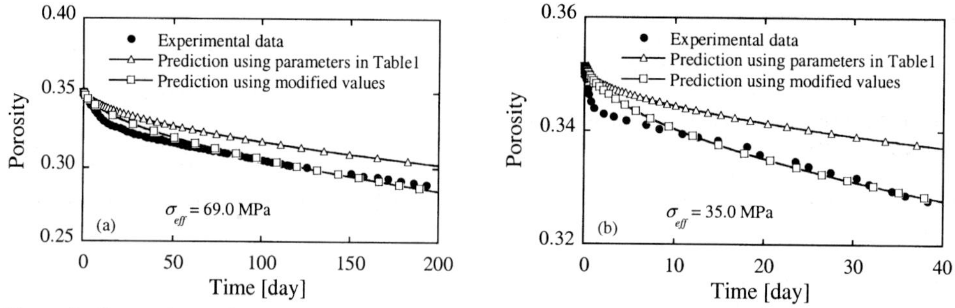

Figure 2. Comparison of porosity reduction with time between the experimental data of Elias and Hajash (1992) and the predictions of the model. Open triangles represent the prediction using parameters of $D_b = 1.12 \times 10^{-9}\ m^2\ s^{-1}$, $k_+ = 2.51 \times 10^{-9}\ mol\ m^{-2}\ s^{-1}$, and $k_- = 1.42 \times 10^{-7}\ s^{-1}$ and open squares represent the prediction using modified values, (a) with $D_b = 1.0 \times (1.12 \times 10^{-9})\ m^2\ s^{-1}$, $k_+ = 1.8 \times (2.51 \times 10^{-9})\ mol\ m^{-2}\ s^{-1}$, and $k_- = 13.0 \times (1.42 \times 10^{-7})\ s^{-1}$ for $\sigma_{eff} = 69.0$ MPa and T = 150 °C, and (b) with $D_b = 1.0 \times (1.12 \times 10^{-9})\ m^2\ s^{-1}$, $k_+ = 3.0 \times (2.51 \times 10^{-9})\ mol\ m^{-2}\ s^{-1}$, and $k_- = 60.0 \times (1.42 \times 10^{-7})\ s^{-1}$ for $\sigma_{eff} = 35.0$ MPa.

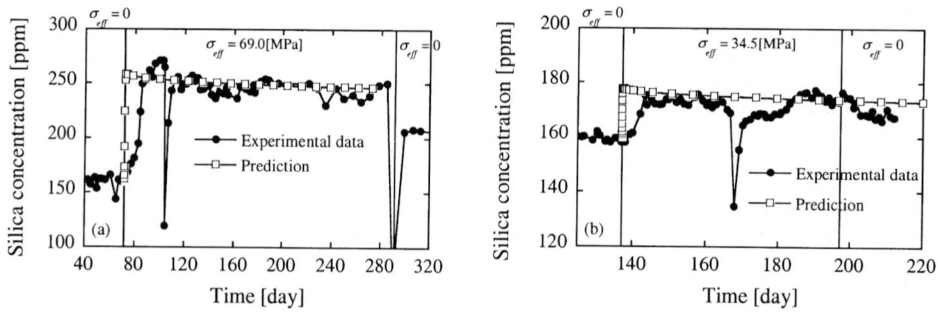

Figure 3. Comparison of silica concentration in the pore fluid between the experimental data of Elias and Hajash (1992) and the predictions of the model. (a) For $\sigma_{eff} = 69.0$ MPa and T = 150 °C, open squares represent the prediction using modified values of $D_b = 1.0 \times (1.12 \times 10^{-9})\ m^2\ s^{-1}$, $k_+ = 1.8 \times (2.51 \times 10^{-9})\ mol\ m^{-2}\ s^{-1}$, and $k_- = 13.0 \times (1.42 \times 10^{-7})\ s^{-1}$. (b) For $\sigma_{eff} = 35.0$ MPa and T = 150 °C, open squares represent the prediction using modified values of $D_b = 1.0 \times (1.12 \times 10^{-9})\ m^2\ s^{-1}$, $k_+ = 3.0 \times (2.51 \times 10^{-9})\ mol\ m^{-2}\ s^{-1}$, and $k_- = 60.0 \times (1.42 \times 10^{-7})\ s^{-1}$.

stress on silica solubility and compaction rate. In our work we attempt to follow the compaction process by predicting both rates of porosity reduction, and evolving concentrations of silica in the pore fluid.

Predicted rates of porosity reduction are shown in Figure 2 together with the data measured within the experiments. To more closely match the porosity and silica concentration histories obtained in the experiments, the principal controlling parameters of diffusion coefficient D_b, and dissolution and precipitation rate constants (k_+, and k_-) are modified with dissolution rate constant increased by factors of 1.8 and 3.0, and precipitation rate constant increased by factors of 13.0 and 60.0 for the effective stresses of 69.0 and 35.0 MPa, respectively. The predictions of porosity and silica concentration obtained using the modified magnitudes of D_b, k_+, and k_- are in remarkably good agreements with the actual data, as shown in Figures 2 and 3. In the early period of porosity reduction with time (Figure 2), each experiment has a steeper slope than predictions. This is due to effect of mechanical consolidation, neglected in the current model. The influence of consolidation can also be recognized from the silica concentration in the pore fluid (Figure 3). Model derived concentrations increase abruptly to a maximum value after loading, while for the experiments a more gradual rise results. In reality, early compaction due to rearrangement of grains is likely to dominate over the effect of pressure solution. Stresses acting on grain-to-grain contacts are insufficiently high to enable sufficient dissolution. Accordingly, silica concentration gradually increases for experiments, while the rapid increase for predictions is observed because the effect of consolidation is not taken into account.

3.2 Dependence on stress, temperature, and grain size

Compaction rates by pressure solution are well-known to be dependent on applied stresses, temperatures, and grain sizes. To examine these effects, we conduct several evaluations. Predicted rates of porosity reduction are shown in Figure 4. The results show that rates of porosity reduction increase as effective stress or temperature of the system increases and as grain size diminishes. These results are consistent with existing experimental results (e.g., Dewers and Hajash, 1995).

4. CONCLUSIONS

A mechanistic model is proposed to describe the compaction of porous aggregates that incorporates the important effects of grain interpenetration that accompany the redistribution of mass due to the three serial processes of dissolution, diffusion, and precipitation. The incorporation of these processes, at a micromechanical level allows the compaction of porous aggregates to develop in a natural manner at any particular effective stress or temperature regime. Although relatively simple, the model is capable of representing the compaction process, including defining the important contribution of mass to the evolving pore fluid mass concentration. The calculation results indicate that porosity reduction increases, in both rate and ultimate magnitude as effective stress or temperature of the system increases, and its rate increases as grain size diminishes.

ACKNOWLEDGEMENTS

The work is a result of partial support from the U.S. Department of Energy under grants DOE-BES-DE-FG02-00ER15111 and DOE-DE-PS26-01NT41048. This support is gratefully acknowledged. This work has benefited from helpful discussions with Susan Brantley, Abraham Grader, and Philip Halleck.

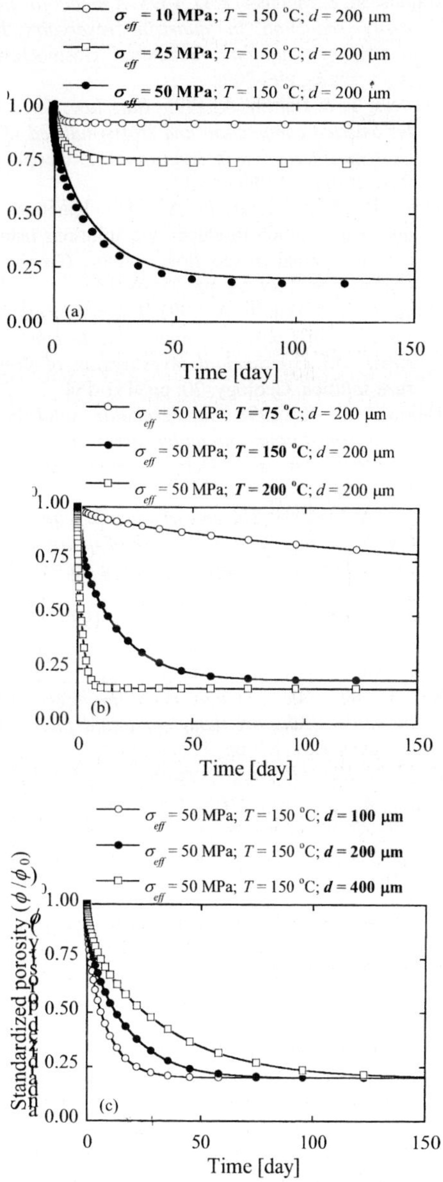

Figure 4. Predictions of porosity reduction with time for various conditions of system. Parametric study is conducted to examine the effects of the compaction process on (a) applied stress, (b) system temperature, and (c) grain size.

REFERENCES

Canals, M. & Meunier, J. D. 1995. *A model for porosity reduction in quartzite reservoirs by quartz cementation.* Geochim. Cosmochim. Acta, 59: pp. 699-709.

Dewers, T. & Hajash, A. 1995. *Rate laws for water-assisted compaction and stress-induced water-rock interaction in sandstones.* J. Geophys. Res., 100: pp. 13,093-13,112.

Dove, P. M. & Crerar, D. A. 1990. Kinetics of quartz dissolution in electrolyte solutions using a hydrothermal mixed flow reactor, *Geochim. Cosmochim. Acta,* 54: pp. 955-969.

Elias, B. P., & Hajash, A. 1992. *Change in quartz solubility and porosity change due to effective stress: An experimental investigation of pressure solution.* Geology, 20: pp. 451-454.

Heidug, W. K. 1995. *Intergranular solid-fluid phase transformations under stress: The effect of surface forces.* J. Geophys. Res., 100: pp. 5931-5940.

Mullis, A. M. 1991. *The role of silica precipitation kinetics in determining the rate of quartz pressure solution,* J. Geophys. Res., 96: pp. 10,007-10,013.

Palciauskas, V. V. & Domenico, P. A. 1989. *Fluid pressures in deforming porous rocks.* Water Resour. Res.: pp. 32, 3041-3049.

Raj, R. 1982. *Creep in polycrystalline aggregates by matter transport though a liquid phase.* J. Geophys. Res., 87: pp. 4731-4739.

Revil, A. 1999. *Pervasive pressure-solution transfer: a poro-visco-plastic model.* Geophys. Res. Let., 26: pp. 255-258.

Revil, A. 2001. *Pervasive pressure solution transfer in a quartz sand.* J. Geophys. Res., 106: pp. 8665-8686.

Rimstidt, J. D. & Barnes, H. L. 1980. *The kinetics of silica-water reactions.* Geochim. Cosmochim. Acta, 44: pp. 1683-1699.

Rutter, E.H. 1976. *The kinetics of rock deformation by pressure solution.* Philos. Trans., R. Soc. London, Ser. A, 283: pp. 203-219.

Stephenson, L. P., Plumley, W. J. & Palciauskas, V. V. 1992. *A model for sandstone compaction by grain interpenetration.* J. Sediment. Petrol., 62: pp. 11-22.

Yasuhara, H., Elsworth, D., & Polak, A. 2003. *A mechanistic model for compaction of granular aggregates moderated by pressure solution,* J. Geophys. Res. (under review).

MEASUREMENT AND 2D MODELING OF FLUID CONTROL ON THE HYDROMECHANICAL BEHAVIOR OF A FRACTURED RESERVOIR

I. Kadiri[1], V. Merrien-Soukatchoff[1], Y. Guglielmi[2], K. Su[3]
[1]) Laboratoire Environnement Géomécanique Ouvrages, Ecole des Mines, Nancy, France
[2]) Geosciences Azur, Université de Nice-Sophia Antipolis, France
[3]) ANDRA, Châtenay-Malabry, France

Abstract: In order to enhance understanding of hydromechanical rock mass processes, original *in situ* field experiments have been carried out since 1997 on a small fractured calcareous rock mass in southeastern France, at a site called Coaraze. On this site, repeated hydraulic loadings are produced by closing a floodgate located on a downstream spring. During the experiment, simultaneous measurements of normal displacement and pressure are recorded both on several joints and in the rock mass. The present paper first describes both the site and conceptual interpretation of the measurements performed. We showed that accurate measurement reproduction requires incorporating certain features: a variable normal joint stiffness, the joint vicinity, and the initial stress state. The paper then develops an original analytical interpretation of the coupled pressure-displacement measurements carried out, which reveal the high level of measurement suitability to the non-linear behaviours examined. 2D modelling, using the discontinuous code (UDEC) was used to proceed with the calibration step.

1. INTRODUCTION

The hydromechanical behavior of fractured media can be described by numerical models based either on a representation of an equivalent continuous medium or on an explicit representation of the fracture network. The continuous equivalent medium approach requires the fracture network to be dense and very well-connected, with a narrow distribution of openings to ensure the absence of preferential channels, Bour (1996). A solid mass generally features several fracture scales, which serve to prevent against adopting a scale beyond which the permeability tensor remains constant, Vuillot (1995). Discrete modeling techniques are therefore frequently applied to model the thermo-hydromechanical behavior of fractured media. Such an approach however requires either an accurate localization of fractures or good knowledge of their statistical parameters, as well as a characterization of the hydraulic and mechanical properties of these fractures. The aim of the present study is to test the compliance of normal fracture behavior with *in situ* measurements and to explore the potential for determining mechanical characteristics by means of calibration with these measurements.

We will begin by describing the Coaraze experimental site, as well as the experiments for measuring fracture opening or closing, coupled with pressure variations, Guglielmi (1999). These results will then be analyzed first to show the importance of taking variable normal joint stiffness into account and then to calculate these stiffnesses. From the parameter sets derived herein, a calibration procedure will be carried out using the UDEC computer code, ITASCA (1996).

2. EXPERIMENTAL PROCEDURE

2.1 Experimental site: Hydrogeology and geometry of the fractures

The Coaraze site, located in the coastal part of the French Alps, consists of a small limestone reservoir with a volume of approximately 19,000 m^3 drained by a spring running at an annual mean flow rate of 10 l/sec (see Fig. 1). This spring has been equipped with a valve that, when closed, causes water to rise in the rock mass until reaching a height of approximately 10 m. The evolution in hydromechanical behavior of the fractures within the saturated zone is recorded by variations in both water pressure and time, Guglielmi (1999).

The fracture network consists of faults oriented N140° with a dip of 75°NE, bedding joints oriented N40° with a dip of 45°E and faults oriented from N50° to N90° with a dip of 70° to 80° SE, Fénart (2002).

Figure 2 shows a transverse section with the small valley situated near the floodgate.

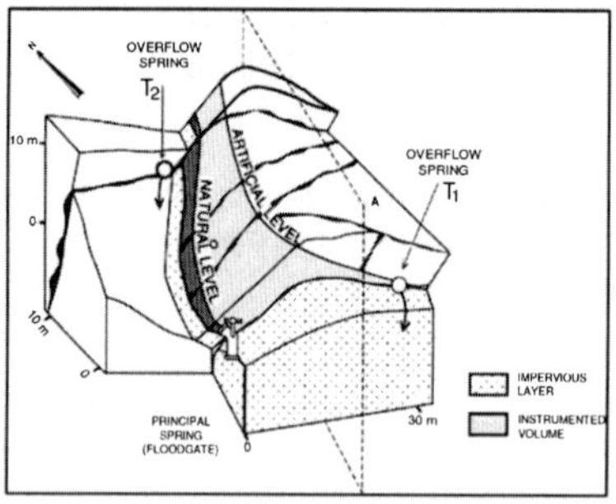

Figure 1: Block-diagram of the site (according to Guglielmi, modified in 1999)

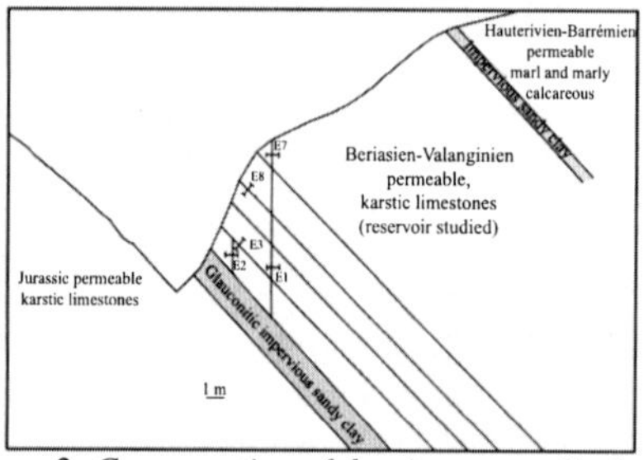

Figure 2: Cross-section of the site perpendicular to the valley, indicating the positions of extensometers E1, E2, E3, E7 and E8

2.2 Instrumentation

Measurements are taken on the three principal joint directions and in 6-cm diameter boreholes. Each measurement point corresponds to two boreholes, both perpendicular to the joints and 20 cm apart (see Fig. 3). A pressure gauge for measuring hydrostatic pressure is placed in the first borehole and a vibrating wire extensometer is installed in the second in order to monitor deformation normal to the joint. This extensometer is positioned within a concrete cylinder. The deformation measurement is thus that of the cylinder with a 6-cm diameter and 50-cm length. The recorded deformations are then transformed into normal displacements by referring to the basic measurement length, i.e. the 15-cm extensometer length.

The eight simultaneous pressure measurements (using pressure gauge M) and deformation measurements (using extensometer E) are conducted near the floodgate, where pressure variations are highest. The instrumented volume corresponds to a parallelepiped of 1,500 m^3.

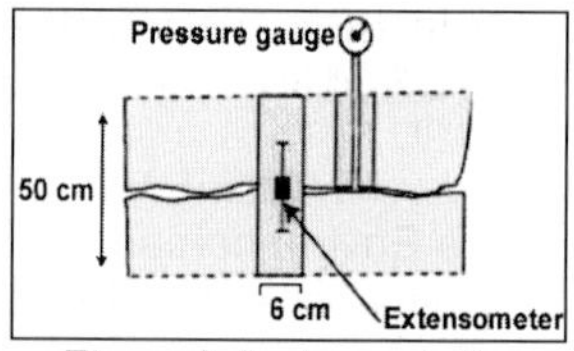

Figure 3: Instrumentation

The experiment itself consists of closing the floodgate during 60 minutes. The deformation vs. time curves show a signal characterized by deformations varying from 1 to 500 μm/m, depending on the point being examined. These deformations may be either shortenings or extensions. We have observed two types of joint deformation behavior over time:

- The first type (in the drains) features significant deformation values (20 and 250 μm/m) at the end of the experiment. Such is the case for measurement couples (E1, M3) and (E7, M9), which are localized on the subvertical N140° joint and a subvertical N50° fault, respectively.
- The second type of behavior occurs in the slightly-conducting joints (mainly bedding joints), which correspond to weak deformations (on the order of one μm). In the case of measurement couple (E3, M4), corresponding to a bedding plane, the joint closes even though water pressure is increasing.

3. ANALYSIS OF MEASUREMENTS

The analysis herein was carried out on the experimental data obtained from the 1-hour floodgate closing. The experimental results were analyzed by applying the BB model (Barton-Bandis model), Bandis *et al.* (1983), along with the continuously-yielding joint model (CYJM), developed by ITASCA (1996). These two models propose a variation in normal stiffness as a function of normal stress, yet they express this behavior in different ways. All experimental studies have demonstrated that joint closure and joint stiffness are remarkably nonlinear with respect to effective normal stress, Goodman (1976), Bandis *et al.* (1983). The BB model is frequently applied in the literature to interpret laboratory tests; for this reason, its use on the Coaraze site seemed appropriate to us.

3.1 The BB model (1983)

Bandis *et al.* (1983) expressed the normal

stiffness of joint K_n as a function of normal displacement U_n in the following equation, where U_m is the maximum joint closure and K_{ni} the initial normal stiffness (at $\sigma'_n = 0$):

$$K_n = \frac{\partial \sigma'_n}{\partial U_n} = \frac{K_{ni}}{(1 - \frac{U_n}{U_m})^2} \quad (1)$$

In assuming the transition from a normal effective stress σ'_{n1} to normal effective stress σ'_{n2} (with $\sigma'_{n2} < \sigma'_{n1}$ since we are studying the stress variation due to pressure in the joint), we can write the effective stresses as a function of displacements starting with a zero effective stress, by means of the following relationships:

$$\sigma'_{n1} = \int_0^{U_{n1}} \frac{K_{ni}}{(1 - \frac{U_n}{U_m})^2} dU_n = \frac{U_{n1}}{\frac{1}{K_{ni}} - \frac{U_{n1}}{K_{ni} U_m}} \quad (2)$$

$$\sigma'_{n2} = \int_0^{U_{n2}} \frac{K_{ni}}{(1 - \frac{U_n}{U_m})^2} dU_n = \frac{U_{n2}}{\frac{1}{K_{ni}} - \frac{U_{n2}}{K_{ni} U_m}} \quad (3)$$

The normal displacement due to the transition of stress σ'_{n1} to stress σ'_{n2} is expressed by:

$$U_n = \frac{\frac{\sigma'_{n1}}{K_{ni}}}{1 + \frac{\sigma'_{n1}}{K_{ni} U_m}} - \frac{\frac{\sigma'_{n2}}{K_{ni}}}{1 + \frac{\sigma'_{n2}}{K_{ni} U_m}} \quad (4)$$

In order to determine a linear relationship between displacement and stress variation, we have carried out the following transformation (this facilitates a lot the approximation of the unknown parameters of equation (4) which can be done easily by linear regression).

$$\frac{U_n}{(\sigma'_{n1} - \sigma'_{n2})} = \frac{U_m^2 K_{ni}}{(\sigma'_{n1} + U_m K_{ni})^2} + \frac{U_n}{(\sigma'_{n1} + U_m K_{ni})} \quad (5)$$

Coupling the normal displacement with pressure inside a fracture thus necessitates knowing the variation in normal stress due to pressure increase. When a joint saturates with water, we state *a priori* that the variation in normal effective stress is merely the pressure inside this joint. However, such an assumption presupposes that the normal total stress within the joint is not undergoing variation. Simple modeling performed using the UDEC code has indicated the presence of variation in normal total stress within the joint during saturation, except for very special cases where joint displacement is not constrained by the rigidity of the surrounding matrix. This model has shown the existence of a Z coefficient that serves to define the variation in normal total stress within the joint as: $\Delta\sigma_{total} = (1\text{-}Z)\,P$, where Z depends on the boundary conditions, the mechanical characteristics of the rock matrix and joints, and the geometrical configuration of fractures around the examined joint (see Table 1 and Fig. 4). This coefficient should not be confused with a Biot coefficient, which remains an intrinsic property of the material (and resembles a Skempton coefficient).

Table 1: Various evaluations of Z for different joint stiffness (K_n), Young's modulus of the matrix (E) and boundary conditions

	Boundary conditions	K_n (Pa/m)	E (MPa)	Z
Test 1	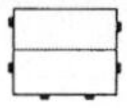	10^{10}	50,000	1
Test 2		10^{10}	50,000	0.62
Test 3	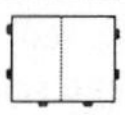	10^{10}	5,000	0.88
Test 4	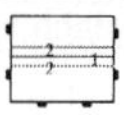	10^{10} for 1 10^9 for 2	50,000	0.58

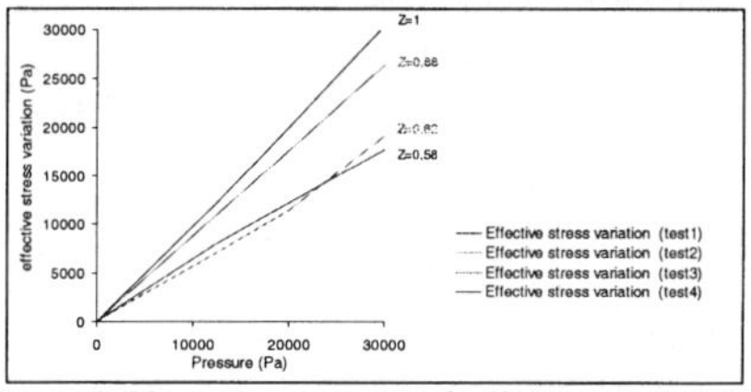

Figure 4: Variation in normal effective stiffness vs. pressure for different Z coefficients

On the basis of these results, the relationship between pressure and normal displacement for each fracture is as follows:

$$\frac{U_n}{P} = Z\left[\frac{U_m^2 K_{ni}}{(\sigma'_{nl} + U_m K_{ni})^2} + \frac{U_n}{(\sigma'_{nl} + U_m K_{ni})}\right] \quad (6)$$

where U_n is the normal displacement recorded in the joint during the experiment, σ_{ni} the normal stress in the joint before saturation of the site, and P the pressure in the joint. We have thus identified a linear relationship between normal displacement and pressure as a function of both the mechanical characteristics of the joints and the initial normal stress inside the joints. The measurement couple (E1,M3), transformed into the form $(U_n, U_n/P)$, was then expressed graphically (Fig. 5).

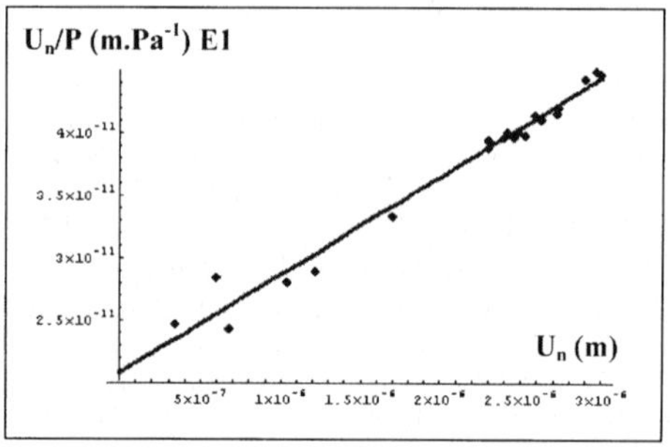

Figure 5: U_n/P vs. U_n (example of point E1)

We can note that this curve is actually linear. Such linearity is also observed for other measurement points. A linear regression makes it possible to calculate $U_m^2 K_{ni}$ and account for a maximum normal stress value at each measurement point.

3.2 *Continuously-yielding joint model (CYJM), ITASCA, 1996*

According to the CYJM model, the variation in normal stiffness as a function of normal effective stress is expressed as:

$$K_n = \frac{\partial \sigma'_n}{\partial U_n} = a_n \; \sigma'^{\,e_n}_n \quad (7)$$

Using the same procedure as for the analysis conducted in the previous section and considering $e_n = 2$, we can then determine a linear relationship between the pressure and normal displacement in the fractures, as follows:

$$\frac{U_n}{P} = Z\left[\frac{1}{a_n \sigma'^2_{ni}} + \frac{U_n}{\sigma'_{ni}}\right] = A + B \cdot U_n \quad (8)$$

where U_n is the normal displacement in the joint during the experiment, σ_{ni} the normal stress in the joint before saturation of the site, and P the pressure in the joint. This linear regression allows us to compute the results provided in Table 2:

Table 2: Hydromechanical parameters of measurement points E1 and E2 given by CYJM (with a_n in $Pa^{-1}.m^{-1}$ and σ'_{ni} in Pa)

	E1	E2
$A = \frac{Z}{a_n \sigma'^2_{ni}}$	2.07×10^{-11}	2.07×10^{-11}
$B = \frac{Z}{\sigma'_{ni}}$	7.86×10^{-6}	1.22×10^{-5}
$\frac{B^2}{A} = Z \cdot a_n$	2.97	16.69
$\frac{1}{B} = \frac{\sigma'_{ni}}{Z}$	1.27×10^5	8.14×10^4

Since Z is less than 1 for drains, this linear regression allows estimating a maximum value of σ'_{ni}, the normal effective stress without pressure.

Both the BB model and CYJM explain the Coaraze measurements relatively well. We should note the negative aspect of the analytical tests carried out: it cannot be affirmed that the models applied are indeed appropriate, yet it can be guaranteed that they do not contradict recorded measurements.

4. TWO-DIMENSIONAL MODELING OF THE COARAZE SITE

4.1 *Modeling principle*

A transverse section of the small valley located near the floodgate has been modeled (see Fig. 2). The choice of this cross-section was dictated by the fact that measurement points are positioned close to this section and can thus be projected onto it. The rock matrix is considered as an elastic, linear and isotropic material. The input characteristics have been derived from laboratory tests (E = 50,731 MPa, ν = 0.3). Joint behavior is also regarded as elastic due to the reversibility of the experiments.

The influence of model size was tested by producing increasingly wider and higher models. This influence becomes negligible beyond a distance corresponding to a cross-section 400 m wide and 650 m high, Kadiri (2002). Several simulations have been run in order to observe the influence of both boundary conditions and the ratio of horizontal to vertical initial stresses ($k = \sigma_{ih0}/\sigma_{iv0}$) on the values and distribution of stresses in the rock mass.

The normal stresses obtained at measurement points E1 and E2 with displacement boundary conditions are ten times weaker than those obtained with stress boundary conditions. Equivalent results have been obtained for variable ranges of normal and shear fracture stiffnesses. Without on-site stress measurements, choosing between these various boundary conditions is difficult. The simultaneous pressure-deformation measurements at the Coaraze site have however enabled carrying out an analysis on the measurement points as well as determining a threshold for normal stress in the joints (see Table 2). The selected boundary conditions are those that more accurately reproduce the initial stresses at Coaraze (i.e. displacement boundary conditions).

While the methodology used has provided a comparative base for choosing Coaraze site boundary conditions, it remains more general and can be applied to various modeling problems.

Hydraulic modeling consists of both applying successive initial hydraulic loads extending from 1 to 10 m and then carrying out a hydraulic calculation in the steady state (once hydromechanical equilibrium has been established, only the final state is analyzed). All model boundaries are impervious (see Fig. 6).

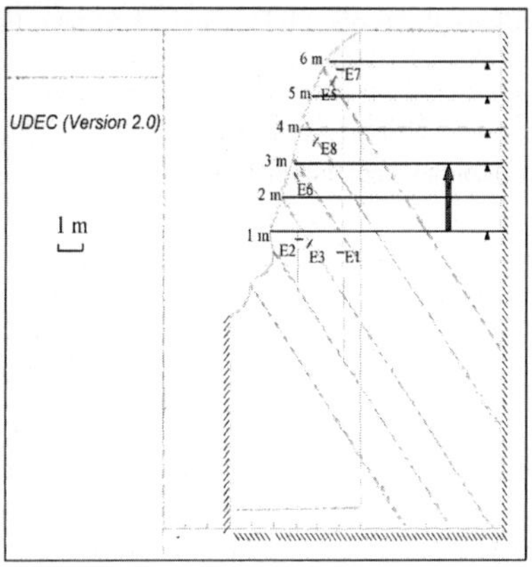

Figure 6: Modeled cross-section

The purpose of this model is to simulate the successive phases of water rising within the rock mass due to floodgate closing.

First of all, a direct analysis was conducted by supposing that the Z coefficient is equal to 1. This assumption amounts to neglecting the effect of fracture interaction within the medium. Each fracture is thus seen as isolated from the site as a whole. For each fracture, parameter a_n is given by the results in Table 1. The hyperbolic shape of the (P,U_n) measurement curves has been well reproduced; nevertheless, comparisons between experimental measurements and modeling results show significant differences. A calibration procedure using variable stiffnesses was carried out with UDEC in order to characterize the fractures and move results closer towards measurements. This procedure consisted, for fractures corresponding to the E1 and E2 measurements, of setting a starting value of a_n (determined by assuming Z = 1) and then gradually increasing this value for each joint until equality between a measurement and a modeled couple (P,U_n) has been obtained. In the model therefore, we have imposed the head corresponding to pressure P at the target calibration point. A steady-state calculation is then performed and the (P_{model},a_{nmodel}) couple is compared with the $(P_{measurement},a_{nmeasurement})$ couple. Figure 7 displays a comparison between numerical and measured couples (E1,M3) and (E3,M4). Results are qualitatively similar for the other measurement points. For point E3, we can note that joint closure, despite the pressure increase, has been well reproduced (Fig. 7). This closure is explained by the strong interaction between the joint and the adjacent fractures.

Calibration may be performed on fractures containing at least one measurement point. On the Coaraze site, no measurements are actually able to provide a preliminary a_n coefficient (under the Z = 1 assumption) on the bedding joints. In the absence of an initial a_n value, calibration of these joints proves to be less straightforward and robust.

5. DISCUSSION

By using joints with a variable normal stiffness and introducing the Z coefficient into computations, we have been able to simulate nonlinear pressure-displacement curves close to site measurements. In order to set up normal joint behavior, it was necessary to determine a Z coefficient for each fracture (given that each fracture had a different surrounding structure). The *a priori* determination of Z however was not possible in the case of the Coaraze site. A Z-

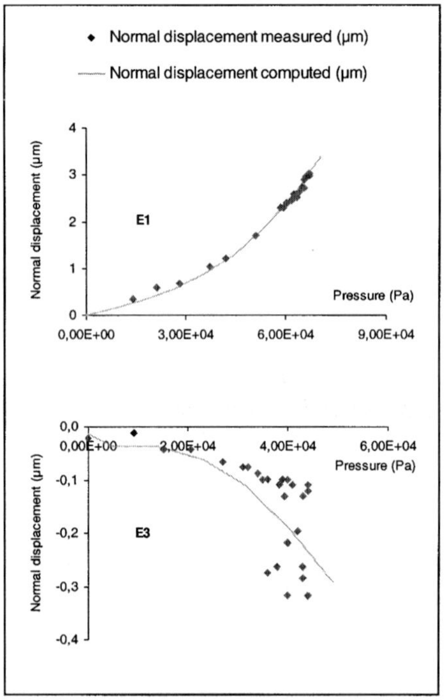

Figure 7: Comparison between measurements and numerical results

coefficient calibration method was thus required to model this site. Deriving a more precise interpretation of the phenomena recorded on site and analyzing the observed scale effects will necessitate further laboratory testing on joints.

Steady state modelling has allowed restoring the global shape of pressure-displacement curves, yet does not explain the heterogeneity of hydraulic behaviour: for two close points, pressure can be quite different at a given time. These phenomena are only partially taken into account by modelling transient flow. 3D modelling is currently being carried out in order both to confirm our methodology and to better incorporate flow paths.

We currently do not have any measurement concerning shear stiffness of the discontinuities of Coaraze. Measurements of shear displacement will be undertaken and analyzed in order to characterize the shear in situ behaviour of discontinuities and to quantify its influence on hydraulic conductivity of fracture network.

ACKNOWLEDGEMENT

The author would like to thank ANDRA and the FORPRO programme for their financial support.

The technical contributions by a number of colleagues working on the project are also gratefully acknowledged.

REFERENCES

Bandis S.C., Lumsdena C., Barton N., 1983. *Fundamentals of rock joint deformation*, Int. J. Rock Min. Sci., Vol. 20 (6), pp. 249-268.

Bour O., 1996. Transferts de fluides dans les milieux fracturés. Effets d'échelle. Ph.D. thesis, Université de Rennes I, 1996.

Fénart P., 2002. *Caractérisation du comportement hydromécanique des massifs rocheux fissures*, Ph.D. thesis, Université de Montpellier II, December 3, 2002, 308 p.

Goodman R., 1976. Methods of geological engineering in discontinuous rock, New York, West Publishing Company.

Guglielmi Y., 1999. Apport de la mesure des couplages hydromécaniques à la connaissance hydrogéologique des réservoirs fissurés. Approche sur site expérimental. HDR, Université de Franche-Comté, Besançon, 158 p.

ITASCA Consulting Group, Inc., 1996. *UDEC, Universal Distinct Element Code, Version 3.0.*

Kadiri I., 2002. *Modélisation hydromécanique des milieux fracturés*, Ph.D. thesis, LAEGO, Ecole des Mines, INPL, France, October 28, 2002, 143 p.

Vuillot E., 1995. Modélisation thermo-hydro-mécanique de massifs rocheux discontinus. Thèse INPL Nancy.

Theme 6
Coupled T-H-M-C Processes in Geotechnical and Environmental Engineering

SIMULATION OF COUPLED THERMAL AND SOLUTE CONCENTRATION EFFECTS ON DENSE RADIOACTIVE WASTE MIGRATION IN DEEP AQUIFERS

Alexander I. Zinin [1], Galina A. Zinina [1], Vitaly M. Kurochkin [2], Andrey I. Rybalchenko [2], Andrey A, Zubkov [3], Sergey P. Pozdniakov [4]
[1]) State Scientific Center Institute of Physics and Power Engineering
[2]) VNIPIPpromtexnologii,
[3]) Siberian Chemical Combine
[4]) Moscow State University

Abstract: The focus of our investigation is simulating local thermal and solute concentration effects on liquid nuclear waste spreading in natural groundwater flow. To study these effects 3-D non-isothermal non-constant density flow and transport model simulated conditions of the radioactive waste injection into a multiple layer system at Siberian Chemical Combine was developed. The used mathematical model includes: equation of continuity, general form of Darcy's law, equation of heat generation and transfer, transport equation for neutral species affected by liquid density and transport equation for radioactive species. The simulation scenario includes 50 years waste injection into a sand aquifer that isolated from top and bottom by semipermeable clay units and after long-term wastes migration in groundwater. Modeling results show that thermal and density effects influence vertical waste spreading. The thermal effect is more significant for radioactive species, while the density effect is more significant for neutral species.

1. INTRODUCTION

Radioactive waste has been injected in deep artesian aquifers at the Siberian Chemical Combine in western Siberia since 1963. The detailed description of hydrogeology of the site and injection setting are given in the books Rybalchenko et al., 1998 and Bradley, 1997. The total volume of waste injected within disposal areas 18 and 18a into aquifers 300 to 400 m below land surface is about $4{\cdot}10^7$ m^3. Most of this waste injected at area 18 is low-level unprocessed waste with total specific activity ranged from 10^{-8} up to 10^{-6} Ci/l. Processed dense intermediate-level waste (MLW) is injected at area 18a. The volume of injection is much less than at area 18. This waste is composed of mostly weakly alkaline solution containing up to 300 g/l total dissolved solids (TDS). The radionuclide composition of this waste includes strontium, cesium, ruthenium and cerium. Total specific activity of waste injected at area 18a ranges from 0.01 -0.1 Ci/l. A few wells placed within the central part of this area periodically injected with HLW waste with activity up to first Ci/l. During injection of this waste radioactive heating occurred with temperature rising in the injection zone up to 180°C (Rybalchenko et al. 1998). Development of a regional flow and long-term transport model for Tomsk site (Shestakov et al., 2002) needs assessment of the influence of local scale effects related with density and temperature of injected wastes at area 18a on transport of this waste by regional groundwater flow. This paper summarizes the results of numerical studies of density and temperature effects on regional waste transport. The simulation runs were performed using the original numerical code GEON-3DM, allowing simulation of coupled thermo-hydro-transport processes for non-constant density/viscosity compressible fluid flow in a porous medium.

The main goal of the work was estimating only the effects of density and temperature on subsurface waste spreading; that is why all simulations were made for a simplified 3-D hydrogeologic setting of the injection site.

2. GEOMETRY OF MODELLED DOMAIN AND HYDRAULIC PARAMETERS

The modeled domain is a rectangular parallelepiped in Cartesian (x,y,z) coordinates. The size of the domain is 5000 · 3000 · 230 m. The plane $y = 0$ is a symmetry plane for the model. In vertical section the domain consists of seven layers with different hydraulic properties.

Thicknesses and properties of all 7 layers are constant in space. The injection zone is placed in the center of the modeled domain. This zone consists of two subzones (zone 1 and zone 2). The plane geometry of two subzones is a rectangle. External subzone (zone 1) is the zone of injection of MLW and internal subzone (zone 2) is the zone of injection of HLW.

Figure 1 shows the geometry of the modeled domain in *xz* section and used values of hydraulic conductivity and porosity of layers. Labeling of layers and its hydraulic properties are given in accordance with work Rybalchenko et al. (1998).

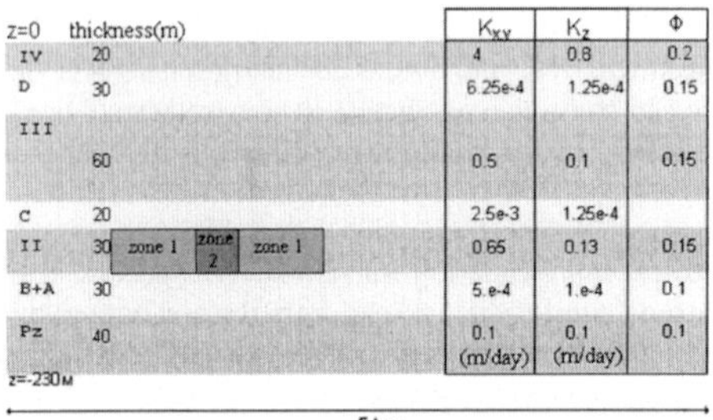

Figure 1. Vertical section of the modelled domain. Grey layers labelled by numbers are aquifers, white layers labelled by letters are semi-permeable units.

3. SOURCE MODEL

For modeling following nuclide composition of injected wastes is assumed: ^{90}Sr (25% of activity), ^{137}Cs (25% of activity) and "effective" short-lived radionuclide with half-decay period 1 year (50% of activity). It is supposed that ^{90}Sr is a non-sorbed species while ^{137}Cs and short-lived radionuclides are sorbed species. The used parameters of these nuclides are shown in Table 1.

Table 1. The nuclide parameters.

Nuclide	Half-life	Energy release	Retardation factor R	Activity in zone 2	Activity in zone 1
	Year	mW/Ci	-	Ci/l	Ci/l
^{90}Sr	28.4	6.67	1-5	1	0.02
^{137}Cs	30	4.43	∞	1	0.02
Short-lived	1	8.85	∞	2	0.04

It is supposed that injection into the aquifer II (fig. 1) is performed during 50 years with average rate 16.7 m³/d within the zone 2 and 483.3 m³/day within the zone 1. The injection rate for each subzone is uniformly distributed over the area and time of injection. The initial temperature of injected waste is 20°C. For TDS three variants were considered: a) small TDS concentration: 20 g/l; b) medium TDS concentration: 100 g/l and c) high concentration of TDS: 200 g/l for HLW and 250 g/l for MLW. The length of the injection period is 50 years, and evolution of the waste in natural groundwater flow is modeled during 250 years after termination of the injection.

4. MATHEMATICAL MODEL OF COUPLED THERMO-HYDRO-TRANSPORT PROCESSES

The coupled system of equations includes equation of continuity, Darcy's law for non-constant density of fluid, balance of thermal energy equation, transport and mass balance equations for TDS and nuclides and closure equations for density and viscosity of fluid (Pruess, 1991). It is supposed that thermal properties do not depend on temperature.

The mass balance of fluid in porous rock can be written as:

$$\frac{\partial \Phi \rho}{\partial t} + \nabla(\rho \vec{v}) = Q_{in}\rho_{in} \tag{1}$$

where Φ is porosity, ρ is density of fluid, kg/m³; Q_{in}, ρ_{in} are volumetric distribution of sources and density of fluid in sources, 1/day, kg/m³;

$\vec{v}$ is Darcy's velocity:

$$\vec{v} = -\frac{\hat{k}}{\mu}(\nabla P + \rho g \vec{e}_z) \tag{2}$$

where P is pressure, Pa, g is gravity acceleration, $\vec{e}_z$ is unit vector of z axes, μ is dynamic viscosity, $\hat{k}$ is permeability tensor.

Using the precondition of local geochemical equilibrium, the mass balance equation for concentration TDS C can be written as:

$$\frac{\partial(R\Phi\rho C)}{\partial t}+\nabla(C\rho\vec{v})=\nabla(\rho\hat{D}\nabla C)+Q_{in}\rho_{in}C_{in} \qquad (3)$$

where C_{in} is concentration of TDS in sources;

$\hat{D}$ is hydrodynamic dispersion tensor, R is retardation factor.

Transport equation for radionuclide i (i=1,2,3) takes into account radioactive decay and retardation:

$$\frac{\partial(R_i\Phi\rho a_i)}{\partial t}+R_i\Phi\rho\lambda_i a_i+\nabla(\rho a_i\vec{v})= \qquad (4)$$
$$Q_{in}\rho_{in}a^i_{in}+\nabla(\rho\hat{D}_i\nabla a_i)$$

where a_i is activity of species i in porous fluid; a^i_{in} is activity of species i in injected wastes; λ_i is the constant of decay.

For nuclides that are fully sorbed, i.e., $R_i=\infty$ eq. (4) is modified into equation of activity balance in sorbed phase:

$$\frac{\partial((1-\Phi)\rho_r a_i)}{\partial t}+(1-\Phi)\rho_r\lambda_i a_i=Q_{in}\rho_{in}a^i_{in} \qquad (5)$$

where ρ_r is bulk density.

Equation of heat energy conservation assumes local temperature equilibrium between fluid and rock in each moment of time:

$$\frac{\partial[(1-\Phi)\rho_r h_r+\Phi\rho h]}{\partial t}+\nabla(h\rho\vec{v})= \qquad (6)$$
$$\nabla(\lambda_m\nabla T)+Q_{in}\rho_{in}h_{in}+\sum_i\sigma_i\rho_m a_i,$$

$$\lambda_m=(1-\Phi)\lambda_r+\Phi\lambda,$$

$$\rho_m=\begin{cases}R_i\Phi\rho, & R_i<\infty\\ (1-\Phi)\rho_r, & R_i=\infty\end{cases}$$

where λ, λ_r are coefficients of thermal conductivity for fluid and rock respectively; h, h_r are enthalpy of fluid and rock respectively, J/kg, σ_i is intensity of heat generation of i- nuclide, (W/Ci). Enthalpy and temperature relates as:

$$dh=C_p dT,\quad dh_r=C^r_p dT \qquad (7)$$

where C_p, C^r_p - specific heat capacities of fluid and rock respectively.

For modeling closure, equations for the physical properties of fluid were found by approximation of experimental data (Kirillov et al., 1984) for solution of sodium nitrate in water. In the equations given below the following common notations are used:

c is TDS concentration in fluid, kg/kg; $\tau=T-20$ is the difference between actual temperature and reference temperature 20°C. $p=P-P_a$ is difference between actual pressure and reference atmospheric pressure, MPa.

Dynamic viscosity is calculated as a function of temperature and concentration as:

$$\frac{\mu}{\mu_0}=\left(0.01002+0.02406\cdot\exp\left(\frac{570.58}{\tau+153.5}\right)\right) \qquad (8)$$
$$\cdot(1+c\cdot(0.7929+c\cdot 10.339))$$

where μ is dynamic viscosity, cps; μ_0 = 1.002 cps is reference viscosity for T= 20°C, and c = 0.

The specific volume of fluid is defined as:

$$\frac{\rho_0}{\rho}= \qquad (9)$$
$$\frac{(1-0.625\cdot c)\cdot(1-4.6\cdot 10^{-4}p)}{1+\tau\cdot(-a_0+\tau\cdot(-a_1+\tau\cdot(a_2+\tau\cdot(-a_3+\tau\cdot a_4))))}$$

$$a_0=2.05\cdot 10^{-4};a_1=5.34\cdot 10^{-6};a_2=3.17\cdot 10^{-8};$$
$$a_3=1.58\cdot 10^{-10};a_4=3\cdot 10^{-13}.$$

where ρ is density of fluid, kg/m^3; ρ_0 = 998.23 is reference density of pure water kg/m^3.

Developed modeling tools permit turning on/off of the influence of solute concentration and/or temperature in the eqs. (8) - (9). It allows simulation of studied effects for fully coupled (T-C-H model), partially coupled (T-H model and C-H model) and non-coupled models.

5. BOUNDARY AND INITIAL CONDITIONS

Initial conditions for pressure assumes steady-state 1-D horizontal constant density flow along x

direction with gradient 0.002. Initial TDS concentration and activities of nuclides are zero. Initial temperature of fluid and rocks is 20°C. The plane $y = 0$ is a symmetry plane, so the all normal derivatives for all modeled fields equal zero at this plane.

Plane $x = X_{left}$ is considered as the input boundary. For this plane the following thermal boundary condition is used:

$$C_p \rho T(\vec{v},\vec{n}) - \lambda_m \frac{\partial T}{\partial n} = C_p \rho T_{env}(\vec{v},\vec{n}) + \alpha_{left}(T - T_{env}) \tag{10}$$

where α_{left} is the parameter of heat exchange, T_{env} is constant temperature of the far field assumed 20°C. For the pressure at the input boundary hydrostatic distribution is assumed:

$$P(X_{left}, y, z, t) = P(X_{left}, y, z_{top}, 0) + g\int_z^{z_{top}} \rho(X_{left}, y, \theta, t)d\theta \tag{11}$$

where z_{top} is the elevation of the upper boundary of the domain.

Plane x $=X_{right}$ is considered as the output boundary. It is assumed that advective fluxes via this boundary are much more than dispersive fluxes, so it is zero normal derivatives are assumed for the concentration of TDS and activities at this plane. For the temperature field the following boundary condition is used:

$$-\lambda_m \frac{\partial T}{\partial n} = \alpha_{right}(T - T_{env}) \tag{12}$$

where α_{right} is the parameter of heat exchange. For the pressure boundary condition similar to eq. (11) is used:

$$P(X_{right}, y, z, t) = P(X_{right}, y, z_{top}, 0) + g\int_z^{z_{top}} \rho(X_{right}, y, \theta, t)d\theta. \tag{13}$$

Closed boundary conditions for concentration and activities are used at other boundaries of the domain: $y = Y_{left}$, $z = z_{top}$, $z = z_{bottom}$:

$$(\vec{v},\vec{n}) = 0, \quad \frac{\partial C}{\partial n} = 0, \quad \frac{\partial a_i}{\partial n} = 0 \tag{14}$$

For the temperature field at these boundaries eq. (12) with parameters $\alpha_{left}, \alpha_{top}, \alpha_{bottom}$ respectively is used.

6. RESULTS OF NUMERICAL SIMULATIONS

In all simulation runs the parameters used of heat exchange at all boundaries are supposed to be zero. The coefficient of thermal conductivity is 2.2 W/m°C and the specific volume heat capacity is $2.8 \cdot 10^6$ J/m^3°C. The retardation coefficient for ^{90}Sr in the layers Pz, II, III, IV is assumed to be 1, while in the layers B+A, C, D it equals 5. Simulation was performed for an anisotropic field of hydraulic conductivity: (Fig. 1). Three different block sizes were used: 50·50·10 m; 25·25·5 m and 50·50·2.5 m to estimate the grid effect on the numerical solution. For the mesh with block size 50·50·10 m the modeled domain has a size in the x, y plane of 5 ·3 km; for finer meshes the modeled domain was reduced to 3.5·3 km. Simulation runs were performed for as given in section 3 different concentrations of TDS in the injected wastes. Below are some results of simulation for the anisotropic medium case - TDS concentration in the waste 100 g/l, and for the mesh with step 25·25·5 are considered.

Figure 2 shows the simulated temperature field in the horizontal section $z = -145$ m for 100 years after start of the injection. This figure demonstrates that the temperature field has two fronts: a) the front from the MLW that gives increasing temperature during 100 years in 3 degrees and b) the front from the HLW that gives maximum temperature values: 124°C for 50 years, 113°C for 100 years and 65°C for 300 years. That means the main temperature effects are related to the HLW and the influence of MLW is small.

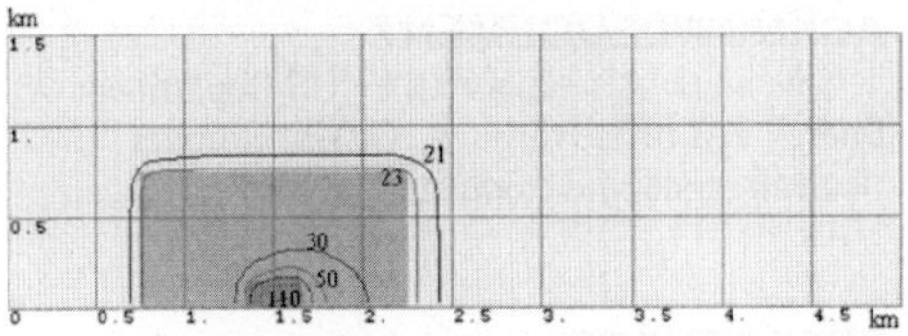

Figure 2. Horizontal section of the modelled domain in the centres of injection aquifer II. Dark grey field is HLW waste and surrounded grey field is MLW. Numbers of isolines are temperature °C.

Increasing the temperature due to radioactive heating decreases the viscosity of fluid. Within the zone of maximum temperature viscosity decreases up to four times in comparison with its reference value. This temperature effect on viscosity exists during all 300 years of simulation.

Figure 3 shows simulated spatial distributions of TDS concentration and log of fluid total activity for 300 years of simulation obtained by the fully coupled model. This figure indicates that decreasing of fluid viscosity in the heat-generating zone makes potential for upward transport of small concentration of TDS and activity. The influence of coupled effects on waste transport is shown in table 2 and 3. These tables summarize the results of overall activity balance calculation for 100 years and TDS balance for 300 years for non-coupled and fully coupled models.

Table 2. Overall total activity balance % for 100 years. First row is non-coupled model and second row is fully coupled model.

Source	Decay	III	C	II	B+A
100	82.93	$2.55 \cdot 10^{-5}$	1.71	14.86	0.50
100	82.93	$4.43 \cdot 10^{-3}$	3.45	13.21	0.40

Table 3. Overall TDS balance % for 300 years. First row is non-coupled model and second row is fully coupled model.

Source	Boundary	III	C	II	B+A
100	0.25	0.12	12.08	81.86	5.69
100	1.79	0.62	7.00	79.89	10.69

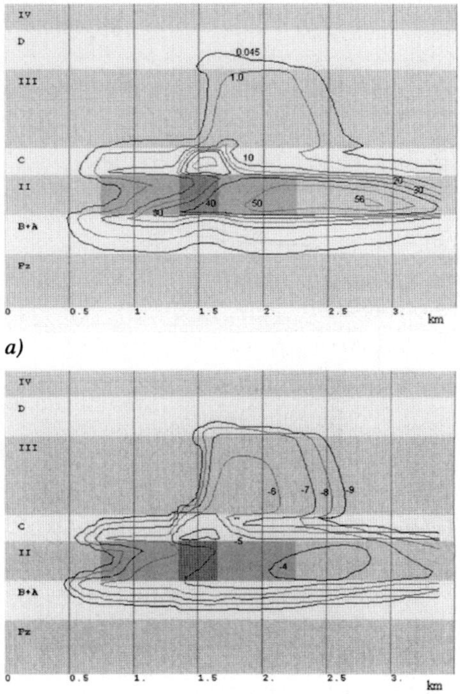

Figure 3. Vertical section of the modelled domain for the plane y=0. a) Numbers of isolines are TDS concentration for 300 years. b) Numbers of izolines are log of total activity.

These tables indicate that thermo-hydro-transport processes coupling leads to increasing wastes spatial spreading. For the activity it is mostly upward spreading due to temperature effect, while for TDS concentration it is more downward spreading due to density effect.

To estimate the influence of partial coupling of the processes the simulation was also performed using T-H and C-H models. Figure 4 shows evolution in time of vertical positions of TDS and activity plumes mass centers. This figure shows the different influences of the modeled processes on the TDS and activity. These plumes are stratified in the vertical direction due to the influence of thermal viscosity effects on nuclides transport and influence of density-concentration relation on TDS transport.

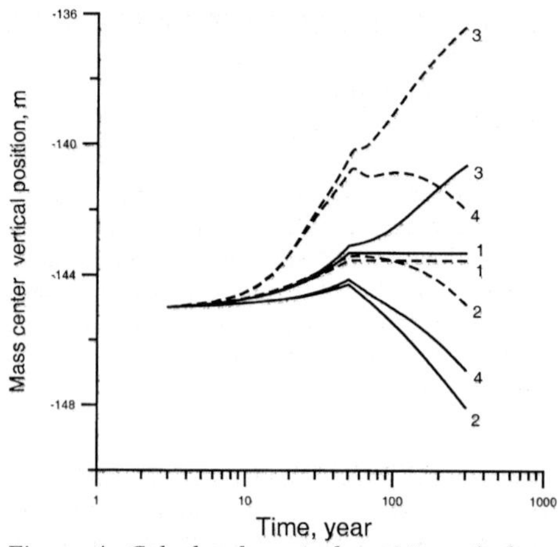

Figure 4. Calculated vertical position of plume mass centers for different models. Dashed line is total activity and solid line is TDS concentration. Labeling of curves: 1 is non-coupled model; 2 is C-H model; 3 is T-H model and 4 fully coupled T-C-H model.

7. CONCLUSIONS

The simulation results indicate:

1. MLW does not practically affect the temperature field in the injection aquifer and neighboring layers. The main temperature effects are governed by radioactive heating of HLW.
2. Radioactive heating leads to decreasing of fluid dynamic viscosity up to 4 times. Because the significant part of heat, generated nuclides, are sorbed within local zones near the injection wells, this viscosity effect can be introduced into the regional-scale model by artificial increasing of hydraulic conductivity surrounding the wells injected with HLW.
3. Thermal and concentration dense convection has a visible effect on vertical migration of waste components. The thermal effect is more significant for the activity field, while the density effect is more significant for the TDS concentration field.
4. To incorporate coupled effects into regional scale flow and transport, artificial "thermal" and "dense" dispersion terms can be introduced into the coupled transport model. The thermal term should be isotropic and the dense term should coincide with gravity acceleration vector.

ACKNOWLEDGEMENT

This work was supported by US Department of Energy EM office through JCCEM contract "Tomsk model development"

REFERENCES

Bradley, Don J., 1997. *Behind the Nuclear Curtain: Radioactive Waste Management in the Former Soviet Union.* Battelle Press,Richland,WA.

Kirillov P.L.,Yuriev Yu.S,.Bobkov, V.P. 1984. *Reference book on thermal hydraulics calculations (nuclear reactors, heat exchangers, steam generators).* Moscow. Energoatomizdat Press. 296p.

Pruess K., 1991 *TOUCH2 – A general-purpose numerical simulator for multiphasefluid and heat flow.* Earth Science Division, Lawrence Berkeley Laboratory. Report No LBL-29400, Berkeley, 1991.

Rybalchenko A.I., Pimenov M.K., Kostin P.P., Balukova V.D., Nosuckhin A.V., Mikerin E.I., Egorov N.N., Kaimin E.P., Kosareva I.M., Kurochkin V.M. 1998. *Deep injection of liquid radioactive waste in Russia.* Battelle Pres, Columbus Richland, Washington. 205p.

Shestakov V.M., Kuvaev A.A., Lekhov A.V., Pozdniakov S.P., Rybalchenko A.I., Zubkov A.V., Davis P.A., Kalinina E.A. 2002. *Flow and transport modeling of liquid radioactive waste injection using data from Siberian Chemical Plan Injection Site.* Environmental Geology. 42(2-3): pp.214-221

STUDY ON COUPLING INFLUENCES OF CONCRETE DAM FOUNDATION SEEPAGE, STRESS, AND CREEP ON STRUCTURE BEHAVIORS OF DAM BODY

Guo Haiqing, Xu Weiya*, Wu Zhongru
HoHai University, 210098, Nanjing, P. R. China

Abstract: Accidents of some arch dams show that dam failures are mainly caused by crack or failure of their foundation rocks, which are directly related to water seepage in the rock. This kind of fluid-rock interaction has an important influence on deformation and stress characters of the dam-rock system. In this paper, the stress and flow fields of dams and their foundation rocks are studied as a coupled system, using visco-elastic constitutive models and finite element solution method. The developed models and FEM technique were applied for analysing the continuous displacement of the 13th dam section of the Longyangxia Dam, and the calculated results agree well with the measured ones.

1. INTRODUCTION

The Longyangxia hydropower project is one of the major hydropower generation and water resources utilization systems built in Yellow River China. Continuous displacement towards the left bank at the 13th dam section has been observed and monitored. This has become a major concern to both the dam safety administration and engineers. The paper represents the results of a research effort devoted to investigate its causes.

Recognizing the importance of the coupled hydro-mechanical effects on the performance of civil engineering structures involving fractured rocks, the stress-flow coupling mechanism of the dam-foundation system at Longyangxia site was simulated using a three-dimensional Finite Element code, supported by two visco-elastic constitutive models to represent the time-dependent material behaviour of the dam concrete and the foundation rock. The calculated results were concord with the measured ones and helped to interpret the causes of this continuous displacement at the 13th dam section of the Longyangxia hydropower project, towards the left bank.

The theoretical foundations of the theory and the FEM formulations presented in this paper are based on works in Zienkiewicz and Cormeau (1974), Oda (1986), Ohnishi et al. (1985), Ohnishi and Kobayashi (1993), Shen et al. (2000) and Wu et al. (2001).

2. THE MATHEMATICAL THEORY

2.1 Basic equations of stress and seepage coupling

The main hypothesis of dam concrete and its foundation rock are as follows: a) the dam and its rock foundation are isotropic continuum media in different areas. b) The seepage follows the Darcy's law. c) The grain skeletons' deformations of dam concrete and foundation rock are ignored. d) The deformations of dam concrete and its foundation rock are mainly caused by deformations of void spaces and cracks between grain skeletons mentioned above.

If the dam and its base rock are affected only by gravity and seepage force, which are considered as body forces, and denoting that the tensile stress is positive, the balance differential equations expressed by displacements and general water heads can be given as

$$\left.\begin{aligned} &G\nabla^2\delta_x + \frac{G}{1-2v}\cdot\frac{\partial\varepsilon_v}{\partial x} - r_w\frac{\partial h}{\partial x} + X_0 = 0 \\ &G\nabla^2\delta_y + \frac{G}{1-2v}\cdot\frac{\partial\varepsilon_v}{\partial y} - r_w\frac{\partial h}{\partial y} + Y_0 = 0 \\ &G\nabla^2\delta_z + \frac{G}{1-2v}\cdot\frac{\partial\varepsilon_v}{\partial z} - r_w\frac{\partial h}{\partial z} + Z_0 + r_w - r_c = 0 \end{aligned}\right\} \qquad (1)$$

where, G is the shear module, E is the Young's modulus, v is the Poisson's ratio, r_w is the density of water, r_c is the saturated density of the concrete or foundation rock, h is the water head. In addition, ∇^2 is the Laplace operator, ε_v is the volume strain, $\delta_i (i = x, y, z)$ are displacement components and X_0, Y_0, Z_0 are equivalent body force components caused by initial strain $\{\varepsilon_0\}$.

According to the law of mass conservation, the continuity equation of the water are derived as given in eq. (2). For simplify, the equation ie expressed along the main seepage directions denoted as coordinate axes x, y, z.

$$\frac{\partial}{\partial x}(k_x\frac{\partial h}{\partial x})+\frac{\partial}{\partial y}(k_y\frac{\partial h}{\partial y})+\frac{\partial}{\partial z}(k_z\frac{\partial h}{\partial z})=\frac{\partial \varepsilon_v}{\partial t}+r_w n\beta\frac{\partial h}{\partial t} \quad (2)$$

where $[k]$ is the permeability tensor, whose main components are k_x, k_y, and k_z; n is the void ratio, and β is the compression parameter of the water.

The basic equations for the coupling analysis of stress and seepage fields are composed of equations (1) and (2), whose boundary conditions include (a) displacement boundary condition $\{\delta\}=\{\delta_0\}$, (b) stress boundary condition $\sigma_{kl}m_l=\sigma_{k_0}$ (where $k,l=1,2,3$), (c) water head boundary condition $h=h_0$, (d) seepage boundary condition $-k_n\frac{\partial h}{\partial n}=q_0$, which should be specified according to the site conditions.

2.2. The constitutive models

For studying the viscous deformation caused by the creep of dam and its base rock under time-dependent loading, different visco-elastic constitutive models are developed to identify the most suitable models and parameters for more accurate simulation of the time-dependent deformation of the dam-foundation system.

2.2.1 The visco-elastic constitutive model of the dam concrete

The creep of dam concrete includes two parts: instantaneous elastic deformation and viscous deformation. A generalized Kelvin model consisting of two standard Kelvin model in series is used to describe the time-dependent deformation of the dam concrete as shown in Figure 1.

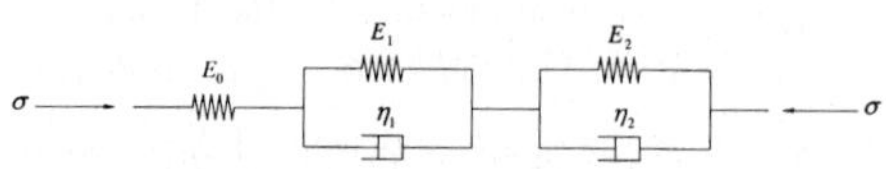

Figure 1. Sketch map of the generalized Kelvin model

The model's strain is the sum of the initial elastic strain and the strains of the two Kelvin models, so that the partial strain e_{ij} of the generalized Kelvin model is expressed as

$$e_{ij}=\frac{s_{ij}}{2E_0}+\sum_{k=1}^{n}\left\{e_{ij,0k}+\frac{1}{2\eta_k}\int_0^t s_{ij}\exp[\frac{E_k}{\eta_k}(t-t_0)]dt\right\} \exp[-\frac{E_k}{\eta_k}(t-t_0)] \quad (3)$$

where E_K, η_K, η_M are the tension (compress, or shear) module and viscous parameters, respectively, and s_{ij} is the deviatoric stress tensor, $s_{ij}=\sigma_{ij}-\frac{1}{3}\delta_{ij}\sigma_{kk}$.

For a special case, when $s_{ij}=s_{ij0}$, equation (3) is changed to

$$e_{ij}=\frac{s_{ij0}}{2E_0}+\sum_{k=1}^{n}\left\{e_{ij,0k}\exp\left[-\frac{E_k}{\eta_k}(t-t_0)\right] +\frac{s_{ij0}}{2E_k}\left[1-\exp[-\frac{E_k}{\eta_k}(t-t_0)]\right]\right\} \quad (4)$$

If the viscous strain of Kelvin model is $\{\varepsilon_{v,K}\}_{t_0}$ at t_0, $t=t_0+\Delta t$, and the stress remains constant during the Δt time increment, the viscous strain increment of the generalized Kelvin model during Δt can be derived from equation (4) as

$$\{\Delta\varepsilon_v\}_t=\sum_{k=1}^{n}\left[1-\exp(-\frac{E_k}{\eta_k}\Delta t)\right]\left(\frac{1}{E_k}[C]\{\sigma\}-\{\varepsilon_{v,k}\}_{t_0}\right) \quad (5)$$

where $[C]$ is the Poisson's ratio matrix.

2.2.2 The visco-elastic constitutive model of the foundation rock

The time-dependent deformation of the foundation rock, caused by loadings, is described by a Burgers model, which is composed of a Kelvin model and Maxwell model in series, as showed in Figure 2. The partial strain expression of Burgers model is

$$e_{ij,K}=\left\{e_{ij,0K}+\frac{1}{2\eta_K}\int_{t_0}^{t} s_{ij}\exp[\frac{E_K}{\eta_K}(t-t_0)]dt\right\} \quad (6)$$

$$\exp\left[-\frac{E_K}{\eta_K}(t-t_0)\right]+e_{ij,0M}+\frac{1}{2E_M}\int_{t_0}^{t}\dot{s}_{ij}dt$$

$$+\frac{1}{2\eta_M}\int_{t_0}^{t}s_{ij}dt$$

Specially, when $s_{ij} = s_{ij0}$ is a constant value, and $\dot{s}_{ij} = 0$, the equation (6) becomes

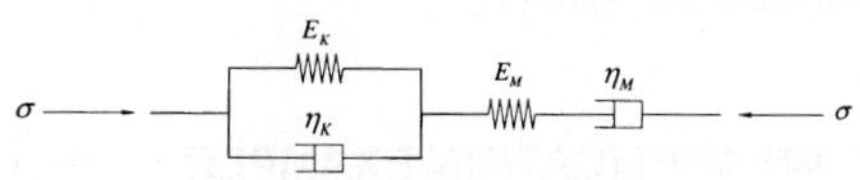

Figure 2. Sketch map of the Burgers model

$$e_{ij} = e_{ij,0K} \exp[-\frac{E_K}{\eta_K}(t-t_0)] + \frac{s_{ij0}}{2E_K} \left\{1-\exp[-\frac{E_K}{\eta_K}(t-t_0)]\right\} + e_{ij,0M} + \frac{s_{ij0}}{2\eta_M}(t-t_0) \tag{7}$$

The visco-elastic strain of the Maxwell component will change into $\{\varepsilon_{v,M}\}_{t_0}$ at time t_0. If $t = t_0 + \Delta t$, and the stress remain the same during the increment of Δt, the viscoelastic strain increment of Burgers model can be derived from equation (7) as given by

$$\{\Delta\varepsilon_v\}_t = \left[1-\exp(-\frac{E_K}{\eta_K}\Delta t)\right]\left(\frac{1}{E_K}[C]\{\sigma\} - \{\varepsilon_{v,K}\}_{t_0}\right) + \frac{\Delta t}{\eta_M}[C]\{\sigma\} \tag{8}$$

where E_K, η_K, η_M is the stretch (compression or shear) modulus and viscous parameters and $[C]$ is the Poisson's ratio matrix, respectively.

2.3 The finite element method

For the basic equations of coupled stress-flow analysis mentioned above, it is very difficult to solve them in closed-form. The transposition method of progression and integration can only be applied for problems of boundary value problems of simple geometry and boundary conditions. Therefore the finite element method (FEM) is used to solve the coupled partial differential equations in this paper.

2.3.1 The basics of the FEM formulation

The FEM solution scheme for the coupled equations of displacement and seepage fields is

$$\left.\begin{aligned} &[\bar{K}]^T\{\Delta\delta\} - [K']\{h\}_{m+1} = \{\Delta F_1\} - [K']\{h\}_m \\ &[K']^T\{\Delta\delta\} + ([S]+\theta\Delta t[\tilde{K}])\{h\}_{m+1} = \Delta t(\theta\{\Delta F_2\}+\{F_2\}_m) \\ &\qquad + ([S]-(1-\theta)\Delta t[\tilde{K}])\{h\}_m \end{aligned}\right\} \tag{9}$$

where θ is the integration parameter, $[\bar{K}]$ is the general stiffness matrix, $[K']$ is the general coupling matrix, $[S]$ is the general compression matrix, $[\tilde{K}]$ is the general seepage matrix, $\{F_0\}$ is the equivalent nodal load vector by the initial strain, $\{F_1\}$ is the equivalent nodal load vector, $\{F_2\}$ is the equivalent nodal discharge vector, and $\{h\}_{m+1}$ is the general water head vector at time t_{m+1}, respectively.

If displacement increment $\{\Delta\delta\}$ and super-static water pressure $\{\Delta p\}$ are taken as the unknown quantities, and the full Hermit differences are adopted with $\theta = 1$, the equation (9) can be re-written as

$$\left.\begin{aligned} &[\bar{K}_t]\{\Delta\delta\} - [K']\{\Delta p\} = \{R\} - \{R_t\} \\ &[K']^T\{\Delta\delta\} - ([S]+\Delta t[\tilde{K}])\{\Delta p\} = 0 \end{aligned}\right\} \tag{10}$$

where $\{R_t\}$ refers to the part of loads balanced by the stresses related to the displacement happened before the time t_m, and are calculated by the following equation

$$\{R_t\} = \sum_{l=0}^{t_m} [\bar{K}_t]_l \{\Delta\delta\}_l \tag{11}$$

where l is the number of calculating periods of time before the time t_m.

2.3.2 The calculation process

According to the incremental initial-strain method used in equation (9), the initial stress field and seepage field of dam-foundation system need to be created by iterations starting from the beginning of the time-marching process. The element stress remains unchanged during the subsequent time step Δt, so that the seepage coefficients, which are decided by the state of stress at the beginning of the time step, remain the same value. They change step by step with loads increasing gradually at each time step. The water heads at the end of every time step are the initial ones of the next time step.

In terms of the magnitude of the time step Δt, two factors need to be considered. One is to guarantee the numerical stability of the viscous initial strain iteration by choosing a small value of Δt and the other is to ensure that the coefficients matrix not to be morbid by making the Δt value adequately large. It be estimated by the following formula

$$\Delta t = \frac{L^2}{\frac{k}{r_w}(K + \frac{4}{3}G)} \quad (12)$$

where L is the linear dimension of the elements, k is the seepage coefficient, and K, G are bulk modulus and shear modulus, respectively.

The appropriate time step Δt should be determined by trial and error. Generally speaking, the state of stress field varies mostly at the beginning of a loading step, so that the time step should be small at the beginning of the loading, and gradually increases in order to speed up the calculations.

2.4 The mechanisms of the dam-foundation interaction for creep analysis

A dam and its rock foundation are the two integral parts of a system. Different types of dams have different interaction behaviours between the two parts. Project practices show that for concrete gravity dams, the rock foundations have significant influence on the stress behaviours of the concrete dams up to one-third of the dam height from the base. The effect is small in other parts. For arch dams, on the other hand, the rock foundations have significant influences on the stress and behaviour of the whole dam body. For both gravity and arch dams, their rock foundations affect significantly the deformations of the concrete dams. Therefore, for stress analysis of the dam-foundation system of gravity dams, the series combination creep model is applied for the first one-third of 1/3 dam height from the base and the parallel combination creep model is applied to the rest. In terms of strain, this relationship can be expressed as

$$\frac{1}{\varepsilon} = \frac{1}{\varepsilon_c} + \frac{1}{\varepsilon_r} \quad (13)$$

for the first one-third part of the dam body and

$$\varepsilon = \varepsilon_c + \varepsilon_r \quad (14)$$

for the rest, where ε_c and ε_r are the strains of dam concrete and foundation rock, respectively. The ε_c can be obtained from equation (3) or (4) and ε_r from equation (6) or (7).

3. AN APPLICATION EXAMPLE

The above theory and FEM code were applied to analyse the continuous displacements of the 13th dam section of Longyangxia dam from July 1989, as an example. The main part of Longyangxia dam is a gravity arch dam of 396 meters in length, 29.2/80 meters in width (at the top/base), 2610 meters for DCL and 178 meters in height.

3.1 The FEM model and conditions

3.1.1 The FEM model and material parameters

The dimension of the FEM model is 540 m in both length and width and 360 m in height. When dividing the FEM mesh, the geological structures in the foundation rock and their engineering treatment measures were taken into account. While laying out the element nodes, the locations of the in situ measuring points of displacement, temperature and stresses are considered. The FEM mesh, which is consisted of 21,189 eight-node-hexahedron elements with 24,873 nodes, is illustrated in Figure 3. The material parameters are listed in Table 1.

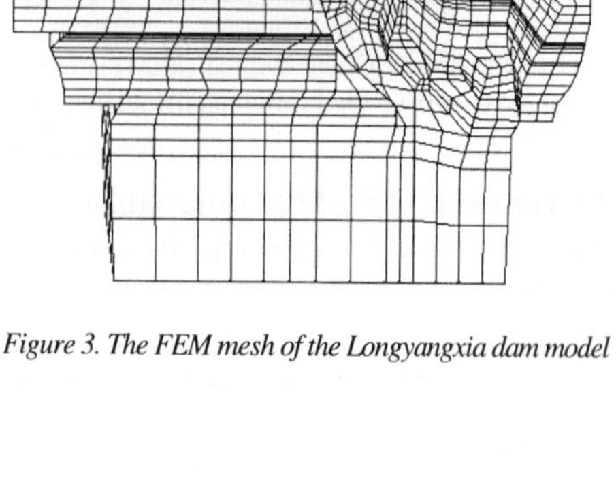

Figure 3. The FEM mesh of the Longyangxia dam model

Table 1. Parameters of Longyangxia dam and foundation materials

Location	Classification	Density kg/m³	Elastic constants: Modulus (GPa)	Elastic constants: Poisson 's ratio μ	Viscous constants: E_K GPa	Viscous constants: η_K $GPa \cdot S$	Viscous constants: E_M GPa	Viscous constants: η_M $GPa \cdot S$	Seepage parameters	cm/s
Dam body (concrete)		2400	20	0.18	250	2.3×10^5	66.7	1.15×10^9	Dam	1.0×10^{-7}
Rock foundation	2580 m ~ 2560 m	2650	8	0.25	50	4.0×10^4	15.0	8.5×10^9	Rock	1.0×10^{-9}
	2560 m ~ 2540 m	2700	12	0.23	200	3.6×10^4	20.0	5.4×10^9		
	2540 m ~ 2500 m	2750	16	0.22	200	2.8×10^4	40.0	5.0×10^9	Crack	1.0×10^{-4}
	Below 2400 m	2755	22	0.22	300	2.0×10^5	51.0	1.8×10^9		
	G_4*	2600	3	0.25	30	5.4×10^4	10.0	9.5×10^9	Curtain	1.0×10^{-8}
	F_{18}	2600	4.5	0.25	40	4.0×10^4	15.0	9.0×10^9		
	F_{71}, F_{73}, F_{32}, F_{67}	2600	3.2	0.25	30	5.4×10^4	10.0	9.5×10^9	Drainage	1.0×10^{-2}
	A_2+F_{120}	2600	5.4	0.25	40	4.0×10^4	15.0	9.0×10^9		

*G_4, F_{18}, F_{71}, F_{73}, F_{32}, F_{67}, and A_2+F_{120} are geological structures and cracks in Longyangxia dam foundation.

3.1.2 Water level variations as time-dependent loading cases in the calculation

From April 16, 1990 to May 1992, the water level of Longyangxia reservoir dropped from 2575.04 m to 2533.15 m. From May 1992 to December 1994, the water level rose from 2533.15 m to 2577.58 m. From January 1995 to July 1998, the water level dropped from 2577.58 m to 2533.54 m again. After July 1998, the water level rose from 2533.54 m to 2581.08 m. These water level variations were used for deriving the water head loading conditions.

3.2 Calculation and analyses of tangential displacements of typical days

The FEM modelling was applied for the simulating the dam behaviour with coupled stress-flow effects under different loading cases according to the water level variations. The tangential displacements of Longyangxia dam's typical dam section on April 16, 1989, April 16, 1990, and April 16, 1996, and December 31, 1999 were calculated and compared with measured data. The results are listed in Table 2 and illustrated in Figure 4.

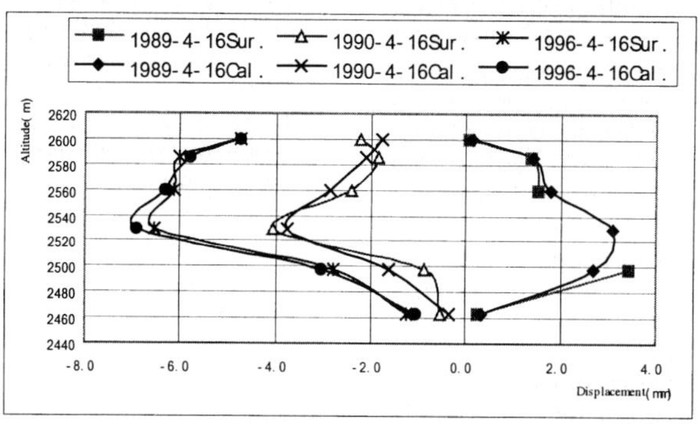

Figure 4. Survey values and calculation values of general displacement of the 13th dam section

Table 2. Measured and calculated values of time-dependent displacements (mm)

Time / Points of measurments	1989-4-16		1990-4-16		1996-4-16		1999-12-31	
	Measured Values	Calculated Values*	Measured values	Calculated Values	Measured Values	Calculated Values	Measured Values	Calculated Values
2463.3 m	--	0.12	-0.39	-0.28	-1.15	-1.01	-1.79	-1.75
2497 m	--	1.05	-0.33	-0.44	-2.70	-2.85	-3.95	-3.82
2530 m	--	1.24	--	-0.79	-4.81	-4.44	-5.86	-5.97
2560 m	--	0.89	-0.68	-0.61	-3.88	-3.69	-5.27	-5.48
2585 m	--	0.57	-0.49	-0.36	-1.93	-1.84	-2.17	-2.26
2600 m	--	0.06	-0.55	-0.31	-4.24	-4.14	-5.60	-5.49

It can be seen from Table 2 and Fig. 4 that: (1) the calculated values are close to the measured ones, and their respective trends of change are very similar as well. (2) The calculated tangential displacements above the 2500m level on April 16, 1996 are as follows: -3.04mm, -6.91mm, -6.33mm, -5.77mm, and -4.68mm, respectively. These data show that the 13th dam section moved towards the left bank. This was caused by the continuing high water level, about 100m ~ 110m higher than adjacent uplift pressure, and the creep displacements of dam body and rock foundation.

4. CONCLUSIONS

(1) The time-dependent constitutive models of the dam concrete and foundation rocks for the Longyangxia hydropower complex were established. (2) The three-dimensional FEM model with the established constitutive models and realistic time-variations of reservoir water level as loading cases were successfully applied to simulate the coupled stress-flow behaviour of the dam-foundation system. (3) The calculated results, which agreed well with the measured ones, clearly indicated that the main reason for the continuous displacement towards the left bank after July 1989, was caused by the influence of the seepage-stress coupling of Longyangxia gravity arch dam on its typical dam section's tangential displacements.

ACKNOWLEDGEMENT

The authors would like to thank the Natural Science Foundation of China for their financial supports Fund 50128908.

REFERENCES

Oda, M. 1986. *An equivalent continuum model for coupled stress and fluid flow analysis in jointed rock massed.* Water Resource Research 22(13): 1845 - 1856

Ohnishi, Y. & Kabayashi, A. 1993. *Thermal-hydraulic-mechanical coupling analysis of rock mass.* In Hudson J. A. (ed), Comprehensive Rock Engineering, Pergamon Press. 191 - 208

Ohnishi, Y. Shibata, H. & Kabayashi, A. 1985. *Development of finite element code for analysis of coupled thermo-hydro-mechanical behaviours of saturated-unsaturated medium.* Int. Symp. On Coupled Process Affecting the Performance of a Nuclear Waste Repository.

Shen Zhenzhong. Xu Zhiying. & Luo Cui. 2000. *Coupled analysis of viscoelasticity stress field and seepage field for the Three Gorges dam's foundation.* Engineering Mechanics. 17(1): 105 - 113.

Wu Zhongru. Gu Chongshi. & Wu Xianghao. 2001. *Theory and its applications of safety monitoring of roller concrete dam.* Science Press.

Zienkiewicz, O. C. & Cormeau, I. C. 1974. *Visco-plasticity, plasticity and creep in elastic solids.* Int. J. Numer Methods, Eng. 8: 821 - 845.

CONSOLIDATION SETTLEMENTS ABOVE DEEP TUNNELS IN FRACTURED CRYSTALLINE ROCK: NUMERICAL ANALYSIS OF COUPLED HYDROMECHANICAL MECHANISMS

E. Eberhardt[1], K. Evans[1], C. Zangerl[1], S. Loew[1]

[1]) Engineering Geology, Swiss Federal Institute of Technology (ETH Zürich), Switzerland

Abstract: Recent measurements of surface displacements 800 m above the Gotthard (N2) highway tunnel in central Switzerland have shown up to 12 cm of subsidence since the tunnel was constructed. Subsidence of this magnitude in a fractured crystalline rock mass is unexpected and appears to be related to large-scale consolidation resulting from groundwater drainage and pore pressure changes around the tunnel. With the recently initiated construction of the 57 km long Gotthard Base Tunnel, which passes through similar geological conditions as the Gotthard highway tunnel and underneath several important dams (the integrity of which could be adversely affected by differential settlements), understanding these processes and being able to predict the magnitude and extent of surface displacement becomes highly relevant. This paper focuses on the use of 2-D finite-element models to analyse the measured subsidence magnitudes recorded over the Gotthard highway tunnel, treating the fractured rock mass as an equivalent poro-elastic medium. The provisional results reported in this paper address the hydrogeological controls and their affect on the extent of the diffusion front and the resulting shape and magnitude of the subsidence profile.

1. INTRODUCTION

Surface subsidence due to consolidation of rock in which the pore pressure has been reduced, through fluid extraction, is well known in the fields of petroleum engineering and groundwater where the problem has received wide attention. However, the phenomenon is not generally recognised as being important in fractured crystalline rock masses, even though large reductions in pore pressure can occur when driving a deep tunnel.

In tunnelling, Schmidt (1989) notes that the tendency for tunnel engineers and researchers to pay little attention to consolidation settlement has often led to surprises, but that such cases have been reported in sufficient detail to identify the underlying cause. These cases almost exclusively involve tunnels excavated in soft soils. Consolidation settlement problems associated with tunnelling in fractured rock have been reported in Oslo, Norway (Karlsrud & Sander 1979), but in this case soft marine clays lying above the bedrock were still considered as being the sole consolidating material and that the fractured bedrock was reported as only providing the drainage network into the tunnel excavation.

Precedence for rock mass consolidation in relation to tunnel drainage has been reported by Lombardi (1992) in relation to the driving of an investigation adit through a confined, fractured, marly-limestone aquifer near the Zeuzier dam in Switzerland. Recorded settlements at the dam site of ~13 cm were measured even though the adit was 1.5 km distant. Although these settlements are small compared to those associated with some fluid extraction schemes, they can have serious consequences since only minor differential displacements are necessary to induce damage in concrete structures. In the Zeuzier case, cracks in the dam appeared which required it to be emptied and repaired over a period of several years.

Zangerl (2003) describes an example of surface subsidence associated with drainage-promoted consolidation in crystalline rock. Recent high-precision levelling measurements over the Gotthard pass road in central Switzerland, have revealed up to 12 cm of subsidence over a 10 km section that passes almost a kilometre above the Gotthard highway tunnel. The levelling surveys were performed by the Swiss Federal Office of Topography prior to and after tunnel construction (i.e. 1970 and 1993/98, respectively). Earlier surveys between 1918 and 1970 showed only a steady alpine uplift rate of approximately 1 mm/year. The presence of the trough was recently confirmed by geodetic triangulation measurements supplemented with Global Positioning System data (Salvini 2002).

The rock mass penetrated by the tunnel primarily consists of paragneisses and granitic gneisses (Figure 1). The majority of fractures and fracture zones are sub-vertical and strike ENE-WSW. Semi-quantitative measurements of tunnel inflow during construction indicated a 3-km wide section where transient inflows were markedly high, with one fracture zone producing an estimated inflow of 300 l/s. The location of this section corresponds closely to the broad trough seen in the subsidence profile, and the peak-inflow coincides with the sharp maximum displacement peak (Figure 1). This spatial correlation, and the temporal relationship between tunnel construction and settlement, points to causality between water drainage into the tunnel and surface deformation. Alternative explanations such as localized surface processes (e.g. landslides) could be excluded given the absence of local indicators and the 10 km extent over which the settlements were measured.

In light of the new 57 km AlpTransit Gotthard base tunnel, which is currently under construction and whose trajectory passes close to several dams, the assessment of possible surface subsidence is of great interest. In this paper we present the results of a two-dimensional finite-element study which includes coupled fluid pressure diffusion and consolidation with the rock mass treated as an equivalent isotropic porous medium.

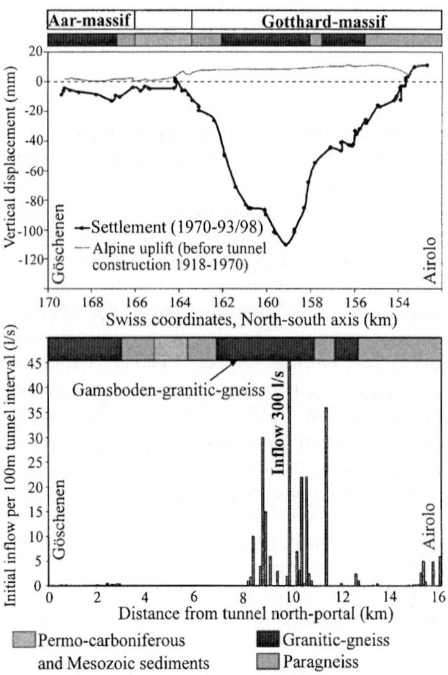

Figure 1. Surface settlement and measured initial water inflow rates into the Gotthard highway tunnel.

2. CONSOLIDATION MECHANISMS IN FRACTURED CRYSTALLINE ROCK

Zangerl *et al.* (2003) recognised several mechanisms that might contribute to the subsidence measured over the Gotthard highway tunnel. All are triggered by reductions in pore pressure within the rock mass due to drainage into the tunnel. The mechanisms are illustrated in Figure 2 and involve: the closure of sub-horizontal joints; the deformation of sub-vertical joints and brittle fault zones through changes in the localized horizontal stress state; and the consolidation of the intact, permeable rock blocks acting as poroelastic bodies.

The approach taken was to include these mechanisms in coupled hydro-mechanical numerical models and attempt to reproduce the observed subsidence. The disturbance to the model is the pore pressure depletion that reflects drainage of the rock mass at the points where major conductive fractures are intersected by the tunnel. To provide data for the modelling, a program of extensive field mapping of geological structures and laboratory tests on core samples was undertaken (Zangerl 2003).

Intuitively it can be expected that closure of horizontal fractures in response to reduced internal pore pressure would be the single largest contributor to subsidence. However, the major conductive structures and brittle fault zones in the Gotthard region are sub-vertical (Zangerl 2003). Numerical modelling using discontinuum techniques (i.e. distinct-element) showed that these features can contribute to subsidence in a complex manner that involve closure and shear of fractures, and vertical strains of intact block induced through a Poisson's ratio effect (Zangerl *et al.* 2003). The asymmetric nature of the subsidence trough and some of the local irregularity on its flanks were reproduced by these models. However, these models did not include consolidation of the intact block material and were unable to account for the observed magnitude of subsidence unless fracture compliance values were adopted which were much higher than seem reasonable. Thus, the results suggested consolidation of the intact blocks contributed significantly to the subsidence.

3. CONTINUUM MODELLING OF CONSOLIDATION SETTLEMENTS

Ideally, the problem requires a fully coupled hydromechanical numerical code that incorporates flow and deformation in both the explicitly-defined discrete fractures (e.g. fault zones) and the intact rock block material bounded by the fractures. Another approach that implicitly includes each of the key contributing factors outlined in Figure 2 (i.e. including matrix consolidation) and requires less expenditure of resources is to represent the discontinuities and intact blocks of the rock mass as an equivalent poro-elastic continuum. This is the approach used in producing the results reported in the remainder of the paper.

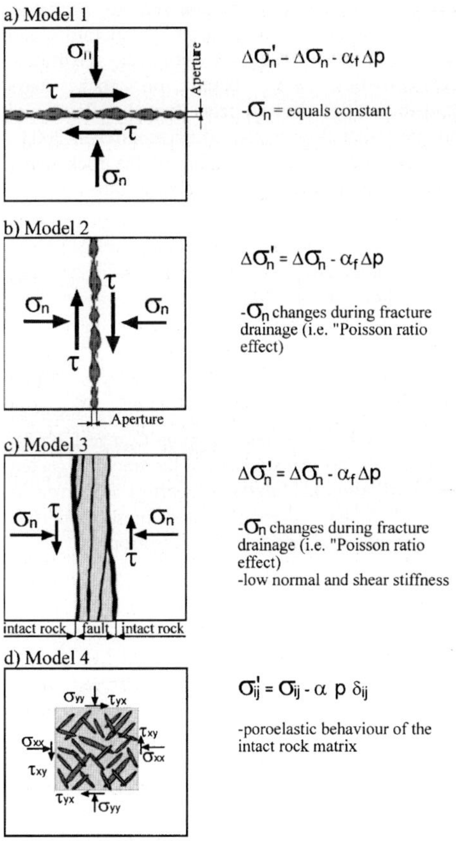

Figure 2. Conceptual consolidation mechanisms for fractured crystalline rock masses involving: (a) horizontal joints, (b) vertical joints, (c) vertical faults, and (d) intact rock. After Zangerl (2003).

3.1 Poroelastic Formulation

The pore pressure within fractures that are members of a connected network will drain when penetrated during tunnel excavation. The reduction in pore pressure will then diffuse out into the rock mass. The extent to which drainage penetrates the rock mass on the long term depends largely upon both the geometry and conductivity of the network. Generally, high conductivity and low divergence of the network, such as tends to occur in brittle fault zones, promotes deep penetration. Saturated, low-permeability intact rock blocks that are bounded by fractures in the network are similarly obliged to adjust their internal pore pressure to maintain equilibrium at their boundaries. These pore pressure changes within the rock mass induce consolidation strains in accord with the theory of linear elasticity.

Biot (1941) derived the 3-D consolidation theory, which describes the coupled hydraulic and mechanical transient response of a linear elastic, isotropic, homogeneous porous medium. One aspect of his theory is the effective stress law for elastic deformation:

$$\sigma'_{ij} = \sigma_{ij} - \alpha p \delta_{ij} \tag{1}$$

where σ'_{ij} and σ_{ij} are the effective and total stress matrices respectively, p is the fluid pore pressure, α is Biot's coefficient, and δ_{ij} is the Kroenecker's delta (Nur & Byerlee 1971). Biot's constant describes the fraction of pore pressure change that is felt by the solid skeleton as a deforming volumetric body force, and takes values ranging from 0 to 1.

Analysis of time-dependent consolidation requires the solution of Biot's consolidation equations coupled to the equations describing flow. The transient hydro-mechanical coupling between pore pressure and volumetric strain for a linear elastic, mechanically isotropic porous medium and fully saturated with a single fluid phase (i.e. water), is given by the fluid continuity equation:

$$\underbrace{\alpha\frac{\partial\varepsilon_v}{\partial t}}_{\text{volumetric strain}} = \underbrace{-S_\varepsilon\frac{\partial p}{\partial t} + \frac{k}{\mu}\nabla^2 p + Q}_{\text{transient-flow-equation}} \tag{2}$$

where: ε_v = volumetric strain, Q = explicit fluid source, k = permeability, μ = dynamic fluid viscosity, p = pore pressure and t = time. The parameter S_ε refers to constrained specific storage (i.e. that which applies under zero macroscopic strain conditions: $\varepsilon_V = 0$) and is given by,

$$S_\varepsilon = \frac{\gamma_w \alpha (1 - \alpha B)}{KB} \tag{3}$$

Here γ_w = specific weight of formation water, K = the drained bulk modulus, ν = the drained Poisson's ratio and B is Skempton's coefficient.

Simulations were performed using the commercial finite-element program VISAGE (VIPS 2003), which incorporates flow and elastic field solutions coupled through poro-elasticity. Each simulation requires as input seven independent parameters: water density, rock density, drained Young's modulus, drained Poisson's ratio, Biot's constant, Skempton's constant and the hydraulic permeability. To date, only 2-D simulations have been performed with future plans involving the analysis of 3-D models currently under development based on experiences gained through the 2-D study. The 2-D mesh was designed to allow inclusion of local topographical, geological and hydrological conditions in the study area.

3.2 Model Geometry and Material Properties

The model geometry used in the present studies is shown in Figure 3. The profile represents an approximate North-South section taken parallel to the tunnel axis, and is thus sub-parallel to the levelling profile shown in Figure 1. The model geometry assumes a laterally extended drainage point along strike of the major permeable fault zone (i.e. out-of-plane) coinciding with the centre of the subsidence trough. This fault zone would act as a sub-vertical drain to the adjacent rock masses to the north and south.

The form of the subsidence profile suggests diffusion of pore pressure drawdown both north and south from the major inflow point at the tunnel. Most certainly there is also diffusion in the East-West direction within the conductive faults intersected by the tunnel. However, there is little data on the pattern of subsidence in these directions. Nonetheless, it seems reasonable to assume that the penetration of drainage along the major fracture zone that produced the largest inflow was extensive. In the 2-D models, it was assumed that the fracture is sufficiently conductive that is can be taken as constant potential feature at the pressure of the tunnel. Thus, at zero time, the pressure along the set of nodes representing this fault is dropped to atmospheric.

Zero displacement boundaries were applied normal to the bottom and sides of the model (shown as 'rollers' in Figure 3). Pore pressures were initialised assuming a water table coincident with the surface topography. Flow boundaries along the side were set as impermeable (i.e. no flow) and the bottom boundary was set to a constant pore pressure of 30 MPa. The surface boundary condition was set assuming a fixed pore pressure condition corresponding to recharge. Material properties were based on laboratory and field estimates whenever possible. Hydraulic conductivities were based on rock mass transmissivities derived from tunnel inflows and simple radial flow models (Luetzenkirchen 2003). The drained Young's modulus of the rock mass, E_{rm}, was calculated using the relationship:

$$E_{rm} = \left(\frac{1}{E} - \frac{1}{Xk_n} \right)^{-1} \tag{4}$$

where E is the drained Young's modulus of intact rock determined from laboratory tests as 45.5 GPa; X is the mean fracture spacing which was taken as 1 m on the basis of field mapping (Zangerl 2003); and k_n is the normal stiffness of the fractures (taken as 38 MPa/mm). Material properties are given in Table 1. Where otherwise stated, these properties were assumed to be constant and isotropic.

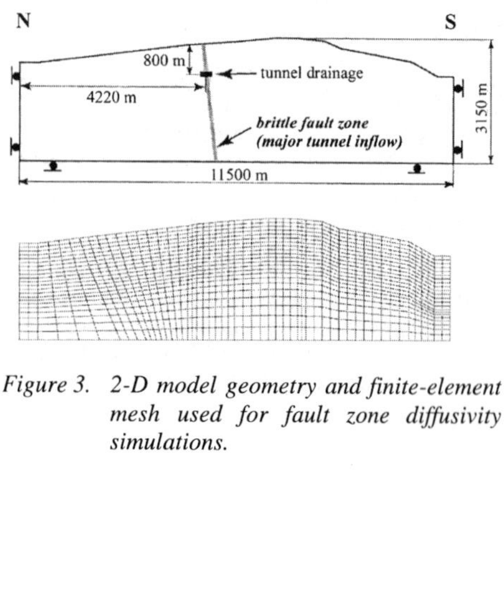

Figure 3. 2-D model geometry and finite-element mesh used for fault zone diffusivity simulations.

Table 1. Model input parameters and material properties.

Parameter	Value
Unit weight of rock, γ	27 kN/m^3
Unit weight of water, γ_w	10 kN/m^3
Young's modulus of rock mass, E_{rm}	20 GPa
Poisson's ratio, ν_{rm}	0.2
Hydraulic conductivity, K_{rm}	1e-8 m/s
Biot's coefficient, α	0.7
Skempton's coefficient, B	0.9

3.2 Model Results

The material properties given in Table 1 yield a very good fit to the magnitude of subsidence (Figure 4). Parametric analysis indicates that the predicted subsidence magnitude is primarily determined by the drained bulk modulus and Biot's coefficient values, and also by the vertical extent of drainage. Indeed, the Table 1 parameter set yields a marginally better fit to the peak magnitude if the drainage along the fault zone is restricted to the lower 200 m section extending upwards from the tunnel, rather than extending to the surface. Limited vertical extent of fault zone drainage is suggested by field observations, which show that surface springs, streams and rivers were not significantly disturbed following construction of the tunnel (Zangerl 2003).

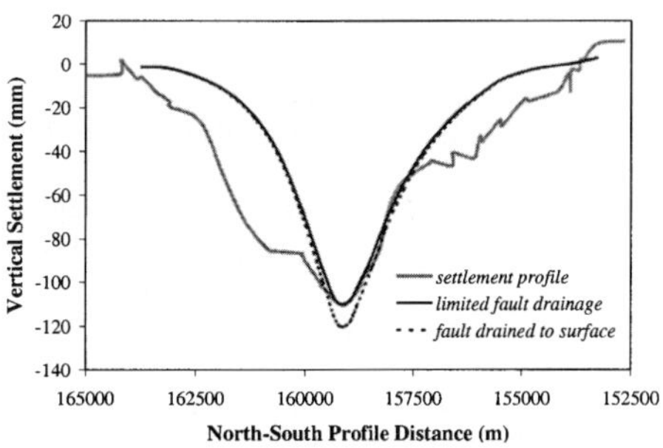

Figure 4. Gotthard settlement profile and those modelled assuming drainage of the fault zone to surface and fault zone drainage limited to a 200 m interval above the tunnel.

Although the magnitude of the subsidence predicted in Figure 4 closely matches the observed, the form of the observed subsidence trough is broader and has detail that is not reproduced by the predicted curves. In part, the detail might reflect the contribution to the subsidence of closure on discrete fractures, as was suggested by the distinct-element modelling (Zangerl 2003). However, within the framework of the poro-elastic continuum model, two possible ways of broadening the predicted trough were examined.

The first approach was to increase the horizontal diffusivity of the medium. This could be accomplished by increasing the horizontal hydraulic conductivity, K_x, or decreasing the specific storage coefficient (through decreasing α and increasing B). Increasing K_x from 1e-8 to 5e-8 m/s resulted in a substantial increase in the width of the subsidence trough, but also increased the magnitude. Thus, α was reduced to 0.45 to yield agreement with the observed (Figure 5). A limitation of the model is that it does not reproduce the marked asymmetry in the observed profile with respect to the northern and southern hinge points where the surface deflections begin. It also tends to under- and over-predict the subsidence along the northern and southern margins of the trough, respectively. Lastly, and perhaps most importantly, geological field observations are not consistent with higher horizontal than vertical hydraulic conductivity. Rather, the sub-vertical, ENE-WSW striking attitude of the major fault zones would suggest if anything the opposite anisotropy, leading to narrower predicted subsidence troughs.

The second approach was to supplement the single drainage fracture zone at the subsidence peak with further drainage zones located 1.2 km to the north where a group of water-bearing fault zones were intersected in the tunnel (Figure 1). That these may also constitute conduits for penetrative drainage is suggested by the co-located inflection in the observed subsidence profile. The results including these extra drainage fractures are also shown in Figure 5. Parameter values were the same as those given in Table 1 except that a lower value of Biot's constant of 0.5 was used to more closely represent values derived from laboratory tests under higher hydrostatic loads (Zangerl 2003). Evidently, the inclusion of the additional drainage zones gives a marked improvement in fitting the asymmetric nature of the settlement trough. It should be noted that this study is still ongoing.

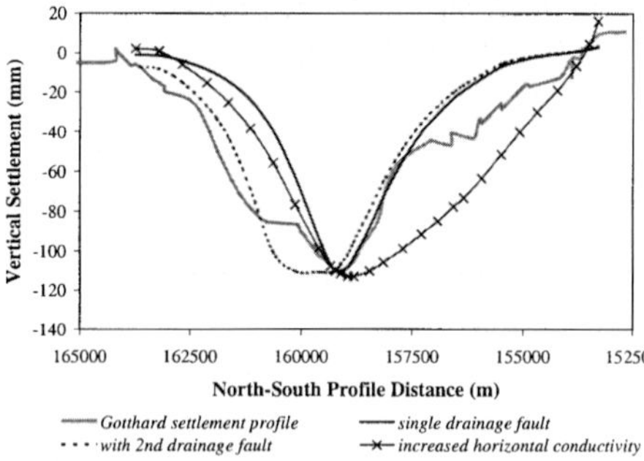

Figure 5. Gotthard settlement profile and those modelled assuming additional key drainage fault structures.

4. CONCLUSIONS

The recent surface geodetic measurements above the Gotthard highway tunnel in Switzerland demonstrate that consolidation of a fractured crystalline rock mass through deep tunnel drainage can result in surface settlements of sufficient magnitude to pose a threat to the integrity of nearby large concrete structures such as arch dams. The preliminary results from this study demonstrate that finite-element modelling based on poro-elastic formulations provides a useful means to quantitatively assess potential consolidation-induced surface settlements in fractured crystalline rock masses

Results of modelling show that an equivalent rock mass deformation modulus scaled from laboratory test results, and other input parameters constrained by field observations provide a good fit to the observed maximum settlement magnitude. However, the fit to the form of the subsidence trough is not so good, particularly with regard to total width and smaller scale variations. A better fit to the width and asymmetry could be obtained by including additional tunnel drainage points in the model at the locations where significant inflows were noted during tunnel construction.

Thus results emphasize the importance of geological anisotropy and asymmetry related to the location of major conductive fault zone structures intersecting the tunnel. Future modelling will concentrate on this anisotropy and asymmetry through the development of additional models incorporating more geological and material heterogeneities constrained by field data and by including tunnel flux drainage conditions based on tunnel inflow measurements. Future modelling will also be directed towards the development of 3-D equivalent continuum models that will allow for the explicit inclusion of important geological and hydrogeological features relative to the path of the Gotthard highway tunnel axis.

REFERENCES

Biot, M.A. 1941. *General theory of three-dimensional consolidation.* Journal of Applied Physics 12: pp. 155-164.

Karlsrud, K. & Sander, L. 1979. *Subsidence problem caused by rock-tunnelling in Oslo.* In S.K. Saxena (ed.), Evaluation and Prediction of Subsidence. New York: American Society of Engineers, pp. 197-213.

Lombardi, G. 1992. *The FES rock mass model - Part 2: Some examples.* Dam Engineering 3(3): pp. 201-221.

Luetzenkirchen, V. 2003. *Structural geology and hydrogeology of brittle fault zones in the central and eastern Gotthard massif, Switzerland.* Department of Earth Sciences, Swiss Federal Institute of Technology (ETH Zurich), Switzerland.

Nur, A. & Byerlee, J.D. 1971. *An exact effective stress law for elastic deformation of rock with fluids.* Journal of Geophysical Research 76(26): pp. 6414-6419.

Salvini, D. 2002. *Deformationsanalyse im Gotthardgebiet.* Zurich: Institute of Geodesy and Photogrammetry, ETH Zurich.

Schmidt, B. 1989. *Consolidation settlement due to soft ground tunneling.* In Proceedings of the Twelfth International Conference on Soil Mechanics and Foundation Engineering, Rio de Janeiro. Rotterdam: A.A. Balkema, pp. 797-800.

VIPS 2003. *VISAGE - Vectorial Implementation of Structural Analysis and Geotechnical Engineering, Version 8.7.* Bracknell, UK: Vector International Processing Systems Limited.

Zangerl, C. 2003. *Analysis of surface subsidence in crystalline rocks above the Gotthard highway tunnel, Switzerland.* Department of Earth Sciences, Swiss Federal Institute of Technology (ETH Zurich), Switzerland.

Zangerl, C., Eberhardt, E. & Loew, S. 2003. *Ground settlements above tunnels in fractured crystalline rock: numerical analysis of coupled hydromechanical mechanisms.* Hydrogeology Journal 11(1): pp. 162-173.

COUPLED DAMAGE-SEEPAGE CONSTITUTIVE MODEL OF JOINTED ROCK MASSES AND ITS ENGINEERING APPLICATION

Zhu Zhende[1], Xu Weiya[1], Zhang Aijun[1], Wang Sijing[2]

[1)] Institute of Geotechnical Engineering, Hohai Unversity, Nanjing 210098, China

[2)] Institute of Geology and Geophysics, The Chinese Academy of Sciences, Beijing 100029, China

Abstract: Based on the principle of energy superposition of liquid diffusion, the permeability tensor of a jointed rock mass is established. By means of the fracture mechanics and damage theory, a constitutive model of jointed rock mass under complex stress and damage states is studied. The damage evolution laws of jointed rock mass under both compression-shear and tension-shear are considered. Furthermore, coupled damage-seepage relationship is derived and programmed into a FEM code. The developed model and FEM code are applied to analyze the stability of high slope of the permanent lock of Three Gorges Project (TGP). Comparing the numerical results with the monitoring results, it can be seen that the model is creditable.

1. INTRODUCTION

The redistribution of stresses around excavations makes the fractures opening, closing, propagating and connecting together. Connected fractures are the main channel of groundwater flow. The size, distribution and hydraulic properties of fractures will have high influence on the flow path of groundwater. According to water pumping test results in a dam site, Louis (1974) set up a empirical equation between normal stress and coefficient of permeability, and this is helpful for establishing the coupled hydro-mechanical constitutive model of jointed rock mass. Similar approach was applied by Tao (1989) to study earthquakes caused by reservoirs through coupled hydro-mechanical simulation of jointed rock mass.

Along with the construction of a series of great hydropower stations, the problems of landslides caused by the rise of water level, and the stability of high slopes under the sharp rise of water pressure by rainfall are in critical needs in China. Consequently, the coupled problem of unsteady flow and damage of jointed rock mass are related to all above problems and is addressed in this paper. Some basic research works had been done by Yang (1995). The authors of this paper set up the coupled constitutive model of unsteady flow and damage, and the model was applied to analyze the stability of high slope of the permanent lock of TGP.

2. THE PERMEABILITY TENSOR OF JOINTED ROCK MASS

Oda (1986) deduced the permeability tensor of jointed rock mass by the theory of statistics, which not only considered the connectivity of fractures, but also the stochastic distributions of the geometrical parameters. Whereas probabilistic density is quite complex, therefore it is quite difficult for practical use. Zhu (1997) deduced the mathematic expression of a permeable tensor for an equivalent continuum given by

$$[K]=\frac{g\pi}{12\upsilon V_P}\sum_{i=1}^{m}\omega_i a_i^2 b_i^3 [M][M]^T \qquad (1)$$

where, ω_i is the weight function of fractures on seepage, a_i is the half trace-length of fractures, b_i is the aperture of fractures, V_P is the representative element volume, υ is the dynamical viscosity coefficients of water, and g is the acceleration of gravity, respectively. The matrix $[M]$ is expressed as

$$[M]=\begin{bmatrix} 1-n_1^2 & -n_1n_2 & -n_1n_3 \\ -n_2n_1 & 1-n_2^2 & -n_2n_3 \\ -n_3n_1 & -n_2n_1 & 1-n_3^2 \end{bmatrix}$$

where, n_i is the cosine of the angle between the common normal of fractures surfaces and the coordinate axes.

3 CONSTITUTIVE MODELS OF THE JOINTED ROCK MASSES

Practically, rock masses are in compression and shear state because of the in situ stresses. The unloading of the rock masses near excavated slopes causes tension-shear stress state in part of the slopes. Consequently, it is necessary to set up constitutive model of jointed rock masses considering both compression-shear and tension-shear mechanisms.

3.1 A constitutive model of the jointed rock masses under compression and shear

Assume C_n is a coefficient for transmitting compression of fractures and C_s is coefficient for transmitting shear of fractures. Denote again K_n as the normal stiffness and K_s as the shear stiffness of fractures. According to the Betti theory (Zhu, 1997), the following equation can be deduced

$$[C]=[C^0]+\sum_{i=1}^{n}[G_i]^T-[\Delta C_i][G_i] \tag{2}$$

Where, $$[\Delta C_i]=\begin{bmatrix} 0 & 0 & 0 & 0 & 0 & 0 \\ 0 & 0 & 0 & 0 & 0 & 0 \\ 0 & 0 & \frac{C_n^i S_0^i}{K_n^i V_i} & 0 & 0 & 0 \\ 0 & 0 & 0 & 0 & 0 & 0 \\ 0 & 0 & 0 & 0 & \frac{C_s^i S_0^i}{K_s^i V_i} & 0 \\ 0 & 0 & 0 & 0 & 0 & \frac{C_s^i S_0^i}{K_s^i V_i} \end{bmatrix},$$

Where, $[C^0]$ is undamaged elastic compliance matrix of rock, $[G_i]$ is the inverse matrix of coordinate-transfer matrix, V_i is the element volume, S_0^i is the area of fractures of the i-th fracture, respectively.

3.2 A constitutive model of the jointed rock masses under tension and shear

According to the principle of self-consistency (Kemeny, 1986), the compliance tensor of a jointed rock mass under tension and shear stress state can be deduced as

$$C_{ijkl}^{0-d}=C_{ijkl}^{0}+C_{ijkl}^{d} \tag{3}$$

Where, $C_{ijkl}^0=\frac{1+\mu}{E}\delta_{ik}\delta_{jl}-\frac{\mu}{E}\delta_{ij}\delta_{kl}$,

$$C_{ijkl}^d=\frac{8(1-\nu^2)}{3E}\sum_{m=1}^{n}[2a_m^3\rho_m n_i^{(m)}n_j^{(m)}n_k^{(m)}n_l^{(m)}+\frac{1}{2-\mu}a_m^3\rho_m (\delta_{jl}n_i^{(m)}n_k^{(m)}+\delta_{jk}n_l^{(m)}n_i^{(m)}+\delta_{il}n_j^{(m)}n_k^{(m)} +\delta_{ik}n_j^{(m)}n_l^{(m)}-4n_i^{(m)}n_j^{(m)}n_k^{(m)}n_l^{(m)})],$$

E is elastic modulus of rock and μ is Poisson's ratio, respectively.

4 DAMAGE EVOLUTION LAW OF THE JOINTED ROCK MASS

4.1 Damage evolution equation of jointed rock mass under compression and shear

Using the same principle of self-consistency (Kemeny, 1986), if there are $K(K\le n)$ sets of fractures in rock mass, the total compliancy tensor of the rock masses is expressed as

$$C_{ijkl}=C_{ijkl}^0+C_{ijkl}^d+dC_{ijkl}^d \tag{4}$$

where, C_{ijkl}^0 is the compliance tensor of the undamaged rock mass, C_{ijkl}^d is the compliance tensor with initial damage and dC_{ijkl}^d is the compliance tensor of jointed rock mass with subsequent incremental damage, given by

$$dC_{ijkl}^d=\sum_{j=1}^{k}[\frac{F_2^{(j)}\tau^{(j)}+F_3^{(j)}\sigma^{(j)}+F_5^{(j)}}{\sigma^{(j)}}n_i^{(j)}n_j^{(j)}n_k^{(j)}n_l^{(j)} +\frac{2F_1^{(j)}\tau^{(j)}+F_2^{(j)}\sigma^{(j)}+F_4^{(j)}}{4\tau^{(j)}}(\delta_{jl}n_i^{(j)}n_k^{(j)}+\delta_{jk}n_i^{(j)}n_l^{(j)} +\delta_{il}n_j^{(j)}n_k^{(j)}+\delta_{ik}n_j^{(j)}n_l^{(j)}-4n_i^{(j)}n_j^{(j)}n_k^{(j)}n_l^{(j)})] \tag{5}$$

The terms in equation (5) are expressed in reference (Zhu, 1997).

4.2 Damage evolution equation of jointed rock mass under tension and shear

The following equation is deduced based on fracture toughness experiment of rock (Segall,1984).

$$\frac{R'}{R}=\left(\frac{a'}{a}\right)^n \tag{6}$$

where, $n\,(n>0)$ is a growth exponent of fracture resistance, R is the critical fracture resistance, and a' and R' are the half-trace length and fracture resistance after fracture growth.

The damage evolution equation of rock masses are usually expressed as

$$C_{ijkl} = C^0_{ijkl} + C^d_{ijkl} + dC^{d(ex)}_{ijkl}$$

(7)

where, $dC^{d(ex)}_{ijkl} = dC^d_{ijkl}(\Delta a)$.

The length of fracture after growing is

$$a' = a^{\frac{n}{n-1}}\left\{\frac{4(1-\mu^2)}{\pi ER}[\sigma'^2 + \frac{2}{2-\mu}\tau'^2]\right\}^{\frac{1}{n-1}}$$

(8)

Fractures at different stress states have different growth model and a damage evolution law is also established for this case based on the arguments described above. Because the more growth of fractures indicates the more degradation of the rock masses, the damage tensor can be expressed as the following

$$\Omega = C^{d(1)}_{ijmn} : K^{0-d}_{mnkl}$$

(9)

where, $C^{d(1)}_{ijkl} = C^d_{ijkl} + dC^{d(ex)}_{ijkl}$, $C^{0-d}_{ijkl} = C^0_{ijkl} + C^{d(ex)}_{ijkl}$ and $C^{0-d}_{ijkl} : K^{0-d}_{klij} = (\delta_{mi}\delta_{nj} + \delta_{mj}\delta_{ni})/2$.

5 EVOLUTION OF PERMEABILITY TENSOR WITH PROGATATION OF FRACTURES

The permeability tensor depends on the stress states. The apertures of fractures will change with stress states. Based on this concept, Zhu (1997) investigated the following aspects:

(1) When fractures are in the tension but with no growth of fractures, permeability tensor can be given as follows:

$$[K] = \frac{g\pi}{12\upsilon v_p}\sum_{i=1}^{m} w_i a_i^2 [b_0 + \frac{8(1-\mu^2)}{\pi E}\sigma_n^i a_i + \frac{3.28\gamma_w H_i(1-\mu^2)}{E} a_i]^3 [M][M]^T \quad (10)$$

(2) When fractures open with tensile stresses and the length of fracture extends to a', the permeability tensor can be given as follows:

$$[K] = \frac{g\pi}{12\upsilon v_p}\sum_{i=1}^{m} w_i (a_i')^2 [b_0 + \frac{8(1-\mu^2)}{\pi E}\sigma_n^i a_i' + \frac{3.28\gamma_w H_i(1-\mu)}{E} a_i']^3 [M][M]^T$$

(11)

where, $\sigma_n = \sigma_{ij} n_i n_j$, $\Delta a_i = (a_i' - a_i) << a_i$.

(3) When the shear stress of the fractures satisfies condition $\tau_s = \tau - \sigma_n f - c > 0$, the fracture surface begins to slide. In the meantime, expansivity appears due to the surface roughness of the fractures. The permeable tensor can then be given as

$$[K] = \frac{g\pi}{12\upsilon v_p}\sum_{i=1}^{m} w_i a_i^2 [b_0 - \frac{\sigma_n^i}{K_n} + \frac{\tau_s^i}{K_s} tgd_m + \frac{3.28\gamma_w H(1-\mu^2)}{E}\cdot a_i]^3 [M][M]^T$$

(12)

where, d_m is the angle of dilatancy.

(4) When the shear stress τ_s is large enough, fractures will propagate and the permeability tensor can be given as

$$[K] = \frac{g\pi}{12\upsilon v_p}\sum_{i=1}^{m} w_i a_i^2 [b_0 - \frac{\sigma_n^i}{K_n} + \frac{\tau_s^i}{K_s} tgd_m + \frac{3.28\gamma_w H(1-\mu^2)}{E}\cdot a_i]^3 \cdot [M][M]^T + \frac{g\pi}{12\upsilon v_p}\sum_{i=1}^{k} w_i a_i L_i [\bar{V}_i + \frac{3.28\gamma_w H(1-\mu^2)}{E}\cdot L_i]^3 [M'][M']^T$$

(13)

with

$$\bar{V}_i = \frac{2(1-\mu^2)}{E}\left[\frac{4a_i\tau_s\cos\theta}{l_i\pi}\int_0^{l_i}\ln\left(\frac{l_i}{x} + \sqrt{(\frac{l_i}{x})^2 - 1}\right)dx - \frac{\pi}{2}\sigma_3 l_i\right]$$

where $\bar{V}_i$ is the average aperture of embranchment opening crack, θ is the angle between normal of fracture surface and maximum principal compressive stress , and

$$[M'] = \begin{bmatrix} 1-(n_1')^2 & -n_1'n_2' & -n_1'n_3' \\ -n_1'n_2' & 1-(n_2')^2 & -n_2'n_3' \\ -n_1'n_3' & -n_2'n_3' & 1-(n_3')^2 \end{bmatrix}.$$

Where, l_i is the length of subcrack (Ashby and Hallam, 1986) and is calculated from

$$\frac{\sigma_1\sqrt{\pi a}}{K_{IC}}=\frac{-(1+l)^{3/2}}{[1-\lambda-f(1+\lambda)-4.3\lambda L][0.23L+\dfrac{1}{\sqrt{3(1+L)}}]} \quad (14)$$

where, $\lambda=\sigma_3/\sigma_1$, $L=l/a$ and f is the friction coefficient of fractures.

6. DAMAGE MODEL COUPLED WITH HYDRO-MECHANICAL BEHAVIOUR OF JOINTED ROCK MASS

Jointed rock mass can be regarded as equivalent continuum media if the distribution of fractures are in uniform style, a RVE (representative volume element) can be established, and the fracture flow controls the seepage of rock masses. A hydro-mechanical model of jointed rock mass coupled with damage is developed in this paper, based on the equivalent continuum assumption. From eqations (1)~(14), a FEM formulation can be established as below

$$\left.\begin{array}{l}
\nabla(K\nabla H)+Q=S_s\dfrac{\partial H}{\partial t}\\
p=\gamma_w H\\
\sigma_{ij}=K_{ijkl}^{0-d}:\varepsilon_{kl}+K_{ijkl}^{0-d}:C_{kl}^{d}:p\\
[\tilde{K}]\{U\}=\{F\}+\{p\}\\
K_{ij}=K(b,a,n)\\
\underset{\sim}{\Omega}_{ijkl}=K_{ijpq}^{0-d}:C_{pqkl}^{d(1)}\\
b=b(\sigma_{ij},p)\\
a=a(\sigma_{ij},p)\\
n=n(\sigma_{ij},p)
\end{array}\right\} \quad (15)$$

The terms in equation (15) are expressed in reference (Zhu, 1997).

7. APPLICATION TO THE TGP SLOPE

The developed models and a FEM code (RPZDFRZ) are applied to investigate the stability of the high slope of the permanent ship lock of the TGP. Mechanical parameters of completely weathered, seriously weathered, slightly weathered and fresh rock mass in TGP's ship lock region are listed in reference (Zhu & Zhang, 1996). Mechanical parameters of rock fractures in TGP's ship lock region are listed in Table 1 (Zhu, 1997).

The in situ stress field is considered as a self-weight stress field in weathered granite and is calculated by the following equation

$$\sigma_x=-4.3982-0.01168H$$
$$\sigma_y=-1.6628-0.03039H$$
$$\tau_{xy}=0.0472-0.00008H$$

(16)

where, stress unit is MPa, H is overburden (m) and x, y are the horizontal and vertical coordinate.

Table 2 shows statistical data of the fracture parameters for the high slope of the permanent ship lock of TGP.

It was estimated that the permeability coefficient $K=(1.4\sim4.6)\times10^{-5}m/s$ in the seriously weathered and fractured zone. The distribution of fractures dominates the penetrability of rock mass in the slightly weathered zone, and excavation caused unloading also plays an important role. Parameters concerned (Zhu and Zhang, 1996) are as follows: the initial aperture of fractures $b=0.1mm$, the dynamical viscosity coefficients of underground water $\upsilon=10\times10^3 m/s$, the unit volume discharge $S=6.4\times10^{-3}m^3/(m^3\cdot s)$ and the bulk dense of water $\gamma_w=1.0\times10^3 kg/m^3$.

8. NUMERICAL RESULTS

A 2D FEM model with 588 elements and 644 nodes was used to study the slope behaviour, considering the geological and geometrical conditions along the No.20 vertical cross section of the slope. According to the design, seven steps were taken in the excavation. Figure 1 shows the positions of some key points on the section No.20 for monitoring results.

From the numerical results at the 7th step, the left slope will move towards up-right, while the right slope will move up-left and the pier will move towards the slopes. Comparing the numerical results between the damage coupled hydro-mechanical model and the classical elasto-plastic model, it shows that the damage

evolution of fractures will

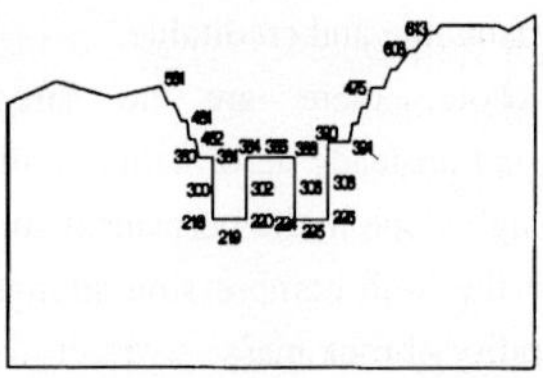

Fig.1 Positions of some key points on the excavation border

Table 1. Mechanical parameters of rock fractures in TGP's ship lock region.

Fracture type		Shear strength		Residual shear strength		Shear stiffness	Normal stiffness
		f'	C/MPa	f'	C'/MPa	$K_s/MPa\cdot m^{-1}$	$K_n/MPa\cdot m^{-1}$
Rigid fracture		0.7	0.30	0.60	0.15	3000	7500
Weak fractures	Slightly weathered	0.6	0.20	0.50	0.15	1000	2500
	Seriously weathered	0.4	0.15	0.35	0.10	1000	2500

Table 2 Statistical data of parameters of fractures.

Trace length / m		Angle of dip / (°)		Spacing / m		Wide of fracture / mm	
Mean	variance	Mean	variance	Mean	variance	Earth's surface	Earth's deep
3.82	0.52	72.8	10.0	1.2	0.02	0.2	0.1

Table 3 Displacements of several key points on excavation border (mm)

Conditions	581	462	381	219	384	388	225	390	475	603
Damage coupled model	27.95	41.43	27.38	21.28	48.13	42.72	31.45	45.47	17.10	14.04
Elasto-plastic model	27.66	7.86	11.65	11.59	34.84	36.00	26.94	34.64	16.70	13.35

lead to additional 5~15mm displacements and in some areas will have 15~30mm displacements. The total displacements of key points are listed in Table 3.

Numerical results show that the maximum principal stress are nearly normal to the excavation surfaces, and in some areas stresses are changed from compressive to tensile, such as the triangle region behind the plumb wall and under the flat roof of left-right slope. The stress state varies from twin-direction compressive stress to nearly zero stress at the top of middle pier. The stress concentration occurs at the corner of brake chamberFigure 2 shows the region of damage evolution that is calculated according to the model proposed in this paper. The evolution regions in fresh and slightly weathered rock masses do not change much, and there are no prodigious regions of damage evolution in middle pier.

The numerical results indicate that the damage evolution of the fractures will increase the permeability coefficient of rock mass. The evolution region of seepage in the right slope is about 10% bigger than that of left. The damage evolution zone in fresh granite will decrease with the increase of depth as shown in Fig. 3.

Fig.2 Evolution damage at the cross-section No.20

Fig.3 Evolution of seepage-damage at the cross-section No.20

It can be seen from Figs.2~3 that the stability of the surrounding rock mass of the brake chamber is very important for the damage evolution regions and seepage caused damage are mostly likely to congregate there. The numerical results obtained from the damage coupled model are bigger than the results obtained using only the Zhu's original approach (Zhu & Zhang, 1996), which just explains the effect of groundwater seepage on the structural characteristic of rock mass. Fortunately, the numerical results are so close to the monitoring results (Wu, 2002).

9. CONCLUSIONS

(1)The coupled model of damage and unsteady-states seepage field for jointed rock mass is put forward for the first time. The FEM program (RPZDFRZ) is based on the model and applied to the calculation of the stability of high slope in the permanent ship lock of Three Gorges Project (TGP). Comparing the numerical results with the monitoring results (Wu, 2002), it can be seen that the model is reasonable and creditable.

(2)As a whole, there are no large-scale demolishment and unsteady deformation in the rock masses of the high slope in the permanent ship lock of TGP due to the high compression strength and the good integrality of rock mass.

(3)Taking account into the damage evolution of jointed rock mass, the resilient displacement increases slightly during the excavation of the ship lock. Displacement apparently increases in the region of damage evolution and the maximum is in the middle of the left slope, the left of plumb wall and middle frusta.

(4)By calculation on the basis of damage coupled hydro-mechanical model, it is clear that the stress of slope increases slightly. The mode of the fractures propagation and coalescence is mainly in tension, but sometimes is also shearing or tension-shearing.

(5)There are more large regions of damage evolution and seepage evolution in the left slope than that of right slope. Therefore, bolting, seepage prevention and drainage measures should be considered.

REFERENCES

Ashby M T, Hallam S D. *The Failure of Brittle Solids Containing Small Cracks under Compressive Stress States.* Acta Metall, 1986, 34(3): 497~510.

Kemeny I, Cook N G W. 1986. *Effective Moduli, Non-linear Deformation and Strength of a Cracked Elastic Solid.* Int. J. Rock. Mech. Sci. &Geomech. Ahstra. 1986,(2) : 107~118.

Louis C. 1974. *Rock Hydraulics in Rock Mechanics.* Verlay wien: New York.

Oda M.1986. *An Equivalent Continuum Model for Coupled Stress and Fluid Analysis in Jointed Rock Masse.* Water Resources Research, 1986, 22(13).

Segall P. 1984. *Formation and Growth Of Extensional Fracture Sets* . Geological Society of America

Bulletin. 1984, 95:454~462.

Tao Zhenyu, Tang Wanfu, Zhang Liming etc. 1989. *The new Forecast of earthquake due to Reservoir.* The second Symposium on Rock Mechanics and Engineering in China. Bei Jing：Reading Press.

Wu Haibin. 2002. *Stabilization Technology on the High-Steep Rock Slope in the Three Gorges Project Shiplock.* Chinese J. of Rock Mechanics and Engineering, 2002, 21(2): 261~267.

Zhu Weishen, Zhang Yujun. 1996. *Primary Research on Stability Analysis and Reinforcement Schemes for Jointed Rock Mass of High Slope of Three Gorges Flight Lock.* Chinese Journal of Rock Mechanics and Engineering, 1996(4): 305~311.

Zhu Zhende. 1997. *Research on the Coupled Model of Unsteady-states Seepage field and Damage field. [Doctoral Dissertation].* Shang Hai: Tongji University.

MATHEMATICAL MODELLING OF BOREHOLE GROUTING IN PERMAFROST

Sergei Fomin[1], Vladimir Chugunov[2] and Toshiyuki Hashida[1]

[1])Fracture Research Institute, Tohoku University, Sendai, Japan

[2])Department of Applied Mathematics, Kazan State University, Kazan, Russia

Abstract: The process of borehole grouting in the regions of permafrost is more complicated than in normal climatic conditions due to the possible thawing of the frozen rock, which can be induced by the heat produced during cement hydration. In order to prevent thawing of the surrounding formation, the main conditions that should be imposed on the grouting off technology and properties of the cementing slurries are rapid cement hydration and low heat generation during chemical reaction. Mathematical model of the grouting off process in permafrost, presented in this study, accounts for heat transferred from the borehole circulating mud and heat generated due to cement hydration. The problem is solved analytically by the approximate integral-heat-balance method. As a result, the close-form solutions for the temperature field in the cemented region and surrounding formation are obtained. On the basis of these analytical solutions, analytical criteria for the borehole grouting off regimes that take place without thawing of the frozen rock are proposed. These criteria can be readily used for assessing physical and chemical properties of the cement slurries and their suitability for grouting off the borehole in permafrost.

1. INTRODUCTION

The heat transfer process plays an important role in a number of industrial applications related to drilling technologies and borehole exploitation. For instance, the process of borehole grouting by the cement slurries is accompanied by generation of significant amount of heat as a result to an exothermic reaction of cement hydration. Assessment of this heat and of the heat transmitted from the circulating mud to the rock formation is of crucial importance, especially for the borehole completion in the region of permafrost where excessive heating is undesirable and may lead to different engineering complications. It is well known that the mud temperature and chemical composition of the cement compounds are the main factors that influence the down-hole performance of cement slurries (Kutasov, 1999). The down-hole temperature controls the pace of chemical reactions during cement hydration resulting in cement setting and strength development. As a mud temperature increases, the cement slurry hydrates and sets faster and develops strength more rapidly. In Arctic wells the cement is set under unfavourable conditions since, on the one hand, external heating of the cement slurry is required for hydration enhancement and, on the other hand, heating of the slurry supplemented by heat produced due to chemical reaction may cause thawing of the frozen rock formation. The latter in turn may lead to deformation and failure of the cemented column behind the casing pipe and further leakage of the cement slurry in thawed area. To this end, sustaining the optimal temperature regime is of primarily importance for the borehole grouting-off in the regions of permafrost. In the present study, interaction of the main parameters that control the grouting process, namely temperature field and thermo-physical and chemical properties of the cement, are modelled by the conjugate heat conduction problem in the cemented column and formation with exponential heat source term due to chemical reaction of cement hydration.

2. SYSTEM MODEL AND ANALYSIS

The axial symmetry of the process is assumed and, therefore, the cylindrical coordinates (r^*, z^*) are employed. In the borehole grouting technology the casing pipe of external radius r_0^* is placed in the borehole of the radius r_w and the gap between the casing tube and the wall of the well is filled by the cement slurry. Fluid in the borehole intensively circulates along the drilling tube (inflow) and annulus between the drilling and casing tubes (outflow). This intensive circulation of the fluid of the appropriate temperature t_L is undertaken in order to provide the proper conditions for cement hardening in the gap behind the casing. Since the radial temperature gradients are much greater than temperature gradients in the vertical direction, the temperature distribution in the cement slurry and surrounding rock can be described by the following non-dimensional mathematical model:

$$\frac{\partial T_1}{\partial \tau} = \frac{D}{r}\frac{\partial}{\partial r}\left(r\frac{\partial T_1}{\partial r}\right) + qbe^{-b\tau}, \; r_0 < r < 1, \quad (1)$$

$$\frac{\partial T_2}{\partial \tau} = \frac{1}{r}\frac{\partial}{\partial r}\left(r\frac{\partial T_2}{\partial r}\right), \; r > 1, \; \tau > 0, \quad (2)$$

$$\tau = 0, \quad T_1 = T_C, \quad T_2 = 0; \quad (3)$$

$$r = r_0, \quad T_1 = T_L; \quad (4)$$

$$r = 1, \quad K\partial T_1/\partial r = \partial T_2/\partial r, \quad T_1 = T_2; \quad (5)$$

$$\lim_{r\to\infty} T_2 < \infty, \quad (6)$$

where

$$r = \frac{r^*}{r_w}, \; r_0 = \frac{r_0^*}{r_w}, \; b = \frac{b^* r_w^2}{d_r}, \; K = \frac{k_c}{k_r},$$

$$\tau = \frac{\tau^* d_r}{r_w^2}, \; q = \frac{wq^*}{c_c(t_f - t_r)}, \; T_{1,2} = \frac{t_{1,2} - t_r}{t_f - t_r}, \quad (7)$$

$$D = \frac{d_c}{d_r}, \; T_C = \frac{t_c - t_r}{t_f - t_r}, \; T_L = \frac{t_L - t_r}{t_f - t_r},$$

t_1, t_2 and t_f are temperatures of cement, formation and freezing point for water in the rock, respectively; t_r and t_c are initial temperature of the rock and cement, respectively; k_c and k_r are thermal conductivities of the cement and rock, respectively; d_c and d_r are thermal diffusivities of the cement and rock, respectively; c_c – specific heat of cement; τ^* - time; w – mass fraction of cement within the slurry; q^* - amount of heat generated due to hydration of cement slurry mass unit; b^* - parameter characterizing the speed of heat generation.

Equations (1) and (2) describe the temperature distribution in the hydrating cement and frozen rock surrounding the borehole, respectively; equation (3) is the initial condition at τ=0, equations (4), (5) and (6) are the boundary conditions at the wall of the casing tube ($r=r_0$), on the wall of the well (r=1) and in infinity, respectively. The term in the right-hand side of equation (1) represent the heat source due to chemical reaction of cement hydration. The problem is formulated under condition that $t_r<t_f$. Since the main target of the present analysis is to define conditions that secure cement hardening without thawing of the surrounding permafrost, it is natural to concentrate on the assessment of temperature variation on the wall of the borehole r=1. When at some moment, $\tau = \tau_f$, this temperature reaches the freezing temperature of the saturated soil (in non-dimensional variables it corresponds to $T_2|_{r=1} = 1$) then formation thawing may occur. Owing to linearity of the boundary-value problem (1)-(6), its solution T_i can be presented as a sum of solutions of three auxiliary problems

$$T_i = T_L X_i + T_C Y_i + qZ_i, \; (i = 1,2); \quad (8)$$

where X_i, Y_i, and Z_i are the following particular solutions of the problem (1)-(6):

1. $T_i = X_i$ for T_L=1 and $T_C=q$=0,
2. $T_i = Y_i$ for $T_L=q$=0 and T_C=1,
3. $T_i = Z_i$ for $T_L=T_C$=0 and q=1.

Each of these particular solutions depends of variables r and τ, and parameters r_0, b, K, and D and can be found separately.

2.1 Approximate solution for X_i

Governing equations for X_1 and X_2 follow from equations (1)-(6). Assuming that T_L=1 and $T_C=q$=0, yields

$$\frac{\partial X_1}{\partial \tau} = \frac{D}{r}\frac{\partial}{\partial r}\left(r\frac{\partial X_1}{\partial r}\right), \; r_0 < r < 1, \quad (9)$$

$$\frac{\partial X_2}{\partial \tau} = \frac{1}{r}\frac{\partial}{\partial r}\left(r\frac{\partial X_2}{\partial r}\right), \; r > 1, \; \tau > 0, \quad (10)$$

$$\tau = 0, \quad X_1 = X_2 = 0; \quad (11)$$

$$r = r_0, \quad X_1 = 1; \quad (12)$$

$$r = 1, \quad K\partial X_1/\partial r = \partial X_2/\partial r, \quad X_1 = X_2; \quad (13)$$

$$\lim_{r\to\infty} X_2 < \infty, \quad (14)$$

For solving this problem the analytical approximate method of integral correlations proposed by Goodman (1958) and improved by Volkov *et al* (1988) can be employed. This method was successfully used by Fomin *et al* (1994) for solving more complex problem of moving heat source within the borehole in application to melting of the paraffin deposition in the annulus.

It is convenient to introduce the following function

$$X(r,\tau) = \begin{cases} X_1, & r_0 \le r < 1, \\ X_2, & r \ge 1. \end{cases} \quad (15)$$

According to Volkov *et al* (1988), the approximate solution of the problem (9)-(14) can be sought in the form:

$$X(r,\tau) = 1 + \sum_{i=1}^{m} a_i(\tau) f_i(r), \quad (16)$$

where $r < l(\tau)$. Here function $l(\tau)$ can be defined as radius of thermal influence, so that for $r \ge l(\tau)$ the temperature is assumed to be undisturbed and $X(r,\tau) = 0$, $\partial X / \partial r = 0$. In the further analysis only first two terms in the sum on the right-hand side of (16) will be retained (which is referred to as a second approximation, m=2). This has been proved to be accurate enough for approximation of the exact solutions for the similar heat conduction problems (Goodman, 1958; Volkov et al, 1988). Functions a_i and f_i in solution (16) are sought to satisfy the integral heat balance correlations, which can be readily obtained by integrating equations (9) and (10) over the interval $[r_0, l]$, and boundary conditions (11)-(14). As a result of this straightforward procedure it can be easily shown that

$$a_1(\tau) = -f_2'(l)/[f_1(l)f_2'(l) - f_2(l)f_1'(l)], \tag{17}$$

$$a_2(\tau) = f_1'(l)/[f_1(l)f_2'(l) - f_2(l)f_1'(l)], \tag{18}$$

$$f_1(r) = \begin{cases} \ln(r/r_0), & r_0 \le r < 1, \\ K \ln r - \ln r_0, & 1 \le r < \infty, \end{cases} \tag{19}$$

$$f_2(r) = \begin{cases} \dfrac{r^2}{4D}[\dfrac{r^2}{r_0^2}(\ln\dfrac{r^2}{r_0^2} - 1) + 1], & r_0 \le r < 1, \\ \dfrac{r^2}{4r_0^2}[K \ln\dfrac{r}{e} + \ln r_0] + \alpha_1 \ln r + \beta_1, & r \ge 1, \end{cases} \tag{20}$$

$$f_3(r) = \begin{cases} \dfrac{1}{64D^2}[\dfrac{r^4}{r_0^4}(\ln\dfrac{r}{r_0} r - \dfrac{3}{2}) + \dfrac{4r^2}{r_0^2} - \dfrac{5}{2}], & r_0 \le r < 1, \\ \dfrac{r^4}{64r_0^4}[K(\ln r - \dfrac{3}{2}) - \ln r_0] + \dfrac{r^2}{4r_0^2}\left[\alpha_1 \ln\dfrac{r}{e} + \beta_1\right] \\ + \alpha_2 \ln r + \beta_2, & 1 \le r < \infty, \end{cases} \tag{21}$$

where

$$\alpha_1 = \frac{K}{4Dr_0^2}\left[D - 1 - 2(1 - \frac{D}{K})\ln r_0\right], \tag{22}$$

$$\beta_1 = \frac{1}{4r_0^2}(\ln r_0 + K) - \frac{1}{4D}\left[\frac{1}{r_0^2}(\ln r_0 + 1) - 1\right], \tag{23}$$

$$\alpha_2 = \frac{K}{64D^2 r_0^4}\left[5(D^2 - 1) - 4(1 - \frac{D^2}{K})\ln r_0\right] + \frac{K}{64D^2 r_0^2}\left[1 - \frac{2D^2}{K}(2\beta_1 - \alpha_1)\right] \tag{24}$$

$$\beta_2 = \frac{1}{64Dr_0^4}\left[D^2\left(\ln r_0 - \frac{3K}{2}\right) - \ln r_0 - \frac{3}{2}\right] + \frac{1}{16D^2 r_0^2}\left[1 - 4D^2(\beta_1 - \alpha_1)\right] - \frac{5}{128D^2}. \tag{25}$$

The radius of thermal influence l is defined implicitly as a solution of algebraic equation

$$-\tau + r_0^2\left[\frac{f_1(l)f_3'(l) - f_1'(l)f_3(l)}{f_1(l)f_2'(l) - f_2(l)f_1'(l)}\frac{l}{\tilde{K}(l)} + \frac{1}{4D}\right] = 0, \tag{26}$$

where $\tilde{K}(l) = 1$ for $r_0 < l < 1$ and $\tilde{K}(l) = K$ for l>1. In the above equations and below $f_k'(l)$ denote the derivatives, df_k / dr, at r=l (k=1,..., m).

Setting in solution (16) for m=2, r=1 and accounting for equations(17)-(21), yields

$$X\big|_{r=1} = 1 + \frac{f_2'(l)\ln r_0 + \dfrac{f_1'(l)(r_0^2 - 1 - \ln r_0)}{4Dr_0^2}}{f_1(l)f_2'(l) - f_2(l)f_1'(l)} \tag{27}$$

The latter expression is valid for, whereas for $r_0 < l(\tau) < 1$, $X\big|_{r=1} = 0$. Obviously, according to notation (15), $X\big|_{r=1} = X_2$.

2.2 Approximate solution for the functions Y_i and Z_i

Assuming $T_L = 0$ in equations (8) and (1)-(6) we obtain the auxiliary problem for the function $\theta_i = T_C Y_i + qZ_i$, where index i=1 denotes temperature in cemented area and i=2 - temperature in the rock formation. Introducing the mean temperature of the cemented ring with internal radius r=r_0 and external radius r=1 as

$$\tilde{\theta}_1 = \frac{2}{1 - r_0^2}\int_{r_0}^{1} r\theta_1 dr \tag{28}$$

and integrating equation (1) with respect to r in the interval $[r_0,1]$, yields

$$\frac{d\tilde{\theta}_1}{d\tau} = \frac{2D}{1 - r_0^2}\left(\frac{\partial\theta_1}{\partial r}\bigg|_{r=1} - r_0\frac{\partial\theta_1}{\partial r}\bigg|_{r=r_0}\right) + qbe^{-b\tau}. \tag{29}$$

Approximately we can assume the parabolic temperature distribution within the narrow cemented ring

$$\theta_1 \approx c_0 + c_1 \ln r + c_2 r^2, \; r_0 \le r \le 1, \tag{30}$$

where the unknown coefficients are determined by boundary conditions $\theta_1|_{r=r_0}=0$, $\theta_1|_{r=1}=\theta_2|_{r=1}$ and equations (28):

$$c_0=[2\alpha\tilde{\theta}_1+(\alpha+r_0^2)\theta_2|_{r=1}]/(1-r_0^2) \quad (31)$$

$$c_1=\alpha(\theta_2|_{r=1}-2\tilde{\theta}_1)/\ln r_0 \quad (32)$$

$$c_2=[(\alpha+1)\theta_2|_{r=1}-2\alpha\tilde{\theta}_1]/(1-r_0^2) \quad (33)$$

where

$$\alpha=\frac{(1-r_0^2)\ln r_0}{(1-r_0^2)(1+\ln r_0)+2r_0^2\ln r_0}. \quad (34)$$

Furthermore, from equation (30) follows that

$$\left.\frac{\partial\theta_1}{\partial r}\right|_{r=1}=2c_2+c_1;\ r_0\left.\frac{\partial\theta_1}{\partial r}\right|_{r=r_0}=2c_2r_0^2+c_1. \quad (35)$$

Hence, equation for the mean temperature (29) can be converted to

$$\frac{d\tilde{\theta}_1}{d\tau}=\frac{4D}{1-r_0^2}\left((\alpha+1)\theta_2|_{r=1}-2\alpha\tilde{\theta}_1\right)+qbe^{-b\tau}. \quad (36)$$

The initial condition at $\tau=0$ for equation (36) follows from (3)

$$\tilde{\theta}_1=\theta_c. \quad (37)$$

Utilizing equations (35) and (31)-(33) for transforming the boundary condition (5) at r=1, the boundary value problem for θ_2, which corresponds to T_2 for $T_L=0$ can be formulated as

$$\frac{\partial\theta_2}{\partial\tau}=\frac{1}{r}\frac{\partial}{\partial r}\left(r\frac{\partial\theta_2}{\partial r}\right),\ r>1,\ \tau>0, \quad (38)$$

$$\tau=0,\qquad \theta_2=0; \quad (49)$$

$$r=1,\qquad \partial\theta_2/\partial r=B(\theta_2-\beta\tilde{\theta}_1); \quad (40)$$

$$\lim_{r\to\infty}\theta_2<\infty, \quad (41)$$

where

$$\beta=\frac{2\alpha(1+2\ln r_0-r_0^2)}{\alpha(1+2\ln r_0-r_0^2)+2\ln r_0} \quad (42)$$

$$B=K\frac{\alpha(1+2\ln r_0-r_0^2)+2\ln r_0}{(1-r_0^2)\ln r_0}. \quad (43)$$

The problem, similar to the above formulated (38)-(41), has been solved by the integral balance method by Volkov *et al* (1988). Following their approach, and denoting by Q the heat flux at r=1 for the case when it is assumed that $\beta\tilde{\theta}_1=1$, the boundary condition (40) can be reduced to

$$\theta_2|_{r=1,\beta\tilde{\theta}_1=1}=1-Q(\tau). \quad (44)$$

Fomin *et al* (2002) have proved that an approximate formula,

$$Q(\tau)=1/[1+B\ln(1+\frac{2.084+0.704B}{1.554+0.407B}\sqrt{\tau})], \quad (45)$$

is accurate enough for describing this function. Due to the Duhamel theorem (Carslaw and Jaeger, 1959), solution of the problem (38)-(41) for arbitrary $\beta\tilde{\theta}_1$ can be presented in the following form

$$\theta_2|_{r=1}(\tau)=-\beta\int_0^\tau\tilde{\theta}_1(z,p)\frac{\partial}{\partial\tau}Q(\tau-p)dp. \quad (46)$$

Substituting (46) into equation (36) and integrating the latter with account to condition (37), yields

$$\tilde{\theta}_1=\theta_c+q(1-e^{-b\tau})+ \\ +\frac{4D}{1-r_0^2}\int_0^\tau\tilde{\theta}_1(p)[\beta(1+\alpha)(1-Q(\tau-p))-2\alpha]dp \quad (47)$$

Equations (46) and (47) constitute the closed system for computing the mean temperature in the cemented ring $\tilde{\theta}_1$ and temperature on the rock-cement boundary $\theta_2|_{r=1}(\tau)$, which characterize the heat interaction of the hydrating cement and frozen formation. Assuming in (46), (47) q=0 and $T_C=1$, leads to equations for computing $\tilde{Y}_1$ and $Y_2|_{r=1}$. In the case of $T_C=0$ and q=1, equations (46), (47) can be used for computing $\tilde{Z}_1$ and $Z_2|_{r=1}$. Obviously, these simple equations can be readily solved numerically by the finite-difference method. Discretization of equations (46), (47) with uniform grid (constant time-step s) is straightforward. Denoting $\tau_i=is$, where $i=1,2,...,n$, and $U_i=\tilde{\theta}_1(\tau_i)$, $V_i=\theta_2|_{r=1}(\tau_i)$, $Q_i=Q(\tau_i)$, $E_i=\beta(\alpha+1)(1-Q(\tau_i))-2\alpha$, $\varphi_i=\theta_c+q(1-e^{-b\tau_i})$, $\gamma=2sD/(1-r_0^2)$ the finite difference approximation of equations (47) and (46) can be presented as follows

$$U_i=\frac{\gamma E_1U_{i-1}+\lambda\sum_{k=0}^{i-2}(U_{k+1}E_{i-k-1}+U_kE_{i-k})+\varphi_i}{1+2\alpha\gamma}, \quad (48)$$

$$V_i = 0.5\beta \sum_{k=0}^{i-1} (U_{k+1} + U_k)(Q_{i-k-1} - Q_{i-k}). \tag{49}$$

As a result, computation of $\theta_2\big|_{r=1}$ and $\tilde{\theta}_1$ is reduced to calculation of U_i and V_i from the recursion relations (48), (49).

3. NUMERICAL RESULTS AND DISCUSSION

For completing the borehole cementation within the time interval $0<\tau<\tau_f$, when thawing of the surrounded permafrost has not yet began, the following necessary and sufficient condition should be satisfied

$$\max_{0\le\tau\le\tau_f} T_2\big|_{r=1} \le 1. \tag{50}$$

The inequality (50) in general is a function parameters T_L, T_C, and r_0, which determine the external and initial conditions of the borehole grouting, and also is dependent of parameters that characterize the properties of the cement slurry, namely, parameters q, b, D, and K. Hence, inequality (50) represents the criteria that allows for the specified conditions (T_L, T_C, and r_0,) to choose the appropriate recipes of the cement slurries (with different q, b, D, and K) that may prevent from overheating and thawing of the frozen formation. Due to suggested decomposition of the general solution T_i of the heat conduction problem (1)-(6) into 3 auxiliary problems for X_i, Y_i and Z_i (see equation (8)), condition (50) can be presented in the following form:

$$\max_{0\le\tau\le\tau_f} (T_L X_2\big|_{r=1} + T_C Y_2\big|_{r=1} + qZ_2\big|_{r=1}) \le 1 \tag{51}$$

Equations (26), (27) that define function $X_2\big|_{r=1}$ are rather simple and do not present any difficulties for numerical calculations. Results of these computations for K=0.1 and D=0.2 and different r_0 are illustrated in Fig. 1.

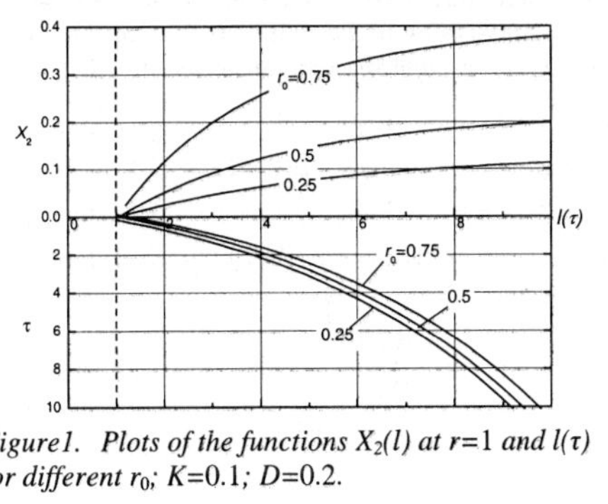

Figure1. Plots of the functions $X_2(l)$ at r=1 and $l(\tau)$ for different r_0; K=0.1; D=0.2.

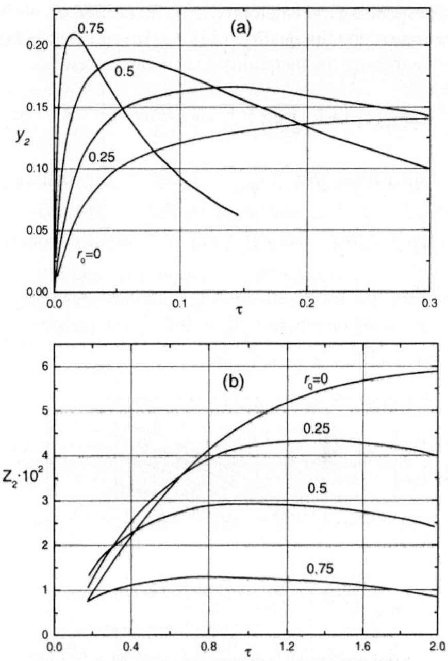

Figure 2. Variation (a) of $Y_2(\tau)$ and (b) of $Z_2(\tau)$ with time at $r=1$ for different r_0; $K=0.1$; $D=0.2$.

For computing $Y_2\big|_{r=1}$ and $Z_2\big|_{r=1}$ equations (48) and (49) are used: the former is computed while assuming in (48) and (49) $T_L = q = 0$ and $T_C = 1$, and the latter is obtained letting $T_L = T_C = 0$ and q=1. The results of computations are presented in Fig. 2. As it can be seen, unlike apparently observed local maximum of functions $Y_2\big|_{r=1}$ and $Z_2\big|_{r=1}$, function $X_2\big|_{r=1}$ exhibits a monotonous growth with time and $\lim_{\tau\to\infty} X_2\big|_{r=1} = 1$. The latter implies that for relatively slow setting of the cement slurry (high values of $\tau=\tau_f$) and $T_L>1$ the inequality (51) will not be satisfied regardless to the values of q, b, D, and K. Physically it means that continued heating of cement slurry by the warm circulating fluid with temperature $t_L>t_f$ inevitably will lead to overheating of the formation above the point of freezing. Hence, in the case of external heating of the cement slurry by the warm circulating mud it is of crucial importance to choose the proper recipe for cement compounds accounting for the time of heating and cement setting τ_f. These recipes should allow for the mud temperatures T_L >1 (though these temperatures should be relatively low) to define the corresponding parameters τ_f, q, b, D, and K which guarantee the inequality (51) to be satisfied. Only strengthening the inequality (51) it can be rewritten as

$$T_L X_2\big|_{r=1}(\tau_f) + T_C \max_{0\le\tau\le\tau_f} Y_2\bigg|_{r=1} + q \max_{0\le\tau\le\tau_f} Z_2\bigg|_{r=1} \le 1. \quad (52)$$

Calculations for $X_2\big|_{r=1}(\tau_f)$ can readily made by the (27) as it is illustrated in Fig. 1. The maximal values for $Y_2\big|_{r=1}$ and $Z_2\big|_{r=1}$ can be computed from the solution of corresponding equations (48) and (49). For instance, the results illustrating solutions sensitivity to variation of parameters K, b and r_0 are presented in Figs. 3 and 4.

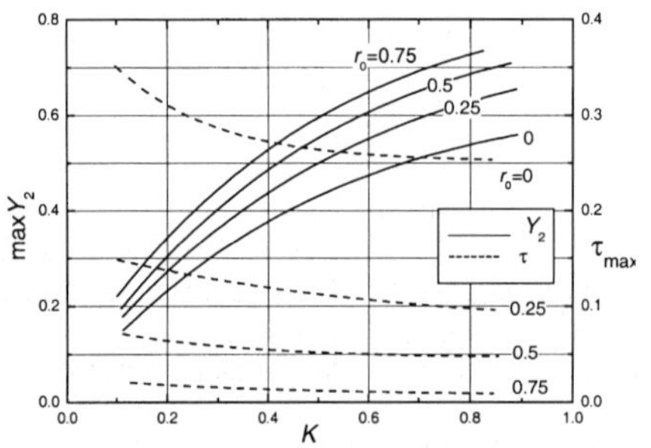

Figure 3. Variation of $\max Y_2$ *at* r=1 and τ_{max} *vs parameter K for different* r_0; *D*=0.2.

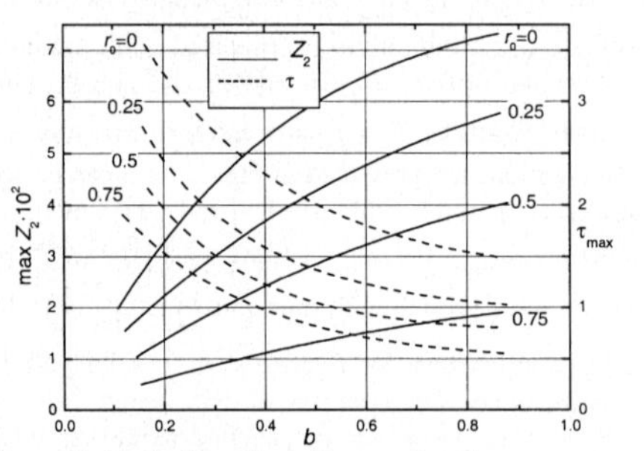

Figure 4. Variation of $\max Z_2$ *at* r=1 and τ_{max} *vs parameter b for different* r_0; *D*=0.2, *K*=0.1.

The described approach allows to select effectively the proper properties of cement solutes and to determine the optimal conditions for borehole grouting in the regions of permafrost. However, it worthwhile mentioning that inequality (52) is rather rough and may not be satisfied even in the case when correlation (50) is valid. The latter may occur primarily due to the fact that functions $Y_2\big|_{r=1}$ and $Z_2\big|_{r=1}$ reach their maximal values at the different moments of time. For instance, this fact can be observed comparing Figs. 2(a) and 2(b). In order to consider this specific feature of the heat transfer process, a series of additional computations were carried out for the values of τ_{max} (time required for $Y_2\big|_{r=1}$ and $Z_2\big|_{r=1}$ to reach their maximal values). These results are presented in Figs. 3 and 4 by the dashed lines.

4. CONCLUSIONS

The following conclusions can be drawn:

The mathematical model for the borehole grouting in the regions of permafrost is proposed and solved analytically by the approximate integral-heat-balance method.

The analytical criteria for the borehole grouting off regime, that will take place without thawing of the frozen rock, are proposed. These criteria can be readily used for assessing physical and chemical properties of the cement slurries and their suitability for grouting off the borehole in permafrost.

The simplified methodology for computing the optimal condition of the borehole grouting is suggested and illustrated by numerical results.

REFERENCES

Carslaw, H.C. & Jaeger, J.C. 1959. *Conduction of Heat in Solids.* Clarendon Press: Oxford, UK.

Fomin, S.A., Wei, P. S., & Chugunov V. A. 1994. *Melting Paraffin Plug Between Two Co-axial Pipes by a Moving Heat Source in the Inner Pipe.* ASME Journal of Heat Transfer 116: pp.1028-1033.

Fomin, S., Chugunov, V., Hashida, T., Saito, S. & Suto, Y. 2002. *Heat Flux on the Bore-Face and Temperature Distribution in the Formation.* GRC Transactions 26: pp.803-807.

Goodman, T.R. 1958. *The Heat Balance Integral and its Application to the Problems Involving a Change of Phase.* ASME Journal of Heat Transfer 80: pp.335-342.

Kutasov, I.M. 1999. *Applied Geothermics for Petroleum Engineers.* Elsevier.

Volkov, I.K., Chugunov, V.A. & Salamatin, A.N. 1988. *Generalization of the Method of Integral Correlations and its Application to Some Heat Transfer Problems.* Journal of Soviet Mathematics 41: pp. 1013-1029.

THERMO-HYDROLOGICAL ANALYSIS TO PREDICT THE TEMPERATURE DISTRIBUTION AROUND A COLD STORAGE CAVERN

Gyu-Sang Lee[1], Chung-In Lee[2]

[1]) Korea Agriculture & Rural Infrastructure Corporation, Ansan 426-160, Korea.

[2]) School of Civil, Urban & Geosystem Engineering, Seoul National University, Seoul 151-742 Korea

Abstract: The first commercial food storage cavern in Korea was constructed for conservation of energy and environment at Gonjiam, Gyunggi district in 1997. The prediction of temperature distribution around cavern was made in earlier design stage for the first ten years of operation. However, the measured temperature during the past five years revealed higher temperature than predicted value. The difference in temperature distribution between the predicted and measured values seems to be caused by the effect of groundwater flow around the cavern, which was not taken into account when the prediction was made.

In this study, the thermal properties of the rock mass were modified to take the effect of groundwater and latent heat of freezing into account. The result of the analysis with these modified thermal properties shows good agreement with the measured temperature distribution.

1. INTRODUCTION

In Korea, 70% of country is covered with mountains and population density is very high. Therefore it is very difficult to find a suitable site to construct a facility occupying extensive area near the capital. Also, energy consumption of cold storage facility is high during the hot summer. The mean temperature was about 25°C in summer and -2.6°C in winter in 2001. The first commercial scale underground cold food storage cavern was constructed at Gonjiam in 1997, to reduce the energy consumption during the daytime in summer.

In this study, briefly explained is the site investigation made in earlier design stage to determine the dimension of cavern. The predicted temperature distribution around cavern was compared with measured values during the 5 years of operation. Also a practical method to evaluate the thermal properties of jointed rock mass was suggested and verified.

2. SITE CHARACTERIZATION

2.1 Location of underground food storage cavern

Gonjiam underground food storage cavern is located at middle part of Korea - Gyunggi distract and is about 50 km away from Seoul – the capital city of Korea. The main rock type is Gyunggi Gneiss complex formed in Precambrian, which is predominant geological formation in middle part of Korea. Gonjiam storage cavern is constructed horizontally at depth of 15m to 50 m below the surface.

Although Gonjiam storage facility was planned to excavate 3 storage caverns having a size of 15×15×100m, the layout of cavern was changed into both 2 caverns having 12×10×60m and 2 caverns having 12×8×50m because of the many small fault zones revealed by the detailed geological investigation (Figure 1). The storage caverns are separated into 5 parts that can be cooled individually.

Figure 1. Layout of the underground food storage cavern

2.2 Thermal properties of rock

To evaluate the thermal properties of rock, NX sized boreholes were drilled and rock specimens were sampled at the same level of rock cavern.

Thermal conductivity was measured by the hot wire method. The principle involved in the measurement has been well explained by Carlslaw and Jaeger (1959). Specific heat was measured by DSC (Differential Scanning Calorimeter) System TA 2910 (DuPont, U.S.A.) with heat flux type. The temperature differences between the reference material and the target specimen were measured during heating. Measurements were conducted twice at a specified temperature and temperature is varied from room temperature to 100 °C. Relationship between specific heat and temperature was linear and the result is summarized in table 1 and figure 2.

Table 1. Thermal properties of the specimen

Properties	Unit	Measured value
Thermal conductivity	W/m ° K	4.99±0.16
Specific heat	J/kg°K	720+1.412T

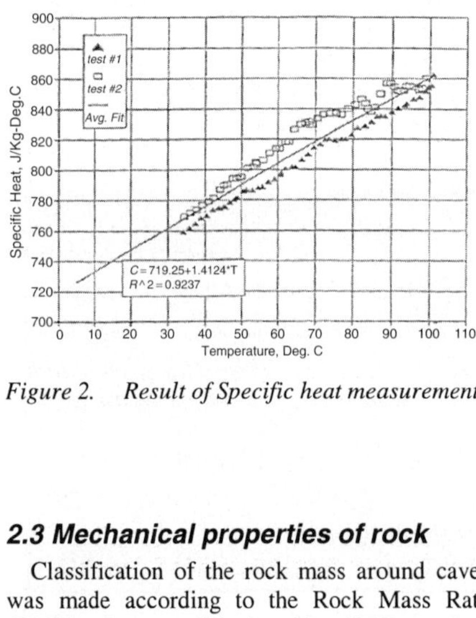

Figure 2. Result of Specific heat measurement

2.3 Mechanical properties of rock

Classification of the rock mass around caverns was made according to the Rock Mass Rating (RMR) System (Bieniawski, 1989) and the Tunnelling Quality Index (Barton et al, 1974). Parameters were selected by the geological site investigation, the characteristic data for the intact rock, the joint sets and the in-situ stresses.

Although both the RMR and Q-system estimated the quality of rock mass at the same level of cavern is fair to good, rock mass quality above the cavern level was very poor to poor in RMR and extremely poor to poor in Q value.

3. TEMPERATURE MONITORING SYSTEM

3.1 Monitoring system

Monitoring of temperature was made using high-precision thermistor string (by Geokon Inc. Canada). Figure 3 shows the location of thermistor strings installed at eight drill holes, and each string has 6 thermistors placed at 0.5, 1.0, 2.0, 4.0, 6.0, 10.0m from the surface of cavern wall. T-1 measuring point was selected to measure the effect on the temperature distribution from ambient temperature and T-2 and T-3 measuring point to measure the temperature distribution of outer rock mass of Storage Room #1. T-4 to T-8 measuring point selected to measure the temperature of pillar between the storage caverns. In this study, the temperature variation at the T-2 and T-3 measuring points are analyzed.

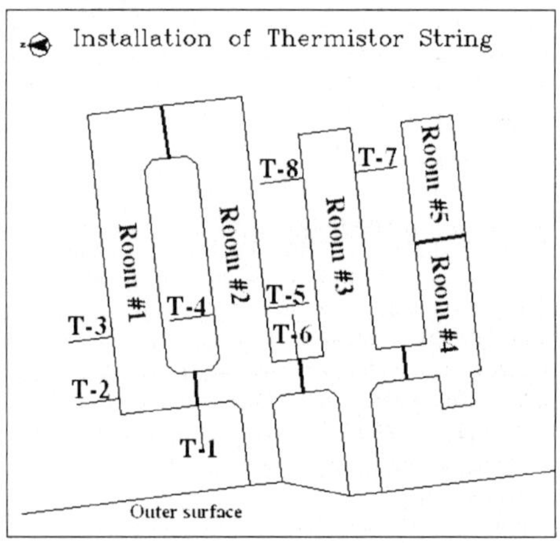

Figure 3. Installation of thermistor strings. Each string has 6 thermistors placed at 0.5, 1.0, 2.0, 4.0, 6.0 10.0m from the surface of cavern wall

3.2 Controlling the temperature of storage cavern

Main storing items at Gonjiam facility are frozen meat and fish, and the target operation temperature is -20°C. Since the construction of storage cavern in December 1997, the pre-cooling procedure to prepare storing of food was began from Room #5 whose floor area was the smallest and then followed by #4 to #1 (see table 2).

Table 2. Pre-cooling procedure

Room #	Pre-cooling Start	Pre-cooling End	Elapsed time (day)	Floor area (m^2)	(Day/ m^2)
#1	99.1.4	99.4.7	93	953	0.1
#2	98.12.8	99.2.6	60	953	0.06
#3	98.7.21	98.10.27	98	791	0.12
#4	98.6.19	98.8.8	50	429	0.12
#5	98.1.4	98.5.26	142	300	0.47

The first pre-cooling at Room #5 was started at 4th Jan. 1998 and lasted for 142 days. Even though it includes time for testing of refrigerators, the time elapsed to reach the target temperature is much longer than other rooms. This is because it was the first room being cooled, so the surrounding rock mass has to be cooled down from 12 to -20°C. The elapsed time to be cooled down at other rooms was shorter because adjacent rock mass was already cooled down. The pre-cooling time per unit area of Room #5 is 0.47 and that of Room #1 is 0.1.

3.3 Measured Temperature

Figure 4 is the results of measuring point T-2. The pre-cooling of Room #1 was started at Jan. 4th 1999 and the temperatures at 0.5m, 1.0m from the surface of cavern wall went down rapidly for the first 200 days, and then decreased slowly afterwards. After 4 years of operation, 0°C temperature was reached at 6.0m from the surface of cavern wall.

Figure 5 shows the results of measuring point T-3 installed at 15m inside of T-2. Although the draw down pattern of temperature is the same as that of T-2, temperature at 6m from the surface of cavern wall is -2°C after 4 years operation and that is 2 °C lower than the result of T-2 due to the influence of ambient temperature.

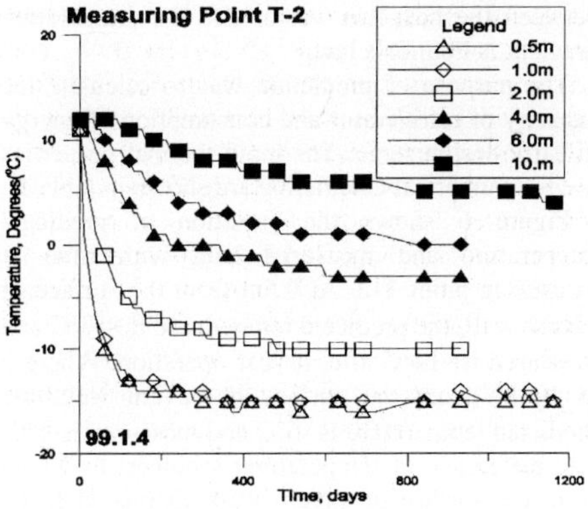

Figure 4. Measured temperature at each thermistors of measuring point T-2

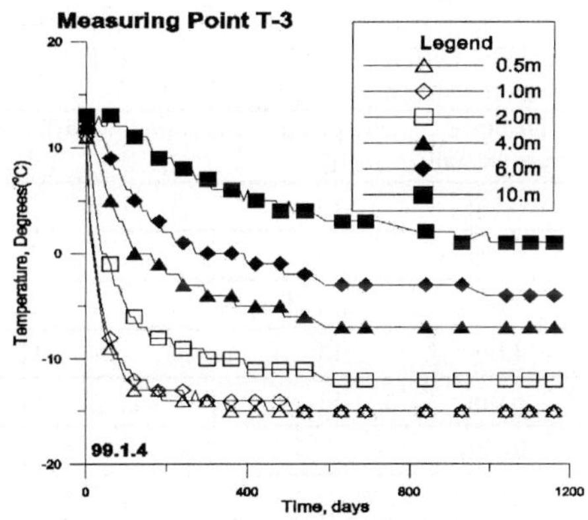

Figure 5. Measured temperature at each thermistors of measuring point T-3

4. PREDICTION OF TEMPERATURE DISTRIBUTION AROUND CAVERN

4.1 Prediction of temperature distribution made in earlier design stage

The numerical code used in the prediction of temperature distribution is FLAC (Fast Lagrangian Analysis of Continua), which is based on the finite difference method (ITASCA Consulting Inc.). Basic law used in FLAC that defines the relation between the heat-flux vector and the temperature gradient is Fourier's law.

The purpose of prediction was to calculate the capacity of refrigerator and consumption of energy in early design stage. The input thermal properties used in numerical calculation are shown in table 1.

Figure 6 shows the variation of predicted temperature and measured data with time at measuring point T-2. At 0.5m from the surface of cavern wall, the predicted temperature is -17°C and measured is -14°C after 4-year operation. Whereas at 10.0m from the surface of cavern wall, the predicted temperature is -6°C and measured is 4°C. The difference of temperature is highest at 10.0m from the surface of cavern wall. Estimated RMR and Q system value of Gonjiam is poor to very poor, so we could estimate that this difference between predicted and measured temperatures is due to the effect of groundwater and many joint sets (see Table 3).

Table 3. Difference of temperature predicted and measured after 4-year operation

Distance from the wall	Predicted (°C)	Measured (°C)	Difference (°C)
0.5m	-18	-15	3
1.0m	-18	-15	3
2.0m	-16	-11	5
4.0m	-14	-3	11
6.0m	-11	0	11
10.0m	-8	4	12

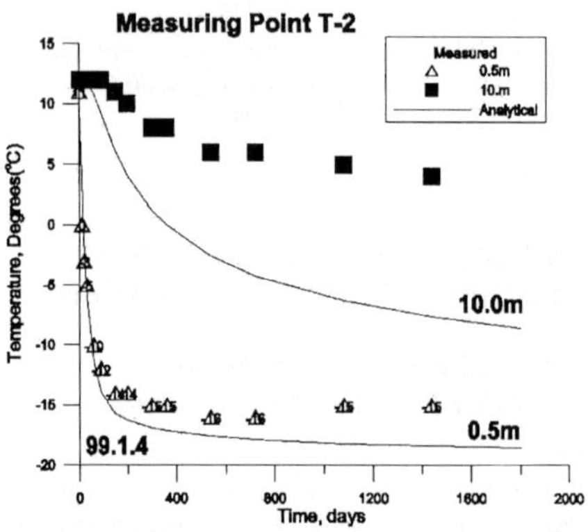

Figure 6. A comparison of numerical and experimental results at T-2 point (made in earlier design stage)

4.2 Temperature prediction considering the groundwater effect

Park et al (1999) conducted the research project on the prediction of temperature distribution around rock cavern on pilot scale in Korea. They proposed that the specific heat increased 25% and thermal conductivity is decreased 23% that of intact value from comparison of the predicted and measured temperature. They also proposed that change of thermal properties are caused from the effect of groundwater and joint for the reason that discontinuity makes thermal conductivity lower and water content makes thermal conductivity and specific heat higher.

Glamheden et al (2002) increased the thermal diffusivity as 42.5% than predicted from the laboratory results to achieve agreement between predicted and recorded temperatures during the cooling period. They explained the possible source of the observed difference is that the laboratory values were determined on unloaded, dry rock samples, whereas the natural rock mass is water saturated and under pressure.

Neaupane et al (1999) suggested the numerical code that could take into account the phase change of pore water during freezing by considering the latent heat of freezing. They showed that predicted temperature considering latent heat is much closer to measured compared to without latent heat case.

Several authors (Woodside & Messmer, 1961; Sass et al., 1971; Pribnow et al., 1995) used the modified thermal conductivity by geometric mean of rock and water component.

$$\lambda_G = \lambda_M^{1-\phi} \cdot \lambda_W^{\phi} \tag{1}$$

Where, ϕ is the volume fraction of water in the mixture, λ is thermal conductivity, and index 'G' applied to geometric mean, 'M' to rock matrix, 'W' to water.

Linear mixing law is valid in volume proportional quantities. And this also applies to volume specific heat capacity,

$$C_{tot}\rho_{tot} = (1-\phi)C_{mtx}\rho_{mtx} + \phi C_{por}\rho_{por} \tag{2}$$

Where ϕ is porosity, C is mass-specific heat capacity and ρ is density, and the index 'tot' applies to total rock, 'mtx' to solid rock matrix and 'por' to pore filling fluid.

In this study, we conducted the prediction of temperature distribution using modified thermal properties considering the effect of groundwater and latent heat of freezing.

We estimated from the maintenance results that the thermal loss in Gonjiam underground cold storage cavern is mostly induced from the groundwater flow and latent heat of freezing due to the poor quality of rock mass of surrounding storage cavern.

Therefore, we conducted the analysis to predict the temperature distribution using modified thermal properties that is calculated by considering the effect of groundwater and latent heat of freezing. Initial thermal properties were estimated considering 20% of volume fraction of groundwater.

During the numerical calculation, thermal properties were changed into modified values considering the latent heat of freezing when temperature of rock mass is reached 0°C. The modification was made by using the numerical formula (1) and (2). And latent heat term is considered to evaluate the energy need to freezing the water in unit volume of rock mass.

After the rock mass temperature reached -1 °C, thermal properties were changed into that of thermal properties calculated from 80% of rock mass and 20% of ice. Initial thermal properties and modified thermal properties are shown in table 4.

Table 4. Thermal properties used in numerical prediction

Properties	Unit	Initial thermal prop.	At 0 °C	Below −1°C
Thermal cond.	W/m° K	3.22	3.22	4.11
Specific heat	J/ kg°K	1013	24,856	882

Figure 7 shows predicted temperature distribution at measuring point T-2 using modified thermal properties after 1-year operation. The measured and predicted temperatures at 0.5m from the wall in measuring point T-2 are same as -15°C, and the temperatures at 10.0m are 10°C and 9°C respectively.

Figure 8 shows that the prediction range extended up to 4 years at each thermistor in measuring point T-2. As the results of variation graph at each thermistor, the predicted temperature distribution matched well with measured data near the surface of the cavern wall, whereas produced a great discrepancy at 4.0m, 6.0m, and 10.0m from the surface of cavern wall with time. And the discrepancy is largest at 10m from cavern wall surface. Figure 9 shows better agreement to measured data compared to figure 8, only that the measuring point is 15m inside from the measuring point T-2.

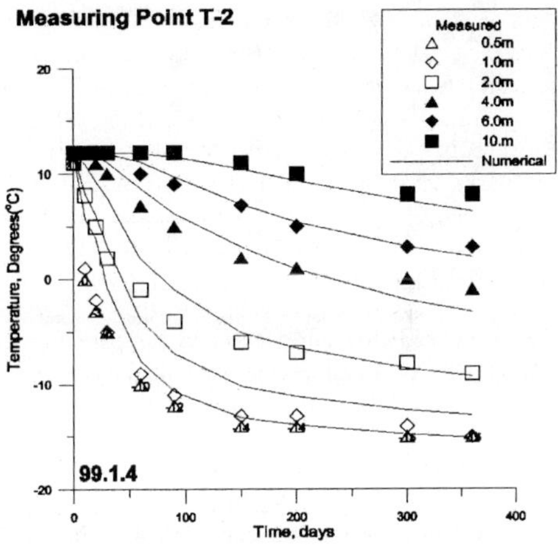

Figure 7. A Comparison of numerical and experimental results at T-2 point (after 1 year operation)

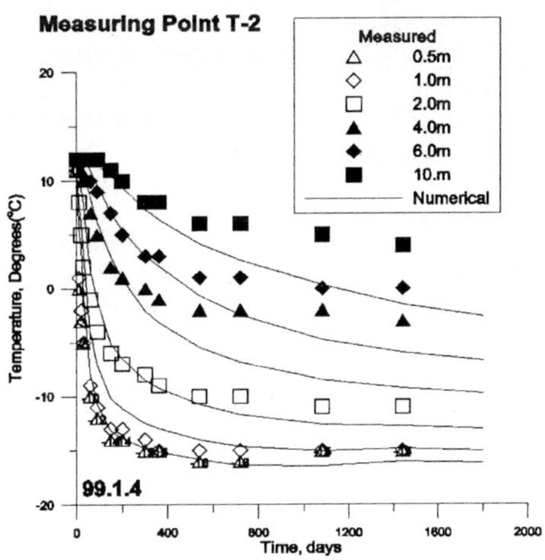

Figure 8. A Comparison of numerical and experimental results at T-2 point (after 4 year operation)

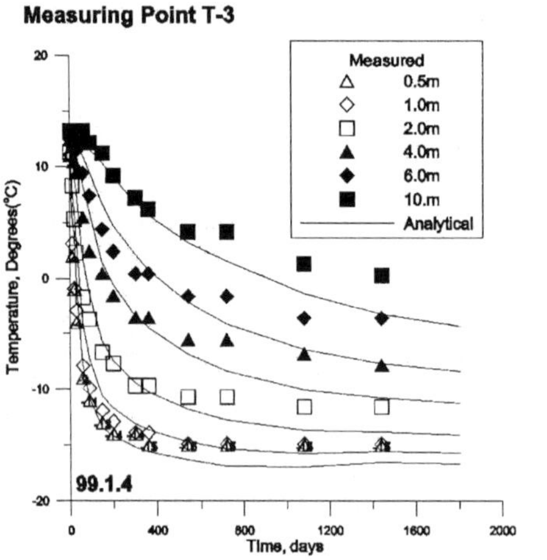

Figure 9. A Comparison of numerical and experimental results at T-3 point (after 4 year operation)

From these results, we can conclude that the ambient temperature influence on the temperature distribution around storage cavern after 1 year operation because rock storage cavern is constructed near the surface.

5. CONCLUSIONS

The comparison of the predicted and measured values at point T-2 with time shows that 3°C difference at 0.5m and 12°C at 10.0m from the surface of cavern wall. Difference between predicted and measured temperatures is due to the effects of groundwater and latent heat of freezing.

We started the analysis to predict the temperature distribution using modified thermal properties that is calculated by considering the 20% volume fraction of groundwater. During the numerical calculation, thermal properties were changed into modified values when temperature of rock mass is reached 0°C. The modified values were calculated using the numerical formula (1) and (2). And latent heat term is considered to evaluate the energy need to freezing the water in unit volume of rock mass. After the rock mass temperature reached -1 °C, thermal properties were changed into that of thermal properties calculated by taking the values from 80% of rock mass and 20% of ice value.

Predicted temperature distribution at measuring point T-2 using modified thermal properties after 1-year operation shows good agreement with measured values at corresponding depths. But the prediction range extended up to 4 year, the predicted temperature distribution matched well with measured data near the surface of the cavern wall, whereas produced a large discrepancy at 4.0m, 6.0m, and 10.0m from the surface of cavern wall with time. However, measuring point T-3 that is installed 15m inside from the T-2 show better agreement compared to T-2. From these results, we can conclude that the ambient temperature influence on temperature distribution around storage cavern as time goes on. Therefore it was found that a 3D numerical analysis is necessary to resolve this problem.

REFERENCE

Troschke, B. and H. Burkhardt. 1998. *Thermal conductivity Models fro Two-Phase Systems*. Phys. Chem. Earth, Vol. 23, No. 3: pp 351-355.

Unver, B. and C. Agan. 2003. *Application of heat transfer analysis for frozen food storage caverns.* Tunnelling and Underground Technology 18: pp 7-17.

Buntebarth G. and J.R. Schopper, 1998. *Experimental and Theoretical Investigations on the Influence of Fluids, Solids and Interactions between them on Thermal Properties of Porous Rocks*. Phys. Chem. Earth. Vol. 23, No.9-10; pp 1141-1146.

Neaupane K., M., T. Yamabe, R. Yoshinaka. 1999. *Simulation of a fully coupled thermo-hydro-mechanical system in freezing and thawing rock.* Int. J Rock Mech Min Sci 36: pp 563-580

Park H.D., Synn, J.H., Park, Y.J., Kim, H.Y., 1999. *A pilot study on the design of an underground food storage cavern in Korea.* Tunnelling and Underground Space Technology. 14: pp 67-73.

Park Y., Synn, J.H., Park, C. & Kim, H.Y. 1997. *Analysis of thermo-mechanical behaviour of underground cold storage cavern.* Environmental and Safety Concerns in Underground Construction: 1st Asian Rock mechanics Symp., Seoul: pp 731-736.

Glamheden R. and U. Lindblom. 2002. *Thermal and mechanical behaviour of refrigerated caverns in hard rock.* Tunnelling and Underground Space Technology 17: pp 341-353.

DEEP WEATHERING AND ALTERATION IN GRANITES – A PRODUCT OF COUPLED PROCESSES

K. Thuro[1], M. Scholz[2]
[1)] Engineering Geology, Swiss Federal Institute of Technology, Zurich, Switzerland
[2)] ISP Scholz Consulting Engineers, Munich, Germany

Abstract: Weathering and alternation in granite has a deep inpact on both geotechnical properties of the rock as well as of the rock mass. In a granite rock mass, the discontinuity pattern together with joint cohesion and friction plays a major role in tunnel construction and road cut slopes. These pheneomena and their coupling could be exclusively studied at the Königshainer Berge Tunnel Project (Germany) where two tubes provided an unique cross section through zones of intensive weathering and hydrothermal alteration within two-mica granites. Some parts of the granites underwent two stages of hydrothermal alteration before beeing under surface conditions. Therefore it was necessary to distinguish between the effect of hydrothermal alteration and weathering on mineralogy as well as on geotechnical rock properties. Only after the endogene alteration processes had finished, the rock has been exposed to the exogene processes of weathering. The change of mineral content and the increase of pore volume promotes the action of chemical decomposition.

1. INTRODUCTION

In the area of the Königshainer Berge (100 km east of Dresden, Saxonia, Germany), two 3.3 km roadway tunnels of have been built in connection with the extension of the German Federal Freeway A4 to the Polish border. The two tubes provided an unique cross section through zones of intensive weathering and hydrothermal alteration within different biotite granites. Although no major problems were encountered during excavation, the pattern and the distribution of the weathered zones was somehow unusual. At the same time, observations have been made concerning zones of unusual altered, "greenish" granites of up to several meters thickness following steep-dipping faults.

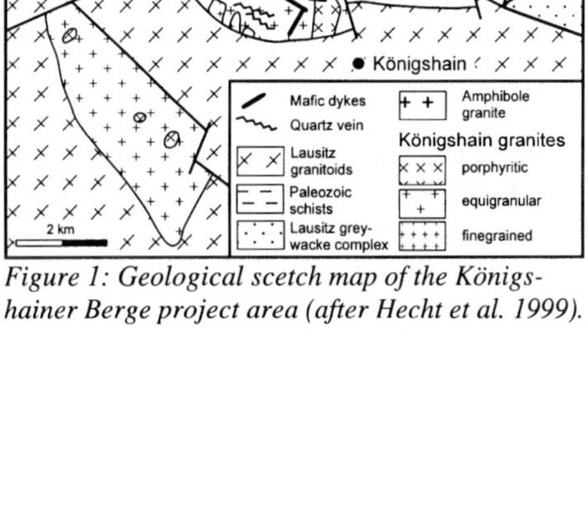

Figure 1: Geological scetch map of the Königshainer Berge project area (after Hecht et al. 1999).

Accordingly, a series of weathered and altered zones have been studied intensively by means of a detailed engineering geological investigation and subsequent rock and soil sampling, testing and monitoring. The main topics of the extensive field studies and laboratory work were:

(1) To characterize each weathering and alteration stage according to the rock or soil properties (e.g. mineral composition, microfabric, compressive and tensile strength, Young´s modulus, specific destruction work, cohesion, friction angle, grain size distribution, etc.)

(2) To correlate the pattern of weathering and alteration in the rock mass along the tunnel section with the discontinuity system.

(3) To distinguish the effect of weathering and hydrothermal alteration on mineralogy as well as on geotechnical rock properties.

(4) To derive relationships between weathering grade and several technical properties (e.g. drilling velocity for blastholes and the consumption of explosives at the tunnel face).

In this paper, only few selected geotechnical aspects of the different disintegration stages of the Königshain granites are presented. Further details are described in the PhD thesis of Scholz 2003 and the Habilitation thesis of Thuro 2002.

2. HYDROTHERMAL ALTERATION OF THE GRANITES

Situated in the most eastern part of Germany, the tunnel cuts through two mica granites (age app. 300 million years) of the Lausitz granodiorite complex. After a complex geologic history, the region was exposed to errosion and weathering following the late Cretaceous/early Tertiary (app. 65 my old, Eidam & Götze 1991, Krauss et al. 1992).

Three types of granites are distinguished in the Königshain area (Möbus, & Lindert 1967), of which the equigranular (light-grey to red in colour) and the porphyric type (coloured blue-grey to white-dotted) have been encountered during tunnelling works (Figure 1). Only some dikes of fine-grained granite (white-yellow to blue-grey) occur within the equigranular type.

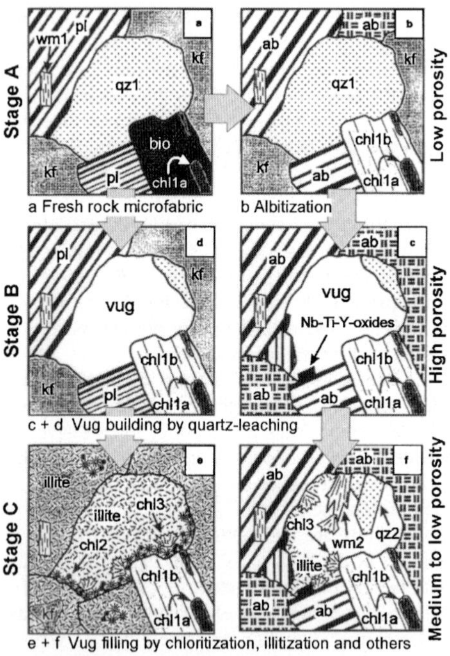

Figure 2: Hydrothermal alteration paths, stages and indication of their pore volume. kf – K-feldspar, pl – plagioclase, ab – albite, qz – quartz, wm – white mica, bio – biotite chl – chlorite (modified after Hcht et al.z 1999).

Some parts of the Königshain granites underwent two stages of hydrothermal alteration before burial, as suggested by (Hecht et al. 1999). The altered granites (described as episyenites) show a variable composition due to the alteration paths they followed (Figure 2). In comparison, the unaltered granites typically contain pink to reddish feldspars, the altered granites show white feldspars due to albitization or sericitization. With progressive chloritization, the granites obtains a green-spotted to dark green colour. More significant, however, is the increase of pore volume due to quartz-leaching in stage B. In stage C, the produced pores cannot be filled completely by blastesis of new minerals such as clay minerals, micas and quartz (Figure 3). Therefore, the maximum porosity is obtained in stage B. Unfortunately, due to predominant weathering of the altered granite, only little material could be gained for rock testing in the laboratory.

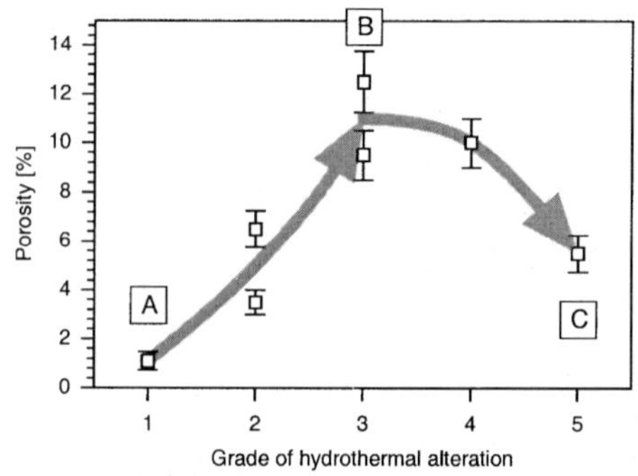

Figure 3: Development of porosity with the grade of hydrothermal alteration. Stages A – B – C refer to Figure 2.

The unconfined compressive strength – determined by point load tests – changes significantly with porosity and the grade of alteration (Figure 4). Although porosity is increasing again in Stage C, compressive strength decreases due to the weak sericite and clay minerals.

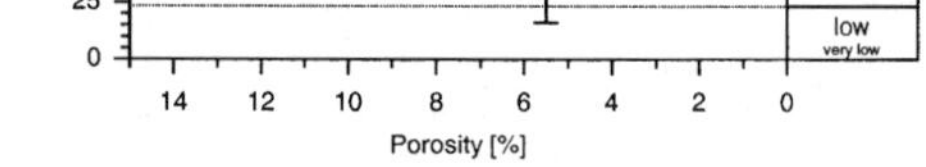

Figure 4: Development of unconfined compressive strength with porosity and alteration.

3. WEATHERING GRADES OF THE GRANITE

It is remarkable, that only after the endogene alteration processes had finished, the rock has been exposed to the exogene processes of weathering. The change of mineral content and the increase of pore volume promotes the action of mechanical disintegration and chemical decomposition.

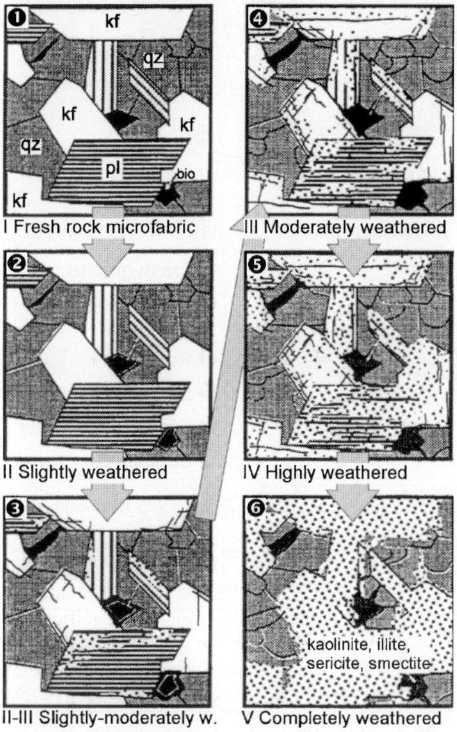

Figure 5: Weathering grades in the microfabric (modified after Scholz 1999).

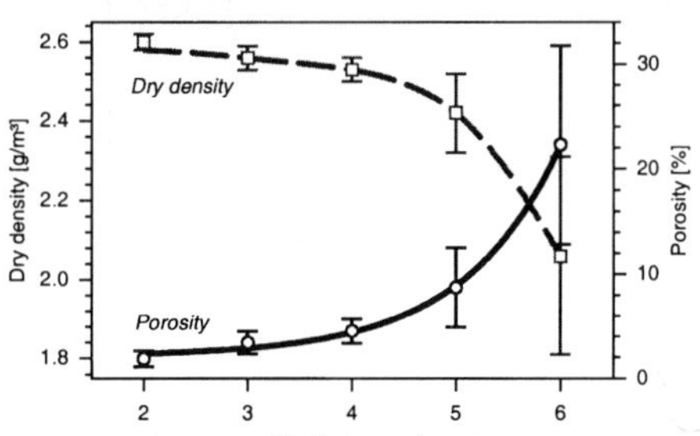

Figure 6: Connection of the weathering grade with dry density and porosity (pore volume). High/mean/low values are plotted for each grade.

Figure 5 shows the stages of weathering in the microfabric of the granite by disintegration and decomposition into a clayey-silty soil material. Grades (I – VI) refer to the ISRM weathering grades (0). An additional grade called "II-III slightly to moderately weathered" could be distinguished. In the following diagrams, the grades therefore are counted in arabic numerals from 1 to 7 (with VI = 7 residual soil, V = 6 completely weathered and so on).

Apart from mineralogical changes, porosity is increasing considerably and dry density of the rock material is decreasing simultaneously (Figure 6).

At the same time, the rock mass undergoes changes (Figure 7) due to weathering. The process begins with the fresh granite ❶ (grade I = 1, which has not been encountered during tunnelling works). Subsequently, a typically reddish-brown rust front ❷ and a zone of microscopic weathered granite ❸ moves from the wall rock into the rock mass (grade II = 2). A bleached light-brown to yellow-white zone ❹ then marks the end of the solid rock (grade II–III = 3). The disintegrated zone ❺ in grade III (= 4) consists mainly of clay and silt as the material of the wall rock decomposes. This proved to be the most dangerous material encountered, because of the low friction angle of the material and its tendency to shear along existing discontinuities. In grade IV (= 5) the granite disintegrates into a mixture of ❻ sandy and/or clayey and silty material. In the completely weathered material (grade V = 6) the rust colours have disappeared and the residual soil (grade VI = 7) only holds remnants of the disintegrated granite.

Accordingly, the physical properties of the rock change in a similar manor. Figure 8 shows the unconfined compressive strength correlated to the different weathering grades. Since the weathering classification is always determined in the rock mass, the correlation is poor with a large standard deviation between values. Nevertheless, a trend can be seen. Only when the UCS is plotted against dry density or porosity (Figure 9) is a good correlation found with a fitted curve.

In addition to the compressive strength, tensile strength, and Young´s modulus, the specific destruction work (integral of the complete stress-strain curve (Thuro 1997, Figure 10) shows a close correlation with porosity (Thuro et al. 2002).

Also technical properties are significantly influenced as can be seen in Figures 11 and 12. Mean values of blasthole drilling rates and consumption of explosives with standard deviations are given.

Several soil properties are also affected in the stages of grade V (6) and VI (7). For example, the grain distribution is shifted from coarse to fine grain size with increasing weathering grade.

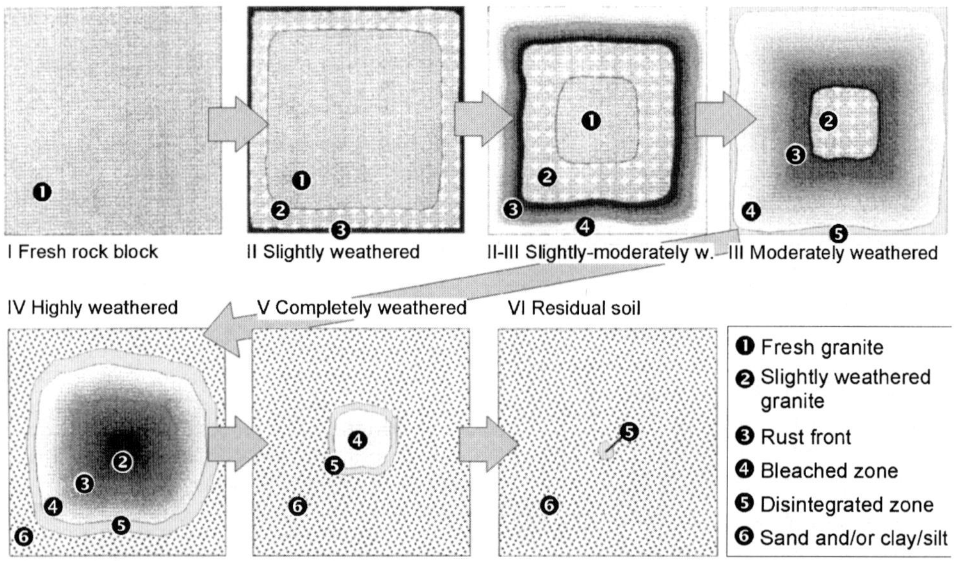

Figure 7: Grades of weathering in a Königshain granite block according to the ISRM 1978 weathering grades (modified after Scholz 1999).

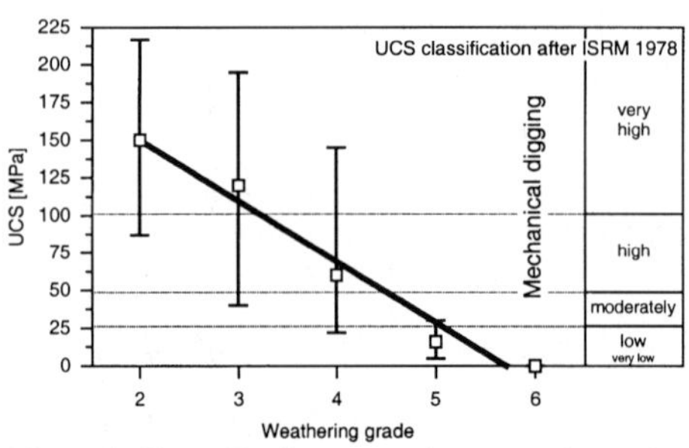

Figure 8: Unconfined compressive strength versus weathering grade (max/mean/min values).

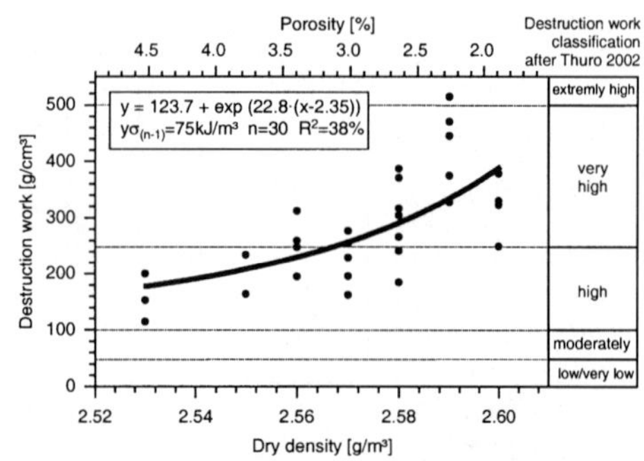

Figure 10: Specific destruction work versus dry density and porosity.

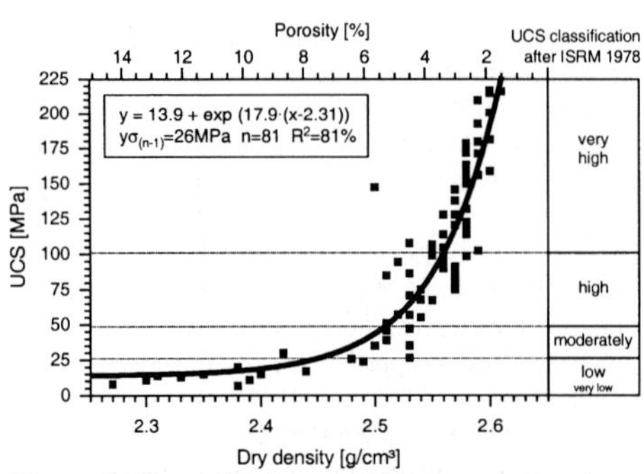

Figure 9: Unconfined compressive strength versus dry density and prosity (single values).

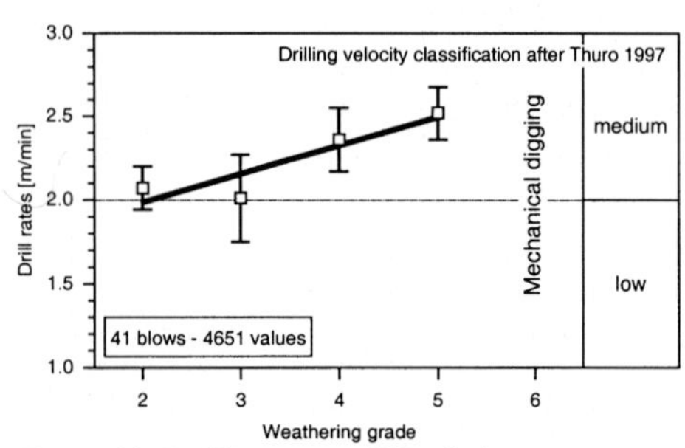

Figure 11: Drilling rates of blastholes versus weathering grade.

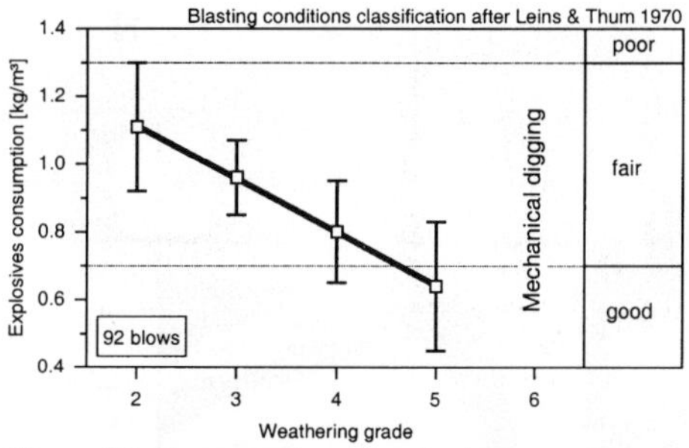

Figure 12: Consumption of explosives versus weathering grade.

4. CONCLUSION

Figure 13 gives an overview of the rock and soil properties determined for each weathering grade. Compared to (Heitfeld 1985) some differences have been encountered concerning description and characterization of the weathered rock mass. In the Königshainer Berge tunnel project, hydrothermal alteration has induced increased porosity in the rock mass. The hydrothermal solutions ascended in dikes following faults and the joint system and produced a finger-like pattern of altered material which easily could be exposed to weathering from the surface. Figure 14 shows the derived model of the weathered rock mass with respect to the scale of the tunnel.

REFERENCES

Eidam, J., Götze. J., 1991. *The granitic massif of Königshain-Arnsdorf (Lusatian Anticlinal Zone) - An example of a reversely zoned pluton.* Chem. der Erde 51, pp 55-71.

Hecht, L., Thuro, K., Plinninger, R.J., Cuney, M., 1999. *Mineralogical and geochemical characteristics of hydrothermal alteration and episyenitization in the Königshain granites, northern Bohemian Massif, Germany.* Int. Journ. Earth Sciences 88, pp 236-252.

Heitfeld, K.-H. 1985. *Ingenieurgeologische Probleme im Grenzbereich zwischen Locker- und Festgesteinen.* Berlin, Heidelberg, New York etc.: Springer, 695 p.

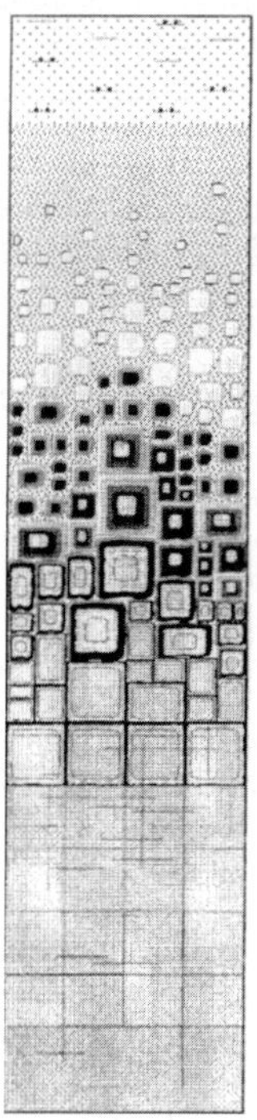

Weathering grades	UCS [MPa]	TS [MPa]	E [GPa]	UCS/SPZ [-]	W_d [kJ/m³]	D_d [g/cm³]	P [%]	φ [°]	c [kN/m²]	
VI - 7 Residual soil	Not possible to obtain data					1.3 - 1.8	50 - 30	40° - 30°	35 - 15	
V - 6 Completely weathered						1.8 - 2.3	30 - 13	30° - 20°	20 - 10	
IV - 5 Highly weathered	0.5 - 25	0.05 - 2	10 - 20	5 - 12	10 - 100	2.3 - 2.52	13 - 5	20° - 10°	50 - 25	Joint only!
		estimated								
III - 4 Moderately weathered	25 - 50	2 - 5	20 - 30	8 - 10	50 - 100	2.50 - 2.56	6 - 3.5			
II-III - 3 Slightly - moderately weathered	50 - 120	5 - 15	30 - 40	10 - 12	100 - 150	2.54 - 2.58	4.5 - 2.5	Not possible to obtain data		
II - 2 Slightly weathered	120 - 250	10 - 25	40 - 60	12 - 25	150 - 550	2.58 - 2.62	2.5 - 1			
I - 1 Fresh rock mass	No obtained data					2.62 - 2.64	1 - 0.5			
						estimated				

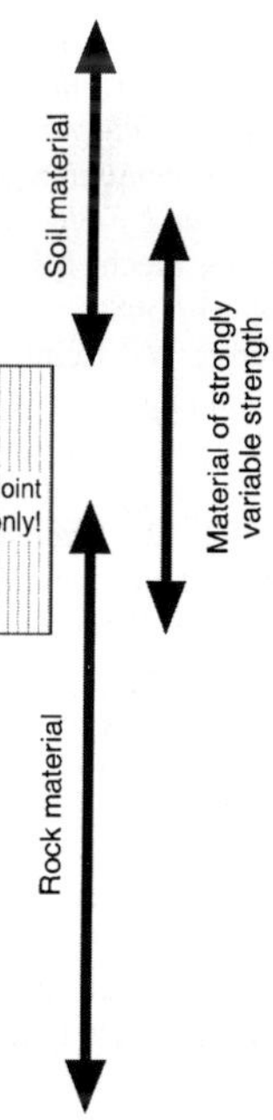

Figure 13: Weathering profile and some physical rock properties. UCS – unconfined compressive strength, TS – tensile strength, E – Young´s modulus, W_d – specific destruction work, D_d – dry density, P – porosity, φ – friction angle, c – cohesion (determined through direct shear tests).

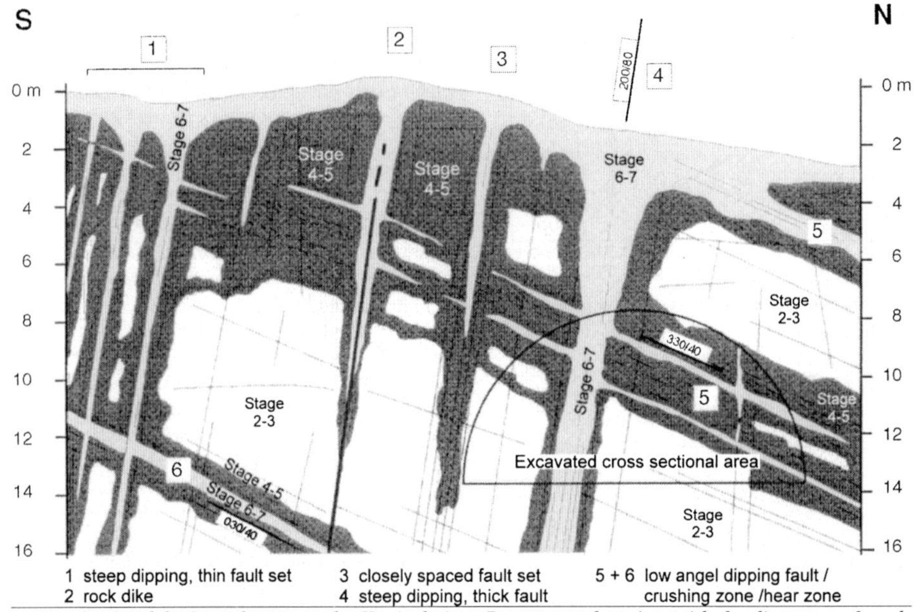

Figure 14: Model of weathering in the Königshainer Berge tunnel project with the dimenson of top heading (excavated cross section area). Stages in arabic numerals refer to Figure 7 (modified after Scholz 1999).

ISRM, 1978. *Suggested methods for the quantitative description of discontinuities in rock masses. International Society for Rock Mechanics, Commission on Standardization of Laboratory and Field Tests, Document No. 4.* Int. J. Rock Mech. Min. Sci. & Geomech. Abstr. 15, pp 319-368.

Kagerer, W., Reintjes, K.-H. & Voyé, R., 1997. *Planning and Construction of the Königshainer Berge tunnel.* Tunnel 1997 (7), pp 20-31.

Krauss, M., Eidam, J., Hammer, J., Korich, D., 1992. *Die cadomisch-variszische Entwicklung des Lausitzer Granodiorit-Komplexes.* Zbl. Geol. Paläont. Teil I II (1/2), pp 71-85.

Leins, W., Thum, W., 1970. *Ermittlung und Beurteilung der Sprengbarkeit von Gestein auf der Grundlage des spezifischen Sprengenergieaufwandes.* Forsch.-ber. d. Landes Nordrhein-Westfalen, 2118.

Möbus, G., Lindert, W., 1967. *Das Granitmassiv von Königshain bei Görlitz (Oberlausitz).* Abh. Dt. Akad. Wissensch. Kl. Bergbau, Hüttenwesen und Montangeologie 1, pp 81-160.

Scholz, M. 1999. *Über die Verwitterung im Königshainer Granitmassiv und ihre Auswirkungen auf den Vortrieb beim Bau der Tunnelanlage Königshainer Berge. Petrographische, fels- und bodenmechanische Untersuchungen.* – 135 S., Diploma thesis, Technical University of Munich (unpublished.).

Scholz, M. 2003. *Geomechanische Eigenschaften verwitterter Granite und ihr Einfluß auf den Vortrieb beim Stollen- und Tunnelbau.* Münchner Geologische Hefte, Reihe B, Angewandte Geologie, B20, I-XI,1-116, in press.

Thuro, K. 1997. *Drillability prediction - geological influences in hard rock drill and blast tunnelling.* Geol. Rundsch. 86, pp 426 - 437.

Thuro, K. 2002. *Geologisch-felsmechanische Grundlagen der Gebirgslösung im Tunnelbau.* – Münchner Geologische Hefte, Reihe B, Angewandte Geologie, B18, I-XIV, 1-160.

Thuro, K. Plinninger, R.J. & Spaun, G. 2002. *Drilling, blasting and cutting – is it possible to quantify geological parameters of excavation? – In* VAN ROY & *JERMY eds. Engineering geology for developing countries.* Proceedings of the 9th Congress of the Intern. Ass. for Eng. Geol. and the Environment, Durban, South Africa, 16-20 Sept. 2002, Ext. Abstracts: 484, CD-ROM: 2853-2861.

MODELING THE THERMO-MECHANICAL PROCESSES OF A TYPICAL THREE-GORGES DAM SECTION DURING AND AFTER CONSTRUCTION

Jian Liu[1,2], Xia-Ting Feng[1], Xiu-Li Ding[2], Chun-Fang Dai[2]
[1])Institute of Rock and Soil Mechanics, the Chinese Academy of Sciences
[2]) Yangtze River Scientific Research Institute

Abstract: Thermal induced stresses are developed as a result of concrete hydration heat in mass concrete dam, which can be significant enough to induce cracking of the dam. The evaluation of the risk of short and long-term thermal reduced cracking is among the key issues in the design of the Three-Gorges Dam. Based on a series of laboratory tests on the thermo-mechanical properties of the concrete materials and the statistics of environmental temperature variations, the thermo-mechanical processes of a typical Three-Gorges Dam section were numerically modelled using a 3D FEM code developed in this study. The detailed construction procedures, concrete creep and water-cooling were simulated. The distribution and evolution of temperature and thermal stress during the duration from the construction, reservoir water filling to the reach of steady interior temperature field were derived and analyzed; some significant suggestions for preventing the dam from thermal induced cracking are provided.

1. INTRODUCTION

Exothermic chemical reaction would occur when concrete being poured, John (2002). Due to hydrate heat and environmental temperature variation, a non-uniform and unsteady temperature field can be developed within mass concrete dam during and after construction. Great thermal gradient and tensile stress can be resulted from the temperature differences between concrete and foundation rock, between concrete core and surface, and among the different inside parts of dam, which is significant enough to induce cracking of the dam, Zhu (1999).

In this study, the thermo-mechanical processes of a typical massive concrete section of the Three-Gorges Dam, the left guide-wall dam section, were analyzed. A series of laboratory tests on the concrete thermo-mechanical properties, including the time-dependent elastic modulus, creep, the rate of hydration heat, etc. were firstly conducted; the statistics of environmental temperature variations with time at the site were performed. Then, a 3D FEM code was developed, in which the material properties, detailed construction procedures, variations of climatic and water temperature, as well as water-cooling were taken into account. Based on the test results and the 3D FEM code, the thermo-mechanical processes of the dam section were numerically modeled. The distribution and evolution of temperature and thermal stresses of the dam section during the duration from the construction, the reservoir water filling to the reach of steady interior temperature were derived; the potential cracking zones thermally induced are analyzed; some significant suggestions for preventing the dam from thermal induced cracking are also provided.

2. STRUCTURE DESCRIPTION AND CONSTRUCTION PROCEDURE

As illustrated in Fig.1, The left guide-wall dam section has a size of 175 m in height, 32 m in thickness and 118 m in width. The dam section is

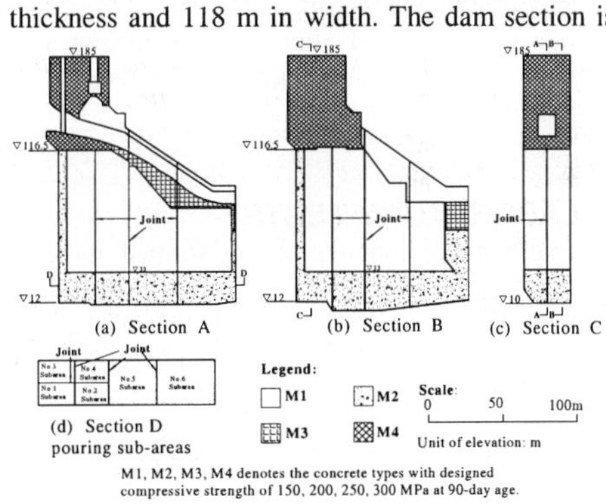

Figure 1. Structural section planes and concrete pouring sub-areas of the left guide-wall dam section.

constructed on a rock foundation of plagioclase and four different concrete materials are used. To control concrete temperature and thermal stress, during the construction, the dam section is separated into six monolith blocks by one temporary transverse joint and three longitudinal joints; correspondingly, there are six pouring sub-areas. Each block is divided into a number of concrete pouring layers with a thickness of 1.5 m or 2.0 m. The total construction time for the left guide-wall dam section is planned to be 37 months (from Jan.1, 1999 to Feb.1, 2002). The construction and operation procedure can be divided as the following phases (refer to Fig.1):

Phase 1: The concrete placement commences in the no.1 pouring sub-area. Then, the construction continues according to the following placing sequence: No.1 sub-area → No.2 and No.3 sub-area→No.4 sub-area→No.5 sub-area→No.6 sub-area. Meanwhile, the height differences between the two neighbouring concrete blocks must be kept within a scope of 6-8 m and the intervals between the successive lifts is planned to be 10-12 days. The first-term water-cooling would be carried out for the placing layers during high air temperature period from Jun. to Aug.; the temperate of cooling water is 8°C; and for each layer, the duration of water-cooling is 15 days.

Phase 2: When the concrete placements in the No.1-6 sub-areas go up to the elevation of 116.5 m, the second-term water-cooling would be conducted; with duration of two months and the river water is adopted as the cooling water.

Phase 3: Concrete placements from the elevation of 116.5 m to 185 m, the No.1-4 pouring sub-areas are combined to one pouring sub-area.

Phase 4: The period after the construction and before the reservoir water filling, 16 months.

Phase 5: Reservoir water filling.

3. LABORATORY TESTS

A trial batch of concrete was performed, thereby producing the satisfactory mixture proportions recommended using in the construction of the left guide-wall dam section. The concrete thermal properties were tested according to the standard methods of the Chinese Test Code for Hydraulic Concrete (1982). Table 1 gives the test results.

Table 1. Thermal properties of concretes

Concrete type	M1	M2	M3	M4
Thermal conductivity (kJ/m·h·°C)	2.26	2.61	2.19	2.15
Thermal diffusivity (m²/h)	0.003 2	0.003 5	0.003 2	0.003 1
Specific heat (J/kg·°C)	959	958	963	982
Dilatation coefficient (10^{-5}/°C)	88	85	86	85
Density (kN/m³)	24.5	24.5	24.5	24.5
Poisson's ratio	0.167	0.167	0.167	0.167

A series of tests on the concrete temperature rise under insulating condition were carried out. The relation between the adiabatic temperature rise and the age of concrete was fitted into the following Equation (1).

$$\theta(\tau) = \theta_0 (1 - e^{-A\tau^B}) \tag{1}$$

where, $\theta(\tau)$ is the adiabatic temperature rise of concrete; τ denotes the age of concrete; A, B and θ_0 is the fitting coefficients.

The concrete creep was tested; the testing data were fitted into the following equation:

$$C(t,\tau) = C(\tau)[1 - e^{-k_1(t-\tau)}] + D(\tau)[1 - e^{-k_2(t-\tau)}] \tag{2}$$

where,

$$C(\tau) = A_1 + \frac{B_1}{\tau^m}; \; D(\tau) = A_2 + \frac{B_2}{\tau^n}$$

$C(\tau)$ is creep; t is the loading age of concrete; τ is the time of strain calculation; k_1, k_2, A_1, A_2, B_1, B_2, m, n are the fitting coefficients.

The time-dependent elastic moduli of the concretes were obtained from the creep stress-strain curves and were fitted into Eq.(3):

$$E(\tau) = E_0 (1 - Ae^{-B\tau}) \tag{3}$$

where, $E(\tau)$ is the elastic modulus of concrete; τ is the age of concrete; E_0, A, B is the fitting coefficients.

Table 2 lists the thermo-mechanical properties of the foundation rock determined.

Table 2. Thermal properties of concretes

Thermal conductivity (kJ/m·h·°C)	2.51
Thermal diffusivity (m²/h)	0.003471
Specific heat (J/kg·°C)	959
Dilatation coefficient (10^{-5}/°C)	85
Density (kN/m³) (kJ/m·h·°C)	27.0
Poisson's ratio	0.2
Elastic modulus (GPa)	35.0

4. ENVIRONMENTAL TEMPERATURE

According to the statistic data from the weather station at the Three-Gorges Dam site, the mean yearly air and water temperature is 17.3°C and 18.0 °C, respectively. The variations of air and water temperature with time were fitted into the following equation:

$$T = T_0 + T_B \mathrm{Sin}\frac{2\pi(t - t_0)}{365} + T_S \tag{4}$$

where, T_0 represents the mean yearly air or water temperature; T_B is the variable range of the yearly air or water temperature; t is time difference between the calculation time and Jan. 1 of the year; t_0 is time difference between Jan. 1 of the year and the origin of the yearly variable temperature curve; T_S represents the additional temperature caused by the solar radiation heat of the concrete surface.

5. A 3D FEM CODE DEVELOPED

5.1 Basic formulations and solution approach of temperature field

The three-dimensional heat transfer within the concrete is governed by the differential equation:

$$\frac{\partial}{\partial x}\left(\alpha_x \frac{\partial T}{\partial x}\right) + \frac{\partial}{\partial y}\left(\alpha_y \frac{\partial T}{\partial y}\right) + \frac{\partial}{\partial z}\left(\alpha_z \frac{\partial T}{\partial z}\right) + \frac{\partial \theta}{\partial \tau} - \frac{\partial T}{\partial \tau} = 0 \tag{5}$$

where, T is the temperature of concrete; α_x, α_y and α_z are the thermal diffusivity coefficients in the x, y and z axial directions; τ is the time; θ is the adiabatic temperature rise.

There are four types of boundary condition as follows, Wu (2001):

(i) When the concrete contacts with water, the surface temperature equals to the known temperature of water.

(ii) When the adiabatic condition is satisfied, it can be expressed by equation (6):

$$\partial T / \partial n = 0 \tag{6}$$

(iii) When the concrete surface is exposed to air, it can be written as equation (7):

$$-\lambda(\partial T / \partial n) = \beta(T - T_a) \tag{7}$$

(iv) When the concrete contacts tightly with a solid medium, the temperature and heat flux of the interface are continuous, that is:

$$T_1 = T_2 \quad \text{and} \quad \lambda_1(\partial T / \partial n) = \lambda_2(\partial T / \partial n) \tag{8}$$

where, T is the surface temperature of concrete; n indicates normal direction of the surface. λ is thermal conductivity coefficient; β is surface exothermal coefficient; T_1 and T_2 are the surface temperatures of concrete and solid; λ_1, λ_2 are the thermal conductivity coefficients of the concrete and solid medium.

The temperature field is solved using a backward implicit difference approximation. The nodal temperature and variable rate of temperature of element are given by the interpolation of shape function. In the solution of temperature field, the equation of heat transfer, initial conditions and boundary conditions have to be satisfied. Based on the variational principle, the problem can be converted into solving the extremum of a functional. The implicit difference equation is written as:

$$\left[H + \frac{2}{\Delta\tau}P\right]\{T\}_\tau + \left[H - \frac{2}{\Delta\tau}P\right]\{T\}_{\tau-\Delta\tau} + \{Q\}_{\tau-\Delta\tau} + \{Q\}_\tau = 0 \tag{9}$$

where,

$$H_{ij} = \sum_e h_{ij}^e = \sum_e \iiint_{\Delta R}\left(\alpha_x \frac{\partial N_i}{\partial x}\cdot\frac{\partial N_j}{\partial x} + \alpha_y \frac{\partial N_j}{\partial y}\cdot\frac{\partial N_j}{\partial y} + \alpha_z \frac{\partial N_i}{\partial z}\cdot\frac{\partial N_j}{\partial z}\right)dxdydz$$

$$P_{ij} = \sum_e p_{ij} = \sum_e \iiint_{\Delta R} N_i N_j dxdydz$$

$$Q_{ij} = \sum_e q_i^e = \sum_e \left\{-\iiint_{\Delta R}\frac{\partial\theta}{\partial\tau}N_i dxdydz - \iint_{\Delta C}\beta T_a N_i ds + \left(\iint_{\Delta C}\beta N_i [N] ds\right)\{T\}^e\right\}$$

$$\bar{\beta} = \frac{\beta}{c\rho}$$

Thus, the temperature at any time can be calculated if the temperature at the initial time is known.

5.2 Formulation of thermal stress

Concrete is considered as linear elastic creep material, Vandewalle (2000). The stress-strain relation is given by the following equation:

$$\{\Delta\sigma\} = [D](\{\Delta\varepsilon\} - \{\Delta\varepsilon^0\}) \tag{10}$$

where, $\{\Delta\sigma\}$ is the stress increment; [D] is the elastic matrix; $\{\Delta\varepsilon\}$ is the strain increment; $\{\Delta\varepsilon^0\}$

is the initial strain increment.

The initial strain increment includes thermal strain increment and creep increment, that is:

$$\left\{\Delta\varepsilon^{0}\right\}=\left\{\Delta\varepsilon^{c}\right\}+\left\{\Delta\varepsilon^{T}\right\} \tag{11}$$

where, $\{\Delta\varepsilon^{c}\}$ is creep increment; $\{\Delta\varepsilon^{T}\}$ is thermal strain increment.

Additionally, there is:

$$\left\{\Delta\varepsilon^{T}\right\}=a\cdot\Delta T\begin{bmatrix}1 & 1 & 1 & 0 & 0 & 0\end{bmatrix}^{T} \tag{12}$$

and,

$$\left\{\Delta\varepsilon^{c}\right\}=\left(1-e^{-k_1\cdot\Delta\tau}\right)\{\omega\}_{\tau+\Delta\tau}+\left(1-e^{-k_2\cdot\Delta\tau}\right)\{\gamma\}_{\tau+\Delta\tau} \tag{13}$$

where, a is linear dilatation coefficient; ΔT is temperature increment;

$$\{\omega\}_{\tau+\Delta\tau}=e^{-k_1\cdot\Delta\tau}\{\omega\}_{\tau}+C_1(\tau)\big|_{\tau}\left\{\Delta\bar{\sigma}\right\}_{\tau};$$

$$\{\gamma\}_{\tau+\Delta\tau}=e^{-k_2\cdot\Delta\tau}\{\gamma\}_{\tau}+C_2(\tau)\big|_{\tau}\left\{\Delta\bar{\sigma}\right\}_{\tau};$$

$$\{\omega\}_{1}=C_1(\tau)\big|_{\tau=0}\left\{\Delta\bar{\sigma}\right\}_{\tau=0};$$

$$\{\gamma\}_{1}=C_2(\tau)\big|_{\tau=0}\left\{\Delta\bar{\sigma}\right\}_{\tau=0};$$

$$\left\{\Delta\bar{\sigma}\right\}_{\tau}=\left[\bar{D}\right]\{\Delta\sigma\}_{\tau};$$

μ is Poisson's ratio;

$\{\Delta\sigma\}_{\tau}$ is the stress increment.

After the creep strain increment is calculated, the corresponding creep stress increment can be solved using finite element method, Song (2002).

5.3 Simulation of water-cooling

The effect of water-cooling is simulated by the following equation, Kim (2001):

$$T_{\tau+\Delta\tau}=T_{w}+(T_{\tau}-T_{w})X \tag{14}$$

where, $T_{\tau+\Delta\tau}$ is the concrete temperature when the water-cooling is implemented for a time increment of $\Delta\tau$; T_w is the temperature of cooling water; T_τ is the concrete temperature at the beginning of water-cooling; X is a coefficient indicating the reduce rate of temperature due to water-cooling.

6. NUMERICAL MODELING

6.1 Numerical implementation

The thermo-mechanical processes of the left guide-wall dam section were numerically modeled using the developed 3D FEM code. The geometry of structure, pouring sub-areas, material sub-areas and concrete placing layers were taken into account in the numerical model, and thereby large numbers of elements and nodes were resulted in. The model was divided into 56133 elements and 64542 nodes. All the construction and operation procedures of the dam section stated in Section 3 were simulated. The thermo-mechanical properties of the concrete and foundation rock stated in Section 3 were used as the input data of the numerical model. The variations of air and water temperature with time were modelled based on the data given in Section 4.

6.2 Temperature evolution

Figure 2 gives the cross-section temperature distributions of the dam section at two stages.

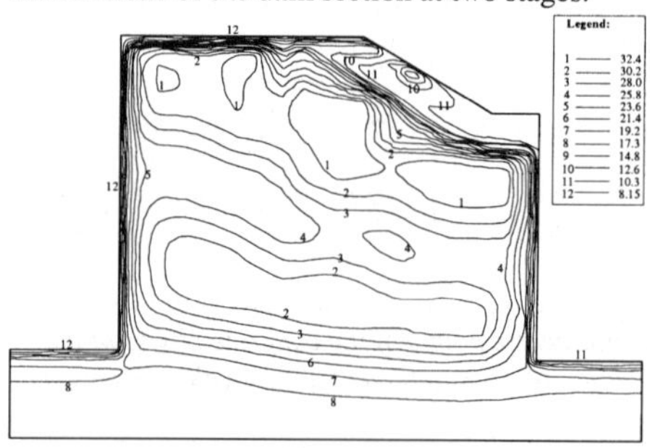

(a) The temperature distribution in Section B before the second-term water cooling in the construction of the left guide-wall dam section of the Three-Gorges Dam

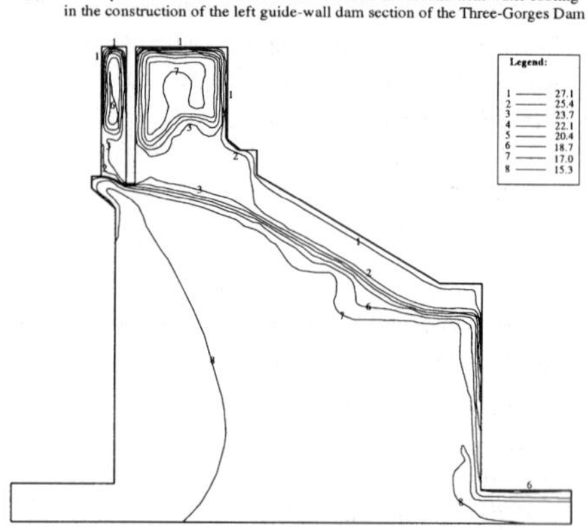

(b) The temperature distribution in Section B after the reservior water filling and the reach of interior steady temperature of the left guide-wall dam section (in summer)

Figure 2. Cross-section temperature distributions at several critical times of the left guide-wall dam section.

During the construction, due to the difference between the thermal properties of the concrete and foundation rock, a great temperature gradient appears in the interior centre close to ground surface of the concrete blocks. Meanwhile, as a result of the differences between the concrete and air temperatures, great temperature gradients occur in the surface layer of the dam section. However, after the construction and reservoir water filling, the temperature gradients disappear or become less.

Figure 3 illustrates the temperature evolutions at the interior centre points of the concrete blocks at the elevation of 22m. From these curves, the temperature evolutions can be divided into five stages:

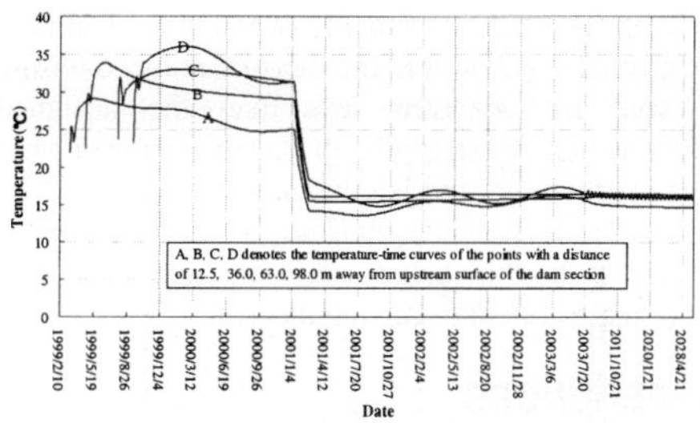

Figure 3. Temperature evolutions with time at the interior centre points of the concrete blocks at the elevation of 22 m of the left guide-wall dam section.

Stage 1: The temperatures rise rapidly due to the generations of the hydrate heat within the concrete. The highest temperature reaches 36.35℃, which appears at the interior centre point D of the concrete block in the No.4 pouring sub-area.

Stage 2: The temperatures gradually descend from their peak values because the thermal conductivity and first-term water cooling make the release of heat within the concrete after the temperatures reach their peak values.

Stage 3: As a result of the second-term water cooling, the temperatures steeply descend with a range from 11℃ to 15℃.

Stage 4: Affected by the residual hydrate heat of concrete, the slight fluctuate and adjustment of the temperatures continues.

Stage 5: The evolutions of temperatures enter into a steady stage.

As for the surface temperature difference of the dam section, Fig.4 depicts the evolutions of the upstream surface temperature differences at the different elevations. At the majority of the curves, the highest peak values of the surface temperature difference appear for two times, respectively on January 15, 2000 and January 15, 2001. This is because of low external temperature in the winter and high interior temperature rise. Nevertheless, affected by the second-term water cooling, the later two peak values, occurring on January 15, 2002 and January 15, 2003, apparently decrease. After the reservoir water filling, the upstream surface temperature differences quickly tend to be steady, the steady value is about 2℃.

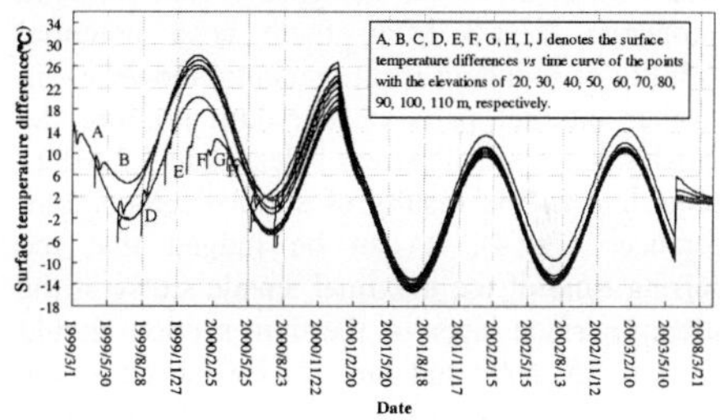

Figure 4. Evolution of the upstream surface temperature difference of left guide-wall section.

6.3 Thermal stress evolution

The results of thermal stress demonstrate that due to the setup of one temporary transverse joint and three longitudinal joints, the horizontal tensile stresses within the concrete are controlled so that the tensile stresses in the vertical direction are much higher than those in horizontal directions. Figs.5 illustrates the thermal creep stress evolutions at the interior centre points of the concrete blocks close to ground surface.

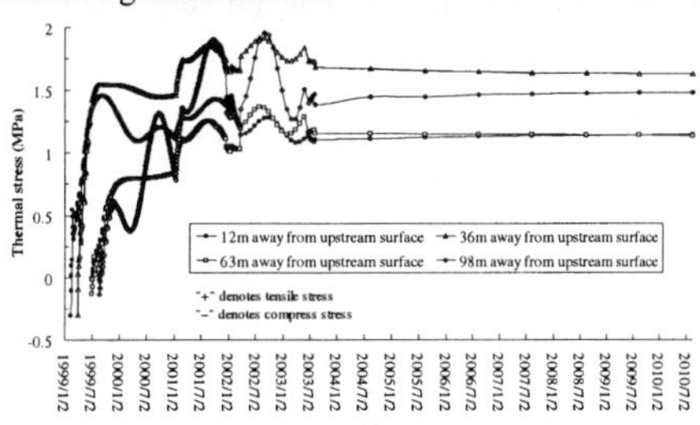

Figure 5. Evolution of the thermal stress at the elevation of 22.0 m of the left guide-wall dam section.

Comparing the above curves with the curves of temperature evolutions depicted in Fig.3, it can be found that the increasing and decreasing of tensile stresses with time are adverse to that of ascending and descending of temperatures. Correspondingly, there are five evolutionary stages of thermal tensile stresses: at the first stage, the compressive stresses with a small value occur. Then, the first peak values of tensile stresses appear. Following this, the tensile stresses reach their second peak values. At the fourth stage, the third peak values of tensile stresses happen. Finally, the tensile stresses tend to be constant. As a whole, the values of tensile stress at the interior centre points of the blocks are more than 1.3 MPa. Since the designed tensile strength of concrete is 1.3 MPa, there is a potential possibility of inducing local cracking. However, in the other internal parts of the dam section, the thermal induced tensile stress is less than 1.2 MPa.

According to the results of surface temperature differences (Fig.4), it can be judged that the occurring time of the maximal tensile stress in the upstream surface layer of the dam section should be on Jan. 15, 2000 and Jan. 15, 2001 because the highest peak values of the surface temperature difference appear at that time. Figure 6 illustrates the distributions of the upstream surface tensile stresses. The following conclusions about the surface tensile stress can be drawn:

On January 15, 2001, in the upstream surface at both the elevation of 38 m and 103m, there appear large tensile stresses, of which the components in the vertical and dam axial directions are 1.25 MPa and 1.10 MPa at the elevation of 38 m, and 2.12 MPa and 1.51 MPa at the elevation of 103 m, respectively. And, the zones of tensile and compressive stress alternatively appear along the vertical direction. In addition, after the reservoir water filling, the tensile stresses are very small or disappear.

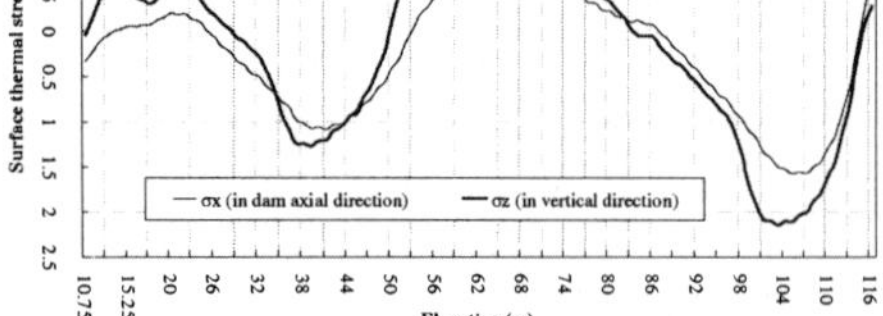

Figure 6. Distribution of the upstream surface thermal stress of the left guide-wall dam section on Jan.15 2001.

7. CONCLUSIONS

In this study, the thermo-mechanical processes of a typical Three-Gorges Dam section were modeled as realistically and reasonably as possible, some conclusions and suggestions about the left guide-wall dam section reached during this study are:

(1) The tensile stresses would be thermally induced in the interior close to the ground surface and the upstream surface (in winter) of the left guide-wall dam section with magnitudes of more than the designed tensile strength of concrete. This would possibly result in concrete cracking in the corresponding positions. Therefore, water-cooling and low initial temperature for the interior placing layers and heat insulation measure for the dam surface were recommended and implemented during the construction of the dam section.

(2) In the core of the dam section away from the surface and foundation, the thermal induced tensile stresses are generally less than the designed concrete tensile strength, thus, suggesting a very small potential of thermal induced cracking.

(3) The systematic philosophy of approaches presented in this paper could be expanded into other mass concrete dam engineering.

REFERENCES

John, J.O, Eugene J.O. 2002.
A new methodology for determining thermal properties and modelling temperature development in hydrating concrete. Construction and Building Materials 17(2003):pp.189-202

Zhu, B.F. 1999.
Thermal stress and temperature control of mass concrete structure. Beijing: Chinese Electric Power Press. [In Chinese]

Chinese Ministry of Water Conservancy and Hydroelectric Power. 1982.
Test Criterion of Concrete Used in Hydraulic Engineering (SD105-82), Beijing: Chinese Water Conservancy and Hydro-electric Power Publishers. [In Chinese]

Wu, Y. & Luna, R. 2001.
Numerical implementation of temperature and creep in mass concrete, Finite elements in analysis and design 37(2001): pp. 97-106

Vandewalle, L. 2000.
Concrete creep and shrinkage at cyclic ambient conditions. Cement & Concrete Composites, 22(2000): pp. 201-208

Song, H.W. et al. 2002.
Creep prediction of concrete for reactor containment structures. Nuclear Engineering and Design 217(2002): pp. 225-236

Kim, J.K, Kim, H.K. & Yang JK. 2001.
Thermal analysis of hydration heat in concrete structures with pipe-cooling system. Computers and Structures 79(2001): pp. 163 – 171

MODELLING OF THE DILATANCY – SATURATION COUPLING DURING EXCAVATION AND CONSOLIDATION OF AN UNDERGROUND STRUCTURE

Philippe Kolmayer[1], Clément Chavant[2]
[1])Electricté de France - Hydro Engineering Centre
[2]) Electricité de France - Research and Development Department

Abstract: In addition to THM coupling,, taking into account the important phenomena of the mechanical behaviour of the rock, like dilatancy, softening and damage, remains fundamental. This article presents an applications of two elastoplastic models which complies with traditional theories of rock mechanics and which are fully coupled with a biphasic unsaturated fluid approach, integrated in code_Aster, the FEM code developed by Electricité de France. In order to estimate the influence of mechanical and hydraulic non linearities on the short term behaviour of underground structures during excavation and consolidation, parametric calculations on a axisymmetrical mono-dimensional tunnel are first performed. Full simulations have been done on the "Mont Terri" tunnel. The importance of the relation between induced damage and permeability is pointed out.

1. INTRODUCTION

The dimensioning and design studies of underground caverns for radioactive waste disposal have to take into consideration the impact of THMC coupling. In order to predict the global stability of underground structures and the extension of the damaged zone, the thermal and non-saturated aspects should not restrict the performance of the mechanical models. Thus, suitable models can work on a large scale of stresses, simulate brittle fracture of the rock and the dilatancy effects.

This article describes non-saturated applications of two elastoplastic models respectful of the above specifications with particular attention given to the behaviour at the end of excavation and after a short resaturation process. The influence of dilatancy, softening and damage has been studied on a simplified model. At last an application on a real tunnel is presented.

2. THE NON LINEAR ELASTOPLASTIC MODELS AND THE NON SATURATED APPROACH

One can regard a non-saturated material as a polyphasic medium made of four homogeneous constituents: a solid skeleton, a liquid phase (superscript l), its vapour and an inert gas. Aiming to predict stresses' and strains' evolutions in usual structures, priority is given to the skeleton motion. Therefore a non-saturation modelling which can easily be coupled to any mechanical law (provided that the assumptions of Bishop effective stresses is done) has been chosen.

$$d\sigma = d\sigma' + d\sigma_p 1 \quad (1)$$

$$d\sigma_p = -b\left(dp_{gz} - S_l dp_c\right) \quad (2)$$

The main assumptions for non-saturation modelling are listed below. See Chavant (1999) for further explanations.

- Two fluids, the liquid water and a gas mixture saturate the porous medium.
- The gas mixture is composed of dry air and water vapour.
- Phase change between liquid water and its vapour is possible.
- Variation of relative permeability to water and gas with water saturation degree are given by equations 3 and 4.
- A unique relation (equation 5) between capillary pressure and liquid saturation is supposed.

$$k_{rel}^{l} = S^{3} \quad (3)$$

$$k_{rel}^{gaq} = (1-S)^2 \cdot \left(1 - S^{\frac{2}{5}}\right) \quad (4)$$

$$S_l = \left(1 + \left(\frac{P_c}{A}\right)^{\frac{1}{1-B}}\right)^{-B} \quad (5)$$

2.1 Notations

$\boldsymbol{\sigma}$ is the stress tensor, $\mathbf{s}$ is the deviatoric stress tensor, I_1 is the trace s_{II} the second invariant of $\mathbf{s}$

We use Lode angle $\cos(3\theta) = 2^{1/2}3^{3/2}\frac{\det(\mathbf{s})}{s_{II}^3}$

$\boldsymbol{\varepsilon}$ is the total deformation tensor, $\mathbf{e}$ is its deviatoric part, ε_v is the volume change. $\boldsymbol{\varepsilon}^p$ is the plastic deformation tensor, $\mathbf{e}^p$ is its deviatoric part, ε_v^p is the plastic volume change. γ^p is the equivalent plastic shear strain (see equation 8).

$$\mathbf{e} = \boldsymbol{\varepsilon} - \frac{Tr(\boldsymbol{\varepsilon})}{3}\mathbf{I} \quad (6)$$

$$\varepsilon_v = Tr(\boldsymbol{\varepsilon}) \quad (7)$$

$$\gamma^p = \int \sqrt{\frac{2}{3}\dot{e}_{ij}^p \dot{e}_{ij}^p}\, dt \quad (8)$$

where K_o is the elastic drained bulk modulus, μ_0 the elastic drained shear modulus.

2.2 The CJS1 model

This model is based on the mean features of the Mohr-Coulomb model and is expressed with stress invariants [Maleki (1999)] instead of principal stresses. Until plasticity is reached, a linear elastic behaviour is assumed. It is fully described by the drained elastic bulk and shear moduli. The yield surface of the perfectly plastic model is given by equation 7. Function $h(\theta)$ is chosen so that the shape of the criterion in the principal stress space is close to the Lade criterion.

$$f = s_{II}h(\theta) - R_m(I_1 - Q) \quad (9)$$

$$h(\theta) = (1 + \gamma_{cjs}\cos(3\theta))^{1/6} \quad (10)$$

The yielding parameters Rm and Q can easily be related to the friction angle ϕ and the cohesion C as shown by equations 11 and 12.

$$R_m = 2\sqrt{\frac{2}{3}} \cdot \frac{\sin\phi}{3 - \sin\phi}(1-\gamma)^{1/6} \quad (11)$$

$$Q = -3 \cdot C \cdot \cot\phi \quad (12)$$

In order to respect volumetric strain, a non-associated plastic theory is used. As suggested by Cambou et al. (1989), the unit vector n is calculated in such a way that equation 13 is equivalent to the definition of plastic volume change of equation 15. One can combine the equations to obtain the expression of vector n:

$$\dot{\boldsymbol{\varepsilon}}^p = \dot{\lambda}\left(\frac{\partial f}{\partial \boldsymbol{\sigma}} - \left(\frac{\partial f}{\partial \boldsymbol{\sigma}}\mathbf{n}\right)\mathbf{n}\right) \quad (13)$$

$$\dot{\boldsymbol{\varepsilon}}^p.\mathbf{n} = 0 \quad (14)$$

$$\dot{\varepsilon}_v^p = -\beta' \frac{\mathbf{s}.\dot{\boldsymbol{\varepsilon}}^p}{s_{II}} \quad (15)$$

$$\mathbf{n} = \frac{\beta' \frac{\mathbf{s}}{s_{II}} + \mathbf{I}}{\sqrt{\beta'^2 + 3}} \quad (16)$$

Parameter β' manages the volume change for a plastic behaviour. It is related to the dilatancy angle ψ according to the equation 17:

$$\beta' = -\frac{2\sqrt{6}\sin(\psi)}{3 - \sin(\psi)} \quad (17)$$

2.3 The "Laigle" model

Perfectly plastic models like Mohr-Coulomb or CJS1 models are not representative of the whole complex behaviour of brittle rocks. In order to improve the usefulness of the numerical models normally used for the dimensioning and design studies for underground caverns and to guarantee the safety of geological underground waste disposals, EDF has developed an improved rheological model [Laigle (2002)]. It focuses essentially on the behaviour of the material beyond its peak of strength, in the domain of medium and large strain.

The main features of this model are:

- reproducing the rock's brittle behaviour (post-peak softening).
- reflecting the influence of the average stress on the rock strength (suitable failure criterion, and evolution of the brittle behaviour towards a ductile behaviour).
- reproducing the increase in volume due to shear load (dilatancy of the rocks).

As for the CJS1 model, a linear elasticity is assumed until plasticity is reached. The yield surface is a "generalized Hoek and Brown criterion" using stress invariants as shown below. k, σ_c, h_0^c and h_0^t are coefficients depending on the material parameters.

$$f=\left(\frac{s_{II}}{\sigma_c}\right)^{\frac{1}{a}}+\frac{mk}{\sqrt{6}\sigma_c}\left(\frac{3}{2}-\frac{1}{2}\frac{2h(\theta)-\left(h_c^0+h_t^0\right)}{h_c^0-h_t^0}\right)s_{II}$$
$$+\frac{mk}{3\sigma_c}I_1-s\cdot k \qquad (18)$$

The volumetric behaviour is given by equation 19. But in order to reproduce brittle behaviour at low confinement and ductility at high confinement, the dilatancy angle depends on the stress state according to equation 20.

$$\sin\psi=\gamma\frac{\alpha'-\alpha_{res}}{\zeta\alpha'+\alpha_{res}} \qquad (19)$$

$$\alpha'=\frac{\sigma_1-\sigma_{t0}}{\sigma_3-\sigma_{t0}} \qquad (20)$$

where γ ζ, α_{res}, σ_{t0} are coefficients depending on materials parameters.

The post-peak behaviour is driven by the equivalent plastic shear strain γ^p which influences the evolution of the parameters m, s and a. The behaviour is broken down into 3 phases correlated to different physical phenomena. At first, softening is governed to a progressive deterioration of the cohesion and an increase in dilatancy. Then the evolution is linked to the development of fractures that induces a progressive discontinuity. Friction and dilatancy angles decrease from their peak values to residual values. Finally, the global behaviour is defined by the residual behaviour of a continuous fracture i.e. a purely frictional behaviour without any volumetric change.

3. APPLICATION ON A SIMPLIFIED STRUCTURE

In this section, an application of the proposed models is presented. The case studied here is in connection with the future construction of an underground research laboratory for nuclear waste disposal in a claystone formation. The main purpose is to evaluate the desaturation-resaturation process around the excavation during the excavation and the consolidation phases with different rheological laws. Special attention is given to the dilatancy phenomenon that influences the saturation.

A horizontal circular tunnel is considered. It is assumed that the rock is initially saturated and the stress state is supposed to be isotropic. This assumption allows us to simplify the geometric configuration and to make an axisymmetrical mono-dimensional calculation as shown in figure 2. The initial values of stress and pressure are given by:

p_l^0=5 MPa, p_g^0=, $\sigma_v^0=\sigma_h^0=\sigma_H^0$=6.79 MPa

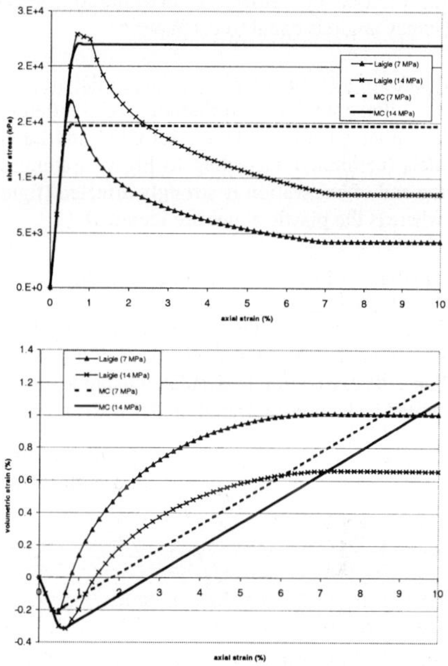

Figure 1. Simulation of the drained triaxial behaviour with CJS1 and Laigle models

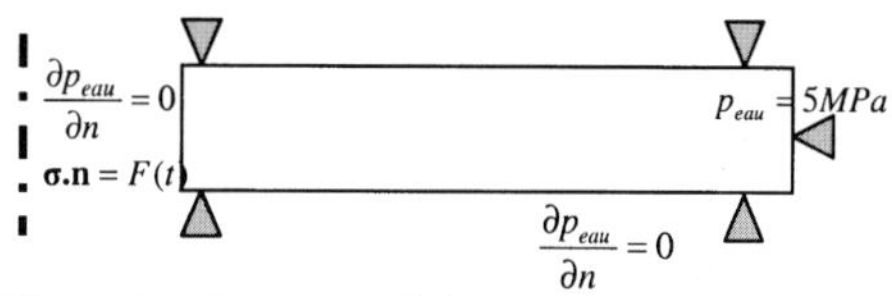

Figure 2. Geometrical domain and boundary conditions

To simulate the excavation of the gallery, a force distribution F(t) is applied at the left boundary. It decreases from 1 to 0 as excavation is completed after approximately 2 days. $\lambda(t)=F(t)/F_0$ is called the confinement.

Importance of dilatancy

Comparative calculations have been carried out using an increasing dilatancy angle. The inner (wall) and the outer plastic zones are affected in a different way by this angle. An increase of the dilatancy induces more suction close to the wall (figure 3) and a stronger evolution in the rock so that the plastic crown is reduced (figure 4). From a mechanical point of view, desaturation has got a beneficial aspect on short-term behaviour. In the studied case, the slope of desaturation versus dilatancy angle is equal to 0.1% per °.

Influence of softening and associated dilatancy

To illustrate the influence of softening, calculations have been done with CJS1 and Laigle models (calibrated according to Fig.1). Saturation at the end of excavation is strongly affected (figure 7) whereas the plastic zones are identical.

Comparison saturated (HM) – unsaturated approach (HHM)

In order to evaluate the influence of the triphasic approach on the short-term behaviour of the gallery, the excavation is simulated using a saturated and a non-saturated model with undrained conditions. Radial displacement and the extension of the plastic zone are identical. Positive suction exists in a crown of 40 cm around the gallery with both methods. The saturated one leads to a linear evolution of the suction whereas the non-saturated one leads to a non-linear evolution of suction that reaches 1.4 MPa. Close to the wall, one can observe a surprising different evolution of the radial stress: it increases with HM assumptions and decreases with HHM ones.

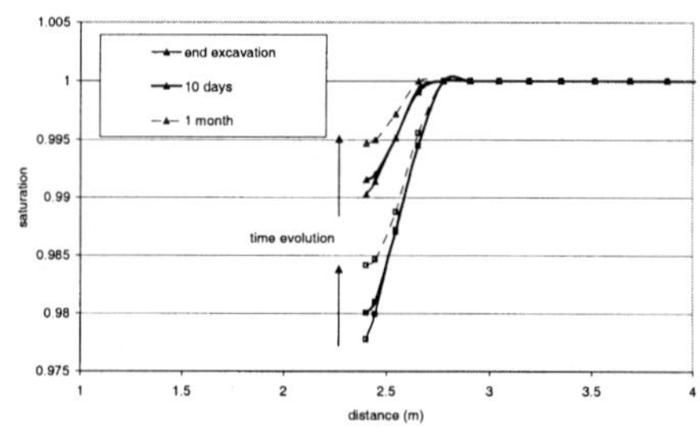

Figure 5. Evolution of saturation profiles during resaturation (1 day, 10 day and 1 month)

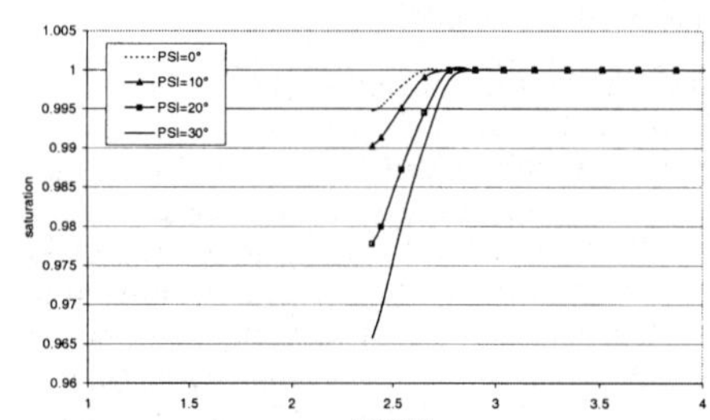

Figure 3. Evolution of saturation with dilatancy angle

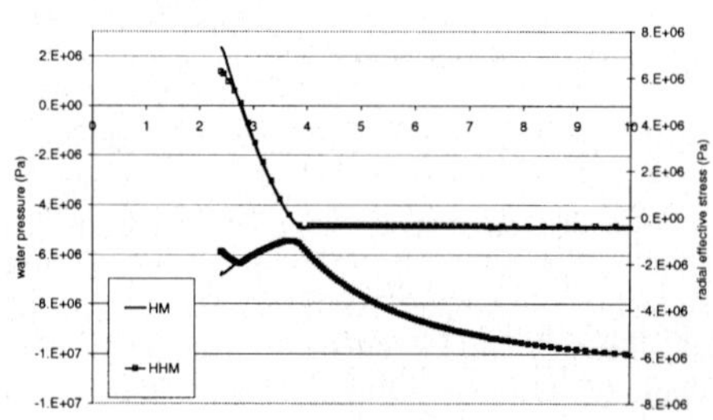

Figure 6. Comparison of saturated and unsaturated approaches

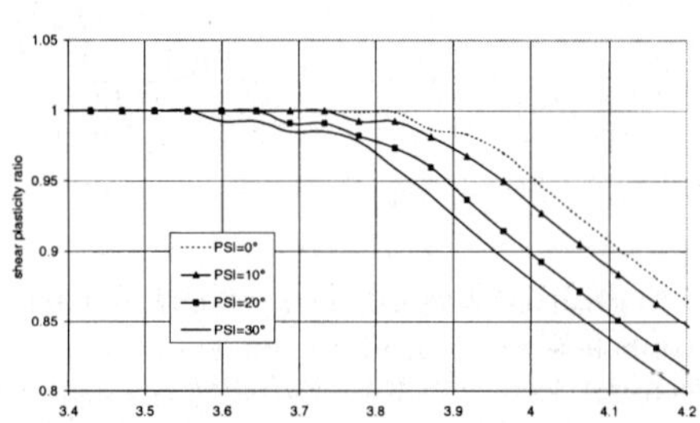

Figure 4. Evolution of plastic zone with dilatancy angle

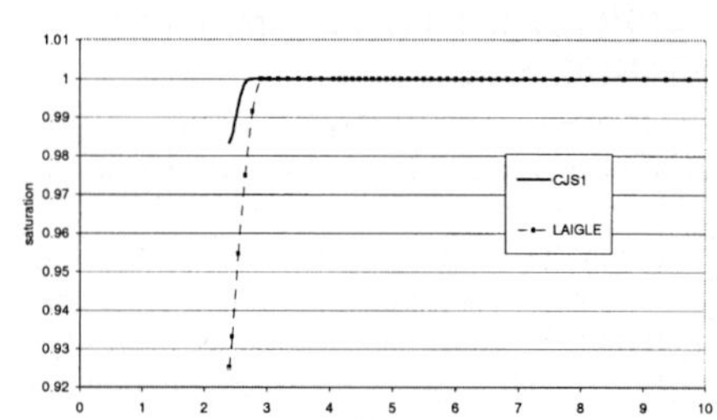

Figure 7. Evolution of saturation with CJS1 and Laigle models

Influence of boundary condition

Two kinds of hydraulic boundary conditions are tested. The first one supposes an impermeable tunnel wall, the second a prescribed relative humidity (HR). Figure 8 shows that the assumption is very influent close to the wall (first 15 cm). The mechanical consequence of this low saturation level is an important increase of suction so that plasticity never occurs. In conclusion, imposed HR is not suitable for elastoplastic model using Bishop's effective stresses. Thus, no flow condition has been adopted.

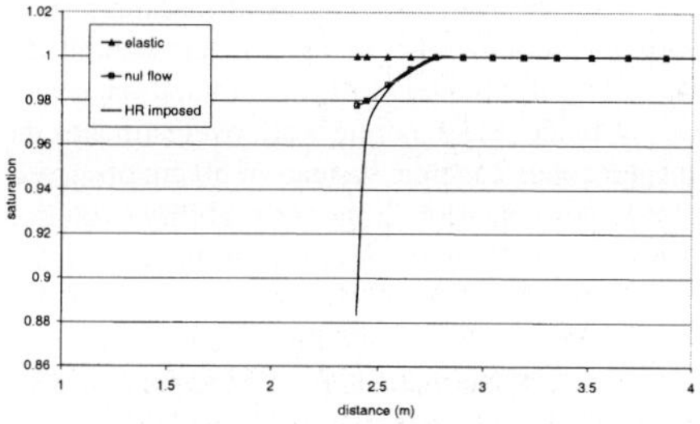

Figure 8. Influence of boundary condition (no flow or imposed relative humidity)

Influence of the retention curve

For this purpose, calculations have been performed using different retention curves. There is a very low influence on the global mechanical behaviour around the tunnel. Displacements are identical and the plastic zone gets longer from 8% (10 cm). The increase of parameter A induces less suction (a smaller desaturated zone in figure 10) and an increase of the non-linear evolution of suction close to the gallery in figure 9. The saturated approach gives similar results as the unsaturated one with A tending to 0.

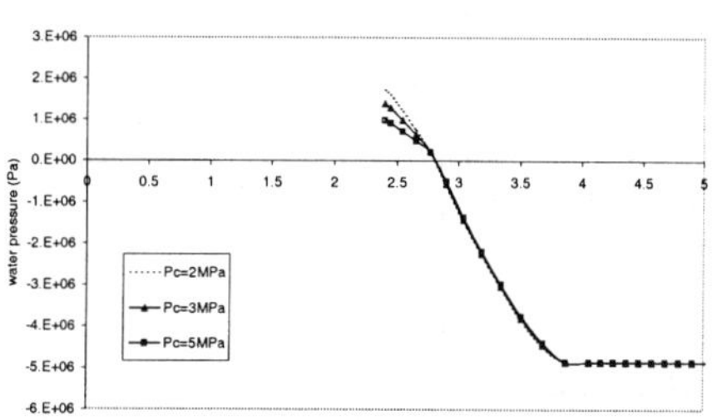

Figure 9. Evolution of suction with the retention curve

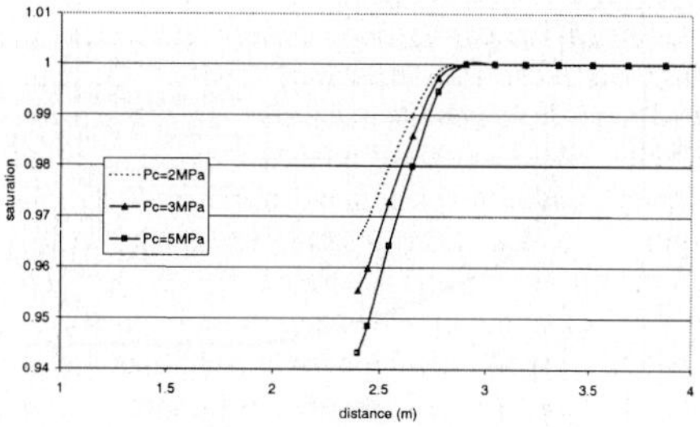

Figure 10. Evolution of saturation with the retention curve

Influence of damage

For primary calculation a constant intrinsic permeability might be suitable. However, many experimental data have shown that rock permeability might increase significantly with the growth of microcracks and fractures. According to results obtained on the Canadian granite, a reasonable relationship linking the volumetric plastic strain to this increase of permeability has been proposed: by Laigle (2003).

$$\frac{k_w}{k_w^0} = 1 + 2.10^{12} \cdot \left(\varepsilon_v^p\right)^3 \tag{20}$$

Taking into account permeability evolution with damage generates a crown of constant water pressure around the tunnel. The size of this crown increases during the resaturation process. Close to the wall, the suction with damage is always smaller than without damage. But 50 cm behind the wall, damage generates more suction after 1 month of resaturation. Thus, it cannot be concluded that damage is always pessimistic for the global stability. Indeed, the radial displacement remains equivalent with and without damage.

The extension of the damaged zone (EDZ) can be estimated in figure 8. The size of the zone is equal with or without damage, but the level of damage is more important with damage. During the resaturation, the gap between the two approaches still increases.

4. APPLICATION ON THE MONT TERRI GALLERY

In order to verify the ability of this complex mechanical model coupled with non-saturation, an application is performed on more representative case. The studied case is inspired from the Mont

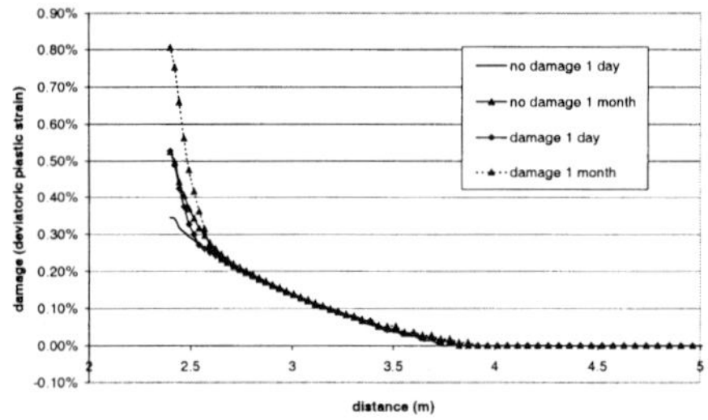

Figure 11. Evolution of damage during resaturation (with and without damage)

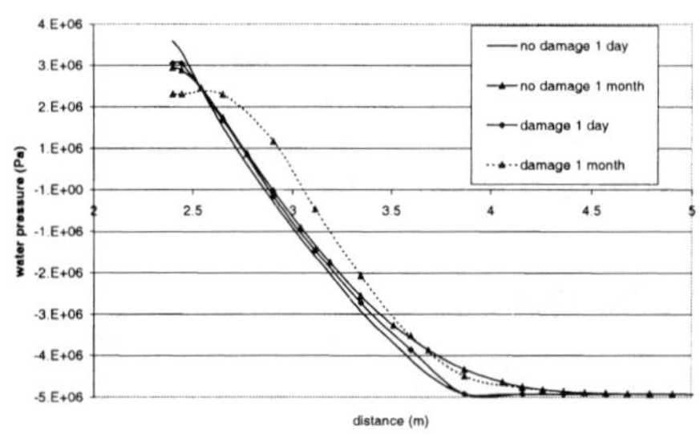

Figure 12. Evolution of suction during resaturation (with and without damage)

Terri gallery in Switzerland. It is located in the Opalinus clay made of argilites at a depth of 250 meters. A "horseshoe" shape is assumed in this case. The initial state of the stress and pressure is given by:

p_l^0=2.0 MPa $\sigma_v^{0'}$=13 MPa $\sigma_h^{0'}$=4 MPa
p_g^0=0.1MPa, $\sigma_H^{0'}$=4 MPa

The mechanical properties listed in table 1are given by Thury (1999). The methodology of identification for the set of parameters of the softening model is developed by Laigle (2003). The hydraulic permeability is taken equal to 2.10^{-13} ms^{-1}. Equations 3 to 5 are used for the retention curve and the relative permeabilities of gas and water with A=15 MPa and B=0.36.

A basic reference calculation is first made using drained conditions (no water). Without lining, excavation couldn't be achieved because the tunnel collapsed when the confinement ratio reached 18%. Hydro-mechanical conditions allow us to complete the excavation. Because water pressure becomes negative in a wide area around the tunnel, an artificial cohesion prevents the damage of the soil.

Table 1.Mechanical properties of the Opalinus clay.

	Low properties
E	4000 MPa
ν	0.20
ϕ'	33 °
C'	2 MPa
Porosity	6%
γ_h	25 kN/m^3
k_0	2 10^{-13} m/s

In Figure 13, one can observe suction as far as 1 meter from the wall whereas local measures show desaturated cracks 50 cm to 1 meter behind the wall. Along the tunnel wall, the simulation, which does not model the concrete wall, overestimates the damaged zone: 2 meters instead of 50 cm observed. According to equation 20, a permeability evolution can be estimated. After excavation, the ration k/k_0 (current and initial permeabilities) reaches 10^4 and after 10 months of consolidation, its value is equal to 10^6. This second value is in good agreement with local measures.

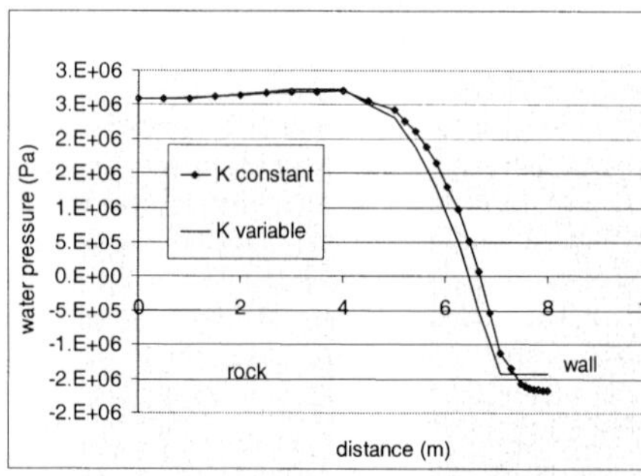

Figure 13. Water pressure evolution with and without damage along the horizontal line after excavation*

Calculations with constant permeability and with variable permeability depending on damage have been performed. At the end of excavation no significant difference is noted between the two approaches. The behaviours differ after 10 months of consolidation:

- In the damaged zone, suction (figure 14) is constant around the tunnel with variable permeability and not concentrated in the lower wedge (constant permeability)..
- The fracture (figure 15) spreads further with variable permeability. Its extension is closed

to that observed just before collapse on a drained calculation. With damage, the fracture seems to be more vertical than without damage.

- The peak value of plastic shear strain (figure 16) is five times greater than with constant permeability. It is concentrated in a thinner crack with variable permeability than with constant permeability.

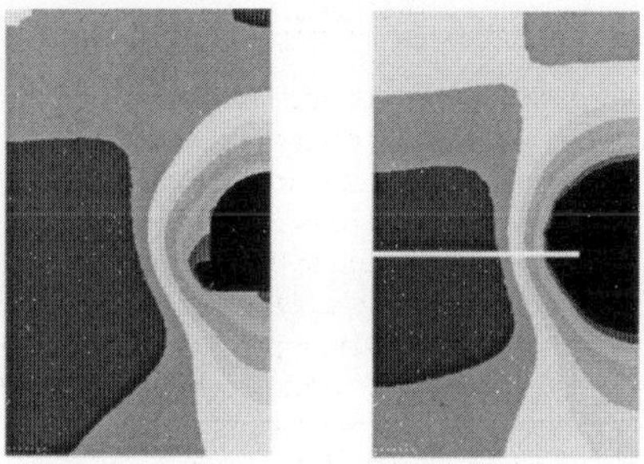

Figure 14. Water pressure with constant (left) and variable (right) permeability after consolidation

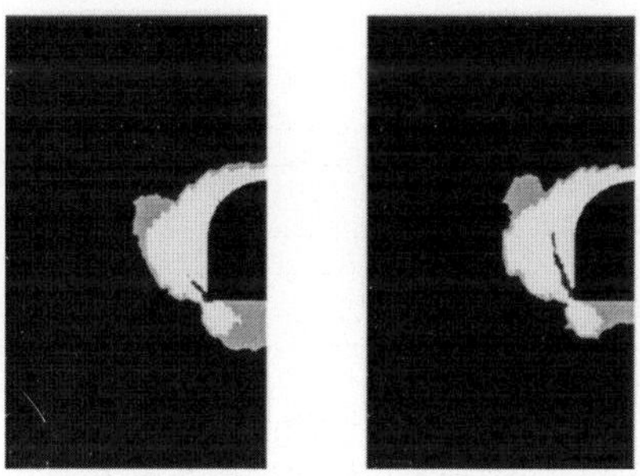

Figure 15. Damage with constant (left) and variable (right) permeability (blue: non damage, green pre-peak damage, yellow: post-peak damage, red: total loss of cohesion)

5. CONCLUSION

In this article, two improved elastoplastic models taking into account dilatancy and softening, and coupled with the non-saturated approach of Bishop's effective stresses have been successfully tested. The influence of dilatancy on saturation degree and on the short-term stability of a tunnel is shown. The consequences of damage (plastic

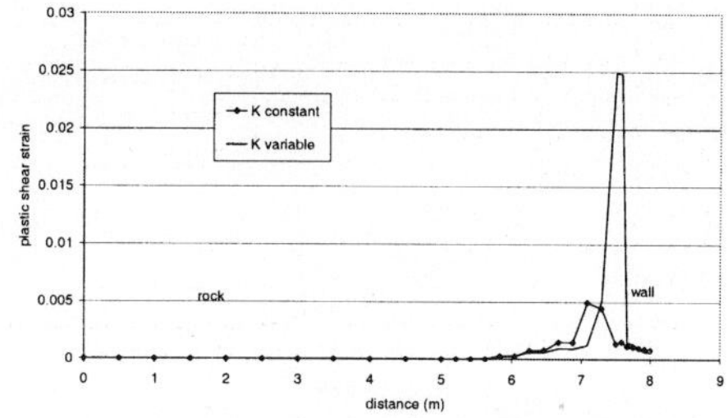

Figure 16. Plastic shear strain evolution with and without damage along the horizontal line (* the location of the horizontal line is given in figure 14.*

volumetric strain) induced by the excavation or by the resaturation process may be very significant as illustrated on the Mont Terri tunnel.

REFERENCES

Cambou B., Jafari k., Elamrani K., 1989 *An elastoplastic model for granular material using tree yielding mechanisms.* Num. Models in geomech. Numog III. Ed Pretruszizak, Pande-Elsevier, pp. 1 – 8.

Chavant C, Charles P., Duforestel T , Voldoire F., 1999, *Thermo-hydro-mechanics of unsaturated porous media in code_Aster*, technical note, EDF R&D, HI-74/99/011/A.

Laigle, F. 2003. *Proposition d'une modélisation rhéologique des roches adaptée à la conception des ouvrages souterrains.* Ecole Centrale de Lyon. Ph. Thesis (à paraître).

Laigle, F.2002. *Numerical simulation for analysis and support design of underground structures.* Int. Rock Mechanics Toronto.

Maleki M, 1998 *Modélisation hiérarchisée du comportement des sols* Ph. Thesis, Ecole Centrale de Lyon.

Thury M, Bossart P, 1999, *Mont-Terri Rock Laboratory: Results of Hydrogeological, Geomechanical and geotechnical Experiments Performed in 1996 and 1997.* Service hydrologique et géologique national. Berne.

AUTHOR INDEX

SUBJECT INDEX